Radiopharmaceuticals in Nuclear Pharmacy and Nuclear Medicine

Third Edition

Notice

The authors and the publisher have made every effort to ensure the accuracy and completeness of the information presented in this book. However, the authors and the publisher cannot be held responsible for the continued currency of the information, any inadvertent errors or omissions, or the application of this information. Therefore, the authors and the publisher shall have no liability to any person or entity with regard to claims, loss, or damage caused or alleged to be caused, directly or indirectly, by the use of information contained herein.

Radiopharmaceuticals in Nuclear Pharmacy and Nuclear Medicine

Third Edition

RICHARD J. KOWALSKY, PharmD, FAPhA
Associate Professor, Pharmacy and Radiology
University of North Carolina
Chapel Hill, North Carolina

STEVEN W. FALEN, MD, PhD
Medical Director
Northern California PET Imaging Center
Sacramento, California

WASHINGTON, D.C.

EDITOR: Nancy Tarleton Landis
ACQUIRING EDITOR: Sandra J. Cannon
COVER DESIGN: Mariam Safi, APhA Creative Services
COMPOSITION: Circle Graphics
PROOFREADING: Kathleen K. Wolter
INDEXING: Jen Burton, Columbia Indexing Group

© 2011 by the American Pharmacists Association

Published by the American Pharmacists Association
2215 Constitution Avenue, N.W., Washington, DC 20037-2985
www.pharmacist.com www.pharmacylibrary.com

APhA was founded in 1852 as the American Pharmaceutical Association.

To comment on this book via e-mail, send your message to the publisher at aphabooks@aphanet.org.

All rights reserved

No part of this book may be reproduced, stored in a retrieval system, or transmitted in any form or by any means, electronic, mechanical, photocopying, recording, or otherwise, without written permission from the publisher.

LIBRARY OF CONGRESS CATALOGING-IN-PUBLICATION DATA
Kowalsky, Richard J., author.
  Radiopharmaceuticals in nuclear pharmacy and nuclear medicine / Richard J. Kowalsky, PharmD, FAPhA, Associate Professor, Pharmacy and Radiology, University of North Carolina, Chapel Hill, North Carolina, Steven W. Falen, MD, PhD, Medical Director, Northern California PET Imaging Center, Sacramento, California. — Third Edition.
      p. ; cm.
  Includes bibliographical references and index.
  ISBN 978-1-58212-118-5
   1. Radioisotope scanning. 2. Radiopharmaceuticals. 3. Radioisotopes in pharmacology. I. Falen, Steven W., author. II. American Pharmacists Association, issuing body. III. Title.
   [DNLM: 1. Radiopharmaceuticals—diagnostic use. 2. Radiopharmaceuticals—therapeutic use. 3. Nuclear Medicine. 4. Radiochemistry—methods. WN 440]
  RC78.7.R4K69 2011
  616.07'575—dc22
                                                                                                2010052206

HOW TO ORDER THIS BOOK

Online: www.pharmacist.com/shop_apha
By phone: 800-878-0729 (770-280-0085 from outside the United States)
VISA®, MasterCard®, and American Express® cards accepted

Contents

Preface — vii

Acknowledgments — xi

Contributors — xiii

1 Radiopharmaceuticals, Nuclear Medicine, and Nuclear Pharmacy: An Overview — 1

2 Radioactive Decay — 13

3 Radionuclide Production — 31

4 Interactions of Radiation with Matter — 47

5 Radiation Detection and Instrumentation — 59

6 Radiation Measurement, Protection, and Risk — 87

7 Radiation Safety — 109

8 Radiation Biology — 127

9 Radiopharmaceutical Chemistry: General Topics — 147

10 Radiopharmaceutical Chemistry: Technetium Agents — 171

11 Radiopharmaceutical Chemistry: Nontechnetium Agents — 207

12 Radiopharmaceutical Chemistry: PET Agents — 229
Stephen M. Moerlein

13	The Nuclear Pharmacy *Kristina M. Wittstrom*	257
14	Quality Control in Nuclear Pharmacy *Joseph C. Hung*	273
15	Microbiologic Control for Radiopharmaceuticals *James F. Cooper*	313
16	Radiopharmaceutical Special Topics *James A. Ponto*	319
17	Molecular Imaging and the Development of New Radiopharmaceuticals *Stephen M. Moerlein*	343
18	Licensing and Regulatory Control *Neil A. Petry*	373
19	Brain	411
20	Thyroid	445
21	Heart	459
22	Lung	499
23	Liver, Spleen, and Gastrointestinal System	517
24	Kidney	547
25	Bone	571
26	Total-Body SPECT Procedures	589
27	Clinical PET Procedures	617
28	Antibodies and Peptides	655
29	Therapeutic Applications of Radioactive Agents	673
30	In Vivo Function Studies	695
	Index	00

Preface

Nuclear pharmacy and nuclear medicine continue to experience extensive change and growth with the development of new radiopharmaceuticals and new technologies for their diagnostic and therapeutic applications. *Radiopharmaceuticals in Nuclear Pharmacy and Nuclear Medicine, Third Edition,* updates the key topics covered in the Second Edition. The book is designed as a comprehensive introductory textbook on the chemical, physical, and biologic properties of radiopharmaceuticals and their applications in nuclear medicine. It is intended for use in courses taught in the disciplines of nuclear pharmacy, nuclear medicine technology, and nuclear medicine. The chapter topics are of moderate depth and breadth and are referenced to the primary literature. As such, this textbook has become a useful resource for professional practitioners in these disciplines and for those preparing for specialty board examination.

The book contains essential information required by state and federal radiation licensing organizations for specialty practitioners preparing to become authorized nuclear pharmacists, authorized nuclear medicine physicians, and nuclear medicine technologists. Accordingly, the chapters cover key information on radiation physics and instrumentation; radiation protection; radiation biology; radioactive decay; radioactivity calculations; radiopharmaceutical chemistry of single-photon emission computed tomography (SPECT) and positron emission tomography (PET) agents; licensing and regulatory controls; shipping and receipt of radioactive material; radiation safety; radiopharmaceutical preparation, measurement, and quality control; and the diagnostic and therapeutic applications of radiopharmaceuticals.

A critical review of the Second Edition by experienced practitioners and educators was solicited before preparation of the Third Edition. This review resulted in reorganization of chapter topics and addition of new information not covered in the Second Edition. A key point in the revision was the division of the Second Edition's radiopharmaceutical chemistry chapter into three chapters to facilitate location of information on this extensive topic. Four new chapters have been added in the Third Edition, covering microbiologic control of radiopharmaceuticals, with emphasis on the guidelines of *U.S. Pharmacopeia (USP)* General Chapter 797; radiopharmaceutical special topics, with a focus on formulation problems, pediatric dosing, breast milk excretion, and adverse reactions; molecular imaging and new radiopharmaceutical development, including discussion of novel approaches and technologies used in the development of new radiopharmaceuticals; and clinical PET procedures, covering a breadth of fundamental diagnostic information provided in PET scans.

The Third Edition also has a new design, featuring a larger page size with two-column format, and four-color, high-resolution illustrations. These changes were instituted to accommodate the new material added to the book and to enhance the quality of illustrations, particularly contemporary nuclear medicine images.

The textbook is organized in three sections. The first section, Chapters 1 through 12, discusses fundamental concepts of radiation physics, radiation safety, radiation biology, and radiopharmaceutical chemistry. The next section, Chapters 13 through 18, covers topics in nuclear pharmacy practice, radiopharmaceutical preparation and quality control, molecular imaging and the development of new radiopharmaceuticals, and the regulatory control of radiopharmaceuticals. The concluding section, Chapters 19 through 30, discusses the diagnostic and therapeutic applications of radiopharmaceuticals used in nuclear medicine practice.

Chapter 1 is an overview of radiopharmaceuticals, their general properties, and the types of procedures performed in nuclear medicine. It also addresses the historical development of nuclear pharmacy as a specialty practice, covering key elements associated with its evolution, the development of training programs in nuclear pharmacy, and development of the nuclear pharmacist certification process that established nuclear pharmacy as the first specialty area recognized by the Board of Pharmaceutical Specialties.

Chapter 2 discusses atomic physics, nuclear force and energy, the modes of radioactive decay, and radioactivity. The topics in this chapter are illustrated with example calculations.

Chapter 3 covers the production of radionuclides used in nuclear medicine, with a discussion of those radionuclides that are produced in nuclear reactors, particle accelerators, and radionuclide generators. Emphasis is placed on the ^{99m}Tc generator.

Chapter 4 discusses the interactions of alpha particles, beta particles, neutrons, and high-energy electromagnetic radiation with matter. The text and figures provide a detailed explanation and illustration of the specific mechanisms involved in ionizing radiation interactions.

Chapter 5 describes radiation detection instrumentation, including radionuclide dose calibrators, Geiger-Müller counters, liquid and solid crystal scintillation counters, SPECT gamma cameras, and PET scanners. The chapter describes instrument operation, calibration, and quality control. Gamma scintillation spectrometry is discussed in detail, as are the fundamental principles involved in counting radioactive samples.

Chapter 6 covers radiation measurement, protection, and risk. Radiation measurement units and radiation dosimetry are discussed. The basic radiation protection elements of time, distance, and shielding are discussed, with an emphasis on practical application to the radiation worker. The chapter concludes with a discussion of the potential carcinogenic and hereditary risks associated with exposure to ionizing radiation and includes a description of radiation risk models and the factors that determine the level of risk. A comparison of radiation risk with other types of risk is presented.

Chapter 7 covers radiation safety. It describes the goals of a radiation safety program for radiation workers and the general public, radiation dose limits established by regulatory bodies, the sources of radiation exposure, and the monitoring of radiation exposure of individual workers in the work environment. Radiation protection regulations and procedures are discussed with regard to notices, instructions, and reports to radiation workers; the receipt and shipment of radioactive material; the disposal of radioactive waste; and emergency procedures for handling accidental spills of radioactive material.

Chapter 8 covers the biologic effects of ionizing radiation. The chapter begins with a discussion of the effects of radiation on cellular biologic systems and includes genetic and cell cycle effects, radiosensitivity, and dose fractionation in radiation treatment procedures. Also covered are the biologic effects of whole-body irradiation, the carcinogenic and hereditary effects of radiation exposure, radiation effects on the embryo and fetus, and radiation-induced cataracts.

Chapters 9 through 12 cover radiopharmaceutical chemistry. Chapter 9 discusses general topics, with a review of chemical properties of the elements, chemical bonding, the solution chemistry of metallic elements, and oxidation–reduction reactions. Examples are given relating these topics to medical radionuclides. This is followed by a description of the ideal properties of diagnostic and therapeutic radiopharmaceuticals, radiolabeling methods, and the processes involved in the in vivo localization of radiopharmaceuticals.

Chapter 10 describes the chemistry of ^{99m}Tc radiopharmaceuticals. It discusses technetium's oxidation states, ^{99m}Tc radiopharmaceutical kit chemistry, and the labeling methods for ^{99m}Tc radiopharmaceuticals, and it presents a detailed description of the preparation and properties of ^{99m}Tc radiopharmaceuticals used in nuclear medicine.

Chapter 11 describes the chemistry of nontechnetium radiopharmaceuticals. This chapter covers the radiolabeling methods and properties of radiopharmaceuticals labeled with ^{14}C, ^{32}P, ^{51}Cr, ^{67}Ga, ^{90}Y, ^{111}In, ^{123}I, ^{125}I, ^{131}I, ^{133}Xe, ^{201}Tl, and ^{153}Sm.

Chapter 12, on the radiochemistry of PET radiopharmaceuticals, is authored by Stephen M. Moerlein, PharmD, PhD, BCNP, of the Mallinckrodt Institute of Radiology in St. Louis, Missouri. The first half of this chapter describes basic concepts of PET, PET scanner design, and data acquisition. This is followed by a description of PET radiopharmaceutical applications for cardiovascular, metabolic, and neurotransmission imaging and a discussion of updated reimbursement for PET procedures by the Centers for Medicare and Medicaid Services. The second half of the chapter covers the radiochemistry and control of PET radiopharmaceuticals, including radionuclide production, radiopharmaceutical synthesis, and radiopharmaceutical production systems. The chapter concludes with a discussion of issues related to radiopharmaceutical formulation, quality assurance, and regulation.

Chapter 13, describing the nuclear pharmacy facility, is authored by Kristina M. Wittstrom, BS, MEd, BCNP, FAPhA, of the University of New Mexico in Albuquerque. The chapter begins with a brief history of how nuclear pharmacies became established as a result of the growth of nuclear medicine. The discussion expands to describe the facilities, equipment, radiation detection instrumentation, and personnel necessary to operate a nuclear pharmacy. The chapter also describes the responsibilities of a nuclear pharmacist, which include radiopharmaceutical procurement, compounding, quality assurance, dispensing, and distribution. The chapter concludes with discussion of ancillary areas of nuclear pharmacy operation that involve the nuclear pharmacist, including health and safety issues, the provision of drug information to the nuclear medicine community, and record keeping.

Chapter 14 is an in-depth discussion of radiopharmaceutical quality control, authored by Joseph C. Hung, PhD, BCNP, FASHP, FAPhA, of the Mayo Clinic in Rochester, Minnesota. It begins with an overview of quality control, followed by a discussion of specific topics that include radionuclide, radiochemical, pharmaceutical, and biologic considerations. The chapter comprehensively covers the instrumentation and methods used to assess radiopharmaceutical identity, quantity, and purity. A discussion of instrumentation quality control includes dose calibrators and survey instruments. The chapter concludes with a discussion of quality control issues specific to PET drug products.

Chapter 15 is a new chapter on microbiologic control for radiopharmaceuticals, authored by James F. Cooper, PharmD, FAPhA, of Greensboro, North Carolina. This chapter expertly describes the operational requirements specified in *USP* Chapter 797 for compounded sterile products (CSPs), with a radiopharmaceutical focus. The discussion centers on the microbial contamination risk levels described in Chapter 797, the control of surface contamination to protect critical sites during radiopharmaceutical preparation, control of airborne contamination of CSPs, microbiologic tests for high-risk radiopharmaceuticals, and standard operating procedures for aseptic compounding of radiopharmaceuticals in a nuclear pharmacy.

Chapter 16 is a new chapter, authored by James A. Ponto, MS, BCNP, FAPhA, of the University of Iowa Hospitals and Clinics in Iowa City. It covers special topics related to radiopharmaceutical preparation and use in nuclear medicine, including formulation problems of radiopharmaceuticals, pediatric dosing, breast milk excretion, and adverse reactions to radiopharmaceuticals. The discussion is strengthened by practical information obtained from the author's own experience and gleaned from reports in the literature. Common problems associated with the preparation and stability of radiopharmaceuticals, particularly ^{99m}Tc agents, are discussed. Methods of determining radiopharmaceutical doses for pediatric patients are reviewed. Breast milk excretion of radiopharmaceuticals is discussed in detail, with tables of radiopharmaceutical excretion data and recommended times for cessation of breast-feeding that practitioners will find useful. The special characteristics that define an adverse reaction to a radiopharmaceutical are discussed, along with the reporting systems used to gain information and the statistical indications that such adverse reactions are infrequent. Examples of reactions are presented, and concerns associated with specific groups of radiopharmaceuticals are discussed. Case studies illustrating pediatric dosing, breast milk excretion, and adverse reactions are presented.

Chapter 17 is a new chapter on molecular imaging and the development of new radiopharmaceuticals, authored by Stephen M. Moerlein. It provides an introduction to the concept of molecular imaging and describes the role it plays in identifying disease at an early stage. The chapter goes on to discuss the radiopharmaceutical discovery and development process, focusing on the selection of an imaging target in the body, identification of a biomarker for this target, development of a radiolabeling strategy, purification and reformulation of the radiotracer, in vitro screening experiments, in vivo validation, and optimization for human application.

Chapter 18, authored by Neil A. Petry, MS, BCNP, FAPhA, of the Duke University Medical Center in Durham, North Carolina, is a revision of the Second Edition chapter on licensing and regulatory control of radiopharmaceuticals. It begins with a review of nuclear pharmacy practice guidelines, followed by a historical review of regulation by the Food and Drug Administration and the Nuclear Regulatory Commission (NRC). An excellent review of investigational new drug (IND) and new drug application (NDA) processes is presented. This is followed by an extensive discussion of the NRC regulations applicable to the medical use of radiopharmaceuticals.

The remaining 12 chapters in the book cover the diagnostic and therapeutic use of radiopharmaceuticals. Chapters 19 through 25, which focus on specific body systems (brain; thyroid; heart; lung; liver, spleen, and gastrointestinal tract; kidney; and bone) are followed by Chapter 26 on total-body SPECT procedures. These chapters follow a similar format, presenting anatomic and physiologic information relevant to the organ system being studied and then discussing the radiopharmaceuticals associated with specific nuclear medicine procedures.

Chapter 27 is a new chapter on clinical PET procedures. It begins with an overview of the value of functional imaging with PET in the diagnostic work-up of a patient. Patient preparation for a PET study and the parameters associated with acquiring the study are then discussed. The primary information depicted in normal FDG PET/CT studies of the brain, neck, heart, kidneys and bladder, and digestive and musculoskeletal systems is described. This is followed by a description of PET/CT studies of various types of cancer. The chapter also discusses nononcologic PET/CT studies to evaluate heart disease, epilepsy, dementias, and infectious and inflammatory processes. The chapter concludes with a discussion of various factors that influence the outcome of a PET/CT study.

Chapter 28, on monoclonal antibodies, has been updated to include radiolabeled peptides. The chapter begins with a review of the immune system and proceeds to discussion of antibody structure, classification, development, and modification; antibody–antigen interactions; and nomenclature. Radiolabeling methods and antibody properties are discussed for specific diagnostic antibodies. The chapter concludes with a discussion of radiolabeled peptides in clinical use and those in various stages of investigation.

Chapter 29 covers therapeutic radiopharmaceuticals. It begins with a discussion of radioimmunotherapy principles, radionuclide and antibody requirements for treating tumors, and dosing methods. Specific therapeutic antibodies are discussed for treating non-Hodgkin's lymphoma, including

^{90}Y-ibritumomab tiuxetan, ^{131}I-tositumomab, ^{131}I-LYM-1, and ^{90}Y-epratuzumab. Also covered are radiotherapy of metastatic bone pain, radiotherapy of polycythemia vera, effusion therapy, radiation synovectomy, brachytherapy of liver cancer, and investigations of targeted therapy with alpha particle emitters. The chapter concludes with a discussion of promising new radiotherapeutic agents.

Chapter 30 covers in vivo function studies in which radiopharmaceutical agents are used to assess body functions. Blood volume measurement, red blood cell survival, measurement of glomerular filtration rate, and miscellaneous in vivo procedures are described. Underlying principles involved in each study and the radiopharmaceuticals used are discussed.

Richard J. Kowalsky
Steven W. Falen
January 2011

Acknowledgments

We owe a debt of gratitude to several individuals who have given us support and guidance during the preparation of the Third Edition. Foremost are our wives, Louise and Susan, whose patience and encouragement have sustained our efforts during the update of the textbook. We are deeply grateful for their care and unlimited support. We would also like to recognize our editor, Nancy Landis, for her expert and thorough review of the text and constructive suggestions for its improvement. Additionally, we recognize the support of Julian Graubart, Senior Director, Books and Electronic Products Department at APhA, Sandy Cannon, Associate Director/Acquisitions Editor, and Kathy Anderson, Senior Manager; and Mariam Safi, Graphic Designer at APhA, for their assistance with publishing details and design elements.

Contributors

James F. Cooper, PharmD, FAPhA
Endotoxin Consultant
Greensboro, NC

Joseph C. Hung, PhD, BCNP, FASHP, FAPhA
Professor of Pharmacy
Professor of Radiology
Mayo Clinic College of Medicine
Director of Nuclear Pharmacy Laboratories and PET Radiochemistry Facility
Mayo Clinic
Rochester, MN

Stephen M. Moerlein, PharmD, PhD, BCNP
Associate Professor of Radiology and Biochemistry
Mallinckrodt Institute of Radiology
Washington University School of Medicine
Adjunct Clinical Associate Professor
Saint Louis College of Pharmacy
St. Louis, MO

Neil A. Petry, BPharm, MS, BCNP, FAPhA
Clinical Assistant Professor, Radiology
Director, Radiopharmacy and Nuclear Medicine Laboratory
Faculty Member, Medical Physics Graduate Program
Duke University Medical Center
Durham, NC

James A. Ponto, BPharm, MS, BCNP
Chief Nuclear Pharmacist
University of Iowa Hospitals and Clinics
Professor (Clinical)
College of Pharmacy
University of Iowa
Iowa City, IA

Kristina M. Wittstrom, BS, MEd, BCNP, FAPhA
Research Lecturer
Radiopharmacy
College of Pharmacy
University of New Mexico
Albuquerque, NM

CHAPTER 1

Radiopharmaceuticals, Nuclear Medicine, and Nuclear Pharmacy: An Overview

Identifying the beginning of nuclear medicine and defining the field depend upon one's perspective on the application of radiation to human disease. If the focus is on the use of natural radioactive material, then nuclear medicine implicitly started in 1901 when the French physician Henri Danlos used radium (a natural element) to treat a tuberculous skin lesion.[1] If the focus is on artificial radioisotopes, then nuclear medicine started after 1934, when the French radiochemists Frederic Joliot and his wife, Irene Curie Joliot, produced the first artificial radioisotope, phosphorus 30 (^{30}P). In the latter case, nuclear medicine began either with George Hevesy's successful use of radiophosphorus in healthy animals in 1935 or with Joseph Hamilton's attempts to treat leukemic patients with sodium 24 (^{24}Na) in 1936. The cyclotron was introduced around that time, and the resulting investigative ferment produced numerous radionuclides that were applied to diagnosis and therapy. Nuclear medicine as officially defined in 1967 was "the specialty of the practice of medicine dealing with the diagnostic, therapeutic (exclusive of sealed radiation sources) and investigative use of radionuclides." Since that time, however, the field has changed extensively.

The application of magnetic resonance imaging (MRI) methods to allow diagnosis without the use of radioactive material inspired a new definition of nuclear medicine that reflected the use of nuclear properties from stable nuclides. In February 1983 the Society of Nuclear Medicine board of trustees adopted the following definition: "the medical specialty which utilizes the nuclear properties of radioactive and stable nuclides for diagnostic evaluation of the anatomic and/or physiologic conditions of the body and provides therapy with unsealed radioactive sources."

Subsequently, the concept of molecular imaging has emerged, with the goal of noninvasive investigation into the cellular and molecular events involved in normal and pathologic processes. Although molecular imaging has always been the intrinsic basis of nuclear medicine, scientists in other disciplines, such as molecular biology, chemistry, and biomedical engineering, are becoming involved. Major areas of focus include the development of new molecular probes and instrumentation for radiation imaging, optical imaging, and magnetic resonance imaging, as well as in vivo spectroscopy. Nanotechnology is playing a significant role in molecular imaging, particularly with the development of new probes for medical diagnosis, drug development, and gene therapy.

While broad goals for molecular imaging are being investigated, the current practice of nuclear medicine continues to be the application of radiopharmaceuticals in diagnosis and therapy, which is the principal focus of this book.

THE RADIOPHARMACEUTICAL

A radiopharmaceutical can be defined as a chemical substance that contains radioactive atoms within its structure and is suitable for administration to humans for diagnosis or treatment of disease. In short, it is a radioactive drug. Radiopharmaceuticals are formulated in various chemical and physical forms to target radioactivity to particular parts of the body. Gamma radiation emitted from diagnostic radiopharmaceuticals readily escapes from the body, permitting external detection and measurement. The pattern of distribution of radiation in an organ system over time permits the nuclear medicine physician to make a diagnostic evaluation of system morphology and function. A therapeutic radiopharmaceutical emits particulate radiation (beta particles) that deposits energy within the organ being treated for disease. Some radionuclides, such as ^{32}P, emit only beta radiation, while other radionuclides, such as ^{131}I, emit beta and gamma radiation simultaneously and therefore possess both therapeutic and diagnostic usefulness.

NUCLEAR MEDICINE PROCEDURES

Examining the types of procedures routinely performed in nuclear medicine is helpful in understanding how radiopharmaceuticals are used. These procedures can be divided into three categories: (1) imaging procedures, (2) in vivo function studies, and (3) therapeutic procedures. The first two categories are diagnostic in nature and account for most of the studies performed in nuclear medicine.

Imaging Procedures

Imaging procedures provide diagnostic information based on the distribution pattern of radioactivity in the body. The procedures are either dynamic or static. *Dynamic studies* provide functional information through measurement of the rate of accumulation and removal of the radiopharmaceutical by the organ. *Static studies* provide morphologic information regarding organ size, shape, and position or the presence of space-occupying lesions, and in some cases relative function.

Detection and measurement of organ radioactivity is usually done with a gamma camera, an electronic device with a radiation detector large enough to visualize, in most cases, the entire organ of interest (Figure 1-1). Before the days of gamma cameras, images were made with rectilinear scanners. The rectilinear scanner detector was 3 to 5 inches in diameter and required multiple passes or scans over the area of interest in a rectangular and linear fashion to obtain an image of the entire organ. Because of this technique, imaging procedures are still referred to as "scans."

Dynamic imaging studies require that the camera detector be positioned over the organ of interest before injection of the radiopharmaceutical, so that the camera is able to capture the radioactivity as it enters and leaves the organ. Information collected can be stored in a computer for further analysis or permanently recorded on photographic film. An example of a dynamic study is the renogram. A kidney-localizing radiopharmaceutical, such as ^{99m}Tc-mertiatide, is injected intravenously, and the time course of its transport and excretion by the renal tubular cells is measured, providing an assessment of kidney function.

Static imaging studies are performed after a radiopharmaceutical is allowed to accumulate in the organ of interest. Images or "pictures" of the organ are acquired as the camera detector is rotated about the body to obtain multiple-angle views of the organ of interest. Static images provide information regarding organ morphology, such as organ size, shape, and position, and help detect the presence of space-occupying lesions, such as tumors.

The pattern of radiopharmaceutical distribution in an organ depends on the particular organ studied and the presence or absence of disease. In some studies, the normal organ readily concentrates the radiopharmaceutical and appears "hot" with uniformly distributed radioactivity. Diseased tissue in these organs excludes the radiopharmaceutical, and lesions appear as "cold" spots within a "hot" organ. An example is a liver colloid scan. A normal liver would appear "hot" after injection of radioactive colloidal particles that localize in the liver's phagocytic cells. If a tumor is present, however, the normal colloid-localizing cells are displaced and an area of decreased or absent radioactivity appears. This is in contrast to other types of organ studies in which the normal organ excludes the radiopharmaceutical and appears "cold," whereas

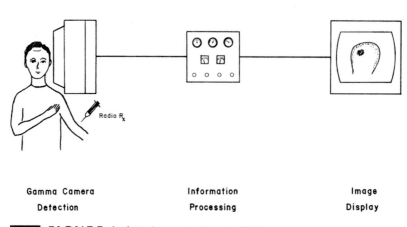

| Gamma Camera | Information | Image |
| Detection | Processing | Display |

FIGURE 1-1 Schematic of a scintillation gamma camera system demonstrating radiopharmaceutical injection, detection of radioactivity, electronic processing, and image display.

diseased tissue concentrates it and appears as a "hot" spot within a "cold" organ. An example is a brain scan obtained with an agent normally excluded by the blood–brain barrier. In disease states where the blood–brain barrier is disrupted, such as with a tumor, radioactivity can leave the vascular space to localize in the tumor.

In still other types of studies a normal organ may accumulate the radiopharmaceutical, but diseased tissue may concentrate it either to a greater degree because of increased function or to a lesser degree because of decreased function. An example is thyroid gland imaging with radioactive iodine. The thyroid gland readily accumulates iodine through normal function, but a diseased gland with either hyperfunctioning or hypofunctioning thyroid tissue demonstrates increased or decreased concentration of radioiodine. Examples of these static studies are illustrated in Figure 1-2. They are planar images presented in two dimensions. A disadvantage of planar imaging is that lesion detection may be impaired, especially when target-to-background ratios are low or there are overlying structures that obscure a clear view of the lesion. Single-photon emission computed tomography (SPECT) or positron emission tomography (PET) cameras are able to construct computer-generated slice images through an organ in transverse, sagittal, and coronal planes and make possible the visualization of an organ in three dimensions (Figure 1-3). Tomographic imaging, therefore, provides greater depth resolution and delineation of the structural and functional information present.

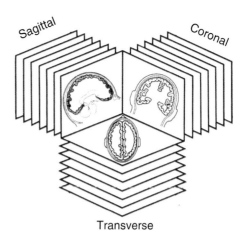

FIGURE 1-3 Diagram illustrating the three-dimensional computer-generated slice images through an organ in transverse, sagittal, and coronal planes.

In Vivo Function Studies

In vivo function studies measure the function of an organ or system on the basis of the absorption, dilution, concentration, or excretion of radioactivity after administration of a radiopharmaceutical. These studies do not require imaging, but analysis and interpretation is based on counting radioactivity emanating either directly from organs within the body or from blood or urine samples counted in vitro. Some examples of in vivo function studies are (1) the radioactive iodine uptake study to assess thyroid gland function as determined by external measurement of the percentage of a dose of radioiodine taken up by the gland over time, (2) determination of whole blood volume by measuring the dilution of a known amount of intravenously injected ^{51}Cr-labeled red blood cells to determine the red cell volume, and (3) assessment of the glomerular filtration rate (GFR) from the plasma clearance of an agent for measuring GFR, such as ^{125}I-iothalamate. An important requirement for in vivo function studies is that the radiopharmaceutical should not alter, in any way, the function of the organ system being measured.

Therapeutic Procedures

Therapeutic procedures in nuclear medicine are on the rise. These procedures are intended to be either curative or palliative and typically rely on the absorption of beta radiation to destroy diseased tissue. The classic therapeutic procedure is the use of ^{131}I-sodium iodide to treat hyperthyroidism and thyroid cancer. Because ^{131}I is a beta and gamma emitter, it can be used both diagnostically and therapeutically in thyroid disease; however, the therapeutic dosage of radioactivity administered is on average 1,000 to 20,000 times larger than the diagnostic dosage used in measuring thyroid function. Radioimmunotherapy (RIT) employing radiolabeled antibodies and peptides has achieved some success in treating tumors. Examples of RIT agents are ^{131}I-tositumomab and ^{90}Y-ibritumomab tiuxetan

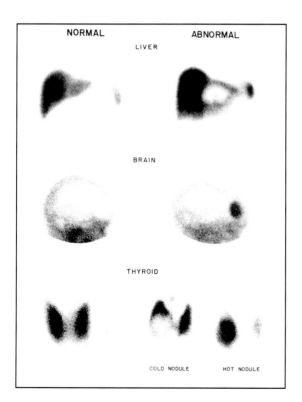

FIGURE 1-2 Typical normal and abnormal static images of three different organs obtained with a conventional planar imaging gamma camera: anterior view of the liver, lateral view of the brain, and anterior view of the thyroid gland.

for treating non-Hodgkin's lymphoma. Another significant area of nuclear medicine therapy is palliative procedures for treating pain associated with bone cancer. In this regard agents such as ^{89}Sr-chloride and ^{153}Sm-lexidronam, which are selectively localized in bone, have been used successfully. The therapeutic application of radiation to target tumor-specific tissue will likely become more successful as advances in molecular targeting are made.

PERSPECTIVE ON RADIOPHARMACEUTICAL USE

The amount of radioactivity administered to a patient in a nuclear medicine procedure is termed the *dosage*, which is measured in units of millicuries (mCi, or 10^{-3} Ci) or microcuries (µCi, or 10^{-6} Ci). The curie (Ci) is equal to 3.7×10^{10} disintegrations (atoms decaying) per second. In the International System of Units, radioactivity is measured in becquerels (Bq). One Bq is equal to 1 disintegration per second; therefore, 1 mCi = 37 MBq. The amount of radiation absorbed by the body from a radioactive substance is termed the radiation *dose* and is traditionally measured in rad (radiation absorbed dose). One rad is equal to 100 ergs of energy absorbed in 1 gram of tissue. The International Unit (IU) of absorbed dose, the gray (Gy), is equal to 1 joule of energy absorbed in 1 kg of tissue (1 Gy = 100 rad).

A goal in diagnostic nuclear medicine is to administer the optimum dosage of radioactivity to acquire the desired information with the lowest radiation dose to the patient, that is, to keep the radiation absorbed dose as low as reasonably achievable. This is accomplished, in part, by the use of short-lived radionuclides that decay quickly. Short-lived radionuclides permit larger amounts of radioactivity to be administered, which improves detection sensitivity.

The radionuclide used in most nuclear medicine studies is ^{99m}Tc. Because of its 6 hour half-life, about 90% of ^{99m}Tc's radioactivity is lost in 1 day. For this reason, it is difficult for drug manufacturers to supply ^{99m}Tc radiopharmaceuticals to nuclear medicine facilities from distant locations. Therefore, ^{99m}Tc and other short-lived radiopharmaceuticals must be prepared on site or supplied by a nearby nuclear pharmacy. The daily preparation of ^{99m}Tc agents involves radiopharmaceutical compounding, measurement of activity, and assessment of radionuclidic purity and radiochemical purity. Other radiopharmaceuticals compounded daily may require sterility and pyrogen testing. Aseptic conditions must be maintained during the preparation of radiopharmaceuticals to be administered by injection. The personnel performing these functions include radiochemists, radiopharmacists, or nuclear medicine technologists. Radiochemists and nuclear pharmacists are usually employed at large medical center hospitals and are instrumental in developing new agents and procedures. Most nuclear pharmacists practice in centralized nuclear pharmacies located in metropolitan areas, which supply radiopharmaceuticals to nearby nuclear medicine facilities. Nuclear medicine technologists sometimes have the responsibility of preparing radiopharmaceuticals in smaller rural hospitals.

Radiopharmaceuticals with longer half-lives are also used in nuclear medicine because of their desirable biochemical properties. A good example is ^{131}I. Its 8 day half-life permits commercial manufacture of ^{131}I radiopharmaceuticals, which can be shipped long distances and stored in the nuclear medicine laboratory for use when needed. ^{131}I is quite useful in nuclear medicine because it is a beta–gamma emitter and can be used for diagnostic imaging and radiation therapy applications. As the iodide ion, it has principal applications for thyroid work. In addition, iodine's diverse chemistry permits its radioisotopes to be compounded into a variety of different chemical forms.

Table 1-1 lists the general classification of chemical and physical forms of radiopharmaceuticals used in nuclear medicine. The diverse forms of these agents include simple elements such as the inert gases, inorganic ions, radiolabeled molecules, and specialized forms such as labeled particles and blood cells.

Table 1-2 lists the usual routes of administration of radiopharmaceuticals. It is noteworthy that radiopharmaceuticals

TABLE 1-1 Physicochemical Forms of Radiopharmaceuticals

Form	Example
Elemental	Xenon 133 (^{133}Xe)
	Krypton 81m (^{81m}Kr)
Simple ions	^{131}I$^-$ (iodide)
	^{99m}TcO$_4^-$ (pertechnetate)
Labeled small molecules	^{131}I-MIBG (covalently bonded)
	^{99m}Tc-DTPA (chelation compound)
Labeled macromolecules	^{125}I-human serum albumin (protein)
	^{111}In-capromab pendetide (antibody)
Labeled particles	^{99m}Tc-sulfur colloid
	^{99m}Tc-macroaggregated albumin
Labeled cells	^{51}Cr- or ^{99m}Tc-erythrocytes
	^{111}In- or ^{99m}Tc-leukocytes

TABLE 1-2 Radiopharmaceutical Routes of Administration and Dosage Forms

Route	Form
Oral	Capsules and solutions
Intravenous injection	Solutions, colloidal dispersions, suspensions
Intrathecal injection	Solutions
Inhalation	Gases and aerosols
Instillation via	Sterile solutions
Eye drops	
Urethral catheter	
Intraperitoneal catheter	
Shunts	

given by intravenous injection may be true solutions, or they may be colloidal dispersions or suspensions of radiolabeled particles, which is sometimes necessary to achieve site-specific localization of radioactivity. Thus, for example, the principal organs of the reticuloendothelial system (liver, spleen, and bone marrow) can be imaged with radiolabeled colloidal particles, the cardiac blood pool with radiolabeled red blood cells, and lung perfusion with albumin aggregates. Table 1-3 lists radiopharmaceuticals used in nuclear medicine.

Radiopharmaceuticals have other unique properties when compared with conventional therapeutic drugs. Intrinsically, they are radioactive and have an associated radiation risk. Therefore, before radiopharmaceuticals are marketed for use in humans, tissue distribution studies are performed in animals to identify the critical organs (those that receive the highest radiation absorbed dose) and to estimate the radiation dose. The magnitude of this dose estimate sets a limit on the amount of radioactivity that can be safely administered to humans in diagnostic studies.

The mass of administered radiopharmaceutical is extremely small, so chemical toxicity is not as great a concern as with traditional pharmaceuticals. In fact, the amount of radiopharmaceutical administered in a standard dosage is not enough to produce a pharmacologic effect. For example, a typical

TABLE 1-3 Radiopharmaceuticals Used in Nuclear Medicine

Radionuclide	Chemical Form and Dosage Form	Use	Typical Dosage (Adult[a])	Route[b]
Carbon C 11	Carbon monoxide	Cardiac: Blood volume measurement	60–100 mCi	Inhalation
Carbon C 11	Flumazenil injection	Brain: Benzodiazepine receptor imaging	20–30 mCi	IV
Carbon C 11	Methionine injection	Neoplastic disease evaluation in brain	10–20 mCi	IV
Carbon C 11	Raclopride injection	Brain: Dopamine D_2 receptor imaging	10–15 mCi	IV
Carbon C 11	Sodium acetate injection	Cardiac: Marker of oxidative metabolism	12–40 mCi	IV
Carbon C 14	Urea	Diagnosis of *Helicobacter pylori* infection	1 μCi	PO
Chromium Cr 51	Sodium chromate injection	Labeling red blood cells (RBCs) for measuring RBC volume, survival, and splenic sequestration	10–80 μCi	IV
Cobalt Co 57	Cyanocobalamin capsules	Diagnosis of pernicious anemia and defects of intestinal absorption	0.5 μCi	PO
Fluorine F 18	Fludeoxyglucose injection	Glucose utilization in brain, cardiac, and neoplastic disease	10–15 mCi	IV
Fluorine F 18	Fluorodopa injection	Dopamine neuronal decarboxylase activity in brain	4–6 mCi	IV
Fluorine F 18	Sodium fluoride injection	Bone imaging	10 mCi	IV
Gallium Ga 67	Gallium citrate injection	Hodgkin's disease, lymphoma Acute inflammatory lesions	8–10 mCi 5 mCi	IV IV
Indium In 111	Capromab pendetide injection	Metastatic imaging in patients with biopsy-proven prostate cancer	5 mCi	IV
Indium In 111	Indium chloride sterile solution	Radiolabeling various ^{111}In radiopharmaceuticals	Various	
Indium In 111	Indium oxine sterile solution	Labeling autologous leukocytes	500 μCi	IV
Indium In 111	Pentetate injection	Cisternography	500 μCi	Intrathecal
Indium In 111	Pentetreotide injection	Neuroendocrine tumors	3 mCi (planar) 6 mCi (SPECT[c])	IV
Indium In 111	Ibritumomab tiuxetan injection	Biodistribution imaging prior to therapeutic dosing with ^{90}Y Zevalin (Biogen Idec) in the treatment of non-Hodgkin's lymphoma	5 mCi	IV
Iodine I 123	Sodium iodide capsules and solution	Thyroid gland imaging Thyroid metastases (total body)	400–600 μCi 2 mCi	PO PO

(continued)

TABLE 1-3 Radiopharmaceuticals Used in Nuclear Medicine (Continued)

Radionuclide	Chemical Form and Dosage Form	Use	Typical Dosage (Adult[a])	Route[b]
Iodine I 123	Iobenguane injection	Pheochromocytoma, carcinoid tumors, nonsecreting paragangliomas, neuroblastoma	0.14 mCi/kg (child) 10 mCi (adult)	IV
Iodine I 125	Albumin injection	Plasma volume determination	5–10 µCi	IV
Iodine I 125	Iothalamate sodium injection	Glomerular filtration rate (GFR) determination	30 µCi	IV
Iodine I 131	Albumin injection	Blood volume/plasma volume determination	5–50 µCi	IV
Iodine I 131	Iobenguane injection	Pheochromocytoma, carcinoid tumors, nonsecreting paragangliomas, neuroblastoma	0.5 mCi/1.7m²	IV
Iodine I 131	Sodium iodide capsules and solution	Thyroid function Thyroid imaging (neck) Thyroid imaging (substernal) Thyroid metastases (total body) Hyperthyroidism Carcinoma	5–10 µCi 50–100 µCi 100 µCi 2 mCi 5–33 mCi 150–200 mCi	PO
Iodine I 131	Iodohippurate sodium injection	Recoverable renal function	200 µCi (2 kidneys) 75 µCi (1 kidney)	IV
Iodine I 131	Tositumomab	Treatment of refractory low-grade non-Hodgkin's lymphoma	Patient-specific dosing; not >75 cGy whole body	IV
Nitrogen N 13	Ammonia injection	Myocardial perfusion studies	10–20 mCi	IV
Oxygen O 15	Water injection	Cardiac perfusion	30–100 mCi	IV
Phosphorus P 32	Chromic phosphate suspension	Peritoneal and pleural effusions	10–20 mCi	Intraperitoneal or intrapleural (Not for IV use)
Phosphorus P 32	Sodium phosphate injection	Polycythemia	1–8 mCi	IV
Rubidium Rb 82	Rubidium chloride injection	Myocardial perfusion studies	30–60 mCi	IV
Samarium Sm 153	Lexidronam injection	Bone pain palliation in confirmed osteoblastic metastatic bone lesions	1.0 mCi/kg	IV
Strontium Sr 89	Strontium chloride injection	Bone pain palliation in confirmed osteoblastic metastatic bone lesions	4 mCi	IV
Technetium Tc 99m	Albumin injection	Heart blood pool imaging	20 mCi	IV
Technetium Tc 99m	Albumin aggregated injection	Perfusion lung imaging	3 mCi	IV
Technetium Tc 99m	Arcitumomab	Recurrent or metastatic colorectal carcinoma	20 mCi	IV
Technetium Tc 99m	Bicisate injection	Adjunct to CT/MRI[d] in patients with confirmed stroke	20 mCi	IV
Technetium Tc 99m	Disofenin injection	Hepatobiliary imaging	5 mCi	IV
Technetium Tc 99m	Exametazime injection	With or without methylene blue for regional cerebral perfusion in stroke Without methylene blue for leukocyte labeling	20 mCi 10 mCi	IV IV
Technetium Tc 99m	Gluceptate injection	Brain imaging Renal perfusion imaging	20 mCi 10 mCi	IV IV

TABLE 1-3 Radiopharmaceuticals Used in Nuclear Medicine (Continued)

Radionuclide	Chemical Form and Dosage Form	Use	Typical Dosage (Adult[a])	Route[b]
Technetium Tc 99m	Mebrofenin injection	Hepatobiliary imaging	5 mCi	IV
Technetium Tc 99m	Medronate injection	Bone imaging	20–30 mCi	IV
Technetium Tc 99m	Mertiatide injection	Kidney imaging	5 mCi	IV
		Renogram—renal transplant	1–3 mCi	IV
		Renogram—captopril	1–3 mCi	IV
Technetium Tc 99m	Oxidronate injection	Bone imaging	20–30 mCi	IV
Technetium Tc 99m	Pentetate injection	GFR (quantitative)	3 mCi	IV
		Renogram (diuretic)	3 mCi	IV
		Renal perfusion imaging	10 mCi	IV
Technetium Tc 99m	Pyrophosphate injection	Infarct-avid scan	15 mCi	IV
Technetium Tc 99m	Red blood cells injection	GI bleed (intermittent)	15 mCi	IV
Technetium Tc 99m	Sestamibi injection	Myocardial perfusion and function, parathyroid imaging	8–40 mCi	IV
Technetium Tc 99m	Sodium pertechnetate injection	Brain imaging	20 mCi	IV
		Thyroid imaging	10 mCi	IV
		Radionuclide ventriculogram	20 mCi	IV
		Radionuclide cystography	1 mCi	Urethral
		Dacryocystography	0.1 mCi	Eye drops
		Meckel's diverticulum	5 mCi	IV
Technetium Tc 99m	Succimer injection	Renal scan—differential renal function	5 mCi	IV
		Renal scan—cortical anatomy	5 mCi	IV
Technetium Tc 99m	Sulfur colloid injection	Liver–spleen scan	5 mCi	IV
		Lymphoscintigraphy (breast)	0.4–0.6 mCi	Interstitial
		Lymphoscintigraphy (melanoma)	0.5–0.8 mCi	Intradermal
		Gastric emptying (scrambled egg)	1 mCi	PO
		GI bleed (acute)	10 mCi	IV
		Lung aspiration	5 mCi	PO
		Gastroesophageal reflux	0.2 mCi	PO
Technetium Tc 99m	Tetrofosmin injection	Myocardial perfusion and function	8–40 mCi	IV
Thallium Tl 201	Thallous chloride injection	Myocardial perfusion imaging	3–4 mCi	IV
		Parathyroid imaging	2 mCi	IV
Xenon Xe 133	Xenon	Lung ventilation imaging	10–20 mCi	Inhalation
Yttrium Y 90	Ibritumomab tiuxetan	Treatment of refractory low-grade non-Hodgkin's lymphoma	0.3–0.4 mCi/kg	IV

[a] Except where otherwise noted.
[b] IV = intravenous; PO = oral.
[c] SPECT = single-photon emission computed tomography.
[d] CT = computed tomography; MRI = magnetic resonance imaging.

10 µCi diagnostic dose of ^{131}I-sodium iodide contains only 8×10^{-11} gram of iodine, one eighty-millionth of the normal total-body iodine stores and about one two-millionth of the daily dietary intake of iodine. It is easy to see that this amount of radioiodine should pose no threat to patients, even to those who may be allergic to iodine-containing substances. Therefore, the premarket testing required to identify acute and chronic toxic effects of traditional drugs is usually not as extensive for radiopharmaceuticals.

Because of the radioactive nature of radiopharmaceuticals, there are specific requirements for the safe use of these agents. Procedures are needed to protect patients, radiation workers, and the general public from unnecessary exposure to radiation. Radioactive material is closely controlled by state and federal agencies. Because radiopharmaceuticals are drugs, they are regulated by the U.S. Food and Drug Administration in regard to their chemical toxicity and efficacy and by the U.S. Nuclear Regulatory Commission (NRC) or NRC agreement state (state approved by NRC to license use of radioactive material) in regard to radiation safety. The use of radioactive material in human subjects requires that physicians and paramedical personnel be properly trained and experienced in the handling of such materials and be recognized in this regard by specific licensure.

HISTORICAL PERSPECTIVES IN NUCLEAR PHARMACY: PRACTICE, EDUCATION, AND SPECIALTY CERTIFICATION

Nuclear Pharmacy Practice

In 1950, John Christian[2] published an article encouraging hospital pharmacists to become informed about radioisotopes in medical practice and to take the initiative to establish facilities for handling radioactive material. He also put forth a plan for a laboratory design with a separate "hot element" room for storage and handling operations and a measurement room for low-level radioactive counting. Christian could advise and counsel with authority, because in 1946 he received the first shipment of radioactive isotopes for biochemical research from Oak Ridge National Laboratory and used them in pharmaceutical development at Purdue University School of Pharmacy and Pharmacal Sciences. Soon thereafter, in 1947, he initiated the first formal lecture and laboratory courses in the United States for teaching the basic principles of radioisotope methodology.

The roots of a pharmacist-run nuclear pharmacy (radiopharmacy) service can be traced to the University of Chicago Clinics, where a radioisotope laboratory was established by Chief Pharmacist Paul Parker in the early 1950s[3] and continued to be operated by Chief Pharmacist Peter Solyom.[4] The staff pharmacist in charge of the radioisotope laboratory where radioactive medications were procured and dispensed for patient use was Larry Summers. In those days, decay tables and a slide rule were used to make calculations. After the establishment of this laboratory, a report on radioisotopes in hospital pharmacy was published.[3] The report was the work of the Committee on Isotopes appointed by George Archambault, then president of the American Society of Hospital Pharmacists, to study the role of the hospital pharmacist in handling radioisotopes. The committee had specific charges to develop special courses in isotope handling and to assess the feasibility of an isotope section in a pharmacy department and determine its layout and design.

In 1958 William Briner[5] informed hospital pharmacists about radiopharmacy and introduced them to pertinent terminology and basic considerations of radiologic health. This was followed in 1960 by another article strongly promoting the role hospital pharmacists should have in the preparation and handling of radiopharmaceuticals for patient use, citing the U.S. Atomic Energy Commission (AEC) requirement that "byproduct material shall not be used in humans until its pharmaceutical quality and assay have been established."[6] Although the program at the University of Chicago Clinics predated the program at the National Institutes of Health (NIH), it is well recognized in the profession that Briner, at NIH, established and maintained a long-standing, active practice of nuclear pharmacy, stalwartly promoted the involvement of pharmacists in nuclear medicine, and was directly responsible for training many of those pharmacists. Early radiopharmacist colleagues who worked with Briner were Edgar Adams, Robert Chandler, and Raymond Farkas. Briner preached the value of pharmacist involvement in nuclear medicine through his professional presentations and publications for more than 40 years. For his pioneering efforts he is affectionately remembered as the "father of radiopharmacy." Briner's leadership helped to create the regulatory and practice environment within which nuclear pharmacists and nuclear medicine professionals work today.

The introduction of ^{99m}Tc-sodium pertechnetate into nuclear medicine practice changed the face of practice dramatically. Its clinical use began in 1961 at the University of Chicago, where several ^{99m}Tc-labeled radiopharmaceuticals were developed by Paul Harper and Kathryn Lathrop.[7] Shortly thereafter, the ^{99}Mo–^{99m}Tc generator became available commercially and the national growth of nuclear medicine procedures escalated. The short half-life of ^{99m}Tc demanded local preparation of radiopharmaceutical agents labeled with this nuclide and increased the demand for nuclear pharmacy services.

The shortage of pharmacists trained in radioisotope methodology stimulated the establishment of a Master of Science program in radiopharmacy at the University of Southern California School of Pharmacy in 1968, under the direction of Walter Wolf and Manuel Tubis. This program ran until 1986 and graduated more than 210 students.

The dearth of nuclear pharmacists also inspired the concept of a shared nuclear pharmacy service, the progenitor of today's centralized or commercial nuclear pharmacy practice. The idea was conceived, established, and evaluated in 1969 by Thomas Gnau in the Nuclear Medicine Division at Bowman Gray School of Medicine in Winston-Salem, North Carolina.[8,9] The intent was to reduce cost, improve staffing efficiency, and ensure a high level of quality control in the delivery of radiopharmaceuticals to several nuclear medicine facilities in a metropolitan area.

The centralized nuclear pharmacy concept proved successful, and such pharmacies developed in the early 1970s. Of note were those established at the University of Washington by David Allen, the University of Tennessee (UT) by James Cooper, the University of Nebraska Medical Center by J. William Dirksen, and the Indiana University Medical Center by Michael Kavula, and the radiopharmacy program at the University of New Mexico (UNM) directed by Richard Keesee.

William Baker, who interned with Gnau at Bowman Gray, joined Keesee and set up the UNM radiopharmacy record-keeping system following the Bowman Gray model. Soon thereafter, in 1972, Keesee established the first centralized radiopharmacy to be licensed by a state board of pharmacy and the AEC. In 1973 Baker moved to the University of Utah Medical Center to establish the Intermountain Radiopharmacy program. He was later joined there by Robert Beightol, who had trained under Cooper at UT.

The UNM program graduated several nuclear pharmacists trained in the centralized pharmacy model who themselves went on to establish private centralized radiopharmacies. Notable among these were Nuclear Pharmacy Inc. (Robert Sanchez and Richard Sakasitz), Pharmatopes (Mark Hebner and Monty Fu), and Texatopes (Nunzio Desantis and Larry Oliver). In 1975 Richard Keesee and David Hurwitz opened Pharmaco Nuclear, which eventually merged with Syncor in 1981. Pharmatopes merged with this company in 1982, followed by Nuclear Pharmacy Inc. in 1984, to become Syncor International Corporation, which is now a part of Cardinal Health. Other large commercial nuclear pharmacy companies in operation today include Tyco Healthcare/Mallinckrodt, GE Healthcare, IBA Group–Eastern Isotopes, and PETNET Solutions. While the bulk of nuclear pharmacy service throughout the United States is provided by these operations, there are a few dozen smaller-scale independent nuclear pharmacies that operate in several states. Over 400 centralized nuclear pharmacies staffed by 800 to 1000 nuclear pharmacists provide more than 80% of the radiopharmaceutical dosage forms used in nuclear medicine in the United States today.

Nuclear Pharmacy Education Programs

The initial lecture and laboratory courses created by Christian at Purdue University led eventually to the creation of the Department of Bionucleonics. The educational efforts of this group resulted in the training of many nuclear pharmacy researchers and practitioners, several of whom are recognized as nuclear pharmacy pioneers (Table 1-4). The Purdue program was in operation for many years under the apt leadership of Stanley Shaw, former head of the Division of Nuclear Pharmacy and now retired. The Purdue program was established in 1972 by Shaw and Gordon Born and has produced hundreds of nuclear pharmacists. It is now under the direction of Kara Duncan Weatherman.

In the 1950s and early 1960s, there were few radioisotope programs where pharmacists could receive training.

One program available at the time was at the Division of Radiological Health, U.S. Public Health Service, Robert A. Taft Sanitary Engineering Center, Cincinnati, Ohio; another was the course on radioisotope techniques at the Oak Ridge Institute of Nuclear Studies in Oak Ridge, Tennessee.

The tremendous growth of nuclear medicine in the 1960s demanded more nuclear pharmacists and therefore more training programs. The earliest sites (before 1970) for nuclear pharmacy education and training were at the University of Southern California, Purdue University, and the NIH Radiopharmaceutical Service. Other programs that eventually developed educational programs for pharmacists were the University of Arkansas, the University of New Mexico, the University of Utah, the University of Nebraska, the University of Pittsburgh, William Beaumont Hospital in Michigan, the University of Minnesota, the University of Wisconsin, the University of Tennessee, the Massachusetts College of Pharmacy, the University of North Carolina, the Medical University of South Carolina, the University of Indiana, the University of Toronto, the University of Cincinnati, Mercer University, Temple University, the University of Michigan, and the University of Oklahoma. These sites offered a variety of programs, including radiopharmacy residencies, short courses, semester courses, condensed (200 hour) authorized nuclear pharmacist programs, and nuclear pharmacy certificate programs.

Despite the apparent large number of institutions offering nuclear pharmacy training, only a few of these programs were turning out pharmacists with sufficient training to obtain licensure as authorized nuclear pharmacists. Thus, the supply of adequately trained nuclear pharmacists could not meet the demand of nuclear pharmacy practice. As a consequence, one nuclear pharmacy company (Syncor, now Cardinal Health) established its own university-associated authorized nuclear pharmacist training program to meet its needs.

Nuclear pharmacy companies have also used certificate programs at institutions to meet the special educational needs of their nuclear pharmacists. Certificate programs are available at Purdue University and Ohio State University. In addition,

TABLE 1-4 Nuclear Pharmacy Pioneers

David R. Allen	John E. Christian	Mark T. Hebner	Larry Oliver
William J. Baker	Clyde N. Cole	Kenneth R. Hetzel	William C. Porter
Robert W. Beightol	James F. Cooper	Dennis R. Hoogland	Richard Sakasitz
Gordon S. Born	Nunzio Desantis	David Hurwitz	Robert Sanchez
Barry M. Bowen	J. William Dirksen	Rodney D. Ice	Stanley M. Shaw
Kenneth Breslow	Raymond J. Farkas	Michael P. Kavula	Anne C. Smith
William H. Briner	Monty Fu	Tom K. Kawada	Arthur C. Soloman
Ronald J. Callahan	Thomas R. Gnau	Richard Keesee	Dennis P. Swanson
Robert P. Chandler	Robert F. Gutkowski	Richard J. Kowalsky	Walter Wolf
Henry M. Chilton	Donald R. Hamilton	Geoffrey Levine	A. Michael Zimmer

TABLE 1-5 Nuclear Pharmacy Specialty Petition Committee			
Ronald J. Callahan	James F. Cooper	Kenneth R. Hetzel	Geoffrey Levine
Robert P. Chandler	John Coupal	Michael P. Kavula	Susan G. Rowles
Henry M. Chilton	Robert Gutkowski	Tom K. Kawada	Stanley M. Shaw
William J. Christopherson	Donald R. Hamilton	Alan S. Kirschner	Arthur C. Soloman

a joint program between the University of Arkansas Medical Center (UAMS) and UNM, called Nuclear Education Online (NEO), was developed to train nuclear pharmacists via distance learning on the Internet. Students completing this program are eligible for NRC certification. A description of nuclear pharmacy education at colleges of pharmacy was published in the *American Journal of Pharmaceutical Education*.[10]

A Syllabus for Nuclear Pharmacy Training, assembled by the Educational Affairs Committee of the Section on Nuclear Pharmacy, Academy of Pharmacy Practice and Management of the American Pharmaceutical Association (APhA; now the American Pharmacists Association), was published in 1995. The syllabus was based on the NRC requirements for authorized nuclear pharmacist training: 200 didactic hours in the basic areas of radiation physics and instrumentation, radiation protection, math related to radioactivity, radiation biology, and radiopharmaceutical chemistry, and 500 hours of off-campus practical training in a nuclear pharmacy under the direction of an authorized nuclear pharmacist. The syllabus is intended as guidance for pharmacy school faculty and nuclear pharmacy preceptors who are involved in the education and training of nuclear pharmacists. Currently, the Code of Federal Regulations (10 CFR 35.55) specifies that training for an authorized nuclear pharmacist should consist of a structured educational program consisting of 700 hours of didactic and experiential training in the areas noted above.

Specialty Certification of Nuclear Pharmacists

Specialty certification in nuclear pharmacy became a reality only after nuclear pharmacists organized as a section within APhA. The process began in Chicago in August 1974 during the Nuclear Pharmacy Symposium at the APhA annual meeting. A petition submitted by these nuclear pharmacists was accepted by APhA, and the Section on Nuclear Pharmacy was officially established in 1975 as the first section within the Academy of General Practice, with James Cooper as chairman pro tem.

The section soon established an education committee, directed by Ronald Callahan, which initiated the comprehensive Task Analysis of Nuclear Pharmacy Practice to identify the types and extent of activities in which nuclear pharmacists were involved. These were organized into the following domains: procurement, compounding, quality assurance, dispensing, distribution, health and safety, and provision of information and consultation. This analysis culminated in practice standards that delineated the recognized duties and responsibilities of nuclear pharmacists. These standards were revised and reissued by APhA in 1995 as *Nuclear Pharmacy Practice Guidelines*.

The guidelines contain the original practice domains and two more: monitoring patient outcome, and research and development. Each domain identifies a list of tasks and a knowledge statement related to each task.

Shortly after the establishment of the Section on Nuclear Pharmacy, the Board of Pharmaceutical Specialties (BPS) was established in 1976 by APhA. BPS immediately recognized nuclear pharmacy as a likely candidate for specialization. Chairman Cooper assembled 16 volunteers to serve as the nuclear pharmacy specialty petition committee (Table 1-5). This committee used the previously identified nuclear pharmacy practice standards as its guide in the petition to BPS. The petition was approved by BPS in 1978, and nuclear pharmacy became the first recognized specialty within the profession of pharmacy.

After approval of the nuclear pharmacy specialty petition, BPS established a Specialty Council on Nuclear Pharmacy (composed of six nuclear pharmacists, three non–nuclear pharmacist generalists, and a test developer/consultant) and charged it with the task of developing a certification program for the new specialty. The first specialty council members are listed in Table 1-6. The standards were again used in this process, serving as the basis for determining the areas of knowledge and skills to be tested via the certification exam. The process involved item writing, test assembly, test administration, and score interpretation.[11,12] The first nuclear pharmacy specialty certification exam was administered simultaneously in Las Vegas and Atlanta on April 24, 1982. The test resulted in

TABLE 1-6 First Nuclear Pharmacy Specialty Council			
David R. Allen	Paul G. Grussing	Sam H. Kalman	Stanley M. Shaw
Ronald J. Callahan	Rodney D. Ice	Richard D. Penna	Arthur C. Soloman
James F. Cooper, Chair			

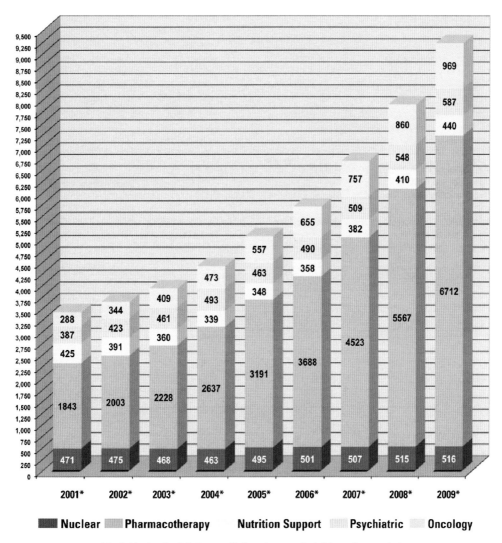

FIGURE 1-4 Numbers of pharmacist specialists holding certification by the Board of Pharmaceutical Specialties, 2001–2009, showing growth in specialization in the five recognized areas for which testing programs have been implemented. (Used with permission of the Board of Pharmaceutical Specialties.)

63 practitioners becoming board certified in nuclear pharmacy (BCNP). Today the *Nuclear Pharmacy Practice Guidelines* are used as the basis for constructing the specialty examination. At the time of this writing there are 516 BCNPs (Figure 1-4).

To ensure the competence of board-certified practitioners, BPS instituted a program of recertification every 7 years after initial certification. Recertification assures the public and the profession that certified practitioners undergo periodic evaluation. A BCNP is recertified by a three-step process: self-assessment, peer review, and formal assessment. Self-assessment involves annual review of the BCNP's nuclear pharmacy practice activities since initial certification or last recertification. Peer review involves the review of documented nuclear pharmacy practice and continuing education activities over the 7 year certification period by the Specialty Council on Nuclear Pharmacy. Formal assessment involves either achieving a passing score on a 100 item recertification exam or completing 70 hours of continuing education in a BPS-approved professional development program. Information on nuclear pharmacy certification and recertification can be obtained from the Board of Pharmaceutical Specialties, 2215 Constitution Ave. N.W., Washington, DC 20037-2985.

As James Cooper[13] remarked in his description of the nuclear pharmacy specialization process, "the unexpected benefit of the process was that it allowed scores of Section members to participate as item writers for the exam and take pride and ownership for the specialty process." The process demonstrated the powerful results that the collaborative effort of a group of dedicated professionals can have. The ultimate benefit, however, is improved pharmaceutical care of patients through the services provided by nuclear pharmacists.

REFERENCES

1. Grigg ERN. The beginnings of nuclear medicine. In: Gottschalk A, Potchen EJ, eds. *Diagnostic Nuclear Medicine.* Baltimore: Williams & Wilkins; 1976:1–13.
2. Christian JE. Radioactive isotopes in hospital pharmacy. *Bull Am Soc Hosp Pharm.* July–August 1950;7:178–83.
3. Latiolais CJ, Parker PF, Hutchinson GB, et al. Radioisotopes in hospital pharmacy. *Bull Am Soc Hosp Pharm.* 1955;12:372–9.
4. Solyom P. Pharmacy service: University of Chicago Clinics. *Am J Hosp Pharm.* 1958;15:52–7.
5. Briner WH. Certain aspects of radiological health. *Am J Hosp Pharm.* 1958;15:44–51.
6. Briner WH. Nuclear medicine has come of age. *Am J Hosp Pharm.* 1960;17:333–8.
7. Harper PV, Lathrop KA, Gottschalk A. Pharmacodynamics of some technetium-99m preparations. In: Andrews GA, Kniseley RM, Wagner HN, eds. *Radioactive Pharmaceuticals.* Oak Ridge, TN: US Atomic Energy Commission; 1966:335–58.
8. Gnau TR, Maynard CD, Finley CW. Shared radiopharmacy services: a community study. *J Nucl Med.* 1971;12:358–9.
9. Gnau TR, Maynard CD. Reducing the cost of nuclear medicine: sharing radiopharmaceuticals. *Radiology.* 1973;108:641–5.
10. Heske SM, Hladik WB 3rd, Laven DL, et al. Status of radiologic pharmacy education at colleges of pharmacy. *Am J Pharm Educ.* 1996;60:152–61.
11. Grussing PG, Allen DR, Callahan RJ, et al. Development of pharmacy's first specialty certification examination: nuclear pharmacy. *Am J Pharm Educ.* 1983;47:11–8.
12. Ponto JA. Nuclear pharmacy and the Board of Pharmaceutical Specialties (BPS). *J Pharm Pract.* 1989;2:299–301.
13. Cooper JF. The nuclear pharmacy specialization process. In: Cooper JF, ed. *25th Anniversary—Nuclear Pioneers.* Washington, DC: American Pharmaceutical Association; 2000:15–9.

CHAPTER 2
Radioactive Decay

The chemical and physical properties of a radiopharmaceutical are responsible for its localization in the body, whereas its radioactive decay properties determine its method of detection and whether it can be used for diagnostic or therapeutic applications. This chapter considers the radionuclide decay properties associated with radiopharmaceuticals.

NUCLIDES

An atom is the smallest particle of an element possessing the properties of the element. It is made up of a nucleus consisting of protons and neutrons surrounded by an electron cloud. The electrons reside in energy shells designated with the principal quantum numbers 1, 2, 3, and so forth or with the letters K, L, or M, respectively, with the K shell closest to the nucleus. Electrons fill the shells in order, with a specified number per shell. A neutral atom has the same number of electrons as protons (Figure 2-1).

A nuclide is an atom characterized by the number of protons and neutrons in its nucleus. Nuclides are designated by the following notation, where X represents the elemental symbol, Z is the number of protons, and N is the number of neutrons. The mass number A is the sum of protons and neutrons:

$$^{A}_{Z}X_{N}$$

Nuclides are classified according to their A, Z, and N values, illustrated by the following examples.

- Isotopes are nuclides with the same Z but different A and N:

$$^{1}_{1}H_{0} \qquad ^{2}_{1}H_{1} \qquad ^{3}_{1}H_{2}$$

- Isobars are nuclides with the same A but different Z and N:

$$^{64}_{28}Ni_{36} \qquad ^{64}_{30}Zn_{34}$$

- Isotones are nuclides with the same N but different Z and A:

$$^{41}_{19}K_{22} \qquad ^{42}_{20}Ca_{22}$$

- Isomers are nuclides with the same A, Z, and N but different nuclear energy states:

$$^{99}_{43}Tc_{56} \qquad ^{99m}_{43}Tc_{56}$$

The lowercase "m" in the mass number denotes the metastable state, an excited nuclear condition that occurs for a measurable period of time.

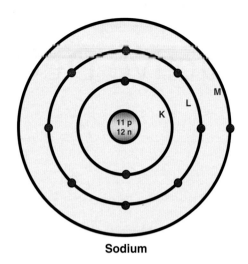

FIGURE 2-1 Bohr model of the sodium atom.

The periodic table lists 116 elements with defined atomic properties. If the isotopes of each element are included, the total number of nuclides is more than 2500, and 266 of them are stable. The remaining unstable species are called radionuclides. Figure 2-2 illustrates a portion of the chart of the nuclides in the region of the light elements.[1]

ELECTRON ENERGY LEVELS

Electrons are bound in their shells by an electron-binding energy, which is the energy that must be applied to remove an electron from its shell. The K-shell electrons have the highest binding energy because they reside closest to the positive attractive force of the nucleus. In general, inner shells fill with electrons before outer shells. If an inner-shell electron is removed, an outer-shell electron will fill the vacancy. When this occurs, energy is released from the atom equal to the difference between the binding energies of the two shells. A diagram of the tungsten (W) atom is shown in Figure 2-3. If its K-shell electron is removed and the vacancy filled by an

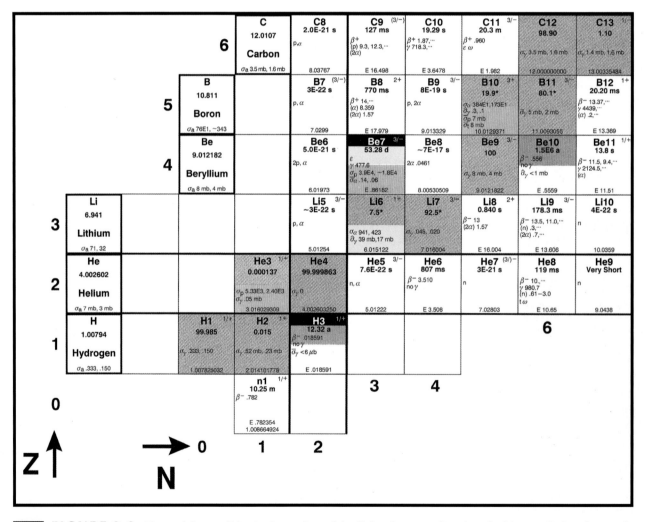

FIGURE 2-2 Chart of the nuclides in the region of the light elements. Reprinted with permission from reference 1. The complete chart is available from Lockheed Martin at http://www.chartofthenuclides.com.

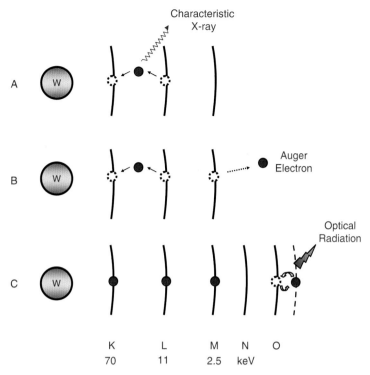

FIGURE 2-3 Perturbations of the tungsten electron shells: (A) characteristic x-ray production after ejection of a K-shell electron following fill-in by M-shell electron, (B) M-shell Auger electron production following ejection of a K-shell electron (as an alternative to x-ray production), (C) optical radiation production after an excited valence electron falls back to its ground state. Numbers are energy in kiloelectron volts, explained in text.

L-shell electron, a characteristic x-ray is produced with an energy of 59 kiloelectron volts (keV), the difference between the K- and L-shell binding energies (Figure 2-3A). The characteristic x-rays released by some radionuclides used in nuclear medicine (e.g., ^{125}I and ^{201}Tl) are the principal radiations used for their detection.

An alternative process to characteristic x-ray production is the *Auger effect* (Figure 2-3B). When this occurs, the energy made available from an electron shell transition is transferred to an outer-shell electron. The electron released from the atom is called an Auger electron. When the Auger effect occurs, no characteristic x-ray is produced. The kinetic energy of the Auger electron is equal to the total energy available minus the electron's binding energy. In the case of an L-shell to K-shell transition in tungsten, an M-shell Auger electron would have 56.5 keV of kinetic energy (59 keV minus 2.5 keV).

Auger electrons have low energy and are completely absorbed over a short range in tissue. Therefore, they are factored into radiation absorbed dose calculations as a nonpenetrating radiation from the decay of radionuclides in the body.

Optical radiation may also be produced from an atom (Figure 2-3C). This occurs when outer-shell valence electrons are excited to higher-energy suborbits. When the electrons return to their ground state, visible light is emitted. An example is the excitation of minerals with ultraviolet light, which causes them to fluoresce.

NUCLEAR ENERGY LEVELS

Neutrons and protons (collectively known as nucleons) reside in the nucleus of an atom in discrete energy levels. In a stable atom, nucleons are in their ground state. They may, however, be excited to higher-energy states during radioactive decay or by interaction with a high-speed particle. When these excited nucleons return to their ground state, energy is emitted from the nucleus as a gamma ray. The energy may be released as one discrete gamma ray or in a cascade as multiple gamma rays of different energies (Figure 2-4). The nuclear de-excitation process is analogous to the electron shell changes discussed previously, but certain differences exist. First, the electron shell de-excitation process occurs immediately after atomic excitation, whereas nuclear de-excitation may be immediate or delayed. When nuclear de-excitation is delayed, the excited nucleus is said to be in the metastable state. Also, although characteristic x-rays and gamma rays are high-energy electromagnetic radiation, they differ in that gamma rays typically have much higher energy and originate from nuclear energy changes, while characteristic

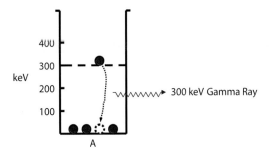

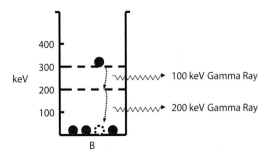

FIGURE 2-4 Nuclear de-excitation by emission of gamma rays after the return of excited nucleons to their ground state. Energy is emitted as a single gamma ray or as a cascade of multiple gamma rays.

x-rays are lower in energy and arise from electron shell energy changes.

NUCLEAR MASS AND ENERGY

In 1905, Albert Einstein proposed his famous equation relating mass to energy: $E = mc^2$, where E is energy in erg units, m is mass in gram units, and c is the velocity of light (2.998×10^{10} cm/sec). The standard atomic mass unit (AMU) is defined as one-twelfth the mass of a ^{12}C atom. The mass of 1 AMU is calculated as

$$\frac{12 \text{ grams}}{6.023 \times 10^{23} \text{ atoms}} \times \frac{1 \text{ atom}}{12 \text{ AMU}} = 1.6603 \times 10^{-24} \text{ gram/AMU}$$

The energy equivalent of this small mass can be calculated from Einstein's equation, where the erg has units of $cm^2 g \cdot sec^{-2}$:

$$E = \left(1.6603 \times 10^{-24} \text{ gram/AMU}\right)\left(2.998 \times 10^{10} \text{ cm/sec}\right)^2$$

$$E = 1.49228 \times 10^{-3} \text{ erg/AMU}$$

In the radiation sciences, the basic energy unit used is the electron volt (eV) and its multiples, the kiloelectron volt (keV) and the megaelectron volt (MeV). An electron volt is a unit of energy equal to the energy acquired by 1 electron falling through a potential difference of 1 volt (V). The energy equivalent of 1 AMU is

$$\frac{1.49228 \times 10^{-3} \text{ erg/AMU}}{1.602 \times 10^{-6} \text{ erg/MeV}} = 931.5 \text{ MeV/AMU} \quad (2\text{-}1)$$

TABLE 2-1 Mass–Energy Relationship of Atomic Particles

Particle	Mass (AMU)	Energy (MeV)
Electron	5.48597×10^{-4}	000.511
Proton	1.0072764	938.278
Neutron	1.0086649	939.571
Hydrogen atom	1.0078250	938.789

Source: www.nndc.bnl.gov/masses/mass.mas03round.

This simple relationship between mass and energy allows calculation of the energy released in a nuclear reaction or in a radioactive decay process. Some useful atomic mass–energy relationships are listed in Table 2-1.

NUCLEAR FORCES

An electrostatic repulsive force occurs between two bodies with the same electrical charge. Thus, when protons come close together they repel each other. However, in the nucleus, protons exist in close proximity to each other. This closeness is possible because of a powerful attractive force known as the nuclear force. The nuclear force is about 100 times greater than the electrostatic repulsive force. The nuclear force occurs between all nucleons in the nucleus and is responsible for holding the neutrons and protons in the nucleus despite the repulsiveness between protons.

The nuclear force is greatest between unlike and uncharged particles (i.e., the attraction between n and p [n,p] > n,n > p,p). The nuclear attractive force is appreciable only over a finite range and is strongest when nucleons are 1×10^{-13} cm apart.[2] The force becomes repulsive at distances less than 0.4×10^{-13} cm and negligible at 2.4×10^{-13} cm.

NUCLEAR BINDING ENERGY

Atoms are composed of neutrons, protons, and electrons. The creation of a ^{12}C atom requires six neutrons, six protons, and six electrons. The atomic masses of these components and their sum are as follows:

$$\begin{aligned}
n &= 6\,(1.0086654 \text{ AMU}) = 6.0519924 \text{ AMU} \\
p &= 6\,(1.0072766 \text{ AMU}) = 6.0436596 \text{ AMU} \\
e &= 6\,(0.0005486 \text{ AMU}) = 0.0032916 \text{ AMU} \\
\text{Sum of components} &= 12.0989436 \text{ AMU}
\end{aligned}$$

However, the nuclidic mass of a formed ^{12}C atom is 12.0000000 AMU, which is 0.0989436 AMU less than the sum of its individual components. A small amount of mass is lost from each component in the formation of an atom. The difference between the mass of the formed atom and the total mass of its individual parts is called the *mass defect*. During an atom's formation this amount of mass is converted into an equivalent amount of energy called the *binding energy*. Since most of an atom's mass resides in the nucleus, most of the binding energy also resides in the nucleus. This nuclear binding energy is the source of the nuclear force. A very small amount of this energy is used to hold the electrons in their orbits. If a formed nucleus is deconstructed back into its individual neutrons and protons, an amount of energy equal to the nuclear binding energy must be provided to the nucleus. Thus, the *nuclear binding energy* is defined as the energy created when a nucleus is produced from its component parts, or conversely as the energy required to separate a nucleus into its individual components. The amount of this energy for ^{12}C is calculated as follows:

$$(0.0989436 \text{ AMU}) \times (931.5 \text{ MeV/AMU}) = 92.166 \text{ MeV}$$

The average binding energy (BE_{avg}) per nucleon for ^{12}C is calculated as

$$BE_{avg} = \frac{92.166 \text{ MeV}}{12 \text{ nucleons}} = 7.68 \text{ MeV/nucleon}$$

Thus, it would take at least 7.68 MeV of energy to remove a neutron or proton from the nucleus of ^{12}C.

The average binding energy per nucleon has been determined for all stable nuclides and is plotted as a function of mass number in Figure 2-5.[2] For mass numbers greater than 11, the BE_{avg} per nucleon is between 7.4 and 8.8 MeV throughout the table of elements. Maximum values of about 8.8 MeV occur in the vicinity of $A = 60$, for iron and nickel, elements that represent a high percentage of the earth's crust. Higher BE_{avg} nuclides are more stable because it takes more energy to break apart their nuclei. The trend in nature is for elements to achieve the greatest nuclear stability. This is evident from the fission of heavy nuclei to form lighter and more stable ones and from the fusion of light nuclei (occurring in the stars) to form heavier ones with higher nuclear stability. Radioactive decay is another way for nuclides to achieve a more stable nuclear state. The BE_{avg} per nucleon of a radioactive nuclide is always smaller than the BE_{avg} per nucleon of the nuclide it decays to.

RADIOACTIVE DECAY

Radioactive decay is the spontaneous emission of radiation from an atom as a result of a transformation within its nucleus. Radiation can be particulate or electromagnetic. The principal forms of radiation are alpha particles, beta particles, gamma rays, and x-rays. An alpha particle is a helium nucleus, $^4He^{2+}$, or a helium atom stripped of its two orbital electrons. Most alpha particle emitters are heavy elements such as uranium, thorium, plutonium, and radium, but a few lighter nuclides are also alpha emitters. Beta particles are electrons emitted from the nuclei of unstable atoms. Negatively charged beta particles are known as *negatrons;* positively charged beta particles are called *positrons.* Since beta particles are electrons, they have the same rest mass as orbital electrons. The energy equivalent of a rest-mass electron is 0.511 MeV. Gamma rays are high-energy electromagnetic radiation. They have no mass or charge and are emitted from unstable nuclei secondary to particle decay.

The ratio of neutrons to protons in a nucleus determines whether a nuclide is stable or radioactive. For light nuclei, stability is achieved when the n/p ratio is 1. Above atomic number 20, however, the n/p ratio must be greater than 1 for stability because the repulsive force of additional protons becomes more prominent and extra neutrons are required to "buffer" this proton interaction. Figure 2-6 illustrates a plot of the stable nuclides,[2] which fall on the *line of nuclear stability.* Nuclides that have too many protons for

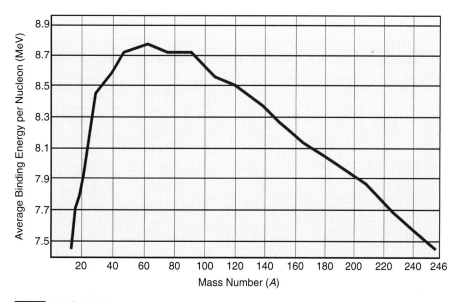

FIGURE 2-5 The average binding energy per nucleon as a function of mass number, A. The line drawn connects the odd A points. (Adapted from reference 2.)

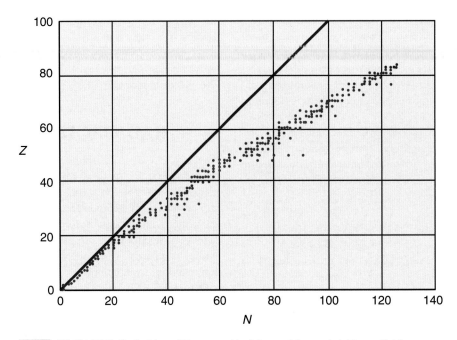

FIGURE 2-6 Plot of Z versus N of the stable nuclei. The solid line represents a neutron-to-proton ratio of unity. Note the increase in this ratio for nuclides of Z greater than 20, reflecting the need for more neutrons to buffer the repulsive force of an increased number of protons. (Adapted from reference 2.)

stability fall in the region above the line of nuclear stability, whereas nuclides with too many neutrons fall in the region below the line of nuclear stability. In either situation, nuclides in these regions are unstable and will undergo radioactive decay until a stable n/p ratio is achieved. Nuclides farther away from the line of nuclear stability, in general, have shorter half-lives, indicating their tendency toward greater instability than those closer to the line. This is illustrated in Table 2-2, which lists the isotopes of carbon. It demonstrates that proton-rich nuclides with n/p ratios less than 1 undergo positron decay, whereas neutron-rich nuclides with n/p ratios greater than 1 undergo negatron decay. In positron decay, an excess positive charge in the nucleus is reduced by ejection of a positive electron created by the transformation of a proton into a neutron and a positive electron. In negatron decay, an excess of neutrons is reduced by ejection of a negative electron created by the transformation of a neutron into a proton and a negative electron.

TABLE 2-2 Carbon Isotopes

Isotope	Neutron:Proton Ratio	Radiation	Half-life	% Isotopic Abundance[a]
^9C	0.50	β^+	0.127 sec	–
^{10}C	0.67	β^+	19.29 sec	–
^{11}C	0.83	β^+	20.3 min	–
^{12}C	1.00	None	Stable	98.90
^{13}C	1.17	None	Stable	1.10
^{14}C	1.33	β^-	5700 yr	Trace amounts
^{15}C	1.50	β^-	2.45 sec	–
^{16}C	1.67	β^-	0.75 sec	–
^{17}C	1.83	β^-	0.19 sec	–
^{18}C	2.00	β^-	0.092 sec	–
^{19}C	2.17	β^-	0.05 sec	–
^{20}C	2.33	β^-	0.01 sec	–

[a] Amounts present in Earth's crust; dash indicates isotope is not present in Earth's crust.

Negatron Decay

Neutron-rich nuclides undergo negatron or beta-minus decay. Negatron decay begins when a neutron is transformed into a proton, an electron, and an antineutrino ($\bar{\nu}$) as follows:

$$n \longrightarrow p^+ + e^- + \bar{\nu}$$

Because the electron is not part of the nucleus, it is ejected as beta radiation. (The antineutrino is discussed later.) In negatron decay the neutron number is decreased by 1 and the atomic number is increased by 1. The mass number remains unchanged and, therefore, parent and daughter nuclides are isobars. Since the atomic number changes, the parent (P) and daughter (D) nuclides are distinctly different chemical entities. A general equation for negatron decay is as follows:

$$^A_Z P_N \longrightarrow \,^A_{Z+1} D_{N-1} + E(\beta^-, \bar{\nu}, \gamma)$$

Consider, for example, the decay of ^{14}C shown in the following decay equation:

$$\underset{14.003241989 \text{ AMU}}{^{14}_{6}C_8} \longrightarrow \underset{14.0030740048 \text{ AMU}}{^{14}_{7}N_7} + E(\beta^-, \bar{\nu})$$

^{14}C is the parent nuclide and nitrogen is the daughter nuclide. We will examine this decay and account for all particles and energy. A ^{14}C atom contains 6 protons and 8 neutrons in its nucleus, and it has 6 orbital electrons. The instant after it decays it no longer is carbon; it becomes nitrogen. The atom now has 7 protons (atomic number of nitrogen), 7 neutrons, and 7 orbital electrons. The beta particle escapes from the ^{14}C nucleus as it transforms to ^{14}N. Before decay, the ^{14}C atom has a nuclidic mass of 14.003241989 AMU, but after decay the ^{14}N atom has a nuclidic mass of 14.0030740048 AMU. The difference is a mass defect of 0.00016798 AMU and is equivalent to 0.156 MeV of energy (0.00016798 AMU × 931.5 MeV/AMU). This is known as the transition energy (E). The transition energy is the total energy released when a parent nuclide decays to its daughter. This energy comes from the small amount of mass lost by the parent. Parent nuclides are always more massive than their daughter nuclides.

In negatron decay the transition energy is dissipated as the kinetic energy of the beta particle and the antineutrino. The maximum energy a ^{14}C beta particle can have is 0.156 MeV. Measurements made by nuclear scientists, however, have demonstrated that on average only about one-third of the transition energy is associated with the beta particle during negatron decay. Because this contradicts the law of conservation of mass and energy, scientists postulated that another particle must carry off the remaining two-thirds of the energy. This particle was subsequently found, and named the neutrino. (It is a general rule that a matter–antimatter pair is formed whenever energy is converted to mass. Since the negatron is a member of the matter system, we write $\bar{\nu}$ for the antineutrino. The neutrino, ν, is associated with the emission of a positron, which is antimatter, so that again a matter–antimatter pair is created. Typically, both are referred to simply as neutrinos.)

The neutrino is a chargeless particle of extremely small mass emitted from the nucleus in all beta decay processes, and it carries away the energy not used by the beta particle.[3] If the energy of each particle from thousands of ^{14}C atoms were measured and their frequency of occurrence versus energy plotted, the beta energy spectrum would be similar to the one in Figure 2-7. If a decaying ^{14}C atom emits a 0.05 MeV beta particle, the neutrino carries away 0.106 MeV. The average energy carried away by beta particles is approximately one-third of the maximum beta energy. The average beta energy varies with Z and beta energy, and ranges from about 0.25 to 0.45.[4] Table 2-3 shows the relationship of beta energies for some common beta emitters.

In negatron decay, one need not account for the electron mass lost from the nucleus of the parent nuclide because an equivalent electron mass is acquired in the electron shell of the daughter nuclide to offset it.

Decay schemes are often used to illustrate the decay process. Radiation emissions in decay schemes are indicated by diagonal arrows drawn from the parent to the daughter nuclide, either to the right or to the left. Gamma ray emissions are designated by vertically drawn arrows. By convention, when the daughter nuclide has a higher atomic number than the parent (e.g., in negatron decay), the diagonal arrow depicting transition from parent to daughter is drawn to the right. When the daughter nuclide is of lower atomic number, as occurs in alpha particle decay, positron decay, or electron capture (EC) decay, the arrow is drawn to the left.

The decay scheme for ^{14}C is shown in Figure 2-8. ^{14}C is known as a pure beta emitter because all the transition energy is distributed between the beta particle and the antineutrino. The nucleus does not receive any of this energy and is therefore not raised to an excited state that would lead to gamma emission. The lack of gamma emission means that ^{14}C cannot be used for diagnostic applications in nuclear medicine. Other examples of pure beta emitters are ^{3}H, ^{32}P, ^{35}S, and ^{90}Y. ^{32}P and ^{90}Y have therapy applications in nuclear medicine.

Some radionuclides are beta and gamma emitters. Examples of those that have been used in nuclear medicine are ^{131}I, ^{99}Mo, ^{133}Xe, ^{153}Sm, ^{198}Au, and ^{203}Hg. The latter two are no longer used routinely; however, the decay of ^{203}Hg will be illustrated because of its simplicity. The decay equation for ^{203}Hg is as follows:

$$\underset{202.9728725 \text{ AMU}}{^{203}_{80}Hg_{123}} \longrightarrow \underset{202.9723442 \text{ AMU}}{^{203}_{81}Tl_{122}} + E(\beta^-, \bar{\nu}, \gamma)$$

The mass defect for this decay is 0.0005283 AMU, equivalent to a transition energy of 0.4921 MeV. The decay scheme is shown in Figure 2-9. ^{203}Hg does not decay directly to the ground state of ^{203}Tl but to its excited state of 0.279 MeV. That is, for every atom of ^{203}Hg that decays, 0.213 MeV of the transition energy is distributed between the beta particle and antineutrino, and 0.279 MeV is released as a gamma ray when the excited ^{203}Tl nucleus de-excites to its ground state. The gamma ray is emitted simultaneously with the beta particle.

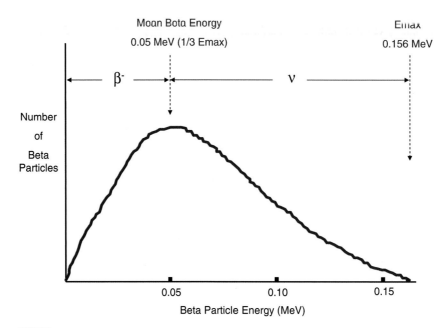

FIGURE 2-7 Beta energy spectrum for ^{14}C decay. The vertical coordinate gives the relative number of beta particles that are emitted at each energy on the horizontal coordinate up to the maximum beta energy of 0.156 MeV. The mean energy of all beta particles is about one-third of the maximum decay energy, with the remaining energy given to the neutrino (ν).

TABLE 2-3 Beta-Emitter Energy Profile

Radionuclide	Atomic Number	Beta Energy (MeV)		
		Maximum E (E max)	Average E (E avg)	E avg/E max
^{3}H	1	0.018	0.006	0.33
^{14}C	6	0.156	0.049	0.31
^{32}P	15	1.71	0.695	0.41
^{89}Sr	38	1.495	0.585	0.39
^{90}Y	39	2.28	0.934	0.41
^{131}I	53	0.606	0.182	0.30
^{137}Cs	55	0.514	0.187	0.36

Source: www.nndc.bnl.gov/chart/reCenter.jsp?z=39&n=51.

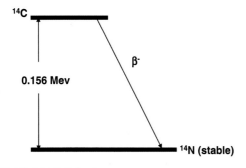

FIGURE 2-8 Simplified decay scheme for ^{14}C.

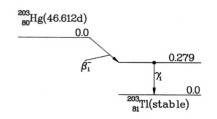

FIGURE 2-9 Decay scheme for ^{203}Hg. (Reprinted with permission of the Society of Nuclear Medicine from reference 5.)

Positron Decay

Positron decay occurs when the n/p ratio is too low for stability. These proton-rich nuclides decay by transforming a proton into a neutron and a positron–neutrino pair ejected from the nucleus. The nuclear transformation is

$$p^+ \longrightarrow n + e^+ + \nu$$

In positron decay the neutron number is increased by 1 and the atomic number is decreased by 1. The mass number remains unchanged and, therefore, parent and daughter nuclides are isobars. Since the atomic number changes, the parent and daughter nuclides are distinctly different chemical entities. A general equation for positron decay is as follows:

$$^A_Z P_N \longrightarrow\ ^A_{Z-1} D_{N+1} + E(\beta^+, \nu, \gamma)$$

Because positron emission decreases the atomic number by one unit, one orbital electron must be lost as well from the daughter nuclide to maintain neutrality. The atomic mass of the daughter nuclide will thus be at least two electron masses less than the parent nuclide. The loss of two electron masses in positron decay requires that at least 1.022 MeV (2 × 0.511 MeV/electron) of transition energy be available for the process to occur. ^{18}F-fluorine is a positron emitter whose decay equation is

$$\underset{18.0009380 \text{ AMU}}{^{18}_9 F_9} \longrightarrow \underset{17.9991610 \text{ AMU}}{^{18}_8 O_{10}} + E(\beta^+, \nu, \gamma)$$

The mass defect for this decay is 0.001777 AMU, equivalent to a transition energy of 1.65528 MeV. Two electron masses are lost in positron decay; one from the nucleus as the positron and one from the electron orbitals because the daughter nuclide has one less proton in its nucleus. Thus, the energy

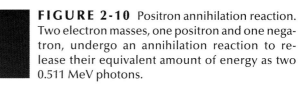

FIGURE 2-10 Positron annihilation reaction. Two electron masses, one positron and one negatron, undergo an annihilation reaction to release their equivalent amount of energy as two 0.511 MeV photons.

equivalent to two electron masses (1.022 MeV) must be subtracted from the transition energy to determine the kinetic energy of the positron and the neutrino. For ^{18}F this energy is 1.65528 MeV minus 1.022 MeV, or 0.63328 MeV. The positron can have all of this energy, but on average only about one-third is dissipated as kinetic energy of the positron and two-thirds as neutrino energy. The mean positron energy of ^{18}F is 0.250 MeV.

Positrons are considered antimatter, existing only for very short periods of time. After being ejected from the nucleus, the positron traverses a distance of a few millimeters in tissue in about 1 microsecond, after which it has lost most of its energy and will combine with a negative electron. The two electron masses annihilate into two 0.511 MeV photons (annihilation radiation), which are emitted in opposite directions (Figure 2-10). Positron emitters always produce 0.511 MeV photons. Several positron emitters have potential usefulness in nuclear medicine, including ^{11}C, ^{13}N, ^{15}O, ^{18}F, ^{68}Ga, and ^{82}Rb. (Figures 2-11 and 2-12). Positron-emitting radiopharmaceuticals are discussed more thoroughly in Chapter 12. A positron emitter may also decay by EC to excited daughter states, which will emit gamma rays. ^{68}Ga and ^{82}Rb are examples of dual-process decay.

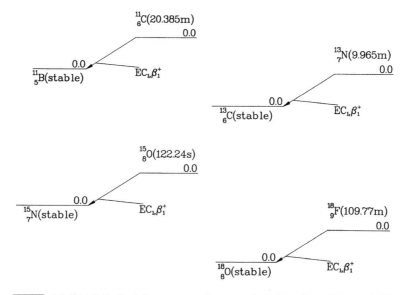

FIGURE 2-11 Decay schemes for ^{11}C, ^{13}N, ^{15}O, and ^{18}F. (Reprinted with permission of the Society of Nuclear Medicine from reference 5.)

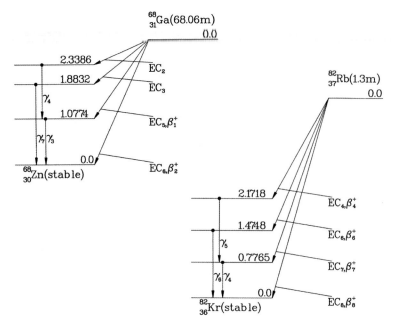

FIGURE 2-12 Decay schemes for ^{68}Ga and ^{82}Rb. (Reprinted with permission of the Society of Nuclear Medicine from reference 5.)

Electron Capture Decay

EC is the second way in which proton-rich nuclides decrease an excess positive nuclear charge. The change in nuclear composition is similar to that occurring in positron decay, but the mechanism is different. Therefore, EC and positron decay are competing processes. In fact, both processes of decay can occur in radionuclides with transition energies greater than 1.022 MeV. If a proton-rich nuclide does not have at least 1.022 MeV of transition energy, a positron cannot be created and EC decay will occur.

In EC decay, an orbital electron, usually the K-shell electron, is captured by the nucleus and combines with a proton to produce a neutron and a neutrino. This reduces the positive nuclear charge by one unit. The nuclear transformation that occurs is

$$p^+ + e^- \longrightarrow n + \nu$$

The general equation for EC decay is

$$^A_Z P_N \longrightarrow\ ^A_{Z-1} D_{N+1} + E(\nu, \gamma)$$

For most EC decay processes a K-shell electron is captured unless the transition energy is less than the K-shell binding energy, in which case L-shell capture occurs.

The neutrino carries off all the transition energy released in the EC decay process unless an excited daughter nuclide is produced, in which case the energy is shared between the neutrino and the gamma ray emitted by the daughter.

Several radionuclides used in nuclear medicine decay by EC: ^{51}Cr, ^{57}Co, ^{67}Ga, ^{111}In, ^{123}I, ^{125}I, and ^{201}Tl. EC is a desirable decay mode for diagnostic radiopharmaceuticals because no particulate radiation is produced, lowering the patient's radiation absorbed dose. The decay equation for ^{51}Cr is

$$\underset{50.9447674\ \text{AMU}}{^{51}_{24}\text{Cr}_{27}} \longrightarrow \underset{50.9439595\ \text{AMU}}{^{51}_{23}\text{V}_{28}} + E(\nu, \gamma)$$

The mass defect for this transition is 0.000808 AMU and is equivalent to a transition energy of 0.753 MeV. The decay scheme is shown in Figure 2-13, where 90% of ^{51}Cr atoms decay directly to the ground state of ^{51}V via the EC$_2$ route by emitting a 0.753 MeV neutrino. The remaining 10% of decays occur by EC$_1$ to an excited state of ^{51}V. In this route the neutrino carries only 0.433 MeV, and a gamma ray of 0.320 MeV is emitted from the excited daughter. Secondary radiations, such as characteristic x-rays and Auger electrons, are also produced during EC decay.

Each of the decay processes previously discussed is called an isobaric transition because in every case of negatron, positron, or EC decay the parent and daughter nuclides have

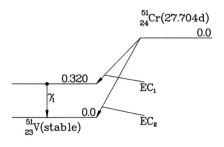

FIGURE 2-13 Decay scheme for ^{51}Cr. (Reprinted with permission of the Society of Nuclear Medicine from reference 5.)

the same mass number. Only the numbers of protons and neutrons change.

Isomeric Transition and Metastable States

When radioactive decay occurs, the available transition energy is released in a variety of ways. Some of it is released as kinetic energy of an emitted particle and some of it may remain with the daughter nucleus, which is left in an excited state. Release of this excess energy by emission of a gamma ray is known as isomeric transition. In this type of decay process there is no change in the atomic number or neutron number of the nucleus, only a change in nuclear energy state. The residual nuclear energy can be released promptly or over an extended period of time. Prompt de-excitation occurs when gamma rays are emitted immediately after the initial decay event, usually within 10^{-13} seconds. Many of the radionuclides used in nuclear medicine are prompt gamma emitters. Some examples are ^{123}I, ^{131}I, ^{51}Cr, ^{133}Xe, ^{111}In, and ^{67}Ga. Delayed de-excitation occurs when the excited nucleus persists for a measurable period of time, with a half-life in the range of 10^{-9} seconds to several months. These nuclei are called metastable isomers and are designated by writing a lowercase "m" after the mass number.

A useful radionuclide in nuclear medicine that decays to a metastable daughter is ^{137}Cs (Figure 2-14). The decay equation for ^{137}Cs is as follows:

$$^{137}_{55}\text{Cs}_{82} \longrightarrow \,^{137}_{56}\text{Ba}_{81} + E(\beta^-, \bar{\nu}, \gamma)$$
$$\text{136.9070895 AMU} \qquad \text{136.9058274 AMU}$$

The mass defect for this decay is 0.001262 AMU, equivalent to a transition energy of 1.176 MeV. ^{137}Cs decays 94.7% by the β_1^- route (β_{max} 0.514 MeV) to the metastable isomer ^{137m}Ba and 5.3% by the β_2^- route (β_{max} 1.176 MeV) directly to the stable daughter ^{137}Ba. The metastable isomer ^{137m}Ba de-excites by isomeric transition to the stable isomer with a half-life of 2.55 minutes, emitting a 0.662 MeV gamma ray. Of these gamma rays, 9.57% undergo electron conversion (discussed below): 7.79% in the K shell, 1.4% in the L shell, 0.3% in the M shell, 0.0646% in the N shell, and 0.00965% in the O shell. Loss of these photons to conversion electrons lowers the photon abundance of ^{137}Cs to 85.1% (94.7% minus 9.57%). When the ^{137}Ba K-shell vacancy is filled in by an outer shell electron, a 32 keV x-ray is released. This x-ray displays a prominent peak in the pulse-height energy spectrum of ^{137}Cs (see Figure 5-12 in Chapter 5). ^{137}Cs is a long-lived source useful for calibrating scintillation counters and dose calibrators.

Some metastable isomers have a long enough half-life that they can be chemically separated from the parent radionuclide. An example is ^{99m}Tc, which is produced by the decay of ^{99}Mo in the ^{99m}Tc generator.

$$^{99}_{42}\text{Mo}_{57} \longrightarrow \,^{99}_{43}\text{Tc}_{56} + E(\beta^-, \bar{\nu}, \gamma)$$
$$\text{98.9077119 AMU} \qquad \text{98.9062547 AMU}$$

A simplified decay scheme is shown in Figure 2-15. The mass defect for this decay is 0.0014572 AMU, equivalent to a transition energy of 1.357 MeV. About 14% of the ^{99}Mo atoms decay to the ground state of ^{99}Tc and 86% decay to the metastable isomer ^{99m}Tc, which is 0.140 MeV above ground state. ^{99m}Tc undergoes de-excitation to its ^{99}Tc isomer, emitting a monoenergetic gamma ray of 0.140 MeV. About 12% conversion electrons are produced in this transition, leaving a photon abundance of 88%.

Internal Conversion

When isomeric transition occurs, the energy released from an excited nucleus may be emitted as a gamma ray or transferred to an inner-shell electron. According to quantum mechanics, some of the orbital electrons, particularly the K-shell electrons, spend an appreciable amount of time near or actually within the nucleus. If an electron absorbs the nuclear energy that is available, it is ejected from the atom, with a kinetic energy equal to the difference between the energy available in the nucleus and the binding energy of the electron. Such an electron is called a *conversion electron,* and the process is referred to as *internal conversion.* Nuclides in the isomeric state may therefore emit either gamma rays or conversion electrons during their de-excitation. This is illustrated in the modified decay scheme for ^{203}Hg (Figure 2-16). ^{203}Hg decays 100% by beta minus decay to an excited state of ^{203}Tl, which has an energy of 0.279 MeV. The excited ^{203}Tl nucleus immediately emits 0.279 MeV gamma rays and internal conversion electrons. Internal conversion occurs

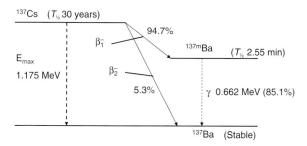

FIGURE 2-14 Simplified decay scheme for ^{137}Cs.

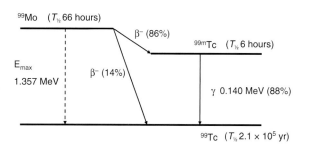

FIGURE 2-15 Simplified decay scheme for ^{99}Mo.

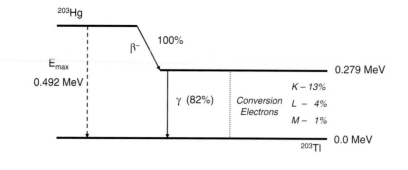

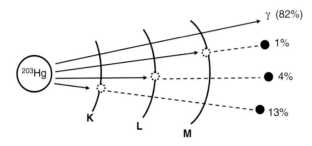

FIGURE 2-16 Modified decay scheme for ^{203}Hg illustrating the percentage of internal conversion in the K, L, and M shells.

13% in the K shell, 4% in the L shell, and 1% in the M shell. Gamma rays of 0.279 MeV are emitted in 82% of decays. The number of detectable photons emitted per 100 disintegrations of a radionuclide is known as the photon intensity or abundance. For ^{203}Hg it is 82%. High photon abundance is desirable for imaging and counting procedures.

Alpha Particle Decay

In alpha particle decay a helium nucleus is lost from the nucleus during decay of the parent nuclide, according to the general decay equation:

$$^{A}_{Z}P_{N} \longrightarrow ^{A-4}_{Z-2}D_{N-2} + ^{4}_{2}He_{2} + E(\gamma)$$

Gamma rays may or may not be emitted during alpha particle decay. Most alpha emitters have massive nuclei with Z greater than 83, although a few light nuclei are alpha emitters. Although alpha radiation has no current clinical usefulness, some alpha-emitting radionuclides have merit, and some may be used in the future for therapeutic application. Soon after radium's discovery in 1898 by Marie and Pierre Curie, ^{226}Ra and its radioactive daughter, ^{222}Rn, were used in medicine. The most frequent use was radiation therapy with sources sealed in glass or platinum seeds that were implanted in or near cancerous tissue for treatment and removed at the end of treatment. Currently, ^{192}Ir and ^{125}I are used for such purposes. ^{103}Pd seeds have been successfully used for treating prostate cancer.

Alpha emitters have no application in diagnostic nuclear medicine because alpha particles produce dense ionization within tissue, accompanied by severe radiation damage. However, the decay of ^{226}Ra is discussed here because of historical interest. The decay equation is

$$^{226}_{88}Ra_{138} \longrightarrow ^{222}_{86}Rn_{136} + ^{4}_{2}He_{2} + E(\gamma)$$
$$226.0254098 \text{ AMU} \qquad 222.0175777 \text{ AMU} \quad 4.002603 \text{ AMU}$$

Alpha decay usually occurs in heavy nuclei where four nucleons (two protons plus two neutrons) can achieve an energy that exceeds the nuclear binding energy to escape from the nucleus. In the decay equation for ^{226}Ra the alpha particle is shown as a neutral helium atom with its orbital electrons, but alpha particles do not assume this state until they have almost exhausted their kinetic energy. The sum of the ^{222}Rn and ^{4}He masses is 0.005229 AMU less than the mass of the ^{226}Ra atom. This mass defect is equivalent to a transition energy of 4.87 MeV.

In contrast to beta emitters, where beta particles are emitted in a spectrum of different energies, alpha particles are emitted with discrete energies. A simplified decay scheme depicting the principal decay routes for ^{226}Ra is shown in Figure 2-17. A branching decay occurs, with 94.5% of the atoms decaying directly to ^{222}Rn and 5.6% to the excited state of ^{222}Rn, which subsequently de-excites by emission of a 0.186 MeV gamma ray. The difference between the alpha-1 energy of 4.78 MeV and the transition energy of 4.87 MeV is 0.09 MeV. This difference is the *recoil energy*, which is given to the nucleus as the massive alpha particle is ejected. This occurs with all alpha emitters and is required for the conservation of energy and momentum. Recoil energy is usually on the order of 0.1 MeV.

Table 2-4 lists some radionuclides used in nuclear medicine and their properties.

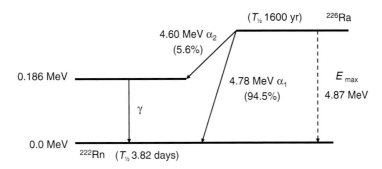

FIGURE 2-17 Simplified decay scheme for ^{226}Ra.

RADIOACTIVITY

Up to this point we have discussed some basic properties of all atoms and considered the ways in which unstable atoms transform into stable nuclei. Soon after the discovery of radioactivity by Henri Becquerel in 1896, it was observed that some elements lost their radioactive properties in a consistent fashion that varied from one element to another. In 1900, before atomic structure was understood, Ernest Rutherford used the terms *decay* and *disintegrate* to describe this process. It is now known that radioactivity involves the transformation of a parent nucleus into a daughter nucleus by radioactive decay.

Activity is the rate of disintegration or decay of a radioactive source. For a given number of radioactive atoms, N, a change in the number of atoms, dN, over an infinitely small period of time, dt, is proportional to N and a proportionality constant, λ, which is the fraction of radioactive atoms that decay per unit of time. This is described by the *radioactive decay law* as

$$\frac{-dN}{dt} = \lambda N \quad (2\text{-}2)$$

The negative sign denotes a decreasing number of atoms with time. Integration of this first-order differential equation yields the following relationship:

$$N = N_0 e^{-\lambda t} \quad (2\text{-}3)$$

where:

N_0 = original number of atoms at $t = 0$
N = number of atoms remaining after decay time $t = t$
$N_0 - N$ = number of atoms that decayed in time $t = t$
λ = decay constant in reciprocal time t^{-1}
t = time of decay

The activity, A, is the number of disintegrations (decays) or transformations occurring per unit of time, $-dN/dt$, and we may write

$$A = \lambda N \quad (2\text{-}4)$$

where:

A = number of disintegrations per unit of time
N = number of radioactive atoms
λ = decay constant

The *decay constant* describes the fraction of atoms that decay per unit of time. For example, a decay constant of 0.01 sec^{-1} means that approximately 1% of the atoms decay per second. A decay constant of 0.15 sec^{-1} means approximately 15% decay per second. This interpretation is only approximate, because radioactive decay is a logarithmic function. The decay constant is peculiar to the radionuclide in question. No two radionuclides are identical.

Since the activity of a radioactive substance is directly proportional to the number of radioactive atoms present, Equation 2-4 can be used to calculate the mass of a radioactive material from its activity. An example of this calculation is given in the section on Radioactivity Calculations. Additional details of the mass/activity relationship are discussed in Chapter 9 under the topic Specific Activity.

In working with radioactive material it is convenient to know the amount of activity present in a sample at various periods of time. Because activity is directly proportional to the number of atoms, we can substitute the expression A/λ for N in Equation 2-3 and arrive at an exponential expression for radioactive decay in terms of activity:

$$A = A_0 e^{-\lambda t} \quad (2\text{-}5)$$

where A_0 is the original activity and A is the activity remaining after time of decay, t.

Units of Activity

Radioactivity can be expressed in three ways: (1) as nuclear transformations per second, typically referred to as decays or disintegrations per second (dps); (2) as curies, millicuries, microcuries, or nanocuries; and (3) as becquerels. The curie was originally defined as the number of disintegrations per second occurring in 1 gram of ^{226}Ra. In the early days of radiochemistry the decay rate of 1 gram of ^{226}Ra was subject to slight variations because it was affected by the purity of a radium sample. Because 1 gram of "pure" radium had a disintegration

TABLE 2-4 Radionuclides in Nuclear Medicine

Nuclide	Decay Mode[a]	Half-life[b]	β$_{max}$ in MeV (% Intensity)[b,c]	β$_{mean}$ (MeV)[b]	Photon MeV (γ and x-rays) (% Intensity)[b]	Half-Value Layer (mm Pb)	Gamma Ray Dose Constant (R/mCi-hr/cm)[e]
^{11}C	β+	20.3 min	0.960 (100)	0.386	0.511 (200)	4.0	5.91*
^{13}N	β+	9.97 min	1.198 (100)	0.492	0.511 (200)	4.0	5.91*
^{14}C	β-	5700 yr	0.156 (100)	0.049	None	None	None
^{15}O	β+	122 sec	1.732 (100)	0.735	0.511 (200)	4.0	5.9*
^{18}F	β+	109.8 min	0.633 (97)	0.250	0.511 (193)	4.0	5.73*
^{32}P	β-	14.26 days	1.710 (100)	0.695	None	None	None
^{51}Cr	EC	27.7 days	None	None	0.320 (10)	1.7	0.18
^{57}Co	EC	271.7 days	None	None	0.122 (86) 0.136 (11)	0.2	1.0
^{58}Co	EC, β+	70.9 days	0.475 (15)	0.201	0.811 (99)	9.0	5.5
^{64}Cu	EC, β+, β-	12.7 hr	β+ 0.653 (17.6) β- 0.579 (38.5)	β+ 0.278 β- 0.191	0.511 (35.2)	0.4	1.15
^{67}Cu	β-	61.8 hr	0.168 (1) 0.377 (57) 0.468 (22) 0.562 (20)	0.141	0.091 (7) 0.093 (16) 0.185 (49)	7.0	0.52*
^{67}Ga	EC	3.26 days	None	None	0.093 (39) 0.185 (21) 0.300 (17) 0.394 (5)	0.66	0.8
^{68}Ga	β+	67.7 min	0.822 (1) 1.899 (88)	0.830	0.511 (178)	4.0	5.37*
^{82}Rb	β+	76.4 sec	2.60 (13) 3.38 (82)	1.48	0.511 (191) 0.777 (15)	7.0	6.1
^{81m}Kr	IT	13.1 sec	None	None	0.190 (68)	0.019	1.6
^{89}Sr	β-	50.5 days	1.4926 (100)	0.585	None	None	None
^{90}Y	β-	64 hr	2.28 (100)	0.934	None	None	None
^{99}Mo	β-	65.94 hr	0.437 (16) 1.215 (82)	0.389	0.740 (12) 0.778 (4)	6.5	0.18
^{99m}Tc	IT	6.02 hr	None	None	0.1405 (89)	0.17	0.78
^{99}Tc	β-	2.11E+5 yr	0.294 (100)	0.085	None	None	None

Radionuclide	Decay mode[a]	Half-life	β$_{max}$[c] (MeV)	γ energies (MeV)		
^{111}In	EC	2.8 days	None	0.023 (69), x-rays 0.026 (13), x-rays 0.171 (91) 0.245 (94)	0.23	3.21
^{123}I	EC	13.22 hr	None	0.027 (70), x-rays 0.030–0.032 (15), x-rays 0.159 (83)	0.05	1.6
^{124}I	EC, β$^+$	4.18 days	1.534 (12) 2.137 (11)	0.027 (47), x-rays 0.511 (46) 0.603 (63) 0.723 (10) 1.691 (11)	8.0[d]	7.6[d]
^{125}I	EC	59.4 days	None	0.027 (114), x-rays 0.031–0.032 (24), x-rays 0.035 (7)	0.017	1.43
^{131}I	β$^-$	8.02 days	0.248 (2) 0.339 (7) 0.606 (90)	0.030 (5), x-rays 0.080 (3) 0.364 (82) 0.637 (7)	2.4	2.27
^{133}Xe	β$^-$	5.24 days	0.346	0.031–0.036 (49), x-rays 0.081 (38)	0.035	0.51
^{137}Cs	β$^-$	30.08 yr	0.514 (95) 1.175 (5)	0.032 (6), x-rays 0.662 (85)	6.0	3.32
^{153}Sm	β$^-$	46.5 hr	0.635 (31) 0.704 (49) 0.808 (18)	0.041 (48), x-rays 0.047 (9), x-rays 0.103 (29)	0.1	0.46
^{186}Re	β$^-$	3.72 days	0.932 (22) 1.070 (71)	0.137 (9)	2.5	0.2
^{201}Tl	EC	3.04 days	None	0.069–0.082 (75), x-rays 0.135 (2.6) 0.167 (10)	0.006	4.7

[a] EC = electron capture, IT = isomeric transition.
[b] Values obtained from National Nuclear Data Center, Brookhaven National Laboratory (www.nndc.bnl.gov/chart/reCenter.jsp?z=39&n=51).
[c] β$_{max}$ is the endpoint energy for the decay.
[d] Source: http://hpschapters.org/northcarolina/NSDS/124IPDF.pdf.
[e] R = roentgen. Values with asterisk are from reference 3, pp. 737–41; except where indicated, other values are from package inserts.

rate close to 3.7×10^{10} dps, this value was officially adopted in 1950. Currently, the International System of Units has adopted the becquerel (Bq) as the official SI unit of radioactivity. One Bq is defined as one nuclear transformation or disintegration per second. Therefore, the following expressions are considered to be equivalent:

1 becquerel (Bq) = 1 dps
1 curie (Ci) = 3.7×10^{10} dps (Bq) or 37 gigabecquerels (GBq)
1 millicurie (mCi) = 3.7×10^{7} dps (Bq) or 37 megabecquerels (MBq)
1 microcurie (μCi) = 3.7×10^{4} dps (Bq) or 37 kilobecquerels (KBq)
1 nanocurie (nCi) = 37 dps (Bq) or 37 Bq

By definition we also have the following equivalent expression:

$$1\,\text{Bq} = 2.7 \times 10^{-11}\,\text{Ci}$$

Although the becquerel is the official unit of radioactivity, the traditional curie units are routinely used in practice. By using the above expressions, conversions can be made readily.

Half-life

The spontaneous or random nature of radioactive decay makes it impossible to predict when a single radioactive atom will disintegrate. However, given a large number of radioactive atoms, some of them will decay immediately, some at intermediate times, and some very late. On average, a certain fraction of the total number of atoms will decay within a definite time period. In 1902 Rutherford noted in his measurements of ^{234}Th that half of any quantity was gone in 24 days. He coined the term *half-life*, which is the time for any quantity of radionuclide to decrease to half its original quantity. Mathematically it is expressed as

$$T_{1/2} = \frac{0.693}{\lambda} \quad (2\text{-}6)$$

Thus, half-life and the decay constant are inversely proportional. For the derivation of this expression, assume that the time of decay (T) in the expression $A = A_0 e^{-\lambda T}$ is that of the half-life or $T_{1/2}$ so that A is one-half A_0; thus, $A = A_0/2$ and

$$\frac{A_0}{2} = A_0 e^{-\lambda T_{1/2}}$$

$$\frac{1}{2} = e^{-\lambda T_{1/2}}$$

$$2 = e^{+\lambda T_{1/2}}$$

$$\ln 2 = \lambda T_{1/2}$$

$$\frac{0.693}{\lambda} = T_{1/2} \quad (2\text{-}6)$$

The half-life of a radionuclide can be determined experimentally by measuring the activity of a sample over time, assuming that the half-life is reasonably short. Radionuclides with very short or very long half-lives require special techniques. Figure 2-18 shows plots of activity versus time on linear and log-linear coordinates. Because radioactive decay is a first-order rate process, the log-linear plot is a straight line from which the half-life is easily determined.

Radioactivity Calculations

Sample calculations 1A and 1B illustrate the random nature of radioactive decay.

Example 1A. The half-lives of ^{99m}Tc and ^{123}I are 6.01 hours and 13.1 hours, respectively. Which of these two radionuclides has the higher probability for radioactive decay?

The probability for decay is related to the nuclide's decay constant, λ, which is defined as the fraction of radioactivity (and atoms) that decays per unit of time. Thus, the decay constant of ^{99m}Tc is $0.693/6.01\,\text{hr} = 0.1153\,\text{hr}^{-1}$, or about 11.5% per hour, and the decay constant for ^{123}I is $0.693/13.1\,\text{hr} = 0.0529\,\text{hr}^{-1}$, or about 5.29% per hour. Therefore, given 100 radioactive atoms of each nuclide, ^{99m}Tc has the probability of decaying about 11 atoms in 1 hour, whereas ^{123}I decays only about 5 atoms in 1 hour.

Example 1B. How many atoms of each radionuclide are in 1 mCi (37 MBq) of ^{99m}Tc and ^{123}I?

The relationship between radioactive atoms and radioactivity is expressed in the formula $A = \lambda N$. Rearranging, we have the following:

$$N = \frac{A}{\lambda} = \frac{A \cdot T_{1/2}}{0.693} = A \cdot T_{1/2} \cdot 1.443$$

$$N_{^{99m}\text{Tc}} = (1\,\text{mCi})(3.7 \times 10^{7}\,\text{dps/mCi})(6.01\,\text{hr} \times 3600\,\text{sec/hr})(1.443)$$

$$= 1.16 \times 10^{12}\,\text{d(atoms)}$$

$$N_{^{123}\text{I}} = (1\,\text{mCi})(3.7 \times 10^{7}\,\text{dps/mCi})(13.1\,\text{hr} \times 3600\,\text{sec/hr})(1.443)$$

$$= 2.52 \times 10^{12}\,\text{d(atoms)}$$

We can see from these relationships that the number of radioactive atoms is directly proportional to the radionuclide's

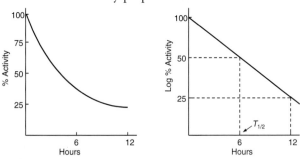

FIGURE 2-18 Half-life determination illustrated by a plot of the decay of ^{99m}Tc over time on linear coordinates (left) and log-linear coordinates (right) that yields a straight line characteristic of a first-order rate process.

half-life and inversely proportional to the decay constant. Because 1 mCi (37 MBq) of activity is equal to a defined number of atoms decaying per second, a slower-decaying nuclide must have proportionally more atoms present to produce the same number of atoms decaying per unit of time as the faster-decaying nuclide. In this case the ratio of atoms (and half-lives) of ^{123}I to ^{99m}Tc is 2.18, meaning that ^{123}I must have slightly more than twice the number of atoms as ^{99m}Tc to have the same amount of activity.

Mean Life

The expression $N = (A)(1/\lambda)$ describes the total number of atoms in a radioactive source. The term $1/\lambda$ is known as the *mean life* of a radioactive source. The relationship between the mean life and the half-life of a radioactive source is expressed as $1/\lambda = T_{1/2}/0.693 = (T_{1/2})(1.443)$. Consider a sample decaying at a rate of 1 mCi (37 MBq). During the first second, 3.7×10^7 atoms decay, but during the next second less than this number decay, and during the third second even less, and so forth (i.e., the decay rate decreases with time). Even though the number of atoms decaying per second decreases, the fraction of radioactive atoms decaying per second remains constant, as defined by λ.

Consider an alternative hypothetical situation in which we assume the sample decay rate does not decrease with time but continues at its initial rate until all the atoms have decayed. The time for this to occur is the mean life. Mean life is a useful term in calculating the radiation absorbed dose to the body, because the dosimetrist must know the total energy that will eventually be deposited after complete decay. This can be calculated from the energy released per atom per decay and the total number of atoms that decay. In the preceding example 1 mCi (37 MBq) of ^{99m}Tc is equal to 1.16×10^{12} atoms. This means that after complete radioactive decay this many atoms of ^{99m}Tc will have transformed into ^{99}Tc. If the gamma ray energy per decay is 140 keV and the photon abundance is 88%, then (140 keV)(0.88)(1.16×10^{12}) or 1.43×10^{14} keV will be emitted as gamma radiation. The remaining 12% of energy will be released as conversion electrons.

Example 2. A vial contains 100 mCi (3700 MBq) of ^{131}I-sodium iodide in a 10 mL volume on Monday at 12 noon. Calculate the volume of solution required for a 12 mCi (444 MBq) dose on Friday at noon.

This problem requires use of exponential decay Equation 2-5. First calculate the radioactive concentration at Monday noon.

$$100 \text{ mCi (3700 MBq)}/10 \text{ mL} = 10 \text{ mCi (370 MBq)/mL}$$

Next, calculate the new concentration at Friday noon (4 days later).

$$A = Ae^{-\lambda t}$$

$$A = 10 \text{ mCi/mL } e^{-\frac{0.693}{8.05 \text{ d}}(4 \text{ d})}$$

$$A = 10 \text{ mCi/mL }(0.7087)$$

$$A = 7.09 \text{ mCi/mL}$$

$$\text{A dose of 12 mCi requires} \left[\frac{12 \text{ mCi}}{7.09 \text{ mCi/mL}}\right] \text{ or 1.7 mL}$$

Example 3. A vial contains 50 µCi (1850 kBq) of ^{131}I-sodium iodide capsules. How many days are required for the capsules to decay to 5 µCi (185 kBq)?

The half-life can be used to estimate the time required.

Number of days	0	8	16	24	32
Number of half-lives	0	1	2	3	4
Capsule activity (µCi)	50	25	12.5	6.25	3.125

Thus, between 24 and 32 days is required. To arrive at the exact answer, use Equation 2-5 in logarithmic form.

$$A = A_0 e^{-\lambda t}$$

$$\ln A = \ln A_0 - \lambda t$$

$$\ln(5 \text{ µCi}) = \ln(5 \text{ µCi}) - \frac{0.693}{8.05 \text{ days}}(t)$$

$$1.61 = 3.91 - 0.0861 t$$

$$t = \frac{2.3}{0.0861 \text{ days}^{-1}} = 26.7 \text{ days}$$

Decay Tables

In making decay calculations on a routine basis in a nuclear pharmacy, it is more convenient to use decay tables instead of the exponential decay equation. A decay table is a tabulation of specified times and the respective fraction of activity remaining at those times. It is prepared by using the exponential decay equation. Rearrangement of Equation 2-5 yields the following expression:

$$\frac{A}{A_0} = e^{-\lambda t} \qquad (2\text{-}7)$$

If A is the activity remaining after an original amount of activity (A_0) decays for a period of time (t), then A/A_0 is the fraction of the original amount remaining. For example, a decay table for ^{131}I can be generated by substituting the values of 1, 2, or 3 days and so forth for t in Equation 2-7 to yield the following:

t (days)	$e^{-\lambda t}$
0	1.0000
1	0.9175
2	0.8418
3	0.7724
4	0.7087
5	0.6502
6	0.5966
7	0.5474
8	0.5022

Example 4. Calculate the activity in a 10 µCi (370 kBq) capsule of ^{131}I-sodium iodide.

a. What is the activity after 5 days' decay?
   Answer: 10 µCi (0.6502) = 6.5 µCi (240.5 kBq)
b. What is the activity after 9 days' decay?
   Answer: 10 µCi (0.5022) (0.9175) = 4.6 µCi (170.2 kBq)

Radiopharmaceuticals are labeled with the total amount of activity at a specified date and time, referred to as the *calibration time*. Only at this time will the vial contain the labeled activity. After the calibration time it will contain less activity because of radioactive decay, and before the calibration time it will contain proportionately more activity. Radiopharmaceuticals are often received in the nuclear pharmacy before the calibration time.

Example 5. Calculate the activity in a capsule of ^{131}I-sodium iodide at noon on January 1 if the label states "10 µCi (370 kBq) per capsule as of 12 noon January 6."

Since the capsule must obviously contain more than 10 µCi (370 kBq) on January 1, the calibration time activity should be divided by the decay factor for 5 days. Thus, the activity in the capsule on January 1 is (10 µCi)/(0.6502) = 15.38 µCi (569 kBq).

The same answer can be obtained by using the reciprocal of the postcalibration decay factor, 1/0.6502 or 1.538, which is the precalibration decay factor. The answer to the previous question is calculated as follows: 10 µCi × 1.538 = 15.38 µCi (569 kBq). Precalibration factors can be readily calculated for any times before the calibration time using the following rearrangement of Equation 2-5:

$$\frac{A_0}{A} = e^{+\lambda t} \quad (2\text{-}8)$$

A precalibration decay table for ^{131}I is.

t (days)	$e^{+\lambda t}$
8	1.9912
7	1.8268
6	1.6762
5	1.5380
4	1.4110
3	1.2947
2	1.1870
1	1.0899
0	1.0000

REFERENCES

1. National Nuclear Data Center, Brookhaven National Laboratory. www.nndc.bnl.gov/chart/reCenter.jsp?z=39&n=51. Accessed August 16, 2010.
2. Friedlander C, Kennedy JW, Miller JM. *Nuclear and Radiochemistry.* 2nd ed. New York: Wiley; 1966:28.
3. Johns HE, Cunningham JR. *The Physics of Radiology.* 3rd ed. Springfield, IL: Charles C Thomas; 1969.
4. Quimby EH, Feitelberg S. *Radioactive Isotopes in Medicine and Biology.* 2nd ed. Philadelphia: Lea & Febiger; 1963:92–3.
5. Weber DA, Eckerman KF, Dillman LT, et al. *MIRD: Radionuclide Data and Decay Schemes.* Reston, VA: Society of Nuclear Medicine; 1989.

CHAPTER 3
Radionuclide Production

All radionuclides used in nuclear medicine are produced by artificial means, in either a nuclear reactor or a particle accelerator. They are available from commercial suppliers either as radiochemicals that must be compounded into desired pharmaceutical dosage forms or as ready-to-use radiopharmaceuticals. Some radionuclides are available from radionuclide generators that can be purchased and kept in the nuclear pharmacy to supply the desired radionuclide at the time it is needed.

The production of medically useful radionuclides involves a nuclear reaction between stable target nuclei and bombarding high-energy particles. The nuclear reaction occurs in two stages. In the first stage, the bombarding particle is captured by the target nucleus, and the particle's kinetic energy and nuclear binding energy are added to the nucleus. The extra allotment of energy in the newly formed "intermediate nucleus" is transferred among its nucleons. In the second stage, by chance, one or more nucleons in the "intermediate nucleus" overcome the nuclear binding energy and escape, or simply a gamma ray escapes. The escaping particle(s) carry off part or all of the available energy. Any energy not carried off by escaping particles is released as gamma radiation from the new nucleus. Note that this release of particles and energy is not radioactivity but is simply the second stage of the nuclear reaction. A variety of particles can escape; typically, they include protons ($^1H^+$), deuterons ($^2H^+$), neutrons (n), and alpha particles ($^4He^{2+}$). The new nuclei formed in the reaction have a different neutron:proton ratio from the stable target nuclei and are radioactive. The radioactive target then undergoes radiochemical processing and purification to isolate the desired radionuclide.

In 1934, Frederick Joliot and Irene Curie Joliot performed a nuclear reaction by bombarding a piece of aluminum foil with alpha particles emitted from a polonium source. When the polonium source was removed, radiation still emanated from the target material. They recognized that the aluminum target had undergone a transformation into some new elemental form that was radioactive. This experiment produced one of the first artificially made radionuclides, ^{30}P, according to the following reaction:

$$^{27}_{13}Al + ^{4}_{2}He \rightarrow ^{1}_{0}n + ^{30}_{15}P$$

The shorthand notation for this reaction is $^{27}Al(\alpha, n)^{30}P$.

NUCLEAR REACTORS

A nuclear reactor contains fuel rods of enriched fissionable ^{235}U positioned in the reactor core (Figure 3-1). The fuel rods are surrounded by a moderator such as heavy water (D_2O). Each fissioning uranium atom releases fast neutrons that are slowed to thermal energy by their interactions with D_2O. A thermal neutron is one that has the same average kinetic energy as the atoms of the surrounding medium. This energy, which is only a fraction of an electron volt at ordinary temperatures, is referred to as the thermal energy, since it depends on the temperature. Thermal neutrons are easily captured by other uranium atoms. When

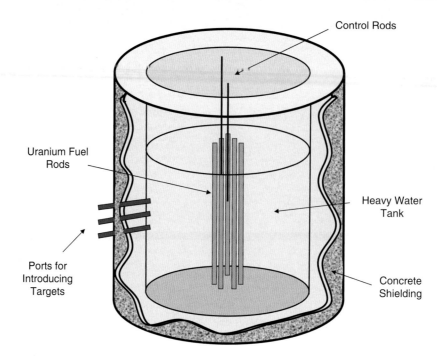

FIGURE 3-1 Simplified schematic diagram of a nuclear reactor for producing medical radionuclides. The uranium fuel rods contain enriched ^{235}U that undergoes fission, releasing neutrons. Fuel rods are placed into a tank of heavy water (D_2O), which serves as a moderator for controlling neutron energy. The rate of nuclear fission is controlled by the relative position of the cadmium or boron control rods. Control rods have a high cross-section for absorbing neutrons. Full insertion of control rods into the reactor core results in absorption of all neutrons, causing the reactor to shut down. External ports provide access of target material to the neutron flux within the reactor for the production of neutron-activated radionuclides. The entire reactor is shielded with concrete.

the uranium atoms fission, they release more neutrons, which sustain the chain reaction. The rate of thermal neutron capture by the uranium nuclei determines the fission rate in the reactor. This rate is controlled by other materials in the reactor that have a high neutron capture cross-section, which is a measure of the probability of neutron capture. Boron or cadmium control rods typically serve this purpose. The fission process generates heat that is carried off by water or other coolants through heat exchangers.

Nuclear reactors are designed for different purposes. Power reactors convert the heat generated from the fission process into electricity. Isotope production reactors have specialized ports (Figure 3-1) where target material may be introduced into the neutron flux, causing neutron activation of stable nuclides into radioactive nuclides.

TYPES OF NUCLEAR REACTIONS

The (n,γ) Reaction

This is a common type of reaction with neutrons. It has the following characteristics: (1) the reaction requires low-energy thermal neutrons (approximately 0.025 eV) and (2) the product nuclide is an isotope of the target nuclide, and chemical separation of target and product nuclides is not possible. This results in a low specific activity product. Example reactions include $^{152}Sm(n,\gamma)^{153}Sm$ and $^{50}Cr(n,\gamma)^{51}Cr$.

The A(n,γ)A* → B Reaction

In some instances radionuclide product separation from stable target material may be possible with (n,γ) reactions if the primary product nuclide (A*) has a short half-life and decays to a longer-lived radionuclide (B) that can be isolated. One example is as follows:

$$^{130}Te(n,\gamma)^{131}Te \xrightarrow{\beta^-, 25 \text{ min}} {}^{131}I(8 \text{ days})$$

The A(n,p)B Reaction

If fast neutrons are captured by a target, an (n,p) reaction is possible. This nuclear reaction imparts an extra allotment of energy to the intermediate nucleus, enabling a proton to escape. Escape of a proton changes the atomic number of the target nuclide; therefore, the radionuclide produced is not an isotope of the target, but an isobar. A benefit of this reaction is that chemical separation of target and product nuclides is possible, and high specific activities can be achieved. The following reaction is an example:

$$^{32}S(n,p)^{32}P$$

The ^{235}U(n,f)Byproducts Reaction

When ^{235}U captures a thermal neutron, the intermediate nucleus is very unstable and fissions into radioactive fragment nuclides as follows:

$$^{235}U + {}^{1}n \rightarrow {}^{236}U$$

$$\downarrow (\text{fission})$$

$$^{131}Sn + {}^{103}Mo + 2\,{}^{1}n$$

$$\downarrow \quad\quad \downarrow$$

$$^{131}Sb \quad {}^{103}Te$$

$$\downarrow \quad\quad \downarrow$$

$$^{131}Te \quad {}^{103}Ru$$

$$\downarrow \quad\quad \downarrow$$

$$^{131}I \quad {}^{103}Rh$$

$$\downarrow$$

$$^{131}Xe$$

The initial fission fragments and the nuclides to which they decay are not isotopes and can be separated chemically to achieve high specific activities. Many radionuclides are made by this process. Some examples are ^{99}Mo (used in the production of the ^{99m}Tc generator), ^{133}Xe, and ^{131}I.

CYCLOTRONS AND LINEAR ACCELERATORS

The basic construction of a cyclotron consists of two hollow, semicircular chambers called Dees placed in a magnetic field. The Dees are coupled to a high-frequency electrical system that alternates the electrical potential on each Dee during cyclotron operation, changing sign about 10^7 times per second. The Dees are heavily shielded and configured so that they can be evacuated to low pressure (Figure 3-2). An arrangement is made at the center of the space between the Dees for releasing protons or deuterons. When a proton is generated, it is attracted into the negatively charged Dee and repelled by the positive Dee. This causes the proton to accelerate into the negative Dee. The magnetic field causes the proton path to bend as it moves through the Dee. When the proton again reaches the gap, the charge on each Dee is reversed. This causes the proton to be accelerated across the gap into the opposite Dee, where the radius of its circular path will increase as a result of its increased kinetic energy. Because of the proton's higher velocity, it will again arrive at the gap precisely when the Dee polarity is reversed, causing further acceleration of the proton. This process is repeated until the proton acquires great energy. At this point the proton exits the Dee and is deflected onto a target where the desired nuclear reaction takes place.

Two types of cyclotrons are used: positive-ion machines, which accelerate protons, and negative-ion machines, which accelerate a proton associated with two electrons, called a positronium. When the positronium ion exits the cyclotron it passes through a carbon foil, which strips away the electrons, allowing a free proton to bombard the target. Modern-day cyclotrons are negative-ion machines and have the advantage of less activation of cyclotron components. Cyclotrons for

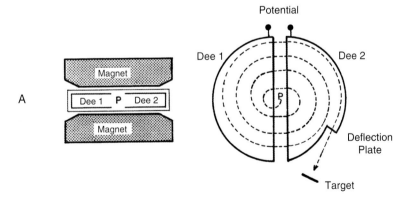

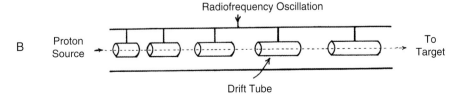

FIGURE 3-2 Schematic diagram of a cyclotron (A) and a linear accelerator (B).

producing medically useful radionuclides operate in the 11 MeV to 17 MeV range. Chapter 12 covers cyclotron applications in more detail.

When a proton attains a velocity approaching the speed of light, its mass becomes relativistic, increasing with energy. As a result, the proton slows somewhat and begins to arrive at the gap late, after the Dee polarity has changed. This limits the energy a proton can achieve in a cyclotron. Because of this problem, alternative methods of producing very high energy particles were conceived. One of these methods is the linear accelerator, or linac (Figure 3-2). The linac is a series of cylindrical drift tubes through which electromagnetic waves pass, along with their associated oscillating electric and magnetic fields. A charged particle such as a proton injected into the drift tube will be carried forward by the traveling wave. The lengths of the drift tubes are designed to accommodate relativistic changes in the accelerated particle. In this way the particle arrives at the gap between drift tubes always in phase, and very high energy particles can be achieved, upwards of 200 MeV.

Accelerator Methods of Radionuclide Production

Most radionuclides used in nuclear medicine are generated in a cyclotron. Radionuclides produced in this manner are not isotopes of the target nuclide. Consequently, chemical separation of product and target nuclides is possible and a high specific activity product can be achieved. Some typical accelerator-produced radionuclides used in nuclear medicine are as follows:

$$^{111}Cd(p,n)^{111}In$$

$$^{18}O(p,n)^{18}F$$

$$^{124}Xe(p,2n)^{123}Cs \xrightarrow{5.8\ min} {}^{123}Xe \xrightarrow{2\ hr} {}^{123}I$$

$$^{124}Xe(p,pn)^{123}Xe \xrightarrow{2\ hr} {}^{123}I$$

$$^{203}Tl(p,3n)^{201}Pb \xrightarrow{9.4\ hr} {}^{201}Tl$$

Many radionuclides produced in nuclear reactors and particle accelerators have half-lives long enough to allow time for processing and shipment by a manufacturer to nuclear medicine facilities across the country. However, some radionuclides with very short half-lives, such as ^{11}C, ^{13}N, and ^{15}O, require fabrication into radiopharmaceuticals at the site of use. This necessitates an on-site cyclotron. For positron emission tomography (PET) studies that use radiopharmaceuticals labeled with longer-lived nuclides such as ^{18}F-fludeoxyglucose (^{18}F-FDG), a cyclotron facility located near the site of use is workable. Many PET nuclear pharmacies are operational across the country at strategic sites to supply PET agents to nearby hospitals and clinics. PET radiopharmaceuticals are discussed in Chapter 12.

Radionuclide Impurities

Radionuclide impurities can be formed in the process of radionuclide production. Undesirable side reactions in a nuclear reactor or particle accelerator due to target impurities can give rise to unwanted radionuclides that are difficult to separate from the primary nuclide of interest. Sometimes a change in the target material and type of nuclear reaction will improve the system. For example, the production of ^{123}I by the reaction $^{124}Te(p,2n)^{123}I$ yields an ^{124}I impurity, which causes poor image resolution due to septal penetration of its 603 keV gamma. The alternative reaction $^{127}I(p,5n)^{123}Xe \rightarrow {}^{123}I$ yields a more desirable product with a ^{125}I impurity, whose low-energy photons (27 keV) do not interfere with image resolution. Another desirable reaction is the p,2n reaction on a ^{124}Xe target shown on this page, which produces very low impurities.

Radionuclide impurities are significant in that they may not only affect imaging resolution but can also increase the radiation dose without adding useful diagnostic information. Of greatest significance are impurities with half-lives longer than that of the desired nuclide, because the percentage of impurity will increase with time. Some examples of this are 66 hour ^{99}Mo impurity in 6 hour ^{99m}Tc and 12 day ^{202}Tl impurity in 73 hour ^{201}Tl.

RADIONUCLIDE GENERATORS

Some radionuclides can be produced from generator systems that make the production of short-lived radionuclides possible in the hospital. Most radiopharmaceuticals used in nuclear medicine are labeled with ^{99}Mo generator-produced ^{99m}Tc. Currently, about 6000 curies of ^{99}Mo is required each week to meet the North American demand for ^{99m}Tc in nuclear medicine. Several large-scale producers located in Canada, the Netherlands, Belgium, France, South Africa, Australia, and Poland supply ^{99}Mo for generator manufacture.

Radionuclide generators are a convenient means of supplying large amounts of short-lived radionuclides for imaging studies. A generator consists of a long-lived parent that decays to a short-lived daughter. The daughter nuclide can be chemically separated from the parent nuclide because they are not isotopes. After separation, fresh daughter activity is generated by the decay of parent nuclide in the generator until the parent activity is depleted. The useful life of a generator depends on the parent half-life. Table 3-1 lists several parent–daughter generator systems.[1] Currently, only the ^{99}Mo–^{99m}Tc and ^{82}Sr–^{82}Rb generators are available commercially. The ^{99}Mo–^{99m}Tc generator is supplied in activities ranging from 1 to 19 curies (Ultra-TechneKow DTE [Covidien]), shielded with lead, tungsten, or depleted uranium, or ranging from 1 to 18 curies (TechneLite [Lantheus]). The lead-shielded ^{82}Sr–^{82}Rb generator (CardioGen-82 [Bracco]) has 90 to 150 mCi of ^{82}Sr at calibration time.

The ^{99m}Tc generator was developed in 1957 at the Brookhaven National Laboratory and was first used clinically in 1961 at the University of Chicago.[2,3] With a ^{99}Mo parent half-life of 65.94 hours, the generator has a useful life of about 2 weeks. Generators are available with different calibration times during the week to meet the activity needs of a hospital or nuclear pharmacy.

TABLE 3-1 Radionuclide Generator Systems

Parent (half-life)	Parent Decay Mode	Daughter Product (half-life)	Column Type	Eluant
^{99}Mo (65.94 hours)	Beta minus	Na^{99m}TcO$_4$ (6.01 hours)	Aluminum oxide	0.9% NaCl
^{82}Sr (25.55 days)	Electron capture	^{82}RbCl (76.4 seconds)	Stannic oxide	0.9% NaCl
^{68}Ge (271 days)	Electron capture	^{68}GaCl$_3$ (67.7 minutes)	Stannic oxide	0.1 M HCl
^{81}Rb (4.57 hours)	Electron capture	^{81m}Kr gas (13.1 seconds)	Cation exchange resin	Water Air/oxygen
^{90}Sr (28.9 years)	Beta minus	^{90}YCl$_3$ (64 hours)	^{90}Sr solution	Solvent extraction
^{113}Sn (115.09 days)	Electron capture	^{113m}InCl$_3$ (1.7 hours)	Zirconium oxide	0.05 M HCl

Production of the ^{99m}Tc Generator

Figure 3-3 illustrates a production scheme for a ^{99m}Tc generator. The ^{99}Mo in contemporary generators is obtained as a uranium fission byproduct. Radiochemical separation and purification isolate the ^{99}Mo for generator preparation. In one method, ^{99}Mo is adjusted to an acidic pH, forming various anionic species such as molybdate (MoO$_4^{2-}$) and paramolybdate (Mo$_7$O$_{24}^{6-}$).[4] The anionic molybdate solution is then loaded onto the generator alumina (Al$_2$O$_3$) column previously washed in pH 5 saline. The positively charged alumina firmly adsorbs the molybdate ions, with a loading capacity of approximately 2 mg of molybdenum per gram of alumina.[4] After loading, generators are sterilized by autoclaving and assembled under aseptic conditions into their lead-shielded container. Each generator is eluted with normal saline (0.9% sodium chloride solution), and the eluate is subjected to several tests before release of the generator.

Generators typically are tested by the manufacturer for elution efficiency, eluate volume, radionuclidic purity to detect the presence of ^{99}Mo and other radionuclide contaminants, radiochemical purity to ensure the proper chemical form of ^{99m}Tc as pertechnetate, aluminum ion concentration in the eluate, pH of the eluate, and, finally, pyrogenicity and sterility. Figure 3-4 shows commercial ^{99m}Tc generators.

^{99m}Tc Generator Elution

Figure 3-3 illustrates the simplified decay scheme for ^{99}Mo to ^{99m}Tc and ^{99}Tc in the generator. Figure 3-5 shows the relative amounts of ^{99}Mo and ^{99m}Tc activity in the generator over time, with the gradual decay of ^{99}Mo and subsequent buildup of ^{99m}Tc. Maximum buildup of ^{99m}Tc activity is achieved in about 23 hours following elution. About 50% of the maximum activity is reached in 4.5 hours and 75% by 8.5 hours after generator elution.[5] Accumulated activity is eluted by washing normal saline solution through the column. The ^{99}Mo activity remains firmly bound to the alumina, but the ^{99m}Tc activity, as the pertechnetate ion (TcO$_4^-$), is easily displaced by the chloride ion (Cl$^-$) in the saline solution. Typically, 80% to 90% of the available ^{99m}Tc activity is removed in one 6 mL elution (Figure 3-6). The ^{99}Mo activity remaining on the column continues to decay, generating more ^{99m}Tc activity. The daily amount of ^{99m}Tc activity in the generator declines with the decay of ^{99}Mo. ^{99m}Tc activity builds up rapidly after generator elution, and the generator may be eluted several times during the same day to obtain more activity. This is more likely to occur at week's end. Weekly replacement of the generator is the norm; however, a generator is useful for about 2 weeks.

There are two basic types of generator systems: the wet system and the dry system (Figure 3-7). In the wet system, a large reservoir of saline is connected to the generator column. Technetium activity is eluted after attachment of a sterile evacuated vial to the elution port. The vacuum draws the saline–pertechnetate solution into the vial. At the end of elution, the generator column remains bathed in saline. In the dry system, a 5 to 20 mL saline charge is attached to the system before elution. An evacuated vial then draws the 5 to 20 mL of saline through the generator column, eluting the ^{99m}Tc activity, followed by a volume of air to "dry" the column.

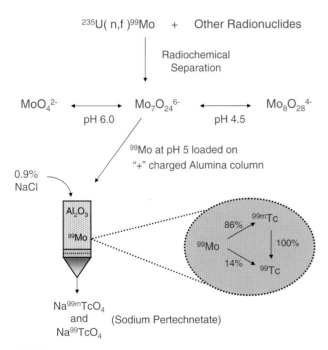

FIGURE 3-3 Schematic diagram of the steps involved in the production of a ^{99m}Tc generator. Fission-produced ^{99}Mo is radiochemically isolated and purified to the anionic molybdate and paramolybdate species, which are loaded on the positively charged alumina (Al$_2$O$_3$) generator column previously washed in pH 5 saline.

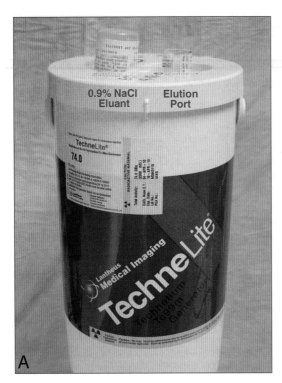
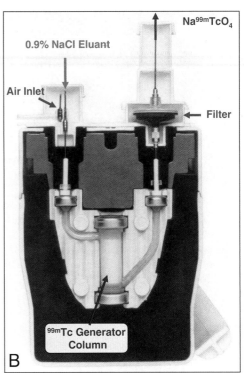

FIGURE 3-4 ^{99m}Tc generators from two different manufacturers. (A) Exterior view. (B) Cutaway view showing the functional parts of a ^{99m}Tc generator.

In both systems the elution vial will end up containing ^{99m}Tc and ^{99}Tc as sodium pertechnetate in normal saline.

^{99m}Tc Generator Yield

Early generator systems were of the wet type. The rapid growth of nuclear medicine over the years required larger amounts of ^{99m}Tc activity and generators with increased amounts of ^{99}Mo activity. However, reduced yields of ^{99m}Tc activity often occurred, particularly during the first elution of a new generator.[6,7] The principal reason for this was the production of radiolysis products from the decay of ^{99}Mo in the generator column solution. This resulted in a reduction of technetium's oxidation state, causing it to bind more firmly to the alumina in the column. Such problems tended to occur with new high-

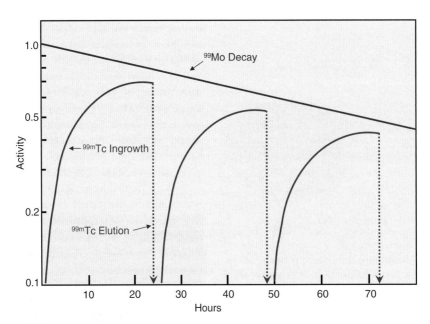

FIGURE 3-5 Decay of ^{99}Mo and ingrowth and elution of ^{99m}Tc activity over time in a ^{99m}Tc generator.

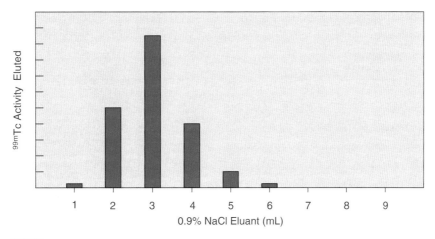

FIGURE 3-6 Elution profile of a ^{99m}Tc generator.

activity generators and when the generator had not been eluted for 2 to 3 days. Upon the first elution, only 10% or so of the total activity could be eluted. However, the noneluted activity in the column could subsequently be removed within an hour or so because the first elution removed the radiolytic contaminants and the fresh saline introduced enough oxygen to revert the reduced technetium back to pertechnetate.

Cifka and Vesely[8] studied the factors that influence elution yield in ^{99m}Tc generators. They found that technetium could not be eluted with pure water and showed that the reduction of pertechnetate on the column begins after the consumption of oxygen from the solution present on the column. One technique used in some generators at the time to improve elution efficiency was to increase the titer of dissolved oxygen in the saline eluant.

Dry system generators were developed to alleviate the poor elution problem of wet systems. Removing saline from the column to introduce air after elution significantly reduced the formation of radiolysis products in order to maintain an oxidized pertechnetate species. Today the problem of incomplete elution rarely occurs with dry generator systems. However, elution problems may still occur if residual eluant remains on the generator column. This might happen if the evacuated vial is removed early before all the eluant has been drawn through the column or if the evacuated vial has a vacuum strength that is insufficient to complete the elution. Therefore, it is important to ensure that the entire eluant volume is drawn through the column and that a sufficient amount of air is introduced after elution.

Quality Control of the Generator Eluate

Three tests should be performed on each generator eluate. The first test is a *radioactivity calibration* (Figure 3-8). This is done

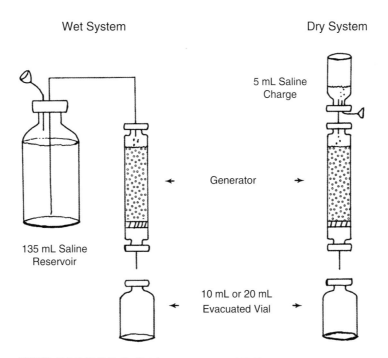

FIGURE 3-7 The two types of ^{99m}Tc generator systems.

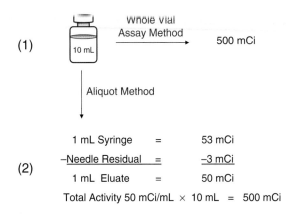

FIGURE 3-8 ^{99m}Tc generator eluate activity measured in a dose calibrator by (1) the whole-vial method or (2) the aliquot method of assay.

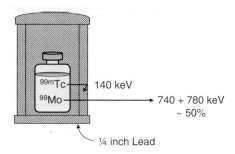

FIGURE 3-9 Generator eluate test for molybdenum breakthrough by the lead shield method in a dose calibrator. The weaker gamma rays from ^{99m}Tc are absorbed by the lead shield, while the higher-energy gamma rays from ^{99}Mo are able to penetrate the shield and be measured in the dose calibrator during the test.

with a dose calibrator to determine the activity eluted and the radioactive concentration. This can be accomplished in two ways. The first is assaying the whole elution vial to obtain the total activity and dividing by the total volume eluted to obtain the concentration. The second is drawing a 1 mL aliquot into a syringe and assaying its contents, then returning the 1 mL volume to the eluate vial and reassaying the syringe for the needle activity; the difference between these two measurements provides the concentration in millicuries per milliliter, and the total activity in the vial is obtained by multiplying the activity in millicuries per milliliter by the total volume in the vial. The aliquot method has the advantages of reduced operator exposure and more accurate assessment of the radioactive concentration.

The second test of a generator eluate is the *radionuclidic purity test* for the presence of ^{99}Mo contamination (Figure 3-9). This so-called moly-99 breakthrough test is accomplished with a specialized lead shield and a dose calibrator.[9] With this technique, the whole vial of generator eluate is placed into a tightly closed ¼ inch thick lead shield, which is then placed into a dose calibrator adjusted to read ^{99}Mo activity. Because of the large difference in lead half-value layers for ^{99m}Tc and ^{99}Mo photons (0.17 mm versus 6.5 mm), the ^{99m}Tc photons will be completely absorbed by the shield while about 50% of ^{99}Mo photons will penetrate the shield and be detected in the dose calibrator ion chamber. The microcuries of ^{99}Mo are read directly from the readout or after application of a correction factor. A drawback with this method is that the dose calibrator has no energy-discriminating capability. Thus, the lead shield method cannot distinguish ^{99}Mo from other radionuclides with high photon energies. An example of this situation is the reported interference of ^{132}I contamination (668–1398 keV gammas) in a generator eluate during a ^{99}Mo test.[10]

The limit for ^{99}Mo contamination in a generator eluate has been set by the Nuclear Regulatory Commission (NRC) and the United States Pharmacopeia (USP) to be not more than 0.15 µCi (5.55 kBq) ^{99}Mo per mCi (37 MBq) ^{99m}Tc in the administered dose. Accordingly, a 20 mCi (740 MBq) dose of ^{99m}Tc may contain no more than 3 µCi (111 kBq) of ^{99}Mo. A convenient method for determining the expiration time for ^{99m}Tc eluates is based on these limits.[11] Table 3-2 lists the expiration time in hours for ^{99m}Tc based on the initial ratio of microcuries of ^{99}Mo per millicurie of ^{99m}Tc at generator elution time. For example, if the initial ratio is 0.058, the ^{99m}Tc eluate cannot be used more than 9 hours after the time of elution. The package inserts of all generator manufacturers indicate that the default expiration time is 12 hours for any ^{99m}Tc generator eluate regardless of the ^{99}Mo:^{99m}Tc ratio.

Results of the moly-99 breakthrough test are typically negative, indicating no significant ^{99}Mo contamination. However, a positive test with significant levels (7 mCi [259 MBq] of ^{99}Mo) in the generator eluate has been reported.[12] In this instance the moly-99 breakthrough was caused by an improperly assembled generator.

The NRC regulations in Title 10 of the Code of Federal Regulations (10 CFR 35.204) stipulate that the ^{99}Mo test must be performed on the first elution of a new generator. Agreement states typically follow NRC requirements but may require this test to be performed more frequently at each generator elution. In addition, the generator manufacturer's package insert specifies that a ^{99}Mo breakthrough test be conducted at each generator elution.

The third test of a generator eluate is a *chemical purity test* for the concentration of aluminum ion present. A colorimetric spot test is typically used. A small drop of generator eluate is placed on a strip of filter paper impregnated with aluminon (the ammonium salt of aurintricarboxylic acid), an aluminum-specific indicator (Figure 3-10). A spot of a standard aluminum ion solution of known concentration (10 µg Al^{3+}/mL) is placed next to the eluate spot. The aluminum ion present reacts with the indicator, producing a pink color with an intensity proportional to the amount of Al^{3+} present. If the color of the eluate spot is less intense than the aluminum standard spot, the eluate passes the test. The USP limit on

TABLE 3-2 Expiration Times for ^{99m}Tc-Sodium Pertechnetate after Generator Elution

Initial Ratio (microcuries of ^{99}Mo/millicurie of ^{99m}Tc)	Expiration Time (hr)	Initial Ratio (microcuries of ^{99}Mo/millicurie of ^{99m}Tc)	Expiration Time (hr)
0.135	1	0.072	7
0.122	2	0.065	8
0.109	3	0.058	9
0.098	4	0.052	10
0.089	5	0.047	11
0.080	6	0.042	12

Source: Reference 11.

aluminum concentration is ≤ 10 μg Al^{3+} per milliliter of eluate. The generator manufacturer's package insert specifies that an aluminum test be conducted at each generator elution to meet the USP limit.

GENERATOR PHYSICS

In Chapter 2 we considered single-step radioactive decay processes in which the daughter products were stable nuclides. With generators, we must consider the situation in which daughter atoms are radioactive, represented by the following decay sequence:

$$N_1 \xrightarrow{\lambda_1} N_2 \xrightarrow{\lambda_2} N_3$$

FIGURE 3-10 Colorimetric spot test for measuring aluminum ion in a ^{99m}Tc generator eluate. Test paper containing a chemical indicator for aluminum ion is spotted with one drop of generator eluate (A) and one drop of standard aluminum solution (B) containing the upper limit of aluminum ion concentration (10 μg Al^{+3}/mL). To pass the test, the color intensity of the eluate spot cannot exceed that of the aluminum standard.

N_1 and N_2 are parent and daughter radionuclides, respectively, and N_3 is stable or very long-lived. Because we are interested in the daughter radionuclide, its decay rate (dN_2/dt) is described by the following expression:

$$\frac{dN_2}{dt} = \lambda_1 N_1 - \lambda_2 N_2 \quad (3\text{-}1)$$

The net rate at which daughter atoms build up is the difference between the rate of their formation by the parent, $\lambda_1 N_1$, and the rate of their own decay, $\lambda_2 N_2$. For the ^{99m}Tc generator, N_1 is the number of ^{99}Mo atoms and λ_1 is its decay constant; N_2 is the number of ^{99m}Tc atoms and λ_2 is its decay constant. Even though the third product, ^{99}Tc (N_3), is radioactive, its half-life is so long (2.11×10^5 years) that it is essentially stable and does not need to be considered in the decay calculations.

After appropriate rearrangement of Equation 3-1, solution of the first-order differential equation yields the following relationship:[13]

$$N_2 = \frac{\lambda_1}{\lambda_2 - \lambda_1} N_1^0 \left(e^{-\lambda_1 t} - e^{-\lambda_2 t} \right) + N_2^0 e^{-\lambda_2 t} \quad (3\text{-}2)$$

$$[\text{-----A-----}] \quad [\text{--B--}]$$

The expressions in bracket A describe the rate of production and decay of daughter ^{99m}Tc atoms, and the expression in bracket B describes the contribution to N_2 from any daughter ^{99m}Tc atoms present initially or remaining after generator elution. The last term is significant only if generator elution efficiency is low or if the generator is re-eluted within a few hours after the previous elution. Recalling the expression $N = A/\lambda$, one can substitute the appropriate activity expression into Equation 3-2 and derive the following activity equation:

$$A_2 = \frac{\lambda_2}{\lambda_2 - \lambda_1} A_1^0 \left(e^{-\lambda_1 t} - e^{-\lambda_2 t} \right) + A_2^0 e^{-\lambda_2 t} \quad (3\text{-}3)$$

Equation 3-3 presumes that 100% of the parent decays to the daughter. For the ^{99}Mo–^{99m}Tc generator, only 86% of ^{99}Mo decays to ^{99m}Tc, as follows:

$$^{99}\text{Mo} \xrightarrow{86\%} {^{99m}\text{Tc}} \longrightarrow {^{99}\text{Tc}}$$
$$\cdots\cdots 14\% \cdots\cdots\cdots\cdots \triangleright$$

Therefore, Equation 3-3 is modified as follows:

$$A_2 = \frac{0.86\lambda_2}{\lambda_2 - \lambda_1} A_1^0 \left(e^{-\lambda_1 t} - e^{-\lambda_2 t}\right) + A_2^0 e^{-\lambda_2 t} \quad (3\text{-}4)$$

Equation 3-4 permits calculation of the theoretical ^{99m}Tc activity (A_2) present in the generator at any time (t) after the previous elution if one knows the ^{99}Mo activity A_1^0 present at the time of the previous elution.

Transient Equilibrium Generators

If the half-life of the parent radionuclide is significantly longer (say 10 to 100 times) than the daughter half-life, and a sufficient period of time is allowed to elapse before the generator is eluted, a condition of transient equilibrium is established between the parent and daughter. This is illustrated in Figure 3-11 for a ^{99m}Tc generator, where line A_2 represents daughter (^{99m}Tc) ingrowth and line A_1 is parent (^{99}Mo) decay. Daughter ingrowth is determined from Equation 3-4 for ^{99m}Tc. Immediately after generator elution, the rate of daughter production is greater than its rate of decay, and daughter activity increases rapidly with time. As daughter atoms accumulate, they begin to decay. Eventually, the rates of daughter production and decay become equal and a maximum activity is reached (point X on the graph). At this point the activities of parent and daughter are equal ($A_1 = A_2$), assuming that 100% of the parent decays to the daughter. The broken line A_2 represents the situation in the ^{99m}Tc generator, where only 86% of the ^{99}Mo decays to ^{99m}Tc (Equation 3-4).

The time required to reach the maximum daughter activity in a generator is derived from Equation 3-2 and is given by the following relationship:

$$t_{max} = \frac{1}{\lambda_2 - \lambda_1} \ln \frac{\lambda_2}{\lambda_1} \quad (3\text{-}5)$$

In the ^{99m}Tc generator, the time to reach the maximum ^{99m}Tc activity is 23 hours.[5] If the generator is undisturbed and sufficient time has elapsed (~60 hours), the ratio of daughter and parent activities is constant (point Y in Figure 3-11). This is the point of *transient equilibrium*, where the daughter appears to decay with the half-life of the parent (65.94 hours). When the daughter is separated from the parent, its activity over time is shown as line B, whose slope is determined by the half-life of the daughter (6.01 hours for ^{99m}Tc). At the point of transient equilibrium, the time of decay since the previous elution has become so large that the value of the exponential term $e^{-\lambda_2 t}$ in Equations 3-3 and 3-4 becomes very small compared with $e^{-\lambda_1 t}$, because λ_2 is much greater than λ_1. Equation 3-4 therefore simplifies to the following expression at transient equilibrium:

$$A_2 = \frac{0.86\,\lambda_2}{\lambda_2 - \lambda_1} A_1^0 e^{-\lambda_1 t} \quad (3\text{-}6)$$

The ratio $\lambda_2/(\lambda_2 - \lambda_1)$ numerically is 1.1, making the constant term in Equation 3-6 equal to 0.946 (0.86 × 1.1). Furthermore, the term $A_1^0 e^{-\lambda_1 t}$ is equal to the activity of ^{99}Mo present in the generator after transient equilibrium is established. The actual ^{99m}Tc activity present in the generator at transient equilibrium

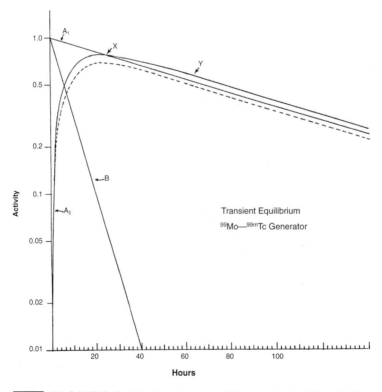

FIGURE 3-11 Transient equilibrium in the ^{99}Mo–^{99m}Tc-generator (see text for description).

is therefore given as 0.946 times the ^{99}Mo activity present. It should be kept in mind that Equation 3-6 is valid for the calculation of ^{99m}Tc activity only after extended decay time (48–72 hours) since the previous generator elution. For times less than 48 hours, Equation 3-4 should be used, although ^{99m}Tc activities calculated with Equation 3-6 at 24 hours will be about 90% of those calculated with Equation 3-4. Table 3-3 lists the relationship between ^{99m}Tc and ^{99}Mo activities in the generator at various times after elution.

Example 1: A ^{99m}Tc generator is manufactured on a Friday, calibrated for 2.5 Ci (92,500 MBq) of ^{99}Mo at 8:00 pm. Calculate the theoretical ^{99m}Tc activity in the generator on the following Monday at 8:00 am if no previous elutions have been made. The decay constants are as follows:

$$\lambda_1(^{99}Mo) = \frac{0.693}{65.94 \text{ hr}} = 0.0105 \text{ hr}^{-1}$$

$$\lambda_2(^{99m}Tc) = \frac{0.693}{6.01 \text{ hr}} = 0.1153 \text{ hr}^{-1}$$

The time of decay is 60 hours, so Equation 3-6 can be used to make the calculation. Thus,

$$A(^{99m}Tc) = \frac{(0.86)0.1153}{0.1153 - 0.0105} 2.5 \text{ Ci } e^{-0.0105 \text{ hr}^{-1}(60 \text{ hr})}$$

$$A(^{99m}Tc) = (0.86)(1.1)(2.5 \text{ Ci})(0.533) = 1.26 \text{ Ci}(46{,}620 \text{ MBq})$$

Alternatively, the data in Table 3-3 can be used to solve this problem. ^{99}Mo activity after 60 hours of decay is

$$2.5 \text{ Ci}(0.533) = 1.33 \text{ Ci}(49{,}210 \text{ MBq})$$

TABLE 3-3 Relationship between ^{99m}Tc and ^{99}Mo in the Generator at Various Times after Elution

Time (hr)	Curies ^{99}Mo ×	Ratio ^{99m}Tc:^{99}Mo =	Curies ^{99m}Tc[a]
0	1.000	—	0
1	0.990	0.094	0.093
2	0.979	0.179	0.175
3	0.969	0.255	0.247
4	0.959	0.324	0.311
5	0.949	0.386	0.366
6	0.940	0.441	0.414
12	0.883	0.677	0.598
18	0.829	0.803	0.666
24	0.779	0.870	0.678
36	0.688	0.924	0.636
48	0.607	0.940	0.571
60	0.533	0.944	0.506
72	0.473	0.946	0.448
78	0.445	0.946	0.421

[a] Actual ^{99m}Tc present based on 86.05% ^{99}Mo decay to ^{99m}Tc.

This value is multiplied by the ^{99m}Tc–^{99}Mo ratio at 60 hours; thus,

$$1.33 \text{ Ci}(0.944) = 1.26 \text{ Ci}(46{,}620 \text{ MBq})$$

Example 2. If the activity actually eluted from the generator in Example 1 is 1.07 Ci (39,590 MBq), what is the elution efficiency?

$$\text{Percent elution efficiency} = \frac{\text{Measured activity} \times 100}{\text{Theoretical activity}}$$

$$= \frac{1.07 \text{ Ci} \times 100}{1.26 \text{ Ci}} = 85\%$$

Example 3. If the generator above is re-eluted at 1:00 pm, what is the expected ^{99m}Tc activity?

Because transient equilibrium is not reestablished after the previous elution, Equation 3-4 must be used for this calculation. The residual ^{99m}Tc activity remaining on the column after the 8:00 am elution is

$$(1.26 \text{ Ci available}) - (1.07 \text{ Ci eluted}) = 0.19 \text{ Ci}(7{,}030 \text{ MBq})$$
retained

From Equation 3-4, the ^{99m}Tc activity present on the column at 1:00 pm is

$$A(^{99m}Tc) = (0.86)(1.1)(1.33 \text{ Ci})\left(e^{-0.0105(5)} - e^{-0.1153(5)}\right)$$
$$+ 0.19 \text{ Ci } e^{-0.1153(5)}$$

$$A(^{99m}Tc) = 0.487 \text{ Ci} + 0.107 \text{ Ci} = 0.594 \text{ Ci}(21{,}978 \text{ MBq})$$

Alternatively, Table 3-3 can be used to solve the problem. ^{99m}Tc activity from 5 hours of ^{99}Mo decay is

$$(1.33 \text{ Ci})(0.949)(0.386) = 0.487 \text{ Ci}(18{,}019 \text{ MBq})$$

^{99m}Tc activity from 5 hours of residual ^{99m}Tc decay is

$$(0.19 \text{ Ci})(0.562) = 0.107 \text{ Ci}(3{,}959 \text{ MBq})$$

Total ^{99m}Tc activity present in the generator at 1:00 pm is

$$(0.487 \text{ Ci} + 0.107 \text{ Ci}) = 0.594 \text{ Ci}(21{,}978 \text{ MBq})$$

Because the elution efficiency is 85%, the expected ^{99m}Tc activity in the generator eluate is

$$(0.594 \text{ Ci})(0.85) = 0.505 \text{ Ci}(18{,}685 \text{ MBq})$$

Secular Equilibrium Generators

If the half-life of the parent is much longer (say 1000 times or more) than the daughter half-life, the parent does not decay appreciably during several daughter half-lives, and a condition of secular equilibrium is established. Figure 3-12 illustrates a secular equilibrium generator in which lines A_1 and A_2 are ^{82}Sr and ^{82}Rb activities, respectively. The activity of ^{82}Rb in

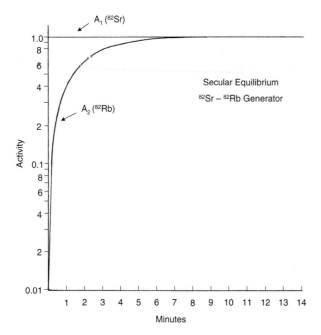

FIGURE 3-12 Secular equilibrium in the ^{82}Sr–^{82}Rb generator (see text for description).

Figure 3-12 (A_2) was calculated from Equation 3-3. As in the case with transient equilibrium generators, the rate of daughter production initially is greater than its rate of decay, and daughter activity increases rapidly over time. When the rate of production equals the rate of decay, secular equilibrium is established (intersection of lines A_1 and A_2). In this condition the daughter appears to decay with the parent half-life. Because of the large difference in decay constants ($\lambda_2 \gg \lambda_1$), the value of λ_1 is insignificant compared with λ_2, and the constant term $\lambda_2/(\lambda_2 - \lambda_1)$ in Equation 3-3 approaches unity. In addition, at secular equilibrium the $e^{-\lambda_2 t}$ terms approach zero; therefore, Equation 3-3 is simplified to

$$A_2 = A_1^0 e^{-\lambda_1 t} \qquad (3\text{-}7)$$

That is to say, at secular equilibrium the daughter activity is equal to the parent activity present in the generator.

All of the generators listed in Table 3-1 are secular equilibrium systems except for the ^{99}Mo–^{99m}Tc generator. The only secular equilibrium generator available commercially in the United States is the ^{82}Sr–^{82}Rb system (Figure 3-12). The ^{68}Ge–^{68}Ga generator is available in Europe and is being developed for use in the United States. The ^{82}Sr parent is produced by a proton spallation reaction on molybdenum, purified, and loaded onto a hydrous stannic oxide column with an activity in the range of 90 to 150 mCi (3330–5550 MBq). ^{82}Sr decays by electron capture (EC) with a half-life of 25.55 days to the daughter nuclide ^{82}Rb that is eluted with normal saline as rubidium chloride.[14] ^{82}Rb decays by positron emission with a half-life of 76.4 seconds. The system regenerates a full charge of ^{82}Rb activity within 10 minutes of elution. The monovalent cationic ^{82}Rb$^+$ is used primarily for cardiac PET imaging. This generator is discussed in more detail in Chapter 12.

Technetium Content in Generator Eluates

The half-lives of the nuclides in a ^{99m}Tc generator are 65.94 hours for ^{99}Mo, 6.01 hours for ^{99m}Tc, and 2.11×10^5 years for ^{99}Tc.[15] Because of its long half-life, ^{99}Tc accumulates over time in the generator. Thus, eluates from all ^{99m}Tc generators contain both ^{99m}Tc and ^{99}Tc atoms. The relative amounts of ^{99}Mo, ^{99m}Tc, and ^{99}Tc in a generator are shown in Figure 3-13. The mole fraction of the ^{99m}Tc isomer in the eluate is given by the following relationship, where N represents the number of atoms:[16]

$$\frac{N_{^{99m}Tc}}{N_{(total)}} = \frac{N_{^{99m}Tc}}{N_{^{99m}Tc} + N_{^{99}Tc}} \qquad (3\text{-}8)$$

The mole fractions, listed in Table 3-4, can be determined at various times from the following equation:[16]

$$\frac{N_{^{99m}Tc}}{N_{(total)}} = \frac{0.86\lambda_1\left(e^{-\lambda_1 t} - e^{-\lambda_2 t}\right)}{\lambda_2 - \lambda_1\left(1 - e^{-\lambda_1 t}\right)} \qquad (3\text{-}9)$$

In this equation, λ_1 and λ_2 are the decay constants for ^{99}Mo and ^{99m}Tc, respectively. With increasing time of decay, the mole fraction of ^{99m}Tc decreases because of the buildup of ^{99}Tc atoms. For example, after 1 day of decay the mole fraction of ^{99m}Tc in the generator is 0.2769, or about 28% of the total number of technetium atoms in the generator eluate; the remaining atoms (approximately 72%) are ^{99}Tc. The amount of ^{99m}Tc in generator eluates is therefore not "carrier-free," and its specific activity continuously decreases as the period of time between generator elutions increases. This has the potential of decreasing labeling efficiency in ^{99m}Tc radiopharmaceutical kits that contain small amounts of reducing agent (Sn^{2+} ion). This becomes most critical when a generator is not eluted for several days. The effect of carrier technetium on the preparation of ^{99m}Tc-labeled human serum albumin is a note-

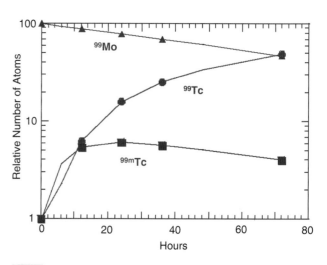

FIGURE 3-13 Relative number of atoms over time of ^{99}Mo, ^{99m}Tc, and ^{99}Tc in a ^{99m}Tc generator.

TABLE 3-4 Mole Fractions of ⁹⁹ᵐTc and ⁹⁹Tc in Generator Eluates

Time Since Prior Generator Elution	Technetium Mole Fraction		Molar Ratio ^{99}Tc/^{99m}Tc
	[^{99m}Tc/^{99m}Tc + ^{99}Tc]	[^{99}Tc/^{99m}Tc + ^{99}Tc]	
3 hr	0.727	0.273	0.38
6 hr	0.619	0.381	0.62
12 hr	0.460	0.540	1.17
24 hr	0.277	0.723	2.61
48 hr	0.131	0.869	6.63
72 hr	0.077	0.923	11.99

Source: Reference 5.

worthy example (Figure 3-14).[17] Kits with limited reducing power that may be affected by buildup of ^{99}Tc in the generator eluate are the Ceretec kit (GE Healthcare) and the Ultra-Tag RBC and TechneScan MAG3 kits (Covidien).

The total number of technetium atoms present in the generator eluate can be determined from the ^{99m}Tc activity and its mole fraction by the following equation:[15]

$$N(\text{total}) = \frac{A(^{99m}\text{Tc})(T_{1/2})(1.443)}{^{99m}\text{Tc mole fraction}} \quad (3\text{-}10)$$

Thus, the total number of technetium atoms, N (total), per millicurie of ^{99m}Tc is

$$N(\text{total}) = \frac{(1\,\text{mCi})(3.7 \times 10^7\,\text{dps/mCi})(6.01\,\text{hr} \times 3600\,\text{sec/hr})(1.443)}{^{99m}\text{Tc mole fraction}}$$

or

$$N(\text{total}) = \frac{(\text{mCi}\,^{99m}\text{Tc eluted})(1.16 \times 10^{12}\,\text{disintegration/mCi})}{^{99m}\text{Tc mole fraction}}$$

(3-11)

Example: A generator is eluted 6 hours after the previous elution and yields 500 mCi (18,500 MBq). How many atoms of ^{99m}Tc and ^{99}Tc, total, are contained in the eluate?

$$N(\text{total}) = \frac{(500\,\text{mCi})(1.16 \times 10^{12}\,\text{disintegration(atoms)/mCi})}{0.619}$$

$$= 9.37 \times 10^{14}\,\text{atoms}$$

Note that each disintegration per second is equivalent to one atom decaying per second.

Table 3-5 lists the total number of technetium atoms per millicurie of ^{99m}Tc eluted from a generator for various time periods, calculated from Equation 3-11 and the mole fraction values listed in Table 3-4. Table 3-5 simplifies the determination of how much ^{99m}Tc eluate should be used for kit preparation in situations in which the technetium atoms should be limited. Note that Table 3-4 indicates that the highest mole fraction of ^{99m}Tc is achieved when the generator elution interval is short. Thus, in preparing kits requiring small amounts of technetium, a practical method is to re-elute the generator within a few hours of the first elution and use this eluate for kit preparation.

Example: Efficient labeling of a red blood cell kit requires that no more than 1.48×10^{14} technetium atoms be used because of the small amount of stannous ion in the kit.[18] What is the maximum volume of generator eluate that can be used to label the kit if the generator was eluted 24 hours after the previous elution and the eluate contained 800 mCi (29,600 MBq) ^{99m}Tc in a 20 mL volume?

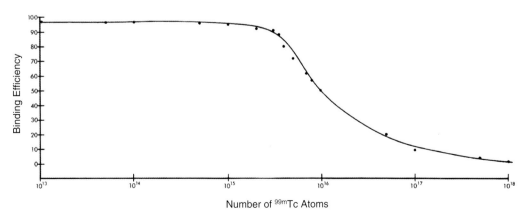

FIGURE 3-14 Effect of carrier technetium on the labeling efficiency of ^{99m}Tc-labeled human serum albumin. (Reprinted with permission from reference 17.)

TABLE 3-5 Total Technetium Atoms (^{99}Tc + ^{99m}Tc) × 10^{12} per Millicurie ^{99m}Tc Eluted from a ^{99m}Tc Generator

Days since Prior Elution	Hours since Prior Elution						
	0	3	6	9	12	15	18	21
0	—	1.59	1.87	2.18	2.52	2.89	3.29	3.72
1	4.18	4.67	5.18	5.73	6.30	6.89	7.51	8.16
2	8.83	9.52	10.25	10.99	11.76	12.56	13.38	14.23
3	15.11	16.03	16.97	17.91	18.91	19.95	21.00	22.12
4	23.23	24.41	25.60	26.85	28.15	29.44	30.86	32.23

From Table 3-5, the total amount of technetium eluted per 800 mCi (29,600 MBq) ^{99m}Tc in 24 hours is

$$4.18 \times 10^{12} \text{ atoms/mCi} \times 800 \text{ mCi} = 3.34 \times 10^{15} \text{ atoms}$$

The amount of eluate that can be used is

$$\frac{1.48 \times 10^{14}}{3.34 \times 10^{15}} \times 20 \text{ mL} = 0.9 \text{ mL}$$

DISPOSAL OF THE ^{99m}TC GENERATOR

Generator manufacturers provide a return program for used ^{99m}Tc generators. Typically, spent generators are returned in an empty generator shipping carton to the manufacturer for credit. Shipping papers must comply with Department of Transportation regulations. Alternatively, spent generators can be decayed to background levels, dismantled, and discarded. Generators may contain long-lived radiocontaminants such as ^{103}Ru (39.5 day half-life) and therefore should be held for 14 weeks before discarding. After that time the generator column can be removed, surveyed with a Geiger-Müller counter, and disposed of in regular trash if the surface exposure rate is no higher than background. The survey should be conducted in an area of low background (e.g., 0.05–0.1 mR/hr).

OTHER GENERATOR SYSTEMS

While the ^{99m}Tc generator system has widespread application in routine nuclear medicine, other generator systems have been used for biomedical applications.[19] Several of these generators have played an important part in the development of imaging agents over the years, some are no longer available, and some may find more application in the future. The properties of several of these generators are briefly discussed in this section. All of them are secular equilibrium generators except for ^{188}W–^{188}Re, which is a transient equilibrium generator. Only the ^{82}Sr–^{82}Rb generator has been approved by the Food and Drug Administration and is commercially available for routine use in cardiac imaging. The other generators described are available only for nonhuman use or are in investigational new drug status.

^{113}Sn–^{113m}In Generator

The ^{113}Sn–^{113m}In generator was used mostly in the 1960s and 1970s for preparing ^{113m}In-labeled radiopharmaceuticals for imaging many of the major organ systems, including the brain, lung, liver, kidney, and blood pool. ^{113}Sn is produced in a nuclear reactor by neutron activation of stable enriched ^{112}Sn through the nuclear reaction ^{112}Sn(n,γ)^{113}Sn. It can also be produced in a cyclotron. The ^{113}Sn is adsorbed on a hydrous zirconium oxide column and the ^{113m}InCl$_3$ is eluted with 0.05 M hydrochloric acid. The trivalent cationic indium (^{113m}In^{3+}) readily forms chelates. At the time, this was an advantage over ^{99m}Tc, which has to be reduced to a cationic species before chelation. When injected intravenously, indium quickly binds to plasma transferrin, forming a blood pool imaging agent. ^{113m}In is also incorporated into a variety of other compounds, including diethylenetriaminepentaacetic acid (DTPA), hydroxide and phosphate colloids, and albumin aggregates.

^{113}Sn decays with a half-life of 115 days by EC to ^{113m}In. ^{113m}In decays by isomeric transition to stable ^{113}In with a half-life of 1.7 hours, emitting a gamma ray of 393 keV. The long half-life of the parent permits a long shelf life of the generator, but the short half-life of the daughter requires radiopharmaceuticals to be prepared more than once daily.

^{82}Sr–^{82}Rb Generator

^{82}Sr is produced by a proton spallation reaction on molybdenum or by ^{85}Rb(p,4n)^{82}Sr and is purified to a final form in dilute hydrochloric acid.[20] It is loaded onto a hydrous stannic oxide column with an activity in the range of 90 to 150 mCi (3330–5550 MBq). ^{82}Sr decays by EC with a half-life of 25 days to ^{82}Rb. The rubidium daughter is eluted as ^{82}Rb-chloride from the generator with normal saline.[14] ^{82}Rb decays by positron emission with a half-life of 75 seconds. Because of this short half-life, the system regenerates a full charge of ^{82}Rb activity within 10 minutes of elution, and studies can be repeated frequently. The monovalent cationic ^{82}Rb$^+$ is used primarily for cardiac PET imaging to assess myocardial blood flow. The generator has a useful life of 1 month.

⁶⁸Ge–⁶⁸Ga Generator

After its production in a cyclotron by the proton bombardment of stable gallium, ^{68}Ge is processed and loaded on various columns, namely, alumina, titanium oxide, or stannic oxide. ^{68}Ge decays with a 275 day half-life by EC to ^{68}Ga, which itself decays by positron emission with a half-life of 68 minutes to stable ^{68}Zn. The accumulated ^{68}Ga activity in the generator can be eluted with 0.1 M to 1.0 M hydrochloric acid as gallium chloride. It is then processed into the desired radiotracer that possesses an appropriate ligand to firmly coordinate ^{68}Ga, such as triethylenetetramine (TETA), 1,4,7,10-tetraazacyclodecane-1,4,7,10-tetraacetic acid (DOTA), or 2,2′,2″-(1,4,7-triazacyclononane-1,4,7-triyl)triacetic acid (NOTA).

⁸¹Rb–⁸¹ᵐKr Generator

The parent radionuclide, ^{81}Rb, can be produced by several nuclear reactions in a cyclotron by the alpha-particle bombardment of a bromine target, as a bromide salt. This secular equilibrium generator is prepared by adsorbing ^{81}Rb on columns of zirconium phosphate, Bio-Rad Ag50 (Bio-Rad, Richmond, CA), or Dowex 50-X8 (Dow Chemical, Midland, MI), from which ^{81m}Kr gas is eluted with water or with an air–oxygen mixture. ^{81}Rb decays by EC with a half-life of 4.5 hours, which greatly limits its utility. The generator is available in Europe but not in the United States, and it has an expiration time of 12 hours. ^{81m}Kr decays by isomeric transition with a 13-second half-life, emitting a 191 keV gamma ray. ^{81m}Kr has been used in pulmonary ventilation imaging and to measure tissue blood flow.

⁹⁰Sr–⁹⁰Y Generator

^{90}Sr is produced as a fission byproduct of ^{235}U in a nuclear reactor. A convenient radiochemical procedure for separation and purification involves precipitation of strontium as the nitrate and subsequent purification by anion exchange chromatography.[19] ^{90}Y-yttrium chloride is a sterile, aqueous solution of yttrium chloride in 0.05 M hydrochloric acid. It is produced by solvent extraction of ^{90}Y from a ^{90}Sr generator solution and purified into the chloride form, ^{90}YCl$_3$. It is available from MDS Nordion (Ontario, Canada) in 5, 10, 20, and 50 mCi amounts per vial with 20 μCi or less of ^{90}Sr radionuclide impurity per curie of ^{90}Y at expiration.

¹⁸⁸W–¹⁸⁸Re Generator

The ^{188}W parent is produced in a high-flux nuclear reactor by double neutron capture on ^{186}W as the trioxide or metal as follows: ^{186}W(n,γ)^{187}W(n,γ)^{188}W.[19,21] ^{188}W decays by beta decay (0.349 MeV E_{max}) with a half-life of 69.4 days to ^{188}Re. ^{188}Re decays also by beta decay (2.12 MeV E_{max}) and gamma emission (0.155 MeV, 15%) with a half-life of 16.98 hours. This transient equilibrium generator has been prepared by packing zirconyl tungstate ^{188}W gel into a column similar to that used for ^{99m}Tc generators. The ^{188}Re is eluted from the generator as perrhenate with water or normal saline in 3 mL fractions with approximately 90% eluted in the first fraction.[19] ^{188}Re has been used to label antibody fragments with diamide dithiolate technology in yields and purity virtually identical to those with ^{99m}Tc.[22]

REFERENCES

1. Finn RD, Molinski VJ, Hupf HB, et al. *Radionuclide Generators for Biomedical Applications*. Oak Ridge, TN: US Department of Energy; 1983. NAS-NS-3202.
2. Stang LG Jr, Tucker WD, Doering RF, et al. Development of methods for the production of certain short-lived radioisotopes. In: *Proceedings of the First UNESCO Conference*. Vol 1. London: Pergamon Press; 1958:50–70.
3. Harper PV, Andros G, Lathrop K. *Preliminary Observations on the Use of Six-hour Tc-99m as a Tracer in Biology and Medicine*. Washington, DC: US Atomic Energy Commission; 1962:77–88. Report ACRH-18.
4. Arino H, Kramer HH. Fission product Tc-99m generator. *Int J Appl Radiat Isot*. 1975;26:301–3.
5. Lamson M III, Hotte CE, Ice RD. Practical generator kinetics. *J Nucl Med Technol*. 1976;4:21–7.
6. Molinski VJ. A review of Tc-99m generator technology. *Int J Appl Radiat Isot*. 1982;33:811–9.
7. Steigman J. Chemistry of the alumina column. *Int J Appl Radiat Isot*. 1982;33:829–34.
8. Vesely P, Cifka J. Some chemical and analytical problems connected with technetium-99m generators. In: *Radiopharmaceuticals from Generator-Produced Radionuclides*. Vienna, Austria: International Atomic Energy Agency; 1971:71–81.
9. Richards P, O'Brien MJ. Rapid determination of Mo-99 in separated Tc-99m. *J Nucl Med*. 1969;10:517.
10. Briner WH, Harris CC. Radionuclidic contamination of eluates from fission product molybdenum-technetium generators [letter]. *J Nucl Med*. 1974;15:466–7.
11. Ponto JA. Expiration times for Tc-99m. *J Nucl Med Technol*. 1981;9:40.
12. Kowalsky RJ, Preslar J. Report of a positive Mo-99 breakthrough test. *J Nucl Med Technol*. 1979;2:108.
13. Friedlander G, Kennedy JW, Miller JM. *Nuclear and Radiochemistry*. 2nd ed. New York: Wiley; 1966:71–5.
14. Gennaro GP, Neirinckx RD, Bergner B, et al. A radionuclide generator and infusion system for pharmaceutical quality Rb-82. In: Knapp FF Jr, Butler TA, eds. *Radionuclide Generators: New Systems for Nuclear Medicine Applications*. ACS Symposium Series No. 241. Washington, DC: American Chemical Society; 1984:135–50.
15. National Nuclear Data Center, Brookhaven National Laboratory. www.nndc.bnl.gov/chart/reCenter.jsp?z=39&n=51. Accessed August 16, 2010.
16. Lamson ML III, Kirschner AS, Hotte CE, et al. Generator-produced 99m TcO4−: carrier free? *J Nucl Med*. 1975;16:639–41.
17. Porter WC, Dworkin HJ, Gutkowski RF. The effect of carrier technetium in the preparation of Tc-99m-human serum albumin. *J Nucl Med*. 1976;17:704–6.
18. Smith TD, Richards P. A simple kit for the preparation of Tc-99m-labeled red blood cells. *J Nucl Med*. 1976;17:126–32.
19. Finn RD, Molinski VJ, Hupf HB, et al. *Radionuclide Generators for Biomedical Applications*. Oak Ridge, TN: US Department of Energy; 1983. NAS-NS-3202.
20. MDS Nordion. Sr-82 Fact Sheet. Strontium-82 radiochemical strontium chloride solution. www.mds.nordion.com/documents/products/Sr-82_Can.pdf. Accessed August 24, 2010.
21. Ehrhardt GJ, Ketring AR, Turpin TA, et al. A convenient tungsten-188/rhenium-188 generator for radiotherapeutic applications using low specific activity tungsten-188. In: Nicolini M, Bandoli G, Mazzi U, eds. *Technetium and Rhenium in Chemistry and Nuclear Medicine 3*. Verona, Italy: Cortina International; 1990:631–4.
22. Vanderheyden JL, Fritzberg AR, Rao TN, et al. Rhenium labeling of antibodies for radioimmunotherapy [abstract]. *J Nucl Med*. 1987; 28:656.

CHAPTER 4
Interactions of Radiation with Matter

The interaction of ionizing radiation with matter is the basis for radiation detection and measurement. Radiation is either particulate or electromagnetic. Particulate radiation includes charged particles, such as alpha particles (He^{2+}) and beta particles (e^+ and e^-); and neutral particles, such as neutrons and neutrinos. Electromagnetic radiation includes x-rays and gamma rays, which are high-energy photons that interact with matter in the same manner as particles. The most important types of ionizing radiation in nuclear pharmacy and nuclear medicine are beta particles, x-rays, and gamma rays, which are the principal radiations emitted from radiopharmaceuticals. There are no radiopharmaceuticals in routine use that emit alpha particles or neutrons. Neutrinos are emitted during the beta decay process, but they have no significant interactions in matter. The properties of these radiations are listed in Table 4-1.

The absorption of ionizing radiation in matter involves interactions with orbital electrons and the nuclei of atoms and molecules. During the interactive process a radiation's energy is partially or completely released in the absorber. The energy can be dissipated as atomic or molecular excitations and ionizations, or via the emission of low-energy electromagnetic radiation.

CHARGED-PARTICLE INTERACTIONS

The interactions of charged particles occur via electrostatic forces of repulsion or attraction between particles of like or unlike charge. Repulsive forces occur when negatively charged orbital electrons in atoms and molecules interact with other negatively charged electrons, such as negatrons. Attractive forces prevail when orbital electrons interact with positively charged particles, such as alpha particles, protons, and positrons. Excitation interactions occur when a charged-particle radiation passes near an atom. During the interaction the atom absorbs a small amount of energy from the passing radiation, causing an orbital electron to be excited to a higher-energy suborbital. The energy absorbed by the electron is eventually released in some form, such as visible or ultraviolet (UV) light, when the electron returns to its ground state. Ionization interactions occur when the radiation comes very close to an atom so that the electrical forces of attraction or repulsion become strong. In an ionization event, an orbital electron absorbs energy in excess of its binding energy and is ejected from the atom. An ion pair is formed, consisting of the negatively charged electron and the remaining positively charged atom. The ejected electron acquires kinetic energy equal to the energy imparted to it minus its binding energy.

Alpha Particle Interactions

Alpha particles are high-energy helium nuclei carrying two positive charges. The positive charge has a great electrostatic influence on the atoms of an absorber. As it passes through

TABLE 4-1 Properties of Radiation Emitted from Radioactive Sources

Type of Radiation	Symbol	Mass (AMU)	Charge	Comments
Alpha	α	4.0015	2+	Nucleus of a helium atom
Negatron	β−	0.0005486	1−	Electron released from the nucleus of neutron-rich nuclides during decay
Positron	β+	0.0005486	1+	Antimatter electron released from the nucleus of proton-rich nuclides during decay
Neutrino	ν	0	0	Emitted during beta (β+ and β−) and electron capture decay
Neutron	n	1.0086654	0	Emitted during nuclear fission
Photon	γ	0	0	A quantum of electromagnetic energy

matter, an alpha particle causes excitations and ionizations of atoms by electrostatic attraction of electrons (Figure 4-1). An alpha particle has its highest energy and velocity at the beginning of its path. Through its interactions, the alpha particle loses energy and slows. At lower speeds an alpha particle spends more time in the vicinity of atoms, where it has a greater opportunity to extract electrons. Hence, fewer ionizations are produced by an alpha particle early in its path of travel than at the end of its path when ionization density increases because the particle's rate of speed is decreased (Figure 4-2). When an ion pair is formed, the electron can receive sufficient energy to send it off at a high rate of speed, at which it will cause ionization of other atoms. This high-speed electron is called a *delta ray*. About 60% to 80% of ionization produced by alpha particles is due to secondary ionization by delta rays.[1]

An alpha particle is about 7300 times more massive than an electron. Consequently, alpha particles are not easily deflected by interactions with electrons; think of a bowling ball interacting with Ping-Pong balls. Alpha particles travel in straight-line paths over a definite range. Occasionally, an alpha particle's path can be altered by an encounter with a nucleus, which has a large mass relative to the alpha particle.[1] After thousands of interactions and loss of its kinetic energy, an alpha particle will acquire two orbital electrons and come to rest as a neutral helium atom.

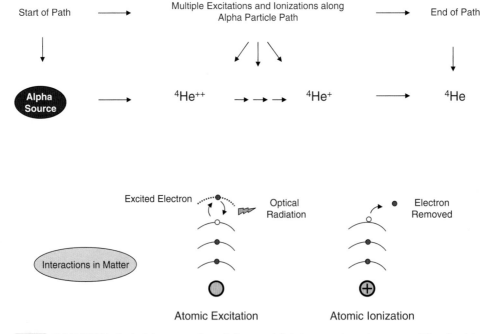

FIGURE 4-1 Diagram of an alpha particle's interactions in matter. The double positive charge on an alpha particle causes attraction of electrons from atoms in the absorber, resulting in excitation and ionization of atoms along its path of travel, shown at the bottom of the figure. As a result of multiple interactions, the alpha particle's velocity slows and it eventually acquires one electron and then two electrons near the end of its path, to become a neutral helium atom, shown at the top of the figure.

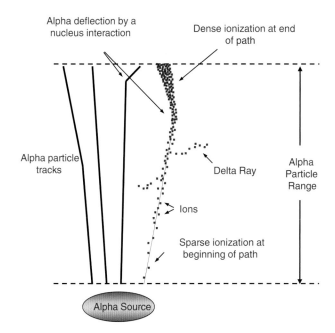

FIGURE 4-2 Diagram of an alpha particle's interactions over its range in an absorber. An alpha particle moves in a straight-line path over a definite range and exhibits no deflection in its path following interactions with orbital electrons. An interaction with a nucleus can cause a deflection of the alpha particle's straight-line path because of the large mass of the nucleus. There will be some recoil of the nucleus, depending on its mass relative to the alpha particle, which is not shown here. The density of ionizations caused by alpha particle interactions is smaller initially because of the alpha particle's high velocity, but ionization density increases greatly near the end of the alpha particle's path, where its velocity is slowed significantly.

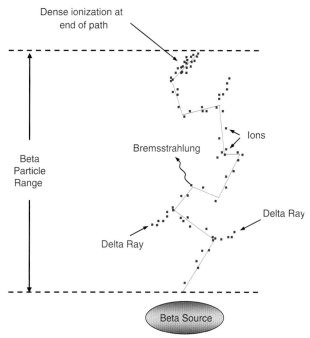

FIGURE 4-3 Diagram of a beta particle's interactions over its range in an absorber. Beta particles have a tortuous path in an absorber because they are easily deflected by interactions with orbital electrons, which have equal mass. Along its path a beta particle loses energy by repulsive ejection of orbital electrons in atoms and by nuclear interactions, which cause emission of bremsstrahlung radiation. The delta rays produced along the path are ejected electrons with sufficient energy to cause secondary ionizations along a spur. Denser ionization occurs near the end of a beta particle's path, where it moves more slowly. The density of ionization of a beta particle is considerably less than that of an alpha particle because of its higher velocity and smaller charge. A beta particle's range in matter can vary widely and depends on the energy it acquires during beta decay. The average range of a beta particle is approximately one-fourth its maximum range in an absorber.

Beta Particle (Positron and Negatron) Interactions

Positrons are positively charged beta particles (β^+) released from the nuclei of proton-rich radionuclides. Negatrons are negatively charged beta particles (β^-) released from the nuclei of neutron-rich radionuclides. Positrons produce excitation and ionization by electrostatic attraction of orbital electrons, similar to alpha particles. Negatrons produce excitation and ionization by electrostatic repulsion of orbital electrons. When excitation interactions occur, positrons and negatrons cause a loosely bound valence electron to be energized into an optical suborbital. When the excited electron returns to its ground state, low-energy radiation, such as visible or ultraviolet light, is emitted. When ionizations occur, positrons and negatrons impart sufficient energy to eject an electron from its orbital. Some ejected electrons have sufficient energy to cause secondary ionizations along a spur track near the atom. These electrons are called delta rays. If an inner-shell electron in an atom is ejected, characteristic x-rays and Auger electrons will be produced when an outer-shell electron fills in the inner-shell deficit (see Figure 2-3, Chapter 2). Because a beta particle interaction involves particles of equal mass, the interactive path of a beta particle in an absorber is tortuous (Figure 4-3). Thousands of excitations and ionizations occur along the beta particle's path before it comes to rest. Near the end of its path, the density of ionization increases because the beta particle's velocity is slowed. Eventually, the beta particle (negatron or positron) loses all of its energy and ceases to produce excitation and ionization. At the end of its path, a negatron comes to rest by simply combining with an atom that needs an electron. In contrast, a positron comes to rest by association with a negative electron and undergoes an annihilation reaction, converting the two electron masses into two 0.511 MeV photons according to

$$E_{\text{MeV}}(e^+ + e^-) = 2(mc^2)(6.24 \times 10^5 \text{ MeV/erg})$$

Immediately before the annihilation reaction, the positron may momentarily form a positronium, e⁺e⁻, which can be thought of as an atom with the positron acting as the nucleus. (See discussion of positron decay in Chapter 2.)

Positrons and negatrons can also interact with an atom in the vicinity of the nucleus, where the positive electrostatic force of the nucleus can alter the path of an electron. A positron, having a charge similar to the nucleus, is repulsed from it, while a negatron, having an opposite charge to the nucleus, is attracted to it. In either case, the change in direction of the beta particle causes it to lose energy. This energy is released as electromagnetic radiation called *bremsstrahlung* (braking radiation). The probability of bremsstrahlung production increases directly with beta particle energy and the atomic number of the absorber. The ratio of beta particle energy loss by bremsstrahlung radiation to energy loss by ionization in an absorber with atomic number Z is approximately $EZ/800$, where E is in MeV. For example, radiation and ionization energy losses are about equal for a 10 MeV beta particle in lead ($Z = 82$), whereas for a 1 MeV beta particle the radiation loss will be about one-tenth the ionization loss. In nuclear pharmacy a few radionuclides, such as ^{32}P and ^{90}Y (maximum beta particle energy [beta max] 1.7 MeV and 2.28 MeV, respectively), produce sufficient amounts of bremsstrahlung to permit their measurement in a dose calibrator. However, because bremsstrahlung radiation is fairly weak and its generation is affected by Z, the geometry of the container used for measuring these nuclides must be carefully specified so that the measurement will be accurate. From a radiation protection point of view, it is best to shield high-energy beta emitters with an inner shield of low Z material, such as glass or plastic, and an outer shield of high Z material, such as lead, to absorb the bremsstrahlung. The processes of beta particle (electron) interactions with matter are summarized in Figure 4-4.

Specific Ionization and Linear Energy Transfer

When ionizations occur in air, the average energy required to produce an ion pair (W) is about 34 eV. Much of this energy is expended as excitations, since only about 12 to 15 eV is required to actually remove an electron from oxygen and nitrogen atoms in air.[1] Thousands of excitations and ionizations will occur, however, before the radiation's energy is expended. For example, a 340 keV (340,000 eV) beta particle will produce about 10,000 ion pairs before it comes to rest.

The number of ion pairs produced per unit path length traveled by radiation is termed *specific ionization* (SI). It is expressed as the number of ion pairs per millimeter. The SI of charged particles is directly proportional to the square of the charge and inversely proportional to particle velocity. The SI for alpha particles, shown in Figure 4-5, is caused by the large mass and dual positive charge of the particles. Most of an alpha particle's ionization occurs near the end of its path of travel in an absorber, where its velocity is slowed. An alpha particle has a higher SI than a proton because it moves more slowly and has twice the charge. In contrast, an electron, having unit negative charge and very small mass, moves through matter at high velocity, creating a much lower SI. The kinetic energy of a particle is given by

$$KE = \tfrac{1}{2} mv^2 \qquad (4\text{-}1)$$

where energy in ergs equals one-half the mass in grams multiplied by the square of the velocity in centimeters per second. Hence, a 1 MeV (1.603×10^{-6} erg) alpha particle, with $m = 6.65 \times 10^{-24}$ g, would have a velocity of

$$v = \sqrt{\frac{2 \times (1.603 \times 10^{-6}\,\text{erg})}{6.65 \times 10^{-24}\,\text{gram}}} = 0.69 \times 10^9\,\text{cm/sec}$$

The velocity of a 1 MeV proton ($m = 1.673 \times 10^{-24}$ g) would be twice that of the alpha particle:

$$v = \sqrt{\frac{2 \times (1.603 \times 10^{-6}\,\text{erg})}{1.673 \times 10^{-24}\,\text{gram}}} = 1.38 \times 10^9\,\text{cm/sec}$$

The relationship between the SI and velocity of electrons, protons, and alpha particles in air is compared in Table 4-2.[1]

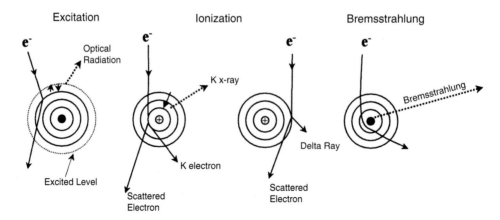

FIGURE 4-4 Beta particle interactions in matter, demonstrating excitation, ionization, and bremsstrahlung production.

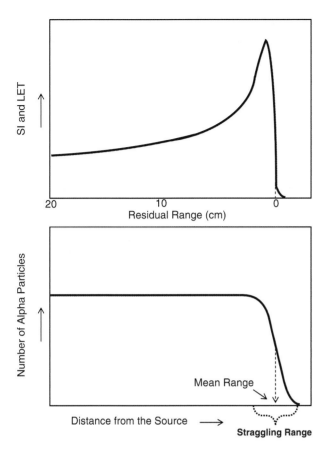

FIGURE 4-5 Composite diagram of alpha particle specific ionization (SI) (upper graph) and range (lower graph) in air. The solid line in the lower graph represents the number of alpha particles from a point source as a function of distance from the source. Monoenergetic alpha particles from a source travel about the same distance into an absorber. There is a slight spread of particle range of about 3% near the end of the path, known as straggling. The mean range is the distance traveled by 50% of the particles. The upper graph, known as a Bragg curve, indicates that SI and linear energy transfer (LET), which are directly related, increase toward the end of an alpha particle's path, where its velocity is slowed and its coulombic interaction with electrons is increased. (Adapted from reference 1.)

Observe that as a particle's velocity is slowed, its specific ionization increases. Note also in Table 4-2 that specific ionization has units of density thickness (mg/cm²). Density thickness is the weight of 1 cm² of absorber with thickness equal to the range of the radiation in the absorber. The thickness of absorber equal to the range of the radiation is determined by dividing density thickness by the absorber density. The relationship between density thickness and absorber density is

$$\frac{\text{gram}}{\text{cm}^3} \times \text{cm} = \frac{\text{gram}}{\text{cm}^2} \quad (4\text{-}2)$$

Equation 4-2 permits calculation of the linear range, in centimeters, over which a radiation will travel in any absorber of known density. Thus, in Table 4-2, the linear range (thickness) over which the ion pairs of each particle are produced in air, with $d = 1.239$ mg/cm³, is 0.77 cm (1.0 mg/cm² ÷ 1.239 mg/cm³). If the absorber in Table 4-2 was water ($d = 1$ g/cm³) the linear range over which ionization would occur would be 0.001 cm (1.0 mg/cm² ÷ 1000 mg/cm³).

The energy dissipated per unit path length is termed the *linear energy transfer* (LET) and is expressed as kiloelectron volts per micrometer (keV per μm). SI and LET are related to W as follows:

$$\text{LET} = \text{SI} \times W \quad (4\text{-}3)$$

LET is directly related to SI. Thus, densely ionizing radiation will release large amounts of energy in a smaller volume, compared with sparsely ionizing radiation. This has great significance for the biologic effects caused by different types of radiation, as will be discussed in Chapter 6. The LET of electrons, protons, and alpha particles in water is compared in Table 4-3.[2]

NEUTRON INTERACTIONS

The neutron was proposed by Ernest Rutherford in 1920 as a particle of unit mass and charge of zero that could easily penetrate a nucleus and unite with it. In 1932 James Chadwick demonstrated that the neutron could not be deflected in electric and magnetic fields, attesting to its neutral charge.[3]

It is known from negatron decay that a neutron is converted into a proton and an electron in the nucleus. Energy is liberated in this process, since the combined mass of a proton and electron is less than the mass of a neutron. Using the example of a hydrogen atom, the neutron mass (1.0086654 AMU) exceeds the combined proton and electron mass (1.0078252 AMU) by 0.0008402 AMU, so an energy equivalent to 0.782 MeV is liberated in the process; therefore, a free neutron should decay spontaneously as follows:

$$^1_0 n \longrightarrow\, ^1_1 H +\, ^{\ 0}_{-1} e + \nu$$

Experiments have demonstrated that a neutron indeed decays spontaneously with a half-life of 700 seconds.[3]

Although neutron emitters are not used in nuclear medicine, their interactions with matter are of interest. Neutrons can be classified by their energies as being thermal (approximately 0.025 eV), slow (0.03 eV–100 eV), intermediate (100 eV–10 keV), and fast (10 keV–20 MeV). Neutrons interact with matter in three ways: (1) scatter by nuclear collision, (2) capture by a nucleus, and (3) decay of a free neutron. Typical scatter and capture reactions that may occur in biologic systems are as follows:[4]

Scatter reactions

$$^1_0 n +\, ^1_1 H \longrightarrow\, ^1_1 H +\, ^1_0 n' \quad \text{(elastic scatter)}$$

$$^1_0 n +\, ^{12}_{\ 6} C \longrightarrow\, ^{12}_{\ 6} C +\, ^1_0 n' + \gamma \quad \text{(inelastic scatter)}$$

Capture reactions

$$^1_0 n +\, ^{14}_{\ 7} N \longrightarrow\, ^{14}_{\ 6} C +\, ^1_1 H$$

TABLE 4-2 Relationship of Radiation Energy, Velocity, and Specific Ionization in Air

Energy (MeV)	Particle Velocity (cm/sec × 10⁹)			Specific Ionization (ion pairs per 1.0 mg/cm²)		
	Beta Particle	Proton	Alpha Particle	Beta Particle	Proton	Alpha Particle
10	29.94	4.38	2.20	53	1,100	13,000
5	29.85	3.10	1.55	49	2,100	22,000
4	29.78	2.77	1.39	48	2,400	26,000
3	29.66	2.40	1.20	47	3,100	32,000
2	29.35	1.96	0.98	46	4,200	41,000
1	28.21	1.38	0.69	46	6,800	54,000
0.5	25.87	0.98	0.49	50	11,200	56,000
0.1	16.44	0.44	–	116	18,000	–
0.05	12.37	–	–	154	–	–
0.01	5.85	–	–	850	–	–
0.000146	0.72	–	–	5950 (maximum)	–	–

Source: Reference 1.

Scatter reactions are nonionizing events, typically occurring in water or graphite-moderated nuclear reactors, in which neutrons interact by collisions with the nuclei of hydrogen or carbon atoms. *Elastic scatter* collisions are those in which the total kinetic energy and momentum of the neutron and nucleus remain constant. This is most likely to occur between two bodies of equal mass, as in a billiard ball collision. Paraffin, water, and other materials rich in hydrogen are good moderators of neutrons by elastic scatter. In the reaction with hydrogen, a portion of the neutron's energy is transferred to the nucleus and the remaining energy is retained by the scattered neutron. *Inelastic collisions* result in a loss of energy from the system. The reaction is (n,n'γ) where n is the incident neutron and n' is the slower neutron released by the nucleus, the energy difference being emitted as a gamma ray. Moderation of fast neutrons by carbon occurs primarily by this process.

Neutron capture reactions are ionizing events because a proton is ejected from the collision with a nucleus. This type of reaction has more biologic consequence than a scatter reaction because a proton is set in motion, which is highly ionizing. Thermal and slow neutrons can more easily be captured by a nucleus because the probability of capture is inversely related to neutron speed. Elements with high capture cross-sections, such as boron and cadmium, have a high potential for capturing neutrons. Control rods made of these elements are used to control the chain reaction in nuclear reactors.

The capture of neutrons by the nuclei of stable elements is a method for producing radioisotopes (see Chapter 3). This process is known as neutron activation. Neutron activation has also been used in a variety of analytical applications. The composition of moon rocks obtained in lunar excursions by astronauts in the 1960s was determined partially by neutron activation analysis. The rocks were exposed to a flux of neutrons from a source, such as ^{252}Cf, and the gamma energy spectrum of radioisotopes produced in rock minerals was analyzed for elemental composition.[5] Another application is in the pharmaceutical industry, for evaluating solid dosage forms in vivo. For example, capsules and compressed tablets can be formulated with stable nuclides of high neutron-capture cross sections, such as ^{152}Sm, ^{138}Ba, and ^{170}Er. The dosage forms are irradiated with neutrons in a nuclear reactor to produce the radioisotopes ^{153}Sm, ^{139}Ba, and ^{171}Er. After administration to human subjects, the dosage form can be imaged in vivo by gamma camera scintigraphy to evaluate its dissolution and gastrointestinal transit.[6]

TABLE 4-3 Linear Energy Transfer (LET) of Radiation in Water

Energy (MeV)	LET (keV/μm)		
	Beta Particle	Proton	Alpha Particle
5.0	0.18	9	110
1.0	0.18	27	300
0.5	0.21	47	350
0.1	0.45	90	170
0.01	2.3	–	–
0.001	19.0	–	–

Adapted from reference 2, p. 86.

ELECTROMAGNETIC RADIATION INTERACTIONS

Gamma rays and x-rays are electromagnetic radiation with properties characterized by frequency, wavelength, and energy. This is shown in Table 4-4 and described by the following equations:

$$v = \frac{c}{\lambda} \qquad (4\text{-}4)$$

$$E = h v = \frac{hc}{\lambda} \qquad (4\text{-}5)$$

TABLE 4-4 Properties of the Electromagnetic Wave Spectrum

Wave Type	Frequency[a]	Wavelength[b]	Photon Energy
Radio	1×10^5	3×10^5 cm	4.13×10^{-10} eV
	1×10^9	30.0 cm	4.13×10^{-6} eV
Microwave	1×10^9	30.0 cm	4.13×10^{-6} eV
	1×10^{11}	0.3 cm	4.13×10^{-4} eV
Infrared	3×10^{12}	0.01 cm	0.0124 eV
	3×10^{14}	0.0001 cm (10,000 Å)	1.24 eV
Visible light	4.3×10^{14}	7000 Å	1.77 eV
	7.5×10^{14}	4000 Å	3.1 eV
Ultraviolet	7.5×10^{14}	4000 Å	3.1 eV
	3×10^{16}	100 Å	124 eV
Soft x-rays	3×10^{16}	100 Å	124 eV
	3×10^{18}	1 Å	12.4 keV
Diagnostic x-rays, gamma rays	3×10^{18}	1 Å	12.4 keV
	3×10^{20}	0.01 Å	1.24 MeV
Cosmic rays	3×10^{20}	0.01 Å	1.24 MeV
	3×10^{23}	0.00001 Å	1240 MeV

[a] Waves/second.
[b] Photon emitters in nuclear medicine generally range from 0.155 Å (80 keV) to 0.031 Å (400 keV).

where
- ν = frequency (waves/sec)
- c = speed of light (2.998×10^{10} cm/sec)
- λ = wavelength (in angstroms $\times 10^{-8}$ cm)
- E = energy in ergs
- h = Planck's constant (6.626×10^{-27} erg-sec)

Thus, the energy of a 1 angstrom photon is

$$E = \frac{(6.626 \times 10^{-27} \text{ erg-sec})(2.998 \times 10^{10} \text{ cm/sec})}{1 \times 10^{-8} \text{ cm}}$$

$$= 1.986 \times 10^{-8} \text{ erg}$$

and its equivalent energy in electron volt units is

$$E = \frac{1.986 \times 10^{-8} \text{ erg}}{1.602 \times 10^{-9} \text{ erg/keV}} = 12.4 \text{ keV}$$

Thus,

$$E_{keV} = \frac{12.4}{\lambda} \quad (4\text{-}6)$$

Radio waves, microwaves, and visible light, which have long wavelength and low energy, exhibit wavelike properties, interacting with matter differently than short-wavelength, high-energy radiation such as x-rays and gamma rays. High-energy electromagnetic radiations, called photons, travel with the speed of light. Although a photon has a wavelength, it propagates through space not as a wave but as a discrete packet or quantum of light energy. A photon, therefore, interacts with matter like a particle of energy. The three processes of photon interaction with matter are the photoelectric effect, Compton scatter, and pair production. Each of these interactive processes results in the ionization of matter. Most of the ionization produced by photons is caused by the electrons set into motion by the interaction. In general, the average SI of photons is about one-tenth to one-hundredth that of beta particles of equal energy.[1] Thus, the paucity of photon interactions in matter causes photons to have ranges considerably greater than the ranges of beta particles.

Photoelectric Effect

Absorption of photons by the photoelectric effect is illustrated in Figure 4-6. This process involves a relatively low-energy photon interacting with an inner-shell electron, usually a K-shell electron. The total energy of the incident photon is lost, being transferred to the ejected electron, which is called a photo-

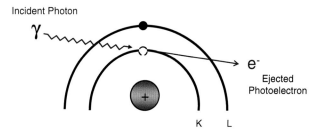

FIGURE 4-6 Photoelectric absorption. The incident photon's energy is entirely transferred to an orbital electron, such as a K-shell electron. The kinetic energy of the ejected electron (KE_K) is equal to the photon's initial energy minus the electron's binding energy.

electron. If a K-shell electron is ejected, the kinetic energy of the photoelectron (KE_k) is equal to the incident photon energy (E_i) minus the K-shell binding energy (BE_k).

$$KE_k = E_i - BE_k$$

For example, a 40 keV photon interacting with the K-shell electron of an iodine atom in a sodium iodide detector (BE_k = 33.17 keV) would eject a photoelectron with 6.83 keV of kinetic energy. The low-energy photoelectron would not travel far and would release its energy close to its site of origin by causing secondary ionization of other atoms within its vicinity. The principal ionization in matter caused by photon interactions is due to the ionizations produced by the electrons set in motion. Because an inner-shell electron is removed in a photoelectric interaction, a characteristic x-ray or Auger electron will also be produced as secondary radiations when the K-shell vacancy is filled by an outer-shell electron.

The probability of a photoelectric interaction increases with the atomic number of an absorber. The probability of photoelectric interaction becomes greatest when the photon's energy approximates the electron's binding energy. This is evidenced by an abrupt increase in photon attenuation at electron binding energies of the absorber. In Figure 4-7, photon attenuation in lead is increased about 5-fold when its energy is equal to the K-shell binding energy (88 keV). This increased probability occurs because all of the K-shell electrons become available for photoelectric interaction. The increased cross-section for photoelectric interaction has significance in the response of ionization detectors where detection efficiency of weak photons increases. In soft tissue, the photoelectric effect is a significant interactive process for photon

TABLE 4-5 Relative Importance of Different Types of Absorption in Water

	Percent Absorption	
Photon Energy	Photoelectric	Compton
10 keV	95	5
25 keV	50	50
50 keV	10	90
100 keV	1	99
150 keV–3 MeV	0	>99

Source: Reference 7.

energies up to 50 keV (Table 4-5). Therefore, radionuclides with photon energies below 50 keV, such as ^{125}I (27–35 keV), are unsatisfactory for diagnostic imaging because of high tissue attenuation by the photoelectric effect.

Compton Scatter

The process of Compton scatter involves the interaction of a medium-energy photon with a loosely bound outer-shell electron. A recoil electron is ejected from the atom and a secondary Compton-scattered photon is produced, having a longer wavelength and reduced energy. Thus, an ion pair is produced in the Compton interaction as shown in Figure 4-8. The energy of the scattered photon E_s can be calculated from the following relationship, where photon energy E is in keV:

$$E_s = \frac{E_i}{1 + \frac{E_i}{511}(1 - \cos\theta)} \quad (4\text{-}7)$$

The kinetic energy given to the Compton electron depends on the angle of interaction (θ). A direct hit with a scatter angle of 180° transfers the largest amount of energy to the Compton electron, whereas a grazing hit at near 0° transfers the least amount to the Compton electron. In any event, the kinetic energy of the Compton electron (KE_{ce}) is the difference between the energies of the incident photon (E_i) and the scattered photon (E_s) given by

$$KE_{ce} = E_i - E_s \quad (4\text{-}8)$$

The energy of the Compton-scattered photon also varies with the angle of scatter. While a backscatter Compton interaction at 180° transfers the maximum energy to the Compton electron, it also confers the minimum energy to the Compton-scattered photon. Thus, substituting 180° for θ in Equation 4-7 gives the minimum energy of the scattered photon:

$$E_s^{min} = E_i / [1 + (2E_i / 511)] \quad (4\text{-}9)$$

Thus, the maximum kinetic energy of the Compton electron is

$$KE_{ce}^{max} = E_i - E_s^{min} \quad (4\text{-}10)$$

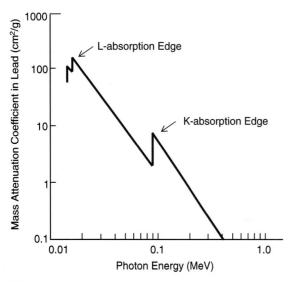

FIGURE 4-7 Photoelectric attenuation coefficient in lead as a function of photon energy. An absorption edge occurs at the binding energies of the K shell (K_{BE} = 88 keV) and L shell (L_{BE} = 13–15 keV), where the probability of interaction is highest at photon energies just above the binding energy.

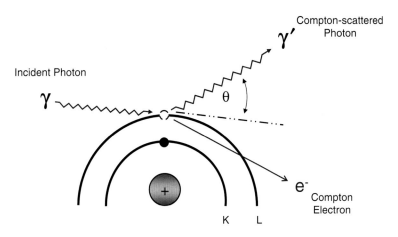

FIGURE 4-8 Compton scatter. An incident photon transfers part of its energy to a loosely bound outer-shell electron that is ejected from the atom. The remaining energy is associated with the scattered photon. The electron's binding energy is insignificant relative to the photon's energy and is disregarded. The kinetic energy of the ejected electron (KE_L) is equal to the difference between the incident and scattered photon energies, which depends on the angle of scatter θ. The ejected electron acquires maximum kinetic energy when θ is 180°.

Substituting Equation 4-9 into Equation 4-10 gives

$$KE_{ce}^{max} = E_i^2 / (E_i + 255.5) \qquad (4\text{-}11)$$

The relationships between the energies of the incident photon, scattered photon, and Compton electron at various angles of scatter for ^{125}I, ^{99m}Tc, ^{131}I, and ^{137}Cs photons are given in Table 4-6. The very small energy required to overcome electron-binding energy has been ignored in these calculations. The Compton electron set in motion will go on to produce secondary ionizations of other atoms. The Compton-scattered photon can continue to undergo additional Compton interactions until the photon is eventually absorbed by a photoelectric interaction. This series of interactions in matter is summarized in Figure 4-9.

The probability for photon interaction in tissue by the photoelectric effect and Compton scatter is about equal at photon energies of 25 keV. Compton scatter predominates from 100 keV to 3 MeV (Table 4-5, Figure 4-10).[7] Photons emitted from most radionuclides in nuclear medicine interact in tissue initially by Compton scatter. The probability of a Compton interaction depends upon electron density of the absorber. High-density materials provide more stopping power for photons undergoing photoelectric and Compton interactions, because more atoms (and electrons) are present per unit volume of absorber compared with low-density material. For this reason, lead ($d = 11.35$) is a good absorber of gamma radiation.

Pair Production

The pair production process involves the interaction of a very high energy photon within the nuclear force field with conversion of the photon into two electronic particles: one positron and one negatron (Figure 4-11). Because two electron

TABLE 4-6 Energy of Compton-Scattered Photons and Compton Electrons[a]

Radionuclide	Photon Energy (keV)	Angle of Scattered Photon, θ							
		45°		90°		135°		180°	
		E_s	KE_{ce}	E_s	KE_{ce}	E_s	KE_{ce}	E_s	KE_{ce}
^{125}I	27.5	27.1	0.4	26.1	1.4	25.2	2.3	24.8	2.7
^{99m}Tc	140.5	130.0	10.5	110.2	30.3	95.6	44.9	90.7	49.8
^{131}I	364	301.1	62.9	212.6	151.4	164.3	199.7	150.1	213.9
^{137}Cs	662	479.9	182.1	288.4	373.6	206.1	455.9	184.3	477.7

[a] E_s = energy of the scattered photon, KE_{ce} = kinetic energy of the Compton electron.

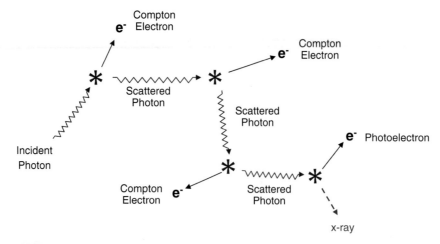

FIGURE 4-9 A photon of moderate energy (0.1–3 MeV) undergoes multiple Compton interactions in matter. Each Compton interaction produces a Compton electron and a scattered photon of diminished energy. The process ends with a photoelectric interaction, releasing an inner-shell photoelectron and a characteristic x-ray when the vacant electron shell is filled.

masses are created, the minimum energy required for this process to occur is 1.022 MeV. Any photon energy above 1.022 MeV is distributed to the particles as kinetic energy. The probability for pair production increases with increasing atomic number of the absorber because of the increased nuclear force field present with high Z material. Pair production begins to occur in soft tissue with photon energies above 3 MeV (Figure 4-10) and therefore is not an important mode of photon absorption with the radionuclides used in nuclear medicine.

Table 4-7 summarizes the products of photon interactions in matter.

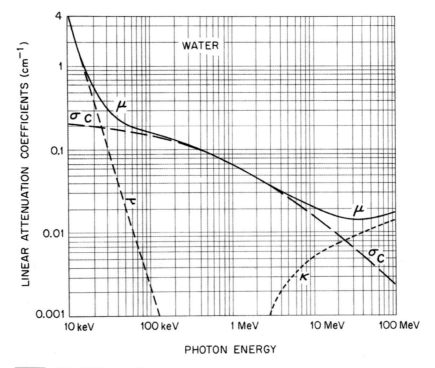

FIGURE 4-10 The behavior of the linear attenuation coefficients for photon absorption in water for photoelectric effect, τ, Compton scatter, σ, and pair production, κ, as contributors to the total attenuation coefficient, μ. (Reprinted from reference 7.)

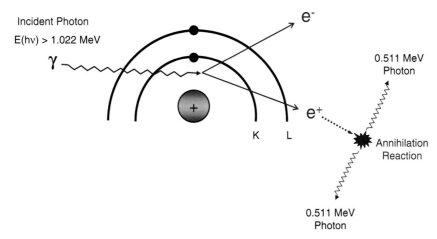

FIGURE 4-11 Pair production. An incident photon of at least 1.022 MeV in the vicinity of the nuclear force field is transformed into two electrons, one negative and one positive. The positron eventually is annihilated outside the atom, producing two 0.511 MeV photons.

TABLE 4-7 Products of Photon Interactions in Matter

Photon Interaction	Electrons	Secondary Photons
Photoelectric effect	Photoelectron, Auger electrons	Characteristic x-rays
Compton scatter	Recoil electron	Scattered photon
Pair production	Positive and negative electron pair	Annihilation photons

REFERENCES

1. Friedlander G, Kennedy JW, Miller JM. *Nuclear and Radiochemistry.* 2nd ed. New York: Wiley; 1964:86–130.
2. Rohrer RH. Nuclear physics and radiation. In: Wagner HN Jr, Szabo Z, Buchanan JW, eds. *Principles of Nuclear Medicine.* 2nd ed. Philadelphia: WB Saunders; 1995.
3. Glasstone S. *Sourcebook on Atomic Energy.* 3rd ed. New York: Van Nostrand Reinhold; 1967:395–431.
4. Chase CD, Rabinowitz JL. *Principles of Radioisotope Methodology.* 3rd ed. Minneapolis: Burgess; 1967:244–8.
5. Ehmann WD, Vance DE. *Radiochemistry and Nuclear Methods of Analysis.* New York: Wiley; 1991:253–311.
6. Parr A, Jay M, Digenis GA, et al. Radiolabeling of intact tablets by neutron activation for in vivo scintigraphic studies. *J Pharm Sci.* 1985; 74:590–1.
7. Grodstein GU. National Bureau of Standards Circular 583; U.S. Department of Commerce: 1957.

CHAPTER 5
Radiation Detection and Instrumentation

The detection and measurement of radiation in nuclear pharmacy and nuclear medicine are important for radiation protection purposes, for accurate assessment of radiopharmaceutical activity, and for imaging procedures. Proper use of radiation detection equipment requires an understanding of instrument construction and operation. This chapter covers the principal methods of detecting ionizing radiation, namely, ion collection and scintillation; the instruments used for radiation detection and measurement; imaging systems; and the counting of radioactive samples.

ION COLLECTION METHODS OF RADIATION DETECTION

Ion collection methods of radiation detection are based on the ability of radiation to ionize atoms of gas in an ionization detector. Typical gases used in ionization detectors are air, helium, and argon. The gas is contained in a sealed chamber configured with a positive and negative electrode pair (Figure 5-1). A power supply creates a voltage potential across the electrodes. An ammeter measures the current, which is directly proportional to the number of ion pairs produced by the radiation source.[1]

Figure 5-2 illustrates the relationship between applied voltage and current in a gas-filled ionization detector. The change in current as voltage is increased can be understood by considering a source of activity exposed to an ionization detector containing a finite number of gas molecules. Under these conditions the radiation enters the chamber and ionizes a certain number of the gas molecules. When no voltage is applied, the ion pairs recombine (*recombination region*) and no current is generated. However, as the voltage is increased, some of the electrons in the ion pairs are collected at the anode, generating a current. A greater proportion of the total number of electrons initially generated is collected with each proportional increase in voltage, producing a proportional increase in current. At a certain voltage, all of the primary electrons are collected and a current plateau is reached, known as the *ionization region*. Additional increases in voltage do not cause an appreciable increase in current over the plateau; the primary electrons are simply collected at a faster rate as voltage is increased over the plateau region. Over the voltage range of this plateau simple ionization occurs. However, additional increases in voltage beyond the plateau produce proportional increases in current caused by the generation of secondary ionizations from primary electrons moving at high speed toward the anode. This is known as the *proportional region*. Toward the end of the proportional region most of the gas molecules in the chamber are ionized and a region of nonproportionality is reached. As voltage is increased above this point, another current saturation plateau is reached, known as the *Geiger region*. The voltage in this region is high enough that an initial ionizing event in the detector will result in an avalanche of ion pairs being produced as all available gas molecules in the detector are

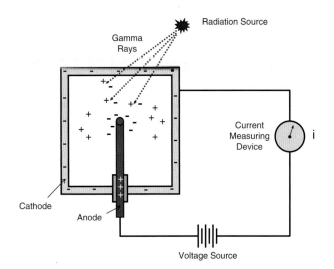

FIGURE 5-1 A simple gas-filled ionization detector. Electrons released by ionization of gas molecules by gamma ray interactions are collected at the central anode, producing a current proportional to the amount of ionization; i = current.

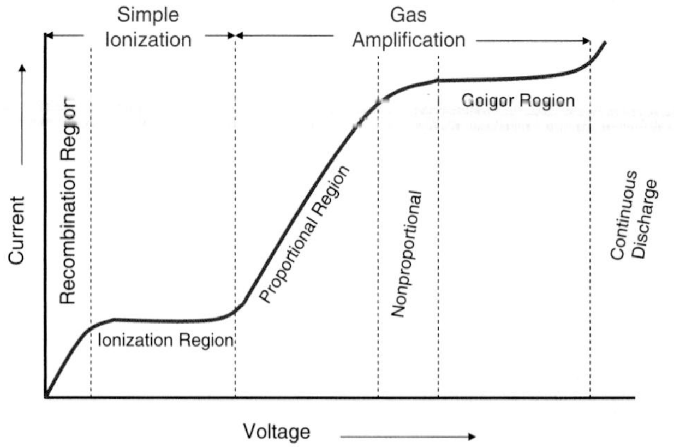

FIGURE 5-2 Current–voltage curve for gas-filled ionization detectors. See text for explanation.

ionized by high-speed electrons moving toward the anode. At still higher voltages above the plateau, a continuous discharge or arcing of current across the electrodes occurs as electrons are extracted from the cathode to the anode by the high voltage potential. This voltage range should be avoided because the detector will be damaged.

Three types of ion collection instruments are represented in this current/voltage response curve: ionization chambers, proportional counters, and Geiger-Müller (GM) counters. *Ionization chambers* are useful for measuring medium- to high-intensity sources of radioactivity and have operating voltages in the range of 50 to 300 volts (saturation current plateau region), depending on the design of the detector. Examples of instruments include the hand-held "Cutie Pie" ionization chamber survey meter, which is useful for measuring output from high-activity sources such as x-ray machines, and the dose calibrator, which is used to measure the activity of radiopharmaceuticals in the microcurie to curie range. GM detectors are capable of measuring low levels of radiation and are useful for surveying the work environment for radioactive contamination. Their operating voltages are usually set near 1000 volts (Geiger plateau region).

Radionuclide Dose Calibrator

Figure 5-3 is a block diagram of a dose calibrator's essential components.[2] The dose calibrator is used to measure the activity of radiopharmaceuticals. Its major components are an ionization chamber constructed with a central well for accepting radiopharmaceutical vials and syringes, and electronic circuitry to convert the ionizations generated in the chamber

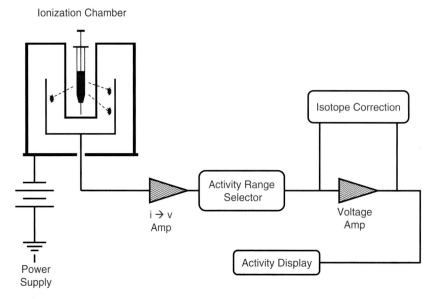

FIGURE 5-3 Block diagram of a dose calibrator; i = current, v = voltage.

into an activity value on the readout. The readout, in curie units or becquerels, is proportional to the amount of activity placed in the ionization chamber well. Operating voltage applied between the anode and cathode in the detector is about 150 volts. Typically, the ionization chamber contains several atmospheres of argon gas to increase detection sensitivity. The range selector is a circuit that adjusts the instrument for the activity range (microcuries, millicuries, curies) being measured. Modern instruments employ automatic ranging. The useful activity range of a dose calibrator is 0.1 mCi to approximately 10 Ci (3.7 MBq to 3.7×10^5 MBq), although activities below 0.1 mCi (3.7 MBq) can be accurately measured if background activity is low. The response, R_i (current output), of the ionization detector to a radionuclide is related to the sum total of its photon energy, E_i, of given intensity, I_i, and the sensitivity, S_i, of the detector to the photon energy, according to the following relationship:[3]

$$R_i = \sum E_i I_i S_i$$

The sensitivity of the detector is defined as its response to a unit of activity of a particular radionuclide (e.g., 1 Ci) divided by its response to the same amount of activity of a reference nuclide (e.g., ^{60}Co). When these measurements are made for several different radionuclides, a sensitivity curve can be generated (Figure 5-4), which shows that the sensitivity of the ionization chamber is not linear with photon energy. The nonlinearity is due to several factors; above 200 keV, ionizations are mainly due to electrons generated by Compton scattering of photons by the filling gas (e.g., argon) and the chamber walls; as photon energy decreases below 200 keV, chamber sensitivity increases as a result of photoelectric interactions, peaking around 60 keV; below 60 keV, sensitivity begins to decrease because of photon attenuation in the ion chamber wall, and photons below about 13 keV are significantly absorbed by the chamber wall and do not enter the chamber gas. The response of the ionization detector to any radionuclide can be calculated from the sensitivity curve and the decay properties of the radionuclide.

The output response of the ionization chamber per unit of activity will not be the same for all radionuclides because of differences in their decay properties (i.e., photon energy and intensity). Therefore, a gain adjustment is required in the electrical circuit to standardize the output so that equal activities of different radionuclides will display the same value on the instrument's readout.[3] The gain adjustment is accomplished by selecting a radionuclide calibration setting for the radionuclide being measured. The instrument is adjusted for background activity, and the source is then placed in the ionization well and its activity is displayed on the readout.

Radionuclide calibration settings on a dose calibrator are established by the manufacturer and confirmed with certified radionuclide standards. The standard is measured in the dose calibrator, and a calibration setting is selected that displays the known activity of the standard.

Standards may be primary standards or secondary standards (reference standards). A primary standard has its absolute disintegrations determined in a detector with 4-π geometry. A typical primary standard supplied by the National Institute of Standards and Technology (NIST) is 5 grams of radioactive solution in a borosilicate glass ampul with a wall thickness of about 0.6 mm. Primary standards supplied by NIST have a nominal uncertainty of 1% to 2%. Secondary standards are produced by comparing a source and a primary standard; they have an uncertainty in the range of 3% to 4%. Secondary standard sources, such as ^{133}Ba and ^{137}Cs, in sealed vials are typically used to check the accuracy and precision of dose calibrators in a nuclear pharmacy. The full complement of quality control tests required for dose calibrators is discussed in Chapter 14.

Adjustment of Dose Calibrator Measurements

The dose calibrator will produce accurate readings of most radionuclides contained in the standard plastic syringes and glass vials used in routine practice. The accuracy of a measurement, however, will depend on the radionuclide's decay properties and the source geometry (i.e., container density and volume). Beta emitters and gamma emitters with low-energy photons (<50 keV) have the greatest potential for errors in measurement caused by attenuation of radiation in the container. High-energy beta particles generate bremsstrahlung, a weak electromagnetic radiation. Beta particles and bremsstrahlung are easily attenuated in container materials; therefore, correction factors or specific calibration settings may need to be determined for accurate measurements. In general, gamma emitters with photon energies greater than 100 keV are not significantly attenuated by standard containers, and manufacturer calibration settings on the dose calibrator will yield accurate measurements. However, some gamma emitters may also emit low-energy photons with high intensity that can have significant and variable attenuation in containers. In these situations, correction factors must be applied to obtain a correct measurement of activity.[4] For example, ^{125}I may require a correction factor

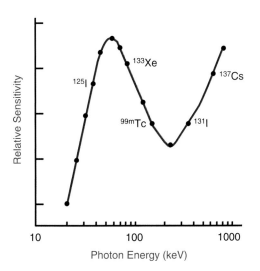

FIGURE 5-4 Relative change in dose calibrator ion chamber sensitivity with photon energy. See text for explanation.

TABLE 5-1 Photon-Emitting Radionuclides Subject to Attenuation of Low-Energy Photons in Dose Calibrator Measurements

Radionuclide	Principal Gamma Rays		K-Shell x-Rays	
	Energy (keV)	% Abundance	Energy (keV)	% Abundance
^{111}In	171	91	23	69
	245	94	26	13
^{123}I	159	83	27	70
			30–32	15
^{125}I	35	7	27	114
			31–32	24
^{133}Xe	81	38	31–36	49
^{201}Tl	135	2.7	69–82	75
	167	10		

of 2 between measurements made in plastic syringes and glass vials[2] and ^{123}I and ^{111}In may need correction for low-energy x-ray components in their decay schemes.[5,6]

Instead of applying correction factors to the calibration setting assigned by the dose calibrator manufacturer, another approach is to establish a new calibration setting for a specific radionuclide and container configuration. A known amount of activity of a standard source of the radionuclide in question is placed into the desired container and assayed in the dose calibrator. A new calibration setting is then chosen to display the activity in the container. This technique has been employed for beta-emitting radionuclides, such as ^{153}Sm and ^{90}Y, where container type and volume may affect the measurement.[7]

An alternative to using a standard source for establishing a new calibration setting relies on the manufacturer's calibration information on the radiopharmaceutical vial label and the difference in vial measurements before and after a dose is removed and placed into a syringe. In this method, the activity in a radiopharmaceutical vial is first measured in the dose calibrator and recorded. Based on the calibration information on the radiopharmaceutical label, a volume of solution to contain a desired amount of activity is removed from the vial into a syringe. The vial is measured again and the remaining activity recorded. The difference in the vial measurements is the calculated amount of activity placed into the syringe. The syringe is then assayed in the dose calibrator and the calibration setting is adjusted to display the calculated activity transferred into the syringe. The specific syringe configuration and calibration setting are then used for all future measurements of the radionuclide.

Several radionuclides that are subject to variations in dose calibrator measurement due to attenuation of low-energy photons are listed in Table 5-1. Of particular interest are ^{123}I and ^{111}In. While these nuclides have high-energy gamma rays, they also have low-energy characteristic x-rays that are easily attenuated; thus, they require correction factors in order to obtain an accurate measurement. An alternative technique to the use of correction factors is the use of a copper filter during dose calibrator measurement.[4] Copper will significantly absorb low-energy x-rays with little attenuation of high-energy gamma photons (Figures 5-5 and 5-6). A copper well liner of 0.6 mm will absorb 99.5% of photons in the 20 to 50 keV range and thus will significantly reduce the low-energy x-ray component of ^{123}I and ^{111}In, eliminating the need for correction factors

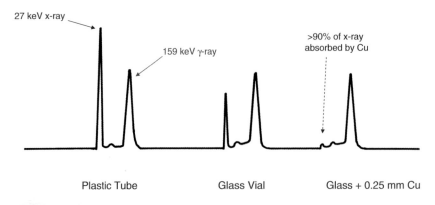

FIGURE 5-5 Sodium iodide-generated photon energy spectra of ^{123}I in different source configurations. Note that a glass container attenuates the characteristic x-ray to a greater extent than the gamma ray and that copper filtration removes more than 90% of low-energy x-rays that can contribute to inaccurate measurements of ^{123}I in a dose calibrator. There is only slight attenuation of the gamma ray by copper.

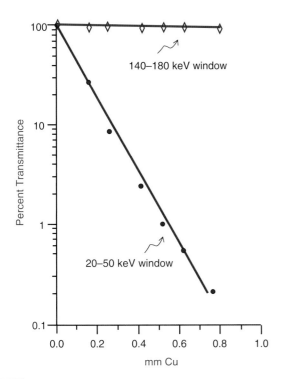

FIGURE 5-6 Attenuation of ^{123}I in copper of 27–32 keV characteristic x-rays (20–50 keV window) and 159 keV gamma rays (140–180 keV window) in a sodium iodide well counter.

for different container configurations with these nuclides (Table 5-2). When a copper well liner is used, new calibration settings must be established on the dose calibrator.

Syringe volume may affect the measurement of a radionuclide, particularly with beta emitters. Zimmerman et al.[7] demonstrated that the calibration setting required for an accurate measurement of ^{90}Y in a dose calibrator varies with syringe volume (Table 5-3) and model of dose calibrator. The magnitude of the volume effect over the range of 3 to 9 mL was 0.84% to 2.5% for Capintec dose calibrators (models CRC-12, CRC-15R, and CRC-35R), 1.2% for the AtomLab dose calibrator, and 5.2% for the PTW Curiementor 3 dose calibrator. The investigators concluded that a single calibration factor for measuring ^{90}Y solutions in 10 mL Becton Dickinson syringes can be applied over a volume range of 3 to 9 mL for the Capintec and AtomLab dose calibrators and that correction factors should be applied for the PTW Curiementor 3 unit.

Geiger-Müller (GM) Survey Meters

A GM survey meter is a very sensitive device for detecting small amounts of radiation. A typical GM meter reports radiation measurement in units of roentgens per hour (R/hr) or mR/hr. Some meters also have a scale for counts per minute. Other meters may use the International System of Units, measuring in grays per hour (Gy/hr) or mGy/hr. The high operating voltage of the GM detector generates a voltage pulse when a single ionization event occurs in the detector probe; therefore, the GM survey meter is suitable for low-level radiation detection. The GM survey meter is commonly used in radiation protection surveys to monitor work areas for ambient radiation exposure levels and radioactive contamination.

Detector probes used with GM survey meters can have different configurations to measure gamma radiation and beta radiation. GM probes are available with end-window or side-window configurations. A typical end-window tube is shown in Figure 5-7. The thin mica window allows passage of beta particles and weak gamma rays that normally would be stopped by the metal casing of the tube. Because of the high operating voltage in GM probes, radiation entering the probe produces primary ionizations that proceed to ionize the entire gas, generating an electric pulse. A quenching agent in the gas absorbs energy and momentarily stops discharge of the probe between ionizing events. This alternating ionization–quenching sequence produces electrical pulses that are audible as ticking sounds and drive an exposure meter.

GM detectors have no energy-discriminating ability, but after calibration they are useful for measuring exposure rate in roentgens per hour from gamma ray sources. GM counters are typically calibrated for exposure response in air with a high-energy ^{137}Cs source with a photopeak energy of 662 keV. The predominant interaction in the gas and wall of the detector probe is Compton scatter, liberating Compton electrons that generate ionizations in the gas. The ionizations produce an electrical pulse that drives a meter. The response of a GM detector over the energy range of a few hundred keV to 1 MeV is relatively flat, and exposure measurements are proportional to the amount of activity in a radioactive source. However, as photon energies decrease below 100 keV, photoelectric interaction increases. The mass attenuation coefficient τ for photoelectric interaction increases with the atomic number (Z) of the absorber and lower photon energies ($\tau \sim Z^3/E^3$). The higher Z of the GM detector wall and gas compared with air or soft tissue causes the detector to overrespond with low-energy photons, causing falsely high exposure readings. This overresponse can be corrected partially with an energy-compensated GM probe that uses a thin metallic shield to remove a significant portion of low-energy photons. In essence, energy compensation normalizes the detector's response over a wider range of photon energies. However, energy compensation causes some loss of detector sensitivity as a result of shielding of low-energy photons. Thus, uncompensated probes are better for detecting low levels of radiation because of their higher sensitivity, whereas energy-compensated probes are more accurate for exposure measurements of low-energy photons.

SCINTILLATION METHODS OF RADIATION DETECTION

Two basic types of instruments employ scintillation methods of radiation detection in nuclear pharmacy and nuclear medicine: solid-crystal scintillation counters and liquid scintillation counters. These instruments have a detector where radiation interacts coupled to electronic circuits for analyzing and recording

TABLE 5-2 Effect of Container Type and Volume on Dose Calibrator Measurement of ^{123}I and ^{111}In with and without Copper Filtration

Volume in Container (mL)	^{123}I Activity (mCi)[a]		^{111}In Activity (mCi)[b]	
	With Copper	Without Copper	With Copper	Without Copper
1-mL Syringe				
0.25	1.23[c]	1.74	1.18[c]	1.54
0.50	1.23	1.76	1.18	1.55
0.75	1.23	1.77	1.18	1.56
1.00	1.23	1.79	1.18	1.55
Mean	1.23	1.77 (+44%)	1.18	1.55 (+31%)
3-mL Syringe				
0.25	1.23	1.62	1.18	1.49
0.50	1.23	1.64	1.18	1.51
1.00	1.23	1.68	1.18	1.51
2.00	1.23	1.71	1.18	1.51
3.00	1.23	1.73	1.18	1.51
Mean	1.23	1.68 (+37%)	1.18	1.51 (+28%)
10-mL Syringe				
1.00	1.23	1.57	1.18	1.44
2.00	1.23	1.59	1.18	1.44
4.00	1.23	1.63	1.18	1.45
8.00	1.23	1.66	1.18	1.45
10.00	1.23	1.67	1.18	1.45
Mean	1.23	1.62 (+32%)	1.18	1.45 (+23%)
10-mL Vial				
1.00	1.20	1.30	1.18	1.28
2.00	1.20	1.28	1.18	1.27
4.00	1.20	1.25	1.18	1.26
8.00	1.19	1.22	1.18	1.24
10.00	1.19	1.21	1.18	1.23
Mean	1.20	1.25 (+4%)	1.18	1.26 (+7%)

[a] ^{123}I calibration setting: 052 with 0.64 mm copper; 277 without copper (manufacturer setting).
[b] ^{111}In calibration setting: 193 with 0.64 mm copper; 303 without copper (manufacturer setting).
[c] True activity of the source obtained from a National Institute of Standards and Technology primary standard.

TABLE 5-3 Dose Calibrator Settings for Measuring ^{90}Y Solutions as a Function of Syringe Volume

Syringe Volume (mL)	Calibrator				
	Capintec CRC-12	Capintec CRC-15R	Capintec CRC-35R	AtomLab	PTW Curiementor 3
3	58	56	55	391	1.031
4	58	56	54	392	1.032
5	58	56	55	393	1.024
6	58	56	54	394	1.020
7	57	56	55	394	1.004
8	57	52	54	394	0.993
9	57	52	53	395	0.980

Source: Reference 7.

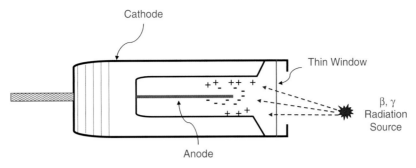

FIGURE 5-7 Cutaway diagram of an end-window Geiger-Müller detector. Radiation produces electrons from primary ionizations of the gas. The electrons move at a high rate of speed toward the anode, creating additional ionizations and ultraviolet radiation, which generate an exponential avalanche of electrons. The collection of electrons at the anode generate a current pulse.

the amount of radiation in a source. Solid-crystal detectors typically are sodium iodide (NaI) doped with a small amount of thallium (Tl). NaI(Tl) crystals scintillate more efficiently at ambient temperatures, whereas pure NaI requires impractical temperatures of minus 80°C. The NaI(Tl) crystal is hygroscopic, requiring a hermetically sealed metal casing. The NaI(Tl) detector is primarily used to detect electromagnetic radiation that is able to penetrate into the crystal. Beta emitters, such as ^{3}H and ^{14}C, are best counted by liquid scintillation detection because beta particles are not able to penetrate into the NaI(Tl) detector. With liquid scintillation the radioactive sample is dissolved or suspended in a scintillation "cocktail" composed of solvent and fluor compounds. The intimate admixture of beta-emitting molecules with fluor compounds in the cocktail provides efficient detection of beta particle radiation. Except for differences in detector composition, the operating principle of liquid and solid-crystal scintillation detectors is similar.

Solid-Crystal Scintillation Spectrometer

A basic scintillation spectrometer for detecting gamma radiation consists of a sodium iodide crystal detector optically coupled to a photomultiplier (PM) tube. An electrical pulse or output signal produced in the PM tube can be amplified and transmitted to a pulse-height analyzer (PHA) where pulses can be rejected or accepted as counts. Various devices are used to record output from the spectrometer. Typically, single-channel PHAs record counts in a preselected region (energy window) of the gamma energy spectrum as counts on a scaler or as counts per minute on a rate meter, whereas multichannel PHAs record data output on a monitor screen where the entire gamma energy spectrum of the radionuclide is displayed and sample counts are recorded in a selected energy window in the spectrum. The data can be stored in a computer (Figure 5-8).

Figure 5-9 illustrates a sodium iodide detector and PM tube. The NaI(Tl) crystal can be configured as a flat-surfaced

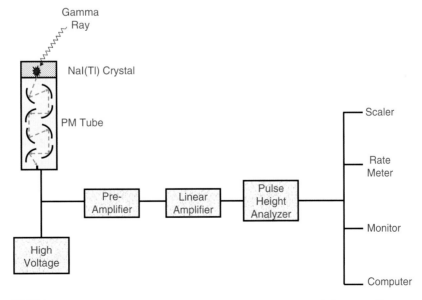

FIGURE 5-8 Block diagram of a sodium iodide scintillation counter. See text for explanation.

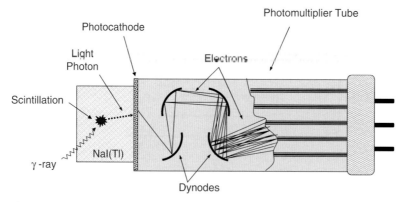

FIGURE 5-9 Sodium iodide crystal–PM tube scintillation detector. Total light output is proportional to gamma ray energy deposited in the crystal. If one electron ejects four electrons from each dynode, 10 dynodes will result in 4^{10} or about 1 million electrons collected at the anode of the PM tube.

probe or with a well for counting test-tube samples. High-energy photons interacting within the crystal transfer energy to the sodium iodide molecules by Compton scatter and photoelectric interactions. The energy of electrons released from the ionization process is mostly absorbed as heat. However, when a NaI(Tl) crystal is activated with 0.1% thallium, some of the excited electrons become trapped in the vicinity of the thallium atoms and are subsequently released as light photons of about 3 eV. This is the scintillation event. About 20 to 30 of these 3 eV light photons are produced per keV of energy transferred to the crystal.[8] There is a reflective surface on the inside of the crystal to increase capture of the light photons. Light photons cause electrons to be ejected from the photosensitive cathode of the PM tube that is optically coupled to the NaI(Tl) crystal. The electrons are then attracted to a series of dynodes, each about 100 V more positive than the previous one. On average, about four electrons are ejected from each dynode for each incident electron on the dynode, so that electron multiplication occurs. A series of 10 dynodes will result in 4^{10} or about 1 million electrons, creating a small electrical pulse at the collecting anode. There is a proportional relationship between the magnitude of the electrical pulse generated and the amount of gamma energy deposited in the crystal. Thus, a 200 keV gamma ray will produce a voltage pulse that is twice the height of that produced by a 100 keV gamma ray. Selection of pulses for counting and analysis is accomplished by a PHA circuit.

Pulse-Height Analyzer

The electrical pulse from the PM tube is fed into a preamplifier located at the base of the detector. The preamplifier serves to amplify weak pulses, match impedance between the detector and other components, and shape the pulse for optimal processing. From the preamplifier the pulse enters the PHA, which has a gain control for modifying pulse amplitude for acceptance or rejection of a pulse for counting purposes and for rejecting voltage pulses created by background and scattered radiation. The PHA is adjusted by potentiometers (variable resisters) that control a lower-level energy discriminator (LLD) and an upper-level energy discriminator (ULD). The LLD on some instruments is called the *threshold* or *base level*. The LLD sets a lower or threshold voltage for counting, whereas the ULD sets an upper voltage for acceptance. The opening or space between the LLD and the ULD is called the window. An input pulse from the detector must fall within this window to be counted, as determined by an anticoincidence circuit. An anticoincidence circuit has two input terminals that transmit an output pulse (a count) only if one terminal receives a pulse but not if both terminals receive a pulse.[9] Thus, if a pulse from the detector exceeds the voltage of the LLD but not the ULD, it will fall within the counting window. This will cause a single input pulse to go to the anticoincidence circuit, and an output pulse will be generated to produce a count (Figure 5-10). If an input pulse from the detector falls below the threshold voltage of the LLD, there will be no input pulse to the anticoincidence circuit. No output pulse is generated and no count is produced. Likewise, if an input pulse exceeds both LLD and ULD voltages, in each case generating an input pulse, no count will occur because the two coincident input pulses sent to the anticoincidence circuit will be rejected.

The potentiometers of the LLD and ULD typically range from 0 to 1000 divisions (0–10 volts) each. The ULD typically rides on top of the LLD. For example, if the LLD is set at 100 divisions and the ULD at 50 divisions, the counting window will be between 100 and 150 divisions. If the LLD is raised to 150 divisions, the counting window would then be between 150 and 200 divisions. Thus, the PHA permits independent counting of different photon energies through adjustment of the counting window (Figure 5-11). The ULD can be disengaged so that all energies that exceed the LLD will be counted.

Spectrometer Calibration and Linearity

Calibration of the spectrometer converts the potentiometer units or divisions into energy units. This is accomplished by

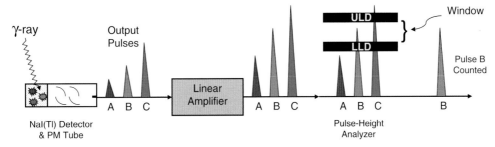

FIGURE 5-10 Schematic of pulse-height analysis in a scintillation counter. Pulses A, B, and C are amplified linearly before processing by the pulse-height analyzer. Pulse B is counted because it falls within the "window" created by the lower-level discriminator (LLD) and upper level discriminator (ULD). Pulse A and pulse C are rejected and do not produce a count. See text for explanation.

setting a narrow window for the PHA centered on the photopeak energy of a radionuclide of known energy. Typically, ^{137}Cs is used. Thus, a 10 keV window is established with the LLD at 657 divisions and the ULD at 667 divisions, so that the photopeak energy of 662 keV is centered in the window. The high voltage is adjusted until a maximum count rate is achieved in the counting window. During the calibration procedure, increases in the high voltage applied to the PM tube will cause proportional increases in the pulse heights generated in the detector until the pulses appear in the counting window. When the counting rate in the window reaches a maximum, the pulse heights of the 662 keV gamma rays will be centrally positioned in the counting window. At this voltage, known as the operating voltage, the spectrometer is calibrated (i.e., 662 keV per 662 divisions or 1 keV per division). Alternatively, the high voltage can be adjusted with a counting window centered at 331 divisions, in which case the spectrometer will be calibrated at 2 keV per division (662 keV/331 divisions).

The amplifier of a scintillation counter ideally should exhibit a linear response, meaning that the amplitude of an output pulse should be proportional to the energy absorbed in the detector over a wide range of energies. Linear amplification should permit a maximal count rate of any other radionuclide when a PHA window is centered about its photopeak energy. Thus, ^{137}Cs (photopeak = 662 keV) will be found at 662 divisions, ^{131}I (photopeak = 364 keV) will be found at 364 divisions, ^{99m}Tc (photopeak = 140 keV) will be found at 140 divisions, and so forth. However, amplifiers exhibit some nonlinearity, which typically occurs at energies below 150 keV when calibration is done with a high-energy nuclide such as ^{137}Cs. For example, with a spectrometer calibrated with ^{137}Cs at 662 keV, the best counting rate for ^{99m}Tc may be found to occur with a window centered at 145 keV instead of 140 keV. Thus, it is often best to recalibrate or "peak in" a counting system with the radionuclide of interest at its photopeak energy. This will create the optimal PHA window for counting the radionuclide.

Pulse-Height Energy Spectrum

A pulse-height energy spectrum of a gamma-emitting radionuclide can be created after a source is counted in a calibrated

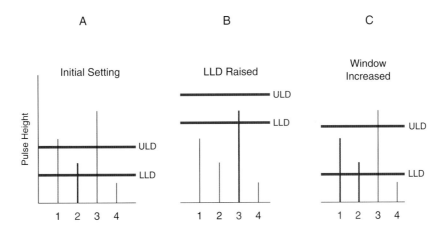

FIGURE 5-11 Effect of pulse height analyzer settings on acceptance or rejection of various pulse heights. (A) Initially only pulse 2 is counted; (B) raising the lower-level discriminator (LLD) adjusts the window so that only pulse 3 is counted; (C) the window is increased by raising the upper level discriminator (ULD) so that pulses 1 and 2 are counted.

scintillation counter. Photopeak energies in the spectrum can then be identified. A scintillation counter may be equipped with a single-channel PHA or a multichannel PHA. A single-channel analyzer (SCA) requires manual adjustment of the PHA window for counting a radioactive source. There are two operational types of PHA discriminators; in one type the ULD is independent of the LLD, and in the other the ULD is dependent on or "rides on top of" the LLD. For example, with the first type, a 10 keV window centered on 605 keV would require setting the LLD at 600 keV and the ULD at 610 keV. With the second type, the LLD would be set at 600 keV and the ULD at 10 keV. In the latter case, resetting the LLD to 610 would give a 10 keV window between 610 and 620, and so forth.

With a SCA, a plot of the ^{137}Cs photopeak can be obtained by counting a ^{137}Cs source in a 10 keV window for 1 minute time periods at various LLD settings ranging from 560 keV to 720 keV. When the count per minute (cpm) per energy interval is plotted on the y-axis versus energy on the x-axis, a pulse-height energy histogram is created (Figure 5-12). A curve drawn through the midpoint of each energy interval outlines the ^{137}Cs photopeak. This same process can be repeated over a wider range of LLD settings from 0 keV to 1000 keV to generate the entire energy spectrum of ^{137}Cs.

A pulse-height energy spectrum for any radionuclide is best established with a multichannel analyzer (MCA). The MCA has electronic circuitry that assigns the various voltage pulses generated in the NaI(Tl) detector into "energy channels" (Figure 5-13), producing a gamma energy spectrum on the monitor with energy in kiloelectron volts on the x-axis versus counts per minute per kiloelectron volt on the y-axis. The gamma energy spectrum for ^{137}Cs is shown in Figure 5-14. The count in a specific channel can be determined by placing an electronic cursor in that channel. Alternatively, the total count in the area under the curve of a photopeak can be determined, which is typically how the amount of activity in a sample is measured. In Figure 5-14 this is shown as the gray-shaded area from 600 keV to 720 keV. Since each radionuclide has a unique pulse-height energy spectrum, the MCA is useful in identifying which radionuclides are present in an unknown sample (Figures 5-15 and 5-16).

Pulse-Height Spectrometry

Pulse-height spectrometry involves analysis of the photon energy peaks in a gamma energy spectrum. The energy peaks may arise from several different types of interactions in the detector. The amplitude of a voltage pulse created by a photon interaction is proportional to the energy released in the NaI(Tl) detector. The full amount of a photon's incident energy may or may not be released in the detector. For example, if a photon interacts by a photoelectric interaction, all of the photon's energy will be released in the detector. The photoelectron produced in the interaction will generate a characteristic x-ray. If the x-ray is absorbed in the detector, an energy pulse will appear in the photopeak region of the pulse-height energy spectrum. If the x-ray escapes from the detector, the pulse will appear in the spectrum at a position equal to the photopeak energy minus the x-ray energy. A photon may interact in the detector by Compton scatter with escape of the scattered photon. In this situation, the amount of energy deposited in the detector will be equal to the energy imparted by the Compton electron and the pulse will appear in the Compton region of the pulse-height energy spectrum. Thus, the information appearing in the pulse-height energy spectrum is influenced by a variety of events that may or may not occur in the detector. Pulse-height spectrometry involves a consideration of the possible events that may contribute to the energy spectrum during scintillation counting.

Some unique features of the pulse-height energy spectrum should be considered in pulse-height spectrometry. A good start would be to consider an idealized spectrum of ^{137}Cs, shown in Figure 5-17. If we consider the 662 keV gamma ray produced in the decay of ^{137}Cs, one possibility is complete absorption of the photon in the detector. This could occur, for example, by a combination of Compton scatter and photoelectric interactions within the NaI(Tl) crystal. In this situa-

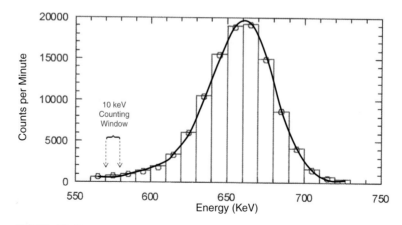

FIGURE 5-12 Pulse-height energy histogram of the ^{137}Cs photopeak generated with a single-channel analyzer by counting the source for 1 minute in a 10 keV window at various baseline discriminator settings from 560 keV to 720 keV.

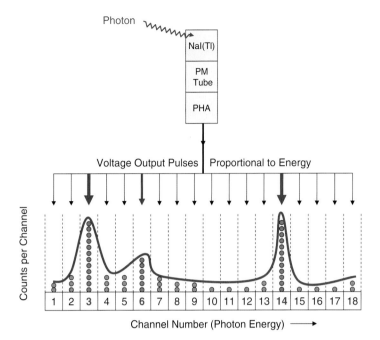

FIGURE 5-13 Diagram of multichannel analyzer operation. Gamma rays of different energies emitted from a radionuclide source deposit different amounts of energy in the NaI(Tl) detector, generating a variety of output voltage pulses that are proportional to energy. These pulses, which are recorded as counts, are sorted into channel numbers according to their energy. The buildup of counts of similar energy in specific channels generates a pulse-height energy spectrum for the radionuclide.

tion, the amplitude of the pulse would appear in the *photopeak* region, shown as a vertical line at 662 keV in Figure 5-17. Another possible interaction could be a 180° Compton scatter of the 662 keV photon with escape of the backscattered photon from the detector. In this case, the amplitude of the pulse generated in the detector would reflect the maximum possible amount of energy that could be given to a Compton electron (KE_{ce}^{max}), as described in Equation 4-11 in Chapter 4. The pulse generated in the detector by this interaction would appear in the region of the spectrum called the *Compton edge*. For ^{137}Cs the pulse would appear at 478 keV in the spectrum (see Table 4-6 in Chapter 4). If the 180° backscattered

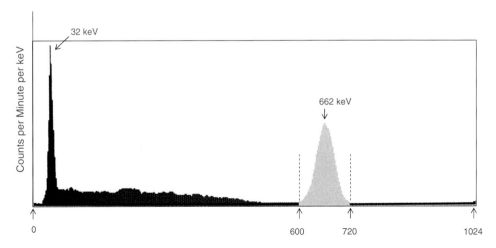

FIGURE 5-14 Complete pulse-height energy spectrum of ^{137}Cs generated with a multichannel analyzer, showing the 32 keV x-ray peak of the ^{137}Ba daughter and the 662 keV gamma ray photopeak emitted by the metastable ^{137m}Ba daughter.

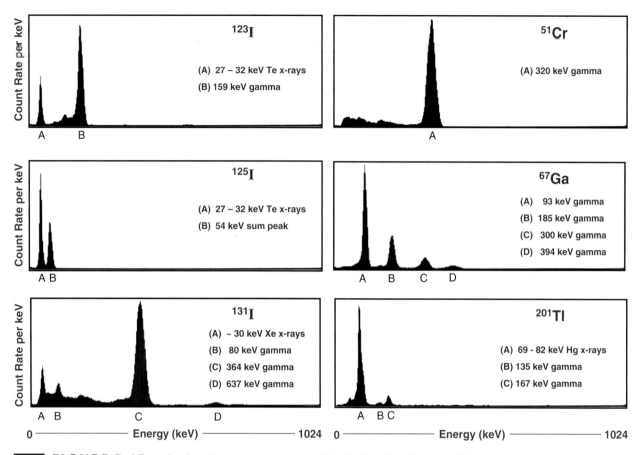

FIGURE 5-15 Pulse-height energy spectra for ^{123}I, ^{125}I, ^{131}I, ^{51}Cr, ^{67}Ga, and ^{201}Tl in a 2 × 2 inch sodium iodide crystal and multichannel analyzer.

Compton photon does not escape but undergoes additional interactions in the detector, the pulse generated will appear in the region between the Compton edge and the photopeak energy. If complete absorption of the Compton-scattered photons occurs in the detector, a pulse will appear at the photopeak energy of 662 keV. A Compton interaction that occurs at a scatter angle less than 180° with escape of the Compton photon would generate pulses in the Compton region of the spectrum below the Compton edge equivalent to the energy given to the Compton electron.

Another feature that may appear in the pulse-height energy spectrum is a *coincidence sum peak*. A coincidence sum peak can occur when two or more gamma rays are produced per disintegration of the radionuclide, which causes them to be recorded as a single peak equal to the sum of the two gamma ray energies. Sum peaks can also result from the addition of gamma ray and x-ray peaks. In nuclear medicine, three single-photon emission computed tomography (SPECT) radionuclides that can exhibit sum peaks in their spectra are ^{125}I, ^{123}I, and ^{111}In. Electron capture by these radionuclides causes the generation of characteristic x-rays by their respective daughter nuclides. ^{125}I decay generates two ^{125}Te K-shell x-rays during its decay, one at 27.2 keV (40%) and one at 27.5 keV (74%), and its spectrum displays a photopeak at 27 keV and a sum peak at 54 keV (Figure 5-15). A high abundance of K-shell x-rays is also produced by ^{123}I (27.2 keV, 24% and 27.5 keV, 46%) and ^{111}In (23 keV, 24% and 23.2 keV, 45%) (Table 5-1). Sum peaks are more prevalent in the spectrum when a source is counted within the well detector (Figure 5-18). If the source is moved out of the well, counting efficiency declines and sum peaks are less evident or are eliminated. Counting a source out of the well may generate a *lead x-ray peak* in the spectrum. Such peaks are caused by photoelectric interactions of γ-rays in the lead shielding, producing characteristic lead x-rays (73–88 keV) that are then detected in the sodium iodide crystal.

The pulse-height energy spectrum may contain other features. A *characteristic x-ray peak* due to internal conversion can appear in the pulse-height spectrum. A good example of this is the decay of ^{137}Cs. ^{137}Cs decays by beta particle emission to an excited state of ^{137m}Ba. ^{137m}Ba de-excites by emission of a 662 keV gamma ray. A fraction of these gamma rays produce a K-shell conversion electron. When the K-shell vacancy is filled by an outer-shell electron, a 32 keV ^{137}Ba x-ray is produced (Figure 5-14). *Iodine escape peaks* can occur following a photoelectric interaction with an iodine atom in the NaI(Tl) crystal, generating a 28 keV iodine x-ray. If the x-ray escapes from the detector, a small peak appears in the spectrum 28 keV below the photopeak energy. Iodine escape peaks are more likely to be seen with low-energy (100–200 keV) photon emitters, which tend to have photoelectric interactions near the surface of a sodium iodide crystal where the x-ray is likely to escape. These types of effects are discussed in more detail by Cherry et al.[10]

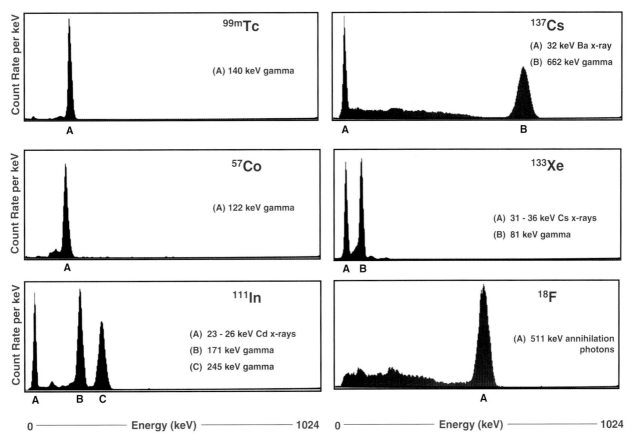

FIGURE 5-16 Pulse-height energy spectra for ^{99m}Tc, ^{57}Co, ^{111}In, ^{137}Cs, ^{133}Xe, and ^{18}F in a 2 × 2 inch sodium iodide crystal and multichannel analyzer.

Photopeak Energy Resolution

The earlier discussion of pulse-height spectrometry made reference to an idealized gamma-energy spectrum with photopeak energies appearing as sharp lines in the spectrum. In reality, spectral photopeaks typically appear as broadened peaks rather than sharp lines. This occurs for several reasons. For example, for a given amount of energy deposited in the Na(Tl) crystal, variable quantities of light photons may be produced. In addition, variable numbers of photoelectrons may be released at the photocathode initially and in the series of dynodes in the PM tube. Other causes could be nonuniform light production or collection in the crystal and electronic noise caused by high-voltage fluctuations and electrical noise in the PM tube.[10] Thus, the photopeak that appears in a pulse-height spectrum has more of a Gaussian-shaped appearance rather than a sharp line. The width of the photopeak curve at one-half its maximum height, known as the full width at half maximum (FWHM), divided by the photopeak energy is known as the *energy resolution* of the detector. A tall, narrow peak yields higher energy resolution than a broad peak. Typical energy resolution of ^{137}Cs in a NaI(Tl) detector is shown in Figure 5-19. The energy resolution of a sodium iodide detector can range from about 6% for 1 MeV gamma rays to about 15% for 0.1 MeV gamma rays.[10]

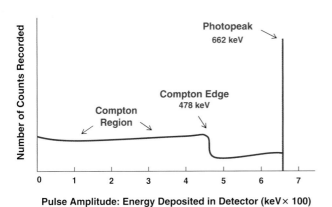

FIGURE 5-17 Idealized gamma energy spectrum of ^{137}Cs. See text for complete explanation.

RADIATION COUNTING INSTRUMENTATION

The instruments typically used in nuclear medicine for counting gamma radiation include the scintillation well counter, scintillation probe, and liquid scintillation counter and the SPECT camera and positron emission tomography (PET) scanner for imaging.

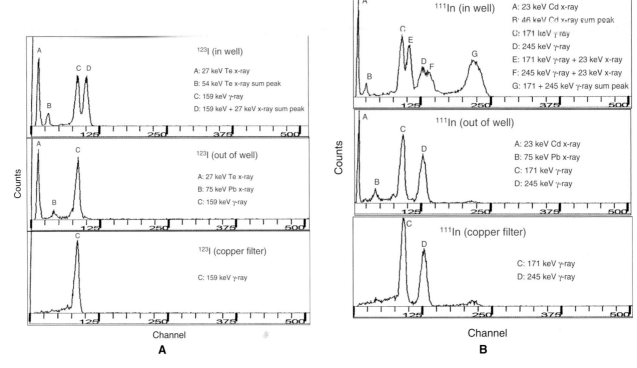

FIGURE 5-18 (A) Pulse-height energy spectrum of ^{123}I. Counting the ^{123}I source in the sodium iodide well (top panel) yields a spectrum with a 27 keV Te x-ray and a 54 keV sum peak of these x-rays, a 159 keV gamma ray and a sum peak of the gamma ray and x-ray. Counting the source out of the well (middle panel) eliminates the sum peaks but introduces a lead x-ray peak. Shielding the ^{123}I source with a copper filter (bottom panel) removes the 27 keV x-rays, leaving only the 159 keV gamma photopeak. (B) Pulse-height energy spectrum of ^{111}In. Counting the ^{111}In source in configurations similar to those for ^{123}I yields spectra with multiple energy peaks as noted in the figure. See text for additional discussion.

Scintillation Well Counter

The scintillation well counter is a solid crystal NaI(Tl) spectrometer designed for counting test-tube samples. The sodium iodide crystal may vary in size, but typically it is 4.5 cm in diameter and 5 cm deep, with a well 1.6 cm in diameter and 3.8 cm deep (Figure 5-20). It is a sensitive device, typically used for counting samples containing less than 1 µCi (37 kBq). Increased counting efficiency is the major advantage of a well design; the sample is almost completely surrounded by the detector. The detector is shielded with lead to reduce background radiation.

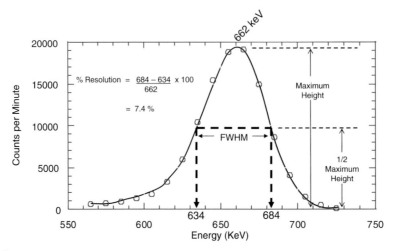

FIGURE 5-19 Energy resolution of a sodium iodide detector for ^{137}Cs gamma rays of 662 keV.

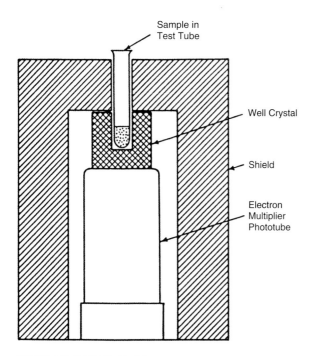

FIGURE 5-20 Cross-section of a scintillation well counter. (Reprinted from *A Manual of Radioactivity Procedures*, Handbook 80, U.S. Department of Commerce, National Bureau of Standards, 1963.)

Scintillation Probe

A scintillation probe is similar to the well counter except that the sodium iodide detector is flat-faced and has no well (Figure 5-21). A typical probe for thyroid uptake measurements has a crystal 2 inches in diameter and 2 inches deep. Smaller hand-held portable probes 1 inch in diameter and 1 inch deep are used to monitor radioactivity over various body parts and as survey instruments in radiation safety operations. An instrument employing a probe detector used in thyroid uptake studies is shown in Figure 5-22.

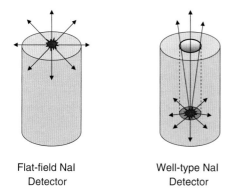

FIGURE 5-21 The counting geometry of a flat-field NaI(Tl) detector is less efficient than that of a well-type detector, which almost completely surrounds the radioactive source.

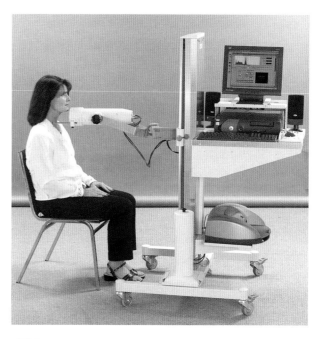

FIGURE 5-22 Stationary scintillation thyroid uptake probe and counter. (Photo courtesy of Biodex Medical Systems, Inc.)

Liquid Scintillation Counter

Liquid scintillation counting is illustrated in Figure 5-23. Liquid scintillation can be used for counting sources that emit any type of radiation, but it is most useful for counting beta particles. Since beta particle radiation cannot penetrate into solid crystal NaI(Tl) detectors, liquid scintillation counting provides an efficient way to measure the activity of beta-emitting nuclides such as ^3H, ^{14}C, ^{32}P, ^{35}S, ^{89}Sr, and ^{90}Y. The sample to be counted is dispersed in a liquid scintillation "cocktail" that contains solvent and fluorescent compounds (fluors) to detect the beta radiation. The close molecular admixture of radionuclide molecules and fluor molecules provides an environment for highly efficient detection of beta particles. The sample to be counted is transferred into a glass or plastic scintillation vial along with the scintillation cocktail and is placed into the counting well of the liquid scintillation counter (Figure 5-24). Beta particles emitted from the sample transfer their energy to solvent molecules and fluor molecules, as shown in Figure 5-23. The energy transferred is converted into light photons that cause the release of electrons in the PM tubes of the scintillation counter. Two opposing PM tubes are used to increase counting efficiency and to employ a coincidence gate that excludes randomly generated electrical pulses due to background and electronic noise. The electrical pulses generated in the PM tubes travel to the amplifier and PHA system adjusted to accept or reject counts. Counts are recorded by a scaler device.

Quenching is a primary consideration in liquid scintillation counting. Quenching refers to any process that will decrease counting efficiency. Counting efficiency for beta emitters is defined as the counts per minute divided by the disintegrations per minute (cpm/dpm) emitted by the source. There are two principal types of quenching: chemical quenching and

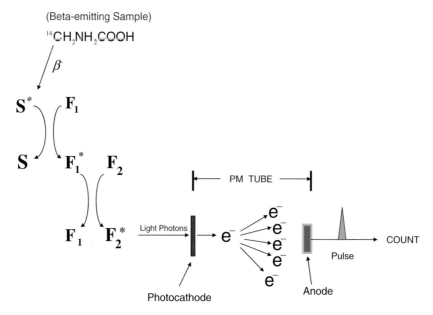

FIGURE 5-23 Liquid scintillation counting. Decay of a beta-emitting radiochemical in the scintillation "cocktail" releases beta particles (electrons) which, through a series of interactions, transfer energy to the solvent molecules. Excited solvent molecules (S*) transfer their energy to the fluor molecules. Excited fluor molecules (F*) release their energy as visible light photons (scintillations). The light photons strike the photocathode of the PM tube, releasing electrons. These electrons are multiplied, generating an electrical pulse that is registered as a count. The number of pulses counted per minute is proportional to the amount of radioactivity in the sample.

color quenching (Figure 5-25). Chemical quenching reduces counting efficiency by interfering with energy transfer from excited solvent molecules to fluor molecules. Oxygen, hydrogen ion, and other chemicals in the admixture can cause chemical quenching. Color quenching reduces the number of light photons that reach the PM tube, through attenuation by colored substances. Typical color quenchers in biological samples are hemoglobin and bile. Thus, some method of quench correction must be employed to determine counting efficiency with liquid scintillation. Quench correction methods are usually provided with the instrument. Additional information can be found in standard texts on liquid scintillation counting.[10–12]

Imaging Systems

The principal types of imaging systems used in nuclear medicine are SPECT cameras and positron emission tomography (PET) scanners. Hybrid systems that combine these with computed tomography (CT) are also available (SPECT/CT and PET/CT).

The earliest imaging system used in nuclear medicine for diagnostic procedures was the rectilinear scanner. A typical scanner had a sodium iodide crystal 3 to 5 inches in diameter and 2 to 3 inches deep. Images were made by moving the detector down and across the region of interest, recording information line by line as the detector scanned across the organ (Figure 5-26). The images obtained were called "scans."

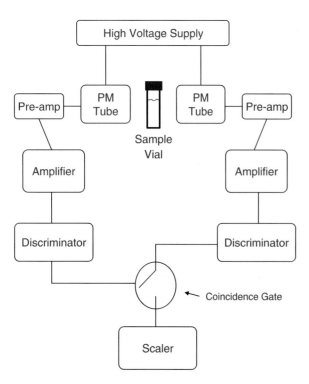

FIGURE 5-24 Component diagram of a liquid scintillation counter.

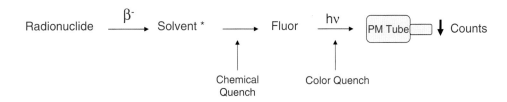
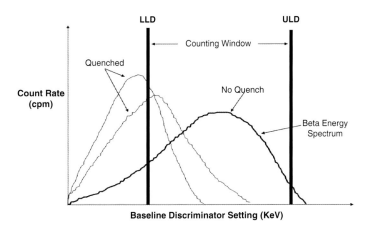

FIGURE 5-25 Quenching in a liquid scintillation sample reduces counting efficiency. Reduced counting efficiency results in a lower count per minute (cpm) recorded per disintegration per minute (dpm) released into the counting fluid by beta decay. Upper panel: Chemical quenching interferes with transfer of energy from excited solvent molecules to fluor molecules; color quenching interferes with transfer of light photons (hv) to the photomultiplier (PM) tube, through absorption by colored substances. Lower panel: Sample count rate (cpm) is proportional to the area under the beta energy spectrum curve. Quenching causes the beta energy spectrum to shift downward, out of the counting window of the pulse-height analyzer, which results in a reduced count rate.

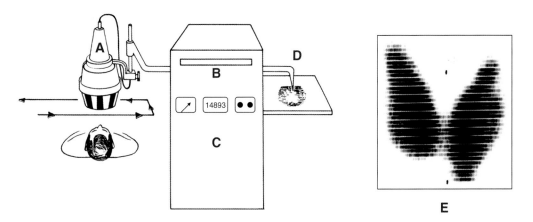

FIGURE 5-26 Diagram of a rectilinear scanner, an early imaging device used in nuclear medicine prior to the gamma camera. (A) Sodium iodide detector, consisting of a collimator, NaI crystal, and photomultiplier tube assembly that moved laterally back-and-forth and vertically over the area to be scanned, creating an image line-by-line; (B) photographic film recording device; (C) electronic processing unit; (D) "dot" image recording device; (E) example of a thyroid gland scan made with a rectilinear scanner.

Modern-day imaging employs gamma cameras whose detectors are large enough to view an entire organ of interest in most instances. The images obtained are often referred to as "scans," even though the method of acquisition no longer involves scanning.

The planar camera was the first type of gamma camera to replace the rectilinear scanner. It acquired a two-dimensional image of a three-dimensional distribution of radioactivity in an organ by acquiring images in different planes around the patient, typically in anterior, posterior, oblique, and right and left lateral projections. However, with this technique of image acquisition, a deep-seated lesion within an organ could be obscured by overlying normal tissue, which caused difficulties in scan interpretation.

A typical gamma camera employs a ⅜ inch thick NaI(Tl) detector of various diameters. The detector is faced with a lead collimator. The detector is wired to the camera's electronic processing unit, which consists of a PHA, a scaler–timer, a positioning logic system, a monitor, and a computer. The sodium iodide crystal detects radioactivity in the patient's body and produces the primary scintillation events used to generate an image. The PHA permits energy discrimination of pulses and is used to set appropriate windows for acquiring information. The scaler–timer records the number of counts and sets the desired time required for image acquisition.

A collimator is a lead disk with an array of holes drilled through it. It is positioned immediately in front of the sodium iodide crystal and limits detection to only those photons emitted from the patient within the angle of view of the collimator's holes. Any photon emitted from the patient outside the angle of view (greater than or less than 90°) will be absorbed by the lead septa between the holes and will not be counted. Without the collimator, a photon could hit the crystal at a position unrelated to its point of origin in the patient, which would result in an image with incorrect activity distribution. In short, a collimator permits spatial resolution by limiting the field of view of the crystal so that activity distribution within an organ can be resolved. The basic types of collimators are shown in Figure 5-27. In general, high-energy nuclides require collimators with thicker lead septa to achieve acceptable image resolution.

A positioning logic system in the gamma camera produces x- and y-coordinates of the pulses generated in the crystal. In essence, it determines the location of each gamma ray interaction within the crystal that in turn is related to its site of origin within the organ. The acquired image is stored in computer memory for data analysis and can be displayed on a monitor for viewing. The monitor display is useful for patient positioning during image acquisition.

SPECT Cameras

A typical SPECT camera employs rectangular NaI(Tl) detectors (17 inches × 23 inches) ⅜ inch thick and backed by an array of 59 PM tubes. In a SPECT camera system, the detector rotates around the patient to acquire two-dimensional images over 360°, similar to a planar imaging camera. Computer reconstruction algorithms permit display of stored information in three orthogonal planes: transverse, sagittal, and coronal. In essence, electronic "slices" are made through the organ in each plane so that the activity distribution can be seen in three dimensions (see Figure 1-3, Chapter 1). A SPECT camera can be used for both planar and SPECT imaging. It is used mostly to detect single photons emitted from radionuclides, as opposed to the coincidence detection of dual annihilation photons in PET imaging. SPECT cameras use sodium iodide crystals and are typically configured with one or two detecting heads. Sensitivity increases in proportion to the number of heads. The detector typically rotates 360° about the patient in a continuous or stepwise manner. A typical dual-head SPECT camera system is shown in Figure 5-28.

A significant advantage of SPECT over planar imaging is an improved target-to-nontarget ratio.[9] In planar imaging, nontarget foreground and background activity around the target is recorded with the target activity, degrading image quality. In SPECT imaging, this nontarget activity is reduced significantly by the image reconstruction process.

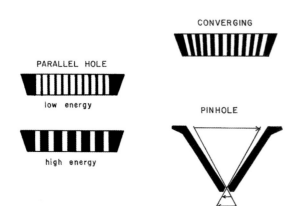

FIGURE 5-27 Diagram of several types of gamma camera collimators.

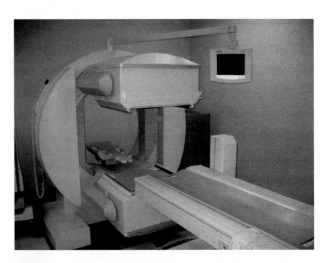

FIGURE 5-28 Dual-head single-photon emission computed tomography (SPECT) gamma camera.

PET Scanners

The emission of a positron from a radionuclide distributed within the body produces two 511 keV annihilation photons that travel in opposite directions 180° apart in accordance with the momentum conservation principle. This colinear property of the photons allows them to be detected simultaneously by opposing detectors coupled to a coincidence device (Figure 5-29). Because only coincident photons are detected, there is no need for a lead collimator to determine the line of origin, and that is why this process is referred to as electronic collimation. Lines of response (coincidence lines) for coincidence events are stored in computer memory and processed by an image reconstruction algorithm to generate cross-sectional images of the activity distribution. Theoretically, the two detectors will detect an event at precisely the same time only if annihilation has occurred precisely in the middle of the line between the two detectors. Because events occur at other points along the coincidence line, the time of the coincidence window must be adjusted to accommodate these events. To accomplish this, the average time for the coincidence window to accept both counts in PET scanners is 6 to 12 nanoseconds.[10,13] This time is long enough to record a significant number of true coincidence events that occur "off-center" on the coincidence line but short enough to limit the number of random coincidence events that strike the crystal detector coincidently with true events.

A number of possible coincidence events can be recorded with PET (Figure 5-30). A true coincidence event is one in which the photons recorded by opposing detectors belong to the same annihilation event. Unwanted events, such as ran-

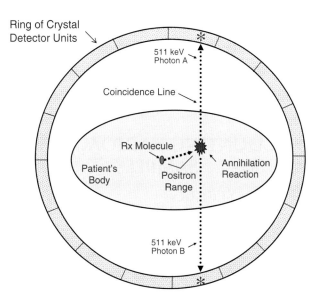

FIGURE 5-29 Diagram of a PET detector demonstrating position of positron release (Rx molecule), its travel to a point of annihilation, and the coincident detection of the two 511 keV photons along the coincidence line. The coincidence line is also referred to as the line of response (LOR) or line of origin.

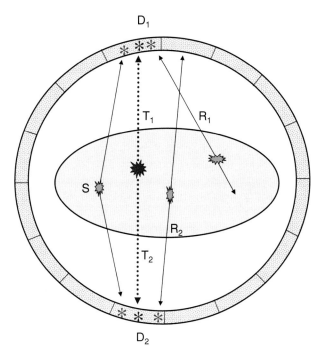

FIGURE 5-30 A true coincidence event produces annihilation photons (T_1 and T_2) within the field-of-view (FOV) of detector pair D_1 and D_2. They are recorded because they originate from the same annihilation event and arrive within the coincidence timing window. Random (R) and scattered (S) coincidence events can also occur. Random events R_1 and R_2 occur from two unrelated annihilation events outside the FOV but can arrive at the detectors within the coincidence window and be recorded. Likewise, a scattered coincidence event (S) can also occur outside the FOV and be detected within the coincidence window. Such events produce incorrect positional information. The probability of such undesirable events increases with increased activity. Random or scattered events that do not arrive within the coincidence window are not recorded.

dom and scattered photons, may also be recorded simultaneously with true events, resulting in incorrect information. Random coincidence events are those in which photons from unrelated annihilation events are detected simultaneously. Scattered coincidence events are those in which one or both photons from the same annihilation event are scattered prior to detection. In random and scattered coincidence events, the recorded line of response does not correspond to the true line of response. Such events degrade image quality, and correction factors must be applied to diminish their effect.

One advantage of PET imaging over SPECT cameras is that no collimator is required because coincidence detection is used. The absence of a collimator significantly increases the detection sensitivity of PET scanners, by a factor of 10 to 100.[13] Another important factor with coincidence detection is that sensitivity is not dependent on depth, because two photons must always cross a total path length to be detected in coincidence. Thus, if one 511 keV photon travels a short distance to one detector, then the other photon must travel the long distance to the opposing detector. Along a given coincidence line the total distance two photons travel to be detected is the same regardless of where along the line the annihilation occurs. Thus, the attenuation of annihilation radiation is a function only of the total length of travel, independent of the depth of its origin.[14]

Another major advantage of coincidence detection is that precise photon attenuation correction is possible. This is typically accomplished by x-rays from a CT scanner to determine tissue density maps for attenuation correction. Attenuation correction permits an accurate quantitative measurement of the amount of activity at a given location and is a major advantage of PET, in addition to its higher image resolution. However, SPECT/CT cameras, which can provide quantitative information, are now available.

Spatial resolution in PET is affected by detector size and design. Modern scanners have a resolution of 4 to 6 mm, which is near the maximum resolution physically possible.[13] Resolution is limited by the distance positrons travel in tissue from their point of origin to their point of annihilation. Positron emitters that have a short mean linear range in tissue have better resolution than those with a longer range (Table 5-4). In addition, because the positron may possess some residual kinetic energy when it undergoes an annihilation reaction with an electron, the two 511 keV photons emitted will not travel in precisely opposite directions, that is, at exactly 180°. This results in a small angular deviation from colinearity that degrades spatial resolution.

A PET scanner consists of a gantry with a patient portal and detector system, a patient bed, and electronic components to control the unit. Different types of detectors are used. Common detector materials in modern PET cameras are bismuth germanate (BGO), lutetium oxyorthosilicate (LSO), and gadolinium silicate (GSO), which have high detection efficiency of annihilation photons and high signal-to-noise ratio. Sodium iodide detectors offer high sensitivity but poor detection efficiency. PET imaging employs computer reconstruction algorithms to display images in transverse, sagittal, and coronal planes, providing three-dimensional imaging capability similar to SPECT.

A PET/CT scanner is shown in Figure 5-31. It has the advantage of being able to acquire an anatomic image with

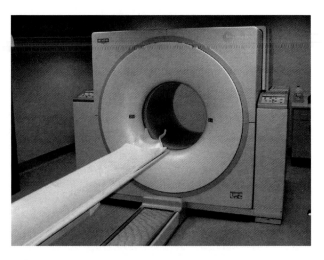

FIGURE 5-31 A PET/CT scanner.

CT and a functional image with PET sequentially. Computer programs permit exact coregistration or overlaying of the CT image with the PET image to confirm whether functional activity is associated with a suspected lesion. This combined imaging modality provides a powerful diagnostic means of determining, for example, whether a previously treated tumor is still viable.

Some SPECT cameras can also be used to image 511 keV positron annihilation photons, either singly, with the use of special collimators, or in coincidence mode, employing dual heads without collimators, similar to PET cameras. In coincidence mode they differ from PET cameras in that the dual-head SPECT camera must rotate around the patient and this requires more time for image acquisition. In summary, the advantages of PET over SPECT imaging are increased sensitivity; precise attenuation correction, which permits quantitative measurements; higher resolution; and the use of isotopically labeled compounds that permit physiologic tracer studies.

Table 5-5 compares the characteristics of SPECT and PET imaging systems.

COUNTING RADIOACTIVE SAMPLES

Counting Statistics

Two types of errors are associated with counting radioactive samples: determinate errors and indeterminate errors.[15] Determinate (systematic) errors are errors determined not by chance but by bias. They are errors caused by, for example, malfunctioning equipment or inconsistent methods of making measurements (e.g., variation in geometry between samples). Such errors are avoidable and can be controlled by the experimenter. Indeterminate (random) errors are errors not under the control of the experimenter that are caused by the random variability of the system being studied. Determining the activity of a radioactive sample is associated with random error because of the spontaneous nature of radioactive decay.

TABLE 5-4 Ranges of Positrons in Water

Radionuclide	Beta E_{max} (MeV)	Range (mm) Maximum	Mean
^{18}F	0.635	2.3	0.64
^{11}C	0.961	3.9	1.03
^{13}N	1.19	5.1	1.32
^{15}O	1.72	8.0	2.01
^{82}Rb	3.38	16.5	4.29

Adapted from reference 7.

TABLE 5-5 Imaging System Properties				
System Type	**Property**			
	Images	Detector[a]	Collimation	Utility
SPECT	2-D, 3-D Tomographic slices (transverse, coronal, sagittal)	NaI Single, dual, triple head	Yes	Detector rotates continuously or stepped 360° around patient[b]
PET	2-D, 3-D Tomographic slices (transverse, coronal, sagittal)	BGO, LSO(Ce), GSO(Ce), NaI, BaF2, CsF Circular detector	None Electronic collimation Coincidence detection 10–100-fold ↑ sensitivity[c]	Patient moves slowly through a stationary detector[b]

[a]NaI = sodium iodide; BGO = $Bi_3Ge_4O_{12}$, bismuth germanate; LSO(Ce) = $Lu_2(SiO_4)O$:Ce, lutetium oxyorthosilicate; GSO(Ce) = $Gd_2(SiO_4)O$:Ce, gadolinium silicate; BaF_2 = barium fluoride; CsF = cesium fluoride.
[b]Uses computer-based image reconstruction algorithms.
[c]Short positron range = higher resolution.

Determinate errors in counting radioactive samples can be minimized by the worker's awareness of good laboratory practices. The first step in minimizing errors is to confirm that all counting equipment is in good working order. This is accomplished by calibrating equipment with standard radioactive sources and applying statistical tests (e.g., chi square) to ensure proper operation. In addition, the experimenter must be aware of the limitations of counting equipment, such as the maximum and minimum activity levels that can be counted accurately. The second step in minimizing determinate errors is to confirm that all samples to be counted are configured in identical geometry, that is, they must have the same volume and container type.

Indeterminate errors in counting radioactive samples can be minimized by being familiar with the random nature of radioactive decay and applying statistical parameters and tests to ensure accuracy within acceptable limits of error.

Accuracy is defined as the closeness of a measurement to the true value. Precision is defined as reproducibility of measurement. A measurement may be precise but not accurate. For example, you may count a sample multiple times and obtain a similar count each time, and thus be precise, but if the sample was pipetted incorrectly, the activity of the sample will be inaccurate. The goal of good scientific measurement is to be both precise and accurate.

It is not possible to predict the exact time a single radioactive atom will undergo decay, but it is possible to determine the fraction of a large quantity of atoms that will decay in a specified period of time. The decay of radioactive atoms is a spontaneous or random process, and therefore multiple independent counts of a source in the same time period will typically yield a different count. Thus, the random nature of radioactive decay makes it difficult to identify the true count rate of the source. Confidence can be gained in estimating the true value by making several 1 minute counts of a source and finding an average count per minute. Consider the counting data in Table 5-6, which shows 10 1 minute counts of a source with nominal activity of 108 cpm. The average or mean cpm is expressed by

$$\bar{n}_{(cpm)} = \frac{\sum n_{i(cpm)}}{N} \quad (5\text{-}1)$$

where $\bar{n}$ is the mean cpm, Σn_i is the sum of individual 1 minute counts (n_i), and N is the number of observations. In the example in Table 5-6, $\bar{n}$ is 108 cpm. If the experiment were repeated, $\bar{n}$ might be 106 cpm. One could then estimate from the two sets of data that the true mean count rate was between 106 cpm and 108 cpm, or 107 cpm. Still another series of counts might produce a different mean cpm. It is obvious from such experimentation that the true count lies somewhere near the mean count and that the individual counts are equal to the mean count plus or minus a few counts. Such results exemplify the nature of a random variable. That is, for a very large number of sample counts, a plot of each count on the x-axis versus the probability of that count on the y-axis would give a Poisson distribution. The Poisson distribution, however, is not symmetric about the mean. Therefore, a normal or Gaussian distribution is typically used to describe counting statistics because it is symmetric about the mean and is very close to the Poisson distribution.[15] A normal distribution curve is shown in Figure 5-32 and is described by the Gaussian probability equation as

$$G_n = \frac{1}{\sqrt{2\pi\mu}} e^{-\frac{(n_i-\mu)^2}{2\mu}} \quad (5\text{-}2)$$

where μ is the true mean value of the source count, n_i is the experimental value of the source count, and G_n is the probability of an individual count n_i occurring. If μ is 30, the probability of obtaining a count of 28 is 0.0681, or 6.81%.

$$G_n = \frac{1}{\sqrt{2\pi(30)}} e^{-\frac{(28-30)^2}{2(30)}} = 0.0681$$

The further away a count is from μ, the lower is its probability of occurring. The probability of a count of 20 occurring is 0.0138, or 1.38%. Since the Gaussian distribution is

TABLE 5-6 Counting Data for a Nominal Source Count Rate of 108 Counts per Minute (cpm)

Observation	Sample Count per Minute (n_i)	Deviation of Sample Count per Minute from Mean Count per Minute ($n_i - \bar{n}$)	Square of Sample Deviation $(n_i - \bar{n})^2$
1	118	+10	100
2	113	+5	25
3	103	−5	25
4	96	−12	144
5	130	+22	484
6	110	+2	4
7	106	−2	4
8	90	−18	324
9	95	−13	169
10	119	+11	121
Total	1080	0	1400

symmetric, the probability of a count of 32 is also 6.81% and the probability of a count of 40 is 1.38%.

The Gaussian distribution is defined by the parameters μ, which defines the center of the distribution, and σ, the standard deviation, which defines the spread or dispersion of data about the mean. The experimental or sample mean ($\bar{n}$) and the sample standard deviation (s) are used to estimate μ and σ.

The usefulness of statistics in dealing with random events is that it describes the magnitude of error introduced by randomness and helps establish conditions that will minimize the error between μ and its estimator $\bar{n}$. Figure 5-32 can be interpreted to mean that for multiple counts of a single source, 68.3% of all observed counts are expected to occur within ± one standard deviation of the mean count and 95.5%, within ± two standard deviations of the mean count.

The sample standard deviation, s in cpm, for multiple counts is expressed as

$$s_{(cpm)} = \sqrt{\frac{\sum(n_i - \bar{n})^2}{N-1}} \quad (5\text{-}3)$$

For the data in Table 5-6

$$\bar{n}_{(cpm)} = \frac{1080}{10} = 108 \text{ and } s_{(cpm)} = \sqrt{\frac{1400}{10-1}} = 12.5$$

Thus, the sample count is expressed as $\bar{n} \pm s$, or 108 ± 13 cpm. In terms of the Gaussian distribution, 68.3% of the 1 minute counts would be expected to fall within the range of 95 to 121 cpm.

Counting Error

The standard deviation gives the numerical (absolute) spread of counts about the mean count. A useful parameter for expressing the error associated with the standard deviation is the coefficient of variation. The coefficient of variation (CV) is the ratio of the sample standard deviation to the sample mean, or CV = $s/\bar{n}$. The percent error (%CV) in a count is

$$\%CV = \frac{s_{(cpm)} \times 100}{\bar{n}_{(cpm)}} \quad (5\text{-}4)$$

In the example above, the percent error is

$$\frac{13 \text{ cpm} \times 100}{108 \text{ cpm}} = 12\%$$

In a single 10 minute count of the source, the mean count rate $\bar{n}_{(cpm)}$ is 108 cpm (1080 counts/10 minutes). For a single

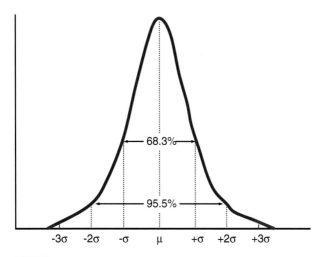

FIGURE 5-32 Normal (Gaussian) distribution curve displaying the mean count and random distribution of counts about the mean from a radioactive source.

TABLE 5-7 Sample Counting Error

Sample Count Time (min)	Total Sample Count	Sample Count per Minute (cpm)	Sample Standard Deviation (cpm)	Error in Standard Deviation (%CV), by Confidence Level		
				68.3	95.5	99.7
1	1000	1000	31.6	3.2	6.3	9.5
5	5000	1000	14.1	1.4	2.8	4.2
10	10,000	1000	10.0	1.0	2.0	3.0
100	100,000	1000	3.2	0.32	0.63	0.95

count n_i, the standard deviation is $\sqrt{n}$ and in terms of the count rate is

$$s_{(cpm)} = \frac{\sqrt{n_i}}{t} \quad (5\text{-}5)$$

In the example above, $s_{(cpm)}$ is

$$s_{(cpm)} = \frac{\sqrt{1080 \text{ counts}}}{10 \text{ min}} = 3.3 \text{ cpm}$$

The sample count rate is expressed, therefore, as the mean count rate $\overline{n}_{(cpm)} \pm s_{(cpm)}$ or in this case 108 cpm ± 3 cpm. The percent error of standard deviation in the single sample count rate is

$$\%CV = \frac{s_{(cpm)} \times 100}{\overline{n}} = \frac{3.3 \text{ cpm} \times 100}{108 \text{ cpm}} = 3\%$$

It is obvious that the error in the standard deviation of a single count is much smaller than that of taking multiple counts on the same sample. Consider, for example, counting a source with a nominal count rate of 1000 cpm for a single count of 1, 5, 10, and 100 minutes. The data shown in Table 5-7 demonstrate that a longer single sample counting time yields a smaller error and therefore a mean count that is closer to the true count rate of the sample. As a general rule in nuclear medicine, samples should be counted for a minimum of 10,000 counts. This produces a 3% error at the 99.7% confidence level, which is quite acceptable for clinical work. The error is actually a measure of the range of counts about the mean caused by the randomness of radioactive decay, and the greater the number of counts in the sample, the smaller is the effect of randomness in the count. At this confidence level there is only a 0.3% chance that the true count rate of the sample falls outside the mean count plus or minus three standard deviations. Another way of saying this is that if the source were counted 1000 times, 3 times out of 1000 the count rate would be expected to fall outside this range because of the normal random nature of radioactive decay, and 997 times it would be expected to fall within this range.

Counting Efficiency

In its simplest form, counting efficiency is counts per minute, recorded by an instrument, divided by the disintegrations per minute (dpm) occurring in the sample being counted (efficiency = cpm/dpm). Counting efficiency is typically below 100%; for a number of reasons, the detector may not be able to capture all of the disintegrations occurring in a radioactive sample. The main factors that affect counting efficiency are the intrinsic efficiency of the detector and geometry factors.

Intrinsic efficiency is the number of radiations interacting within the detector divided by the number of radiations incident on the detector.[10] Intrinsic efficiency depends on the type of radiation, the size and composition of the detector, and the attenuation coefficient of the detector for the radiation's energy. For example, the intrinsic efficiency of solid-crystal sodium iodide detectors for pure beta emitters, such as ^{3}H and ^{14}C, is zero because the weak energy of beta particles cannot penetrate into and interact with the sodium iodide crystal. However, if a beta emitter is dispersed within a liquid scintillation fluid, most of the emitted radiation is absorbed by the scintillation fluid and counting efficiency is quite high, depending on the energy of the beta particles. For example, the counting efficiency in liquid scintillation fluid of low-energy beta particles from ^{3}H, having a maximum beta particle energy (beta max) of 12.3 keV, is approximately 60%, whereas the efficiency of higher-energy beta particles from ^{14}C (beta max 156 keV) is closer to 90%. The counting efficiency of a beta emitter is easily determined with liquid scintillation by counting an accurate aliquot of a calibrated standard of the beta source and dividing the net counts per minute observed by the known disintegrations per minute in the sample.

The counting efficiency of a gamma emitter in a sodium iodide detector depends on the intrinsic efficiency of the detector and the geometry of the source. Efficiency must be adjusted for the photon abundance, since each disintegration may not produce a photon. The counting efficiency of a gamma emitter in a particular geometry is determined by counting an aliquot of a calibrated sample of the gamma source and dividing the net counts per minute observed by the known disintegrations per minute in the sample and the photon abundance or PA (mean number of emitted photons per disintegration). Thus

$$\text{Efficiency} = \frac{\text{Net cpm}}{(\text{Source } \mu\text{Ci})(2.22 \times 10^6 \text{ dpm}/\mu\text{Ci})(\text{PA})} \quad (5\text{-}6)$$

Photon energy and geometry factors, such as detector size, distance of the source from the detector, and absorption and scatter of radiation within the source itself and in any material between the source and the detector, can affect counting efficiency.[10] The attenuation coefficient of photons in sodium iodide is inversely related to photon energy. Thus, for a sodium iodide detector of a given thickness, higher gamma energy sources will be less efficiently detected than lower-energy sources, and for a given photon energy large-diameter crystals will be more efficient detectors than small-diameter crystals. A typical sodium iodide well counter crystal is 1.75 inches in diameter and 2 inches deep and contains a well that is 0.7 inch in diameter and 1.6 inches deep. Crystal detectors without wells (flat-field detectors) range in size from 0.5 to 3 inches in diameter and 1 to 3 inches deep. Higher counting efficiencies are achieved using a well counter because the source is almost completely surrounded by the detector, which maximizes capture of radiation, compared with counting the source using a flat-field detector, where a greater fraction of disintegrations escape detection (Figure 5-21).

If a radioactive source is moved farther away from the detector, fewer emissions will reach the detector and counting efficiency will fall. This can be a useful technique for counting sources that are "too hot" and would exceed the dead time of the detector.

The configuration of the source container can also affect counting efficiency. Counts may escape detection when the volume of a sample placed in a scintillation well detector is nearly equal to the volume of the well. In this situation, disintegrations occurring near the surface of the sample at the top of the well are more apt to escape detection by the crystal. It is best to keep the sample volume small and near the bottom of the well so that most of the sample is surrounded by the crystal detector. The source container is important as well. A source counted in a plastic tube will count with a different efficiency compared with the same source counted in a glass tube, because of differences in attenuation by the container. This is especially important if the gamma energy of the source is weak (<50 keV). In nuclear medicine it is particularly significant for ^{125}I, which emits 27 keV x-rays.

The most important consideration in radioactive counting is keeping source geometry and counting instrument settings consistent when making relative counts between unknown samples and standards.

Example: A 1.0 µCi (37 kBq) source of ^{133}Xe gas in a 3 mL glass vial is counted in a scintillation counter to yield 486,508 net cpm. The photon abundance of the 81 keV gamma ray for ^{133}Xe is 36%. Calculate the counting efficiency for ^{133}Xe in this configuration.

$$\text{Efficiency} = \frac{486{,}508 \text{ cpm}}{(1.0 \text{ µCi})(2.22 \times 10^6 \text{ dpm/µCi})(0.36)} = 0.61$$

The activity of an unknown source of ^{133}Xe counted in the same geometry is

$$\text{Activity (µCi)} = \frac{^{133}Xe \text{ sample net cpm}}{(\text{Efficiency})(2.22 \times 10^6 \text{ dpm/µCi})(PA)} \quad (5\text{-}7)$$

Example: A "grab-sample" of exhaust gas from a charcoal trap in a ^{133}Xe lung ventilation machine yields 350 net cpm in a 3 mL vial. How many microcuries of ^{133}Xe are in the sample?

$$\text{Activity (µCi)} = \frac{350 \text{ cpm}}{(0.61)(2.22 \times 10^6 \text{ dpm/µCi})(0.36)}$$

$$= 7.19 \times 10^{-4} \text{ µCi}$$

If the maximum permissible concentration (MPC) of ^{133}Xe in the work environment is 1×10^{-4} µCi/mL, does this charcoal trap need to be changed?

The concentration of ^{133}Xe in the sample is (7.19×10^{-4} µCi in 3 mL or 2.4×10^{-4} µCi/mL). Since this is 2.4 times higher than the MPC, the trap must be changed.

The counting efficiency must be determined experimentally for each radionuclide to be counted with a particular instrument.

Resolving Time and Maximum Detectable Activity

The resolving time (dead time) of a detector is the time required, between two successive interactions in the detector, for the interactions to be recorded as separate events. It is also known as dead time, because during this time the instrument is unable to record an interaction occurring in the detector. For example, a detector whose resolving time is 10 microseconds theoretically can resolve 100,000 counts per second ($1/1 \times 10^{-7}$ sec) or 6×10^6 cpm. If the detector's efficiency for ^{133}Xe is 61%, the theoretical maximum amount of ^{133}Xe activity it could count accurately is

$$\frac{6 \times 10^6 \text{ cpm}}{0.61(2.22 \times 10^6 \text{ dpm/µCi})(0.36)} = 12.3 \text{ µCi}$$

In this example, any sample containing more than 12.3 µCi (455 kBq) would record a count rate that is erroneously low, because the number of photons per minute reaching the detector would exceed the detector's capability to resolve them as individual events. This phenomenon is known as coincidence loss, because two or more incident photons coincidently hitting the detector would be unresolved. In practice, each instrument must be challenged to determine the actual maximum detectable activity it can count for each radionuclide.

If a source activity is larger than the maximum detectable activity, adjustments must be made to lower the count rate to minimize coincidence events and not exceed instrument resolving time. Techniques that can be used to prevent resolving-time loss are (1) increasing the source distance from the detector, (2) partially shielding the source, and (3) diluting the source. In the practice of nuclear pharmacy, coincidence losses can possibly occur during chromatographic analysis of radiopharmaceuticals, because a chromatogram may contain an excessive amount of activity. For example, a ^{99m}Tc-oxidronate kit containing 200 mCi (7400 MBq) in 10 mL will contain 20 µCi (0.74 MBq) in a 1 µL spot on the chro-

TABLE 5-8 Radiochromatographic Analysis of ^{99m}Tc-Oxidronate by Three Different Techniques

Detector	Chromatographic Strip	Net Strip Activity	Percent Pertechnetate[a]
Dose calibrator	Origin	15 µCi	
	Solvent front	1 µCi	6.3
Above well counter	Origin	6,757 counts	
	Solvent front	570 counts	7.8
In well counter	Origin	188,355 counts[b]	
	Solvent front	324,334 counts	63.3

[a] See text for detailed discussion.
[b] This count is falsely low because strip activity exceeded resolving time of the detector.

matography paper. This amount of activity placed into the well of a scintillation counter can exceed the dead time of the instrument. In situations like this, a counting geometry should be selected that will reduce the counting efficiency (e.g., placing the chromatogram strips several inches above the detector in a defined geometry). An example illustrating this point is shown in Table 5-8, where a 1 µL sample (16 µCi) of ^{99m}Tc-oxidronate was spotted on chromatography paper and developed in acetone. In this system, the ^{99m}Tc-oxidronate remains at the origin and free pertechnetate impurity travels to the solvent front. The developed strip was cut into two pieces (origin half and solvent front half). Each strip half was counted in a dose calibrator and in a well counter by two different techniques: (1) counting the strips placed in the scintillation well and (2) counting the strips 6 inches above the well. The percent pertechnetate impurity was determined by dividing the net solvent front counts by the origin plus solvent front counts multiplied by 100. The strips counted in the scintillation well yielded an erroneously high amount of pertechnetate impurity. This technical error occurred because the origin strip contained an amount of activity (15 µCi) that exceeded the resolving time of the detector (1 µCi), resulting in a falsely low count. When the strips were counted above the well, the count rate was reduced and did not exceed the resolving time of the detector; the result with this technique is more accurate. To not exceed resolving time with a scintillation counter, a general rule of thumb is to place not more than 1 µCi into the well.

The *maximum detectable activity* for a scintillation counter for a particular radionuclide can be determined by counting a series of sources of increasing activity and plotting source counts per minute versus activity. The result of such a determination is shown in Figure 5-33. The point on the graph where source counts per minute deviates from linearity is the maximum detectable activity.

Minimum Detectable Activity

In some circumstances it is necessary to measure low levels of activity, for example, in assessing the concentration of radioactive material released into restricted or unrestricted environments. An example is the release of ^{133}Xe into the workroom or the outside environment during a nuclear medicine procedure. Such releases should not exceed the MPC for ^{133}Xe defined in the Code of Federal Regulations (10 CFR Part 20). The MPC for ^{133}Xe is 1×10^{-4} µCi/mL for restricted areas (work environment) and 5×10^{-7} µCi/mL for unrestricted areas (effluent air). In monitoring released concentrations of ^{133}Xe the question becomes, "Is the scintillation counter sensitive enough to detect these low concentrations?" This is where determination of the minimum detectable activity (MDA) is important. In monitoring ^{133}Xe release, for example from a lung ventilation machine, one method is to collect a sample of gas effluent from the charcoal trap into a 3 mL vial and count it in a scintillation counter to determine whether it exceeds the MPC. The following discussion and examples illustrate the calculations involved.

Minimum sensitivity (MS) is defined as the net count rate above background that must be exceeded before a sample is said to contain any measurable radioactivity. MS is considered to be three standard deviations above the mean background count rate. It is calculated as follows:

$$\text{MS} = \frac{3\sqrt{\text{Background count}}}{\text{Count time}} \quad (5\text{-}8)$$

Example: A background of 400 counts in 5 minutes gives a standard deviation of 4 cpm ($\sqrt{400}$ counts/5 min). The MS is therefore 12 cpm. From Gaussian statistics, the mean count rate is 80 cpm, and 99.7% of background counts would be expected to fall within ± three standard deviations of the mean count rate, or 80 ± 12 cpm. Thus, the upper limit of background count that could be expected to occur randomly is 92 cpm (80 cpm + 12 cpm). There would be only a 0.3% chance that the background count would exceed 92 cpm. Thus, a sample whose gross count is 88 cpm would not be considered to contain activity because its count rate is within the expected range of the background count. Only samples whose gross count is greater than 92 cpm would be expected to contain radioactivity.

The *minimum detectable activity* (MDA) is defined as the smallest quantity of radioactivity that can be measured under

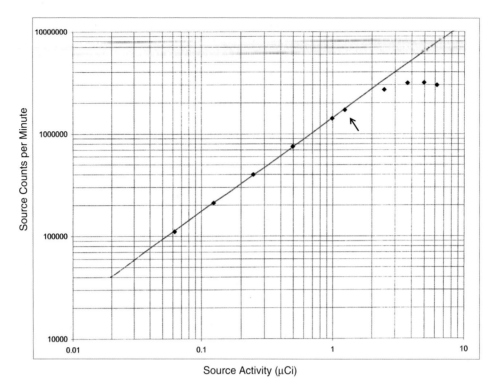

FIGURE 5-33 Maximum detectable activity determination for ^{99m}Tc in a gamma scintillation well counter with a pulse-height analyzer window of 80 keV centered on 140 keV photopeak. Counts per minute for sources (♦) are linearly proportional to source activity up to 1.25 μCi. Sample counts per minute above 1.25 μCi are spuriously low because of coincidence loss. A safe maximum detectable activity for this counter would be 1 μCi of ^{99m}Tc.

specific conditions of MS and counting efficiency. It is calculated as follows:

$$\text{MDA} = \frac{\text{MS cpm}}{(\text{Efficiency, cpm/dpm})(2.22 \times 10^6 \text{ dpm/μCi})(\text{PA})} \quad (5\text{-}9)$$

Example: A 3 mL vial is counted for background in a scintillation counter with a baseline set at 60 keV and a window of 40 keV to center on the ^{133}Xe photopeak of 81 keV. The background count was 523 counts in 5 minutes. The MS is

$$\text{MS} = \frac{3\sqrt{523} \text{ counts}}{5 \text{ min}} = \frac{3(23)}{5} = 3(4.57) = 14 \text{ cpm}$$

The MDA for ^{133}Xe in this counter is

$$\text{MDA} = \frac{14 \text{ cpm}}{(0.61 \text{ cpm/dpm})(2.22 \times 10^6 \text{ dpm/μCi})(0.36)}$$
$$= 2.87 \times 10^{-5} \text{ μCi}$$

If this activity was acquired in a 3 mL sample vial, the minimum detectable concentration of ^{133}Xe would be 9.57×10^{-6} μCi/mL. This value is 9.6% of the MPC for a restricted area; therefore, the scintillation counter can easily measure ^{133}Xe activity below the MPC and can be used to monitor effluent from the lung ventilation machine. Efficiency for ^{133}Xe can be obtained by placing a known number of microcuries in a 3 mL vial measured in a dose calibrator and allowing it to decay down to an amount that can be counted without exceeding the instrument dead time.

Chi-Square Test

The chi-square test is used to determine the "goodness of fit" of measured data to a Poisson distribution. The value of χ^2 is computed as follows:

$$\chi^2 = \frac{\sum_{i=1}^{N}(n_i - \bar{n})^2}{\bar{n}} \quad (5\text{-}10)$$

where n_i represents each of N individual measurements and $\bar{n}$ is the sample mean. Chi-square analysis is used to estimate whether a group of random variable data fall within the expected range of variability of a Poisson distribution. Radioactive decay is known to follow a Poisson distribution; that is, there is an expected spread of individual counts above or below the mean count. An instrument used to count a radioactive source multiple times should yield counts that fall within a Poisson distribution. Thus, the chi-square test may be used to evaluate instrument reliability for counting ra-

TABLE 5-9 Table of Chi-Square

Degrees of Freedom ($N-1$)	There is a probability of				
	0.95	0.90	0.50	0.10	0.05
	That the computed value of chi-square will be $\geq$				
2	0.103	0.211	1.386	4.605	5.991
3	0.352	0.584	2.366	6.251	7.815
4	0.711	1.064	3.357	7.779	9.488
5	1.145	1.610	4.351	9.236	11.070
6	1.635	2.204	5.348	10.645	12.592
7	2.167	2.833	6.346	12.017	14.067
8	2.733	3.490	7.344	13.362	15.507
9	3.325	4.168	8.343	14.684	16.919
10	3.940	4.865	9.342	15.987	18.307

dioactive samples. From the computed value of χ^2, the number of degrees of freedom ($N-1$), and the table of chi-square, the probability, P, of a set of data falling within the expected distribution can be determined. The chi-square distribution is known and is listed in Table 5-9. For a typical chi-square test of instrument reliability the computed value of χ^2 should fall within the probability range of P 0.9 to P 0.1 for ($N-1$) degrees of freedom.

Example: Chi-square for the data in Table 5-6 is computed as follows:

$$\chi^2 = \frac{1400}{108} = 12.96$$

For 9 degrees of freedom the computed χ^2 value of 12.96 falls between P values of 0.5 and 0.1, and therefore the instrument is considered to be functioning properly. There is an 80% chance of the computed χ^2 value for 9 degrees of freedom falling within the range of 4.168 and 14.684. There is also a 10% chance of the χ^2 value being less than or greater than these values, respectively. That is, there is a 20% chance of the computed χ^2 value falling outside the P 0.9 to 0.1 range. Thus, if an instrument is determined to have an abnormal chi-square value outside this range, the chi-square test should be repeated several times. If the χ^2 value is found to be abnormal more than 20% of the time, malfunction of the instrument would be suspected.

Values of χ^2 greater than the P 0.1 value suggest that the data are distributed over a wider range than that expected by a Poisson distribution (i.e., the data have too much variability), whereas values of χ^2 smaller than the P 0.9 value suggest that the data are confined to a narrower range than that predicted by the Poisson distribution (i.e., the data are not variable enough).

REFERENCES

1. Chase CD, Rabinowitz JL. *Principles of Radioisotope Methodology.* 3rd ed. Minneapolis: Burgess; 1967:244–8.
2. Kowalsky RJ, Johnston RE, Chan FH. Dose calibrator performance and quality control. *J Nucl Med Technol.* 1977;5:35–40.
3. Suzuki A, Suzuki MN, Weis AM. Analysis of a radioisotope calibrator. *J Nucl Med Technol.* 1976;4:193–8.
4. Kowalsky RJ, Johnston RE. Dose calibrator assay of iodine-123 and indium-111 with a copper filter. *J Nucl Med Technol.* 1998;26:94–8.
5. Dose calibrator correction factors for assaying indium-111 in various geometries and volumes. Technical bulletin. St. Louis: Mallinckrodt Medical; March 29, 1994.
6. Ceccatelli A, Benassi M, D Andrea M, et al. Experimental determination of calibration settings of a commercially available radionuclide calibrator for various clinical measurement geometries and radionuclides. *Appl Radiat Isot.* 2007;65:120–5.
7. Zimmerman BE, Cessna JT, Millican MA. Experimental determination of calibration settings or plastic syringes containing solutions of ^{90}Y using commercial radionuclide calibrators. *Appl Radiat Isot.* 2004;60:511–7.
8. Hine GJ. Sodium iodide scintillators. In: Hine GJ, ed. *Instrumentation in Nuclear Medicine.* Vol 1, chap 6. New York: Academic Press; 1967.
9. Early PJ. SPECT imaging: single photon emission computed tomography. In: Early PJ, Sodee BD, eds. *Principles and Practice of Nuclear Medicine.* 2nd ed. St Louis: Mosby; 1995:291–313.
10. Cherry SR, Sorenson JA, Phelps ME. *Physics in Nuclear Medicine.* 3rd ed. Philadelphia: WB Saunders; 2003.
11. Thompson J. Liquid scintillation counting application note: use and preparation of quench curves. Perkin Elmer Life Sciences. www.perkinelmer.com/. Accessed September 3, 2010.
12. Knoche HW. *Radioisotopic Methods for Biological and Medical Research.* New York: Oxford University Press; 1992.
13. Newiger H, Hamisch Y, Oehr P, et al. Physical principles. In: Ruhlmann J, Oehr P, Biersack HJ, eds. *PET in Oncology: Basics and Clinical Applications.* Berlin: Springer-Verlag; 1999:1–34.
14. Phelps ME, Hoffman EJ, Nizar AM, et al. Application of annihilation coincidence detection to transaxial reconstruction tomography. *J Nucl Med.* 1975;16:210–24.
15. Chase CD, Rabinowitz JL. *Principles of Radioisotope Methodology.* 3rd ed. Minneapolis: Burgess; 1967:75–108.

CHAPTER 6

Radiation Measurement, Protection, and Risk

Evidence of the biologic effects of ionizing radiation was experienced soon after the discovery of radioactivity. Skin erythema was first observed by Henri Becquerel, who kept a radioactive source in his side coat pocket on his lecture circuit. Pierre Curie found that an ulcer formed on his hand after it was exposed to radium radiation. As experience was gained with ionizing radiation, safe handling and radiation protection techniques were eventually developed. Today, exposure to ionizing radiation still carries the potential for biologic harm. This chapter focuses on radiation measurement and protection, radiation dosimetry, and radiation risk assessment. Knowledge of these topics enables radiation workers to understand the techniques for safe handling of radioactive sources and appreciate the level of risk associated with exposure to ionizing radiation.

RADIATION MEASUREMENT UNITS

Several different units are used in the measurement of radiation: the curie (Ci) and the becquerel (Bq) are units for measuring the activity of a source; the roentgen (R) is a dose unit related to the amount of exposure to x- and γ-radiation; the radiation absorbed dose (rad) and the gray (Gy) are dose units that express the amount of energy absorbed from exposure to all types of ionizing radiation; and the roentgen equivalent man (rem) and the sievert (Sv) are dose units used in radiation safety and dosimetry, which take into account the relative biologic hazard of different types of radiation and the risk associated with exposure of different areas of the body.

The Curie and Becquerel

The curie and becquerel were defined in Chapter 2. They are the units for measuring the amount of radioactivity present in a source. The activity term is related to a defined number of atoms decaying per unit of time. The older term, curie, is equivalent to 3.7×10^{10} disintegrations or atoms decaying per second. The becquerel is the SI unit of radioactivity and is equal to 1 disintegration per second or 2.7×10^{-11} Ci. The amount of activity in radiopharmaceuticals is typically measured in subunits of the curie: the millicurie and microcurie, and in becquerels.

The Roentgen

The roentgen (R) is a unit of exposure dose related to the amount of ionization in air produced by x- or γ-radiation. By current definition, 1 R equals 2.58×10^{-4} coulombs per kilogram of air or 2.082×10^{9} ion pairs per cubic centimeter of air at standard temperature and pressure. It is important to note that the roentgen does not measure exposure to particulate radiation such as alpha and beta particles. Furthermore, the roentgen is only an exposure quantity, with no qualification of the time of exposure or the amount of radiation absorbed by the exposure. Roentgen units are typically used with Geiger-Müller (GM) counters in measuring radiation exposure from packages of radioactive material or during surveys of the work environment.

Radiation Absorbed Dose

The radiation absorbed dose (rad) is a quantitative measure of radiation dose equivalent to 100 ergs of energy absorbed per gram of absorber:

$$\text{Absorbed dose } (D) \text{ in rad} = \frac{\text{Total energy absorbed (ergs)}}{\text{Mass of absorber (grams)}}$$

In contrast to the roentgen, the rad relates to all types of radiation, not just x- or γ-radiation. Although the roentgen is a measure of exposure dose only in air, it can be converted to an absorbed dose in rad by the following expression (IP = ion pairs):

$$1\,R = \frac{2.082 \times 10^9 \text{IP}}{0.001293 \text{ gram air}} \cdot \frac{33.7 \text{ eV}}{1\,\text{IP}} \cdot \frac{1.602 \times 10^{12} \text{erg}}{1\,\text{eV}}$$

$$\cdot \frac{1\,\text{rad}}{100\,\text{ergs/gram}} = 0.869 \text{ rad} \qquad (6\text{-}1)$$

Because tissue is denser than air by a factor of 1.108, the absorbed dose in tissue of 1 R is equivalent to 1.108 × 0.869 rad, or 0.96 rad. Thus, bodily exposure to 1 R of x- or γ-radiation is approximately equal to 1 rad of absorbed dose. The SI unit of absorbed dose is the gray (Gy), which has units of J kg^{-1} and is equal to 100 rad. Therefore, 1 rad is equal to 1 centigray (cGy).

Relative Biologic Effectiveness

The biologic consequences of radiation exposure relate not only to how much energy is absorbed but also to how it is distributed within the absorber. It is fairly easy to understand that if 100 ergs of energy is deposited within 1 gram of tissue, the biologic harm that may occur within the cells of that tissue will potentially be greater if the 100 ergs is localized in a small portion of the 1 gram than if that energy is spread uniformly throughout the 1 gram. A simple analogy illustrates this concept: If you expose your hand to the noonday sunlight for a few minutes, you will notice a feeling of warmth on your skin. However, if you interpose a magnifying glass between the incident sunlight and your hand so that the photons of sunlight are focused on a point on your skin, you will feel a different effect, even though the same amount of sunlight interacts with your hand. The biologic effect differs because of the distribution of energy.

Relative biologic effectiveness (RBE) is a term used to describe the degree of biologic effect produced by different types of radiation at the same absorbed dose. RBE is defined as the dose in rad of x- or γ-radiation required to produce a given biologic effect divided by the dose in rad of any ionizing radiation required to produce the same effect. Gamma rays of ^{60}Co (average energy 1.25 MeV) and 200 to 300 keV x-rays have been used as the reference radiation in determining RBE. The RBE depends on the type of radiation and its energy, and differences in RBE are principally due to differences in linear energy transfer (LET) of the radiation. High-LET radiation, such as alpha particles, has the potential to cause a greater biologic effect from a given absorbed dose than low-LET radiation such as x-rays or beta particles. From RBE measurements it has been found that 0.05 rad of alpha radiation in tissue will produce the same biologic effect as 1 rad of x- or γ-radiation. The RBE for alpha particles therefore is 20. One rad of beta particles produces the same biologic effect as 1 rad of x- or γ-radiation and therefore has an RBE of 1.

Equivalent Dose

The equivalent dose is the radiation dose absorbed by an organ or tissue weighted for the type of radiation exposure. It is a measure of the biologic consequences produced by the type of radiation causing the exposure. Different types of radiation produce different biologic effects, even for the same absorbed dose. To address this issue, the International Commission on Radiological Protection, in its 1991 report, referred to as ICRP 60, introduced the term equivalent dose, which modifies the absorbed dose in rad by a radiation-weighting factor.[1] The radiation-weighting factor is a value assigned to a specific type and energy of radiation to represent the RBE of that radiation to induce stochastic (cancer or hereditary) effects in tissue at low doses.[1] The *equivalent dose* (H_T) is the product of the absorbed dose (D) in rad and the radiation-weighting factor (W_R) (Table 6-1):

$$H_T = D \cdot W_R \qquad (6\text{-}2)$$

The units of equivalent dose are the rem (roentgen equivalent man) or the SI term sievert (Sv). One Sv, which has SI units of J kg^{-1}, is equivalent to 100 rem; one rem is equivalent to 1 centisievert (cSv).

Effective Dose

The effective dose is the sum of the equivalent doses in all organs and tissues of the body weighted for the sensitivity

TABLE 6-1 Radiation Weighting Factors

Radiation Type and Energy	Weighting Factor (W_R)
Photons, all energies	1
Electrons, all energies	1
Neutrons < 10 keV	5
Neutrons 10 keV to 100 keV	10
Neutrons > 100 keV to 2 Mev	20
Protons > 2 MeV	5
Alpha particles	20

Adapted from reference 1.

of those organs and tissues. It is important to note that the same equivalent dose in different tissues can produce different degrees of risk. For example, the risk of detrimental hereditary effects to an individual is greater from 1 rem gonadal irradiation than from 1 rem hand or skin irradiation. Likewise, the risk of developing thyroid cancer is greater from thyroid gland exposure than from exposure of other organs. Modification of the equivalent dose by a tissue-weighting factor (W_T) yields the effective dose (E):

$$E = \sum H_T \cdot W_T \qquad (6\text{-}3)$$

The modification is necessary because an individual's risk of stochastic effects from exposure to ionizing radiation depends not only on the absorbed dose and RBE (i.e., the equivalent dose) but also on the radiosensitivity of the particular organ or tissue exposed. The tissue-weighting factor represents the relative contribution of that organ or tissue to the total body detriment when the total body is irradiated uniformly. Thus, the effective dose provides one number that reflects the total body risk from nonuniform radiation exposure. The values and relative distributions of the weighting factors have changed over the years. Factors for ICRP 26, 60, and 103 are shown in Table 6-2.[1–3]

For an example of how these weighting factors are used, consider the data in Table 6-3 for calculation of the effective dose for ^{99m}Tc-medronate (^{99m}Tc-MDP) based on ICRP 60 weighting factors. The table lists the equivalent dose (H_T) to various organs, the tissue-weighting factors (W_T), and their product, which gives the contribution of each organ dose to the overall effective dose to the body. The sum of these is the effective dose for ^{99m}Tc-MDP (1.93×10^{-2} rem(cSv)/mCi). Thus, a 30 mCi (1110 MBq) bone scan dose would have an effective dose of 0.579 rem(cSv).

The effective dose concept has evolved over time. Terminology has changed, as have the data evaluation methods for radiation risk assessment. ICRP 26 (1977) applied radiation quality factors (Q) to the absorbed dose, and the product of the two was called the dose equivalent (H). Application of tissue-weighting factors (W_T) to the dose equivalent resulted in the effective dose equivalent (H_E). The values assigned to W_T were based on the risk of fatal cancer and serious hereditary disease in the first two generations. ICRP 60 (1991) changed the names dose equivalent to equivalent dose and effective dose equivalent to effective dose when it expanded the list of ICRP 26 tissue-weighting factors and changed their values. There is a conceptual difference between the terms dose equivalent and equivalent dose. Dose equivalent (H) was based on the absorbed dose at a "point" in tissue, weighted by a distribution of quality factors that are related to the LET distribution of the radiation at that point. The equivalent dose (H_T) is based on an average absorbed dose in the tissue and weighted by the radiation-weighting factor for the type of radiation impinging on the body or emitted by an internal source.[1] ICRP 60 also expanded the detriment from fatal cancer and hereditary disease from the first two generations to all future generations and took into account the severity of disease and years

TABLE 6-2 Tissue-Weighting Factors (W_T) for Calculating Effective Dose

Organ	ICRP 26 (1977)	ICRP 60 (1991)	ICRP 103 (2007)
Red marrow	0.12	0.12	0.12
Colon	—	0.12	0.12
Lung	0.12	0.12	0.12
Stomach	—	0.12	0.12
Breasts	0.15	0.05	0.12
Gonads	0.25	0.20	0.08
Urinary bladder	—	0.05	0.04
Liver	—	0.05	0.04
Esophagus	—	0.05	0.04
Thyroid	0.03	0.05	0.04
Skin	—	0.01	0.01
Bone surfaces	0.03	0.01	0.01
Brain	—	—	0.01
Salivary glands	—	—	0.01
Remainder	0.30[a]	0.05[b]	0.12[c]
Total	1.00	1.00	1.00

[a] Remainder is equally divided between the five organs or tissues with the highest doses.
[b] Remainder = adrenals, brain, upper large intestine, small intestine, kidney, muscle, pancreas, spleen, thymus, and uterus. If a remainder organ dose is greater than any organ listed, use W_T of 0.025 for that organ and 0.025 to average the dose of the rest of the remainder.
[c] Remainder = adrenals, extrathoracic region, gallbladder, heart, kidneys, lymphatic nodes, muscle, oral mucosa, pancreas, prostate, small intestine, spleen, thymus, and uterus/cervix.

TABLE 6-3 Effective Dose Estimate for ^{99m}Tc-Medronate

Organ	H_T (rem/mCi)[a]	W_T	$H_T \cdot W_T$ (rem/mCi)
Gonads (ovaries)	1.2E-02	0.20	2.40E-03
Breast	3.4E-03	0.05	1.70E-04
Red marrow	3.3E-02	0.12	3.96E-03
Lung	5.7E-03	0.12	6.84E-04
Thyroid	6.0E-03	0.05	3.0E-04
Bone surfaces	2.2E-01	0.01	2.2E-03
Colon (lower large intestine wall)	1.3E-02	0.12	1.56E-03
Stomach	5.6E-03	0.12	6.72E-04
Liver	5.9E-03	0.05	2.95E-04
Esophagus	5.13E-03	0.05	2.56E-04
Urinary bladder	1.2E-01	0.05	6.0E-03
Skin	4.2E-03	0.01	4.2E-05
Uterus[b]	1.9E-02	0.025	4.75E-04
Adrenals	9.0E-03	0.0028	2.52E-05
Brain	7.0E-03	0.0028	1.96E-05
Small intestine	8.7E-03	0.0028	2.44E-05
Upper large intestine	7.7E-03	0.0028	2.16E-05
Kidneys	3.2E-02	0.0028	8.96E-05
Muscle	7.1E-03	0.0028	1.99E-05
Pancreas	7.4E-03	0.0028	2.07E-05
Spleen	6.5E-03	0.0028	1.82E-05
Thymus	5.0E-03	0.0028	1.40E-05
Total		1.00	1.93E-02[c]

[a] Equivalent dose (H_T) data from reference 4.
[b] Remainder organs: In the ICRP 60 10 organs (uterus, adrenals, brain, small intestine, upper large intestine, kidneys, muscle, pancreas, spleen, thymus) are assigned weighting factors of 0.005 each (total weight 0.05 for 10 organs). In cases in which one remainder organ receives a dose that is much higher than the others, that organ is assigned one-half of the 0.05 weight, and the remaining 9 organs share the rest (0.025/9 = 0.0028).
[c] Effective dose (E) = $\Sigma H_T \cdot W_T$ = 1.93 E-02 rem/mCi.

of life lost. This was done because in the 1980s new epidemiological information became available from the Life Span Study in Japan. The data available from a study of extended follow-up of survivors of the atomic bombings indicated that new risk estimates would allow improvements in dosimetry calculations. A more recent update of radiation risk factors is now available from ICRP 103.[3] Table 6-4 provides a summary of radiation dose terminology.

In summary, the curie is a dose unit for measuring the quantity of radioactivity in a source, such as a radioactive drug. It is the unit that is measured in dose calibrators. The roentgen is a unit of exposure dose for measuring electromagnetic radiation in the workplace. Radiation exposure levels, in milliroentgens per hour, are typically measured with a GM survey meter. The gray is a unit of absorbed dose equal to a defined amount of energy deposited in a given quantity of matter. It is the unit used in radiation dosimetry. The sievert is a unit of effective dose, which takes into consideration the absorbed dose, radiation quality, and organ or tissue sensitivity. The sievert is used in radiation protection to report radiation worker exposure as monitored by film badges and dosimeters. Some useful conversions are given in Table 6-5, and a scheme reviewing the steps from absorbed dose to effective dose is shown in Figure 6-1.

Other Dose Terms

In addition to the absorbed dose, equivalent dose, and effective dose there are other radiation dose terms used in the area of radiation protection. Some of these, as defined in Title 10 of the Code of Federal Regulations (10 CFR 20.1003), are as follows:

- Committed dose equivalent (CDE, or $H_{T,50}$). The dose equivalent to organs or tissues of reference that will be received from an intake of radioactive material by an individual during the 50 year period after the intake. This

TABLE 6-4 Radiation Dose Terminology

Dose Equations	Conventional Unit	SI Unit
$D = \tilde{A} \cdot S$	rad	gray
$H = D \cdot Q$	rem	sievert
$H_T = D \cdot W_R$	rem	sievert
$H_E = \Sigma H \cdot W_T$	rem	sievert
$E = \Sigma H_T \cdot W_T$	rem	sievert

Symbol	Description
$\tilde{A}$	Cumulated activity in organ (µCi hr)
S	Mean dose/unit cumulated activity (rad/µCi hr)
D	Absorbed dose
H	Dose equivalent (ICRP 26)
H_T	Equivalent dose (ICRP 60)
H_E	Effective dose equivalent (ICRP 26)
E	Effective dose (ICRP 60)
Q	Radiation quality factor (ICRP 26)
W_R	Radiation-weighting factor (ICRP 60)
W_T	Tissue-weighting factor (ICRP 26, 60)

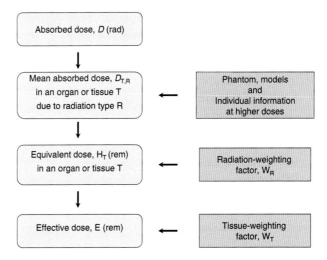

FIGURE 6-1 System of dose quantities used in radiologic protection. (Adapted from reference 3.)

term considers internally deposited radionuclides whose dose will be determined by the time course of radiation residing in the body. It differs from an external dose received over a short exposure time. ICRP considers what the committed dose equivalent will be over a person's working lifetime of 50 years. Radionuclides with effective half-lives of 2.5 months or less are essentially decayed in 1 year, and the committed dose equivalent is essentially equal to the annual dose equivalent in the year of the intake.

- Committed effective dose equivalent (CEDE or $H_{E,50}$). The sum of the products of the weighting factors applicable to each of the body organs or tissues that is irradiated and the committed dose equivalent to these organs or tissues.

$$H_{E,50} = \sum W_T \cdot H_{T,50}$$

TABLE 6-5 Radiation Dosimetric Conversions

1 R (tissue) = 0.96 rad
1 Gy = 100 rad
1 rad = 1 cGy or 10 mGy
1 Sv = 100 rem
1 rem = 1 cSv or 10 mSv
1 Bq = 1 dps
1 mCi = 3.7 × 10^7 dps (Bq)
1 mCi = 37 MBq
1 Bq = 2.7 × 10^{-11} Ci

- Deep-dose equivalent (DDE or Hd). The dose equivalent at a tissue depth of 1 cm from whole-body external exposure.
- Lens dose equivalent (LDE). The dose equivalent from external exposure of the lens of the eye at a tissue depth of 0.3 cm (300 mg/cm^2).
- Shallow-dose equivalent (SDE or Hs). The dose equivalent from external exposure of the skin of the whole body or the skin of an extremity at a tissue depth of 0.007 cm (7 mg/cm^2) averaged over an area of 10 cm^2.
- Total effective dose equivalent (TEDE). The sum of the deep-dose equivalent (for external exposures) and the committed effective dose equivalent (for internal exposures).

RADIATION DOSIMETRY AND DOSE CALCULATION

The radiation dose to individual organs and to the whole body from an administered radiopharmaceutical is important for assessing the risk to a patient. It sets the upper limit on administered activity for a radiologic procedure, and it is useful for comparing the risk of a radiologic procedure to other types of risky activities.

A radiopharmaceutical distributes throughout the body, but not necessarily in a uniform manner. Consequently, different organs absorb different amounts of radiation. The organ receiving the highest radiation dose is termed the *critical organ*. In some instances it is the target organ, the one being imaged, but sometimes it is not. For example, ^{99m}Tc-exametazime (^{99m}Tc-HMPAO) is used for brain imaging, but the critical organ is the lacrimal glands.

The radiation dose to an organ from an internally administered radionuclide is given by the expression

$$\bar{D}(r_t \leftarrow r_s) = \tilde{A}_s \cdot S(r_t \leftarrow r_s) \qquad (6\text{-}4)$$

where $\bar{D}$ is the mean absorbed dose in rad to a target organ (r_t) from a radionuclide distributed uniformly in a source organ (r_s). The absorbed dose to an organ depends on several factors; those of significance are the

- Amount of radioactivity in the organ
- Type and energy of the radiation
- Amount of energy absorbed by the organ
- Residence time of radiation in the organ
- Distribution of radiation in the organ and
- Organ mass

It must be kept in mind that the target organ will receive the radiation dose from radioactivity within it and from neighboring organs.

In Equation 6-4 $\tilde{A}_s$ (pronounced "A tilde") is the cumulated activity, in units of microcurie-hour (µCi hr), in the source region (r_s). It is the sum, or accumulation, of all the nuclear transitions occurring in the organ s during the time interval of interest, usually taken as infinity when complete decay has occurred. For complete nuclide decay, $\tilde{A}_s$ is determined by the amount of activity in the organ and its effective half-life, T_{eff}, as follows:

$$\tilde{A}_s (\mu Ci\ hr) = \frac{A_0 (\mu Ci)}{\lambda_{eff}} = A_0 (\mu Ci) \cdot 1.443 \cdot T_{eff} (hr) \quad (6\text{-}5)$$

The value of $\tilde{A}_s$ is influenced by the fraction of administered activity taken up by the organ, which is governed by normal physiologic factors and any alterations due to organ pathology.

The S in Equation 6-4 relates to physical data regarding the radionuclide and the organ mass because the dose will be expressed in rad. It is given by the expression

$$S(r_t \leftarrow r_s) = \frac{\Sigma \Delta_i \Phi_i (r_t \leftarrow r_s)}{m_t} \quad (6\text{-}6)$$

where

$$\Delta_i = 2.13 \cdot n_i \cdot E_i \quad (6\text{-}7)$$

In Equation 6-7, the value 2.13 is a unit conversion constant, n_i is the mean number of particles or photons per nuclear transformation, and E_i is the mean energy of the radiation in megaelectron volts. The unit for Δ_i is gram-rad/µCi hr. In Equation 6-6, the term m_t is the mass in grams of the target organ, making the S unit rad/µCi hr. The term Φ_i is the absorbed fraction of radiation in the target organ and is unitless. For nonpenetrating radiations such as beta particles, the fraction absorbed is 1. For photons the fraction absorbed is usually less than 1 and depends on photon energy.

The Medical Internal Radiation Dose (MIRD) Committee of the Society of Nuclear Medicine has tabulated values of S for several radionuclides, greatly facilitating radiation dose calculations using Equation 6-4. Table 6-6 lists the S values for ^{99m}Tc.[5]

Example: An investigational ^{99m}Tc radiopharmaceutical for spleen imaging has the following distribution after intravenous administration: 80% spleen, 15% liver, and 5% total body. Estimate the radiation dose to the spleen from a 1 mCi (37 MBq) dose. Assume very slow biologic elimination, that is, $T_{eff} = T_p$ (physical half-life) or 6 hours.

The first step in the process is to calculate the cumulated activities in the source organs (spleen, liver, total body). Thus

$$\tilde{A}\ spl = (1000\ \mu Ci)(0.80)(1.443)(6\ hr) = 6926\ \mu Ci\ hr$$
$$\tilde{A}\ liv = (1000\ \mu Ci)(0.15)(1.443)(6\ hr) = 1299\ \mu Ci\ hr$$
$$\tilde{A}\ tb = (1000\ \mu Ci)(0.05)(1.443)(6\ hr) = 433\ \mu Ci\ hr$$

The second step is to multiply these cumulated activities by the appropriate S values from Table 6-6 to calculate the total dose to the spleen:

$$D\ spl = \tilde{A}\ spl \cdot S(spl \leftarrow spl) + \tilde{A}\ liv \cdot S(spl \leftarrow liv)$$
$$\quad + \tilde{A}\ tb \cdot S(spl \leftarrow tb)$$
$$= (6226\ \mu Ci\ hr)(3.3 \times 10^{-4}\ rad/\mu Ci\ hr)$$
$$\quad + (1299\ \mu Ci\ hr)(9.2 \times 10^{-7}\ rad/\mu Ci\ hr)$$
$$\quad + (433\ \mu Ci\ hr)(2.2 \times 10^{-6}\ rad/\mu Ci\ hr)$$
$$= 2.286\ rad + 0.001\ rad + 0.001\ rad$$
$$D\ spl = 2.288\ rad$$

Example: Estimate the radiation dose to the lungs from a ^{99m}Tc-DTPA aerosol used for lung ventilation imaging. Assume instantaneous uptake in the lungs of 1 mCi (37 MBq) with biologic removal from the lungs into blood of 1.5% per minute.

Because there is a biologic clearance component, the effective half-life will need to be calculated as a first step. Thus, if λ_b, λ_p, and λ_{eff} are, respectively, the biologic, physical, and effective decay constants:

$$\lambda_b = 0.015\ min^{-1} \cdot 60\ min/hr = 0.900\ hr^{-1}$$
$$\lambda_p = 0.693/6.02\ hr = 0.1151\ hr^{-1}$$
$$\lambda_{eff} = 0.9000 + 0.1151 = 1.015\ hr^{-1}$$

The cumulated activity and dose to the lung is as follows:

$$\tilde{A}_{lu} = \frac{1000\ \mu Ci}{1.015\ hr^{-1}} = 985\ \mu Ci\ hr$$
$$\bar{D}_{lu} = \tilde{A}_{lu} \cdot S(lu \leftarrow lu)$$
$$= (985\ \mu Ci\ hr)(5.2 \times 10^{-5}\ rad/\mu Ci\ hr)$$
$$= 0.051\ rad$$

RADIATION PROTECTION

In nuclear pharmacy the principal concern in radiation protection is external exposure from gamma and x-ray emissions. In addition, some consideration needs to be given to high-energy beta particle emitters, but they are not a major external exposure threat because they are easily attenuated by low-density

TABLE 6-6 MIRD Values of S, Absorbed Dose per Unit Cumulated Activity (rad/μCi hr), for ^{99m}Tc ($T_{½}$ 6.03 hr)

	Source Organs[a]								
				Intestinal Tract					
Target Organs	Adrenals	Bladder Contents	Stomach Contents	SI Contents	ULI Contents	LLI Contents	Kidneys	Liver	Lungs	Other Tissue (Muscle)
Adrenals	3.1E-03	1.5E-07	2.7E-06	1.0E-06	9.1E-07	3.6E-07	1.1E-05	4.5E-06	2.7E-06	1.4E-06
Bladder wall	1.3E-07	1.6E-04	2.7E-07	2.6E-06	2.2E-06	6.9E-06	2.8E-07	1.6E-07	3.6E-08	1.8E-06
Bone (total)	2.0E-06	9.2E-07	9.0E-07	1.3E-06	1.1E-06	1.6E-06	1.4E-06	1.1E-06	1.5E-06	9.8E-07
GI (stomach wall)	2.9E-06	2.7E-07	1.3E-04	3.7E-06	3.8E-06	1.8E-06	3.6E-06	1.9E-06	1.8E-06	1.3E-06
GI (SI)	8.3E-07	3.0E-06	2.7E-06	7.8E-05	1.7E-05	9.4E-06	2.9E-06	1.6E-06	1.9E-07	1.5E-06
GI (ULI wall)	9.3E-07	2.2E-06	3.5E-06	2.4E-05	1.3E-04	4.2E-06	2.9E-06	2.5E-06	2.2E-07	1.6E-06
GI (LLI wall)	2.2E-07	7.4E-06	1.2E-06	7.3E-06	3.2E-06	1.9E-04	7.2E-07	2.3E-07	7.1E-08	1.7E-06
Kidneys	1.1E-05	2.6E-07	3.5E-06	3.2E-06	2.8E-06	8.6E-07	1.9E-04	3.9E-06	8.4E-07	1.3E-06
Liver	4.9E-06	1.7E-07	2.0E-06	1.8E-06	2.6E-06	2.5E-07	3.9E-06	4.6E-05	2.5E-06	1.1E-06
Lungs	2.4E-06	2.4E-08	1.7E-06	2.2E-07	2.6E-07	7.9E-08	8.5E-07	2.5E-06	5.2E-05	1.3E-06
Marrow (red)	3.6E-06	2.2E-06	1.6E-06	4.3E-06	3.7E-06	5.1E-06	3.8E-06	1.6E-06	1.9E-06	2.0E-06
Other tissue (muscle)	1.4E-06	1.8E-06	1.4E-06	1.5E-06	1.5E-06	1.7E-06	1.3E-06	1.1E-06	1.3E-06	2.7E-06
Ovaries	6.1E-07	7.3E-06	5.0E-07	1.1E-05	1.2E-05	1.8E-05	1.1E-06	4.5E-07	9.4E-08	2.0E-06
Pancreas	9.0E-06	2.3E-07	1.8E-05	2.1E-06	2.3E-06	7.4E-07	6.6E-06	4.2E-06	2.6E-06	1.8E-06
Skin	5.1E-07	5.5E-07	4.4E-07	4.1E-07	4.1E-07	4.8E-07	5.3E-07	4.9E-07	5.3E-07	7.2E-07
Spleen	6.3E-06	6.6E-07	1.0E-05	1.5E-06	1.4E-06	8.0E-07	8.6E-06	9.2E-07	2.3E-06	1.4E-06
Testes	3.2E-08	4.7E-06	5.1E-08	3.1E-07	2.7E-07	1.8E-06	8.8E-08	6.2E-08	7.9E-09	1.1E-06
Thyroid	1.3E-07	2.1E-09	8.7E-08	1.5E-08	1.6E-08	5.4E-09	4.8E-08	1.5E-07	9.2E-07	1.3E-06
Uterus (nongravid)	1.1E-06	1.6E-05	7.7E-07	9.6E-06	5.4E-06	7.1E-06	9.4E-07	3.9E-07	8.2E-08	2.3E-06
Total body	2.2E-06	1.9E-06	1.9E-06	2.4E-06	2.2E-06	2.3E-06	2.2E-06	2.2E-06	2.0E-06	1.9E-06

(continued)

TABLE 6-6 MIRD Values of S, Absorbed Dose per Unit Cumulated Activity (rad/μCi hr), for ^{99m}Tc ($T_{1/2}$ 6.03 hr) (Continued)

Target Organs	Source Organs[a]								
	Ovaries	Pancreas	R Marrow	Skeleton Cort Bone	Tra Bone	Skin	Spleen	Testes	Thyroid	Total Body
Adrenals	3.3E-07	9.1E-06	2.3E-06	1.1E-06	1.1E-06	6.8E-07	6.3E-06	3.2E-08	1.3E-07	2.3E-06
Bladder wall	7.2E-06	1.4E-07	9.9E-07	5.1E-07	5.1E-07	4.9E-07	1.2E-07	4.8E-06	2.1E-09	2.3E-06
Bone (total)	1.5E-06	1.5E-06	4.0E-06	1.2E-05	1.0E-05	9.9E-07	1.1E-06	9.2E-07	1.0E-06	2.5E-06
GI (stomach wall)	8.1E-07	1.8E-05	9.5E-07	5.5E-07	5.5E-07	5.4E-07	1.0E-05	3.2E-08	4.5E-08	2.2E-06
GI (SI)	1.2E-05	1.8E-06	2.6E-06	7.3E-07	7.3E-07	4.5E-07	1.4E-06	3.6E-07	9.3E-09	2.5E-06
GI (ULI wall)	1.1E-05	2.1E-06	2.1E-06	6.9E-07	6.9E-07	4.6E-07	1.4E-06	3.1E-07	1.1E-08	2.4E-06
GI (LLI wall)	1.5E-05	5.7E-07	2.9E-06	1.0E-06	1.0E-06	4.8E-07	6.1E-07	2.7E-06	4.3E-09	2.3E-06
Kidneys	9.2E-07	6.6E-06	2.2E-06	8.2E-07	8.2E-07	5.7E-07	9.1E-06	4.0E-08	3.4E-08	2.2E-06
Liver	5.4E-07	4.4E-06	9.2E-07	6.6E-07	6.6E-07	5.3E-07	9.8E-07	3.1E-08	9.3E-08	2.2E-06
Lungs	6.0E-08	2.5E-06	1.2E-06	9.4E-07	9.4E-07	5.8E-07	2.3E-06	6.6E-09	9.4E-07	2.0E-06
Marrow (red)	5.5E-06	2.8E-06	3.1E-05	4.1E-06	9.1E-06	9.5E-07	1.7E-06	7.3E-07	1.1E-06	2.9E-06
Other tissue (muscle)	2.0E-06	1.8E-06	1.2E-06	9.8E-07	9.8E-07	7.2E-07	1.4E-06	1.1E-06	1.3E-06	1.9E-06
Ovaries	4.2E-03	4.1E-07	3.2E-06	7.1E-07	7.1E-07	3.8E-07	4.0E-07	0.0	4.9E-09	2.4E-06
Pancreas	5.0E-07	5.8E-04	1.7E-06	8.5E-07	8.5E-07	4.4E-07	1.9E-05	5.5E-08	7.2E-08	2.4E-06
Skin	4.1E-07	4.0E-07	5.9E-07	6.5E-07	6.5E-07	1.6E-05	4.7E-07	1.4E-06	7.3E-07	1.3E-06
Spleen	4.9E-07	1.9E-05	9.2E-07	5.8E-07	5.8E-07	5.4E-07	3.3E-04	1.7E-08	1.1E-07	2.2E-06
Testes	0.0	5.5E-08	4.5E-07	6.4E-07	6.4E-07	9.1E-07	4.8E-08	1.4E-03	5.0E-10	1.7E-06
Thyroid	4.9E-09	1.2E-07	6.8E-07	7.9E-07	7.9E-07	6.9E-07	8.7E-08	5.0E-10	2.3E-03	1.5E-06
Uterus (nongravid)	2.1E-05	5.3E-07	2.2E-06	5.7E-07	5.7E-07	4.0E-07	4.0E-07	0.0	4.6E-09	2.6E-06
Total body	2.6E-06	2.6E-06	2.2E-06	2.0E-06	2.0E-06	1.3E-06	2.2E-06	1.9E-06	1.8E-06	2.0E-06

[a] SI = small intestine; ULI = upper large intestine; LLI = lower large intestine; R Marrow = red marrow; Cort Bone = cortical bone; Tra Bone = trabecular bone.

Source: Society of Nuclear Medicine. MIRD pamphlet no. 10. May 13, 1975.

TABLE 6-7 Radiation Absorption in Air and Water

Energy (MeV)	Maximum Range into Absorber (cm)				% Photons Absorbed per Centimeter[a]	
	Electrons		Alpha Particles			
	Air	Water	Air	Water	Air	Water
0.025	1.2	0.0013	0.06	0.00006	<0.1	45
0.050	4.1	0.004	0.09	0.00009	<0.1	20
0.1	13.5	0.014	0.14	0.00014	<0.1	16
0.2	42.2	0.045	0.21	0.00022	<0.1	13
0.3	79.1	0.084	0.26	0.00027	<0.1	11
0.5	165.6	0.177	0.35	0.00037	<0.1	9
1.0	407.6	0.437	0.56	0.00059	<0.1	7

[a] % Absorbed = $100(1 - e^{-\mu x})$, where x = 1 cm.
Source: www.nist.gov/physlab/data/radiation.cfm.

materials. Table 6-7 compares the absorption ranges of electrons and alpha particles in air and water with the attenuation of photons of similar energy.[6] The data demonstrate that low-energy photon emitters, such as ^{125}I (27–35 keV), are absorbed close to 50% per centimeter in water (similar to soft tissue) and therefore could deposit a significant radiation dose to tissue. The high level of tissue absorption by weak gamma emitters is the primary reason they are not useful for diagnostic imaging. The data also show that the short range of particulate radiation in water would be similar in tissue; therefore, ingested radionuclides would produce a significant radiation bioburden.

There appears to be little hazard to the body from external exposure to particulate radiation, because electrons and alpha particles are readily absorbed by air or a few millimeters of skin thickness. However, a few high-energy beta emitters such as ^{32}P (1.7 MeV), ^{90}Y (2.28 MeV), and ^{89}Sr (1.46 MeV) can pose some external threat because of the range these betas have in air and tissue. The maximum range in air, water, and plastic of some typical beta emitters used in nuclear medicine is listed in Table 6-8. The range or distance that a beta particle travels into an absorber is of some interest. The range (R) can be expressed in centimeters and grams per centimeter squared, both being related to the density (d) of an absorber as follows: R (cm) = R (grams/cm^2)/d (grams/cm^3). Since the density of water is 1.0 gram/cm^3, the range in centimeters that a beta particle travels in water is the same as its range in grams/cm^2. Therefore, the range of a similar-energy beta particle in a different absorber can be found by dividing its range in water in grams/cm^2 by the density of the absorber. Thus, for example, the maximum range of a 2.28 MeV ^{90}Y beta particle in Lucite is 1.13 grams/cm^2 divided by the density of Lucite (1.19 grams/cm^3), or 0.95 cm. ^{90}Y beta particles have a maximum range in air of about 8.7 m, which creates a potential external exposure threat from open vessels of ^{90}Y. However, these beta particles are completely absorbed in about 1 cm of Lucite. Typical shielding of beta emitters employs low Z material such as Lucite to minimize the formation of bremsstrahlung radiation, with an overwrap of lead to absorb any bremsstrahlung that is produced.

Potential sources of internal radiation exposure are the ingestion of radiation-contaminated food or water or inhalation of airborne radionuclides. The most common threat in nuclear medicine is inhalation of radioiodine vapor during activities

TABLE 6-8 Beta Particle Range in Matter

Nuclide	Beta Max (MeV)	Beta Range in Water (g/cm^2)	Maximum Beta Particle Range into Absorber (cm)		
			Water d = 1 g/cm^3	Air d = 1.293 × 10^{-3} g/cm^3	Plastic[a] d = 1.19 g/cm^3
^{3}H	0.018	0.0007	0.0007	0.58	0.0006
^{14}C	0.155	0.03	0.03	23	0.03
^{32}P	1.71	0.83	0.83	642	0.70
^{89}Sr	1.46	0.69	0.69	534	0.58
^{90}Y	2.28	1.13	1.13	874	0.95
^{153}Sm	0.81	0.34	0.34	263	0.29

[a] Polymethylmethacrylate (Lucite, Plexiglass, Perspex).
Source of beta particle range in water: www.nist.gov/physlab/data/radiation.cfm.

such as radiopharmaceutical dose preparation, administration of therapeutic radioiodine solution to patients, and radioiodination procedures. Other examples are radioaerosols and radioactive gases used in lung imaging studies. For the most part, however, exposure of radiation workers in nuclear pharmacy and nuclear medicine occurs primarily from handling unshielded sources in the lab and from patients who have received radiopharmaceuticals. Protection from all these sources requires vigilance and the use of various techniques. Airborne contamination can be controlled by using exhaust hoods during dose preparation and radioiodination procedures. Imaging rooms for lung ventilation studies with radioactive xenon gas should have dedicated exhaust to the outside. In addition, functional charcoal traps should be used on lung ventilation machines to limit room contamination from radioactive xenon during lung ventilation studies. In general, the three most important considerations for protection from external exposure to gamma radiation are time, distance, and shielding.

Time

The shorter the time of exposure, the lower will be the radiation dose. This means that work with radioactive material must be planned well and performed as quickly as possible, especially if the task requires handling unshielded sources.

Nuclear Regulatory Commission (NRC) regulations (10 CFR 20.1301) state that the total effective dose equivalent to individual members of the public must not exceed 0.1 rem (0.001 Sv) in a year and that the dose in any unrestricted area from external sources must not exceed 2 mrem in any 1 hour. These dose rate limits are intended only for short-term, nonoccupational exposures over periods of not more than 50 hours (i.e., 100 mrem divided by 50 hours = 2 mrem/hour). These limits apply particularly to "nonoccupational personnel"—persons such as hospital nurses, visitors, and non–radiation-treated patients who may be exposed to a patient treated with radioactive material. Table 6-9[7] lists approximate times for an exposure of 100 mrem from 100 mCi of various radionuclides at specific distances.

In particular circumstances, such as in some NRC agreement states (described in Chapter 7), the total dose to a non-treated patient near a treated patient may exceed 100 mrem. In North Carolina, for example, the regulations permit a 125 mrem dose limit for the duration of a brachytherapy procedure. The National Council on Radiation Protection and Measurements (NCRP Report 37) provides guidance on precautions in managing patients who have received therapeutic amounts of radionuclides.[7] If nonoccupational personnel will have chronic exposure longer than 50 hours, the hourly dose rate must be reduced below 2 mrem so that the total exposure does not exceed 100 mrem.

Regarding exposure from patients released from an institution after receiving radioactive materials, NRC regulations (10 CFR 35.75 and Regulatory Guide 8.39) state that a licensee may "release from its control any individual who has been administered radiopharmaceuticals or permanent implants containing radioactive material if the total effective dose equivalent (TEDE) to any other individual from exposure to the released individual is not likely to exceed 500 mrem (0.005 Sv)." Since the TEDE in this situation can exceed 100 mrem (0.001 Sv), the licensee must provide the released patient with oral and written instructions on how to maintain doses as low as reasonably achievable to other individuals. The instructions should contain guidance on limiting the time other individuals are exposed to the patient. The licensee must apply to NRC for exposure of members of the public up to 500 mrem (0.005 Sv), according to 10 CFR 20.1301.

Hospital personnel may also be at risk for chronic radiation exposure if their workplace is adjacent to a radiation therapy department that uses a linear accelerator for patient treatment. Adequate shielding of floors, walls, and ceiling around an accelerator or high-radiation area must be provided so that exposure does not exceed 100 mrem (0.001 Sv) per year to nonradiation workers. Another example of chronic radiation exposure is a nuclear pharmacy where the business office is adjacent to the radiation preparation or storage area.

Distance

Maintaining as much distance as practical from a radiation source is an effective method for reducing exposure because of the *inverse square law*. This law, which applies only to x- and γ-radiation, states that the amount of radiation from a point source is inversely proportional to the square of the distance from the source. Simply stated, doubling the distance from a source reduces the exposure to one-fourth. This principle of exposure reduction works only if the source is small relative to the exposed body.

The specific gamma ray constant (Γ) of a radionuclide must be known to apply the inverse square law. This constant is the exposure rate in R/hour at a distance of 1 cm from a 1 mCi (37 MBq) source. The units of Γ are R-cm^2/mCi hour. Table 6-10 lists the specific gamma ray constants for several radionuclides used in nuclear medicine. For any given number of millicuries, N, the dose rate at distance d from the source is given by the following equation:

TABLE 6-9 External Exposure from Various Radionuclides

Radionuclide	Approximate Time (hr) for 100 mrem per 100 mCi	
	At 2 ft (0.61 m)	At 6 ft (1.83 m)
^{137}Cs	1	10
^{60}Co	0.33	3
^{125}I	12	115
^{131}I	1.5	15
^{192}Ir	0.75	7

TABLE 6-10 Radionuclide-Specific Gamma Ray Constants (Γ) and Half-Value Layers in Lead

Radionuclide	Half-Value Layer (mm Pb)	Γ (R-cm²/mCi hr)
^{18}F	4.0	5.73
^{51}Cr	1.7	0.18
^{57}Co	0.2	1.0
^{58}Co	9.0	5.5
^{67}Ga	0.66	0.8
^{99m}Tc	0.17	0.78
^{111}In	0.23	3.21
^{123}I	0.05	1.6
^{125}I	0.017	1.43
^{131}I	2.4	2.27
^{133}Xe	0.035	0.51
^{137}Cs	6.0	3.32
^{201}Tl	0.006	4.7

$$R/hr = \frac{N\Gamma}{d^2} \quad (6\text{-}8)$$

Example 1: What is the dose rate from a 100 mCi (3700 MBq) ^{131}I source at 1 cm and at 2 feet (61 cm)?

$$R/hr\,@1\,cm = \frac{N\Gamma}{d^2} = \frac{(100\,mCi)(2.2\,R\cdot cm^2/mCi\cdot hr)}{(1\,cm)^2}$$
$$= 220\,R/hr$$

$$R/hr\,@61\,cm = \frac{N\Gamma}{d^2} = \frac{(100\,mCi)(2.2\,R\cdot cm^2/mCi\cdot hr)}{(61\,cm)^2}$$
$$= 0.059\,R/hr$$

Example 2: How much time would it take to accumulate a 100 mR (0.1 R) exposure dose from 100 mCi (3700 MBq) of ^{131}I at the distance of 2 feet?

$$\text{Time to accumulate }0.1\,R = \frac{0.1\,R}{0.059\,R/hr} = 1.7\,hr$$

Example 3: What distance would lower the dose rate to 2 mR/hr from the 100 mCi (3700 MBq) ^{131}I source?

$$\frac{N\Gamma}{d^2} = 2\,mR/hr$$

$$d\,(cm) = \sqrt{\frac{100\,mCi \times 2.2\,R/hr/mCi \times 1000\,mR/R}{2\,mR/hr}}$$
$$= 332\,cm\text{ or about }11\text{ feet}$$

Maintaining distance from a source reduces exposure significantly, but this alone does not provide adequate safety in the handling of high-activity sources. These sources must be shielded. Practical applications of the inverse square law in the handling of radioactive sources in nuclear pharmacy are discussed in Chapter 7.

Shielding

The effectiveness of any shielding material depends upon its atomic number, density, and thickness. Material of high density and high Z has many atoms (and electrons) packed into a small volume, producing high stopping power. As the energy of gamma photons increases, thicker shields are required to stop them. If one interposes an absorber between a radiation source and a radiation counter, the fraction of the original intensity transmitted through the shield will be a function of the absorber thickness, x, and the linear attenuation coefficient, μ. The attenuation coefficient depends on the atomic number (Z) of the absorber and the photon energy (E), but for given values of Z and E, μ has a constant value. The linear attenuation coefficients of various photon energies in several absorbers are listed in Table 6-11. The following formula shows the relationship between original intensity and transmitted intensity after shielding:

$$I = I_0\,e^{-\mu x} \quad (6\text{-}9)$$

where I is the transmitted intensity after shielding, I_0 is the original intensity before shielding, and μ is the linear attenuation coefficient (mm^{-1}).

If one plots transmitted intensity (I) values for various absorber thicknesses, a linear relationship is obtained on semilog graph paper as shown in Figure 6-2. The absorber thickness required to reduce the original intensity to half its value is known as the *half-value layer* (HVL). HVL values for several radionuclides are listed in Table 6-10. Mathematically, the HVL is inversely related to the linear attenuation coefficient as follows:

$$\mu = \frac{0.693}{HVL} \quad (6\text{-}10)$$

For example, the thickness of lead required to reduce the radiation intensity from a 100 mCi (3700 MBq) point source of ^{99m}Tc from its original intensity to 2 mR/hour can be calculated. From Table 6-10, Γ (^{99m}Tc) = 0.78 R-cm²/mCi hour. Using the natural log form of Equation 6-9 we have

$$I_0 = \frac{(0.78\,R\cdot cm^2/mCi\cdot hr)(100\,mCi)(1000\,mR/R)}{1\,cm^2}$$
$$= 78,000\,mR/hr$$

$$\ln I = \ln I_0 - \mu x$$

$$\ln 2\,mR/hr = \ln 78,000\,mR/hr - \frac{0.693}{0.17\,mm}(x)$$

$$0.693 = 11.26 - 4.08x$$

$$x = 2.59\,mm$$

| **TABLE 6-11** Photon Attenuation Coefficients |||||
| | Linear Attenuation Coefficients, μ (cm⁻¹)[a] ||||
Energy (keV)	Soft Tissue $d = 1.06$	Bone $d = 1.92$	Copper $d = 8.94$	Lead $d = 11.35$
20	0.87	7.68	302.1	980.2
30	0.40	2.56	97.6	374.8
50	0.24	0.81	23.4	91.3
100	0.18	0.36	4.1	63.0
150	0.16	0.28	2.0	22.9
200	0.14	0.25	1.4	11.3
300	0.12	0.21	1.0	4.6
500	0.10	0.17	0.7	1.8
1000	0.07	0.13	0.5	0.8

[a] μ (cm⁻¹) = mass attenuation coefficient (cm²/g) × absorber density (g/cm³).
Source: http://physics.nist.gov/PhysRefData/XrayMassCoef/tab4.html.

RADIATION RISK

Risk is defined as the possibility of loss or injury. With regard to radiation it refers to the probability of a defined deleterious outcome from radiation exposure. Key questions that can be asked about the risks of ionizing radiation are, What physical harm can result from exposure to radiation? and What is the risk of getting cancer or causing a genetic mutation? These questions cannot be answered precisely, mainly because there is no ethical way of experimentally exposing humans to radiation to measure its effects. However, data on radiation-induced biologic effects are available from cell and animal irradiation experiments, accidental human exposure incidents, patients with adverse effects of exposure to radiation for medical treatment, and Japanese survivors of atomic bombs dropped during World War II. Even with these data, in most instances estimations and extrapolations must be made about the risks of radiation at the low levels typical of occupational and medical exposure.

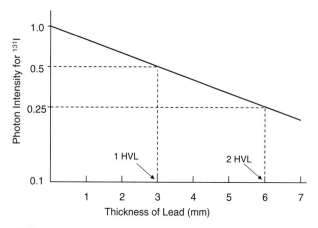

FIGURE 6-2 Plot of log-transmitted gamma ray intensity (ordinate) versus absorber thickness (abscissa) for ^{131}I in lead. HVL = half-value layer.

Stochastic and Deterministic Effects of Radiation

Stochastic effects of ionizing radiation exposure are those effects whose probability of occurrence increases with dose but whose severity is independent of dose. Examples of stochastic effects are cancer (a somatic risk of ionizing radiation exposure) and hereditary disorders (a genetic risk of ionizing radiation exposure). A stochastic effect is an all-or-none effect that can arise from damage to the DNA of a single cell; there is no threshold dose required to produce it. This is because there is a finite probability for the occurrence of a stochastic event even at very small doses so that, unless all such events can be repaired up to some level of dose, there can be no threshold. As radiation dose increases, the frequency of stochastic effects increases but the severity of the effect is not expected to increase.[3] Thus, the severity of cancer caused by 100 rad (1 Gy) of radiation is no different from that caused by 10 rad (0.1 Gy), but the chance of developing cancer is increased at 100 rad (1 Gy).

Deterministic effects of ionizing radiation exposure are somatic and genetic effects whose severity does increase with dose because of a proportional increase in damage to or death of cells. Radiation-induced cell killing plays a crucial role in the pathogenesis of tissue injury. Nonlethal damage can also contribute significant detriment by, for example, interfering with the inflammatory response of cells or with the natural migration of cells in developing organs. A threshold dose exists for deterministic effects. The threshold dose is the dose below which no measurable effect is detected and above which an effect is observed because tissue damage exceeds repair. Examples of deterministic effects from radiation exposure are lens opacification, bone marrow depression, decrease in sperm count, skin erythema, epilation, and mental retardation. Previously, these effects were termed *nonstochastic* effects. However, the meaning of the term deterministic is "causally determined by preceding events," and ICRP 60 changed the term, considering it a more appropriate description of the events that occur.

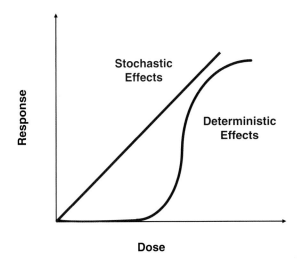

FIGURE 6-3 Idealized plot of stochastic and deterministic responses to ionizing radiation dose.

Thus, the death of an individual cell may be considered a random or stochastic effect, but the composite effect of killing a high proportion of cells in an organ, altering its structure or function, is deterministic.[1] Figure 6-3 shows idealized dose–response curves for stochastic and deterministic effects of ionizing radiation.

Risk Assessment

As stated earlier, the primary concern about exposure to ionizing radiation is the risk of producing stochastic and deterministic effects. For acute whole-body exposures above 3 Gy of low-LET radiation, early effects occur primarily because of cell killing. This can give rise to organ and tissue damage and, in extreme cases, death. These are deterministic effects. A second type of effect can occur at later times after exposure. This type is from damage to cellular nuclear material (DNA), resulting in radiation-induced cancer in exposed individuals (somatic effects) or hereditary disease in their descendants (genetic effects). These effects are stochastic.

Much of the data used to estimate the cancer risk from ionizing radiation comes from epidemiologic studies of Japanese survivors of exposure from atomic bombs. This is considered a good model for risk estimation because exposure of the survivors was uniform over the total body and people of all ages were exposed. Although this population experienced a wide range of exposure from low dose to high dose, all exposures were at high dose rate. Therefore, for radiation protection purposes the effects of this type of exposure must be extrapolated down to the low dose and low dose rate conditions typical of occupational exposures.

For radiation-induced hereditary disease, risk estimates are made primarily on the basis of animal studies, mostly in mice, but only after exposures at intermediate to high dose levels. Limited human data are available from studies carried out on the children of Japanese survivors who were irradiated.

Risk Models for Radiation-Induced Cancer

Epidemiologic studies of Japanese survivors of the atomic bombings and patients treated with radiation have shown that there is a minimum period of time between irradiation and the appearance of a radiation-induced tumor, although this latency period varies with age and from one tumor type to another. For example, the latency period for acute myeloid leukemia is about 2 years, compared with 5 to 10 years for other cancers. Many solid tissue tumors (e.g., liver and lung) have latency periods of 10 years or more.[1]

Because of the unknown effects over time, and the effects of other factors such as age and sex, empiric models have been postulated that extrapolate data based on only a limited portion of the lives of exposed individuals. Two models have been proposed.[1,8] The first is the *additive* or *absolute risk model.* This model postulates that radiation will induce cancer independently of the spontaneous rate and that variations in risk due to age and sex at exposure may occur. It predicts a constant excess of induced cancer throughout life, unrelated to the natural spontaneous rate of cancer. The second model is the *multiplicative* or *relative risk model.* It postulates that radiation exposure will increase the natural incidence of cancer by a constant factor and will consequently increase with age. Both forms of response occur after a minimum latency period.

Most organizations in the 1970s used the absolute risk model for risk assessment. This model produces predictions of eventual probability of death that are about half the values predicted by the relative risk model. ICRP 60 has since favored the relative risk model and a modified relative risk model proposed by the Biologic Effects of Ionizing Radiation committee (BEIR V) that considers sex, age, age at exposure, and time since exposure.[9] The United Nations Scientific Committee on the Effects of Atomic Radiation (UNSCEAR) used both models, absolute and relative, for its estimates of lifetime probability of cancer death.[10]

ICRP 60 selected data from the atomic bomb survivors as the most complete set of information on which to base quantitative risk estimates, following the lead of BEIR V[9] and UNSCEAR.[10] The Japanese study group is large (75,991 with dosimetry data) and includes both sexes, all ages, and an extensive dose range with uniform whole-body exposure. Of the total survivors, 34,272 serve as an internal control group; their radiation doses were negligible (<0.5 rad) because they were so far from the hypocenter. The remaining 41,719 survivors had doses greater than 0.5 rad and, of these, 3,435 died of some form of cancer between 1950 and 1985. Compared with previous ICRP 26 estimates, the ICRP 60 estimates of the probability of death from cancer for the period 1950 to 1985 were higher because of several factors, including an increase in the number of excess solid tumors observed between 1975 and 1985, new dosimetry estimates for survivors, and preference for the relative risk model rather than the absolute model for projecting the observed numbers of solid cancers to lifetime values.[1]

Dose–Response Relationship

Experimental information on dose–response relationships and the influence of dose rate on radiation-induced cancer incidence has been comprehensively reviewed by NCRP.[11] The general conclusion was that the shape of the dose–response relationship for low-LET radiation, in most biologic systems, was curvilinear with dose, that is, linear–quadratic in form given by the relationship $E = \alpha D + \beta D^2$. This is shown as curve A in Figure 6-4. This relationship had its origins in the 1930s, when it was used to fit data for radiation-induced chromosome aberrations. At low doses, the slope of the dose–response curve is less than at higher doses. At low doses and low dose rates, it is unlikely that more than one ionizing event will occur in the critical parts of a cell (DNA) while repair mechanisms in the cell are operational. Under these conditions the effect per unit dose is constant ($E/D = \alpha$). At higher doses and dose rates, however, the effect increases more rapidly; it increases linearly with dose squared as the quadratic term becomes operative, that is, $E = \beta D^2$. This response is consistent with two or more events combining to produce an enhanced effect. At still higher doses, the effect often declines because cell killing reduces the number of cells at risk.

The ratio α/β, which is the dose at which the linear and quadratic contributions to the biologic effect are equal, can vary from about 100 rad (1 Gy) to more than 1000 rad (10 Gy). Fitted dashed line B in Figure 6-4 is a high dose and high dose rate linear response derived from the available data from Japanese survivors (data points shown). The low dose and low dose rate linear response (dotted line C) is an extrapolation of the low dose portion of the sigmoid curve A.

From a radiation protection perspective, most exposures of the general public, patients undergoing radiologic procedures, and radiation workers involve low dose, low dose rate radiation. Estimates of risk for these groups have been obtained by direct extrapolation from epidemiologic studies of populations exposed at high doses and dose rates. To obtain risk estimates for radiation-induced cancer when exposures are at low doses or low dose rates, most organizations have recommended the use of a reduction factor. Suggested reduction factors have ranged from 2 to 10, meaning that the risk of radiation-induced cancer from low dose, low dose rate radiation should be reduced from that of high dose, high dose rate radiation by one-half to one-tenth. NCRP termed the reduction factor the *dose rate effectiveness factor* (DREF) and defined it as the ratio of the slopes of curves B to C shown in Figure 6-4.[11] ICRP prefers to call this reduction factor the *dose and dose rate effectiveness factor* (DDREF). To provide a conservative risk coefficient for radiation protection purposes, ICRP 60 has applied a DDREF of 2 for doses below 20 rad (0.2 Gy) at any dose rate and for higher doses if the dose rate is less than 10 rad (0.1 Gy) per hour. These dose rates apply to curve C, which is, in effect, an extrapolation of the linear portion of the actual dose–response relationship expected at a low dose and low dose rate.

Curve C is known as the linear nonthreshold (LNT) relationship for radiation-induced cancer. It is a conservative model, which assumes that radiation-induced cancer is possible at all doses of radiation, even very low doses. In contradiction to this model is the concept of radiation hormesis, which hypothesizes that chronic low doses of ionizing radiation can stimulate repair mechanisms that protect against disease. In other words, that low-dose radiation promotes health. While there is some experimental evidence from animal and laboratory investigations to support radiation hormesis, the evidence is not conclusive and there remains considerable controversy on this topic. This is because it is very difficult to detect and measure with certainty the effects caused by low-dose radiation on biologic systems. The BEIR VII report[12] considered the topic of radiation hormesis and the literature evidence in support of this concept and included the following statement: "The committee concludes that the assumption that any stimulatory hormetic effects from low doses of ionizing radiation will have a significant health benefit to humans that exceeds potential detrimental effects from the radiation exposure is unwarranted at this time."

Cancer Risk Estimates

Studies have shown that the risk of cancer depends on the particular kind of cancer, the age and sex of the person exposed, the magnitude of the dose to a particular organ, the quality of the radiation, the nature of the exposure, whether brief or chronic, the presence of other factors such as exposure to carcinogens and promoters that may interact with the radiation, and individual characteristics of the person.[9] Since nearly 20% of all deaths in the United States result from cancer, the estimated number of cancers attributable to low-level radiation is only a small fraction of the total number of deaths that occur from all causes. Furthermore, the cancers that result from radiation have no special features by which they can be distinguished from those having other causes. Thus, the probability of cancer resulting from a small dose of radiation is difficult

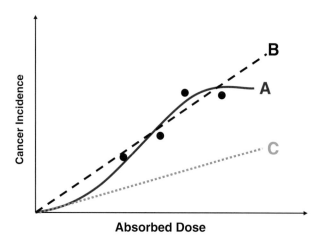

FIGURE 6-4 Schematic curves of cancer incidence versus absorbed dose. Curve A represents the linear-quadratic response to radiation dose observed in studies of biologic systems exposed to radiation; curve B represents the linear nonthreshold high dose/dose rate response observed in Japanese survivors of atomic bomb exposure; curve C represents the linear nonthreshold low dose/dose rate response to radiation dose arising from an extrapolation of curve A at low radiation dose. (Adapted from reference 11.)

to quantify and can be estimated only by extrapolation from the increased rates of cancer that have been observed in individuals after larger doses. Following the relative risk model of radiation-induced cancer, where radiation exposure increases the spontaneous rate of cancer, the approach used to estimate the risk of cancer from radiation exposure is to examine an exposed population over time to see if its cancer rate exceeds the spontaneous rate.

The BEIR V report based its cancer risk estimates on the data gathered from Japanese atomic bomb survivors. It used a time-dependent relative risk model that considered not only how risk increases with dose but also how it varies as a function of time for persons exposed at various ages. Only the atomic bomb survivor cohort contains persons of all ages at exposure. Survivors who were young when exposed are just now reaching the age at which cancer becomes an appreciable cause of death in the general population. The number of excess cancer deaths in this group to date has been small. Estimates of how the radiation-induced excess changes over time for those exposed as children introduce a great deal of uncertainty into attempts to project lifetime risks for the population as a whole.

Although the number of excess cases of cancer has increased as exposed groups have been followed for longer periods, the data are not strong when stratified into different dose, age, and time categories for all cancers at specific sites in the body. Reliance on data for all types of cancer has been limited, and attention has been focused on estimating the risk for leukemia, breast cancer, thyroid cancer, and cancers of the respiratory and digestive systems, of which the numbers of excess cases are substantial. However, to obtain an estimate of the total risk of mortality from all cancers, the BEIR V committee also modeled cancers other than those in Japanese survivors. The committee used epidemiologic data from patients treated with radiation for ankylosing spondylitis, cervical cancer, and postpartum mastitis and women who received multiple fluoroscopies in conjunction with artificial pneumothorax treatment for tuberculosis.

Risks were calculated for the following patterns of exposure to low-LET radiation: (1) instantaneous exposure causing an equivalent dose to all body organs of 10 rem (0.1 Sv), varying the age at exposure by 10-year intervals, (2) continuous lifetime exposure causing an equivalent dose in all body organs of 0.1 rem (0.001 Sv) per year, and (3) continuous exposure from age 18 to 65 causing an equivalent dose to all body organs of 1 rem (0.01 Sv) per year. The excess cancer risks in 100,000 exposed persons associated with these exposure assumptions are summarized in Table 6-12. Note that the natural prevalence of cancer is 20,000 per 100,000 population. Thus, Table 6-12 estimates that in males exposed to a single dose of 10 rem (0.1 Sv) there will be 770 additional cancers above the natural rate of 20,000. The BEIR V committee stratified by age at exposure (in 10 year intervals) the excess cancer mortality for males and females by site for a single exposure of 10 rem (0.1 Sv), as summarized in Tables 6-13 and 6-14. In general, individuals exposed at a younger age are more susceptible to radiation-induced cancer. This is especially significant for breast cancer in females exposed before the age of 15; those exposed at age 45 or older show little or no excess. Susceptibility to radiation-induced leukemia is relatively constant throughout life, whereas susceptibility to respiratory cancers increases in middle age. The nonleukemia cancers listed in the tables are simply the sum of respiratory, digestive, and other risks.

From the data in Tables 6-13 and 6-14 it is estimated that if 100,000 persons of all ages received a whole-body dose of 10 rem (0.1 Sv) of gamma radiation in a single brief expo-

TABLE 6-12 Excess Cancer Mortality Estimates per 100,000 Exposed Persons

Type of Exposure	Male	Female
Single exposure to 100 mSv (10 rem)	770	810
Continuous lifetime exposure to 1 mSv/yr (0.1 rem/yr)	520	600
Continuous exposure to 10 mSv/yr (1.0 rem/yr) from age 18 to 65	2880	3070

Adapted from reference 9, pp. 172–3.

TABLE 6-13 Excess Cancer Mortality by Age at Exposure and Site for 100,000 Males of Each Age Exposed to 0.1 Sv (10 rem)

Age at Exposure	Total	Leukemia	Non-Leukemia	Respiratory	Digestive	Other
5	1276	111	1165	17	361	787
15	1144	109	1035	54	369	612
25	921	36	885	124	389	372
35	566	62	504	243	28	233
45	600	108	492	353	22	117
55	616	166	450	393	15	42
65	481	191	290	272	11	7
75	258	165	93	90	5	—
85	110	96	14	17	—	—
Average	770	110	660	190	170	300

Adapted from reference 9, p. 175.

TABLE 6-14 Excess Cancer Mortality by Age at Exposure and Site for 100,000 Females of Each Age Exposed to 0.1 Sv (10 rem)

Age at Exposure	Total	Leukemia	Non-Leukemia	Breast	Respiratory	Digestive	Other
5	1532	75	1457	129	48	655	625
15	1566	72	1494	295	70	653	476
25	1178	29	1149	52	125	679	293
35	557	46	511	43	208	73	187
45	541	73	468	20	277	71	100
55	505	117	388	6	273	64	45
65	386	146	240	—	172	52	16
75	227	127	100	—	72	26	3
85	90	73	17	—	15	4	—
Average	810	80	730	70	150	290	220

Adapted from reference 9, p. 175.

sure (high dose rate), about 800 extra cancer deaths would be expected to occur during these persons' remaining lifetimes in addition to the nearly 20,000 cancer deaths that would occur naturally in the absence of the radiation. Therefore, these data indicate that 10 rem (0.1 Sv) radiation exposure increases the cancer mortality risk to 20.8% or an excess of 0.8%. Because exposures of radiation workers and patients are considered low dose, low dose rate, the DDREF would reduce the excess risk by one-half, to 0.4%. Thus, according to BEIR V estimates, the excess risk of death from radiation-induced cancer would be about 0.04% per 1 rem effective dose for low-LET radiation exposure such as with radiologic procedures.

ICRP 60 considered all the risk estimates of its committees and others, namely BEIR and UNSCEAR, and developed overall risk coefficients for radiation-induced cancer. Table 6-15 summarizes the coefficients for all ages and for the working population (ages 25–64) at high dose, high dose rate and at low dose, low dose rate where a DDREF of 2 is applied.[13] These coefficients indicate, for example, that for the whole population exposed to low dose, low dose rate radiation the risk of cancer above the spontaneous incidence is 5×10^{-2} excess deaths per person exposed to 100 rem (1 Sv). Another way of expressing this risk coefficient is to say that there will be an excess of 1 death per 2000 persons exposed to 1 rem (0.01 Sv) effective dose of low dose, low dose rate radiation. This is an excess risk of 0.05% above the natural incidence of cancer, similar to the BEIR V estimate discussed above, and would increase a person's risk of developing cancer from 20% to 20.05%. This total risk coefficient was derived from the number of radiation-induced cancers that developed in each organ in the body. Table 6-16 compares the fatal probability coefficients for organ cancers developed by ICRP 26 and ICRP 60. Note that the total risk coefficient of ICRP 60 (5×10^{-2} Sv^{-1}) is four times higher than that of ICRP 26 (1.25×10^{-2} Sv^{-1}). The difference is due principally to new data on the increased probability of cancer acquired from Japanese survivors, new dosimetry methods used, and use of the relative rather than the absolute risk model for projecting the observed number of solid cancers to lifetime values.

Risk Models for Radiation-Induced Hereditary Effects

Carcinogenic effects of radiation exposure arise from transformational changes in the DNA of somatic cells in the person irradiated, whereas hereditary effects arise from transformations in the DNA of germs cells of the person irradiated but manifest themselves in their offspring of later generations.

Two models are used to estimate hereditary risks of radiation, the doubling dose method and the direct method. These are similar to the relative and absolute carcinogenic risk mod-

TABLE 6-15 Nominal Probability Coefficients (10^{-2} Sv^{-1}) for Stochastic Effects[a]

	Fatal Cancer				
Exposed Group	High Dose, High Dose Rate	Low Dose, Low Dose Rate	Nonfatal Cancer	Severe Hereditary Effects	Total[c]
Working population[b]	8	4	0.8	0.8	5.6
Whole population	10	5	1.0	1.3	7.3

[a] Fatal cancer or hereditary effects in excess of spontaneous or naturally occurring effects.
[b] Adult workers (ages 25–64 years) of both sexes.
[c] Total excludes the coefficients for high dose, high dose rate.
Source: Reference 1, p. 22, and reference 13, page 31.

TABLE 6-16 Nominal Somatic Risks in a Population[a] from Fatal Radiation-Induced Cancer in Specific Organs after Exposure to Low Dose, Low Dose Rate Low-LET Radiation

Organ	Nominal Risk of Excess Cancer (Fatal Probability Coefficient × 10^{-2} Sv^{-1})	
	ICRP 26	ICRP 60
Bladder	—	0.30
Bone marrow	0.20	0.50
Bone surface	0.05	0.05
Breast	0.25	0.20
Colon	—	0.85
Liver	—	0.15
Lung	0.20	0.85
Esophagus	—	0.30
Ovary	—	0.10
Skin	—	0.02
Stomach	—	1.10
Thyroid	0.05	0.08
Remainder[b]	0.50	0.50
Total	1.25[c]	5.00[d]

[a] Nominal risks are average values for a population of equal numbers of males and females of all ages; breast and ovary are for females only.
[b] The composition of the remainder is quite different in the two cases.
[c] This total was used for both workers and the general public.
[d] General public only. The total fatal cancer risk for a working population only is taken to be 4.00 × 10^{-2} Sv^{-1}.

els, respectively. The *doubling dose model* compares radiation-induced mutations with those that occur spontaneously. The model expresses the result in terms of the doubling dose, that is, the dose of radiation that will double the spontaneous mutation rate. The *direct model* considers the incidence of disorders that occur in the first generation exclusive of the spontaneous rate. It ignores the natural mutation rate and looks for new mutations. ICRP 60 prefers the doubling dose model. The estimate of doubling dose is 100 rad (1 Gy) and is based on mouse data and low dose rate, low-LET exposure.[3] This is a calculated rather than measured quantity, based on the measured mutation rate per gene locus in mice, adjusted for the estimated comparable number of loci in humans.[13]

The induced hereditary burden from radiation exposure based on the doubling dose method is estimated from the following equation:[9]

$$\text{Induced burden} = \text{Spontaneous burden} \times (\text{Doubling dose})^{-1}$$
$$\times \text{Mutation component} \times \text{Dose}$$

For example, if the spontaneous burden is 20,000 per million liveborn humans for some class of genetic disease, and the doubling dose is 100 rem (1 Sv), and the average mutational component for the disease is 50%, then for parents in each generation exposed to 100 rem (i.e., 100 rem/30 years) the induced burden will be 10,000 cases/10^6 liveborn/generation, or 100 additional cases above the spontaneous rate per rem exposure.[9]

On the assumption of a doubling dose of 100 rem (1 Sv), the BEIR V committee has estimated the genetic effects of 1 rem (0.01 Sv) exposure per generation. Table 6-17 lists the type and number of mutations that occur spontaneously, together with those that are produced by radiation per million live births. The total number of spontaneous defects averages 42,300 (37,300–47,300), or about 4.2% of all live births. The right columns in the table list the expected increase in the respective spontaneous mutations resulting from 1 rem (0.01 Sv) of radiation exposure per 30 year generation. The number of mutations in the first generation is small because only dominant mutations are manifested at this time. At equilibrium, after several generations have been irradiated and there is sufficient time for all types of mutations to become manifest, the number of mutations is larger, giving the full measure of the radiation-induced burden. Thus, for example, the number of clinically severe autosomal dominant mutations per million live births at equilibrium is 25 per rem,

TABLE 6-17 Estimated Genetic Effects of 1 rem Exposure per Generation

Type of Exposure	Current Spontaneous Incidence/10^6 Liveborn Offspring	Additional Cases/10^6 Liveborn Offspring/rem/Generation	
		First Generation	Equilibrium
Autosomal dominant			
Severe	2,500	5–20	25
Mild	7,500	1–15	75
X-linked	400	<1	<5
Recessive	2,500	<1	Very slow increase
Chromosomal			
Translocations	600	<5	Very little increase
Trisomies	3,800	<1	<1
Congenital abnormalities	20,000–30,000	10	10–100
Totals	37,300–47,300	20–40	115–200

Adapted from reference 9.

which is predicted by the induced burden formula previously described. That is, a doubling dose of 100 rem (1 Sv) will induce a mutation rate at equilibrium equal to the spontaneous rate of 2500 mutations.

Hereditary Risk Estimates

Estimation of the probability of radiation-induced hereditary effects in humans is based primarily on genetic studies in animals, mainly mice, exposed to ionizing radiation. Animal studies provide information for estimating mutation rates; these data and certain assumptions are used to estimate the probability of radiation-induced hereditary disorders in humans. A hereditary disorder is a pathologic condition arising as a consequence of a mutation or chromosomal aberration transmitted from one human generation to the next. The disorders are classified into three groups: (1) X-chromosome-linked gene mutations and autosomal dominant and recessive gene mutations occurring on all other chromosomes, (2) chromosome number or structural abnormalities such as deletions, duplications, and translocations, and (3) multifactorial effects resulting from a combination of genetic and environmental factors, including congenital abnormalities present at birth and common disorders of adult life.

Heritable effects of radiation have yet to be clearly demonstrated in humans, but the absence of a statistically significant increase in genetically related disease in the children of atomic bomb survivors is not inconsistent with animal data.[1] The only data available are from Japanese offspring of atomic bomb survivors; parents of these offspring exposed to low-dose (1–9 rad) and high-dose (≥100 rad) radiation were studied.[13] In this study the following genetic risk factors were considered: untoward pregnancy outcomes (stillbirths, congenital defects, death in the first week), childhood mortality, and frequency of sex chromosome abnormality. Analysis of the data showed a difference in these two groups in the direction of higher radiation, but the difference was not statistically significant. However, an average doubling dose calculated from these data was found to be 156 rem (1.56 Sv).

From the results of all types of genetic studies conducted, ICRP 60 considers the nominal hereditary probability coefficients for severe genetic effects for the whole population to be 1×10^{-2} per sievert per individual and for workers to be 0.6×10^{-2} per sievert. When weighted further for years of life lost if harm occurs, the corresponding numbers are 1.3×10^{-2} per sievert and 0.8×10^{-2} per sievert.[1] Another way of expressing this is to say that in the working population there would be 60 excess genetic disorders and in the whole population 100 excess disorders above the normal incidence per million people exposed to 1 rem (0.01 Sv) effective dose.

Tissue-Weighting Factors and Detriment

When the whole body is uniformly irradiated, the probability of the occurrence of cancer and genetic effects is assumed to be proportional to the dose equivalent (now called equivalent dose) to the whole body, and the risk can be represented by a single value.[14] Because irradiation from internally administered radionuclides is not uniform, the concept of effective dose equivalent (now called effective dose) was developed by a scientific committee of NCRP in 1967 and subsequently adopted by ICRP. Tissue-weighting factors, recommended by ICRP 26, were derived from organ risk coefficients.[2] These factors and their risk coefficient appear in Table 6-18. For example, the contribution from breast exposure is 15% of the total body risk based on a tissue-weighting factor (W_T) of 0.15 (0.25/1.65). ICRP 60 (1991) and ICRP 103 (2007) revised organ risk coefficients, weighting factors, and detriment. The revised factors are listed in Table 6-2.

ICRP noted that detriment is a measure of the total harm that would eventually be experienced by an exposed group and its descendants as a result of the group's exposure to ionizing radiation. ICRP 60 considers four main components of detriment: (1) risk of fatal cancer, (2) risk of serious hereditary disease in future generations, (3) morbidity from nonfatal cancer, and (4) life lost due to fatal cancer. The probability coefficients for all of these risks are listed in Table 6-15, which include stochastic effects from fatal cancer and hereditary effects, as discussed previously. These values have also been endorsed by NCRP.[15]

Comparison of Radiation Risk with Other Risks

The concept of effective dose permits comparison of radiation risk with other risks that people are exposed to in their daily lives. This can be particularly helpful with consent forms, which require disclosure of the radiation risk associated with an investigational study. The effective dose is additive, so that if a person undergoes different radiologic procedures the effective doses can be summed so as not to exceed the annual limit on radiation exposure. Another useful aspect of effective

TABLE 6-18 Somatic and Genetic (Gonadal) Risk Coefficients[a] and Recommended Values of Tissue-Weighting Factors Derived by ICRP 26

Tissue (T)	Risk Coefficient 10^{-2} Sv^{-1}	W_T
Gonads	0.40	0.25
Breast	0.25	0.15
Red marrow	0.20	0.12
Lung	0.20	0.12
Thyroid	0.05	0.03
Bone surfaces	0.05	0.03
Remainder[b]	0.50	0.30
Total[c]	1.65	1.00

[a] The risk coefficient is the probability of developing fatal cancer per person for 100 rem (1 Sv) of exposure.
[b] A W_T of 0.06 is assigned to each of the five remainder tissues receiving the highest dose equivalents, and the other remainder tissues are to be neglected.
[c] The total somatic risk alone is 1.25×10^{-2} Sv^{-1}. The genetic risk is 0.4×10^{-2} Sv^{-1} making the total risk 1.65×10^{-2} Sv^{-1}.

dose is that it can be compared with other types of risky activities to facilitate a person's understanding of the degree of risk involved in a radiologic procedure. For example, the estimated risk of smoking is 1.37×10^{-7} deaths per cigarette; that is, one death is expected to occur for every 7,299,270 cigarettes smoked. The risk of dying in an automobile accident in North America is 5.6×10^{-8} deaths per mile driven; that is, one death is expected to occur for every 17,857,143 miles driven.[16] The stochastic risk of death from radiation-induced cancer (Table 6-15) is 5×10^{-2} per Sv (100 rem), equivalent to 5×10^{-4} per rem effective dose; that is, one extra death above natural causes of cancer death is expected to occur for every 2000 people exposed to 1 rem (0.01 Sv). The death rate from natural causes of cancer is 20%. Thus, 1 rem effective dose will increase the death rate from 400 people to 401 people out of 2000; that is, from 20% to 20.05%. Compared with other risky activities, 1 rem exposure carries the same risk of dying as driving 8929 miles (5×10^{-4} deaths/rem ÷ 5.6×10^{-8} deaths/mile) or smoking 3650 cigarettes (5×10^{-4} deaths/rem ÷ 1.37×10^{-7} deaths/cigarette). The same analogy can be used to explain risk to patients undergoing a nuclear medicine procedure. As an example, the effective dose for a 30 mCi ^{99m}Tc-MDP bone scan is estimated to be about 0.5 rem (0.005 Sv) (Table 6-3). Thus, the procedure would carry the same risk for death as smoking about 1800 cigarettes or driving about 4500 miles in an automobile.

Other ways of discussing radiation risk with patients are to compare the effective dose for a procedure with the average annual amount of natural background radiation, which is about 0.3 rem (0.003 Sv), or with the annual allowable whole-body exposure for a radiation worker of 5 rem (0.05 Sv), or with the effective dose of another radiologic procedure.[17,18] Common radiologic procedures and their average effective doses are listed in Tables 6-19 and 6-20.[4]

TABLE 6-19 Average Effective Doses of Radiologic Exposures

Radiation Source	Effective Dose (mrem)
Annual occupational exposure limit	5000
Average annual natural background radiation in the United States	300
Diagnostic X-ray Procedures	
Chest (PA and lateral of chest)	10
Chest (PA)	2
Cervical spine	20
Thoracic spine	100
Pelvis	60
Lumbar spine	150
Intravenous urography	300
Upper gastrointestinal[a]	600
Barium enema[a]	800
Mammography	400
Abdomen	700
Dual x-ray absorptiometry (without CT)	0.1
Dual x-ray absorptiometry (with CT)	4
Small-bowel series	500
CT Procedures	
Head	200
Neck	300
Chest	700
Chest for pulmonary embolism	1500
Abdomen	800
Pelvis	600
Three-phase liver study	1500
Spine	600
Coronary angiography	1600
Calcium scoring	300
Virtual colonoscopy	1000

CT = computed tomography.
[a]Includes fluoroscopy.
Source: References 4, 16, and 17.

TABLE 6-20 Average Effective Doses of Nuclear Medicine Procedures

Scan Procedure	Agent	Administered Activity (mCi)	Effective Dose (mrem)
Brain	^{99m}Tc-exametazime	20	690
Brain	^{99m}Tc-bicisate	20	570
Brain	^{18}F-fludeoxyglucose	20	1410
Thyroid	^{123}I-sodium iodide	0.25	190
Thyroid	^{99m}Tc-sodium pertechnetate	10	480
Parathyroid	^{99m}Tc-sestamibi	20	670
Cardiac (stress)	^{201}Tl-thallous chloride	5	407
Cardiac (rest–stress)	^{99m}Tc-sestamibi	40	1280
Cardiac (rest–stress)	^{99m}Tc-tetrofosmin	40	1140
Cardiac ventriculography	^{99m}Tc-red blood cells	30	780
Cardiac viability	^{18}F-fludeoxyglucose	20	1410
Lung perfusion	^{99m}Tc-albumin aggregated	5	200
Lung ventilation	^{133}Xe-xenon	20	50
Lung ventilation	^{99m}Tc-pentetate aerosol	1.1	200
Liver–spleen	^{99m}Tc-sulfur colloid	6	210
Hepatobiliary	^{99m}Tc-disofenin	5	310
Gastrointestinal bleed	^{99m}Tc-red blood cells	30	780
Gastric emptying	^{99m}Tc-solids	0.4	40
Renal	^{99m}Tc-pentetate	10	180
Renal	^{99m}Tc-mertiatide	10	260
Renal	^{99m}Tc-succimer	10	330
Renal	^{99m}Tc-gluceptate	10	200
Bone	^{99m}Tc-medronate	30	630
Total body	^{67}Ga-gallium citrate	4	1500
Total body	^{111}In-pentetreotide	6	120
Total body	^{99m}Tc-white blood cells	20	810
Total body	^{111}In-white blood cells	0.5	670
Total body (tumor)	^{18}F-fludeoxyglucose	20	1410

Source: Reference 4.

REFERENCES

1. International Commission on Radiological Protection. *1990 Recommendations of the International Commission on Radiological Protection.* ICRP Publication 60. New York: Pergamon Press; 1991.
2. International Commission on Radiological Protection. *Limits for Intakes of Radionuclides by Workers.* ICRP Publication 26. New York: Pergamon Press; 1977.
3. International Commission on Radiological Protection. Recommendations of the ICRP. ICRP Publication 103. *Ann ICRP.* 2007 (2-4);247–332 (annex B). www.sciencedirect.com/science?_ob=PublicationURL&_tockey=%23TOC%234965%232003%2399966 9995%23469832%23FLA%23&_cdi=4965&_pubType=J&_auth=y&_acct=C000047720&_version=1&_urlVersion=0&_userid=7922 358&md5=32e5accbb89435cc46d41e1a5b05279c. Accessed September 7, 2010.
4. Mettler FA Jr, Huda W, Yoshizumi TY, et al. Effective doses in radiology and diagnostic nuclear medicine: a catalog. *Radiology.* 2008;248:254–63.
5. Snyder WS, Ford MR, Warner GG, et al. *"S," Absorbed Dose per Unit Cumulated Activity for Selected Radionuclides and Organs.* MIRD Pamphlet No. 11. New York: Society of Nuclear Medicine; 1975.
6. Quimby EH, Feitelberg S. *Radioactive Isotopes in Medicine and Biology.* 2nd ed. Philadelphia: Lea & Febiger; 1963:85.
7. National Council on Radiation Protection and Measurements. *Precautions in the Management of Patients Who Have Received Therapeutic Amounts of Radionuclides.* NCRP Report No. 37. Washington, DC: National Council on Radiation Protection and Measurements; 1970.
8. United Nations Scientific Committee on the Effects of Atomic Radiation. *Sources and Effects of Ionizing Radiation. Vol II: Effects.* UNSCEAR 2000 Report to the General Assembly. New York: United Nations; 2000.
9. National Academy of Sciences. *Health Effects of Exposure to Low Levels of Ionizing Radiation.* BEIR V Report. Washington, DC: National Academies Press; 1990.
10. United Nations Scientific Committee on the Effects of Atomic Radiation. Radiation carcinogenesis in man. In: *Sources, Effects and Risks of Ionizing Radiation.* Annex F. New York: United Nations; 1988.
11. National Council on Radiation Protection and Measurements. *Influence of Dose and Its Distribution in Time on Dose-Response*

Relationships for Low-LET Radiations. NCRP Report No. 64. Bethesda, MD: National Council on Radiation Protection and Measurements; 1980.
12. National Academy of Sciences. *Health Risks from Exposure to Low Levels of Ionizing Radiation.* BEIR VII Phase 2. Washington, DC: National Academies Press; 2006.
13. Hall EJ. *Radiobiology for the Radiologist.* 4th ed. Philadelphia: JB Lippincott; 1994.
14. National Council on Radiation Protection and Measurements. *Recommendations on Limits for Exposure to Ionizing Radiation.* NCRP Report No. 91. Bethesda, MD: National Council on Radiation Protection and Measurements; 1987.
15. National Council on Radiation Protection and Measurements. *Limitation of Exposure to Ionizing Radiation.* NCRP Report No. 116. Bethesda, MD: National Council on Radiation Protection and Measurements; 1993.
16. Huda W. Nuclear medicine dose equivalent: a method for determination of radiation risk. *J Nucl Med Technol.* 1986;14:199–201.
17. Standards for Protection Against Radiation, Occupational Dose Limits for Adults. 10 CFR 20.1201 (2003).
18. National Council of Radiation Protection and Measurements. *Ionizing Radiation Exposure of the Population of the United States.* NCRP Report No. 93. Bethesda, MD: National Council of Radiation Protection and Measurements; 1987.

CHAPTER 7
Radiation Safety

Radiation safety refers to the activities and control measures that limit the amount of radiation exposure received by radiation workers, members of the general public, and patients undergoing radiologic procedures. The radiation protection techniques described in Chapter 6 apply to workplace radiation safety practice. Guidance on radiation safety issues is found in the Code of Federal Regulations under Title 10: Energy, Part 19: Notices, Instructions and Reports to Workers: Inspections and Investigations (10 CFR 19); Part 20: Standards for Protection Against Radiation (10 CFR 20); and Part 35: Medical Use of Byproduct Material (10 CFR 35). Additional information regarding transport of radioactive material is found in U.S. Department of Transportation (DOT) regulations. All of these documents can be accessed at www.gpoaccess.gov/cfr/index.html. Helpful information related specifically to nuclear pharmacies is located in NUREG 1556, Volume 13, Program Specific Guidance about Commercial Radiopharmacy Licenses, available on the web at www.nrc.gov/reading-rm/doc-collections/nuregs/staff/sr1556/v13/r1/.

This chapter discusses important points related to the safe handling of radioactive material in nuclear medicine and nuclear pharmacy.

RADIATION PROTECTION ORGANIZATIONS

The population at risk from exposure to ionizing radiation is divided into two groups: (1) the general public (nonoccupational exposure group) and (2) radiation workers (occupational exposure group). A number of organizations are involved in studying the effects of ionizing radiation on biologic systems. These groups monitor and analyze investigational studies and reports to assess the risks associated with radiation exposure and make recommendations for setting radiation dose limits.

Two principal scientific committees conduct this type of assessment. The first is the United Nations Scientific Committee on the Effects of Atomic Radiation (UNSCEAR), whose mandate is to assess and report levels and effects of exposure to ionizing radiation to the General Assembly of the United Nations. The last report by this international organization appeared in 2000.[1] The second committee is the Biologic Effects of Ionizing Radiations (BEIR) Committee, appointed in the United States by the National Academy of Sciences. This committee advises the U.S. government, through periodic reports, on the health consequences of radiation exposure. Its last report (BEIR VII) appeared in 2006.[2] The report can be read online at www.nap.edu/catalog.php?record_id=11340. These committees analyze and summarize data and suggest risk estimates for radiation-induced cancer and genetic effects, but they are not obligated to make recommendations on dose limits.

Two principal committees make recommendations regarding radiation dose limits. The first is the International Commission on Radiological Protection (ICRP). This committee is well respected by the scientific community and often takes a leadership role in formulating concepts in radiation protection and in recommending dose limits. Its most recent report is ICRP 60, published in 1991.[3] ICRP 92 in draft form is available on the web at www.icrp.org/docs/2005_recs_CONSULTATION_Draft1a.pdf. The second committee is the National Council on Radiation Protection and Measurements (NCRP), a U.S. organization. NCRP often follows the recommendations of ICRP. Its most recent report on dose limits is NCRP 116, published in 1993.[4]

The principal regulatory body in the United States is the Nuclear Regulatory Commission (NRC). It was established in 1974 by the Energy Reorganization Act and replaced the Atomic Energy Commission, the organization that previously regulated byproduct material in the United States. NRC is composed of five members and is responsible for licensing and regulating nuclear facilities and materials. It establishes and enforces rules and regulations concerning the safe use of radiation and radioactive byproduct material in the United States. Its regulations are published in 10 CFR.

Naturally occurring and accelerator-produced radionuclides are regulated by individual states. NRC may grant regulatory authority for byproduct material to individual states, in which case these so-called agreement states regulate all forms of radioactive material: naturally occurring material, byproduct material, and accelerator-produced material. A state's regulation of byproduct material must be compatible with NRC regulations. Shortly after September 2001, the U.S. Congress granted the NRC jurisdiction over all sources of nuclear material, including accelerator-produced isotopes.

AS LOW AS REASONABLY ACHIEVABLE CONCEPT

The goal of any radiation protection program is to limit radiation exposure to levels that are considered safe. This goal is met by (1) preventing or limiting the development of deterministic (nonstochastic) effects of radiation by setting equivalent dose limits well below the threshold limits for a person's working lifetime and (2) limiting the risk of stochastic effects to a frequency no greater than the risks seen in nonradiation occupations.[4] Deterministic effects of radiation exposure are those effects that become more severe with increasing dose but manifest themselves only above a threshold dose. Some examples are skin erythema, cataract formation, and reduction in sperm count. Stochastic effects are those that occur without a threshold dose and whose probability of occurrence increases with dose. Such effects include the development of cancer and genetic defects.

As stated in 10 CFR 20.1101, a licensee must establish a radiation protection program so that occupational doses and doses to members of the public are as low as reasonably achievable (ALARA). Additionally, the program must establish air emission limits for the general public, and the program must be reviewed annually to ensure compliance with ALARA. With such radiation protection programs in place, the public is assured that radiation will be used in a safe manner.

To keep radiation exposure ALARA, access to areas that house radiation sources must be controlled. A restricted area is one to which access is limited for the purpose of protecting individuals against undue risks from exposure to radiation and radioactive materials, such as the areas in a nuclear pharmacy or nuclear medicine laboratory where sources are stored, a radiopharmaceutical compounding and dispensing area, a nuclear medicine imaging room, or a hospital room where a patient is being treated with radionuclide therapy. Restricted areas are typically posted with signs to warn people of the potential for exposure to ionizing radiation. An unrestricted area is one to which access is neither limited nor controlled by the licensee, such as a nuclear pharmacy office, a nuclear medicine waiting room, or a nuclear medicine reading room.

RADIATION SAFETY PROGRAM

The authority and responsibilities for a radiation protection program are outlined in 10 CFR 35.24. The regulation states that a licensee's management must

(A) Approve in writing (1) license applications, renewals, or amendments before submission to NRC, (2) individuals who work as authorized users, authorized nuclear pharmacists, or authorized medical physicists, and (3) any radiation protection program changes permitted under 35.26.
(B) Appoint a radiation safety officer (RSO) who agrees, in writing, to be responsible for implementing the radiation protection program.
(C) Permit an authorized user or qualified individual to serve as a temporary RSO.
(D) Appoint more than one temporary RSO.
(E) Establish, in writing, the authority, duties, and responsibilities of the RSO.
(F) Establish a radiation safety committee if two or more different types of uses of byproduct material or types of units are authorized; this committee must include an authorized user of each type of use permitted in the license, the RSO, a representative of the nursing service, and a representative of management.
(G) Provide the RSO with authority and resources to conduct the radiation safety program to (1) identify radiation safety problems, (2) initiate, recommend, or provide corrective actions, (3) stop unsafe operations, and (4) verify implementation of corrective actions.
(H) Retain records of action taken in (A), (B), and (E) of part 35.24 for 5 years.

In a nuclear pharmacy the RSO must have a level of basic technical knowledge sufficient to understand the work to be performed with byproduct material and be qualified by training and experience to perform the duties noted above. Any individual who has sufficient training and experience to be named as an authorized nuclear pharmacist is also considered qualified to serve as the facility RSO.[5] Typical duties and responsibilities of an RSO in a nuclear pharmacy are outlined in Appendix H of NUREG-1556.

OCCUPATIONAL DOSE LIMITS AND RISK

An occupational dose is a radiation dose received by an individual in the course of employment in which the individual's assigned duties involve exposure to ionizing radiation. An occupational dose does not include doses received from background radiation, from any medical administration, or from voluntary participation in medical research investigations. Occupational dose limits and dose limits for the public are enforced by the NRC and agreement states and are found in 10 CFR 20. These limits are derived from recommendations made by ICRP and NCRP. Table 7-1 lists the annual dose limits for exposure to ionizing radiation. The occupational dose limits have changed considerably over the years. The annual limit for whole-body dose was reduced by a factor of about 3 between 1934 and 1950, and by another factor of 3, to the equivalent of 50 mSv (5 rem), by 1958.[1] The lower dose limits became feasible mainly because radiation protection techniques and methods were developed that workers could use in their practices to keep exposures ALARA.

The philosophy of NCRP is that for occupational exposure, the level of protection provided should ensure that the risk of developing fatal cancer from exposure to radiation be no greater than that of fatal accidents in safe industries.[4] Quantitatively, this can be viewed as follows. It is estimated that the average nominal lifetime excess risk from a single, uniform whole-body equivalent dose of 1 Sv (100 rem) is 4×10^{-2} for fatal cancer, 0.8×10^{-2} for severe genetic effects, and 0.8×10^{-2} for nonfatal cancer, for a total detriment of 5.6×10^{-2} (see Table 6-15, Chapter 6). The average fatal accident rate in all industries is 1×10^{-4} per year or 1 death in 10,000, with a range of 0.2×10^{-4} to 5×10^{-4}.[4] NCRP data from 1980 indicate that the average equivalent dose of monitored workers was approximately 2.1 mSv (0.21 rem) per year, which would suggest a total detriment of about 1.1×10^{-4} per year (2.1×10^{-3} Sv per year $\times 5.6 \times 10^{-2}$ detriment Sv^{-1}).[4] This suggests, therefore, that the occupational risk of the average radiation worker is roughly comparable to the average risk of accidental death for all industries.

NCRP considers further what the risks might be for the radiation worker who is exposed to maximum permissible doses of radiation. The NCRP limits on occupational radiation exposure are listed in Table 7-1: not more than 50 mSv (5 rem) per year, and a cumulative exposure not to exceed a person's age $\times$ 10 mSv (1 rem). Using these limits and a hypothetical worst-case scenario of acceptable maximal exposure, the lifetime fatal cancer risk is approximately 3×10^{-2}.[4] The worst-case scenario for accidental death in a safe industry is 5×10^{-4} per year $\times$ 50 years, which results in a lifetime fatal accident risk of 2.5×10^{-2}. Thus, the risk of a fatal outcome from maximal allowable exposure for a radiation worker is consistent with the maximal risk of accidental death for all industries.

DOSE LIMITS FOR VOLUNTEER SUBJECTS

Radiation safety concerns typically involve monitoring the exposure of radiation workers (occupational exposure), the general public (nonoccupational exposure), and patients undergoing diagnostic or therapeutic procedures. A fourth exposure group consists of human subjects who volunteer to undergo experimental procedures that involve exposure to ionizing radiation during research studies approved by the local radioactive drug research committee (RDRC). Such studies are typically conducted at universities or research centers and monitored by institutional review boards (IRBs) in compliance with Food and Drug Administration regulations. The main purpose of an IRB is to ensure that the research is conducted in an ethical manner and that the volunteer subject has been informed of the benefits and risks of participating in the study and has consented to do so. If the subject will receive a radioactive drug or be exposed to external beam radiation from a source such as an x-ray or computed tomography (CT) machine, a human-use radioisotope committee must also approve the research. The

TABLE 7-1 Annual Radiation Dose Limits

A. Occupational dose limits for adults (10 CFR 20.1201)	
A licensee must establish controls to not exceed the following:	
1. An annual limit, which is the more limiting of	
a. Total effective dose equivalent	5 rem (50 mSv)
Or	
b. Sum of deep-dose equivalent and committed dose equivalent to any organ or tissue except lens of the eye	50 rem (500 mSv)
2. The annual limits to lens of the eye, to the skin, and to the extremities, which are	
a. Eye dose equivalent	15 rem (150 mSv)
b. Shallow-dose equivalent to skin or any extremity	50 rem (500 mSv)
3. Cumulative exposure (NCRP limit)	1 rem $\times$ age in yr
B. Occupational dose limits for minors (10 CFR 20.1207)	10% of adult limits
C. Embryo or fetus of occupational worker (10 CFR 20.1208)	0.5 rem (5 mSv)
D. Total effective dose equivalent (public) 10 CFR 20.1301	0.1 rem (1 mSv)

two main purposes of this committee are to ensure that the radiation dose received by the subject is within the acceptable guidelines for radiation exposure and that no female subject is or might be pregnant during the study.

NCRP Report 70, Nuclear Medicine Factors Influencing the Choice and Use of Radionuclides in Diagnosis and Therapy (1982), contains radiation limits for exposure to adult volunteer subjects who receive radionuclides for investigative purposes. These guidelines, which are listed in 21 CFR 361 (April 1, 2000), state that the limits for radiation exposure of the whole body, active blood-forming organs, lens of the eye, and gonads are 3 rem per year for a single dose and 5 rem per year for an annual and total dose commitment; for other organs the limits are 5 rem and 15 rem for single dose and annual and total dose commitment, respectively. For research subjects under 18 years of age, the radiation dose is not to exceed 10% of these limits. These limits can be used by a human-use radioisotope committee in evaluating the radiation safety of proposed research and by the principal investigator of the study in drafting the consent form that informs the subject of the radiation risk.

SOURCES OF RADIATION EXPOSURE

Every individual is exposed to some form of ionizing radiation, whether or not he or she works with radiation. In the United States, the average annual effective dose equivalent from natural background radiation is 3 mSv (300 mrem). About one-third of this exposure comes from radiation in the cosmos and terrestrial sources, and about two-thirds comes from radon. Additional sources of ionizing radiation exposure, principally from medical procedures, increase the annual exposure to about 3.6 mSv (360 mrem) (Table 7-2).[6]

Background Radiation

The natural sources of background radiation include cosmic, terrestrial, and human body sources. Cosmic radiation, mostly from high-energy protons, varies with altitude and latitude. The average annual dose in the United States is 26 mrem. It doubles with every 6562 feet of altitude. At 39,000 feet, which is representative of airline altitude, it is 0.5 mrem/hour.[6] Cosmic radiation is higher at the North and South Poles, where charged particles are attracted, and lower at the equator. Terrestrial radiation is from the earth and building materials and varies geographically. In the Rocky Mountains, where rocks are rich in thorium and uranium, the average annual effective dose equivalent is 63 mrem/year; in the Atlantic and Gulf Coast regions it is 16 mrem/year; and for the remainder of the country it is 30 mrem/year.[6] Natural bodily exposure comes mainly from inhaled radon gas, with a small component from ingested nuclides, principally ^{14}C, ^{210}Po, and ^{40}K.[6]

Manufactured sources of background radiation include medical procedures in diagnostic radiology and nuclear medicine and nonmedical sources of exposure such as consumer products (e.g., smoke detectors) and radioactive fallout from the atmosphere.

The average natural background radiation exposure varies around the world. For example, in the United States it is 300 mrem, in the Brazilian coastal region it is 500 mrem, in Niue Island in the Pacific it is 1000 mrem, and in Kerala, India, it is 1300 mrem. Interestingly, despite the wide variation in these background dose rates, there appears to be no significant difference in the rate of stochastic effects of radiation between these populations.

Radiation Exposure in Nuclear Medicine and Nuclear Pharmacy

The primary exposure to ionizing radiation in professional practice occurs while handling radioactive material during transport and receipt, during radiopharmaceutical compounding and dispensing operations, during patient dose administration, and during patient imaging procedures. Patient procedures pose a radiation risk to both nuclear medicine personnel and the patients themselves. Radiation workers wear whole-body film badges and extremity dosimeters that are processed monthly to assess exposure. The results are reviewed by the RSO for compliance with regulations. In general, radiation worker exposure is well below the occupational dose limits. A few principal areas are reviewed here to identify the sources and magnitude of worker exposures from procedures in nuclear medicine and nuclear pharmacy, as described in NCRP Report 124.[7]

Transport and Receipt of Radioactive Material

As a general rule, radiation sources should be shielded so that the external exposure does not exceed 20 μGy (2 mrem)/hour. For example, a 3700 MBq (100 mCi) source of ^{131}I requires 5 cm (2 inches) of lead to achieve 20 μGy/hour. Thus, for example, it is important that the cart used to transport doses of ^{131}I-sodium iodide to a patient's room provide adequate shielding to the transporter to keep exposure ALARA. After the patient receives a dose of ^{131}I-sodium iodide, exposure increases significantly because the patient is not shielded. For example,

TABLE 7-2 Annual Effective Dose Equivalent in the U.S. Population, 1980–1982

Source	Average Annual H_E (mrem)
Natural sources	
Radon	200
Cosmic, terrestrial, body	100
Medical	
Diagnostic x-rays, nuclear medicine	53
Nonmedical	7
Total	360

Source: Reference 6.

the dose rate at 1 m from a patient immediately after receiving 100 mCi of ^{131}I is approximately 0.2 mGy (20 mrem)/hour. A typical dose rate from a shielded radiopharmaceutical package containing 100 mCi of ^{131}I is about 60 mrem/hour at the surface and 0.2 mrem/hour at 1 m.[7] This dose rate, of course, varies depending on the amount of lead used to shield the source.

Hospital inpatients are sometimes injected with a radiopharmaceutical in their rooms when the nuclear medicine procedure requires a waiting period between dose administration and imaging. This typically occurs with bone imaging. In this circumstance the radiopharmaceutical dose should be transported to the patient room in a lead-lined syringe carrier to minimize exposure of the technologist and others along the way. This ensures that ALARA exposure is maintained in nonoccupational areas.

Dose Preparation and Patient Injection

NRC regulations require that a syringe shield be used in compounding radiopharmaceutical kits, because hand exposure can be quite high when large amounts of radioactivity are handled in such operations. Figure 7-1 illustrates the exposure from ^{99m}Tc with and without lead shielding. It demonstrates that 0.3 cm (⅛ inch) of lead reduces the exposure rate by a factor of 1000.

Exposure rate to the hands has been assessed by several investigators. The equivalent doses reported vary because of the differing conditions of exposure and monitoring, but they provide a guide to the potential magnitude of the exposure. Anderson et al.,[8] using thermoluminescent dosimeters (TLDs) taped to the fingers, found that the average dose to the index finger was 1600 mrem per curie of ^{99m}Tc injected using unshielded syringes, with a significantly smaller dose to the other fingers. For personnel who performed only generator elution, calibration, and dose preparation (no patient injection), the average dose was only 200 mrem per curie eluted. This report indicates that the majority of exposure comes from patient injection and averages about 1400 mrem per curie or about 28 mrem per 20 mCi (740 MBq) dose injected.

Use of syringe shields during patient injection is considered optional by NRC regulations, but dose rates, as just noted, can be high if precautions are not taken. Henson,[9] using a computer analysis and calculation method, assessed the exposure rate from unshielded radiopharmaceutical syringes at various positions on the syringe. Figure 7-2 summarizes the dose rates from syringes containing ^{99m}Tc and illustrates that there is a significant difference in dose depending on how close the finger is to the volume of activity. Figure 7-2 indicates that the dose rate decreases considerably when the finger is moved away from the active volume in the syringe. For example, if the average dose administered is 0.5 mL in a 2 mL syringe, the dose rate is 55 mrem per millicurie per minute at position A. Considering that finger contact time with the syringe during patient injection averages 15 seconds,[8] a 20 mCi (740 MBq) dose would deliver 275 mrem to the finger at position A, 6.25 mrem at position B, and 1.5 mrem at position C. The occupational dose limit for extremity exposure is 50 rem per year or an average of 1 rem per week. From the example just cited it is clear that one could easily exceed this limit after administering only 4 unshielded doses with the fingers held at position A and after 160 doses at position B. Based on an average dose rate from radiopharmaceuticals, noted previously by Anderson et al., of 28 mrem per 20 mCi (740 MBq) dose injected, 35 doses per week (1000 mrem per week limit divided by 28 mrem per dose) would be the maximum allowed in order to avoid exceeding the 50 rem per year limit of extremity exposure.

It is obvious that the position of the fingers on the syringe relative to the radioactive volume is an important factor in reducing exposure of the hands. This points out the advantage of using the inverse square law (see Chapter 6) when handling

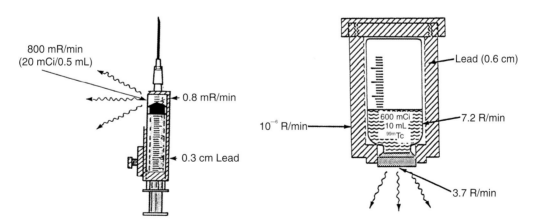

FIGURE 7-1 Rates of film exposure from a ^{99m}Tc source in a syringe and a generator elution vial. The syringe contains 20 mCi of ^{99m}Tc activity emitting a dose rate of 800 mR/minute (unshielded) and 0.8 mR/minute (shielded with 0.3 mm of lead). The generator elution vial contains 600 mCi of ^{99m}Tc emitting 3.7 R/minute at the unshielded entrance port of the vial and 10^{-6} R/minute at the surface of a lead shield 0.6 cm thick. (Courtesy of Dr. John Howley, Radiation Safety Branch, National Institutes of Health, Bethesda, MD.)

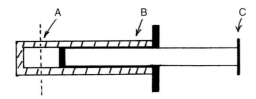

Volume (mL)	A	B	C
0.5	55	1.25	0.30
1.0	40	1.75	0.25
2.0	25	5.95	0.20

FIGURE 7-2 Exposure rate (in mrem per mCi per minute) of fingers at various positions on an unshielded 2 mL plastic syringe containing ^{99m}Tc. (Compiled from reference 9.)

syringes, especially if they are not shielded. It is important also to recognize that hand radiation badges worn on the ring finger will not give a true estimate of exposure at the fingertips. On the basis of these data, technologists who inject patients and nuclear pharmacists who routinely prepare patient doses should monitor hand and finger exposure rates during these procedures. Syringe shields may be necessary to keep exposures within the acceptable dose limit. The design of syringe shields has improved over the years to make them less bulky and easier to handle. Figure 7-3 illustrates a shield constructed of tungsten, which is 50% denser than lead.

Skin contamination can be a potential problem. Kereiakes[10] reported that complete decay of 1 μCi of ^{99m}Tc on the skin can range from 200 Gy from a point source to 0.07 Gy if it is spread out over 1 cm^2. The potential for such a high dose rate emphasizes the importance of wearing disposable gloves while handling radioactive material.

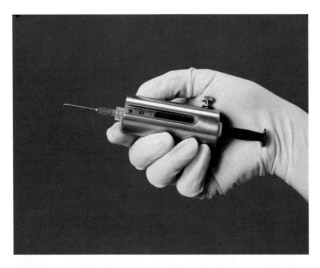

FIGURE 7-3 Tungsten syringe shield with leaded-glass viewing window. (Photo courtesy of Biodex Medical Systems, Shirley, NY.)

Beta-emitting sources can also produce significant exposure of the hands. It has been estimated that a dose of 4.5 rad is delivered to a finger held for 30 seconds in contact with a syringe containing 10 mCi (370 MBq) of ^{32}P in a 5 mL volume.[9] This high dose rate is possible because of the penetrating power of high-energy beta particles. With the increased use of beta emitters for radionuclide therapy, the handling of these doses should be given careful consideration. See Table 6-8, Chapter 6, for physical data on beta emitters used in nuclear medicine.

Imaging Procedures

Several studies have reported on exposures of nurses and nuclear medicine technologists from patients who have received a radiopharmaceutical, and these have been summarized.[7] Some of the highest exposure doses to technologists arise from patients injected with a ^{99m}Tc bone agent. For example, the dose rates from a 20 mCi (740 MBq) dose of ^{99m}Tc-medronate are summarized in Table 7-3. NCRP Report 124 states that the exposure from diagnostic nuclear medicine patients is typically less than 1 mR/hour at 1 m. Benedetto et al.[11] estimated that the dose per year to nuclear medicine personnel exposed to patients is about 400 mrem.

Summary

In summary, the following recommendations will reduce radiation dose to the hands and body.

- Time: Work quickly and efficiently.
- Distance:
  - Hold syringes at a safe distance from the active volume.
  - Make venipuncture with a butterfly set and stopcock for difficult injections to reduce handling time of the dose.
  - Handle "hot" sources with tongs.
- Shielding: Use syringe shields, especially with large-activity sources.
- Protectives: Wear disposable gloves.
- Monitors:
  - Monitor hands with a Geiger–Müller (GM) counter frequently to detect contamination.
  - Wear ring badges on the index finger or finger likely to receive the highest dose on the basis of syringe handling technique.

PERSONNEL MONITORING

Radiation workers are monitored for exposure with whole-body badges and with extremity badges such as finger dosimeters. Persons who should be badged are outlined in 10 CFR 20.1502 as follows.

External exposure monitoring is required for

1. Adults likely to receive an annual dose in excess of any of the following:
   - 5 mSv (0.5 rem) deep-dose equivalent
   - 15 mSv (1.5 rem) lens-dose equivalent

TABLE 7-3 Dose Rates from Patients Injected with 20 mCi ^{99m}Tc-Medronate

Time after Injection	Patient Condition	Position of Measurement	Dose Rate in mR/hr (patient's waist)	
			At surface	At 1 meter
5 min	With and without bone metastases	Anterior	9	0.9
		Posterior	9	1.0
4 hr	Without metastases	Anterior	3	0.3
		Posterior	5	0.6
	With metastases	Anterior	14	2
		Posterior	49	4

Source: Reference 7, p. 21.

- 50 mSv (5 rem) shallow-dose equivalent to the skin or any extremity
2. Minors likely to receive an annual dose in excess of any of the following:
   - 1 mSv (0.1 rem) deep-dose equivalent
   - 1.5 mSv (0.15 rem) lens-dose equivalent
   - 5 mSv (0.5 rem) shallow-dose equivalent to the skin or any extremity
3. Declared pregnant women likely to receive an annual dose during their entire pregnancy of 1.0 mSv (0.1 rem) deep-dose equivalent

Internal exposure monitoring is required for

1. Adults likely to receive in 1 year in excess of 10% of the applicable annual limit on intake (ALI) for ingestion and inhalation. The ALI is the intake of a given radionuclide per year that results in a committed effective dose equivalent of 5 rem or a committed dose equivalent to an organ or tissue of 50 rem.
2. Minors and declared pregnant women likely to receive in 1 year a committed effective dose equivalent in excess of 1.0 mSv (0.1 rem).

Beyond these regulations, a person may be badged at any time for any reason if he or she is concerned about risk from radiation exposure during work involvement.

Whole-body badges, which monitor for deep-dose and shallow-dose equivalent exposure, must be worn on the body where they are likely to receive the highest exposure (10 CFR 20.1201c). Typically, this is the chest area or the waist.

Investigational Levels

During implementation of a radiation safety program, the RSO establishes personnel exposure levels above which an investigation is initiated to identify potential causes of excessive radiation exposure. Investigational level 1 exposures are typically set at about one-fourth the occupational exposure limits to identify exposure trends in the work environment and to provide a level of control that keeps doses ALARA. Table 7-4 gives an example of investigational levels at an institution. If a level 1 exposure occurs, the RSO may send the person involved a written report with a request to review radiation safety procedures to keep exposure ALARA. If a level 2 exposure occurs, the RSO may conduct a direct investigation and an interview with personnel involved and prepare a written report with suggested corrective actions. Figure 7-4 illustrates a typical radiation dosimetry report for personnel exposures that lists current-month exposure and exposure history by quarter, year, and lifetime for each radiation worker.

Monitoring Devices

Exposure to ionizing radiation is monitored with whole-body badges and extremity (finger) badges. Monitors typically employ either photographic film or luminescent chemicals that are exposed or activated by electrons released by Compton or photoelectric interactions that occur in the monitor.

Film Badges

A whole-body film badge uses photographic film that is housed in a plastic holder. It is worn on the chest or waist regions of the body to monitor exposure during the workday. Ionizing radiation generates electrons that expose the film. The badge may also employ plastic, aluminum, or copper filters that selectively absorb radiation to distinguish the type and energy of radiation exposure. An area with no filter permits direct access to the film by weak betas and gamma photons, whereas an area with metallic filters monitors more energetic photon exposure. Badges are usually worn for 1 month and then sent to a laboratory that

TABLE 7-4 Investigational Levels for Radiation Exposure Monitoring

Body Part	Investigational Level (mrem per monthly monitoring period)	
	Level 1	Level 2
Whole body: head, trunk, gonads, or eye lens	100	400
Extremities: elbow, arm below elbow, foot, knee, leg below knee, or skin	1000	3000
Conceptus	30	40

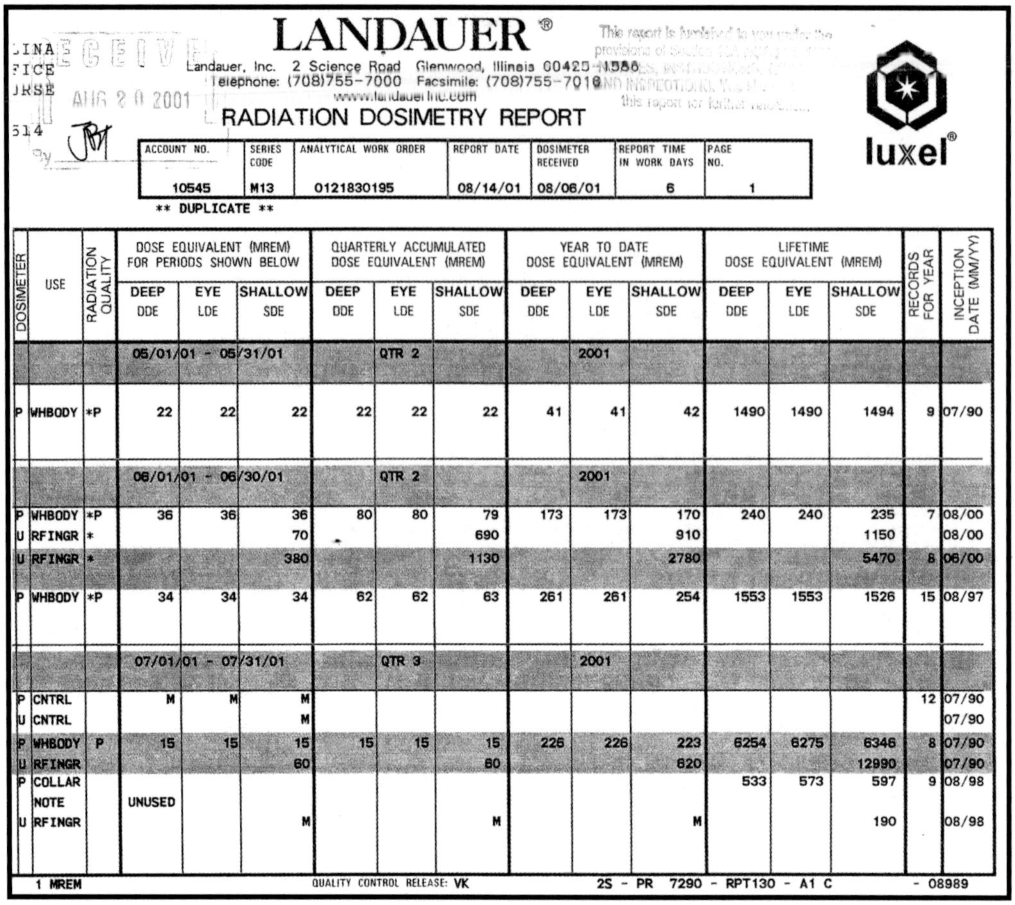

LANDAUER®

Landauer, Inc. 2 Science Road Glenwood, Illinois 60425-1586
Telephone: (708)755-7000 Facsimile: (708)755-7016
www.landauerinc.com

RADIATION DOSIMETRY REPORT

luxel®

AUG 20 2001

ACCOUNT NO.	SERIES CODE	ANALYTICAL WORK ORDER	REPORT DATE	DOSIMETER RECEIVED	REPORT TIME IN WORK DAYS	PAGE NO.
10545	M13	0121830195	08/14/01	08/06/01	6	1

** DUPLICATE **

DOSIMETER	USE	RADIATION QUALITY	DOSE EQUIVALENT (MREM) FOR PERIODS SHOWN BELOW			QUARTERLY ACCUMULATED DOSE EQUIVALENT (MREM)			YEAR TO DATE DOSE EQUIVALENT (MREM)			LIFETIME DOSE EQUIVALENT (MREM)			RECORDS FOR YEAR	INCEPTION DATE (MM/YY)
			DEEP DDE	EYE LDE	SHALLOW SDE	DEEP DDE	EYE LDE	SHALLOW SDE	DEEP DDE	EYE LDE	SHALLOW SDE	DEEP DDE	EYE LDE	SHALLOW SDE		
			05/01/01 - 05/31/01			QTR 2			2001							
P	WHBODY	*P	22	22	22	22	22	22	41	41	42	1490	1490	1494	9	07/90
			06/01/01 - 06/30/01			QTR 2			2001							
P	WHBODY	*P	36	36	36	80	80	79	173	173	170	240	240	235	7	08/00
U	RFINGR	*			70			690			910			1150		08/00
U	RFINGR	*			380			1130			2780			5470	8	06/00
P	WHBODY	*P	34	34	34	62	62	63	261	261	254	1553	1553	1526	15	08/97
			07/01/01 - 07/31/01			QTR 3			2001							
P	CNTRL		M	M	M										12	07/90
U	CNTRL				M											07/90
P	WHBODY	P	15	15	15	15	15	15	226	226	223	6254	6275	6346	8	07/90
U	RFINGR				60			60			620			12990		07/90
P	COLLAR											533	573	597	9	08/98
	NOTE		UNUSED													
U	RFINGR				M			M			M			190		08/98

1 MREM QUALITY CONTROL RELEASE: VK 2S - PR 7290 - RPT130 - A1 C - 08989

FIGURE 7-4 Personnel radiation dosimetry report.

develops the film and measures the exposure. The greater the exposure, the darker is the film. The radiation worker's film badge is compared with control films that have been exposed to known amounts of radiation to quantitate the worker's exposure level. The typical sensitivity range for film badges is 0.3 mSv to 10 mSv (30 mrem to 1 rem).

Optically Stimulated Luminescent Dosimeters

Optically stimulated luminescent dosimetry (OSLD) is an advanced technology for measuring exposure to ionizing radiation.[12,13] The technique employs aluminum oxide crystals doped with carbon (Al_2O_3:C) capable of trapping and storing electron energy generated by ionizing radiation in the crystals. The amount of radiation exposure is measured by exposing the crystals to a stimulating light source, typically a green light (540 nm) from a laser or light-emitting diode source. Stimulation causes the crystals to release the trapped energy as blue light (420 nm) that is proportional to the amount of radiation exposure. Both high- and low-energy photons and beta particles can be measured by OSLD. The technique offers the advantages of greater control and accuracy, with a sensitivity from 10 μSv to 100 Sv (1 mrem to 10,000 rem). In addition, multiple readouts from the dosimeter can be made because only a small fraction of the stored energy is released following stimulation with the green light. OSLD may be used in whole-body badges or finger badges.

Thermoluminescent Dosimeters

TLDs are devices that contain chemical crystals such as lithium fluoride doped with magnesium or titanium (LiF:Mg,Ti) or calcium fluoride doped with manganese (CaF:Mn). Ionizing radiation produces electrons that become trapped in crystal defects and cause the crystals to become excited to a metastable state. Upon heating under controlled conditions, the crystals release their "stored" energy in the form of visible light, which is proportional to the amount of radiation exposure. The sensitivity range is 0.1 mSv to 1 Sv (10 mrem to 100 rem). TLDs are typically used in finger badges, but they are also used in some body badges. Figure 7-5 shows examples of a film badge and a TLD ring badge.

Bioassays

In accordance with 10 CFR 20.1502, a licensee is required to monitor occupational intake of radioactive material by and assess the committed effective dose equivalent to (1) adults who are likely to receive in 1 year greater than 10% of ALI in 10 CFR 20, Appendix B, Table 1, columns 1 and 2, and (2) minors and declared pregnant women likely to receive a committed effective dose equivalent in excess of 0.05 rem in 1 year. A recommended time of monitoring is within 24 hours of exposure.

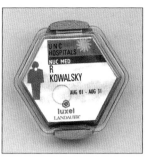

FIGURE 7-5 Radiation monitoring devices: finger ring badge (left) and body badge (right).

A critical area of concern for nuclear medicine and nuclear pharmacy personnel is the handling of radioiodine, particularly ^{131}I. The performance of thyroid bioburden measurements for all occupationally exposed individuals involved in the preparation or administration of therapeutic dosages of ^{131}I (capsules and solutions) is addressed in 10 CFR 20.1204 and 20.1501. Each licensee must monitor (20.1204) the occupational intake of radioactive material by and assess the committed effective dose equivalent to adults likely to receive, in 1 year, an intake in excess of 10% of the applicable ALI(s) in 10 CFR 20.1001–20.2402, Appendix B, Table 1, columns 1 and 2. In addition, Regulatory Guide 8.9 (Acceptable Concepts, Models, Equations and Assumptions for a Bioassay Program) provides guidance for evaluating a thyroid burden of radioiodine. The guide recommends an evaluation level when any thyroid measurement exceeds 2% of the ALI and an investigational level when it exceeds 10% of the ALI. The ALI values for a few radionuclides are given in Table 7-5. An example of how the evaluation procedure works follows, taken from Appendix A of NRC Regulatory Guide 8.9.

A technologist prepared a dose of ^{131}I-sodium iodide for thyroid therapy. A bioassay of the technologist's thyroid indicated a thyroid content of 0.08 µCi 24 hours after dosage preparation. Does this thyroid bioburden require further evaluation?

Determine whether the thyroid intake of ^{131}I exceeds 2% of the ALI.

$$\text{Intake} = \frac{A \, \mu\text{Ci}}{\text{IRF for } ^{131}\text{I}}$$

where:
$A \, \mu\text{Ci}$ = thyroid content at time of measurement
IRF for ^{131}I = intake retention fraction for ^{131}I at time interval after intake

The IRF for ^{131}I after 24 hours is 0.133 (from the table of IRF thyroid values in NUREG/CR 4884, Interpretation of Bioassay Measurements). The evaluation level for ^{131}I is 2% (1 µCi) of the ALI (50 µCi). Since the intake for this technologist (0.6 µCi) does not exceed this level, no further evaluation is needed. If a person's thyroid content is greater than 0.133 µCi, the intake value will be greater than 1 µCi, and the RSO is required to evaluate the situation by, for example, repeating measurements to verify the results. If a person's thyroid uptake exceeds 0.665 µCi, the intake value will exceed 10% of the 50 µCi ALI, and the RSO needs to conduct a more thorough investigation (level 2) with multiple measurements, air sampling, and so on. A thyroid content of 6.65 µCi reaches the ALI, and the person should be excluded from further exposure to radioiodine for the year. Preferably, the person should be removed from radioiodine exposure before the ALI is reached.

Pregnant Radiation Workers

A licensee must develop a pregnancy policy to ensure that the dose equivalent to the embryo or fetus during the entire pregnancy of a declared pregnant radiation worker does not exceed 0.5 rem (10 CFR 20.1208). NCRP 91 (1987) also recommends a total dose limit of 0.5 rem and no more than 0.05 rem in any month once a pregnancy becomes known. The policy should require a woman to voluntarily declare her pregnancy in writing and include the estimated date of conception (10 CFR 20.1003). The records may be maintained in a separate file for privacy (10 CFR 20.2106e). Without this declaration, the licensee cannot be held responsible for the radiation safety of the fetus. If the dose equivalent to the embryo or fetus is found to have exceeded the 0.5 rem limit, or to be within 0.05 rem

TABLE 7-5 Annual Limits on Intake (ALI), Derived Air Concentrations (DAC), and Maximum Permissible Concentrations (MPC) for Selected Radionuclides, from 10 CFR 20, Appendix B

	Table 1 Occupational Values			Table 2 Effluent MPC		Table 3 Release to Sewers
	Oral Ingestion	Inhalation (Restricted Areas)		Unrestricted Areas		
Radio-nuclide	Column 1 ALI (µCi)	Column 1 ALI (µCi)	Column 2 DAC (µCi/mL)	Column 1 Air (µCi/mL)	Column 2 Water (µCi/mL)	Avg Conc/Month (µCi/mL)
^{123}I	3000	6000	3×10^{-6}	2×10^{-8}	1×10^{-4}	1×10^{-3}
^{125}I	40	60	3×10^{-8}	$3 \text{ v } 10^{-10}$	2×10^{-6}	2×10^{-5}
^{131}I	30	50	2×10^{-8}	2×10^{-10}	1×10^{-6}	1×10^{-5}
^{133}Xe			1×10^{-4}	5×10^{-7}		

of this dose by the time of declaration, the licensee must take measures to ensure that any additional dose does not exceed 0.05 rem for the remainder of the pregnancy.

AREA MONITORING

Surveys

In accordance with 10 CFR 35.70, at the end of each day of use a licensee must survey with a radiation detection instrument all areas where unsealed byproduct material requiring a written directive was prepared for use or administered. Records of the survey must be kept for 3 years. The regulation gives no guidance regarding survey limits, but good laboratory practice suggests use of a calibrated GM survey meter to assess whether radiation levels exceed an upper limit of 2 mR/hr at 30 cm from a work surface.

Effluent Monitoring

In accordance with regulations stated in 10 CFR 20.1301, a licensee's operation must be conducted so that (1) the total effective dose equivalent to the public does not exceed 0.1 rem per year, and (2) the dose rate from an external source to an unrestricted area does not exceed 2 mrem in any 1 hour. Concerning compliance with these limits, 10 CFR 20.1302 states that the licensee must make measurements or calculations to demonstrate that the total effective dose equivalent to any member of the public does not exceed 0.1 rem per year or that (1) the average concentrations of radioactive material released in gaseous or liquid effluents to an unrestricted area do not exceed the values in 10 CFR 20, Appendix B, Table 2, and (2) if an individual is continuously present, the dose from external sources does not exceed 2 mrem in any 1 hour and 0.05 rem in a year.

An example of demonstrating compliance with these regulations is a calculation to determine that the exhaust of ^{133}Xe gas to the outside of a facility does not exceed the maximum permissible concentration (MPC), as follows.

A nuclear medicine facility conducts lung ventilation studies with ^{133}Xe gas. On average, 10 doses of 20 mCi each are administered per week. Determine whether the expected amount of ^{133}Xe released outside the facility will exceed the MPC level given in 10 CFR 20, Appendix B, Table 2, Column 1, or $5 \times 10^{-7} \mu Ci/mL$ (see Table 7-5).

The room exhaust rate is $1714 \text{ ft}^3/\text{minute} \times 2.83 \times 10^4 \text{ mL/ft}^3 = 4.85 \times 10^7$ mL/minute. Assume 20% spillage of ^{133}Xe.

$$^{133}\text{Xe } \mu\text{Ci/mL exhausted/wk} = \frac{20{,}000 \text{ }\mu\text{Ci/dose} \times 10 \text{ doses/wk} \times 0.20}{4.85 \times 10^7 \text{ mL/min} \times 10{,}080 \text{ min/wk}}$$

$$= 8.09 \times 10^{-8} \text{ }\mu\text{Ci/mL}$$

In this example the concentration of ^{133}Xe released is 16% of the MPC. Regulatory Guide 8.37 (ALARA Levels for Effluents) recommends aiming for not more than 20% of the MPC in effluents as a conservative limit for radioactive material release.

Clearance Time for Spilled Radioactive Gas

If a radioactive gas is spilled in the work environment, it is recommended that the time be determined for workers to leave the room so that occupational exposure does not exceed the derived air concentration (DAC). The DAC is the concentration of radionuclide breathed for 2000 working hours per year that results in an intake of 1 ALI; 1 DAC = ALI/2.4×10^9 mL. Note: 2.4×10^9 mL = (2000 hours) × (60 minutes/hour) × (2×10^4 mL/minute). The following example illustrates how to calculate the evacuation time.

What is the evacuation time for a radiation worker in the imaging room if a 20 mCi (740 MBq) dose of ^{133}Xe is spilled in the room (restricted area)?

Regulatory Guide 10.8 (Guide for Preparation of Applications for Medical Use) includes in Appendix 0.4 a model procedure for calculating spilled gas clearance times (t), where:

A = highest activity of gas in a single container, in microcuries
 = 20,000 μCi
V = volume of room, in milliliters
 = (5628 ft^3) × (2.83 10^4 mL/ft^3) = 1.59×10^8 mL
Q = exhaust rate, in mL/minute
 = (1714 ft^3/minute) × (2.83 × 10^4 mL/ft^3) = 4.85×10^7 mL/minute
C = DAC ^{133}Xe restricted area = 1×10^{-4} μCi/mL (Table 7-5)

$$t \text{ (clearance time)} = -\frac{V}{Q} \times \ln\left(C \times \frac{V}{A}\right)$$

$$= -\frac{1.59 \times 10^8 \text{ mL}}{4.85 \times 10^7 \text{ mL/min}}$$

$$\times \ln\left(1 \times 10^{-4} \text{ }\mu\text{Ci/mL} \frac{1.59 \times 10^8 \text{ mL}}{20{,}000 \text{ }\mu\text{Ci}}\right)$$

$$= 0.75 \text{ min}$$

Note that this is the time until the concentration reaches the DAC for ^{133}Xe in a restricted area, and workers must leave the room during this period.

In a comprehensive radiation safety program, dose equivalents are assessed from surveys or other measuring devices to ensure compliance. DAC and ALI values from 10 CFR 20, Appendix B, Table 1 can be used to demonstrate compliance. This is particularly useful for volatile nuclides such as radioiodines and radioxenons. A partial list of radionuclides given in 10 CFR 20, Appendix B, is shown in Table 7-5.

Sealed-Source Monitoring

A licensee must wipe test a sealed source every 6 months to ensure that not more than 0.005 μCi (185 Bq) of removable

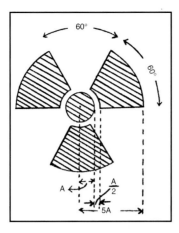

FIGURE 7-6 Standard radiation symbol. The cross-hatched area is magenta or purple; the background is yellow. (Reprinted from 10 CFR 20.1901.)

contamination is present (10 CFR 35.67 and 35.2067). This procedure does not apply to sources with half-lives of less than 30 days, gases, and 100 µCi (3.7 MBq) or less of beta- or gamma-emitting radionuclides. Typical sources that apply in nuclear pharmacy would be reference standards, such as ^{137}Cs, ^{133}Ba, and ^{57}Co, that are used to check the dose calibrator. Records of the leak test must be retained for 3 years.

CAUTION SIGNS

Signs are used to warn individuals that radiation is present in a container, work area, or room. In general, all signs and labels bear the radiation symbol shown in Figure 7-6. Caution signs for various exposure levels of radiation, as listed in 10 CFR 20.1902, are as follows:

- Caution, Radiation Area: >5 mrem in 1 hour at 30 cm
- Caution, High Radiation Area: >100 mrem in 1 hour at 30 cm
- Grave Danger, Very High Radiation Area: >500 rem in 1 hour at 1 m

Because of the types and amounts of radioactive material used and stored in nuclear medicine and nuclear pharmacy laboratories, these areas are likely to require no more than the Caution, Radiation Area sign.

NOTICES, INSTRUCTIONS, AND REPORTS TO RADIATION WORKERS

The regulations in 10 CFR 19 list the requirements for information transfer by the licensee to workers who participate in radiation-associated activities, as well as the rights of workers during NRC or agreement state inspections.

Posting Requirements (10 CFR 19.11)

1. Each licensee must post the following documents or a notice stating where they may be examined:
   a. Current copies of the regulations in 10 CFR 19 and 20.
   b. The radioactive materials license.
   c. The operating procedures applicable to licensed activities.
2. A licensee must post, for 5 days or until the violation is corrected, any notice of violation involving working conditions within 2 days of receipt of the violation notice.
3. A licensee must post NRC Form 3, "Notice to Employees," in a sufficient number of places to be observed by workers.

Instruction to Workers (10 CFR 19.12)

Any person likely to receive an occupational dose higher than 100 mrem per year must be

1. Informed of storage, transfer, or use of radioactive material.
2. Instructed in radiation safety procedures.
3. Instructed to report any violations to the licensee.
4. Instructed on how to respond to warnings in case of radiation exposure occurrences.
5. Instructed about requesting radiation exposure reports.

Notification and Reports to Individuals (10 CFR 19.13)

A licensee must provide each worker with the following reports:

1. A written report of radiation exposure data, analyses, and calculations of an individual's exposure.
2. An annual report of a worker's radiation dose.
3. A radiation exposure record upon a former worker's request.
4. A report of any radiation exposure incident report sent to NRC.
5. A radiation exposure report upon termination of employment.

Presence of Individuals during Inspections (10 CFR 19.14)

A licensee must comply with the following conditions of inspection:

1. A licensee must allow NRC to inspect facilities annually.
2. A licensee must notify NRC of any individual authorized to represent workers at inspection.
3. The workers' representative must be actively engaged in radiation work.
4. A consultant may accompany the inspection.

Consultation with Workers during Inspection (10 CFR 19.15)

The following conditions must prevail during inspections:

1. NRC may consult with workers privately.
2. Workers may bring any perceived violations of regulations, overexposures, and so on to NRC's attention.

Request for Inspections (10 CFR 19.16)

The following conditions must prevail regarding requests for inspection:

1. Any worker may submit a written request to the NRC regional office to inspect a licensee regarding any perceived violation. The complainant may request anonymity.
2. NRC will determine whether inspection is warranted.

RECEIPT OF RADIOACTIVE MATERIAL

Regulations describing the procedures and monitoring limits for receiving and opening packages of radioactive material are found in 10 CFR 20.1906, 10 CFR 71.87 and 71.47, and 49 CFR 173. The regulations for some agreement states may differ and may be more stringent than NRC regulations. Part 20 requires a survey only for packages containing greater than Type A quantities (which are typically >10 Ci), but all packages bearing White-I, Yellow-II, or Yellow-III labels must be wipe tested for surface contamination (see Warning Labels section). Table 7-6 summarizes the package receipt requirements for NRC. The GM survey for external radiation levels is made at the package surface and at 1 m distance. The 1 m survey is the transport index (TI). Wipe testing of the external surface is typically done with a 1 inch filter paper disk rubbed over at least 300 cm^2 of multiple surfaces. The disk is counted in a scintillation counter, corrected for counter efficiency to convert counts per minute (cpm) to disintegrations per minute (dpm), and the result reported as dpm per cm^2. Monitoring of packages must be performed as soon as practicable but not later than 3 hours after receipt. If measured levels exceed the limits stated in Table 7-6, the licensee must immediately notify (1) the final delivery carrier and (2) the NRC operations center by telephone. Note that efficiency of the wipe procedure must be taken into account in recording test results for removable contamination. If wipe efficiency is 100%, the limit for surface contamination is 220 dpm/cm^2. However, if wipe efficiency is 10%, then the limit for a wipe test is 22 dpm/cm^2. When wipe efficiency is not known, the regulation assumes that it is 10%.

Each licensee must establish and maintain written procedures for safely opening radioactive material packages, ensure that they are followed, and give any special instructions for the type of package being opened. In addition to the monitoring requirements, the package-opening procedures should include the following good practice guides listed in NUREG 1556:

1. Open the outer package and remove the packing slip; open the inner package to verify contents, comparing requisition, packing slip, and label on the container.
2. Check integrity of container, inspecting for breaks in seals, loss of liquid, and high count rates on wipes.
3. Survey packing material and packages for contamination before discarding, treating any contaminated materials as radioactive waste.
4. Obliterate any radiation labels prior to discard of materials.
5. Maintain records of receipt, survey, and wipe tests for 3 years.

SHIPMENT OF RADIOACTIVE MATERIAL

Many regulations apply to pharmaceutical manufacturers and nuclear pharmacies that ship radiopharmaceuticals to hospitals and clinics. Requirements for packaging and transportation of radioactive material are described in DOT regulations (49 CFR) and NRC regulations (10 CFR 71), and summarized in two additional publications.[14,15] Specifically,

TABLE 7-6 Nuclear Regulatory Commission Monitoring Requirements for Radioactive Material Packages upon Receipt

Limit Type	Qualifications	Requirements
External radiation limits (GM survey)	Only radioactive materials with Radioactive White-I, Yellow-II, or Yellow-III label containing > Type A quantities or if crushed, wet, or damaged. Most radionuclide Type A quantities are >10 Ci (see Table 7-7)	Package surface levels (not >200 mR/hr) Transport index (not >10 mR/hr at 1 m)[a]
Surface contamination limits (wipe test)	All packages with Radioactive White-I, Yellow-II, or Yellow-III label except if a gas or special form. All nuclear medicine radioactive materials are normal form.	Averaged over 300 cm^2 Must be not greater than 220 dpm (0.0001 µCi) per cm^2 of package surface[b]

[a]From 49 CFR 173.441 and 10 CFR 71.47.
[b]From 49 CFR 173.443.

10 CFR 71.5 states that "each licensee who transports licensed material outside the site of usage, as specified in the NRC license, or where transport is on public highways, or who delivers licensed material to a carrier for transport, shall comply with the applicable requirements of the DOT regulations in 49 CFR parts 107, 171 through 180, and 390 through 397." The parts that particularly relate to packaging and shipment of radioactive materials are

1. Packaging: 49 CFR 173: subparts A, B, and I
2. Marking and labeling: 49 CFR 172: subpart D, 172.400 through 172.407, 172.436 through 172.440, and 172.436 through 172.441 of subpart E
3. Placarding: 49 CFR 172: subpart F, 172.500 through 172.519 and 172.556, and Appendices B and C
4. Accident reporting: 49 CFR 171.15 and 171.16
5. Shipping papers and emergency information: 49 CFR 172: subparts C and G
6. Hazardous material employee training: 49 CFR 172: subpart H
7. Security plans: 49 CFR 172: subpart I
8. Hazardous material shipper/carrier registration: 49 CFR 107: subpart G
9. Modes of transport, via public highway: 49 CFR 177 and 390 through 397, and via air: 49 CFR 175

10 CFR 71.13 exempts physicians from the transport regulations in 10 CFR 71.5 when the physician transports licensed material for use in the practice of medicine. However, the physician must be licensed under 10 CFR 35 or an agreement state to exercise this exemption.

Radioactive Material Definition

For purposes of transportation, a substance is considered to be radioactive if it exceeds both a specific activity concentration threshold and a consignment activity threshold listed in the table of 49 CFR 173.436. Table 7-7 lists threshold values for radionuclides routinely shipped from nuclear pharmacies.

Packaging Requirements

Packaging requirements for radioactive material are determined by the form, type, and quantity being shipped. The two radionuclide forms are special form and normal form. Special form radioactive material is material that may present a direct radiation hazard if released from the package but causes little hazard resulting from contamination because it is in a nondispersible solid form or sealed in a durable capsule. Special form materials are much less likely to spread contamination in the event of package failure. The regulations therefore generally allow substantially larger quantities of special form materials than of normal form materials to be placed in a given package. Normal form radioactive materials may be solid, liquid, or gaseous and include any material that has not been qualified as special form. All radiopharmaceuticals are normal form materials.

TABLE 7-7 Specific Activity and Consignment Activity Thresholds for Radionuclides to Be Considered Radioactive

Radionuclide	Specific Activity Threshold (µCi/g)	Consignment Activity Threshold (µCi)
^{18}F	2.7×10^{-4}	27
^{32}P	2.7×10^{-2}	2.7
^{51}Cr	2.7×10^{-2}	270
^{67}Ga	2.7×10^{-3}	27
^{90}Y	2.7×10^{-2}	2.7
^{99}Mo	2.7×10^{-3}	27
^{99m}Tc	2.7×10^{-3}	270
^{111}In	2.7×10^{-3}	27
^{123}I	2.7×10^{-3}	270
^{125}I	2.7×10^{-2}	27
^{131}I	2.7×10^{-3}	27
^{133}Xe	2.7×10^{-2}	0.27
^{153}Sm	2.7×10^{-3}	27
^{201}Tl	2.7×10^{-3}	27

Source: 49 CFR 173.436.

The principal types of package used to ship radioactive material are Type A, Type B, and "excepted" packages. The criteria for testing the integrity of Type A packages are found in 49 CFR 173.465 and those for excepted packages are in 49 CFR 173.410. Shippers are not required to personally test the package, only to ensure that the testing was performed before use.

The quantity of a particular radionuclide that can be shipped in a Type A package is listed in Table 7-8. Every radionuclide is assigned an A_1 and an A_2 value. These values are the maximum activity (in curies) of that radionuclide that may be shipped in a Type A package. The A_1 values are for special form and the A_2 values for normal form radionuclides. Quantities exceeding Type A package limits require Type B packages. Radiopharmaceuticals are normal form and are shipped in Type A packages and excepted packages. The activity limits for these packages are shown in Table 7-8.

Excepted packages typically contain limited amounts of radioactivity, such as in spent patient syringes returned to a nuclear pharmacy. The excepted quantities are listed in Table 7-8. Excepted packages are excepted from specification packaging, marking (except for the United Nations [UN] identification number), and labeling, and shipping-paper requirements if they do not contain a hazardous substance, but they must meet the following requirements as outlined in 49 CFR 173.421:

1. The package must meet the requirements of 49 CFR 173.410.
2. External radiation at any point on the external surface cannot exceed 0.5 mrem/hour.
3. Removable external contamination cannot exceed 220 dpm/cm^2.
4. For limited quantities, the outside of the inner packaging, or if there is no inner packaging, the outside of the

TABLE 7-8 Type A Package Activity Limits for Radionuclides in Nuclear Medicine

Radionuclide	Form	A_2 (Normal Form)[a] (Ci)	Excepted Quantity Package Limits[b] (mCi)
^{11}C	Liquid	16	1.6
^{57}Co	Liquid	270	27
^{57}Co	Solid	27	270
^{51}Cr	Liquid	810	8.1
^{137}Cs	Liquid	16	1.6
^{18}F	Liquid	16	1.6
^{67}Ga	Liquid	81	8.1
^{123}I	Liquid	81	8.1
^{123}I	Solid	81	81
^{125}I	Liquid	81	8.1
^{131}I	Liquid	19	1.9
^{131}I	Solid	19	19
^{111}In	Liquid	81	8.1
^{99}Mo	Liquid	20	2.0
^{13}N	Liquid	16	1.6
^{32}P	Liquid	14	1.4
^{186}Re	Liquid	16	1.6
^{188}Re	Liquid	11	1.1
^{81}Rb	Liquid	22	2.2
^{153}Sm	Liquid	16	1.6
^{89}Sr	Liquid	16	1.6
^{99m}Tc	Liquid	110	11
^{99}Tc	Liquid	24	2.4
^{201}Tl	Liquid	110	11
^{117m}Sn	Liquid	11	1.1
^{133}Xe	Gas	270	270
^{90}Y	Liquid	8.1	0.81

[a]From 49 CFR 173.435.
[b]From 49 CFR 173.425 (liquids = $A_2 \times 10^{-4}$; solids and gases = $A_2 \times 10^{-3}$).

packaging itself, bears the marking "Radioactive." An empty package must bear the marking "EMPTY" and meet the requirements of 49 CFR 173.428.
5. The outside of the package must be marked with UN 2910 for limited quantities and UN 2908 for empty packages.

Package Markings

Package marking regulations are provided in 49 CFR 172.301, 172.302, and 172.310. Durable markings for nonbulk radioactive packages are as follows.

Basic package-based markings:
- Proper shipping name: Radioactive Material, Type A Package
- UN ID number: UN 2915.
- International Country Code: U.S.A. DOT 7A Type A.
- RQ (reportable quantity), if a "hazardous substance": most radionuclides in nuclear pharmacy shipments are in amounts less than the hazardous materials limit; however, ^{131}I has a limit of 370 MBq (10 mCi) and "RQ" must appear on the package marking and shipping papers (e.g., RQ Radioactive Material, Type A Package, UN 2915).

Additional marking requirements:
- Packages greater than 110 pounds must have the gross weight marked on the outside of the package.
- The words "Type A" must be 0.5 inch high.
- Excepted packages must be marked with UN ID number: UN 2910 and the word "radioactive."
- Empty packages must be marked: EMPTY UN 2908.
- Packages containing liquids must have on opposite sides a marking with arrows to indicate the upward position of the inside package.
- Name and address of the consignor and consignee must be marked.
- All markings must be isolated from other marks, unobscured by labels or attachments, displayed on a background of sharply contrasting color, and durable.

Package Labeling

Warning Labels

Each package of radioactive material, unless excepted, must be labeled on two opposite sides with a distinctive warning label (Figure 7-7). Each of the three label types bears the unique trefoil symbol recommended by NCRP. The labels alert persons that the package contains radioactive materials and that the package may require special handling. A label with an all-white background (White-I) indicates that the external radiation level is low and no special handling is required. If the upper half of the label is yellow (Yellow-II and Yellow-III), it signifies higher radiation levels and the need for more precautions. Radiation level limits for these labels are given in Table 7-9.

The following items must be entered in the blank spaces on the warning label with durable, water-resistant marking:

1. Contents: The name of the radionuclide or its symbol (e.g., molybdenum 99, ^{99}Mo, or Mo-99).
2. Activity: The total activity of all radionuclides contained in the package; these must be expressed in SI units (e.g., Bq, GBq) and may be followed by customary units (e.g., mCi) in parentheses.
3. Transport index (TI): The value measured with a calibrated survey meter at 1 m from the external surface, rounded up to the next higher tenth (e.g., 0.12 mR/hr at 1 m = TI of 0.2). Note: TI is a unitless number required only on Yellow-II and Yellow-III labels. It is not required on White-I labels.

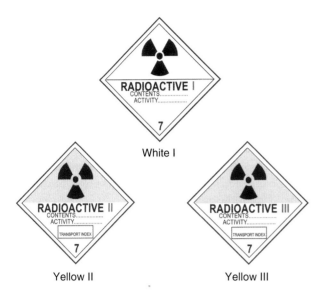

FIGURE 7-7 U.S. Department of Transportation (DOT) labels for radioactive material packages. The I, II, or III referring to level of radiation emanating from the package is in red. Each label is diamond shaped, 4 inches on each side, and has a black solid-line border ¼ inch from the edge. The background color of the upper half is white for the I label and yellow for the II and III labels. The 7 is the DOT hazardous material classification number for radioactive material.

Most packages shipped from nuclear pharmacies are either White-I or Yellow-II. Yellow-III is rarely required but typically is found on ^{99m}Tc generators and ^{131}I-sodium iodide therapy packages.

Transport Index

The TI is the maximum dose rate in mR/hr at 1 m from the exterior surface of the package. The TI limits are listed in Table 7-9. No package offered for transport may have a TI greater than 10. A package surface dose rate less than 0.5 mR/hr is considered to have a TI of 0; therefore, no space is allotted on the Radioactive I (White-I) label for TI. The total TI of all packages in any single transport vehicle or storage location may not exceed 50. The TI system provides control over radiation exposure of personnel handling packages. It prescribes, for example, the minimum separation distance between radioactive materials and crew or passengers (Table 7-10).

Package Monitoring

Any package of radioactive material offered for transport must be wipe tested for removable contamination on its external surface in accordance with 49 CFR 173.443. Common sites for radioactive contamination on nuclear pharmacy shipping containers are the handles and closure clasps of transport boxes. The wipe must occur over at least 300 cm^2 of surface, and the limit of removable contamination is 220 dpm per 1 cm^2. The test is for removable (nonfixed) contamination, and it is assumed that the removal technique is 10% efficient unless proven differently. Therefore, shippers should multiply the wipe dpm by 10 before comparing it with the limit. In other words, the wipe must not have any more than 22 dpm/cm^2 to meet the limit if wipe efficiency is 10%.

Example: A wipe is made over 300 cm^2 of a ^{99m}Tc radiopharmaceutical package surface to yield 3000 cpm in a scintillation well counter. The well counter efficiency for ^{99m}Tc is 50%. If the wipe technique is 10% efficient, what is the removable contamination in dpm/cm^2 on the package surface?

$$\frac{(3000 \text{ cpm})(2)(10)}{300 \text{ cm}^2} = 200 \text{ dpm}/\text{cm}^2$$

Note that a scintillation counter efficiency of 50% requires multiplying cpm by 2 to determine the dpm on the wipe.

Placarding

The shipper of radioactive material packages, by rail or highway, must apply a "radioactive" placard (Figure 7-8) to the transport vehicle if any package on board bears a radioactive Yellow-III label. The placard must appear on four sides of the transport vehicle. Placarding is usually not of concern for

TABLE 7-9 Labeling, Transport Index, and Radiation Limits for Radioactive Packages

Label Required	Transport Index	At Package Surface
		Limits of Radiation Exposure from Package
White-I	Not applicable	≤0.5 mR/hr
Yellow-II	>0 and <1.0	>0.5 mR/hr and ≤50 mR/hr
Yellow-III	>1.0 and ≤10	>50 mR/hr and ≤200 mR/hr

TABLE 7-10 Separation Distances between People and Radioactive Material (RAM) Packages for Highway Vehicles, Based on Transport Index Values

Sum of Transport Indexes on All RAM Packages	Minimum Separation Distance in m (ft) between People and RAM Packages
0.1–1.0	0.3 (1)
1.1–5.0	0.6 (2)
5.1–10.0	0.9 (3)
10.1–20.0	1.2 (4)
20.1–30.0	1.5 (5)
30.1–40.0	1.8 (6)
40.1–50.0	2.1 (7)

Source: 49 CFR 177.842.

FIGURE 7-8 Radioactive placard for transport vehicles if any radioactive material package on board bears a radioactive Yellow-III label. The background color of the black trefoil in the upper half of this 12 × 12 inch placard is yellow.

nuclear pharmacies because they rarely offer Yellow-III packages for transport.

Other Package Requirements

If a package contains liquid radioactive material, it must be packed with enough absorbent material to absorb at least twice the volume of liquid. In addition, the outside of each package must contain a security seal that is not readily breakable and that will give evidence that the package has not been illicitly opened.

Shipping Papers

Shipping papers must be included with transported radioactive material as described in 49 CFR 172.101 and 172.200 through 172.204. Items included are

- Proper shipping name (49 CFR 172.101) (e.g., Radioactive Material, Type A Package nonspecial form nonfissile or fissile-excepted)
- Hazard class: For radiopharmaceuticals, Class 7
- ID number: UN 2915 (for greater than excepted quantities and less than A_2 quantities)
- Package type: Type A
- Name or abbreviation of each radionuclide in the shipment
- Physical and chemical form of the radioactive material
- Activity of each radionuclide in the shipment in SI units (e.g., Bq) followed by customary units (e.g., mCi) in parentheses
- Designation "RQ" if amount is equal to or greater than the reportable quantity for a hazardous material listed in 49 CFR 172.101, Table 2, Appendix A (e.g., this would apply for ^{125}I or ^{131}I whose RQ is 10 mCi; in this case the proper shipping name for ^{125}I or ^{131}I is "RQ Radioactive Material Type A package")
- Category of label applied to each package (e.g., White-I, Yellow-II)
- TI of each package
- 24 hour emergency response telephone number
- Shipper's certification ("This is to certify that the above-named materials are properly classified, described, packaged, marked, and labeled, and are in proper condition for transportation according to the applicable regulations of the Department of Transportation")
- Shipper's signature

Accident Reporting

According to 49 CFR 171.15 and 171.16, the carrier of radioactive material must ensure that DOT and the shipper are notified in the event of fire, breakage, spillage, or suspected radioactive contamination involving the shipment. Carriers must also ensure that vehicles, areas, or equipment in which radioactive material may have spilled are not placed in service again until they have been surveyed and decontaminated. An emergency response telephone number must be included in the shipping papers and the number must be monitored at all times the material is being transported by a person knowledgeable about the hazardous material being shipped. Emergency response information must be included with the shipping papers.

Personnel Training

Subpart H of DOT regulations, 49 CFR 172.700, 172.702, and 172.704, describes the purpose and scope, applicability, and responsibility for training of employees involved in the transport of radioactive material. The training must include

1. Familiarization with DOT requirements and ability to identify hazardous materials
2. Training for each specific function the employee performs
3. Safety training concerning emergency response information, including measures to protect the employee, and methods to avoid accidents, such as proper procedures for handling packages containing radioactive material

Quality Control Measures

The following are important measures the shipper must attend to before and during a shipment of radioactive material.

1. Ensure that the package is proper for the contents.
2. Determine that the package is in good physical condition.
3. Check that the external radiation and contamination levels are within allowable limits.
4. Conduct vehicle inspections for proper operation.
5. Review requirements pertaining to vehicle attendance and incident reporting.
6. Establish a procedure for loading and unloading radioactive material, including securing packages by blocking or bracing so that packages will not move during transport.
7. Maintain shipping papers within accessible reach while driving.

Summary of Radioactive Material Shipment

Packages that contain less than the limited quantities for excepted packages shown in Table 7-8 and that have less than 0.5 mR/hour at the surface with no significant external contamination may be shipped in an excepted package. The container must be capable of preventing leakage during normal shipment. No outer label is required, but the inner container must be labeled "Radioactive." The package must contain the marking "Radioactive material, excepted package—limited quantity of material UN 2910."

Packages that contain more than a limited quantity or no more than the A_2 quantity must be shipped in a Type A package. A surface wipe test must be performed and the package surveyed at the surface and at 1 m to determine the TI. The package should be labeled with two warning labels (White-I, Yellow-II, or Yellow-III), one each on opposite sides of the package. The labels must contain the name or symbol of the radionuclide, its activity, and the TI. The package must contain the marking "Radioactive Material, Type A package, UN 2915." It must have a security seal and be accompanied by shipping papers.

Empty, reusable packages such as those used to deliver radiopharmaceutical doses must be surveyed inside and out before they are returned to the shipper and should have any external "Radioactive" signs removed if no activity is present. The package must contain the marking "Radioactive material, excepted package—empty packaging."

DISPOSAL OF RADIOACTIVE WASTE

The following methods are acceptable for the disposal of radioactive waste according to 10 CFR 20.2001: (1) transfer to an authorized recipient, (2) decay in storage, (3) effluent release, (4) release to sanitary sewer, (5) incineration, and (6) burial in the soil. Of these, the first four methods are the most practical for nuclear medicine and nuclear pharmacy practice.

Transfer to an Authorized Recipient

Radioactive material may be transferred to an authorized commercial radiation waste dump, but this is very costly and is rarely, if ever, used for nuclear medicine waste because of the relatively short half-lives of the material. Transfer of waste applies mostly in situations in which a hospital participates in a return shipment program with a radiopharmaceutical manufacturer that reuses the lead shielding material. A typical example is the ^{99m}Tc generator return program. Another example is a nuclear medicine lab returning spent syringes to the nuclear pharmacy that supplied the patient doses. Of course, DOT regulations for shipment must be followed if radioactive material is returned.

Decay in Storage

Since most radionuclides used in nuclear medicine have short half-lives, a licensee may allow a radioactive material with a physical half-life of less than 120 days to decay in storage before disposal (10 CFR 35.92). If this option is chosen, the licensee must survey the vial to confirm background levels of radioactivity and obliterate all radiation labels before discarding into ordinary trash. This method is particularly suitable if sufficient storage space for waste is available.

Effluent Release

This method applies to the release to an unrestricted area of volatile radionuclide waste such as ^{131}I and ^{133}Xe. The requirements under 10 CFR 20.1301 must be followed. See the previous section on Effluent Monitoring.

Release to Sanitary Sewer

In accordance with 10 CFR 20.2003, the following conditions must be met in order to dispose of radioactive waste to a sanitary sewer: (1) The radioactive material must be readily soluble or dispersible in water, (2) the monthly release must not exceed the concentration listed in 10 CFR 20, Appendix B, Table 3, and (3) the total quantity released in a year cannot exceed 5 Ci of ^{3}H, 1 Ci of ^{14}C, and 1 Ci of all other radioactive material combined. All radioactive excreta are exempt from this regulation.

Example: How much ^{131}I can be disposed of in the sewer each month from a 200 bed hospital?

The monthly limit for ^{131}I in 10 CFR 20, Appendix B, Table 3 is 1×10^{-5} µCi/mL (see Table 7-5). Sewer release, based on 10^6 mL/bed/day,[16] is $(1 \times 10^6$ mL/bed/day) (30 days/month) (200 beds) = 6×10^9 mL/month. So the allowed release is $(1 \times 10^{-5}$ µCi/mL) $(6 \times 10^9$ mL/month) = 60,000 µCi (2220 MBq)/month.

Incineration and Burial in Soil

Each of these methods requires a special license and is not generally applicable to nuclear medicine or nuclear pharmacy.

RADIATION EMERGENCY PROCEDURES

The most common radiation emergency in nuclear medicine or nuclear pharmacy is accidental spill of radioactive material. Spills may be minor or major depending on the amount of activity released. The general guide to follow in the event of any spill is confinement and shielding of the spilled activity to protect personnel from exposure. NRC Regulatory Guide 10.8 provides a model spill procedure in Appendix J and lists items that can be used to assemble a spill kit.

TABLE 7-11 Radionuclide Activities above Which Major Spill Procedure Is Used

Radionuclide	Activity (mCi)	Radionuclide	Activity (mCi)
^{32}P	10	^{111}In	10
^{51}Cr	100	^{123}I	10
^{57}Co	100	^{125}I	1
^{67}Ga	100	^{131}I	1
^{99m}Tc	100	^{201}Tl	100

Source: NRC Regulatory Guide 10.8.

The decision to implement a major spill procedure instead of a minor spill procedure depends on several variables, including the number of individuals affected, the radiation energy and exposure rate, the likelihood of spread of contamination, and the radiotoxicity of the spilled material. For some short-lived radionuclides the best spill procedure is to simply restrict access to the area until complete decay occurs. Shielding spilled material that has a low gamma energy, such as ^{99m}Tc, is an effective way to limit exposure of workers. In any event, the RSO should be consulted. Regulatory Guide 10.8 lists several radionuclides and the amounts above which a major spill procedure should be instituted; these are reproduced in Table 7-11. These amounts can be used as a guide for developing any laboratory spill procedure guidelines. The model procedures for minor and major spills recommended by Regulatory Guide 10.8 are as follows.

Minor Spill Procedure (Liquids and Solids)

1. Notify persons in the area that a spill has occurred.
2. Prevent spread by covering with absorbent material.
3. Clean spill using disposable gloves and absorbent paper, placing material into a plastic bag.
4. Survey area with a survey meter, also checking hands, clothing, and shoes for contamination. Shield non-removable contamination.
5. Report the incident to the RSO.
6. The RSO will follow up and complete the necessary reports.

Major Spill Procedure (Liquids and Solids)

1. Clear the area. Notify all persons not involved in the spill to vacate the room.
2. Prevent spread by covering with absorbent material, but do not attempt to clean up the spilled material. Limit the movement of any personnel who may be contaminated.
3. Shield the source if possible.
4. Close the room and lock or secure the area to prevent entry.
5. Notify the RSO immediately.
6. Decontaminate personnel by removing contaminated clothing, flushing contaminated skin with water, and washing with mild soap.
7. The RSO will supervise cleanup of the spill and completion of necessary reports.

REFERENCES

1. United Nations Scientific Committee on the Effects of Atomic Radiation. *Sources and Effects of Ionizing Radiation.* Vol I: Sources and Vol II: Effects. UNSCEAR 2000 Report to the General Assembly. New York: United Nations; 2000.
2. National Academy of Sciences. *Health Risks from Exposure to Low Levels of Ionizing Radiation.* BEIR VII Phase 2 Report. Washington, DC; National Academies Press; 2006
3. International Commission on Radiological Protection. *1990 Recommendations of the International Commission on Radiological Protection.* ICRP Publication 60. New York: Pergamon Press; 1991.
4. National Council on Radiation Protection and Measurements. *Limitation of Exposure to Ionizing Radiation.* NCRP Report No. 116. Bethesda, MD: National Council on Radiation Protection and Measurements; 1993.
5. US Nuclear Regulatory Commission. *Consolidated Guidance about Materials Licenses: Program-Specific Guidance about Commercial Radiopharmacy Licenses.* NUREG-1556, vol 13. Washington, DC: US Nuclear Regulatory Commission; 1999.
6. National Council on Radiation Protection and Measurements. *Ionizing Radiation Exposure of the Population of the United States.* NCRP Report No. 93. Bethesda, MD: National Council on Radiation Protection and Measurements; 1987.
7. National Council on Radiation Protection and Measurements. *Sources and Magnitude of Occupational and Public Exposures from Nuclear Medicine Procedures.* NCRP Report No. 124. Bethesda, MD: National Council on Radiation Protection and Measurements; 1996.
8. Anderson DW, Richter CW, Ficken VJ, et al. Use of TLD for measurement of dose to the hands of nuclear medicine technicians. *J Nucl Med.* 1972;13:627–9.
9. Henson PW. Radiation dose to the skin in contact with unshielded syringes containing radioactive substances. *Br J Radiol.* 1973;46:972–6.
10. Kereiakes JG. *Biophysical Aspects: Medical Uses of Technetium-99m.* American Association of Physicists in Medicine Topical Review Series. New York: American Institute of Physics Inc; 1992.
11. Benedetto AR, Dziuk TW, Nusynowitz ML. Population exposure from nuclear medicine procedures: measurement data. *Health Phys.* 1989;57:725–31.
12. Jursinic PA. Characterization of optically stimulated luminescent dosimeters, OSLDs, for clinical dosimetric measurements. *Med Phys.* 2007;34:4594–604.
13. Schembri V, Heijmen BJM. Optically stimulated luminescence (OSL) of carbon-doped aluminum oxide (Al_2O_3:C) for film dosimetry in radiotherapy. *Med Phys.* 2007;34:2113–8.
14. US Department of Transportation Pipeline and Hazardous Materials Safety Administration. *Radioactive Materials Regulations Review (Preliminary Draft).* Washington, DC: US Department of Transportation; February 2008. http://hazmat.dot.gov/HMpubsreview/docs/RAMreg.pdf. Accessed August 30, 2010.
15. Vanderslice SD. *Update on Department of Transportation Regulations.* Continuing Education for Nuclear Pharmacists and Nuclear Medicine Professionals. Albuquerque: University of New Mexico Health Sciences Center; 2006.
16. *A Manual of Radioactivity Procedures.* Handbook 80. Washington, DC: US Department of Commerce, National Bureau of Standards; 1961:133.

CHAPTER 8
Radiation Biology

Scientists learned in their earliest work with ionizing radiation that it had the capacity to produce biologic effects. A radium source that Henri Becquerel carried in his vest pocket produced skin erythema. Pierre Curie's exposure of his hand to radium caused an ulcer that was slow to heal. We now know from radiation biology research that ionizing radiation releases energy in the body that can potentially cause biologic damage. The severity of damage depends on the type and amount of radiation exposure to the body and the length and extent of exposure. When ionizing radiation interacts in the body it releases energy. The average amount of energy released per ionization is about 34 eV, more than enough to break a covalent bond between the elements that compose body tissues. The body can repair broken bonds, but if misrepair occurs deleterious effects may result. Scientists have discovered that a variety of effects can occur in cells and tissues, depending on the amount of radiation exposure. Survival studies of mammalian cells irradiated at low doses have demonstrated that cells have the ability to repair damage from radiation but that cells can also be killed by radiation if the dose is large enough. The effects of large-dose radiation exposure may be manifested soon after exposure, whereas the effects from smaller doses of radiation, if not repaired, may show up at a later time. Studies have also shown that different types of cells in the body have different sensitivities to radiation; therefore, exposure to different parts of the body poses different degrees of risk. Fortunately, we also know that ionizing radiation can be beneficial. For example, cancer can be cured by the application of ionizing radiation under controlled conditions in radiation therapy, because rapidly dividing cancer cells are sensitive to radiation.

This chapter covers the basic principles of radiation biology, including the effects of ionizing radiation on normal cells and on cancer.

BIOLOGIC EFFECTS OF IONIZING RADIATION

Figure 8-1 is a flow diagram of the significant elements to be considered in a discussion of the biologic effects of ionizing radiation (i.e., alpha, beta, gamma, and neutron radiation). Although ultraviolet radiation has potential for biologic damage, particularly to the skin and eyes, it is not an ionizing radiation and will not be included in this discussion.

The primary ionizing event in the body releases a high-speed electron that can cause further cellular interactions. During these subsequent interactions, energy is transferred by direct or indirect action to critical biologic molecules within cells. The energy deposited may break the bonds of molecules that are necessary for cellular function or reproduction. The most critical target is DNA. If the radiation dose is small and the damage slight, the cell may be able to repair the damage and return to health. Higher doses and dose rates may cause harm that exceeds the cell's ability to repair, and permanent damage will become evident within days or weeks of exposure. These are known as deterministic effects. If the dose is high enough, deterministic effects can be life threatening, as in the deaths caused by the

Ionizing Radiation
(excitation & ionization)
⇩
Cellular Interactions
(direct & indirect)
⇩
Chromosome Breaks & DNA Modification
⇩
Repair or Permanent Damage
⇩
Deterministic Effects & Stochastic Effects
- Effect increases   - Probability of effect
with dose   increases with dose

FIGURE 8-1 Overview of biological effects of ionizing radiation.

Chernobyl nuclear reactor accident in 1986, where large doses of whole-body exposure occurred. Some examples of non–life-threatening deterministic effects are skin erythema caused by x-ray exposure, loss of hair (epilation) in a patient being treated with radiation therapy for a brain tumor, and the development of cataracts several years after high-dose radiation exposure of the eye. Deterministic effects have a threshold dose below which no effect is observed but above which the severity of effect increases in proportion to the dose. Radiation protection practices to keep exposure as low as reasonably achievable are designed to prevent the occurrence of deterministic effects.

If any repair process from radiation-induced damage is incomplete or defective, the cell may be altered in such a way that it will become cancerous or develop a mutation that is heritable. These are stochastic effects, which typically appear long after the radiation exposure. Stochastic effects do not have a threshold dose, that is, they either occur or do not occur. Thus, a particular cancer may occur after a low dose of radiation exposure or a high dose, but the likelihood of cancer developing is greater after a high dose. Stochastic effects can be somatic or genetic. The development of cancer is a somatic stochastic effect. Genetic stochastic effects occur when germinal cells (oocytes or spermatogonia) sustain radiation-induced damage resulting in a mutation that is passed on to future generations.

The International Commission on Radiological Protection (ICRP) has estimated that the excess risk of fatal cancer to the general population, above the natural spontaneous rate of cancer (20%) from uniform whole-body exposure, is 5% per sievert or one extra cancer fatality out of 2000 persons exposed to 1 rem.[1] Without radiation exposure 400 people out of 2000 will die of cancer from natural causes. Their exposure to 1 rem will raise the death toll to 401. The chance of anyone receiving this magnitude of radiation exposure outside of medical treatment is low, so the risk to the general population and radiation workers from typical radiation exposure in the environment or on the job is small (see Chapter 7).

DIRECT AND INDIRECT INTERACTIONS OF RADIATION

Experiments have shown that the critical radiosensitive target within cells is DNA and that interactions within DNA or any other tissue occur by both direct and indirect effects. Direct effects occur when a high-speed electron is set in motion following a primary ionization event caused by the photoelectric effect or Compton scatter. If this occurs within the vicinity of a chromosome, the electron can then interact directly with a DNA molecule, causing a bond to break. The critical points are the bonds between base, sugar, and phosphate molecules in the DNA chain. Direct effects are more probable with high linear energy transfer (LET) radiation, such as alpha particles, protons, and neutrons, because the high ionization density of their tracks releases many electrons as the radiation passes through the DNA molecule.

Indirect effects of radiation are mediated by free radicals produced secondarily in the interaction of radiation with water molecules. The free radical, which carries an unpaired electron, is highly reactive and may interact with DNA, causing bond breakage. Indirect effects are more probable with low-LET radiation, such as beta particles, gamma rays, and x-rays. Since water constitutes such a high percentage of biologic matter, about two-thirds of x-ray damage in mammalian cells is due to the hydroxyl radical (OH·).[2]

The direct and indirect processes of radiation interaction are illustrated in Figure 8-2. In each situation the primary event is the same: release of a high-speed electron from an ionizing event. With the direct effect, this primary electron deposits its energy directly within the DNA molecule, for example by removal of an electron from a covalent bond by an ionization interaction. This can lead to cellular damage. With the indirect effect, the primary electron interacts first

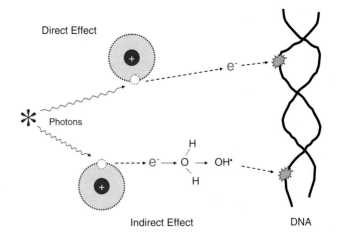

FIGURE 8-2 Mechanisms of direct and indirect effects of ionizing radiation on biologic systems. Gamma ray photons from an ionizing radiation source release free electrons by photoelectric or Compton interactions. The electrons then interact directly with DNA or indirectly via free radicals.

with water, producing a free radical that then goes on to deposit its energy within DNA.

The reactions between radiation and water molecules to produce free radicals are

$$H_2O \leftrightarrow H^+ + OH^- \quad (8\text{-}1a)$$

$$H_2O \xrightarrow{h\nu} H^\cdot + OH^\cdot \quad (8\text{-}1b)$$

$$RH \xrightarrow{h\nu} R^\cdot + H^\cdot \quad (8\text{-}1c)$$

Equation 8-1a describes the normal dissociation of water into hydrogen ions and hydroxyl ions, for comparison with free radical formation. This process does not involve any interaction with radiation. When the water molecule dissociates, all the electrons involved in the covalent bond between oxygen and hydrogen are possessed by the hydroxyl group. This leaves it with one more electron than its number of protons and, therefore, with a single negative charge. The hydrogen atom, having no electrons, remains as a positively charged proton.

Equation 8-1b describes the production of hydrogen and hydroxyl free radicals. The radiation interaction with a water molecule separates the atoms so that hydrogen retains its lone unshared electron, leaving the hydroxyl group with an unshared electron. Both species are now free radicals and have neutral ionic charges. Equation 8-1c shows a similar reaction with a biologic molecule (represented by RH).

The stepwise mechanism for the radiolysis of water into hydrogen and hydroxyl free radicals is shown in Equations 8-2a, b, and c.

$$H_2O \xrightarrow{h\nu} H_2O^+ + e^- \quad (8\text{-}2a)$$

$$H_2O^+ + H_2O \rightarrow H_3O^+ + OH^\cdot \quad (8\text{-}2b)$$

$$e^- + H_2O \rightarrow OH^- + H^\cdot \quad (8\text{-}2c)$$

In the first step, a photon ($h\nu$) ionizes a water molecule, producing a fast electron and an ion radical. In the second step, the ion radical combines with another water molecule to produce the hydronium ion and a hydroxyl radical. Finally, the fast electron combines with a water molecule to produce a hydroxide ion and the hydrogen radical.

FREE RADICAL REACTIONS

The radiolysis of water is a significant event in radiobiology because of free-radical formation. Electrons not only orbit around the nucleus of an atom but also spin on their own axes. The most stable atoms have electrons paired and spinning in opposite directions. If one electron is removed from the pair, for example by a radiation interaction, an unpaired electron remains, creating a free radical. The free radical is much less stable, making it very reactive chemically as it seeks to pair with another electron of opposite spin. If the free radical is produced in the vicinity of a cell's DNA, it may interact with the DNA helix. The most effective free radical is the oxidizing agent $OH^\cdot$, which can abstract a hydrogen atom from DNA, yielding a highly reactive site in the form of a DNA radical. If the site is not repaired, the cell may not be able to replicate, may fail to produce a critical protein required for cell viability, or may produce a mutation.

A number of free-radical reactions are possible. Some of the more significant ones are as follows:

$$R^\cdot + H^\cdot \rightarrow RH \quad \text{(restored organic molecule)} \quad (8\text{-}3a)$$

$$R^\cdot + O_2 \rightarrow RO_2^\cdot \quad \text{(restoration blocked)} \quad (8\text{-}3b)$$

$$H^\cdot + O_2 \rightarrow HO_2^\cdot \quad \text{(hydroperoxy radical)} \quad (8\text{-}3c)$$

$$RH + HO_2^\cdot \rightarrow R^\cdot + H_2O_2 \quad \text{(molecule inactivated)} \quad (8\text{-}3d)$$

$$HO_2^\cdot + H^\cdot \rightarrow H_2O_2 \quad (8\text{-}3e)$$
$$\text{(hydrogen peroxide)}$$

$$OH^\cdot + OH^\cdot \rightarrow H_2O_2 \quad (8\text{-}3f)$$

Equation 8-3a demonstrates the possibility of restoration of a biologic molecule after it interacts to form free radicals, and Equation 8-3b shows how the presence of oxygen can combine with a free-radical molecule to block its restoration. Oxygen can also combine with a hydrogen radical to produce the reactive hydroperoxy radical shown in Equation 8-3c. The hydroperoxy radical can in turn inactivate a biologic molecule, as shown in Equation 8-3d. If the inactivated molecule plays a significant role in the cell's metabolism, the cell's viability may be threatened. The hydroperoxy radical can combine with a hydrogen radical (Equation 8-3e) or two hydroxyl radicals can combine (Equation 8-3f) to produce hydrogen peroxide, a known cell toxicant. Thus, multiple mechanisms are possible for causing biologic effects.

THE CELL CYCLE

Before we discuss the effects of radiation on biologic systems, a review of the chemical composition and structure of DNA and the functions occurring during the cell cycle is in order. DNA appears in the nucleus of the cell as a coiled double helix of the sugar molecule deoxyribose, a phosphate, and the bases adenine, thymine, guanine, and cytosine. If the DNA helix is uncoiled and flattened out, it resembles a ladder (Figure 8-3). The covalently bonded sugar–phosphate groups form the rails or backbone of the ladder, and the rungs are formed by hydrogen bonding of the base pairs attached to the sugar molecules in each rail. The bases bond in specific pairs: adenine with thymine and guanine with cytosine. The DNA molecule is replicated in the nucleus of every cell that undergoes mitosis. If it is damaged in any way, by either chemicals or radiation, it may not be able to replicate or may replicate in a way that creates a mutant gene. Extensive research has been

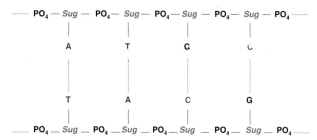

FIGURE 8-3 Ladder structure of a portion of the DNA molecule illustrating the two phosphate–sugar backbone rails covalently bonded to the bases: adenine, thymine, guanine, and cytosine. The two rails are connected by hydrogen bonds (dashed lines) between the base pairs.

done to study the effect of radiation on cell growth, DNA, and the cell cycle.

The cell cycle is divided into two main segments: interphase and mitosis, or M phase (Figure 8-4). These, in turn, are divided into additional phases. Interphase is the period from the end of cell division to the beginning of a new mitosis. During this time the cell appears to be quiescent, in part because the chromosomes in the nucleus are not readily visible because the DNA strands are uncoiled and filamentous. However, DNA synthesis occurs during the S phase of interphase. This is the event that allows each new daughter cell formed during mitosis to acquire exactly the same genetic makeup. During DNA synthesis the hydrogen bonds holding the base pairs together break, allowing the filaments to separate. Each filament serves as a template. Available "free" bases in the cell cytoplasm then match up with their complementary bases attached to the DNA filament (adenine with thymine and guanine with cytosine). New phosphate and deoxyribose molecules follow base pairing and the DNA strand is duplicated. The nucleus now contains two identical chromatids, one for each daughter cell to be formed during mitosis. The periods just before and immediately after the S phase, G1 and G2 respectively, are gaps during interphase when DNA is not replicating.

The cell cycle enters mitosis at prophase (Figure 8-5). During this phase the DNA filaments of each chromatid shorten and condense into coils, thickening the chromatin material and making it visible after staining procedures. The nuclear membrane disappears, and a centromere appears between the two chromatids. Spindle fibers begin to form at the opposite ends of the cell. The beginning of chromatid movement toward the center of the cell initiates metaphase. During this phase the centromeres divide, the chromatids align themselves at the center of the cell, and each centromere attaches itself to a spindle fiber. During anaphase the chromatids repel each other, with half the number migrating along the spindle toward one end of the cell and the other half migrating toward the opposite end of the cell. When they reach the opposite poles of the cell, telophase begins. During telophase the chromosomes uncoil and elongate into filamentous strands once again, and a nuclear membrane is formed around them. The center of the cell indents, separating the cytoplasm into two distinct cells after the formation of a new cell membrane. The two new daughter cells, each with a complete set of identical chromosomes, enter interphase and the cell cycle is complete.

The length of the cell cycle and the various phases for a particular type of cell can be determined by labeling techniques. One method uses a radioactive substrate (tritiated thymidine) that is taken up during DNA synthesis. Cells labeled with tritiated thymidine are allowed to grow for a period of time and then are fixed, stained, and subjected to autoradiography. If the cells are analyzed immediately after incorporation of the label, those cells that incorporate

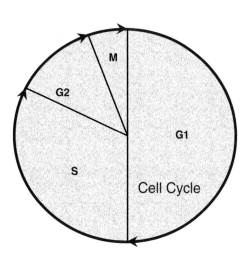

FIGURE 8-4 Phases of the cell cycle. See text for description.

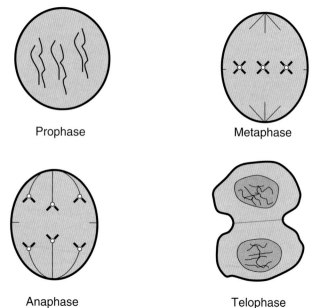

FIGURE 8-5 Four phases of mitosis in the cell cycle, illustrating movement of the chromosomes.

the radioactive thymidine are in the S phase of DNA synthesis. If staining and autoradiography is delayed for several hours after labeling, some cells will move on to mitosis, in which cells incorporate radioactivity at the various stages of cell division depending on how much time has passed between labeling and the analysis. By using this technique, the lengths of various phases of the cell cycle can be determined.

EFFECTS OF RADIATION ON GENETIC MATERIAL

Ample evidence from radiobiologic experiments demonstrates that DNA is the critical target for the biologic effects of radiation in cellular systems. Other molecules and organelles in the cell may also exhibit detrimental effects from radiation, such as structural changes in proteins, alterations in enzyme activity, and increased permeability of membranes, but the most critical target is DNA. When radiation interacts with DNA, breaks can occur in its backbone. If a break occurs in a single rail of the backbone, repair enzymes can repair it easily, using the opposite rail as a template. Such a break in DNA is called a single-strand break (Figure 8-6A). However, the effectiveness of enzymes in repairing a break may be thwarted if a free radical binds to the broken strand of DNA, blocking the enzyme's access to the break. If both rails of the DNA molecule break in such a way that base pairing holds them together, the strands may still be easily repaired. This is called a separated double-strand break (Figure 8-6B). If the rails break directly opposite each other, the DNA backbone may separate. This is called an opposite double-strand break (Figure 8-6C). This type of break is more difficult to repair and can lead to misrepair and chromosomal aberrations that may cause lethal effects in the cell.

Radiobiologic studies on irradiated cells have shown that various types of chromosome abnormalities can result from opposite double-strand breaks in DNA in which the chromosome breaks in two. These abnormalities typically result from breaks occurring in two separate chromosomes, followed by misrepair of the breaks. Examples of aberrations that cause cell death are shown in Figure 8-7. A *dicentric chromosome* is formed when a double-strand break occurs in two different chromosomes during the G1 phase. A misrepair can occur when the broken fragment from one chromosome is exchanged with the fragment from the other chromosome, resulting in one chromosome with two centromeres (a dicentric chromosome) and one with no centromere (an acentric chromosome). Subsequent duplication of these chromosomes in the S phase yields an aberrant dicentric chromosome and two acentric chromosomes (Figure 8-7A). The acentric chromosomes lack a centromere, cannot attach to a spindle fiber, and will not be transmitted to the daughter cell. A *ring chromosome* is formed when a double-strand break occurs in the same chromosome in the G1 phase (Figure 8-7B). Misrepair occurs when the two broken ends attach to each other, forming a ring chromosome with one centromere and an acentric chromosome. When these replicate during the S phase, a ring chromosome is transmitted to each daughter cell minus the pieces lost in the acentric chromosome. A *dicentric chromatid* forms when a cell is irradiated immediately after DNA synthesis during the G2 phase. Breaks occurring in each chromatid of the same chromosome are misrepaired by sister union of the two ends (Figure 8-7C). After centromere duplication, a dicentric chromatid attempts to migrate toward opposite poles at anaphase and the chromatid stretches across the cell, forming an anaphase bridge, preventing individual daughter cells from forming.

An important distinction should be made regarding chromosome breaks induced by radiation. During the S phase of the cell cycle DNA replicates, producing an exact duplicate of itself. These two identical sister chromosomes are called chromatids. If cells are irradiated at this point of the cell cycle, one of the chromatids may be damaged. If it is not repaired, the chromatid aberration produced will be passed on to the daughter cell receiving that chromatid. The other daughter cell will receive the normal chromatid. Chromatid aberrations, therefore, are produced in individual chromatids when irradiation occurs after DNA synthesis. Chromosome aberrations are produced when cells are irradiated prior to DNA synthesis in the G1-phase. If successful repair of the damage is completed before DNA synthesis occurs, each daughter cell receives normal chromosomes. If the repair is not completed or if there is a misrepair, it will be duplicated in the S phase and the damaged chromatids will be passed on to both daughter cells.

Chromosome breaks that lead to nonlethal aberrations are shown in Figure 8-8. A *translocation* aberration involves a double-strand break in two G1-phase chromosomes, with exchange of the broken ends between the two chromosomes.

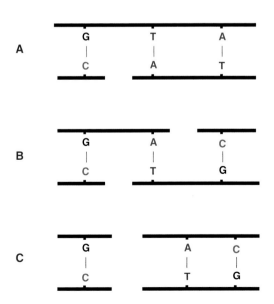

FIGURE 8-6 Radiation-induced breaks in DNA molecule. (A) single-strand break, (B) double-strand separated break, (C) double-strand opposite break.

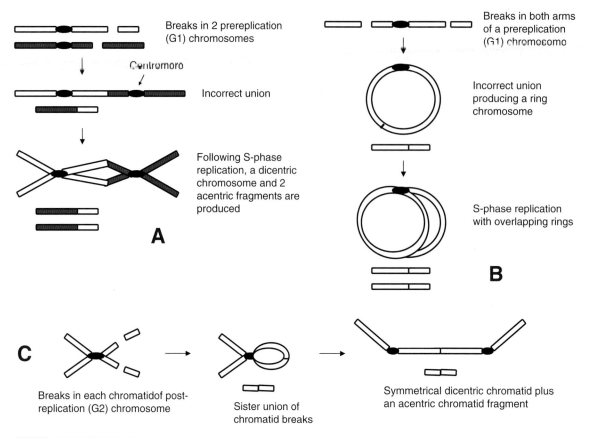

FIGURE 8-7 Chromosome aberrations induced by radiation. (A) Breaks in two irradiated G1-phase chromosomes before replication followed by illegitimate union of the broken fragments. After S-phase replication a dicentric chromosome and two acentric fragments are formed. (B) Breaks in both arms of the same irradiated G1-phase prereplication chromosome followed by incorrect union forming a chromosome ring. After replication, overlapping ring chromosomes form. (C) Breaks in each chromatid of a G2-phase irradiated postreplication chromosome followed by sister union of the broken ends. During anaphase a dicentric chromatid is formed and each centromere migrates to a pole, stretching the chromatid between the two poles to form an anaphase bridge.

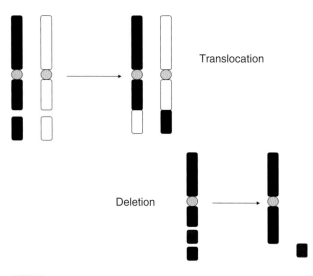

FIGURE 8-8 Double-strand breaks in adjacent chromosomes. Translocation misrepair results in exchange of broken pieces between chromosomes. Deletion misrepair causes loss of a piece of genetic material after a double-strand break in the same chromosome.

This type of break and misrepair has been shown to cause activation of an oncogene, which leads to malignancy. Burkitt's lymphoma is an example. A *deletion* aberration can result when two double-strand breaks occur in the same arm of a chromosome, causing a piece of genetic material to be removed. If the piece of genetic material that is lost is associated with the production of a suppressor gene in the cell, cancer can develop.

CELL SURVIVAL CURVES

Cell killing by radiation is defined as a loss of proliferative ability. It may result from any number of causes, such as point mutations, chromosome breakage, and rupture of vital cellular membranes. The magnitude of cell killing is measured by the fraction of cells that survive after radiation exposure, the endpoint being the ability of cells to form colonies in growth media. Therefore, survival curve analysis is limited to cells that undergo mitosis.

Much of the information known about the effects of ionizing radiation on biologic systems comes from studies in which viruses, bacteria, yeasts, and mammalian cells are irradiated and the fraction of cells that survive as a function of dose is measured. A survival curve is generated by irradiating cells at various doses, plating them on growth media, and counting the number of cell colonies that survive after a given time. Unirradiated cells are also plated as a baseline to determine the fraction of untreated cells that will grow. This is the plating efficiency (PE). The survival fraction of treated cells at each dose of radiation is determined by the following formula:

$$\text{Survival fraction} = \frac{\text{Number of colonies}}{\text{Number of cells plated} \times \text{PE}}$$

The first survival curve for mammalian cells was from measurements after x-ray irradiation of cells derived from human squamous cell carcinoma of the cervix[3] (HeLa cells, named after Henrietta Lacks, the woman from whom the cell line was originally obtained). The curve had a shoulder region at low doses up to 150 rad and became exponential at higher doses. Survival curves, in general, are of two types: exponential and sigmoid (Figure 8-9). An exponential curve indicates that there is a single target in the cell that must be hit or inactivated to kill the cell. With this type of survival, the fraction of cells killed is constant as the dose increases, but the number of cells killed per unit of dose diminishes because the starting number of live cells at higher doses is smaller, similar to exponential decay of a radionuclide. This type of curve has been observed after irradiation of viruses and certain bacteria and yeasts. A sigmoid curve indicates that multiple targets must each be hit (inactivated) to kill the cell. This is the multitarget, single-hit theory of cell survival kinetics. In the sigmoid curve a shoulder region is evident at lower doses (cell killing is not proportional to dose), signifying that damage must be accumulated before a cell is killed (i.e., sublethal damage occurs in this region). However, as the radiation dose increases, a point is reached at which the amount of sublethal damage is eventually maximized. At this point, cell killing becomes proportional to dose, and the survival curve becomes exponential. A sigmoid curve is typically observed after irradiation of mammalian cells.

Many studies conducted over the years have yielded survival curves typified by those shown in Figure 8-10, which demonstrate the response of mammalian cells to high- and low-LET radiation at high dose rates (>10 rad [0.1 Gy]/minute).[1] Looking at curve A, it is evident that for all doses of densely ionizing radiation of high LET (alpha particles and neutrons) and for high doses of low-LET radiation (x-rays or gamma rays), the dose–response curve is exponential, that is, linear on a semilog plot. Survival of cells under these conditions is given by the equation

$$\frac{N}{N_0} = e^{-D/D_0} \qquad (8\text{-}4)$$

where N is the number of cells surviving, N_0 is the number of cells initially, N/N_0 is the surviving fraction, D is the applied dose of radiation, and D_0 is the mean lethal dose. D_0 is the dose considered to provide, on average, one inactivating event per cell. According to this equation, when the applied dose is equal to the mean lethal dose (i.e., $D = D_0$), then $N = 0.37 N_0$. Therefore, the D_0 dose is the dose at which 37% of cells survive (i.e., 63% of cells are killed).

The question then is, if the applied dose is sufficient to kill all cells, why do 37% survive? The answer is related to the random nature of radiation interaction with matter. When the D_0 dose is applied, some cells will sustain a lethal event more than once and some will escape being hit at all. Another way of saying this is that the random nature of radiation interaction dictates that

- Some cells will sustain hits in all targets (lethal damage).
- Some cells will sustain hits in a few targets (sublethal damage).
- Some cells will sustain no hits in a cell (cells not affected).

The survival curve in Figure 8-10A for low-LET radiation is characterized by the parameters n (the extrapolation number), D_0 (the D_{37} dose), and D_q (the quasi-threshold dose). The extrapolation number is obtained by extrapolating the

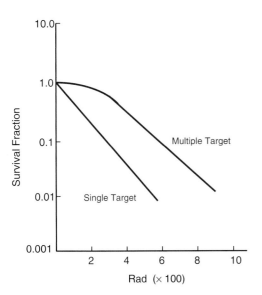

FIGURE 8-9 Survival curves resulting from radiation interaction in cells. The exponential curve shows that cell killing increases proportionally with dose, implying that a single target must be inactivated to kill the cell. The sigmoid curve implies that there are multiple targets that must be hit (inactivated) to kill the cell. A shoulder region is evident at lower doses (cell killing not proportional to dose), signifying that damage must be accumulated before a cell is killed or that repair processes are in effect. At higher doses cell killing is exponential, signifying that all targets in a cell that is hit are inactivated and repair is not possible. As dose increases, the probability of more cells being inactivated increases, and the survival fraction declines.

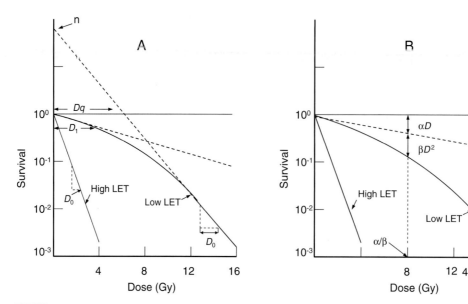

FIGURE 8-10 Survival curves for mammalian cells exposed to high-LET and low-LET ionizing radiation at high dose rates (>0.1 Gy min^{-1}). D is the applied dose of radiation, D_0 is the mean lethal dose, and D_q is the quasi-threshold dose. (Adapted from reference 1, p. 100.)

linear portion of the high-dose low-LET curve to the ordinate. This was originally regarded as the number of targets required to be hit to inactivate a cell, but it is now called simply the extrapolation number. For mammalian cells, n ranges from 2 to 10. The D_0 dose was previously defined. The quasi-threshold dose is the dose at which the straight portion of the low-LET survival curve, extrapolated backward, intersects the dose axis drawn through a survival fraction of unity. It can be viewed as the dose during which most of the sublethal damage occurs after a large dose of radiation. The D_q and n terms are measures of the size of the shoulder region typically seen in the early part of the curve and are related by the following equation:[2]

$$\frac{D_q}{D_0} = \log_e n \qquad (8\text{-}5)$$

The survival curve in Figure 8-10A for low-LET radiation shows a shallower initial slope (shoulder) that increases with dose. This response is explained by the multitarget theory, which states that only sublethal damage occurs in the cell, allowing the cell to repair the damage. The repair of cellular damage has been demonstrated in mammalian-cell experiments involving two doses of radiation separated by intervals of time.[4] It was shown that when the same total dose of radiation is administered in fractions, separated by a period of time, the number of cells surviving increases with the time between fractions and the survival curve after the second fraction exhibits the same D_0, n, and D_q as the previous survival curve (Figure 8-11). Cells surviving the first dose fraction respond as unirradiated cells to the second fraction. This was interpreted to mean that sublethal radiation damage had been repaired during the time between the first and second dose.

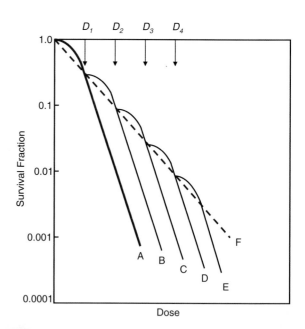

FIGURE 8-11 Effect of dose fractionation on the survival of cells irradiated with successive fractions of ionizing radiation separated by a set time interval. Each curve exhibits the same shoulder, slope, and extrapolation number, indicating that repair of damage has occurred between doses. Curve A represents survival from a single large dose. Curves B–E represent survival after equal fractions of the large dose. The dotted line F connecting single points on curves B–E has a shallower slope than line A and therefore a larger D_0 compared with curve A, demonstrating repair. (Adapted from reference 4, p. 287.)

Another explanation for what is happening in cell survival curves is shown in Figure 8-10B. The low-LET curve in B is similar to the one in curve A, but it is continuously bending, with no final straight portion. This curve is fitted to a linear–quadratic function that assumes there are two components to cell killing, where the frequency of lethal events is given by $F(D) = \alpha D + \beta D^2$, indicating that cellular effects are proportional to dose and dose squared, depending on the magnitude of the dose. This concept is similar to the induction of cancer and hereditary effects caused by low-dose and high-dose radiation, discussed in Chapter 6. Survival of cells is given by the equation

$$S = e^{-(\alpha D + \beta D^2)}$$

where S is the surviving fraction (N/N_0) from applied dose (D), and α and β are constants of proportionality. The dose at which the linear and quadratic components are equal is the ratio of α/β. This ratio has significance in radiotherapy.

According to the linear–quadratic model, radiation response in cells is related, in a simplified way, to radiation-induced chromosome aberrations.[2] Recall that lethal aberrations are likely the result of misrepair of double-strand breaks occurring in two separate chromosomes. A double-strand break can be caused by a single track of radiation passing through DNA of each chromosome, such as an electron released from a single ionizing event. A double-strand break in each chromosome can also be caused by two electrons produced from two separate ionizing events, one electron causing a double-strand break in one chromosome and the other electron causing a double-strand break in the second chromosome (Figure 8-12).

At low doses, the two breaks may result from the passage of a single electron set in motion by an x-ray, and the probability of an interaction between the two breaks is proportional to dose. This makes the dose–response curve for chromosomal aberrations linear at low doses. At higher doses, the two chromosome breaks may result from two separate electrons, and the probability is then proportional to dose squared. The quadratic effect causes the dose–response line to become curved. It is therefore reasonable to link the linear–quadratic relationship characteristic of the induction of chromosome aberrations to the cell survival curve.[2]

RADIOSENSITIVITY

Not too many years after the discovery of x-rays and radioactivity, physicians observed that rapidly growing tissue such as tumors appears to be more readily affected by radiation than nearby normal tissue. To gain more objective evidence of cellular sensitivity, Bergonie and Tribondeau[5] exposed rodent testicles to radiation, reporting their results in 1906. Testicles contain cells at various stages of maturity, and Bergonie and Tribondeau's experiments demonstrated that ionizing radiation is more effective against immature spermatogonia undergoing mitosis than against mature differentiated sperm cells. Their conclusions were that dividing cells are more sensitive than cells that do not divide and that a cell's sensitivity is determined by the cell's characteristics rather than the radiation.

In 1925 Ancel and Vitemberger[6] modified the law of Bergonie and Tribondeau; they proposed that the damage to any cell from radiation is the same but the appearance of damage is influenced by two factors. They suggested that mitosis is an important factor because radiation-induced damage is expressed only when the cell attempts to divide, and that damage is more apparent when various conditions are present during and after irradiation, changing the cell's sensitivity.

In 1968 Casarett[7] classified cells into five categories of radiosensitivity, ranging from vegetative, undifferentiated stem cells, which are the most radiosensitive, to differentiated mature cells, which are the most radioresistant. These categories are summarized in Table 8-1. It is clear from this classification that cells actively undergoing division are the most radiosensitive and are at greatest risk from excessive radiation exposure.

One of the first clinical symptoms to arise from high-dose whole-body exposure to ionizing radiation, such as occurred in the Chernobyl nuclear reactor accident, is nausea and vomiting caused by the effects of radiation on the gastrointestinal cells. In calculating effective dose from radiopharmaceuticals, the cells of the body given the greatest risk weighting factor are the gonadal cells because of their potential for passing on radiation-induced genetic defects when they divide.

FACTORS AFFECTING RADIOSENSITIVITY

As mentioned previously, several factors have a role in determining the sensitivity of cells to radiation. The significant factors are LET, dose rate, the presence of oxygen, and the phase of the cell cycle during irradiation.

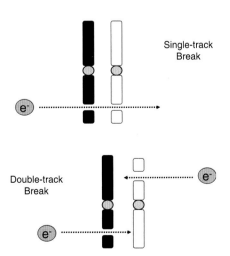

FIGURE 8-12 Double-strand breaks in two adjacent chromosomes by one electron from a single ionizing event (single-track break) and by two electrons from separate ionizing events (double-track break).

TABLE 8-1 Relative Radiosensitivity of Cell Populations

Radiosensitivity	Cell Type
High sensitivity	Lymphocytes Erythroblasts Spermatogonia
Relative sensitivity	Myelocytes Intestinal crypt cells Epidermal basal cells
Intermediate	Endothelial cells Osteoblasts
Relatively resistant	Granulocytes Osteocytes Erythrocytes Spermatozoa
Highly resistant	Fibroblasts Muscle cells Nerve cells

Source: Reference 7.

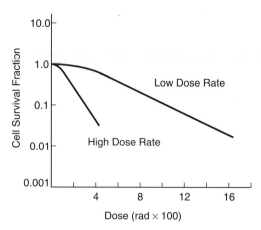

FIGURE 8-13 Effect of dose rate on cell survival from low LET radiation. Low dose rate (10 rad/min) produces a broad shoulder and shallow slope compared with high dose rate (100 rad/min), which produces a smaller shoulder and steeper slope.

Linear Energy Transfer

High-LET radiation, such as neutrons and alpha particles, has a high specific ionization and therefore produces dense ionization along its tracks, in contrast to x-rays or gamma rays. The high density of ion pairs increases the probability for interaction within the critical target centers of a cell. Rad for rad, biologic damage is greater than with sparsely ionizing radiation. The shoulder region seen on survival curves for low-LET radiation, which is associated with cellular repair of sublethal damage, is reduced or absent with high-LET radiation because of the increased efficiency for inflicting lethal damage (more double-strand breaks) (see Figure 8-10). The quasi-threshold dose, D_q, is absent on the survival curve, resulting in an extrapolation number of 1.

Dose Rate

For low-LET x-rays or gamma radiation, a high dose rate is more damaging to cells than a low dose rate because low dose rates produce more single-strand breaks, which allow for repair of sublethal damage. Therefore, the shoulder of the survival curve is broader with radiation at a low dose rate (Figure 8-13). High dose rates deliver more ionizing tracks per unit of time, thus producing more double-strand breaks, and also allow less time for repair of sublethal damage. There is no dose-rate effect with high-LET radiation because of its dense ionization.

Oxygen Effect

Oxygen has been shown to be an effective radiosensitizer; its presence during irradiation of cells and tissue can enhance the killing effect of radiation. The magnitude of oxygen's effect can be measured and is known as the *oxygen enhancement ratio* (OER). The OER is the ratio of the radiation dose required to produce a given effect without oxygen to that required to produce the same effect with oxygen. Experiments have shown that administration of oxygen during irradiation enhances the effect more than administration before or after irradiation. Oxygen is most effective when it is present over a range of 0 to 20 mm Hg tension, but effectiveness falls off at oxygen tensions in the 20 to 40 mm Hg range. Because the effect is most pronounced during irradiation, it is postulated that the mechanism of action is related to free radical formation of biologic molecules that combine with oxygen, preventing the molecules from restoring themselves (see Equation 8-3b).[8] This effectively increases cell damage and lethality at a lower dose than if oxygen were absent.

For mammalian cells the OER averages approximately 2.5. This means that a 100 rad (1 Gy) dose in the presence of oxygen will produce the same killing effect as 250 rad (2.5 Gy) in the absence of oxygen. The oxygen effect is most pronounced with x-rays or gamma rays; it is absent or diminished with high-LET radiation (alpha and neutron) because the damage produced by densely ionizing radiation is not repairable.

Tumors have a distinct architectural pattern consisting of a central region of necrosis (dead cells) surrounded by a rim of viable well-oxygenated cells. In between are viable cells that have a relative deficiency in oxygen (hypoxic cells). Hypoxic viable cells pose the biggest problem in radiotherapy because they are relatively radioresistant compared with well-oxygenated cells. The presence of these cells will result in tumor regrowth and treatment failure. Exogenous methods of increasing the oxygen titer in hypoxic tumor cells to increase their radiosensitivity have not met with much success. There is evidence, however, that the proportion of hypoxic cells decreases after a dose of radiation because as oxygenated cells are killed their oxygen supply is made available to the hypoxic cells. Thus, these cells become more radiosensitive at subsequent doses of radiation (dose fractions), and this phenomenon is believed to play a role in the effectiveness of fractionated radiation treatment plans.

Cell Cycle Effects

Experiments have shown that the radiosensitivity of cells is different at different phases of the cell cycle. The most sensitive phases are the G2 phase immediately after DNA synthesis and the M phase. The G1 phase is less sensitive to radiation, and the S phase is the most resistant.

When mammalian cells grown in culture are irradiated at 37°C, more cells in the G2 and M phases are killed than cells in the S phase. The surviving cells tend to become partly synchronized in the cell cycle. In this situation, most surviving cells will be in the S phase of the cell cycle. If the length of the cell cycle for the particular cell type is known, radiobiologists can irradiate a group of synchronized cells when they reach another phase of the cycle. In this way, the relative radiosensitivities of the cell cycle can be determined.

The total cell cycle time for Chinese hamster cells in vivo is 11 hours. The times for the phases of the cycle are M phase, 1 hour; S phase, 6 hours; G2 phase, 3 hours; and G1 phase, 1 hour. Elkind et al.[9] studied the growth of hamster cells in vitro, which have a shorter cycle time (about 9 hours). These cells were irradiated at various times during the cell cycle with a total dose of 1551 rad of x-rays fractionated into two doses (747 rad and 804 rad) (Figure 8-14). After the first dose, cells in the more sensitive phases have been killed but cells in the S phase, being more resistant, are spared. The remaining cell population, then, is synchronized in the S phase of the cycle and can undergo repair before the second dose is given. If 6 hours elapses before the second dose, this cohort of cells cycle around (reassortment) to the G2 and M phases, where they become more radiosensitive. If the second dose is given at this time in the cycle, the increased radiosensitivity exceeds the effect of repair, and the surviving fraction will decrease. If the second dose is not given until 10 to 12 hours after the first dose (i.e., more than the length of the cell cycle), the cells will have cycled through mitosis and increased in number (repopulation). This triad of repair, reassortment, and repopulation, coupled with reoxygenation, is known as the "4 Rs" of radiobiology and plays a significant role in the treatment of tumors with radiation.

DOSE FRACTIONATION IN TUMOR TREATMENT

Soon after its discovery, radiation was applied to treat cancer. The treatment schedules that were developed fall into two main groups. Single-dose therapy involves large radiation doses at one time, and multiple-dose therapy involves smaller (fractionated) doses over a longer time.[10] The information gained from clinical experience and radiobiology research has shown that dose fractionation has several inherent advantages that enhance the treatment of cancer with radiation.

A key factor in treating a cancerous tumor with radiation is the effect of the radiation on surrounding normal tissue. The amount of radiation used to treat a tumor is limited by the tolerance of normal tissue. The dose of radiation that will cause total destruction of tumor cells is called the tumor lethal dose (TLD), while the dose that begins to cause normal tissue necrosis is called the normal tolerance dose (NTD). The ratio of NTD to TLD is called the therapeutic ratio. It is desirable that this ratio be large in order to spare normal tissue and reduce morbidity. Therapeutic ratios vary by tumor type because of the differences in tumor and normal cell radiosensitivities. The total tumor dose and fractionation schedule are selected to maximize tumor killing and minimize normal tissue damage.

Dose fractionation is an important factor in planning radiation treatment. Ultimately, the benefit of radiotherapy depends on a therapeutic gain between the responses of the tumor and that of the normal tissue. A gain from fractionation is realized because a number of factors are operational during therapy:

- Repair of sublethal and potentially lethal damage; normal cells often have a greater capacity to repair intracellular effects than tumor cells do;
- Repopulation of cells between fractions, allowing regrowth of normal cells;
- Redistribution of cells throughout the cell cycle, which tends to sensitize the more rapidly dividing cells in tumors; and

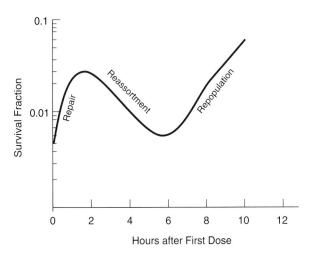

FIGURE 8-14 Survival of hamster cells exposed to two fractions of x-radiation separated by various time intervals. Repair of cells is evident in the first phase of the curve between 0 and 2 hours. Cells synchronized in the S phase from the first dose cycle to more sensitive G2–M phases by 6 hours, causing a fall in survival when the second dose is given here. If the second dose is given at 10–12 hours, cells show an increase in survival, having exceeded their cycle time and repopulated by mitosis. (Adapted from reference 9.)

- Reoxygenation effects; repeated fractions permit reduction of the number of hypoxic cells in tumors while little change in oxygenation occurs in normal tissues, permitting a gradual increase in the tumor's sensitivity to radiation.[11]

Survival curves of normal and tumor cells have provided insight into the effects of different fractionation schemes on tumor destruction. In general, more rapidly dividing tissues (skin, mucosal cells, rapidly dividing tumors) have relatively greater α/β ratios, about 1000 rad (10 Gy), while more slowly dividing tissues (kidney, brain) have smaller α/β ratios, about 200 rad (2 Gy), representing their greater capacity to repair at lower doses.[11] The greater the α/β value in tumor cells and the lower the value in adjacent normal tissue cells, the more fractionation will help the therapeutic ratio. If the radiation dose is divided into several fractions, each with a smaller dose, more slowly dividing tissues like normal organ tissues can be preferentially spared from the effects of radiation. More rapidly dividing tissues like tumors show a relatively greater radiation effect during fractionation because they have a greater chance of being irradiated during sensitive phases of the cell cycle. More rapidly dividing normal cells like skin and mucosa show a greater effect as well, but these tissues have an excellent capacity to replenish themselves. After each treatment both normal cells and tumor cells recover, but if normal cells recover more, a differential is achieved and eventually a dose can be reached at which normal cells can recover but tumor cells cannot. Thus, sometimes a greater tumor cell kill with relatively fewer late normal tissue effects can be achieved by reducing the fraction size and increasing the total number of fractions to create a greater therapeutic gain.[11]

ACUTE EFFECTS OF WHOLE-BODY IRRADIATION

The acute effects that occur after whole-body irradiation are well described.[2,12] These effects are deterministic in that they result in a measurable loss of cell and tissue function as a direct consequence of radiation's inactivation of vital elements within cells. Information on these effects is derived mostly from Japanese atomic bomb survivors, patients who undergo radiotherapy, and victims of nuclear reactor accidents. The effects that occur after acute whole-body radiation exposure in excess of 100 to 200 rad (1–2 Gy) are known as the acute radiation syndrome.

The three principal syndromes that may occur, based on the organ system principally involved in causing death, are the hematopoietic, gastrointestinal, and cerebrovascular or acute incapacitation syndromes. Each of these is preceded by a prodromal phase and a latency phase. The prodromal phase lasts about 24 hours, with peak symptoms occurring 6 to 8 hours after exposure. The most frequent prodromal symptoms are anorexia, nausea, and vomiting. The time of onset and the severity of prodromal symptoms are dose dependent. The frequency of prodromal symptoms is about 25% in persons exposed to 150 rad (1.5 Gy) and about 95% in those exposed to 300 rad (3 Gy).[12] The latency phase is an asymptomatic period following the prodromal phase. Its length is inversely proportional to the magnitude of exposure. This phase may not be evident at doses in excess of 1000 rad (10 Gy) because of the severity of prodromal symptoms. At lower doses the latency period is the result of the time required for the deleterious consequences of cell depletion in mitotically active tissues to become clinically evident. This is most important for cells in the gastrointestinal tract and bone marrow. Thus, after an exposure event, the stem cells of these systems may be killed but the mature cells that are present will continue to function for a time until they die. When this occurs, clinical symptoms and signs of stem cell deficiency will become evident. The symptoms that appear in the latent phase include chills and fever due to neutropenia and infection and the development of petechiae and hemorrhage, which are consequential signs of thrombocytopenia.

Hematopoietic Syndrome

This syndrome results from an exposure in the range of 200 to 800 rad (2–8 Gy). At higher doses in this range, stem cells in the bone marrow that produce the blood cellular elements are killed and are no longer available to replenish the circulating cells as they are lost to senescence. As a result, the victim experiences lymphopenia, neutropenia, thrombocytopenia, and anemia. The characteristic signs are petechiae, bleeding from mucous membranes, and infection. If death occurs, it will be within about 2 to 4 weeks after exposure. The cause of death is bone marrow suppression and hematopoietic failure.

The likelihood of death during the hematopoietic syndrome depends on the dose received. Experiments in monkeys exposed to a single dose of x-rays showed that no animals died with doses up to 200 rad (2 Gy), whereas a dose of about 800 rad (8 Gy) killed all the exposed animals.[2] The $LD_{50/30}$, which is the dose that kills 50% in 30 days, is about 530 rad (5.3 Gy) for monkeys. For humans the peak time of death from hematopoietic failure is also 30 days, but it can be as long as 60 days, and the LD_{50} is expressed as $LD_{50/60}$. The $LD_{50/60}$ of Japanese atomic bomb survivors averages 310 rad (3.1 Gy), ranging from 270 to 400 rad (2.7–4.0 Gy) depending on the circumstances of exposure.[12] The available data indicate that the $LD_{50/60}$ for healthy humans after an acute total-body exposure ranges from 250 to 350 rad (2.5–3.5 Gy) without medical intervention, increasing to 450 rad (4.5 Gy) with supportive medical care and 1000 to 1100 rad (10–11 Gy) with a successful bone marrow transplant.[12] After the nuclear reactor accident at Chernobyl in 1986, there were 32 immediate deaths due to thermal and radiation burns to plant workers attempting to put out the fire. A total of 143 workers and firefighters developed acute radiation syndrome, of whom 31 died within 3 months (28 died from bone marrow suppression or gastrointestinal damage).[13]

Gastrointestinal Syndrome

This syndrome results from an exposure of 1000 to 5000 rad (10–50 Gy). The most frequent symptoms of gastrointestinal damage are nausea, vomiting, and bloody diarrhea. Without specific therapy, the loss of body fluids and electrolytes results in dehydration, with death occurring 5 to 12 days after exposure. The cause of death is necrosis and mitotic arrest of mucosal stem cells. The cells most affected are the stem cells in the crypts of the small intestine. As the mature cells lining the intestine become senescent and slough, no cells are available to replace them. Thus, the intestinal wall becomes denuded and unprotected, permitting invasion of bacteria into the body. Hemorrhage and fluid loss ensue. Similar effects occur in the stomach, colon, and rectum, but at a slower rate because these cells turn over more slowly.[12]

Cerebrovascular Syndrome

This syndrome results from an exposure of more than 5000 rad (50 Gy). Soon after exposure the principal symptoms are apathy, lethargy, somnolence, tremors, convulsions, and coma. Death occurs within 10 to 36 hours after exposure.[12] This syndrome is so named because the most consistent lesions involve the arterioles and venules. Hemorrhage and leakage of fluid occur from the vascular space. The exact cause of death is unknown, but it is speculated to be cerebral edema from small vessel leakage, which is typically found at autopsy.[12] The high-dose exposure in this syndrome is also sufficient to cause severe damage to the hematopoietic and gastrointestinal systems, so that if the victim lives long enough death occurs because of failure of one or both of these systems.

Prognosis and Treatment

Recovery from acute radiation exposure depends on the dose received, the health of the victim, and the medical care provided. If the prodromal phase is short and the latency period long with milder symptoms, the prognosis for recovery is better than if the latency period is short or absent and symptoms are severe. Supportive medical treatment can improve the outcome. Typical measures include reverse isolation to protect against infection and aggressive replacement of fluids, electrolytes, and blood products. In some instances bone marrow transplantation may be indicated. In general, if the dose received is less than 400 to 500 rad (4–5 Gy), treatment of symptoms should be instituted. Antibiotics for infection and platelets for bleeding are recommended, but blood transfusions should be withheld because they suppress regeneration of new blood cells.[2] Animal experimentation has shown that the use of antibiotics to control infection raises the LD_{50} by a factor of 2. The patient should be isolated in a sterile environment. After higher doses of 800 to 1000 rad (8–10 Gy), bone marrow transplantation may be useful.[2]

The magnitude of radiation exposure may not be known in some cases. The blood levels of cellular elements can provide some measure of exposure, because lymphopenia, neutropenia, thrombocytopenia, and anemia occur in that order. The absolute lymphocyte count is especially useful in this context.

Despite the severity of effects from acute radiation exposure, heavily irradiated survivors of radiation accidents in the nuclear industry who have been followed for as long as 30 years have shown no remarkable difference from the normal aging population in terms of shortened life span, early malignancies, or rapidly growing cataracts.[2] Although this is encouraging information for victims who survive acute exposure, it cannot be considered definitive of the actual level of risk because of the small study population.

CARCINOGENIC EFFECTS OF RADIATION

The preceding section discussed the deterministic effects of acute large-dose radiation exposure in which radiation terminates the reproductive and metabolic functions of cells and causes the victim to die. The carcinogenic and hereditary effects of radiation exposure (discussed in the next section) are known as stochastic effects because they arise from random modifications (mutations) in somatic or germ cells. Because the cells are not killed, the mutation produced may eventually cause somatic cells to become cancerous or germ cells to produce genetic defects in offspring from the parent cell. As was discussed in Chapter 6, the probability of stochastic effects increases with radiation dose, and the chance for repair of radiation-induced alterations improves as dose and dose rate decrease.

Literature reports from the early years of radiation use document the increased incidence of cancer in persons exposed to radiation in the work environment and from medical procedures. Some of these reports are anecdotal and do not provide accurate measures of the radiation exposures; however, there is a reasonable correlation between radiation exposure and cancer development. For example, skin cancer was common among dentists who held x-ray film in their patients' mouths while exposures were being made. Lung cancer was a frequent occurrence among miners exposed to ore containing radium; the cancer was believed to be caused by intense alpha-particle irradiation of lung tissue from inhaled radon gas, the principal daughter product of radium decay. An excess of liver tumors was reported in patients who received an x-ray contrast material known as Thorotrast, which contained radioactive thorium, an alpha-particle emitter.

The risk factors for cancer induction and hereditary effects caused by radiation were discussed in Chapter 6. This section will briefly describe the radiation-induced cancers for which data are available and the mechanisms involved in radiation carcinogenesis.

Tumors Induced by Radiation

The development of cancer after radiation exposure occurs after a latent period, that is, from the time of exposure to the appearance of cancer. The latent period varies with the type of cancer. For leukemias the latent period is short; the average for all types is 2 years. In contrast, the minimum latency for solid tumors varies by tumor type, ranging from 5 years for carcinoma of the thyroid to 20 years for multiple myeloma.[12]

A variety of cancers have been identified whose incidence is increased by exposure to radiation.[2,14] The principal data sources are Japanese atomic bomb survivors and patients treated with radiation.

- *Leukemia.* There is an increased incidence of acute and chronic myeloid leukemia in adults, whereas acute lymphatic or stem cell leukemia is highest in children. The mean excess mortality above other causes of leukemia for all ages in males and females is 110 and 80, respectively, per 100,000 persons exposed to 10 rem (0.1 Sv) of uniform whole-body low-LET radiation.[14] The risk is relatively constant at all ages.
- *Thyroid cancer.* The incidence of radiation-induced thyroid cancer is dose related, with the following characteristics:[14] (1) Susceptibility is greatest early in childhood, and for those exposed before puberty the tumors usually do not become apparent until after sexual maturation. (2) Females are two to three times more susceptible than males. (3) It is frequently preceded or accompanied by benign thyroid nodules. (4) It is generally papillary rather than of follicular or mixed histopathology. (5) It is slow growing. Thyroid cancer is amenable to surgery and treatment with radioiodine. About 5% of patients with radiation-induced thyroid cancer die as a result of their disease. The principal data on thyroid cancer come from atomic bomb survivors, fallout victims from Marshall Islands nuclear tests, and children treated with x-rays for tonsil disease, enlarged thymus gland, and tinea capitis. A significant finding has been the rapid development of thyroid cancer following the Chernobyl accident from radioiodine, particularly in children; the incidence of pediatric thyroid cancer increased by a factor up to 50 times the preexposure rate.[13]
- *Breast cancer.* The incidence increases with radiation dose and is higher in American and Canadian women and lower in Japanese women. The data are derived from atomic bomb survivors, female patients in a Nova Scotia sanitorium who were fluoroscoped multiple times during artificial pneumothorax for pulmonary tuberculosis, and women treated with x-rays for postpartum mastitis. The excess mortality for breast cancer is highest in younger women exposed before age 15; women age 50 or older show little or no excess mortality (see Table 6-14, Chapter 6).[14]
- *Lung cancer.* Risk estimates come from three sources: atomic bomb survivors, patients exposed to x-rays during treatment of ankylosing spondylitis, and uranium mine workers, for whom the potential risk is from inhalation of ^{222}Rn, which decays to long-lived alpha-particle emitters of high LET.[14] Even though there is a clearly increased incidence of lung cancer among mine workers, it is difficult to separate the effects of radiation from those of other contributing factors, such as cigarette smoke. The excess cancer mortality is slightly higher for males than females and is greatest between the ages of 25 and 65 for both sexes (see Tables 6-13 and 6-14, Chapter 6).
- *Bone cancer.* The ingestion of long-lived, alpha-emitting radium isotopes, which have metabolism similar to calcium and localize in bone, has been associated with the development of bone cancer. One of the two known populations affected was the painters of luminous dials on clocks and watches, who licked their radium paint brushes to achieve a fine point. Small amounts of radium ingestion (^{226}Ra and ^{228}Ra) over several years produced a radiation bone burden. Bone sarcomas and carcinomas of the epithelial cells lining the paranasal sinuses and nasopharynx developed in these workers. The other affected population was patients given injections of ^{224}Ra for treatment of tuberculosis and ankylosing spondylitis. These cancers were induced by the slow turnover of high-LET radiation in bone.
- *Skin cancer.* The first person to die of radiation-induced cancer in the United States was Clarence Madison Dally, Thomas Edison's assistant, whose hand was exposed to x-rays from a fluoroscope.[14] He developed radiation dermatitis and finally died from metastatic epidermoid carcinoma. In the years that followed, many such cases occurred among physicists, physicians, x-ray technologists, and dentists before safety standards were instituted. Squamous cell and basal cell carcinomas were observed frequently. Radiation-induced skin cancers are readily diagnosed and treated, and there is a large difference between incidence and mortality.[2]

Mechanisms of Carcinogenesis

Radiation exposure does not cause a unique type of tumor; it simply increases the incidence of tumors that will form spontaneously. In Chapter 6, Figure 6-4, it was shown that the incidence of radiation-induced cancer follows a linear–quadratic relationship with dose. This relationship is based on data from the Japanese atomic bomb survivors, who were exposed at high dose and dose rates. The linear–quadratic relationship demonstrates that at low dose and dose rate cancer induction is linearly related to dose, whereas at high dose and dose rate cancer induction is exponential, being proportional to dose squared. The incidence of cancer induction falls off at very high dose because cells are killed.

A plot of cell survival and neoplastic transformation in mammalian cells as a function of radiation dose from ^{60}Co gamma-ray exposure is shown in Figure 8-15. The plot shows that the frequency of neoplastic transformation per surviving cell increases with dose up to a few grays and reaches a plateau at higher doses, where there is a balance between transformation and cell killing. Cell survival declines as the number of neoplastic transformations increases with dose. At high doses the survival of cells destined to become transformed declines exponentially, paralleling the survival curve of normal untransformed cells.

The dose–response relationship for cancer induction is similar to that for the induction of chromosome aberrations and therefore links the induction of cancer to radiation's alteration of genetic material. It is known from experimental

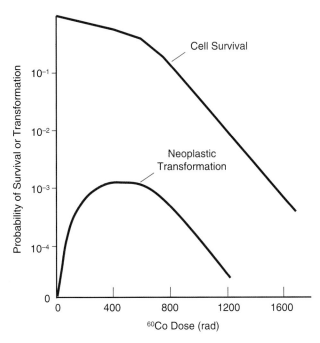

FIGURE 8-15 Probability of survival of 10T1/2 mammalian cells (top curve) and neoplastic transformation per irradiated cell (bottom curve) as a function of radiation dose from ^{60}Co gamma radiation at 100 rad/min. See text for explanation. (Adapted from Han A, Hill CK, Elkind MM. Repair of cell killing and neoplastic transformation at reduced rates of Co-60 gamma rays. *Cancer Res.* 1980;40:3329.)

studies that control of cell proliferation is the consequence of signals affecting cell division and differentiation, and if a cell turns cancerous it is because of a change in the signaling system. The conversion of a normal cell into a malignant state may result from activation of an oncogene, loss of a suppressor gene, or a combination of these two mechanisms.[2,14]

Oncogenes

Cells in their normal state contain proto-oncogenes that function to regulate cell growth. Studies have shown that proto-oncogenes may be activated to the corresponding oncogene by various means, causing the cell to become cancerous. Activation can be induced by chemicals, viruses, and radiation, which each can produce tumors that are indistinguishable from one another. The three principal mechanisms by which proto-oncogenes are activated to produce a malignant cell are as follows:[2,14]

- *Point mutation.* The change of a single base pair, resulting in a protein with a single amino acid change. An example is the point mutation in the oncogene N-*ras* found in cancer cells of acute leukemia.
- *Chromosome translocation.* Radiation breaks occurring in two separate chromosomes with translocation of the pieces. For example, breaks in chromosomes 2 and 8 with translocation activate the *myc* oncogene that is responsible for Burkitt's lymphoma.
- *Gene amplification.* Extra copies (overexpression) of a proto-oncogene can lead to activation of the oncogene. For example, gene amplification in the N-*myc* oncogene produces neuroblastoma.

Suppressor Genes

Normal cells may contain tumor suppressor genes that are associated with various chromosomes. Hybridization studies have demonstrated that when normal cells are combined with certain tumor cells, tumorigenicity of the tumor cells is suppressed.[15] One example is chromosome 11 in normal human fibroblasts, which contains a suppressor gene for the malignant phenotype of HeLa cells (a human tumor cell line). Studies have shown that when HeLa cells are fused with human fibroblasts, the hybrid cells do not produce the malignant phenotype of HeLa cells. However, when chromosome 11 is removed from the hybrid cells, the tumor is expressed again.

Radiation may inactivate suppressor genes in chromosomes of human cells, resulting in cancer. Some examples of suppressor genes whose inactivation or loss is associated with human cancer are the p105-RB gene on chromosome 13, which is associated with retinoblastoma, and the p53 gene on chromosome 17, which is associated with breast cancer, small cell lung cancer, cervical cancer, and bladder cancer.[2] The risks of radiation-induced cancer are discussed in Chapter 6.

HEREDITARY EFFECTS OF RADIATION

Radiation does not produce unique genetic effects; it simply increases the probability of effects that would occur naturally, similar to the induction of carcinogenic effects by radiation. While the carcinogenic effects of radiation arise from transformational changes in the DNA of somatic cells, thereby affecting the person who received the radiation exposure, the hereditary effects of radiation arise from transformations in the DNA of germ cells, which will manifest themselves in the offspring of those irradiated. If a radiation-induced mutation affects a dominant gene, the defect will be expressed in the first generation. If, however, the mutation occurs in a recessive gene of one parent, the defect will be expressed only if there is no dominant complementary gene from the other parent. For example, a mutation occurring on the X chromosome of a mother will not be expressed in a daughter who receives a normal X chromosome from her father. However, the defective gene will remain with the daughter as a sex-linked recessive defect, and it may be expressed in her son if the Y chromosome from his father does not carry a dominant complementary gene. Examples of sex-linked recessive defects are hemophilia and color blindness. The hereditary risks associated with exposure to ionizing radiation are discussed in Chapter 6.

Genetic Effects

The effects of radiation on the DNA of reproductive cells are similar to those in somatic cells. Unrepaired gene mutations that are not lethal to the cell can be passed on to offspring, whereas lethal chromosome breaks will not be passed on. However, some chromosome translocations resulting from breaks are not lethal and will be passed on. Offspring will have a defective chromosome from one parent and a normal complementary chromosome from the unaffected parent. If the defective chromosome does not cause embryonic death, the offspring will likely have some physical or mental abnormality as a result of the defective chromosome.

Radiation Effects on Testes and Ovaries

The male produces sperm from puberty to death. Mature sperm are the result of several developmental stages. Spermatogonial stem cells progress to become primary and secondary spermatocytes, then spermatids, and finally mature spermatozoa. The average time from the immature stem cell stage to mature spermatozoa is 10 weeks. Since the first studies by Bergonie and Tribondeau, exposure of male germ cells to radiation has shown that spermatogonia are most sensitive to radiation and mature spermatozoa are most radioresistant. Temporary and permanent sterility can be caused by gonadal exposure to radiation. Because these cells undergo continual cell division and are at various stages of maturity, a moderate dose of radiation may kill the more sensitive immature spermatocytes. Reproductive potency will remain as long as the unaffected mature sperm are viable; however, a risk of mutation will be present in these cells. When these sperm die, a period of temporary sterility will ensue until the viable stem cells can repopulate by division.

The ovary has its full complement of oocytes at birth; by 3 days of age the oocytes are in a resting phase and no further cell division occurs. Thus, there is no possibility of temporary sterility as with spermatogonia. The radiation exposures necessary to confer temporary and permanent sterility to human reproductive cells are shown in Table 8-2.

Some interesting and useful findings have come from what is known as the "megamouse project." In this study millions of specially bred mice were irradiated under various conditions and the effects were measured. Hall[2] summarized the findings as follows:

- Radiosensitivities of mice for different mutations vary widely.
- Dose fractionation results in a reduced mutation rate from acute exposure.
- The male mouse is much more radiosensitive than the female mouse, so much so that at low dose rate almost all the genetic burden in a population is carried by males.
- The genetic consequences of a given dose can be greatly reduced if a time interval is allowed between irradiation and conception. The decrease in mutation rate with time after irradiation is likely due to some repair process.

The last point is relevant in the genetic counseling of persons who receive gonadal exposure to radiation. In general, it is recommended that persons exposed to 10 rad or more allow a period of 6 months to elapse between exposure to radiation and a planned conception, to permit repair processes to minimize the risk of genetic mutation.

RADIATION EFFECTS ON THE EMBRYO AND FETUS

The principal effects of irradiation of the mammalian fetus are lethality of the embryo, malformations, mental retardation, and induction of malignancy.[1] The most important factors causing these effects are dose, dose rate, and stage of gestation during irradiation.

Russell and Russell[16,17] divided the developmental period in utero into three stages. (1) *Preimplantation* is the time from fertilization to when the embryo attaches to the uterine wall. This period is about 9 days in human development. During this time the fertilized ovum repeatedly divides, forming a ball of cells that are highly undifferentiated. (2) *Organogenesis* begins after the embryo implants in the uterine wall at around day 9 of gestation. During this stage the cells of the embryo differentiate into the various stem cells that will eventually form the major organs of the body. This process continues for about 6 weeks, at which point the embryo is termed a fetus. (3) The *fetal stage* is the period during which growth of the formed structures takes place. At this stage the central nervous system (CNS) is developing. The CNS in adults consists of nondividing highly differentiated cells, but in the fetus the cells forming the CNS are continually dividing, migrating, and differentiating. The neuroblasts appear by the 18th day of gestation in the human. As the fetus develops, the neuroblasts disperse throughout the body and become more differentiated. Neuroblasts continue to exist throughout fetal development of the CNS and until at least 2 weeks after birth. The CNS continues to develop until 10 to 12 years of age.[8] Much of the evidence for radiation effects on the embryo and fetus has been obtained from experiments in animals, notably mice. Therefore, it is useful to compare the time spans of gestational stages in mice and humans:

- Preimplantation: Mouse, 0 to 5 days; human, 0 to 9 days
- Organogenesis: Mouse, 5 to 13 days; human, 10 days to 6 weeks

TABLE 8-2 Threshold Doses for Deterministic Effects in Human Germ Cells

Tissue and Effect	Acute Single Dose (rad)	Prolonged Exposure[a] (rad/yr)
Testes		
Temporary sterility	15	40
Permanent sterility	350–600	200
Ovaries		
Permanent sterility	250–600	20

[a] Annual dose rate if received in highly fractionated or protracted exposures.
Source: Reference 1.

- Fetal period: Mouse, 13 days to full term (20 days); human, 6 weeks to 9 months

Animal Studies

The principal effects observed from irradiation of mice and rats are growth retardation, congenital malformations, and embryonic, neonatal, or fetal death.

Irradiation during the preimplantation period results in either embryonic death or normal development if the embryo survives. At this stage there are few cells in the conceptus and they are not yet specialized. Damage to one cell, being the progenitor of many descendant cells, has a high probability of fatality; growth retardation or malformations are not seen at this stage of development. In mice, doses as low as 10 rad (0.1 Gy) can kill a fertilized egg.[2]

Irradiation during early organogenesis produces severe growth retardation, seen as low birth weight at term. Animals can recover, however, and go on to attain full growth as adults. A dose of 100 rad (1 Gy) will produce growth retardation. Another principal effect during this sensitive stage is the development of congenital anomalies. Some anomalies observed in mice are exencephaly (imperfect cranium with protrusion of the brain outside the skull), anencephaly (absence of a cranial vault and cerebral hemispheres), stunted development, and evisceration. Since various organs form on specific gestational days, irradiation on those days produces specific abnormalities. For example, radiation exposure of the mouse embryo on the 9th day results in a high frequency of ear and nose abnormality, whereas exposure on the 10th day results in bone abnormalities.[8] The neuroblasts in the fetus are highly undifferentiated, mitotically active, and highly radiosensitive if irradiated at this time. Some of the common abnormalities observed in mice after in utero irradiation are microcephaly (small brain), hydrocephaly (water on the brain), and eye deformities such as microphthalmia (small eyes).[8] A dose of 200 rad in mice during the period of maximum sensitivity 8 to 12 days after conception can result in a nearly 100% rate of malformations at birth. A 200 rad dose on day 10 after conception carries a 70% death rate because of gross fetal abnormalities.[2]

The fetal growth stage is less sensitive to radiation, because the organs are formed and cells are more differentiated. Higher doses are required to produce effects, mostly on formed organ systems. Any growth retardation at this stage, however, is permanent. Irradiation during this period may result in stochastic effects such as cancer later in life.

Human Studies

Information about irradiation of humans in utero comes primarily from studies of Japanese atomic bomb survivors and exposure of pregnant women during medical procedures.

In Japanese offspring, no birth defects were found as a result of irradiation before 15 days of gestation; this is consistent with animal data. That is, any damage that occurs during this stage will likely result in embryonic death. The embryos that escape damage develop normally.[2] The principal effects of irradiation in utero are microcephaly and mental retardation. The most pronounced microcephaly was seen with 150 rad (1.5 Gy) of in utero exposure, but effects were also seen with maternal exposure as low as 10 to 19 rad (0.1 to 0.19 Gy).[2] Mental retardation was not observed to be induced by radiation before 8 weeks from conception or after 25 weeks.[1] During the most sensitive period, 8 to 15 weeks after conception, the fraction of those exposed who became severely mentally retarded increased by approximately 0.4 per sievert; approximately 40% of women exposed to 100 rem (1 Sv) produced retarded offspring. For exposure during weeks 16 to 25, the risk is only one-fourth as great (0.1 per sievert).[1] Mental retardation is thought to be associated with decreased migration of cells from their place of origin to their site of function in the brain; the highest risk is during the gestational stage when the brain cortex is being formed.[2] Pooled data from Hiroshima and Nagasaki for children exposed at 8 to 15 weeks of gestation demonstrate that the dose–response relationship for mental retardation is linear, with a threshold of 12 to 20 rad (0.12–0.2 Gy) (Figure 8-16).[17] This is consistent with the deterministic nature of retardation, which requires the killing of a minimum number of cells to be manifest.

Less severe mental retardation in children exposed to ionizing radiation before birth has been shown by intelligence test scores.[18,19] With exposure during the sensitive period 8 to 15 weeks after conception, the observed shift in IQ scores is about 30 IQ points per 100 rad (1 Gy).

One of the first studies of medical exposure of pregnant women, demonstrating the adverse effects of x-irradiation in utero, was performed by Goldstein and Murphy,[20] who reported microcephaly and mental retardation and other defects. Dekaban[21] surveyed the literature on abnormalities in children exposed to x-irradiation during various stages of gestation and reported several findings:

- Doses higher than 250 rad (2.5 Gy) to the embryo before 2 to 3 weeks of gestation may cause embryos to abort but are not likely to produce severe abnormalities in fetuses that survive to term.
- Irradiation between 4 and 11 weeks of gestation may produce severe abnormalities of many organs, particularly in the skeleton and CNS.
- Irradiation between weeks 11 and 16 of gestation may cause a few eye, skeletal, and genital abnormalities and frequently causes microcephaly, retardation, and stunting of growth.
- Mild microcephaly, mental retardation, and growth stunting may result from irradiation during weeks 16 to 20 of gestation.
- After 20 weeks, the fetus is more radioresistant, but irradiation during this time may produce some functional defects.

Children who were exposed to radiation in utero seem to be susceptible to childhood leukemias and other cancers that are expressed during approximately the first decade of life.[1] Studies in England[22] and the United States[23,24] of x-ray

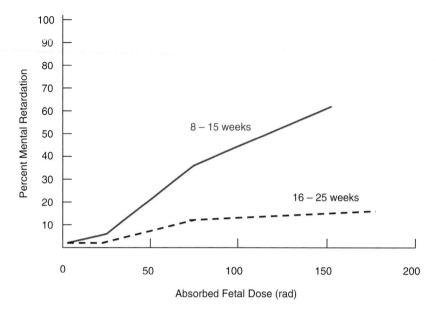

FIGURE 8-16 Frequency of mental retardation as a function of dose and gestational age from atomic bomb radiation exposure in utero. (Adapted from reference 18.)

exposure in utero and subsequent malignancies show an association between exposure and the subsequent development of childhood malignancies. Stewart and Kneale[22] performed an analysis of the Oxford Survey of Childhood Cancers. The survey analyzed 15,298 children. Fifty percent (7,649) died of malignancy before age 10 (case group); of these, 1,141 had been x-rayed in utero. There were 7,649 children who did not die and were cancer free (control group), and of these 774 had been x-rayed before birth. The case/control ratio of those who were x-rayed (1.48) was significantly greater than the ratio of those who were not x-rayed (0.95). The case/control ratio increased with fetal radiation dose from 1.26 with one x-ray exam to 2.24 for more than five exams. When the groups were stratified by trimester of irradiation, it was found that the case/control ratio was 8.25 in the first trimester, 1.49 in the second, and 1.43 in the third. Thus, the excess of cancer risk from x-ray examination in utero was directly related to fetal dose, and the risk of cancer was greatest when exposure occurred during the first trimester. Another way to analyze these data is to look at the rate of death from cancer of those x-rayed versus those not x-rayed. Of the 1,915 children x-rayed, 1,141 (59.3%) died of cancer, whereas of the 13,383 not x-rayed, 6,508 (48.6%) died of cancer. This demonstrates a 22% increase in cancer death after x-ray exposure in utero.

Additional evidence that in utero irradiation causes childhood malignancy comes from a study that showed the same incidence of leukemia and other cancers in twins of irradiated women as in single children of such women, with a clear excess over children who were not exposed to radiation.[25] ICRP estimates that the excess risk of fatal childhood cancer due to prenatal exposure throughout pregnancy is 2.8×10^{-2} per sievert.[1] This corresponds to an excess over the spontaneous rate of 280 malignancies per 10,000 person-Sv (2.8% per 100 rem exposure).

NCRP recommends that the total dose to the fetus during gestation not exceed 0.5 rem (0.5 cSv), with a monthly limit of 0.05 rem (0.05 cSv).[26] Once pregnancy is declared, the duties of a radiation worker should be reviewed to ensure that this limit is not exceeded.

The most critical stage of gestational development for radiation-induced congenital malformations, including microcephaly and mental retardation, extends from 10 days to 26 weeks. The data from Japan suggest a threshold of 12 to 20 rad (0.12–0.2 Gy) for retardation. On the basis of this threshold, 10 rad is often considered the cutoff point above which an anomaly may occur in a child irradiated in utero.[2]

RADIATION-INDUCED CATARACT FORMATION

A cataract is an opacification of the normally transparent lens of the eye. Cataracts may be caused by aging, family history (genetic), medical problems such as diabetes, injury, medication such as steroids, and ionizing radiation. The lens contains cells that continue to divide and replenish the lens tissue. Radiation injures the dividing cells, making them nontranslucent. Accumulation of these injured cells leads to cataract development, which is a deterministic effect.

Studies in humans, principally radiation therapy patients, and in animals have shown that ionizing radiation can cause cataracts.[27,28] A cataract may be stationary at a defined locus, or it may progress to cloud a larger portion of the lens (progressive cataract).

The minimum dose required to produce a progressive cataract is about 200 rad (2 Gy) in a single exposure, with larger doses necessary in a fractionated regimen.[27] The latent period between irradiation and the appearance of lens opacity is dose

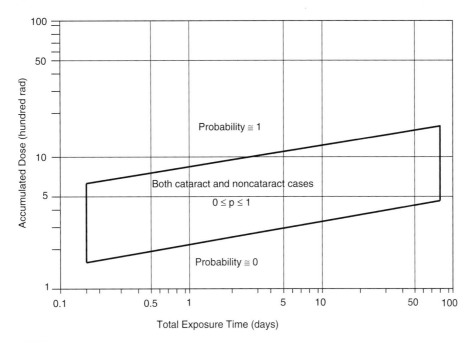

FIGURE 8-17 Time–dose relationship for cataract formation in humans. (Adapted from reference 27.)

related.² The latency period is about 8 years after exposure to a dose in the range of 250 to 600 rad (2.5–6.5 Gy). In one study of radiotherapy patients who had received low doses to the eye (220–650 rad [2.2–6.5 Gy]), progressive cataracts developed in only about 12%, compared with 88% in those who received high doses (650–1150 rad [6.5–11.5 Gy]).² Radiation-induced cataracts are a deterministic late effect. There is a practical threshold dose below which cataracts are not produced and above which the severity of the biologic response is dose related (Figure 8-17).

What are the potential risks for those who handle large amounts of activity, such as ^{99m}Tc generator eluates? The following simple example provides some guidance on the amount of exposure possible and demonstrates the small probability of cataract induction from this type of exposure.

What time of exposure is required to accumulate 200 rad (2 Gy) from the unshielded injection port on a ^{99m}Tc elution vial containing 7 Ci (259 GBq) of activity?

$$\frac{7000 \text{ mCi} \times 0.8 \text{ R/hr/mCi/cm}}{60 \text{ min/hr}} = 93 \text{ R/min at 1 cm}$$

$$\frac{200 \text{ rad}}{93 \text{ R/min}} = 2.15 \text{ min at 1 cm from the unshielded end of the vial}$$

Applying the inverse square law, we have

$$\frac{93 \text{ R/min}}{(30 \text{ cm})^2} = 0.1 \text{ R/min at 1 foot}$$

$$\frac{200 \text{ rad}}{0.1 \text{ R/min}} = 2000 \text{ min (\textasciitilde 33 hr exposure required at 1 foot from the unshielded end of the vial)}$$

REFERENCES

1. International Commission on Radiological Protection. *1990 Recommendations of the International Commission on Radiological Protection.* ICRP Publication 60. New York: Pergamon Press; 1991.
2. Hall EJ. *Radiobiology for the Radiologist.* 4th ed. Philadelphia: JB Lippincott; 1994.
3. Puck TT, Marcus TI. Action of x-rays on mammalian cells. *J Exp Med.* 1956;103:653–66.
4. Elkind MM, Whitmore GF. *The Radiobiology of Cultured Mammalian Cells.* New York: Gordon and Breach; 1967.
5. Bergonie J, Tribondeau L. De quelques resultats de la radiotherapie et essai de fixation d'une technique rationelle. *C R Acad Sci* (Paris). 1906;143:983.
6. Ancel P, Vitemberger P. Sur la radiosensibilite cellulaire. *C R Soc Biol.* 1925;92:517.
7. Casarett AP. *Radiation Biology.* Englewood Cliffs, NJ: Prentice Hall; 1968.
8. Travis EL. *Primer of Medical Radiobiology.* Chicago: Year Book; 1975.
9. Elkind MM, Sutton-Gilbert H, Moses WB, et al. Radiation response of mammalian cells in culture. V: Temperature dependence of the repair of x-ray damage in surviving cells (aerobic and hypoxic). *Radiat Res.* 1965;25:359–76.
10. Thames HD. On the origin of dose fractionation regimens in radiotherapy. *Semin Radiat Oncol.* 1992;2:3–9.
11. Meyer JL. Techniques and technology in the practice of radiotherapy. In: Fajardo L-G LF, Berthrong M, Anderson RE, eds. *Radiation Pathology.* New York: Oxford University Press; 2000:143–54.
12. Fajardo L-G LF, Berthrong M, Anderson RE, eds. *Radiation Pathology.* New York: Oxford University Press; 2000:43–51.
13. Tuttle RM, Becker DV. The Chernobyl accident and its consequences: update at the millennium. *Semin Nucl Med.* 2000;30:133–40.
14. National Academy of Sciences. *Health Effects of Exposure to Low Levels of Ionizing Radiation.* BEIR V Report. Washington, DC: National Academies Press; 1990.
15. Stanbridge EJ. Suppression of malignancy in human cells. *Nature.* 1976;260:17–20.
16. Russell LB, Russell WL. Radiation hazards to the embryo and fetus. *Radiology.* 1952;58:369–76.
17. Russell LB, Russell WL. An analysis of the changing radiation response of the developing mouse embryo. *J Cell Physiol.* 1954;43(suppl 1):1030–1149.

18. Otake M, Schull WJ. In utero exposure to A-bomb radiation and mental retardation: a reassessment. *Br J Radiol.* 1984;57:409–11.
19. Otake M, Schull WJ, Fujukoshi Y, et al. *Effect on School Performance of Prenatal Exposure to Ionizing Radiation in Hiroshima.* RERF TR2-88. Hiroshima, Japan: Radiation Effects Research Foundation; 1988.
20. Goldstein L, Murphy DP. Microcephalic idiocy following radium therapy for uterine cancer during pregnancy. *Am J Obstet Gynecol.* 1929;18:189–95, 281–2.
21. Dekaban AS. Abnormalities in children exposed to x-irradiation during various stages of gestation: tentative timetable of radiation to the human fetus. *J Nucl Med.* 1968;9:471–7.
22. Stewart A, Kneale GW. Radiation dose effects in relation to obstetric x-rays and childhood cancers. *Lancet.* 1970;1:1185–8.
23. MacMahon B. Prenatal x-ray exposure and childhood cancer. *J Natl Cancer Inst.* 1962;28:1173–91.
24. MacMahon B, Hutchinson GB. Prenatal x-ray and childhood cancer: a review. *Acta Univ Int Contra Cancrum.* 1964;20:1172–4.
25. Harvey EB, Boice JD, Honeyman M, et al. Prenatal x-ray exposure and childhood cancer in twins. *N Engl J Med.* 1985;312:541–5.
26. National Council on Radiation Protection and Measurements. *Recommendations on Limits for Exposure to Ionizing Radiation.* NCRP Report No. 91. Bethesda, MD: National Council on Radiation Protection and Measurements; 1987.
27. Merriam GR, Szechter A, Focht EF. The effects of ionizing radiations on the eye. *Front Radiat Ther Oncol.* 1972;6:346–85.
28. Bateman JL, Bond VP. Lens opacification in mice exposed to fast neutrons. *Radiat Res.* 1967;7(suppl):239–49.

CHAPTER 9

Radiopharmaceutical Chemistry: General Topics

What is a radiopharmaceutical? How is it different from a radiochemical? The U.S. Food and Drug Administration (FDA) defines a radiopharmaceutical as "any substance defined as a drug that exhibits spontaneous disintegration of unstable nuclei with the emission of nuclear particles or photons." A radiochemical has basically the same definition except that it is not a drug and therefore is not approved for direct human use. In essence, a radiopharmaceutical is a radiochemical that has undergone a specific series of tests in animals and humans and has been shown to be safe and effective for a particular diagnostic or therapeutic application.

The physicochemical forms of radiopharmaceuticals are diverse. They include radiolabeled elemental forms, simple ions, small molecules, macromolecules, particles, and blood cells. Radiopharmaceuticals are administered mostly by intravenous injection, but also as oral dosage forms, such as capsules or solutions, and by inhalation, as gases and aerosols.

Radiopharmaceuticals possess a few unique characteristics. Unlike traditional drugs, radiopharmaceuticals do not produce a pharmacologic effect in the body because only trace amounts are administered. Consequently, their risk of chemical toxicity is essentially nil. However, radiopharmaceuticals possess an inherent radiation risk, which limits the amount of radioactivity that can be administered. A radiopharmaceutical is used in nuclear medicine for the diagnosis or treatment of disease. Thus, it must possess specific biochemical properties to effect localization in the desired target organ and appropriate radiation properties for detection and measurement in vivo.

Radiopharmaceuticals can be prepared by a drug manufacturer, at a nuclear pharmacy, or at the site of use. Radiopharmaceuticals with long physical half-lives are typically prepared by a drug manufacturer and can be purchased and stored in the nuclear medicine department until needed. An example of a long-lived radiopharmaceutical is one labeled with ^{131}I (half-life = 8 days). Many radiopharmaceuticals have short physical half-lives and must be prepared daily in the nuclear medicine clinic or at a nearby nuclear pharmacy. Examples include radiopharmaceuticals labeled with ^{99m}Tc (half-life = 6 hr). ^{99m}Tc-labeled radiopharmaceuticals are prepared extemporaneously from radiopharmaceutical kits, following simple stepwise procedures.

This chapter discusses some of the basic physicochemical properties of radiopharmaceuticals important to their preparation and use.

PERIODIC TABLE OF THE ELEMENTS

Atomic Weights of Elements

There are 116 elements with unique chemical properties (www.nist.gov/physlab/data/upload/periodic-table.pdf). Each element is composed of atoms whose principal mass resides in a nucleus, composed of protons and neutrons. The nucleus is surrounded by a cloud of electrons. The general term nuclide is used to describe an atom characterized by the number of protons and neutrons in its nucleus. The number of protons is the atomic number of the element. In a neutral atom the number of electrons surrounding the nucleus is equal to the number of nuclear protons. The total number of protons and neutrons in the nucleus is the mass number of the element. Each element has different isotopes. Isotopes have the same number of protons but a different number of neutrons. For example, hydrogen, with atomic number 1, has three isotopes: ^{1}H (protium) has one proton and no neutrons; ^{2}H (deuterium) has one proton and one neutron; and ^{3}H (tritium) has one proton and two neutrons. Protium and deuterium are stable and tritium is radioactive. All of the isotopes of all 116 elements together total more than 2500 nuclides. Of this number, 266 nuclides are stable and the remainder are radioactive (radionuclides).

Each element has an atomic weight that indicates how heavy, on average, one of its atoms is relative to an atom of another element. Thus, hydrogen, with atomic weight 1.00794, is one-twelfth as heavy as carbon, with atomic weight 12.011, and one-sixteenth as heavy as oxygen, with atomic weight 15.9994. The atomic weight of an element is an average because it consists of a mixture of naturally occurring isotopes. For example, oxygen occurs in nature as three stable isotopes, ^{16}O, ^{17}O, and ^{18}O, which have natural abundances of 99.759%, 0.0374%, and 0.2039%, respectively. The atomic weight of oxygen is the sum of the atomic masses of each isotope multiplied by its natural abundance:

$$^{16}O(15.9949149 \times 0.99759) + {}^{17}O(16.999133 \times 0.000374)$$
$$+ {}^{18}O(17.9991598 \times 0.002039)$$

By definition, the gram atomic weight (GAW) of an element, or one mole, is the atomic weight expressed in grams. One mole is equivalent to Avogadro's number (6.023×10^{23}) of atoms of the element. Thus, one mole each of carbon (12.011 grams), oxygen (15.9994 grams), and hydrogen (1.00794 grams) will contain 6.023×10^{23} atoms.

Chemical Properties of Elements

From a physics perspective, the ratio of neutrons to protons in the nucleus of a nuclide determines whether it is stable or radioactive, the types of radiation it will emit during radioactive decay, and whether it has potential for diagnostic or therapeutic use. From a chemistry perspective, a nuclide's chemical properties determine the type of bond it will form when it is incorporated into a chemical compound such as a radiopharmaceutical.

The isotopes of a given element have different nuclear properties, but their chemical properties are similar. The atomic number of an element is equal to the number of positively charged protons in its nucleus. A neutral atom has a number of electrons equivalent to the atomic number in its energy levels or shells surrounding the nucleus. Each electron shell can have a maximum number of electrons. The electrons in the outermost shell of an element are called the valence electrons. They are involved in chemical interactions between elements to form chemical compounds. These chemical interactions between elements serve to fill the elements' valence shell electron requirements. Valence electrons determine an element's chemical properties, such as oxidation state, ionization potential, electronegativity, and type of bond formed with other elements.

When all of the elements are arranged in increasing order by atomic number, similar chemical properties are found to occur periodically among groups of elements. The similar chemical properties of elements in a group are associated with the number of valence electrons in these elements. The periodicity of chemical properties is known as the periodic law of the elements. Observation of this phenomenon led to establishment of the periodic table of the elements (Table 9-1).

In the periodic table, elements with similar properties are arranged under each other in vertical columns called groups. There are 18 main groups of elements. The horizontal rows of elements formed as the groups advance across the table are called periods. One additional electron is added to each successive element in a period. A new period is begun after the formation of a chemically inert element. Inertness is associated with a filled electron energy level or sublevel. The contrast of chemical properties between different elemental groups can be illustrated by comparing the elements of groups 1 and 17. Group 1 elements, known as the alkali metals, have one electron in the outermost electron (valence) shell. They have large atomic volumes, meaning their valence electrons are relatively far removed from the positive attractive force of the nucleus. They have low ionization potentials, meaning that little work is required to remove the valence electrons. They are good conductors of electricity and heat. Each group 1 element has a metallic luster and reacts vigorously with water to liberate hydrogen gas, forming basic solutions as metallic hydroxides. By contrast, group 17 elements, known as the halogens, have seven electrons in their valence shells. Their atomic volumes are small relative to their atomic numbers and their ionization potentials are high. They are nonmetals, being poor conductors of electricity and heat. Halogens react with hydrogen to form compounds producing acidic solutions in water. The distinct differences in properties between the elements in groups 1 and 17 are related primarily to the number of electrons in their outermost energy levels and the distance of these electrons from the nucleus. These differences play a significant role in the formation of chemical bonds between reacting atoms.

CHEMICAL BONDS

The attraction between two atoms within a molecule is called a chemical bond. The nature of the chemical bond is determined by the electronic properties of the atoms involved. The principal electron energy levels around the nucleus of an atom are numbered consecutively: 1, 2, 3, and so forth. Within each principal energy level are sublevels designated by the letters s, p, d, and f. The number of sublevels in a given principal energy level is equal to the number of that level. Thus, the first energy level has one sublevel s, the second energy level has two sublevels s and p, the third energy level has three sublevels, s, p, and d, and so forth.

The electrons in each energy level are confined to a certain space termed the atomic orbital. No more than two electrons may occupy an orbital. The total number of atomic orbitals in a principal energy level is equal to the square of the number of that level. Thus, the first principal level has one sublevel (s), one (i.e., 1^2) atomic orbital (s), and a maximum capacity of two electrons. The second principal level has two sublevels (s and p), four (i.e., 2^2) atomic orbitals (s, p_x, p_y, and p_z), and a capacity of eight electrons, and so forth (Table 9-2).

The s atomic orbital has the form of a spherical cloud about the nucleus, and the s electrons can be randomly located anywhere within this cloud. The p orbitals are designated with subscript letters x, y, and z to indicate that they have direction about the nucleus as well as form. The p-orbital electrons travel in a figure 8 pattern about the nucleus in three planes along x-, y-, and z-coordinates that pass directly through the nucleus. These forms are illustrated in Figure 9-1.

The first 10 elements in the periodic table and their electronic configurations are listed in Table 9-3. As a general rule, an atom seeks to achieve its lowest energy state, and it does this by completing its outermost energy level with electrons through chemical reactions with other atoms. Table 9-3 indicates that, of the 10 elements listed, only helium and neon have a full complement of electrons in their principal energy levels, making these elements inert. Inertness is achieved because these elements cannot increase their stabilities further by combining with other elements. The remaining elements, however, have unfilled energy levels and must combine with other elements to satisfy their electron deficiencies and achieve a lower energy state. This can be accomplished in three ways: by an atom donating electrons, acquiring electrons, or sharing electrons with another atom. These three processes result in two basic types of chemical bonds: electrovalent (ionic) bonds and covalent bonds.

Electrovalent Bonds

As the name implies, an electrovalent bond is associated with an electrical charge on the reacting atoms. A neutral atom acquires a positive charge if it loses one or more of its electrons and a negative charge if it gains one or more electrons. For example, in the formation of lithium fluoride (LiF) a lithium atom readily donates its single 2s electron to a fluorine atom, resulting in a positively charged lithium ion and a negatively charged fluoride ion. In this situation, lithium is left with its only remaining principal energy level ($1s^2$) filled with electrons, and fluorine is left with its outer energy level ($2s^2$, $2p^6$) filled with electrons. As a consequence of the electron transfer between lithium and fluorine, an electrovalent bond forms between two oppositely charged species, creating the more stable lithium fluoride molecule.

The electrovalent bond tends to form between atoms that have large differences in electronegativity. Electronegativity is the tendency of an atom to attract electrons. In the example of lithium fluoride above, lithium has weak electronegativity (low ionization potential), tending to lose its electron, and fluorine has strong electronegativity (high ionization potential), tending to acquire an electron (Table 9-4). The electrovalent bond is a relatively weak bond, however, and in aqueous solution lithium fluoride readily dissociates to form lithium and fluoride ions. Many chemical compounds that are routinely used in chemistry and pharmacy are characterized by ionic bonding. Some examples are sodium and potassium salts as acetates, chlorides, and phosphates. Examples of radiopharmaceuticals with ionic bonding are ^{131}I-sodium iodide and ^{99m}Tc-sodium pertechnetate.

Covalent Bonds

A covalent bond is characterized by the sharing of a pair of electrons between atoms, one electron donated from each atom. In a covalent bond an electron orbital of one atom overlaps with an orbital in another atom. Covalent bonds are stronger than ionic bonds and are favored between atoms with similar electronegativities. When the electronegativities of bonding atoms are equal, electron sharing is equal, and the bond is nonpolar covalent. The bond between two hydrogen atoms in a hydrogen molecule (H_2) is a good example, being characterized by overlapping of the 1s electron clouds of each H atom. When the electronegativities between bonding atoms are slightly different, electron sharing is unequal, and the bond is polar covalent. That is, the electrons spend more time in the region of the molecule with the more electronegative atom, making it slightly negative, designated as $\delta-$. The region of the molecule with the less electronegative atom is slightly positive, designated as $\delta+$. Water (H_2O) is a good example of a polar covalent molecule, in which the covalent bond between oxygen and hydrogen is characterized by an overlap of the s-orbital electrons of hydrogen and the p-orbital electrons of oxygen. The electronegativities of hydrogen and oxygen are 2.1 and 3.5, respectively. Consequently, the electrons in the water molecule spend more time in the vicinity of oxygen compared with hydrogen because of oxygen's higher electronegativity. This makes oxygen $\delta-$ and the two hydrogens $\delta+$.

Table 9-5 gives examples of the types of bonds that form between elements of differing electronegativities. Note that there is no sharp distinction between ionic and covalent bonds. In general, if the electronegativity difference between bonding atoms is 0.0 the bond is nonpolar covalent; if it is greater than 2.0 it is more ionic; and if it is between 0.0 and 2.0 it is polar covalent.

The shared electrons in a covalent bond are in molecular orbitals encompassing two positive nuclei. Because the electrons are attracted by two nuclei, the bond is stronger and the molecule more stable (lower energy state). In the representation of the chemical structures of organic molecules, composed principally of carbon, hydrogen, and oxygen, a single covalent bond is conventionally written as a dash or single line (representing 2 electrons) between bonding atoms.

Carbon is a basic constituent of all organic molecules and is an interesting case in point regarding covalent bonding. In

TABLE 9-1 Periodic Table of the Elements[a]

Group	1 IA	2 IIA	3 IIIB	4 IVB	5 VB	6 VIB	7 VIIB	8 VIII	9 VIII
Period 1	1 **H** 1.00794 **1** (2.1)								
2	3 **Li** 6.941 **1** (1.0)	4 **Be** 9.0122 **2** (1.5)							
3	11 **Na** 22.9898 **1** (0.9)	12 **Mg** 24.3050 **2** (1.2)							
4	19 **K** 39.0983 **1** (0.8)	20 **Ca** 40.078 **2** (1.0)	21 **Sc** 44.9559 **3** (1.3)	22 **Ti** 47.867 **4** (1.5)	23 **V** 50.9415 **5** (1.6)	24 **Cr** 51.9961 **6,3** (1.6)	25 **Mn** 54.9380 **7,4,2** (1.5)	26 **Fe** 55.845 **3,2** (1.8)	27 **Co** 58.9332 **3,2** (1.8)
5	37 **Rb** 85.4678 **1** (0.8)	38 **Sr** 87.62 **2** (1.0)	39 **Y** 88.9059 **3** (1.2)	40 **Zr** 91.224 **4** (1.4)	41 **Nb** 92.9064 **5** (1.6)	42 **Mo** 95.94 **6** (1.8)	43 **Tc** [98] **7** (1.9)	44 **Ru** 101.07 **3** (2.2)	45 **Rh** 102.9055 **3** (2.2)
6	55 **Cs** 132.905 **1** (0.7)	56 **Ba** 137.327 **2** (0.9)	57–71*	72 **Hf** 178.49 **4** (1.3)	73 **Ta** 180.948 **5** (1.5)	74 **W** 183.84 **6** (1.7)	75 **Re** 186.207 **7** (1.9)	76 **Os** 190.233 **8,4** (2.2)	77 **Ir** 192.217 **4,3** (2.2)
7	87 **Fr** [223] **1** (0.7)	88 **Ra** [226] **2** (0.9)	89–103#	104 **Rf** [261]	105 **Db** [262]	106 **Sg** [266]	107 **Bh** [264]	108 **Hs** [277]	109 **Mt** [268]
*	57 **La*** 138.9055 **3**	58 **Ce** 140.116 **3**	59 **Pr** 140.9077 **3**	60 **Nd** 144.242 **3**	61 **Pm** [145] **3**	62 **Sm** 150.36 **3**	63 **Eu** 151.964 **3**	64 **Gd** 157.25 **3**	65 **Tb** 158.925 **3**
#	89 **Ac#** [227] **3**	90 **Th** 232.038 **3**	91 **Pa** 231.036 **5**	92 **U** 238.029 **6**	93 **Np** [237] **5**	94 **Pu** [244] **4**	95 **Am** [243] **3**	96 **Cm** [247] **3**	97 **Bk** [247] **3**

[a] Order of Information, top to bottom: atomic number and elemental symbol, atomic weight, oxidation state (in bold), and electronegativity (in parentheses). Numbers in brackets are atomic weights of longest-lived isotope. Elements 57–71(*) and 89–103(#) are displayed individually below elements 1–116.

10	11 IB	12 IIB	13 IIIA	14 IVA	15 VA	16 VIA	17 VIIA	18 VIIIA
								2 **He** 4.002602
			5 **B** 10.811 3 (2.0)	6 **C** 12.011 4 (2.5)	7 **N** 14.0067 5,−3 (3.0)	8 **O** 15.9994 −2 (3.5)	9 **F** 18.99840 −1 (4.0)	10 **Ne** 20.1797
			13 **Al** 26.98154 3 (1.5)	14 **Si** 28.0855 4 (1.8)	15 **P** 30.97376 5 (2.1)	16 **S** 32.065 6,−2 (2.5)	17 **Cl** 35.453 −1 (3.0)	18 **Ar** 39.948
28 **Ni** 58.6934 2 (1.8)	29 **Cu** 63.546 2 (1.9)	30 **Zn** 65.409 2 (1.6)	31 **Ga** 69.723 3 (1.6)	32 **Ge** 72.64 4 (1.8)	33 **As** 74.9216 3 (2.0)	34 **Se** 78.96 6,−2 (2.4)	35 **Br** 79.904 −1 (2.8)	36 **Kr** 83.798
46 **Pd** 106.42 2 (2.2)	47 **Ag** 107.868 1 (1.9)	48 **Cd** 112.411 2 (1.7)	49 **In** 114.818 3 (1.7)	50 **Sn** 118.710 4,2 (1.8)	51 **Sb** 121.760 3 (1.9)	52 **Te** 127.60 6,−2 (2.1)	53 **I** 126.9045 −1 (2.5)	54 **Xe** 131.293
78 **Pt** 195.084 4,2 (2.2)	79 **Au** 196.967 3 (2.4)	80 **Hg** 200.59 2 (1.9)	81 **Tl** 204.383 3 (1.8)	82 **Pb** 207.2 2 (1.8)	83 **Bi** 208.980 3 (1.9)	84 **Po** [209] 2 (2.0)	85 **At** [210] −1 (2.2)	86 **Rn** [222]
110 **Uun** [281]	111 **Uuu** [272]	112 **Uub** [285]		114 **Uuq** [289]		116 **Uuh** [292]		

66 **Dy** 162.500 3	67 **Ho** 164.930 3	68 **Er** 167.259 3	69 **Tm** 168.934 3	70 **Yb** 173.043 3	71 **Lu** 174.967 3
98 **Cf** [251] 3	99 **Es** [252] 3	100 **Fm** [257] 3	101 **Md** [258] 3	102 **No** [259] 2	103 **Lr** [262] 3

TABLE 9-2 Properties of Electron Energy Levels

Property	Principal Energy Level, n			
	1	2	3	4
Number of sublevels	1 (s)	2 (s,p)	3 (s,p,d)	4 (s,p,d,f)
Number of orbitals, n^2	1	4	9	16
Number of electrons (two per orbital)	2	8	18	32

TABLE 9-3 Electronic Configuration of Elements 1 through 10

Element	Atomic Number	Electronic Configuration
H	1	$1s^1$
He	2	$1s^2$
Li	3	$1s^2, 2s^1$
Be	4	$1s^2, 2s^2$
B	5	$1s^2, 2s^2, 2p_x^1$
C	6	$1s^2, 2s^2, 2p_x^1, 2p_y^1$
N	7	$1s^2, 2s^2, 2p_x^1, 2p_y^1, 2p_z^1$
O	8	$1s^2, 2s^2, 2p_x^2, 2p_y^1, 2p_z^1$
F	9	$1s^2, 2s^2, 2p_x^2, 2p_y^2, 2p_z^1$
Ne	10	$1s^2, 2s^2, 2p_x^2, 2p_y^2, 2p_z^2$

the normal state of elemental carbon, having the electronic configuration $1s^2, 2s^2, 2p_x^1, 2p_y^1$, only two covalent bonds may form, since only the two p electrons, p_x and p_y, are unpaired. However, when carbon enters into covalent bonding with other atoms, one of the 2s electrons is promoted to the $2p_z$ orbital, making four unshared electrons available for bonding, that is, one electron each in the 2s, $2p_x$, $2p_y$, and $2p_z$ orbitals. The formation of single covalent bonds via these four sp^3 hybridized orbitals, known as sigma (σ) bonds, is more profitable in terms of energy and results in a molecule that has greater stability than if carbon bonded through only two p orbitals.

When two carbon atoms bond via a double bond, one of the p electrons in each carbon atom does not become hybridized but remains in a separate orbital. The remaining three electrons (one s and 2 p) become hybridized to form three sp^2 orbitals. These sp^2 orbitals form sigma bonds that are in one plane, as compared with the x, y, and z planes of sp^3 orbitals. In this situation, the unhybridized p electron occupies a separate orbital at right angles to the plane of the sp^2 orbitals. The double bond in ethylene ($CH_2=CH_2$) is an example. The double bond between the two carbon atoms in

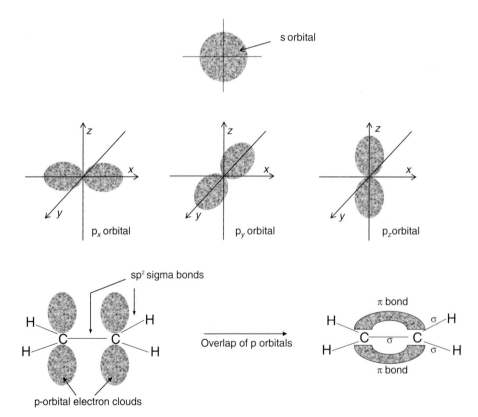

FIGURE 9-1 Bonding orbital diagrams. Top, spherical electron orbital present in the s subshell; middle, figure-eight shaped electron orbitals in the p subshell; lower, single pi bond formed by two overlapping p electron orbitals from two carbon atoms.

TABLE 9-4 Properties of Group I and Group VII Elements

Element	Atomic Number	Gram Atomic Volume (cm³)	Ionization Potential (eV)	Electronegativity
Group I				
H	1	13	13.6	2.1
Li	3	13	5.4	1.0
Na	11	24	5.1	0.9
K	19	46	4.3	0.8
Rb	37	56	4.2	0.8
Cs	55	71	3.9	0.7
Group VII				
F	9	15	17.4	4.0
Cl	17	19	13.0	3.0
Br	35	23	11.8	2.8
I	53	26	10.4	2.5

ethylene is made up of one hybrid sp^2 sigma bond and one pi bond (shown in Figure 9-1). When an sp^2 sigma bond forms in ethylene, two p orbital electrons, one from each carbon atom, orient at right angles to the sp^2 orbitals, forming a molecular orbital known as the pi (π) orbital. These orbitals overlap, forming another bond between the carbon atoms known as the pi bond. Thus, the double bond between two carbon atoms is made up of one covalent sp^2 sigma bond and one covalent pi bond. A triple bond between two carbon atoms is made up of one sp^2 bond and two pi bonds.

Pi bonds are important in radiopharmaceutical coordination compounds. ^{99m}Tc-sestamibi is one example where the d orbital electrons in the reduced technetium atom are stabilized by pi bonding orbitals in isonitrile ligands.

Coordinate Covalent Bonds

Coordinate covalent bonds are often present in organic molecules and are characteristically present in organometallic chelates, such as technetium, indium, and gallium compounds. The coordinate covalent bond is characterized by the donation of a pair of unshared electrons by one of the atoms involved in the covalent bond. Nitrogen is frequently involved in coordinate covalent bonding. Nitrogen's electronic configuration $1s^2$, $2s^2$, $2p_x^1$, $2p_y^1$, $2p_z^1$ shows that it has five valence electrons in its second principal energy level: a pair of 2s electrons and three unshared p orbital electrons. When these three p-orbital electrons are used in covalent bonding, the 2s-electrons remain and are available for coordinate covalent bonding, being donated as a pair of electrons. The coordinate covalent bond is sometimes written in chemical structures as an arrow to distinguish it from a standard covalent bond.

Coordination Compounds

A coordination compound is a complex molecular species formed by the association of two or more simpler species, each capable of independent existence. When one of the species is a metal ion, the resulting entity is known as a metal complex. In coordination chemistry the metal atom acts as an electron acceptor and is referred to as the central atom. Electron-donor atoms that bond with the metal are called coordinating groups, or ligands. Ligands ordinarily have an unshared pair of electrons that are donated to the central atom, forming a coordinate covalent bond. The number of bonds formed by a metal with a ligand is known as the coordination number. It is typically 4, 5, or 6. If the complexing agent coordinating with a metal does so through one donor atom, donating one electron pair (i.e., is monodentate), a simple metal–ligand complex is formed. If the complexing agent contains two or more donor atoms (i.e., is multidentate), forming a ring structure with the metal, the complex is called a metal chelate and the complexing agent is called a chelating agent. There is often coordination between a radionuclide metal and an organic ligand, the result being an organometallic complex. Typical donor atoms in these complexes are nitrogen, oxygen, and sulfur. The reactions shown in Figure 9-2 illustrate these complexation processes between a metal (M) and a ligand (L).

Metal chelates may involve coordinate covalent bonding between neutral and charged ligands. An example is the hexadentate chelating agent ethylenediaminetetraacetic acid (EDTA), which possesses two nitrogen donor atoms and four

TABLE 9-5 Bond Types Relative to Electronegativity Differences

Compound	Electronegativity Difference	Bond Type
Molecular hydrogen (H–H)	0	Nonpolar covalent
Water (H₂O)	1.4	Polar covalent
Lithium fluoride (LiF)	3.0	Ionic

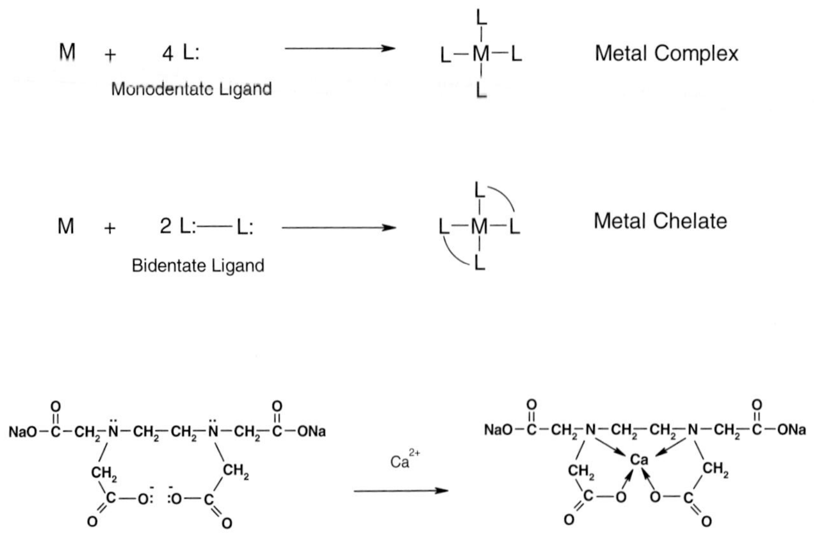

FIGURE 9-2 Generalized complexation reactions between a metal and monodentate, bidentate, and multidentate ligands.

oxygen donor atoms. Figure 9-2 shows the coordination of calcium with Na_2EDTA by coordinate covalent bonds via a pair of electrons donated from each of two nitrogen atoms and two oxygen atoms. The two positive charges on the calcium ion are neutralized by the negative charges on each of the bonding oxygens. A coordination compound may be a neutral complex or have a net ionic charge. Many natural substances are chelates: chlorophyll, hemoglobin, vitamin B_{12}, and insulin, for example. Table 9-6 gives a partial list of the types of coordination geometries found in technetium coordination compounds that are illustrated in Figure 9-3.[1] The electron configurations d^0, d^2, and d^6 given in the table refer to the number of valence electrons involved in bonding in the d sublevel of technetium. Many radiopharmaceuticals are coordination compounds, and more detailed information regarding technetium chemistry is given in Chapter 10.

SOLUTION CHEMISTRY OF METAL IONS

In general, metal ions are soluble in aqueous solution when the pH is quite low or they are complexed with a ligand. Higher pH typically results in hydrolysis of metals that are not complexed. If the pH of a radiolabeling reaction is likely to result in metal ion hydrolysis, some adjustment in the formulation must be made. For example, stannous chloride is often used in the preparation of technetium radiopharmaceuticals, which typically occurs at slightly acidic or neutral pH. However, a fairly low pH is required to prevent tin hydrolysis. For example, when stannous chloride is dissolved in hydrochloric acid, a soluble hexacoordinate tin–chloro coordination complex is formed, which keeps tin soluble:

$$SnCl_2 + 4Cl^- \rightarrow [SnCl_6]^{4-}$$

When the pH is raised to between 1.2 and 4.5, the insoluble hydrolysis product tetratin (II) hexahydroxide dichloride, $Sn_4(OH)_6Cl_2$, is formed. At pH higher than 5.5, an amorphous hydrous oxide, $Sn_5O_3(OH)_4$, is formed.[2] Formation of these insoluble hydroxide complexes can cause co-precipitation of reduced technetium, resulting in low labeling efficiency. Hydrolysis can be prevented during the preparation of technetium radiopharmaceuticals by inclusion of a chelating agent in the formulation to sequester stannous ion as a soluble metal chelate.

Many radiopharmaceuticals are chelates of radionuclide metals, which are prone to hydrolysis reactions similar to those with tin. This fact was used to advantage in the early years of radiopharmaceutical development, when the hydrolysis reac-

TABLE 9-6 Coordination Geometries of Representative Technetium Compounds

Technetium Oxidation State	Electron Configuration	Coordination Number	Tc-Complex Geometry	Example
7+	d^0	4	Tetrahedron	Tc-pertechnetate
5+	d^2	5	Square pyramid	Tc-mertiatide
1+	d^6	6	Octahedron	Tc-sestamibi

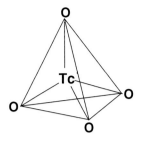
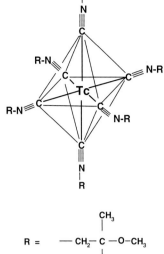

FIGURE 9-3 Coordination geometries of technetium complexes. Tetrahedron (^{99m}Tc-pertechnetate), square pyramid (^{99m}Tc-mertiatide), octahedron (^{99m}Tc-sestamibi).

tion was deliberately employed to produce particulate radiotracers for diagnostic procedures. Some examples are the hydrolysis of ^{113m}In-chloride with sodium hydroxide or sodium phosphate to produce insoluble ^{113m}In-hydroxide or ^{113m}In-phosphate colloids for liver imaging.[3,4] Another example is the production of ^{113m}In-iron hydroxide macroaggregates for lung imaging by raising the pH of ^{113m}In-chloride with sodium hydroxide in the presence of ferric chloride.[4]

Chelation suppresses certain chemical reactions of a metal ion without its removal from the system. Trace metal ion contamination of drug products may contribute to oxidative degradation of a drug molecule. EDTA and sodium citrate are often used in nonradioactive pharmaceuticals to sequester metal ions that can enhance drug degradation. An example is the use of EDTA in epinephrine injection to retard metal ion–induced auto-oxidation of epinephrine to an inactive product, adrenochrome. An example in nuclear pharmacy is the chelation of aluminum ion by EDTA in ^{99m}Tc-sulfur colloid injection. The chelation effectively retards the potential reaction of any aluminum ion present in ^{99m}Tc generator eluates with phosphate buffers in the sulfur colloid kit. Formation of insoluble aluminum phosphate would co-precipitate the sulfur colloid particles in ^{99m}Tc sulfur colloid, making this radiopharmaceutical an ineffective radiotracer.

Another example, shown here diagrammatically, is the use of a transfer ligand (TL) in technetium kits to form a weak intermediate complex with reduced technetium (Tc), preventing the formation of the insoluble hydrolysis product $TcO_2 \cdot 2H_2O$. The Tc–TL complex maintains reduced technetium in a soluble form that can readily complex with the coordinating ligand (Rx). This technique is employed in the ^{99m}Tc-mertiatide kit, with tartrate functioning as the transfer ligand.

$$TcO_4^- + TL \xrightarrow{Sn^{2+}} Tc-TL \xrightarrow{Rx} Tc-Rx$$

Still another example of preventing a metal hydrolysis reaction is in the preparation of ^{111}In-labeled antibodies. Antibody labeling with indium occurs near pH 5; however, ^{111}In-indium chloride is available only in hydrochloric acid at pH 1. Typically the indium chloride is mixed with sodium acetate buffer before addition to the antibody. This forms indium acetate, which keeps indium in a soluble form while it coordinates with the antibody. Inadequate labeling and antibody degradation can occur if the acidic indium chloride is added directly to the antibody.

In addition to technetium coordination compounds, several other radiopharmaceuticals are metal chelates. These include ^{111}In-pentetate, ^{111}In-oxine, ^{111}In-pentetreotide, ^{67}Ga-citrate, and ^{153}Sm-lexidronam. Understanding the chemistry involved in radionuclide metals is important to radiolabeling success.

OXIDATION–REDUCTION CHEMISTRY

Oxidation–reduction (redox) reactions are important in the preparation of certain radiopharmaceuticals, particularly those labeled with technetium and iodine. Redox reactions are also important because they can be involved in the degradation of many substances. Redox reactions involve an exchange of electrons between reacting species in solution, producing a change in the oxidation state of reacting atoms. *Oxidation* is an increase in the oxidation state of a substance through loss of electrons, gain of oxygen, or loss of hydrogen. *Reduction* is a decrease in the oxidation state of a substance through gain of electrons or hydrogen or loss of oxygen. When one substance is oxidized, a complementary substance must be reduced. There is no net loss or gain of electrons in a redox reaction. Table 9-7 summarizes oxidation–reduction terminology. The oxidation state or valence of an element in a redox reaction is the charge it acquires when it gains or loses one or more electrons. Oxidation states of the chemical elements are given in Table 9-1.

The formation of an electrovalent bond in lithium fluoride was discussed previously. When a neutral lithium atom reacts with a neutral fluorine atom under appropriate conditions, a redox reaction occurs in the formation of lithium fluoride. The lithium atom, with atomic number 3, has three protons in its nucleus. With its full complement of electrons, lithium has the electron configuration $1s^2 2s^1$. In this state it has zero charge and a single electron in its outermost electron shell. Being weakly electronegative, lithium can readily lose that electron through oxidation, becoming a lithium ion with net charge of 1+. The fluorine atom, having atomic number 9 and a full complement of electrons, has the electron configuration $1s^2 2s^2 2p^5$. In this state it has zero charge with seven electrons in its second principal electron shell. Being strongly electronegative, a fluorine atom readily acquires the one electron from lithium and becomes reduced to form a fluoride ion with a net charge of 1−. The two oppositely charged ions attract each other to form LiF by an electrovalent bond. In the process a redox reaction occurs, with a change in valence or oxidation state of each atom. Lithium is oxidized and fluorine is reduced. Lithium is the reducing agent and fluorine is the oxidizing agent. The reactions are represented in Figure 9-4 as electronic dot configurations showing the valence electrons in the outermost shell of each reactant.

In a manner similar to LiF, hydrogen fluoride (HF) can form by the reaction of hydrogen and fluorine. However, the bond in HF is covalent because of sharing of electrons between hydrogen and fluorine. The sharing of electrons is unequal,

TABLE 9-7 Oxidation–Reduction Terminology

Term	Oxidation State	Electron Change
Oxidation	Increases	Loss of electrons
Reduction	Decreases	Gain of electrons
Oxidizing agent	Decreases	Acquires electrons
Reducing agent	Increases	Supplies electrons
Substance oxidized	Increases	Loses electrons
Substance reduced	Decreases	Gains electrons

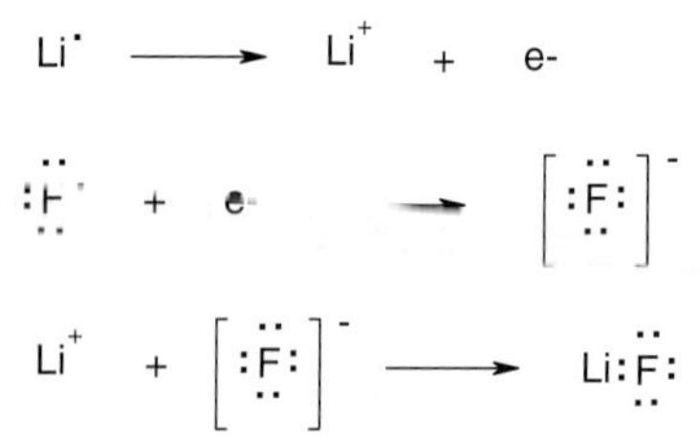

FIGURE 9-4 Oxidation–reduction reaction in the formation of an ionic bond between lithium and fluorine in lithium fluoride.

however, because the electronegativity of fluorine (4.0) is greater than that of hydrogen (2.1). Therefore, the bond in HF is polar covalent. A redox reaction also occurs in the formation of HF even though there is no actual exchange of electrons. A change in oxidation state of the reacting atoms occurs because of unequal sharing of electrons. Thus, the oxidation number assigned to hydrogen is 1+ and that assigned to fluorine is 1−.

Technetium radiopharmaceuticals are often used in nuclear medicine. The starting material for preparing technetium radiopharmaceuticals is sodium pertechnetate. This compound is characterized by both ionic bonding and covalent bonding. Technetium forms a polar covalent bond with four oxygen atoms, and this technetium–oxo complex becomes negatively charged when it acquires the $3s^1$ valence electron from a sodium atom. An ionic bond is formed between a sodium cation and a pertechnetate anion. Pertechnetate is a complex ion in which technetium is oxidized by oxygen. Technetium acquires an oxidation state of 7+ because of unequal sharing of electrons with oxygen, which is more electronegative than technetium.

Technetium Redox Reactions

Redox reactions important in nuclear pharmacy involve primarily technetium and iodine compounds. With technetium compounds the general redox reactions involve the reduction of technetium pertechnetate by stannous chloride. The technetium atom in pertechnetate is in its highest oxidation state of 7+. The 7+ charges on the technetium atom, combined with the eight negative charges of oxygen (each oxygen having a valence of 2−), give pertechnetate a net charge of 1−. The oxidation state of technetium found in most technetium radiopharmaceuticals is 5+, 4+, 3+, or 1+, depending on the coordinating ligand and reaction conditions. The following half reactions illustrate typical redox reactions in technetium chemistry.

$$Sn^{2+} \rightarrow Sn^{4+} + 2e^- \qquad (9\text{-}1)$$

$$TcO_4^- + 2e^- + 6H^+ \rightarrow TcO^{3+} + 3H_2O \qquad (9\text{-}2)$$

The sum of reactions 9-1 and 9-2 is as follows:

$$TcO_4^- + Sn^{2+} + 6H^+ \rightarrow TcO^{3+} + Sn^{4+} + 3H_2O \qquad (9\text{-}3)$$

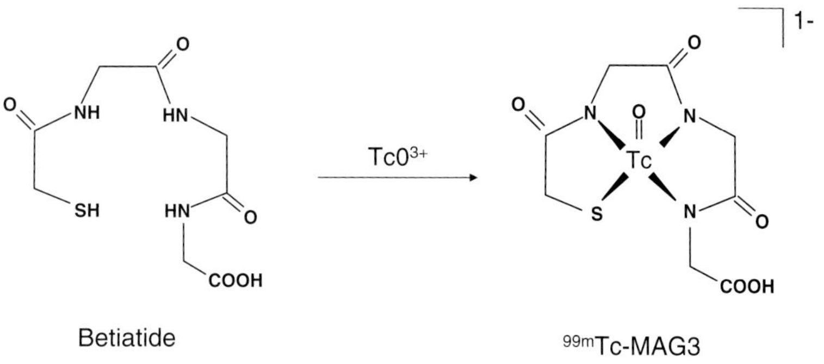

FIGURE 9-5 Coordination of a technetium–oxo core TcO^{3+} with the betiatide ligand in the ^{99m}Tc-mertiatide (^{99m}Tc-MAG3) kit.

In this reaction sequence, each atom of stannous tin (Sn^{2+}) is oxidized to stannic tin (Sn^{4+}) by losing two electrons. These two electrons reduce the technetium atom from the 7+ oxidation state in pertechnetate to the 5+ oxidation state in TcO^{3+}. The net charge of 3+ on this reduced technetium–oxo species is the sum of 2− on the oxygen atom and 5+ on the technetium atom. There is a net exchange of two electrons between the reducing agent (stannous tin) and the oxidizing agent (pertechnetate). When this reaction occurs in a radiopharmaceutical kit, the reduced technetium–oxo core TcO^{3+} is complexed by the coordinating ligand in the kit, forming the desired radiopharmaceutical. An example is the preparation of ^{99m}Tc-mertiatide (^{99m}Tc-MAG3) illustrated in Figure 9-5, where betiatide is the coordinating ligand in the kit.

Iodine Redox Reactions

Radioiodination reactions are another example of redox reactions. Iodine is a member of the group VII elements, the halogens, which have seven valence electrons. Iodine's electron configuration is $[Kr]4d^{10}5s^25p^5$. The halogens have high electronegativity and form negative halide ions typically found in ionic salts. All halogens have positive oxidation states achieved by a loss of electrons. In contrast to technetium chemistry, which requires technetium reduction prior to complexation reactions, radioiodination reactions require iodide oxidation. Radioiodine is usually obtained as sodium iodide. Under certain conditions, iodide can lose electrons, becoming oxidized as follows:

$$2I^- \rightarrow I_2 + 2e^- \rightarrow 2I^+ + 2e^- \quad (9\text{-}4)$$

The significant oxidative reactions that can occur in radioiodide solutions are as follows:[5–9]

$$4I^- + O_2 + 4H^+ \leftrightarrow 2I_2 + 2H_2O \quad (9\text{-}5)$$

$$2HI + H_2O_2 \leftrightarrow I_2 + 2H_2O \quad (9\text{-}6)$$

$$2I^- + 2OH^{\cdot} \leftrightarrow I_2 + 2OH^- \quad (9\text{-}7)$$

Oxidation of iodide to molecular iodine (I_2) in aqueous solution is favored in acidic pH and in the presence of hydrogen peroxide or hydroxyl radicals, which may be present as radiolysis byproducts. Molecular iodine is volatile and poses a potential radiologic biohazard. Thus, steps must be taken to reduce the formation of volatile radioiodine in aqueous solutions. Reaction 9-5 can be effectively retarded by buffering the radioiodide solution to an alkaline pH. In addition, using reducing agents such as thiosulfate and bisulfites will reverse these reactions to maintain the iodine in the nonvolatile iodide state. The reduction reaction with thiosulfate is as follows:

$$I_2 + 2[S_2O_3]^{2-} \rightarrow 2I^- + [S_4O_6]^{2-} \quad (9\text{-}8)$$

The indirect oxidation of iodide by free radicals and peroxides is more likely to occur at high radioactive concentrations and in ^{131}I solutions because of the significant amount of beta radiation. These radiation-induced reactions are difficult to prevent entirely, but they can be minimized by lowering the radioactive concentration and using radical scavengers and antioxidants.[10,11] Oral solutions of ^{131}I-sodium iodide for clinical use can be stabilized in various ways, which include adjusting solution pH to 7.5 to 9.0 with disodium phosphate, adding antioxidants such as sodium bisulfite or sodium thiosulfate, and adding a chelating agent such as disodium EDTA. Chelating agents can retard oxidative catalysis induced by trace metal ions in solution.[5] Unstabilized solutions of ^{123}I-, ^{125}I-, and ^{131}I-sodium iodide as radiochemicals are also available in sodium hydroxide solution for radioiodination. Radiopharmaceutical radioiodination is discussed in Chapter 11.

Oxidative reactions are accelerated by heat and light, and since iodine has a low solubility in water (0.03% at 20°C), refrigerated storage will lower its volatility. The chlorine in tap water will oxidize radioiodide to volatile radioiodine; therefore, dilutions of radioiodide solutions should be made with distilled water.[12] Solutions prepared for long-term storage can be stabilized with 0.2% to 1.0% sodium thiosulfate, a reducing agent that reverts volatile iodine to stable iodide according to reaction 9-8.

Oral ^{131}I-sodium iodide solution for therapy procedures in nuclear medicine should be handled with precautions

TABLE 9-8 Recommendations for Safe Handling of Radioiodide Solutions

1. Wear disposable gloves when handling radioiodide solutions
2. Wipe test solution vials for removable contamination upon receipt
3. Open solution vials and perform all radioiodinations in an exhaust hood
4. Dilute solutions for immediate use with distilled water
5. For prolonged storage, buffer solutions to pH 7.0–8.5 and add 0.25% sodium thiosulfate antioxidant
6. Store solutions in a cool environment to reduce volatility
7. Perform bioassays on personnel who handle 33 mCi or more of radioiodine

TABLE 9-9 Ideal Properties of Radionuclides for Diagnostic Imaging

1. *Decay mode:* Electron capture or isomeric transition from metastable isomers; no particulate radiation; gamma or x-rays only
2. *Photon energy:* 100–200 keV is ideal
   Below 100 keV = tissue absorption and scatter (decreases resolution)
   Above 200 keV = lower detection efficiency (decreases sensitivity)
3. *Half-life:* Effective half-life equal to 1 to 1.5 times the imaging time
4. *Chemical properties:* Can be compounded into different chemical forms

against external contamination. It is recommended that the screw-cap bottles of oral iodide solution be wipe tested directly with a cotton-tipped applicator to detect any contamination on the external surface of the bottle. In rare instances bottles with loose caps and secure caps have been found to be grossly contaminated on their external surfaces. Wearing disposable rubber gloves is a necessity to prevent potential skin contamination. Table 9-8 lists some recommendations for safe handling of radioiodide solutions.

IDEAL PROPERTIES OF RADIOPHARMACEUTICALS

A radiopharmaceutical has either diagnostic or therapeutic application. Diagnostic radiopharmaceuticals are used to gain information about disease processes and can be divided into two categories of use: in vivo function agents and imaging agents. Therapeutic radiopharmaceuticals are used to treat disease with radiation.

Diagnostic Radiopharmaceuticals

By definition, a tracer is something that follows something else. A tracer is a labeled substance that can be measured and acts the same as the substance being followed. The tracer can be identical to the substance being followed, or it can simply act in the same manner. To be effective, the tracer must not interfere with the system being studied. Radiopharmaceuticals are often called radiotracers; however, only a few are true tracers of a physiologic process. A radiopharmaceutical for assessing in vivo function traces a physiologic process or space to measure its function or capacity. Noteworthy examples are the measurement of thyroid gland function with ^{131}I sodium iodide, measurement of glomerular filtration rate (GFR) with ^{99m}Tc-diethylenetriaminepentaacetic acid (^{99m}Tc-DTPA or ^{99m}Tc-pentetate) or ^{125}I-iothalamate, the determination of red blood cell volume with ^{51}Cr-labeled red blood cells, and the measurement of plasma volume with ^{125}I-labeled human serum albumin. With in vivo function studies, the radioactive agent is administered to a patient and the physiologic process is assessed by counting radiation emitted from the organ of interest, for example, with thyroid gland function studies, or by counting blood samples, such as with GFR studies or blood volume studies.

Diagnostic imaging accounts for a majority of nuclear medicine studies. Radiopharmaceuticals used for this purpose localize in specific organs of interest and emit gamma radiation. Gamma camera images of radiotracer distribution within the organ can then be obtained to assess organ morphology (size, shape, position, or presence of space-occupying lesions) and function. An ideal imaging agent should rapidly and avidly localize in the organ of interest, remain there for the duration of study, and be quickly excreted from the body thereafter. No single agent currently meets these stringent requirements, and a judicious selection of both radionuclide and chemical form must be combined to achieve the best compromise. Table 9-9 lists several properties of an ideal radionuclide for diagnostic imaging.

Decay Mode and Energy

Electromagnetic radiations (gamma rays and x-rays) are the most suitable forms of radiation for external detection. Particulate radiation, being completely absorbed by tissue, cannot be detected externally and only increases radiation burden to the patient. The most desirable decay modes for diagnostic imaging, therefore, are electron capture and isomeric transition, since these modes have no primary particulate emissions associated with them. It is also desirable that the decay mode produce a high photon abundance. The characteristics of the decay process may make a radionuclide undesirable for diagnostic imaging even though the decay mode is ideal. A good example is ^{51}Cr. ^{51}Cr decays to ^{51}V by two electron capture (EC) routes, EC_1 and EC_2 (Figure 2-13, Chapter 2). The EC_2 route occurs in 90% of decays to the ground state of ^{51}V, with no photon emission. The EC_1 route occurs in 10% of decays to the excited state of ^{51}V, which subsequently de-excites to

TABLE 9-10 Photon Detection Efficiency in a Half-Inch Sodium Iodide Crystal

Radionuclide	Photon Energy (keV)	Detector Efficiency (%)
^{133}Xe	81	92
^{99m}Tc	140	86
^{111}In	172	73
	247	45
^{131}I	364	23
Positron emitters	511	13

Source: Anger HO. Radioisotope cameras. In: Hine GJ, ed. *Instrumentation in Nuclear Medicine.* New York: Academic Press; 1967: 485.

release 0.320 MeV photons. Thus, only 10 photons are produced in 100 decays of ^{51}Cr. This low photon abundance makes ^{51}Cr unsatisfactory for imaging procedures. Another imaging consideration is internal conversion. Internal conversion should be minimal to maximize the yield of detectable photons. For example, the 38% photon abundance of the 0.081 MeV gamma ray from ^{133}Xe makes it a less desirable imaging agent than ^{99m}Tc, which has an 88% abundance of its 0.140 MeV gamma ray. There is a higher percentage of photon loss due to the production of conversion electrons with ^{133}Xe than with ^{99m}Tc.

The energy of gamma rays should be high enough to readily penetrate and escape from the body with minimal scatter, yet be low enough for efficient detection by the gamma camera detector. Table 9-10 lists the detection efficiencies of various radionuclide photon energies in a sodium iodide detector. Comparing ^{133}Xe with ^{99m}Tc, it is evident that the photons from ^{133}Xe have higher detection efficiency than those from ^{99m}Tc. However, the lower-energy photons from ^{133}Xe are more easily attenuated in tissue, offsetting their higher detection efficiency.

Half-life

The effective rate of loss (R_{eff}) of radioactivity from an organ or the body is directly proportional to the rates of physical decay (R_p) of the radionuclide and of biologic excretion (R_b) of the radiopharmaceutical as shown in Equation 9-9:

$$R_{eff} = R_p + R_b \quad (9\text{-}9)$$

In accordance with Equation 9-10, the rate of removal by either process is inversely proportional to the half-life of the process:

$$R(\text{removal rate}) \propto \frac{1}{T_{1/2}} \quad (9\text{-}10)$$

Combining Equations 9-9 and 9-10, we have the following relationships:[13]

$$\frac{1}{T_{eff}} = \frac{1}{T_p} + \frac{1}{T_b} \quad (9\text{-}11)$$

or

$$T_{eff} = \frac{T_p \times T_b}{T_p + T_b} \quad (9\text{-}12)$$

The effective half-life is therefore the time required to remove half of the radioactivity from an organ by a combination of physical decay and biologic elimination. The T_{eff} is always less than either the T_p or T_b, but it will be nearly equal to the smaller function when the other is very large, as illustrated in the following example, when (1) T_p is 1 hour and T_b is 10 hours, (2) T_p and T_b are both 10 hours, and (3) T_p is 10 hours and T_b is 1 hour:

1. $T_{eff} = \dfrac{1 \times 10}{1 + 10} = \dfrac{10}{11} = 0.91 \text{ hour}$

2. $T_{eff} = \dfrac{10 \times 10}{10 + 10} = \dfrac{100}{20} = 5.0 \text{ hours}$

3. $T_{eff} = \dfrac{10 \times 1}{10 + 1} = \dfrac{10}{11} = 0.91 \text{ hour}$

From an imaging standpoint, the optimum effective half-life should be about 1 to 1.5 times the period of observation or study time. This time provides enough radioactivity for acceptable counting statistics and a removal rate that diminishes the radiation dose to the patient. In practice, however, it is difficult to achieve an optimum balance between imaging time and T_{eff}. Typically, many radiopharmaceuticals have a prolonged biologic retention, and use of a short-lived radionuclide is the best way to limit radiation dose. Fortunately, most of the clinical procedures performed in nuclear medicine employ ^{99m}Tc, which has a short physical half-life of 6 hours.

The type of study sometimes dictates the half-life of the radionuclide selected for use. For example, cisternography studies are performed to assess cerebrospinal fluid dynamics. A standard cisternogram to evaluate hydrocephalus may require 2 to 3 days for completion, in which case the agent of choice would be ^{111}In-DTPA, which has a 2.8 day T_p. Six-hour ^{99m}Tc agents are generally unsatisfactory for studies beyond 1 day, although ^{99m}Tc-DTPA may be a satisfactory choice for cerebrospinal fluid leak studies, which require only a few hours to complete.

Radiopharmaceuticals with a long T_{eff} cause radioactivity to linger in the body, which may interfere with subsequent diagnostic studies using radioactivity. For example, when performing in vivo function studies that require blood or urine samples, it is standard practice to obtain background blood or urine samples before administration of the radiopharmaceutical. This will allow any residual radioactivity in the body that may be present from a previous study to be accounted for during the analysis.

Radionuclide Chemistry

A radionuclide ideally should have chemical properties that allow it to be compounded into a variety of chemical forms useful for imaging studies. Radionuclides such as radioiodine have been quite useful because of iodine's diverse chemistry. Free radioiodide ion has been useful for thyroid gland studies, but radioiodine can be covalently labeled to molecules for other applications. For example, radioiodinated meta-iodobenzylguanidine (^{123}I- or ^{131}I-MIBG), a metabolic analogue of norepinephrine, is useful for imaging neuroendocrine tumors, whereas radioiodinated proteins and antibodies are useful for other applications. Technetium has been incorporated into a large number of chemical compounds that are used in a majority of nuclear medicine studies. Indium's complexation chemistry has yielded several useful radiopharmaceuticals. Indium-111, with a half-life of 2.8 days, is particularly suited for labeling antibodies that require imaging over several days.

Therapeutic Radiopharmaceuticals

Some radiopharmaceuticals are used to deliver a therapeutic dose of radiation to diseased tissue. Ideally, they must be designed to localize in the organ to be treated, with limited distribution to surrounding tissues, so that the radiation dose is maximized in the target organ and minimized in normal organs. The principal radiation for therapeutic applications is beta radiation, which has a short range in tissue. This topic is discussed in more detail in Chapter 29.

RADIOLABELING

Figure 9-6 illustrates the elements in the periodic table that have isotopes with suitable radioactive decay and chemical properties for the preparation of diagnostic or therapeutic radiopharmaceuticals. The most widely used radionuclides in nuclear medicine are ^{18}F, ^{32}P, ^{51}Cr, ^{57}Co, ^{67}Ga, ^{82}Rb, ^{90}Y, ^{99m}Tc, ^{111}In, ^{123}I, ^{125}I, ^{131}I, ^{133}Xe, ^{201}Tl, and ^{153}Sm. Incorporation of several of these nuclides into radiopharmaceutical compounds involves two types of radiolabeling: isotopic labeling and nonisotopic labeling.

Isotopic Labeling

Isotopic labeling involves the substitution of a stable atom in a compound with its radioisotope, yielding a radioactive analogue. The radioactive analogue has similar chemical and biologic properties to the parent compound, and thus it is a true physiologic tracer. A good example of this in nuclear medicine is radioactive iodide (^{131}I⁻ or ^{123}I⁻) as a radioactive analogue of stable iodide ^{127}I for thyroid gland studies. Radioactive iodide is indistinguishable from stable iodide in its biochemical reactions in the thyroid gland. Other examples are the substitution of stable phosphorus ^{31}P with radioactive phosphorus ^{32}P in ^{32}P-sodium phosphate and ^{32}P-chromic phosphate for use in therapeutic procedures in nuclear medicine.

Although isotopic labeling is an ideal approach to radiopharmaceutical preparation, it has practical limitations. Biologic molecules and drugs of interest are composed mostly of the elements carbon, hydrogen, oxygen, nitrogen, phosphorus, and sulfur. The main limitation of these elements is that most of their radioisotopes have undesirable radioactive properties for clinical use, such as unsatisfactory half-life, decay mode, or photon energy. However, the development of positron emission tomography (PET) has led to useful radiopharmaceuticals labeled with the positron emitters ^{11}C, ^{13}N, and ^{15}O. Although the number of potential molecules that can be synthesized with these isotopes is virtually unlimited, a practical limitation is the need for an on-site cyclotron and radiosynthesis facilities because of the very short half-lives of these nuclides. Table 9-11 lists the physical properties of radionuclides for isotopic labeling.

Another potential issue with isotopic labeling is that the substitution of one isotope for another in a molecule may produce isotopic effects; that is, reaction rates may differ because of mass differences of the isotopes. The isotopic effect is more likely to occur with small molecules and if the radioisotopic label is located at the reaction site in the molecule. However,

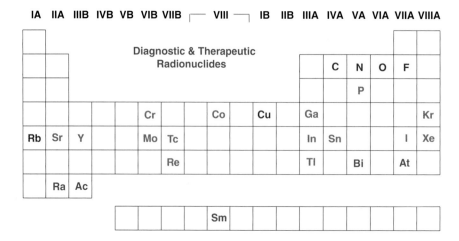

FIGURE 9-6 Elements with radioisotopes useful for diagnostic and therapeutic applications. Elements in red undergo positron decay, those in blue undergo alpha decay, and those in green undergo negatron decay, electron capture decay, or isomeric transition.

TABLE 9-11 Properties of Isotopic Labels for Biologic Molecules

Nuclide	Decay Mode	Half-life	Photon Energy
^{3}H	β–	12.32 yr	No photons
^{14}C	β–	5715 yr	No photons
^{32}P	β–	14.28 days	No photons
^{35}S	β–	87.2 days	No photons
^{11}C	β+	20.3 min	511 keV
^{13}N	β+	9.97 min	511 keV
^{15}O	β+	2.04 min	511 keV

the isotopic effect does not appear to be a concern with radiopharmaceuticals used in nuclear medicine studies.

Nonisotopic Labeling

Nonisotopic or "foreign" labeling involves incorporating a radioactive atom that is not native to the compound being labeled. Nonisotopic labeling is not ideal because the presence of a foreign atom in a molecule may alter its biochemical properties. Small molecules may be affected more than large molecules. In vivo localization of the radiolabeled molecule can be affected if the radionuclide is located at a site in the molecule involved with in vivo receptor interaction. Despite this potential problem, however, many useful radiopharmaceuticals with foreign labels have been developed. Some examples are simple metallic coordination compounds, such as ^{99m}Tc- and ^{111}In-labeled DTPA, in which the chelating agent is primarily responsible for nuclide localization in vivo. More sophisticated coordination compounds have also been developed in which the radionuclide metal is remotely coordinated to a bifunctional chelating agent attached to a targeting molecule such as a peptide or antibody. Another design approach is to build the coordination compound around the nuclide of interest. In this situation the radionuclide becomes an integral component in the molecular structure of the coordination compound and essential for its in vivo localization. Examples include technetium-essential compounds such as ^{99m}Tc-mebrofenin and ^{99m}Tc-mertiatide.

Some nonisotopically labeled compounds can act as if they were isotopically labeled. For example, the biochemical properties of a large molecular weight protein may not be significantly affected if the number of radioactive atoms incorporated in the protein is small. An example is ^{125}I-human serum albumin (^{125}I-HSA), which has no more than one radioiodine atom per molecule of albumin. ^{125}I-HSA is useful for measuring plasma volume in vivo because its metabolism is similar to that of native albumin. Another example is the radiofluorinated analogue of deoxyglucose, ^{18}F-fludeoxyglucose (^{18}F-FDG). ^{18}F-FDG participates in the first-step metabolism of glucose in vivo, which makes it useful for assessing disease processes that involve alterations in glucose metabolism. These two compounds do not behave identically to the native unlabeled species, but they have similar enough properties to be useful as biologic markers.

Specific Activity

The amount of a radioactive substance can be expressed in activity units or in concentration units. Activity units are typically given in curies (Ci), millicuries (mCi), and microcuries (µCi), or in becquerels (Bq). Units of activity are discussed in Chapter 2. Specific activity is the ratio of the radioactivity to the mass of a radionuclide or a radioactive compound. Typical units might be microcuries per microgram (µCi/µg) or curies per millimole (Ci/mmol). The mass of a radionuclide in a given amount of activity can be determined from the number of radioactive atoms present. By applying Avogadro's number and the GAW of the nuclide, the number of atoms can be converted into grams. In the formula $A = \lambda N$, A is activity, in Ci, mCi, or µCi; N is the number of atoms or disintegrations; and λ is 0.693 divided by the half-life. Recall that there are Avogadro's number (6.023×10^{23}) of atoms or disintegrations per GAW of an element.

A question sometimes asked is "What mass of iodine is present in a 10 µCi (370 kBq) dose of ^{131}I-sodium iodide for a thyroid uptake study?" The calculation is as follows:

$$T_{\frac{1}{2}} \text{ of } ^{131}\text{I} = (8.02 \text{ days})(24 \text{ hr/day})(3600 \text{ sec/hr}) = 6.93 \times 10^5 \text{ sec}$$

$$N = \frac{A}{\lambda} = \frac{(A)(T_{\frac{1}{2}})}{0.693} = (A)(T_{\frac{1}{2}})(1.443)$$

$$N = (10 \mu Ci)(3.7 \times 10^4 \text{ dps}/\mu Ci)(6.93 \times 10^5 \text{ sec})(1.443)$$

$$N = 3.70 \times 10^{11} \text{ disintegrations or atoms}$$

The mass in grams of this many atoms is calculated from Avogadro's number and the GAW of ^{131}I, as follows:

$$\frac{3.70 \times 10^{11} \text{ atoms} \times 131 \text{ grams/GAW}}{6.023 \times 10^{23} \text{ atoms/GAW}} = 8.05 \times 10^{-11} \text{ gram}$$

This trace amount of radioiodine represents only about one four-millionth of the average daily dietary intake of iodine (300 µg) and about one eighty-millionth of the body's iodine stores of 6.5 mg.

The theoretical specific activity (SA), in millicuries per microgram, of any nuclide can be determined from the formula $A = \lambda N$, which is modified to yield Equation 9-13.

$$SA (mCi/\mu g) = \frac{(0.693)(6.023 \times 10^{23} \text{ d/mole})}{(3.7 \times 10^7 \text{ dps/mCi})(T_{\frac{1}{2}} \text{sec})} \quad (9\text{-}13)$$
$$(GAW \text{ grams/mole})(10^6 \mu g/\text{gram})$$

Example: Calculate the specific activity of isotopically pure ^{131}I in millicuries per microgram. The half-life of ^{131}I is 8.02 days.

$$\text{SA }^{131}\text{I (mCi/µg)} = \frac{(0.693)(6.023 \times 10^{23} \text{ d/mole})}{(3.7 \times 10^7 \text{ dps/mCi})(T_{1/2} \text{ sec})}$$
$$(\text{GAW grams/mole})(10^6 \text{ µg/gram})$$

$$= \frac{(0.693)(6.023 \times 10^{23} \text{ d/mole})}{(3.7 \times 10^7 \text{ dps/mCi})(6.93 \times 10^5 \text{ sec})}$$
$$(131 \text{ grams/mole})(10^6 \text{ µg/gram})$$

$$= 124 \text{ mCi}(4.6 \text{ GBq})/\text{µg }^{131}\text{I}$$

This high level of purity of ^{131}I is difficult to achieve because of the limitations of chemical methods.

In addition to the radionuclide of interest, radioactive samples may contain stable isotopes of the element. The presence of other isotopes will increase the sample mass and decrease the specific activity of the radionuclide. Thus, a pure sample of ^{131}I will have a specific activity of 124 mCi/µg, but a sample containing 50% stable ^{127}I will have a specific activity of only 62 mCi/µg. The decay constant of a radionuclide varies inversely with half-life. Thus, radionuclides with long half-lives (small decay constants) have a greater number of atoms (more mass) per unit of activity compared with short half-life nuclides, and therefore a lower specific activity. Table 9-12 lists the maximum specific activities of iodine radioisotopes and their mass per millicurie, which is the inverse of their specific activities.

An example of when specific activity is an important consideration in nuclear medicine is in the dosing of ^{99m}Tc-labeled macroaggregated albumin (^{99m}Tc-MAA) for perfusion lung imaging. The number of ^{99m}Tc-MAA particles administered is an important consideration in certain situations, such as with pediatric patients, who have underdeveloped lungs; patients with pulmonary hypertension, who have compromised lung function; and patients who have a right-to-left cardiac shunt, where particles will be shunted directly to the systemic arterial circulation. In these situations, a minimum number of particles should be administered for the nuclear medicine procedure. The number of particles administered to a patient from a vial of ^{99m}Tc-MAA will vary with time because of the decrease in specific activity over time. For example, if a kit of ^{99m}Tc-MAA containing 6 million particles (2 mg MAA) is reconstituted with 60 mCi of ^{99m}Tc-sodium pertechnetate, the initial specific activity will be 30 mCi/mg of MAA and a 3 mCi dose will contain 300,000 particles in 0.5 mL. Table 9-13 shows that a decline in activity over time due to radioactive decay will cause a proportionate decrease in specific activity. Consequently, a proportionately larger volume

TABLE 9-12 Specific Activities (SA) of Radioiodines

Isotope	Half-life	SA (mCi/µg)	Mass (µg/mCi)
^{123}I	13.2 hours	1930.0	5.2×10^{-4}
^{131}I	8.02 days	124.0	8.1×10^{-3}
^{125}I	59.4 days	17.6	5.7×10^{-2}

TABLE 9-13 Effect of Time on Specific Activity and Dosage Amounts of ^{99m}Tc-MAA[a]

		3 mCi Dose of ^{99m}Tc-MAA	
Time	Specific Activity	No. MAA Particles	Volume
0 hr	30.0 mCi/mg	300,000	0.50 mL
3 hr	21.2 mCi/mg	424,233	0.71 mL
6 hr	15.0 mCi/mg	600,000	1.00 mL

[a] When kit containing 2 mg of MAA equivalent to 6 million MAA particles is prepared with 60 mCi ^{99m}Tc-sodium pertechnetate.

and number of ^{99m}Tc-MAA particles will be needed to obtain the same 3 mCi dose of activity at any time after initial radiopharmaceutical preparation.

The label on most radiopharmaceutical vials provides information on specific activity and radioactive concentration. From this information one can determine the amount of labeled compound in a unit volume as follows:

$$\frac{\text{Radioactive concentration (mCi/mL)}}{\text{Specific activity (mCi/mg)}} = \frac{\text{Drug concentration}}{\text{(mg/mL)}}$$

In the field of radiochemistry, the small chemical amount of a radioisotope in a sample can create a problem with its recovery during chemical processing, primarily because of adsorption losses on glassware surfaces. To mitigate this problem, radiochemists add an amount of stable isotope of the radionuclide being analyzed to the radioactive sample to minimize adsorption and "carry" it through the chemical process. The stable isotope added is called a "carrier." The term "carrier-free" originally indicated radioactive preparations with no isotopic carrier intentionally added and no isotopic material detectable by chemical or spectrographic means.[14] Over time, the term carrier-free became misunderstood and misused in relation to its original definition. It came to mean that a radionuclide sample contained only the radionuclide of interest, one of absolute theoretical specific activity. However, because it is questionable whether such samples can actually be achieved given the limitations of radiochemical methods, new terminology was proposed.[14] *Carrier-free* (CF) now indicates a radionuclide or stable nuclide that is not contaminated with any other stable or radioactive nuclide of the same element. *No carrier added* (NCA) indicates an element or compound to which no carrier of the same element has been intentionally or otherwise added during its preparation. *Carrier added* (CA) indicates any element or compound to which a known amount of carrier has been added. Thus, a preparation that is termed CF is, by definition, NCA. However, one that is designated NCA may not necessarily be CF, because it may contain stable or radioactive isotopic contaminants that were not intentionally added but are present because of the limitations of production and purification processes. Most radiotracer nuclide preparations are NCA.

IN VIVO LOCALIZATION OF RADIOPHARMACEUTICALS

Diagnostic and therapeutic procedures in nuclear medicine require radiopharmaceutical (radiotracer) localization in target organs. Localization necessitates that radiotracers cross biologic membranes. The plasma membrane of cells in an organ is the primary barrier that drugs must traverse to reach their site of action; this same barrier must be traversed for localization of a radionuclide in the target organ. As with all drugs, radiotracers have certain properties that affect membrane transport, the most important being plasma protein binding, molecular size and shape, ionic charge, and lipid solubility.

Membrane Structure and Transport Processes

The plasma membrane that surrounds a single cell in the body is composed of a phospholipid bilayer interspersed with protein molecules (Figures 9-7 and 9-8).[15,16] The individual phospholipid molecules making up each layer of the membrane are composed of a hydrophilic "head" (phosphate or carboxylate portion) and a hydrophobic "tail" (fatty acid portion). The two layers of phospholipid molecules are aligned so that the hydrophilic (head) portion of one layer is oriented toward the outside (extracellular side) of the membrane and the hydrophilic (head) portion of the other layer is oriented toward the inside (intracellular side) of the membrane. The hydrophobic (tail) portions of each layer are oriented toward each other on the inside portion of the membrane. Hence, the interior of the membrane is hydrophobic, having an attraction for nonpolar solutes, and the exterior of the membrane is hydrophilic, having an attraction for polar solutes. In addition, imbedded within the phospholipid bilayer are membrane proteins. Proteins function to transport solutes (molecules in solution) across the membrane. Most protein transporters are enzymes, which transport only specific solutes known as substrates. For example, a particular enzyme can transport an amino acid substrate but not a sugar substrate. Some proteins function as ion channels, permitting specific ions to diffuse across the membrane. Small ionic substances and water diffuse through minute pores in the membrane, which are believed to be intramolecular channels within protein molecules that penetrate all the way through the membrane. The driving force for charged molecule diffusion is the electrochemical potential across a membrane (i.e., diffusion is affected by the difference in chemical concentration of the molecule and the charge concentration on each side of the membrane).

Transport of solutes across the cell membrane occurs by two main processes: passive diffusion and mediated transport. The process of *passive diffusion* is controlled by the relative concentrations of a solute on each side of the membrane and occurs from higher concentration to lower concentration. The net flux by passive diffusion is also influenced by the membrane surface area and the permeability constant of the solute being transported. For a given membrane concentration gradient and surface area, flux will be higher for solutes with larger permeability constants. An additional requirement for membrane transport is that a drug must be freely available in blood (i.e., must not have a high degree of binding to plasma proteins).

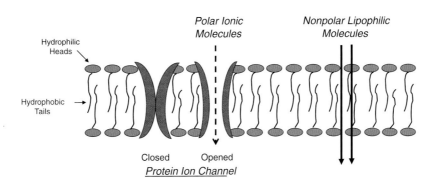

FIGURE 9-7 Basic structural elements of the plasma cell membrane, illustrating diffusion of molecules through the membrane from the extracellular space to the intracellular space. Outer portions of the membrane have hydrophilic properties and inner portions have hydrophobic properties. Polar ionic molecules undergo transcellular transport through open selective protein ion channels; nonpolar molecules undergo passive diffusion directly across the membrane because of their lipid solubility.

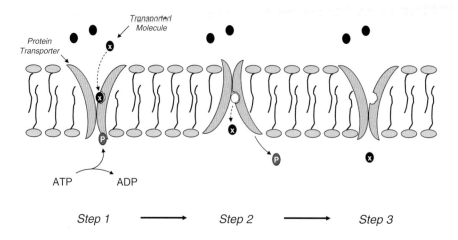

FIGURE 9-8 Active transport of a substance across a plasma cell membrane from the extracellular space to the intracellular space. Release of a phosphate group following the enzymatic breakdown of ATP to ADP by ATPase increases the exposure and affinity of the binding site for the transported molecule (Step 1). Attachment of the molecule to the binding site changes the conformation of the transporter, resulting in loss of the phosphate group (Step 2). This decreases the molecule's affinity for the binding site and releases it into the intracellular space (Step 3).

The ionic charge on a molecule influences its ability to be transported across membranes. In general, nonpolar (un-ionized) solutes that are lipid soluble readily diffuse into the nonpolar regions of the membrane and have higher transport than polar (ionized) solutes. Lipid-soluble compounds, therefore, have higher flux rates across membranes. Most drugs are weak acids or weak bases, and their ionized or un-ionized form is determined by the drug's pKa and the local pH according to the Henderson-Hasselbalch equation, as follows:

$$pH = pKa + \log \frac{\text{ionized drug}}{\text{unionized drug}}$$

A modification of this equation permits calculation of the percentage of ionization of a weak acid (WA) or weak base (WB) drug, as follows:

$$\% \text{ ionized WA} = \frac{100}{\left[1+10^{(pKa-pH)}\right]}$$

$$\% \text{ ionized WB} = \frac{100}{\left[1+10^{(pH-pKa)}\right]}$$

Thus, the direction of diffusion of weak acids or weak bases across a membrane is influenced by a difference in pH on each side of the membrane. For example, for a WA or a WB in extracellular fluid, intracellular diffusion of a WA will be favored with higher intracellular pH; intracellular diffusion of a WB will be favored with lower intracellular pH. This pH-gradient hypothesis has been used to predict the absorption of weak acids from gastric juice (pH 1.4) into plasma (pH 7.4) and the excretion of weak bases from plasma into urine (pH 5.4).

The influence of ionic charge, lipophilicity, and protein binding on membrane transport and radiotracer localization can be illustrated by the development of diffusible brain imaging agents. Figure 9-9 illustrates the processes of drug uptake and retention in the brain, which requires a small molecular weight molecule that is un-ionized and lipid soluble. In 1978 Oldendorf[17] pointed out that the lipophilic agent ^{123}I-iodoantipyrine had brain uptake proportional to cerebral blood flow because it readily crossed the blood–brain barrier (BBB). However, it also diffused back out and therefore had limited usefulness without a brain trapping mechanism.[17,18] Subsequently, nuclear medicine scientists directed their efforts toward the development of lipophilic radioiodinated and ^{99m}Tc tracers that could cross the BBB, so that cerebral blood flow could be measured. To this end, Loberg et al.[19] prepared a series of ^{99m}Tc-labeled iminodiacetic acid analogues with varying lipophilicity assessed by octanol-to-water partition coefficients. They demonstrated that brain uptake was proportional to lipophilicity; however, none of the agents tested had clinical potential because of high plasma protein binding. At about the same time, Kung and Blau[20] developed two diamine compounds, di-β-(piperidinoethyl)-selenide (PIPSE) and di-β-(morpholinoethyl)-selenide (MOSE), labeled with ^{75}Se for brain localization (Figure 9-10). Their

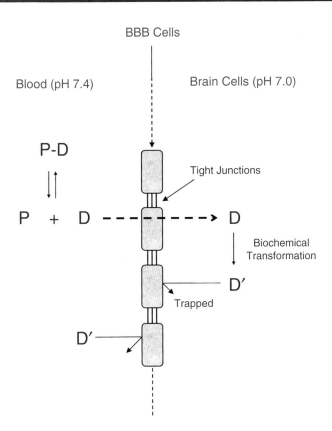

FIGURE 9-9 Mechanism of radioactive drug uptake in the brain. Ionized drug in the blood (D′) does not diffuse across an intact capillary endothelium into brain tissue. A small molecular weight un-ionized, lipophilic drug (D) dissociated from its protein-bound form (P-D) in blood can undergo transcellular passive diffusion across the neural capillary endothelium (blood–brain barrier [BBB] cells) into brain tissue. Drug that undergoes biochemical transformation in the brain is unable to diffuse back out and is trapped in the brain. Biochemical transformation can occur by different mechanisms, such as through protonation of a neutral amine (R-NH$_2$) to its ionized form (R-NH$_3^+$) because of lower intracellular pH in the brain or through a metabolic process such as hydrolysis of an ester (RCOOR) to its corresponding acid (RCOO$^-$). Brain localization of radiolabeled amine compounds is an example of the former mechanism, and localization of ^{99m}Tc-bicisate is an example of the latter mechanism.

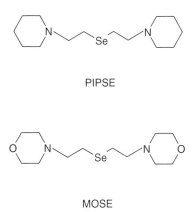

FIGURE 9-10 Chemical structures of selenium-labeled amine-type first-generation diffusible tracers for brain imaging. See text for discussion.

work was based on the pH-gradient hypothesis, which assumes that an un-ionized amine species in the blood at pH 7.4 is lipophilic and will freely diffuse across the BBB into the more acidic intracellular fluid (pH 7.0) in the brain cells. Once the amine molecule gains access to the brain, it is protonated and ionized and unable to diffuse back out.[20] The seminal work of Kung and Blau led the way toward development of useful cerebral blood flow imaging agents and suggested a mechanism for how agents could become trapped in the brain, a necessary requirement for tomographic imaging with single-photon emission computed tomography (SPECT). Eventually, two lipophilic tracers were developed that are ionized and trapped in the brain, ^{99m}Tc-exametazime and ^{99m}Tc-bicisate; these became the principal radiotracers for assessing regional cerebral blood flow.

Mediated transport across membranes can occur by facilitated diffusion or by active transport. These processes involve a membrane carrier (transporter) to translocate the substance across the membrane. *Facilitated diffusion* is a downhill transport process (high to low concentration) and is stereospecific and saturable. Substances that are not lipid soluble but are necessary for cell viability, such as sugars and amino acids, cross the cell membrane by facilitated diffusion. The transport of glucose into cells occurs by facilitated diffusion. The effect of the carrier is to make glucose soluble in the membrane, in which it would otherwise not be soluble because of its hydrophilicity.

Active transport across the cell membrane is an energy-dependent, uphill movement against a substance's electrochemical gradient, from a lower concentration to a higher concentration if the substance is a nonelectrolyte and from a lower to a higher electrochemical potential if it is an electrolyte. Two types of active transport are possible, depending on the source of energy to drive the process. Primary active transport uses ATP to drive the process, and secondary active transport uses an ion concentration difference across the membrane to drive the process. In general, active transport involves combining a substance with a carrier protein (enzyme) at the outer membrane, diffusion through the membrane, and release of the substance from the carrier at the inner membrane–cytoplasm interface (Figure 9-8). The freed substance then diffuses into the cytoplasm. In primary active transport, ATPase catalyzes conversion of ATP to ADP, releasing phosphate. The major difference between facilitated diffusion and active transport is that active transport can move substances from low to high concentration, whereas facilitated diffusion can move substances only from high to low concentration. The principal difference between passive diffusion and facilitated diffusion is that the latter is limited by the amount of carrier present to transport molecules. A prime example of an active transport process in the body is the sodium–potassium pump, which is present in all cell membranes of the body. It carries potassium into the cell and sodium out of the cell at a net transfer rate of three sodium ions out for two potassium ions in. The carrier protein is sodium–potassium (Na-K) ATPase, which is capable of transporting ions and splitting ATP to provide energy for the pump. The pump can transport sodium ions and potassium ions against concentration gradients of 20 to 1 and 30 to 1, respectively.

A fourth method of transport across cell membranes is endocytosis.[15] It varies among cell types and is very active in gastrointestinal absorption. Endocytosis involves contact of a substance such as a fluid, molecule, or particle with a surface protein on the cell. The cell then invaginates to engulf the substance and releases it into the cell cytoplasm. Endocytosis of fluids or molecules in fluid is called *pinocytosis* (cell drinking), whereas endocytosis of solids, such as bacteria or colloidal particles, is called *phagocytosis* (cell eating).

Stereochemistry

It is well known from structure–activity relationship (SAR) studies that the localization and pharmacologic action of a drug molecule in the body are affected by many factors and that the stereochemical configuration (isomeric form) of the molecule can play an important part in the effectiveness of drug therapy. During the design and testing of second-generation technetium radiopharmaceuticals it became evident that the stereoisomeric form of certain radiopharmaceuticals was important for distribution and localization. Since a radiopharmaceutical does not elicit a pharmacologic response upon interaction with a receptor, its interaction with the receptor is referred to as a structure–distribution relationship (SDR).[21]

Isomers

Compounds that have the same molecular formula with identical numbers and types of atoms but distinctly different arrangement of the atoms are called *isomers*.[22] Isomers can be of two types: constitutional isomers or stereoisomers. Isomers that have different bonding connectivities of their atoms are called *constitutional isomers*. For example, ethanol (CH_3CH_2OH) and acetone (CH_3COCH_3) each have the molecular formula C_3H_6O, but they differ in the nature of the functional group. Likewise, 1-propanol ($CH_3CH_2CH_2OH$) and 2-propanol ($CH_3CHOHCH_3$) have the molecular formula C_3H_8O but differ in the location of the OH group. Isomers that differ only in the relative spatial orientation of atoms or groups are called *stereoisomers*. There are several types of stereoisomers, but those of greatest interest regarding the SDR of radiopharmaceuticals are known as enantiomers.

Enantiomers are pairs of molecular species that are mirror images of each other. In addition, they are nonsuperimposable, which means that their three-dimensional configuration cannot be arranged so that one enantiomer can be overlaid upon the other. Enantiomers are nonsuperimposable because they have chirality. A chiral molecule is one that lacks symmetry. In organic compounds, a tetrahedral carbon atom becomes chiral when the four groups attached to it are different. The different possible spatial arrangement of groups on a chiral carbon atom creates the enantiomers. Although enantiomers have identical physicochemical properties, their different molecular configurations may cause them to have different biologic properties. The molecular configuration of an enantiomer is designated by letter descriptors, either with a small capital (D) or (L) or with a large, italicized capital (*R*) or (*S*), as described in a later section.

One property of chiral molecules is that they are optically active. When enantiomers are separated and their solutions are exposed to plane-polarized light, each enantiomer will cause the light to rotate by opposite but equal amounts. The enantiomer that rotates the plane of light clockwise, to the right, is said to be dextrorotatory and is designated with a plus sign (+) before its name. The enantiomer that rotates the plane of light counterclockwise, to the left, is said to be levorotatory and designated with a minus sign (−) before its name. An equimolar mixture of enantiomers, designated as (±), is called a racemic mixture or a racemate. Racemates are devoid of optical activity. Although enantiomers have identical physicochemical properties, their different optical properties provide a means of identifying the enantiomer that has a desired biologic activity.

Each chiral atom in a molecule can give rise to two optical isomers. If n is the number of asymmetric carbons in a molecule, the number of optical isomers is 2^n unless the molecule as a whole is symmetrical. Figure 9-11 illustrates the relationship between the possible isomers of tartaric acid. The figure shows that D-(−)-tartaric acid is the mirror image of L-(+)-tartaric acid. Therefore, they are enantiomers and optically active. *Meso*-tartaric acid, however, is not a mirror image of the D and L forms and is therefore a diastereomer of these forms. *Diastereomers* are stereoisomers that are not mirror images. They have completely different properties. *Meso*-tartaric acid melts at 140°C (D and L forms melt at 170°C), is less dense, and is less soluble in water than the D and L forms. Also, the *meso* molecule is symmetrical as a whole; that is, even though it contains two asymmetric carbon atoms, the top half and bottom half of the molecule are identical. This plane of symmetry in the molecule makes the *meso* form optically inactive.

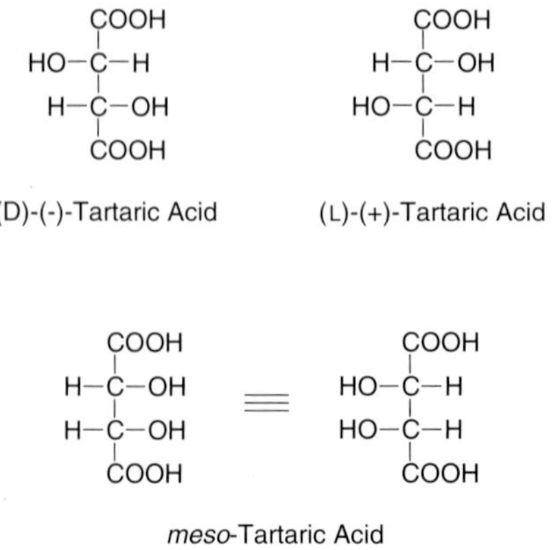

FIGURE 9-11 Enantiomers of tartaric acid, (D) and (L) forms, and their diastereomer (*meso*-tartaric acid). See text for details.

Since the only difference in enantiomers is their optical properties, their physicochemical properties such as melting point, boiling point, and solubility are identical and cannot be used to separate them. However, high-performance liquid chromatography (HPLC) using a chiral column packing material has been used to resolve enantiomers on both an analytical and a preparative scale.

Configuration

The spatial arrangement of atoms in a chiral molecule that distinguishes it from its mirror image is known as the absolute configuration. Specific conventions are used to assign configuration. The convention established by Fischer for assigning the configuration of amino acids and carbohydrates is that the longest carbon chain is written vertically with the most oxidized end placed at the top of the chain (e.g., CHO in a carbohydrate). Figure 9-12 illustrates this assignment for carbohydrates and amino acids. If the OH group at the bottom-most (highest numbered) chiral center is on the right-hand side, the molecule is given a D or dextro configuration, and if it is on the left-hand side, it is given an L or levo configuration. With amino acids, the L designation is used for those whose α-amino group is on the left-hand side of the carbon chain and the D designation is used if the α-amino group is on the right-hand side. This convention has been applied to other molecules that are closely related to amino acids and carbohydrates. It must be remembered, however, that the D and L configurational descriptors have nothing to do with the signs of optical rotation. For example, naturally occurring glucose has a D configuration and just happens to be dextrorotatory. Therefore, it is written as D-(+)-glucose. However, naturally occurring tartaric acid has an L configuration and happens to also be dextrorotatory. It is written as L-(+)-tartaric acid as shown in Figure 9-11.

Molecules are three dimensional, and Fischer developed a two-dimensional projection formula to help in visualizing the three-dimensional structure. This is illustrated in the bottom panel of Figure 9-12 for D-glyceraldehyde. By convention, the projection formula, shown as structure (2) in Figure 9-12, illustrates that the solid-wedge groups bonded to the chiral carbon atom project out of the plane of paper and the dashed-wedge groups project behind the plane of paper. The three-dimensional pyramidal structure of D-glyceraldehyde is shown for comparison.

In drug development, it is often found that only one enantiomeric form of a drug is active biologically, because it exactly complements the receptor configuration whereas its mirror image entantiomer may not (Figure 9-13). Several second-generation radiopharmaceuticals are labeled peptides. Peptides are short-chain amino acid sequences that may contain the D- or L-enantiomer of a particular amino acid. Natural amino acids are of the L-form, which may undergo enzymatic degradation in vivo. Therefore, the corresponding D-enantiomer of the amino acid is sometimes substituted in a peptide to increase its in vivo stability. A prime example is the somatostatin analogue Indium In-111 pentetreotide (OctreoScan [Covidien]), in which the peptide has (D)-phenylalanine and (D)-tryptophan in its structure to inhibit metabolism by amino- and carboxy-peptidases.

A more universal system of configurational nomenclature is the Cahn-Ingold-Prelog (CIP) convention, which is based on the three-dimensional structures of molecules.[22] According to this system, the configuration of a molecule is described as either *R* (from *rectus,* Latin for right) or as *S* (from *sinister,* Latin

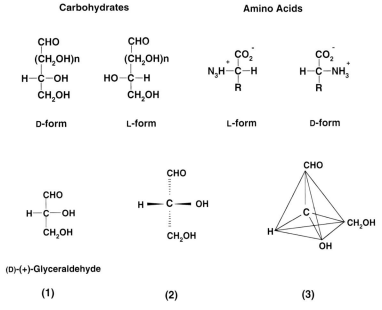

FIGURE 9-12 Upper panel shows the Fischer (D) and (L) configurations for carbohydrates and amino acids. Lower panel shows the structure of (D)-(+)-glyceraldehyde (1), its Fischer projection formula (2), and its three-dimensional pyramidal structure (3). See text for detailed explanation.

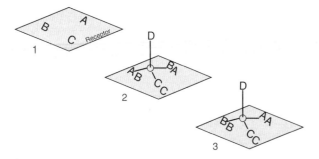

FIGURE 9-13 Stereochemical interaction of a drug molecule (D-ABC) and its receptor (ABC) demonstrating drug–receptor mismatch (2) and match (3).

for left). The assignment is made by considering a hierarchy of assigned values of various ligand groups bonded to a chiral carbon. Priority is given to groups with highest atomic number. The chiral center in a three-dimensional structure is viewed from the side opposite the lowest-ranked ligand. If, from this view, the arrangement of the three remaining ligands (in order of highest to lowest ranking) appears in a clockwise direction (right-handed), the configuration is *R*, and if the arrangement of ligands is counterclockwise (left-handed), the configuration is *S*. This is known as the sequence rule. The CIP descriptors for D-glucose are shown in Figure 9-14. The benefit of this system is that it permits a compound's structure to be drawn correctly by anyone who understands the sequence rule. As with D and L descriptors, *R* and *S* descriptors have nothing to do with signs of optical rotation. Thus, an *R*- or *S*-configured compound may be dextrorotatory or levorotatory.

In some instances, when an organic molecule contains several different functional groups, it is desirable to describe the spatial configuration of one group relative to another group. *Syn* (synperiplanar) and *anti* (antiperiplanar) designations denote the relative configuration of any two stereogenic centers in a chain. A stereogenic center is a focus in a molecule at which the interchange of two ligands attached to an atom leads to a stereoisomer. A stereogenic center may not be chiral, but all chiral centers are stereogenic. If ligands on the stereogenic centers are on opposite sides of the plane, the relative configuration is *anti*. If they are on the same side of the plane, they are *syn*. The *syn* and *anti* configurations for hydroxyl groups in D-glucose are shown in Figure 9-14. The *syn/anti* designation is used in technetium radiopharmaceuticals to describe the configuration of functional groups relative to the technetium–oxo core; an example is the *syn* and *anti* forms of ^{99m}Tc-N,N′-bis(mercaptoacetyl)-2,3-diaminopropanoate (^{99m}Tc-CO$_2$–DADS).

Another type of isomerism found in octahedral complexes having three identical groups coordinated to a metal atom is meridional/facial isomerism. The meridional (*mer*) isomer has the three identical groups bound to the metal in the same plane, whereas the facial (*fac*) isomer has the three groups occupying the same face (Figure 9-14). An example of this type of isomerism in technetium chemistry is the technetium tricarbonyl compounds, which have three carbonyl groups coordinated in a facial configuration.

RADIOPHARMACEUTICAL DEVELOPMENT

A major focus in nuclear medicine has been the design of target-specific radiopharmaceuticals. Most of the first-generation agents had a radionuclide that was labeled or "tagged" to a

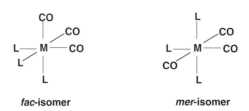

FIGURE 9-14 Upper panel left shows the (*R*) and (*S*) configuration for asymmetric carbon atoms in (D)-glucose. To its right is illustrated the zig-zag projection of (D)-glucose indicating the use of *syn* and *anti* descriptors for designating the spatial configuration of one group relative to another. Lower panel illustrates the facial (*fac*) and meridional (*mer*) isomers of an octahedral metal complex. See text for complete description.

molecular species that localized in target organs of interest on the basis of substrate-nonspecific mechanisms; that is, no specific biochemical interaction between the radiolabeled compound and a particular receptor site in the target organ was involved. Localization occurred by nonspecific in vivo processes. Typically, the radionuclide was a passive component and had little influence on the radiopharmaceutical's localization. Examples include radiocolloids for liver, spleen, and bone marrow imaging, which localized by phagocytosis; radiolabeled macroaggregated albumin particles for lung perfusion imaging, which localized by capillary blockade; heat-damaged radiolabeled red blood cells for spleen imaging, which localized by cell sequestration; radiometal chelates for assessing renal blood flow and excretion, which localized by diffusion; and radioactive xenon gas for lung ventilation imaging, which localized by diffusion.

Over time, radiotracers were developed that localized by a specific biochemical interaction between the radiotracer and a biologic receptor. Agents that localize by this process are known as substrate-specific tracers. Examples include ^{18}F-fludeoxyglucose, a substrate for the enzyme hexokinase, which localizes by metabolic trapping; ^{11}C-palmitic acid, a substrate for oxidative metabolism in the heart; ^{131}I-sodium iodide, which is actively co-transported with sodium ions across the follicular-cell plasma membrane in the thyroid gland; ^{201}Tl-thallium chloride, a biologic analogue of potassium, which is taken up by the sodium–potassium ATPase membrane pump in metabolically active cells such as myocytes and tumor cells; radiolabeled antibodies that undergo specific immunologic reactions with tumor-associated antigens; and ^{123}I-iobenguane sulfate, a biologic analogue of norepinephrine, which can be used to image tissues rich in adrenergic innervation such as the adrenal glands, heart, and neuroendocrine tumors.

Many radiopharmaceuticals used in nuclear medicine are labeled with ^{99m}Tc, and most of these agents are substrate nonspecific. With the early agents, technetium was a passive component tagged to a molecule that was responsible for technetium's in vivo localization. Technetium was not essential for localization. Examples include technetium-labeled complexes with pentetate, medronate, oxidronate, and gluceptate. As knowledge and experience grew in the area of technetium chemistry, new ^{99m}Tc complexes were developed with technetium as a core atom essential for complex formation and a molecular partner in the determination of biologic localization. Those agents that localize via specific cell transporters can be classified as substrate-specific radiopharmaceuticals. Examples in this group include ^{99m}Tc-mebrofenin and disofenin for hepatobiliary imaging and ^{99m}Tc-mertiatide for kidney imaging. Further advances in technetium chemistry led to the design of second-generation technetium-tagged radiopharmaceuticals. These agents can be classified as substrate specific because they have technetium labeled to a bifunctional chelate. One end of the molecule binds technetium and the other end is functionalized with a moiety to target a specific receptor in the body. Among these agents are technetium-labeled antibodies and peptides. Chapter 10 discusses these compounds in detail.

Radiotracer development for molecular imaging is discussed in detail in Chapter 17.

REFERENCES

1. Liu S, Edwards DS. ^{99m}Tc-labeled small peptides as diagnostic radiopharmaceuticals. *Chemgl Rev.* 1999;99:2235–68.
2. Steigman J, Eckelman WC. *The Chemistry of Technetium in Medicine.* Nuclear Science Series NAS-NS-3204 Nuclear Medicine. Washington, DC. National Academies Press; 1992:16.
3. Alvarez J. Preparation of ^{113m}In radiopharmaceuticals. In: Subramanian G, Rhodes BA, Cooper JF, Sodd VJ, eds. *Radiopharmaceuticals.* New York: Society of Nuclear Medicine; 1975:102–6.
4. Cooper JF, Wagner HN Jr. Preparation and control of ^{113m}In radiopharmaceuticals. In: *Radiopharmaceuticals from Generator-Produced Radionuclides.* Vienna, Austria: International Atomic Energy Agency; 1971:83–9.
5. Wolfangel RL. Accumulation of radioiodine in staff members [reply]. *J Nucl Med.* 1979;20:995.
6. Howard BY. Safe handling of radioiodinated solutions. *J Nucl Med Technol.* 1976;4:28–30.
7. Miller KL, Bott SM, Velkley DE, et al. Review of contamination and exposure hazards associated with therapeutic doses of radioiodine. *J Nucl Med Technol.* 1979;7:163–6.
8. Pollock RW, Meyer RD. Concentration of I-131 in the air during thyroid therapies. *Health Phys.* 1979;36:68–9.
9. Rubin LM, Miller KL, Schadt WW. A solution to the radioiodine volatilization problem. *Health Phys.* 1977;32:307–9.
10. Burgess JS, Partington EJ. Radiation decomposition effects in aqueous solutions of carrier-free sodium iodide I-131. Amersham, England: United Kingdom Atomic Energy Authority, Radiochemical Centre; March 1960. Publication RCC/R-98.
11. Shubnyakova LP, Kharlamov VT, Pikaev AK. Radiolysis of dilute aqueous solutions of Na^{131}I. *Khimiya Vysokikh Energii.* 1976;10:41–5.
12. Maguire WJ. A precaution for minimizing radiation exposure from iodine vaporization. *J Nucl Med Technol.* 1980;8:90–3.
13. Pizzarello DJ, Witcofski RL. *Basic Radiation Biology.* Philadelphia: Lea & Febiger; 1967:61
14. Wolf AP. Terminology concerning specific activity of radiopharmaceuticals [reply]. *J Nucl Med.* 1981;22:392–3.
15. Widmaier EP, Raff H, Strang KT. *Vander, Sherman, and Luciano's Human Physiology: The Mechanisms of Body Function.* 9th ed. Boston: McGraw Hill; 2004:109–33.
16. Buxton ILO. Pharmacokinetics and pharmacodynamics: The dynamics of drug absorption, distribution, action, and elimination. In: Brunton LL, Lazo JS, Parker KL, eds. *Goodman & Gilman's The Pharmacological Basis of Therapeutics.* 11th ed. New York: McGraw-Hill Medical Publishing Division; 2006.
17. Oldendorf WH. Need for new radiopharmaceuticals [letter]. *J Nucl Med.* 1978;19:1182.
18. Uzler JM, Bennett LR, Mena I, et al. Human CNS perfusion scanning with ^{123}I-iodoantipyrine. *Radiology.* 1975;115:197–200.
19. Loberg MD, Corder EH, Fields AT, et al. Membrane transport of Tc-99m-labeled radiopharmaceuticals. I. Brain uptake by passive transport. *J Nucl Med.* 1979;20:1181–8.
20. Kung HF, Blau M. Regional intracellular pH shift: a new mechanism for radiopharmaceutical uptake in brain and other tissues. *J Nucl Med.* 1980;21:147–52.
21. Volkert WA. Stereoreactivity of ^{99m}Tc-chelates at chemical and physiological levels, In: Nicolini M, Bondoli G, Mazzi U, eds. *Technetium and Rhenium in Chemistry and Nuclear Medicine 3.* Verona, Italy: Cortina International; 1990:343–52.
22. Eliel EL, Wilen SH. *Stereochemistry of Organic Compounds.* New York: John Wiley & Sons; 1994.

CHAPTER 10

Radiopharmaceutical Chemistry: Technetium Agents

Technetium, as element 43, was discovered in 1937 by Perrier and Segrè in a sample of molybdenum that had been irradiated by deuterons.[1] The new element received its name from the Greek word *technetos,* meaning artificial, because technetium was the first element previously unknown on earth and made artificially.[2] In 1939, Seaborg and Segrè observed that molybdenum-98 irradiated with slow neutrons gave rise to ^{99}Tc through decay of the metastable isomer, ^{99m}Tc.[3] Eventually, 21 isotopes of technetium were discovered, ranging from ^{90}Tc to ^{110}Tc, with ^{110}Tc having the shortest half-life (0.86 sec) and ^{97}Tc the longest (2.6×10^6 yr). All technetium isotopes are radioactive.

The technetium isotope that has achieved widespread application in diagnostic nuclear medicine is ^{99m}Tc. This metastable isomer decays with a half-life of 6.01 hr to ^{99}Tc, shown in the decay scheme in Figure 10-1.

^{99}Tc has a half-life of 2.11×10^5 years, so it is essentially stable. ^{99}Tc has been useful in characterizing the chemistry of technetium radiopharmaceuticals. The metastable state of ^{99m}Tc is 0.1427 MeV above the ground state of ^{99}Tc. Three gamma photons (γ_1, γ_2, and γ_3), of 0.0022, 0.1405, and 0.1427 MeV, respectively, are released in the decay of ^{99m}Tc to ^{99}Tc. The most abundant of these is γ_2 (0.1405 MeV), occurring in 89.1% of all nuclear transitions. It is the principal photon detected in nuclear medicine imaging studies.

In the 1950s, purification work on the ^{132}Te–^{132}I generator at Brookhaven National Laboratory (BNL) turned up a contaminant that proved to be technetium. The technetium contaminant was due to ^{99}Mo, which was also present because it had followed tellurium in the chemical separation process.[4] The discovery eventually led to the production of the ^{99}Mo–^{99m}Tc generator in 1957 at BNL. Final improvements were made by Powell Richards.[4] The ^{99}Mo used in present-day generators is obtained as a fission byproduct of ^{235}U. It is chemically processed into molybdate and loaded onto an aluminum oxide chromatography column. The column is then sterilized and assembled into the generator. A simplified decay scheme for ^{99m}Tc production is shown below.

$$^{99}\text{Mo} \xrightarrow{86\%} {}^{99m}\text{Tc} \longrightarrow {}^{99}\text{Tc}$$
$$\cdots\cdots 14\% \cdots\cdots \triangleright \cdots$$

The first technetium generator for investigational use was purchased from BNL by the University of Chicago.[5] Elution of the generator with 0.9% sodium chloride solution yielded technetium as sodium pertechnetate, Na^{99m}TcO$_4$. Following intravenous injection into mice, pertechnetate ion (^{99m}TcO$_4^-$) activity was observed to localize in the thyroid gland, salivary glands, stomach, and urinary bladder, similar to iodide ion (Figure 10-2). These first observations of pertechnetate's in vivo behavior and subsequent developmental studies with early technetium-labeled compounds in animals and humans demonstrated the potential that ^{99m}Tc had for diagnostic imaging.[5]

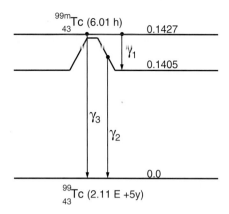

FIGURE 10-1 Decay scheme for ^{99m}Tc.

The following advantages were identified for ^{99m}Tc as a diagnostic radionuclide:

1. Low radiation dose due to short half-life and absent beta radiation
2. High photon yield (89%) of 140 keV gamma rays; good tissue penetration; easily collimated and efficiently detected by the gamma camera
3. Local availability from an on-site ^{99m}Tc generator.
4. Capability of being compounded into a variety of chemical forms

TECHNETIUM CHEMISTRY

The chemical forms of early technetium compounds for clinical use were ^{99m}Tc-sodium pertechnetate (for brain and thyroid imaging) and ^{99m}Tc-sulfur colloid (for liver, spleen, and bone marrow imaging). Different ^{99m}Tc-labeled compounds were soon developed for imaging other organs, namely the kidneys, lungs, and placenta.[5] An essential requirement for preparing technetium-labeled compounds was changing technetium's valence or oxidation state. This was achieved by using a reducing agent in the radiolabeling procedure. The early technetium compounds were prepared extemporaneously following a stepwise admixture of coordinating ligand, reducing agent, pertechnetate, and adjuvants and any necessary pH or temperature adjustments. Labeling yields were often not quantitative, so a final purification step was needed to remove unbound pertechnetate. The final product was sterilized by membrane filtration or by autoclaving. Eventually, sterile radiopharmaceutical kits were developed that simplified the compounding procedure. The kit formulations yielded high-purity technetium compounds and eliminated the need to remove unreacted pertechnetate. This compounding method continues to be the standard of practice today.

Technetium Oxidation States

Technetium is positioned in the periodic table along with manganese and rhenium, but its chemistry is more similar to that of rhenium. Table 10-1 compares the electronic configuration of technetium with that of other elements. As a transition metal in group VIIB, technetium has seven electrons beyond the noble gas configuration of krypton, and these electrons reside in the 4d and 5s subshells (d^7 configuration) (Table 10-2). Loss of these seven electrons yields the 7+ oxidation state (d^0 configuration) as pertechnetate Tc(VII), which is the most stable oxidation state in aqueous solution. A reduced form of technetium, at a lower oxidation state, is

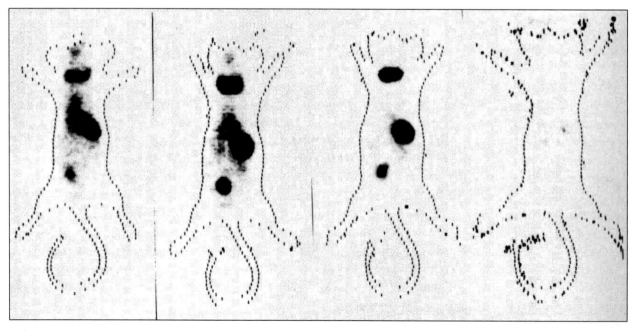

FIGURE 10-2 Serial scans of a mouse at 10 minutes and 1, 6, and 24 hours (left to right) after an intravenous injection of ^{99m}Tc-sodium pertechnetate, showing concentration in the thyroid gland, salivary glands, stomach, and urinary bladder. (Reprinted from reference 5.)

TABLE 10-1 Electron Shell Configuration of Selected Elements

		Principal Energy Level and Sublevel												
		1	2		3			4				5		
Z	Element	s	s	p	s	p	d	s	p	d	f	s	p	d	f
1	H	1													
2	He	2													
3	Li	2	1												
4	Be	2	2												
5	B	2	2	1											
6	C	2	2	2											
7	N	2	2	3											
8	O	2	2	4											
10	Ne	2	2	6											
15	P	2	2	6	2	3									
16	S	2	2	6	2	4									
36	Kr	2	2	6	2	6	10	2	6						
43	**Tc**	**2**	**2**	**6**	**2**	**6**	**10**	**2**	**6**	**6**		**1**			
53	I	2	2	6	2	6	10	2	6	10		2	5		

Z = element atomic number.

required for the preparation of technetium radiopharmaceuticals. One problem encountered with reduced technetium is that Tc(IV) is readily hydrolyzed to insoluble $TcO_2 \cdot H_2O$, as follows:

$$TcO^{2+} + 2\ OH^- \longrightarrow TcO_2 \cdot H_2O \downarrow$$

The amount of technetium (^{99m}Tc and ^{99}Tc) in a generator eluate, if reduced to Tc(IV), is sufficient to exceed the solubility product of $TcO_2 \cdot H_2O$.[6] The insolubility of this species over a wide pH range creates a thermodynamic trap, making reduced technetium unavailable for complexation with the coordinating ligand.[7] Therefore, limiting its formation is critical during technetium radiopharmaceutical preparation. This hydrolysis can be minimized by providing a sufficient concentration of complexing agent in the reaction mixture. Radiochemical investigations during the development of technetium-labeled compounds of specific oxidation states demonstrated that in cold concentrated hydrochloric acid (HCl) technetium is reduced to Tc(V) as $[TcOCl_4]^-$ and in hot HCl it is reduced to Tc(IV) as $[TcCl_6]^{2-}$.[8] Hydrolysis and disproportionation were prevented if the HCl concentration was ≥2M. The halide ligands in these chlorocomplexes could be displaced by an appropriate coordinating ligand that

TABLE 10-2 Electron Configuration and Oxidation States of Technetium

Electron Configuration							
	Principal Level and Sublevel						
	4		5				
Core[a]	d	f	s	p	d	f	Oxidation State	Configuration Descriptor
[Kr]	6		1				Tc^0 (metal)	d^7
[Kr]	6						Tc^{+1}	d^6
[Kr]	5						Tc^{+2}	d^5
[Kr]	4						Tc^{+3}	d^4
[Kr]	3						Tc^{+4}	d^3
[Kr]	2						Tc^{+5}	d^2
[Kr]	1						Tc^{+6}	d^1
[Kr][b]	0						Tc^{+7}	d^0

[a] [Kr] core = $1s^2 2s^2 2p^6 3s^2 3p^6 3d^{10} 4s^2 4p^6$.
[b] This level represents pertechnetate $[Tc^{+7}O_4]^-$ with technetium at its highest oxidation state.

stabilized reduced technetium. The result was a variety of specific technetium-labeled compounds having biologic properties potentially suitable for diagnostic imaging.[8]

Ascorbic acid and ferrous iron are reducing agents that were used in early radiolabeling studies, but they often led to incomplete reduction, requiring removal of unreacted pertechnetate. More powerful reducing agents were subsequently introduced. Sodium borohydride ($NaBH_4$) and sodium dithionite ($Na_2S_2O_4$) were found to be effective in alkaline pH, and stannous chloride ($SnCl_2$) was effective in acidic pH. Stannous chloride is capable of producing high yields of technetium-labeled compounds, eliminating the need to remove free pertechnetate. This led to the introduction of "instant kits" for the preparation of ^{99m}Tc radiopharmaceuticals.[9] Other stannous salts, such as stannous fluoride and stannous tartrate, also became useful in kit formulations.

The oxidation state of technetium and the stability of its compounds are controlled by several factors, including pH, the type of reducing system, the chemical properties of the coordinating ligand, and the use of adjunctive ingredients in the kit. The most stable states in water are Tc(VII) as TcO_4^- and Tc(IV) as the insoluble hydrolyzed reduction product, $TcO_2 \cdot H_2O$.[10] Reduction/titration experiments have shown that some ligands can produce complexes with technetium in a specific oxidation state and other ligands yield complexes with technetium with mixed oxidation states. The oxidation state is determined by the number of electrons, n, acquired by pertechnetate. Thus, when n = 2, 3, 4, and 6, technetium is reduced to the (V), (IV), (III), and (I) oxidation states, respectively. Once technetium is reduced, it can disproportionate unless sufficient coordinating ligand is present.[7] Disproportionation occurs when technetium is both oxidized and reduced:

$$3 \text{Tc(V)} \longrightarrow 2 \text{Tc(IV)} + \text{Tc(VII)} \quad (10\text{-}1)$$

$$3 \text{Tc(VI)} \longrightarrow \text{Tc(IV)} + 2 \text{Tc(VII)} \quad (10\text{-}2)$$

For example, during the synthesis of a ^{99m}Tc radiopharmaceutical, initial reduction of pertechnetate ($^{99m}TcO_4^-$) may lead rapidly to the formation of the Tc(VI) intermediate $^{99m}TcO_4^{2-}$, which is unstable and can disproportionate to $^{99m}TcO_4^-$ (Tc-VII) and $^{99m}TcO_2$ (Tc-IV) according to Equation 10-2. These unwanted reactions may compromise labeling yields of the desired technetium complex. The presence of chelating agents can stabilize intermediate oxidation states of technetium. Some technetium complexes are quite stable to oxidation (e.g., ^{99m}Tc-DTPA, ^{99m}Tc-HIDA derivatives, ^{99m}Tc-gluceptate), whereas others are more labile (e.g., ^{99m}Tc-DMSA, ^{99m}Tc bone agents). Labile complexes may require the addition of an antioxidant to the kit formulation.

The oxidation states of technetium in several compounds are listed in Table 10-3.[7] It should be noted, however, that electron transfer studies to identify the oxidation state of technetium in first-generation complexes, such as ^{99m}Tc-succimer (DMSA), ^{99m}Tc-pentetate (DTPA), ^{99m}Tc-pyrophosphate (PPi or PYP), and ^{99m}Tc-diphosphonates have not been conclusive and depend on the reducing conditions.[7,10] On the other hand, technetium's oxidation state in many second-generation complexes has been well characterized. Some examples include ^{99m}Tc-sestamibi (Cardiolite [Lantheus]), in which technetium attains the Tc(I) oxidation state (d^6 configuration) by gaining six electrons; ^{99m}Tc-mebrofenin (Choletec [Bracco]), in which technetium attains the Tc(III) oxidation state (d^4 configuration) by gaining four electrons; and ^{99m}Tc-mertiatide (TechneScan MAG3 [Covidien]), in which technetium attains the Tc(V) oxidation state (d^2 configuration) by gaining two electrons. An exception to the requirement for chemical reduction is ^{99m}Tc-sulfur colloid, in which technetium retains the Tc(VII) oxidation state by virtue of its stability as insoluble technetium heptasulfide, Tc_2S_7.[10] Thus, technetium exhibits a diverse chemistry that allows it to be incorporated into a variety of chemical compounds for diagnostic use in nuclear medicine. Table 10-3 lists the usual oxidation states present in technetium radiopharmaceuticals prepared from kits. Specific technetium compounds are discussed later in this chapter.

TABLE 10-3 Oxidation State of Technetium in Various Compounds

Oxidation State	Chemical Form
Tc(VII)	Pertechnetate, sulfur colloid
Tc(V)	Citrate, DMSA (high pH), ECD, gluceptate, gluconate, HMPAO, MAG3, tetrofosmin
Tc(IV)	DTPA, HDP, MDP, PPi (PYP), $TcO_2 \, H_2O$
Tc(III)	DMSA (low pH), HIDA analogues, furifosmin, teboroxime
Tc(I)	Sestamibi

Kit Chemistry

Technetium kits are typically formulated with a coordinating ligand, a reducing agent, and adjuvants such as ancillary chelating agents, buffers, and antioxidants. The optimum formulation is determined for each type of kit.[6,11,12] The pH of a radiolabeling mixture is important for reactions to proceed appropriately, and buffers are used to provide optimal pH.

Coordinating Ligand

The coordinating ligand in the kit reacts with reduced technetium to form the desired technetium radiopharmaceutical. Many technetium radiolabeling reactions occur at pH 4 to 6 where hydrolysis of metal ions can occur. Therefore, sufficient ligand must be available in kits to ensure complexation of reduced ^{99m}Tc and ^{99}Tc, and in some kits the Sn(II) and Sn(IV) as well.[11] For example, a DTPA-to-tin molar ratio of 890 produced higher labeling yields with technetium than a molar ratio of 9.[6] The excess ligand is necessary to ensure complexation of reduced technetium and tin so that tin–technetium side reactions do not compete with ligand–technetium com-

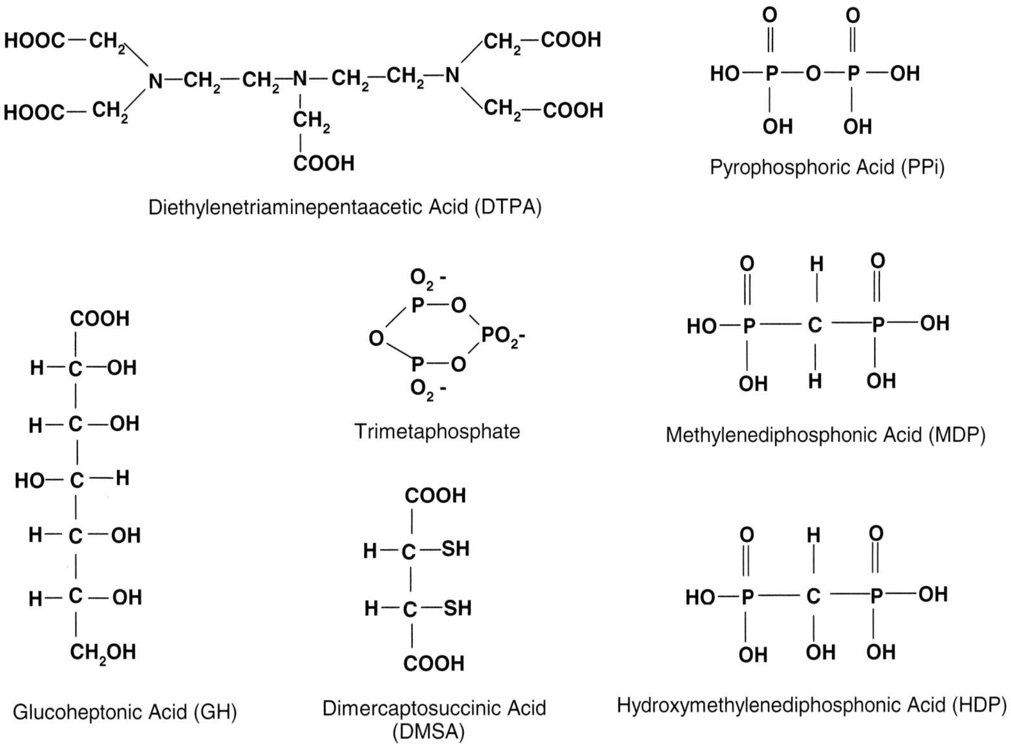

FIGURE 10-3 Chemical structures of several coordinating ligands used in ^{99m}Tc kits.

plexation. Commercial DTPA kits have a DTPA-to-tin molar ratio of 33 to 1. Some examples of coordinating ligands in first-generation kits are shown in Figure 10-3. Most of them are still in use today.

Ancillary Chelating Agents

In many kits the coordinating ligand complexes all species of tin and reduced technetium. Some examples are shown in Figure 10-3. In several kits, however, the principal coordinating ligand for ^{99m}Tc cannot act as a complexing agent for tin, and ancillary chelating agents must be included in the kit for this purpose. Ancillary chelating agents form soluble complexes with tin and reduced technetium species, preventing their hydrolysis.[10] Ancillary chelating agents may also function as transfer ligands. A transfer ligand, otherwise known as a donor ligand or exchange ligand, forms a weak complex with reduced technetium. The complex temporarily stabilizes reduced technetium against disproportionation and hydrolysis when technetium's reaction rate with the coordinating ligand is slow. During the radiolabeling reaction the stronger, coordinating ligand displaces the weaker ligand from the technetium donor complex. Examples of transfer ligands in technetium radiopharmaceuticals are sodium tartrate in the mertiatide kit, sodium gluconate in the tetrofosmin kit, sodium glucoheptonate in the apcitide kit, and sodium citrate in the sestamibi kit.

Reducing Agents

The most common reducing agent in ^{99m}Tc kits is Sn(II) as stannous chloride dihydrate, $SnCl_2 \cdot 2H_2O$. Typically very little of the Sn(II) in kits is present as free ions in solution, most of it being complexed with ligand or chelating agent and some of it as colloidal tin aggregates.[7] Being a powerful reducing agent, Sn(II) is easily oxidized to Sn(IV) by oxygen in the air or in solution (Figure 10-4). Usually, a large excess of stannous chloride is present with respect to pertechnetate in radiopharmaceutical solutions, with the ratio of $SnCl_2$ to $^{99m}TcO_4^-$ being as high as 10^8 to 10^9.[10] Consequently, very little of the Sn(II) during radiolabeling reactions is oxidized by pertechnetate per se. Most of its reducing power is lost through oxidation by oxygen and free radicals, according to the following reactions:

$$2\ Sn^{2+} + 4\ H^+ + O_2 \longleftrightarrow 2\ Sn^{4+} + 2\ H_2O \qquad (10\text{-}3)$$

$$Sn^{2+} + 2\ H^+ + 2\ OH^\bullet \longleftrightarrow Sn^{4+} + 2\ H_2O \qquad (10\text{-}4)$$

$$Sn^{2+} + 2\ H^+ + 2\ HO_2^\bullet \longleftrightarrow Sn^{4+} + 2\ H_2O_2 \qquad (10\text{-}5)$$

Oxygen has been shown to have a greater effect during the radiolabeling reaction than after the reaction is complete, because it competes with pertechnetate for Sn(II).[13] Therefore, except for a few special situations, it is important to exclude air from technetium radiopharmaceuticals during and after preparation. To limit the oxidation of tin, kits are typically lyophilized and sealed under an oxygen-free atmosphere such as nitrogen or argon. In addition, it has been shown that radiation-induced decomposition of the Tc complex can occur, catalyzed by the presence of dissolved oxygen, and that the presence of excess stannous ion acts as an inhibitor of this

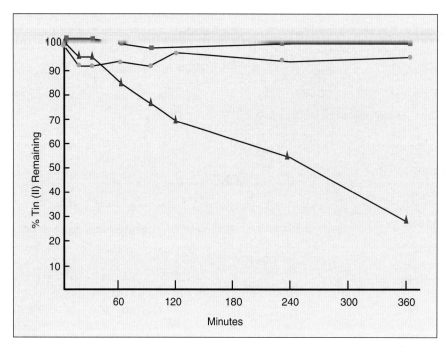

FIGURE 10-4 Influence of preparation and storage conditions on stability of stannous pyrophosphate solutions. ● = oxygenated water, nitrogen atmosphere; ■ = nitrogen-purged water, nitrogen atmosphere; ▲ = nitrogen-purged water, room air.

reaction.[14] Therefore, with few exceptions, technetium radiopharmaceuticals require a sufficient level of stannous ion throughout the useful life of the prepared kit. However, the complexing ligand to Sn(II) ratio must be high to keep Sn(II) in a usable form and to prevent side reactions of hydrolyzed tin coprecipitating reduced technetium as a colloidal impurity.[6]

Antioxidant Stabilizers

Antioxidants are agents that have a higher oxidation potential than the drug in question, and therefore are preferentially oxidized, protecting the drug from oxidation. They are reducing agents. After radioactivity is added to a kit, radiolytic degradation of the solution may occur, producing free radicals (see Chapter 6 for free-radical production reactions). A free radical contains a single unpaired reactive electron. If that electron is donated to another species, the free radical is a reducing agent. If the free radical accepts an electron from another species, it is an oxidizing agent. The principal radiolytic species produced in aqueous solution are the hydrated electron e^-_{aq} and the hydrogen radical H·, which are powerful reducing agents, and the hydroxy radical OH· and hydroperoxy radical HO·$_2$, which are powerful oxidizing agents.[15,16] The concentration of the hydroperoxy radical, in particular, increases in the presence of oxygen. Because an antioxidant can protect a drug from being oxidized by oxidizing free radicals, it is also called a radical scavenger.

Free radicals may be present in the sodium pertechnetate solution added to a kit and may generate further in the radiopharmaceutical kit after it is reconstituted. Free-radical production is promoted in high-activity concentrations, when oxygen is present, and when low amounts of Sn(II) are in the kit. Molinski[17] showed that free-radical reactions in pertechnetate solutions generate hydrogen peroxide at a constant rate of 33×10^{-5} μg mCi^{-1} h^{-1} and that this is promoted by increased amounts of radioactivity and oxygen. The presence of ascorbic acid will reduce peroxide generation to insignificant levels. Excess stannous ion in kits has also been shown to act as an inhibitor of technetium complex decomposition mediated by oxygen.[14]

The oxygen-catalyzed radiolytic reaction probably occurs through the generation of two free-radical oxidation intermediates.[12] The first intermediates are alkoxy or hydroxy radicals produced by the scission of the oxygen–oxygen bond of peroxides according to Equation 10-6. The other intermediate is the peroxy radical formed from oxygen and radiolysis byproducts according to Equation 10-7.

$$RO - OH \longrightarrow RO^\bullet + OH^\bullet \quad (10\text{-}6)$$

$$R^\bullet + O_2 \longrightarrow RO_2^\bullet \quad (10\text{-}7)$$

These free-radical species may then degrade the Tc-complex with the generation of free pertechnetate according to the following generalized reaction:

$$Tc - Ligand \xrightarrow{RO_2^\bullet} Ligand + TcO_4^- \quad (10\text{-}8)$$

Some technetium radiopharmaceuticals, such as bone imaging agents, are particularly prone to oxidation. Kits with low levels of Sn(II) can be stabilized against oxidation by free-

TABLE 10-4 Standard Oxidation Potentials for Various Antioxidants

Antioxidant	Oxidation Potential E° (volts)	pH	Temperature (°C)
Sodium thiosulfate	+0.50	7.0	30
Ascorbic acid	+0.003	7.0	25
	−0.115	5.2	30
	−0.136	4.6	30
Methylene blue	−0.111	7.0	30
Sodium methabisulfite	−0.114	7.0	25
Sodium bisulfite	−0.117	7.0	25

radical scavenging antioxidants such as ascorbic acid, gentisic acid, paraaminobenzoic acid (PABA), and methylene blue (Table 10-4). Antioxidants function by donating reactive hydrogen atoms to the free-radical intermediates to yield a resonance-stabilized and nonreactive molecule, RO_2H (Figure 10-5).[12] An example of bone kits stabilized by sodium ascorbate and gentisic acid is shown in Figure 10-6.[12,18]

It should be noted that there are some exceptions to excluding air from technetium kits. With certain radiopharmaceuticals, air is deliberately added to the kit to minimize the formation of radiochemical impurities. For example, air is added to the ^{99m}Tc-MAG3 kit to oxidize excess stannous ion, which might otherwise further reduce the oxidation state of technetium in the desired complex. Another example is the addition of air to the ^{99m}Tc-tetrofosmin kit, which has been shown to minimize autoradiolysis of the technetium complex by reducing free radicals.[19]

Carrier Technetium

A ^{99m}Tc generator eluate will contain both ^{99m}Tc and ^{99}Tc isomers. Since ^{99m}Tc decays to ^{99}Tc, which is essentially stable, the total number of technetium atoms will remain constant over time. However, the ratio of ^{99}Tc to ^{99m}Tc atoms will increase over time. For example, the ratio of ^{99}Tc to ^{99m}Tc atoms is 2.6 in the eluate obtained from a generator that has not been eluted for 24 hours. This ratio will increase to 10.4 in 12 hours at the expiration time of the generator eluate. If a generator has not been eluted for 72 hours the ratio of ^{99}Tc to ^{99m}Tc atoms will be 12.1 at the time of elution, increasing to 48.4 in 12 hours (see Chapter 3). Because the ^{99m}Tc activity declines over time, its specific activity will also decline. Therefore, ^{99m}Tc is never "carrier free." This can be important in technetium radiopharmaceutical chemistry, because a number of technetium kits are formulated with low amounts of reducing agent, limiting the amount of ^{99m}Tc-pertechnetate that can be added to the kit. Kits with limited reducing power (e.g., the Ceretec kit [GE Healthcare], the UltraTag RBC and TechneScan MAG3 kits [Covidien]) that are reconstituted with pertechnetate from generators with long ingrowth will be most vulnerable to reduced ^{99m}Tc labeling yields because of the large amounts of ^{99}Tc. In such situations, labeling problems can be minimized by eluting the generator twice,

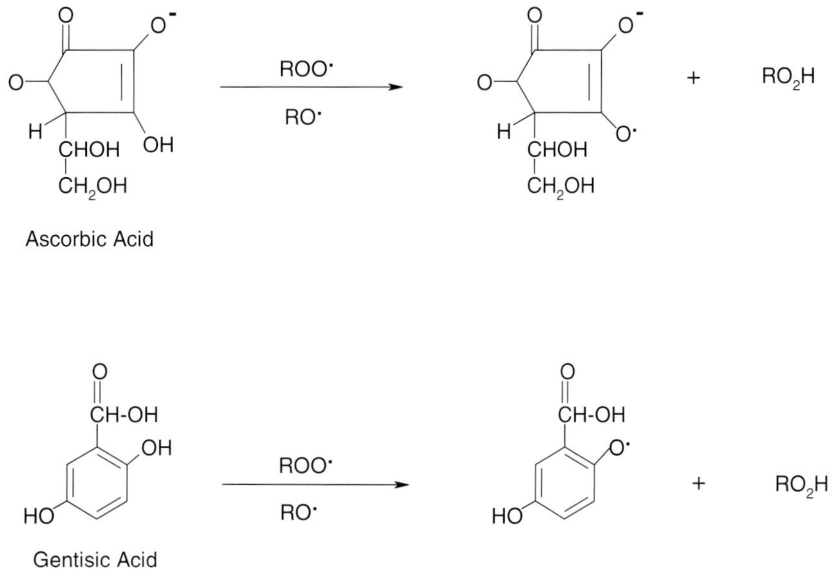

FIGURE 10-5 Mechanism of action of free-radical scavengers ascorbic acid and gentisic acid.

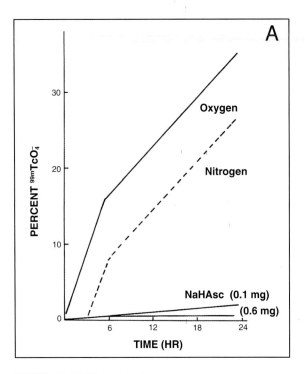 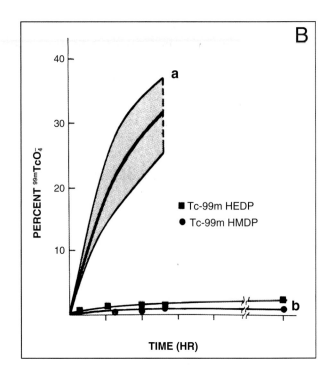

FIGURE 10-6 (A) In vitro stability of ^{99m}Tc-Sn-HEDP (^{99m}Tc-Sn-EHDP) without stabilizer, in vials under oxygen (air) and nitrogen atmosphere (top two curves); and with stabilizer (0.1 mg and 0.6 mg sodium ascorbate) under either oxygen (air) or nitrogen atmosphere (bottom two curves). (Reprinted with permission of the Society of Nuclear Medicine from reference 12.) (B) In vitro stability of ^{99m}Tc-HEDP and ^{99m}Tc-HMDP under (a) oxygen (no stabilizer) or (b) nitrogen or oxygen (with stabilizer). Stabilizer was 0.56 mg of either ascorbic acid or gentisic acid. (Reprinted with permission of the Society of Nuclear Medicine from reference 12.)

with the second elution made a few hours after the first elution. Pertechnetate from the second elution will have a more favorable ratio of ^{99m}Tc to ^{99}Tc and have less likelihood of causing inefficient labeling.

TECHNETIUM RADIOPHARMACEUTICALS

A major goal in nuclear medicine has been the design and development of target-specific radiopharmaceuticals. In first-generation agents, technetium is labeled or "tagged" to a molecular species that localizes technetium in a target organ through a substrate-nonspecific mechanism; that is, no specific biochemical interaction occurs between the technetium compound and a particular receptor site in the target organ. Localization of technetium in a target organ occurs, for example, by an in vivo physiologic process, such as simple diffusion, phagocytosis, entrapment, or cell sequestration, that is attributable to the properties of the compound that binds technetium. Thus, with first-generation technetium-labeled compounds, technetium is a passive atom attached to the localizing molecule.

Many of the early technetium complexes were not well characterized chemically because their technetium concentration (approximately 10^{-8} to 10^{-9} M, e.g., 10 mCi ^{99m}Tc = 1.9 ng) was below that required for conventional methods of analysis. Identification of the properties of technetium compounds was limited to analysis of tracer concentrations of technetium solutions. Properties such as oxidation state, formation constant, and electrical charge were assessed by polarography, chromatography, solvent extraction, and electrophoresis. As investigative studies progressed, however, macroscopic amounts of ^{99}Tc as ammonium pertechnetate were used to prepare carrier added (CA) technetium compounds. This allowed structural characterization by conventional methods such as infrared spectroscopy, mass spectroscopy, nuclear magnetic resonance spectroscopy, and x-ray crystallography.[7] An important advantage of using standard characterization methods with second-generation technetium complexes was that it permitted the chemical and biologic equivalence of CA (macroscopic) and NCA (no carrier added; tracer) quantities to be demonstrated.[20]

The development of second-generation technetium compounds began in the 1970s. Loberg and Fields,[21] in their quest to develop a technetium heart imaging agent, serendipitously discovered the substituted iminodiacetic acid (IDA) analogues for hepatobiliary imaging (HIDA compounds) (Figure 10-7). Their elegant work in characterizing the structure of the first hepatobiliary imaging complex (^{99m}Tc-lidofenin, or ^{99m}Tc-HIDA) led to two important contributions to the future development of technetium radiopharmaceuticals. The first was the discovery that the technetium atom in the ^{99m}Tc-HIDA complex was essential for uptake and excretion of the com-

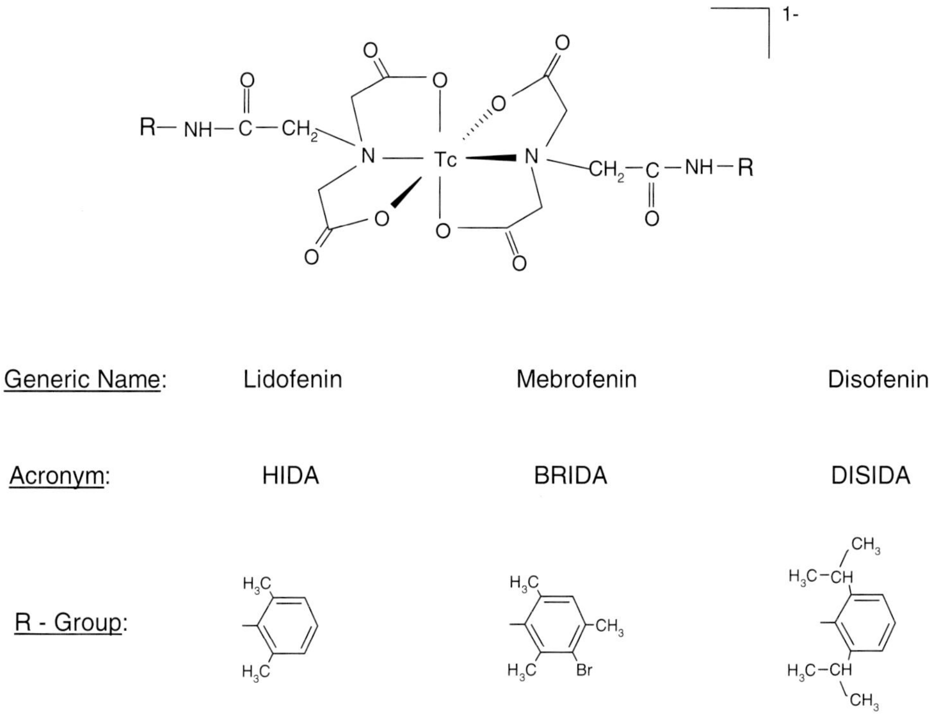

FIGURE 10-7 Chemical structure of iminodiacetic analogues.

plex by the liver.[22,23] In the absence of technetium, the IDA analogues were renally excreted following intravenous injection. However, when technetium was coordinated with IDA ligands, the principal excretion pathway was hepatobiliary.[23] The second important contribution of this work was the introduction of the concept of bifunctional chelating agents (BFCAs), ligands that not only chelate technetium but can be modified with functional groups to control biodistribution of the technetium complex.

With the ability to structurally characterize technetium compounds, research was eventually directed toward examining ligands that could stabilize technetium in lower oxidation states, previously thought to be unstable. Davison, Jones, et al.[24] at the Massachusetts Institute of Technology demonstrated that Tc(V) oxo complexes with bisdithiolate (S_4) and diamidedithiolate (N_2S_2) type ligands could produce oxidation-stable square pyramidal complexes. In these complexes the sulfur and nitrogen atoms form the basal plane of the pyramid, with the technetium atom displaced toward the apex and bonded to an oxygen atom at the apex (Figure 10-8). The presence of the ligating groups on the technetium–oxygen core makes the configuration behave electronically as a closed shell and renders the complex kinetically inert.[25] Work with these ligands laid the foundation for introducing ligand backbone substitutions with noncoordinating functional groups that could influence in vivo localization. This permitted the development of a new generation of technetium-labeled radiopharmaceuticals that were technetium-essential. Important among these technetium-essential compounds are ^{99m}Tc-bicisate, ^{99m}Tc-exametazime, and ^{99m}Tc-mertiatide with the technetium(V)–oxo core; and ^{99m}Tc-tetrofosmin with a technetium(V)–dioxo core.[26–30]

During the development of these agents a number of technetium cores were identified around which a radiopharmaceutical could be designed.[31–33] The concept of using a technetium core coordinated to a BFCA in the design of technetium radiopharmaceuticals was developed further in the 1990s by the addition of a reactive site on the BFCA that enables conjugation of the technetium complex to another molecule for targeting purposes, creating what are known as second-generation technetium-tagged radiopharmaceuticals.

TECHNETIUM CORES

Technetium compounds have a core where the technetium atom exists in a specific oxidation state, determined by the number of electrons in its d orbitals. In some cores the technetium atom exists alone ("naked" technetium atom) and in other cores it is associated with another functional group such as oxygen or nitrogen. The different oxidation states of technetium are stabilized in these cores by a variety of coordinating ligands.[25] Technetium cores found in a number of diagnostic radiopharmaceuticals are shown in Figure 10-9. Cores associated with other technetium complexes have also been identified.[32,34]

Tc(I) Compounds

Reduction of Tc(VII), as pertechnetate, to Tc(I) creates a technetium atom with six additional electrons (d^6 configuration) that must be stabilized by pi-acceptor ligands.[31] Some

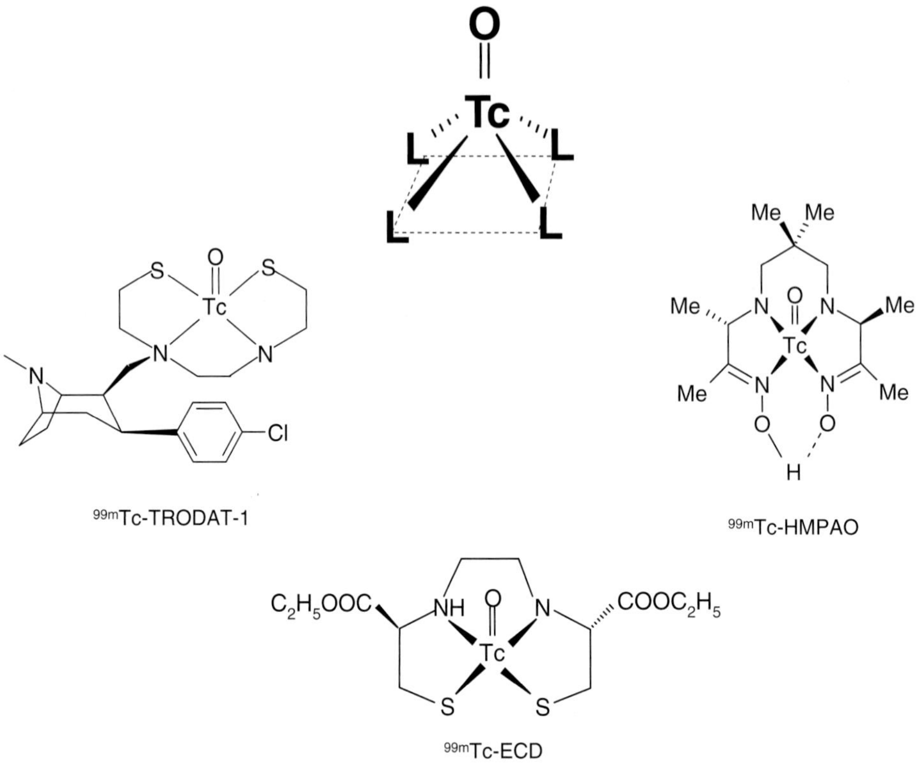

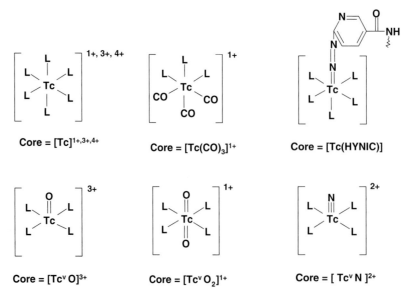

FIGURE 10-8 Tc(V) in a five-coordinate square pyramidal complex (top structure) with the oxygen atom at the apex and the coordinating ligand atoms forming the basal plane with the technetium atom displaced toward the apex. Other structures are examples of radiopharmaceuticals that incorporate this Tc(V) complex.

FIGURE 10-9 Technetium cores and the general structures of technetium complexes with ligands. L = neutral donor ligand.

of the coordinating ligands that will stabilize Tc(I) are the phosphines (P), diphosphines (P-P), isonitriles (CNR), and carbonyl (CO). Technetium is stabilized by these ligands because they have pi-bonding orbitals devoid of electrons, which can be filled by the d-orbital electrons from the Tc(I) atom. The cores most frequently explored are the naked core $[Tc]^+$ and the tricarbonyl core $[Tc(CO)_3]^+$. Tc(I) complexes are very stable when six coordination sites are occupied, forming an octahedral complex. This is readily accomplished under NCA labeling conditions in radiopharmaceutical kits, because the excess ligand available forces complete coordination around the metal to maintain reducing conditions. In addition, the ligands in Tc(I) complexes can be functionalized with groups that alter chemical properties, such as lipophilicity, without affecting complex stability. The functional groups chosen can also influence biologic localization in vivo.

$[Tc]^+$ Core

A prime example of a radiopharmaceutical with this naked core is the lipophilic heart imaging agent ^{99m}Tc-sestamibi, in which the Tc(I) atom is coordinated by six monodentate 2-methoxy-isobutyl isonitrile (MIBI) ligands, forming a stable octahedral complex (Figure 10-10). Since the MIBI ligands are neutral, the sestamibi complex retains the single positive charge of the Tc$^+$ core.

$[Tc(CO)_3]^+$ Core

A variety of technetium complexes can be made starting with the tricarbonyl core. This core can be coordinated with a BFCA having residual functional groups to couple technetium to receptor-avid molecules. At this time, no technetium radiopharmaceuticals with this core have been approved for routine use, but it offers a unique method of creating target-specific radiotracers.

Radiopharmaceuticals with the $[Tc(CO)_3]^+$ core form stable octahedral organometallic complexes of two subtypes: (1) $[fac\text{-}Tc(CO)_3]^+$, in which Tc(I) can accommodate a variety of ligands besides the three carbonyls to complete the octahedral sphere, and (2) $[CpTc(CO)_3]^+$, in which, in addition to the three carbonyls, Tc(I) is coordinated to a functionalized cyclopentadiene ligand that can attach the complex to a targeting molecule.[35] A novel synthon of $[fac\text{-}Tc(CO)_3]^+$ is the water- and air-stable organometallic aqua complex $[fac\text{-}Tc(CO)_3(H_2O)_3]^+$ (Figure 10-10).[36] (A synthon is a molecular unit designed to facilitate the synthesis of a radiopharmaceutical.) The labile water ligands in this synthon can be readily substituted with donor ligands from a functionalized targeting molecule. The $[fac\text{-}Tc(CO)_3(H_2O)_3]^+$ synthon can be prepared by adding sodium pertechnetate to an Isolink kit (Covidien) containing sodium tartrate 8.5 mg, sodium tetraborate decahydrate 2.85 mg, sodium carbonate 7.15 mg, and sodium boranocarbonate 4.5 mg and heating the mixture for 40 minutes at 100°C.

The synthon $[fac\text{-}Tc(CO)_3Cl(H_2O)_2]$ has been used to prepare a neutral lipophilic complex, TROTEC-1 (Figure 10-10), wherein two water ligands are displaced by the sulfurs in a dithioether-derivatized tropane analogue. The complex is neutral because of the presence of the chloride ligand. TROTEC-1

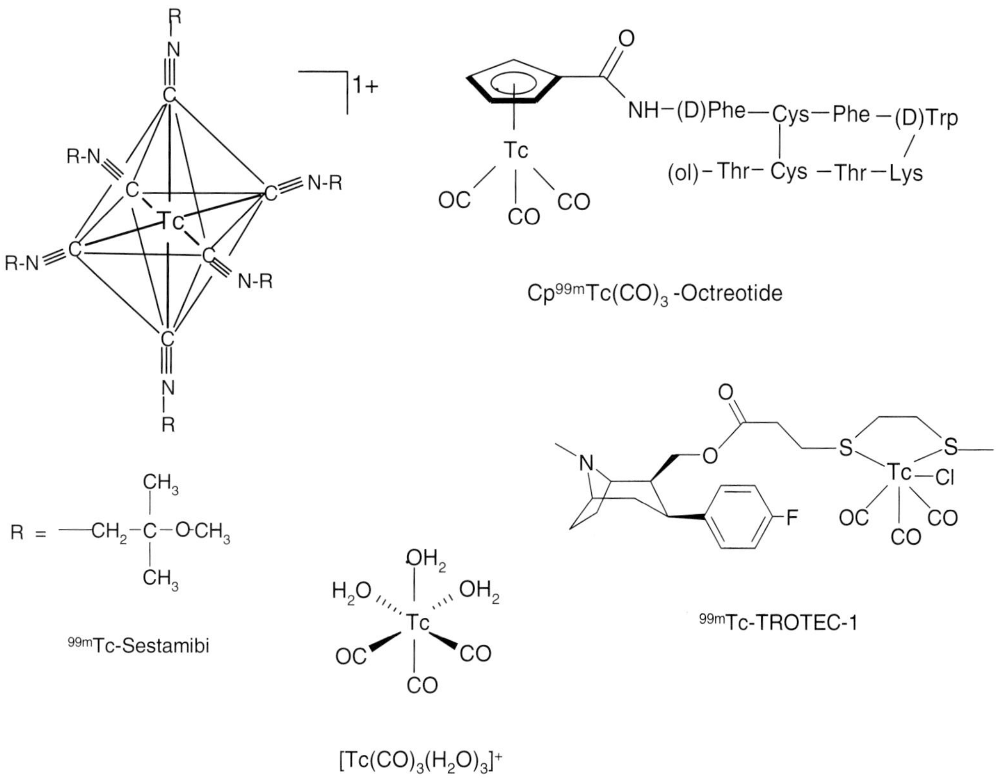

FIGURE 10-10 Chemical structures of various coordination compounds with technetium in the 1+ oxidation state.

has been shown to have high affinity for the dopamine transporter (DAT) in the brain; however, brain uptake is low.[37]

The cyclopentadienyl ($C_5H_4COCH_3$ or Cp) core complex [fac-$Tc(CO)_3Cp$]$^+$ is interesting because it is highly stable, lipophilic, and can be readily derivatized for conjugation to bioactive molecules. This core complex has been used to prepare diagnostic technetium or therapeutic rhenium compounds because the Cp moiety is readily functionalized to introduce a targeting molecule. The method was originally designed to accomplish the reduction, carbonylation, and cyclopentadienylation of pertechnetate in a relatively mild, one-pot reaction termed a "double ligand transfer" reaction because two ligands (Cp and CO) are transferred together from two different metals (Fe and Mn) to a third metal, Tc.[38,39] For example, this method was applied to produce a $Tc(CO)_3Cp$–octreotide conjugate (Figure 10-10), which demonstrated receptor-mediated uptake in the adrenal glands and pancreas.[40]

A novel approach has been developed for preparing bifunctional chelates with the [$Tc(CO)_3$]$^+$ core.[41] This synthetic approach is based on a single amino acid chelate (SAAC). The design of a bifunctional SAAC exploits modification of a natural or synthetic amino acid to incorporate a tridentate chelation moiety to bind the [$Tc(CO)_3$]$^+$ core and a terminus for conjugation to a small peptide via solid-phase synthesis. Figure 10-11 illustrates the general approach for labeling a SAAC with ^{99m}Tc.

Tc(III) and Tc(IV) Compounds

The "naked" technetium atom with intermediate oxidation states, Tc(III) ([Tc]$^{3+}$ core) and Tc(IV) ([Tc]$^{4+}$ core), typically is six-coordinate but may have coordination numbers of five, six, or seven.[32] In addition, complexes of intermediate oxidation states may undergo redox reactions, exhibiting different oxidation states with a coordinating ligand. Evidence for this has been demonstrated in technetium complexes with DTPA, citrate, phosphonates, and other ligands.[10] A mixture of different ligands is sometimes necessary to stabilize an oxidation state, as exemplified by the boronic acid adducts of technetium oxime (BATO compounds), namely, ^{99m}Tc-teboroxime; the final complex contains a heptacoordinate Tc(III) atom bound to a chlorine atom and to the six nitrogens of three dioxime ligands.[42]

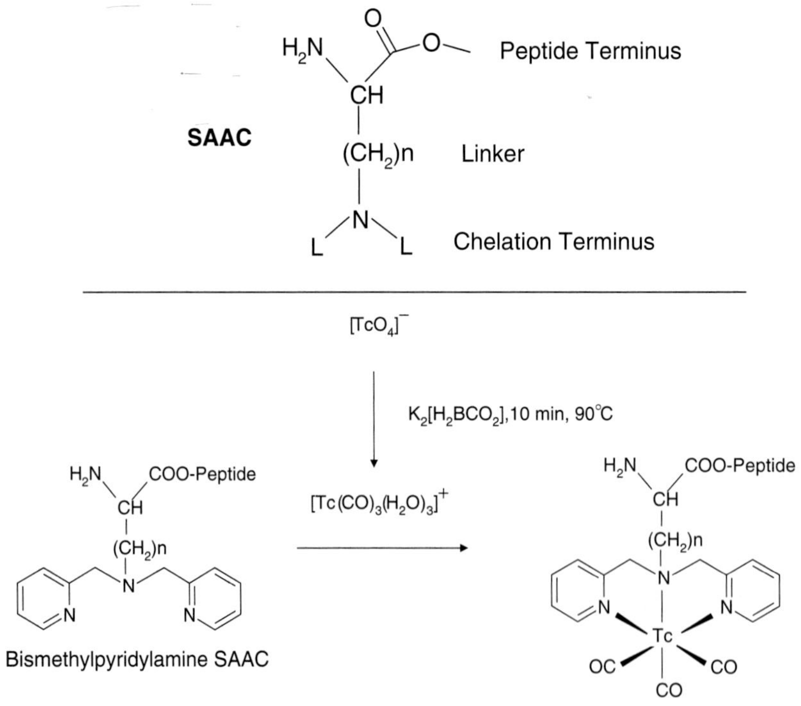

FIGURE 10-11 Single amino acid chelate (SAAC) design of technetium radiopharmaceuticals. The upper panel illustrates a SAAC molecule showing termini for peptide conjugation and radionuclide chelation. The linker can be adjusted to provide optimal separation between the termini, and the amino group can be functionalized to optimize radiotracer properties. The lower panel illustrates radiolabeling of a preformed SAAC with ^{99m}Tc by chemical reduction of pertechnetate with potassium boranocarbonate and coordination of the [$Tc(CO)_3(H_2O)_3$]$^+$ core by the bismethylpyridylamine chelator with displacement of the three labile water ligands in the technetium core by three nitrogen ligands in the chelator. (Source: References 36 and 41.)

[Tc]³⁺ Core

Reduction of Tc(VII) as pertechnetate to Tc(III) creates a technetium atom with four additional electrons (d^4 configuration). The Tc(III) state is sometimes reached in multiple reduction steps, initially to Tc(V) and then to Tc(III).[32] Several technetium compounds with a Tc^{3+} core have been developed for use in nuclear medicine. These include the ⁹⁹ᵐTc-iminodiacetic acid (IDA) analogues, ⁹⁹ᵐTc-succimer, ⁹⁹ᵐTc-teboroxime, and ⁹⁹ᵐTc-furifosmin.

[Tc]⁴⁺ Core

Reduction of Tc(VII) to Tc(IV) creates a technetium atom with three additional electrons (d^3 configuration). Several first-generation technetium compounds have technetium in the Tc(IV) oxidation state. Tc(IV) radiopharmaceuticals prepared from kits are formed by reduction of pertechnetate with stannous tin in the presence of a coordinating ligand. These include technetium complexes with pentetate (DTPA), the phosphonates (MDP, HDP), and pyrophosphate (PPi). Although these agents are included under the Tc(IV) oxidation state, investigations on oxidation state chemistry have not demonstrated this conclusively in every case; the Tc(III) state may also occur, depending on the reducing conditions.[7] For example, the likely oxidation state of technetium in ⁹⁹ᵐTc-DTPA prepared from kits is Tc(IV);[43] however, other oxidation states, such as Tc(III) and Tc(V) have also been shown to exist, depending on the reduction conditions.[43,44] An important compound in this oxidation state is hydrolyzed reduced technetium ($TcO_2·H_2O$), which exists in the Tc(IV) state, most likely because of its insolubility over a wide range of pH.[7] Although $TcO_2·H_2O$ is not useful in nuclear medicine per se, it is problematic as a radiochemical impurity in technetium kits.

Tc(V) Compounds

Somewhat opposite the Tc(I) oxidation state is Tc(V), d^2 configuration, which is two electrons reduced from Tc(VII) in pertechnetate. As such, Tc(V) has a high electron deficiency (5−) and requires good electron-donating ligands to confer stability to its complexes.[32] Typical donor atoms coordinating with Tc(V) are N, O, S, and P, which are present in several ligands used to prepare technetium radiopharmaceuticals. Two common cores present in Tc(V) radiopharmaceuticals are $[Tc=O]^{3+}$ and $[O=Tc=O]^+$. The net charge on the core is determined by balancing the total charge of the oxo groups with that of technetium. For example, the 3+ charge on the $[Tc=O]^{3+}$ core results from the sum of 5+ on technetium and 2− on oxygen. Likewise, the net charge on a technetium complex with one of these cores is a balance of the total charge of the ligating groups and that of the technetium core.

[Tc=O]³⁺ Core

Technetium compounds containing this core are five-coordinate, forming square pyramidal complexes with the technetium atom bonded to oxygen oriented apically and four coordinating ligands forming the basal plane (Figure 10-8). Important ligands that have produced stable in vivo complexes with the $[Tc=O]^{3+}$ core include the triamidethiol (N_3S) ligand (N-[mercaptoacetyl]glycylglycylglycine) in ⁹⁹ᵐTc-MAG3, the PnAO (N_4) ligand (hexamethylpropyleneamineoxime) in ⁹⁹ᵐTc-HMPAO, and the diaminedithiol (N_2S_2) ligands (N,N′-1,2-ethenediylbis-L-cysteine diethylester, otherwise known as ethylcysteinate dimer) in ⁹⁹ᵐTc-ECD and ⁹⁹ᵐTc-TRODAT-1, where TRODAT is [2-[[2-[[[3-(4-chlorophenyl)-8-methyl-8-azabicyclo[3.2.1]oct-2-yl]methyl](2-mercaptoethyl)amino]ethyl]amino] ethanethiolato(3-)-N2,N2′,S2,S2′]oxo-[1R-(exo-exo)], a diaminedithiol complex of the TcO^{3+} core with a tropane analogue derivatized from one nitrogen molecule. The N_2S_2 and N_3S ligands form very stable technetium complexes and can be fitted with functional groups to alter biodistribution.

[O=Tc=O]⁺ Core

The dioxo core of Tc(V) has two trans-oxygen atoms that neutralize four of the five positive charges on the technetium atom, creating an overall core charge of 1+. The $[O=Tc=O]^+$ core forms six-coordinate octahedral complexes with technetium, with oxygen atoms occupying the two apical sites and coordinating ligands occupying the four equatorial sites. Cyclam (N-N-N-N) and diphosphine (P-P) ligands have been successfully used to produce stable complexes with the $[O=Tc=O]^+$ core.[32] The most important technetium dioxo compound to date is ⁹⁹ᵐTc-tetrofosmin (Figure 10-12), which is the cationic complex [⁹⁹ᵐTc-(tetrofosmin)₂O₂]⁺ in which tetrofosmin is the ether-functionalized diphosphine ligand 1,2-bis [bis(2-ethoxyethyl) phosphino]ethane.

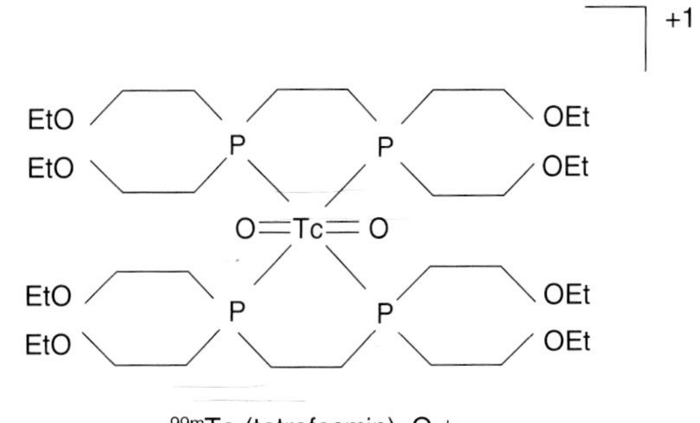

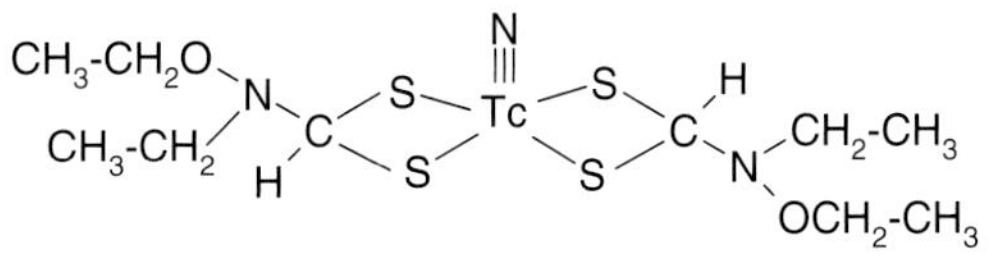

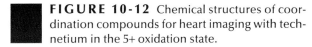

FIGURE 10-12 Chemical structures of coordination compounds for heart imaging with technetium in the 5+ oxidation state.

[Tc≡N]²⁺ Core

The [Tc≡N]²⁺ core can be produced in stable form at the NCA level.[32] A ⁹⁹ᵐTc-nitrido compound, [bis (N-ethyl-N-ethoxydithiocarbamato)nitrido ⁹⁹ᵐTc(V)], was developed for myocardial perfusion imaging.[45–47] It is a neutral lipophilic complex with a [Tc≡N]²⁺ core with a Tc(V) atom triple-bonded to a strong pi-electron donor nitride atom (N³⁻) and four donor sulfur atoms (Figure 10-12). A lyophilized kit has been developed [⁹⁹ᵐTc-N-(NOEt)₂, from CIS Bio International, France] using stannous chloride that permits labeling at neutral pH.[47] After intravenous injection, ⁹⁹ᵐTc-N-(NOEt)₂ localizes in the myocardium proportional to blood flow. Following myocardial uptake, ⁹⁹ᵐTc-N-(NOEt)₂ redistributes from the heart and has been compared with ²⁰¹Tl for myocardial perfusion imaging.[48,49]

[Tc-HYNIC] Core

HYNIC (6-hydrazinonicotinamide) in the [Tc-HYNIC] core was first utilized by Abrams et al.[50] to label polyclonal immunoglobulin G. Since then it has been used to label chemotactic peptides, somatostatin analogues, and other biologically interesting molecules.[34] This core forms complexes of the general form [⁹⁹ᵐTc(HYNIC–peptide)(L)₂], in which the HYNIC group satisfies one coordination site on technetium and the remaining sites are completed by various co-ligands. A ternary ligand system with the general form [⁹⁹ᵐTc(HYNIC–peptide)(tricine)(L)] has been used to prepare technetium complexes. The system contains three different ligands: the bifunctional coupling group (HYNIC), a monodentate triphosphine co-ligand (L), which is trisodium triphenylphosphine-3,3′,3″-trisulfonate (TPPTS), and a tetradentate co-ligand, tris(hydroxymethyl)methylglycine (tricine). Following this idea, a technetium complex that targets the glycoprotein (GP) IIb/IIIa platelet receptor (Figure 10-13) was developed using the postlabeling approach, where the peptide–HYNIC conjugate is formed prior to coordination with technetium and the co-ligands.[51] The components in the GP IIb/IIIa ternary ligand complex exist in a 1:1:1:1 ratio of Tc:HYNIC–peptide:tricine:phosphine. This complex demonstrated arterial and venous thrombi in a canine model with thrombus-to-muscle ratios of ~10:1 at 2 hours.[52]

The oxidation state of technetium in HYNIC complexes is not clear and may depend on the type of co-ligand, taking into account that some ligands (e.g., phosphines) have reducing capability.[34,51]

Tc(VII) Compounds

The Tc(VII) oxidation state is characterized by technetium in the d⁰ configuration, its highest, most stable oxidation state, in which technetium has lost all seven valence electrons. There are two principal compounds used in nuclear medicine with technetium in the 7+ oxidation state: ⁹⁹ᵐTc-sodium pertechnetate and ⁹⁹ᵐTc-sulfur colloid.

TECHNETIUM-ESSENTIAL RADIOPHARMACEUTICALS

Technetium-labeled compounds are considered to be of two types: technetium essential and technetium tagged.[20] *Technetium-essential* compounds have technetium as a neces-

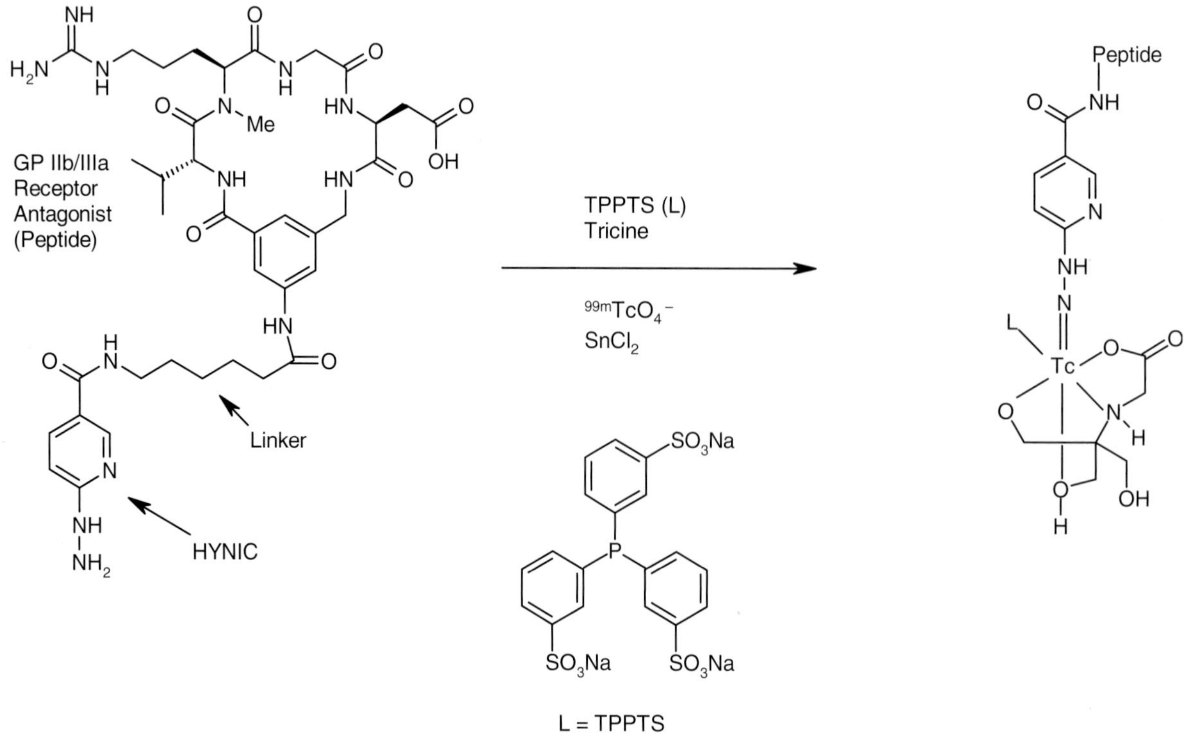

FIGURE 10-13 General scheme for the production of ⁹⁹ᵐTc-HYNIC–peptide conjugates.

sary core atom around which other components are arranged. Neither of the individual components alone (coordinating ligand or technetium core) localize in the same way that the integrated molecule does. The ligands that coordinate with the core may be monodentate or multidentate and are designed to stabilize technetium in its oxidation state and provide desirable pharmacokinetic properties to the final complex. Technetium's coordination number may be satisfied by multiple monodentate ligands, such as the six individual isonitrile ligands in ^{99m}Tc-sestamibi; one or two multidentate ligands, such as those in ^{99m}Tc-bicisate and ^{99m}Tc-tetrofosmin, respectively; or a combination of a multidentate ligand and individual monodentate ligands, such as that found in ^{99m}Tc-furifosmin. The functional groups on the ligands are chosen to confer certain properties to the final complex, such as lipophilicity, ionic charge, and molecular size. Such modifications alter the pharmacokinetic properties of the technetium complex to enhance its localization and excretion.[32] For example, the design of ^{99m}Tc-mertiatide (^{99m}Tc-MAG3) not only involved careful selection of the N_3S coordinating ligand but also required the strategic location of a carboxylate substituent in the peptide sequence to give the complex the necessary renal excretion properties, which mimic *o*-iodohippurate (OIH).[31] Another example is the rearrangement of two methyl groups on the propyleneamine oxime (PnAO) ligand in ^{99m}Tc-PnAO, producing the configuration found in ^{99m}Tc-HMPAO, in order to increase the brain retention properties of ^{99m}Tc-HMPAO.[28]

TECHNETIUM-NONESSENTIAL RADIOPHARMACEUTICALS

Technetium-nonessential compounds have a technetium atom bound (tagged) to a substance that is responsible for technetium's in vivo localization. The technetium atom is passive and not required for in vivo localization. First-generation technetium-nonessential compounds have technetium bound to diverse types of substances that include complexing agents (e.g., DTPA), particles (e.g., sulfur colloid or macroaggregated albumin), blood cellular elements (e.g., red blood cells), and proteins (e.g., human serum albumin). These technetium compounds localize by simple physiologic processes such as diffusion, phagocytosis, capillary blockade, and compartmental confinement. Second-generation technetium-nonessential compounds have technetium bound to a receptor-binding moiety to localize technetium via biochemical interaction at a specific receptor site. The design and labeling methods of these compounds are more sophisticated than those of first-generation technetium-nonessential radiopharmaceuticals.

Two approaches have been used in the design of second-generation technetium-nonessential radiopharmaceuticals: the integrated approach and the bifunctional chelate approach.[33–35,53] The *integrated approach* incorporates technetium into a binding site within a molecule so that technetium becomes an integral part of the molecule without affecting its conformation and receptor-binding affinity in vivo. This approach has been applied, for example, to the design of radiotracers that mimic the three-dimensional configuration of biologically important molecules such as steroids (testosterone, progesterone, and estradiol) that can bind to specific receptors.[33,35] Integrating a technetium complex into a receptor-binding motif without affecting receptor specificity presents a great challenge in this approach.

In the *bifunctional chelate approach*, technetium is coordinated to a molecule that has a technetium-binding moiety and a target-localizing moiety. The extensive experience gained from the development of technetium-essential radiopharmaceuticals with BFCAs, particularly with the tripeptide MAG3, made the BFCA approach to peptide labeling a natural extension of that work. Furthermore, it instilled the idea of incorporating a technetium-binding amino acid sequence into an active peptide molecule. The peptide sequence allows the introduction of coordinating donor groups to form a stable complex with a technetium core and an activation site that can conjugate with a targeting molecule such as a peptide or antibody.

Second-generation technetium-tagged compounds have the following general components: *Targeting molecule–Linker–BFCA–^{99m}Tc*.[34] The targeting molecule is often a peptide, antibody, or some other small molecule designed to target a specific receptor in vivo. The linker is a simple hydrocarbon chain of variable length for modifying pharmacokinetics or distancing the technetium chelate region from the receptor-binding region of the targeting molecule. A linker molecule may or may not be necessary. The BFCA serves two main purposes: (1) to coordinate technetium and (2) to provide a molecular backbone that can be modified with functional groups for attachment to the targeting molecule. Some examples of BFCAs are the N_2S_2 ligands diaminedithiol, diamidedithiol, and monoaminemonoamide dithiol; triamidethiol (N_3S); and PnAO (N_4), all of which form five-coordinate square pyramidal technetium complexes, and hydrazine nicotinamide (HYNIC). The functional group on the BFCA is the conjugation site where it covalently attaches to the targeting molecule, either directly or through the linker molecule. With this design the technetium chelate is often far removed from the receptor-binding motif to minimize possible interference with binding at the biologic receptor site. In such complexes, technetium is a passive component nonessential for localization. However, in some technetium complexes with small peptides, biodistribution and uptake at the receptor may be influenced by the metal chelate as a whole because the technetium atom may contribute greatly to the overall molecular size and configuration of the radiopharmaceutical.[34] In such cases technetium is not entirely passive and the radiopharmaceutical may be considered to be technetium essential.

The principal targeting molecules employed in technetium-nonessential radiopharmaceuticals are antibodies and peptides.[34] They differ primarily in molecular weight and structure. Antibodies are analogous to large and small proteins in size. Whole antibodies have molecular weights on the order of 150 kDa and antibody fragments about 50 kDa for Fab and 100 kDa for F(ab')$_2$. By contrast, peptides usually contain fewer than 100 amino acids and have molecular weights of about 10 kDa or less. Although antibodies exhibit high receptor-binding affinity and specificity, their lack of effectiveness has been attributed to the inaccessibility of the cells in solid tumors and the heterogeneous distribution of tumor-associated antigens on the tumor surface. By contrast, the affinities of many peptides for their receptors are significantly greater than that of antibodies or their fragments. In most cases, the receptors for peptides are readily accessible on the external surface of cell membranes. Peptides are often modified to improve stability and targeting. Replacing an L-enantiomer with a D-enantiomer minimizes degradation by peptidases,[54] and changing amino acids can improve receptor selectivity.[55]

RADIOLABELING APPROACHES FOR BIFUNCTIONAL CHELATES

Conjugation of the peptide, protein, or antibody targeting molecule (TM) with the BFCA is often accomplished through a reaction between a primary amino group on the TM and an activated ester group or an isothiocyanate group on the BFCA, or between a sulfhydryl group on the TM and a maleimide group on the BFCA (Figure 10-14).[34] Conjugation can occur either after coordination with technetium (prelabeling approach) or before coordination with technetium (postlabeling approach) (Figure 10-15).

The *prelabeling approach* involves the following sequence of reactions: chelation of technetium with the BFCA, activation of the BFCA, and conjugation with the TM. The advantage of prior chelation of technetium with the BFCA is that the TM is not subjected to the sometimes harsh labeling conditions (e.g., low pH, high temperature) necessary for coordination of technetium with the BFCA. The disadvantage is that the prelabeling approach is not particularly amenable to simple kit formulation. The sequence of reactions with the *postlabeling approach* involves activation of the BFCA, conjugation with the TM, and chelation with technetium. The advantage of this approach is that it permits a carefully worked out chemistry for conjugation of the TM with the BFCA. It also has particular appeal for kit formulation, provided that the chelation reaction conditions with technetium are not detrimental to the TM. Radiolabeling can be accomplished with either approach by direct reduction of pertechnetate in the presence of the BFCA–TM conjugate or via ligand exchange with a technetium donor complex such as ^{99m}Tc-glucoheptonate. The postlabeling approach has been used in the preparation of technetium-labeled peptides, such as ^{99m}Tc-apcitide (AcuTect [Diatide]) and ^{99m}Tc-depreotide (NeoTect [Diatide]).

A similar labeling approach is used for some non-technetium radiopharmaceuticals labeled with radionuclide metals. For example, capromab pendetide (ProstaScint [Covidien]) is an antibody covalently conjugated with a short peptide–DTPA linker (glycyl-tyrosyl-lysyl-DTPA or GYK-DTPA) to complex the antibody with ^{111}In-indium by the postlabeling approach. Similarly, ibritumomab tiuxetan (Zevalin [Biogen Idec]) is an antibody conjugated with tiuxetan, an isothiocyanatobenzyl-derivatized DTPA linker, bound covalently by a thiourea bond to the antibody to form a site for chelation with ^{111}In-indium or ^{90}Y-yttrium by the postlabeling approach.

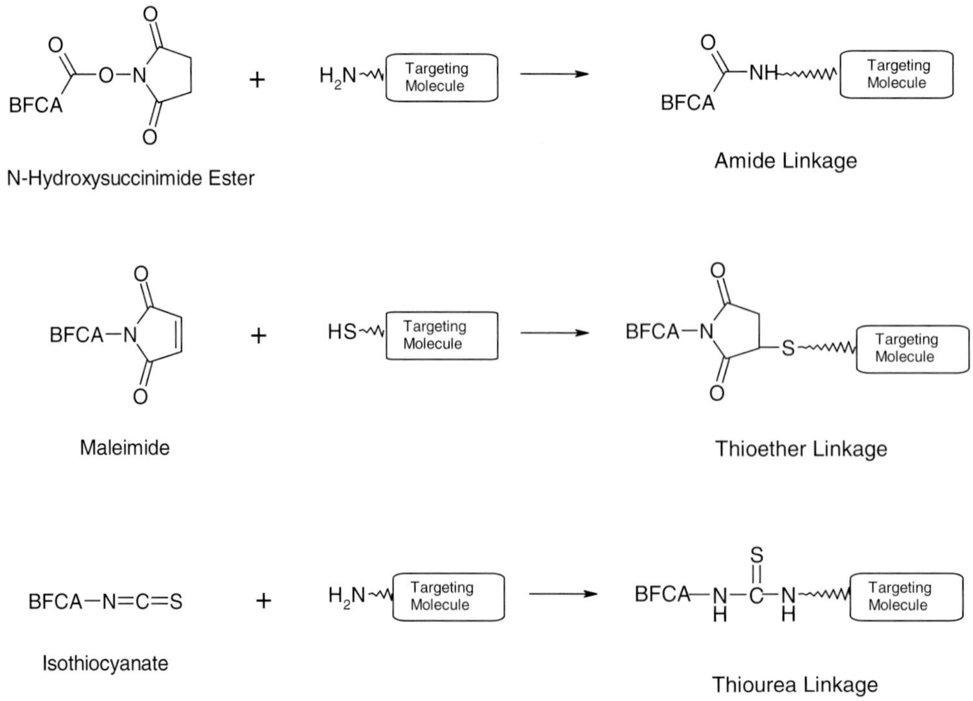

FIGURE 10-14 Typical methods of conjugating bifunctional chelating agents (BFCAs) with targeting molecules.

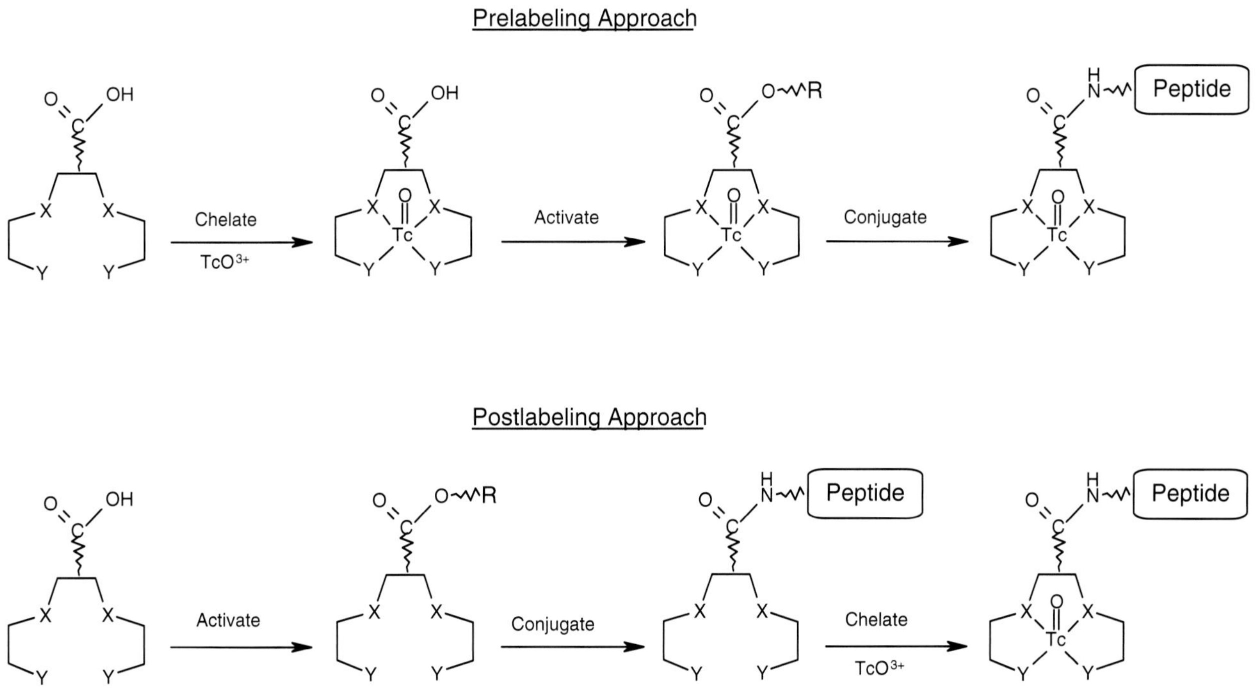

FIGURE 10-15 Prelabeling and postlabeling approaches for coordinating technetium to a targeting molecule (peptide). See text for details.

TECHNETIUM RADIOPHARMACEUTICAL PREPARATION AND PROPERTIES

Technetium can exist in a number of oxidation states. It is obtained from the ^{99m}Tc generator as sodium pertechnetate with technetium in the Tc(VII) oxidation state. The preparation of a technetium radiopharmaceutical requires the addition of pertechnetate to a kit, whereupon technetium is reduced to a lower oxidation state, typically Tc(I), Tc(III), Tc(IV), or Tc(V), before it will complex with a coordinating ligand to form the desired technetium radiopharmaceutical complex. Chemical reduction is usually accomplished with stannous tin. In addition to the technetium complex, various undesirable radiochemical impurities can be produced during radiolabeling reactions, the principal ones being insoluble hydrolyzed–reduced technetium (TcO$_2$·H$_2$O) and pertechnetate. The amount of each technetium species present in technetium radiopharmaceuticals can be determined by simple radiochromatography procedures.

Sodium Pertechnetate Tc 99m Injection

Sodium pertechnetate Tc 99m injection is a sterile aqueous solution of ^{99m}Tc and ^{99}Tc as sodium pertechnetate in 0.9% sodium chloride injection obtained by elution of the ^{99}Mo–^{99m}Tc generator (Ultra-TechneKow DTE [Covidien], TechneLite [Lantheus]). It is the source material for reconstituting radiopharmaceutical kits. Specifications and quality control requirements for the generator eluate are stated in the manufacturer's package insert, in NRC regulations in 10 CFR 35.204 and 10 CFR 35.2204, and in the *United States Pharmacopeia and the National Formulary (USP 28/NF 23)*.[56] The radioactivity concentration of ^{99m}Tc and ^{99}Mo in the eluate must be determined by a suitable calibrated instrument, such as a dose calibrator, to determine the ^{99}Mo to ^{99m}Tc activity ratio. The amount of ^{99}Mo activity must not exceed 0.15 μCi ^{99}Mo per mCi ^{99m}Tc at the time of patient dose administration. NRC regulations require this test only on the first eluate after receipt of a generator. NRC agreement-state regulations may require this test to be done with each elution. NCR regulations in 10 CFR 35.2204 require that a record, listing the ^{99}Mo concentration, the time and date of the measurement, and the name of the individual who made the measurement, be maintained for 3 years.

According to *USP 28/NF 23*, the presence of aluminum ion in the eluate must not exceed 10 μg/mL. The NRC does not require a test for aluminum ion. The manufacturer's instructions in the generator package insert state that the aluminum ion concentration and the ratio of ^{99}Mo microcuries to ^{99m}Tc millicuries in each eluate be determined prior to administration.

The radiochemical purity must be not less than 95% as ^{99m}Tc-sodium pertechnetate and not more than 5% other chemical forms of technetium.[56] The pH range of the generator eluate is between 4.5 and 7.5. The generator package insert states that the expiration time for ^{99m}Tc-sodium pertechnetate injection is not later than 12 hours from the time of generator elution, because the eluate does not contain an antimicrobial preservative. A generator expires 14 days from the calibration date printed on the generator label.

TABLE 10-5 Composition of ^{99m}Tc-MAA Kits

DraxImage MAA		Pharmalucence (Pulmolite)	
Human albumin aggregated	2.5 mg	Human albumin aggregated	1.0 mg
Human albumin	5.0 mg	Human albumin	10.0 mg
SnCl$_2$·H$_2$O	0.06–0.11 mg	SnCl$_2$	2.4–7.0 µg
Sodium chloride	1.2 mg		

Technetium Tc 99m Albumin Injection

Technetium Tc 99m albumin injection (^{99m}Tc-albumin or ^{99m}Tc-HSA) is a sterile aqueous solution of ^{99m}Tc complexed to human serum albumin. A kit for preparing ^{99m}Tc-HSA is no longer on the market, but one can be prepared extemporaneously. A patent disclosure gives the following kit formulation: human albumin (21 mg) and stannous tartrate (0.23 mg), at a pH between 2 and 4, sealed under nitrogen gas.[57] Without bacteriostatic preservative the kit must be stored before and after labeling at 2°C to 8°C. The ^{99m}Tc-HSA complex is prepared by adding up to 3.0 mL containing 100 mCi (3700 MBq) maximum of ^{99m}Tc-sodium pertechnetate to the kit, followed by gentle mixing to avoid foaming, and allowing the preparation to stand for 20 minutes to achieve maximum tagging. The product is stable for 6 hours after preparation. The radiochemical purity is not less than 90%. Possible radiochemical impurities are free pertechnetate and hydrolyzed–reduced technetium,[58] each of which should not exceed 5%.[57]

^{99m}Tc-HSA is confined to the vascular space after intravenous injection and is indicated for imaging the heart blood pool. For this application the suggested administered activity of ^{99m}Tc-HSA is 5 mCi (185 MBq) by intravenous injection. The critical organ is the urinary bladder wall, with an absorbed radiation dose of 0.033 rad(cGy)/mCi.[59]

Several methods have been developed for preparing ^{99m}Tc-HSA, and labeling mechanisms have been proposed;[60] however, no technique has been found to characterize the exact oxidation state of technetium in the complex.[10] It has been proposed that technetium complexes with albumin at reduced sulfhydryl (SH) groups in the protein.

Technetium Tc 99m Albumin Aggregated Injection

Technetium Tc 99m albumin aggregated injection (^{99m}Tc-albumin aggregated or ^{99m}Tc-MAA; DraxImage MAA [DraxImage], Pulmolite [Pharmalucence]) is a sterile aqueous suspension of ^{99m}Tc labeled to human albumin aggregate particles in the pH range of 3.8 to 8.0. Kits for preparing ^{99m}Tc-MAA are available from several manufacturers as a sterile lyophilized powder of nonradioactive ingredients sealed under nitrogen. Table 10-5 lists the components of various kits, and Table 10-6 lists kit properties. The stannous albumin aggregates in the kits are produced by heating human albumin in the presence of stannous chloride under controlled conditions of temperature, pH, and mixing. Not less than 90% of the aggregate particles have a diameter between 10 µm and 90 µm, and none can exceed 150 µm.[56] Figure 10-16 is a photomicrograph of MAA particles.

The labeled product is prepared by adding the required amount of ^{99m}Tc-sodium pertechnetate to the kit to reconstitute and disperse the particles. The kit is allowed to stand at room temperature for up to 15 minutes to ensure maximum tagging. The amount of activity added to the kit must be at a concentration that will permit removal of a 3 mCi (111 MBq) dose containing typically between 100,000 and 600,000 particles, as recommended by the manufacturer. The number of particles per dose may need to be lowered for pediatric patients and for those who have right-to-left cardiac shunts. ^{99m}Tc-MAA is contraindicated in patients with severe pulmonary hypertension. (See Chapter 22 for details on potential particle toxicity.) Accordingly, a kit may be reconstituted with sterile saline and an aliquot removed to reduce the ultimate number of particles to be labeled with technetium. The number of particles per dose will increase with time because of constantly declining specific activity. Thus, for example, adding 50 mCi (1850 MBq) to a kit containing 5 million particles will provide 300,000 particles in a 3 mCi (111 MBq) dose initially, but 6 hours later the same dose will contain 600,000 particles. See Table 9-13 in Chapter 9.

In general, ^{99m}Tc-MAA kits are stored at 2°C to 8°C before and after labeling. The expiration time of a labeled kit is 6 hours from the time of kit preparation. The radiochem-

TABLE 10-6 Properties of ^{99m}Tc-MAA Kits

Kit Manufacturer	Particles per Kit	Max Activity (Labeling Time)	Lung Half-life	Kit Storage Before Labeling	Kit Storage After Labeling	Expiration Time
Draximage	4–8 × 10^6	100 mCi (15 min)	2–3 hr	2–8°C	2–8°C	6 hr
Pharmalucence	3.6–6.5 × 10^6	50 mCi (1–2 min)	5 hr	15–30°C	2–8°C	6 hr

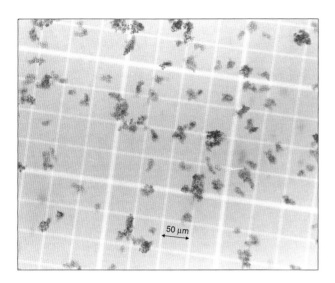

FIGURE 10-16 Photomicrograph of ^{99m}Tc-MAA particles.

ical purity is not less than 90% of the total radioactivity tagged to albumin particles, as determined by chromatography. Not more than 10% soluble supernatant impurities may be present, as determined by centrifugation.[56]

Upon intravenous injection, ^{99m}Tc-MAA particles become trapped in obstructed blood vessels and in any vessel whose diameter is smaller than the particle size. ^{99m}Tc-MAA is therefore indicated for perfusion lung imaging to assess the presence of pulmonary emboli, isotope venography to identify lower-extremity venous thrombosis, and assessment of peritoneovenous (LeVeen) shunt patency. The usual adult administered activity of ^{99m}Tc-MAA for perfusion lung imaging is 3 to 4 mCi (111–148 MBq) by intravenous injection. The critical organ is the lung, with an absorbed radiation dose of 0.22 rad(cGy)/mCi.[61]

Technetium Tc 99m Apcitide Injection

Technetium Tc 99m apcitide injection (^{99m}Tc-apcitide; AcuTect [Diatide]) is a sterile aqueous solution prepared from a kit containing a lyophilized mixture of the peptide bibapcitide (100 μg), sodium glucoheptonate dihydrate (75 mg), and stannous chloride dihydrate (89 μg), adjusted to a pH of 7.4. The kit must be stored at 2°C to 8°C before labeling with ^{99m}Tc-sodium pertechnetate. Although this kit is no longer marketed, its preparation method illustrates the unique way in which technetium can be incorporated into a peptide molecule.

In the preparation of ^{99m}Tc-apcitide, ^{99m}Tc-sodium pertechnetate is mixed with the kit components and reacted in a boiling water bath for 15 minutes to cause the following labeling reaction:

$$TcO_4^- + \text{Sn-Glucoheptonate} \longrightarrow \text{Tc-Glucoheptonate}$$

$$\text{Bibapcitide} + \text{Tc-Glucoheptonate} \xrightarrow{\text{Heat}} \text{Tc-Apcitide}$$

During this time the bibapcitide molecule (Figure 10-17), composed of two apcitide monomers, is split to release apcitide. Apcitide is labeled with ^{99m}Tc by ligand exchange from ^{99m}Tc-glucoheptonate as the transfer agent.[62]

^{99m}Tc-apcitide was indicated for the detection of acute venous thrombosis in the lower extremities[63,64] and localized by the following process. After blood vessel injury, platelets become activated and express a GP IIb/IIIa receptor on their membrane surface. The platelet receptor binds fibrinogen, resulting in platelet aggregation and blood clotting. The peptide arginine-glycine-aspartate (Arg-Gly-Asp, or RGD in single-letter amino acid code) is the amino acid sequence identified as the platelet attachment site on fibrinogen.[65] Synthetic peptides containing the RGD sequence can effectively compete with endogenous fibrinogen during the clotting process. Therapeutic drugs have been developed to control clotting by this mechanism.[66] ^{99m}Tc-apcitide is a thrombus-localizing radiopharmaceutical designed to mimic the RGD peptide sequence. Apcitide contains the mimetic sequence Apc-Gly-Asp. The synthetic amino acid Apc (S-aminopropyl-L-cysteine) is an arginine surrogate. It replaces arginine in the receptor-binding sequence and confers additional selectivity for the platelet receptor.[65] The chemical structures of apcitide and ^{99m}Tc-apcitide have been characterized as shown in Figure 10-18.[65] Apcitide is a BFCA that contains a platelet receptor-binding region composed of the peptide Apc-Gly-Asp and a technetium-binding region composed of the peptide Gly-Gly-Cys-NH$_2$.

Technetium Tc 99m Bicisate Injection

Technetium Tc 99m bicisate injection (^{99m}Tc-bicisate or ^{99m}Tc-ECD; Neurolite [Lantheus]) is a sterile aqueous solution prepared from a kit containing a lyophilized mixture of bicisate dihydrochloride (0.9 mg) (coordinating ligand), disodium edetate dihydrate (0.36 mg) (auxiliary complexing agent and transfer ligand), mannitol (24 mg)(bulking agent), and stannous chloride dihydrate (72 μg) (reducing agent), adjusted to pH 2.7 and sealed under nitrogen. The kit also contains a vial of phosphate buffer at pH 7.6. The kit should be stored at 15°C to 25°C and protected from light. Bicisate is also known as ethyl cysteinate dimer (ECD). The chemical structure of ^{99m}Tc-bicisate complex is shown in Figure 10-19.

^{99m}Tc-bicisate is prepared in a three-step process by (1) adding 100 mCi of ^{99m}Tc-sodium pertechnetate in a 2 mL volume to the phosphate buffer vial (this raises the pH to optimum for labeling reaction), (2) adding 3 mL of 0.9% sodium chloride injection to the lyophilized bicisate vial, and (3) within 30 seconds, removing 1 mL from the bicisate vial and adding it to the pertechnetate/buffer vial. This final mixture is incubated for 30 minutes at room temperature to effect labeling. Note that no less than 50 mCi should be used; otherwise, incomplete labeling may occur. If necessary, the ^{99m}Tc-bicisate may be diluted with normal saline only after the product is labeled. The labeled product is stored at 15°C to 25°C and must be used within 6 hours of preparation. The radiochemical purity must be 90% or higher, as determined by instant thin-layer chromatography.[56,67]

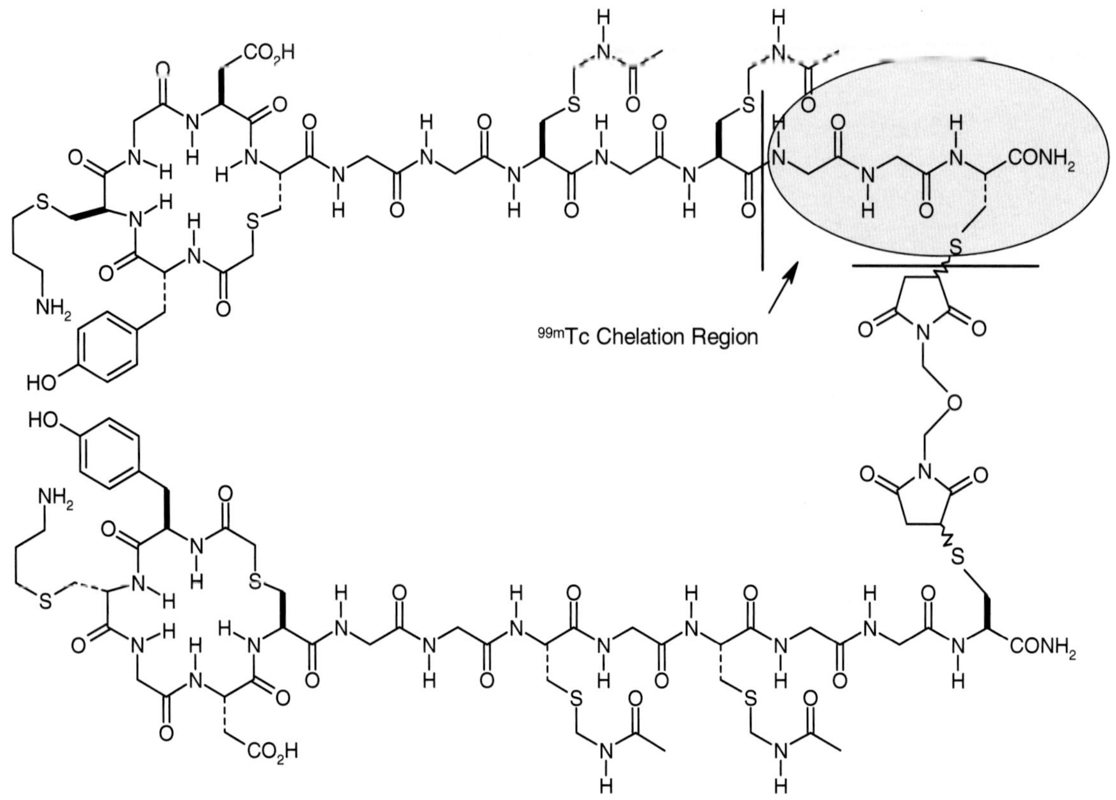

FIGURE 10-17 Chemical structure of the bibapcitide peptide dimer molecule from which the apcitide ligand is derived in the ^{99m}Tc-apcitide kit.

The use of ^{99m}Tc-bicisate is indicated as an adjunctive procedure to computed tomography (CT) and magnetic resonance imaging (MRI) in the localization of stroke in patients in whom stroke has already been diagnosed. The usual adult administered activity for brain imaging is 10 to 30 mCi (370–1110 MBq) by intravenous injection. The critical organ is the urinary bladder wall, with an absorbed radiation dose of 0.11 rad(cGy)/mCi.[67]

During the preparation of ^{99m}Tc-bicisate, the diamine-dithiol ligand ECD loses one hydrogen ion from an amine group and two hydrogen ions from the sulfhydryl groups. The resulting three negative charges on these donor atoms neutralize the three positive charges on the Tc $=$ O^{3+} core. This creates a neutral complex, ^{99m}Tc-bicisate, which is lipophilic and stable in aqueous solution (Figure 10-19).

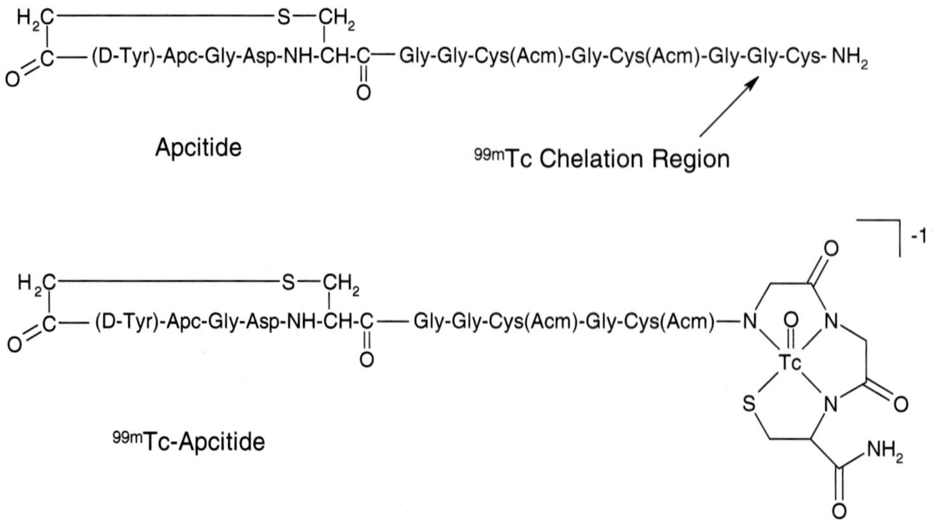

FIGURE 10-18 Chemical structures of the apcitide ligand and ^{99m}Tc-apcitide.

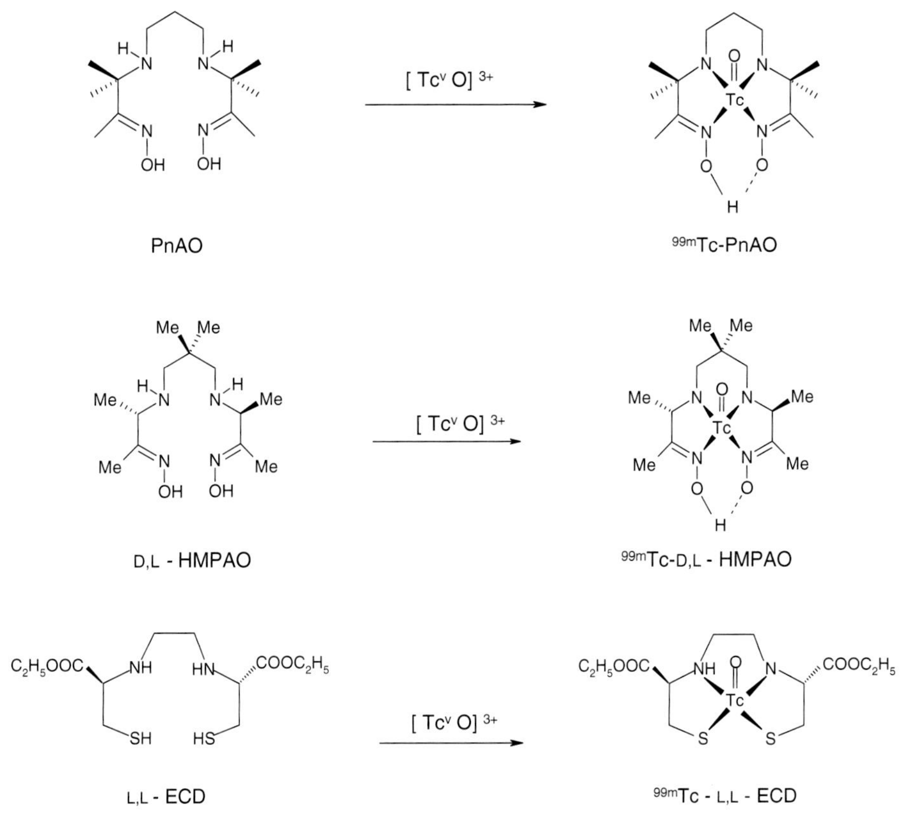

FIGURE 10-19 Chemical structures of brain imaging ligands and their technetium coordination compounds: ^{99m}Tc-PnAO, ^{99m}Tc-HMPAO, ^{99m}Tc-ECD.

The ECD ligand exists as the L,L and D,D isomers; both isomers demonstrate brain uptake but only the L,L isomer exhibits brain retention,[68] and only in primates (monkeys and humans). While the carbon backbone of the ligand system is quite stable, substitution on this backbone with two ester functionalities makes it labile to enzymatic hydrolysis. After intravenous injection, ^{99m}Tc-bicisate localizes in the brain by passive diffusion of the un-ionized, lipid-soluble complex. Slow hydrolysis in blood and rapid hydrolysis in brain tissue to the more hydrophilic metabolite results in high brain uptake and retention.

Technetium Tc 99m Disofenin Injection

Technetium Tc 99m disofenin injection (^{99m}Tc-disofenin or ^{99m}Tc-DISIDA; Hepatolite [Pharmalucence]) is a sterile aqueous solution prepared from a lyophilized kit containing the complexing ligand disofenin (20 mg) and stannous chloride dihydrate (0.6 mg) at pH 4 to 5, sealed under nitrogen. The chemical name of disofenin is diisopropylacetanilido iminodiacetic acid; its structure is shown in Figure 10-7. The kit is stored at 15°C to 25°C before and after labeling with ^{99m}Tc.

^{99m}Tc-disofenin is prepared by adding 12 to 100 mCi (444–3700 MBq) of ^{99m}Tc-sodium pertechnetate in a 4 to 5 mL volume to the kit. The vial is swirled for 1 minute to dissolve the powder and allowed to incubate at room temperature for 4 minutes to effect complexation of ^{99m}Tc with disofenin. ^{99m}Tc-disofenin has a useful life of 6 hours. The radiochemical purity of the ^{99m}Tc-disofenin complex is not less than 90%.[69]

^{99m}Tc-disofenin is indicated as a hepatobiliary agent in the diagnosis of acute cholecystitis. The usual adult administered activity is 1 to 5 mCi (37–185 MBq) by intravenous injection in nonjaundiced patients and 3 to 8 mCi (111–296 MBq) in patients whose bilirubin level is greater than 5 mg/dL. The critical organ is the upper large intestinal wall, with an absorbed radiation dose of 0.35 rad(cGy)/mCi.[69]

The ^{99m}Tc-IDA analogues are hexacoordinate complexes having the general form [Tc-(IDA)$_2$]$^-$, with the Tc(III) atom stabilized by two nitrogens and four oxygens from two IDA ligands. The four negatively charged oxygens neutralize the 3+ charge on the Tc^{3+} core to give the complex a net charge of 1−. The complex is kinetically inert. ^{99m}Tc-disofenin is extracted from blood by active transport at the anionic site on the hepatocyte membrane, where it competes with bilirubin.

Technetium Tc 99m Exametazime Injection

Technetium Tc 99m exametazime injection (^{99m}Tc-exametazime or ^{99m}Tc-HMPAO; Ceretec [GE Healthcare]) is a sterile aqueous solution prepared from a kit containing a sterile lyophilized mixture of exametazime (0.5 mg), stannous chloride dihydrate (7.6 µg), and sodium chloride (4.5 mg)

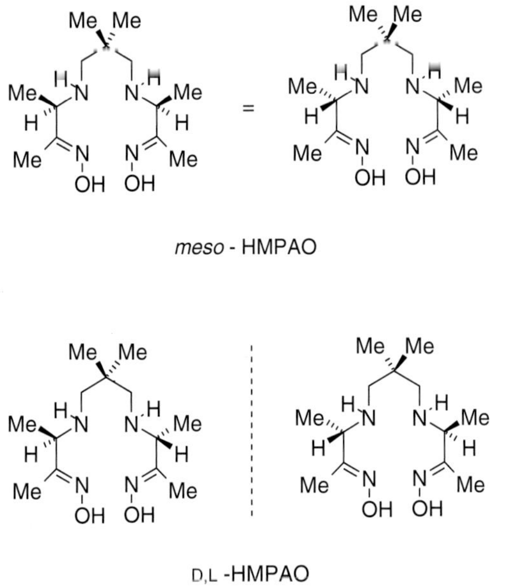

FIGURE 10-20 Chemical structures of the D,L-HMPAO enantiomers and their *meso* diastereomer.

sealed under nitrogen. Exametazime is also known as hexamethylpropyleneamine oxime, or HMPAO (Figure 10-20). The structural formula of the technetium complex is shown in Figure 10-19. The kit also contains one vial each of 1% methylene blue and a phosphate buffer that, when mixed together, act as a stabilizer. The kit is stored at 15°C to 25°C before labeling and at 20°C to 25°C after labeling with technetium.

^{99m}Tc-exametazime is prepared in one of two ways: (1) with or without stabilizer for brain imaging and (2) without stabilizer for labeling leukocytes. The stabilized product is prepared by mixing 0.5 mL of methylene blue 1% with 4.5 mL of phosphate buffer. The lyophilized kit is reconstituted with 10 to 54 mCi (370–1998 MBq) of ^{99m}Tc-sodium pertechnetate, followed within 2 minutes by the addition of 2 mL of the methylene blue/buffer mixture, which confers product stability for 4 hours. Because the product is blue and opaque to visualization, it must be injected through a 0.45 μm membrane filter supplied with the kit to remove any particulate matter that may be present in the solution. Unstabilized ^{99m}Tc-exametazime, which is used for labeling leukocytes, is prepared by reconstituting the lyophilized powder with 10 to 54 mCi (370–1998 MBq) of ^{99m}Tc-sodium pertechnetate in a volume of 5 mL. Because the methylene blue stabilizer is not mixed with this product, it has a useful life of only 30 minutes and therefore must be prepared close to the time of need.

Because the kit contains such a small amount of stannous chloride, it is recommended that to obtain a product with the highest radiochemical purity, the ^{99m}Tc-sodium pertechnetate solution used be obtained from a generator previously eluted within 24 hours and that eluate not more than 30 minutes old be used for brain imaging and not more than 2 hours old for leukocyte labeling. These precautions will reduce the amount of radiolytic products (free radicals) and "carrier" ^{99}Tc added to the kit. The radiochemical purity of ^{99m}Tc-exametazime must be not less than 80%, as determined by instant thin-layer chromatography.[56,70]

^{99m}Tc-exametazime, with or without stabilization, is indicated for the detection of altered cerebral perfusion in stroke. The usual adult administered activity is 10 to 30 mCi (370–1110 MBq). The critical organ is the lacrimal glands, with an absorbed dose of 0.258 rad(cGy)/mCi. Unstabilized ^{99m}Tc-exametazime labeled to autologous leukocytes is indicated for the localization of intraabdominal infection and inflammatory bowel disease. The usual adult administered activity is 7 to 25 mCi (259–925 MBq), and the critical organ is the spleen, with an absorbed dose of 0.556 rad(cGy)/mCi.[70]

During radiolabeling of ^{99m}Tc-exametazime, the ^{99m}Tc=O^{3+} core coordinates with two amine and two oxime nitrogen groups. The two amine nitrogens and one of the oxime nitrogens lose hydrogen ions to create three negative charges on the HMPAO ligand, which neutralize the three positive charges on the ^{99m}Tc=O^{3+} core. Thus, the final ^{99m}Tc-exametazime complex is neutral. In addition, an intermolecular hydrogen bond forms with the ionized oxime group (Figure 10-19). Because ^{99m}Tc-exametazime has two chiral carbons, it can form up to four stereoisomers; however, only three forms exist since the *meso* isomers are identical and the D,L isomers are enantiomers. The commercial kit contains the D,L racemate since the *meso* isomers have poor brain uptake. Studies with the D,L-HMPAO complex demonstrated that it was neutral and lipophilic, but unstable in aqueous solution. The instability was found to be a conversion from the primary lipophilic complex to a secondary hydrophilic complex, which was mediated by reducing agents.[71,72] Although the D,L isomers had poor in vitro stability, they demonstrated high brain retention. Brain uptake was believed to be associated with the lipophilic complex, and brain retention to be caused by its intracellular conversion to a nondiffusible hydrophilic complex. The brain conversion was shown to be mediated by the intracellular reducing agent glutathione.[71] Neirinckx et al.[73] demonstrated that the ^{14}C-labeled D,L-HMPAO isomer without technetium did not cross the blood–brain barrier, in contrast to the identical compound labeled with technetium. Thus, the technetium complex with D,L-HMPAO is considered to be a technetium-essential radiopharmaceutical.

Decomposition of the lipophilic ^{99m}Tc-HMPAO complex in vitro is mediated by reducing agents. Consequently, the kit formulation contains a small amount of stannous ion, which limits the amount of ^{99m}Tc-sodium pertechnetate that can be added to the kit. Excess stannous ion not used in technetium reduction will promote complex decomposition, which limits the shelf life of the reconstituted kit without stabilizer to 30 minutes. The shelf life of the ^{99m}Tc-HMPAO kit in the United States was extended to 4 hours by incorporating a stabilizing buffer and an antioxidant/radical scavenger (methylene blue). European kits contain cobaltous chloride (CoCl$_2$) as the stabilizer, which forms a Co(II)-HMPAO complex that is oxidized to Co(III)-HMPAO by oxygen.[19] Co(III)-HMPAO rapidly oxidizes stannous ion.

Co-(III) amines and methylene are also effective free-radical scavengers.

Technetium Tc 99m Gluceptate Injection

Technetium Tc 99m gluceptate injection (^{99m}Tc-gluceptate or ^{99m}Tc-GH [DraxImage]) is a sterile aqueous solution prepared from a lyophilized kit containing gluceptate calcium (50 mg) and stannous chloride dihydrate (0.7–1.1 mg) adjusted to pH 6.9 to 7.1 and sealed under nitrogen. The kit is stored at 15°C to 30°C before labeling and at 2°C to 8°C after labeling with technetium. This kit is available in Canada and Europe but not in the United States.

The ^{99m}Tc-gluceptate complex is prepared by adding up to 300 mCi (11,100 MBq) in 2 to 10 mL of ^{99m}Tc-sodium pertechnetate solution to the kit and letting it stand at room temperature for 15 minutes to effect labeling. ^{99m}Tc-gluceptate is stable for 6 hours after preparation. The radiochemical purity, determined by instant thin-layer chromatography, is not less than 90% as the ^{99m}Tc-gluceptate complex.[56]

^{99m}Tc-gluceptate is indicated for use in brain and kidney imaging. The usual adult administered activity is 15 to 20 mCi (555–740 MBq) for brain imaging and 10 to 15 mCi (370–555 MBq) for kidney imaging by intravenous injection. The critical organ is the renal cortex, with an absorbed radiation dose of 0.24 rad(cGy)/mCi.[74]

^{99m}Tc-gluceptate is a Tc(V) complex with the seven-carbon carboxylic acid sugar glucoheptonate (Figure 10-21). Studies on the formation of ^{99m}Tc-gluceptate with millimolar ^{99}Tc-pertechnetate and nanomolar ^{99m}Tc-pertechnetate demonstrate that the complex is formed rapidly in high yield when the ligand-to-technetium molar ratio is 25 or greater.[75] No significant in vitro and in vivo differences were found between the ^{99m}Tc- and ^{99}Tc-glucoheptonate complexes. The chemical stability of the complex is high over a long period in the presence of no other ligands. Electrophoretic measurements have determined the net charge on the complex as 1–. Similar results were found when a sodium borohydride (NaBH$_4$) reductant was used in place of stannous chloride. The complex has been characterized as an oxo-bis(glucoheptonato)technetate(V) anion, composed of a TcO^{3+} core and two glucoheptonate ligands (Figure 10-21).[75] The stoichiometry of the complexation reaction is described as follows:

$$TcO_4^- + 2(C_7H_{13}O_8)^- + Sn^{2+} + H_2O$$
$$\longrightarrow [TcO(C_7H_{12}O_8)_2]^- + 2H^+ + [SnO_2(OH)_2]^{2-}$$

Technetium Tc 99m Mebrofenin Injection

Technetium Tc 99m mebrofenin injection (^{99m}Tc-mebrofenin or ^{99m}Tc-BRIDA; Choletec [Bracco], generic [Pharmalucence]) is a sterile aqueous solution prepared from a lyophilized kit sealed under nitrogen containing the complexing ligand mebrofenin (45 mg), stannous fluoride dihydrate (1.03 mg) as reducing agent, and methylparaben (5.2 mg) and propylparaben (0.58 mg) as preservatives. The pH of the reconstituted product is 4.2 to 5.7. Mebrofenin is a methyl- and bromine-substituted acetanilido iminodiacetic acid analogue; its chemical structure is shown in Figure 10-7. The kit is stored before and after labeling with technetium at 20°C to 25°C.

^{99m}Tc-mebrofenin is prepared by adding up to 100 mCi (3700 MBq) in 1 to 5 mL of ^{99m}Tc-sodium pertechnetate solution to the kit and allowing it to stand at room temperature for 15 minutes to effect labeling. The labeled product is stable for 18 hours (Bracco) and 6 hours (Pharmalucence). The radiochemical purity of the ^{99m}Tc-mebrofenin complex is not less than 90%.[56]

^{99m}Tc-mebrofenin is indicated for use as a hepatobiliary imaging agent. The usual adult administered activity is 2 to 5 mCi (74–185 MBq) in nonjaundiced patients and 3 to 10 mCi (111–370 MBq) in patients with bilirubin levels greater than 1.5 mg/dL. The critical organ is the upper large intestinal wall, with an absorbed radiation dose of 0.248 rad(cGy)/mCi.[76]

The IDA analogues are hexacoordinate complexes having the general form [Tc-(IDA)$_2$]$^-$, with the Tc(III) atom stabilized

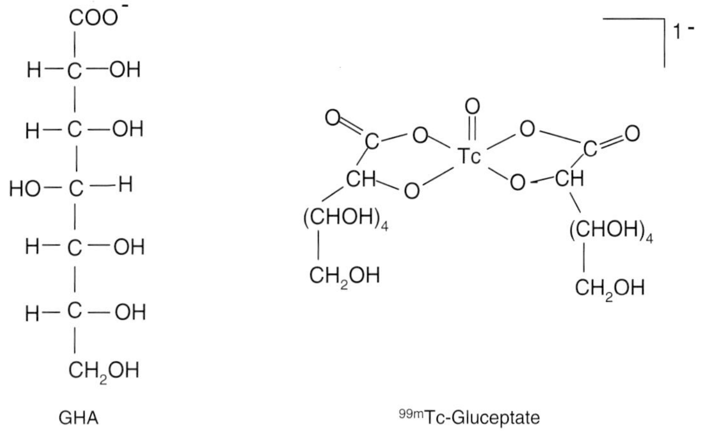

FIGURE 10-21 Chemical structures of the glucoheptonic acid (GHA) ligand and the ^{99m}Tc-gluceptate complex.

by two nitrogens and four oxygens from two IDA ligands. The four negatively charged oxygens neutralize the 3+ charge on the Tc³⁺ core, to give the complex a net charge of 1−. The complex is kinetically inert. ^{99m}Tc-mebrofenin is extracted from blood by active transport at the anionic site on the hepatocyte membrane, where it competes with bilirubin.

Technetium Tc 99m Medronate Injection

Technetium Tc 99m medronate injection (^{99m}Tc-medronate or ^{99m}Tc-MDP [DraxImage and Pharmalucence]) is a sterile aqueous solution prepared from a lyophilized kit containing medronic acid and stannous chloride. Medronate is otherwise known as methylene diphosphonic acid or MDP; its chemical structure is shown in Figure 10-3. Kits are available from several manufacturers; kit properties are shown in Table 10-7.

^{99m}Tc-medronate is prepared by adding the specified amount of activity and volume of ^{99m}Tc-sodium pertechnetate solution to the kit and mixing for a maximum of 1 to 2 minutes to effect labeling. Several kits are formulated with antioxidant stabilizer (gentisic acid or ascorbic acid) to protect the relatively labile complex from degradation by oxygen and radiolytically generated free radicals. The radiochemical purity should be no less than 90%, as determined by instant thin-layer chromatography.[56]

^{99m}Tc-medronate is indicated as a bone-imaging agent to delineate areas of altered osteogenesis due to various causes. Frequently it is used to identify metastatic bone lesions from breast and prostate cancer. The usual adult administered activity is 10 to 20 mCi (370–740 MBq) by intravenous injection. The critical organ is the urinary bladder wall, with an absorbed radiation dose of 0.13 rad(cGy)/mCi (2 hour void).[77]

Technetium Tc 99m Mertiatide Injection

Technetium Tc 99m mertiatide injection (^{99m}Tc-mertiatide or ^{99m}Tc-MAG3; TechneScan MAG3 [Covidien]) is a sterile aqueous solution prepared from a kit containing a lyophilized mixture of betiatide (1 mg), stannous chloride dihydrate (0.2 mg), sodium tartrate dihydrate (40 mg), and lactose monohydrate (20 mg) sealed under argon. The chemical structures of the betiatide ligand and ^{99m}Tc-mertiatide are shown in Figure 10-22. The kit should be stored at 15°C to 30°C and protected from light before labeling with technetium.

^{99m}Tc-mertiatide is prepared by inserting a venting needle into the kit followed by injection of 20 to 100 mCi (740–3700 MBq) in 4 to 10 mL of ^{99m}Tc-sodium pertechnetate. The ^{99m}Tc-sodium pertechnetate should not be more than 6 hours old. The plunger on the ^{99m}Tc syringe is pulled back to remove 2 mL of argon gas and introduce air into the vial. Within 5 minutes of adding the ^{99m}Tc-sodium pertechnetate solution, the vial is placed into a shielded boiling water bath for 10 minutes. The vial is then cooled for 15 minutes before use. The pH of the reconstituted product is between 5 and 6. The labeled product is stored at 15°C to 30°C and must be used within 6 hours of preparation. The radiochemical purity must be not less than 90%, as determined with Sep-Pak C18 (Waters) reverse-phase mini-column chromatography.[56,78]

^{99m}Tc-mertiatide is indicated for renal imaging for the diagnosis of congenital and acquired abnormalities, renal failure, urinary tract obstruction, and calculi. It is a diagnostic aid for creating renal angiograms to visualize renal blood flow and for creating renogram curves to assess overall kidney function, individual kidney function, and the function of kidney segments such as the renal cortex and pelvis. The usual adult administered activity is 5 to 10 mCi (185–370 MBq) by intravenous injection. The critical organ is the urinary bladder wall, with an absorbed radiation dose of 0.48 rad(cGy)/mCi.[79]

The kit formulation for preparing ^{99m}Tc-MAG3 contains an S-benzoyl mercaptoacetyltriglycine coordinating ligand (betiatide), stannous chloride as the reducing agent, and sodium tartrate as transfer ligand.[80] The reactive thiol (SH) in betiatide is protected with a benzoyl group. Adding ^{99m}Tc-sodium pertechnetate and heating releases the protective group, and the Tc=O³⁺ core transfers quantitatively from the tartrate

TABLE 10-7 Properties of Technetium Medronate (MDP) and Oxidronate (HDP) Kits

Manufacturer	Composition		Max Activity/Vol (Labeling Time)	Storage Conditions Before Labeling	Storage Conditions After Labeling	Expiration Time
Bracco and Pharmalucence (MDP)	Medronic acid SnF$_2$ Ascorbic acid	20 mg 0.13–0.38 mg 1.0 mg	500 mCi 0.5–5 mL (1 to 2 min)	20–25°C	2–8°C	6 hr
DraxImage (MDP)	Medronic acid SnCl$_2$·2H$_2$O PABA	10 mg 0.8–1.21 mg 2.0 mg	500 mCi (immediate)	2–30°C	2–30°C	6 hr
GE Healthcare (MDP Multidose)	Medronic acid SnCl$_2$·2H$_2$O Ascorbic acid	10 mg 0.17–0.29 mg 2.0 mg	No max listed 2–8 mL (1 to 2 min)	≤25°C	≤25°C	6 hr
Covidien (TechneScan HDP)	Oxidronate Na SnCl$_2$·2H$_2$O Gentisic acid	3.15 mg 0.258–0.343 mg 0.84 mg	300 mCi 3–6 mL (30 sec)	20–25°C	20–25°C	8 hr

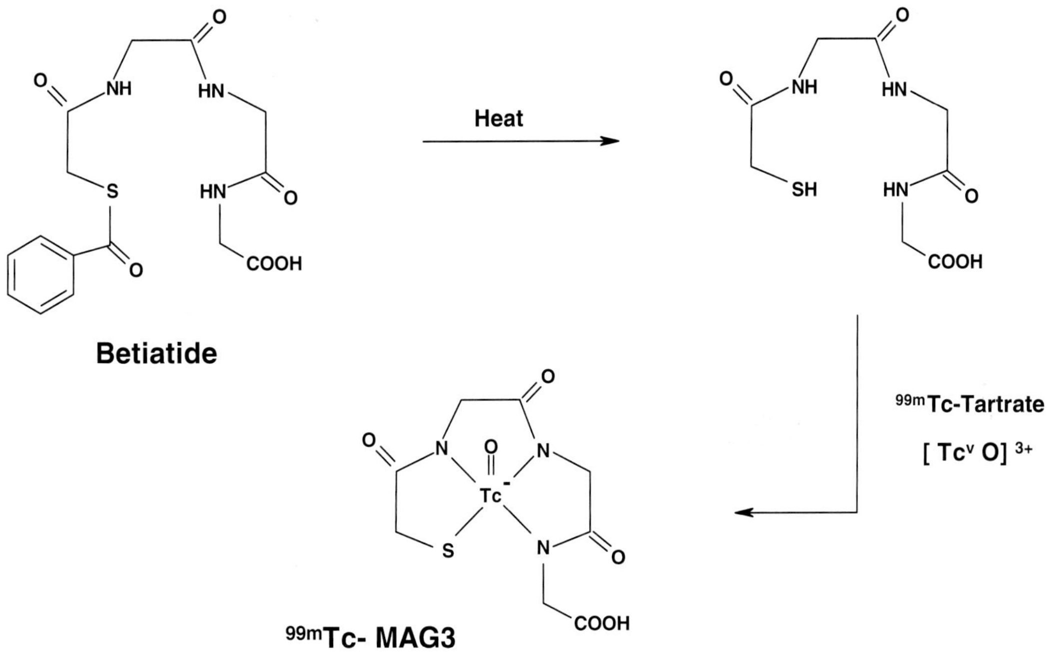

FIGURE 10-22 Synthetic pathway for labeling ^{99m}Tc-mertiatide (^{99m}Tc-MAG3).

ligand to mercaptoacetyltriglycine, to form ^{99m}Tc-MAG3. A negative charge of 1− on the ^{99m}Tc-MAG3 complex results from neutralization of the 3+ charge in the Tc=O^{3+} core with four negative charges created on the coordinating donor atoms during the radiolabeling reaction: three on nitrogen atoms through loss of hydrogen ions and one on the sulfur atom by release from its protective group. Factors that may cause radiochemical impurities to form during ^{99m}Tc-MAG3 preparation are using more than 100 mCi (3700 MBq) and less than 4 mL to reconstitute the kit, waiting longer than 5 minutes to place the vial into the boiling water bath, and not adding air to the reaction vial during radiolabeling.[80] The main impurities are ^{99m}Tc-pertechnetate, ^{99m}Tc-tartrate, and hydrolyzed–reduced technetium. Air is needed to oxidize excess Sn(II) in the kit, which could possibly reduce Tc(V) to Tc(IV).[19]

Technetium Tc 99m Oxidronate Injection

Technetium Tc 99m oxidronate injection (^{99m}Tc-oxidronate or ^{99m}Tc-HDP; TechneScan HDP [Covidien]) is a sterile aqueous solution prepared from a kit containing oxidronate sodium (3.15 mg), stannous chloride dihydrate (0.297 mg), gentisic acid (0.84 mg) as antioxidant stabilizer, and sodium chloride (30 mg), sealed under nitrogen (Table 10-7). The kit should be stored at 20°C to 25°C before and after labeling with ^{99m}Tc-sodium pertechnetate. The chemical structure of oxidronate (HDP) is shown in Figure 10-3.

^{99m}Tc-oxidronate is prepared by adding up to 300 mCi (11,100 MBq) in 3 to 6 mL of ^{99m}Tc-sodium pertechnetate solution and mixing for 30 seconds. The pH of the reconstituted product is between 4 and 5.5. ^{99m}Tc-oxidronate is stable for 8 hours after preparation. The radiochemical purity must be not less than 90%, as determined by instant thin-layer chromatography.[56]

^{99m}Tc-oxidronate is indicated as a diagnostic skeletal imaging agent to demonstrate areas of altered osteogenesis. The usual adult administered activity is 10 to 20 mCi (370–740 MBq) by intravenous injection. The critical organ is the bone surface, with an absorbed radiation dose of 0.322 rad(cGy)/mCi.[81]

Technetium Tc 99m Pentetate Injection

Technetium Tc 99m pentetate injection (^{99m}Tc-pentetate or ^{99m}Tc-DTPA [DraxImage]) is a sterile aqueous solution prepared from a kit containing a lyophilized mixture of the pentetate (dietheylenetriaminepentaacetic acid, or DTPA), stannous chloride, and other adjuvants sealed under nitrogen (Table 10-8). Early formulations of some DTPA kits included calcium ion to lessen the chance of depleting this ion in the cerebrospinal fluid (CSF) when ^{99m}Tc-DTPA was used for certain CSF studies and to avoid depleting calcium ion in the blood in studies with ^{99m}Tc-DTPA given intravenously. This is the reason for including calcium chloride in the current kit formulation. The DTPA kit also contains paraaminobenzoic acid (PABA) as a free-radical scavenger, which increases the shelf life of the kit. The chemical structure of the DTPA ligand is shown in Figure 10-3.

^{99m}Tc-pentetate is prepared by simply adding the required amount of ^{99m}Tc-sodium pertechnetate solution and allowing the kit to stand for a few minutes to effect labeling. The pH of the reconstituted product is between 3.8 and 7.5. The complex is stable for 12 hours. The radiochemical purity must be 90% or higher.[56]

TABLE 10-8 Properties of a Technetium Pentetate (DTPA) Kit

Manufacturer	Composition		Max Activity/Vol (Labeling Time)	Storage Before Labeling	Storage After Labeling	Expiration Time
DraxImage (DTPA)	Pentetic acid	20 mg	500 mCi	2–25°C	2–25°C	12 hr
	SnCl$_2$·2H$_2$O	0.25–0.385 mg	2–10 mL			
	PABA	5.0 mg	(15 min)			
	CaCl$_2$·2H$_2$O	3.73 mg				

^{99m}Tc-pentetate is indicated for use in brain and kidney imaging, to assess renal perfusion, and to estimate glomerular filtration rate (GFR). ^{99m}Tc-DTPA prepared from early kit formulations had to be used within 1 hour of preparation for GFR studies, but this is not a requirement with the current kit. ^{99m}Tc-pentetate is also used for other applications, as an aerosol for lung ventilation studies (FDA approved) and CSF leak studies (not FDA approved). Typical administered activities are 3 mCi for GFR measurement, 5 to 10 mCi (185–370 MBq) for renal perfusion, and 10 to 20 mCi (370–740 MBq) for brain imaging. The critical organ is the bladder wall, with an absorbed radiation dose of 0.115 rad(cGy)/mCi (2 hour void) and 0.27 rad(cGy)/mCi (4.8 hour void).[82]

Technetium Tc 99m Pyrophosphate Injection

Technetium Tc 99m pyrophosphate injection (^{99m}Tc-pyrophosphate or ^{99m}Tc-PPi; TechneScan PYP [Covidien], generic [Pharmalucence]) is a sterile aqueous solution prepared from a kit containing a sterile lyophilized mixture of sodium pyrophosphate and stannous chloride sealed under nitrogen. Kit composition is shown in Table 10-9.

^{99m}Tc-pyrophosphate is prepared by simply adding the required amount of ^{99m}Tc-sodium pertechnetate solution to the kit and mixing for a few minutes at ambient temperature to effect labeling. The pH of the reconstituted product is between 4.0 and 7.5. ^{99m}Tc-pyrophosphate is stable for 6 hours. The radiochemical purity must be 90% or higher.[56]

Stannous pyrophosphate (without ^{99m}Tc) is used as a source of tin for labeling red blood cells (RBCs). For this application the kit is reconstituted with 0.9% sodium chloride injection and stannous pyrophosphate is injected into the patient 5 to 60 minutes prior to administration of ^{99m}Tc-sodium pertechnetate for in vivo RBC labeling.

^{99m}Tc-pyrophosphate is indicated for bone and cardiac (infarct avid) imaging with an administered activity of 15 mCi (555 MBq). The critical organ is the bladder wall, with an absorbed radiation dose of 0.23 rad(cGy)/mCi (4.8 hour void).[83]

Technetium Tc 99m Red Blood Cells

In nuclear medicine, ^{99m}Tc-labeled red blood cells (^{99m}Tc-RBCs) are used primarily for cardiac blood pool studies and gastrointestinal bleeding studies. Heat-denatured ^{99m}Tc-RBCs are used for spleen imaging. The three principal methods for labeling RBCs with technetium are the in vitro method, the in vivo method, and the modified in vivo method.

In Vitro Method

With the in vitro method, RBCs are labeled in whole blood using the UltraTag RBC kit (Covidien). The kit consists of three components:

1. 10 mL reaction vial: lyophilized mixture of stannous chloride dihydrate 105 μg, sodium citrate dihydrate 3.67 mg, and dextrose anhydrous 5.5 mg at pH 7.1 to 7.2.
2. Syringe I: sodium hypochlorite 0.6 mg in 0.6 mL at pH 11 to 13.
3. Syringe II: citric acid monohydrate 8.7 mg, sodium citrate dihydrate 32.5 mg, and dextrose anhydrous 12 mg in 1.0 mL at pH 4.5 to 5.5.

A sample of whole blood in one to three milliliters, anticoagulated with ACD or heparin, is incubated for 5 minutes

TABLE 10-9 Properties of Stannous Pyrophosphate Kits

Manufacturer	Composition		Max Activity/Vol (Labeling Time)	Storage Composition Before Labeling	Storage Composition After Labeling	Expiration Time
Pharmalucence	Sodium pyrophosphate	12 mg	100 mCi/10 mL	15–30°C	15–30°C	6 hr
	SnCl$_2$·2H$_2$O	2.8–4.9 mg	(10 minutes)			
Covidien (TechneScan PYP)	Sodium pyrophosphate	11.9 mg	100 mCi/10 mL	28°C	20–25°C	6 hr
	SnCl$_2$·2H$_2$O	3.2–4.4 mg	(5 minutes)			

with the stannous citrate/dextrose mixture in the reaction vial. The contents of syringes I and II are then added sequentially, followed by 10 to 100 mCi (370–3700 MBq) of ^{99m}Tc-sodium pertechnetate and incubation at ambient temperature for 20 minutes. During the labeling procedure, a portion of the stannous ion complexed with citrate enters the RBCs, becoming associated with intracellular hemoglobin.[84] The sodium hypochlorite in syringe I is added to oxidize the extracellular stannous ion to stannic ion. This prevents extracellular reduction of ^{99m}Tc-sodium pertechnetate when it is added to the cells. Extracellular reduced technetium does not penetrate the RBC membrane, and it will result in poor labeling efficiency. Stannous ion within RBCs is not oxidized by hypochlorite, because hypochlorite cannot cross the cell membrane.[84] The citrate solution, which is added from syringe II, sequesters extracellular stannous ion, enhancing its oxidation by hypochlorite. After its addition to the pre-tinned cells, ^{99m}Tc-sodium pertechnetate diffuses into the cells and becomes reduced and bound to hemoglobin. The labeling efficiency is typically greater than 95%.[85] After intravenous injection, the ^{99m}Tc activity associated with RBCs in blood has an estimated biologic half-life of approximately 29 hours.[85,86]

In Vivo Method

The in vivo method of labeling RBCs with ^{99m}Tc requires intravenous injection of stannous pyrophosphate (Sn-PPi), optimally between 10 and 20 μg Sn(II)/kg body weight, 20 to 30 minutes prior to intravenous administration of 15 to 25 mCi (555–925 MBq) of ^{99m}Tc-sodium pertechnetate.[84] Labeling efficiency of the tinned RBCs (fraction of the total administered activity incorporated into RBCs) in vivo is variable, ranging from 60% to 90%.[84] Labeling is not quantitative, because of competition for ^{99m}Tc-pertechnetate between RBCs, extracellular fluid, excretory processes, and pertechnetate-avid tissues, such as the thyroid gland, the stomach, and the gastrointestinal tract. Variable amounts of gastric and urinary activity may be evident on scans; this is attributable to ^{99m}Tc-pertechnetate not bound to RBCs in the bloodstream. The activity incorporated into RBCs labeled by the in vivo method has a clearance half-time of 50 ± 4 hours and total urinary excretion of 21% by 24 hours.[84]

In vivo labeling of RBCs with ^{99m}Tc-sodium pertechnetate can occur for a prolonged period after administration of Sn-PPi. From a practical point of view, this means that a few hours' delay can occur before administration of pertechnetate for a blood pool study. Indeed, RBCs can be labeled with pertechnetate several weeks after administration of Sn-PPi.[87] When pertechnetate was administered 30 minutes, 7 days, 21 days, and 42 days after Sn-PPi, the labeling efficiencies were >90%, 41%, 27%, and 25%, respectively.[87]

Modified In Vivo Method

The modified in vivo method for labeling RBCs with ^{99m}Tc, also known as the in vivitro or in vivo in vitro method, was developed to increase labeling efficiency.[88] With this technique, the patient receives approximately 500 μg of stannous ion intravenously as Sn-PPi. Twenty minutes later, 3 mL of tinned RBCs is withdrawn from the patient through a heparinized infusion set into a shielded syringe containing 20 mCi (740 MBq) of ^{99m}Tc-sodium pertechnetate. The mixture is incubated for 10 minutes with gentle agitation and then reinjected into the patient. Labeling yields greater than 90% are achieved because RBCs compete for ^{99m}Tc-sodium pertechnetate only with plasma in the syringe.

Labeling Mechanism of ^{99m}Tc-Red Blood Cells

After incubation of RBCs with stannous pyrophosphate or stannous citrate, stannous ion is transported across the cell membrane,[88] where it is believed to be associated with an intracellular protein.[89] RBCs become labeled with technetium because the pertechnetate ion ^{99m}TcO$_4^-$ readily diffuses into the cell, where it becomes reduced by stannous ion and bound to hemoglobin.[89] Approximately 20% of ^{99m}Tc is associated with heme and 80% with globin in the hemoglobin molecule.[90,91] The pertechnetate ion ^{99m}TcO$_4^-$ is believed to be transported across the membrane via the band-3 protein transport system in exchange for chloride ion or bicarbonate ion.[92]

Technetium Tc 99m Sestamibi Injection

Technetium Tc 99m sestamibi injection (^{99m}Tc-sestamibi; Cardiolite and Miraluma [Lantheus], generic [Covidien, DraxImage, GE Healthcare, Pharmalucence]) is a sterile aqueous solution prepared from a kit consisting of a lyophilized mixture of tetrakis (2-methoxy isobutyl isonitrile) copper (I) tetrafluoroborate (1.0 mg), sodium citrate dihydrate (2.6 mg), L-cysteine hydrochloric acid monohydrate (1.0 mg), mannitol (20 mg), and stannous chloride dihydrate (0.025–0.075 mg) sealed under nitrogen. ^{99m}Tc-sestamibi is prepared by adding 25 to 150 mCi (925–5550 MBq) of ^{99m}Tc-sodium pertechnetate in 1 to 3 mL to the kit and mixing vigorously to dissolve the powder. The vial is then placed into a boiling water bath for 10 minutes. The vial is allowed to cool for 15 minutes before use. ^{99m}Tc-sestamibi is stored at 15°C to 25°C and is stable for 6 hours. Its radiochemical purity is 90% or higher.[56] The chemical structure of ^{99m}Tc-sestamibi is shown in Figure 10-10.

The composition and labeling of the available sestamibi kits are identical. ^{99m}Tc-sestamibi is indicated for myocardial perfusion studies in a dosage range of 10 to 30 mCi (370–1110 MBq) and for breast imaging to confirm the presence or absence of malignancy in a dosage range of 20 to 30 mCi (740–1110 MBq). Because of its lipophilicity, the principal route of excretion is hepatobiliary, and the critical organ is the upper large intestinal wall, with a radiation absorbed dose of 0.15 rad(cGy)/mCi after stress injection and 0.18 rad(cGy)/mCi after rest injection.[93]

In the sestamibi kits the methoxy isobutyl isonitrile (MIBI) ligand is complexed into a copper/boron fluoride complex to facilitate lyophilization, since MIBI alone is a volatile liquid.[19] Cysteine and stannous chloride are reducing agents.[94] Sodium citrate and cysteine can form complexes with tin and reduced technetium to prevent hydrolysis of these metals. Mannitol is a bulking agent for lyophilized injections.

During the heating step of the radiolabeling process, the copper–MIBI complex is broken, releasing MIBI ligands. The MIBI ligands then displace citrate or cysteine from the preformed ^{99m}Tc-citrate/cysteine donor complexes, to form ^{99m}Tc-sestamibi. The stepwise process with Tc-citrate is shown below:[19]

Step 1. $TcO_4^- + Citrate \xrightarrow{Sn^{2+}} Tc\text{-}Citrate$

Step 2. $[Cu(I)(MIBI)_4]^+ + Tc\text{-}Citrate \xrightarrow{heat} [Tc(I)(MIBI)_6]^+$

In the ^{99m}Tc-sestamibi complex, a Tc$^+$ core is coordinated with six monodentate MIBI ligands, forming a stable octahedral complex. Since the MIBI ligands are neutral, the ^{99m}Tc-sestamibi complex retains the single positive charge of the Tc$^+$ core.

Technetium Tc 99m Succimer Injection

Technetium Tc 99m succimer injection (^{99m}Tc-succimer or ^{99m}Tc-DMSA [GE Healthcare]) is a sterile aqueous solution prepared from a kit consisting of a lyophilized mixture of DMSA (1.0 mg) and stannous chloride dihydrate (0.42 mg), ascorbic acid (0.70 mg), and inositol (50 mg) as a bulking agent, sealed under nitrogen. The kit is stored at 2°C to 8°C.

Kit labeling is accomplished by adding up to 40 mCi (1480 MBq) of ^{99m}Tc-sodium pertechnetate in 1 to 6 mL to the vial and incubating for 10 minutes. During this time the following reactions take place:

$Sn(II)DMSA + TcO_4^- \xrightarrow{fast} Tc\text{-}DMSA\,(Complex\ I)$

$Tc\text{-}DMSA\,(Complex\ I) \xrightarrow{10\ min} Tc\text{-}DMSA\,(Complex\ II)$

The resulting product has a pH between 2 and 3 and is stable for 4 hours. Its radiochemical purity must be 85% or higher.[56] The chemical structure of DMSA and proposed structure of ^{99m}Tc-DMSA are shown in Figure 10-23.[95]

^{99m}Tc-DMSA is indicated for kidney imaging for evaluation of renal parenchymal disorders. The usual adult administered activity is 5 mCi (185 MBq). The critical organ is the renal cortex, with an absorbed radiation dose of 0.85 rad(cGy)/mCi.[96]

Ikeda and colleagues[97,98] demonstrated that four possible ^{99m}Tc-DMSA complexes can form when ^{99m}Tc-sodium per-

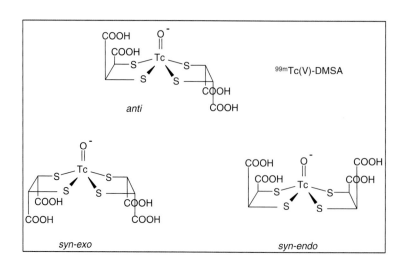

FIGURE 10-23 The chemical structure of 2,3-dimercaptosuccinic acid (DMSA); the proposed structure of ^{99m}Tc(III)DMSA; and the geometric isomers of Tc(V)O(DMSA)$_2$. (Reprinted with permission from reference 100.)

technetate and Sn(II) DMSA are reacted. Complex II has the highest kidney uptake. The maximum yield of Complex II is achieved at pH 2.5 in the absence of oxygen. The labeling reaction of ^{99m}Tc-DMSA proceeds in two steps: rapid formation of Complex I, followed by a slower rate-determining step from Complex I to Complex II, with the latter being greatly affected by oxygen.[97] This is the reason for a 10 minute incubation period. Once Complex II is formed, it may revert back to Complex I by oxidation. This is promoted by the oxidation of Sn(II) to Sn(IV), which lowers the reduction potential of the system. Diminished kidney uptake will occur because Complex I is readily excreted. The inclusion of ascorbic acid in present-day kits retards this oxidation. Ikeda's group[97] believed that Complex I is Tc(IV)DMSA and Complex II is Tc(III)DMSA (Figure 10-23), although this has not been confirmed.

When ^{99m}Tc-sodium pertechnetate is reduced by dithionite in the presence of DMSA at alkaline pH, a stable complex of ^{99m}Tc(V)DMSA is formed, having the formula [^{99m}TcO(DMSA)$_2$]$^-$.[99] The complex contains the Tc=O^{3+} core coordinated by the four thiol groups of two DMSA ligands. Three geometric isomers of this complex have been characterized (Figure 10-23).[100] The isomers shown in Figure 10-23 have a 1– charge; however, at physiologic pH the four carboxylate groups are nearly ionized, giving the complex a 5– charge. A method for preparing high-purity ^{99m}Tc(V)DMSA, using the GE Healthcare DMSA kit, requires reconstituting the kit with 1.0 mL of 4.2% sodium bicarbonate followed by rapid addition of 20 to 40 mCi (740–1480 MBq) of ^{99m}Tc-sodium pertechnetate diluted to 3 mL with normal saline.[101] Following incubation for 10 minutes at room temperature, sterile oxygen is bubbled through the solution for 10 minutes, followed by sterile filtration. The oxygen oxidizes excess Sn(II), which would reduce ^{99m}Tc(V)DMSA to ^{99m}Tc(III)DMSA, resulting in kidney localization. Compared with the ^{99m}Tc(III)DMSA kidney imaging agent, ^{99m}Tc(V)DMSA localizes differently, with little kidney uptake, and has an affinity for a variety of tumors, in particular, medullary thyroid carcinoma.[102,103]

Technetium Tc 99m Sulfur Colloid Injection

Technetium Tc 99m sulfur colloid injection (^{99m}Tc-sulfur colloid or ^{99m}Tc-SC) is a sterile colloidal dispersion of sulfur particles labeled with ^{99m}Tc and prepared from a kit (Pharmalucence). The kit consists of three components: (1) a reaction vial containing a lyophilized mixture of 2.0 mg anhydrous sodium thiosulfate (the source of sulfur), 2.3 mg disodium edetate (Al^{3+} ion chelator), and 18.1 mg gelatin (protective colloid); (2) a solution A vial with 1.8 mL of 0.148 M hydrochloric acid; and (3) a solution B vial with 1.8 mL of 24.6 mg/mL anhydrous sodium biphosphate and 7.9 mg/mL sodium hydroxide. The kit should be stored at 15°C to 30°C.

^{99m}Tc-sulfur colloid is prepared by adding 1 to 3 mL of ^{99m}Tc-sodium pertechnetate (not more than 500 mCi [18,500 MBq] in each mL) to the reaction vial to dissolve the powder. After the addition of 1.5 mL of solution A (acid), the vial is placed into a boiling water bath for 5 minutes. At the end of boiling, the vial is cooled, and 1.5 mL of solution B (buffer) is added. The pH of the final mixture is between 4.5 and 7.5 and the radiochemical purity must be 92% or higher.[56] The labeled product is stable for 6 hours stored at 15°C to 30°C.

The chemistry of ^{99m}Tc-sulfur colloid has been extensively reviewed.[7] During the boiling incubation step of the acidified mixture, thiosulfate is hydrolyzed, releasing elemental sulfur. The sulfur atoms aggregate to form colloid-size particles. Gelatin, as a protective colloid, controls particle size and aggregation by coating the sulfur particles with a protective charged protein sheath that causes the particles to repel each other. Also during the incubation step, technetium heptasulfide is formed and becomes incorporated into the sulfur particles. ^{99m}Tc-sulfur colloid has technetium in the nonreduced 7+ oxidation state. It is able to maintain this state because of its stability as insoluble technetium heptasulfide, Tc$_2$S$_7$.[7] The chemical reactions are summarized below:

$$S_2O_3^{2-} + H^+ \xrightarrow{Heat} S + HSO_3^-$$

$$2\,TcO_4^- + 7\,S_2O_3^{2-} \xrightarrow{Heat} Tc_2S_7 + 7\,SO_4^{2-} + H_2O$$

The EDTA (edetate) in the formulation chelates any aluminum ion that may be present in the ^{99m}Tc-sodium pertechnetate solution. Any free aluminum ion will react with the phosphate buffer to form insoluble aluminum phosphate, which precipitates from solution and carries ^{99m}Tc-sulfur colloid with it. Such a product would localize in the lungs. Figure 10-24 shows a microscopic view of a normal ^{99m}Tc-sulfur colloid preparation and the results of formulation changes with gelatin omitted and EDTA omitted. Figure 10-25 illustrates the stabilizing effect of EDTA against aluminum phosphate flocculation.

^{99m}Tc-sulfur colloid is indicated for several uses in various dosages: imaging the reticuloendothelial system (liver and spleen), 1 to 8 mCi [37–296 MBq]; bone marrow studies, 3 to 12 mCi [111–444 MBq]; to evaluate the patency of peritoneovenous (LeVeen) shunts; and in oral preparations for esophageal transit studies, gastroesophageal reflux studies, and pulmonary aspiration studies. ^{99m}Tc-sulfur colloid, filtered through a 0.1 μm or 0.22 μm membrane filter, is used in lymphoscintigraphy to localize the sentinel lymph node in melanoma and breast cancer. The critical organ after intravenous injection is the liver, with a radiation absorbed dose of 0.34 rad(cGy)/mCi.[104]

Technetium Tc 99m Tetrofosmin Injection

Technetium Tc 99m tetrofosmin (^{99m}Tc-tetrofosmin; Myoview [GE Healthcare]) injection is a sterile aqueous solution prepared from a kit containing a lyophilized mixture of tetrofosmin (0.23 mg), stannous chloride dihydrate (30 μg), disodium

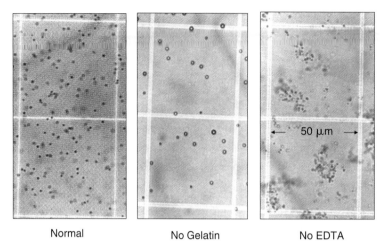

FIGURE 10-24 Photomicrographs of sulfur colloid prepared with gelatin, without gelatin, and without EDTA. Product prepared without gelatin has larger particle size. Product prepared without EDTA shows sulfur colloid particles co-precipitated with flocculated particles of aluminum phosphate. Magnification ×450.

sulfosalicylate (0.32 mg), sodium D-gluconate (1.0 mg), and sodium bicarbonate (1.8 mg) sealed under nitrogen. The kit is stored at 2°C to 8°C before reconstitution.

Labeling of tetrofosmin is accomplished by introducing a venting needle to the kit and adding up to 240 mCi (8880 MBq) of ^{99m}Tc-sodium pertechnetate (in 4 to 8 mL and not more than 30 mCi/mL concentration). This is followed by removal of 2 mL of gas from the vial and incubation at room temperature for 15 minutes. Figure 10-26 shows the reaction sequence that occurs during this time.[19] This reaction employs a gluconate transfer ligand to facilitate the complexation reaction between the technetium oxo core and tetrofosmin. Sulfosalicylate is a metal-chelating agent and sodium bicarbonate adjusts the pH to the slightly alkaline range for labeling and stability. Since the tetrofosmin ligand is neutral, the ^{99m}Tc-tetrofosmin complex has an overall charge of 1+ from the $O = Tc = O^{1+}$ core. ^{99m}Tc-tetrofosmin is stored at 2°C to 25°C after technetium labeling and is stable for 12 hours. Its radiochemical purity must be no less than 90%.[56]

^{99m}Tc-tetrofosmin is indicated for imaging the heart to assess myocardial perfusion. For a cardiac stress–rest study the recommended stress dose is 5 to 8 mCi (185–296 MBq), followed in 4 hours by a rest dose of 15 to 24 mCi (555–888 MBq). The critical organ is the gallbladder wall, with an absorbed radiation dose of 0.123 rad(cGy)/mCi (stress) and 0.180 rad(cGy)/mCi (rest).[105]

The cationic complex [^{99m}Tc-(tetrofosmin)$_2$O$_2$]$^+$ is the most important technetium dioxo compound to date, in which tetrofosmin is the ether-functionalized diphosphine ligand 1,2-bis[bis(2-ethoxyethyl)phosphino]ethane. Structural characterization of this dimeric complex has shown that the ^{99m}Tc and ^{99}Tc complexes are identical and possess the $O = Tc = O^+$ core (Figure 10-12).[30,106] The original formulation had a pH range of 8.3 to 9.1 and was prepared without admission of air to the reaction vial. It had a shelf life of 8 hours. Stability studies by the manufacturer revealed that the complex was sensitive to autoradiolytic decomposition and that admission of 2 mL of air at the time of ^{99m}Tc-sodium pertechnetate addition, and a change to a final pH range of 7.5 to 9.0, would result in a product that was stable for 12 hours.[19] The increased stability is attributed to the ability of oxygen to scavenge reducing species (the hydrated electron e^-_{aq} and the hydrogen radical H$^\bullet$), to form less active moieties, shown by the reactions below:[19]

$$H^\bullet + O_2 \longrightarrow HO_2^\bullet \qquad (10\text{-}9)$$

$$e^-_{aq} + O_2 \longrightarrow O_2^{-\bullet} \qquad (10\text{-}10)$$

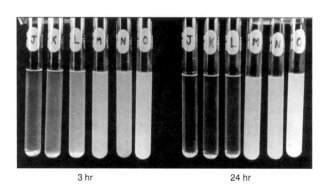

FIGURE 10-25 Stabilizing effect of EDTA against aluminum phosphate flocculation in sulfur colloid. EDTA-to-Al ion molar ratio: J, 0.125; K, 0.25; L, 0.5; M, 1.0; N, 2.0; O, 4.0. Note gradual improved stabilization in the 3 hour samples as the ratio increases.

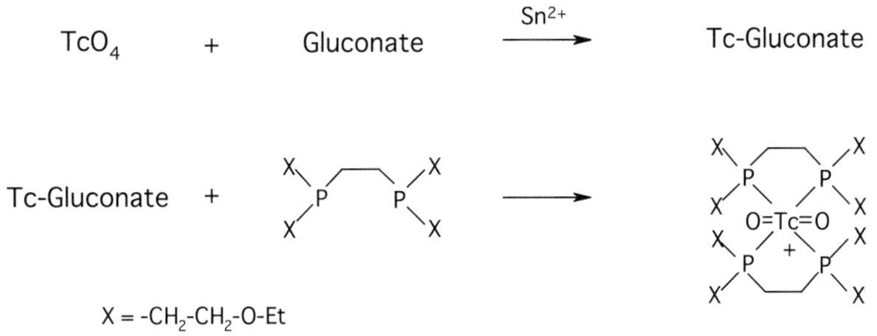

$$TcO_4 \;+\; \text{Gluconate} \;\xrightarrow{Sn^{2+}}\; \text{Tc-Gluconate}$$

Tc-Gluconate + [diphosphine ligand] → [Tc(diphosphine)$_2$ complex with O=Tc=O core]

X = -CH$_2$-CH$_2$-O-Et

FIGURE 10-26 Reaction sequence for labeling ^{99m}Tc-tetrofosmin.

Technetium 99m White Blood Cells

In early developmental work it was found that efficient radiolabeling of leukocytes occurs with a neutral lipophilic radionuclide chelate, such as ^{111}In-oxine or ^{111}In-tropolone.[107–109] Peters et al.[109] demonstrated that leukocytes labeled with the lipophilic complex ^{99m}Tc-exametazime (^{99m}Tc-HMPAO) compared favorably with ^{111}In-tropolone-labeled leukocytes. Labeling efficiency of this method is about 50%, with 78% of the activity associated with granulocytes.[109,110] The advantages of labeling leukocytes with ^{99m}Tc rather than ^{111}In are convenience, since ^{99m}Tc is readily available; better count density and resolution; and the ability to label leukocytes in the presence of plasma,[110,111] which is an important factor for maintaining leukocyte viability. Labeling efficiency with ^{111}In-oxine can be reduced in the presence of plasma transferrin, which competes with leukocytes for indium. ^{111}In-tropolone is a more efficient label of leukocytes in plasma than ^{111}In-oxine.[112]

The routine method for labeling leukocytes with technetium requires the separation of leukocytes from whole blood. Technique is important for cell viability. In general, low speed centrifugation is used to obtain the leukocyte button. A routine method, illustrated in Figure 10-27, involves collection of 43 mL of whole blood in 7 mL of ACD, addition of 10 mL of 6% hetastarch, and incubation for about 45 minutes. In children, the recommended minimum volume of blood is 10 to 15 mL. The leukocyte-rich plasma (LRP) supernatant is removed and centrifuged at 450g to pellet the leukocytes.

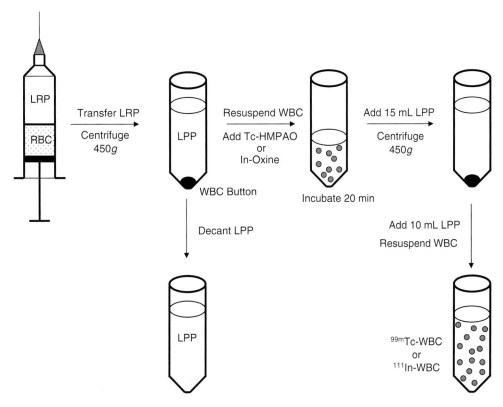

FIGURE 10-27 Procedure for labeling leukocytes (WBCs) with technetium or indium.

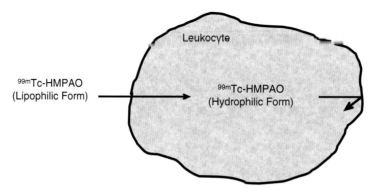

FIGURE 10-28 Proposed mechanism for labeling leukocytes with ^{99m}Tc-HMPAO.

The leukocyte-poor plasma (LPP) supernatant is removed and saved. The leukocyte pellet is mixed with 30 mCi of freshly prepared ^{99m}Tc-HMPAO in 5 mL and incubated for 15 to 20 minutes to label the cells. The mixture is then centrifuged at 450g and the supernatant containing unbound technetium is removed. The labeled leukocytes are resuspended in LPP for reinjection into the patient.

Hetastarch, added to the anticoagulated whole blood, facilitates the aggregation and settling rate of RBCs. Hetastarch can be omitted in patients with leukocytosis, who inherently have a high RBC sedimentation rate. RBCs do not separate well in patients with sickle cell anemia. Blood with sickle cells can be separated into two tubes and up to 50% hetastarch added to settle the RBCs. After settling of RBCs, the top layer (buffy coat) of LRP contains most of the platelets and about 70% of the leukocytes.[112] The LRP may be slightly pink in color because of the presence of 2% to 4% RBCs that are not considered to interfere with leukocyte labeling.[113]

The recommended minimum number of granulocytes in the blood to be labeled is 2×10^6/mL.[114] The mean number of granulocytes in normal blood is 4.15×10^6/mL, which provides a sufficient number for radiolabeling. The granulocyte count should be considered in patients with bone marrow suppression, such as those receiving chemotherapy.

During the incubation period, after ^{99m}Tc-HMPAO is added to the leukocyte button, the lipophilic ^{99m}Tc-HMPAO complex diffuses into the leukocytes and is converted to a nondiffusible hydrophilic form that becomes trapped within the cell (Figure 10-28).[115] ^{99m}Tc-HMPAO preferentially labels granulocytes,[115,116] and differential binding studies indicate that the label is associated mostly with eosinophils.[117,118] After intravenous injection, the plasma half-life of ^{99m}Tc-HMPAO-labeled leukocytes is approximately 4 hours, compared with 6 hours for ^{111}In-labeled leukocytes.[115] The shorter half-life of ^{99m}Tc-HMPAO-labeled leukocytes in vivo is attributed to elution of the technetium label from the leukocytes. The eluted form of ^{99m}Tc is believed to be hydrophilic, having genitourinary and gastrointestinal excretion. The absence of thyroid accumulation indicates that it is not pertechnetate.[115]

The recommended maximum amount of time between drawing blood and reinjection should not exceed 5 hours, because of loss of granulocyte chemotaxis during storage. The temperature of blood and labeled leukocytes should not exceed 70°F to avoid cell damage. Radiolabeled cells should be reinjected within 1 to 3 hours after preparation.[114,119] The usual administered activity is 0.1 to 0.2 mCi/kg (3.7–7.4 MBq/kg) for pediatric patients, with a minimum activity of 0.5 to 1.0 mCi (18.5–37 MBq), and 5 to 10 mCi (185–370 MBq) for adults.[119] Patients who are on hemodialysis should have their ^{99m}Tc-leukocyte study performed on a nondialysis day.

REFERENCES

1. Perrier AC, Segrè E. Some chemical properties of element 43. *J Chem Physiol.* 1937;5:712.
2. Glasstone S. *Sourcebook on Atomic Energy.* 3rd ed. New York: Van Nostrand Reinhold; 1967:657–8.
3. Seaborg GR, Segrè E. Nuclear isomerism of element 43. *Phys Rev.* 1939;55:808–14.
4. Richards P. Nuclide generators. In: Andrews GA, Knisely RM, Wagner HN Jr, eds. *Radioactive Pharmaceuticals.* Oak Ridge, TN: US Atomic Energy Commission; 1965:155–63.
5. Harper PV, Lathrop KA, Gottschalk A. Pharmacodynamics of some technetium-99m preparations. In: Andrews GA, Knisely RM, Wagner HN Jr, eds. *Radioactive Pharmaceuticals.* Oak Ridge, TN: US Atomic Energy Commission; 1965:335–58.
6. Srivastava SC, Meinken G, Smith TD, et al. Problems associated with stannous ^{99m}Tc-radiopharmaceuticals. *Int J Appl Radiat Isot.* 1977; 28:83–95.
7. Steigman J, Eckelman WC. *The Chemistry of Technetium in Medicine.* Nuclear Science Series NAS-NS-3204 Nuclear Medicine. Washington, DC. National Academies Press; 1992.
8. Jones AG, Davison A. The chemistry of technetium I, II, III, and IV. *Int J Appl Radiat Isot.* 1982;33:867–74.
9. Eckelman W, Richards P. Instant Tc-99m DTPA. *J Nucl Med.* 1970; 11:761.
10. Richards P, Steigman J. Chemistry of technetium as applied to radiopharmaceuticals. In: Subramanian G, Rhodes BA, Cooper JF, et al., eds. *Radiopharmaceuticals.* New York: Society of Nuclear Medicine; 1975:23–35.
11. Eckelman WC, Levenson SM. Radiopharmaceuticals labeled with technetium. *Int J Appl Radiat Isot.* 1977;28:67–82.
12. Tofe AJ, Francis MD. In vitro stabilization of a low-tin bone imaging agent (Tc-99m-Sn-HEDP) by ascorbic acid. *J Nucl Med.* 1976;17: 820–5.

13. Owanwanne A, Church LB, Blau M. Effect of oxygen on the reduction of pertechnetate by stannous ion. *J Nucl Med.* 1977;18:822–6.
14. Billinghurst MW, Rempel S, Westendorf BA. Radiation decomposition of technetium-99m radiopharmaceuticals. *J Nucl Med.* 1979;20:138–43.
15. Steigman J. Chemistry of the alumina column. *Int J Appl Radiat Isot.* 1982;33:829–34.
16. Chase GD, Rabinowitz JL. *Principles of Radioisotope Methodology.* 3rd ed. Minneapolis: Burgess; 1967:442–7.
17. Molinski VJ. A review of Tc-99m generator technology. *Int J Appl Radiat Isot.* 1982;33:811–9.
18. Tofe AJ, Bevan JA, Fawzi MD, et al. Gentisic acid: a new stabilizer for low tin skeletal imaging agents [concise communication]. *J Nucl Med.* 1980;21:366–70.
19. Burke J. Advances in ^{99m}Tc Radiopharmaceutical Chemistry. Paper presented at: 148th Annual Meeting of the American Pharmaceutical Association; March 2001; San Francisco, CA.
20. Deutsch E, Glavan KA, Sodd VJ, et al. Cationic Tc-99m complexes as potential myocardial imaging agents. *J Nucl Med.* 1981;22:897–907.
21. Loberg MD, Fields AT. Chemical structure of technetium-99m-labeled N-(2,6-dimethylphenylcarbamoylmethyl)iminodiacetic acid (Tc-HIDA). *Int J Appl Radiat Isot.* 1978;29:167–73.
22. Loberg MD, Fields AT. Stability of ^{99m}Tc-labeled N-substituted iminodiacetic acids: ligand exchange reaction between ^{99m}Tc-HIDA and EDTA. *Int J Appl Radiat Isot.* 1977;28:687–92.
23. Ryan J, Cooper M, Loberg M, et al. Technetium-99m-labeled N-(2,6-dimethylphenyl carbamoylmethyl) iminodiacetic acid (Tc-99m HIDA): a new radiopharmaceutical for hepatobiliary imaging studies. *J Nucl Med.* 1977;18:995–1004.
24. Davison A, Jones AG, Orvig C, et al. A new class of oxotechnetium(5+) chelate complexes containing a TcON$_2$S$_2$ core. *Inorg Chem.* 1981;20:1629–32.
25. Jones AG, Davison A. The relevance of basic technetium chemistry to nuclear medicine. *J Nucl Med.* 1982;23:1041–3.
26. Jones AG, Abrams MJ, Davison A, et al. Biological studies of a new class of technetium complexes: the hexakis(alkylisonitrile)technetium (I) complexes. *Int J Nucl Med Biol.* 1984;11:225–34.
27. Cheesman EH, Blanchette MA, Ganey MV, et al. Technetium-99m ECD: ester-derivatized diamine-dithiol Tc complexes for imaging brain perfusion [abstract]. *J Nucl Med.* 1988; 29:788.
28. Nowotnik DP, Canning LR, Cumming SA, et al. Development of a Tc-99m-labelled radiopharmaceutical for cerebral blood flow imaging. *Nuc Med Commun.* 1985;6:499–506.
29. Fritzberg AR, Kasina S, Eshima D, et al. Synthesis and biological evaluation of technetium-99m MAG3 as a hippuran replacement. *J Nucl Med.* 27;1986:111–6.
30. Forster AM, Storey AE, Archer CM, et al. Structural characterization of the new myocardial imaging agent technetium-99m tetrofosmin (^{99m}Tc-P 53) [abstract]. *J Nucl Med.* 1992; 33:850.
31. Jones AG, Davison A, Trop HS, et al. Oxotechnetium (V) complexes [abstract]. *J Nucl Med.* 1979;20:641.
32. Mazzi U, Nicolini M, Bandoli G, et al. Technetium coordination chemistry: development of new backbones for ^{99m}Tc radiopharmaceuticals. In: Nicolini M, Bandoli G, Mazzi U, eds. *Technetium and Rhenium in Chemistry and Nuclear Medicine 3.* Verona, Italy: Cortina International; 1990:39–50.
33. Mease RC, Lambert C. Newer methods of labeling diagnostic agents with Tc-99m. *Semin Nucl Med.* 2001;31:278–85.
34. Liu S, Edwards DS. ^{99m}Tc-labeled small peptides as diagnostic radiopharmaceuticals. *Chem Rev.* 1999;99:2235–68.
35. Jurisson SS, Lydon JD. Potential technetium small molecule radiopharmaceuticals. *Chem Rev.* 1999;99:2205–18.
36. Alberto R, Ortner K, Wheatley N, et al. Synthesis and properties of boranocarbonate: a convenient in situ CO source for the aqueous preparation of [^{99m}Tc(OH$_2$)$_3$(CO)$_3$]$^+$. *J Am Chem Soc.* 2001;123:3135–6.
37. Hoepping A, Reisgys M, Brust P, et al. TROTEC-1: a new high-affinity ligand for labeling of the dopamine transporter. *J Med Chem.* 1998;41:4429–32.
38. Spradau TW, Katzenellenbogen JA. Preparation of cyclopentadienyl-tricarbonylrhenium complexes using a double ligand-transfer reaction. *Organometallics.* 1998;17:2009–17.
39. Spradau TW, Katzenellenbogen JA. Protein and peptide labeling with (cyclopentadienyl)tricarbonyl rhenium and technetium. *Bioconjugate Chem.* 1998;9:765–72.
40. Spradau TW, Edwards WB, Anderson CJ, et al. Synthesis and biological evaluation of Tc-99m cyclopentadienyl tricarbonyl technetium-labeled octreotide. *Nucl Med Biol.* 1999;26:1–7.
41. Bartholoma M, Valliant J, Maresca KP, et al. Single amino acid chelates (SAAC): a strategy for the design of technetium and rhenium radiopharmaceuticals. *Chem Commun.* 2009;5:493–512.
42. Treher EN, Francesconi LC, Gougoutas JZ, et al. Monocapped tris (dioxime) complexes of technetium(III): synthesis and structural characterization of TcX(dioxime)$_3$ B-R (x = Cl, Br; dioxime = dimethylglyoxime, cyclohexanedione dioxime; R = CH$_3$, C$_4$H$_9$). *Inorg Chem.* 1989;28:3411–6.
43. Eckelman WC, Meinken G, Richards P. The chemical state of ^{99m}Tc in biomedical products. II. The chelation of reduced technetium with DTPA. *J Nucl Med.* 1972;13:577–81.
44. Steigman J, Meinken G, Richards P. The reduction of pertechnetate-99 by stannous chloride—I. The stoichiometry of the reaction in HCl, in a citrate buffer and in a DTPA buffer. *Int J Appl Radiat Isot.* 1975;26:601–9.
45. Baldas J, Bonnyman J. Substitution reactions of ^{99m}TcNCl$_4^-$: a route to a new class of ^{99m}Tc-radiopharmaceuticals. *Int J Appl Radiat Isot.* 1985;36:133–9.
46. Baldas J, Bonnyman J. Effect of ^{99m}Tc-nitrido group on the behavior of ^{99m}Tc-radiopharmaceuticals. *Int J Appl Radiat Isot.* 1985;36:919–23.
47. Pasqualini R, Duatti A, Bellande E, et al. Bis(dithiocarbamato) nitrido technetium-99m radiopharmaceuticals: a class of neutral myocardial imaging agents. *J Nucl Med.* 1994;35:334–41.
48. Fagret D, Pierre-Yves M, Brunotte F, et al. Myocardial perfusion imaging with technetium-99m-Tc NOET: comparison with thallium-201 and coronary angiography. *J Nucl Med.* 1995;36:936–43.
49. Calnon DA, Ruiz M, Vanzetto G, et al. Myocardial uptake of 99mTc-N-NOet and 201Tl during dobutamine infusion: comparison with adenosine stress. *Circulation.* 1999;100:1653–9.
50. Abrams MJ, Juweid M, tenKate CI, et al. Technetium-99m-human polyclonal IgG radiolabeled via the hydrazine nicotinamide derivative for imaging focal sites of infection in rats. *J Nucl Med.* 1990;31:2022–8.
51. Edwards DS, Liu S, Barrett JA, et al. New and versatile ternary ligand system for technetium radiopharmaceuticals: water soluble phosphines and tricine as coligands in labeling a hydrazinonicotinamide-modified cyclic glycoprotein IIb/IIIa receptor antagonist with ^{99m}Tc. *Bioconjugate Chem.* 1997;8:146–54.
52. Barrett JA, Crocker AC, Damphouse DJ, et al. Biological evaluation of thrombus imaging agents utilizing water soluble phosphines and tricine as coligands when used to label a hydrazinonicotinamide-modified cyclic glycoprotein IIb/IIIa receptor antagonist with ^{99m}Tc. *Bioconjugate Chem.* 1997;8:155–60.
53. Banerjee S, Pillai MRA. Evolution of Tc-99m in diagnostic radiopharmaceuticals. *Semin Nucl Med.* 2001;31:260–77.
54. Reddy JA, Xu LC, Parker N, et al. Preclinical evaluation of ^{99m}Tc-EC-20 for imaging folate receptor-positive tumors. *J Nucl Med.* 2004;45:857–66.
55. Lister-James J, Knight LC, Maurer AH, et al. Thrombus imaging with a technetium-99m-labeled, activated platelet receptor-binding peptide. *J Nucl Med.* 1996;37:775–81.
56. *The United States Pharmacopeia, 28th rev, and The National Formulary, 23rd ed.* Rockville, MD: United States Pharmacopeial Convention, Inc; 2005.

57. Technetium-99m labeled radiodiagnostic agents and method of preparation. United States Patent 4042677. August 8, 1977. www.freepatentsonline.com/4042677.html
58. Robbins PJ. *Chromatography of Technetium-99m Radiopharmaceuticals—A Practical Guide.* New York: Society of Nuclear Medicine, Inc; 1984.
59. Technetium Tc-99m HSA Multidose Kit [package insert]. Arlington Heights, IL: Medi-Physics Inc/Amersham Healthcare; 1993.
60. Williams MJ, Deegan T. The processes involved in the binding of technetium-99m to human serum albumin. *Int J Appl Radiat Isot.* 1971;22:767–74.
61. TechneScan MAA Kit [package insert], St Louis: Mallinckrodt Inc; 2000.
62. Lister-James J, Dean RT. Technetium-99m-labeled receptor-specific small synthetic peptides: practical imaging agents of biochemical markers. In: Nicolini M, Mazzi U, eds. *Technetium, Rhenium, and Other Metals in Chemistry and Nuclear Medicine 5.* Padova, Italy: SGE Ditorioli; 1999:401–7.
63. Tc-99m Apcitide [package insert]. Londonderry, NH: Diatide Inc; 1998.
64. Taillefer R, Edell S, Innes G, et al. Acute thromboscintigraphy with Tc-99m-apcitide: results of the phase-3 multicenter clinical trial comparing Tc-99m apcitide scintigraphy with contrast venography for imaging acute DVT. *J Nucl Med.* 2000;41:1214–23.
65. Andrieux A, Hudry-Clergeon G, Ryckewaert J, et al. Amino acid sequences in fibrinogen mediating its interaction with its platelet receptor, GP IIb/IIIa. *J Biol Chem.* 1989;264:9258–65.
66. Lefkovits J, Plow EF, Topol EJ. Platelet glycoprotein IIb/IIIa receptors in cardiovascular medicine. *N Engl J Med.* 1995;332:1553–9.
67. Tc-99m Bicisate [package insert]. Billerica, MA: Dupont Pharmaceuticals; 1998.
68. Walovitch RC, Hill TC, Garrity ST, et al. Characterization of technetium-99m-L,L-ECD for brain perfusion imaging, part 1: pharmacology of technetium-99m ECD in nonhuman primates. *J Nucl Med.* 1989;30:1892–901.
69. Tc-99m Disofenin [package insert]. Bedford, MA: CIS-US Inc; 1999.
70. Tc-99m Exametazime [package insert]. Arlington Heights, IL: Medi-Physics Inc/Amersham Healthcare; 1996.
71. Neirinckx RD, Burke JF, Harrison RC, et al. The retention mechanism of technetium-99m-HM-PAO: intracellular reaction with glutathione. *J Cereb Blood Flow Metab.* 1988;8:S4–S12.
72. Hung JC, Corlija M, Volkert WA, et al. Kinetic analysis of technetium-99m d,l-HM-PAO decomposition in aqueous media. *J Nucl Med.* 1988;29:1568–76.
73. Neirinckx RD, Canning LR, Piper IM, et al. Technetium-99m d,l-HM-PAO: a new radiopharmaceutical for SPECT imaging of regional cerebral blood perfusion. *J Nucl Med.* 1987;28:191–202.
74. Tc-99m Gluceptate [package insert]. Kirkland, Quebec, Canada: DraxImage Inc; 1998.
75. de Kieviet W: Technetium radiopharmaceuticals: chemical characterization and tissue distribution of Tc-glucoheptonate using Tc-99m and carrier Tc-99. *J Nucl Med.* 1981;22:703–9.
76. Tc-99m Mebrofenin [package insert]. Princeton, NJ: Bracco Diagnostics Inc; 2000.
77. Tc-99m Medronate [package insert]. Princeton, NJ: Bracco Diagnostics Inc; 1999.
78. Sattuck LA, Eshima D, Taylor AT Jr, et al. Evaluation of the hepatobiliary excretion of technetium-99m-MAG3 and reconstitution factors affecting radiochemical purity. *J Nucl Med.* 1994;35:349–55.
79. Tc-99m Mertiatide [package insert]. St Louis: Mallinckrodt Inc; 1995.
80. Nosco DL, Wolfangel RG, Bushman MJ, et al. Technetium-99m MAG3: labeling conditions and quality control. *J Nucl Med Technol.* 1993;21:69–74.
81. Tc-99m Oxidronate [package insert]. St Louis: Mallinckrodt Inc; 2000.
82. Tc-99m Pentetate [package insert]. Bedford, MA: CIS-US, Inc; 1993.
83. Tc-99m Pyrophosphate [package insert]. St Louis: Mallinckrodt Inc; 2000.
84. Srivastava SG, Chervu LR. Radiolabeled red blood cells: current status and future prospects. *Semin Nucl Med.* 1984;14:68–82.
85. UltraTag RBC [package insert]. St Louis: Mallinckrodt Inc; 2000.
86. Larson SM, Hamilton GW, Richards P, et al. Kit-labeled technetium-99m red blood cells (Tc-99m RBCs) for clinical cardiac chamber imaging. *Eur J Nucl Med.* 1978;3:227–31.
87. Srivastava SC, Richards P, Yonekura Y, et al. Long-term retention of tin following in-vivo RBC labeling. *J Nucl Med.* 1982;23:P91.
88. Callahan RJ, Froelich JW, McKusick KA, et al. A modified method for the in vivo labeling of red blood cells with Tc-99m [concise communication]. *J Nucl Med.* 1982;23:315–8.
89. Dewanjee MK, Rao SA, Penniston GT. Mechanism of red blood cell labeling with ^{99m}Tc-pertechnetate. The role of the cation pump at RBC membrane on the distribution and binding of Sn^{2+} and ^{99m}Tc with the membrene proteins and hemoglobin. *J Labelled Comp Radiopharm.* 1982;19:1464–5.
90. Dewanjee MK. Binding of Tc-99m ion to hemoglobin. *J Nucl Med.* 1974;15:703–6.
91. Straub RF, Srivastava SC, Meinken GE, et al. Transport, binding and uptake kinetics of tin and technetium in the in-vitro Tc-99m labeling of red blood cells. *J Nucl Med.* 1985;26:P130.
92. Callahan RJ, Rabito CA. Radiolabeling of erythrocytes with technetium-99m: role of band-3 protein in the transport of pertechnetate across the cell membrane. *J Nucl Med.* 1990;31:2004–10.
93. Cardiolite/Miraluma [package insert]. Billerica, MA: Bristol Myers Squibb; 2000.
94. Johannsen B, Syhre R, Spies H, et al. Chemical and biological characterization of different Tc complexes of cysteine and cysteine derivatives. *J Nucl Med.* 1978;19:816–24.
95. Moretti JL, Rapin JR, Saccavina JC, et al. 2,3-Dimercaptosuccinic-acid chelates—1. Structure and pharmacokinetic studies. *Int J Nucl Med Biol.* 1984;11:270–4.
96. DMSA kit for the preparation of Tc-99m succimer injection [package insert]. Arlington Heights, IL: Medi-Physics Inc/Amersham Healthcare; 1993.
97. Ikeda I, Inoue O, Kurata K. Chemical and biological studies on Tc-99m DMS-II: effect of Sn(II) on the formation of various Tc-DMS complexes. *Int J Appl Radiat Isot.* 1976;27:681–8.
98. Ikeda I, Inoue O, Kurata K; Preparation of various Tc-99m dimercaptosuccinate complexes and their evaluation as radiotracers. *J Nucl Med.* 1977;18:1222–9.
99. Sampson C. Preparation of ^{99m}Tc-(V)DMSA. *Nucl Med Comm.* 1987;8:184–5.
100. Blower PJ, Singh J, Clarke SEM. The chemical identity of pentavalent technetium-99m-dimercaptosuccinic acid. *J Nucl Med.* 1991;32:845–9.
101. Washburn LC, Biniakiewica DS, Maxon HR. Reliable kit preparation of Tc-99m pentavalent dimercaptosuccinic acid [Tc-99m(V)DMSA]. *J Nucl Med.* 1994;35:263P.
102. Arslan N, Ilgan S, Yuksel D, et al. Comparison of In-111 octreotide and Tc-99m(V)DMSA scintigraphy in the detection of medullary thyroid tumor foci in patients with elevated levels of tumor markers after surgery. *Clin Nucl Med.* 2001;26:683–8.
103. Adams S, Acker P, Lorenz M, et al. Radioisotope-guided surgery in patients with pheochromocytoma and recurrent medullary thyroid carcinoma: a comparison of preoperative and intraoperative tumor localization with histopathologic findings. *Cancer.* 2001;92:263–70.
104. CIS-Sulfur Colloid [package insert]. Bedford, MA: CIS-US Inc; 1998.
105. Myoview [package insert]. Arlington Heights, IL: GE Healthcare; 1996.
106. Kelly JD, Forster AM, Higley B, et al. Technetium-99m-tetrofosmin as a new radiopharmaceutical for myocardial perfusion imaging. *J Nucl Med.* 1993;34:222–7.
107. McAfee JG, Thakur ML. Survey of radioactive agents for in vitro labeling of phagocytic leukocytes. I. Soluble agents. *J Nucl Med.* 1976;17:480–7.
108. Baker WJ, Datz FL. Preparation and clinical utility of In-111 labeled leukocytes. *J Nucl Med Technol.* 1984;12:131–8.

109. Peters AM, Saverymuttu SH, Reavy HJ, et al. Imaging inflammation with 111-indium-tropolonate labeled leukocytes. *J Nucl Med.* 1983;24:39–44.
110. Peters AM, Danpure HJ, Osman S, et al. Clinical experience with ^{99m}Tc-hexamethylpropylene-amineoxime for labeling leucocytes and imaging inflammation. *Lancet.* 1986;2:946–9.
111. Roddie ME, Peters AM, Danpure HJ, et al. Inflammation: imaging with Tc-99m HMPAO-labeled leukocytes. *Radiology.* 1988;166:767–72.
112. McAfee JG, Subramanian G, Gagne G. Technique of leukocyte harvesting and labeling: problems and perspectives. *Semin Nucl Med.* 1984;14:83–106.
113. Gobuty AH. Technologic, clinical, and basic science considerations for In-111 labeled leukocytes. *J Nucl Med Technol.* 1984;12:131–8.
114. Palestro CJ, Brown ML, Forstrom LA, et al. Procedure guideline for ^{99m}Tc-exametazime (HMPAO)-labeled leukocyte scintigraphy for suspected infection/inflammation 3.0, June 2, 2004. In: *Procedure Guidelines Manual 2007.* Reston, VA: Society of Nuclear Medicine.
115. Peters AM. The utility of ^{99m}Tc-HMPAO-leukocytes for imaging infection. *Semin Nucl Med.* 1994;24:110–27.
116. Danpure HJ, Osman S, Carroll MJ. Development of a clinical protocol for radiolabeling of mixed leukocytes with technetium-99m hexamethylpropyleneamine oxime. *Nucl Med Commun.* 1988;9:465–75.
117. Puncher MR, Blower PJ. Autoradiography and density gradient separation of technetium-99m-exametazime (HMPAO) labelled leucocytes reveals selectivity for eosinophils. *Eur J Nucl Med.* 1994;21:1175–82.
118. Moberg L, Karawajczyk M, Venge P. ^{99m}Tc-HMPAO (Ceretec) is stored in and released from the granules of eosinophil granulocytes. *Br J Haematol.* 2001;114:185–90.
119. Indium In-111 Oxine [package insert]. Arlington Heights, IL: Medi-Physics Inc/GE Healthcare; 1996.

CHAPTER 11
Radiopharmaceutical Chemistry: Nontechnetium Agents

More than 2500 nuclides have been discovered, most of which are radioactive. Unfortunately, a majority of these radionuclides have half-lives that are too short for practical application in medicine. Only a few possess the nuclear and chemical properties essential for diagnostic or therapeutic application in human disease. Among the most useful radionuclides are ^{14}C, ^{32}P, ^{51}Cr, ^{67}Ga, ^{90}Y, ^{99m}Tc, ^{111}In, ^{123}I, ^{125}I, ^{131}I, ^{133}Xe, ^{201}Tl, and ^{153}Sm. Of these, ^{32}P, ^{90}Y, and ^{153}Sm have established applications in radiation therapy procedures. They are discussed in Chapter 29. Radiopharmaceuticals labeled with ^{14}C, ^{51}Cr, and ^{125}I have application in diagnostic in vivo function studies and are discussed in Chapter 30. The remainder are used in diagnostic single-photon emission computed tomography (SPECT) imaging procedures. An additional group of radionuclides is used for diagnostic imaging by positron emission tomography (PET). They include the positron emitters ^{11}C, ^{13}N, ^{15}O, ^{18}F, and ^{82}Rb. The first three, ^{11}C, ^{13}N, and ^{15}O, have very short half-lives of approximately 20, 10, and 2 minutes, respectively. Radiopharmaceuticals labeled with these nuclides require on-site cyclotron production and a radiochemistry laboratory. Radiopharmaceuticals labeled with cyclotron-produced ^{18}F, which has a 110-minute half-life, can be prepared in a nuclear pharmacy facility that is located within a few hours' transport of hospitals and clinics. ^{82}Rb, which has a 76-second half-life, is obtained from a commercially available ^{82}Sr–^{82}Rb generator, making ^{82}Rb readily accessible in a nuclear medicine PET imaging facility.

This chapter describes the chemistry of nontechnetium radiopharmaceuticals labeled with one of the following radionuclides: ^{32}P, ^{51}Cr, ^{67}Ga, ^{90}Y, ^{111}In, ^{123}I, ^{125}I, ^{131}I, ^{133}Xe, ^{201}Tl, and ^{153}Sm. Chapter 10 focuses separately on the chemistry of ^{99m}Tc-labeled radiopharmaceuticals, and Chapter 12 covers the chemistry of PET agents.

IODINE CHEMISTRY

The most useful radioisotopes of iodine for nuclear medicine are ^{123}I, ^{125}I, and ^{131}I. In addition, there has been some interest in the positron emitter ^{124}I. Their physical properties are shown in Table 11-1. The type of nuclear medicine procedure dictates which radioiodine isotope is used. For SPECT imaging, only ^{123}I and ^{131}I are suitable. ^{123}I has favorable imaging properties because its gamma ray energy is efficiently detected by the sodium iodide crystal in the gamma camera. However, its 13.2 hour half-life requires daily purchase, which makes it less convenient to use. ^{131}I has a long shelf life, but its high gamma ray energies yield lower detection efficiency and require heavy collimation on gamma cameras. In addition, because of its beta particle emission, ^{131}I has a radiation absorbed dose per microcurie 100-fold higher than ^{123}I. This limits the amount of activity that can be administered for imaging studies but is an advantage for radiation therapy procedures to treat hyperthyroidism and thyroid cancer. ^{131}I has also been used as a radiolabel for therapeutic antibodies, such as ^{131}I-tositumomab (Bexxar [Corixa]), for the treatment of non-Hodgkin's lymphoma.[1,2]

TABLE 11-1 Physical Properties of Radioiodine Isotopes

Nuclide	Half-life	Decay Mode/Product	Photon MeV	Photon Abundance(%)	SA[a] mCi/μg
^{123}I	13.22 hours	EC/^{123}Te	0.159	83	1930.0
			0.027 (Te x-rays)	70	
^{124}I	4.18 days	EC, β+/^{124}Te	0.027 (Te x-rays)	47	251.9
			0.511	46	
			0.603	63	
			0.723	10	
			1.691	11	
^{125}I	59.4 days	EC/^{125}Te	0.035	7	17.5
			0.027 (Te x-rays)	114	
^{131}I	8.02 days	β–/^{131}Xe	0.364	82	124.0
			637	7	

[a] Theoretical maximum specific activity.

The photons emitted from ^{125}I have very low energies that are readily absorbed in the body. For this reason, ^{125}I is excluded from diagnostic imaging procedures, but its activity can be readily counted with a scintillation well counter. The principal application of ^{125}I-labeled radiopharmaceuticals is for in vivo function studies. In this regard, ^{125}I-iothalamate is used to measure glomerular filtration rate (GFR), and ^{125}I-human serum albumin is used to measure plasma volume. ^{125}I is a useful radiolabel for drugs, peptides, and proteins associated with in vitro laboratory test methods. The low-energy radiation and lack of beta emission make ^{125}I less damaging to labeled molecules, and its 60 day half-life gives it a long shelf life. Although its photon energies are too low for diagnostic imaging studies, ^{125}I is sometimes used in brachytherapy devices for radiation treatment of tumors in situ, because its radiation is effectively absorbed over a short distance. One agent with this application is Iotrex (Proxima Therapeutics, Alpharetta, GA). It is a ^{125}I intracavitary radiation therapy product for treating malignant brain tumors following tumor resection.[3] ^{125}I is also used in the form of implantable seeds in the treatment of prostate cancer in lieu of surgery.[4,5]

Radioiodine Production

^{131}I

^{131}I is obtained as a byproduct of uranium fission but can also be produced by the neutron activation of tellurium. Both methods produce ^{131}Te, which decays to ^{131}I:

$$^{235}U(n,f) \, ^{131}Te \xrightarrow{\beta^-, 30\,hr} \, ^{131}I$$

$$^{130}Te(n,\gamma) \, ^{131}Te \xrightarrow{\beta^-, 30\,hr} \, ^{131}I$$

The nuclear material is processed to yield sodium iodide as the final chemical form.

The nucleus of ^{131}I is characterized by 53 protons and 78 neutrons, 4 neutrons more than the stable isotope ^{127}I. Neutron-rich ^{131}I undergoes negatron (beta-minus) decay with a half-life of 8.02 days to stable ^{131}Xe according to the following decay equation:

$$^{131}_{53}I_{78} \xrightarrow{n \to p^+ + e^- + \nu} \, ^{131}_{54}Xe_{77} + 0.971\,MeV(\beta, \nu, \gamma)$$

130.9061246 AMU 130.9050824 AMU

The transition energy between ^{131}I and the ^{131}Xe ground state is 0.971 MeV (0.0010422 AMU × 931.5 MeV/AMU). Several beta transitions are possible in the decay scheme of ^{131}I (Figure 11-1), but the most frequent transition (γ_{14}) releases a 0.364 MeV gamma ray with 82% abundance.[6] It is

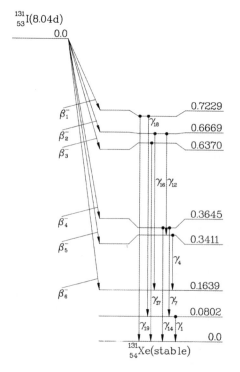

FIGURE 11-1 Decay scheme for ^{131}I. (Reprinted with permission from reference 6.)

the principal photon used in gamma-camera imaging. The remaining 0.607 MeV of transition energy is given to the beta particle and the neutrino.

^{125}I

^{125}I is produced indirectly by the neutron irradiation of ^{124}Xe to yield a short half-lived ^{125}Xe that quickly decays to ^{125}I:

$$^{124}\text{Xe}(n,\gamma)\,^{125}\text{Xe} \xrightarrow{EC,\,17\,hr} {}^{125}\text{I}$$

The nucleus of ^{125}I is characterized by 53 protons and 72 neutrons, 2 neutrons less than ^{127}I. This proton-rich nucleus undergoes electron capture (EC) decay with a half-life of 59.4 days to stable ^{125}Te according to the following decay equation:

$$^{125}_{53}\text{I}_{72} \xrightarrow{p^+ + e^- \to n + \nu} {}^{125}_{52}\text{Te}_{73} + 0.1858\,\text{MeV}(\nu,\gamma)$$

124.9046302 AMU   124.9044307 AMU

The transition energy for this decay is 0.1858 MeV. Of this amount, the neutrino carries away 0.151 MeV to an excited state of ^{125}Te, which promptly de-excites to the ground state, releasing the remaining 0.035 MeV as a gamma ray (Figure 11-2). For every 100 atoms of ^{125}I that decay, 93 of the 0.035 MeV gammas undergo electron conversion (K-, L-, and M-shell electrons are removed), so only 7 gamma rays are detectable (7% abundance). However, the result of electron conversion is emission of a high percentage of ^{125}Te x-rays (0.027 MeV [114%] and 0.031 MeV [24%]). These weak photons are easily absorbed. In fact, about 50% of the photons will be absorbed by the glass vial containing an ^{125}I source. For this reason, container geometry must be considered in order to make an accurate measurement of ^{125}I with the dose calibrator. See Chapter 5 and Table 14-12 in Chapter 14 for more details on ^{125}I assay.

^{123}I

^{123}I has nearly ideal properties for imaging, with a 13.2 hour half-life and a 0.159 MeV gamma ray that is efficiently detected by the sodium iodide crystal of the gamma camera. ^{123}I can be produced in a cyclotron by several different methods. As discussed in Chapter 3, the type of radionuclidic impurities formed in the production of ^{123}I is determined by the nuclear reaction employed. A method that yields high-purity ^{123}I is the proton bombardment of a ^{124}Xe target. The short-lived ^{123}Xe product quickly decays to ^{123}I:

$$^{124}\text{Xe}(p,2n)\,^{123}\text{Cs} \xrightarrow{5.8\,min} {}^{123}\text{Xe} \xrightarrow{2\,hr} {}^{123}\text{I}$$

The ^{123}I nucleus is characterized by 53 protons and 70 neutrons, making it 4 neutrons less than ^{127}I. It decays by EC to ^{123}Te, which is essentially stable ($T_{1/2} = 1.2 \times 10^{13}$ yr) (Figure 11-3).

$$^{123}_{53}\text{I}_{70} \xrightarrow{p^+ + e^- \to n + \nu} {}^{123}_{52}\text{Te}_{71} + 1.2286\,\text{MeV}(\nu,\gamma)$$

122.905589 AMU   122.904270 AMU

The transition energy for this decay is 1.2286 MeV, which is dissipated by several available EC transitions. The principal route (EC_{14}) proceeds to the excited state of ^{123}Te, which emits the 0.159 MeV gamma ray (γ_2). A partial loss of the 0.159 MeV gamma ray to K-, L-, and M-conversion electrons yields a photon abundance of 83%.

An important point to consider with dose calibrator measurement of ^{123}I is the emission of a high abundance of ^{123}Te x-rays (0.027 MeV [70%] and 0.031 MeV [15%]), which are subject to significant absorption by the ^{123}I source container. This can affect its measurement in the dose calibrator by more than 10%. Therefore, care must be taken to develop geometry correction factors for the different types of containers. A facile technique for removing the low-energy x-rays during dose calibrator measurement has been published;

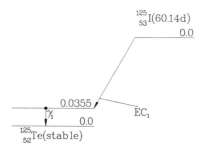

FIGURE 11-2 Decay scheme for ^{125}I. (Reprinted with permission from reference 6.)

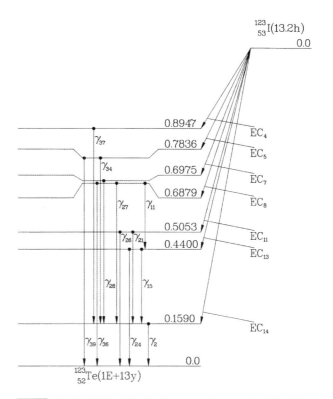

FIGURE 11-3 Decay scheme for ^{123}I. (Reprinted with permission from reference 6.)

it obviates the necessity for applying geometry correction factors for different container configurations.[7]

Labeling with Radioiodine

Radioiodine's importance in nuclear medicine stems from its diverse chemistry and the availability of several isotopes with different physical properties. Iodinated compounds can be prepared by several methods, including isotope exchange, nucleophilic substitution, electrophilic substitution, addition to double bonds, iododemetallation, and conjugation labeling with prosthetic groups. These methods have been reviewed by Lever.[8] The most widely used methods for preparing radiopharmaceuticals are electrophilic substitution, isotope exchange, and conjugation labeling.

Electrophilic Substitution

With the exception of nucleophilic substitution reactions that involve substitution of I$^-$ for a leaving group such as bromide, many radioiodinations involve substituting an electrophilic I$^+$ for hydrogen in an aromatic compound activated by an electron-donating group such as OH or NH$_2$ in the aromatic ring. The *p*-hydroxyl group in tyrosine is a good example (Figure 11-4). Radioiodination is facilitated by the presence of iodine monochloride, chloramine-T, iodogen, or iodobeads. Direct iodination of proteins and antibodies into the tyrosyl moiety is accomplished by electrophilic substitution. The general reaction is as follows, with asterisk indicating radioactive isotope:

$$R - H + I_2^* \leftrightarrow R - I^* + HI \qquad (11-1)$$

In molecular iodine (I$_2$), the structure (I$^-$ – I$^+$) is assumed. The I$^+$ ion does not exist alone but usually forms a complex with a nucleophilic species in aqueous solution. The following reactions are possible:

$$I_2 + H_2O \leftrightarrow H_2OI^+ + I^- \qquad (11-2)$$

$$I_2 + OH^- \leftrightarrow HOI + I^- \qquad (11-3)$$

It is believed that the iodinating species in labeling reactions is either the hydrated complex (H$_2$OI$^+$)[9] or hypoiodous acid (HOI).[10]

Protein iodination is one of the most important radiolabeling techniques in nuclear medicine. Typical iodination sites in protein molecules include the 3 and 5 positions in the aromatic ring of tyrosine as the primary site or the imidazole ring of histidine as a secondary site. Because the anionic form of the molecule to be labeled seems to be the reactive species with I$^+$, close attention must be paid to the protein's pKa and the pH of the reaction mixture. At low pH, tyrosine is protonated and labeling yields are low. Basic pH, however, promotes dissociation of tyrosine hydrogens to form the desired tyrosinate anion and also promotes the hydrolysis of I$_2$ according to reaction 11-2.[11] This greatly facilitates the rate of iodination into the tyrosine ring (Figures 11-3 and 11-5). Solutions of pH 7 to 9 are typical in protein iodination. One must avoid higher pH values (>10) because of the irreversible disproportionation of HOI to iodate according to the following reaction. Some iodate, however, will still form at pH 7 to 9:

$$3\,HOI + 3\,OH^- \leftrightarrow 2\,I^- + IO_3^- + 3\,H_2O \qquad (11-4)$$

The order of mixing reagents is important in achieving high iodination yield. This depends on the method of radioiodination, but in general the molecule to be labeled is mixed first with buffer followed by addition of radioiodide and iodinating agent. To retard protein damage, mild iodinating conditions must be used. More care must be taken in using iodinating agents that are fairly strong oxidizing agents, because they may attack the protein. It is recommended that not more than one atom of iodine per molecule of protein, on average, be introduced, in order to preserve protein integrity. However, this too depends on the protein, the iodinating method, and the ultimate use or application of the iodinated protein.[12] For example, iodinated albumin prepared by the iodine monochloride method should contain not more than one mole of iodine per mole of protein.

Isotope Exchange

Isotope exchange involves exchanging a radioactive iodine atom with a stable iodine atom present in the molecule.

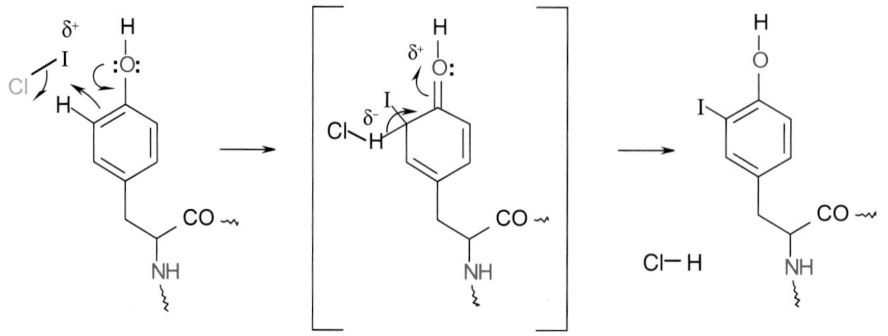

FIGURE 11-4 Mechanism of iodination of tyrosyl residues in proteins by electrophilic substitution.

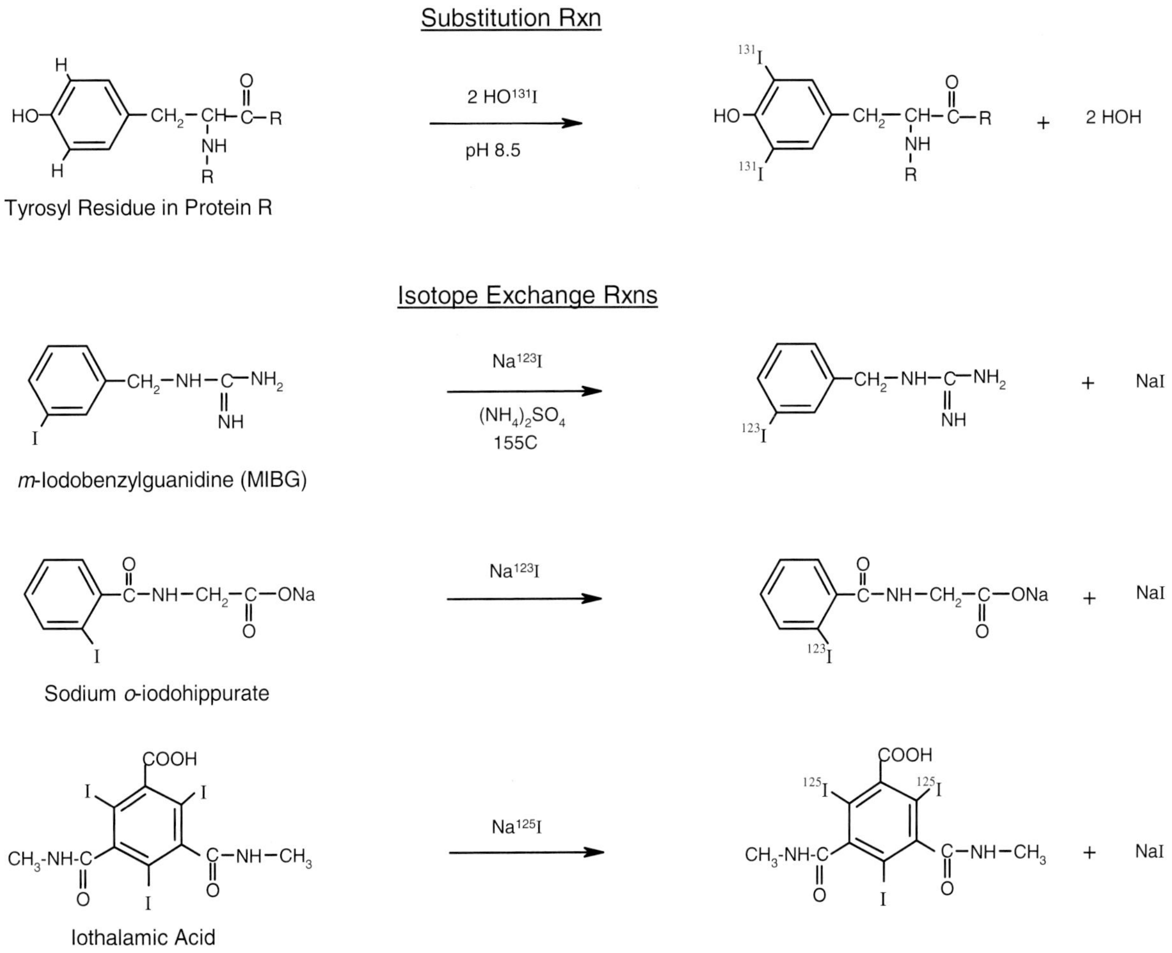

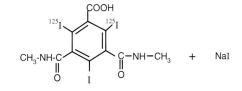
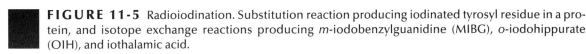

FIGURE 11-5 Radioiodination. Substitution reaction producing iodinated tyrosyl residue in a protein, and isotope exchange reactions producing m-iodobenzylguanidine (MIBG), o-iodohippurate (OIH), and iothalamic acid.

Theoretically, a 1:1 substrate-to-radioiodine molar ratio will produce a 50% labeling yield, whereas a 20:1 ratio will increase the yield to more than 95%, although specific activity will be lower.[8] Labeling reactions are typically conducted with the application of heat, either in solution or in the solid phase. Compounds that have been labeled in this manner include ^{123}I-iodoamphetamine (IMP), ^{123}I-N,N,N′-trimethyl-N′-[2-hydroxy-3-methyl-5-iodobenzyl]-1,3-propanediamine (^{123}I-HIPDM), ^{123}I- or ^{131}I-OIH, and ^{123}I- or ^{131}I-MIBG. The general exchange reaction is as follows:

$$R-I + I_2^* \leftrightarrow R I^* + I_2 \quad (11\text{-}5)$$

Iodination Methods

Different reagents have been used to facilitate radioiodination of compounds. These include iodine monochloride, chloramine-T, lactoperoxidase, electrolysis, Iodo-Gen (Pierce, Rockford, IL), Iodo-Beads (Pierce), and the Bolton-Hunter reagent. A few of the methods are described here. All of them except the Bolton-Hunter method involve covalent attachment of cationic iodine (I$^+$) to the aromatic ring of tyrosine residues in the protein.

Iodine Monochloride

The iodine in iodine monochloride is in the I$^+$ state because of chlorine's greater electronegativity. Iodine monochloride is formed by the oxidation of iodide with iodate in strong acid according to the following reaction:

$$2\,NaI + NaIO_3 + 6\,HCl \leftrightarrow 3\,ICl + H_2O + 3\,NaCl \quad (11\text{-}6)$$

When mixed with iodine monochloride, radioiodide undergoes isotope exchange. Because the iodine monochloride is in excess and all of its iodine is in the I$^+$ form, essentially all of the radioiodide is converted to I$^+$ accordingly:

$$ICl + NaI^* \leftrightarrow I^*Cl + NaI \quad (11\text{-}7)$$

Subsequently, the I*Cl hydrolyzes to HOI*, which iodinates the compound.[11] Radiolabeling yields are about 75%, but specific activity of the product is low because stable iodine is also incorporated into the compound because of the excess stable iodine monochloride present. In this labeling technique radioiodide is added to the buffered compound, and iodine monochloride is jetted into the mixture. If iodine monochloride is added before the radioiodide, labeling yields are lowered considerably because the isotope exchange reaction is impaired by the reaction of iodine monochloride directly with the compound. The iodine monochloride is added rapidly to disperse the mixture quickly in order to reduce the degree of multiple labeling within the same molecule created by localized concentrations of reactive iodine. An advantage of this method is that the amount of iodine incorporated into the compound is controlled by the amount of iodine monochloride used. This is an advantage if excessive substitution must be avoided.

Chloramine-T

This method uses the sodium salt of N-chloro-4-methyl benzene sulfonamide (chloramine-T) as the iodinating agent.[12,13] Chloramine-T undergoes hydrolysis at pH 7 to 8, liberating sodium hypochlorite, which oxidizes radioiodine to hypoiodous acid according to the following reactions:

$$CH_3 - C_6H_4SO_2NaNCl + H_2O \leftrightarrow$$
$$CH_3 - C_6H_4SO_2NH_2 + NaOCl \qquad (11\text{-}8)$$

$$NaOCl + HI^* \leftrightarrow HOI^* + NaCl \qquad (11\text{-}9)$$

The general technique is to mix the compound to be labeled with buffer and radioiodide and then to add the fresh chloramine-T solution. A period of incubation is required, and the reaction is stopped by adding a reducing agent. The typical labeling conditions for immunoglobulin G (IgG) with ^{131}I in a 1:1 molar ratio of antibody to iodine are shown in Table 11-2.

High labeling yields can be obtained, but labeling conditions must be carefully controlled because chloramine-T is a powerful oxidizing agent and may damage proteins.[14] The advantage of chloramine-T is that no carrier iodide is needed; therefore, high specific activities can be obtained. In addition, virtually complete utilization of the isotope can be achieved. This method has been effectively used to label the LYM-1 antibody for the treatment of non-Hodgkin's lymphoma.[15]

Iodo-Gen and Iodo-Beads

Iodo-Gen (Pierce) and Iodo-Beads (Pierce) are solid-phase oxidants in a technology developed to permit iodinations to proceed in a two-phase system, with the oxidant material bound in a solid phase from which reactants can be separated by simple aspiration (Figure 11-6). The Iodo-Gen method involves dissolving the Iodo-Gen oxidizing reagent (1,3,4,6-tetrachloro-3α,6α-diphenylglycouril) in an organic solvent such as chloroform and plating the reagent onto the sides of a glass tube after solvent evaporation. The buffered protein and radioiodide are added to the Iodo-Gen-coated tube and incubated for 10 minutes, whereupon the Iodo-Gen oxidizes the iodide. The reaction is stopped by aspirating the mixture from the tube.

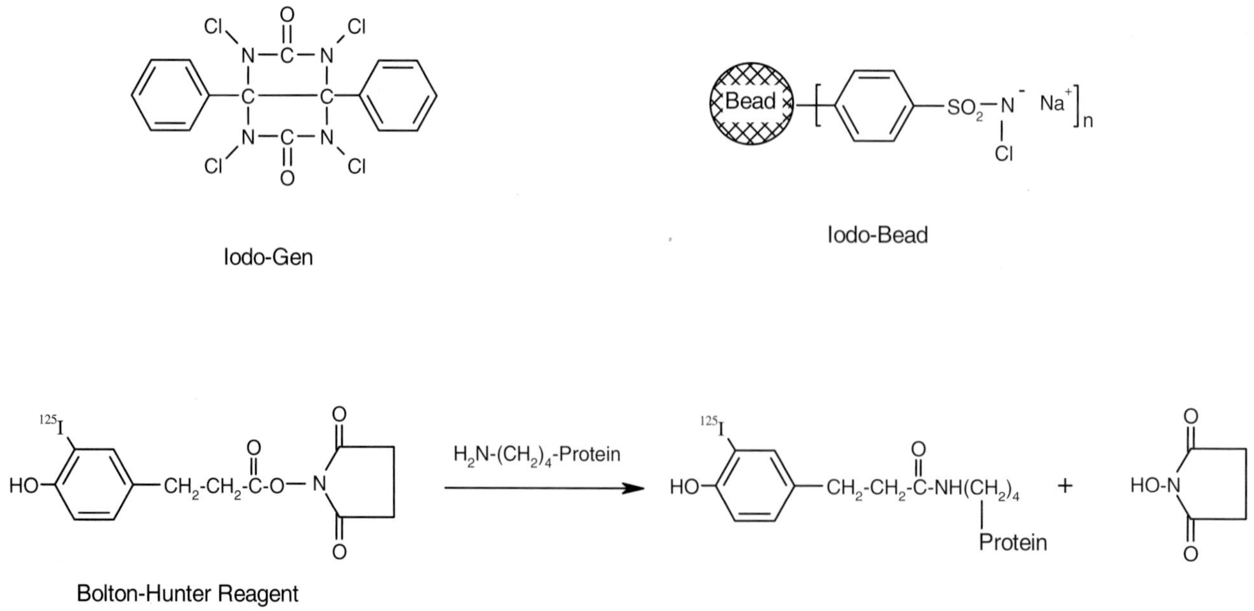

TABLE 11-2 Radioiodination of a Protein

Mixing Sequence	Typical Amounts
1. Protein	100 µg
2. Phosphate buffer	Adjust to pH range 7 to 8
3. NaI*	Na^{131}I 0.087 µg
4. Iodinating agent	Chloramine-T 1.25 µg
React for 10 min on ice	
5. Reducing agent	Sodium metabisulfite 2.5 µg

Asterisk indicates radioactive isotope.

FIGURE 11-6 Chemical structures of two mild oxidizing iodinating agents: Iodo-Gen and Iodo-Beads. Radioiodination of non-tyrosine-containing proteins via the Bolton-Hunter reagent.

Iodo-Beads are nonporous polystyrene beads to which is bonded the sodium salt of N-chloro-benzenesulfonamide as the oxidant. Labeling involves mixing radioiodide with several beads, incubating for 5 minutes to allow oxidation of iodide, then adding the protein in buffer, incubating for 10 minutes more, and aspirating the mixture to stop the reaction.

These solid-phase iodination techniques are claimed to provide milder oxidation conditions for the protein compared with chloramine-T and do not require the addition of reducing agent to stop the reaction. Labeling yields are good, and high specific activity labeling can be achieved.[16]

Bolton-Hunter Reagent

To circumvent the problem of protein damage by direct iodination and to be able to label proteins that lack the tyrosine moiety, indirect radioiodination with a prosthetic group can be used. An example is the Bolton-Hunter reagent.[17] This reagent is a reactive conjugate prelabeled with radioiodine that is reacted with the protein, eliminating contact with oxidizing and reducing agents. The reagent is an ^{125}I-labeled acylating agent, iodinated 3-(4-hydroxy-phenyl) propionic acid N-hydroxysuccinimide ester, which reacts with lysine amino groups in the protein (Figure 11-6). This method is mild, producing proteins that retain immunoreactivity.

IODINATED RADIOPHARMACEUTICALS

Radiopharmaceuticals labeled with ^{123}I, ^{125}I, and ^{131}I can be obtained from several radiopharmaceutical companies or can be compounded at the time of use.

Sodium Iodide I 123, I 125, and I 131 Capsules or Solution

Sodium iodide ^{131}I is supplied in hard gelatin capsules and in aqueous solution for oral administration to meet United States Pharmacopeia (USP) requirements for strength and purity for clinical use. Capsules of ^{123}I-sodium iodide are also available for diagnostic use in thyroid imaging (Table 11-3). Capsules and solution are available in various amounts of activity for diagnostic and therapeutic procedures involving the thyroid gland. These dosage forms have been formulated to stabilize radioiodide in its nonvolatile reduced form.[18] Adjuvants such as disodium phosphate, sodium bisulfite, and disodium EDTA are often present to stabilize iodide against oxidation. Because of their antioxidant properties, solutions for thyroid studies should not be used for radioiodination procedures, which require oxidation of radioiodide. ^{123}I-, ^{125}I-, and ^{131}I-sodium iodide in sodium hydroxide solution without stabilizers are available as radiochemicals for use in radioiodinations. Dilutions of radioiodide solutions should be made only with distilled water, since chlorine in tap water will oxidize iodide (I$^-$) to its volatile form (I$_2$) according to the following reaction:

$$2\ NaI + Cl_2 \rightarrow 2\ NaCl + I_2$$

This reaction will occur spontaneously because the standard oxidation potential of iodide to iodine (−0.536 V) is more positive than the standard oxidation potential of chloride to chlorine (−1.360 V). Thus, the oxidation of iodide and the complementary reduction of chlorine occur spontaneously,

TABLE 11-3 Radioiodide Capsule and Solution Formulations		
Supplier	**Radionuclide Formulation**	**Activities**
Covidien	Sodium iodide I-131 USP therapeutic capsules (containing sodium radioiodide, inert filler)	0.75–100 mCi
Covidien	Sodium iodide I-131 USP therapeutic solution (containing sodium radioiodide, 0.1% sodium bisulfite, 0.2% disodium edetate, 0.5% sodium phosphate buffer)	5–150 mCi
DraxImage	Sodium iodide I-131 USP therapeutic capsules (containing sodium radioiodide, disodium edetate dihydrate as stabilizer, sodium thiosulfate pentahydrate as reducing agent, dibasic sodium phosphate anhydrous)	2–200 mCi
DraxImage	Sodium iodide I-131 USP kit for therapeutic capsules and solution (*solution* containing sodium radioiodide, disodium edetate dihydrate as stabilizer, sodium thiosulfate pentahydrate as reducing agent, dibasic sodium phosphate anhydrous, pH 7.5–9.0; *capsules* [hard gelatin] containing 300 mg dibasic sodium phosphate)	250, 500, 1000 mCi (solution concentration 1 mCi/μL)
Covidien, DraxImage	Sodium iodide I-131 USP diagnostic capsules (containing sodium radioiodide, disodium edetate dihydrate as stabilizer, sodium thiosulfate pentahydrate as reducing agent, dibasic sodium phosphate anhydrous)	15, 25, 50, 100 μCi
Covidien	Sodium iodide I-123 diagnostic capsules (containing sodium radioiodide, sucrose)	100 and 200 μCi

because the net oxidation potential of the system is positive according to the following reactions:

$$2\,I^- \xrightarrow{\text{oxidation}} I_2 + 2e^- \quad -0.536\ V$$

$$Cl_2 + 2e^- \xrightarrow{\text{reduction}} 2\,Cl^- \quad +1.360\ V$$

$$2\,I^- + Cl_2 \rightarrow I_2 + 2\,Cl^- \quad +0.824\ V$$

There has been some concern over the years regarding the relative bioavailability of radioiodide from capsules versus solution. It has been reported in the literature that radioactive iodine uptake (RAIU) tests can be altered (falsely low) by differences in capsule formulation because of incomplete release of radioiodide from the capsule after oral dosing.[19,20] Incomplete release of iodide has been attributed to incomplete dissolution of capsule filler (e.g., Gelfoam) or possible complexation with adjuvants, such as magnesium stearate lubricant. Fortunately, neither of these agents is present in current formulations of [131]I- or ^{123}I-sodium iodide capsules. Diagnostic and therapeutic ^{131}I-sodium iodide capsules contain dibasic sodium phosphate and ^{123}I-sodium iodide capsules contain sucrose fillers. Dibasic sodium phosphate filler has been shown to release 95% of iodide after 20 minutes using a USP dissolution test, which correlated with RAIU results in human subjects.[21] Unfortunately, the USP monographs for [131]I- and ^{123}I-sodium iodide capsules do not require a dissolution test. This would be a useful test, because therapy dosages of ^{131}I are based on the diagnostic RAIU test. If a RAIU test result is falsely low because of poor bioavailability of radioiodide from diagnostic capsules, potential overdosage of radioiodine therapy with ^{131}I-sodium iodide solution for thyroid gland disease may result.[22]

Iodinated I 125 Albumin Injection and Iodinated I 131 Albumin Injection

Iodinated I 125 albumin injection (^{125}I-HSA; Jeanatope [Iso-Tex Diagnostics]) and iodinated I 131 albumin injection (^{131}I-HSA; Megatope [Iso-Tex Diagnostics]) are prepared by mild iodination of normal human albumin to introduce not more than one atom of iodine per molecule of albumin to minimize denaturation of the protein. The iodine is firmly bound and is released only by metabolism of the protein in vivo. The radiochemical purity of radioiodinated albumin is not less than 97%. Being a biologic product, it is stored at 2°C to 8°C. ^{125}I-HSA is available in 100 μCi (3.7 MBq) multidose vials of 10 mL, and ^{131}I-HSA is available in 0.5 mCi (18.5 MBq) and 1.0 mCi (37 MBq) vials. ^{125}I-HSA is indicated primarily for the measurement of plasma volume, and ^{131}I-HSA is indicated for plasma volume and whole blood volume measurement. The usual adult administered activity for plasma volume measurement is 5 to 10 μCi (185–370 kBq) of ^{125}I-HSA or ^{131}I-HSA. The critical organ for ^{125}I-HSA is the total body, with a radiation absorbed dose of 0.0006 rad(cGy)/μCi.[23] For ^{131}I-HSA the critical organ is the heart wall, with an absorbed dose of 0.011 rad(cGy)/μCi.

Iobenguane I 123 and I 131 Injection

Iobenguane sulfate (MIBG) has the chemical structure shown in Figure 11-5. It is available as a commercially prepared product labeled with [123]I (AdreView [GE Healthcare]) as a 5 mL solution, 2 mCi (74 MBq)/mL, or labeled with [131]I (Iobenguane Sulfate I 131 Injection [Pharmalucence]) supplied as a frozen solution in a concentration of 2.3 mCi (85.1 MBq)/mL. The latter product should be kept frozen until needed and used within 6 hours after thawing. ^{123}I-MIBG and ^{131}I-MIBG as lyophilized products are available from AnazaoHealth. ^{131}I-MIBG received Food and Drug Administration (FDA) approval for marketing in 1994 and ^{123}I-MIBG (AdreView), in 2008.

The radioiodinated product can be prepared by a solid-phase isotope exchange method. [123]I- or ^{131}I-sodium iodide in dilute sodium hydroxide and 1 mL of MIBG exchange solution, containing 2 mg MIBG sulfate and 10 mg ammonium sulfate, is heated at 155°C for 30 minutes. The dried product is redissolved in 1 mL of water and reheated for an additional 30 minutes.[24] The final product is redissolved in water or saline and sterile-filtered before use. Labeling yields are typically 98% or higher, requiring no purification step to remove unbound iodide. During the solid-phase exchange reaction, thermal decomposition of ammonium sulfate drives off ammonia and lowers the pH. The mildly acidic, oxidizing conditions ensure that I^+ will be formed to effect the electrophilic exchange reaction. High labeling yields require an absence of chloride ion; therefore, saline should not be used during the heating step.[25] Once the product is labeled, it can be reconstituted in normal saline for injection. The radiochemical purity of [123]I- and ^{131}I-MIBG is not less than 90%.[26]

[123]I- or ^{131}I-MIBG is indicated for the localization of primary or metastatic pheochromocytomas and neuroblastomas. The usual intravenous adult administered activity of ^{131}I-MIBG is 0.5 mCi (18.5 MBq). In patients weighing over 65 kg, the dose is 0.3 mC(11.1 MBq)/m² up to a maximum of 1 mCi (37 MBq). Children's doses of ^{131}I-MIBG are based on 0.3 mCi (11.1 MBq)/m² up to a maximum dosage of 0.5 mCi (18.5 MBq). Dosing of ^{123}I-MIBG is based on 0.14 mCi/kg, with a suggested maximum dose of 10 mCi (370 MBq) in adults.[27,28] A thyroid-blocking dose of potassium iodide (KI) should be administered 1 day before dosing and daily for 7 days with ^{131}I-MIBG and for 2 days with ^{123}I-MIBG.[29] The FDA-recommended thyroid-protective doses of KI daily are as follows: infants less than 1 month old, 16 mg; children age 1 month to 3 years, 32 mg; children 3 years to 18 years, 65 mg; adults, 130 mg.

The critical organs for ^{131}I-MIBG in the adult are the urinary bladder wall and the liver, each with a radiation absorbed dose of 3 rad(cGy)/mCi.[29] For ^{123}I-MIBG the critical organ is the urinary bladder wall, with a radiation absorbed dose of 0.35 rad(cGy)/mCi.

Iodohippurate Sodium I 131 Injection

Iodohippurate sodium I 131 injection (^{131}I-iodohippurate or ^{131}I-OIH; Hippuran) has been essentially supplanted by

^{99m}Tc-mertiatide (^{99m}Tc-MAG3) for kidney imaging. A brief description is included here because this was the major agent for evaluating renal function for more than 30 years. Its only disadvantage is that it could not be labeled with ^{99m}Tc. The chemical structure of OIH is shown in Figure 11-5. It is prepared by isotope exchange. Its radiochemical purity is not less than 97%, with the major radiochemical impurity being radioiodide. The high purity limits are necessary because OIH's ability to measure renal function is based on its high renal extraction and elimination in the urine. Too much free radioiodide would prolong renal clearance because of radioiodide's high degree of reabsorption (approximately 70%) by the tubular cells. OIH's principal route of elimination is tubular secretion, and it played a prime role in assessing global renal function via the renogram. It is the "gold standard" to which other renal function agents are compared. A typical dosage for renal function assessment is 75 µCi (2.775 MBq) for one kidney and 200 µCi (7.4 MBq) for two kidneys. The critical organ is the urinary bladder wall, with an absorbed radiation dose of 5.7 mrad/µCi.[30] For economic reasons, an OIH product is no longer marketed in the United States.

Iothalamate Sodium I 125 Injection

Iothalamate sodium I 125 injection (^{125}I-iothalamate; Glofil-125 [IsoTex Diagnostics]) is prepared by isotope exchange. Its chemical structure is shown in Figure 11-5. It is not used for renal imaging because of the ^{125}I label. Its renal clearance following intravenous injection closely approximates that of inulin, and its primary indication is for the assessment of GFR. It is supplied in a multidose vial with a radioactivity concentration of 250 to 300 µCi/mL (9.25–11.1 MBq/mL) in a 4 mL volume. Its radiochemical purity is not less than 98%.[26] The product should be stored at 2°C to 8°C. Dosage varies with the method used to assess GFR. A single-dose technique requires a dosage of 10 to 30 µCi (370–1110 kBq) and a continuous infusion technique, 20 to 100 µCi (740–3700 kBq). The critical organ is the thyroid gland, with an absorbed radiation dose of 7.8 mrad/µCi.[31]

GALLIUM, INDIUM, AND THALLIUM CHEMISTRY

Gallium, indium, and thallium are members of the group III metals in the periodic table. The electron configurations are as follows: gallium is [Ar]$3d^{10}4s^24p^1$, indium is [Kr]$4d^{10}5s^25p^1$, and thallium is [Xe]$5d^{10}6s^26p^1$. Each of these elements can assume different oxidation states by losing one, two, or three valence electrons. The most relevant oxidation states in radiopharmaceuticals are 3+ for gallium and indium and 1+ for thallium.

In acidic aqueous solution (typically, hydrochloric acid) below pH 3, gallium and indium exist in a soluble ionic form. As the pH is raised above 3, these metals readily hydrolyze to form hydroxides with very low solubility products; for Ga(OH)$_3$, K$_{sp}$ = 7.1 × 10^{-36}, and for In(OH)$_3$, K$_{sp}$ = 1 × 10^{-33}.[32,33] Thus, above pH 3 these radiometals must be complexed with a suitable ligand to prevent precipitation. Complexes with intermediate stability are formed with ligands such as acetate, citrate, tartrate, or 8-hydroxyquinoline (oxine). Higher-stability complexes are formed with multidentate ligands such as EDTA or DTPA, but their rate of formation may be slow. For example, a slow rate of complexation between indium and DTPA at neutral pH causes the precipitation of indium hydroxide, which forms more rapidly. In the presence of an acetate buffer, however, indium hydroxide precipitation is prevented, because soluble indium acetate forms rapidly at neutral pH. But because the In-acetate stability constant (K = 10$^{3.5}$) is small relative to that for In-DTPA (K = 10$^{28.4}$), ligand exchange will occur favoring In-DTPA.[34] Hydrolysis of gallium is also a potential problem. The degree of formation of soluble gallium citrate (K = 10$^{10.02}$) and insoluble gallium hydroxide (K = 10$^{11.1}$) is a function of pH and citrate concentration. A competition exists between citrate and hydroxyl ions for gallium; when the citrate concentration is low (≤0.2%), successful chelation occurs only at low pH and when the pH is high (greater than 7), chelation occurs only at a high citrate concentration (≥1%).[35] Therefore, ^{67}Ga-citrate is prepared by neutralizing no carrier added (NCA) ^{67}Ga-chloride with sodium hydroxide in the presence of 4% sodium citrate, producing a 1:1 ^{67}Ga:citrate complex.

Another issue of importance is the complexation of gallium and indium radiopharmaceuticals with plasma transferrin because these metals behave similarly to iron, which binds firmly to transferrin. When indium chloride in 0.05 M hydrochloric acid is injected intravenously, the free indium ion binds rapidly to transferrin, creating a blood pool imaging agent. When indium is chelated with DTPA, binding to transferrin does not occur even though the thermodynamic equilibrium for ligand exchange favors In-transferrin (In-DTPA, K = 10^{29} and In-transferrin, K = 10$^{30.5}$). This is because the rate of dissociation of indium from In-DTPA is very low (i.e., In-DTPA is kinetically inert).[36] In the case of In-oxine (K ~ 10^{10}), indium readily translocates to transferrin. Thus, an important consideration in labeling leukocytes with indium oxine is minimizing the amount of plasma in the labeling process to avoid reducing cell labeling efficiency. With gallium citrate (K = 10^{10}) the equilibrium favors gallium transferrin (K = 10$^{23.7}$), and gallium translocates from citrate to plasma transferrin and other proteins after intravenous administration.[35]

Thallium exists typically in the 1+ (thallous) or 3+ (thallic) oxidation states in its compounds. The principal chemical form used as a radiopharmaceutical is thallous chloride, ^{201}TlCl.

Production of Gallium, Indium, and Thallium

Gallium, indium, and thallium are produced in a cyclotron. ^{67}Ga is most commonly produced by the nuclear reaction ^{68}Zn(p,2n)^{67}Ga, and ^{111}In by the nuclear reaction

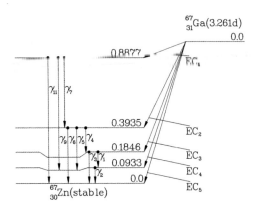

FIGURE 11-7 Decay scheme for ^{67}Ga. (Reprinted with permission from reference 6.)

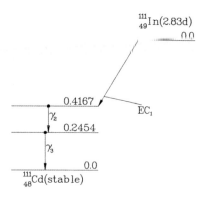

FIGURE 11-8 Decay scheme for ^{111}In. (Reprinted with permission from reference 6.)

^{111}Cd(p,n)^{111}In. The targets are processed by ion exchange and solvent extraction techniques, evaporated to dryness from isopropyl ether, and redissolved in 0.05 M hydrochloric acid to produce the metal chlorides.^{37,38}

The ^{67}Ga nucleus decays by EC with a half-life of 3.26 days to stable ^{67}Zn:

$$^{67}_{31}Ga_{36} \xrightarrow{p^+ + e^- \rightarrow n + \nu} {^{67}_{30}Zn_{37}} + 1.0008 \text{ MeV} (\nu, \gamma)$$

66.9282017 AMU   66.9271273 AMU

The transition energy for this decay is 1.0008 MeV, which is dissipated by several available EC transitions (Figure 11-7). Several gamma photons are emitted; the principal ones used for imaging are 0.093 MeV (γ_2), 0.185 MeV (γ_3), 0.300 MeV (γ_5), and 394 keV (γ_6) (Table 11-4).^{6}

The ^{111}In nucleus decays by EC with a half-life of 2.8 days to stable ^{111}Cd:

$$^{111}_{49}In_{62} \xrightarrow{p^+ + e^- \rightarrow n + \nu} {^{111}_{48}Cd_{63}} + 0.8615 \text{ MeV} (\nu, \gamma)$$

110.905103 AMU   110.9041781 AMU

The transition energy for this decay is 0.8615 MeV, which is dissipated by a neutrino at 0.690 MeV and two gamma rays in cascade at 0.171 MeV (γ_2) and 0.245 MeV (γ_3) (Figure 11-8 and Table 11-4).^{6} As in the decay of ^{123}I, there is a high abundance of low-energy characteristic x-rays (0.023 MeV [69%] and 0.026 MeV [13%]) emitted that are subject to significant attenuation by the source container. Therefore, correction factors must be applied to dose calibrator measurements of ^{111}In. Copper filtration can be used to remove the x-rays to facilitate measurements.^{7}

^{201}Tl is produced in a cyclotron by bombarding a target of pure natural thallium metal with protons:^{39}

$$^{203}_{81}Tl_{122} (p, 3n) {^{201}_{82}Pb_{119}} \xrightarrow{9.4 \text{ hr}} {^{201}_{81}Tl_{120}}$$

After irradiation, the target is dissolved in mineral acid and the ^{201}Pb is separated by ion-exchange chromatography. After decay of the ^{201}Pb, ^{201}Tl is isolated by ion-exchange chromatography, removing lead impurities. The chloride salt is formed in final stages of preparation by dissolution in hydrochloric acid, adjusted to pH 7.0, sterilized, and tested to detect any carrier thallium present. Radiochromatography is performed to differentiate Tl$^+$ and Tl^{3+}.

^{201}Tl decays by EC with a half-life of 3.04 days to stable ^{201}Hg:

$$^{201}_{81}Tl_{120} \xrightarrow{p^+ + e^- \rightarrow n + \nu} {^{201}_{80}Hg_{121}} + 0.4813 \text{ MeV} (\nu, \gamma)$$

200.970819 AMU   200.9703023 AMU

TABLE 11-4 Physical Properties of Gallium, Indium, and Thallium

Nuclide	Half-life	Decay Mode	Photons (MeV)	% Abundance
^{67}Ga	3.26 days	EC	0.093	39
			0.185	21
			0.300	17
			0.394	5
^{111}In	2.80 days	EC	0.171	91
			0.245	94
^{201}Tl	3.04 days	EC	0.135	2.7
			0.167	10
			0.069–0.082	75

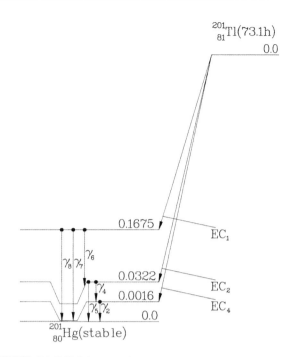

FIGURE 11-9 Decay scheme for ^{201}Tl. (Reprinted with permission from reference 6.)

The transition energy for this decay is 0.4813 MeV, which is dissipated by one of three EC routes to excited levels of mercury (Figure 11-9). The principal photons are gamma rays of 0.135 MeV (γ_6) and 0.167 MeV (γ_8) that have low abundance and 0.069 to 0.082 MeV mercury x-rays at 75% abundance, which are used for imaging (Table 11-4).[6]

Gallium Citrate Ga 67 Injection

Although many potential radiopharmaceuticals can be made with gallium, the only compound that has achieved major use in nuclear medicine is ^{67}Ga-gallium citrate (^{67}Ga-citrate). It is available as a sterile aqueous solution in multidose vials (2 mCi [74 MBq]/mL, 3 to 12 mCi (111–444 MBq) from several manufacturers. It is prepared by neutralizing acidic NCA ^{67}Ga-chloride with sodium hydroxide in the presence of 4% sodium citrate, producing a 1:1 ^{67}Ga:citrate complex. Depending on the pH, several protonated species are possible. The species at around pH 4 is $[C_3H_4OGa(OH)_2(COO)(COOH)_2]^-$ and at pH higher than 8 it is $[C_3H_4OGa(OH)_2(COO)_3]^{3-}$.[40] The injection has a pH between 5.5 to 8.0 and is preserved with 0.9% benzyl alcohol. It is stored at room temperature (20°C–25°C). Its radiochemical purity is not less than 97%.[26] At the time of calibration, it contains not less than 99% ^{67}Ga and not more than 0.02% ^{66}Ga and 0.2% ^{65}Zn radionuclidic impurities.

^{67}Ga-citrate is indicated for diagnostic imaging of Hodgkin's disease, lymphoma, bronchogenic carcinoma, and inflammatory lesions to identify fevers of unknown origin. The standard adult dosage is 3 to 8 mCi (111–296 MBq) administered intravenously. The critical organ is the lower large intestine, with a radiation absorbed dose of 0.9 rad(cGy)/mCi.[41]

Indium In 111 Chloride Solution

Indium In 111 chloride injection (^{111}In-chloride) is available from several manufacturers as a sterile aqueous solution of NCA ^{111}In-chloride in 0.05 M hydrochloric acid, pH 1.1 to 1.4 (Covidien and IsoTex Diagnostics) and in 0.04 M hydrochloric acid, pH approximately 1.4, as Indichlor (GE Healthcare). Both products are available as 5 mCi (185 MBq) in 0.5 mL volume single-use vials, to be stored at room temperature (20°C–25°C). The radiochemical purity is not less than 95% as ionic In^{3+}. Radionuclidic purity at the time of calibration is not less than 99.925% ^{111}In, with not more than 0.075% ^{114m}In and ^{65}Zn, combined, as radionuclidic impurities. ^{111}In-chloride is intended for use in labeling antibodies such as capromab pendetide (^{111}In-capromab pendetide; ProstaScint [Cytogen]).

Indium In 111 Oxyquinoline Solution

Indium In 111 oxyquinoline solution (^{111}In-oxine, GE Healthcare) is a sterile aqueous solution of NCA ^{111}In^{3+} complexed to 8-hydroxyquinoline (oxine). The complex is a 3:1 ratio of oxine:indium (Figure 11-10). The product is available as a single-use vial at a pH range of 6.5 to 7.5. Each milliliter contains at calibration time 1 mCi (37 MBq) ^{111}In, 50 μg oxine, 100 μg polysorbate 80 (detergent stabilizer), and 6 mg N-2-hydroxyethyl-piperazine-N'-2-ethane sulfonic acid (HEPES) buffer in 0.75% sodium chloride solution. It is stored at room temperature. Its radiochemical purity is not less than 90%.[26] The product contains not more than 1 μCi (37 kBq) ^{114m}In impurity per 1 mCi (37 MBq) ^{111}In at the time of calibration. At the time of expiration, it contains not less than 99.75% ^{111}In and not more than 0.25% of $^{114m/114}$In impurities.

^{111}In-oxine is indicated for radiolabeling autologous leukocytes (described below). The usual adult dosage of ^{111}In-labeled leukocytes is 200 to 500 μCi (7.4–18.5 MBq) intravenously. The critical organ is the spleen, with a radiation absorbed dose of 20 rad(cGy)/500 μCi (7.4 MBq) (at the expiration date of the product).[42]

Indium In 111 Pentetate Injection

Indium In 111 pentetate injection (^{111}In-pentetate or ^{111}In-DTPA [GE Healthcare]) is a sterile aqueous solution of NCA ^{111}In^{3+} complexed to disodium pentetate in a 1:1 molar ratio. The eight-coordinate structure for the complex is shown in Figure 11-10.[43] The single-use product, available in a 1.5 mL vial, contains in each milliliter of isotonic solution at the time of calibration 1 mCi (37 MBq) ^{111}In, 20 to 50 μg pentetic acid, and sodium bicarbonate with the pH adjusted to 7 to 8. It is stored at 5°C to 30°C. The radiochemical purity is not less than 90%.[26] The radionuclidic purity at calibration time is not less than 99.88% ^{111}In and less than 0.06% ^{114m}In and ^{65}Zn combined.

^{111}In-DTPA is indicated for use in radionuclide cisternography. The usual intrathecal adult dosage is 500 μCi

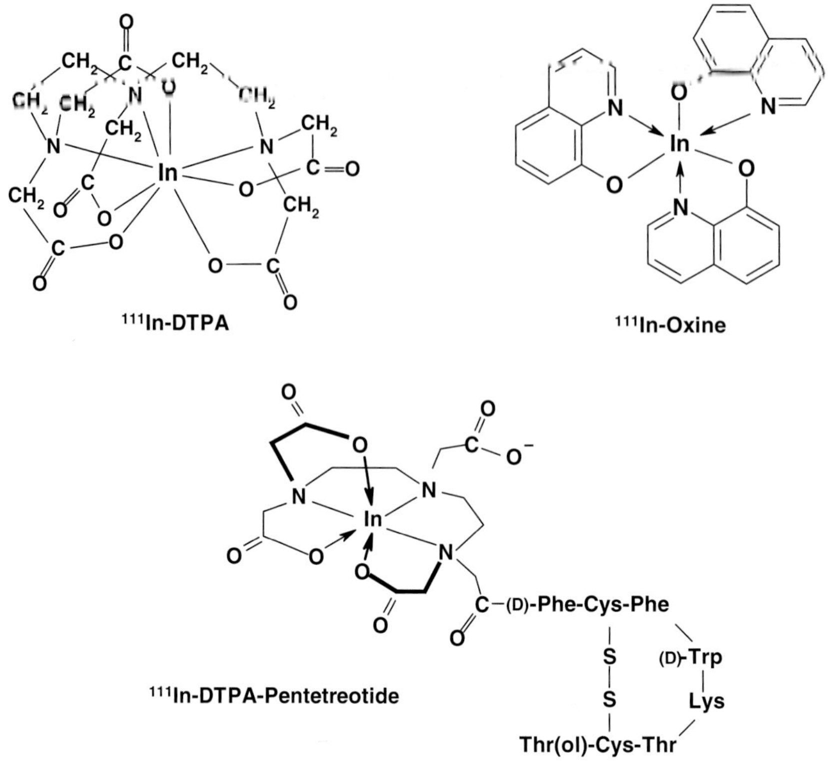

FIGURE 11-10 Chemical structures of ^{111}In-labeled DTPA, oxine, and pentetreotide.

(18.4 MBq) (maximum dosage). The critical organ is the spinal cord surface, with a radiation absorbed dose of 5 rad(cGy)/500 μCi.[44]

Indium In 111 Pentetreotide Injection

Indium In 111 pentetreotide injection (^{111}In-pentetreotide; OctreoScan [Covidien]) is prepared from a kit that contains a lyophilized mixture of 10 μg of pentetreotide, 2.0 mg gentisic acid, 4.9 mg anhydrous trisodium citrate, 0.37 mg anhydrous citric acid, and 10 mg inositol as a bulking agent. The kit also contains a vial of indium In 111 chloride injection that contains, at the calibration date, 1.1 mL of ^{111}In(Cl)$_3$ at 3 mCi(111 MBq)/mL in 0.02 M hydrochloric acid, and 3.5 μg/mL ferric chloride, which is stated by the manufacturer to increase the labeling yield. The kit is stored at refrigerated temperature (2°C–8°C) before labeling. The product is labeled by adding the ^{111}In-chloride to the lyophilized mixture and incubating at room temperature for 30 minutes, during which time there is transchelation from indium citrate to indium pentetreotide. The final product has a pH between 3.8 and 4.3 and is stored at or below 25°C. It must be used within 6 hours of preparation. A proposed chemical structure for indium pentetreotide is shown in Figure 11-10. The labeled product may be diluted with up to 3 mL of normal saline immediately before injection. Dilution should not be done before the labeling reaction. The radiochemical purity must be checked before patient administration and should be not less than 90%.[26,45]

^{111}In-pentetreotide is indicated for localization of primary and metastatic neuroendocrine tumors bearing somatostatin receptors. The usual adult intravenous dosage for SPECT imaging is 6 mCi (222 MBq). The critical organ is the spleen, with a radiation absorbed dose of 14.77 rad(cGy)/6 mCi.[45]

Indium In 111-Labeled Antibodies

Indium-111 has been used to label antibodies for radioimmunodiagnosis. Labeling is made possible by covalent attachment of a linker molecule to the antibody away from the antigen binding site. The linker molecule contains a chelating group to coordinate with ^{111}In. Antibody labeling is accomplished by first mixing ^{111}In-chloride with an acetate buffer, which is then added to the antibody. Antibodies currently available are indium In 111 capromab pendetide (ProstaScint [Cytogen]) for imaging prostate cancer metastases and indium In-111 ibritumomab tiuxetan and yttrium Y-90 ibritumomab tiuxetan (^{111}In- and ^{90}Y-Zevalin [Spectrum Pharmaceuticals]) for use in the treatment of patients with relapsed or refractory follicular non-Hodgkin's lymphoma. These and other radiolabeled antibodies are discussed in Chapter 28.

Indium In 111-Labeled White Blood Cells

The basic process for labeling leukocytes is illustrated in Chapter 10, Figure 10-27. The method involves collection of

43 mL of whole blood in 7 mL of anticoagulant citrate dextrose (ACD), addition of 10 mL of 6% hetastarch, and incubation for about 45 minutes. Hetastarch facilitates the settling rate of red blood cells (RBCs). This agent may be omitted in patients with leukocytosis who inherently have a high RBC sedimentation rate. RBCs do not separate well in patients with sickle cell anemia. In these patients blood should be separated into two tubes and up to 50% hetastarch added to settle the RBCs. After the RBCs have settled, the leukocyte-rich plasma (LRP) supernatant is removed and centrifuged at 450g to pellet the leukocytes. The leukocyte-poor plasma (LPP) supernatant is removed and saved. The leukocyte pellet may have a red color due to the presence of a small quantity of RBCs, which does not inhibit leukocyte labeling. The leukocyte pellet is resuspended in 2 mL of normal saline, and ^{111}In-oxine solution (1 mCi [37 MBq] in 1 mL) is added and the mixture incubated for 15 to 20 minutes to label the cells. At the end of incubation, 15 mL of LPP is added, the mixture is centrifuged at 450g, and the supernatant containing unbound indium is removed. The ^{111}In-labeled leukocytes are resuspended in 10 mL of LPP for reinjection into the patient.

The principal white blood cell (WBC) fraction labeled with ^{111}In-oxine is granulocytes, primarily neutrophils.[46] The recommended minimum number of granulocytes to be labeled is 3×10^6/mL of whole blood.[47] The mean number of granulocytes in normal blood is 4.15×10^6/mL. The granulocyte count should be considered in cases of bone marrow suppression, such as in patients on chemotherapy. Labeling yields are improved with increased numbers of leukocytes. McAfee and Thakur[48] demonstrated that yields increased linearly from approximately 35% to 75% as leukocytes increased from 3.6×10^7 to 1.3×10^8. With an average of 10^8 cells, Goodwin et al.[49] achieved a labeling efficiency of 84% ± 9%; in patients with WBC counts above 15,000/μL in peripheral blood, yields were usually 90% or more.

^{111}In-oxine is a relatively weak complex and is prone to releasing the indium to plasma transferrin.[50] Thus, most of the plasma should be separated from the pelleted leukocytes before addition of ^{111}In-oxine. Cells can be washed once with normal saline to remove plasma and increase labeling yields. Coleman et al.[51] reported 34% labeling efficiency in dilute plasma compared with 87% in saline. Alternatively, layering 1 mL of saline carefully over the white cell button and removing it will remove most but not all of the plasma. Removal of all plasma may affect cell viability. Danpure et al.[52] demonstrated that concentrating leukocytes in a small amount of plasma enables a large number of cells to compete effectively with plasma transferrin during labeling and at the same time sustain leukocyte viability. Another technique is to label leukocytes with ^{111}In-tropolone, which can label cells in the presence of plasma.[48,52]

During the labeling process lipophilic ^{111}In-oxine diffuses into the leukocytes and dissociates. Free intracellular ^{111}In binds to nuclear and cytoplasmic proteins, and the oxine diffuses out (Figure 11-11).[48] There may be a small percentage of red cells (2%–4%) present in the white cell button, but

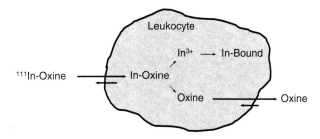

FIGURE 11-11 Proposed mechanism for labeling leukocytes (WBCs) with ^{111}In-oxine.

^{111}In-oxine does not bind to RBCs as it does to WBCs, which have a protein that binds indium. The presence of any ^{111}In-RBCs does not interfere with infection imaging.[53] After intravenous injection, the plasma half-life of ^{111}In-labeled leukocytes is 6 to 7 hours.[49,54]

The recommended maximum amount of time between blood drawing and reinjection is 5 hours because of loss of granulocyte chemotaxis during storage. The temperature of blood and labeled leukocytes should not exceed 70°F to avoid cell damage. Radiolabeled cells should be reinjected within 1 to 3 hours after preparation.[47,55] The maximum administered activity in adults is 0.5 mCi (18.5 MBq), with a useful range of 0.3 to 0.5 mCi (11.1–18.5 MBq); in pediatric patients it is 0.007 to 0.0135 mCi/kg (0.15–0.25 MBq/kg), with a minimum activity of 0.05 to 0.06 mCi (1.85–2.3 MBq) and a maximum activity of 0.5 mCi (18.5 MBq).[47] Patients who are on hemodialysis should have their blood drawn before dialysis begins. Labeled cells can be reinjected after dialysis for imaging the next day.

Indium In 111-Labeled Platelets

Radiolabeled platelets have diagnostic value in the investigation of hematologic conditions such as deep venous thrombosis and diseases that affect platelet survival, such as idiopathic thrombocytopenic purpura. ^{111}In is considered to be the most effective platelet label for diagnostic studies because it has useful imaging photons and a 2.8 day half-life that is necessary for platelet survival studies. ^{111}In forms a lipophilic complex with oxine or tropolone and will label platelets with high efficiency.[56–60] The general labeling procedure involves separating platelets from ACD-anticoagulated whole blood by differential centrifugation, incubating the isolated platelets with a lipophilic indium complex, separating unbound indium, and resuspending the labeled platelets in platelet-poor plasma for reinjection. A detailed method for preparing and labeling platelets with ^{111}In-oxine or ^{111}In-tropolone for platelet survival studies has been recommended by the International Committee for Standardization in Hematology Panel on Diagnostic Applications of Radionuclides.[60] Following intravenous administration, the normal values for ^{111}In-labeled platelet recovery range from 55% to 72% of the injected dose, and normal values for platelet survival in the circulation range

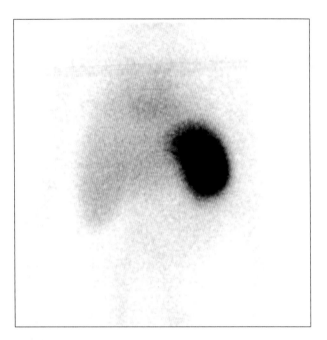

FIGURE 11-12 Anterior abdominal image of a 5 year old female with a history of thrombocytopenia who was being evaluated for altered distribution of platelets. The image was obtained 2 hours after intravenous administration of 307 µCi ^{111}In-oxine-labeled donor platelets and demonstrates intense radiotracer uptake in an enlarged spleen, compatible with splenic sequestration. Activity is also evident within the liver and cardiac blood pool, compatible with normal intravascular distribution of platelets. Delayed images at 3 and 24 hours (not shown) demonstrated a distribution pattern similar to the 2 hour image.

from 7.3 to 9.5 days.[60] This procedure was used to prepare ^{111}In-oxine labeled platelets for a platelet distribution study in a patient with thrombocytopenia (Figure 11-12).

Thallous Chloride Tl 201 Injection

Thallous chloride Tl 201 injection (^{201}Tl-chloride) is a sterile aqueous solution that contains at the time of calibration 1 mCi(37 MBq)/mL of ^{201}Tl-chloride in 0.9% sodium chloride solution, pH adjusted to 4.5 to 7.0 and preserved with 0.9% benzyl alcohol.[61] Multidose vials in 2, 4, 8, and 9 mCi (74, 148, 296, and 333 MBq) sizes are available at different calibration times throughout the week. Its radiochemical purity is not less than 95%.[26] At the time of calibration, it contains no more than 1.0% each of ^{200}Tl and ^{202}Tl and not more than 0.25% ^{203}Pb as radionuclidic impurities, and no less than 98% ^{201}Tl as the desired radionuclide. The product is stored at room temperature.

^{201}Tl-chloride is indicated for evaluation of myocardial perfusion in the diagnosis and localization of myocardial infarction and ischemic heart disease. Its other approved indication is for the localization of parathyroid hyperactivity (adenoma) in patients with elevated serum calcium and parathyroid hormone levels. Thallium has also been used as a tumor marker in brain, breast, and lung cancer. Its mechanism of tumor localization is believed to be related to tumor blood flow and permeability and active uptake via the Na-K ATPase pump. The usual adult dosage for myocardial imaging is 2 to 4 mCi (74–148 MBq) intravenously. The critical organ in males is the testes, with a radiation absorbed dose of 3 rad(cGy)/mCi; in females it is the thyroid gland, with a dose of 2.3 rad(cGy)/mCi.[61]

XENON Xe 133 GAS

Xenon, an inert gas, is used in nuclear medicine in its native elemental state. Its electron configuration is $[Kr]4d^{10}5s^25p^6$. Two isotopes of xenon have been used clinically in nuclear medicine: ^{127}Xe and ^{133}Xe (Table 11-5). ^{127}Xe is cyclotron produced; it was used in the past for lung ventilation studies but is no longer available. The principal isotope used currently is ^{133}Xe. It is produced as a fission byproduct in a nuclear reactor as follows: ^{235}U(n,f)^{133}Xe. The product is available in unit dose vials ready for patient use in amounts of 10 mCi (370 MBq) and 20 mCi (740 MBq) from various suppliers. (The properties described here pertain to vials supplied by Covidien, formerly Tyco-Mallinckrodt.) It may contain small amounts of other radioactive gas impurities, namely, ^{133m}Xe, ^{131m}Xe, and ^{85}Kr, in addition to ^{131}I.[62] It should be stored at room temperature. Xenon Xe 133 gas (^{133}Xe-xenon) is indicated for inhalation studies to evaluate lung function and for the assessment of cerebral blood flow. The usual adult dosage is 10 mCi (370 MBq) by inhalation. The critical organ is the lung. The radiation absorbed dose depends on the volume of the spirometer used to perform the study; for a 5 L spirometer it is 0.11 rad(cGy)/10 mCi and for a 10 L spirometer it is 0.065 rad(cGy)/10 mCi.[62]

^{133}Xe undergoes negatron (beta minus) decay with a half-life of 5.24 days to stable cesium:

TABLE 11-5 Physical Properties of Xenon Radioisotopes

Nuclide	Half-life	Decay Mode	Photons (MeV)	% Abundance
^{127}Xe	36.4 days	EC	0.172	26
			0.203	68
			0.375	17
^{133}Xe	5.24 days	Negatron	0.081	38

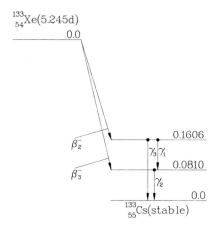

FIGURE 11-13 Decay scheme for ^{133}Xe. (Reprinted with permission from reference 6.)

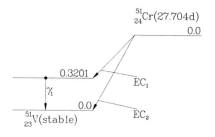

FIGURE 11-14 Decay scheme for ^{51}Cr. (Reprinted with permission from reference 6.)

$$^{133}_{54}Xe_{79} \xrightarrow{n \to p^+ + e^-} {}^{133}_{55}Cs_{78} + 0.4274 \text{ MeV } (\beta, \gamma)$$

132.9059107 AMU   132.9054519 AMU

The transition energy for this decay is 0.4274 MeV, which is dissipated principally by a beta particle of 0.346 MeV and emission of a gamma ray of 0.081 MeV (γ_2) to the ground state of cesium (Figure 11-13). The photon abundance is only 38% because of a high number of conversion electrons.

SODIUM CHROMATE Cr 51 INJECTION

Chromium, with electron configuration [Ar]$3d^54s^1$, is a member of the transition elements, which owe their separate classification in the periodic table to belated filling of the next-to-the-outermost electron energy level. Chromium can exist in several oxidation states, most commonly 6+ as sodium chromate (Na$_2{}^{51}$CrO$_4$), which can be easily reduced to the 3+ oxidation state. ^{51}Cr is useful for labeling RBCs. The chromate ion (^{51}CrO$_4{}^{2-}$) readily diffuses through the erythrocyte membrane and is reduced intracellularly to chromic ion (^{51}Cr^{3+}), which binds to hemoglobin.[63,64]

^{51}Cr is produced by neutron activation of stable chromium metal or chromiuim oxide enriched in ^{50}Cr by the reaction ^{50}Cr(n,γ)^{51}Cr.[65] The irradiated material is dissolved in hydrochloric acid, and the chromic chloride is oxidized to chromate. ^{51}Cr decays by EC with a half-life of 27.7 days to stable ^{51}V as follows:

$$^{51}_{24}Cr_{27} \xrightarrow{p^+ + e^- \to n + \nu^-} {}^{51}_{23}V_{28} + 0.753 \text{ MeV } (\nu, \gamma)$$

50.9447674 AMU   50.9439595 AMU

The transition energy for this decay is 0.753 MeV, which is dissipated by one of two EC routes. The EC$_2$ route occurs in 90% of decays; all of the energy is taken away by the neutrino. The EC$_1$ route occurs in 10% of decays to the excited level of vanadium (Figure 11-14).[6] In this route, the neutrino carries away 0.433 MeV, and a 0.320 MeV gamma ray is emitted from the excited nuclear state of vanadium. ^{51}Cr is unsatisfactory for imaging because of the low abundance of the 0.320 MeV photon, but it is quite suitable for in vitro procedures that involve scintillation counting.

Sodium chromate Cr 51 injection (^{51}Cr-sodium chromate) is a clear, colorless sterile solution. It is available in 250 μCi (9.25 MBq) vials at a concentration of 100 μCi (3.7 MBq)/mL at the time of calibration. Its pH is between 7.5 and 8.5, and its specific activity is not less than 10 mCi (370 MBq)/mg Na$_2$CrO$_4$ at expiration.[26] The limit on specific activity is to prevent potential chromium toxicity to RBCs. The radiochemical purity is not less than 90% as sodium chromate, as determined by paper chromatography.[26] The presence of excess chromic impurity (^{51}Cr^{3+}) must be limited because it does not label erythrocytes. The product is stored at room temperature.

^{51}Cr-sodium chromate is indicated for the determination of RBC volume (mass), RBC survival, and evaluation of blood loss. For determination of RBC volume, the dosage ranges from 10 to 30 μCi (370–1110 kBq). RBC survival studies use a dosage of 150 μCi (5.55 MBq). The critical organ is the spleen, with a radiation absorbed dose of 2.64 rad(cGy)/200 μCi.[65]

^{51}Cr is also available as the radiochemical chromic chloride, ^{51}CrCl$_3$, for other labeling purposes and is sometimes used to label plasma proteins for assessing gastrointestinal protein loss. As ^{51}Cr-EDTA, it has been used to determine GFR.

PHOSPHORUS CHEMISTRY

Phosphorus is a member of the group V elements. Members of this group have properties ranging from nonmetallic (nitrogen and phosphorus) to semimetallic (argon and antimony) to metallic (bismuth).[66] Each of these elements has five valence electrons in its outermost energy level. The electron configuration of phosphorus is [Ne]$3s^23p^3$. Phosphorus can assume valences from 3$^-$ (by acquiring three electrons) to 5$^+$ (by losing five electrons). It is commonly available as orthophosphoric acid (H$_3$PO$_4$), a triprotic acid that dissociates into three possible conjugate base forms depending on the pH of the solution. The predominant forms of phosphate present in the pH range of pharmaceutical injections are the monobasic ($H_2PO_4^-$) and the dibasic (HPO_4^{2-}) anions. The ionic equilibrium equation for these forms is

$$H_2PO_4^- \xleftrightarrow{\leftarrow H^+ OH^- \rightarrow} HPO_4^{2-} \text{ (pKa 7.2)}$$

At a pH higher than 7.2, the dibasic form predominates, and the presence of excess Ca^{2+} ions in solution may cause the precipitation of dibasic calcium phosphate ($CaHPO_4$) because of its low solubility (approximately 30 mg/dL). This can be prevented by keeping the pH below 7. The presence of Al^{3+} ions in solution with phosphate can precipitate aluminum phosphate. When this is a possibility, EDTA is often used to sequester the aluminum ion away from phosphate. EDTA is used in ^{99m}Tc-sulfur colloid kits for this purpose.

Among the other phosphoric acids are pyrophosphoric acid, $H_4P_2O_7$, and metaphosphoric acid, HPO_3. While pyrophosphoric acid, like orthophosphoric acid, is a discrete molecule, metaphosphoric acid is polymeric. Solutions of these acids are not stable; on standing over time, they will convert to orthophosphoric acid. The sodium salts of these acids are useful agents in chemistry because they will complex metal ions. Sodium trimetaphosphate ($Na_3P_3O_9$) is prepared by heating NaH_2PO_4 for several hours at 550°C. It forms soluble chelates with cations, notably Ca^{2+}, and has been used extensively as a water-softening agent under the trade name Calgon. Sodium pyrophosphate, $Na_4P_2O_7$, is used extensively in nuclear medicine as the stannous chelate in kits for tagging RBCs, which include the stannous pyrophosphate kits from Pharmalucence and TechneScan PYP from Covidien.

The radioisotope of phosphorus used in nuclear medicine is ^{32}P. It is prepared in a nuclear reactor by the capture of a fast neutron by stable sulfur, according to the reaction $^{32}S(n,p)^{32}P$. ^{32}P decays by negatron emission with a half-life of 14.26 days to stable ^{32}S as follows:

$$^{32}_{15}P_{17} \xrightarrow{n \rightarrow p^+ + e^-} {}^{32}_{16}S_{16} + 1.71 \text{ MeV}(\beta, \nu)$$

31.97390727 AMU   31.97207100 AMU

^{32}P is a pure beta emitter with no gamma photons emitted during its decay (Figure 11-15). The transition energy for this decay is 1.71 MeV. On average, the beta particle receives 0.695 MeV and the neutrino 1.015 MeV. Bremsstrahlung radiation is produced during the decay of ^{32}P, but this is secondary to the decay process and is caused by the interaction of the high-speed beta particle with matter. ^{32}P is used solely for radiation therapy procedures. The two radiopharmaceuticals labeled with ^{32}P are chromic phosphate P 32 suspension and sodium phosphate P 32 solution.

Chromic Phosphate P 32 Suspension

Chromic phosphate P 32 suspension (^{32}P-chromic phosphate) is a grayish-green insoluble colloidal chromic phosphate ($Cr^{32}PO_4$). It is produced by mixing chromic nitrate solution with radiophosphoric acid. The resulting precipitate of chromic phosphate is dried in an oven and reduced to particles of about 1 μm size in a ball mill.[66,67] A commercial product is available (Phosphocol P-32 [Covidien]) that is a sterile aqueous suspension in 30% dextrose and 0.1% sodium acetate, preserved with 2% benzyl alcohol. Its particle size distribution is as follows: <0.6 μm, 0.7%; 0.7 to 1.3 μm, 74.4%; 1.4 to 2.0 μm, 16.6%; 2.1 to 4.0 μm, 7%; >4.0 μm, 2.9%. Its radiochemical purity is not less than 95%, and it has a pH range of 3.0 to 5.0.[26] The product is stored at room temperature. It is available in 15 mCi (555 MBq) multidose vials at a concentration of 5 mCi (185 MBq)/mL with a specific activity of 3.3 mCi (122.1 MBq)/mg of chromic phosphate at calibration.[68]

^{32}P-chromic phosphate is administered by intracavitary instillation for the treatment of peritoneal or pleural effusions caused by metastatic disease. It has also been used in the interstitial treatment of cancer. An off-label application is radiation synovectomy in diseases involving inflamed synovial lining, such as rheumatoid arthritis and hemophiliac arthropathy.[69] Typical adult dosage for intraperitoneal instillation is 10 to 20 mCi (370–740 MBq); for intrapleural instillation it is 6 to 12 mCi (222–444 MBq); for synovectomy in the knee joint

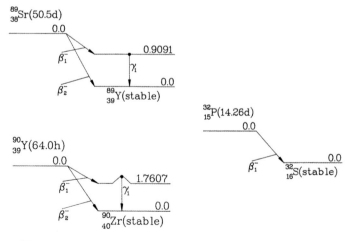

FIGURE 11-15 Decay schemes for ^{89}Sr, ^{32}P, and ^{90}Y. (Reprinted with permission from reference 6.)

TABLE 11-6 Radiation Dose Estimates for Chromic Phosphate P 32 Suspension[a]

Tissue Depth (cm)	Dose Rate (rad/hr)	Tissue Absorbed Dose (rad) per 20 mCi (740 MBq)		
		Pleural (4000 cm^2)	Peritoneal (5000 cm^2)	Prostate (16 grams)
0.004	10.2	23,000	18,000	910,000
0.008	8.58	19,000	15,000	
0.012	7.61	17,000	14,000	
0.016	6.91	15,000	12,000	
0.020	6.36	14,000	11,000	
0.10	2.41	5,400	4,300	
0.20	0.94	2,100	1,700	

[a] For 70 kg patient with 90% retention.
Source: Phosphocol P-32 package insert (Mallinckrodt; November 2000).

it is 6 mCi (222 MBq). Interstitial tumor dose is based on estimated tumor mass and is in the range of 0.1 to 0.5 mCi (3.7–18.5 MBq)/gram. The radiation absorbed dose depends on the tissue surface area exposed and the depth; calculations for an average (70 kg) patient with 90% retention are shown in Table 11-6.[68]

Sodium Phosphate P 32 Solution

Sodium phosphate P 32 solution (^{32}P-sodium phosphate) is a clear, colorless sterile solution of sodium phosphate at pH 5.0 to 6.0. It is suitable for oral or intravenous administration. *USP 28/NF 23*[70] states that the product is dibasic sodium phosphate, but at the final pH range listed in the product monograph it is mostly the monobasic form, since the pKa for the equilibrium of these two forms is 7.2 (see first reaction under Phosphorus Chemistry). The commercial product is available with a radioactive concentration of 0.67 mCi (24.79 MBq)/mL in 5 mCi (185 MBq) vials at the time of calibration. Its radiochemical purity must be 100%.[26] It is stored at room temperature.

^{32}P-sodium phosphate is indicated for the treatment of polycythemia vera, chronic myelocytic leukemia, and chronic lymphocytic leukemia. It has also been used as a palliative treatment for skeletal metastases in the treatment of bone pain. The recommended adult dosage for polycythemia vera, established by the Polycythemia Vera Study Group, is an initial intravenous dose of 2 to 3 mCi (74–111 MBq)/m^2 body surface area, not to exceed 5 mCi (185 MBq).[71] The radiation absorbed dose to the bone marrow has been estimated to be 24 rad(cGy)/mCi divided between marrow, 13 rad(cGy); trabecular bone, 10 rad(cGy); and cortical bone, 1 rad(cGy).[72]

STRONTIUM CHEMISTRY

Strontium is a member of the group II (alkaline-earth) elements, which include beryllium, magnesium, calcium, barium, and radium. Strontium's electron configuration is [Kr]5s^2. Alkaline-earth metals are so named because alchemists referred to any nonmetallic substance that was insoluble and unchanged by fire as an "earth," and the earths of group II give an alkaline reaction.[66] They are moderately strong reducing agents, and their compounds have an oxidation state of 2+. In vivo these elements have a propensity to localize in bone. Several radioisotopes of strontium have been used in nuclear medicine over the years, notably ^{85}Sr and ^{87m}Sr for bone imaging and ^{89}Sr for therapy.

Strontium Chloride Sr 89 Injection

^{89}Sr is produced by neutron activation of enriched ^{88}Sr by the reaction ^{88}Sr(n,γ)^{89}Sr. The target is processed into its final form as strontium chloride, ^{89}SrCl$_2$. ^{89}Sr decays by beta emission with a 50.53 day half-life to stable ^{89}Y as follows:

$$^{89}_{38}Sr_{51} \xrightarrow[n \to p^+ + e^-]{} {}^{89}_{39}Y_{50} + 1.4926 \, MeV(\beta, \nu)$$

88.9074507 AMU   88.9058483 AMU

The transition energy for this decay is 1.4926 MeV, which is the maximum energy of the beta particle. Strontium decays by one of two beta transitions. The β_1 transition occurs with very low intensity (0.009%) to the excited level of ^{89}Y, resulting in a gamma ray emission of 0.091 MeV (γ_1) of the same intensity (Figure 11-15). In the β_2 transition (99.99% intensity), the beta particle carries off on average 0.585 MeV and the neutrino 0.9076 MeV. Although ^{89}Sr is a pure beta emitter, a technique has been developed for its measurement in the dose calibrator.[73]

Strontium chloride Sr 89 injection (^{89}Sr-chloride; Metastron [GE Healthcare]) is a sterile aqueous solution at pH 4.0 to 7.5. It is supplied in 4 mCi (148 MBq) vials at a concentration of 1 mCi (37 MBq)/mL, with a specific activity of 80 to 167 μCi (2.96–6.18 MBq)/mg at the time of calibration. The product is stored at room temperature (15°C–25°C) and expires 28 days after calibration.

^{89}Sr-chloride is indicated for the palliative treatment of bone pain in patients with skeletal metastases. The usual adult dosage is 4 mCi intravenously, and it can be dosed by weight at 40 to 60 µCi (1.48–2.22 MBq)/kg. The critical organ is the bone surface, with a radiation absorbed dose of 63 rad(cGy)/mCi.[74]

YTTRIUM CHEMISTRY

Yttrium is a member of group IIIB transition elements. Its electron configuration is [Kr]$4d^1 5s^2$. It is quite rare naturally, but as a metal it reacts well to produce compounds in which its oxidation state is 3+. Yttrium hydrolyzes below pH 7 and requires buffering with sodium acetate during labeling reactions, similar to indium. To be an effective label for antibodies, yttrium must be chelated firmly with a derivatized DTPA ligand; otherwise, it will dissociate and localize in bone, where it can deliver a high radiation dose to the bone marrow.

^{90}Y is the daughter product of ^{90}Sr decay. ^{90}Y itself decays by beta emission with a half-life of 64.0 hours to stable zirconium ^{90}Zr as follows:

$$^{90}_{39}\text{Y}_{51} \xrightarrow{n \to p^+ + e^-} {}^{90}_{40}\text{Zr}_{50} + 2.2798 \text{ MeV}(\beta, \nu)$$

89.9071519 AMU   89.9047044 AMU

The transition energy for this decay is 2.2798 MeV, which is carried away by two possible beta transitions. The β_1 route occurs with very low intensity (0.0115%) (Figure 11-15). The β_2 route occurs 99.9885% to the ground state of ^{90}Zr with an average beta-particle energy of 0.934 MeV and the remainder carried off by the neutrino.

Yttrium Y 90 Chloride Solution

Yttrium Y 90 chloride (^{90}Y-chloride) is a sterile aqueous solution of yttrium chloride in 0.05 M hydrochloric acid. It is produced by solvent extraction of ^{90}Y from a ^{90}Sr generator solution and purified into the chloride form, ^{90}YCl$_3$. ^{90}Y-chloride is available from MDS Nordion (Ontario, Canada) in 5, 10, 20, and 50 mCi (185, 370, 740, and 1850 MBq) amounts per vial with no more than 20 µCi of ^{90}Sr radionuclide impurity per Ci of ^{90}Y at expiration. ^{90}Y-chloride is also available from Iso-Tex Diagnostics (Friendswood, TX). The product is used for labeling monoclonal antibodies for radioimmunotherapy. A notable example is ibritumomab tiuxetan (^{90}Y-ibritumomab tiuxetan [Spectrum Pharmaceuticals]) for the treatment of patients with relapsed or refractory low-grade, follicular, or transformed B-cell non-Hodgkin's lymphoma. This antibody is discussed in detail in Chapter 28.

SAMARIUM CHEMISTRY

Samarium is a member of the lanthanide transition elements known as the rare earths. Its electron configuration is [Xe] $4f^6 5d^0 6s^2$, and it forms complexes in the 3+ oxidation state. ^{153}Sm has useful properties for therapeutic application in palliation of painful bone metastases. It forms coordination complexes of high thermodynamic stability with phosphonate and aminocarboxylate ligands.[75] ^{153}Sm is produced in high yield and purity by neutron irradiation of isotopically enriched ^{152}Sm oxide (^{152}Sm$_2$O$_3$) by the reaction ^{152}Sm(n,γ)^{153}Sm. The oxide target is dissolved in 1.0 M hydrochloric acid, diluted to 0.1 M hydrochloric acid, added to a lyophilized kit of ethylenediaminetetramethylenephosphonic acid (EDTMP) for complexation, and brought to pH 7.0 to 8.5 with sodium hydroxide.[76] The ionic formula for the complex is ^{153}Sm^{3+}[CH$_2$N(CH$_2$ PO$_3^{2-}$)$_2$]$_2$. The molecular weight is 581.1. The chemical structure of the pentasodium salt is shown in Figure 11-16.

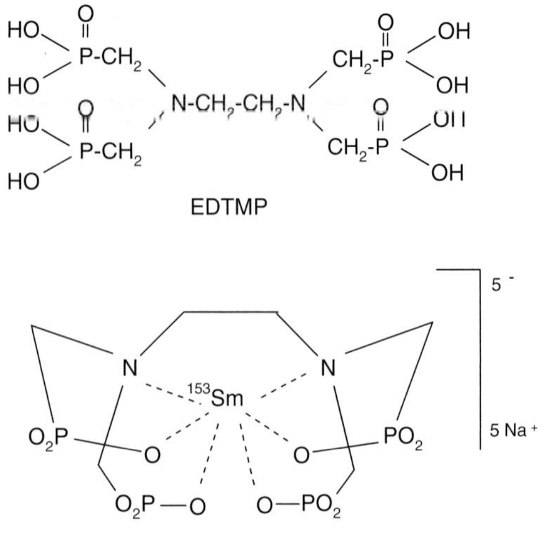

FIGURE 11-16 Chemical structures of EDTMP ligand and its ^{153}Sm coordination complex (^{153}Sm-lexidronam).

^{153}Sm decays by beta emission with a 46.5 hour half-life to stable ^{153}Eu according to the following decay equation:

$$^{153}_{62}\text{Sm}_{91} \xrightarrow{n \to p^+ + e^-} {}^{153}_{63}\text{Eu}_{90} + 0.8077 \text{ MeV}(\beta, \nu, \gamma)$$

152.9220974 AMU   152.9212303 AMU

The transition energy for this decay is 0.8077 MeV, which is dissipated by several beta decay routes (Figure 11-17). The most abundant routes are β_{15} (18%), 0.808 MeV; β_{13} (49%), 0.704 MeV; and β_{11} (31%), 0.635 MeV. The most abundant gamma ray is γ_{11} (29%) at 0.103 MeV. The average energy released as beta particles per decay of ^{153}Sm is 0.224 MeV. This is computed by multiplying each maximum beta particle energy listed above by its abundance and by 0.33 (the average fraction of transition energy per decay dissipated by beta particles) and summing the individual average beta energies. Thus, we have 0.33 [(0.808 MeV) (0.18) + (0.704 MeV) (0.49) + (0.635 MeV) (0.31)] = 0.224 MeV.

Samarium Sm 153 Lexidronam Injection

Samarium Sm 153 lexidronam injection (^{153}Sm-lexidronam; Quadramet [Berlex]) is a sterile, aqueous, clear to light amber solution for intravenous administration. Each milliliter of

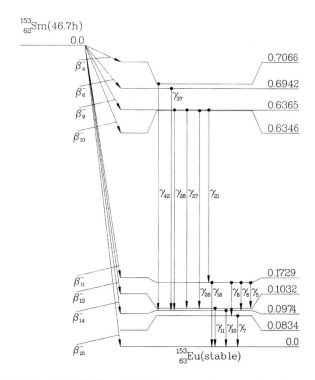

FIGURE 11-17 Decay scheme for ^{153}Sm. (Reprinted with permission from reference 6.)

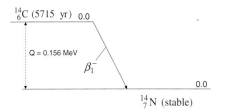

FIGURE 11-18 Simplified decay scheme for ^{14}C.

solution contains, at calibration, 35 mg EDTMP·2H$_2$O; 5.3 mg Ca as Ca(OH)$_2$; 14.1 mg Na as NaOH, equivalent to 44 mg Ca/Na EDTMP (anhydrous); and 50 mCi (1850 MBq) ^{153}Sm, at a specific activity of 1.0 to 11.0 mCi (37 to 407 MBq)/μg Sm. The solution pH is 7.0 to 8.5. The product is available as a frozen solution at a concentration of 50 mCi (1850 MBq)/mL in vial sizes of 100 mCi (3700 MBq) and 150 mCi (5550 MBq). The drug expires 48 hours after the calibration time or within 8 hours of thawing.

Although the thermodynamic stability constant of ^{153}Sm-EDTMP is quite high, the complex in not kinetically inert in plasma. Therefore, ^{153}Sm-lexidronam has a ligand-to-metal ratio of approximately 250:1 to 300:1. The excess ligand prevents any samarium that dissociates from the EDTMP from forming insoluble ^{153}Sm-hydroxyl species that localize in the liver.[75]

^{153}Sm-lexidronam is indicated for relief of pain in patients with confirmed osteoblastic metastatic bone lesions that enhance on radionuclide bone scan.[77] Neoplastic bone disease associated with lytic-type bony metastases, such as multiple myeloma, may not be treatable with ^{153}Sm-lexidronam.[76] The recommended dosage is 1 mCi (37 MBq)/kg intravenously. The critical organ is the bone surface, with a radiation absorbed dose of 25 rad(cGy)/mCi.[77]

CARBON CHEMISTRY

Carbon is a group IV element in the periodic table with an electronic configuration 1s^2 2s^2 2p^2. It has four valence electrons because of promotion of one 2s orbital electron to a p orbital, creating four sp^3 hybridized orbitals. Consequently, carbon does not easily lose or gain electrons when it bonds with other atoms and forms covalent bonds. Chapter 9 contains a more detailed discussion of carbon bonding chemistry.

The isotopes of carbon that are of interest in nuclear medicine are ^{11}C, a positron emitter, and ^{14}C, a negatron emitter. ^{11}C chemistry is discussed in Chapter 12. ^{14}C is produced in a nuclear reactor by neutron irradiation of nitrogen in beryllium nitride or aluminum nitride. The nuclear reaction is ^{14}N(n,p)^{14}C. The target material is processed into a variety of ^{14}C-labeled compounds. ^{14}C-carbon dioxide is a typical reactant for preparing other ^{14}C-labeled compounds.

^{14}C decays by negatron decay with a half-life of 5700 years to stable nitrogen:

$$^{14}_{6}C_8 \xrightarrow[n \to p^+ + \nu]{} {}^{14}_{7}N_7 + E(\beta^-, \nu) + 0.156 \text{ MeV}$$

14.003241989 AMU   14.003074005 AMU

The transition energy for this decay is 0.156 MeV, which is dissipated as kinetic energy of a beta particle and a neutrino. The mean energy distribution is approximately one-third (0.05 MeV) to the beta particle and two-thirds (0.106 MeV) to the neutrino. The decay scheme for ^{14}C is shown in Figure 11-18. Since there is no gamma ray emitted in ^{14}C decay, this isotope of carbon cannot be used for imaging studies. Its use in nuclear medicine is limited to diagnostic studies that employ liquid scintillation counting of ^{14}C.

Carbon C 14 Urea Capsules

Carbon C 14 urea capsules (^{14}C-urea; PYtest [Tri-Med Specialties]) are gelatin capsules that contain 1 μCi (37 kBq) of ^{14}C-urea adsorbed on sugar spheres.[78] A method for synthesizing ^{14}C-urea involves reacting ammonia with ^{14}C-carbon dioxide, forming ammonium carbamate, which is dehydrated to ^{14}C-urea:

$$2\,NH_3 + {}^{14}CO_2 \to H_2N-{}^{14}COONH_4 \quad \text{(ammonium carbamate)}$$

$$H_2N-{}^{14}COONH_4 \to H_2O + H_2N-{}^{14}CO-NH_2 \quad \text{(urea)}$$

^{14}C-urea is indicated for the detection of gastric urease in the diagnosis of *Helicobacter pylori* infection in the human stomach.[79] The dosage is one capsule (1 μCi [37 kBq]) taken orally. The effective dose equivalent is 0.3 mrem/μCi.[78]

REFERENCES

1. DeNardo GL, DeNardo SJ. Treatment of B-lymphocyte malignancies with ^{131}I-LYM-1 and ^{67}Cu-2IT-BAT-LYM-1 and opportunities for improvement. In: Goldenberg DM, ed. *Cancer Therapy with Radiolabeled Antibodies*. Boca Raton, FL: CRC Press; 1995: 217–27.
2. Kaminski MS, Zasadny KR, Francis IR, et al. Iodine-131-anti-B1 radioimmunotherapy for B-cell lymphoma. *J Clin Oncol*. 1996;14:1974–81.
3. Stubbs J, Frankel R, Schultz K, et al. Preclinical evaluation of a novel device for delivering brachytherapy to resected brain tumor cavity margins. *J Neurosurg*. 2002;96:335–43.
4. Blasko JC, Grimm PD, Ragde H. Brachytherapy and organ preservation in the management of carcinoma of the prostate. *Semin Radiat Oncol*. 1993;3:240–9.
5. Storey MR, Landgren RC, Cottone JL, et al. Transperineal 125-iodine implantation for treatment of clinically localized prostate cancer: 5-year tumor control and morbidity. *Int J Radiat Oncol Biol Phys*. 1999;43:565–70.
6. Weber DA, Eckerman KF, Dillman LT, et al. *MIRD: Radionuclide Data and Decay Schemes*. New York: Society of Nuclear Medicine; 1989.
7. Kowalsky RJ, Johnston RE. Dose calibrator assay of 123 and indium-111 with a copper filter. *J Nucl Med Technol*. 1998;26:94–8.
8. Lever JR. Radioiodinated compounds. In: Wagner HN Jr, Szabo Z, Buchanan JW, eds. *Principles of Nuclear Medicine*. 2nd ed. Philadelphia: WB Saunders; 1995:199–213.
9. Hughes WL. The chemistry of iodination. *Ann NY Acad Sci*. 1957;70:3–18.
10. McFarlane AS. Efficient trace-labeling of proteins with iodine. *Nature*. 1958;182:53–5.
11. Helmkemp RW, Contreras MA, Bale WF. I-131 labeling of proteins by the iodine monochloride method. *Int J Appl Radiat Isot*. 1967;18:737–46.
12. Bayly RJ, Anthony E, Evans JS, et al. Synthesis of labeled compounds. In: Tubis M, Wolf W, eds. *Radiopharmacy*. New York: Wiley; 1976: 303–78.
13. Hunter WM, Greenwood FC. Preparation of iodine-131 labeled human growth hormones of high specific activity. *Nature*. 1962;194: 495–6.
14. McConahey PJ, Dixon EJ. A method of trace iodination of proteins for immunological studies. *Int Arch Allergy*. 1966;29:185–9.
15. Mills SL, DeNardo SJ, DeNardo GL, et al. I-123 radiolabeling of monoclonal antibodies for in vivo procedures. *Hybridoma*. 1986;5: 265–75.
16. Markwell MAK. A new solid-state reagent to iodinate proteins. *Anal Biochem*. 1982;125:427–32.
17. Bolton AE, Hunter WM. The labeling of proteins to high specific radioactivities by conjugation to a I-125 containing acylating agent. *Biochem J*. 1973;133:529–39.
18. Haney TA, Wedeking P, Morcos N, et al. A therapeutic and diagnostic I-131 capsule formulation with minimal volatility and maximal bioavailability. *J Nucl Med*. 1981;22:P74.
19. Robertson JS, Verhasselt M, Wahner HW. Use of ^{123}I for thyroid uptake measurements and depression of ^{131}I thyroid uptakes by incomplete dissolution of capsule filler. *J Nucl Med*. 1974;15:770–4.
20. Green JP, Wilcox JR, Marriott JD, et al. Thyroid uptake of I-131: Further comparisons of capsules and liquid preparations. *J Nucl Med*. 1976;17:310–2.
21. Yu MD, Huang WS, Cherng CC, et al. Effect of formulation on reduced radioiodide thyroid uptake. *J Nucl Med*. 2002;43:56–60.
22. Rini JN, Vallabhajosula S, Zanzanico P, et al. Thyroid uptake of liquid versus capsule I-131 tracers in hyperthyroid patients treated with liquid I-131. *Thyroid*. 1999;9:347–52.
23. Kereiakes JG, Rosenstein M. *Handbook of Radiation Doses in Nuclear Medicine and Diagnostic X-ray*. Boca Raton, FL: CRC Press; 1980.
24. Mock BH, Weiner RE. Simplified solid-state labeling of I-123 m-iodobenzylguanidine. *Appl Radiat Isot*. 1988;39:939–42.
25. Mangner TJ, Wu J-L, Weiland DM. Solid-phase exchange radioiodination of aryl iodides. Facilitation by ammonium sulfate. *J Org Chem*. 1982;47:1484–8.
26. *The United States Pharmacopeia 28th rev. and The National Formulary, 23rd ed.* Rockville, MD: United States Pharmacopeial Convention, Inc; 2005.
27. Gelfand MJ, Elgazzar AH, Kriss VM, et al. Iodine-123-MIBG SPECT versus planar imaging in children with neural crest tumors. *J Nucl Med*. 1994;35:1753–7.
28. Shapiro B, Gross MD. Radiochemistry, biochemistry, and kinetics of ^{131}I-metaiodobenzylguanidine (MIBG) and ^{123}I-MIGB: clinical implications of the use of ^{123}I-MIGB. *Med Pediatr Oncol*. 1987;15:170–7.
29. Iobenguane sulfate I-131 [package insert]. Bedford, MA: CIS-US Inc; 1994.
30. Iodohippurate sodium I-131 [package insert]. St Louis: Mallinckrodt Inc; 1995.
31. Iothalamate sodium I-125 [package insert]. Carlsbad, CA: Cypros Pharmaceutical Corp; 1995.
32. Anderson CJ, Welch MJ. Radiometal-labeled agents (non-technetium) for diagnostic imaging. *Chem Rev*. 1999;99:2219–34.
33. Moerlein SM, Welch MJ. The chemistry of gallium and indium as related to radiopharmaceutical production. *Int J Nucl Med Biol*. 1981;8:277–87.
34. Hill T, Welch M, Adatepe M, et al: A simplified method for the preparation of indium DTPA as a brain scanning agent. *J Nucl Med*. 1970;11:28–30.
35. Hnatowich DJ, Kulprathipanja S, Beh R. The importance of pH and citrate concentration on the in vitro and in vivo behaviour of radiogallium. *Int J Appl Radiat Isot*. 1977;28:925–31.
36. Welch MJ, Welch TJ. Solution chemistry of carrier-free indium. In: Subramanian G, Rhodes BA, Cooper JF, Sodd VJ, eds. *Radiopharmaceuticals*. New York: Society of Nuclear Medicine; 1975:73–9.
37. Brown LC. Chemical processing of cyclotron-produced Ga-67. *Int J Appl Radiat Isot*. 1971;22:710–3.
38. Brown LC, Beets AL. Cyclotron production of carrier-free indium-111. *Int J Appl Radiat Isot*. 1972;23:57–63.
39. Lebowitz E, Green MW, Fairchild R, et al. Thallium-201 for medical use I. *J Nucl Med*. 1975;16:151–5.
40. Dymov AM, Savostin AP. *Analytical Chemistry of Gallium*. Ann Arbor: Ann Arbor Science Publishers; 1970.
41. Gallium citrate Ga-67 [package insert]. St Louis: Mallinckrodt Inc; 2000.
42. Indium In-111 oxine [package insert]. Arlington Heights, IL: Medi-Physics Inc/Amersham Healthcare; 1996.
43. Maecke HR, Riesen A, Ritter W. The molecular structure of indium-DTPA. *J Nucl Med*. 1989;30:1235–9.
44. Indium In-111 DTPA [package insert]. Arlington Heights, IL: Medi-Physics Inc/Amersham Healthcare; 1994.
45. OctreoScan kit [package insert]. St Louis: Mallinckrodt Inc; 2000.
46. Thakur ML, Lavender JP, Arnot RN, et al. Indium-111-labeled autologous leukocytes in man. *J Nucl Med*. 1977;18:1012–9.
47. Palestro CJ, Brown ML, Forstrom LA, et al. In-111 leukocyte scintigraphy for suspected infection/inflammation 3.0. In: *Procedure Guidelines Manual*. Reston, VA: Society of Nuclear Medicine; 2004. www.snm.org/guidelines. Accessed September 15, 2010.
48. McAfee JG, Thakur ML. Survey of radioactive agents for in vitro labeling of phagocytic leukocytes. I. Soluble agents. *J Nucl Med*. 1976;17:480–7.
49. Goodwin DA, Doherty PW, McDougall IR. Clinical use of indium-111 labeled white cells: an analysis of 312 cases. In: Thakur ML, Gottschalk A, eds. *Indium-111 Labeled Neutrophils, Platelets, and Lymphocytes*. New York: Trivirum; 1981:131.
50. Thakur ML, Coleman RE, Welch MJ. Indium-111-labeled leukocytes for the localization of abscesses: preparation, analysis, tissue distribution, and comparison with gallium-67 citrate in dogs. *J Lab Clin Med*. 1977;82:217–28.
51. Coleman RE, Welch DM, Baker W, et al. Clinical experience using indium-111 labeled leukocytes. In: Thakur ML, Gottschalk A, eds. *Indium-111 Labeled Neutrophils, Platelets, and Lymphocytes*. New York: Trivirum; 1981:103.

52. Danpure HJ, Osman S, Brady F. The labeling of blood cells in plasma with In-111 tropolonate. *Br J Radiol.* 1982;55:247–9.
53. Gobuty AH. Technologic, clinical, and basic science considerations for In-111-oxine-labeled leukocyte studies. *J Nucl Med Technol.* 1983;11:190–5.
54. Peters AM. The utility of ^{99m}Tc-HMPAO-leukocytes for imaging infection. *Semin Nucl Med.* 1994;24:110–27.
55. Indium In-111 oxine [package insert]. Arlington Heights, IL: Medi-Physics Inc/GE Healthcare; 1996.
56. Thakur ML, Walsh L, Malech HL, et al. Indium-111-labeled platelets: improved method, efficacy, and evaluation. *J Nucl Med.* 1981;22:381–5.
57. Dewanjee MK, Rao SA, Didisheim P. Indium-111 tropolone, a new high-affinity platelet label: preparation and evaluation of labeling parameters. *J Nucl Med.* 1981;22:981–7.
58. Mathias CJ, Welch KJ. Radiolabeling of platelets. *Semin Nucl Med.* 1984;14:118–27.
59. Kotze HF, Heyns A du P, Lotter MG, et al. Comparison of oxine and tropolone methods for labeling human platelets with indium-111. *J Nucl Med.* 1991;32:62–6.
60. International Committee for Standardization in Hematology Panel on Diagnostic Applications of Radionuclides. Recommended Method for Indium-111 Platelet Survival Studies. *J Nucl Med.* 1988;29:564–6.
61. Thallous chloride Tl-201 [package insert]. St Louis: Tyco-Mallinckrodt Inc; 2006.
62. Xenon Xe 133 gas [package insert]. St Louis: Tyco-Mallinckrodt Inc; 2006.
63. Person HA. The binding of Cr-51 to hemoglobin. I. In vitro studies. *Blood.* 1963;22:218–20.
64. Gray SJ, Sterling K. The tagging of red cells and plasma proteins with radioactive chromium. *J Clin Invest.* 1950;29:1604–10.
65. Sodium chromate Cr-51 [package insert]. St Louis: Mallinckrodt Inc; 2000.
66. Sienko MJ, Plane RA. *Chemistry.* 2nd ed. New York: McGraw-Hill; 1961.
67. Morton ME. Colloidal chromic radiophosphate in high yields for radiotherapy. *Nucleonics.* 1952;10:92–7.
68. Phosphocol P-32 [package insert]. St Louis: Mallinckrodt Inc; 2005.
69. Siegel BA. Radiation synovectomy revisited. *Eur J Nucl Med.* 1993;20:113–27.
70. *The United States Pharmacopeia, 28th rev, and The National Formulary, 23rd ed.* Rockville, MD: United States Pharmacopeial Convention, Inc; 2005.
71. Wasserman LR. The treatment of polycythemia vera. *Semin Hematol.* 1976;13:57–78.
72. Spiers FW, Beddoe AH, King SD, et al. The absorbed dose to bone marrow in the treatment of polycythemia vera by P-32. *Br J Radiol.* 1976;49:133–40.
73. Herold TJ, Gross GP, Hung JC. A technique for measurement of strontium-89 in a dose calibrator. *J Nucl Med Technol.* 1995;23:26–8.
74. Metastron [package insert]. Arlington Heights, IL: GE Healthcare; 1998.
75. Volkert WA, Hoffman TJ. Therapeutic radiopharmaceuticals. *Chem Rev.* 1999;99:2269–92.
76. Singh A, Holmes RA, Farhangi M, et al. Human pharmacoknetics of samarium-153 EDTMP in metastatic cancer. *J Nucl Med.* 1989;30:1814–8.
77. Quadramet [package insert]. Richmond, CA: Berlex; 1999.
78. PYtest [package insert]. Charlottesville, VA: TRI-MED Specialties; 1997.
79. Marshall BJ, Surveyor I. Carbon-14 urea breath test for the diagnosis of Campylobacter pylori associated gastritis. *J Nucl Med.* 1988;29:11–6.

CHAPTER 12
Radiopharmaceutical Chemistry: PET Agents

Stephen M. Moerlein

Clinical applications of positron emission tomography (PET) have grown remarkably in recent years. Like other nuclear medicine procedures, PET imaging detects the distribution of tracers within the body. Whereas magnetic resonance imaging (MRI) and computed tomography (CT) imaging methods provide structural or anatomic information, PET imaging is intrinsically linked to function because the biologic fate of PET radiotracers is determined by the in vivo rate of the physiologic processes of the organism.

PET differs from conventional single-photon emission computed tomography (SPECT) imaging in that short-lived positron-emitting radionuclides are used as radiolabels. This has several important implications. First, because of the physics of positron decay, coincidence circuitry can be used, which facilitates acquisition of high-resolution images with accurate attenuation correction. In addition, most of the elements used as positron-emitting labels (Table 12-1) are attached via covalent bonds and are commonly found in several biochemical and drug structures. This leads to much greater versatility in the development of novel PET radiopharmaceuticals, compared with the application of ^{99m}Tc (attached via bulky chelating linker groups) or ^{123}I (which forms a biologically unstable bond with carbon) as radiopharmaceutical labels.

PET imaging has unique applications that have proven useful for routine procedures in nuclear medicine. Given the special characteristics of PET methods and the wide range of tracer design and production, further growth in this area of nuclear medicine is anticipated, despite its higher costs. This chapter discusses the current status of PET radiopharmaceuticals used routinely for clinical imaging procedures and presents examples of clinical research applications of this powerful imaging technique.

POSITRON EMISSION TOMOGRAPHY

Positron Decay

Proton-rich nuclei decay either by positron emission or by electron capture (EC).[1] EC is a prevalent decay process for nuclides used in conventional nuclear medicine, since it generates photons that are well suited for imaging by Anger-type thallium-activated sodium iodide [NaI(Tl)] cameras. Examples of such clinically useful radiopharmaceutical labels are ^{123}I and ^{111}In. The photons emitted by these nuclides are collimated by lead collimators to determine the directionality of the photons detected by NaI(Tl) crystals used for image reconstruction.

Positron decay transforms proton-rich nuclei closer to the line of beta stability by converting the unstable nucleus to a daughter nuclide with one less proton and one more neutron, together with the emission of a positron and a neutrino. The elementary process can be written as follows:

$$p \rightarrow n + \beta^+ + \nu$$

TABLE 12-1 Radionuclides Used in Positron Emission Tomography

Nuclide	Half-Life	E_{max} (MeV)
^{15}O	2 min	1.7
^{13}N	10 min	1.2
^{11}C	20 min	1.0
^{19}F	110 min	0.7
^{82}Rb	76 sec	3.4
^{68}Ga	68 min	1.9

Emission of a neutrino (ν) is necessary in this decay process to conserve momentum and angular momentum.[1] The emitted neutrino has very little interaction with matter and is of importance to PET imaging only in that, because of conservation laws, it influences the energy of the corresponding positron. Unlike the photons imaged in conventional nuclear medicine, which are monoenergetic with discrete energy groups, positrons are emitted with a continuous energy spectrum and have a median energy equal to approximately one-third of the maximum positron energy.[1] As discussed later, the range of the emitted positrons has practical impact in that it degrades image resolution in PET.

An emitted positron will lose energy through ionization and excitation of atoms along its pathway until it reaches thermal velocities and combines with an orbital electron in an annihilation reaction.[1] This elementary process is written as follows:

$$\beta^+ + e^- \rightarrow \gamma + \gamma$$

A positron is an antiparticle of an ordinary electron, so the annihilation reaction converts the entire mass of each particle into energy.

Momentum conservation in this process demands that two gamma quanta be produced with equal and opposite momenta. The two quanta are emitted almost exactly 180° apart; any deviation from colinearity derives from residual momentum in the positron at the time of annihilation (Figure 12-1). Because the rest mass of the positron and of the electron is 511 keV, the annihilation photons each have an energy of 511 keV. It is important to note that it is these annihilation photons (and not the positrons themselves) that are detected by PET imaging devices.

Coincidence Detection

The directionality of the annihilation photons after positron detection allows the unique PET imaging method to be achieved. The key unit of PET detection is the coincidence circuit, shown in Figure 12-2, which provides for the colinearity of the two annihilation photons to be discriminated. In this method, only simultaneous events detected within the common field of view of the two detectors are registered as an annihilation event, and hence as a positron emission.

This technique is based on the fact that the annihilation photons always escape at approximately 180° from one another and arrive at opposing detectors nearly simultaneously. Detection of an event at only a single detector would not be registered as a coincidence, so singles or randoms are excluded from the data collection process. This electronic collimation is a unique advantage of PET imaging; there is no need for the heavy lead shielding with collimating septa that is required for image definition in single-photon imaging.

Another major difference between detection in PET and in conventional nuclear medicine cameras is the type of scintillation crystals used. Whereas relatively thin (three-eighths inch standard) thallium-activated sodium iodide [NaI(Tl)] crystals are widely used in single-photon nuclear medicine cameras, these are inadequate for efficient stopping of the 511 keV annihilation photons imaged in PET.

To achieve efficient detection with PET, the characteristics of scintillation crystals must be scrutinized. Crystals can be generally characterized by their physical and optical properties. Physical properties of crystals include density and atomic number, which together determine how efficiently annihilation photons are stopped by the crystal, as well as their hygroscopic

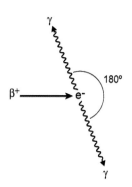

FIGURE 12-1 Positron decay with production of annihilation radiation.

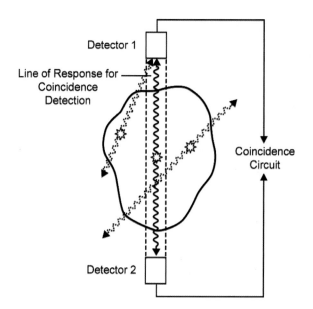

FIGURE 12-2 Coincidence circuitry for detection of annihilation radiation.

TABLE 12-2 Properties of Scintillation Crystals for PET

Property	NaI(Tl)	BGO	GSO	LSO
Density (g/cm³)	3.67	7.13	6.71	7.40
Effective atomic number	51	75	59	65
Relative emission intensity	100	15	30	75
Optical decay constant (nsec)	230	300	60	40

nature and mechanical durability. Physical properties thus affect detector sensitivity and the ease with which the crystals can be commercially manufactured for PET instruments. Optical properties include emission intensity (or light output) and the decay constant (how long the pulse of light persists following a scintillation). Both of these properties directly affect image quality of the PET scanner. Crystals with high emission intensity permit several crystal elements to be mounted on a single photomultiplier tube (PMT), which improves energy resolution and spatial resolution. Faster crystals with rapid decay constants enhance scanner count rate performance and reduce system dead time as well as detection of random events.

Table 12-2 lists pertinent physical and optical properties for the most common crystal types used in the design of PET scanners.[2] These crystals are bismuth germanate [BGO or $Bi_4Ge_3O_{12}$], gadolinium silicate [GSO or $Gd_2SiO_5(Ce)$], and lutetium oxyorthosilicate [LSO or $Lu_2(SiO_4)O$]. Data for NaI(Tl) are included for comparison. Although sodium iodide scintillators have very high light output, for PET application this crystal has poor efficiency in detecting annihilation photons (because of low density and effective Z), as well as slow response (large optical decay constant). The large optical decay constant of NaI(Tl) leads to increased dead time with high count rates, and the relatively low effective atomic number leads to an increased Compton scatter during coincidence detection.

LSO, BGO, and GSO all have characteristics superior to NaI(Tl) for application in PET scanners. Among these three scintillation crystals, GSO and BGO have shortcomings relative to LSO. BGO is a relatively slow scintillator and has low emission intensity. GSO has relatively low light output and low effective Z, and it cleaves easily so it is not mechanically durable. LSO is currently the crystal with the best combination of properties for PET, and most modern PET scanners use LSO crystals for detection elements. LSO is characterized by high density and atomic number, short decay constant, and high light output, which translates to high photon detection efficiency, good timing resolution, and high spatial resolution, respectively. Moreover, LSO crystals are rugged and nonhygroscopic, which facilitates ease of manufacture and maintenance.

Factors Affecting Image Quality

Dedicated PET scanners consist of a series of coincidence circuits assembled into a ring shape to enclose all the angles around the subject to be imaged (Figure 12-3).[3] Several rings of detectors are mounted next to one another in the transaxial direction, so that multislice image acquisition can occur simultaneously.

The PET scanner is based on coincidence detection; if a photon is detected at two detectors within the ring simultaneously, the interaction is recorded as a coincidence event. The number of coincidences for a particular line of response is thus directly proportional to the amount of positron-emitting radioactivity within the field of view of the two detectors.

The recorded coincidence counts between detectors can be expressed as sinograms, which are the collection of coincidence data expressed as an angular function. Sinograms are reconstructed from raw data into cross-sectional images by using computer algorithms similar to those used in SPECT or CT.

This simplified description of PET system design is complicated by several factors that act in concert to degrade image quality.[3] These factors are summarized in Table 12-3.

Scatter is an important effect in PET, especially because electronic collimation is used instead of lead collimation. As shown in Figure 12-4, scatter can be in-plane or out-of-plane and acts to artificially increase counts within a line of response,

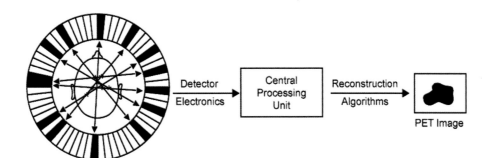

FIGURE 12-3 PET scanner design consisting of circular banks of coincidence circuits.

TABLE 12-3 Factors that Degrade PET Images

Problem	Correction
Scatter	Increase energy resolution (light output) of detectors
Attenuation	Transmission scan
Random events	Fast scintillators and electronics
Dead time	Fast scintillators and electronics
Noise	Increase signal counts by increasing tracer dose and acquisition time
Spatial resolution	Decrease detector size to increase crystal light output

even though no positron decay occurred there. Scatter can be decreased by using lead septa around scintillators to reduce access to scattered photons, by increasing energy resolution (light output) of the detectors, and by applying scatter correction algorithms.

Attenuation is another parameter that strongly influences PET image quality. Attenuation results in the loss of true counts, resulting in increased noise and inaccurate quantification in reconstructed images. The 511 keV photons detected by PET are individually attenuated to a lesser extent than the lower-energy photons used in conventional SPECT imaging, because two attenuated photons must be detected in coincidence detection, but the overall impact of attenuation is actually more important in PET imaging than in single-photon imaging.

Attenuation is proportional to tissue density and is accounted for by a measured transmission attenuation correction. The attenuation correction is determined using a transmission measurement similar to that used in CT imaging. For standalone PET scanners, this approach entails the use of a moving line source, with a photon energy similar to that of annihilation radiation. A common line source is a bar of ^{68}Ge, which decays to ^{68}Ga, which decays by positron emission and release of annihilation photons. The line source projects photons through the patient diagonal to the detectors as it rotates around the ring of scintillators. The resulting transmission scan accurately measures the attenuation of photons for each detector and is used to correct the emission data for this effect. The transmission attenuation correction measurement is ideally made before administration of the PET radiopharmaceutical.

For PET/CT scanners (see below), the need for an external radioactive source is obviated, and the CT component of the dual-modality imaging device is used for attenuation correction. Because CT energies average 70 keV, which is substantially lower than the 511 keV photons from annihilation radiation, attenuation coefficients derived from CT images must be appropriately corrected using energy scaling algorithms.

Random events degrade PET image quality, as they are not true coincidences. The scintillator and its associated electronics require a finite time to detect true annihilation photons and record a coincidence; therefore, a minimum time window is required for operation of the coincidence circuits of the PET scanner. Thus, the time interval for the scanner must be optimized so that there is a window large enough to efficiently detect annihilation photons but not large enough to allow a high proportion of random events to occur. Random events can be decreased by the use of fast scintillators that have small optical decay constants.

Dead time is another parameter affecting PET image quality, especially with high count-rate imaging procedures. When an excessive number of events impinge upon a scintillator, the response of the crystal may become saturated and detection efficiency may decrease. This effect is minimized with the use of faster scintillators and electronics, which permit higher levels of annihilation radiation to be accurately detected.

PET scanner noise diminishes image quality because it increases the level of background counts and thereby decreases the signal-to-background contrast. To increase the signal counts in relation to random noise counts, higher radiopharmaceutical dosages or longer scan intervals can be used. Increasing the efficiency of the scanner for detection of annihilation radiation also increases the signal-to-noise ratio.

The spatial resolution of PET images is affected by several factors, some of which can be controlled through scanner design. Spatial resolution is affected by positron range, which is related to the positron energy, as well as by the slight noncolinearity of the two 511 keV photons from positron annihilation. Resolution is also inversely proportional to detector size; the smaller the detector, the greater is the resolution. Note that this relationship is opposite that of detector efficiency, which decreases as the crystal dimensions decrease. The design of scanners with smaller detectors is ultimately limited by physical constraints, since the PMT and related electronics attached to scintillators are relatively bulky. To circumvent this effect, PET scanners often use detector blocks that have several detector elements etched onto a single scintillator crystal to give a smaller effective detector size. In this way, the effective detector element dimensions are decreased with no change in the actual detector crystal-to-PMT ratio.

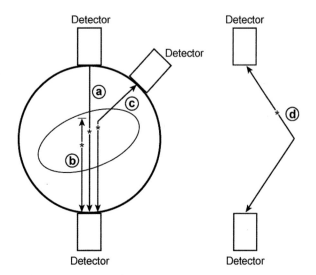

FIGURE 12-4 Factors leading to degradation of PET image resolution. (a) True coincidence; (b) tissue absorption; (c) in-plane scatter; (d) out-of-plane scatter.

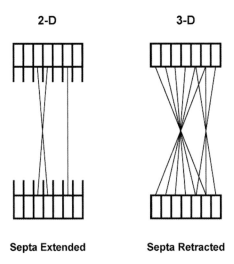

FIGURE 12-5 Two- and three-dimensional data acquisition modes in PET. With 2-D imaging, septa are extended, and only in-plane or simple cross-plane coincidences are detected. With 3-D imaging, septa are retracted, and the number of coincidences is dramatically increased.

Two-Dimensional versus Three-Dimensional Data Acquisition

Emission data from PET can be collected in either two-dimensional (2-D) or 3-D mode. A simplified illustration of the difference between these data-acquisition modes is shown in Figure 12-5. In the 2-D mode, coincidences are collected in the direct plane or in simple cross-plane orientation in which coincidences between scintillators immediately adjacent to the direct plane are included in the database for image reconstruction. In 2-D data acquisition, lead septa are used to surround the sides of the detector elements to reduce the solid angle of detection and thereby minimize counts from scattered photons and randoms arriving from directions outside the line of response for the specific detectors.

In the 3-D mode, the shielding septa are removed, and all possible configurations of detector elements are used as coincidence circuits for image reconstruction. The total number of coincidences analyzed in 3-D mode is clearly much larger than that in 2-D PET imaging. The advantage of applying the 3-D mode for PET data acquisition is that sensitivity is greatly increased. A disadvantage is that because the detector elements are unshielded, scatter is more prevalent and more sophisticated algorithms are needed for image reconstruction. In addition, 3-D PET imaging is especially prone to detector saturation caused by the high number of coincidence and random events that are detected. The short optical decay time of LSO (Table 12-2) improves the count rate performance of PET systems acquiring data in 3-D mode.

PET/CT Scanners

To anatomically locate areas of increased tracer localization in PET images, it is necessary to coregister emission (PET) and transmission (CT) images. Such coregistration facilitates the mapping of functional changes (detected by PET) onto their anatomic sites (detected by CT). When the CT and PET data are obtained on individual scanners, increased scheduling time is required, causing greater patient inconvenience. Moreover, substantial technical difficulties are involved in realigning tomographic images created from nonidentical patient imaging done on two different machines.

A solution to these problems is to combine CT and PET imaging into a single scanner.[4] Such fused PET/CT devices facilitate increased patient throughput in busy imaging clinics, improving patient convenience. The fused devices allow technical improvements in precise positioning of imaging subjects. Most clinical PET imaging facilities today use PET/CT instruments because of their superior diagnostic accuracy and faster acquisition of image data. These devices do not use the same detection instrumentation but consist of a dedicated PET scanner mounted next to a CT scanner. Although different imaging detection rings are used, they are physically mounted in close proximity, so the patient reclines on a bed that can be slid into position for imaging by each portion of the device.

Although the high-resolution CT images generated by this fused device exceed the requirements for attenuation correction and scatter correction of PET images, the clinical-quality anatomic images obtained with the CT component of the scanner are of the caliber that radiologists normally use. This has the benefit of providing radiologists, surgeons, and oncologists with the type of anatomic information they routinely apply in practice, combined with coregistered PET images, data they may not have used in the past. The combined PET/CT delivers precise information about the location and extent of tracer localization to experienced practitioners of PET and also has the valuable effect of introducing nonspecialist clinicians to the role of PET in providing information that supplements more established testing methods for patient care and disease management.[5]

There are several technical issues with PET/CT imaging. Since CT transmission imaging occurs within seconds and PET image acquisition takes place over a longer period (minutes), there is often a positional mismatch between the attenuation and emission scans. This is especially likely to occur in tissues affected by respiratory motion, cardiac motion, and bowel motility. In addition to this complication, attenuation overcorrection occurs when CT images of high-density contrast agents are used for tissue attenuation measurements. Both of these confounders are accounted for in clinical protocols, however, and do not significantly limit the widespread utility of PET/CT imaging.

IMAGING APPLICATIONS

Since its beginnings as a research tool, PET has been used over the years to study the in vivo distribution of numerous positron-emitting tracers. Studies with these tracers have provided valuable insight into the pathophysiology of diseases and the in vivo action of pharmacologic agents. However, only a few of the tracers have come into routine clinical use, partially because the field of PET imaging has not yet fully developed. The primary limiting factor, however, is the stringent requirements a PET protocol must meet in order to be implemented in routine patient care. These requirements include diagnostic efficacy, imaging protocol simplicity, and radiopharmaceutical compounding efficiency. These specific clinical requirements supplement the general safety restrictions for all PET tracers administered to humans, such as limits on absorbed radiation dose and quality assurance for purity, sterility, and apyrogenicity. Its ability to meet the more stringent requirements determines whether a tracer is suitable as a radiopharmaceutical for routine patient care or is appropriate only for specialized clinical research applications.

An important criterion for clinical utility is that the radiopharmaceutical must be used in PET imaging procedures that have validated diagnostic efficacy. If a PET procedure will give clinical information that is not achievable by other, less expensive means, the cost of PET is justified within the overall scope of health care expenses.

Another requirement for routine clinical utility is that the PET imaging protocol be relatively simple. Simple procedures streamline clinical operations, facilitate rapid dissemination of answers to clinical questions, reduce costs, and minimize patient inconvenience.

Finally, clinically useful PET tracers must be prepared in an efficient, reliable manner. This ensures timely delivery of the radiopharmaceutical in quantities adequate for scheduled studies as well as for potential emergency additions to the workday schedule. Efficiency is important in preparing radiopharmaceutical dosages for unscheduled studies, which may become more prevalent as PET imaging becomes as common as other diagnostic tests used in the hospital setting.

To date, despite the many PET tracers that have been developed, only a few radiopharmaceuticals satisfy all three requirements for clinical utility. Although multisite studies over time may show the diagnostic efficacy of several tracers, other radiopharmaceuticals (especially metabolic tracers and receptor ligands) utilize complicated tracer kinetic models that are time consuming and unsuitable for rapid turnaround in the clinical setting. Such PET tracers will probably be limited to clinical research projects unless simplified image analysis techniques can be developed and validated.

PET tracers that satisfy the stringent requirements for routine clinical use are listed in Table 12-4. Both cyclotron-based and generator-available radiopharmaceuticals are listed. The physiologic basis for the localization of these tracers is discussed here; the radiosynthesis of individual radiolabeled drugs is detailed in a later section.

Table 12-4 lists several physiologic functions that can be assessed noninvasively with PET. These are categorized as cardiovascular parameters, metabolic processes, and neurotransmission measurements. Cardiovascular parameters include perfusion and blood volume. These processes affect the delivery of nutrients and other tracers to the tissues under study. PET studies of metabolic processes assess the utilization of nutrients on a regional basis, including glycolytic and oxidative pathways, as well as osteogenic activity. Metabolic changes occur as a direct consequence of human pathology. Neurotransmission is a complicated physiologic process that involves the synthesis and release of neurotransmitters as well as their binding to neuronal receptor sites. Changes in the subtle equilibrium of any of these components of neurotransmission can lead to disease. Pathophysiologic processes often involve changes in more than one of these categories (i.e., cardiovascular, metabolic, and neurotransmission), so multiple PET measurements are often warranted to fully evaluate patient status.

Table 12-4 lists the appropriate radiopharmaceuticals indicated for measurement of each of these physiologic processes by PET. All of these tracers and applications have achieved a sufficient level of clinical acceptance that monographs for each are found in *The United States Pharmacopeia, 32nd Revision, and The National Formulary 27th Edition (USP 32/NF 27)*.[6] The following section discusses details of these clinical PET radiopharmaceuticals, including the mechanism of localization of each radiotracer and its specific clinical imaging indication.

Cardiovascular Parameters

The major cardiovascular parameter is tissue perfusion, and several PET tracers are indicated for clinical use. Blood flow can be measured using the PET radiopharmaceuticals water

TABLE 12-4 PET Radiopharmaceuticals with USP Monographs

Radiopharmaceutical	Physical Half-Life	Physiologic Parameter
Cardiovascular Tracers		
Water O 15 injection	2 min	Perfusion
Ammonia N 13 injection	10 min	Perfusion
Rubidium Rb 82 injection	76 sec	Perfusion
Carbon monoxide C 11	20 min	Blood volume
Metabolic Tracers		
Fludeoxyglucose F 18 injection	110 min	Glucose utilization
Sodium fluoride F 18 injection	110 min	Osteogenic activity
Sodium acetate C 11 injection	20 min	Metabolic oxidation
Neurotransmission Tracers		
Fluorodopa F 18 injection	110 min	Dopa decarboxylase
Raclopride C 11 injection	20 min	D_2 receptors
Flumazenil C 11 injection	20 min	BZD receptors

O 15 injection, ammonia N 13 injection, and rubidium chloride Rb 82 injection. Each of these tracers has its individual advantages for measuring tissue perfusion by PET. Blood volume is another cardiovascular parameter that can be determined by PET imaging, with carbon monoxide C 11.

Blood Flow

Tissue perfusion was the first PET measurement to be instituted for routine clinical use. The impetus for this application is the high-resolution images and sensitivity of PET together with the prevalence of ischemic conditions in the heart and brain. Blood flow is essential for maintenance of adequate tissue oxygenation and nutrition, and suboptimal perfusion is a major cause of pathology and therapeutic failure. From the imaging perspective, perfusion is of fundamental importance because it affects delivery of tracer to the region of interest. Indications for PET perfusion measurements include evaluation of coronary artery disease, cerebrovascular defects, and organ perfusion in conjunction with PET measurement of other parameters.

Necessary characteristics of a PET perfusion tracer are that the radiopharmaceutical be deposited in tissues in proportion to blood flow and that the localized radioactivity within the region be retained for the duration of the scan. Tissue accumulation of the flow tracer can be caused by diverse mechanisms, including passive diffusion and active transport, and retention of radioactivity may be caused by dilutional effects or binding to intracellular constituents.

^{15}O-water is the most versatile perfusion tracer available. This radiopharmaceutical is freely diffusible, and tissue extraction does not substantially decrease at high flow rates. Tissue perfusion in the brain and heart can be quantified by using tracer kinetic modeling of PET data derived with ^{15}O-water. Since the physical half-life of this tracer is only 2 minutes and the turnaround time for radiopharmaceutical production is short, multiple PET measurements can be made on a single patient during a single imaging session. This is a great advantage when repeat studies are required, such as in the evaluation of resting and stressed myocardial blood flow or measurement of cerebral activation. ^{15}O-water has found widespread application in PET measurement of cerebral perfusion and is the only tracer that quantitatively measures myocardial blood flow. A disadvantage of this tracer for clinical perfusion studies of the heart is that clearance of activity from the blood is slow compared with the half-life of ^{15}O. For this reason, when myocardial perfusion measurements are undertaken, it is necessary to correct for activity in the heart chambers. Although this is readily accomplished using ^{15}O-carbon monoxide, it is an inconvenience because it necessitates the administration of a second PET tracer, and the subtraction images used to delineate myocardium tend to be statistically noisy. For this reason, alternative PET tracers may be preferred for clinical assessment of myocardial ischemia.

Ammonia N 13 injection (^{13}N-ammonia) is favored as a tracer for PET measurement of myocardial perfusion. This tracer is partially extracted, meaning that the extraction fraction decreases at high perfusion rates. The trapping mechanism for ^{13}N-ammonia involves incorporation of the radionuclide into glutamine because of the enzymatic action of glutamine synthetase. Myocardial perfusion is difficult to quantify with this tracer because of the tracer's variable extraction. However, since most clinical measurements of myocardial perfusion require only a qualitative assessment of heart perfusion, in which an ischemic zone is compared with healthy myocardium, absolute quantification of myocardial perfusion is not a prerequisite for clinical utility. A major advantage of ^{13}N-ammonia in PET imaging of myocardial perfusion is that the relatively long physical half-life (10 minutes) of the radionuclide facilitates clearance of blood background, and very high quality PET perfusion images are attained without need for an image subtraction procedure. On the other hand, the half-life of ^{13}N also creates difficulties for clinical scheduling, since repeated studies (as in rest–stress perfusion protocols) are not possible in rapid succession, in contrast to the use of perfusion tracers that have shorter half-lives.

^{82}Rb-rubidium chloride was the first radiopharmaceutical applied for clinical PET and is the only PET tracer used clinically that is available from a generator system. Rubidium ion is an analogue of potassium and accumulates in myocardium via active transport. As with ^{15}O-water, rapid sequential perfusion measurements can be made, and imaging protocols that involve rest–stress conditions can be scheduled with minimal difficulty. Like ^{13}N-ammonia, ^{82}Rb-rubidium chloride is partially extracted, which means that quantification of myocardial perfusion is difficult. In addition, the short physical half-life of the nuclide (76 seconds) is inadequate for sufficient clearance of blood background activity from the heart chambers. Attempts have been made to correct for blood background activity for continuous-infusion or bolus-injection protocols, but the resulting data are not as easy to analyze as for longer-lived tracers. Despite these limitations, a major attraction of ^{82}Rb-rubidium chloride for clinical application is that it can be used in the absence of a cyclotron and capital investment costs are thus lower than for cyclotron-produced PET radiopharmaceuticals.

Blood Volume

Assessment of blood volume with PET is often useful because it facilitates correction of other PET data for regional disparities in blood volume. This adjunctive PET measurement can be applied to the raw data for tracers that localize via other mechanisms, providing a cleaner signal for the process of interest. A good example is correction for blood volume in flow or metabolic measurements of the heart, since the heart chambers occupy such a large fraction of the total myocardial volume.

Radiolabeled carbon monoxide is the tracer of choice for accomplishing PET blood-volume measurement. The positron-emitting gas is inhaled through a filtered mouthpiece, and the tracer rapidly binds in vivo to hemoglobin to form carboxyhemoglobin. Carboxyhemoglobin remains trapped within the plasma compartment, so the radioactivity detected by PET within a region of interest represents the blood volume for that region. Carbon monoxide C 11 can be used as a PET tracer for the measurement of blood volume. Although not listed in *USP 32/NF 27*, ^{15}O-carbon monoxide may be a more appropriate tracer for measurement of blood volume in the clinical setting. The shorter half-life of the ^{15}O label facilitates rapid PET measurement of blood volume, creates minimal scheduling difficulty, and results in a decreased radiation burden to the patient. It is thus more prudent to use ^{15}O-carbon monoxide as an adjunctive PET measurement when imaging other PET tracers than to add the longer-lived carbon monoxide C 11 to an imaging protocol.

Metabolic Processes

The radiopharmaceuticals indicated for clinical measurement of metabolic processes are fludeoxyglucose F 18 injection (^{18}F-FDG or ^{18}FDG), sodium fluoride F 18 injection (^{18}F-fluoride), and sodium acetate C 11 injection (^{11}C-acetate). The first is widely used for assessment of glucose utilization, and the third has found predominant application for evaluation of tissue oxidative metabolism, especially in the myocardium. Sodium fluoride F 18 injection is used for PET assessment of osteogenic activity.

Glucose Utilization

The predominant radiopharmaceutical for clinical PET is the metabolic tracer ^{18}F-FDG. Several favorable characteristics promote its widespread use for noninvasive assessment of human pathology. These include multiple and diverse indications for use in PET imaging, the facile manner of drug preparation, and the convenient half-life of the radiolabel.

Glucose is a nutrient used by most tissues of the body, so evaluation of regional glucose utilization in vivo by PET is clinically important for assessing the vitality of organs as well as for monitoring therapeutic response to interventions. ^{18}F-FDG is thus indicated for several different applications, including PET imaging of the brain, heart, and various neoplasms.

As shown in Figure 12-6, the unique biologic property of ^{18}F-FDG that promotes its use as a radiopharmaceutical is its metabolic trapping in vivo. The radiolabeled glucose derivative participates in the initial steps of glucose metabolism (carrier-mediated transport and phosphorylation at the 6 position by the enzyme hexokinase), so the radiopharmaceutical faithfully traces aerobic or anaerobic flux. However, after these initial biochemical steps, the phosphorylated intermediate does not undergo further metabolism to glycogen or to carbon dioxide, nor is it effectively degraded by glucose 6-phosphatase. Thus, tissues accumulate radioactivity in proportion to their glucose utilization rate, and there is no complicating redistribution of radiometabolites that would weaken the image contrast of tracer accumulation. In clinical PET, images of glucose utilization are typically acquired 45 to 60 minutes after the intravenous injection of ^{18}F-FDG.

Osteogenic Activity

Osteogenic activity is measured using sodium fluoride F 18 injection. Bone remodeling involves a balance between osteoclastic and osteoblastic activity. Fluoride ions diffuse from the blood compartment and exchange with hydroxyl groups in hydroxyapatite crystal to form fluoroapatite. Uptake of fluoride into bone may be due to increased regional blood flow and bone turnover. Sodium fluoride F 18 injection is an ideal radiopharmaceutical for measurement of this process, because it is characterized by rapid bone uptake, rapid clearance from the blood compartment (so bone/blood ratios are favorable), and a convenient half-life. Although localization of fluoride to bone may occur in nonmalignant orthopedic conditions and other benign pathologies, the primary clinical interest in sodium fluoride F 18 injection is for evaluation of bone metastases in oncology patients.

Oxidative Metabolism

Another important metabolic measurement made by PET is oxidative metabolism. This is especially pertinent in cardiology, because oxidative metabolism is the major nutritional pathway for myocardium under aerobic conditions. Under these conditions, the primary source of energy for the healthy myocyte derives from fatty acids that enter the Krebs cycle and subsequently undergo beta-oxidation.

Sodium acetate C 11 injection is the PET radiopharmaceutical used to clinically assess oxidative metabolism in the heart. ^{11}C-acetate is the smallest fatty acid, and it enters the Krebs cycle in a substrate-independent manner to be metabolized to ^{11}C-carbon dioxide at a rate dependent on the oxidative capability of the heart (Figure 12-7). The viability of heart tissue can therefore be measured by PET determination of the

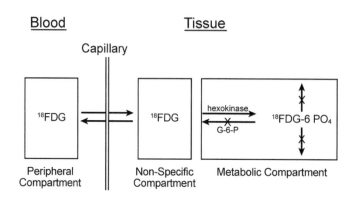

FIGURE 12-6 Metabolic trapping of ^{18}F-FDG. The tracer is phosphorylated by hexokinase but is not further metabolized and does not redistribute.

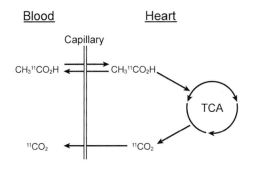

FIGURE 12-7 Oxidative metabolism of ^{11}C-acetate. In viable myocardial tissue, the tracer enters the Krebs (TCA) cycle, is converted to carbon dioxide, and clears from the heart.

clearance half-life of ^{11}C-acetate from myocardium after bolus injection of the radiopharmaceutical.

PET measurement of myocardial viability using ^{11}C-sodium acetate has clinical value. It is frequently used in concert with myocardial perfusion measurements to discriminate infarction (dead myocardium) from ischemia (underperfused tissue that may be rescued through appropriate medical intervention).

More sophisticated viability studies sometimes combine PET measurement of ^{11}C-acetate with a flow measurement as well as with PET measurement of glucose utilization using ^{18}F-FDG. In these studies, ischemic zones (flow tracer) are compared with decreases in oxidative metabolism (^{11}C-sodium acetate) or increases in glucose utilization (^{18}F-FDG), the latter of which is up-regulated as anaerobic glycolysis becomes a more important source of energy for the ischemic myocytes. Note that infarcted myocardium will metabolize neither ^{11}C-acetate nor ^{18}F-FDG.

In addition to the primary indication for ^{11}C-sodium acetate in clinical PET myocardial viability measurements, there are off-label uses for this PET radiopharmaceutical (Table 12-5). These uses were identified through serendipity, and their discovery suggests that there may be other indications for alternative PET radiopharmaceuticals that have not been identified and exploited.

^{11}C-sodium acetate has also proven useful as a myocardial perfusion tracer.[7,8] Soon after bolus injection, the tracer has kinetics similar to that of a partially extracted perfusion tracer. It may streamline clinical PET protocols if the same tracer (^{11}C-sodium acetate) is used for both flow and viability measurements in patients scheduled for myocardial viability studies, rather than adding a second flow tracer to the imaging session. An added benefit of using one tracer for two measurements is that the patient is spared the additional radiation dose from a second tracer.

^{11}C-sodium acetate is also useful for PET imaging of prostate cancer[9] and hepatocellular carcinoma.[10] The development of PET radiopharmaceuticals for assessment of prostate cancer has proven to be challenging; the location of the gland near the bladder makes discrimination difficult for tracers that clear through the kidneys and urinary tract. For similar reasons, tracers that clear by the hepatic route are difficult to use for discrimination of liver tumors. ^{11}C-acetate is eliminated as ^{11}C-carbon dioxide from the lungs, so there is neither bladder nor liver accumulation of activity, in contrast to the use of ^{18}F-FDG. Localization of ^{11}C-acetate within tumor sites is believed to be due to participation in lipid synthesis within lesions.

Neurotransmission Imaging

The third broad category of clinical PET applications is the noninvasive assessment of neurotransmission. Neuronal communication is a complicated process that involves the synthesis and release of neurotransmitters that diffuse across the synaptic cleft to bind to and activate specific receptors on postsynaptic nerve cells. Myriad neurotransmitter pathways in the brain interact in an inhibitory or excitatory fashion. Intensive worldwide research is devoted to gaining insight into the complicated manner in which these pathways interact in health and disease.

Some of the results of PET neuroscience research have been translated to clinical nuclear medicine. Three PET radiopharmaceuticals have demonstrated clinical utility (Table 12-4). Two of these, fluorodopa F 18 injection (^{18}F-FD) and raclopride C 11 injection (^{11}C-RAC), are used in the assessment of the dopaminergic nervous system. Dopamine neurotransmission is affected in degenerative diseases like Parkinson's and Huntington's, and dopamine receptors are the site of action of antipsychotic drugs.

The third PET tracer used for neurotransmission imaging is flumazenil C 11 injection (^{11}C-FMZ). Flumazenil binds to benzodiazepine (BZD) receptors in the brain. Although these drug-binding sites are not bona fide neurotransmitter receptors, BZD receptors are allosterically linked to receptors for the inhibitor neurotransmitter gamma aminobutyric acid (GABA). PET tracers for imaging GABA receptor activity are not currently available, but radiopharmaceuticals that bind to BZD receptors give in vivo data that are indicative of GABA activity.

Enzyme Activity

^{18}F-FD is useful for PET assessment of a number of cerebral dopamine neurons. As illustrated in Figure 12-8, the tracer follows the enzymatic pathway that converts levodopa to dopamine. ^{18}F-FD is used rather than an ^{18}F-labeled derivative of dopamine because dopamine is a charged species that does not partition across the blood–brain barrier to gain access to dopamine neurons in the brain. Within the brain, ^{18}F-FD is acted upon by aromatic amino acid decarboxylase to form

TABLE 12-5 Off-Label Applications of PET Radiopharmaceuticals

Radiopharmaceutical	Application
Sodium acetate C 11 injection	Prostate cancer Hepatocellular carcinoma Myocardial perfusion
Fluorodopa F 18 injection	Pheochromocytoma Thyroid cancer

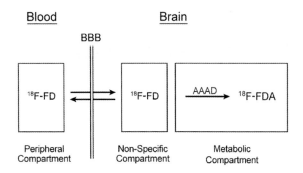

FIGURE 12-8 In vivo localization of 6-^{18}F-fluorodopa (^{18}F-FD) within dopaminergic neurons. The tracer is decarboxylated by cellular aromatic amino acid decarboxylase (AAAD) to form fluorodopamine (^{18}F-FDA), which remains within the neuron.

^{18}F-fluorodopamine and accumulates within dopaminergic neurons. There is selective localization within the basal ganglia of the brain, the area that controls movement. In degenerative diseases like Parkinson's disease, there is loss of dopaminergic neurons, so the level of accumulation of activity in the basal ganglia is less than in healthy, age-matched controls. The decrease in ^{18}F-FD localization in dopamine-rich areas of the brain has been shown to correlate with disease severity.

Like sodium acetate C 11 injection, ^{18}F-FD has off-label applications (Table 12-5). It has been found to accumulate in vivo within tumors and has proven useful for PET evaluation of pheochromocytoma and thyroid carcinoma.[11,12] The localization mechanism is believed to be due to increased amino acid utilization by the cancerous lesions.

Neuroreceptor Binding

Both ^{11}C-RAC and ^{11}C-FMZ are examples of receptor-binding radiopharmaceuticals. Receptor-binding PET radiopharmaceuticals have high affinity and high specific activity so that primarily receptor-bound localization is detected in the PET image (Figure 12-9).

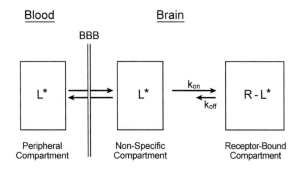

FIGURE 12-9 In vivo localization of radioligands to receptor sites. The radioligand (L*) must pass the blood–brain barrier (BBB) and have high affinity for binding to the receptor (R) and relatively low nonspecific binding.

The in vivo localization of ^{11}C-RAC is anatomically similar to that of ^{18}F-FD, but the two tracers measure two different aspects of the same neurotransmitter system. ^{11}C-RAC binds to postsynaptic dopamine D_2 receptors with moderately high affinity. There are few D_2 receptors, so ^{11}C-RAC must be of very high specific activity (>500 Ci/mmol) to ensure that only a small percentage of the binding sites are occupied. Generally, D_2 receptors are up-regulated when dopamine levels in the synapse are low (as in Parkinson's disease) and down-regulated by high levels of the neurotransmitter (as in psychosis), but this relationship can be altered by receptor dysfunction. The major application of ^{11}C-RAC stems from the fact that the radioligand is displaceable by endogenous dopamine. ^{11}C-RAC binding is measured before and after an intervention that induces dopamine release, and the decrease in D_2 binding after the intervention is an index of the functional capability of the dopaminergic system. These receptor-displacement studies have been used to evaluate drug potency in vivo and to measure the effect of stimuli on dopamine release, as well as to assess the residual function of dopaminergic neurons in degenerative states.

^{11}C-FMZ binds with high affinity to BZD receptors in the brain, which are ubiquitous in cerebral tissue but are especially concentrated in the cortex. The link between the inhibitory neurotransmitter GABA and BZD binding underlies the clinical utility of ^{11}C-FMZ. Because there is decreased inhibitory GABA innervation in epileptic foci, binding of ^{11}C-FMZ is also reduced at these tissue sites. This has clinical utility in imaging the foci of intractable seizures, for which surgical intervention may be considered. Although there is also decreased glucose utilization and thus decreased accumulation of ^{18}F-FDG in the region of the epileptic focus during the interictal period, the size of the region of decreased glucose utilization exceeds the more discriminatory area of decreased ^{11}C-FMZ binding. Thus, the application of ^{11}C-FMZ in PET can identify tissue that is possibly healthy, permitting a more conservative approach to removal of brain tissue.

REIMBURSABLE PET PROCEDURES

Reimbursement for imaging procedures is a very important issue in the advancement of PET as a clinical tool. Only when PET procedures are reimbursed is it financially feasible to apply them on a widespread basis for patient care. The key decision maker about reimbursement is the federal government's Centers for Medicare and Medicaid Services (CMS), since most third-party payers follow CMS reimbursement decisions. Although CMS uses private companies as local contractors to process and pay Medicare claims, its coverage decisions apply nationwide and specify coverage instruction, billing codes, and effective dates for payment.

Of the PET imaging procedures that have been used in research, only a select few have been approved for CMS reimbursement. This approval is an important step in technology transfer, because procedures are approved only after careful

TABLE 12-6 PET Cardiology Procedures with Approval for Coverage by CMS[a]

Pathology	Tracer	Indication
Myocardial viability	^{18}F-FDG	Assessment prior to revascularization following inconclusive SPECT study
Myocardial perfusion	^{13}N-ammonia	In place of, or after inconclusive, SPECT study
	^{82}Rb-rubidium chloride	In place of, or after inconclusive, SPECT study

SPECT = single-photon emission computed tomography.
[a] Procedures for which the Centers for Medicare and Medicaid Services has issued a national coverage decision.

scrutiny of the scientific evidence that evaluates the effectiveness and clinical benefit of the particular imaging procedure. Professional organizations such as the Society of Nuclear Medicine (SNM) and the Academy of Molecular Imaging (AMI) play an important role by lobbying and providing expert information to CMS for approval of new PET procedures. Once a procedure is deemed worthy of CMS reimbursement, CMS issues a national coverage decision (NCD), which describes in detail conditions under which the PET procedure will be covered. After an NCD is issued the procedure can move from the specialized realm of medical research, with its limited number of subjects funded by a granting agency for a specific study, to the health care arena, where PET imaging can be performed routinely to benefit the population at large. Only those PET procedures that have been shown to have advantages over existing clinical test methods are granted NCD status. CMS-covered PET procedures are used in cardiology, neurology, and oncology. These NCD-approved PET studies are summarized in Tables 12-6 through 12-8.

The first procedure to be granted NCD status, in 1995, was myocardial perfusion studies with rubidium chloride Rb 82 injection (Rb-PET) (Table 12-6). Several years later (in 2001), five indications for ^{18}F-FDG in PET imaging (FDG-PET) were added. The devices used in these approved procedures included both dedicated (full- or partial-ring) PET scanners and hybrid (coincidence) gamma cameras, because the data reviewed for approval were acquired using either type

TABLE 12-7 FDG-PET Procedures in Neurology with Approval for Coverage by CMS[a]

Pathology	Indication
Refractory seizures	Presurgical evaluation of seizure foci
Dementia	Differentiation of frontotemporal dementia from Alzheimer's disease

[a] Procedures for which the Centers for Medicare and Medicaid Services has issued a national coverage decision.

TABLE 12-8 FDG-PET Procedures in Oncology with Approval for Coverage by CMS[a]

	Treatment Strategy	
Tumor Type	Initial	Subsequent
Colorectal	C	C
Esophagus	C	C
Head and neck	C	C
Lymphoma	C	C
Non–small-cell lung	C	C
Ovary	C	C
Myeloma	C	C
Breast	C	C
Melanoma	C	C
Cervix	C or CED	C
Brain	C	CED
Thyroid	C	C or CED
Small-cell lung	C	CED
Soft-tissue sarcoma	C	CED
Pancreas	C	CED
Testes	C	CED
Prostate	NC	CED
All other solid tumors	C	CED
All other cancers	CED	CED

C = covered; NC = noncovered; CED = coverage with evidence development.
[a] Procedures for which the Centers for Medicare and Medicaid Services has issued a national coverage decision.

of scanner. The more recent expansion of clinical PET into further indications for FDG-PET and PET imaging with ^{13}N-ammonia (NH$_3$-PET) is based on regional or whole-body imaging data acquired only with dedicated PET scanners. The data show the superiority of these devices for the specific clinical applications and reflect the improvement in PET instrumentation over time. Thus, recent NCDs are for PET procedures that use dedicated PET or PET/CT scanners, since these represent state of-the-art instrumentation in current clinical use.

It is noteworthy that Tables 12-6 through 12-8 do not include the application of PET for imaging of infection. CMS has been petitioned for coverage of FDG-PET imaging of infection, including osteomyelitis, infection of hip arthroplasty, and fever of unknown origin. However, CMS has not found the current data for imaging infection to be compelling and thus has denied coverage. Additional supporting evidence needs to be generated before NCDs for these indications can be issued.

Cardiology

CMS provides coverage for specific PET applications in cardiology (Table 12-6). There are two reimbursable indications for PET imaging of the heart. These NCDs are myocardial viability measurements made using FDG-PET and myocardial perfusion assessment using NH_3-PET or Rb-PET.

Identification of deficient heart muscle movement in hibernating or stunned myocardium is important for the determination of appropriateness for revascularization. FDG-PET distinguishes between dysfunctional but viable myocardial tissue and infarcted scar tissue. This knowledge provides preoperative prognostic information and thereby affects decisions about the management of patients with ischemic cardiomyopathy and left ventricular dysfunction.

FDG-PET is covered for the determination of myocardial viability as a primary or initial assessment prior to revascularization. Reimbursement is also provided if the PET study is performed following an inconclusive SPECT study. Because of the original approval date, imaging with either full or partial ring scanners is reimbursable. Note that whereas an inconclusive SPECT study may be followed with an FDG-PET study, if an FDG-PET study is initially performed and found to be inconclusive, a follow-up SPECT study is not covered.

NCDs also exist for PET myocardial perfusion measurements. Perfusion measurements performed at rest or with pharmacologic stress are covered for diagnosis and management of patients with known or suspected coronary artery disease. The PET radiopharmaceuticals approved for this indication are generator-produced rubidium chloride Rb 82 injection and cyclotron-produced ammonia N 13 injection. Rb-PET was the first PET procedure to be approved by CMS, and the addition of an NCD for NH_3-PET was a benefit to imaging centers with cyclotrons, since it made possible both perfusion and viability studies using cyclotron-produced radiopharmaceuticals only. Thus, PET myocardial perfusion measurements can be performed without the additional capital expense of generator technology for Rb-PET.

PET myocardial perfusion imaging (either rest alone, or rest plus stress) is covered in place of, but not in addition to, SPECT. However, NH_3-PET or Rb-PET is permitted after an inconclusive SPECT study when it is necessary for the determination of appropriate medical or surgical intervention in a specific patient.

Neurology

In the neurological arena, there are two covered indications for FDG-PET (Table 12-7). Medicare reimburses FDG-PET scans for differential diagnosis of frontotemporal dementia (FTD) from Alzheimer's disease (AD). This NCD is approved under specific requirements or as part of a CMS-approved clinical trial. The specific requirements include suspicion of FTD, comprehensive clinical evaluation of the patient that rules out other neurodegenerative causes, and lack of prior SPECT or PET scan for the same indication.

FDG-PET is not approved for diagnosis of either FTD or AD when there is no ambiguity between the two. CMS has also denied coverage for FDG-PET imaging of patients with a presumptive diagnosis of dementia-causing neurogenerative disease, including possible AD, clinically typical FTD, dementia of Lewy bodies, or Creutzfeldt-Jakob disease.

The other neurological application of FDG-PET that has NCD approval is for use in patients with epilepsy. PET imaging of fludeoxyglucose F 18 injection is used to localize the seizure foci in patients with intractable complex seizure disorders. PET studies of refractory seizures are covered for presurgical cases only.

Oncology

The greatest number of CMS-covered PET applications are in the field of oncology. A cursory view of Tables 12-6 through 12-8 shows that the explosive growth of clinical PET is predominantly due to applications of FDG-PET for tumor imaging. Although there are valuable applications of PET in neurology and cardiology, the numbers of clinically relevant procedures in these areas do not compare to the large number of NCDs issued for FDG-PET of oncology cases (Table 12-8).

The rapid growth of PET imaging in oncology has led to a novel approach by CMS to evaluating coverage for tumor imaging. The agency has taken a new approach to gathering evidence for technology assessment. In the past, to evaluate PET procedures for oncology and other specialties, CMS reviewed data acquired through studies with specific designs that examined relatively small populations of similar subjects within academic research centers.

With this approach, CMS was able to issue NCDs for only nine cancer types: breast, cervical, colorectal, esophageal, head, neck (including thyroid), and non–small-cell lung cancers and melanoma and lymphoma. Coverage for these indications was decided on the basis of a cumbersome method in which data were gathered on a cancer-by-cancer basis. These evaluations were of no use in assessing other tumor types, and additional studies were needed before NCDs could be issued for the remaining, noncovered cancers. Such studies were unlikely to take place for cancers that occur with low frequency, since a traditional clinical trial would probably not achieve the sample size needed for statistical power with only a small number of study sites.

Because of these limitations in acquiring the data necessary for NCDs, CMS embarked on a new payment program called coverage with evidence development (CED). With this transition, CMS moved from being a passive consumer of evidence to an active supporter of evidence development. A major reason for this change was the lack of relevance of many clinical trials that enroll subjects who differ substantially from the Medicare population (i.e., elderly patients with multiple co-morbidities). CED enables Medicare to specifically set study requirements so that the design and implementation of clinical research is more relevant to its mission. The ability to develop its own evidence in this manner enables CMS to make appropriate reimbursement decisions.

CED was applied to PET in 2005, with the establishment of the National Oncologic PET Registry (NOPR).[13] The goal of the NOPR was to acquire data on whether FDG-PET changed intended patient management strategies and whether other tests or procedures could be avoided. To facilitate data acquisition, CMS extended coverage of FDG-PET for essentially all cancers and indications as long as the requirements of the NOPR were met. Requirements for reimbursement are that physician-completed prestudy and poststudy forms be submitted together with the FDG-PET scan report.

There are several strengths to the NOPR model for CED. A very large sample population that included virtually every cancer type was rapidly assembled. Multiple practice-based PET imaging sites were used in these studies, so the external validity of the findings was ensured. Moreover, the elderly subjects who were enrolled represented a diverse group of patients with multiple co-morbidities, typical of the Medicare population. Most important, the NOPR is a mechanism whereby physicians and patients can participate in CED and benefit from the use of FDG-PET while data are being acquired for CMS.

Preliminary findings of the NOPR are remarkable. Physicians' self-reported changes in patient management occurred 36.5% of the time across all cancer types and indications.[14,15] In most cases, the extent of cancer was judged to be of a higher stage after FDG-PET scanning. Although the registry reported intended management rather than actual care delivered, and it is not yet known if FDG-PET affects long-term clinical outcomes, the preliminary data convincingly demonstrate the value of PET imaging to clinical oncology. Also, the effect of FDG-PET on clinical outcomes is unknown. The NOPR serves as a model for future CMS assessment of new technology.[16,17]

The evidence gathered by the NOPR has compelled CMS to expand FDG-PET coverage to a broad range of cancers. As shown in Table 12-8, FDG-PET studies are now reimbursable for several types of tumors beyond the initial nine types. CMS decided that the aggregate evidence was sufficient for coverage of FDG-PET imaging for the initial diagnosis and staging of all previously uncovered cancer types. However, the value of FDG-PET for monitoring response to therapy and restaging is still uncertain and will be reimbursed only through CED policy.

In 2009, CMS adopted a coverage framework that differentiates FDG-PET into procedures used to inform the initial antitumor therapy and those that guide subsequent treatment strategies (Table 12-8). This change applies to FDG-PET for all cancer types and replaces the previous four-part categorization into diagnosis, staging, restaging, and monitoring response to treatment. Under the current framework, diagnosis and staging fall under the category of initial treatment strategy, and restaging and monitoring response to treatment are categorized as subsequent treatment strategy.

CMS will cover one FDG-PET scan for patients with solid tumors that are biopsy proven or strongly suspected when the physician needs to determine the location or extent of tumors for one of three reasons. The first is whether or not the patient is a candidate for an invasive diagnostic or therapeutic procedure. The second is to determine the optimal anatomic location for an invasive procedure. The third reason is to determine the anatomic extent of a tumor when it affects the selection of antitumor therapy.

CMS will not automatically cover FDG-PET imaging for subsequent antitumor treatment strategies except for the specific tumor types listed in Table 12-8, most of which were evaluated prior to initiation of CED and NOPR. Coverage of FDG-PET for subsequent treatment strategy in the remaining cancer types is still possible under CED, however. Note that prostate cancer is noncovered for initial treatment strategy but that reimbursement for subsequent treatment strategy is permitted under CED.

RADIONUCLIDE PRODUCTION

Positron-emitting radionuclides lie on the proton-rich side of the line of beta stability.[1] For this reason, they are usually produced by accelerators that bombard targets with charged particles, rather than by using nuclear reactors as a source of neutrons to irradiate targets. The positron-emitting nuclides commonly used as radiolabels in clinical PET are listed in Table 12-1. Four of these cyclotron-produced radionuclides are isotopes of elements (oxygen, nitrogen, carbon, fluorine) that are frequently found in biochemical and drug structures, have useful half-lives that range from 2 minutes (^{15}O) to 110 minutes (^{18}F), and are incorporated into radiopharmaceutical structures via covalent bonds. These physicochemical characteristics explain the widespread use of cyclotron-produced nuclides for labeling PET radiopharmaceuticals.

Two of the positron-emitting nuclides available from generator systems, ^{82}Rb and ^{68}Ga, do not require a cyclotron for production. Compared with the cyclotron-produced nuclides, however, the properties of these nuclides are much less versatile for labeling drug structures for PET imaging. ^{82}Rb is used solely as a perfusion tracer, and ^{68}Ga is most often used in clinical PET simply as a source of photons for transmission scans.

Radionuclide Generators

Compared with cyclotron-prepared radiopharmaceuticals, PET radiopharmaceuticals available from generators (Table 12-9) are much less commonly used in clinical practice. However, for the sake of completeness, their use in clinical PET imaging is discussed briefly.

The ^{82}Rb generator was the first system authorized for reimbursement for clinical PET imaging (Table 12-6). The commercially available generator system consists of the parent isotope ^{82}Sr bound to a hydrous stannic oxide column and is eluted with 0.9% sodium chloride injection to isolate the ^{82}Rb daughter (Figure 12-10).[18] The versatility of the generator system is limited by the short physical half-lives of the parent nuclide and the daughter radioisotope (Table 12-9). ^{82}Sr decays by EC with a half-life of 25.6 days, which limits the useful lifetime of the generator system to only a few weeks. The very short 76 second half-life of the rubidium daughter, although useful for repeat studies in PET, prevents any coordination chemistry from being performed with the nuclide prior to patient administration.

TABLE 12-9 Generator-Produced PET Radionuclides

Nuclide	Half-Life	E_{max} (MeV)	Parent Isotope (half-life)
^{82}Rb	76 sec	3.35	^{82}Sr (25.6 days)
^{68}Ga	68 min	1.90	^{68}Ge (271 days)

The generator system is thus limited to the role of a source of ^{82}Rb infusate for patients undergoing cardiovascular PET measurements. The generator eluate is passed through a dosimeter and sterilizing filter and directly into the patient for myocardial blood flow measurements (Figure 12-10). A microprocessor-controlled syringe pump determines the elution profile so that either bolus or continuous infusion can be used.[18] Because of the short half-life of the radioisotope, the actual batch of ^{82}Rb injection used does not undergo any quality control testing before administration to the patient. However, the generator elution performance is carefully assessed immediately before the PET study to confirm proper functioning of the device; the total radioactivity that is eluted, the elution profile, and potential breakthrough of ^{82}Sr are measured.

Although the physical characteristics of the ^{68}Ge–^{68}Ga generator appear to be ideally suited for clinical applications, to date the use of ^{68}Ga in PET imaging has been limited to research. Nevertheless, there is great potential for future development of this generator as a source of positron-emitting radiopharmaceuticals. ^{68}Ge decays by EC with a half-life of 271 days, which translates to a useful lifetime of approximately 1 year for the generator. In addition, the 68 minute half-life of the daughter confers substantial versatility to radiopharmaceutical labeling or PET imaging procedures with ^{68}Ga. A commercially available ^{68}Ga generator system consists of the ^{68}Ge parent bound to a column of stannous oxide and is eluted with 1 N hydrochloric acid.[19] The generator is characterized by high recovery of ^{68}Ga daughter and low breakthrough of ^{68}Ge parent.

The most common application of the ^{68}Ge–^{68}Ga decay system is not for radiopharmaceutical production but for instrumentation use. The generator system is often used on standalone PET devices as a source of 511 keV photons needed for transmission measurements and attenuation correction of emission scans. In this case, the daughter ^{68}Ga is not eluted from the generator column but instead is merely allowed to decay on the rods in situ. The long half-life of the ^{68}Ge parent is a major advantage for this application, because the source rods have to be replaced infrequently.

Charged Particle Accelerators

Most PET radionuclides are produced with cyclotrons, devices designed by physicists to accelerate charged particles. As illustrated in schematic form in Figure 12-11, the operation of the cyclotron is based on the concept that repeated application of small accelerating voltages to ions will accelerate those ions to high velocities. These accelerated particles will be constrained to circular pathways by the presence of a magnetic field induced by magnetic poles situated above and below the accelerating electrodes. As the particles pass the interface between the hollow electrodes (called "dees" because of their shape), the difference in electrical potential (controlled by a radio-frequency oscillator) imparts additional kinetic energy to the ion. As the ions increase in velocity, their radius of rotation

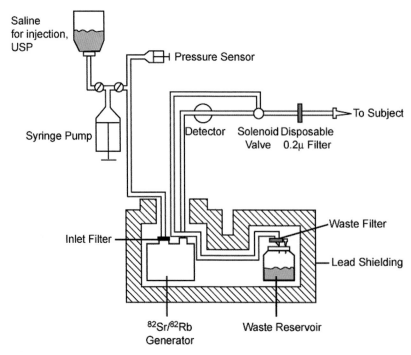

FIGURE 12-10 Diagram of the ^{82}Sr–^{82}Rb generator system. This is the only generator system used for clinical PET imaging.

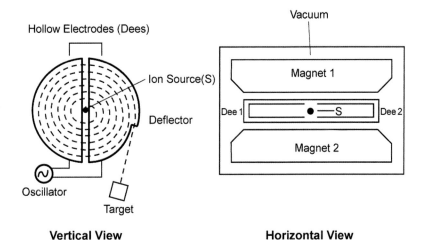

FIGURE 12-11 Diagram of cyclotron acceleration of particle beams for PET radionuclide production. Accelerated particles can be either positive or negative ions.

increases until they achieve their terminal velocity, which is determined by the engineering parameters designed into the cyclotron. The beam of high-energy charged particles is then deflected away from the magnetically constrained pathway of the cyclotron to impinge upon the target. The irradiation of the target material by the cyclotron beam produces, by nuclear reaction, the radionuclides used for labeling PET radiopharmaceuticals.

Clearly, more details are involved in actual radionuclide production than are included in this simplistic description; readers are referred to more comprehensive information.[20–22] Cyclotrons are designed to many different specifications depending on their nuclear applications, and those used for medical radioisotope production generally have lower beam energy and higher beam current than corresponding machines used for physics research. Higher beam currents are desirable because the amount of radionuclide produced is proportional to the number of nuclear reactions that are induced by the beam. Also, the threshold energies for the nuclear reactions used to produce positron-emitting nuclides are relatively low compared with those used in particle physics research.

Accelerators for the production of radionuclides have been classified into four levels (Table 12-10) based on the particles accelerated and their energies.[22] The higher the energy of the accelerator, the more expensive are the capital installation and subsequent operating costs. For most in-hospital production of PET radionuclides, either level I or level II machines are used; the energies of these cyclotrons are adequate for the production of needed quantities of positron emitters for PET. Level III or IV accelerators are used to produce less common radionuclides used in nuclear medicine procedures, but the radionuclides used routinely for labeling PET radiopharmaceuticals (Table 12-1) do not require the use of such large devices.

Most modern PET-based cyclotrons use negative-ion acceleration technology, although several older positive-ion machines are still in current use in the PET community. The advantage of negative-ion machines is that substantially less radiation shielding is required, so installation costs are reduced.

In addition, beam extraction is greatly simplified with negative-ion machines. With negative-ion cyclotrons, a thin graphite foil is placed in the beam pathway, so the ions are stripped of their negative charge and converted to positive ions. According to the laws of electromagnetism, the positively charged particles will rotate in the direction opposite that of the original beam of negative ions, so they will impinge upon targets mounted on appropriately situated exit ports. By placing the graphite foils at different radii of the cyclotron, the energy of the beam hitting the target can be selected. Also, by adjusting the depth of foil insertion, either a fraction or the entire beam can be stripped, enabling adjustment of beam intensity so that it hits a single target or irradiates multiple targets simultaneously. This type of flexibility is a great asset when radionuclide production is scheduled for busy PET clinics.

Cyclotron Targetry

The selection of targets to be irradiated in cyclotrons for isotope production is an area that has undergone intensive research and optimization.[21,23] Cyclotron targets can be gaseous, liquid, or solid, but those used for producing routine PET radionuclides are gaseous or liquid, and their loading or unloading can be achieved by application of a pressure differential. A caution

TABLE 12-10 Classification of Particle Accelerators

Level I	Single particle (usually p or d) $E_p \leq 10$ MeV
Level II	Single or multiple particle (usually p and/or d) $E_p \leq 20$ MeV
Level III	Single or multiple particle (usually p) $E_p \leq 0$ MeV
Level IV	Single or multiple particle (usually p) E_p 70–500 MeV

TABLE 12-11 Cyclotron Production of PET Radionuclides

Nuclide	Half-Life	Reaction	Target	Target Product	Batch Yields
^{15}O	2 min	^{14}N(d,n)^{15}O	0.1% O_2/N_2	[^{15}O]O_2	2,000 mCi
		^{15}N(p,n)^{15}O			2,000 mCi
^{13}N	10 min	^{16}O(p,α)^{13}N	[^{16}O]water	[^{13}N]NH$_3$, [^{13}N]NO$_3^-$, [^{13}N]NO$_2^-$	450 mCi
^{11}C	20 min	^{14}N(p,α)^{11}C	0.5% O_2/N_2	[^{11}C]CO$_2$	1,000–3,000 mCi
^{18}F	110 min	^{18}O(p,n)^{18}F[a]	[^{18}O]water	[^{18}F]F$^-_{aq}$	2,000–10,000 mCi
		^{20}Ne(d,α)^{18}F[b]	0.5% F_2/^{20}Ne	[^{18}F]F$_2$	500–1,000 mCi

[a] ^{18}F produced as high specific activity fluoride ion (F$^-$).
[b] ^{18}F produced as moderate specific activity fluorine gas (F$_2$).

concerning the irradiation of liquid targets is that free radicals may be generated in situ. Such radicals can potentially interfere with the reactivity of the product radionuclide by creating species that interfere with subsequent radiopharmaceutical labeling reactions. However, by careful control of beam and target conditions during cyclotron irradiation, such deleterious effects can be minimized.

The nuclear reactions commonly used for radionuclide production in clinical PET centers are listed in Table 12-11. These reactions use incident particles of either protons or deuterons, which are accelerated to adequate energies in moderate beam intensities by level I or level II cyclotrons. The selection of the nuclear reaction to be used for radioisotope production is determined by the capability of the particle accelerator. More expensive multiparticle machines accelerate both deuterons and protons and have the advantage that inexpensive targets of naturally abundant nuclei can be used for radioisotope production. An exception to this generalization is the production of ^{18}F-fluoride, which requires isotopically enriched ^{18}O-water as a target, regardless of whether level I or level II accelerators are used.

Single-particle machines that produce only proton beams are less expensive than level II cyclotrons but have the disadvantage that they require expensive isotopically enriched target materials for the production of ^{15}O, since the ^{14}N(d,n)^{15}O nuclear reaction is not an option. However, for clinical PET centers that do not require heavy production of ^{15}O tracers, single-particle (level I) cyclotrons offer a relatively inexpensive means of radionuclide production.

Labeling Precursors

It is important to consider the chemical form of the radionuclide as it leaves the target, because this dictates the chemical reactivity and hence the synthetic reactions that can be used for radiopharmaceutical preparation. In some cases, target products can be immediately applied for radiopharmaceutical synthesis. However, it is more usual for the target product to be purified, either to remove chemicals that interfere with the labeling procedure or to convert the target product to alternative chemical forms that can be more easily incorporated into the molecular structure of the desired radiopharmaceutical. The degree of chemical manipulation is thus dictated by the spectrum of radiochemicals and chemicals present in the irradiated target and is further constrained by the half-life of the product radionuclide. The predominant labeling precursors for synthesis of PET radiopharmaceuticals are listed in Table 12-12.

Gaseous ^{15}O-Oxygen

As shown in Table 12-12, the target product for ^{15}O (which has a 2 minute half-life) is gaseous ^{15}O-oxygen ([^{15}O]O_2). Although ^{15}O-oxygen can be used as a PET radiopharmaceutical for measuring tissue oxygen consumption, it is more commonly used to synthesize alternative radiopharmaceutical structures, such as ^{15}O-water or ^{15}O-carbon monoxide, for imaging use. Thus, it is appropriate to consider ^{15}O-oxygen a labeling precursor for these more popular ^{15}O-labeled PET radiopharmaceuticals. Because of the short half-life of ^{15}O, the target product must be rapidly converted to desired chemical forms using on-line techniques. These production techniques are described below.

Oxo Anions of ^{13}N-Nitrogen

The 10 minute half-life of ^{13}N also limits the number of radiosynthetic steps that can be used to incorporate the nuclide into a labeled drug, and to date ammonia N 13 injection is the sole ^{13}N-labeled radiopharmaceutical of clinical significance. ^{13}N is produced in the irradiated target in a spectrum of chemical forms including ^{13}N-ammonia, ^{13}N-nitrate, and ^{13}N-nitrite. Instead of purifying the relatively modest fraction that is produced in the form of ^{13}N-ammonia, it is more efficient to convert all of the chemical forms of ^{13}N produced in the target to ^{13}N-ammonia. In this regard, ^{13}N-nitrate and ^{13}N-nitrite

TABLE 12-12 Radioactive Precursors for Labeling PET Radiopharmaceuticals

Radionuclide	Labeling Precursor
^{15}O	^{15}O-oxygen
^{13}N	^{13}N-nitrate, ^{13}N-nitrite
^{11}C	^{11}C-carbon dioxide
	^{11}C-methyl iodide
^{18}F	Reactive ^{18}F-fluoride
	^{18}F-fluorine

can be considered labeling precursors for ^{13}N-ammonia. Techniques have been developed for the rapid reduction of nitrogen oxides in the target mixture to ^{13}N-ammonia for radiopharmaceutical use, as well as for minimization of their production in the target during irradiation; these are described below.

^{11}C-Methyl Iodide

The relatively long 20 minute half-life of ^{11}C facilitates much greater latitude in chemical manipulation, both for preparation of the labeling precursor and during radiopharmaceutical synthesis. The target product for ^{11}C is ^{11}C-carbon dioxide, which can be either used directly as a labeling reagent or converted to a large number of synthetic precursors.[24,25] Of these diverse ^{11}C labeling reagents, ^{11}C-methyl iodide is especially valuable because of the large number of drug structures that can be labeled by N- or O-methylation.[26]

There are liquid-phase and gas-phase methods for preparing ^{11}C-methyl iodide. Liquid-phase methods were the first to be developed.[27–29] This synthetic approach involves the reduction of ^{11}C-carbon dioxide target to ^{11}C-methanol using lithium aluminum hydride and subsequent treatment of the ^{11}C-methanol intermediate with hydriodic acid to form ^{11}C-methyl iodide (Figure 12-12). The second step of this procedure can take place either by addition of hydriodic acid to the reduction vessel ("one-pot" synthesis) or after distillation of ^{11}C-methanol into a second vessel that contains hydriodic acid ("two-pot" synthesis). With either synthetic approach, the final ^{11}C-methyl iodide must be passed through purifying columns of soda lime and phosphorus pentoxide into a separate reaction vessel for radiopharmaceutical labeling. This purification step is necessary in order to remove any traces of hydrogen iodide or water, which would quench ^{11}C-methylation reactions.

Although recommendations have been proposed for standardizing production conditions for ^{11}C-methyl iodide prepared by liquid-phase methods,[26] the labeling technique is cumbersome and the radiochemical yield and specific activity are extremely sensitive to environmental conditions and reagents. Moreover, the turnaround time for production of ^{11}C-labeled PET radiopharmaceuticals in this manner is unwieldy because of the time needed for apparatus replacement and setup. This is a major impediment to the efficient operation of a clinical production center at a busy PET clinic, especially when repeated imaging scans are required in the same patient.

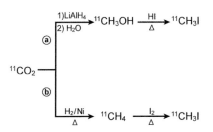

FIGURE 12-12 Radiosynthesis of the labeling precursor ^{11}C-methyl iodide. (a) Liquid-phase method; (b) gas-phase method.

Gas-phase methods provide a better means of preparing ^{11}C-methyl iodide.[30,31] As shown in Figure 12-12, in this synthetic approach ^{11}C-carbon dioxide is first reduced to ^{11}C-methane by heating in the presence of hydrogen gas with nickel catalysis. The ^{11}C-methane intermediate is then converted to ^{11}C-methyl iodide by recirculation through a heated quartz tube that contains iodine crystals. The ^{11}C-methyl iodide is then trapped on a porous polymer support and subsequently released as a gas for radiopharmaceutical labeling applications.

The gas-phase approach to making ^{11}C-methyl iodide has several advantages. The system is easily automated, and commercial sources supply microprocessor-controlled devices that accomplish this radiochemistry over a 12 minute period, completely without operator intervention. The system is also capable of synthesizing multiple batches of the labeling precursor without the need for apparatus replacement or setup. Thus, sequential batches of ^{11}C-methylated radiopharmaceuticals can be prepared without excessive radiation burden to production personnel. Moreover, the turnaround time between successive batches of ^{11}C-methyl iodide is only 20 to 30 minutes, which enables scheduling flexibility. The gas-phase synthesis of ^{11}C-methyl iodide also yields a higher specific activity product than the liquid phase approach, and the production sequence promotes greater consistency of precursor quality, which is important for radiopharmaceutical quality assurance.

Reactive ^{18}F-Fluoride

The 110 minute half-life of ^{18}F is very convenient for radiopharmaceutical synthesis. The radionuclide can be produced in two chemical forms, high specific activity ^{18}F-fluoride ion ($[^{18}F]F^-$) and moderate specific activity gaseous ^{18}F-fluorine ($[^{18}F]F_2$).

The preferred form of ^{18}F for radiopharmaceutical labeling is reactive ^{18}F-fluoride, which is used in nucleophilic substitution reactions. Aqueous ^{18}F-fluoride is produced with high specific activity in multicurie batches by proton irradiation of ^{18}O-water. Because of the high batch yields for cyclotron production of ^{18}F-fluoride, large quantities of PET radiopharmaceuticals can be synthesized from this precursor for use in the imaging clinic throughout the workday. In other situations, these high radiopharmaceutical production yields can be used to advantage because they facilitate distribution of PET tracers to off-site remote imaging locations. An additional advantage of ^{18}F-fluoride as a labeling precursor is its high specific activity, which makes possible the synthesis of PET radiopharmaceuticals for receptor-binding studies. It also minimizes the potential for pharmacologic or toxicologic effects of ^{18}F-labeled tracers, whether receptor-binding or not.

The irradiated ^{18}O-water target that contains ^{18}F-fluoride can be either used directly for nucleophilic fluorination reactions or purified by passage through a resin that separates the aqueous ^{18}F-fluoride from byproducts in the irradiated solution (Figure 12-13).[32] Heat generation during target irradiation causes byproducts to be dissolved in the target solution that can potentially interfere with subsequent labeling reactions. For this reason, resin purification has the dual advantages

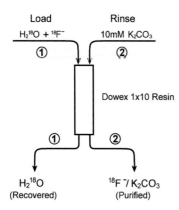

FIGURE 12-13 Resin purification of ^{18}F-fluoride and recovery of irradiated ^{18}O-water. (1) Loading of target water and ^{18}F-fluoride, with recovery of ^{18}O-water for reuse; (2) elution of purified ^{18}F-fluoride with aqueous potassium carbonate.

of facilitating recovery of expensive irradiated ^{18}O-water for reuse and removing chemicals that may impede radiopharmaceutical labeling reactions with ^{18}F-fluoride.

Nucleophilic fluorination reactions are inhibited by water, so it is essential that the aqueous solution of ^{18}F-fluoride be completely dried prior to the substitution reaction that incorporates ^{18}F into the molecular structure of the radiopharmaceutical. Such chemical transformation of aqueous ^{18}F-fluoride into reactive ^{18}F-fluoride is a key aspect of radiopharmaceutical synthesis, and the major steps have been reviewed.[25,33] There are three major steps to this process. First, a cation such as potassium, complexed to an aminopolyether cryptand such as Kryptofix 2.2.2 (Sigma-Aldrich) or a tetraalkylammonium salt, is added to the aqueous solution to serve as a counterion to the fluoride anion. Second, all traces of water are removed from the solution by azeotropic distillation with acetonitrile. Finally, a dipolar, aprotic organic solvent such as dimethylsulfoxide, dimethylformamide, or acetonitrile is added to resolubilize the dried complex of ^{18}F-fluoride. The resulting anhydrous solution of radioactivity contains reactive ^{18}F-fluoride and is added to the appropriate substrate to facilitate the nucleophilic substitution reaction necessary for radiopharmaceutical labeling.

Electrophilic ^{18}F-Fluorine

Although nucleophilic reactions may give high radiochemical yields for some ^{18}F-labeled PET tracers, this is not always the case. Some drug structures cannot be fluorinated in useful quantities with nucleophilic reactions because of the electronic nature (electron-rich rather than electron-poor aromatic rings) of the labeling substrate. An example of such a PET radiopharmaceutical is ^{18}F-fluorodopa (^{18}F-FD).

For these select applications, electrophilic ^{18}F-fluorine is used as a labeling precursor. Several electrophilic ^{18}F-fluorination reagents, including ^{18}F-labeled acetyl hypofluorite and xenon difluoride, have been developed over the years,[25] but ^{18}F-fluorine is the electrophilic fluorination precursor of choice from the perspective of simplicity and efficiency of labeling procedures.

^{18}F-fluorine is produced in hundred-millicurie quantities by the irradiation of a ^{20}Ne gas target with deuterons. It is necessary for the neon target to contain 0.1% to 2% F_2 as a carrier,[34] so the radionuclide is produced in only moderate specific activity (<12 Ci/mmol).[35] Both the batch yield and the specific activity of ^{18}F-fluorine are much lower than for ^{18}F-fluoride. Nevertheless, ^{18}F-fluorine is an extremely valuable precursor for that special category of PET radiopharmaceutical that cannot be labeled via ^{18}F-fluoride and does not require high specific activity to be effective.

RADIOPHARMACEUTICAL SYNTHESIS

The preparation of a short-lived positron-emitting radiopharmaceutical has stringent requirements because of the short half-lives of the relevant radionuclides. Radiosyntheses must be rapid and must produce the desired radiopharmaceutical in high radiochemical yield and purity. Although not always essential, high specific activity is often advantageous, either to avoid saturation of the biologic process under study or to minimize potential pharmacologic and toxicologic effects.

To accomplish this, facile labeling techniques using substrates that direct the radionuclide to a specific molecular site are needed. The preparation of these substrates often requires substantial synthetic expertise, and various commercial sources specialize in supplying these labeling reagents. As explained above, rapid, efficient conversion of the target product (the chemical bearing the product radionuclide) into a labeling precursor (or reactive chemical reagent form) of the radionuclide is frequently used to facilitate radiopharmaceutical labeling efficiency.

^{15}O and ^{13}N have short half-lives that preclude time-consuming radiosynthetic or imaging procedures. Radiopharmaceuticals labeled with these nuclides are rapidly produced on line, with minimal chemical manipulation. By contrast, the longer-lived ^{11}C and ^{18}F labels allow more flexibility with radiopharmaceuticals that have more lengthy preparation or imaging requirements. Generally speaking, however, clinically used PET nuclides have short half-lives and, with the exception of ^{18}F, must be prepared on site to satisfy the needs of the PET imaging suite.

^{15}O-Labeled Gases

Because of the short half-life of ^{15}O, PET radiopharmaceuticals that are labeled with this nuclide are produced on line with little operator intervention. As illustrated in Figure 12-14, target product ^{15}O-oxygen is converted to either ^{15}O-carbon monoxide or ^{15}O-carbon dioxide by recirculation through a charcoal furnace heated to 1000°C or 400°C, respectively.[36–39] For production of ^{15}O-carbon monoxide, careful control of beam and target conditions is necessary to minimize the

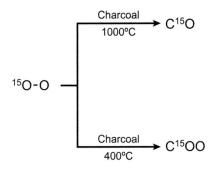

FIGURE 12-14 On-line synthesis of ^{15}O-carbon monoxide and ^{15}O-carbon dioxide from target product ^{15}O-oxygen.

amount of toxic nonradioactive carbon monoxide present in the radiopharmaceutical product.[40]

^{15}O-oxygen is commonly used in PET research protocols for measuring the cerebral metabolic rate of oxygen, whereas ^{15}O-carbon monoxide is used in assessing regional cerebral blood volume. Both of these radioactive gases are administered to the subject by direct inhalation of the target gas through a disposable mouthpiece with a 0.2 μm filter.

Water O 15 Injection

^{15}O-labeled water is produced for perfusion measurements by isotopic exchange or reduction reactions. In the exchange method, ^{15}O-carbon dioxide is used as a labeling precursor that is converted to ^{15}O-water, as shown in Figure 12-15. During cyclotron bombardment of the target, the ^{15}O-carbon dioxide that is produced on line is bubbled though a sterile bag of water for injection mounted in an ionization chamber.[38] The entire apparatus is mounted within a shielded "hot" cell. When adequate exchange labeling has occurred, as monitored by the ionization chamber reading at the cyclotron control console, some of the bag contents is remotely transferred to a 12 mL sterile syringe, which is pneumatically sent to the PET suite for bolus injection into the subject. Routinely, 50 to 125 mCi is produced in this manner, with a turnaround time of only 10 to 12 minutes.[38]

^{15}O-water can also be produced by reduction of ^{15}O-oxygen, as shown in Figure 12-15. Using this production route, conversion of the target product ^{15}O-oxygen into the labeling intermediate ^{15}O-carbon dioxide is unnecessary. Hydrogen gas is instead admixed with the target stream of ^{15}O-oxygen, and the mixture is passed over palladium or platinum catalyst heated to 450°C.[37,39] The reduced ^{15}O-water that results is trapped by bubbling though sterile water.

The reliable yield and short production turnaround time for preparation of ^{15}O-water make this radiopharmaceutical ideally suited for application in cerebral activation studies, in which rapidly repeated PET measurements of brain perfusion in the resting and activated state are performed.[41,42] In addition, the ease of production of this short-lived PET tracer also facilitates ancillary measurement of tissue perfusion to supplement PET imaging protocols that involve longer-lived radiopharmaceuticals.

Ammonia N 13 Injection

As shown in Table 12-11, the proton irradiation of ^{16}O-water results in ^{13}N in a variety of chemical forms.[43,44] The desired form, ^{13}N-ammonia, represents only a fraction of the total ^{13}N radioactivity that is produced. ^{13}N-ammonia (^{13}NH$_3$) is formed when ^{13}N atoms abstract hydrogen atoms from water. However, radiolytic oxidation also occurs during cyclotron irradiation of water, and ^{13}N-ammonia is converted to a large extent to oxo anions such as ^{13}N-nitrate or ^{13}N-nitrite. These radiolytically produced oxo anions can account for up to 85% of the total ^{13}N radioactivity.[45]

The oxo anions of ^{13}N can be removed from the desired ^{13}N-ammonia to yield a radiochemically pure product, but this is an inefficient use of the ^{13}N activity that is produced. Thus, most production sequences for ^{13}N-ammonia involve chemical reduction of the oxo anions to ^{13}N-ammonia (Figure 12-16). This is accomplished by using reducing agents such as DeVarda's alloy in aqueous sodium hydroxide[46] or titanium (III) salts,[47] followed by distillation of the ^{13}N-ammonia into a slightly acidic aqueous solution for product reformulation.

An alternative to chemically reducing oxidized target products from the target solution is to prevent their formation during irradiation (Figure 12-15). This is achieved by adding radical scavengers such as ethanol, acetic acid, methane, or

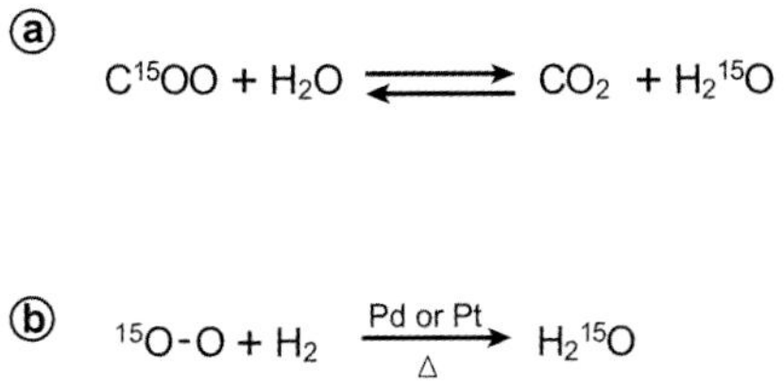

(a) $C^{15}OO + H_2O \rightleftharpoons CO_2 + H_2{}^{15}O$

(b) $^{15}O\text{-}O + H_2 \xrightarrow[\Delta]{\text{Pd or Pt}} H_2{}^{15}O$

FIGURE 12-15 Radiosynthesis of ^{15}O-water. (a) Isotopic exchange method; (b) chemical reduction method.

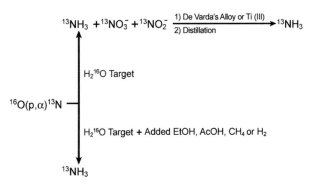

FIGURE 12-16 Radiosynthesis of ^{13}N-ammonia. Oxo anions of ^{13}N can be reduced either after irradiation or in situ during irradiation by the addition of reductants to the ^{16}O-water target.

hydrogen to the target mixture, which results in ^{13}N-ammonia only, and no oxo anions, being produced.[48–51] This slight alteration in production simplifies the overall radiopharmaceutical production process for ^{13}N-ammonia.

Sodium Acetate C 11 Injection

The preparation of sodium acetate C 11 injection for PET studies of myocardial viability involves radiosynthesis of ^{11}C-acetate from target product ^{11}C-carbon dioxide.[52] As shown in Figure 12-17, the radiosynthetic pathway is to first carbonate methylmagnesium Grignard reagent with ^{11}C-carbon dioxide, followed by acid hydrolysis to form ^{11}C-acetic acid. This two-step reaction has been adapted for several radiopharmaceutical production systems. The reaction conditions vary only slightly for these different devices; the main difference between these production systems lies in how the labeled drug is purified.

Sodium acetate C 11 injection can be purified by using solid-phase extraction,[53–56] liquid–liquid extraction,[57,58] or distillation.[59–61] Solid-phase extraction is preferred for clinical production, since it facilitates automated production of the PET radiopharmaceutical via robotics or other dedicated systems. A flexible robotic system has been described that reliably produces the labeled drug in batch yields of 220 to 300 mCi within 25 minutes.[56] Similar production results are accomplished in the clinical setting with alternative dedicated systems, and many of these production devices are available from commercial suppliers.

Fludeoxyglucose F 18 Injection

The workhorse of clinical PET imaging is ^{18}F-FDG. This radiopharmaceutical has several clinical indications and can be produced as a multidose batch suitable for use throughout the workday or for distribution to imaging sites remote from the production location. Although this radiopharmaceutical can be synthesized via electrophilic as well as nucleophilic reactions, the nucleophilic approach is preferred because of the higher production yields and the enantiomeric purity of the drug product.[62]

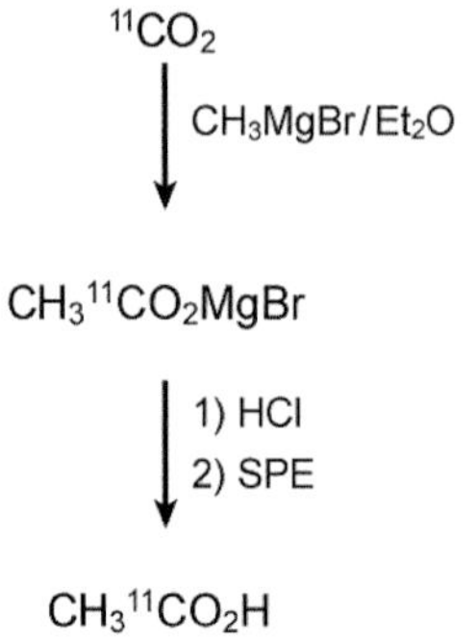

For clinical application, the universal method for production of this PET radiopharmaceutical is nucleophilic ^{18}F-fluoride displacement of a triflate ($-OS_2CF_3$) leaving group on a mannose precursor, followed by hydrolysis of protecting groups (Figure 12-18). Since the original description of this radiosynthetic method,[63] automation of drug production has been accomplished with only minor alteration of the original nucleophilic substitution pathway. Some of these automated systems are based on aminopolyether- or tetrabutylammonium-mediated nucleophilic fluorination,[64,65] whereas other systems utilize resin-supported ^{18}F-fluoride for the nucleophilic substitution reaction.[66] Several commercially available automated systems have been developed for routine production of the tracer and yield several curies of ^{18}F-FDG.

Recent synthetic refinements use basic rather than acidic hydrolysis to deprotect the fluorinated intermediate. The practical advantages of this change are shorter drug preparation times and avoidance of the formation of 2-chloro-2-deoxyglucose as a chemical contaminant in the radiopharmaceutical product.[67] Studies have shown that 2-^{18}F-fluoro-deoxy-D-glucose does not undergo epimerization under these alkaline conditions.[68]

^{18}F-FDG prepared in the above devices is purified via solid-phase extraction and ion exchange in a straightforward manner. The neutralized hydrolysis mixture is simply passed through cartridges bearing exchange resins that efficiently remove any traces of potential chemical and radiochemical byproducts from the final drug solution. The simplicity of these procedures enhances the reliability of automated production systems for clinical PET centers.

Fluorodopa F 18 Injection

[18]F-fluorodopa is an example of a useful PET radiopharmaceutical that is prepared in moderate specific activity using ^{18}F-fluorine as a labeling precursor. The reaction pathway,

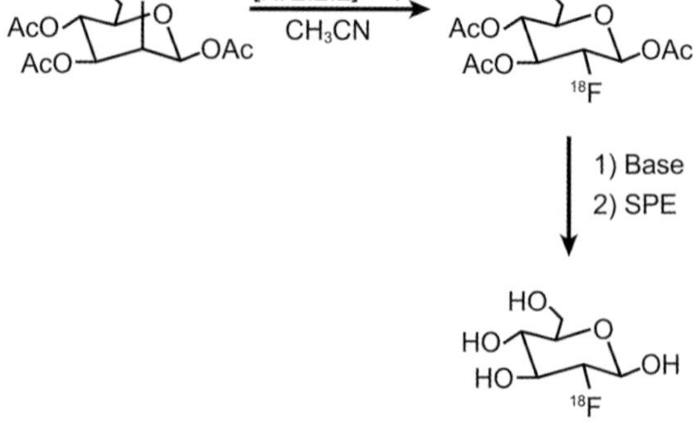

FIGURE 12-17 Radiosynthesis of ^{11}C-acetate from ^{11}C-carbon dioxide gas. This is a three-step procedure involving carbonation of a Grignard reagent, hydrolysis of the intermediate, and purification by solid phase extraction (SPE).

FIGURE 12-18 Radiosynthesis of 2-^{18}F-fluoro-2-deoxy-D-glucose (^{18}F-FDG) from ^{18}F-fluoride ion. This is a three-step procedure involving fluorination of a mannose triflate precursor, deprotection of the intermediate, and purification by solid phase extraction (SPE).

shown in Figure 12-19, involves the region-specific electrophilic ^{18}F-fluorodemetallation of an organometallic precursor by ^{18}F-fluorine gas.[69] The fluorine displaces the metal from the aromatic site of the labeling substrate in an electrophilic substitution reaction, so only the 6-^{18}F-fluoro isomer of fluorodopa is produced. Mercuric[70,71] or stannylated[72,73] labeling precursors are used because they give high fluorodemetallation yields.

Following the fluorination step, the labeled intermediate is deprotected via acidic hydrolysis and the drug is purified using preparative reverse-phase high-performance liquid chromatography (HPLC). The purification of this radiopharmaceutical, as well as of the other tracers of neurotransmission (Table 12-4), requires HPLC, which adds an element of complexity to the overall drug production sequence that is not necessary for the other radiopharmaceuticals used in clinical PET. In the case of ^{18}F-FD, dilute acetic acid is used as the HPLC mobile phase for the purification, which simplifies product reformulation. The isolated HPLC product peak is collected and sterilized by terminal filtration into the final product container, since the HPLC solvent and its pH range are already physiologically compatible.

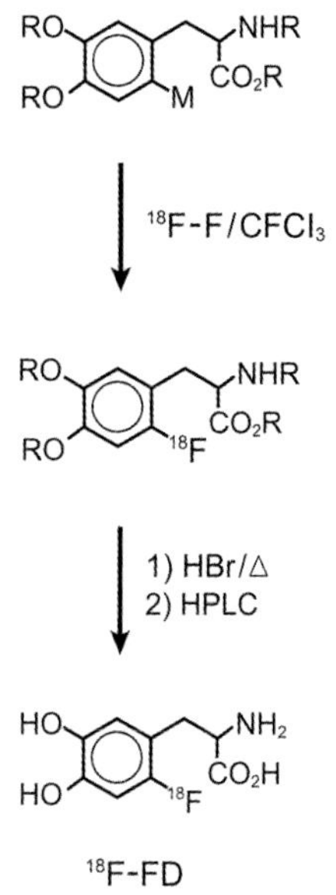

Flumazenil C 11 Injection and Raclopride C 11 Injection

^{11}C-FMZ and ^{11}C-RAC receptor-binding radiopharmaceuticals are examples of high specific activity tracers synthesized using ^{11}C-methyl iodide as a labeling precursor. ^{11}C-FMZ is synthesized via N-^{11}C-methylation,[74–77] whereas ^{11}C-RAC is prepared via O-^{11}C-methylation.[78–80]

The basic procedure used for the preparation of both of these receptor ligands is the same (Figure 12-20). ^{11}C-methyl iodide is bubbled through a solution of the respective normethyl substrate and the ^{11}C-methylation reaction is promoted by heating the reaction mixture for 1 to 3 minutes. The product is then purified by HPLC, which separates the ^{11}C-labeled ligand from the unlabeled substrate and other contaminants.

FIGURE 12-19 Radiosynthesis of 6-^{18}F-fluoro-L-dopa (^{18}F-FD) from ^{18}F-fluorine gas. This is a three-step procedure involving fluorodemetallation of an organotin- or mercury-leaving group (M), deprotection of the intermediate, and purification by HPLC.

The typical preparation time (including HPLC purification) is 30 to 40 minutes from delivery of ^{11}C-methyl iodide to the reaction solution. This time frame is practical for use with ^{11}C, which has a 20 minute half-life.

RADIOPHARMACEUTICAL PRODUCTION SYSTEMS

Preparation of radiopharmaceuticals in multimillicurie quantities in a routine manner requires special production systems to avoid unduly hazardous conditions for production personnel.

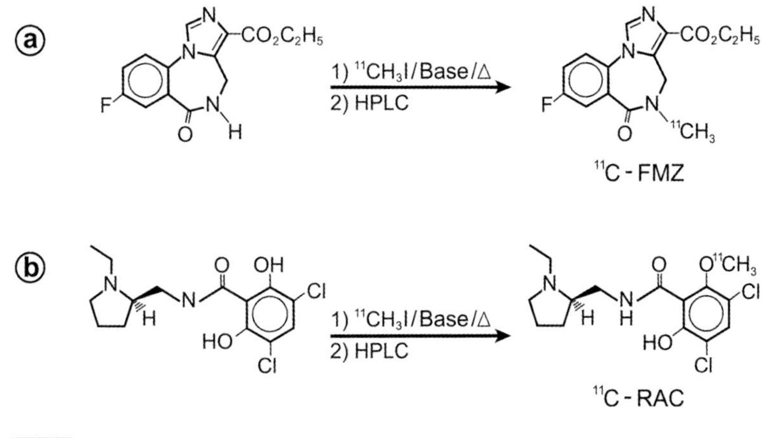

FIGURE 12-20 Radiosynthesis of ^{11}C-flumazenil (^{11}C-FMZ) and ^{11}C-raclopride (^{11}C-RAC) from ^{11}C-methyl iodide. These are two-step procedures involving N- or O-methylation, followed by purification using HPLC.

Special demands associated with the preparation of PET radiopharmaceuticals include a high-radiation environment, bulky lead shielding, short production intervals because of radionuclidic half-lives, and a batch yield that satisfies the needs of the clinical imaging schedule. In addition to these requirements, the radiosynthesis must produce the labeled drug in a pure state on a reliable, efficient basis. Frequently, tracer radiosyntheses must be repeated multiple times within a single workday. Finally, the production system must facilitate in-process documentation of the major steps of the compounding procedure so that labeled drug production is in compliance with federal and state regulations.

To address these challenges, several PET radiopharmaceutical production systems have been developed over the years.[81,82] Although these systems have differences, they all share certain characteristics. They all use the procedural steps of fluid flow, radioactivity measurement, temperature control, solvent dispensing, and product purification, within a shielded hot cell. The systems uniformly emphasize proper selection of equipment, materials, controllers, dispensing devices, extraction hardware, and radioactivity detectors so that each of the synthetic steps is accomplished in a predictable fashion, consistently yielding a high-purity drug product in an efficient manner. Disposable equipment is used wherever possible to minimize the potential introduction of contaminants during radiopharmaceutical compounding procedures.

The various designs of PET radiopharmaceutical production systems can be broadly categorized as remotely operated systems, automated modular systems, and robotic systems. These three categories are discussed below.

Remotely Operated Systems

Remotely operated systems are the most flexible and least expensive means of high-radioactivity radiopharmaceutical syntheses, since there is no computer or microprocessor used for radiopharmaceutical compounding. These systems can thus be considered an intermediate step in the evolution of methods from low-level preclinical synthesis of tracers to optimized high-level production procedures using robotics or automated modular designs.

Remotely operated systems are collections of devices that facilitate real-time operator control over the various steps of radiopharmaceutical production. This control takes place from behind lead shielding so that radiosyntheses are carried out in a safe manner. Components such as motor-driven needles, screw-driven capping devices, and remotely pressurized fluid lines are custom designed to accomplish tasks such as fluid delivery to reaction vessels, sealing of reaction vessels, and transfer of reaction mixtures to HPLC injectors for product purification. Use of long-handled tongs for radioactivity manipulation is also common with this approach.

These systems are easy to implement for radiopharmaceutical production and require a minimum number of optimization studies prior to their application. Remotely operated systems are especially valuable for the initial high-radioactivity syntheses of PET radiopharmaceuticals because of their ease of installation. The flexibility of the system design facilitates rapid configuration of the appropriate devices and components for a given radiopharmaceutical within a shielded hot cell. Also, the same hot cell can accommodate the various components needed for preparation of several different radiopharmaceuticals. After production conditions are optimized for routine production of a given PET tracer, the decision can be made whether to continue with remotely operated production or to make a capital investment in more expensive automated modular or robotic systems.

A major disadvantage of remotely operated systems is that the drug preparation sequence relies totally on operator intervention. During routine application there is potential for great variation in production conditions, with possible variance in radiopharmaceutical product quality.

With appropriate adherence to process guidelines, however, remotely operated systems offer an adequate and inexpensive means for compounding PET tracers, and several systems have been developed for the routine production of clinically used PET radiopharmaceuticals. Systems have been described for the ^{18}F-labeled tracers ^{18}F-FDG[83,84] and ^{18}F-FD.[85,86] Remotely operated systems are also available for preparing ^{11}C-sodium acetate[59–61,87,88] and ^{13}N-ammonia.[45,46]

Automated Modular Systems

Once all of the production parameters have been optimized for preparation of a PET radiopharmaceutical, it is possible to automate the entire drug production process. This is accomplished using modular systems in which the movement of synthetic intermediates through fixed-plumbed devices and equipment is completely controlled by computer software or timing circuits. In the modular approach, all the equipment necessary for the production of a given radiopharmaceutical is mounted together, and the operation from start to finish is preprogrammed according to optimized reaction conditions. Because all of the compounding steps are programmed, deviations in the preparation conditions are minimal and the quality of the final product is thus standardized.

A major advantage of automated modular systems for the routine production of PET radiopharmaceuticals is that product consistency is enhanced. Also, the fact that the system is dedicated to the production of a single tracer simplifies maintenance and troubleshooting. For these reasons, most commercially supplied PET radiopharmaceutical production systems are of the automated modular type. In some cases, commercial systems are marketed in which all reagents are mounted on a single disposable cartridge for each production batch. Use of such disposables tends to minimize the potential for contamination of the final drug product.

A disadvantage of this drug production approach is that it is difficult to do repetitive batch production on the same workday using the same module because of the radioactivity remaining in the system after the initial synthesis. Also, hot-cell space must be allocated for the production of each tracer, since separate modular systems are used for the preparation of the individual radiopharmaceuticals. This limitation has been recognized by some research groups, which have proposed application of modular systems for the production of more than one tracer when radiosynthetic steps are similar. However, this expanded application of automated modules is possible only with research tracers. Each of the clinically applied radiopharmaceuticals listed in Table 12-4 requires a separate automated module for preparation, since the radiosynthetic steps used to prepare the different tracers do not overlap.

Several automated modular systems have been described for the production of clinical PET radiopharmaceuticals. Various modular systems are available for producing ^{18}F-FDG.[89–92] Automated production modules have also been described for ^{11}C-sodium acetate,[57,58] ^{13}N-ammonia,[47,93] ^{18}F-FD,[94] and ^{11}C-FMZ.[77]

Robotic Systems

Robotic production systems use commercially available robots to move radioactive intermediates between fixed work stations to accomplish radiopharmaceutical synthesis, purification, and reformulation. Because the robot arm and gripping device are flexible, they have capabilities similar to a human arm and hand. Because the action of the robot arm is microprocessor controlled, optimized radiopharmaceutical production sequences can be programmed into the control software. Robotic production systems thus combine the flexibility of remotely controlled systems with the standardized production parameters of automated modular production systems.

The flexibility of robotics allows synthesis of a variety of tracers with the same device. This is especially useful in research endeavors, in which the radiosynthesis of novel PET tracers must be scaled up to the radioactivity levels necessary for performance of PET studies. For this reason, the primary promoters of robotic production systems have been PET centers that emphasize clinical and preclinical imaging research.[95–99]

Robotics also has special advantages when applied in the routine production of radiopharmaceuticals for clinical PET. The flexibility of robotic systems facilitates the production of multiple PET tracers within the same shielded hot cell, which is valuable to sites with limited space for radiopharmaceutical production. The microprocessor control over production steps increases the batch-to-batch consistency of radiopharmaceutical quality and aids in the documentation of production procedures. Moreover, robotic systems can be programmed to move radioactive waste remaining from a production session to a shielded area within a hot cell, so setup for subsequent radiopharmaceutical syntheses is less problematic in terms of the radiation dose to personnel.

Robotics has the unique potential for simultaneous synthesis of two or more tracers within a single hot cell. If this potential were realized through creative software, it would be especially valuable for busy PET clinics. With this capability, the PET radiopharmaceutical production schedule would be limited only by the capability of the cyclotron to produce the starting radionuclides.

Robotic synthesis of several radiopharmaceuticals for clinical PET has been achieved. Labeled drugs prepared via robotics include ^{18}F-FDG,[65,99] ^{11}C-sodium acetate,[56] ^{11}C-FMZ,[99] and ^{18}F-FD.[98]

RADIOPHARMACEUTICAL FORMULATION

A major concern for nuclear pharmacists is the reformulation of PET radiopharmaceuticals. A PET tracer may be radiochemically pure, but the isolation process may yield the tracer dissolved in a solvent inappropriate for administration to humans. The solvent used for purification must then be removed and the tracer reconstituted into a physiologically compatible solution.

Liquid chromatography is generally used for purification of radiopharmaceuticals. It typically involves simple solid-phase extraction cartridges or HPLC. In either case, the basic components are the same: the impure reaction solution traverses over a stationary phase through which flows the mobile phase. The eluant solution bears the purified radiopharmaceutical dissolved in the mobile phase. These chromatographic systems are optimized so that retention times are as low as feasible without compromising product purity. This facilitates purification of the radiopharmaceutical within the temporal constraints of the short half-life of the positron-emitting label. In addition, the tracer is isolated in a minimum volume of mobile phase, which makes radiopharmaceutical reformulation less problematic.

In ideal situations, radiopharmaceutical purification involves chromatographic systems in which the mobile phase is a physiologically compatible liquid. Examples are the solid-phase extraction purification of ^{18}F-FDG, which uses sterile water for injection as the mobile phase, and the HPLC purification of ^{18}F-FD, which uses a weakly acidic (acetic acid, pH 5) aqueous solution. These purification systems are possible because the final reaction mixture contains chemical contaminants that are easily resolved from the radiopharmaceutical product.

The production of some PET radiopharmaceuticals generates byproducts that cannot be removed from the product by chromatographic systems with aqueous mobile phases. Notable examples are receptor-binding radioligands such as ^{11}C-FMZ and ^{11}C-RAC. The reaction mixture for these tracers includes the nonreacted, nor-methyl labeling substrate as well as the ^{11}C-labeled tracer. Since these tracers differ from the labeling substrate by only a single methyl substituent, the chromatographic characteristics of these two chemicals are very similar. Typically, reverse- or normal-phase HPLC is required to resolve the radioligand from the labeling substrate. In either case, toxic organic compounds are used in the HPLC mobile phase, so the radiochemically pure radioligand isolated in the product fraction of the HPLC eluant is unsuitable for human administration.

Reformulation of the radiopharmaceutical thus involves removal of the HPLC mobile phase from the purified radioligand. This can be accomplished by one of two means. The first is to remove the solvent using a rotary evaporator. The

HPLC solvent is evaporated by gentle heating under reduced pressure. The radiopharmaceutical, which remains on the wall of the glass vessel after evaporation of volatiles, is then dissolved in a physiologically compatible solvent for further work-up. This approach is cumbersome, subjects the product to heat- or vacuum-related losses, and is limited to HPLC solvents that have relatively low boiling points.

The second reformulation method is more amenable to remote operation and involves solid-phase extraction. In this approach, the product fraction from the HPLC is diluted in a 25 to 250 mL volume of aqueous buffer and passed across a reverse-phase Sep-Pak (Waters) cartridge. The lipophilic radiopharmaceutical is retained on the cartridge, while the diluted organic solvents are eluted to waste. The shielded cartridge is then rinsed with additional sterile water for injection to remove residual contaminants, and the radiopharmaceutical product is subsequently eluted off the cartridge with a small volume of ethanol. The ethanol can be diluted with sterile water for injection or 0.9% sodium chloride injection to complete the reformulation procedure.

Once the radiopharmaceutical is reformulated into an appropriate injectate, it must meet further requirements for suitability for human administration. In addition to radiochemical and chemical purity, sterility is key. Because heat sterilization is impractical for these tracers, terminal sterilization via membrane filtration is the method most commonly used. Passage of the reformulated radiopharmaceutical through a 0.2 μm filter into a final product container effectively removes bacterial contamination from the reformulated tracer solution. There are important caveats with this method of compounding PET tracers, however.

First, terminal sterilization removes bacterial contamination but not pyrogenic compounds, which are capable of passing through the 0.2 μm pores of the filter. Thus, this method ensures against bacterial contamination but does not guarantee a pyrogen-free product. Absence of pyrogens in the radiopharmaceutical is best ensured by eliminating pyrogens in the production procedure. If a radiopharmaceutical product is found to be pyrogenic, it is necessary to replace any component or reagent in the production process that may act as a source of pyrogens, since there is no filter to eliminate these fever-inducing compounds from the drug solution.

Second, it is important to use membrane filters that have compatibility specifications appropriate for the solution being sterilized. Many radiopharmaceuticals require solvents that have pH adjustments or solubilizing agents (such as ethanol) to facilitate the dissolution of otherwise poorly soluble substances. Although the filter membrane itself may be resistant to degradation by these solvents, the plastic housing or the adhesive used to hold the filter assembly together may not withstand the solvent. Commercial suppliers of filter assemblies can often recommend specific filter types for given applications.

Finally, it is essential after each terminal filtration to test for membrane integrity. If a filter has ruptured during the compounding procedure because of overpressurization, there is no guarantee that the radiopharmaceutical product is sterile. The only way to confirm that the filter is still intact after filtration is to perform a bubble test, in which the intact, wet membrane will create resistance to a pressurized syringe. This bubble test can be performed manually immediately after filtration of the radiopharmaceutical into the final product container. Alternatively, some production systems have the membrane integrity test automated as a part of the radiolabeled drug compounding procedure.[56]

QUALITY ASSURANCE

Quality assurance of the final drug product is an important aspect of PET, just as it is in conventional nuclear medicine practice. The quality of the radiopharmaceutical must meet high standards to ensure safety of the patient and to be an effective imaging agent that results in useful diagnostic information.

Because of the greater level of sophistication of PET radiopharmaceutical compounding, however, the instrumentation used in quality control testing is more complicated than that used in conventional nuclear medicine. For example, radiochemical purity testing, which is usually performed on a prerelease basis, can involve gas chromatography (^{15}O tracers) or HPLC (^{11}C or ^{18}F tracers). In some cases (^{18}F-FDG), thin-layer radiochromatographic methods have been developed to simplify radiochemical purity testing.

Because of the synthetic methods used, testing for chemical purity is also a requirement for PET radiopharmaceuticals. This is usually accomplished using gas chromatography or HPLC techniques to quantify the amount of contaminant chemicals in the final drug product. To streamline the clinical production of PET tracers, it is sometimes possible to develop simple colorimetric tests to confirm that contaminant levels fall below official limits. An example of this is the color spot test method that was developed for the detection of Kryptofix 2.2.2 in preparations of ^{18}F-FDG.[100]

Product quality is defined in *USP 32/NF 27*, which is an official compendium of drug standards.[6] The United States Pharmacopeia (USP) publishes monographs for the major PET radiopharmaceuticals used in clinical practice (Table 12-4). These monographs define standards of purity with regard to sterility, apyrogenicity, pH, radiochemical purity, radionuclidic purity, chemical purity, and labeling requirements. USP also sets standards for methods of assessing these aspects of purity, such as for sterility testing and for pyrogen testing with bacterial endotoxin techniques.

USP Chapter 823 establishes mandated requirements for the compounding of PET tracers.[6] These requirements include the control of components, materials, and supplies; verification of compounding procedures; stability testing and expiration dating; and steps to be taken during the compounding of PET radiopharmaceuticals for human use. Standards for quality control, sterilization, and sterility assurance are also given.

USP also publishes guidelines, in Chapter 1015, for the use of automated radiochemical synthesis apparatus for PET radiopharmaceuticals.[6] These guidelines address such important issues as equipment quality assurance, routine quality control testing, reagent audit trail, and documentation of apparatus parameters. The guidelines point out that any changes made in the synthesis method should be validated to confirm that there is no effect on final drug quality.

It is important to be aware that similar quality assurance guidelines exist outside the United States.[101] Many innovative PET studies are performed in foreign PET centers, and it is reassuring to know, when interpreting image results, that the PET tracer being imaged is of high quality.

REGULATORY ISSUES

As with radiopharmaceuticals in general, the oversight of PET radiopharmaceutical production, application, and disposal involves several different regulatory bodies. These include the Food and Drug Administration (FDA), Environmental Protection Agency (EPA), Occupational Safety and Health Administration (OSHA), and Nuclear Regulatory Commission (NRC); organizations responsible for the radiation safety of the public; and organizations and agencies that promulgate additional rules for the safe and effective use of drugs. Examples are The Joint Commission, which accredits health care organizations, and federal agencies that implement laws such as the Health Insurance Portability and Accountability Act of 1996, which mandates control of protected health information. For PET radiopharmaceuticals that are transported off site, regulations of the Department of Transportation must also be followed. Product quality is regulated by USP, which establishes purity standards for PET radiopharmaceuticals. Finally, professional licensing bodies, such as state boards of pharmacy, play key roles in regulating activities related to PET and PET radiopharmaceuticals.

Compliance with all of these regulatory demands is important, not only because it is ethical and beneficial for the public at large but because failure to comply can have unwanted legal consequences. Furthermore, adherence to the pertinent regulations is key to financial reimbursement for PET procedures.

A major hurdle to reimbursement for PET imaging procedures has been FDA approval of the relevant radiopharmaceuticals. CMS and third-party payers are reluctant to reimburse for PET procedures that involve "experimental" imaging of radiochemicals in human subjects. However, as described above, reimbursement will be made for PET imaging procedures that involve drugs with FDA-approved indications for imaging (Tables 12-6 through 12-8). Thus, from a financial perspective, drawing a distinction between a labeled tracer and an FDA-approved radiopharmaceutical is essential.

Selecting PET radiopharmaceuticals for approval from the many existing PET tracers was an arduous task for FDA, given that the agency had no history of regulating this category of drug. Unlike conventional FDA-approved drugs, the drugs used for PET have no pharmacologic effect, disappear rapidly, and have personnel radiation safety issues associated with their manufacture. The evolution of the current FDA involvement with PET and its relationship to reimbursement have been reviewed elsewhere.[102]

The traditional route of drug approval is not applicable to PET radiopharmaceuticals, but the FDA Modernization Act of 1997 (FDAMA) allows modifications of new drug applications (NDAs) and current good manufacturing practices (CGMPs) that are relevant and enforceable for the PET community. The safety and efficacy of clinically used PET radiopharmaceuticals were evaluated by FDA with the assistance of the PET community in reviewing the literature. Although the process of evaluating PET radiopharmaceuticals and their indications is ongoing, precedent has been established for involving the PET community at large rather than requiring individual PET sites to prove the safety and efficacy of tracers.

FDA ensures compliance with manufacturing standards by requiring the filing of NDAs, adherence to CGMPs, and inspection by FDA staff. FDAMA has streamlined these requirements for PET, and FDA will adopt templates that can be used by various PET production sites in a manner that greatly simplifies regulatory compliance and standardizes the manufacture of PET radiopharmaceuticals. FDA has enlisted the assistance of the PET community in developing templates for NDAs and CGMPs for clinically used PET radiopharmaceuticals. Under this regulatory model, PET sites will register as drug manufacturers, submit NDAs based on available templates, follow standardized CGMPs, and be inspected for compliance by FDA.

An alternative model for the production of PET radiopharmaceuticals is based on the traditional right of professional pharmacists to compound drugs upon receipt of a prescription written by a physician in the course of caring for a given patient. This model requires the PET radiopharmaceutical dispensed for the patient to meet all official standards of purity, but it differs in that the site is not registered as a drug manufacturer and an NDA has not been filed for the PET radiopharmaceutical. In this alternative model, the professional activities of the compounding pharmacist are regulated by the state board of pharmacy, and the nuclear pharmacy performing PET radiopharmaceutical compounding is licensed and inspected by the state board of pharmacy.

FUTURE OUTLOOK

Further research into the development, validation, and implementation of new radiopharmaceuticals for PET imaging is integral to the continued growth of this area of nuclear medicine practice. Efforts to discover and evaluate new radiolabeled compounds for clinical use in PET protocols are increasing our understanding of human physiology and pathophysiology. New PET imaging techniques will undoubtedly be added to the list of PET procedures used in health care. New PET radiopharmaceuticals and methods may advance medical practice by promoting the longitudinal assessment of therapy for various disorders. Future PET techniques may also have a role in the development of new therapeutic agents, since PET imaging efficiently evaluates important parameters such as dose–occupancy relationships and could help streamline clinical trials of new drugs.

PET imaging is especially well positioned for future growth because of its versatility in quantifying physiologic parameters. Substantial progress has been made in meeting the challenges associated with the radiochemical synthesis of radiopharmaceutical structures for PET.[103] Moreover, recent advances in the instrumentation of small-animal scanners now permit PET imaging of improved animal models of human disease. The capability to noninvasively validate novel radiotracers in animals yields valuable insight into new imaging techniques for drug development and for clinical diagnostic applications in humans.

Examples of these exciting new developments in PET research have been presented at meetings of the International Isotope Society and the Academy of Molecular Imaging. Abstracts from the International Isotope Society demonstrate the large pipeline of novel PET tracers that have been synthesized for preclinical evaluation.[104] Abstracts from the World Molecular Imaging Congress demonstrate important advances in instrumentation (Institute for Molecular Imaging), preclinical imaging in drug discovery (Society of Non-Invasive Imaging in Drug Development), and state-of-the-art diagnostic PET applications (Institute of Clinical PET).[105] This research suggests that clinical PET applications will continue to grow.

REFERENCES

1. Friedlander G, Kennedy JW, Macias ES, et al. *Nuclear and Radiochemistry.* 3rd ed. New York: Wiley-Interscience; 1981.
2. Melcher CL. Scintillation crystals for PET. *J Nucl Med.* 2000;41:1051–5.
3. Turkington TG. Introduction to PET instrumentation. *J Nucl Med Technol.* 2001;29:1–8.
4. Townsend DW. Positron emission tomography/computed tomography. *Sem Nucl Med.* 2008;38:152–66.
5. Blodgett TM, Meltzer CC, Townsend DW. PET/CT: form and function. *Radiology.* 2007;242:360–85.
6. *The United States Pharmacopeia, 32nd rev, and The National Formulary, 27th ed.* Rockville, MD: United States Pharmacopeial Convention, Inc; 2008.
7. Sciacca RR, Akinboboye O, Chou RL, et al. Measurement of myocardial blood flow with PET using 1-^{11}C-acetate. *J Nucl Med.* 2001;42:63–70.
8. Van den Hoff J, Burchert W, Börner A-R, et al. [1-^{11}C]Acetate as a quantitative perfusion tracer in myocardial PET. *J Nucl Med.* 2001;42:1174–82.
9. Oyama N, Miller TR, Dehdashti F, et al. 11C-Acetate PET imaging of prostate cancer: detection of recurrent disease at PSA relapse. *J Nucl Med.* 2003;44:549–55.
10. Ho C-L, Yu SCH, Yeung DWC. ^{11}C-Acetate PET imaging in hepatocellular carcinoma and other liver masses. *J Nucl Med.* 2003;44:213–21.
11. Timmers HJLM, Chen CC, Carrasquillo JA, et al. Comparison of ^{18}F-fluoro-L-dopa, ^{18}F-fluorodeoxyglucose, and ^{18}F-fluorodopamine PET and ^{123}I-MIBG scintigraphy in the localization of pheochromocytoma and paraganglioma. *J Clin Endocrinol Metab.* 2009;94:4757–67.
12. Beheshti M, Pöcher S, Vali R, et al. The value of ^{18}F-DOPA PET-CT in patients with medullary thyroid carcinoma: comparison with ^{18}F-FDG PET. *Eur Radiol.* 2009; 19:1425–34.
13. Hillner BE, Liu D, Coleman RE, et al. The National Oncology PET Registry (NOPR): design and analysis plan. *J Nucl Med.* 2007;48:1901–8.
14. Hillner BE, Siegel BA, Liu D, et al. Impact of positron emission tomography/computed tomography and positron emission tomography (PET) alone on expected management of patients with cancer: initial results from the National Oncologic PET Registry. *J Clin Oncol.* 2008;26:2155–61.
15. Hillner BE, Siegel BA, Shields AF, et al. Relationship between cancer type and impact of PET and PET/CT on intended management: findings of the National Oncologic PET Registry. *J Nucl Med.* 2008;49:1928–35.
16. Lindsay MJ, Siegel BA, Tunis SR, et al. The National Oncologic PET Registry: expanded Medicare coverage for PET under Coverage with Evidence Development. *AJR.* 2007;188:1109–13.
17. Tunis S, Whitner D. The National Oncologic PET Registry: lessons learned for Coverage with Evidence Development. *J Am Coll Radiol.* 2009;6:360–5.
18. Gennaro GP, Neirinckx RD, Bergner B, et al. A radionuclide generator and infusion system for pharmaceutical quality Rb-82. In: Knapp FF, Butler TA, eds. *Radionuclide Generators. New Systems for Nuclear Medicine Applications.* Washington, DC: American Chemical Society; 1984:135–50.
19. Loc'h C, Mazière B, Comar D. A new generator for ionic gallium-68. *J Nucl Med.* 1980;21:171–3.
20. Ruth TJ. Accelerators available for isotope production. In: Welch MJ, Redvanly CS, eds. *Handbook of Radiopharmaceuticals. Radiochemistry and Applications.* Hoboken, NJ: John Wiley and Sons; 2003:71–85.
21. Schlyer DJ. Production of radionuclides in accelerators. In: Welch MJ, Redvanly CS, eds. *Handbook of Radiopharmaceuticals. Radiochemistry and Applications.* Hoboken, NJ: John Wiley and Sons; 2003:1–70.
22. Wolf AP, Jones WB. Cyclotrons for biomedical isotope production. *Radiochim Acta.* 1983;34:1–7.
23. Qaim SM, Clark JC, Crouzel C, et al. PET radionuclide production. In: Stöcklin G, Pike VW, eds. *Radiopharmaceuticals for Positron Emission Tomography. Methodological Aspects.* Boston: Kluwer Academic Publishers; 1993:1–43.
24. Antoni G, Kihlberg T, Långström B. Aspects on the synthesis of ^{11}C-labelled compounds. In: Welch MJ, Redvanly CS, eds. *Handbook of Radiopharmaceuticals. Radiochemistry and Applications.* Hoboken, NJ: John Wiley and Sons; 2003:14–194.
25. Ferrieri RA. Production and application of synthetic precursors labeled with carbon-11 and fluorine-18. In: Welch MJ, Redvanly CS, eds. *Handbook of Radiopharmaceuticals. Radiochemistry and Applications.* Hoboken, NJ: John Wiley and Sons; 2003:229–82.
26. Crouzel C, Långström B, Pike VW, et al. Recommendations for a practical production of ^{11}C-methyl iodide. *Appl Radiat Isot* 1987; 38:601–4.
27. Långström B, Lundqvist H. The preparation of ^{11}C-methyl iodide and its use in the synthesis of ^{11}C-methyl-L-methionine. *Int J Appl Radiat Isot.* 1976;27:357–63.
28. Marazano C, Maziere M, Berger G, et al. Synthesis of methyl iodide-^{11}C and formaldehyde-^{11}C. *Int J Appl Radiat Isot.* 1977;28:49–54.
29. Iwata R, Ido T, Saji H, et al. A remote-controlled synthesis of ^{11}C-iodomethane for the practical preparation of ^{11}C-labeled radiopharmaceuticals. *Int J Appl Radiat Isot.* 1979;30:194–6.
30. Larsen P, Ulin J, Dahlstrom K, et al. Synthesis of [^{11}C]iodomethane by iodination of [^{11}C]methane. *Appl Radiat Isot.* 1997;48:153–7.
31. Link JM, Krohn KA, Clark JC. Production of [^{11}C]CH$_3$I by single-pass reaction of [^{11}C]CH$_4$ with I$_2$. *Nucl Med Biol.* 1997;24:93–7.
32. Schlyer DJ, Bastos MAV, Alexoff D, et al. Separation of [^{18}F]fluoride from [^{18}O]water using anion exchange resin. *Appl Radiat Isot.* 1990;41:531–3.
33. Snyder SE, Kilbourn MR. Chemistry of fluorine-18 radiopharmaceuticals. In: Welch MJ, Redvanly CS, eds. *Handbook of Radiopharmaceuticals. Radiochemistry and Applications.* Hoboken, NJ: John Wiley and Sons; 2003:195–227.
34. Casella V, Ido T, Wolf AP, et al. Anhydrous ^{18}F labeled elemental fluorine for radiopharmaceutical production. *J Nucl Med.* 1980;21:750–7.
35. Blessing G, Coenen HH, Franken K, et al. Production of [^{18}F]F$_2$, H^{18}F and ^{18}F$_{aq}$ using the ^{20}Ne(d,α)^{18}F process. *Appl Radiat Isot.* 1986;37:1135–40.
36. Strijkmans K, Vandecasteele C, Sambre J. Production and quality control of ^{15}O$_2$ and C^{15}O$_2$ for medical use. *Int J Appl Radiat Isot.* 1985;36:279–83.

37. Clark JC, Buckingham PD. *Short-lived Radioactive Gases for Clinical Use.* London: Butterworths; 1975.
38. Welch MJ, Kilbourn MR. A remote system for the routine production of oxygen-15 radiopharmaceuticals. *J Lab Comp Radiopharm.* 1986; 22:1193–1200.
39. Berridge MS, Terris AH, Cassidy EH. Low-carrier production of [^{15}O]oxygen, water and carbon monoxide. *Appl Radiat Isot.* 1990;41: 1173–5.
40. Clark JC, Crouzel C, Meyer GJ, et al. Current methodology for oxygen-15 production and clinical use. *Appl Radiat Isot.* 1987;38:597–600.
41. Sergent T. Brain-imaging studies of cognitive functions. *Trends in Neurosciences.* 1994;17:221–7.
42. Cowell SF, Code C. Thinking nuclear medicine—PET activation. *J Nucl Med Technol.* 1998;26:17–22.
43. Krizek H, Lembares N, Dinwoodie R, et al. Production of radiochemically pure ^{13}NH$_3$ for biomedical studies using the ^{16}O(p,α)^{13}N reaction. *J Nucl Med.* 1973;14:629–30.
44. Tilbury RS, Dahl JR. ^{13}N species formed by proton irradiation of water. *Radiation Res.* 1979;79:22–3.
45. Parks NJ, Krohn KA. The synthesis of ^{13}N labelled ammonia, dinitrogen, nitrite and nitrate using a single cyclotron target system. *Int J Appl Radiat Isot.* 1978;29: 754–6.
46. Vaalburg W, Kamphusi JAA, Beerling-van der Moles HD, et al. An improved method for the production of ^{13}N-labelled ammonia. *Int J Appl Radiat Isot.* 1975;26:316–8.
47. Ido T, Iwata R. Fully automated synthesis of ^{13}NH$_3$. *J Lab Comp Radiopharm.* 1981;18:244–6.
48. Wieland BW, Bida G, Padgett H, et al. In-target production of ^{13}N-ammonia via proton irradiation of dilute aqueous ethanol and acetic acid mixtures. *Appl Radiat Isot.* 1991;42:1095–8.
49. Berridge MS, Landmeier BJ. In-target production of [N-13]ammonia—target design, products, and operating parameters. *Appl Radiat Isot.* 1993;44:1433–41.
50. Krasikova RN, Fedorova OS, Korsakov MV, et al. Improved [N-13]ammonia yield from the proton irradiation of water using methane gas. *Appl Radiat Isot.* 1999;51:395–401.
51. Mullholland GK, Sutorik A, Jewett DM, et al. Direct in-target synthesis of [^{13}N]NH$_3$ by irradiation of water under hydrogen pressure. *J Nucl Med.* 1989;30:926.
52. Moerlein SM, Gropler RJ, Welch MJ, et al. USP standards for Sodium Acetate C 11 Injection. *Pharm Forum.* 1994;20:7523–5.
53. Meyer G-J, Günther K, Matzke K-H, et al. A modified method for ^{11}C acetate, preventing liquid phase extraction steps. *J Lab Comp Radiopharm.* 1993;32:182–3.
54. Kruijer PS, Ter Lindin T, Mooij R, et al. A practical method for the preparation of [^{11}C]acetate. *Appl Radiat Isot* 1995; 46: 317–21.
55. Iwata R, Ido T, Tada M. Column extraction method for rapid preparation of [^{11}C]acetic and [^{11}C]palmitic acids. *Appl Radiat Isot.* 1995;46: 117–21.
56. Moerlein SM, Gaehle GG, Welch MJ. Robotic preparation of Sodium Acetate C 11 Injection for use in clinical PET. *Nucl Med Biol.* 2002;29: 613–21.
57. Pike VW, Horlock PL, Brown C, et al. The remotely controlled preparation of a ^{11}C-labelled radiopharmaceutical—[1-^{11}C]acetate. *Int J Appl Radiat Isot.* 1984;35:623–7.
58. Del Fiore G, Peters JM, Quaglia L, et al. Automated production of carbon-11 labelled acetate to allow detection of transient myocardial ischemia in man, using positron emission tomography. *J Radioanal Nucl Chem Lett.* 1984;87:1–14.
59. Norenberg JP, Simpson NR, Dunn BB, et al. Remote synthesis of [^{11}C]acetate. *Appl Radiat Isot.* 1992;43:943–5.
60. Ishiwata K, Ishii S-I, Sasaki T, et al. A distillation method of preparing C-11 labeled acetate for routine clinical use. *Appl Radiat Isot.* 1993; 44:761–3.
61. Berridge MS, Cassidy EH, Miraldi F. [^{11}C]Acetate and [^{11}C]methionine: improved syntheses and quality control. *Appl Radiat Isot* 1994; 46: 173–5.
62. Coenen HH, Pike VW, Stöcklin G, et al. Recommendation for a practical production of [2-^{18}F]fluoro-2-deoxy-D-glucose. *Appl Radiat Isot.* 1987; 38:605–10.
63. Hamacher K, Coenen HH, Stöcklin G. Efficient stereospecific synthesis of no-carrier-added 2-[^{18}F]-fluoro-2-deoxy-D-glucose using aminopolyether supported nucleophilic substitution. *J Nucl Med.* 1986;27:235–8.
64. Hamacher K, Blessing G, Nebeling B. Computer-aided synthesis (CAS) of no-carrier-added 2-[^{18}F]fluoro-2-deoxy-D-glucose: an efficient automated system for the aminopolyether-supported nucleophilic fluorination. *Appl Radiat Isot.* 1990;41:49–55.
65. Brodack JW, Dence CS, Kilbourn MR, et al. Robotic production of 2-deoxy-2-[^{18}F]fluoro-D-glucose: a routine method of synthesis using tetrabutylammonium [^{18}F]fluoride. *Appl Radiat Isot.* 1988;39:699–703.
66. Toorongian SA, Mulholland GK, Jewett DM, et al. Routine production of 2-deoxy-2-[^{18}F]fluoro-D-glucose by direct nucleophilic exchange on a quaternary 4-aminopyridium resin. *Nucl Med Biol.* 1990;17:273–9.
67. Füchtner F, Steinbach J, Mäding P, et al. Basic hydrolysis of 2-[^{18}F]fluoro-1,3,4,6-tetra-O-acetyl-D-glucose in the preparation of 2-[^{18}F]fluoro-2-deoxy-D-glucose. *Appl Radiat Isot.* 1996;47:61–6.
68. Meyer G-J, Matzke KH, Hamacher K, et al. The stability of 2-[^{18}F]fluoro-deoxy-D-glucose towards epimerisation under alkaline conditions. *Appl Radiat Isot.* 1999;51:37–41.
69. Luxen A, Guillaume M, Melega WP, et al. Production of 6-[^{18}F]fluoro-L-DOPA and its metabolism in vivo—a critical review. *Nucl Med Biol.* 1992;19:149–58.
70. Luxen A, Perlmutter M, Bida GT, et al. Remote, semiautomated production of 6-[^{18}F]fluoro-L-dopa for human studies with PET. *Appl Radiat Isot.* 1990;41:275–81.
71. Adam MJ, Jivan S. Synthesis and purification of L-6-[^{18}F]fluorodopa. *Appl Radiat Isot.* 1988;39:1203–6.
72. Namavari M, Bishop A, Satyamurthy N, et al. Regioselective radio-fluorodestannylation with [^{18}F]F$_2$ and [^{18}F]CH$_3$COOF: a high yield synthesis of 6-[^{18}F]fluoro-L-dopa. *Appl Radiat Isot.* 1992;43:989–96.
73. de Vries EFJ, Luurtsema G, Brüssermann M, et al. Fully automated synthesis module for the high yield one-pot preparation of 6-[^{18}F]fluoro-L-DOPA. *Appl Radiat Isot.* 1999;51:389–94.
74. Moerlein SM, Mintun MA, Perlmutter JS, et al. USP Standards for flumazenil C 11 injection. *Pharm Forum.* 1998;24:6360–5.
75. Maziere M, Hantraye P, Prenant C, et al. Synthesis of ethyl 8-fluoro-5,6-dihydro-5-[^{11}C]methyl-6-oxo-4H-imidazo[1,5-a][1,4]benzodiazepine-3-carboxylate (RO 15.1788-^{11}C): a specific radioligand for the in vivo study of central benzodiazepine receptors by positron emission tomography. *Int J Appl Radiat Isot.* 1984;35:973–6.
76. Halldin C, Stone-Elander S, Thorell J-O, et al. ^{11}C-Labelling of Ro-15-1788 in two different positions, and also ^{11}C-labelling of its main metabolite Ro 15-3890, for PET studies of the benzodiazepine receptors. *Appl Radiat Isot.* 1988;39:993–7.
77. Suzuki K, Inoue O, Hashimoto K, et al. Computer-controlled large scale production of high specific activity [^{11}C]Ro 15-1788 for PET studies of benzodiazepine receptors. *Int J Appl Radiat Isot.* 1985;36:971–6.
78. Moerlein SM, Perlmutter JS, Welch MJ. USP Standards for Raclopride C 11 Injection. *Pharm Forum.* 1995;21:172–6.
79. Farde L, Pauli S, Hall H, et al. Stereoselective binding of 11C-raclopride in living human brain—a search for extrastriatal central D2-dopamine receptors by PET. *Psychopharmacology.* 1988;94:471–8.
80. Halldin C, Farde L, Högberg T, et al. A comparative PET-study of five carbon-11 or fluorine-18 labelled salicylamides. Preparation and in vitro dopamine D-2 receptor binding. *Nucl Med Biol.* 1991;18:871–81.
81. Crouzel C, Clark JC, Brihaye C, et al. Radiochemistry automation for PET. In: Stöcklin G, Pike VW, eds. *Radiopharmaceuticals for Positron Emission Tomography. Methodological Aspects.* Boston: Kluwer Academic Publishers; 1993:45–89.
82. Alexoff DL. Automation for the synthesis and application of PET radiopharmaceuticals. In: Welch MJ, Redvanly CS, eds. *Handbook of Radiopharmaceuticals. Radiochemistry and Applications.* Hoboken, NJ: John Wiley and Sons; 2003:229–82.
83. Fowler JS, MacGregor RR, Wolf AP, et al. A shielded synthesis system for production of 2-deoxy-2-[^{18}F]fluoro-D-glucose. *J Nucl Med.* 1981; 22:376–80.
84. Barrio JR, MacDonald NS, Robinson GD, et al. Remote, semiautomated production of F-18-labeled 2-deoxy-2-fluor-D-glucose. *J Nucl Med.* 1981;22:372–5.

85. Adam MJ, Ruth TJ, Grierson JR, et al. Routine synthesis of L-[^{18}F]6-fluorodopa with fluorine-18 acetyl hypofluorite. *J Nucl Med.* 1986;27:1462–6.
86. Luxen A, Barrio JR, Van Moffaert G, et al. Remote, semiautomated production of 6-[F-18]fluoro-L-dopa for human studies with PET. *J Lab Comp Radiopharm.* 1989;26:465–6.
87. Pike VW, Eakins MN, Allan RM, et al. Preparation of [1-^{11}C]acetate—an agent for the study of myocardial metabolism by positron emission tomography. *Int J Appl Radiat Isot.* 1982;33:505–12.
88. Welch MJ, Dence CS, Kilbourn MR. Remote systems for the routine production of some carbon-11 radiopharmaceuticals. *J Lab Comp Radiopharm.* 1982;19:1382.
89. Iwata R, Ido T, Takahashi T, et al. Automated synthesis system for production of 2-deoxy-[^{18}F]fluoro-D-glucose with computer control. *Appl Radiat Isot.* 1984;35:445–54.
90. Alexoff DL, Russell JAG, Shiue C-Y, et al. Modular automation in PET tracer manufacturing: application of an autosynthesizer to the production of 2-deoxy-2-[^{18}F]fluoro-D-glucose. *Appl Radiat Isot.* 1986;37:1045–61.
91. Padgett HC, Schmidt DG, Luxen A, et al. Computer-controlled radiochemical synthesis: a chemistry process control unit for the automated production of radiochemicals. *Appl Radiat Isot.* 1989;40:433–45.
92. Hamacher K, Blessing G, Nebeling B. Computer-aided synthesis (CAS) of no-carrier-added 2-[^{18}F]fluoro-2-deoxy-glucose: an efficient automated system for the aminopolyether-supported nucleophilic fluorination. *Appl Radiat Isot.* 1990;41:49–55.
93. Suzuki K, Tamate K, Nakayama T, et al. Development of equipment for the automatic production of ^{13}NH$_3$ and L-(^{13}N)-glutamate. *J Lab Comp Radiopharm.* 1982;19:1374–5.
94. Ruth TJ, Adam MJ, Jivan S, et al. An automated system for the synthesis of L-6[^{18}F]fluorodopa. *J Lab Comp Radiopharm.* 1991;30:304.
95. Brodack JW, Kilbourn MR, Welch MJ, et al. Application of robotics to radiopharmaceutical preparation: controlled synthesis of fluorine-18 16α-fluoroestradiol-17β. *J Nucl Med.* 1986;27:714–21.
96. Brodack JW, Kilbourn MR, Welch MJ. Automated production of several positron-emitting radiopharmaceuticals using a single laboratory robot. *Appl Radiat Isot.* 1988;39:689–97.
97. Brodack JW, Kaiser SL, Welch MJ. Laboratory robotics for the remote synthesis of generator-based positron-emitting radiopharmaceuticals. *Lab Robotics Autom.* 1989;1:285–94.
98. Brihaye C, Lemaire C, Damhaut P, et al. Robot-assisted synthesis of [^{18}F]altanserin, 4-[^{18}F]fluorotropapride, 6-[^{18}F]fluorodopa and 2-[^{18}F]fluoro-L-tyrosine. *J Lab Comp Radiopharm.* 1994;35:160–2.
99. Krasikova RN. Automated synthesis of radiopharmaceuticals for positron emission tomography. *Radiochemistry.* 1998;40:352–61.
100. Mock BH, Winkle W, Vavrek MT. A color test for the detection of kryptofix 2.2.2 in [^{18}F]FDG preparations. *Nucl Med Biol.* 1997;24:193–5.
101. Meyer G-J, Coenen HH, Waters SL, et al. Quality assurance and quality control of short-lived radiopharmaceuticals for PET. In: Stöcklin G, Pike VW, eds. *Radiopharmaceuticals for Positron Emission Tomography. Methodological Aspects.* Boston: Kluwer Academic Publishers; 1993:91–150.
102. Keppler JS. Federal regulations and reimbursement for PET. *J Nucl Med Technol.* 2001;29:173–9.
103. Welch MJ, Redvanly CS, eds. *Handbook of Radiopharmaceuticals. Radiochemistry and Applications.* Hoboken, NJ: John Wiley and Sons; 2003.
104. 18th International Symposium on Radiopharmaceutical Sciences, Edmonton, Canada, 12–17 July 2009. *J Lab Comp Radiopharm.* 2009;52:S1-S539.
105. Proceedings of the 2009 World Molecular Imaging Congress, Montreal, Canada, September 23–26. *Mol Imaging and Biol.* 2010;12(suppl 1):1–493.

CHAPTER 13
The Nuclear Pharmacy

Kristina M. Wittstrom

In a nuclear pharmacy radiopharmaceuticals are procured, prepared, stored, and dispensed primarily for patient administration within a nuclear medicine facility. Large teaching and research facilities may have a complete in-house nuclear pharmacy, while smaller institutions may have a modest space within the nuclear medicine department for minimal handling procedures. It is estimated that 70% to 80% of all radiopharmaceutical doses are dispensed through commercial centralized nuclear pharmacies.[1] This chapter examines some of the unique aspects of a commercial nuclear pharmacy.

DEVELOPMENT OF A NUCLEAR PHARMACY

The development of nuclear pharmacy advanced with the advent of the ^{99m}Tc generator in the late 1960s.[1,2] With the increasing availability of reagent kits, pharmacists became involved in the preparation and dispensing of short-lived radiopharmaceuticals. The first educational program in nuclear pharmacy was established in 1969 at the University of Southern California. Other early educational programs included those at Purdue University, the University of Michigan, the University of Tennessee, and the University of New Mexico.

The 1970s saw tremendous growth in nuclear medicine, new radiopharmaceuticals, and nuclear pharmacy. The first commercial centralized nuclear pharmacy was created in 1972 by Richard Keesee at the University of New Mexico College of Pharmacy in Albuquerque. Keesee is credited with the concept of unit dose radiopharmaceuticals: dispensing a single patient dose of radioactive drug on the prescription order of a physician. Graduates of university educational programs began to establish commercial centralized nuclear pharmacies in 1974. Nuclear pharmacies spread across the country, numbering about 50 by 1980.

Currently, the National Association of Nuclear Pharmacies estimates that about 500 centralized nuclear pharmacies staffed by about 1200 nuclear pharmacists provide radiopharmaceuticals to hospitals and clinics in the United States. Recently, centralized commercial nuclear pharmacies have been developed in other countries. The continued demand for positron emission tomography (PET) imaging has resulted in a new growth area: PET radiopharmaceuticals.

A nuclear pharmacy may serve a variety of customers and radiopharmaceutical needs. A small percentage of nuclear pharmacies are located within university hospitals or imaging centers that prepare and dispense radiopharmaceutical prescriptions for on-site use only. A university hospital nuclear pharmacy is commonly involved in research and development of new radiopharmaceuticals. Centralized commercial nuclear pharmacies offer a full range of radiopharmaceutical products, which may include PET radiopharmaceuticals. Centralized nuclear pharmacies are most commonly found in cities with populations greater than 100,000, where they supply radiopharmaceuticals to many hospitals and clinics within the geographic service area. A PET nuclear pharmacy site prepares and dispenses only PET radiopharmaceuticals. Such sites may be found within a medical center, as free-standing entities, or affiliated with a more traditional nuclear pharmacy to maximize resources. In some areas, nuclear pharmacy services may be combined with a home care pharmacy, a custom compounding pharmacy, or a traditional community pharmacy.

FACILITIES AND EQUIPMENT

The physical facility design must meet criteria of both the state board of pharmacy and the U.S. Nuclear Regulatory Commission (NRC). State boards of pharmacy have specific requirements for the provision, maintenance, and evaluation of environmental quality in facilities used for sterile pharmaceutical preparation and compounding,[3] and facilities must be designed to meet these requirements.

The NRC requires that nuclear pharmacy facilities have adequate controls to protect health and minimize danger to life or property, minimize the likelihood of contamination, and keep exposures to workers and the public as low as reasonably achievable (ALARA). The requirements include documentation that the nuclear pharmacy has sufficient engineering controls and barriers to protect the public and employees. Specifically, the facility and equipment must be designed to effectively keep personnel exposures to radiation and radioactive materials ALARA in order to minimize the risks from handling radioactive materials.[4]

There are many ways to design a nuclear pharmacy to meet regulatory requirements. Ideally, the pharmacy should be a free-standing structure or the end unit of a multitenant building to minimize common walls and reduce potential radiation exposure of the general public. The floor plan should have areas designated for specific functions, with radioactive storage areas located away from other work areas. The physical layout can vary in size and design, but most nuclear pharmacies have common features. Figure 13-1 shows a typical floor plan with work and storage areas.

The nuclear pharmacy is divided into two areas. The *unrestricted area* (shaded area in Figure 13-1) consists of an office, work areas, a conference room, an employee lounge, and the delivery foyer. The general public and employees have

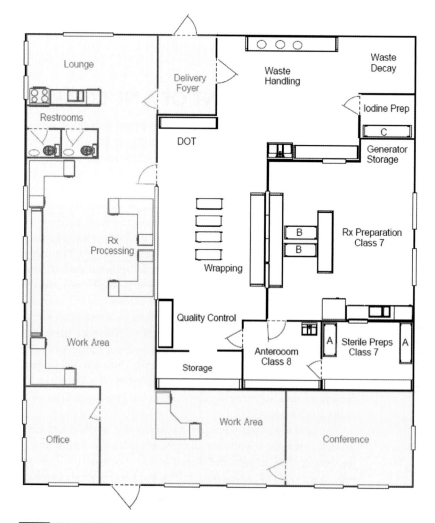

FIGURE 13-1 Nuclear pharmacy floor plan. The shaded area is the unrestricted area (no radioactivity). The unshaded area (where radioactivity is stored and handled) is restricted. (A) Biological safety cabinets, (B) laminar-airflow hoods, (C) isotope fume hood.

unrestricted access to this area. The *restricted area* (unshaded area in Figure 13-1) consists of the storage and work areas for handling radioactive materials. Only trained radiation workers have access to this area.

Unrestricted Area

The unrestricted area contains no radioactive materials. The prescription processing area is where telephone orders are received by the nuclear pharmacist. These are transcribed and entered into networked computers. The electronic prescriptions are accessed as needed from within the restricted area. Near the prescription processing area is the primary entrance to the restricted area.

At the rear of the building is the delivery foyer. This area is used for couriers to securely deliver packages after the pharmacy is closed. Outgoing packages for hospitals move through this foyer to the parking area. Returned containers are placed into the waste-handling area.

Restricted Area

The restricted area contains areas for specific functions. Some are specific for radiation safety reasons; others are specific to pharmacy compounding issues. The compounding area consists of the anteroom, the sterile preparation area, and the prescription preparation area.

Compounding Area

The area used for preparation of sterile radiopharmaceuticals must be physically designed and environmentally controlled to minimize airborne contamination. All surfaces—ceilings, walls, floors, fixtures, counters, and equipment—must be smooth and easily cleanable. Items brought into the compounding areas must be cleaned and disinfected beforehand. Equipment and other items used for compounding should reside permanently within the compounding area and should be removed only for necessary maintenance.

All work surfaces must be smooth and nonpermeable for easy cleaning and disinfecting. Surfaces should be constructed of materials resistant to the absorption of liquid, such as stainless steel, plastic, epoxy resin, or Formica. Surfaces in areas where radioactive materials are routinely handled should be protected with plastic-backed absorbent sheeting. This sheeting absorbs liquid contamination and can be easily removed. Environmental testing is conducted with such paper in place to approximate real conditions.

Sinks with hot and cold water are often found in the compounding area. A sink located in the anteroom is for personnel hygiene before entering the other areas. A sink is often found within the prescription processing area; this is most often used for hand washing before leaving the compounding area. Sinks are not intended for the disposal of radioactive solutions.

Entry into the compounding area is through the anteroom. This is an International Organization for Standardization (ISO) Class 8 or better area where personnel hand hygiene and garbing procedures, supply preparation, and other high-particulate activities are performed.[3] This area is under positive pressure to ensure that air flows from clean to dirty areas.

After preparation activities in the anteroom, the pharmacist may enter the sterile preparation area, which is an ISO Class 7 buffer area. This area contains primary engineering control devices such as compounding aseptic containment isolators (CACIs) that provide an ISO Class 5 environment.[3] Typical CACIs used are vertical laminar-airflow workbenches (LAFWs) or biologic safety cabinets (BSCs).[3] Devices of this type protect the pharmacist from exposure to airborne contaminants and provide an aseptic environment for open-transfer product preparation. The sterile preparation area is isolated from other activities to minimize traffic and to allow uninterrupted concentration on the compounding activity. Here unique, patient-specific radiopharmaceuticals are compounded upon physician order. A BSC that meets the requirements for an ISO Class 5 environment for radioactive labeling procedures is essential for open-container procedures such as blood cell labeling.

From the anteroom, the properly prepared pharmacist enters the buffer zone for prescription preparation. This area is under positive pressure to both the anteroom and the wrapping area and meets the requirements for an ISO Class 7 environment. Each nuclear pharmacy has one or more radiopharmaceutical dose-drawing stations in this area that are located within vertical LAFWs with appropriate shielding and dose calibrators. All closed-system compounding is performed within these hoods in an ISO Class 5 environment. Each drawing station is equipped with a touch-screen monitor linked to the pharmacy computer system (Figure 13-2). This

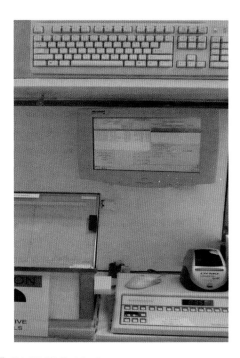

FIGURE 13-2 Drawing station configured with computer monitor for electronic record keeping. Keyboard is located above the hood opening. Calibrator display box is on the bottom right.

FIGURE 13-3 Labeled unit dose syringe being placed into lead syringe shield.

allows for synchronous record keeping to be done electronically and eliminates the need for paper records.

Unit dose radiopharmaceuticals are also prepared within the vertical laminar-airflow hood. The labeled syringe is placed into a lead syringe shield (Figure 13-3). Computer-generated prescription labels are printed near the hood. The shielded dose and its label are passed through a positive-pressure opening to the wrapping area.

A nuclear pharmacy uses many large (multicurie) generators each week. The generators are stored within the prescription preparation area to maintain clean environmental conditions. They are stored in the manufacturers' shielding, some distance away from the compounding stations to minimize employee exposure to radiation.

Many radiopharmaceuticals require refrigeration or freezing to maintain product integrity and to prevent microbial growth and ensure sterility. No food or drink should be stored in the refrigerator located in the prescription preparation area. The refrigerator and freezer compartment temperatures should be monitored and recorded daily.

Outside the Compounding Area

In the wrapping area, unit dose shields are closed and labeled. Security shrink-wrap is often added at this point. The shields are wipe tested for removable contamination before being placed into a designated shipping container. Once each hospital order is complete, the shipping container is moved on a cart to the Department of Transportation (DOT) area.

Doses are confirmed against a packing list, and the shipping container is closed and secured. Each container is monitored for DOT labeling requirements. Two labels are completed and affixed to the container. After shipping papers are completed, the container is wipe tested for removable contamination before release for transport and delivery. DOT requirements are discussed in Chapter 7.

While doses are being wrapped and prepared for shipment, each compounded radiopharmaceutical is tested in the quality control area. Although this is not required by regulation, nuclear pharmacy standards of practice recommend that each radiopharmaceutical be tested for purity before its release for patient administration. Testing includes radionuclidic and radiochemical purity checks. These quality control tests are discussed in Chapter 14.

The iodine preparation room contains an exhausting fume hood for the storage of volatile radioactive materials, usually ^{131}I and ^{133}Xe. The fume hood ducting contains special traps to monitor the release of effluent into the atmosphere (see Chapter 7, Radiation Safety). A glove box unit that exhausts into the main fume hood may be used. Manipulations of volatile substances are performed within the glove box. A second "trap" is located near the glove box to monitor potential employee exposure.

As a service to unit dose users, most nuclear pharmacies offer a waste management system. Spent syringes are returned in shields and DOT shipping containers to the nuclear pharmacy. The containers are opened and the spent syringes are sorted by isotope half-life into large cardboard barrels contained in movable lead barrels (Figure 13-4). The waste is held for decay in the waste storage area. After 10 half-lives, the waste is transferred to a biohazardous-waste disposal service (see Chapter 7).

Storage and Shielding of Radioactive Materials

To minimize employees' exposure to radiation, protective equipment should be used in the storage and handling of radioactive materials. Most protective devices are made of lead in different configurations. Basic shielding equipment is

FIGURE 13-4 Lead barrel used for radioactive waste storage.

TABLE 13-1 Storage and Shielding Equipment for Radioactive Materials

Lead-lined storage areas for
    Radiopharmaceuticals
      Generators
      Radioactive sealed sources
      Radioactive waste
Lead L-block with leaded-glass viewport
Lead bricks
Lead waste barrels
Vial shields (lead or tungsten)
Unit dose syringe delivery shields (lead or tungsten)
Dispensing syringe shields
Lead sheeting of various size and thickness

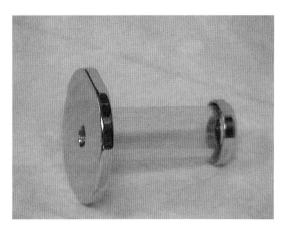

FIGURE 13-6 Dispensing-syringe shield used in drawing doses from radiopharmaceutical vials.

listed in Table 13-1. Storage equipment is available in a variety of sizes and designs to accommodate specific needs.

Lead bricks, usually $2 \times 4 \times 8$ inches, are useful for providing extra shielding in areas where extemporaneous shielding is required, such as in the fume hood. Bricks provide the flexibility to move or change shielding as needed. Lead bricks are available from commercial suppliers.

Vial shields of either lead or tungsten can be purchased or salvaged from manufacturer shields supplied with radiopharmaceuticals. Vial shields are used to shield generator elution vials, prepared radiopharmaceuticals, and multidose vials prepared for customers. The thickness of lead shields should match the gamma ray energy of the source in order to minimize exposure during handling and transit to the nuclear medicine facility. The need for adequate shielding is most important for handling and transporting high gamma energy sources, such as ^{131}I and ^{18}F, which require thicker lead to reduce exposure rates.

There are several styles of syringe shields. Some shields are designed to be used while injecting patients; these are often made of lead with a leaded-glass inset to enable the syringe markings to be read (Figure 13-5). The dispensing-syringe shield (Figure 13-6) is unique to nuclear pharmacy. It is made entirely of leaded glass to eliminate the need to manipulate a syringe to see the markings. The needle end of the shield is threaded to lock the syringe in place during compounding procedures. A chrome-covered lead plate is affixed to the end of the shield to protect the pharmacist's hands from the cone of radiation emitted from a bulk radiopharmaceutical preparation.

A lead L-block (Figure 13-7) is an L-shaped piece of lead with an inset of leaded glass to permit viewing the work area. An L-block is routinely used in dispensing stations to protect the worker's body and face during manipulation of radioactive materials. L-blocks are also used at the quality control station, in the glove-box exhausting fume hood, and in front of devices used to heat radiopharmaceuticals during compounding.

Hoods

As discussed earlier, it is desirable to have several types of hoods within the restricted area. Exhaust hoods are used to contain volatile radiopharmaceuticals such as iodine and xenon (C in Figure 13-1). A traditional chemical exhaust hood is acceptable if the process does not require a sterile air environment. Special monitoring systems are used to assess containment of the volatile radioactivity.

Procedures that involve the labeling of blood cells and proteins or open-container compounding require a sterile air environment and operator biologic safety. A vertical-flow BSC (A in Figure 13-1) is recommended for handling any potentially biohazardous material, including radioactive material. Air entering the hood is forced through a high-efficiency particulate air (HEPA) filter to render it 99.9% free of particles less than 0.3 μm. Air leaving the hood passes through a second HEPA filter to prevent the release of potentially biohazardous material into the workspace. The hood must meet ISO Class 5 sterile environment conditions.

A vertical-flow BSC that exhausts through a HEPA filter into the room is adequate for most procedures except those involving volatile radionuclides, which require a system that exhausts external to the facility. Dispensing stations are usually placed within a vertical-flow hood meeting ISO

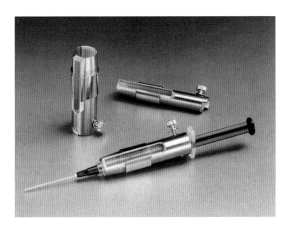

FIGURE 13-5 Syringe shields with leaded-glass viewing area. (Used with permission of Biodex Medical Systems, Inc.)

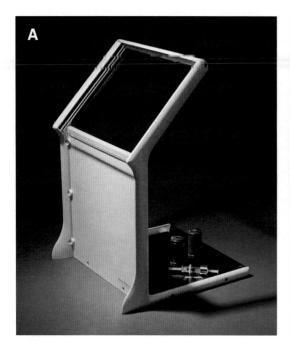

FIGURE 13-7 Lead-shielded L-block drawing stations with leaded-glass viewing area. (A) Standard L-block (½ inch thick lead) with ¼ inch thick leaded glass for single-photon-emitting radiopharmaceuticals; (B) PET 511 L-block (1 inch thick lead) with 4 inch thick leaded glass for positron-emitting radiopharmaceuticals. (Used with permission of Biodex Medical Systems, Inc.)

Class 5 standards (B in Figure 13-1). A hood should be allowed to run for at least 30 minutes at the beginning of a new workday before cleaning, decontamination, and use and should remain operational throughout the workday. Figure 13-8 illustrates the two basic types of laminar-flow hoods, the horizontal-flow hood and the vertical-flow hood. The vertical-flow hood is best suited for nuclear pharmacy procedures.

RADIATION DETECTION INSTRUMENTATION

The devices used to measure or detect radioactivity in a nuclear pharmacy include dose calibrators, portable Geiger-Müller (GM) survey meters, ion chambers or meters with energy-

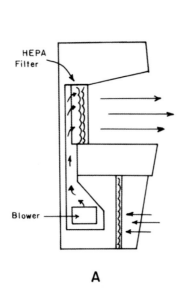
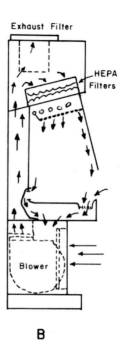

FIGURE 13-8 Laminar-flow hoods. (A) Horizontal-flow hood; (B) vertical-flow hood.

compensated probes, area monitors, and scintillation detectors with single-channel or multichannel analyzers. The principles of operation of these instruments are discussed in Chapter 5.

Dose Calibrator

The dose calibrator is an ionization chamber that is calibrated to measure the radioactivity of different isotopes used in radiopharmaceutical doses (Figure 13-9). It is capable of measuring a range of activities from microcuries to curies. This easy-to-use instrument is adjusted for the radionuclide to be measured, by either selecting a preset button or adjusting a potentiometer to a specific calibration number. The source of radioactivity is placed into the chamber well and the activity is displayed on a digital readout.

The user must be aware of geometric factors that may affect the readout on a dose calibrator. This is particularly important in measuring the activity of low photon energy nuclides, such as ^{125}I (27–35 keV) and nuclides that produce abundant low-energy characteristic x-rays, such as ^{123}I (27–31 keV) and ^{111}In (23–26 keV). The type of container (e.g., plastic syringe, glass vial) can affect the readout; different readings of the same amount of activity may be obtained as a result of differences in absorption of low-energy photons by the container. Correction factors must be applied to accurately measure certain radionuclides in different configurations. Nuclides such as ^{123}I and ^{111}In emit high-energy gamma rays for imaging, but they also emit a high abundance of low-energy x-rays in the 23 to 31 keV range that will affect the dose calibrator readout with different types of containers. As described in Chapter 5, a simple device using a copper filter inserted into the dose calibrator eliminates this problem by absorbing the low-energy x-rays without affecting the high-energy gamma rays.[5]

Dose calibrators are quite rugged and operate satisfactorily for many years. Daily and periodic quality control assessments are required to ensure their accurate operation. These assessments are discussed further in Chapter 14.

Geiger-Müller Survey Meters

As discussed in Chapter 5, the GM detector is a gas-filled portable device for measuring radiation exposure in counts per minute (cpm) or milliroentgens per hour (mR/hr). The meter measures radiation detected by a probe. The preferred probe for use in nuclear pharmacies is the pancake probe. The front of the probe (Figure 13-10) has little shielding, allowing for a very sensitive check for contamination. The back of the probe is covered with a metal plate. Since the meters are calibrated using the back side of the probe, the back side can be used to assess radiation fields. Other types of probes include the end-window probe, suitable for contamination checks, and beta probes, which have a sliding beta window, allowing for monitoring both gamma and beta emissions.

Survey meters are the workhorses of a nuclear pharmacy. They are used in checking packages of radioactive materials, checking for contamination in work areas, monitoring personnel, and measuring radiation fields. However, a GM survey

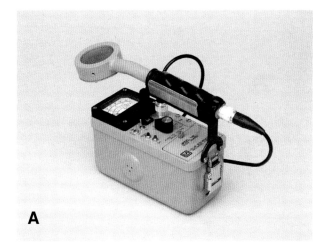

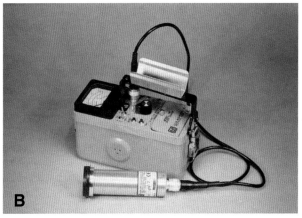

FIGURE 13-10 Geiger-Müller survey meters with (A) pancake probe and (B) end-window probe. (Used with permission of Biodex Medical Systems, Inc.)

FIGURE 13-9 Dose calibrator for assaying radiopharmaceuticals. (Used with permission of Capintec, Inc.)

meter is limited in the amount of radiation exposure it can detect accurately. Most portable survey meters are accurate to 1 or 2 R/hr. Detection of radiation exposure in excess of these levels requires the use of a different instrument.

Ionization Chambers and Energy-Compensated Probes

High levels of radiation exposure such as those from multicurie ^{99m}Tc generators or high-activity ^{131}I shipments may require an instrument that can read above the levels of the GM survey meter. An ionization chamber (Figure 13-11) will read high levels of radiation exposure accurately, but not as rapidly as a GM meter. An alternative to an ionization chamber is a multipurpose survey meter that not only operates with a gas-filled probe but also has an internal energy-compensated solid detector probe. If a nuclear pharmacy does not receive or ship packages containing high activity levels of radioactive materials with high radiation exposure rates, then this additional equipment is not necessary.

Area Monitors

Area monitors are similar to GM survey meters except that they are usually stationary and are plugged into an electrical outlet. Area monitors are placed at the exit points of a restricted area and in other areas where ambient radiation exposure is a concern. Many monitors are equipped with an adjustable alarm that can be set to go off when a specific radiation level is exceeded. A device is sometimes equipped with two probes and is used to monitor hands, feet, and clothing before personnel exit the restricted area. A typical monitor is shown in Figure 13-12.

FIGURE 13-11 Cutie Pie ionization chamber. (Photo courtesy of Biodex Medical Systems, Inc.)

Scintillation Well Counters

A scintillation well counter consists of a sodium iodide crystal detector designed with a well to accept test tubes. The crystal is coupled to a photomultiplier tube. This unit detects gamma radiation and generates voltage pulses proportional to the gamma energy deposited in the crystal. The pulses are sent to a single-channel or multichannel analyzer (spectrometer) that can discriminate between the photon energy pulses. This permits the identification of unknown radionuclide samples and measurement of the amounts of radioactivity in samples containing one or more known nuclides. The well counter is used primarily to count small amounts of radioactivity (<1 µCi) present in room wipes, package wipes, and samples of biologic fluid (e.g., plasma or urine). This sensitive instrument must

FIGURE 13-12 (A) Geiger-Müller alarm rate meter for monitoring ambient radiation levels and (B) foot monitor. (Used with permission of Biodex Medical Systems, Inc.)

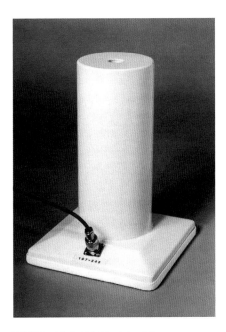
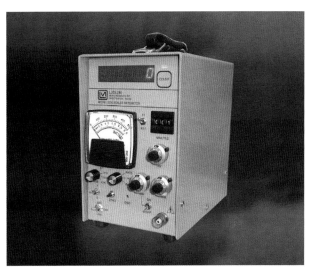

FIGURE 13-13 Sodium iodide scintillation well counter. (Used with permission of Biodex Medical Systems, Inc.)

be located in a low-level background counting area. An example of a scintillation well counter is shown in Figure 13-13.

MISCELLANEOUS EQUIPMENT AND SUPPLIES

A number of other items are necessary in the nuclear pharmacy. A microscope and hemacytometer are used to size particulate-containing radiopharmaceuticals such as radiocolloids, macroaggregated albumin, and labeled cells. A centrifuge is necessary for separating blood cells from plasma for cell-labeling studies. Several radiopharmaceuticals require heating for periods of time; a water bath or heat block is often used to heat these products. A refrigerator and freezer are necessary for proper storage of radiopharmaceuticals. Daily temperature checks should be made on both the refrigerator and the freezer. Various types of chromatography materials and reagents are needed for quality control of prepared radiopharmaceuticals. An in-depth discussion of these supplies can be found in Chapter 14.

PERSONNEL

A nuclear pharmacy is licensed by a state board of pharmacy. Compounding and dispensing operations must be done under the supervision of a licensed pharmacist and in accordance with all pharmacy statutes and regulations. The possession, handling, and dispensing of radiopharmaceuticals is regulated by the NRC or an agreement-state regulatory body. NRC regulations require that the use and handling of radiopharmaceuticals be under the supervision and direction of an authorized nuclear pharmacist (ANP).

Authorized Nuclear Pharmacist

All activities within a nuclear pharmacy must be performed under the supervision and direction of an ANP. An ANP has completed traditional pharmacy education, is licensed as a pharmacist within the state of practice, and has completed an accredited education and training program consisting of 200 contact hours of didactic education and 500 hours of experiential training. A handful of universities offer an undergraduate ANP program. Alternatives to the university classroom programs include university-sponsored distance education programs and on-site programs. Commercial nuclear pharmacy chains may offer an internal course for employees.

The primary responsibility of an ANP is to prepare sterile, efficacious radiopharmaceuticals and provide the right drug in the right dose for the right patient at the right time. The ANP must have a strong knowledge base in nuclear pharmacy methods and techniques and in the scientific principles underlying the practice of pharmacy.

Nuclear Pharmacy Technician

As specified by individual state boards of pharmacy, nonpharmacist personnel may be allowed to perform routine, nonjudgmental dispensing tasks. Nuclear pharmacy technicians (NPTs) may be certified by the Pharmacy Technician Certification Board (CPhT). However, preparation for this national technician certification does not include the training necessary to work with radiopharmaceuticals. Formal instructional programs specific to NPT tasks and responsibilities are available, in addition to in-house training. Tasks commonly delegated to technicians include routine dispensing of diagnostic doses, quality control testing, inventory control, packaging, and record keeping.

Delivery Personnel

Most nuclear pharmacies maintain a staff of delivery personnel and a fleet of vehicles. Delivery personnel are specially trained in the handling of packages containing radioactive material to be transported to the final user (hospital or clinic). All personnel working in a nuclear pharmacy have received training in emergency procedures involving packages containing radioactive materials.

Handling Techniques

Nuclear pharmacy techniques can be divided into two categories: protective techniques and aseptic techniques. Protective techniques are methods that prevent or minimize radioactive contamination and unnecessary exposure of personnel to radiation. These techniques maximize the basic tenets of radiation safety: time, distance, and shielding. Several common safety techniques are listed in Table 13-2.

Aseptic techniques are methods that prevent or minimize the chance of microbial contamination of sterile solutions and devices. The sterile compounding of radiopharmaceuticals is addressed in Chapter 797 of *The United States Pharmacopeia (USP)*. Microbiologic requirements for radiopharmaceuticals are discussed in Chapters 14 and 15. Compounding personnel must be adequately educated and trained to correctly perform specific sterile compounding activities.[3] Minimum requirements for aseptic handling are listed in Table 13-3. Each nuclear pharmacy has an internal policies and procedures manual that specifies procedural detail and outlines the required techniques.

NUCLEAR PHARMACY PRACTICE

The Section on Nuclear Pharmacy of the American Pharmacists Association has developed nuclear pharmacy practice guidelines; these were established initially in 1978 and updated most recently in 2007. These guidelines form the content outline for the nuclear pharmacy specialty certification examination

TABLE 13-2 Protective Techniques in Nuclear Pharmacy

Wear disposable gloves and a street-length lab coat.
Handle radioactive materials behind a lead L-block.
Use syringe shields when preparing and dispensing radioactive doses.
Use tongs to move unshielded radioactive materials to maximize distance.
Cover work area with absorbent plastic-backed paper.
Maintain negative vial pressure to prevent spraying.
Plan ahead and work quickly and efficiently to minimize handling time.
Do not eat or drink in restricted areas.

TABLE 13-3 Aseptic Techniques in Radiopharmacy

Perform aseptic hand cleaning and wear protective clothing.
Perform disinfection of nonsterile compounding surfaces.
Maintain sterility in ISO Class 5 hoods while protecting personnel and environment from radioactive contamination.
Use proper technique to prevent needle coring when entering vials.
Manipulate sterile products aseptically.
Inspect all radiopharmaceutical components, materials, devices, and solutions for accuracy and integrity.
Comply with institutional policies and procedures.

administered by the Board of Pharmaceutical Specialties.[6] The outline is based on a role delineation study of practicing nuclear pharmacists. The 2007 outline consists of domains, subdomains, tasks, and knowledge statements pertinent to the practice of nuclear pharmacy. Table 13-4 summarizes the practice areas.

The guidelines do not consider differences in practice settings, job responsibilities, or other factors. The guidelines may not be applicable to all nuclear pharmacists, nor are they all-inclusive. Pharmacists should use professional judgment in interpreting the guidelines.

Drug Order Provision

Procurement

Procurement of radiopharmaceuticals and the associated activities are learned from experience on the job. Some tasks, such as determining product specifications, initiating purchase orders, receiving shipments, maintaining inventory, and ensuring proper storage of materials, are similar to those performed in other areas of pharmacy practice. The art of procurement involves anticipating daily needs for short-lived radiopharmaceuticals and obtaining amounts sufficient to meet these needs with minimal waste. Procurement involves daily, and often hourly, analysis of inventory against anticipated next-day demands.

Ordering radiopharmaceuticals requires a thorough knowledge of calibration and expiration dates and times and of shipping and delivery schedules. Unlike traditional pharma-

TABLE 13-4 Content Outline for Nuclear Pharmacy Specialty Certification Examination

Domain 1: Drug Order Provision
   Subdomain A: Procurement
   Subdomain B: Compounding
   Subdomain C: Quality Assurance
   Subdomain D: Dispensing
Domain 2: Health and Safety
Domain 3: Drug Information Provision

ceuticals, radiopharmaceuticals are ordered directly from the manufacturer. A nuclear pharmacy may have daily standing orders for some products, but these orders must be constantly assessed and adjusted. Most orders are shipped by air to arrive in the early hours of the morning, often before the nuclear pharmacy opens for the day. The materials must be on hand in the pharmacy in time to get them to hospitals and clinics before their nuclear medicine departments open.

Nuclear pharmacies have a designated delivery area for shipments arriving when the nuclear pharmacy is closed. This may be an exterior lock box or a secured foyer to which the delivery company has a key. Package receipt involves following the regulatory procedures for opening packages, as described in Chapter 7.

Compounding

Although many radiopharmaceuticals are available in ready-to-use form, most radiopharmaceuticals must be compounded on an as-needed basis. Compounding activities can range from the relatively simple task of reconstituting reagent kits with ^{99m}Tc-sodium pertechnetate to complex tasks such as operating a cyclotron and synthesizing PET radiopharmaceuticals. Nuclear pharmacists may also perform extemporaneous compounding of commercially unavailable radiopharmaceuticals, such as ^{123}I-iobenguane sulfate (^{123}I-MIBG). As in other pharmacy practice settings, a valid prescription order is needed. Other considerations include appropriate components, supplies, and equipment; a suitable environment for sterile dosage forms; and appropriate record keeping, including lot-specific information to ensure traceability and validation or verification of the radiopharmaceutical's compounding procedure, storage, and expiration date.

Most radiopharmaceuticals are prepared using ^{99m}Tc from a radionuclide generator. The generator is eluted to provide the radioisotope for labeling the drugs to be dispensed. Physical information about the generator (activity, volume, time) is immediately recorded in an electronic worksheet (Figure 13-14) that is linked to the dose calibrator and to the pharmacy computer system.

The technetium radiopharmaceuticals are prepared by reconstituting reagent kits with ^{99m}Tc-sodium pertechnetate. Each nuclear pharmacy has internal written procedures for the reconstitution of ^{99m}Tc kits. If no written procedure has been tested and validated, the manufacturer's package insert recommendations should be followed.

The preparation of radiopharmaceuticals labeled with technetium and other radionuclides is more complex than most traditional drug compounding because issues of radio-

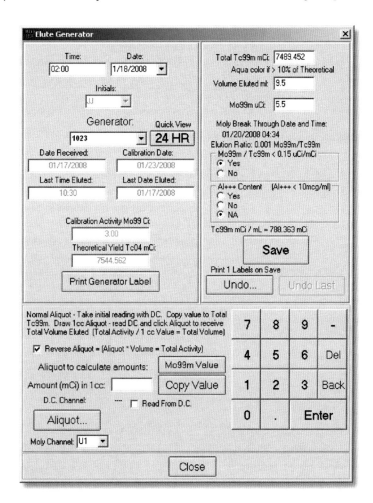

FIGURE 13-14 Generator elution worksheet screen. (Used with permission of BioDose, Las Vegas, NV.)

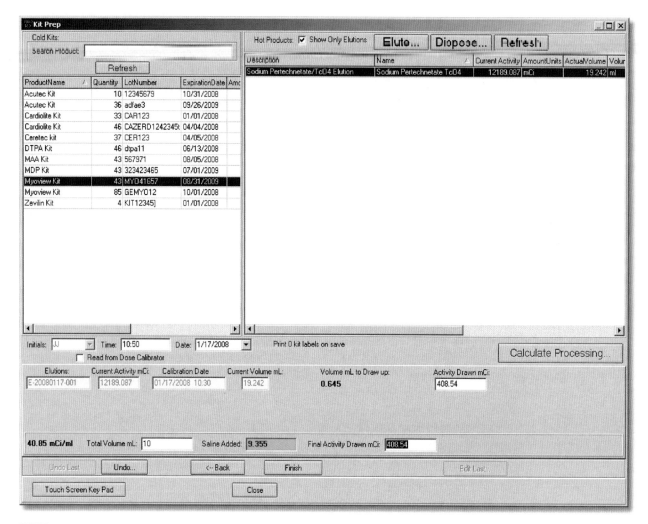

FIGURE 13-15 Kit preparation worksheet screen. (Used with permission of BioDose, Las Vegas, NV.)

active accountability and decay, radiation protection, and radiolabeling conditions must be considered. A kit preparation screen (Figure 13-15) allows the nuclear pharmacist to select the components to be compounded and simultaneously generate the required records, including a lot number. All information about each component used in the radiopharmaceutical preparation, such as the manufacturer's lot number, expiration, time and activity of reconstitution, and volumes used, is recorded as each product is prepared.

Nontechnetium radiopharmaceuticals are commercially available as approved drugs. Some of these products are in ready-to-use form and simply require a decay calculation prior to dispensing a patient dose. Examples include ^{67}Ga-gallium citrate, ^{201}Tl-thallium chloride, ^{131}I-sodium iodide, and ^{111}In-indium pentetate. Other products are in kit form, and the preparation of these products requires following the manufacturer's package insert instructions. Examples include ^{111}In-pentetreotide (OctreoScan [Covidien]) and ^{111}In-capromab pendetide (ProstaScint [Cytogen]). Record keeping for these products is similar to that for technetium radiopharmaceuticals.

Upon a written order from a physician for a specific radiopharmaceutical for a specific patient, a nuclear pharmacist may extemporaneously compound a radiopharmaceutical that is not commercially available. Standards of practice for reagent chemical purity, sterility, and apyrogenicity become the responsibility of the nuclear pharmacist. Such compounding activities must be performed and documented in accordance with local state board of pharmacy and Food and Drug Administration (FDA) regulations.

The compounding of PET radiopharmaceuticals requires more controls, validation procedures, and record keeping than are needed for any other radiopharmaceutical. PET radiopharmaceuticals may require the pharmacist to synthesize the desired radionuclide, purify it, and chemically incorporate it into a biologic tracer form in accordance with good manufacturing practices. Additional information on PET radiopharmaceuticals can be found in Chapter 12.

Quality Assurance

Quality assurance of radiopharmaceuticals involves the performance of appropriate chemical, physical, and biologic tests to ensure that the product is suitable for human use. Certain standards, such as sterility and apyrogenicity, may be guaranteed by the manufacturer of commercially available reagent kits; this minimizes the quality assurance demands on the nuclear pharmacist. When radiopharmaceuticals, including PET products, are compounded extemporaneously, verification of product specifications is the responsibility of the nuclear pharmacist. In such cases the pharmacist's responsibilities include not only performing the test(s) but also interpreting the results, evaluating analytical test methods, performing calibration or functional checks of equipment and instruments used, and keeping appropriate records. Record keeping is more than just ensuring component traceability; records must also document procedure validation, including test results and analysis. Radiopharmaceuticals must meet all specifications described in the appropriate *USP* monograph. These include radionuclidic purity, radiochemical purity, chemical purity, pH, particle size, sterility, apyrogenicity, and specific activity. Quality assurance testing is described in detail in Chapters 14 and 15.

Dispensing

An authorized prescription order is made by a nuclear medicine physician (or delegate) in accordance with state and local requirements. The radiopharmaceutical is dispensed to the authorized nuclear physician at an authorized location and in accordance with the site's radioactive materials license.

Prescription orders can be received electronically into the pharmacy computer system or entered manually after a telephone order. The computer system provides many regulatory checks on the validity of orders. An order entry screen is shown in Figure 13-16.

Radiopharmaceuticals are not dispensed directly to patients, but to those health professionals licensed to administer radiopharmaceuticals. Most radiopharmaceutical prescriptions are dispensed as unit doses ready for administration to a particular patient. Multidose vials of radiopharmaceuticals may be delivered to nuclear medicine departments to cover emergency or unexpected situations "per physician order."

Each state's board of pharmacy and radiation safety agency have specific requirements for the labeling of radiopharmaceuticals. The syringe or vial that contains the radioactive material must be tagged with specific information, including prescription number, patient name, radiopharmaceutical name, activity dispensed, and date and time of calibration. The labeled syringe or vial is placed into a lead shield, which also must be labeled. The shield label is larger and can accommodate additional information, such as directions for storage or administration.

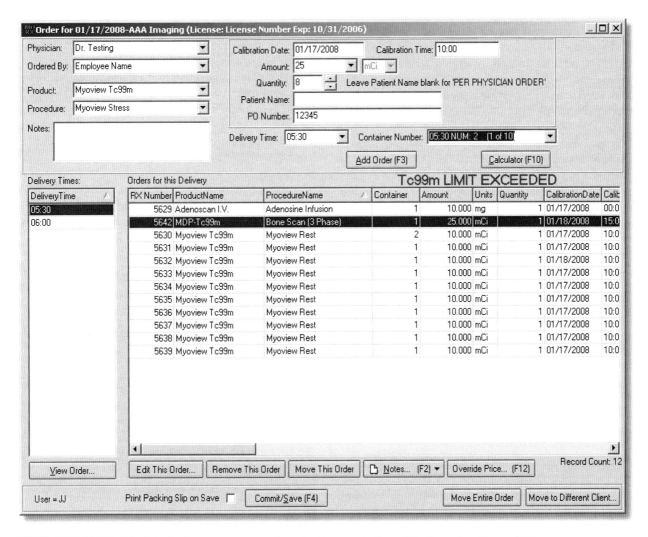

FIGURE 13-16 Order entry screen. (Used with permission of BioDose, Las Vegas, NV.)

The nuclear pharmacist is responsible for ensuring that the radiopharmaceutical dosage is consistent with the prescription order and is appropriate for a particular patient study. Patient considerations may include prior history, age, weight, sex, and disease state. Other considerations include adjustment for radioactive decay between preparation and dispensing times and between dispensing and administration times. For some radiopharmaceuticals, stability concerns may mandate a shorter product expiration time. The nuclear pharmacist must have a working knowledge of product chemistry and pharmacokinetics to optimize patient care.

Most radiopharmaceuticals are injectable products and require the use of aseptic technique during compounding and dispensing. The nuclear pharmacist must be able to ensure maintenance of sterile controls throughout both processes. This involves the proper use of well-maintained laminar-airflow hoods for both compounding and dispensing procedures.

Distribution

The distribution of radiopharmaceuticals is a major responsibility of the nuclear pharmacist. After radiopharmaceuticals have been compounded and dispensed, they must be transported to the final site of use. Radiopharmaceuticals must arrive in a timely manner so as not to disrupt the flow of a nuclear medicine department's operations. There are many regulatory and logistical considerations in the distribution process.

DOT regulates packaging, labeling, shipping papers, employee training, and actual transport. The recipient of radioactive material has regulatory requirements regarding where, when, and how deliveries are to be made. Other regulatory groups that may be involved include the Environmental Protection Agency (EPA; for biohazardous waste transport), hazardous materials agencies, and state law enforcement agencies (for example, the California Highway Patrol requires a special license to transport radioactive materials). A nuclear pharmacist must be knowledgeable about the applicable local, state, and federal regulations governing the transport of radioactive materials.

Nuclear pharmacies routinely offer established deliveries at certain times throughout the day. These are often identified with production runs. For example, the first run is the first compounding session of the day, with doses delivered prior to 0700; the second run provides doses delivered around midday; and the third run provides late afternoon products. Each run is divided into routes of several hospital deliveries in sequence. In addition, deliveries are made outside these regular runs to meet same-day orders and emergency orders. The nuclear pharmacist must have a comprehensive working knowledge of transport regulations, know the location and distance of each hospital, and efficiently juggle delivery personnel and vehicles for timely delivery of radiopharmaceuticals. Larger nuclear pharmacies may use a dispatcher to handle delivery logistics.

The request for radiopharmaceuticals does not cease at 1700. The nuclear pharmacist is available 24 hours a day, 7 days a week to provide imaging materials. After-hours requests often are of an emergency nature and require expeditious delivery. Sometimes the nuclear pharmacist not only compounds and dispenses the prescription but also delivers it.

Health and Safety

Radiation safety requirements and standards have been established and are enforced by the NRC or agreement states. These requirements include limits for radiation doses, area levels of radiation, airborne concentrations of radioactivity, waste disposal, and precautionary procedures to protect the health and safety of the occupationally exposed worker and the general public. Figure 13-17 shows a home page for daily regulatory tasks. The ANP designation signifies that a nuclear pharmacist is not only capable of operating a nuclear pharmacy but is also qualified to function as the site radiation safety officer. Radiation safety issues are discussed in detail in Chapter 7.

In addition to radiation safety concerns, other aspects of health and safety are important. The Occupational Safety and Health Administration regulates chemical safety and other personnel hazards. EPA regulates the release of radioactivity into the air (air monitoring) and the handling of biologic or biohazardous materials. The federal agencies may often have a local or state equivalent with regulations that are equally or more restrictive. To prevent citations and penalties, nuclear pharmacies must be in compliance with all safety standards.

Drug Information Provision

A nuclear pharmacist is part of the nuclear medicine team. Nuclear pharmacists share their expert knowledge by providing the appropriate information to physicians, technologists, patients, and others. This information ranges from regulatory requirements to patient-specific variables. Table 13-5 lists types of information often requested from a nuclear pharmacist.

The nuclear pharmacist provides information in many different settings. Educational information may be presented to the pharmacy staff, a hospital, a local or national professional meeting, or regulatory agencies. Organizational policies

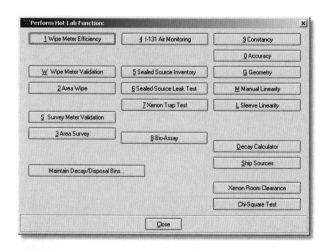

FIGURE 13-17 Hot lab screen showing routine tasks to be completed for regulatory compliance. (Used with permission of BioDose, Las Vegas, NV.)

TABLE 13-5 Information and Consultation Topics

Biological effects of radiation
Radiation physics
Radiopharmaceutical compounding
Quality assurance
Clinical applications of radiopharmaceuticals
Pharmacologic interventions used with radiopharmaceuticals
Drug–radiopharmaceutical interactions
Adverse reactions to radiopharmaceuticals
Patient-specific variables that alter radiopharmaceutical distribution
Radiopharmaceutical product defects
Regulatory issues

TABLE 13-6 Representative List of Required Records

Generator elution records
Compounding records
Dispensing records
Radiopharmaceutical quality assurance testing
Incoming package receiving record
Outgoing package shipping papers
Room wipes and surveys
Air monitoring
Employee bioassay
Employee training documents
Employee dosimetry reports
Equipment testing—daily and periodic
Laminar-airflow hood cleaning
Laminar-airflow integrity testing
Temperature checks of refrigerator/freezer
Waste disposal records—hazardous material and radioactive materials

and procedures are often developed under the guidance of a group of nuclear pharmacists. These may be for a single pharmacy or a corporate structure or for the benefit of the profession. Information pertinent to the care of a specific patient (pharmaceutical care) is requested daily by physicians and technologists.

Although nuclear pharmacists usually do not have direct contact with patients, they provide pharmaceutical care indirectly. Their contributions to patient outcomes may include patient preparation, dose calculation, the use of proper interventional agents and imaging sequence, and the development of institutional standards for the use of radiopharmaceuticals.

The development of new radiopharmaceuticals and clinical applications is vital for the viability and future growth of nuclear medicine and the profession of nuclear pharmacy. Nuclear pharmacists may participate in the development of new radiopharmaceuticals, new compounding procedures, and quality control tests. Many nuclear pharmacists participate in clinical investigations and evaluations of new uses of radiopharmaceuticals. An important contribution of the nuclear pharmacist is the dissemination of information about new products and techniques to practicing nuclear medicine physicians and technologists.

Records in the Nuclear Pharmacy

NRC and agreement-state agencies require an accounting of all transfer, disposal, and decay of radioactive material. Records substantiating daily and periodic testing must be maintained for specific amounts of time. DOT requires that copies of shipping documents be maintained for 1 year. EPA wants "cradle to grave" traceability of biohazardous waste. Compounding and dispensing records must also be kept. Boards of pharmacy expect certain records, such as temperature checks on refrigerators and freezers. Nuclear pharmacy internal protocols and procedures may require additional records not mandated by a regulatory body. A key concept promulgated by all regulatory agencies is that if a written record is not available, the process or procedure did not occur. Table 13-6 provides a representative list of required records for a nuclear pharmacy.

Summary

In summary, a nuclear pharmacist procures, prepares, and dispenses radiopharmaceuticals for use in nuclear medicine procedures. A nuclear pharmacist is uniquely educated in the art of pharmacy and the safe handling of radioactive materials. The nuclear pharmacist is responsible for safe and efficacious radiopharmaceutical preparation and delivery. The nuclear pharmacist is an important member of the nuclear medicine team, offering pharmaceutical and educational information to nuclear medicine practitioners and pharmaceutical care to patients in nuclear medicine.

REFERENCES

1. Shaw SM, Ponto JA. Nuclear pharmacy practice. In: Gennaro AR, ed. *Remington: The Science and Practice of Pharmacy*. 20th ed. Philadelphia: Lippincott Williams & Wilkins; 2000.
2. Brucer MA. *A Chronology of Nuclear Medicine 1600–1989*. St Louis: Heritage; 1990.
3. Pharmaceutical compounding—sterile preparations (revised general chapter 797). In: *The United States Pharmacopeia, 31st rev, and The National Formulary, 26th ed.* Rockville, MD: United States Pharmacopeial Convention, Inc; 2007.
4. US Nuclear Regulatory Commission. *Consolidated Guidance about Materials Licenses: Program-Specific Guidance about Commercial Radiopharmacy Licenses*. NUREG-1556, vol 13. Washington, DC: US Nuclear Regulatory Commission; 1999.
5. Kowalsky RJ, Johnston RE. Dose calibrator assay of iodine-123 and indium-111 with a copper filter. *J Nucl Med Technol*. 1998;26:94–8.
6. Content outline for nuclear pharmacy specialty certification examination. Washington, DC: Board of Pharmaceutical Specialties; 2007.

CHAPTER 14
Quality Control in Nuclear Pharmacy

Joseph C. Hung

The old adage that a chain is only as strong as its weakest link is applicable to nuclear medicine practice. The nuclear medicine physician must be sure that the information obtained from a radiodiagnostic procedure is a true representation of the patient's condition. The strength and conviction of a physician's diagnostic impression are based not only on his or her knowledge and experience but on trust in a nuclear diagnostic system in which all measures have been taken to prevent errors. There are several areas where problems can occur to weaken the system: competency of personnel, data collection, data processing and display systems, instrumentation, and radiopharmaceuticals. Although diagnostic radiopharmaceuticals have historically been the mainstay of nuclear pharmacy practice, during the past decade there has been increased interest in the development and use of therapeutic radiopharmaceuticals. Suboptimal quality of a therapeutic radiopharmaceutical could cause more detriment to the patient, since the higher amount of radioactivity associated with a therapeutic radioisotope can produce damaging effects on nontarget tissues and organs. This chapter focuses on practical operational and quality control methods used in nuclear pharmacy practice to ensure high-quality diagnostic and therapeutic radiopharmaceuticals.

RADIOPHARMACEUTICAL QUALITY CONTROL

Because radiopharmaceuticals are intended for human administration, quality control (QC) procedures are imperative to ensure their safety and effectiveness. QC of radiopharmaceuticals has been defined by Briner[1] as "a series of tests, observations and analyses that will indicate beyond a reasonable doubt the identity, quality, and quantity of all ingredients present in a product and which will demonstrate that the technology employed in its formulation or manufacture will yield a dosage form of highest safety, purity, and efficacy." This definition implies that the QC program must be in operation continuously throughout the radiopharmaceutical preparation process.

Radiopharmaceuticals prepared by drug manufacturers are routinely monitored through a series of vigorous testing procedures. Consequently, the end user is generally not required to conduct any further QC evaluation on such products, except for certain mandated tests such as the ^{99}Mo breakthrough test on a ^{99m}Tc generator's first eluate.[2] However, for extemporaneously prepared ^{99m}Tc-labeled compounds, it is the responsibility of the preparer to ensure that the final product meets acceptable standards of quality and purity. Fortunately, the occurrence of defective or substandard radiopharmaceutical products has reportedly been infrequent.[3-11]

There are three main sources of information on QC methods for radiopharmaceuticals: drug product package inserts, the *United States Pharmacopeia* (*USP*), and the scientific literature. *USP* is the official compendium of drug products manufactured in the United States. It is recognized by the federal Food, Drug, and Cosmetic Act and is referenced in numerous statutes regulating items used in medical practice.[12] USP standards for establishing a drug's identity, strength, quality, and purity, and specifications for packaging and labeling provide a guide for the quality of drug products prepared for medical use, including radiopharmaceuticals. A new *USP* General Chapter 1165, Radiopharmaceutical Quality Assurance and Compounding, is now in the review process and would apply directly to longer half-life radiopharmaceuticals, whereas *USP* General Chapter 823, Radiopharmaceuticals for Positron Emission Tomography—Compounding,[13] provides details on requirements during the compounding process (including QC) for the very short half-lived radiopharmaceuticals used in positron emission tomography (PET). Package inserts, which must be approved by the Food and Drug Administration (FDA), are included with radioactive drug products and provide preparation and QC information for the nuclear pharmacist preparing these products. However, there can be shortcomings in the QC information contained in package inserts.[14] A third source of information is the scientific literature. Although journal articles, textbooks, and technical reports do not have the power of statute associated with a package insert or a *USP* standard, they are sometimes used as alternative sources for QC test methods in nuclear pharmacy practice.

The official test method for QC described in the *USP* monograph for a radiopharmaceutical does not preclude the use of alternative methods.[15] However, alternative methods must be shown to be equivalent to the *USP* method, which is the reference method. It is not clear whether FDA has a similar policy regarding deviations from test procedures described in package inserts. FDA permits physicians to use approved drug products for unapproved or unlisted clinical indications.[16,17] Likewise, nuclear pharmacists should be permitted to use alternative QC testing methods in order to meet production capabilities and constraints, especially if the testing information described in the package insert is incomplete or is listed only as a recommendation.[14] With regard to PET radiopharmaceuticals, *Guidance: PET Drug Products—Current Good Manufacturing Practice (CGMP)*[18] and *Sample Formats—Application to Manufacture Ammonia N 13 Injection, Fludeoxyglucose F 18 Injection (FDG 18), and Sodium Fluoride F 18 Injection: Chemistry, Manufacturing, and Controls Section*,[19] both issued by FDA, offer specific QC requirements for overall PET drug products and for three specific PET radiopharmaceuticals.

Routine radiopharmaceutical QC procedures can be broken down into four categories: radiation considerations, chemical considerations, pharmaceutical considerations, and biologic considerations.

RADIATION CONSIDERATIONS

The safe and efficacious use of radiopharmaceuticals requires that they be of the highest purity with regard to their radionuclide and chemical composition. Any nuclear medicine procedure requires the administration of a particular radionuclide in a particular chemical form. The presence of impurities, such as different radionuclides or unwanted chemical forms of the desired radionuclide, may produce undesirable information from the diagnostic procedure and give unnecessary radiation to the patient. Therefore, it is important to conduct purity tests on all radiopharmaceuticals prior to patient administration.

Radionuclide Identity

A radionuclide can be identified by the type and energy of radiation(s), by half-life, or by both. Radiation types include α-particle, β-particle, γ-ray, and x-ray. Only electromagnetic radiation (i.e., γ-ray and x-ray) can be detected by gamma spectrometry. Determination of the nature and energy of the emitted electromagnetic radiation(s) is usually carried out by gamma spectrometry. This can be readily accomplished with a multichannel analyzer (MCA) equipped with a sodium iodide thallium-activated [NaI(Tl)] detector or a lithium-drifted germanium [Ge(Li)] semiconductor detector (Figure 14-1). The NaI(Tl) detector has a much lower energy resolution than a Ge(Li) detector but is more commonly available.

Half-life is a unique property of each radioisotope (similar to the fingerprint of every individual) and thus can be used to identify a radionuclide. Half-life is determined by successive counting of a radioactive sample at intervals corresponding to half of the estimated half-life for a time period equal to about three half-lives.[20–22] A dose calibrator can be used to measure half-life in radionuclide identity tests. A graph can be

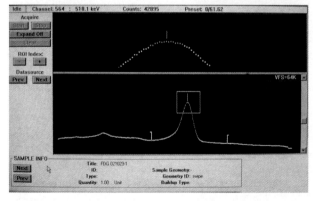

FIGURE 14-1 A multichannel analyzer (sodium iodide thallium-activated detector) with a pulse-height spectrum shown on the monitor.

drawn with time as the abscissa and the logarithm of the measured radioactivity as the ordinate. Half-life determinations are not required in traditional (non-PET) nuclear pharmacy practice but are essential in identifying PET radionuclides. In some situations, such as when two or more positron-emitting radionuclides need to be identified (e.g., ^{18}F impurities in a ^{13}N-labeled preparation), half-life determinations must be carried out in addition to gamma spectrometry, since ^{18}F and ^{13}N demonstrate identical spectra.

Radionuclidic Purity

Radionuclidic purity of a radiopharmaceutical is defined as a ratio, expressed as a percentage, of the radioactivity of the desired radionuclide to the total radioactivity in the preparation. For example, a 100 µCi (3.7 MBq) ^{99m}Tc-sodium pertechnetate preparation containing 99.5 µCi (3.68 MBq) as ^{99m}Tc and 0.5 µCi (18.5 kBq) ^{99}Mo would have a radionuclidic purity of 99.5% with respect to ^{99m}Tc. In this example, ^{99}Mo activity would represent a 0.5% radionuclidic impurity.

Radionuclidic impurities are significant because they can contribute unnecessary radiation dose to the patient without adding to the diagnostic information obtained. Examples of radionuclidic impurities include 66 hour ^{99}Mo impurity in 6 hour ^{99m}Tc, 4 day ^{124}I impurity in 13 hour ^{123}I, and 12 day ^{202}Tl impurity in 73 hour ^{201}Tl. Relevant radionuclidic impurities and acceptable limits in radiopharmaceuticals are usually listed in package inserts or specified in *USP* monographs. Some of these are given in Table 14-1. Radionuclidic purity requirements must be fulfilled throughout the useful life of a radiopharmaceutical.

Radionuclidic impurities in radiopharmaceuticals can arise from several factors. These include the method of radionuclide production, target impurities that contribute to competing nuclear reactions, and incomplete radionuclide separation during radiochemical processing. A change in target material and type of nuclear reaction can reduce the level of radionuclidic impurities. An example is the production of ^{123}I by the (p,2n) reaction on a ^{124}Te target, which produces significantly more radionuclidic impurities (i.e., ^{124}I, ^{125}I, ^{130}I, and ^{131}I) than the (p,5n) reaction on an ^{127}I target, which produces only an ^{125}I impurity.

Radionuclidic purity is constantly changing in a product. If the radionuclidic impurity has a half-life longer than that of the principal radionuclide, the concentration of the impurity will increase with time. Such a situation often forms the basis for establishing a radiopharmaceutical's expiration date. An example is ^{202}Tl ($T_{1/2}$ = 12 days) impurity in thallous chloride ^{201}Tl injection ($T_{1/2}$ = 73 hours). The highly abundant (95%) 493 keV gamma ray of ^{202}Tl may affect image quality when low-energy collimation is used during ^{201}Tl imaging. Another example is the presence of ^{99}Mo ($T_{1/2}$ = 66 hr) in ^{99m}Tc-sodium pertechnetate ($T_{1/2}$ = 6 hr). Figure 14-2 illustrates the expiration time for a ^{99m}Tc generator eluate as a function of the Mo:Tc ratio.

Assessment of radionuclidic purity in radiopharmaceuticals is best accomplished by gamma spectrometry using an MCA equipped with either a Ge(Li) or NaI(Tl) detector (Figure 14-1). The photopeak height and area under the curve on the gamma-ray spectrum are analyzed to assess the amount of radionuclide present in a product. Ge(Li) detectors are preferred for gamma spectrometry because of their superior energy resolution in separating gamma-ray photopeaks. NaI(Tl) detectors are less suitable for analysis because their lower energy resolution can cause photopeak overlap, making it impossible to resolve individual gamma ray energies. The overall MCA system should have sufficient sensitivity and resolution for the intended purpose. Other types of detectors are required if alpha- and beta-emitting impurities need to be detected. In cases where two or more PET radionuclides need to be identified or differentiated (e.g., ^{18}F impurities in a ^{13}N-labeled preparation), a combination of gamma-ray spectrum and half-life determination should be used.

Most radiopharmaceuticals used in nuclear medicine have their radionuclidic purity tested by the manufacturer. The routine test required in non-PET nuclear pharmacy practice is the ^{99}Mo breakthrough test on the first eluate of a ^{99m}Tc generator,[2] which is a relatively simple procedure. Determining the radionuclidic purity of a PET radiopharmaceutical, however, can be time-consuming because a suitable period of decay would be required before the presence of a long half-life radionuclidic impurity could be identified and measured. Consequently, it is difficult to release a batch of a short half-life PET radiopharmaceutical. Nevertheless, when possible, a gamma-ray spectrum should be acquired and analyzed to assess radionuclidic purity prior to release of a PET radiopharmaceutical preparation.

Radiochemical Identity

A radiochemical is best identified by in vitro analytic methods. These methods may include electrophoresis, gas chromatography (GC), liquid chromatography, such as high-pressure liquid chromatography (HPLC; also called high-performance liquid chromatography), paper chromatography, solid-phase extraction, and thin-layer chromatography (TLC). A reference standard (RS), preferably USP grade, should be used in conjunction with the test sample during the radiochemical identification process to accurately identify and differentiate the desired radiochemical from radiochemical impurities. Technical descriptions of these analytic methods are presented in *USP* General Chapter 621, Chromatography.[23] Methods commonly used in radiopharmaceutical analysis are discussed briefly below.

Gas Chromatography

GC is a valuable technique for qualitative and quantitative analysis of organic compounds (Figure 14-3). The GC system allows various components of a sample to be separated on the basis of their volatility and ability to partition between a high–boiling-point liquid stationary phase and a gaseous mobile phase. For complete analysis of complex mixtures of drug molecules, each component separated must possess a relatively high level of partitioning and volatility under the operating temperature range. System suitability tests as described in *USP* General

TABLE 14-1 Radionuclidic Purity of Commonly Used Radionuclides

Radionuclide	Half-Life (Principal Energy)	Purity	Contaminants	Half-Life (Principal Energy)	Acceptable Limits
^{11}C	20 min (511 keV)	90% ^{11}C	^{13}N	9.96 min (491 keV)	<10%
			^{10}C	19 sec (718.3 keV)	
			^{14}O	70 sec (2.3 MeV)	
^{18}F	110 min (511 keV)	99.5% ^{18}F as in fludeoxyglucose F 18 injection and sodium fluoride F 18 injection 99.9% ^{18}F as in fluorodopa F 18 injection			
^{67}Ga	77.9 hr (93 keV, 184 keV, 296 keV, 388 keV)	99% ^{67}Ga			
^{111}In	2.8 days (173 keV, 247 keV)		^{110m}In		<3 µCi/mCi (<3 kBq/MBq) ^{111}In
			^{114m}In	50.0 days (192 keV, 558 keV, 724 keV)	<3 µCi/mCi (<3 kBq/MBq) ^{111}In
			^{65}Zn	243.9 days (1.12 MeV)	<3 µCi/mCi (<3 kBq/MBq) ^{111}In
^{123}I	13.3 hr (159 keV)	85% ^{123}I			
^{13}N	9.96 min (511 keV)	99.5% ^{13}N			
^{15}O	2.03 min (511 keV)	99.5% ^{15}O			
^{82}Rb	76 sec (511 keV, 777 keV)		^{82}Sr	25.5 days (511 keV, 777 keV)	<0.02 µCi/mCi (<0.02 kBq/MBq) and <0.05 µCi/mCi (<0.05 kBq/MBq) ^{82}Rb at the end of elution
			^{83}Rb	86.2 days (530 keV)	
			^{85}Sr	64.8 days (514 keV)	<0.2 µCi/mCi (<0.2 kBq/MBq) ^{82}Rb at the end of elution
^{99m}Tc	6.02 hr (140 keV)		^{99}Mo	65.94 hr (181 keV, 740 keV, 778 keV)	<0.15 µCi/mCi (0.15 kBq/MBq) ^{99m}Tc
^{201}Tl	73.0 hr (60–80 keV, 135 keV, 167 keV)	95% ^{201}Tl	^{200}Tl	26.1 hr (368 keV, 579 keV, 829 keV, 1.21 MeV)	<2%
			^{203}Pb	52.0 hr (279 keV, 401 keV)	<0.3%
			^{202}Tl	12.2 days (439 keV, 522 keV, 961 keV)	<2.7%
^{133}Xe	5.3 days (31 keV, 81 keV)	95% ^{133}Xe			

Source: *The United States Pharmacopeia and The National Formulary.*

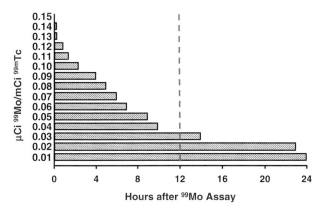

FIGURE 14-2 Expiration time of a ^{99m}Tc generator eluate based on the initial ^{99}Mo/^{99m}Tc activity ratio. The upper limit for this ratio is 0.15 μCi ^{99}Mo/mCi ^{99m}Tc. For example, an initial ratio of 0.09 will become 0.15 in about 4 hours and the eluate will expire. The dotted line indicates the default expiration time for any generator eluate (12 hours).

Chapter 621, Chromatography,[23] should be carried out to ensure that the resolution and reproducibility of the GC system are adequate for the analysis to be performed. At least one injection of the RS or internal standard should be performed in GC before the injection of test samples. The GC system usually consists of a gas supply, an injection port, a thermostat-controlled oven that contains the column, a detector system, and an integrator for recording retention time along with peak area information for the separated chemical species.

Sample Injection Port

The sample solution is introduced into the system through the injection port with a syringe. The sample volume should not be too large, and the sample should be introduced onto the column as a bolus to avoid loss of peak resolution. A silicone rubber septum is normally installed in the injection port to prevent leakage of the sample and mobile phase. The temperature of the sample injection port is usually set at approximately 50°C higher than the boiling point of the least volatile component of the sample, or at least 10°C to 15°C above the column oven temperature to ensure volatilization of sample components.

Mobile Phase Gases

The vaporized sample is transported by the flow of inert gas (mobile phase) through the column, where separation of the sample components occurs. Nitrogen is typically used as the mobile phase, but other gases such as argon or helium are sometimes used. If necessary, the gases are filtered of impurities through special in-line absorbent cartridges. Regulator valves control the overall gas pressure from the tanks.

Columns

Two general types of columns are used for GC analysis, packed columns and capillary (open tubular) columns. A packed column contains a finely divided, inert, solid support material

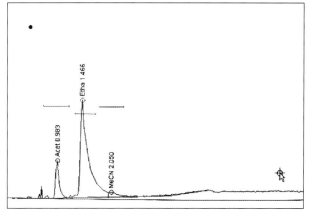

FIGURE 14-3 A gas chromatography system and a spectrum of various residual solvents detected from a preparation of fludeoxyglucose F 18 injection.

coated with liquid stationary phase. A capillary column has the stationary phase coated on the inside wall of a thin glass capillary tube. For precise work, column temperature must be controlled within tenths of a degree. The optimum column temperature depends on the boiling point of the sample. As a rule, a temperature slightly above the average boiling point of the sample results in an elution time of 2 to 30 minutes. Minimal temperatures give good resolution but result in increased elution times. If a sample has a wide boiling point range, then temperature programming can be useful.

Detectors

Many types of detectors can be used in GC analysis. Mass selective detectors (MSD) are common, although electron capture, flame ionization, and thermal conductivity detectors are also used. An MSD has the advantage of allowing identification of chemical components from the mass spectrum. With the flame ionization detector system, hydrogen and air are added to the column eluant at the detector and ignited to maintain a flame just above the column exit. When organic compounds are burned in the detector flame, ions are formed that change the voltage at the collector electrodes. As more molecules enter the detector, more ions are formed, increasing the intensity of the signal recorded by the integrator.

Chromatogram

The area under a chromatographic peak, known as the peak area, represents the total amount of analyte in a sample. The time, in minutes, between sample injection and analyte detection is referred to as retention time. It is a measure of the time an analyte spends on the column and is used qualitatively for analyte identification after comparison with a series of standards.

High-Pressure Liquid Chromatography

HPLC is one of the most versatile tools used in radiopharmaceutical analysis, because it provides high-resolution component separation (Figure 14-4). HPLC can be used with organic or inorganic molecules, including nonvolatile species that cannot be analyzed by GC. As with GC, system suitability tests described in *USP* General Chapter 621, Chromatography,[23] should be performed to make sure that the resolution and reproducibility of the HPLC system are suitable for the QC analysis.

There are many different makes and models of chromatographs. The major components of any HPLC system include the mobile phase reservoir(s), pump, injector, column, detector, and recorder. Both the injector and the column are equipped with overflow reservoirs.

The analytic process begins with a mobile phase that is drawn up from the reservoir and propelled by the pump into the injector. Here, a sample is injected via an injection valve or by a syringe–septum arrangement and then carried through the column under pressure (up to 6000 psi) at a precisely controlled rate. The column is the heart of the system; it is where component separation occurs. Separated components leave the column and enter a detector that measures the concentration of different solutes in the sample. A signal sent to a recording device (an integrator) generates a chromatogram that displays retention times and peak areas for each component. At least one injection of the RS or internal standard should be carried out prior to the injection of the test samples.

Solvent Reservoirs

One or more reservoirs hold the mobile phase (solvent) required for sample elution. Mobile phase is applied to the column (the stationary phase) and acts as a carrier for the sample solution. Particulate matter in the mobile phase, which could clog the system, is removed by a fritted steel filter on the end of the uptake tube. There are two major types of mobile phases: isocratic and gradient. In isocratic elution chromatography, compounds are eluted using mobile phase with constant composition. In gradient elution chromatography, different compounds are eluted by various compositions of mobile phase. Isocratic systems are preferable because they provide more reproducible separation; however, gradient elution chromatography significantly increases the separation power of an HPLC system because of improved peak resolution (decreased peak width and increased peak height) for most compounds.

Pump and Damper

A high-pressure pump draws up the mobile phase and drives it forward at a constant rate via a piston-type pump. (Other pump types can also be used.) The pump operates by withdrawing a piston, which creates a sudden drop in pressure in an internal chamber. To compensate, the mobile phase is drawn into the chamber through a lower one-way check valve. The upper exit valve is closed because of negative pressure. When the piston pushes back into the chamber, the pressure increases and the mobile phase is forced out. The upper check valve opens while the lower check valve blocks backflow.

A pulse damper is placed in line with the pump to reduce the pulsations caused by the action of the piston in the pump head. In this case, the pulse damper consists of a long expandable coil that smoothes out the flow of the mobile phase by stretching with each pulse of pressure from the pump and recoiling when the pressure ceases. Without the damper in place, the pulses would register as bumps on the chromatogram.

The pump should be capable of delivering a constant and pulse-free flow rate at pressures up to 6000 psi. This is achievable if the mobile phase is free of air. It is therefore essential that the solvent(s) be degassed before use. The mobile

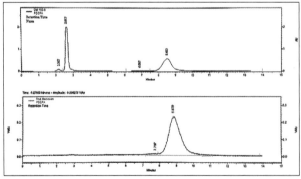

FIGURE 14-4 The upper panel illustrates a high-performance (high-pressure) liquid chromatography (HPLC) system coupled with a scintillation detector for analysis of fluorodopa F 18 injection. The upper spectrum, obtained with an ultraviolet (UV) detector, identifies the fluorodopa peak; the lower spectrum, generated with a radiation detector, identifies the ^{18}F activity associated with the fluorodopa peak.

phase or eluant should be filtered free of particulate matter and should flow through an in-line filter before entering the pump head. Before analysis, the pump should always be bled of any trapped air by priming the pump head. Pump heads should be periodically inspected for leaks and pump seal wear. The check valves and pistons should also be regularly inspected for damage.

Sample Injector Port
From the pulse damper, the mobile phase travels to the injector, where it can either flow directly into the column via an internal loop or be redirected by the switching of a valve through an external loop where it picks up the sample. In the load position of the injector, the external loop is isolated from the high-pressure mobile phase, allowing the sample solution to be injected into the external loop at atmospheric pressure. Excess sample is removed via the injector overflow tube. External loops can vary in size, but the 20 µL size is common. When switched to the inject position, the mobile phase passes through the external loop, carrying the sample solution to the column at high pressure.

Injectors may become blocked, and the internal components (rotor seal, injection port seal) may wear with use. An injector should be visually inspected before use to check for leaks and to ensure that there is liquid flow through the injector loop pathway.

Columns
Columns are available in many different types and sizes. Some are used for analysis of small samples (i.e., analytic HPLC), while others are used for isolating large quantities of a particular component in a mixture (i.e., preparative HPLC).

HPLC analytic methods are of two types: normal-phase and reverse-phase. In normal-phase HPLC, the packing material is polar in nature (e.g., Si–O–Si–R packing), whereas the mobile phase consists of nonpolar solvent(s) (e.g., hexane, acetone, other hydrocarbons). Samples of moderate to strong polarity are well separated by normal-phase HPLC. In reverse-phase HPLC, the column uses a nonpolar stationary phase (e.g., C18 packing) and a relatively polar mobile phase (e.g., water, methanol, acetonitrile). In this system, polar analytes will elute from the column ahead of less polar analytes. When HPLC is used for purification of a radiopharmaceutical, there should be no bleeding of unintended materials (e.g., column material) into the mobile phase.

HPLC columns give little trouble if they have been properly conditioned, cleaned, and stored. End fittings should always be tight so that there are no leaks, and column frits should be replaced in the event of persistent leaks.

Detectors and Recorder
Analytes eluted from the column are monitored by a suitable detector of sufficient sensitivity. A wide range of detectors are available for HPLC systems. The most common is the UV spectrophotometer. For working with radiopharmaceutical samples, a radiation detector is necessary in addition to the UV detector. The typical wavelength in UV analysis is 254 nm. The analyte passes through a flow cell, altering the absorbance of UV light shone through it. The change in absorbance is recorded and printed out as a chromatogram peak. If a radiation detector is used, such as a NaI(Tl) detector, it will measure the concentration of radioactive components in the eluate. Detectors are generally very reliable; however, UV detectors should be periodically checked for cell gasket failure, cracked windows, leaks, and blockages in the flow cell. The source lamp of a UV detector will need to be replaced after it has reached its maximum life.

The *USP* monograph for radiochemical purity (RCP) testing of ammonia N 13 injection[24] specifies an HPLC system with an electrical conductivity detector, which can detect all charged species in the sample. For analysis of ammonia N 13 injection, the chemical species can include $^{13}NH_4^+$, $^{13}NO_2^-$, $^{13}NO_3^-$, and $^{18}F^-$, which can be detected only by this method,

Drawbacks of HPLC
HPLC testing for radiochemical identity has some drawbacks: (1) An HPLC system is more expensive and elaborate than a TLC setup; (2) an HPLC system is not widely available in nuclear medicine or nuclear pharmacy laboratories; and (3) completion of the test requires two HPLC runs, one with the test solution and another using the reference solution, whereas paper chromatography or TLC requires only one run for completion because the sample and RS are spotted on the same strip. Thus, for practical reasons, paper chromatography and TLC (described in later sections) are the simplest and most rapid methods for routine QC testing of radiochemical species in radiopharmaceuticals.

Solid-Phase Extraction

Solid-phase extraction (SPE) is similar in principle to HPLC in that the technique involves a solid support medium coupled with a solvent mobile phase. A convenient device using this method is the Sep-Pak cartridge (Waters Chromatography, Millipore Corporation, Milford, MA). Sep-Pak cartridges are disposable SPE devices commonly used for sample preparation (e.g., purification, trace enrichment or concentration, fractionation, solvent exchange) (Figure 14-5). They can also be used as analytic devices. Generally, the proper use of a Sep-Pak cartridge requires a five-step process: sample preparation, cartridge conditioning and equilibration, sample application, washing, and elution. However, the separation procedure for a given compound is unique, and not all of the five steps may be required for the application.

A Sep-Pak cartridge method for determining the RCP of ^{99m}Tc-mertiatide (^{99m}Tc-MAG3) is described in the ^{99m}Tc-MAG3 package insert;[25] that description is provided here to illustrate the SPE technique. Reverse-phase chromatography with a Sep-Pak C18 cartridge involves a series of steps using solvents of different polarities to separate the different radiochemical species.

1. Conditioning. The cartridge is conditioned with 6 to 10 void volumes of a moderately nonpolar, water-

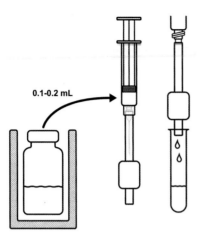

FIGURE 14-5 The Sep-Pak cartridge system for radiopharmaceutical purity determination. Radiopharmaceutical is applied to the cartridge, followed by elution of radioactive components.

miscible solvent, such as methanol, ethanol, or acetonitrile, followed by a polar solvent similar to the sample solution. For ^{99m}Tc-MAG3, 10 mL of ethanol is used to prepare the cartridge, followed by 10 mL of 0.001 N hydrochloric acid. The cartridge is then drained by pushing 5 mL of air through it with the syringe; however, the cartridge should not be allowed to dry out before sample application.

2. Sample application and washing. If the less polar analyte is to be retained by the C18 sorbent and the more polar unwanted species eluted, the cartridge should be washed with a polar solvent in which the analyte has limited solubility. For ^{99m}Tc-MAG3, 0.1 mL of the ^{99m}Tc-MAG3 preparation solution is applied to the cartridge, and then the cartridge is rinsed with 10 mL of 0.001 N hydrochloric acid, whereupon ^{99m}Tc-MAG3, being the less polar analyte, is retained by the C18 sorbent while the more polar ^{99m}Tc-sodium pertechnetate and ^{99m}Tc-tartrate impurities are washed off.

3. Analyte elution. The retained less polar analyte is eluted with a nonpolar or moderately polar solvent. In this example, 10 mL of 1:1 ethanol:saline is used to elute the ^{99m}Tc-MAG3.

4. When all components are recovered, the used cartridge is discarded in an appropriate manner. For ^{99m}Tc-MAG3, the cartridge will contain the retained hydrolyzed–reduced ^{99m}Tc (H-R ^{99m}Tc) impurity.

The separation efficiency of a Sep-Pak cartridge varies with flow rate and sample load. In general, resolution may be poor if the sample flow rate is too high, because components may not interact sufficiently with the sorbent. However, the results of RCP determination of ^{99m}Tc-MAG3 with a C18 solid-phase minicolumn Sep-Pak cartridge (WAT051910 [Waters Corp.]) do not seem to be affected by the faster flow rates (e.g., as high as ~15–20 mL/min for conditioning and elution).[26] Faster flow rates would save time and consequently reduce radiation exposure of the end user. With regard to the relationship of sample load and separation efficiency of a Sep-Pak cartridge, the overloading of a test sample onto the cartridge results in sample breakthrough, which later leads to variability in sample recovery or outcome interpretation.

The Sep-Pak cartridge can provide a complete chromatographic separation of ^{99m}Tc radiochemical species; however, the solvent elution process, costly cartridge, and lengthy procedure make the Sep-Pak cartridge method less than ideal for routine use in a busy nuclear medicine department or nuclear pharmacy. Also, technical errors can occur with this cartridge method. For example, after radiopharmaceutical sample loading, the cartridge must be eluted slowly with an appropriate solvent; otherwise, the analyte bound to the sorbent (e.g., ^{99m}Tc-MAG3) may not be completely removed, resulting in false estimation of radiochemical identity or RCP (described in the next section). Another disadvantage of using Sep-Pak cartridges for either radiochemical identification or RCP determination is the increased radiation exposure of the person performing the Sep-Pak procedure. The Sep-Pak cartridge method requires at least 0.1 mL of ^{99m}Tc radiopharmaceutical preparation to be loaded onto the cartridge. This volume may contain several millicuries (megabecquerels) of activity. For example, a regular ^{99m}Tc-MAG3 kit preparation will contain between 0.5 mCi (18.5 MBq) and 2.5 mCi (92.5 MBq) per 0.1 mL application. An alternative technique, using a two-strip paper chromatography system, requires only two 5 μL samples (50–250 μCi; 1.9–9.3 MBq) for RCP analysis and less than 3 minutes to complete the RCP determination for ^{99m}Tc-MAG3.[27]

Paper and Thin-Layer Chromatography

Paper chromatography and TLC methods are used most frequently to identify the various radiochemical species in a radiopharmaceutical preparation, especially for ^{99m}Tc-labeled radiopharmaceuticals. In each of these techniques, microliter amounts of the radiopharmaceutical are spotted at the origin of a chromatographic strip (the stationary phase). The chromatographic strip is then placed vertically in a chromatographic chamber (usually a vial or glass tube) that contains an appropriate solvent (mobile phase). The strip is placed in the solvent so that the origin of the spot is not immersed.

The stationary phase may be paper, such as Whatman 31ET or Gelman Solvent Saturation Pads. A modified thin-layer support called instant thin-layer chromatography (ITLC) is also used. ITLC is a glass microfiber mesh impregnated with silica gel (ITLC-SG) or polysilicic acid (ITLC-SA), which results in a support resembling paper. ITLC support strips are available from Biodex (Shirley, NY; www.biodex.com). The mobile phase is usually water, saline, or an organic solvent. The chromatographic chamber must be covered tightly to maintain a solvent-saturated atmosphere. The electrostatic forces (adsorption and capillary action) of the stationary phase tend to retard the movement of various radiochemical species, whereas the mobile phase carries each radiochemical component according to its partition between the stationary phase and the mobile phase. The partitioning of each radiochemical species between the stationary and mobile phases is determined by the solubility of the radiochemical species in the mobile phase, which is affected by the polarity of the solvent. Therefore, the electrostatic attractive forces of the stationary phase and the polarity of the mobile phase are the two determining factors in the separation of different radiochemical components in a sample.

Selection of appropriate support and solvent systems permits separation of the different chemical species in a radiopharmaceutical. After chromatogram development, the strip is removed, dried, and analyzed using a radiochromatogram scanner or another method for counting the radioactivity distribution on the strip.

The solvent front (S_f) is the distance that the solvent travels from the origin of the chromatographic strip, whereas the relative front (R_f) of a radiochemical component is the distance the component travels from the origin relative to the S_f (Figure 14-6).

R_f values for chemical components in a radiopharmaceutical are established by using known radiochemical species in a given chromatographic system. The identities of various radiochemical species present in a radiopharmaceutical are determined by comparing their R_f values with known R_f values in the same system. One of the following three methods can be used to determine R_f values and the relative amount of a radiochemical component in the radiopharmaceutical preparation for the purpose of determining RCP:

1. Scanning the strip with a radiochromatogram scanner that traces out various activity peaks (Figures 14-6 and 14-7). Peak position (its R_f) is indicative of the particular species present, and peak area corresponds to the respective amount of activity for each species.
2. Cutting the chromatogram into centimeter segments that are individually counted. Subsequently, a histogram plot of the activity distribution is made and the amount of activity in each species is determined.
3. Using a miniaturized chromatography system that uses a 6 cm strip. This method is used when the species are well separated on the strip so that it can be cut into two pieces and counted for analysis. This method is used routinely for ^{99m}Tc radiopharmaceuticals because it is rapid and easy to perform on a daily basis.

Paper chromatography and TLC methods are generally accurate and reliable; however, with these systems R_f values vary somewhat depending on the brand of solid support, quality of solvent, and operating conditions.[21] Consequently, radiochemical identification is best accomplished when a pure, authentic sample of the compound in question is used as an RS on the same chromatogram. In addition, although both test and RS samples are spotted on the same chromatography paper or TLC strip, the R_f values measured for the test substance may differ from the values obtained for the RS

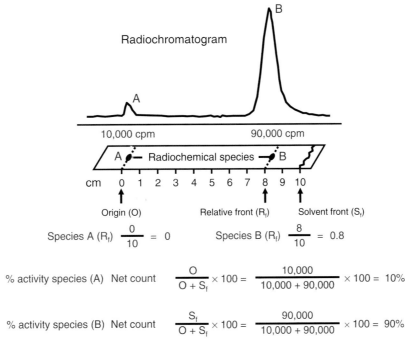

FIGURE 14-6 Radiochromatogram analysis. Species A represents the desired radiopharmaceutical; species B represents radiochemical impurity. See text for explanation of R_f and S_f.

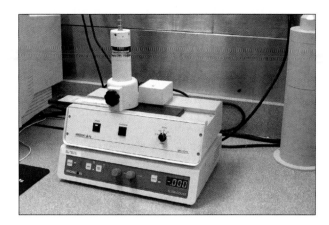

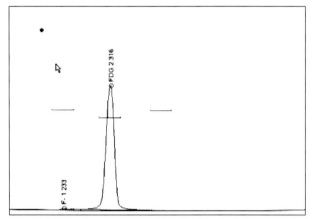

FIGURE 14-7 Radiochromatogram scanner and a spectrum of fludeoxyglucose F 18 injection.

compound.[23] An acceptable range (as determined by system suitability testing, described below) for this difference should be established.

To verify the resolution and reproducibility of any gas or liquid chromatographic system (e.g., GC, HPLC, TLC), suitability testing should be completed before sample analysis.[15,23] The difference in measured R_f values between the test and RS samples should not exceed reliability estimates as determined statistically from replicate assays.[23] These types of tests are necessary to validate a chromatographic system's precision and accuracy.

Technical Precautions in Performing Chromatography Procedures

ITLC and paper chromatography procedures are easy to perform and are usually trouble free; however, several procedural errors and artifactual results can occur. Possible causes of the artifacts and errors are summarized below.[28–31]

1. Strips counted in a dose calibrator should contain at least 100 µCi (3.7 MBq) of activity to keep the counting error to 1% or less. This assumes dose calibrator sensitivity of 1 µCi (37 kBq) and, at minimum, a 1% impurity in a 100 µCi (3.7 MBq) sample. If a scintillation well counter is used for counting, the strips should be counted far enough away from the detector so that the counting rate does not exceed the dead time of the detector.[32,33] This is not a problem if the dose calibrator or radiochromatogram scanner is used for analysis.

2. The radiopharmaceutical sample spot should be placed at least 1 cm from the bottom of the strip so the spot itself does not enter the solvent but is well above its level. In this way the solvent will pass through the spot and cause soluble species to migrate in a normal manner. If the spot is even partly submerged in the solvent, these species will be retarded from migration and erroneous results will be obtained.[34]

3. Fresh chromatography strips and solvents should be used to ensure reproducibility in analysis.

4. To avoid incorrect analysis, care should be taken to not allow the solvent to migrate past the S_f line. If the strip is eluted significantly past the S_f line, the cut line must be changed to maintain the same R_f value.

5. Oxidation reactions or strip interactions can occur on the chromatography paper or ITLC strip. Reduced states of ^{99m}Tc are easily oxidized; therefore, the spot of radiopharmaceutical should not undergo prolonged air drying on the strip before it is placed in the solvent. ^{99m}Tc species may bind with the media or strip, which is often the cause of streaking (e.g., inadequate or no separation of free ^{99m}Tc and H-R ^{99m}Tc from the principal ^{99m}Tc-labeled complex).

6. In spotting the radiopharmaceutical, uneven spotting or splattering should be avoided.

7. Interaction of radiochemical species can occur with preparative compounds or ink markers used to visualize solvent flow.

8. Grease from fingerprints can alter migration patterns during development.

9. The chromatography strip should not be allowed to touch the side of the wet chromatography chamber, or solvent will rise up rapidly on the strip by capillary action, invalidating the analysis.

10. Cross-contamination of chromatography strips with other radiopharmaceuticals can lead to erroneous results.

11. Use of the wrong solvent or improper solvent preparation will alter results.

12. Use of the wrong chromatographic strips for a particular radiopharmaceutical will cause erroneous results.

13. Insufficient solvent in the chromatography chamber will cause solvent evaporation and incomplete development of the strip.

14. Use of contaminated tweezers or scissors in handling chromatographic paper or ITLC media can lead to spurious results.

15. Exposure of solvent to the atmosphere for too long can result in evaporation of one solvent in a mixture or absorption of water vapor in organic solvents, causing altered R_f values.

16. Mechanical factors such as chamber movement and unleveled surfaces may cause erroneous results.

Prolonged air drying of the applied sample spot on the chromatography strip is not desirable because the ^{99m}Tc-labeled complex may be oxidized.[30] Drying should be done quickly with a hot-air dryer or in a stream of nitrogen gas. Drying the spot before development is important if the applied spot is 5 μL or larger, particularly when organic solvents such as acetone are used. Acetone will mix freely with water. Hence, radiochemical species that are soluble in water but not soluble in acetone, such as ^{99m}Tc-labeled complexes, will streak up the strip from the origin if the spot is wet. If streaking extends into the top half of the strip, erroneous results will be obtained. This is more likely to occur with 5 cm ministrips than with standard 10 cm strips. If the applied spot is small, about 1 μL, drying of the spot may not be necessary.

Figure 14-8 shows that the extent of streaking depends on spot size, and Table 14-2 lists the results obtained with several ^{99m}Tc-labeled complexes developed on 5 cm ITLC-SG strips in acetone after application of a 5 μL wet or dried spot. It is evident from these data that significant migration (i.e., streaking) of the ^{99m}Tc-labeled complex occurs into the solvent-front half of the strip if a large wet spot is developed in acetone. During analysis this would be interpreted as a ^{99m}Tc-sodium pertechnetate impurity, but it would be an artifact created by improper technique.[34]

Corrective Actions

If a low RCP value is obtained, the following actions should be taken:

1. Review the RCP testing technique to ensure that the appropriate procedures to avoid errors and artifacts, as described above, have been followed. Repeat the RCP testing procedure.
2. If the RCP is still below the acceptable limit, discard the reconstituted kit and prepare another kit.
3. If several kits from the same lot number have failed RCP testing, notify the manufacturer.

The use of a USP RS in evaluating radiochemical identity is ideal;[15] however, an RS may not be available. For example,

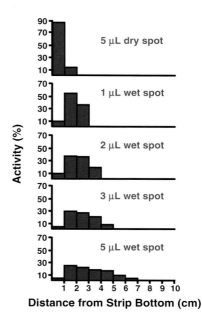

FIGURE 14-8 Demonstration of chromatogram streaking of ^{99m}Tc-DTPA with the instant thin-layer chromatography–silica gel (ITLC-SG)/acetone system as a function of wet spot size.

USP Fludeoxyglucose Related Compound B RS (2-chloro-2-deoxy-D-glucose), for fludeoxyglucose F 18 (^{18}F-FDG) analysis, is identified and listed in the *USP* monograph for ^{18}F-FDG,[35] but it cannot be obtained from USP or any commercial source. If a USP RS is not available, a high-quality source of chemical substance, such as one listed as Food Chemicals Codes (FCC) grade, authentic substance (AS) grade, or analytic reagent (AR) grade, or a substance that has been certified by the American Chemical Society (ACS) can be used.[15] For non-PET radiopharmaceuticals, a RS or high-quality chemical substance corresponding to the nonradioactive compound of interest usually does not exist or is not commercially available. Therefore, it may not be possible to carry out radiochemical identity testing on reconstituted non-PET radiopharmaceuticals.

TABLE 14-2 Effect of Wet versus Dry Spot on Minichromatography Results (% Activity) for ^{99m}Tc-Labeled Compounds in ITLC-SG/Acetone System

Radiopharmaceutical	Wet Spot[a]		Dry Spot[a]	
	Origin	Solvent Front	Origin	Solvent Front
^{99m}Tc-gluceptate (^{99m}Tc-GH)[b]	74.01	25.99	99.81	0.19
^{99m}Tc-pentetate (^{99m}Tc-DTPA)[b]	60.04	39.96	99.70	0.30
^{99m}Tc-pyrophosphate (^{99m}Tc-PYP or ^{99m}Tc-PPi)[b]	85.37	14.63	99.39	0.61
^{99m}Tc-medronate (^{99m}Tc-MDP)[b]	82.60	17.40	99.82	0.18

[a]Results are means of five determinations on 5 cm ITLC-SG strips.
[b]Common chemical abbreviation.
Source: Reference 34.

Counting Instruments for Radiochromatography

Radionuclide Dose Calibrators

The dose calibrator is often used for RCP measurements in the nuclear pharmacy because it is easy to use and results are obtained quickly. However, the accuracy of measurement with this instrument is of major concern when low amounts of radioactivity are used. In general, when a dose calibrator is used for RCP determination, the chromatography strips should contain 100 µCi (3.7 MBq) or more of activity in order to reduce the error to ≤1%.

Well Scintillation Counters

The gamma scintillation well counter is an appropriate instrument for counting chromatography strips; however, to achieve accurate assessment of radioactivity, the dead time of this instrument must not be exceeded. To avoid exceeding the maximum counting capabilities of the well counter, four different methods are recommended: (1) increasing the distance from source to detector; (2) using an attenuator on the well counter, such as an appropriately sized metal disk or coin placed over the opening of the well counter; (3) decreasing the activity of the radioactive sample; and (4) correcting for instrument dead time if the well counter is equipped with a device that automatically compensates for high amounts of radioactivity.[32,36]

Radiochromatogram Scanners

A radiochromatogram scanner (Figure 14-7) detects and measures the distribution of radioactivity along the intact radiochromatography strip. The scanner can analyze samples over a wide range of activities and assess the relative amounts of radiochemical species distributed over the strip; however, the procedure is time-consuming and the instrument is quite expensive. The scanner should have sufficient sensitivity and spatial resolution for the intended analysis.

Radiochemical Purity

The RCP value of a radiopharmaceutical preparation is defined as the ratio, expressed as a percentage, of the radioactivity in the desired chemical form to the total radioactivity in the radiopharmaceutical preparation. For example, a 100 µCi (3.7 MBq) sample of ^{99m}Tc-sulfur colloid (^{99m}Tc-SC) of which 95 µCi (3.5 MBq) is present as ^{99m}Tc bound to sulfur particles and 5 µCi (1.85 kBq) is ^{99m}Tc-sodium pertechnetate would have an RCP of 95%. The radiopharmaceutical product in this case contains a 5% ^{99m}Tc-sodium pertechnetate impurity.

Radiochemical impurities are undesirable because their distribution in the body differs from that of the radiopharmaceutical of interest, making it difficult to obtain useful information in nuclear medicine studies. High background counts that are due to the presence of radiochemical impurities in

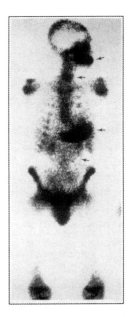

FIGURE 14-9 Whole-body bone scan with ^{99m}Tc-MDP showing ^{99m}Tc-pertechnetate impurity localized in the salivary glands, thyroid, stomach, and intestinal tract (arrows).

areas not of primary interest in the study can degrade image quality, interfere with diagnostic interpretation, and expose the patient to unnecessary absorbed radiation dose. For example, ^{99m}Tc-sodium pertechnetate present in bone imaging radiopharmaceuticals at 5% or greater impurities is readily seen as stomach and thyroid uptake on the scan, as shown in Figure 14-9.

Radiochemical impurities in radiopharmaceuticals can arise from a number of sources, including competing chemical reactions during radiolabeling, problems in preparative techniques, radiolytic decomposition of the product, oxidation–reduction reactions, and chemical changes during storage because of pH or temperature changes and exposure to light. Radiation causes decomposition of water, a major ingredient of most radiopharmaceutical preparations, leading to the production of reactive hydrogen atoms and hydroxyl radicals, hydrated electrons, hydrogen, hydrogen ions, and hydrogen peroxide. Hydrogen peroxide is formed in the presence of oxygen radicals originating from the radiolytic decomposition of dissolved oxygen. Many radiopharmaceuticals show improved stability if oxygen is excluded. Radiation can also affect the radiopharmaceutical itself, giving rise to ions, radicals, and excited states. These species can combine with one another and with the active species formed from water. Radiation decomposition can be minimized through the use of chemical agents that act as electron or radical scavengers.

Relevant radiochemical impurities and their limits are usually listed in the package insert or in the *USP* monograph

(Table 14-3). RCP requirements must be fulfilled throughout the useful life of the radiopharmaceutical. The three basic sources of information on RCP test methods are package inserts, the *USP,* and published literature. However, because the information provided in these sources is not always complete or practical, it can be difficult to select and perform adequate RCP testing on the prepared radiopharmaceutical.[14,37] Table 14-4 lists routine chromatography methods for RCP determination of various single-photon emission computed tomography (SPECT) radiopharmaceuticals and PET radiopharmaceuticals by FDA-approved methods. The methods in Table 14-4 are taken from package inserts and the published literature.[38–62] Selection of the methods listed in Table 14-4 was based on the following: a compilation of procedures published by the American Pharmacists Association,[63] the *USP,* and FDA's *Chemistry, Manufacturing, and Controls Section* for ammonia N 13 injection, FDG F 18 injection, and sodium fluoride F 18 injection.[19] Information on RCP test methods for PET radiopharmaceuticals not listed in Table 14-4 can be found in the current version of *USP.*

Radiopharmaceuticals labeled with ^{99m}Tc are widely used in nuclear medicine studies. There are four principal radiochemical species that can be present in stannous-reduced ^{99m}Tc-labeled radiopharmaceuticals: (1) the ^{99m}Tc-labeled complex, which is the desired radiochemical species; (2) free ^{99m}Tc in the form of ^{99m}Tc-sodium pertechnetate (which was not reduced initially because of low levels of divalent stannous ion [Sn^{2+}] or was reduced initially and regenerated by oxidation reactions); (3) H-R ^{99m}Tc, which includes reduced ^{99m}Tc that reacts with water to form various hydrolyzed species (e.g., ^{99m}TcO$_2$) and ^{99m}Tc-Sn^{2+}colloids formed by the binding of reduced ^{99m}Tc to hydrolyzed stannous ion; and (4) ^{99m}Tc-labeled complexes of transfer ligands such as glucoheptonate and tartrate. Free ^{99m}Tc-sodium pertechnetate increases tissue background activity because of activity localized in the salivary glands, thyroid gland, stomach, and GI tract (Figure 14-9). H-R ^{99m}Tc not only compromises the labeling yield of the ^{99m}Tc-labeled complex but also interferes with imaging interpretation because of its localization in the reticuloendothelial system (see Figure 25-4, Chapter 25).

In general, with technetium chromatography systems, H-R ^{99m}Tc contaminants remain at the origin ($R_f = 0.0$) because they are insoluble particles, and free ^{99m}Tc-sodium pertechnetate travels to the solvent front ($R_f = 1.0$) in most solvents. The ^{99m}Tc-labeled complex may have an $R_f = 0.0$ or 1.0 or somewhere in between, depending on the system used. A simple chromatography system for RCP determination would allow the ^{99m}Tc-labeled complex to migrate to the S_f, with free ^{99m}Tc-sodium pertechnetate and H-R ^{99m}Tc remaining at the origin. This type of system is illustrated for ^{99m}Tc-sestamibi, as an example, in Figure 14-10. The RCP for the ^{99m}Tc-labeled complex is easily calculated by dividing the radioactivity on the top portion of the strip by the total activity in both sections of the chromatography strip. Many other ^{99m}Tc-labeled complexes require a two-solvent dual-strip system to determine RCP, because the ^{99m}Tc-labeled complex is always associated with one of the two radiochemical impurities (i.e., either free ^{99m}Tc or insoluble H-R ^{99m}Tc) and cannot be isolated by itself. This is illustrated in Figure 14-11 for ^{99m}Tc-apcitide, which has three impurities (free ^{99m}Tc-sodium pertechnetate, ^{99m}Tc-glucoheptonate, and insoluble ^{99m}Tc). Thus, the RCP of the ^{99m}Tc-labeled complex is determined indirectly by subtracting the percentage of each radiochemical impurity from 100% to obtain the percentage of the ^{99m}Tc-labeled complex.

Assessment of Radioactivity

The previous version of Part 35 in Title 10 of the Code of Federal Regulations (CFR) required that each radiopharmaceutical dose be assayed in a dose calibrator unless it was a pure beta- or alpha-emitting material (10 CFR §35.53, which has been deleted from 10 CFR). Changes in 10 CFR §35.63, "Determination of dosages of unsealed byproduct material for medical use," are as follows.[64]

- For a unit dose (a radiopharmaceutical dose that is received in a unit form and is not manipulated by the licensee), the dosage may be determined by direct measurement of activity (e.g., in a dose calibrator), or based upon the reported activity by a manufacturer, a preparer licensed under 10 CFR §32.72 or equivalent agreement state (e.g., a licensed nuclear pharmacy),[65] an NRC or agreement state licensee for use in research (i.e., radioactive drug research committee or investigational new drug), or a PET radiopharmaceutical producer licensed under 10 CFR §30.32(j)[66] or equivalent agreement state requirements and accounting for decay.[65]
- For a non–unit dose (anything prepared or manipulated by the licensee), the dosage may be determined by a direct measurement, a combination of measurement of dosage and mathematical calculations, or a combination of volumetric measurements and mathematical calculations based upon the measurement made by a manufacturer or preparer licensed under 10 CFR §32.72[65] or equivalent agreement state, or a PET radiopharmaceutical producer licensed under 10 CFR §30.32(j)[66] or equivalent agreement state requirements.

A dose calibrator is commonly used in a nuclear pharmacy for determination of the amount of radioactivity because it permits rapid and accurate measurement of radiopharmaceutical dosages. However, accurate measurement with a dose calibrator depends on the following factors: (1) a properly calibrated dose calibrator (see "Dose Calibrator" in the Instrument Quality Control section), (2) adherence to the specified measurement range (i.e., maximum and minimum measurable radioactivity), especially for PET radioisotopes, and (3) the type of container (e.g., syringe, vial, test tube) and the filling volume of solution in the container.

TABLE 14-3 Radiochemical Purity Limits of Radiopharmaceuticals from *USP* and Package Inserts

Radiopharmaceutical (Common Chemical Abbreviation and/or Brand Name)	USP Monograph[a]	Package Insert[a,b]
^{11}C-carbon monoxide	98	NA
^{11}C-flumazenil (^{11}C-FMZ)	98	NA
^{11}C-mespiperone	98	NA
^{11}C-methionine	98	NA
^{11}C-raclopride (^{11}C-RAC)	95	NA
^{11}C-sodium acetate	95	NA
^{14}C-urea	90	NL
^{57}Co-cyanocobalamin[c]	90	NL
^{58}Co-cyanocobalamin capsules[c]	90	NL
^{51}Cr-edetate	95	NA
^{51}Cr-sodium chromate	90	NL
^{18}F-fluorodopa (^{18}F-FD)	90	NA
^{18}F-fludeoxyglucose (^{18}F-FDG)	90	90
^{18}F-sodium fluoride	95	NL
^{67}Ga-citrate	97	NL
^{111}In-capromab pendetide (ProstaScint)	90	90
^{111}In-chloride	95	95
^{111}In-ibritumomab tiuxetan (^{111}In Zevalin)	95	95
^{111}In-oxyquinoline (^{111}In-oxine)	90	NL
^{111}In-pentetate (^{111}In-DTPA)	90	NL
^{111}In-pentetreotide (OctreoScan)	90	90
^{111}In-satumomab pendetide (OncoScint)[c]	90	NL
^{123}I-iobenguane (^{123}I-MIBG)	90	NL
^{123}I-iodohippurate sodium	97	NL
^{123}I-sodium iodide	95	NL
^{123}I-sodium iodide capsule	95	NL
^{123}I-sodium iodide solution	95	NL
^{125}I-albumin	97	NL
^{125}I-iothalamate	98	NL
^{131}I-iobenguane (^{131}I-MIBG)	90	NL
^{131}I-albumin (^{131}I-HSA)	97	NL
^{131}I-albumin aggregated (^{131}I-MAA)	NL	NL
^{131}I-iodohippurate sodium	97	NL
^{131}I-iodomethylnorcholesterol (NP59)[d]	NL	NL
^{131}I-rose bengal[c]	90	NL
^{131}I-sodium iodide	95	NL
^{131}I-tositumomab (Bexxar)	NL	NL
^{81m}Kr-gas	NL	NA
^{13}N-ammonia	95	NL
^{15}O water	95	NA
^{32}P-chromic phosphate[c]	95	NL
^{32}P-sodium phosphate[c]	100	NL
^{82}Rb-chloride	NL	NL
^{153}Sm-lexidronam (^{153}Sm-EDTMP)	99	NL

(continued)

TABLE 14-3 Radiochemical Purity Limits of Radiopharmaceuticals from *USP* and Package Inserts (Continued)

Radiopharmaceutical (Common Chemical Abbreviation and/or Brand Name)	USP Monograph[a]	Package Insert[a,b]
^{89}Sr-chloride	NL	NL
^{99m}Tc-albumin[c]	95	NL
^{99m}Tc-albumin aggregated (^{99m}Tc-MAA)	90	NL
^{99m}Tc-albumin colloids[c]	90	NL
^{99m}Tc-apcitide (AcuTect)[c]	90	90
^{99m}Tc-arcitumomab (CEA-Scan)[c]	95	90
^{99m}Tc-bicisate (^{99m}Tc-ECD; Neurolite)	90	90
^{99m}Tc-depreotide (NeoTect)[c]	90	90
^{99m}Tc-disofenin (^{99m}Tc-DISIDA; Hepatolite)	90	NL
^{99m}Tc-etidronate[c]	NL	NL
^{99m}Tc-exametazime (^{99m}Tc-HMPAO; Ceretec)	80	80
^{99m}Tc-fanolesomab (NeutroSpec)[c]	NA	90
^{99m}Tc-gluceptate (^{99m}Tc-GH)	90	NL
^{99m}Tc-lidofenin (^{99m}Tc-HIDA)[c]	90	NL
^{99m}Tc-mebrofenin (^{99m}Tc-BRIDA; Choletec)	90	NL
^{99m}Tc-medronate (^{99m}Tc-MDP)	90	NL
^{99m}Tc-mertiatide (^{99m}Tc-MAG3)	90	90
^{99m}Tc-nofetumomab merpentan (Verluma)[c]	85	NL
^{99m}Tc-oxidronate (^{99m}Tc-HDP)	90	NA
^{99m}Tc-pentetate (^{99m}Tc-DTPA)	90	NL
^{99m}Tc-pyro- and trimeta-pyrophosphate[c]	90	NL
^{99m}Tc-sodium pertechnetate	95	NL
^{99m}Tc-pyrophosphate (^{99m}Tc-PYP or ^{99m}Tc-PPi)	90	NL
^{99m}Tc-red blood cells (UltraTag RBC)	90	95
^{99m}Tc-sestamibi	90	90
^{99m}Tc-succimer (^{99m}Tc-DMSA)	85	NL
^{99m}Tc-sulfur colloid (^{99m}Tc-SC)	92	NL
^{99m}Tc-tetrofosmin (Myoview)	90	90
^{201}Tl-chloride	95	NL
^{127}Xe-gas	NL	NA
^{133}Xe-gas	NL	NL
^{133}Xe injection	NL	NA
^{90}Y-ibritumomab tiuxetan (^{90}Y Zevalin)	95	95

[a] Minimum percent radiochemical purity.
[b] NA = not available; NL = not listed.
[c] Discontinued product.
[d] Investigational new drug application (IND).

Certain radionuclides, such as ^{123}I, ^{111}In, and ^{127}Xe, not only have high-energy photopeaks but also emit substantial amounts of low-energy characteristic x-rays that produce variations in dosage measurements, especially when different container configurations or materials are used.[67,68] With the use of a copper filter (0.6 mm thickness), the low-energy component of ^{123}I and ^{111}In can effectively be reduced, even when these radionuclides are placed in various types of containers and are dispensed in different solution volumes.[69] ^{90}Y has attracted much research and clinical attention in radioimmunotherapy (especially with regard to its role in the Zevalin [ibritumomab tiuxetan] radiotherapeutic regimen) because its 64 hour half-life and pure beta emissions are very useful in cancer therapy. Although the typical dose calibrator has been designed to measure gamma-emitting radionuclides, it may also be quite useful in dosage measurement of pure beta

TABLE 14-4 Chromatographic Systems for Radiochemical Purity Testing of Reconstituted Radiopharmaceuticals[a]

Radiopharmaceutical (Common Chemical Abbreviation and/or Brand Name of Reagent Kit)	Stationary Phase	Mobile Phase	Relative Front (R_f) Values				Minimum RCP	Special Instructions	Ref.
			Labeled Complex	Unlabeled Radioisotope	Other Radiochemical Impurities				
^{18}F-fludeoxyglucose (^{18}F-FDG)	TLC-SG (activated silica gel on aluminum support)	Acetonitrile:water (95:5, v/v)	0.4	NA	NA		90	b	35
^{18}F-sodium fluoride	HPLC equipped with a flow-through gamma ray detector, a conductivity detector, and a 7.5 mm × 30 cm column that contains 10 μm packing L17	0.003 N sulfuric acid in water	c	c	c		95	c	38
^{111}In-capromab pendetide (ProstaScint)	ITLC-SG	0.9% NaCl	0.0–0.5	0.5–1.0	0.5–1.0		90	d	39
^{111}In-ibritumomab tiuxetan (^{111}In Zevalin)	ITLC-SG	0.9% NaCl	0.0–0.5	0.5–1.0	0.5–1.0		95		40
^{111}In-pentetreotide (OctreoScan)	Waters Sep-Pak C18 cartridge	e	e	e	e		90	e	41
^{13}N-ammonia	HPLC equipped with a flow-through gamma ray detector, a conductivity detector, and a 4.1 mm × 25 cm column that contains 10 μm packing L17	0.25 mL of concentrated nitric acid to 1000 mL of a mixture of water and methanol (7:3)	f	f	f		95	f	24
^{99m}Tc-albumin aggregated (^{99m}Tc-MAA)	Whatman 31ET	Acetone	0.0	1.0	0.0		90		42
^{99m}Tc-apcitide (AcuTect)	ITLC-SG	Water	0.25–1.0	0.25–1.0	0.0–0.25, 0.25–1.0[g]		90	h	43
	ITLC-SG	Saturated NaCl solution (SAS)[i]	0.0–0.75	0.75–1.0	0.0–0.75, 0.75–1.0[j]				
^{99m}Tc-arcitumomab (CEA-Scan)	ITLC-SG	Acetone	0.0–0.5	0.5–1.0	0.0–0.5		90/95[k]		44
^{99m}Tc-bicisate (^{99m}Tc-ECD; Neurolite)	1. Bakerflex silica gel 1B-F 2. Whatman 17	Ethyl acetate Ethyl acetate	1.0 1.0	0.0 0.0	0.0 0.0		90 90	l	45 46
^{99m}Tc-depreotide (NeoTect)	ITLC-SG	Methanol:1 M ammonium acetate (1:1, v/v) (MAM)	0.4–1.0	0.4–1.0	0.0–0.4, 0.4–1.0[m]		90	n	47
	ITLC-SG	SAS[i]	0.0–0.75	0.75–1.0	0.0–0.75, 0.75–1.0[o]				

Radiopharmaceutical	Medium	Solvent			%	Ref
99mTc-disofenin (99mTc-DISIDA; Hepatolite)	ITLC-SA	20% NaCl	0.0	1.0	0.0	90	48
99mTc-exametazime (99mTc-HMPAO; Ceretec)	1. ITLC-SG	Methyl ethyl ketone	0.8–1.0	0.8–1.0	0.0[p]	80	49
	ITLC-SG	0.9% NaCl	0.0	0.8–1.0	0.0		
	Whatman 31ET	50% Acetonitrile	0.8–1.0	0.8–1.0	0.8–1.0/0.0[q]		50, 51
	2. Pall Solvent Saturation Pads	Ether	1.0	0.0	0.0	80	
99mTc-gluceptate (99mTc-GH)	Whatman 31ET	Acetone	1.0	1.0	0.0	90	32
	ITLC-SG	Water	0.0	1.0	0.0		
99mTc-mebrofenin (99mTc-BRIDA; Choletec)	ITLC-SA	20% NaCl	0.0	1.0	0.0	90	48
99mTc-medronate (99mTc-MDP)	Whatman 31ET	Acetone	0.0	1.0	0.0	90	42
99mTc-mertiatide (99mTc-MAG3; TechneScan MAG3)	1. Waters Sep-Pak C18 cartridge	r	r	r	r	90	25
	2. Pall Solvent Saturation Pads	Chloroform:acetone: tetrahydrofuran (1:1:2, v/v)	0.0–0.5	0.5–1.0	0.0–0.5	90	27
	Gelman Solvent Saturation Pads	0.9% NaCl	0.5–1.0	0.5–1.0	0.5–1.0/0.0[s]		
99mTc-oxidronate (99mTc-HDP)	Whatman 31ET	Acetone	0.0	1.0	0.0	90	42, 52
99mTc-pentetate (99mTc-DTPA)	Whatman 31ET	Acetone	0.0	1.0	0.0	90	42
99mTc-pyrophosphate (99mTc-PYP or 99mTc-PPi)	Whatman 31ET	Acetone	0.0	1.0	0.0	90	42
99mTc-sodium pertechnetate	ITLC-SG	Acetone	1.0	0.0	0.0	95	32
99mTc-red blood cells (UltraTag RBC)		t	t	t	t	90	53, 54
99mTc-sestamibi (Cardiolite, Miraluma)	1. Baker Flex aluminum oxide coated, plastic TLC plate (#1 B-F)	Ethanol	1.0	0.0	0.0	90	u	55
	2. Pall Solvent Saturation Pads	Chloroform: tetrahydrofuran (1:1, v/v)	0.8–1.0	0.0	0.0	90	56, 57
	3. Waters Sep-Pak alumina N cartridge	v	v	v	v	90	v	58

(continued)

TABLE 14-4 Chromatographic Systems for Radiochemical Purity Testing of Reconstituted Radiopharmaceuticals[a] (Continued)

Radiopharmaceutical (Common Chemical Abbreviation and/or Brand Name of Reagent Kit)	Stationary Phase	Mobile Phase	Relative Front (R_f) Values			Minimum RCP	Special Instructions	Ref.
			Labeled Complex	Unlabeled Radioisotope	Other Radiochemical Impurities			
^{99m}Tc-succimer (^{99m}Tc-DMSA)	ITLC-SA	Acetone	0.0	1.0	0.0	85		42
^{99m}Tc-sulfur colloid (^{99m}Tc-SC)	Whatman 31ET	Acetone	0.0	1.0	1.0	92		42
^{99m}Tc-tetrofosmin (Myoview)	1. ITLC-SG (5 × 20 cm)	Acetone:dichloromethane (36:65, v/v)	0.2–0.8	0.8–1.0	0.0–0.2	90		59
	2. ITLC-SG (2.5 × 10 cm)	2-Butanone	0.55	1.0	0.0	90		60
	3. Waters silica Sep-Pak cartridge	w	w	w	w	90	w	61
	4. Pall Solvent Saturation Pads	Chloroform: tetrahydrofuran (1:1, v/v)	0.8–1.0	0.0	0.0	90		62

[a] HPLC = high-pressure liquid chromatography; ITLC-SA = instant thin-layer chromatography–polysilicic acid; ITLC-SG = instant thin-layer chromatography–silica gel; NA = not available; RCP = radiochemical purity; TLC = thin-layer chromatography; TLC-SG = thin-layer chromatography–silica gel.

[b] Prepare a solution of the *test solution* and the *standard solution*, inject about 50 μL of the combined solution into the chromatograph, maintain the flow rate at about 0.8 mL per minute, record the chromatograms, and measure the areas for both the radioactive and nonradioactive peaks (the injection volume may be adjusted to obtain suitable detection system sensitivity—the relative standard deviation for replicate injections is not more than 5%). The radioactivity of the major peak is not less than 95% of the total radioactivity measured, and the retention time of the *test solution* corresponds to the retention time (about 8 minutes) of the *standard solution*.

[c] Apply about 10 μg of the *standard solution* to the same activated ITLC-SG plate. Develop the chromatogram in a solvent system consisting of a mixture of acetonitrile and water (95:5) until the solvent has moved about three-fourths of the length of the plate. Remove the plate and allow the chromatogram to dry. Determine the radioactivity distribution by scanning the chromatogram with a suitable collimated radiation detector. Determine the location of FDG by spraying the developed chromatographic plate with 2 N sulfuric acid and heating the plate at 110° for 10 minutes; the R_f value of ^{18}F-FDG (determined by radiochromatogram scanning) corresponds to that of the *standard solution* (about 0.4). The radioactivity of ^{18}F-FDG is not less than 90% of the total radioactivity.

[d] Before applying the test sample to the ITLC-SG strip, mix equal parts (several drops of each) of ^{111}In-capromab pendetide with 0.05 M ^{99m}Tc-DTPA solution and allow the mixture to stand at room temperature for 1 minute.

[e] Slowly push 10 mL of methanol through the longer end of a fresh Waters Sep-Pak C18 cartridge. Similarly, rinse the cartridge with 10 mL water and then with another 5 mL water. Discard the eluates. Load 0.05–0.1 mL of ^{111}In-pentetreotide on the longer end of the cartridge column, making sure the test sample migrates onto the column and not into the tube neck. With a disposable syringe, slowly push (in dropwise manner) 5 mL of water through the longer end of the cartridge, using a test tube (tube 1) to collect the eluate. Similarly, elute the cartridge with 5 mL methanol drop by drop and collect the eluate in a test tube (tube 2). Place the Sep-Pak cartridge in a test tube (tube 3). The percentages of ^{111}In-pentetreotide (tube 2), hydrophilic impurities (tube 1), and nonelutable impurities (tube 3) are calculated by dividing the radioactivity in each tube by the total of activity of all three tubes.

[f] Prepare a mixture of the *standard solution* and the *test solution* (1:1) and inject about 20 μL of the mixture into the chromatograph, maintain the flow rate at about 2.0 mL per minute, record the chromatograms, and measure the peak areas (the volume of injection may be adjusted to obtain suitable detection system sensitivity—the relative standard deviation for replicate injections is not more than 5%.) The areas of both the main radioactive and nonradioactive peaks are equal. The radioactivity of the major peak is not less than 95% of the total radioactivity measured. The retention time of the *test solution* corresponds to the retention time of the *standard solution*.

[g] ^{99m}Tc immobile materials (R_f = 0.0–0.25), ^{99m}Tc-glucoheptonate (R_f = 0.25–1.0).

h RCP of ^{99m}Tc-apcitide = 100 − (% activity in the bottom piece of ITLC-SG water strip + % activity in the top piece of ITLC-SG SAS strip).

i Mix 5 grams of NaCl with 5 to 10 mL of water and shake periodically for 10 to 15 minutes. Add more NaCl and shake again for 10 to 15 minutes until a solid residue remains.

j ^{99m}Tc immobile materials (R_f = 0.0–0.75), ^{99m}Tc-glucoheptonate (R_f = 0.75–1.0).

k Package insert: 90%; USP: 95%.

l (1) Pre-equilibrate the chromatographic developing tank with ethyl acetate for 15 to 30 minutes. (2) The sample spot should not be greater than 10 mm; allow the spot to dry for 5 to 10 minutes. (3) The developing time is approximately 15 minutes.

m ^{99m}Tc nonmobiles (R_f = 0.0–0.4), ^{99m}Tc-glucoheptonate and ^{99m}Tc-edetate (R_f = 0.4–1.0).

n RCP of ^{99m}Tc-depreotide = 100 − (% activity in the bottom piece of ITLC-SG MAM strip + % activity in the top piece of ITLC-SG SAS strip).

o ^{99m}Tc nonmobiles (R_f = 0.0–0.75), ^{99m}Tc-glucoheptonate and ^{99m}Tc-edetate (R_f = 0.75–1.0).

p Secondary ^{99m}Tc-exametazime complex and H-R ^{99m}Tc stay at the origin.

q Secondary ^{99m}Tc-exametazime complex migrates to R_f 0.8 to 1.0, whereas H-R ^{99m}Tc remains at the origin.

r (1) Preparation of Sep-Pak C18 cartridge. The Sep-Pak C18 cartridge is first flushed with 10 mL of 200 proof ethanol, followed by pushing 5 mL of air through the cartridge with a syringe. (2) Sample analysis. Apply 0.1 mL of ^{99m}Tc-mertiatide to the long end of the cartridge. The cartridge is then eluted successively with 10 mL of 0.001N hydrochloric acid and 10 mL of 1:1 ethanol/0.9% sodium chloride solution. The two fractions of sample eluates and cartridges are collected in separate culture tubes for counting. (3) Counting. The radioactivity of the first sample elution (hydrophilic ^{99m}Tc impurity plus a fraction of H-R ^{99m}Tc), the second sample elution (^{99m}Tc-mertiatide), and the cartridge (the remaining H-R ^{99m}Tc plus nonelutable impurities) is assayed in a dose calibrator. The percentages of ^{99m}Tc-mertiatide (fraction 2), hydrophilic ^{99m}Tc species (primarily H-R ^{99m}Tc), and nonelutable impurities are calculated by dividing each fraction of radioactivity by the total of activity of both sample liquid fractions and the cartridge.

s Hydrophilic ^{99m}Tc impurity migrates to R_f 0.5 to 1.0, whereas H-R ^{99m}Tc remains at the origin.

t (1) Transfer 0.2 mL of ^{99m}Tc-red blood cells (RBCs) to a centrifuge tube containing 2 mL of 0.9% NaCl. (2) Centrifuge the ^{99m}Tc-RBC sample at 150g for 1 minute. (3) Carefully pipette off the diluted plasma. (4) Measure the radioactivity in the plasma and the RBCs separately in a dose calibrator. (5) Calculate the labeling efficiency of ^{99m}Tc-RBCs as follows:

% RBC labeling = [RBC activity ÷ (RBC activity + Plasma activity)] × 100.

u (1) The TLC plate has to be predried in an oven at 100°C for 1 hour. (2) After two drops of ^{99m}Tc-sestamibi sample are applied side-by-side on top of the ethanol wet spot, the TLC plate is placed in a desiccator to allow the spot to dry before the plate is developed in the TLC tank. (3) This TLC system requires about 30 minutes to complete the drying and development.

v Slowly push 5 mL of ethanol (100% ethanol is preferable) through the longer end of a fresh Waters Sep-Pak alumina N cartridge. Load 0.05 to 0.1 mL of ^{99m}Tc-sestamibi on the longer end of the cartridge column, making sure the test sample gets on the column and not in the tube neck. With a disposable syringe, slowly push (in dropwise manner) 10 mL of ethanol through the longer end of the cartridge, using a test tube (tube 1) to collect the eluate. Follow with a few milliliters of air to collect all of the ethanol. Place the Sep-Pak cartridge in a test tube (tube 2). The percentage of ^{99m}Tc-sestamibi (tube 2) is calculated by dividing the radioactivity in tube 2 by the total activity in tubes 1 and 2.

w Slowly equilibrate the silica Sep-Pak cartridge with 5 mL of normal saline followed by 1 mL of air. Load 50 μL of sample onto the long-necked side of the column, with care taken to place the sample on the column and not in the tube neck. Then, push 10 mL of 70:30 methanol:water dropwise through the column into a test tube at a flow rate of 5 mL/min. Finally, place the cartridge in a second test tube and assay each tube in a dose calibrator. The percentage of ^{99m}Tc-tetrofosmin is calculated by dividing the radioactivity in tube 2 by the total radioactivity in tubes 1 and 2.

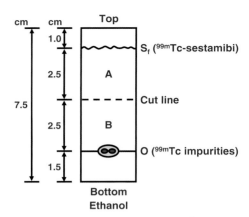

FIGURE 14-10 Single-strip thin-layer chromatography (TLC) system for radiochemical purity (RCP) determination of ^{99m}Tc-sestamibi applied to the bottom of a Baker-Flex aluminum oxide–coated TLC plate and developed in absolute ethanol.

emitters such as ^{90}Y using its readily produced bremsstrahlung radiation.[70–73]

According to 10 CFR §35.63,[64] if the dosage of a radiopharmaceutical preparation needs to be assayed, with regard to all beta-emitting and difficult-to-measure gamma-emitting materials (e.g., ^{153}Sm), it is practical to use volumetric measurements and mathematical calculations. For all other radiopharmaceutical preparations, a dose calibrator can be used to directly measure the radioactivity.

The amount of radioactivity in a radiopharmaceutical is usually determined either by whole-vial assay or by counting an aliquot of the radiopharmaceutical in a dose calibrator. Radioactive concentration is usually expressed in terms of specific activity, which is defined as the activity per unit weight of the labeled compound (e.g., millicuries per gram or megabecquerels per gram), or specific concentration, which is defined as the activity per unit volume (e.g., millicuries per milliliter or megabecquerels per milliliter). The radioactive concentration is usually stated on the vial label along with the calibration date and time; however, 10 CFR §35.69, "Labeling of vials and syringes," requires only that the identity of the radioactive drug be shown on the label.[74] Although curies or millicuries are the units of radioactivity commonly used in the United States, the becquerel is the internationally recognized unit for radioactivity. Therefore, becquerel units are also typically listed on the vial or syringe label. It is common practice that each radiopharmaceutical vial or syringe be radioassayed to confirm that its activity is within ±10% of the labeled activity after decay correction, although 10 CFR §35.63 allows a difference of ±20% between the measured dosage and the prescribed dosage.[64]

It is important to verify by measurement the activity stated on the label of a radiopharmaceutical received by a nuclear medicine department or nuclear pharmacy. The European Association of Nuclear Medicine reported its findings with regard to defects in radiopharmaceuticals and noted instances in which incorrect radioactivities were provided by manufacturers.[3–9] Its 2000 annual report states that the measured activity of one shipment of ^{131}I therapeutic capsules was 20% to 40% higher than that stated on the labels.[8] This illustrates the importance of assaying the radioactivity of any radioactive material received. A concern raised by this finding is that if an assay of some type is not done upon receipt, dosage calculations determined by volumetric measurements and mathematical calculations based on information provided by the manufacturer could result in incorrect dosing of patients. This would be particularly significant for pure beta- or alpha-emitting material, which is difficult or impossible to assay in a dose calibrator. Some method for assessing the activity of all radiopharmaceuticals should be a part of the standard of practice to avoid misadministration to patients. It is also important to ensure that the radioisotope setting selected on the dose calibrator is the correct one for the measured radioisotope. In addition, the worker should carefully read the unit (e.g., millicuries versus microcuries) of the measured radioactivity as displayed on the dose calibrator. A 2007 information notice published by the Nuclear Regulatory Commission indi-

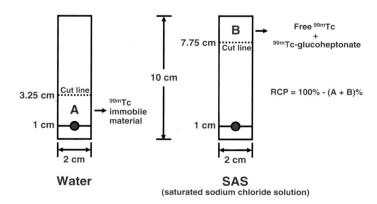

FIGURE 14-11 Dual-strip ITLC-SG systems for radiochemical purity (RCP) determination of ^{99m}Tc-apcitide (AcuTect [Rentschler Biotechnologie]).

cated that a nuclear pharmacy technician misread a dose of "0.915 mCi" ^{131}I sodium iodide displayed on the dose calibrator as "9.15 μCi."[75] This resulted in an administered dosage of 915 μCi rather than the requested dosage of 5 to 20 μCi.

Specific Activity

Specific activity is defined as the radioactivity of the radionuclide per unit mass of the element or chemical form concerned. It is a critical specification for receptor-targeted radiopharmaceuticals whose uptake at receptors in the body is influenced by the mass of radiotracer injected. Specific activity changes with time; therefore, the stated specific activity should include a date and, if necessary, a time.

No testing for specific activity of a radiopharmaceutical preparation is required if the radionuclide of interest is prepared by a no-carrier-added method of synthesis. If carrier-added synthesis is used for production of the radionuclide concerned, the appropriate test and acceptance criteria as well as the testing frequency for specific activity must be defined.

Enantiomeric Purity

The stereoisomeric purity should be verified for applicable compounds. 2-^{18}F-fluoro-2-deoxy-D-mannose (FDM) is an epimer of ^{18}F-FDG, and the enantiomeric (or isomeric) purity test for ^{18}F-FDM is described in the *European Pharmacopeia*.[76–79]

CHEMICAL CONSIDERATIONS
Chemical Purity

Chemical purity is a measurement of the presence of undesirable nonradioactive chemical species in radiopharmaceuticals. This assessment is important to ensure that the presence of any chemical substance with potential toxic, physiologic, or pharmacologic effects is within appropriate limits. Chemical purity testing should be performed on any chemical substance that is either used or formed during the synthesis of a radiopharmaceutical (e.g., chemical impurities, unlabeled ingredients, reagents, and byproducts), even if such substances are not listed in the *USP* monograph or package insert.[1]

Examples of chemical impurities significant to nuclear pharmacy are as follows:

1. Aluminum ion (Al^{3+}) in ^{99}Mo–^{99m}Tc generator eluate. Excessive Al^{3+} can induce flocculation of ^{99m}Tc-SC because Al^{3+} combines with the phosphate buffer in the product to form insoluble aluminum phosphate.[80,81] In bone scanning, liver localization and degradation of image quality have been noted when Al^{3+} concentrations exceed 10 μg per milliliter of ^{99m}Tc eluate.[81] Accordingly, the *USP* limits the concentration of aluminum ion in ^{99m}Tc-sodium pertechnetate to 10 μg/mL.[82] Assessment of aluminum ion concentration in the ^{99m}Tc eluate is usually performed with a colorimetric spot test (discussed in Chapter 3) and should be done on every ^{99m}Tc generator eluate.
2. Carrier iodine in radioiodide solution. Carrier iodine can compete with radioiodine in the radioiodination process, resulting in poor labeling efficiency, and will interfere with uptake of tracer radioiodide in the thyroid gland.
3. Trace metals such as iron in ^{111}In-chloride solution. Trace metals have been known to significantly reduce labeling yields with ^{111}In, especially in platelet labeling with ^{111}In-oxine.

For ^{18}F-FDG injection, potential chemical impurities are aminopolyether 4,7,13,16,21,24-hexaoxa-1,10-diazabicyclo-(8,8,8)-hexacosane (Kryptofix 2.2.2. [Sigma-Aldrich], or K222), 2-chloro-2-deoxy-D-glucose (ClDG), nonradioactive 2-fluoro-2-deoxy-D-glucose (FDG), tetra-alkyl ammonium salts, and 4-(4-methylpiperidino)pyridine. Testing for these chemical impurities can be done by comparison of spot size and intensity shown on a developed TLC strip (K222), with a sophisticated HPLC system equipped with a UV detector (ClDG, FDG, and tetra-alkyl ammonium salts), or with a UV spectrophotometer [4-(4-methylpiperidino)pyridine].[35,76]

For ^{13}N ammonia injection, aluminum is a potential chemical impurity if Devarda's alloy is used to reduce ^{13}N nitrate or ^{13}N nitrite. The chemical purity test for aluminum is carried out by comparing the color intensity of the sample solution and the standard solution (the color of the sample solution should not be more intense than that of the standard solution)[24] or by using a spectrophotometer to determine that the concentration of aluminum ion in the injection is not greater than 10 μg/mL.[83]

Residual Solvents

Similar to the chemical purity assessment, any residual solvents present in the final radiopharmaceutical preparation must be determined and found to be within USP limits.

FDA guidance for the pharmaceutical industry groups residual solvents into three classes.[84] The classification of residual solvents involves assessment of not only their potential toxicity with respect to human health concerns but also their possible deleterious effects on the environment.[84] Class 1 comprises solvents known and strongly suspected to be human carcinogens and environmental hazards. Their use should be avoided in the manufacture of drug substances, excipients, and drug products. Class 2 solvents have inherent toxicity, and their use should be limited in pharmaceutical products.[84,85] Solvents in Class 3 are those that have less toxic potential and thus pose a lower risk to human health.[84,85]

A GC system with flame ionization detection is the instrument of choice in the determination of residual solvents. In a manner similar to an HPLC system, a GC system must be validated for proper analysis of residual solvents.[23,18] According to the *USP* monograph for fludeoxy-

TABLE 14-5 Appearance and Color of Radiopharmaceuticals

Radiopharmaceutical	Appearance	Color
^{99m}Tc-albumin aggregated (^{99m}Tc-MAA)	Turbid	White
^{99m}Tc-sulfur colloid	Slightly turbid	Milky
Other ^{99m}Tc-labeled compounds	Clear	Colorless
^{131}I-sodium iodide	Clear (turns light amber with time)	Colorless
^{32}P-sodium phosphate[a]	Clear	Colorless
^{32}P-chromic phosphate[a]	Turbid	Bluish green
Other non-^{99m}Tc-labeled compounds	Clear	Colorless

[a] Withdrawn from U.S. market.

glucose F 18 injection, the resolution between standard and test solutions should be not less than 1.0, and the relative standard deviation for replicate injections should not be more than 5%.[35]

PHARMACEUTICAL CONSIDERATIONS

The basic considerations regarding the pharmaceutical aspects of a radiopharmaceutical are appearance and color, particle number and size, pH, osmolality, and stabilizers or preservatives.

Appearance and Color

Annual reports of defective radiopharmaceutical products prepared by the European Association of Nuclear Medicine frequently describe the presence of foreign substances (e.g., black particles, glass elements, fiber) in vials, coring of vial stoppers, and vial breakage or cracking before, during, and after reconstitution.[3–8] Although these occurrences are rare, they do underscore the value of inspecting the content of kit vials before and after radiopharmaceutical preparation. Workers should be thoroughly familiar with the normal appearance and color of every radiopharmaceutical (Table 14-5). Formulations used in nuclear medicine may be true solutions, colloidal dispersions, suspensions, or solid materials such as oral capsules; they have defined properties that allow their identification. Gross macroscopic inspection of solutions should be performed to identify any foreign material or change in appearance or color. For example, ^{131}I-sodium iodide solution is clear and colorless when freshly prepared but turns light amber with time because of radiolysis effects; this color change is not deleterious. Another example is the difference between ^{32}P-sodium phosphate, which is a colorless, clear solution, and ^{32}P-chromic phosphate, which is a bluish-green, turbid, insoluble suspension (Table 14-5). Direct visual observation of a radiopharmaceutical should be conducted in accordance with the principle of ALARA (as low as reasonably achievable) (e.g., view the product behind leaded glass) in order to limit acute and chronic radiation exposure of inspecting personnel. The visual inspection should be made through leaded-glass shielding that is not tinted or fogged and under adequate light so that the radiopharmaceutical observed is not obscured.

Particle Number and Size

Particulate radiopharmaceuticals such as ^{99m}Tc-SC and ^{99m}Tc-albumin aggregated (^{99m}Tc-MAA) normally have a cloudy appearance (Table 14-5). Microscopic inspection of these agents, or filtration analysis of ^{99m}Tc-SC, can be performed to confirm that particles are uniformly dispersed and of proper size (Table 14-6).

Microscopic inspection with a light microscope and hemacytometer grid can be performed to estimate particle size and number with ^{99m}Tc-MAA (Figure 14-12). According to the *USP*, at least 90% of ^{99m}Tc-MAA particles should have a diameter between 10 and 90 μm, with none greater than 150 μm.[86]

pH

All radiopharmaceuticals have an optimal pH range for stability, and most radiopharmaceuticals are within a pH range of 4 to 8, with some exceptions:

1. Radioiodine solution should be kept in alkaline pH to prevent volatilization of iodine.

TABLE 14-6 Particle Size of Common Particulate Radiopharmaceuticals

Radiopharmaceutical	Diameter (μm)
^{99m}Tc-sulfur colloid (thiosulfate)	0.1–1.0
Filtered ^{99m}Tc-sulfur colloid (thiosulfate)	<0.1 to 0.22
^{99m}Tc-albumin aggregated (^{99m}Tc-MAA)	10 to 90 (90%) <150 (100%)

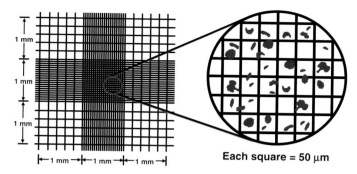

FIGURE 14-12 Estimating size and number of ^{99m}Tc-MAA particles with the hemacytometer grid.

2. Indium chloride solution must be kept quite acid (pH 1 to 3) to remain in solution and to prevent formation of insoluble indium hydroxides. Chelated forms of indium, however, such as ^{111}In-DTPA, are very soluble at neutral pH.
3. The pH of unstabilized ^{99m}Tc-exametazime injection is in the range of 9.0 to 9.8, whereas the pH of the stabilized ^{99m}Tc-exametazime is buffered between 6.5 and 7.5 for stability.[49,87]

According to *USP* General Chapter 791, test paper may be suitable for the measurement of an approximate pH value.[88] A narrow-band pH paper is suitable for pH assessment, because USP pH limits for radiopharmaceutical products are usually quite broad. However, the accuracy and traceability of pH paper should be initially verified with standard buffers. The pH value obtained with pH paper is approximate, and the accuracy is probably no better than ±0.25 pH units.[89]

Osmolality

Because most radiopharmaceutical preparations are formulated in physiologic saline and are therefore isotonic, an osmolality test is usually not necessary for radiopharmaceutical injections. Osmolality and pH are more critical for intrathecal injections than for intravenous injections because of the high dilution and buffering capacity of the blood.

Stabilizers and Preservatives

Stabilizers may be added to radiopharmaceuticals for a number of reasons. Typically they are added to retard oxidation or to reduce radiolytic degradation in preparations with high specific concentration. A preservative is usually added to a radiopharmaceutical formulation to retard the growth of microorganisms that might be introduced into the vial during multiple penetrations. Preservatives can extend the shelf life of a ^{99m}Tc radiopharmaceutical beyond the default expiration time of 12 hours. ^{99m}Tc-mebrofenin is a good example of this. If the added stabilizer or preservative may cause toxic, physiologic, or pharmacologic effects, it must be properly evaluated before release of the drug product for patient use. The testing method, acceptance limits, and testing schedule for assay of the potentially toxic stabilizer or preservative should be established and validated. Such testing is typically the domain of the radiopharmaceutical manufacturer and is not part of the routine practice of nuclear pharmacy.

BIOLOGIC CONSIDERATIONS

Most radiopharmaceuticals are designed for parenteral administration and therefore must be prepared by aseptic processing. The goal of aseptic processing is to prepare a drug product that is free of microorganisms and toxic microbial byproducts, most notably bacterial endotoxins.

Microbiologic control (i.e., sterility testing) and bacterial endotoxin testing (BET) must be performed according to written policies and procedures for injectable radiopharmaceuticals when these tests are required as release criteria. Retrospective sterility testing and BET should be conducted on randomly selected batches of a non-PET radioactive drug product to check the adequacy of aseptic technique. These tests should be conducted at regular intervals, depending on historical results and trends, and should be completed more frequently when new personnel are involved. Sterility testing and BET should be performed according to procedures based on and adapted from those described in *USP* General Chapters 71 and 85, respectively.[90–92]

Sterility

A sterile solution is one that contains no living organisms, pathogenic or nonpathogenic. All injectable products must be sterilized by steam (autoclave) or membrane filtration. Autoclaving, with steam under pressure, is useful only for products that can withstand the severe physical conditions of 121°C under pressure at 15 pounds per square inch gauge (psig). These conditions obviously preclude autoclaving protein and biologic products. For heat-labile products, sterilizing membrane filtration is the method of choice. Membrane porosity should be at 0.22 μm.

If radiopharmaceuticals are formulated from raw materials, sterile glassware, syringes, and other components should be used to lessen the chance of introducing microorganisms and pyrogenic material into the product. Radiopharmaceutical products with a long enough half-life should be subjected to the USP sterility test before use. The official sterility test uses fluid thioglycolate medium to test for bacterial contamination and soybean casein digest medium to test for fungi.[90] Sterility tests can usually be conducted in a hospital's microbiology laboratory. Of course, proper precautions must be taken if the product is radioactive.

The test for sterility should be carried out in accordance with *USP* General Chapter 71.[90] However, special difficulties arise with radiopharmaceutical preparations because of the short half-life of some radionuclides (especially those used for diagnostic purposes), small size batches, low production volume, and radiation hazards.

The USP sterility testing method requires observation of the tubes of media over a 14 day incubation period unless otherwise specified elsewhere in the sterility testing chapter or in the individual monograph.[90] It is common to observe the test samples at days 3, 7, and 14. In any event, the short half-lives of most radiopharmaceuticals used in nuclear medicine studies prohibit completion of the sterility testing before the release of radiopharmaceutical products. In addition, when the half-life of the radionuclide is very short (e.g., less than 20 minutes), administration of a radiopharmaceutical preparation to a patient is generally on-line with a validated production system. It is justifiable to dispense radioactive drug products before completion of the sterility test if the radiopharmaceutical is prepared by a validated aseptic process.

For safety reasons (i.e., high levels of radioactivity), it may not be possible to use the required quantity of a radiopharmaceutical preparation in accordance with the *USP* sterility testing chapter.[90] To limit radiation exposure of personnel, the product should be diluted or allowed to decay to a safe working level before sterility tests are conducted.

In summary, radiopharmaceutical preparations should meet the requirements stated in *USP* General Chapter 71 on sterility tests, with the exception that radioactive drug products may be distributed or dispensed before completion of the tests for sterility.[90] The sterility test for a PET radiopharmaceutical must be started within 30 hours after completion of production.[93]

Membrane Filter Integrity

Since the sterility test is completed retrospectively for most radiopharmaceuticals, the membrane filter integrity test can be considered an indicator of the microbiologic purity of the product. All membranes used for product sterilization must pass an integrity test prior to product release. "Bubble point" measurement is a simple test for membrane filter integrity.[18,19]

Bacterial Endotoxin Testing

Pyrogens are metabolic products of microorganisms that cause a pyretic response upon injection. Endotoxin is the most significant pyrogen. The response in humans is characterized by the onset of chills and fever within 45 to 90 minutes after the injection of pyrogenic material. General malaise and headache may also be present. The severity of response is dependent upon the amount of pyrogen administered.

All injectable products are required to be pyrogen free (ideally) or below the endotoxin limit. For radiopharmaceuticals not administered intrathecally, the endotoxin limit is set at 175 endotoxin units (EUs) per V, where V is the maximum recommended dose in milliliters; for intrathecally administered radiopharmaceuticals the limit is 14 EU/V.[91] Endotoxin is not removed by sterilizing membrane filtration methods; therefore, a solution could be sterile but pyrogenic. The most likely sources of pyrogens have been impure water and chemicals used in product preparation. Pyrogenic contamination is usually prevented by using high-quality chemicals and water for injection. Glassware can be rendered pyrogen free by dry heat at 250°C for 30 minutes, which incinerates the endotoxin. Autoclaving does not completely destroy pyrogens.

The official test for the presence of pyrogens in parenterals has been the USP rabbit test.[92] The USP rabbit pyrogen test requires intravenous administration of the drug to be tested into the marginal ear vein of three rabbits whose body temperature is monitored. Temperature is recorded each hour for 3 hours and compared with the baseline temperature before injection. The product passes the test if no rabbit shows an individual rise in temperature of 0.6°C or more above its respective control temperature and if the sum of the three individual maximum temperature rises does not exceed 1.4°C. Details of the test can be found in the *USP*.[92] Special care and facilities are needed to house the rabbits so that the test will be valid. Rabbits are subject to temperature elevations solely from fright or excitation. Thus, sham testing must be performed to ensure reliability. Although this has been a reliable test for the presence of pyrogens in parenterals, it has limitations for use with radiopharmaceuticals, particularly those containing short-lived radionuclides. In addition, the rabbit test appears to lack the sensitivity required to detect endotoxins in radiopharmaceuticals intended for intrathecal use.[94] Fortunately, an in vitro test has been developed that is far more sensitive to bacterial endotoxin than the rabbit test; this is the limulus amebocyte lysate (LAL) test, now officially known as the bacterial endotoxins test (BET).

The bioassay in rabbits was the first pyrogen test and was used for many years,[92] but the BET was shown to be more sensitive.[94] Development of a viable alternative to the USP rabbit pyrogen test began with the observation by Bang[95] that gram-negative bacteria caused intravascular coagulation in the American horseshoe crab, *Limulus polyphemus*. In collaboration, Levin and Bang[96] later found that this clotting resulted from action between endotoxins and a clottable protein (a proenzyme) in the lysate of the crab's circulating blood cells, which are amebocytes. The need for a more suitable pyrogen test for drugs, particularly radiopharmaceuticals, led Cooper, Levin, and Wagner[97] to develop a simple method of applying this new approach to endotoxin testing of drug products. The BET (then called the LAL test), was introduced in the *United States Pharmacopeia, 20th Revision*, and assigned official status by the USP Committee of Revision in 1993.[97]

The BET for pyrogens is preferred over the rabbit test (especially for radiopharmaceutical preparations) because of simplicity, sensitivity, specificity, rapidity, and cost-effectiveness.[97] Table 14-7 lists characteristics of the BET. It is a qualitative and quantitative test for gram-negative bacterial endotoxin. A proenzyme obtained from a lysate of washed amebocytes of *L. polyphemus* blood is extremely sensitive to the presence of gram-negative bacterial endotoxin. Once the proenzyme is activated by the endotoxin (its rate of activation is determined by the concentration of endotoxin present), the activated enzyme (coagulase) hydrolyzes specific bonds within a clotting protein (i.e., coagulogen) that is also present in LAL (Figure 14-13). The hydrolyzed coagulogen then

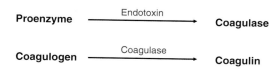

FIGURE 14-13 The activation sequence of a proenzyme in limulus amebocyte lysate catalyzed by a gram-negative bacterial endotoxin.

TABLE 14-7 Characteristics of Bacterial Endotoxin Test (BET)

1. It is an in vitro test
2. It can be performed in-house with minimal equipment and personnel
3. Test volume can be as small as 0.1 mL
4. The test is completed within 1 hour
5. An in-process 20 minute endotoxin limit test can be used to allow for early release of an ultra-short-lived radiopharmaceutical
6. Positive and negative controls and inhibition or enhancement testing can be performed with each test
7. It is relatively inexpensive, and the test materials can be stored until needed
8. The sensitivity of the BET allows for substantial dilution of the test sample (except for intrathecal drug products)
9. It can be used to test drugs not amenable to the rabbit test, such as anesthetics, cancer chemotherapeutic agents, sedatives, narcotics, intrathecal drugs, and drugs that exert potent pharmacologic effects on animal systems[97]

turns into coagulin and forms a gelatinous clot. Lyophilized lysate must be stored at between 2°C and 8°C to maintain effectiveness.

There are two types of BET techniques: the gel-clot technique and the photometric technique (i.e., a turbidimetric method and a chromogenic method).[91] Specifications listed in the *USP* monograph for selecting a testing technique should be followed. Unless otherwise indicated in the *USP* monograph, the gel-clot technique is the method of choice in case of dispute.[91]

The LAL reaction is enzyme mediated and has an optimal pH range for proper performance of the LAL reagent. As with most enzymatic reactions, test samples for LAL testing should be done at neutrality. If necessary, pH should be adjusted with endotoxin-free acid or base. Samples to be tested may be stored at 2°C to 8°C for 24 hours or be kept frozen if the storage period is longer than 24 hours.

The validity of the BET requires adequate demonstration that a test sample does not inhibit or enhance the reaction or otherwise interfere with the test. Interference can be overcome by suitable treatment, such as filtration, neutralization, dialysis, or heating.[91] Many compounds in a drug product, however, are inhibitory, and the easiest way to overcome inhibition is dilution of the test sample. Sample dilution must not exceed the maximum valid dilution at which the endotoxin limit can be determined by the BET technique.[91] In any event, inhibition or enhancement testing should be performed on test sample solutions to determine whether they interfere with the LAL reagent and to ensure that any treatment has effectively eliminated interference without loss of endotoxin. A product is said to be free of product inhibition or enhancement if the geometric mean endpoint of endotoxin in the product is between one-half and 2 times the labeled lysate sensitivity.[91]

Each vial of LAL is labeled with the lysate sensitivity (λ), which is obtained using the USP RS endotoxin (RSE) and is expressed in endotoxin units (EU) per milliliter. One EU (or USP-EU) is equal to 1 international unit (IU) of endotoxin. A sensitivity of not less than 0.15 EU/mL should be met for an LAL reagent to be used for routine BET.[91] Before using a new batch of LAL reagent kits, each user should confirm the labeled LAL reagent sensitivity by using the control standard endotoxin (CSE) supplied with an LAL reagent kit. This should be verified again when there is any change in test conditions that might affect the outcome of testing, such as a change in the formulation.[91] Acceptable variation of the measured

TABLE 14-8 Gel-Clot Limit Test

Test Solution	Number of Replicates	Endotoxin Concentration/ Test Sample or Control	Function	Results or Comments
A	2	None/sample solution	Test sample	May be positive or negative
B	2	2 λ/sample solution	Test for interference	Should be positive
C	2	2 λ/LAL reagent water	Positive control	Should be positive
D	2	None/LAL reagent water	Negative control	Should be negative

sensitivity of the LAL reagent is between one-half and 2 times the stated lysate sensitivity.

According to the *USP*, 5 EU/kg body weight is the endotoxin limit for a drug product administered by any route other than intrathecally (for which the limit is 0.2 EU/kg).[91] The endotoxin limit for radiopharmaceutical products is 175 EU/V, where V is the maximum dose in milliliters, at the expiration time. For intrathecally administered radiopharmaceuticals, the endotoxin limit is 14/V because of increased toxicity of endotoxin by the intrathecal route. Drug products intended for intrathecal administration must be tested at a lesser dilution to meet the limit of 14 EU/V.

Because, at expiration time, the maximum administered total volume of a radiopharmaceutical may be equal to the total volume of the entire vial of the radioactive drug product, it might seem appropriate to calculate the bacterial endotoxin limit (i.e., 175 EU/V) using the total volume in the vial. This approach is simple, but it results in a stringent limit. This is because the total volume of the entire vial of radiopharmaceutical preparation, rather than a partial volume, is used as the denominator for calculation of the acceptable bacterial endotoxin limit.

The USP gel-clot limit test is used when a monograph or a package insert contains a requirement for bacterial endotoxin limit.[98] As shown in Table 14-8, the test is not valid unless both replicates of test solutions B and C are positive and those of test solution D are negative.[91] The test should be repeated when the results of replicates of test solution A are not consistent (i.e., one positive and one negative).[91] Practical procedures for applying the gel-clot limit test in pharmacy compounding have been described.[98]

To quantify the endotoxin concentration of an unknown solution, serial 2-fold dilutions of sample are tested until an endpoint is reached (Table 14-9). The endotoxin concentration (E) is calculated by multiplying the lysate sensitivity (λ) by the reciprocal of the endpoint dilution. For example, a product yielded an endpoint at a 1:8 dilution with LAL reagent water when using an LAL reagent with λ = 0.125 EU/mL (E = λ × 8 = 0.125 EU/mL × 8 = 1 EU/mL).

The commonly used gel-clot technique for determining bacterial endotoxin concentration requires a 60 minute incubation period, as described in *USP* General Chapter 85.[91] This is also the endotoxin testing method recommended in FDA's guidance on CGMP for PET drug products.[18] Because the remainder of the required QC testing for a typical PET radiopharmaceutical (e.g., ^{18}F-FDG), with the exception of the sterility test, can be completed in 20 to 30 minutes, it is not practical and is indeed quite wasteful to delay release of the short-lived ^{18}F-FDG injection for an additional 30 to 40 minutes. Kinetic LAL methods using multitube readers are the quickest ways to complete a BET for PET radiopharmaceuticals.[99]

USP General Chapter 823 on the compounding of PET radiopharmaceuticals indicates that an in-process 20 minute endotoxin "limit test" (i.e., incorporating positive controls in the range of 5 EU/mL to 175 EU/V) can be used to allow for

TABLE 14-9 Gel-Clot Assay

Test Solution	Number of Replicates	Endotoxin Concentration/ Test Sample or Control	Diluent	Dilution Factor	Initial Endotoxin Concentration	Results or Comments
A	2	None/sample solution	LAL reagent water	1	—	See text for calculation of endotoxin concentration of solution A
	2			2	—	
	2			4	—	
	2			8	—	
B	2	2 λ/sample solution	—	1	2 λ	Should be positive
C	2	2 λ/LAL reagent water	LAL reagent water	1	2 λ	Geometric[a] mean endpoint concentration is in the range of 0.5 λ to 2 λ
	2			2	1 λ	
	2			4	0.5 λ	
	2			8	0.25 λ	
D	2	None/LAL reagent water	—	—	—	Should be negative

[a] The endpoint is the last positive test in the series of decreasing concentrations of endotoxin.
Source: Reference 91.

the possibility of early release, for human use, of an injectable PET drug radiolabeled with a radionuclide having a half-life greater than 20 minutes. However, the standard 60 minute BET must be performed and completed.[13] The two-part BET is a sensible approach, especially for the short half-lived PET drug products, because all other required QC testing procedures (with the exception of the sterility test and 60 minute BET testing) are usually completed in 20 to 30 minutes. Nonetheless, it is interesting to note that the detailed procedure of this "limit test" 20 minute BET is mentioned in neither USP General Chapter 85[91] nor USP General Chapter 823.[13] Cooper[100] proposed a test scheme for the 20 minute BET. The proposed 20 minute gel-clot limit test includes use of the standard test tubes as described in USP General Chapter 85, plus one additional positive product control tube that contains a test sample solution mixed with 160 λ endotoxin concentration.

USP General Chapter 823 indicates that BET may also be performed using other recognized procedures as stated in USP General Chapter 85.[13,91] The Endosafe-PTS (PTS [Charles River Laboratories, Wilmington, MA]) is a handheld spectrophotometer that employs FDA-licensed disposable cartridges for a quick (15 to 20 minute) assay of endotoxins. PTS uses LAL kinetic chromogenic methods to measure color intensity in direct relationship to the endotoxin concentration in a sample. Since the chromogenic method is an acceptable method for BET according to USP General Chapter 85,[91] the rapid test system provided by the PTS would allow the end user to complete the two-part BET in one simple and quick test, offering a valuable tool to the PET community.

INSTRUMENT QUALITY CONTROL

A nuclear pharmacy laboratory usually has a dose calibrator, a scintillation well counter, and a Geiger-Müller (GM) survey meter as standard equipment for measuring radioactivity. CGMP for PET drug products requires that a PET drug production facility be equipped with additional instruments (e.g., HPLC, GC, MCA) for QC testing. The purposes and operating principles of these instruments have been addressed previously (the dose calibrator, scintillation well counter or MCA, and GM survey meter are discussed in Chapter 5; HPLC and GC are discussed earlier in this chapter). This section will describe routine QC procedures for the dose calibrator and survey meters.

Dose Calibrator

The dose calibrator is used in nuclear medicine and nuclear pharmacy primarily to measure the activity of radiopharmaceuticals before administration to the patient. A number of QC tests must be conducted to ensure the proper operation of this instrument.

The current 10 CFR §35.60, "Possession, use, and calibration of instruments used to measure the activity of unsealed byproduct material," no longer spells out the procedures for checking accuracy, constancy, geometry, and linearity for dose calibrators.[101] Instead, the regulation states that licensees must follow "nationally accepted standards" or "the manufacturer's instructions" for calibrating dose measurement systems.[101] For reference, the requirements for dose calibrator QC testing described in the previous 10 CFR §35.60 are summarized in Table 14-10. These requirements may still be enforced by some agreement-state licensing agencies, and the requirements as stated in the "old" 10 CFR §35.50 are still sound guidelines for dose calibrator quality control.

Dose calibrator manufacturers' instructions may vary and therefore may not be a practical option for standardized QC testing of dose calibrators. In regard to "nationally accepted standards," two documents are generally recognized as authoritative guidelines: the American National Standard Calibration and Usage of "Dose Calibrator" Ionization Chambers for the Assay of Radionuclides, and NCRP Report 99—Quality Assurance for Diagnostic Imaging, published, respectively, by

TABLE 14-10 Dose Calibrator Quality Control Requirements of 10 CFR §35.50 ("Old" Part 35)

Quality Control Test	Frequency[a]	Sources and Conditions[b]
Constancy	At beginning of each day of use	Not less than 50 μCi of a photon emitter
Accuracy	At installation and annually	Not less than 50 μCi of a photon emitter Two sources used, each within ±5% of its stated activity; one source must have a principal photon energy between 100 and 500 keV
Linearity	At installation and quarterly	Test from highest patient or research subject dosage down to 30 μCi
Geometry	At installation	Test range of volumes and configurations to be used

[a] Each test must be performed after instrument adjustment or repair.
[b] Mathematically correct dosage readings for geometry or linearity errors that exceed 10%. Repair or replace dose calibrator if accuracy or constancy error exceeds 10%.

the American National Standards Institute (ANSI) and the National Council on Radiation Protection and Measurements (NCRP).[102,103]

According to 10 CFR §35.2060, "Records of calibrations of instruments used to measure the activity of unsealed byproduct material," a record of calibration must be kept for 3 years and should include dose calibrator model and serial numbers, date of calibration, results of calibration, and the name of the individual who performed the calibration.[104]

To ensure proper dose calibrator operation, the following QC tests are required: constancy (precision), accuracy, linearity, and geometry.[101] Standard radionuclide sources, otherwise known as standard reference materials, are available for conducting the accuracy and constancy tests. A primary standard reference material is one whose disintegrations have been determined under known geometry conditions, typically 4-pi geometry. Primary standards are prepared by the National Institute of Standards and Technology (NIST) and have an uncertainty of <1%. They are supplied as 5 mL solutions in borosilicate glass ampuls with a wall thickness of approximately 0.6 mm. A secondary standard reference material is one whose activity is determined by comparison with a primary standard. Secondary standards are prepared by commercial firms (e.g., Capintec) and have an uncertainty of 2% to 4%. These sources are typically supplied as a solid epoxy material sealed in a polyethylene bottle.

The following discussion of QC standards and procedures is based primarily on NCRP Report 99.[103]

Constancy

A constancy (precision) test ensures that the dose calibrator can measure a source of constant activity repeatedly within a stated degree of reproducibility over a long period of time. A satisfactory response to this check indicates that the dose calibrator is operating consistently from day to day.

The constancy test is performed on the dose calibrator at the beginning of each day of use. A long-lived reference source, usually ^{137}Cs, is used. Other sources may be used as well for this test. ^{137}Cs is similar in photon energy to ^{99}Mo; ^{57}Co has a principal photon energy similar to that of ^{99m}Tc. If one uses a dose calibrator only for the measurement of ^{131}I dosages, a good reference source would be ^{133}Ba, whose photon energy is similar to that of ^{131}I. A suggested method for the constancy test is as follows:

1. Assay the reference source using the dose calibrator setting for that radionuclide (e.g., use the ^{137}Cs setting to assay a ^{137}Cs source).
2. Measure the background activity at the same setting and subtract this from the measured activity. If an automatic background subtraction circuit is used, confirm that background was correctly subtracted.
3. Record the net activity (source minus background) in the QC logbook for that dose calibrator.
4. It is advantageous to repeat the above procedure, using the same reference source, for all routinely used radioisotope settings on the dose calibrator (e.g., ^{99m}Tc, ^{201}Tl,

^{111}In) to check their response. Readings will differ from the reference source but should consistently follow the characteristic half-life of the source.

Source measurements deviating by more than ±5% of the predicted activity indicate a need for instrument adjustment or repair.

Accuracy

Accuracy is defined as the closeness of a measurement to the true value. The purpose of the dose calibrator is to measure the radioactivity of a radiopharmaceutical with a high degree of accuracy. Dose calibrator accuracy is assessed by measurement of standard reference sources of known activity traceable to NIST. Typical sources used for accuracy assessment are ^{57}Co, ^{133}Ba, and ^{137}Cs. The measured activity of a reference source must agree within ±5% of its certified activity after decay correction. Calibration checks that do not agree within ±5% indicate that the instrument must be either repaired or adjusted. This test must be performed at the time of dose calibrator installation and thereafter on a yearly basis. It must also be performed after repair or adjustment of the dose calibrator. A suggested method for the accuracy test is as follows:

1. Assay one of the reference sources using the appropriate radioisotope setting (e.g., use the ^{57}Co setting to assay ^{57}Co). Allow sufficient time for a stable reading to be obtained.
2. Remove the reference source and measure background activity. Subtract background activity from the measured activity to obtain the net activity. Confirm proper operation of the automatic background subtraction circuit if it is used.
3. Record the net measured activity, the source geometry, and the instrument settings.
4. Repeat the previous steps for a total of three independent determinations and average the net measured activity.
5. The average value must be within ±5% of the certified radioactivity after decay corrections.

Linearity

Dose calibrator linearity implies an accurate instrument response over a wide range of activities. A nonlinear response is more likely to occur at high activities than at low activities.[105] Nonlinearity at high activities is likely due to recombination of ion pairs in the chamber before they are collected at the chamber electrodes, causing a falsely low readout of the true activity present.

A linearity test should be performed at dose calibrator installation and quarterly. For practical reasons, the linearity test need be conducted only over the activity range of patient dosages being measured. ^{99m}Tc is typically used to conduct the linearity test because it is readily available in large amounts of activity. However, if the highest dose administered to a patient is 300 mCi (11.1 GBq) of ^{131}I, the amount of ^{99m}Tc activity must be adjusted upward because of differences in the decay characteristics of these two nuclides. Thus, 430 mCi (15.9 GBq) of ^{99m}Tc should be used in place of

300 mCi (11.1 GBq) of ^{131}I to compensate for the gamma energy differences between ^{131}I and ^{99m}Tc.[106]

Two methods can be used for the linearity test. The most accurate method is the decay method, which consists of multiple measurements of the same radionuclide source over an extended time period, typically 2 to 3 days.[107,108] An alternative method is the attenuation method, which uses a series of lead-lined sleeves or tubes of varying thickness (Figure 14-14). A set of 6 to 8 tubes is designed to progressively increase attenuation of the ^{99m}Tc source to simulate decay from 0 through 50 hours. Tube sets are commercially available from a number of vendors.

Decay Method

1. Before the linearity test, properly zero the dose calibrator. A reliable and accurate clock should be used to record the time (an error of 10 minutes in the recorded time for ^{99m}Tc will result in an apparent error of 2% in the dose calibrator linearity check).
2. Obtain a ^{99m}Tc source that contains the largest activity routinely administered to a patient or routinely measured in the dose calibrator (e.g., the first elution from a new ^{99}Mo–^{99m}Tc generator).
3. Assay the ^{99m}Tc source in the dose calibrator and subtract background activity to obtain the net radioactivity in millicuries or megabecquerels. Record the date, time, and measured activity of the ^{99m}Tc source. Repeat this measurement three times and obtain the average value.
4. Repeat this process at 6, 24, 30, 48, 54, 72, and 78 hours thereafter, until the activity level of the ^{99m}Tc source has dropped to 30 µCi (111 kBq) or less.

   Using the 30 hour activity measurement (A_{30}) as a reference point, calculate the predicted activity (A_t) at the other specified times, t, using the following decay equation:

   $$A_t = A_{30} e^{-0.1151(t-30)}$$

5. On semilog graph paper, label the logarithmic vertical axis in activity (millicuries per megabecquerel) and label the linear horizontal axis in decay time (hours). At the top of the graph, note the dose calibrator information (i.e., manufacturer, model number, and serial number), as well as the date and time of the initial assay.
6. Plot the data points and draw a best-fit straight line through the data points. Figure 14-15 shows a linearity decay graph. The linearity of the dose calibrator is determined by selecting the point that deviates furthest from the best-fit line. The deviation is calculated from the following equation:

$$\% \text{ Deviation} = \frac{\text{Measured activity} - \text{Calculated or fitted activity}}{\text{Calculated activity}} \times 100\%$$

Attenuation Method

The set of lead sleeves must be calibrated before use for each dose calibrator and source configuration to be used routinely for the linearity test. A typical calibration method is as follows:

1. Begin the linearity test as described above in the decay method. After completing the first assay using the decay method, calibrate the lead sleeves as follows. Each measurement must be performed three times and an average value obtained. In addition, the entire calibration must be completed within 6 minutes.
2. Place the ^{99m}Tc source into a base sleeve that is not lead lined. Assay the activity of the ^{99m}Tc source using a dose calibrator and record the results.
3. Place the first lead-lined sleeve (the one with the thinnest lining) over the unlined base sleeve and record the results (Figure 14-16).
4. Remove lead sleeve 1 and place the source in lead sleeve 2. Record the measured radioactivity.
5. Repeat this process until all the lead sleeves have been used.
6. Determine the attenuation (calibration) factors for each lead-lined sleeve by dividing the activity measured in step 1 by the measured activity for the lead sleeve.

FIGURE 14-14 Lead sleeve shielding system for the dose calibrator linearity test.

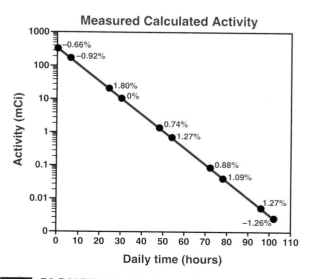

FIGURE 14-15 Dose calibrator linearity decay graph.

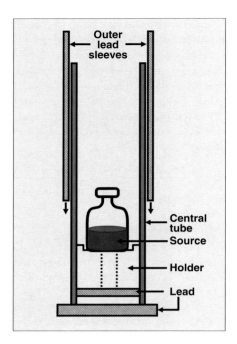

FIGURE 14-16 Diagram illustrating the shielding method for measuring dose calibrator linearity.

7. This procedure need be performed only once on a new set of lead sleeves. The lead-lined sleeves can now be used to test the dose calibrator for linearity.
8. Obtain a ^{99m}Tc source that contains the highest activity routinely administered to a patient or a human research subject.
9. Assay the ^{99m}Tc source in the dose calibrator and subtract background activity to obtain the net radioactivity in millicuries or megabecquerels. Record the date, time, and measured activity of the ^{99m}Tc source. Repeat this process using each of the lead sleeves. This should be done as quickly as possible to minimize errors due to decay. A maximum of 6 minutes (1% decay) is allowed before decay correction must be applied to the results.
10. Repeat step 9 three times and obtain the average value for each lead sleeve.
11. For each average activity measured, multiply the result by the appropriate calibration factor for each sleeve to obtain the factored activity. See the example in Table 14-11.
12. Sum the products from step 11 and divide by the total number of sleeves to obtain the mean factored activity. The percent deviation of each measurement from the mean is determined from the following equation:

$$\% \text{ Deviation} = \frac{\text{Mean factored activity} - \text{Factored activity}}{\text{Mean factored activity}} \times 100\%$$

Deviations greater than ±5% between predicted and measured activity with either the decay method or the attenuation method indicate the need for repair or adjustment of the dose calibrator. Alternatively, a correction factor can be applied for the specific amount of activity being measured.

Geometry

The measured activity of a radioactive source may be significantly different from its true activity, particularly in assays of radionuclides that have weak gamma emissions. In these situations the source measurement can be greatly affected by its volume or container configuration. The assayed activity can also be affected by the volume and size of the dose calibrator chamber; therefore, geometry effects should be established for each type of dose calibrator.[103] The extent of these geometry effects should be ascertained for commonly used radionuclides and appropriate correction factors applied to the measurements if variations are found to be significant (>2%). Dose

TABLE 14-11 Example of Dose Calibrator Linearity Test Report

Tube No.	Net Measured Activity (mCi)			Average Readings	×	Calibration Factor	=	Factored Activity	% Error
	1	2	3						
1	61.700	60.800	60.400	60.967		1.000		60.967	−0.36
2	35.300	34.800	34.500	34.867		1.758		61.296	−0.90
3	19.300	19.000	18.900	19.067		3.144		59.946	1.32
4	6.170	6.030	5.990	6.063		10.089		61.173	−0.70
5	2.590	2.540	2.520	2.550		23.638		60.277	0.78
6	0.441	0.437	0.434	0.437		140.456		61.426	−1.11
7	0.131	0.128	0.128	0.129		474.422		61.137	−0.64
8	0.060	0.060	0.060	0.060		1020.35		61.153	−0.66
9	0.009	0.009	0.009	0.009		6736.32		59.369	2.27

Average factored activity = 60.749.
Mean + 5% = upper limit = 63.787.
Mean − 5% = lower limit = 57.712.
Instrument performance is within the acceptable limits of ±5% error.

calibrator correction factors that are supplied by the manufacturer should be verified. In some instances differences of 200% in dose calibrator readings between glass and plastic syringes have been observed for lower-energy radionuclides such as ^{125}I. This type of situation would require a correction factor for measurements made in different types of containers. Correction factors may also need to be established for various types of syringes and for different volumes in a given syringe. An alternative technique to syringe-correction factors is to assay the stock vial before and after filling the syringe. This method maintains consistent assay geometry, and the activity in the syringe is then the difference between the two stock vial readings, with appropriate correction for any syringe volume effects if required. The geometry effects on source measurement for all types of containers should be ascertained upon installation of a dose calibrator, particularly for weak gamma emitters for which these effects may be large.

The three principal geometry considerations that may affect the accuracy of source measurement in a dose calibrator are container configuration, position of the source in the ion chamber well, and the effect of source volume.

Container Configuration

The type of source container used can greatly affect the accuracy of radionuclide measurements. This effect will be most significant with weak gamma emitters. It is important to be aware that dose calibrator radioisotope calibration factors are established for a particular instrument by the manufacturer using RSs in a particular container configuration, typically 5 mL of reference solution in a glass ampul. Other containers, such as glass serum vials or plastic syringes, will yield somewhat different values because of variations in photon absorption by the container. This is not a major problem with most radionuclides used in nuclear medicine, but a few exceptions do exist, such as ^{133}Xe and ^{125}I. Table 14-12 demonstrates the differences in the assay of ^{99m}Tc, ^{131}I, and ^{125}I with standard containers and indicates a significant influence of container configuration on ^{125}I measurements.[105] For weak gamma emitters, it would be more appropriate to determine calibration factors for the specific container used to assay the radiopharmaceutical. For example, a calibration factor for ^{125}I in a plastic syringe can be determined by transferring a known amount of ^{125}I reference solution into the syringe. This can be accomplished by weighing the syringe empty and then full of the reference solution. The difference will be the weight of the reference solution. The true activity in the syringe is calculated from the known activity per unit weight of the reference solution. Subsequently, the filled syringe is assayed in the dose calibrator, and the calibration factor dial is adjusted to read out the true activity on the display. From that point on, all dose measurements of ^{125}I solution in a similar syringe with that calibration factor will be accurate. This method can be applied to any radionuclide.

Ion Chamber Well Geometry

The position of a source in the ion chamber well will affect the accuracy of measurement. Each chamber has a well position where source detection sensitivity is greatest. Ideally, this position will extend over a reasonably wide range to minimize variation in measurements. Figure 14-17 illustrates variations in dose calibrator activity measurements of a 1.0 mCi (37 MBq) source in a 10 mL serum vial at different distances from the bottom of the ion chamber well. Note that the most accurate readings are obtained between 4 and 8 cm from the well bottom. Readings near the bottom and top of the well are lower because a significant number of photons escape from the detection volume of the chamber in these positions. Well geometry characteristics vary and should be determined for each dose calibrator.

Volume Variation

The measured activity of a radionuclide in a dose calibrator may vary with the volume of solution because of self-absorption. That is, 1 mCi (37 MBq) in a 10 mL volume may produce a different reading than if it were in a 1 mL volume. Experience has shown that this is not a significant problem with most radionuclides used in nuclear medicine except for

TABLE 14-12 Effect of Container Configuration on Dose Calibrator Measurements of ^{99m}Tc, ^{131}I, and ^{125}I

Radionuclide	Activity Concentration in µCi/gram of Solution (% Deviation Relative to Ampul Activity)		
	Ampul	Serum Vial	Plastic Syringe
^{99m}Tc	44.5	44.0 (−1.1)	45.2 (+1.4)
^{131}I	37.1	35.8 (−3.5)	38.6 (+4.0)
^{125}I	39.6	20.4 (−48.4)	57.8 (+45.9)

Source: Reference 105.

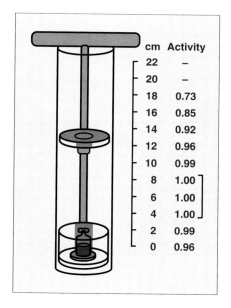

FIGURE 14-17 Ion chamber well geometry effect on activity measurements in a dose calibrator.

weak gamma emitters such as ^{125}I. One can easily test for volume variation by placing 1 mCi (37 MBq) of any radionuclide in a 1 mL volume into a 30 mL vial. An initial reading and subsequent readings after 2 mL additions of water will indicate whether correction factors need to be applied for different volumes. Regulatory Guide 10.8, Revision 2, recommends the use of a 3 mL plastic syringe and a 30 mL glass vial for evaluation of source geometry.[108] If the error in the measured activity at any volume is greater than ±2% of the baseline activity, a correction table must be established so that the measured activity can be converted to "true activity" (i.e., true activity = measured activity × correction factor).

SURVEY INSTRUMENTS

Two types of survey instruments are used in nuclear medicine: the ionization chamber exposure rate meter and the GM survey meter (Figure 14-18). All survey instruments require routine quality control to ensure that they are operating correctly. 10 CFR §35.61, "Calibration of survey instruments," requires instruments to be calibrated before first use, annually, and after any repair that affects the calibration.[109] The regulations require a licensee to calibrate instruments as follows:

1. Calibrate all scales with readings up to 10 mSv (1000 mrem) per hour using a radiation source.
2. Calibrate two separated readings on each scale or decade that will be used to show compliance.
3. Conspicuously note on the instrument the date of calibration.

When calibrating a survey instrument, the licensee should consider a point as calibrated if the indicated exposure rate differs from the calculated exposure rate by not more than 20%, and a correction chart or graph should be conspicuously attached to the instrument. In accordance with 10 CFR §35.2061, "Records of radiation survey instrument calibrations," a licensee must maintain a record of instrument calibration for 3 years.[110] The record must include the model and serial number of the instrument, date of calibration, results of calibration, and name of the individual who performed the calibration. For mobile medical services, verification of proper instrument function must be verified with a dedicated check source before use at each client's address.[111]

Survey Meter Calibration

A model procedure for calibration of ionization chambers and GM survey instruments is as follows:

1. Calibrate each scale that reads below 1 R/hour at two points on the scale located at approximately one-third and two-thirds of full scale.
2. Calibration should be performed with a ^{137}Cs reference source of approximately 100 mCi (3.7 GBq) or greater.
3. The ^{137}Cs reference source activity should have accuracy within ±5% of a NIST calibrated reference source and be decay corrected prior to the calibration procedure. Radiation exposure from the source should be documented in units of milliroentgen per hour (mR/hr) at one meter.
4. To obtain the appropriate scale reading, position the instrument at various distances from a ^{137}Cs source of known exposure rate at one meter (R_1). Record the distance and instrument reading. The expected reading at each distance can be determined with the inverse square law. For example, at a distance of 30 cm (0.3 m), the expected reading ($R_{0.3}$) can be calculated with the following equation:

$$R_{0.3} \times (0.3)^2 = R_1 \times (1)^2$$

A   **B**   **C**   **D**

FIGURE 14-18 Examples of ionization chamber survey instruments (A: Model 5 Geiger-Müller [GM] counter and B: RSO-50E ionization dosimeter) and GM survey meter/probe (C: Model 3 survey meter attached to Model 44-9 pancake GM detector and D: Model 44-25 pancake GM hand probe).

5. If the measured and expected exposure rates differ by more than 10% but less than 20%, a calibration chart, graph, or response factor should be prepared and attached to the instrument.
6. If the measured and expected exposure rates differ by more than 20%, the instrument should be repaired or adjusted until the measured readings are within 10% of the expected reading.

A model procedure for calibration of GM survey meter efficiency is as follows:

1. Place a 0.5 to 1.0 inch disk-type planchette of the desired source on a countertop. This source should contain 0.1 to 0.5 µCi (3.7 to 18.5 kBq) ^{57}Co or any other radionuclide dispersed uniformly throughout the source. Ensure that all other sources of radiation have been removed from the area.
2. Place the GM survey meter either directly on top of the source or at a distance of 5 cm.
3. Measure the exposure rate in counts per minute (cpm). Note the source activity and calibration date and compute its current activity in microcuries (kilobecquerels). Divide the exposure rate by the activity to get the efficiency factor of the meter in counts per minute per microcurie (cpm per kilobecquerel). This factor should be recorded on a label affixed to the detector.

This type of calibration of a thin-window GM counter is useful for assessing removable contamination from radioactive material packages of beta-emitting radionuclides such as ^{32}P, ^{89}Sr, and ^{90}Y, since these nuclides cannot be counted in a gamma scintillation well counter.

Constancy Check

All survey meters should be checked each day before use with a long-lived reference source. The reference source is typically a plastic disk impregnated with 1 to 10 µCi (37 to 370 MBq) of ^{137}Cs, and its reading should be established at the time the survey meter is calibrated. The geometric arrangement of the source and survey meter probe should be noted and reproduced for each daily check of the meter response. Verify that the meter gives a reading within 10% of the source value documented at calibration.

QUALITY ASSURANCE ISSUES RELATED TO PET DRUG PRODUCTS

Section 121 of the Food and Drug Administration Modernization Act of 1997 (FDAMA) requires that FDA establish approval procedures, that is, new drug application (NDA) or abbreviated NDA (ANDA) procedures, and CGMP for PET drugs.[112] A compounded PET drug, compounded pursuant to a valid prescription order and in accordance with state law, is exempt from NDA or ANDA and CGMP requirements. The 1997 FDAMA added section 501(a)(2)(C) to the federal Food, Drug, and Cosmetic Act, which provides that a compounded PET drug is deemed to be adulterated if it is compounded, processed, packaged, or held other than in accordance with PET compounding standards and the official *USP* monograph.[112] Section 501(a)(2)(C) is to expire after the date on which FDA establishes approval procedures and CGMP requirements.[112] In the process of establishing approval procedures for PET drugs, FDA issued a draft chemistry, manufacturing, and controls section, which contains a subsection of regulatory specifications, standard testing procedures, and testing schedules.[19]

Although *USP* stipulates acceptance criteria as well as testing procedures for drugs, it does not specify the frequency for conducting the required QC tests, nor does it indicate whether drug preparations can be released for patient administration before completion of required QC testing. This may be because USP tests are commonly used either in challenges to certain claims made by consumer organizations or in evaluations of marketed drug products conducted by government regulators.[113] However, specifying testing schedules could help end users to ensure that necessary QC tests are initiated within an appropriate time frame and to know whether a drug can be released before completion of QC tests. *USP* General Chapter 823 does give guidelines regarding testing frequency and release criteria for PET drugs.

PET drugs are unique among radiopharmaceuticals because of their short physical half-lives, ranging from a few seconds to several hours. Consequently, PET radiopharmaceuticals are typically prepared using a cyclotron that is located at or near the PET imaging facility. Most PET centers in the United States are part of an academic medical institution that has its own cyclotron and radiochemistry laboratories for producing PET radiotracers for clinical and research applications.

Notices issued by FDA in 1995 and 1997 concerning the regulation of PET radiopharmaceuticals indicate that FDA has considered the production of PET drug products to be significantly different from the production of conventional drugs in a number of important ways, as follows.[18,114–116]

1. Because of the short physical half-lives of PET radioisotopes,
   - Prolonged preparation time significantly erodes the useful clinical life of PET drug products;
   - PET facilities generally produce PET drug products in response to daily demand;
   - PET radiopharmaceuticals must be administered to patients shortly after production because of the short half-lives of the PET radioisotopes.
2. Only a few lots are produced per day, with one lot equaling one multiple-dose vial, for a relatively small number of patients.
3. An entire lot may be administered to one or several patients, depending upon the amount of radioactivity

remaining at the time of administration. Consequently, administration of the entire quantity of a lot to a single patient should be anticipated for every lot prepared.
4. Since each multiple-dose vial contains a homogenous solution of a PET drug product and equals one lot, results from end-product testing of samples drawn from the vial may be representative not only of the entire lot but of all doses administered to patients.
5. The quantities of active ingredients contained in each lot of a PET drug generally vary from microgram to nanogram amounts.
6. PET drugs do not usually enter a general drug distribution chain. Rather, the entire lot (one vial) usually is distributed directly from the PET facility either to a single nuclear medicine department or physician for administration to patients or to a nuclear pharmacy for dispensing. Distribution to other PET centers may occur when geographic proximity allows for distribution and use within the parameters of the PET drug product's half-life.

On September 1, 2005, FDA issued a proposed rule on CGMP for PET drugs.[93] A companion guidance on CGMP for PET drug products was also released on the same day.[18] The final versions of the PET CGMP rule and guidance were eventually issued by FDA on December 10, 2009.[18,93]

The PET CGMP guidance provides details and recommendations on compliance, based on size, scope, and complexity of PET center operations.[18] CGMP is not merely focused on QC issues of the finished PET drug products; CGMP must also use adequate and validated personnel, facilities, equipment, and controls in the preparation, packaging, and holding of PET drugs in order to ensure that the drug products meet safety and quality requirements. Quality assurance (QA) and QC are two important elements in PET drug production. QA is a total package of the various activities that influence the product's identity, strength, purity, and quality. QC is a subset of QA that deals with testing of materials and products to determine whether they meet minimum acceptance specifications; QC should be performed in any PET drug production facility so that the quality of PET drug products is ensured. Unique QA issues related to the preparation, packaging, and holding of PET drugs, as outlined in the final PET CGMP rule and guidance, are summarized as follows.

Personnel Resources

A PET drug production facility must have a sufficient number of qualified personnel to satisfactorily complete all required tasks in a timely manner. Adequate ongoing programs or plans should be established for training employees in new procedures and operations, as well as in areas where deficiencies have been identified. In addition, the various stages of production and test verification must be checked by a second person. However, if a PET center is operated by one person, FDA allows that individual to perform the production and QA functions and self-check his or her own work.

QA Function

Each PET drug product production facility must have a QA process that includes execution and oversight activities.

Execution of the QA function, which involves typical QC tasks, includes the following:

- Examination and evaluation of each lot of incoming material before use to ensure that the material meets its established specifications, and
- Review of production batch records and laboratory control records for accuracy, completeness, and conformance to established specifications before authorization of final release or rejection of a batch or lot of PET drug product.

Oversight of the QA function includes the following:

- Approval of procedures, specifications, process, and methods,
- Assurance that personnel are appropriately trained and qualified,
- Assurance that PET drugs have adequately defined identity, strength, quality, and purity,
- Investigation of errors and assurance that appropriate corrective action is taken to prevent their recurrence, and
- Conduct of periodic audits to monitor compliance with established procedures and practices.

The QA functions of a PET drug production facility producing one or two PET drugs can be performed by the facility's own employees. However, in large PET centers that produce multiple PET drugs, an entity located outside the PET drug production facility (e.g., a corporate QA/QC department, or an independent expert or outside consultant) should be chosen in order to achieve an objective evaluation of the effectiveness of the facility's QA function, especially oversight functions.

Facilities and Equipment

A PET drug production facility should have adequate space and suitable work areas for handling materials and equipment, in order to prevent mix-ups and to allow completion of all production-related tasks in an orderly manner. Critical activities involved in the production and testing of a PET drug product that expose the PET drug product or the sterile surface of a container and closure system to the environment should be conducted within an aseptic work station with a rating of ISO Class 5 (e.g., a laminar-airflow work station or barrier isolator). Examples of typical activities include (1) aseptic assembly of sterile components (e.g., syringe, needle, filter, or vial), (2) storage of sterility samples and finished PET drug products, (3) sterility testing of the finished PET drug product, and (4) dose withdrawal for QC or patient administration. The aseptic processing facility and the aseptic work area should be properly certified, cleaned, and validated in accordance with standard procedures, such as those listed in *USP* General Chapter 797, Pharmaceutical Compounding—Sterile Preparations, and *USP* General Chapter 1208, Sterility Testing—Validation of Isolator Systems.[117,118] Although *USP* General Chapter 823, Radiopharmaceuticals for Positron Emission Tomography—Compounding, supersedes Chapter 797 with regard to proper practice and standards for PET radiopharmaceuticals, Chapter 797 describes in greater depth the required maintenance and evaluation of facilities for compounding sterile drugs.

Similarly, equipment used in the production and QC of a PET drug product should be adequately certified, maintained, and validated. The following reference sources (in descending order) can be used to control the usage and maintenance of such equipment: (1) manufacturer's instructions, (2) relevant *USP* general chapters, and (3) written procedures developed by each individual PET center. The warranty of any piece of equipment is normally valid if the operation procedures and maintenance schedule stated in the owner's manual are properly followed. In addition, the manufacturer may have specific requirements for the calibration, performance, and maintenance of the machine. Hence, it may be prudent to rely primarily on the manufacturer's instructions regarding equipment issues, with the *USP* general chapters and self-established procedures as alternative sources when the manufacturer's instructions are inadequate.

Control of Components, Containers, and Closures

The components used in the production of PET drugs, including containers and closures used to package final products, must be properly controlled to prevent contamination, mix-ups, and deterioration. Only qualified vendors should be used, and it is prudent to have backup vendors for each component. Incoming materials should be carefully examined and should be classified into three categories: quarantined, accepted, and rejected. If a lot of materials is rejected, it should be labeled "rejected," segregated, and properly disposed. A traceability system (e.g., a component logbook) for monitoring each lot of received materials should be established.

Use of a certificate of analysis to replace identity testing (e.g., melting point determination) is permissible for the following materials obtained from reliable vendors:

1. Reagents, solvents, gases, purification columns, and other auxiliary materials.
2. Components that yield an active pharmaceutical ingredient (API), only if it can be verified by the outcome of finished-product testing that these components have been used in the production of the PET drug of interest. If not, specific identity testing needs to be performed. For the production of ^{18}F-FDG injection, the components that yield the APIs are ^{18}O-water, mannose triflate, and ^{18}F-fluoride (if it is obtained from an outside supplier).
3. Commercially available ready-to-use sterile, pyrogen-free, sealed containers and closures.
4. Inactive ingredients (e.g., diluent, stabilizer, preservative). If an inactive ingredient (e.g., 0.9% sodium chloride solution) is prepared on site, an identity test of the components used to prepare the inactive ingredient must be performed before its release for use.

Production and Process Controls

PET drug producers must have adequate production and process controls (i.e., master production and control record, batch production and control record) in place to ensure consistent production of a PET drug product that meets applicable standards for identity, strength, quality, and purity. The records also provide complete traceability and accountability for production and control of each batch. Corrections to paper entries should be dated and signed or initialed, leaving the original entry still readable, and there should be an audit trail to document the changes. Electronic records and signatures should be documented according to 21 CFR Part 11, Electronic Records; Electronic Signatures and the FDA guidance on this mode of documentation.[119,120] Each batch record should be reviewed and approved for final release (with signature or initials and date).

Since most PET drug products are administered to patients by injection, they must be produced by personnel who are trained in aseptic technique. An operator is initially certified to perform aseptic processing by completing three successful media-fill runs and thereafter is requalified by passing an annual media-fill run test.

Aseptic processing of PET drug products should involve microbiologic control of all components as follows: (1) glassware should be washed and rinsed with Water for Injection USP (preferably) or purified water, wrapped in aluminum foil, and terminally sterilized by using a suitable dry-heat oven cycle, (2) transfer lines are suitable for reuse; however, they should be promptly cleaned after each use by organic solvents (e.g., ethanol and acetone), rinsed with Water for Injection USP, flushed with volatile solvents, and finally dried with nitrogen, (3) low-microbial grade resin column material should be used in order to limit bioburden, (4) a sample of sterilizing membrane filter should be tested before use; integrity testing of the membrane filter should be performed after filtration (i.e., bubble-point test), and (5) the environment should be periodically monitored by methods such as the use of swabs or contact plates to test the aseptic work station surfaces and the use of settling plates or dynamic air samples to evaluate air quality of the aseptic facility and work station.

Validation of a new process or of a significant change to an already validated process should be conducted prospectively. However, if a PET center has an established history of producing a particular PET drug, validation of the production process can be conducted retrospectively if the process has been shown to be capable of yielding batches meeting required specifications. In addition, a concurrent validation can be justified if the short half-life of the PET drug prohibits the use of either a prospective or a retrospective validation process. In any event, the respective order of preference with regard to process validation should be prospective, retrospective, and concurrent. However, each PET drug producer should weigh the pros and cons carefully in considering which type of validation scheme to use.

With regard to computer control, PET drug production facilities can rely on a certification by the software or system vendor that the specified software has been verified under its operating conditions.

Laboratory Controls

A PET drug production facility should have written testing procedures that include acceptance standards of identity, strength, quality, and purity for prepared PET drug products. Validation of the testing method is not required if the method is stated in the *USP*; however, any new noncompendial method should be validated to ensure that the analytical parameters (e.g., accuracy, precision, linearity, ruggedness) of the new test method are appropriate.[121,122]

If an RS is obtained from an officially recognized source (e.g., USP), no further testing is required. However, if an RS is purchased from an alternative source, a reference spectrum or other supporting data to confirm its identity and purity must either be obtained from the supplier or established by the PET drug production facility.

Equipment used in a PET drug production facility should be routinely calibrated and maintained. Reagent or solution prepared on site should be adequately controlled and properly labeled. Raw test data (e.g., chromatograms, spectra, printouts) and any calculations performed should be part of the batch production and control record.

Stability Testing

PET CGMP guidance suggests that stability testing be based on at least three production runs of the final drug product. The evaluation of product stability should include the following: (1) setting the radioactive concentration to be at or near the highest level, (2) storage of the whole batch volume in the intended container and closure system, and (3) evaluation for a time period equal to the stated shelf life of the PET drug product. Examples of parameters for stability testing are radiochemical identity and purity (including levels of radiochemical impurities), appearance, pH, stabilizer or preservative effectiveness, and chemical purity.

Finished Drug Product Controls and Acceptance Criteria

The final rule on PET CGMP indicates that PET centers must establish specifications for each batch of a PET drug product, including criteria for identity, strength, quality, purity, and, if appropriate, sterility and pyrogenicity (for injectable drug products).[93] The final rule also states that the accuracy, sensitivity, specificity, and reproducibility of each testing procedure must be established and documented. These controls and acceptance criteria are requirements that must be met before a PET drug production facility may release a finished PET drug product.

However, modifications to these standard guidance principles may be justifiable. For example, the prerelease of PET drug products for commercial distribution before completion of all required QC testing may be appropriate because of transportation deadlines. Reduced testing frequency (e.g., initial validation with annual testing thereafter) of certain QC procedures (e.g., radionuclidic purity, chemical purity for 2-chloro-2-deoxy-D-glucose) may be necessary because of constraints related to the costly QC equipment (e.g., MCA, special HPLC system).[123]

For a PET drug product that has a very short half-life (e.g., ^{13}N-ammonia), production usually involves multiple subbatches on the same day. PET CGMP guidance recommends that the initial subbatch be used as the representative sample of the entire batch, provided that a sufficient number of subbatches (i.e., beginning, middle, and end) have been shown to produce a product meeting the predetermined acceptance criteria. Thus, the release of subsequent subbatches can be qualified for acceptance, provided that the initial subbatch meets all acceptance criteria except sterility. The PET CGMP guidance also indicates that testing each subbatch for certain attributes prior to release may be appropriate in certain cases (e.g., for pH determination in the production of ammonia N 13 injection using Devarda's alloy catalyst).

Sterility testing need not be completed before final release but must be started within 30 hours after the completion of PET drug production. If the test sample is held longer than 30 hours (e.g., weekends, holidays), PET drug producers should demonstrate that the extended period will not adversely affect the viability of a USP indicator organism (e.g., *Escherichia coli*) inoculated in the test sample. Samples should be stored appropriately (e.g., under refrigeration).

The USP Bacterial Endotoxins Test (BET) (General Chapter 85) should be performed for a sterile PET drug product.[91] The BET utilizes gel-clot and rapid photometric methods (i.e., turbidimetric method and chromogenic method) for endotoxin measurement.

If the PET drug product fails the sterility test, pyrogenicity test, or both, the proposed guidance recommends an immediate complete investigation with documentation. Corrective actions based on the outcome of the investigation should be implemented promptly.

The final PET CGMP rule states that final release of a PET drug product is permissible if a particular required QC test (with the exception of radiochemical identity/purity test on the API of a PET drug product) cannot be completed because of a malfunction of the analytical equipment. The prerequisites of such a "conditional release" include (1) documentation showing successful completion of the incomplete test(s) in preceding consecutive batches produced with the same methods, (2) assurance that all other acceptance criteria are met, (3) retention of a reserve sample of the conditionally released batch of drug product, (4) prompt correction of the malfunction and completion of the omitted test after the malfunction is corrected, (5) immediate notification of the receiving facility that an out-of-specification result is obtained, and (6) documentation of all actions regarding the conditional final release of the PET drug product.

According to the final PET CGMP rule, a PET drug product that does not conform to specifications can be reprocessed. Examples of reprocessing could include a second passage through a purification column to remove an impurity or a second passage through a filter if the original filter failed the integrity test. The reprocessed drug product must conform to specifications, except for sterility, before final release. When the option for reprocessing is exercised, PET CGMP guidance recommends that the event be documented and conditions described in a brief deviation report.

Labeling and Packaging

If a sticky label is not suitable for the immediate container (including the vial or syringe) because of a limited surface area or high radiation considerations, PET CGMP guidance allows the use of a string label, provided that a procedure is in place to associate the label with the vial or syringe if the label were to come off.

Distribution

PET centers that distribute PET drug products to affiliated institutions, outside pharmacies, and outside clients should ship PET drug products in accordance with labeled conditions (e.g., temperature) to ensure the identity, purity, and quality of the drug product. In addition, a recall system must be in place to permit any recall notification to promptly reach the receiving facility and pharmacist and the patient's physician, if known. With regard to the distribution of any drug product (including a PET drug), FDA generally considers these activities to be part of the practice of pharmacy.

Complaint Handling

All complaints regarding incidents should be handled by designated individuals. An investigation must be initiated as soon as possible so that all relevant information can be collected in a timely manner. Corrective action should be implemented immediately to avoid a similar incident in the future.

Records

The final rule requires that records be stored at a PET drug production facility or another location that is reasonably accessible to responsible officials of the PET center and to employees of FDA who are designated to perform inspections. An accessible location is one that would enable the PET center to obtain records in a reasonable period of time. All records should be kept for at least 1 year from the date of release of a PET drug product, whereas validation reports should be kept as long as the systems are in use.

REFERENCES

1. Briner WH. Sterile kits for the preparation of radiopharmaceuticals: some basic quality control considerations. In: Subramanian G, Rhodes B, Cooper J, et al., eds. *Radiopharmaceuticals.* New York: Society of Nuclear Medicine; 1975:246–53.
2. US Nuclear Regulatory Commission. Permissible molybdenum-99, strontium-82, and strontium-85 concentrations. 10 CFR §35.204. www.nrc.gov/reading-rm/doc-collections/cfr/part035/part035-0204.html. Accessed February 1, 2010.
3. European system for reporting adverse reactions to and defects in radiopharmaceuticals: annual report 1995. *Eur J Nucl Med.* 1996;23:BP27–31.
4. European system for reporting adverse reactions to and defects in radiopharmaceuticals: annual report 1996. *Eur J Nucl Med.* 1998;25:BP3–8.
5. European system for reporting adverse reactions to and defects in radiopharmaceuticals: annual report 1997. *Eur J Nucl Med.* 1998;25:BP45–50.
6. European system for reporting adverse reactions to and defects in radiopharmaceuticals: annual report 1998. *Eur J Nucl Med.* 1999;26:BP33–8.
7. European system for reporting adverse reactions to and defects in radiopharmaceuticals: annual report 1999. *Eur J Nucl Med.* 2001;28:BP2–8.
8. European system for reporting adverse reactions to and defects in radiopharmaceuticals: annual report 2000. *Eur J Nucl Med.* 2002;29:BP13–9.
9. European system for reporting adverse reactions to and defects in radiopharmaceuticals: annual report 2001. *Eur J Nucl Med.* 2003;30:BP87–94.
10. Ponto JA. Technetium-99m radiopharmaceutical preparation problems: 12 years of experience. *J Nucl Med Technol.* 1998;26:262–4.
11. Decristoforo C, Siller R, Chen F, et al. Radiochemical purity of routinely prepared ^{99m}Tc radiopharmaceuticals: a retrospective study. *Nucl Med Commun.* 2000;21:349–54.
12. Mission and preface. *The United States Pharmacopeia and The National Formulary.* Rockville, MD: United States Pharmacopeial Convention, Inc. www.uspnf.com/uspnf/pub/index?usp=32&nf=27&s=2&officialOn=December%201,%202009. Accessed February 1, 2010 (access available only to subscribers).
13. Radiopharmaceuticals for positron emission tomography—compounding (general chapter 823). *The United States Pharmacopeia and The National Formulary.* Rockville, MD: United States Pharmacopeial Convention, Inc. www.uspnf.com/uspnf/pub/index?usp=32&nf=27&s=2&officialOn=December%201,%202009. Accessed February 1, 2010 (access available only to subscribers).
14. Hung JC, Ponto JA, Gadient KR, et al. Deficiencies of package insert directions for the preparation of radiopharmaceuticals. *J Am Pharm Assoc.* 2004;44:30–5.
15. General notices and requirements. *The United States Pharmacopeia and The National Formulary.* Rockville, MD: United States Pharmacopeial Convention, Inc. www.uspnf.com/uspnf/pub/index?usp=32&nf=27&s=2&officialOn=December%201,%202009. Accessed February 1, 2010 (access available only to subscribers).
16. Unapproved uses of approved drugs: the physician, the package insert, and the FDA. *Pediatrics.* 1978;62:262–4.
17. Kasik JE. Use of approved drugs in a non-approved way. *Iowa Medicine.* October 1987:513–5.
18. US Food and Drug Administration. *Guidance: PET Drug—Current Good Manufacturing Practice (CGMP).* www.fda.gov/downloads/Drugs/GuidanceComplianceRegulatoryInformation/Guidances/UCM070306.pdf. (Accessed February 1, 2010).

19. US Food and Drug Administration. *Sample Formats—Application to Manufacture Ammonia N 13 Injection, Fludeoxyglucose F 18 Injection (FDG 18), and Sodium Fluoride F 18 Injection: Chemistry, Manufacturing, and Controls Section.* www.fda.gov/downloads/Drugs/GuidanceComplianceRegulatoryInformation/Guidances/UCM078740.pdf. Accessed February 1, 2010.
20. Radiopharmaceutical preparation. *European Pharmacopeia.* http://online6.edqm.eu/ep606/#doc_34_of_681. Accessed February 2, 2010 (access available only to subscribers).
21. Personal written communication with technical support department at Capintec, Inc. Pittsburgh, PA; January 2002.
22. Radioactivity (general chapter 821). *The United States Pharmacopeia and The National Formulary.* Rockville, MD: United States Pharmacopeial Convention, Inc. www.uspnf.com/uspnf/pub/index?usp=32&nf=27&s=2&officialOn=December%201,%202009. Accessed February 2, 2010 (access available only to subscribers).
23. Chromatography (general chapter 621). *The United States Pharmacopeia and The National Formulary.* Rockville, MD: United States Pharmacopeial Convention, Inc. www.uspnf.com/uspnf/pub/index?usp=32&nf=27&s=2&officialOn=December%201,%202009. Accessed February 1, 2010 (access available only to subscribers).
24. Ammonia N 13 injection. *The United States Pharmacopeia and The National Formulary.* Rockville, MD: United States Pharmacopeial Convention, Inc. www.uspnf.com/uspnf/pub/index?usp=32&nf=27&s=2&officialOn=December%201,%202009. Accessed February 2, 2010 (access available only to subscribers).
25. TechneScan MAG3 [package insert]. St Louis: Mallinckrodt Inc; 2005.
26. Ponto J. Effect of solvent flow rate in mini-column testing of ^{99m}Tc-mertiatide. *J Nucl Med Technol.* 2005;33:232–3.
27. Hung JC, Wilson ME, Brown ML. Rapid preparation and quality control of technetium-99m MAG3. *J Nucl Med Technol.* 1991;19:176–9.
28. Levit N. *Radiopharmacy Laboratory Manual for Nuclear Medicine Technologists.* Albuquerque, NM: University of New Mexico College of Pharmacy; 1980:69–80.
29. Vivian A, Ice RD, Shen V, et al. *Procedure Manual: Radiochemical Purity of Radiopharmaceuticals Using Gelman Sepachrom (ITLC) Chromatography.* Ann Arbor, MI: Gelman Sciences; 1977:3–56.
30. Williams CC. Radiochemical purity of ^{99m}Tc-oxidronate [letter]. *J Nucl Med.* 1981;22:1015.
31. Zimmer AM, Spies SM. Quality control procedures for newer radiopharmaceuticals. *J Nucl Med Technol.* 1991;19:210–4.
32. Robbins PJ. *Chromatography of Technetium-99m Radiopharmaceuticals: A Practical Guide.* New York: Society of Nuclear Medicine; 1984.
33. Taukulis RA, Zimmer AM, Pavel DG, et al. Technical parameters associated with miniaturized chromatographic systems. *J Nucl Med Technol.* 1979;7:19–22.
34. Kowalsky RJ, Creekmore JR. Technical artifacts in chromatographic analysis of Tc-99m radiopharmaceuticals. *J Nucl Med Technol.* 1982;10:15–9.
35. Fludeoxyglucose F 18 injection. *The United States Pharmacopeia and The National Formulary.* Rockville, MD: United States Pharmacopeial Convention, Inc. www.uspnf.com/uspnf/pub/index?usp=32&nf=27&s=2&officialOn=December%201,%202009. Accessed February 3, 2010 (access available only to subscribers).
36. Martinez E, Study KT. Scintillation well counter: too much activity? *Monthly Scan.* July 1980:1.
37. Hung JC, Budde PA, Wilson ME. Testing the radiochemical purity of technetium Tc 99m-labeled radiopharmaceuticals. *Am J Health Syst Pharm.* 1995;52:310–3.
38. Sodium fluoride F 18 injection. *The United States Pharmacopeia and The National Formulary.* Rockville, MD: United States Pharmacopeial Convention, Inc. www.uspnf.com/uspnf/pub/index?usp=32&nf=27&s=2&officialOn=December%201,%202009. Accessed February 4, 2010 (access available only to subscribers).
39. ProstaScint [package insert]. Princeton, NJ: Cytogen Corporation; 1997.
40. Zevalin [package insert]. San Diego: IDEC Pharmaceuticals; 2003.
41. OctreoScan [package insert]. St Louis: Mallinckrodt Inc; 2006.
42. Zimmer AM, Pavel DG. Rapid miniaturized chromatographic quality-control procedures for Tc-99m radiopharmaceuticals. *J Nucl Med.* 1977;18:1230–3.
43. Acutect [package insert]. Wayne, NJ: Diatide/Rentschler Biotechnologie GmbH; 2001.
44. CEA-Scan [package insert]. Morris Plains, NJ: Immunomedics Inc; 1999.
45. Neurolite [package insert]. North Billerica, MA: Bristol-Myers Squibb Medical Imaging, Inc; 2003.
46. Budde PA, Hung JC, Mahoney DW, et al. Rapid quality control procedure for technetium-99m-bicisate. *J Nucl Med Technol.* 1995;23:190–4.
47. Neotect [package insert]. Londonderry, NH: Diatide/Rentschler Biotechnologie GmbH; 2000.
48. Zimmer AM, Magewski W, Spies SM. Rapid miniaturized chromatography for Tc-99m IDA agents: comparison with gel chromatography. *Eur J Nucl Med.* 1982;7:88–91.
49. Ceretec [package insert]. Arlington Heights, IL: Amersham Health/Medi-Physic, Inc.; 2005.
50. Jurisson S, Schlemper EO, Troutner DE, et al. Synthesis, characterization, and X-ray structural determinations of technetium (V)-oxotetradentate amine oxine complexes. *Inorg Chem.* 1986;25:543–9.
51. Hung JC, Taggart TR, Wilson ME, et al. Radiochemical purity testing for ^{99m}Tc-exametazime: a comparison study for three-paper chromatography. *Nucl Med Commun.* 1994;15:569–74.
52. Zimmer AM. Re: radiochemical purity of Tc-99m oxidronate [reply]. *J Nucl Med.* 1981;22:1016.
53. UltraTag [package insert]. St Louis: Mallinckrodt Inc; 2005.
54. Chowdhury S, Hung JC. Optimal centrifugation parameters for labeling efficiency determination of technetium-99m labeled red blood cells [abstract]. *J Nucl Med Technol.* 1993;21:114–5.
55. Cardiolite [package insert]. North Billerica, MA: Bristol-Myers Squibb Medical Imaging; 2003.
56. Hung JC, Wilson ME, Brown ML, et al. Rapid preparation and quality control method for technetium-99m-2-methoxy isobutyl isonitrile (technetium-99m-sestamibi). *J Nucl Med.* 1991;32:2162–8.
57. Hung JC, Wilson ME, Gebhard MW, et al. Comparison of four alternative radiochemical purity testing methods for ^{99m}Tc-sestamibi. *Nucl Med Commun.* 1995;16:99–104.
58. Hammes R, Kies S, Koblenski D, et al. A better method of quality control for technetium-99m sestamibi. *J Nucl Med Technol.* 1991;19:232–5.
59. Myoview [package insert]. Arlington Heights, IL: GE Healthcare/Medi-Physic, Inc.; 2006.
60. Metaye T, Desmarquet M, Rosenberg T. Rapid quality control for testing the radiochemical purity of ^{99m}Tc-tetrofosmin. *Nucl Med Commun.* 2001;22:1139–44.
61. Hammes R, Joas LA, Kirschling TE, et al. A better method of quality control for ^{99m}Tc-tetrofosmin. *J Nucl Med Technol.* 2004;32:72–8.
62. Egert LA, Dick MD, Mahoney DW, et al. A rapid radiochemical purity testing method for ^{99m}Tc-tetrofosmin. *J Nucl Med Technol.* 2010;38:81–4.
63. American Pharmaceutical Association. *Alternative Radiochemical Purity Testing Procedures for the Compounded Radiopharmaceuticals Approved from 1988–1997.* Washington, DC: American Pharmaceutical Association; 1998.
64. US Nuclear Regulatory Commission. Determination of dosages of unsealed byproduct material for medical use. 10 CFR §35.63. www.nrc.gov/reading-rm/doc-collections/cfr/part035/part035-0063.html. Accessed February 5, 2010.
65. US Nuclear Regulatory Commission. Manufacture, preparation, or transfer for commercial distribution of radioactive drugs containing byproduct material for medical use under part 35. 10 CFR §32.72. www.nrc.gov/reading-rm/doc-collections/cfr/part032/part032-0072.html. Accessed February 5, 2010.
66. US Nuclear Regulatory Commission. Application for specific licenses. 10 CFR §30.32. www.nrc.gov/reading-rm/doc-collections/cfr/part030/part030-0032.html. Accessed February 5, 2010.
67. Wiarda KS. Use of a copper filter for dose calibrator measurements of nuclides emitting K x-rays. *J Nucl Med.* 1984;25:633–4.

68. Harris CC, Jaszczk RJ, Greer KL, et al. Effects of characteristic X-rays on assay of iodine-123 by dose calibrators. *J Nucl Med.* 1984;25:1367–70.
69. Kowalsky RJ, Johnston RE. Dose calibrator assay of iodine-123 and indium-111 with a copper filter. *J Nucl Med Technol.* 1998;26:94–8.
70. Salako QA, DeNardo SJ. Radioassay of yttrium-90 radiation using the radionuclide dose calibrator. *J Nucl Med.* 1997;38:723–6.
71. Zimmerman BE, Coursey BM, Cessna JT. Correct use of dose calibrator values [letter]. *J Nucl Med.* 1998;39:575.
72. Salako QA. Correct use of dose calibrator values [reply]. *J Nucl Med.* 1998;39:575–6.
73. Schultz MK, Cessna JT, Anderson TL, et al. A performance evaluation of ^{90}Y dose calibrator measurements in nuclear pharmacies and clinics in the United States. *Appl Radiat Isot.* 2008;66:252–60.
74. US Nuclear Regulatory Commission. Labeling of vials and syringes. 10 CFR §35.69. www.nrc.gov/reading-rm/doc-collections/cfr/part035/part035-0069.html. Accessed February 5, 2010.
75. US Nuclear Regulatory Commission. NRC Information Notice 2007-25: suggestions from the advisory committee on the medical use of isotopes for consideration to improve compliance with sodium iodide I-131 written directive requirements in 10 CFR 35.40 and supervision requirements in 10 CFR 35.27. July 19, 2007.
76. Fludeoxyglucose (^{18}F) injection. *European Pharmacopeia.* http://online6.edqm.eu/ep606/#doc_7_of_161. Accessed February 5, 2010 (access available only to subscribers).
77. Alexoff DL, Casati R, Fowler JS, et al. Ion chromatographic analysis of high specific activity ^{18}FDG preparations and detection of the chemical impurity 2-deoxy-2-chloro-D-glucose. *Int J Rad Appl Instrum [A].* 1992;43:1313–22.
78. Füchtner F, Steinback J, Mäding P, et al. Basic hydrolysis of 2-[^{18}F]fluoro-1,3,4,6-tetra-O-acetyl-D-glucose in the preparation of 2-[^{18}F]fluoro-2-deoxy-D-glucose. *Appl Radiat Isot.* 1996;47:61–6.
79. Meyer G-J, Matzke KHY, Hamacher K, et al. The stability of 2-[^{18}F]fluoro-2-deoxy-D-glucose towards epimerization under alkaline conditions. *Appl Radiat Isot.* 1999;51:37–41.
80. Staum MM. Incompatibility of phosphate buffer in ^{99m}Tc-sulfur colloid containing aluminum ion. *J Nucl Med.* 1972;13:386–7.
81. Study KT, Hladik WB, Saha GB. Effects of Al^{3+} ion on ^{99m}Tc sulfur colloid preparations with different buffers. *J Nucl Med Technol.* 1984;12:16–8.
82. Sodium pertechnetate Tc 99m injection. *The United States Pharmacopeia and The National Formulary.* Rockville, MD: United States Pharmacopeial Convention, Inc. www.uspnf.com/uspnf/pub/index?usp=32&nf=27&s=2&officialOn=December%201,%202009. Accessed February 8, 2010 (access available only to subscribers).
83. Ammonia (^{13}N) injection. *European Pharmacopeia.* http://online6.edqm.eu/ep600/. Accessed February 2, 2010 (access available only to subscribers).
84. US Food and Drug Administration. Classification of residual solvents by risk assessment. In: *Guidance for Industry Q3C Impurities: Residual Solvents.* Rockville, MD: US Food and Drug Administration; 1997:3. www.fda.gov/downloads/Drugs/GuidanceComplianceRegulatoryInformation/Guidances/ucm073394.pdf. Accessed February 8, 2010.
85. US Food and Drug Administration. *Guidance for Industry Q3C: Tables and List.* Rockville, MD: US Food and Drug Administration; 1997. www.fda.gov/downloads/Drugs/GuidanceComplianceRegulatoryInformation/Guidances/ucm073395.pdf. Accessed February 8, 2010.
86. Technetium Tc 99m albumin aggregated injection. *The United States Pharmacopeia and The National Formulary.* Rockville, MD: United States Pharmacopeial Convention, Inc. www.uspnf.com/uspnf/pub/index?usp=32&nf=27&s=2&officialOn=December%201,%202009. Accessed February 8, 2010 (access available only to subscribers).
87. Technetium Tc 99m exametazime injection. *The United States Pharmacopeia and The National Formulary.* Rockville, MD: United States Pharmacopeial Convention, Inc. www.uspnf.com/uspnf/pub/index?usp=32&nf=27&s=2&officialOn=December%201,%202009. Accessed February 8, 2010 (access available only to subscribers).
88. pH (general chapter 791). *The United States Pharmacopeia and The National Formulary.* Rockville, MD: United States Pharmacopeial Convention, Inc. www.uspnf.com/uspnf/pub/index?usp=32&nf=27&s=2&officialOn=December%201,%202009. Accessed February 8, 2010 (access available only to subscribers).
89. Fisherbrand Alkacid Full-Range pH Kit. Fisher Scientific www.fishersci.com/wps/portal/PRODUCTDETAIL?productId=656640&catalogId=29104&pos=3&catCode=RE_SC&fromCat=yes&keepSessionSearchOutPut=true&brCategoryId=null&hlpi=y&fromSearch=Y. Accessed February 8, 2010.
90. Sterility tests (general chapter 71). *The United States Pharmacopeia and The National Formulary.* Rockville, MD: United States Pharmacopeial Convention, Inc. www.uspnf.com/uspnf/pub/index?usp=32&nf=27&s=2&officialOn=December%201,%202009. Accessed February 8, 2010 (access available only to subscribers).
91. Bacterial endotoxins test (general chapter 85). *The United States Pharmacopeia and The National Formulary.* Rockville, MD: United States Pharmacopeial Convention, Inc. www.uspnf.com/uspnf/pub/index?usp=32&nf=27&s=2&officialOn=December%201,%202009. Accessed February 8, 2010 (access available only to subscribers).
92. Pyrogen test (general chapter 151). *The United States Pharmacopeia and The National Formulary.* Rockville, MD: United States Pharmacopeial Convention, Inc. www.uspnf.com/uspnf/pub/index?usp=32&nf=27&s=2&officialOn=December%201,%202009. Accessed February 8, 2010 (access available only to subscribers).
93. US Food and Drug Administration. *Final Rule: Current Good Manufacturing Practice for Positron Emission Tomography Drugs.* http://frwebgate.access.gpo.gov/cgi-bin/getdoc.cgi?dbname=2009_register&docid=fr10de09-9.pdf. Accessed February 8, 2010.
94. Cooper JF, Harbert JC. Endotoxin as a cause of aseptic meningitis after radionuclide cisternography. *J Nucl Med.* 1975;16:809–13.
95. Bang FB. A bacterial disease of *Limulus polyphemus. Bull Johns Hopkins Hosp.* 1956;98:325–51.
96. Levin J, Bang FB. Clottable protein in *Limulus:* its localization and kinetics of its coagulation by endotoxin. *Thromb Diath Haemorrh.* 1968;19:186–97.
97. Cooper JF. Bacterial endotoxins test. In: Prince R, ed. *Microbiology in Pharmaceutical Manufacturing.* Godalming, UK: Davis Horwood International; 2001:537–67.
98. Cooper JF, Thoma LA. Screening extemporaneously compounded intraspinal injections with the bacterial endotoxins test. *Am J Health Syst Pharm.* 2002;59:2426–33.
99. Cooper JF. BET Methods for PET Radiopharmaceuticals. Paper presented at 148th Annual Meeting of the American Pharmaceutical Association; March 2001; San Francisco, CA.
100. Cooper JF. USP revision for radiopharmaceuticals for positron emission tomography (PET) takes effect June 2001. *LAL Times.* June 2001;8:1–2.
101. US Nuclear Regulatory Commission. Possession, use, and calibration of instruments used to measure the activity of unsealed byproduct material. 10 CFR §35.60. www.nrc.gov/reading-rm/doc-collections/cfr/part035/part035-0060.html. Accessed February 8, 2010.
102. American National Standards Institute. *American National Standard Calibration and Usage of "Dose Calibrator" Ionization Chambers for the Assay of Radionuclides.* Washington, DC: American National Standards Institute; 1986.
103. National Council on Radiation Protection and Measurements. *Quality Assurance for Diagnostic Imaging.* NCRP Report No. 99. Bethesda, MD: National Council on Radiation Protection and Measurements; 1988.
104. US Nuclear Regulatory Commission. Records of calibrations of instruments used to measure the activity of unsealed byproduct material. 10 CFR §35.2060. www.nrc.gov/reading-rm/doc-collections/cfr/part035/part035-2060.html. Accessed February 10, 2010.
105. Kowalsky RJ, Johnston RE, Chan FH. Dose calibrator performance and quality control. *J Nucl Med Technol.* 1977;5:35–40.
106. Oswald WM, Herold TJ, Wilson ME, et al. Dose calibrator linearity testing using an improved attenuator system. *J Nucl Med Technol.* 1992;20:169–72.
107. Chu RY. Accuracy of dose calibrator linearity test [letter]. *Health Phys.* 1988;55:95.

108. US Nuclear Regulatory Commission. Model procedures for dose calibrator calibration. In: *Program-Specific Guidance About Medical Use Licenses (NUREG-1556, Vol 9)*. Washington, DC: US Nuclear Regulatory Commission. www.nrc.gov/reading-rm/doc-collections/nuregs/staff/sr1556/v9/index-old.html. Accessed September 28, 2010.
109. US Nuclear Regulatory Commission. Calibration of survey instruments. 10 CFR §35.61. www.nrc.gov/reading-rm/doc-collections/cfr/part035/part035-0061.html. Accessed February 10, 2010.
110. US Nuclear Regulatory Commission. Records of radiation survey instrument calibration. 10 CFR §35.2061. www.nrc.gov/reading-rm/doc-collections/cfr/part035/part035-2061.html. Accessed February 10, 2010.
111. US Nuclear Regulatory Commission. Records of mobile medical services. 10 CFR §35.2080. www.nrc.gov/reading-rm/doc-collections/cfr/part035/part035-2080.html. Accessed February 10, 2010.
112. Positron emission tomography. US Food and Drug Administration Modernization Act. Bill S. 830, Pub L No. 105-115 §121, 11 Stat 2296 (1997).
113. Briner SJ. USP monographs and tests of radiochemical purity [letter]. *Am J Health Syst Pharm*. 1995;52:1817–8.
114. Draft guidelines on the manufacture of positron emission tomography radiopharmaceutical drug products; availability. 60 *Federal Register* 10593-4 (1995).
115. Guidance for industry: current good manufacturing practices for positron emission tomographic (PET) drug products; availability. 62 *Federal Register* 19580-1 (1997).
116. US Food and Drug Administration. *Proposed Rule: Current Good Manufacturing Practice for Positron Emission Tomography Drugs*. www.fda.gov/OHRMS/DOCKETS/98fr/cd0279.pdf. Accessed February 11, 2010.
117. Pharmaceutical compounding—sterile preparations (general chapter 797). *The United States Pharmacopeia and The National Formulary*. Rockville, MD: United States Pharmacopeial Convention, Inc. www.uspnf.com/uspnf/pub/index?usp=32&nf=27&s=2&officialOn=December%201,%202009. Accessed February 11, 2010 (access available only to subscribers).
118. Sterility testing—validation of isolator systems (general chapter 1208). *The United States Pharmacopeia and The National Formulary*. Rockville, MD: United States Pharmacopeial Convention, Inc. www.uspnf.com/uspnf/pub/index?usp=32&nf=27&s=2&officialOn=December%201,%202009. Accessed February 11, 2010 (access available only to subscribers).
119. Electronic records; electronic signature. 21 CFR §11. www.accessdata.fda.gov/scripts/cdrh/cfdocs/cfCFR/CFRSearch.cfm?CFRPart=11. Accessed February 11, 2010.
120. US Food and Drug Administration. *Guidance for Industry: Part 11, Electronic Records; Electronic Signature—Scopes and Application*. www.fda.gov/downloads/RegulatoryInformation/Guidances/ucm125125.pdf. Accessed February 11, 2010.
121. International Conference on Harmonisation. *ICH Harmonised Tripartite Guideline: Validation of Analytical Procedures: Text and Methodology Q2(R1)*. www.bioforum.org.il/Uploads/Editor/karen/q2_r1_step4.pdf. Accessed February 11, 2010.
122. Validation of compendial procedures (general chapter 1225). *The United States Pharmacopeia and The National Formulary*. Rockville, MD: United States Pharmacopeial Convention, Inc. www.uspnf.com/uspnf/pub/index?usp=32&nf=27&s=2&officialOn=December%201,%202009. Accessed February 11, 2010 (access available only to subscribers).
123. Hung JC. Comparison of various requirements of the quality assurance procedures for 18F-FDG injection. *J Nucl Med*. 2002;43:1495–506.

CHAPTER 15

Microbiologic Control for Radiopharmaceuticals

James F. Cooper

The *U.S. Pharmacopeia* (*USP*) General Chapter 797, Pharmaceutical Compounding—Sterile Preparations first appeared in 2004. After extensive revision, it was republished in *USP 31/NF 26* and became effective in June 2008 as an official, enforceable standard. The revised standard has revolutionized sterile preparation practices, from proper hygiene to air quality. It applies to all health care personnel who conduct preadministration manipulations of compounded sterile products (CSPs), including compounding, dispensing, storage, and transport. In this revised CSP standard, the greatest emphasis is placed on preventing contact contamination.[1] Ready availability of sterile 70% isopropyl alcohol (IPA) is vital. Sterile compounding requires that sterility be maintained during the compounding of CSPs exclusively with sterile ingredients and components and that sterility be achieved during compounding with nonsterile materials. This chapter focuses on practice standards that are relevant to sterile radiopharmaceutical compounding, addressing aseptic technique training and evaluation, standard procedures, and control of surface and airborne contamination.

MICROBIAL CONTAMINATION RISK LEVELS

The determination of microbial risk level dictates the practice behaviors, processes, and equipment used in preparing various radiopharmaceuticals. The preparation of ^{99m}Tc-labeled products from commercial materials is low microbial risk level compounding. Low-risk level CSPs are those that are prepared by the simple aseptic transfer of sterile drugs from one container to another, using sterile devices, in a controlled-air environment. This risk level also applies to the preparation of unit doses of other products, including manufactured positron emission tomography (PET) products. Radiolabeling of autologous cells exemplifies medium-risk level compounding because the procedure's complexity increases microbial risk. Compounding with nonsterile components and subsequent sterilization has a high-risk level, necessitating terminal sterilization and microbial testing. In case of emergency, when a unit dose of a radiopharmaceutical is given to one patient within 1 hour of compounding, the drug may be prepared under the "immediate use" provisions. In emergency compounding, personnel are expected to use suitable aseptic technique, but they are exempted from garbing requirements and air-quality standards. Conditions for a beyond-use-date (BUD) of 12 hours or less potentially apply to ^{99m}Tc drugs.

Radiopharmaceuticals are unique among CSPs. Their radioactive nature renders them readily characterized for potency and radiometric analyses. However, accommodating the radioactive property increases the risk of microbial contamination. For example, shielding materials may alter the direction of laminar airflow in a primary engineering control (PEC). Also, the weight and design of radionuclide generators cause them to be stored in less than Class 5 air quality, even though the elution area is a critical site for compounding.

Special provisions for generators are made wherein elution may be done in a Class 8 or better environment, according to the supplier's directions. A substantial risk arises from necessary reuse of syringe shields; this practice risks a return of microbial contamination to the dispensing area. Definitive disinfection practices are needed to eliminate the hazard of blood and microbial contamination.

PROTECTING CRITICAL SITES THROUGH CONTROL OF SURFACE CONTAMINATION

Continual education, motivation, and oversight are required to produce compounding personnel who work in a suitable environment with impeccable aseptic technique and accuracy. Good compounding is an attitude as well as a set of well-prescribed behaviors. Hands must be properly cleansed. The body must be properly covered with personnel protective equipment (PPE) that is donned correctly. CSPs must be properly handled in a suitably controlled environment.

Hand hygiene and sterile gloves are paramount for protecting the *critical site,* defined as any opening or surface that can provide a pathway between a sterile product and the environment, including vial septa and needle tips. Protection of these critical sites must be given highest priority during compounding. For hand hygiene, the CSP standard specifies pre-gloving hand disinfection with a waterless, alcohol-based, chlorhexidine-containing surgical hand scrub with persistent activity (e.g., Avagard [3M, St. Paul, MN]). Sterile gloves are recommended because they reduce the initial bioburden at critical sites. Frequent and repeated sanitization with sterile IPA maintains low microbial bioburden on gloved hands. Trissel and coworkers[2] reported that the microbial contamination rate in media-fill challenge studies dropped from 5.2% to 0.34% when the use of nonsterile gloves was replaced with a procedure for wearing sterile gloves and repeatedly disinfecting with sterile IPA.

Sterile 70% IPA is the principal disinfectant in CSP compounding. Since bacteria are ubiquitous in nature and are present on every surface, aseptic compounding requires impeccable cleaning and disinfection of work surfaces. *USP* Chapter 1072, Disinfectants and Antiseptics, defines such agents and describes their effectiveness. Sterile 70% IPA is clearly the most useful disinfecting agent because it is inexpensive, is rapid acting against most bacteria, and quickly evaporates without residue. Nonsterile IPA may be used to sanitize supplies in the ante-area. The antimicrobial activity of alcohols is attributed to denaturing of proteins. Alcohols are biocidal against gram-positive and gram-negative bacteria and most fungi. A limitation is that alcohols are not reliably sporicidal against spore-forming bacteria, such as *Bacillus* spp.[2]

The CSP standard details how to don PPE in an order from dirtiest (foot covers) to cleanest (face mask), followed by completing hand-washing tasks and donning a nonshedding gown with snugly fitting sleeves. Sterile gloves are donned before compounding begins, and sterile IPA is applied routinely throughout the compounding process and whenever non-sterile surfaces (e.g., shields, carts) are touched. Therefore, personnel must be trained by an expert on proper disinfection, hand-washing and scrubbing techniques, donning of PPE apparel, and avoiding touch contamination of critical sites. In addition to attentive study of the CSP chapter, use of multimedia instructional sources and professional publications can help personnel achieve competency in cleanroom etiquette. All personnel must complete didactic training, write a competency assessment, be observed for skill assessment, and conduct media-fill tests (see *USP* Chapter 797 Appendices III–V).

The media-fill procedure should simulate basic CSP tasks in a nuclear pharmacy, such as elution of a ^{99m}Tc generator, preparation of a ^{99m}Tc kit, and drawing multiple unit doses. Soybean–casein digest medium (SCDM) is substituted for ^{99m}Tc components to maximize the opportunity for microbial growth from suboptimal technique. The facility's existing procedures for cleaning and disinfection, garbing, and aseptic technique should be applied. A media-fill scheme is presented in Figure 15-1. In this exercise, a shielded 1 mL and a 3 mL aliquot of media simulate kit compounding by ^{99m}Tc eluate rehydration and saline dilution. Finally, six unit doses are drawn from the kit vial and injected into an incubation vial. This protocol is conducted in four replicates in one work period to introduce a fatigue factor. The four kit vials and four incubation vials (each the accumulation of six unit doses) are incubated for at least 14 days at 25°C to 35°C. The media vials are inspected over a 14 day period for turbidity or other signs of microbial growth. The absence of growth and satisfactory completion of other requirements render the operator qualified in aseptic technique. Requalification is an annual requirement. This procedure is a relatively insensitive means for assessing aseptic technique, so it should never be used to validate marginal cleanroom behavior. Rather, this procedure should emulate the spirit and guidance of *USP* Chapter 797. Media may be challenged with certified microbiologic cultures contained in a matrix (e.g., Bioball [bioMérieux, Durham, NC]). Personnel who also conduct medium-risk or high-risk procedures, such as radiolabeling autologous cells, should qualify with a protocol that simulates these specific tasks.

Assessment of CSP competency includes gloved-fingertip sampling in which the evaluator collects a gloved fingertip and thumb sample from both hands onto suitable agar plates by lightly pressing each fingertip onto the agar. No colony-forming units may grow after three fingertip collections. Surface collections are also made during media-fill or other training assessments.

Supplies should be unpackaged before entering the compounding area. However, removal of syringes from their immediate package must be done under laminar-airflow conditions. Unpackaged syringes must be kept in a sanitized

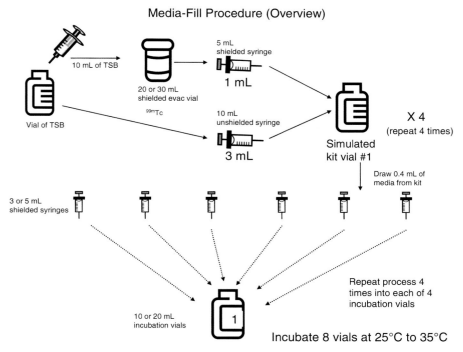

FIGURE 15-1 Simulated media fill for ^{99m}Tc radiopharmaceuticals. Two syringes contain media that simulate reconstitution of a kit with ^{99m}Tc-sodium pertechnetate and saline. Dose-drawing simulation specifies six transfers of media that are collected in an incubation vial; this is done in four replicates. The four kit vials and four incubation vials are incubated and observed after 14 days.

plastic container until transferred to the direct compounding area (DCA) in the PEC for dose-drawing activities.

CONTROL OF AIRBORNE CONTAMINATION FOR CSPS

Airborne contamination is eliminated by a properly certified, maintained, and sanitized laminar-airflow enclosure. Airborne bacteria are carried by dust and dry particulates of bacteria and saliva expelled from the mouth. Bacteria are also rafted in air on skin scales. An operator sheds about 10 million skin scales daily, and most of them carry microbes, making personnel a greater source of airborne bacterial contamination in a DCA than contaminants from the floor or surfaces below the work level. The CSP chapter describes the use of PECs for effective removal of airborne bacteria and other particulates with high-efficiency particulate air (HEPA) filters, which remove 99.97% of particles ≥0.3 μm size. "PEC" is the general name for work stations such as a laminar-airflow workbench (LAFW), biologic safety cabinet (BSC), or compounding aseptic containment isolator (CACI), all of which provide an International Organization for Standardization (ISO) Class 5 air quality work environment (the air quality classification can be found in Table 1 of *USP* Chapter 797). Laminar flow describes air that moves in a confined space with uniform speed along parallel lines. An operator must learn to work with no object between the air source and the critical site.

A basic conflict arises between traditional nuclear pharmacy and the CSP chapter regarding the direction of airflow in cleanrooms. Traditionally, LAFW units and cleanrooms were designed for positive pressure so that bacteria and particulates are swept out of the work area. Contrary to this, good radiologic health practice dictates using negative-pressure work units to avoid exposure of operators and work areas to airborne contamination. Vertical laminar-airflow units (BSCs and CACIs) with filtered exhaust are a compromise between the opposing practices. Another complication is the modification of PEC units to accommodate dose calibrators and shielding requirements. Operators must understand that the PEC is not a storage site; all extraneous or unused components must be removed so that laminar flow is not disrupted or made turbulent.

The "Radiopharmaceuticals as CSPs" section of Chapter 797 specifies location of PECs within restricted-access ISO Class 8 air-quality buffer zones (formerly designated cleanrooms). The minimum number of air changes per hour is 30 for ISO Class 7 buffer zones. This section allows a line of demarcation to define the compounding area for preparing low-risk level CSPs with a BUD of 12 hours or less. However, a 12 hour BUD limits the dispensing of longer-lived drugs, such as ^{201}Tl tracers. A PEC for radiolabeling blood-derived elements must be clearly segregated from other activities to avoid cross-contamination.

Emergence of the CSP chapter prompted substantial redesign of nuclear pharmacies to achieve compliance with facility and air-quality standards. Hung and Anderson[3] reported on the results of costly remodeling for four nuclear pharmacies at Mayo Clinic and the rationale for decisions affecting air quality. The original pharmacies had open-air construction with no air-quality classification. Remodeling brought fixed walls to create a buffer area (formerly cleanroom) that was connected to an anteroom by an interlocking door system. All PECs were placed in ISO Class 7 buffer zones and exhausted 100%. Elution of ^{99m}Tc generators was done in a negative-pressure ISO Class 7 buffer area rather than the minimum Class 8 specified by the CSP standard. Only essential equipment, such as keyboards and intercoms, was placed in the buffer area. Facility design is also discussed in Chapter 13 of this text and elsewhere.[4]

What is ideal cleanroom etiquette in the nuclear pharmacy? At the beginning of a shift, a properly attired operator enters the buffer area and prepares for work in the DCA. With all possible materials removed, the PEC is sanitized according to procedure, from clean to less clean areas, including the dose calibrator. Now, clean and sanitized kit-labeling products, vial shields, a container of syringes, and other necessary items such as shielded vials of ^{99m}Tc are bathed again with sterile IPA as they are brought into the DCA. Vial septa of kit products are sanitized and shielded, and ^{99m}Tc labeling procedures commence. Where applicable, sterile water is placed in the water bath for labeling procedures; the bath is emptied and dried at the end of the shift. After imaging agents are radiolabeled, it is appropriate to don new sterile gloves. As dose drawing begins, the operator selects freshly sanitized syringe shields, optimally handled by a properly attired assistant. Ideally, shielded ^{99m}Tc vials remain in a PEC for later dose-drawing activities. Traffic is kept to a minimum during the dose-drawing period. All PECs are cleaned and sanitized at the end of the shift. This scenario presents a practice to strive for in creating procedures that are compliant with the letter and spirit of USP Chapter 797.

Table 3 of USP Chapter 797 describes the minimum frequency for cleaning and disinfecting the various compounding areas. All cleaning materials must be nonshedding and dedicated to use in the buffer or clean area, ante-area, and segregated compounding areas and must not be removed from these areas until disposal. Cleaning must be done by qualified staff in conformance with written procedures.

All PEC units must be certified for ISO Class 5 air quality by tests for airflow velocity and filter integrity. Certification is required at installation; recertification occurs at 6 month intervals and after any repair or relocation. Environmental monitoring for viable and nonviable particulates should be done and documented during each certification activity. Proper maintenance includes changing prefilters and monitoring airflow gauges for proper performance of PEC units. The respective standard operating procedures (SOPs) must provide clear instructions and appropriate forms for documenting these activities.

MICROBIOLOGIC TESTS FOR HIGH-RISK LEVEL RADIOPHARMACEUTICALS

Radiopharmaceuticals that are prepared from nonsterile materials are usually sterilized by membrane filtration. Examples include ^{123}I-iobenguane, PET products, and investigational drugs. The Food and Drug Administration's current good manufacturing practices guidance for PET drugs is the principal regulation for manufacture of ^{18}F and other PET products.[5]

A successful bubble-point test for membrane integrity is the best way to verify that the filter was free of defects. Devices for filter-integrity tests include a pressure source (nitrogen tank) and calibrated pressure regulator. The observed pressure is compared with the filter manufacturer's specification for pressure resistance.

Aseptically sterilized products also require end-product tests for microbiologic purity, that is, sterility and freedom from pyrogens (bacterial endotoxins). Specific requirements for these tests are described in a continuing-education unit.[6] Successful sterility testing requires aseptic sampling of an aliquot from the final container, sanitization of a certified PEC, and inoculation of media by a qualified operator. The SOP for this task must be consistent with USP Chapter 71, Sterility Tests. Prepared culture media in 8 to 10 mL volumes are convenient tools for inoculation and incubation in controlled environments. Outer surfaces of sterility test media are ideally disinfected by a peroxide-based solution. During the 14 day incubation period, media tubes are examined for cloudiness, precipitation, and other signs of microbial growth. A suitable record must document observation dates, results, and identity of the operator.

The ideal endotoxins test for radiopharmaceuticals is an automated test system that requires a limulus amebocyte lysate (LAL) cartridge containing precalibrated amounts of dried reagents in a polystyrene cartridge.[6,7] No liquid reagents or standards are needed, in contrast to traditional methods. Test results are obtained in 15 minutes by a portable photometric detector with endotoxin-specific software (Portable Test System [Charles River Labs, Charleston, SC]). The new method is FDA approved and adheres to USP Chapter 85, Bacterial Endotoxins Test. A pretest, 1:50 dilution of the CSP reduces personnel exposure and avoids significant interference factors without compromising sensitivity with respect to the USP endotoxin limit.

STANDARD OPERATING PROCEDURES

Failure to follow SOPs resulted in an outbreak of hepatitis C in at least 12 patients who received doses of a cardiac imaging agent prepared at a nuclear pharmacy.[8] Apparently, a vial of saline, used as a diluent for cell labeling, was inadvertently contaminated with sera from a hepatitis patient and then transferred to the hood for ^{99m}Tc kit preparation. The incident illustrates the consequences of a breakdown in the training and oversight of compounding personnel.

SOPs must be tailored to meet the expectations and design of each pharmacy. The CSP standard suggests 23 general SOPs for aseptic compounding. Nuclear pharmacies need procedures for all aspects of practice, including SOPs for the following:

- Creation of standard procedures and document control;
- Description of training and assessment of personnel for aseptic processing;
- Instructions for preparation of all types of radiopharmaceuticals;
- Instructions for hand washing within the ante-area;
- Decontamination of frequently used supplies for the ante-area;
- Instructions for donning attire and sterile gloves in the buffer area;
- Introduction of supply carts and material storage in the buffer area;
- Decontamination of syringe shields and transfer into the buffer area;
- Instructions for handling supplies in the DCA;
- Disinfection of vial septa and shielded radiopharmaceuticals in the DCA;
- Instructions for surface cleaning of the DCA;
- Instructions for use of PECs; and
- Management of traffic flow in the compounding area.

The development and revision of SOPs are more effective when operators have the opportunity to participate in the process. Weekly staff meetings are an occasion to review, revise, and reinforce procedures. Of course, changes must not be implemented until they go through the change-control process described in the document control SOP.

This chapter has superficially reviewed *USP* Chapter 797 as it applies to nuclear pharmacy. There is no substitute for careful study of the CSP chapter in preparation for creating written procedures that apply comprehensively to a specific nuclear pharmacy setting. Once procedures are in place, training, assessment, and encouragement of personnel are the keys to ensuring the safety and efficacy of radiopharmaceuticals for their recipients.

REFERENCES

1. Okeke CC, Newton DW, Kastango ES, et al. Basics of compounding: *USP* <797> Pharmaceutical Compounding—Sterile Preparations, part 10: first revision: the main changes, events and rationale. *Int J Pharm Compound.* 2008;12:530–6.
2. Trissel LA. Quality-control analytical methods: overview of beyond-use dating for compounded sterile preparations. *Int J Pharm Compound.* 2008;12:524–9.
3. Hung JC, Anderson MM. Mayo Clinic approaches to meet *United States Pharmacopeia* <797> requirements for facility design and environmental controls of nuclear pharmacy. *J Nucl Med.* 2009;50:156–64.
4. Cooper JC, Petry NA. Compounding radiopharmaceuticals in compliance with the *USP* Chapter <797> Pharmaceutical Compounding—Sterile Preparations. *Continuing Education for Nuclear Pharmacists.* Vol 13, lesson 6. Albuquerque, NM: University of New Mexico College of Pharmacy; 2007.
5. *Guidance: PET Drugs—Current Good Manufacturing Practice (CGMP).* Rockville, MD: US Food and Drug Administration, Center for Drug Evaluation and Research; December 2009.
6. Cooper JF, Hung JC. Bacterial endotoxins test and sterility test for radiopharmaceuticals. *Continuing Education for Nuclear Pharmacists.* Vol 14, lesson 5. Albuquerque, NM: University of New Mexico College of Pharmacy; 2008.
7. Cooper JF, Latta KS, Smith D. Automated endotoxin testing for high-risk level compounded sterile preparations at an institutional compounding pharmacy. *Am J Health Syst Pharm.* 2010;67:280–6.
8. Patel RP, Larson AK, Castel AD, et al. Hepatitis C virus infections from a contaminated radiopharmaceutical used in myocardial perfusion studies. *JAMA.* 2006;296:2005–11.

CHAPTER 16
Radiopharmaceutical Special Topics

James A. Ponto

FORMULATION PROBLEMS OF RADIOPHARMACEUTICALS

Although a few radiopharmaceuticals are purchased from the manufacturer in a form ready for use, most radiopharmaceuticals require preparation by a nuclear pharmacist or other authorized user. This preparation frequently involves the successful completion of chemical reactions, in contrast to the preparation of conventional drugs, which typically involves simply mixing and dilution. Such chemical reactions, in certain circumstances, may not proceed as anticipated but may instead produce radiochemical impurities. Moreover, the radiation emitted from the radioactive content may cause radiolytic decomposition, which results in radiochemical impurities. Such radiochemical impurities exhibit altered biodistribution and may potentially interfere with the diagnostic or therapeutic effectiveness of the radiopharmaceutical.

Factors involved in the production of radiochemical impurities during the preparation of radiopharmaceuticals have been extensively reviewed previously.[1–4] The purpose of this chapter topic is to highlight common formulation (preparation) problems with radiopharmaceuticals that result in radiochemical impurities. An understanding of potential causes of common radiochemical impurities will enable practitioners to formulate and follow strategies for avoiding or minimizing these impurities.

Common Radiochemical Impurities

The predominant radiochemical impurity associated with ^{99m}Tc radiopharmaceuticals is free, unlabeled ^{99m}Tc in the chemical form of pertechnetate ion (TcO_4^-). Pertechnetate distributes throughout the vascular and interstitial fluid and concentrates primarily in the salivary glands, thyroid, stomach, intestinal tract, and urinary tract (see Figure 19-7, Chapter 19 and Figure 25-5, Chapter 25). A second radiochemical impurity associated with some ^{99m}Tc radiopharmaceuticals is insoluble ^{99m}Tc in the chemical form of technetium hydroxides or technetium-labeled stannous hydroxides (also referred to collectively as hydrolyzed–reduced technetium). These insoluble species, which tend to be in the colloid size range, are phagocytized by reticuloendothelial system cells in the liver and spleen (see Figure 25-4, Chapter 25). A third radiochemical impurity associated with a few ^{99m}Tc radiopharmaceuticals is particle aggregation. Aggregates larger than 10 μm lodge in the first capillary beds that they encounter, which are the lungs after peripheral venous injection.

A variety of other radiochemical impurities may be formed during the preparation of, or result from radiolytic decomposition of, various ^{99m}Tc radiopharmaceuticals. Impurities that are hydrophilic, ionized, non–protein bound, and <5000 molecular weight tend to be excreted in the urine by glomerular filtration. Impurities that are lipophilic, possess both polar and nonpolar groups, and 300–1000 molecular weight tend to be excreted by the liver into the bile. Yet other impurities tend to remain in the blood pool, especially if highly protein bound or labeled to red blood cells.

The predominant concern with radioiodinated radiopharmaceuticals is the radiolytic production of radioactive free iodide (I⁻), which is localized in the thyroid. Stable iodine (SSKI [saturated solution of potassium iodide] or Lugol's solution) is typically administered prior to and for some time after administration of iodinated radiopharmaceuticals in order to block thyroid uptake of free radioiodide.

The predominant concern with ^{111}In radiopharmaceuticals is formation of insoluble indium hydroxides during preparation or as a result of radiolytic decomposition. Nonchelated indium ion (In^{+3}) is soluble at acidic pH but precipitates as indium hydroxide above a pH of about 3. Insoluble indium hydroxide is primarily phagocytized by the liver, spleen, and bone marrow. Nonbound free indium in the blood binds to plasma transferrin, which demonstrates prolonged retention in the blood pool. Other ^{111}In radiochemical impurities include ^{111}In labeled to small molecules (e.g., peptide fragments), which are excreted by the kidneys into the urine, and ^{111}In labeled to very large molecules (e.g., antibody aggregates), which are localized in the liver.[5]

The predominant concern with ^{18}F fluorinated radiopharmaceuticals (e.g., ^{18}F-fludeoxyglucose [^{18}F-FDG], ^{18}F-fluorodopa) is the presence of free fluoride ion (F⁻) produced during the production reaction or resulting from radiolytic decomposition. ^{18}F-fluoride is localized in the bone and excreted in the urine. Depending on the radiopharmaceutical and the particular production method, other radiochemical byproducts may also be produced, which exhibit biodistribution patterns disparate from that intended.

Common Problems Associated with ^{99m}Tc Chelates

Most ^{99m}Tc radiopharmaceuticals are prepared via stannous (Sn^{+2}) ion reduction of pertechnetate (Tc in an oxidation state of +7) to a lower oxidation state (such as +4 or +5), which then allows the reduced technetium atoms to be chelated by multidentate ligands. A number of factors can interfere with this reduction and chelation by altering the intended ratio of Sn^{+2} to ^{99m}Tc atoms.

Excessive ^{99m}Tc

An excessive amount of ^{99m}Tc added to a reagent kit may result in unacceptably high amounts of residual, unreacted free pertechnetate. This problem can occur with a variety of products, especially those reagent kits containing relatively small amounts of Sn^{+2}, such as ^{99m}Tc-exametazime (^{99m}Tc-HMPAO), ^{99m}Tc-mertiatide (^{99m}Tc-MAG3), and ^{99m}Tc-red blood cells (UltraTag [Covidien]). Increased amounts of ^{99m}Tc obviously depress the Sn^{+2} to ^{99m}Tc ratio, which could interfere with the yield of the labeling reaction. However, because Sn^{+2} is nominally in stochiometric excess even in these cases, the decreased radiolabeling yield is more likely caused by, or in combination with, other factors than by ^{99m}Tc mass effects alone.

Excessive ^{99}Tc

^{99}Tc is an ever-present impurity in ^{99m}Tc samples. ^{99}Tc is a decay product of ^{99}Mo (approximately 14% of ^{99}Mo decays directly to ^{99}Tc) in addition to being the transition product following ^{99m}Tc decay (see Figure 2-15, Chapter 2). The long half-life of ^{99}Tc (210,000 years) allows it to build up (in terms of technetium atoms) in a ^{99}Mo–^{99m}Tc generator (see Figure 3-13, Chapter 3). Hence the eluate from a generator that has an ingrowth time more than 10 hours since prior elution will contain many more ^{99}Tc atoms than ^{99m}Tc atoms.

One term used to describe this relationship is mole fraction. The mole fraction of ^{99m}Tc atoms is equal to the number of ^{99m}Tc atoms divided by the total number of technetium atoms (i.e., the sum of ^{99}Tc atoms plus ^{99m}Tc atoms). Mole fractions of ^{99m}Tc in generator eluates are summarized in Table 3-4, Chapter 3. For example, the mole fraction of ^{99m}Tc atoms in the eluate from a generator that has an ingrowth time of 24 hours since prior elution is 0.28. This means only 28% of the technetium atoms are ^{99m}Tc, while the other 72% are ^{99}Tc. The mole fraction at the time of elution further decreases over time because of the continuous decay of ^{99m}Tc atoms into ^{99}Tc atoms. Hence, the ^{99m}Tc mole fraction of the eluate in this example will exponentially decrease over time from its original value of 0.28 to 0.14 at 6 hours of age and 0.07 at 12 hours of age.

^{99}Tc is chemically identical to ^{99m}Tc and will compete in all applicable chemical reactions. Hence it is not surprising that high amounts of ^{99}Tc (i.e., low mole fractions of ^{99m}Tc) can interfere with Sn^{+2} reduction and chelation of ^{99m}Tc. This problem occurs most frequently when eluates containing low mole fractions of ^{99m}Tc (e.g., prolonged ingrowth time before generator elution, prolonged age of eluate) are used to prepare reagent kits containing relatively small amounts of Sn^{+2}, such as ^{99m}Tc-albumin aggregated (^{99m}Tc-MAA), ^{99m}Tc-exametazime (^{99m}Tc-HMPAO), ^{99m}Tc-mertiatide (^{99m}Tc-MAG3), and ^{99m}Tc-red blood cells (UltraTag). An example of this effect is illustrated in Figure 3-14, Chapter 3. However, because Sn^{+2} is nominally in stochiometric excess in all but the most extreme of cases, the decreased radiolabeling yield of a ^{99m}Tc radiopharmaceutical is more likely caused by, or in combination with, other factors than by ^{99}Tc effects alone.

Radiolytic Oxidants

Radiation interacts with water molecules to produce ions, free radicals, and peroxides (see Equations 8-2a, 8-2b, 8-2c, 8-3d, 8-3e, and 8-3f, Chapter 8). Hydroxyl free radicals and peroxides are capable of readily oxidizing Sn^{+2} and thus may interfere with the intended reduction of ^{99m}Tc. The magnitude of these effects is directly related to the radiation levels because the rate of peroxide production increases linearly with increasing activity or radioactive concentration. The presence of oxygen further promotes the production of peroxides.

Because production of these radiolytic oxidants is related to radiation level, they are most abundant in highly concentrated pertechnetate solutions. Peroxides also tend to accumulate over time. Hence, the frequency and severity of problems caused by these radiolytic oxidants is most likely when reagent kits containing relatively small amounts of Sn^{+2} are prepared using highly concentrated ^{99m}Tc solutions, ^{99m}Tc eluates from generators with prolonged ingrowth times, or ^{99m}Tc solutions many hours old.

Inadequate Stannous Ion

Stannous ion is readily oxidized to stannic (Sn^{+4}) ion by atmospheric oxygen, dissolved oxygen, and radiolytic oxidants. One example of this effect is shown in Figure 10-4, Chapter 10. Reagent kits typically contain a stochiometric excess of Sn^{+2} salt lyophilized and sealed in an atmosphere of nitrogen or argon. However, oxidation of Sn^{+2} by entry of air during needle puncture, dissolved oxygen present in aqueous diluent, or radiolytic oxidants in pertechnetate solutions can decrease reductive capacity below the threshold needed for a satisfactory radiolabeling process. Although oxidation of Sn^{+2} is potentially problematic for all Sn^{+2}-containing reagent kits, the frequency and severity of this effect is inherently more pronounced in reagent kits containing relatively small amounts of Sn^{+2}.

Stannous ion can also be oxidized by inadvertent entry of oxidizing antiseptic agents (e.g., hydrogen peroxide) into the reagent vial during needle puncture. Also, many of the chemicals in bacteriostatic sodium chloride injection may cause oxidation of Sn^{+2} if used as a diluent.

In the labeling of red blood cells in vivo with ^{99m}Tc, an adequate amount of Sn^{+2} ion must first be delivered into the bloodstream. Substandard radiolabeling of ^{99m}Tc to red blood cells in vivo may result if the injection of Sn^{+2} ion (e.g., as stannous pyrophosphate) is infiltrated. Substandard radiolabeling may also result if the Sn^{+2} ion product is administered through certain catheters because the Sn^{+2} ions tend to bind to their walls.

Recommendations

In order to minimize the above problems associated with alterations in the intended ratio of Sn^{+2} to ^{99m}Tc atoms, a number of precautions are generally recommended. These recommendations are especially important in preparing reagent kits containing relatively small amounts of Sn^{+2}. Whenever possible,

1. Avoid adding excessive ^{99m}Tc activity to reagent kits.
2. Avoid using ^{99m}Tc eluates from generators that have ingrowth times of more than 24 hours.
3. Avoid the use of aged ^{99m}Tc solutions, especially those more than 12 hours old.
4. Avoid adding air (i.e., oxygen) to reagent vials unless otherwise directed.
5. Avoid maintaining excessive concentrations of ^{99m}Tc solutions (i.e., concentrated ^{99m}Tc solutions should be diluted to lower radioactive concentrations).
6. Avoid use of bacteriostatic sodium chloride injection for preparation or dilution of ^{99m}Tc radiopharmaceuticals.

Common Problems Associated with Preparative Manipulations

Radiopharmaceuticals have been designed by the manufacturer to comply with USP purity specifications when prepared according to manufacturer instructions. Deviations from these instructions, whether purposeful or inadvertent, may result in suboptimal radiochemical purity.

Improper Mixing Order

The preparation of many radiopharmaceuticals involves a series of procedural steps to produce the intended product. Alterations in the prescribed sequence, or mixing order, can drastically interfere with desired outcome. For example, the preparation of ^{99m}Tc-sulfur colloid involves the decomposition of thiosulfate in the presence of heat and acid, which precipitates as a colloid with the coprecipitation of ^{99m}Tc-technetium sulfide. Little if any desired product will be formed if additions of the hydrochloric acid and the phosphate buffer components are reversed. In the preparation of ^{99m}Tc-red blood cells in vitro (UltraTag), sodium hypochlorite is added to pretinned blood to oxidize residual extracellular Sn^{+2}. Negligible radiolabeling will occur if the sodium hypochlorite is added to the reaction vial prior to the addition of blood, because the sodium hypochlorite will prematurely oxidize the Sn^{+2} present in the reaction vial.

Preparation of ^{111}In-capromab pendetide involves addition of buffered ^{111}In-acetate to the antibody solution. Because indium is not soluble at pH >3, the ^{111}In is supplied as $InCl_3$ dissolved in hydrochloric acid. Adding this acidic ^{111}In-chloride directly to the antibody solution will cause poor radiochemical purity and potential damage to the antibody structure. Therefore, the standard procedure is to first add some sodium acetate to the ^{111}In-chloride to produce ^{111}In-acetate at a buffered pH, and then add the buffered ^{111}In-acetate to the antibody solution. Acetate is a weak complexing agent, which maintains ^{111}In in a soluble form ready for chelation by the capromab pendetide, and the buffered pH does not damage the antibody structure. Similarly, in the radiolabeling of ibritumomab tiuxetan, improper mixing of antibody and ^{111}In-chloride or ^{90}Y-chloride (both dissolved in hydrochloric acid) without first buffering with sodium acetate will lead to an unacceptable product.[6]

Improper Heating

Several radiopharmaceuticals require heating as a step in their preparation. Inadequate heating (i.e., insufficient temperature or insufficient time) may not provide acceptable chemical yields. For example, substandard radiochemical purity due to inadequate heating can occur for ^{99m}Tc-mertiatide (^{99m}Tc-MAG3), ^{99m}Tc-sestamibi, and ^{99m}Tc-sulfur colloid. Excessive heating, on the other hand, has much less impact on radiochemical purity but may produce gas pressures inside sealed vials sufficiently high to cause rupture of the septum, ejection of the stopper, or breakage of the glass wall. Therefore, substantial negative pressure (i.e., a partial vacuum) in the vial should be ensured prior to heating.

The particle size distribution of ^{99m}Tc-sulfur colloid is also influenced by the extent of heating. Shorter heating times are associated with smaller particle size (see Table 26-6, Chapter 26), which may be exploited for use in lymphoscintigraphy. Longer heating times, on the other hand, produce larger particles, which will embolize in lung capillaries if they exceed 10 μm in size.

The degree of damage to radiolabeled red blood cells intended for spleen imaging varies directly with the amount of heating. Inadequate heating results in insufficient damage, allowing the radiolabeled erythrocytes to remain in the blood pool rather than being sequestered by the spleen. Excessive heating, on the other hand, results in excessive damage and fragmentation of the radiolabeled erythrocytes, which tend to be phagocytized by the liver rather than sequestered by the spleen.

Incubation/Time Delays

Although many metal chelates are formed very rapidly, some complexation reactions require substantial incubation time because the rate of the coordination reaction between the radionuclide metal and the complexing agent is slow or the diffusion or transport of a reactant is slow.

Incubation times of up to 10 to 20 minutes may be required to achieve maximal radiolabeling of ^{99m}Tc-disofenin (^{99m}Tc-DISIDA), ^{99m}Tc-mebrofenin, ^{99m}Tc-pentetate (^{99m}Tc-DTPA), and ^{99m}Tc-succimer (^{99m}Tc-DMSA) because of the slow progression from the initial rapidly formed mononuclear complex to the final dinuclear (dimeric) complex. For albumin products such as ^{99m}Tc-MAA, a similar incubation time is needed to allow pertechnetate ions to diffuse into the protein's denatured tertiary structure where stannous reduction and chelation take place. An incubation time of 10 to 20 minutes is also required for radiolabeling of red blood cells with ^{99m}Tc, a process that involves the relatively slow transport of pertechnetate ions across the cell membrane. Similarly, for maximal radiolabeling of leukocytes with ^{99m}Tc-exametazime (^{99m}Tc-HMPAO) or ^{111}In-oxine, an incubation time of about 15 to 20 minutes is needed to allow diffusion of the lipophilic chelate across the cell membrane followed by trapping of the radionuclide inside the cell.

Several of the newer radiopharmaceuticals are prepared via rapid initial formation of an intermediate complex from which the final product is formed via exchange reactions. Transfer ligands used to form the intermediate complex are generally weak complexing or chelating agents, such as acetate (used in the preparation of ^{111}In-capromab pendetide and ^{111}In- or ^{90}Y-ibritumomab tiuxetan), citrate (used in the preparation of ^{99m}Tc-sestamibi and ^{111}In-pentetreotide), edetate (used in the preparation of ^{99m}Tc-bicisate [^{99m}Tc-ECD]), gluconate (used in the preparation of ^{99m}Tc-tetrofosmin), and tartrate (used in the preparation of ^{99m}Tc-mertiatide [^{99m}Tc-MAG3]). Because of the relative slowness of the exchange reactions, incubation times of up to 30 minutes are required for the preparation of ^{99m}Tc-bicisate (^{99m}Tc-ECD), ^{99m}Tc-tetrofosmin, ^{111}In-capromab pendetide, ^{111}In-ibritumomab tiuxetan, and ^{111}In-pentetreotide. Heating is used in the preparation of ^{99m}Tc-mertiatide (^{99m}Tc-MAG3) and ^{99m}Tc-sestamibi to cleave off protective side groups and free up complexation binding sites, as well as promote the exchange reaction. Inadequate incubation times may result in larger amounts of radiochemical impurities in the form of residual intermediates.

On the other hand, excessive incubation times or time delays between preparation steps can produce radiochemical impurities. For example, suboptimal radiochemical purity of ^{99m}Tc-mertiatide (^{99m}Tc-MAG3) may result if there is excessive time delay before addition of air to the vial or before the boiling step. An excessive time delay before adding the methylene blue stabilizer to a freshly reconstituted vial of ^{99m}Tc-exametazime (^{99m}Tc-HMPAO) may result in decreased radiochemical purity. In these examples, the excessive delays allow radiolytic effects to proceed and produce radiochemical impurities. In the preparation of ^{99m}Tc-sulfur colloid, an excessive delay before the boiling step may result in ^{99m}Tc-edetate (^{99m}Tc-EDTA) impurity subsequent to reduction of pertechnetate by hydrochloric acid.

Component Concentration

For several radiopharmaceuticals, improper component concentration can result in radiochemical impurities. Most reagents kits are optimized to provide proper component concentrations, but deviation from manufacturer preparation instructions can result in decreased radiochemical purity due to altered component concentration. For example, suboptimal radiochemical purity may occur when an excessive volume of pertechnetate or normal saline is used to prepare ^{99m}Tc-albumin aggregated (^{99m}Tc-MAA), ^{99m}Tc-bicisate (^{99m}Tc-ECD), and ^{99m}Tc-exametazime (^{99m}Tc-HMPAO).

The rate and extent of radiolabeling antibodies with ^{111}In is affected by incubation volume. Use of an excessive volume of buffer during the radiolabeling step can result in suboptimal radiolabeling of ^{111}In-capromab pendetide and ^{111}In-ibritumomab tiuxetan. Similarly, use of an excessive volume of buffer during the radiolabeling step can result in suboptimal radiolabeling of ^{90}Y-ibritumomab tiuxetan. Suboptimal radiolabeling of ^{111}In-pentetreotide can result from excessive incubation volume due to addition of excessive volumes of ^{111}In-chloride or normal saline.

The rate and extent of radiolabeling red blood cells with ^{99m}Tc are affected by erythrocyte concentration. Unexpectedly low labeling efficiencies with the in vivo ^{99m}Tc red blood cell labeling procedure may occur in patients with severely low hematocrits. Substandard labeling efficiencies with the in vitro (UltraTag) red blood cell labeling procedure may also occur, related to low erythrocyte concentration in the reaction vial due to inadequate volume of the blood sample or excessive volume of pertechnetate solution.

Similarly, radiolabeling of white blood cells is affected by leukocyte concentration. Suboptimal radiolabeling with ^{99m}Tc-exametazime (^{99m}Tc-HMPAO) or ^{111}In-oxine can occur when an inadequate number of leukocytes is labeled, either because the patient has an abnormally low white cell count or because an inadequate volume of blood is collected. Decreased radiolabeling can also be caused by low leukocyte concentration in the reaction medium due to addition of a large volume of ^{99m}Tc-exametazime (^{99m}Tc-HMPAO).

Excessive component concentration can lead to radiochemical impurities in ^{99m}Tc-tetrofosmin. In addition to concentration-related radiolytic effects (discussed below), high concentrations of tetrofosmin can produce radiochemical impurities via reactions in which tetrofosmin itself acts as a reducing agent.

Excessive concentration of ^{99m}Tc-disofenin (^{99m}Tc-DISIDA) can result in cloudiness due to precipitation of poorly soluble disofenin.

Radiolytic Effects

Although radiolytic oxidative effects are of primary concern for most traditional ^{99m}Tc radiopharmaceuticals, radiolytic reduction is a key mechanism for radiochemical impurity production in several newer ^{99m}Tc radiopharmaceuticals. Aqueous electrons and hydrogen free radicals formed from radiolysis of water are capable of reducing many metal ions and metal complexes. For example, in the preparation of ^{99m}Tc tetrofosmin, these species can reduce ^{99m}Tc(V) in the desired ^{99m}TcO$_2$(tetrofosmin)$_2$ complex to produce other unwanted complexes such as ^{99m}Tc(IV)tetrofosmin, ^{99m}Tc(III)Cl$_2$(tetrofosmin)$_2$, and ^{99m}Tc(I)tetrofosmin$_3$. This undesired further reduction can be minimized by avoiding high radioactive concentration (i.e., minimizing production of radiolytic reductants) and by purposeful addition of oxygen to interact with these reductive species as they are formed.

Similarly, radiolytic reductants can reduce the desired ^{99m}Tc(V)mertiatide (MAG3) to other unwanted complexes such as ^{99m}Tc(IV)mertiatide. Hence, avoidance of high radioactive concentrations and purposeful addition of air (oxygen) are important methods for limiting the formation of radiochemical impurities.

Recommendations

In order to minimize the above problems associated with preparative manipulations, a number of precautions are generally recommended. Whenever possible,

1. Strictly follow specified mixing order of components.
2. Strictly follow specified heating procedure.
3. Strictly follow specified incubation times and avoid time delays between steps.
4. Strictly follow specified amount/volume/dilution instructions to maintain the appropriate component concentrations.
5. Strictly follow specified activity/volume/dilution instructions to maintain the appropriate radioactivity concentrations and minimize radiolytic effects.

Common Problems Associated with Drug Stability and Delays before Dispensing

Radiopharmaceuticals are rarely dispensed immediately after preparation. Rather, they are typically stored for some period of time before a patient dose is needed. During this storage, even if it is only for a relatively short time, radiochemical impurities may be produced, related to problems with stability. Also, there may be sedimentation of particles or interactions with the container wall over time.

Aluminum Ion

In sufficient concentrations, ionic aluminum can chemically interact with several radiopharmaceuticals. For example, aluminum ion can interact with ^{99m}Tc-diphosphonates to produce a colloidal precipitation that is phagocytized by the liver, with ^{99m}Tc-pentetate (^{99m}Tc-DTPA) to cause dissociation and thereby alter glomerular filtration measurements, with ^{99m}Tc-sodium pertechnetate to cause complexation and thereby alter thyroid uptake and blood pool retention, with ^{99m}Tc-red blood cells to cause agglutination, and with ^{99m}Tc-sulfur colloid to produce flocculent particles that embolize in the lungs. Aluminum ion can also interact with ^{111}In-oxine to result in precipitation and can interfere with the radiolabeling of ^{111}In-pentetreotide.

Aluminum ion contamination from generator breakthrough (i.e., from the alumina column in ^{99}Mo–^{99m}Tc generators) was a frequent problem in the past, but improvements in manufacturing processes have essentially solved generator breakthrough concerns. Another source of aluminum contamination was leaching of aluminum ion from aluminum needle hubs. This problem has been effectively obviated by using needles that have hubs composed of either plastic or stainless steel. Yet another potential source of aluminum ion is leaching from the glass of certain suppliers' vials; hence, only glass vials proven compatible should be used for radiopharmaceuticals that may be affected by aluminum. In the special case of sulfur colloid kits, the manufacturer has included edetate (EDTA) in the formulation to chelate any aluminum ion before it can cause flocculation of the ^{99m}Tc-sulfur colloid particles (see Figures 10-24 and 10-25, Chapter 10).

Radiolytic Effects

Radiolytic decomposition occurs with nearly all radiopharmaceuticals. Most radiolytic effects are caused by free radicals, which are formed as the result of radiolysis of water. Because the production of free radicals is related to energy transfer from ionizing radiation, this problem is more pronounced in products that contain a high amount of activity or a high radioactive concentration. Particulate emissions (e.g., beta or positron decay) also enhance free radical formation because of their high linear energy transfer. Products possessing relatively weak coordination complexation bonds are especially susceptible to radiolytic decomposition.

Oxidative decomposition is a primary concern with most ^{99m}Tc radiopharmaceuticals. Antioxidants, especially ascorbic acid and gentisic acid, have proven useful for inhibiting radiolytic decomposition. For example, Figure 10-6 in Chapter 10 shows the marked stabilization offered by the addition of ascorbic acid to ^{99m}Tc bone agents.

Reductive decomposition, on the other hand, is a greater concern for ^{99m}Tc-exametazime (^{99m}Tc-HMPAO). The primary lipophilic complex is readily converted to a secondary hydrophilic complex via nucleophile-induced polymerization. This process, mediated in large part by radiolytic reductive free radicals, can be minimized by the presence of methylene blue, which functions as an oxidizing agent and free radical scavenger.

High activity concentrations of PET radiopharmaceuticals may demonstrate significant radiolytic decomposition over time, primarily related to free radicals produced from positron interactions in water. Antioxidants or free radical scavengers have demonstrated utility in stabilizing several PET radiopharmaceuticals. For example, ^{18}F-fludexoyglucose (^{18}F-FDG) can be effectively stabilized with ethanol, a free radical scavenger.[7]

The rate of radiolytic decomposition is also affected by temperature. Lower temperatures will slow the rate of free radical diffusion and chemical reactions. Storage in a refrigerator rather than at room temperature may thus improve the stability of many radiopharmaceuticals. Maintenance in a frozen state (e.g., shipping on dry ice and storage in a freezer) is frequently required to ensure stability of susceptible products that contain high amounts of activity and are particulate emitters; examples include the therapeutic products ^{153}Sm-lexidronam (^{153}Sm-EDTMP; Quadramet [EUSA Pharma]) and ^{131}I-tositumomab (Bexxar [GlaxoSmithKline]).

Radiolytic decomposition is generally a time-related process. That is, greater amounts of radiochemical impurities are produced over time as a result of radiolytic decomposition. The manufacturer's recommended expiration time is often based, in part, on radiolytic decomposition effects. It appears obvious, therefore, that prolonged storage should be avoided in order to minimize impurities produced by radiolytic decomposition.

Sedimentation of Particles

Particulate radiopharmaceuticals, notably ^{99m}Tc-albumin aggregated (^{99m}Tc-MAA), tend to settle with time. The rate of sedimentation is variable and depends in large part on the particular manufactured product. Failure to resuspend sedimented particles immediately before withdrawal of a dose may result in unexpectedly low radioactivity in the calculated volume for the dose, a somewhat higher percentage of ^{99m}Tc-pertechnetate impurity in the dose, and potentially an inadequate number of particles for optimal lung imaging. Resuspension is best accomplished by gently inverting the vial several times; vigorous agitation or shaking will result in the formation of foam in the vial.

Sedimentation can similarly occur with radiolabeled cell preparations, such as ^{99m}Tc-red blood cells, ^{99m}Tc-leukocytes, and ^{111}In-leukocytes. Particulate intravascular brachytherapy sources, such as ^{90}Y-microspheres, also demonstrate rapid sedimentation.

Interaction with Container Walls

Some radiopharmaceuticals tend to adsorb over time to the inner surface of glass vials, resulting in somewhat lower radioactivity in the calculated volume for a dose and a slightly higher percentage of ^{99m}Tc-pertechnetate impurity in the dose. Radiopharmaceuticals prone to this phenomenon include ^{99m}Tc-albumin aggregated (^{99m}Tc-MAA), ^{99m}Tc-sestamibi, ^{99m}Tc-succimer (^{99m}Tc-DMSA), and ^{99m}Tc-sulfur colloid. ^{99m}Tc-tetrofosmin also tends to adsorb to the glass walls and rubber stopper of storage vials, with increased adsorption related to storage time, contact between the solution and the rubber stopper, agitation, and low concentration of tetrofosmin.

Recommendations

In order to minimize the above problems associated with drug stability and delays before dispensing, a number of precautions are generally recommended. Whenever possible,

1. Minimize exposure of susceptible radiopharmaceuticals to aluminum ion that may be leached from generator columns, aluminum needle hubs, and certain glass containers.
2. Minimize radiolytic decomposition by using strategies such as diluting unnecessarily high radioactive concentrations to lower concentrations; adding antioxidants, oxidants, or free-radical scavengers as applicable; storing at cold temperatures; and avoiding prolonged storage times.
3. Minimize sedimentation problems inherent to particulate or cellular radiopharmaceuticals by gently resuspending the particles or cells immediately prior to handling.
4. Minimize adsorption of susceptible radiopharmaceuticals to vial walls and stoppers by optimizing the concentration and by avoiding prolonged storage, contact with the rubber vial stopper, and excessive agitation.

Summary

A wide variety of radiopharmaceutical formulation problems have been observed. Common preparation problems and their resulting radiochemical impurities have been highlighted in this section. An understanding of the potential causes of common radiochemical impurities will enable practitioners to formulate and follow strategies for avoiding or minimizing these impurities. Moreover, it may be useful to apply this information in the exploration of potential problems likely to be encountered with new radiopharmaceuticals or when deviating from package insert instructions for preparation of established radiopharmaceuticals. In spite of best efforts, preparation problems that are detected by quality control testing or by altered biodistribution patterns will occasionally occur. These should be monitored closely and documented by the health care professionals involved. Product-related problems should be reported to the manufacturer and to appropriate regulatory agencies (e.g., the Food and Drug Administration [FDA] MedWatch medical products reporting program) and, as appropriate, disseminated in professional communications. The widespread reporting of such problems in a timely manner will contribute to improved safety and efficacy of radiopharmaceuticals.

PEDIATRIC DOSING OF RADIOPHARMACEUTICALS

Children are not small adults! Body constituents, organ masses, and organ function are far different at birth than in adulthood; they change over time with growth and maturation. For example, extracellular water constitutes 45% of a newborn's body weight but 20% of an adult's body weight. Brain organ mass is a much higher proportion of body weight in infants than in adults. Glomerular filtration function is low at birth but rapidly increases over the first few months of life, whereas thyroid function is high in infants and slowly decreases over several years.

Radiation absorbed doses (i.e., absorbed energy per unit mass) for a given amount of radioactivity are higher in children than in adults because the radioactivity is concentrated in a smaller mass (Table 16-1). Radiation absorbed doses per unit of administered radioactivity received by newborns are higher by approximately an order of magnitude compared with adults. Hence, reduction of the administered radioactivity dose (hereafter referred to simply as dose) in pediatric patients is appropriate in order to limit the radiation absorbed dose. Adjustment of pediatric doses as described below typically allows high-quality imaging with radiation absorbed doses at values similar to those received by adults.

Selection of pediatric doses should be based on the general principle of administering the lowest activity that will result in a satisfactory procedure. Larger-activity doses will deliver unnecessarily high radiation absorbed doses without additional diagnostic benefit, whereas lower-activity doses may result in inadequate (i.e., nondiagnostic) procedures in which the delivered radiation absorbed dose was for naught.

Unfortunately, most radiopharmaceutical package inserts do not provide any information on pediatric dosing. Moreover, in the few package inserts that do include some pediatric dosing information, the information is often out of date in regard to changes in imaging equipment, analysis, or use that have occurred over time. Hence, practitioners have relied largely upon experience, anecdotal accounts, and recommendations from colleagues for their selections of pediatric doses. Recently, a work group developed consensus guidelines on pediatric dosing of 11 commonly used radiopharmaceuticals (http://interactive.snm.org/docs/Pediatric_dose_consensus_guidelines_Final_2010.pdf).

Minimum Dose

The minimum dose can be described as the smallest administered activity that is needed to achieve a satisfactory procedure regardless of dose adjustment calculation. This is important because the dose determined by a dose adjustment calculation may be too low for performing a procedure of adequate quality. This issue is most frequently encountered with newborns and infants. In these situations, a preestablished minimum dose should be used.

Minimum doses are quite variable, not only for different radiopharmaceuticals and procedures but in relation to different equipment, analysis, and patient factors.[11-14] For example, dynamic imaging procedures typically require higher doses than do static imaging procedures. Examples of some other imaging-related factors that affect minimum pediatric dose include type of imaging (planar versus SPECT), number of detectors (single versus multiple), type of collimator (sensitivity, resolution), particular reconstructive algorithm, data analysis, and image display. The ability of the infant to remain motionless for the duration of image acquisition also influences the minimum dose. For example, if an infant requires general anesthesia to control motion, a larger minimum dose might be appropriate in order to shorten the time of image acquisition and thus shorten the time and associated risks of anesthesia. Larger minimum doses may also be reasonable in cases where concerns for long-term stochastic radiation effects are moot, such as in infants with terminal diseases or probable brain death.

TABLE 16-1 Radiation Absorbed Doses (mrad/mCi) at Various Ages for Selected Radiopharmaceuticals

Radiopharmaceutical	Organ[a]	Newborn	1 yr	5 yr	10 yr	15 yr	Adult
^{67}Ga-citrate	Spleen	7.7	2.6	1.6	0.95	0.70	0.63
	ED	4.8	2.0	1.0	0.63	0.40	0.33
^{99m}Tc-MAA	Lung	3.1	1.0	0.59	0.35	0.26	0.20
	ED	0.59	0.23	0.13	0.09	0.06	0.04
^{99m}Tc-pyrophosphate	Bone	0.65	0.19	0.14	0.09	0.06	0.05
	ED	0.20	0.08	0.04	0.03	0.02	0.02
^{99m}Tc-sodium pertechnetate	Large intestine	2.0	0.7	0.4	0.3	0.2	0.2
	ED	0.81	0.32	0.17	0.10	0.07	0.05
^{99m}Tc-pentetate	Kidneys	0.39	0.15	0.10	0.07	0.05	0.04
	ED	0.14	0.06	0.03	0.03	0.02	0.02
^{99m}Tc-sulfur colloid	Liver	2.9	1.3	0.82	0.56	0.42	0.34
	ED	0.41	0.18	0.10	0.07	0.04	0.03
^{111}In-pentetreotide	Kidneys	18.5	7.8	4.4	3.1	2.3	1.9
	ED	2.9	1.3	0.74	0.48	0.34	0.27
^{123}I-sodium iodide	Thyroid	160	109	51	30	21	15
	ED	12	8.1	4.4	1.9	1.4	0.85

[a] ED = effective dose.
Source: Data extracted from references 8 through 10. For a more complete listing of radiation absorbed doses from radiopharmaceuticals, see www.doseinfo-radar.com/RADAR-INT-NM.html.

Examples of minimum pediatric doses can be found in various references.[10,11,14–16] Rather large variations in minimum doses, even among premier pediatric nuclear medicine facilities, suggest the need for the nuclear medicine community to establish pediatric dose guidelines.[11] Nonetheless, the particular minimum dose for an individual infant will still depend on the specific imaging-related and patient-related factors described above.

Age-Based Dose Adjustment

In the past, dose adjustments for pediatric patients were often calculated by age. Two common age-based adjustments were Young's rule and the modified Young's rule, which use the following formulas:

$$\text{Pediatric dose} = \frac{\text{Child age (yr)}}{\text{(Child age} + 12)} \times \text{Adult dose} \quad \text{(Young's rule)}$$

$$\text{Pediatric dose} = \frac{\text{Child age (yr)} + 1}{\text{Child age (yr)} + 7} \times \text{Adult dose} \quad \text{(Modified Young's rule)}$$

Young's rule generally follows body weight relationships for children from about 2 years up to about 10 years old, whereas the modified Young's rule gives a closer empiric fit to the body surface area (BSA) relationship.[8,10] Age-based adjustments currently are not recommended because they are susceptible to substantial error in the adjustment of doses for children with abnormal physiques.[10]

Weight-Based Dose Adjustment

For most drugs, pediatric doses are calculated on a milligram per kilogram (mg/kg) basis (i.e., proportional to body weight). Radiopharmaceutical doses can be similarly adjusted for pediatric patients on a weight basis, such as in millicuries per kilogram (mCi/kg). Dose adjustment proportional to weight is the easiest and most commonly used method for determining radiopharmaceutical doses in pediatric patients.[8,11,12,14] The weight-based dose is calculated using the following formula, also referred to as Clark's rule:

$$\text{Pediatric dose} = \frac{\text{Child body weight (kg)}}{\text{Standard adult weight (70 kg)}} \times \text{Adult dose}$$

Weight-based adjustment for pediatric radiopharmaceutical doses is appropriate for most nuclear medicine procedures because, except for the brain and testes, the mass of major organs stays reasonably proportional to body weight during growth.[8,10,11,12]

Weight-based radiopharmaceutical doses calculated for infants may result in quite small doses of administered radioactivity, sometimes falling below the amount needed to perform a satisfactory imaging procedure. Hence, nuclear medicine facilities using this dose adjustment method should establish minimum doses for each radiopharmaceutical and procedure as appropriate.

Dose Adjustment Based on Body Surface Area

In nuclear medicine imaging, emitted photons are detected by the camera in two dimensions. Hence, maintaining a constant activity per unit area can be used as the basis for dose adjustment.[8] Because area is related to a dimension squared (i.e., x^2) and volume is related to a dimension cubed (i.e., x^3), the radioactivity per unit surface area is equal to the radioactivity in the volume raised to the two-thirds power (i.e., $v^{2/3}$ or $v^{0.667}$). Assuming that the density of major organs is approximately 1 (i.e., volume = mass), the pediatric dose can be calculated using the following formula:[8]

$$\text{Pediatric dose} = \frac{(\text{Child organ mass})^{0.667}}{(\text{Standard adult organ mass})^{0.667}} \times \text{Adult dose}$$

Remember those old "fortunately–unfortunately" stories? Unfortunately, specific organ masses in individual patients are not readily known. Fortunately, (organ mass)$^{2/3}$ pretty closely follows BSA, so the dose may be adjusted using the following formula:

$$\text{Pediatric dose} = \frac{\text{Child BSA (m}^2)}{\text{Standard adult BSA (1.73 m}^2)} \times \text{Adult dose}$$

Unfortunately, BSA cannot be readily measured in individual patients. Fortunately, BSA can be closely estimated by using a height–weight nomogram[8] or by calculation using one of the following formulas incorporating height and weight:[17]

DuBois formula: $\quad \text{BSA (m}^2) = \text{kg}^{0.425} \times \text{cm}^{0.725} \times 0.007184$

Mosteller formula: $\quad \text{BSA (m}^2) = \sqrt{\dfrac{\text{cm} \times \text{kg}}{3600}}$

or $\quad \text{BSA (m}^2) = \sqrt{\dfrac{\text{in} \times \text{lb}}{3131}}$

For children with relatively normal height–weight relationships, BSA can also be estimated from weight alone (Table 16-2).

Although somewhat less commonly used than the weight-based method, the BSA-based method for pediatric dose adjustment is appropriate for most nuclear procedures. BSA is closely related to organ size, cardiac output, glomerular filtration rate, plasma volume, and extracellular fluid volume.[12] Establishment of minimum doses is less frequently necessary with the BSA-based method because of the somewhat higher calculated doses, especially in infants, than with the weight-based method.[10]

Height-Based Dose Adjustment

Height-based dose adjustment for pediatric radiopharmaceutical doses has been suggested for two nuclear medicine procedures, namely, dynamic renal imaging and gated blood pool cardiac imaging, because weight-based or BSA-based adjust-

TABLE 16-2 Body Surface Area (BSA) for Children of Normal Height–Weight Relationship and Corresponding Fraction of Adult Dose

Weight (kg)	BSA (m²)	Fraction of Adult Dose[a]
2	0.16	0.09
4	0.25	0.14
6	0.33	0.19
8	0.40	0.23
10	0.47	0.27
12	0.54	0.31
14	0.61	0.35
16	0.67	0.39
18	0.74	0.43
20	0.80	0.46
22	0.85	0.49
24	0.90	0.52
26	0.95	0.55
28	1.00	0.58
30	1.05	0.61
32	1.10	0.64
34	1.15	0.67
36	1.20	0.70
38	1.25	0.72
40	1.30	0.75
42	1.35	0.78
45	1.40	0.81
47	1.45	0.84
50	1.50	0.87
55	1.55	0.90
60	1.60	0.92
64	1.65	0.95
68	1.70	0.98
70	1.73	1.00

[a] Calculated as child BSA (m²) ÷ adult BSA (1.73 m²).
Source: Derived from reference 15 and reference 18 (West nomogram).

ments may yield doses lower than those required to obtain good quality studies.[10,13] For procedures other than these two, however, doses calculated by a height-based method yield inappropriately high values.[10] When deemed appropriate, height-based dose adjustments can be calculated using the following formula:

$$\text{Pediatric dose} = \frac{\text{Child height (cm)}}{\text{Standard adult height (176 cm)}} \times \text{Adult dose}$$

Special Case: ^{99m}Tc-MAA

Microembolization of lung capillaries and small arterioles is a potential safety concern in performing perfusion lung imaging with ^{99m}Tc-MAA. The numbers of pulmonary arterioles and alveoli at birth are only about 1/10 of the numbers in adults, increasing to about 1/3 at 1 year, 1/2 at 3 years, and near adult values at 8 years of age. Hence, it has been recommended that the number of MAA particles injected not exceed 50,000 in newborns, 150,000 in 1-year olds, and 300,000 in 5-year olds.[19,20] Several methods have been described for preparing ^{99m}Tc-MAA doses with these restricted numbers of particles.[21,22]

Special Case: Radiopharmaceuticals Containing Benzyl Alcohol

Benzyl alcohol is an aromatic alcohol often used as a bacteriostatic preservative in multiple-dose vials of products intended for parenteral injection. Benzyl alcohol is normally oxidized rapidly to benzoic acid, conjugated with glycine in the liver, and excreted as hippuric acid. However, this metabolic pathway may not be well developed in newborns. Insufficient conjugation of benzoic acid by an immature liver may lead to its accumulation and thereby result in metabolic acidosis. Fatal "gasping syndrome" has been associated with the use of benzyl alcohol in neonates, particularly in premature infants and infants of low birth weight.[23]

Selected radiopharmaceuticals that contain benzyl alcohol include gallium citrate Ga 67 injection (^{67}Ga-citrate), thallous chloride Tl 201 injection (^{201}Tl-chloride), sodium chromate Cr 51 injection (^{51}Cr-sodium chromate), iodinated I 125 albumin injection (^{125}I-albumin), and iobenguane I 123 injection (^{123}I-iobenguane, or ^{123}I-MIBG). Before use of these radiopharmaceuticals in neonates and infants, benzyl alcohol content should be carefully considered. For example, in a clinical trial performed at the author's institution, the amount of radiopharmaceutical administered to infants was limited on the basis of benzyl alcohol content so that the administered dose would deliver no more than 1 mg of benzyl alcohol per kilogram of body weight. Regardless of the amount, infants injected with products containing benzyl alcohol should be observed for signs and symptoms of benzyl alcohol toxicity, such as gasping respirations, seizures, bradycardia, and hypotension.

Special Case: PET/CT

Pediatric doses of positron emission tomography (PET) radiopharmaceuticals, predominantly ^{18}F-fludeoxyglucose (^{18}F-FDG), are commonly adjusted by weight.[24,25] However, studies of noise-specific count rate density, a surrogate of image quality, show that substantial reductions in the administered dose of ^{18}F-FDG and in image acquisition times are possible without affecting quality, especially for lighter-weight children.[24] Injection rules have been developed for one particular PET/computed tomography (CT) scanner, based on a posteriori noise-specific count rate density curves from one group of pediatric patients;[24] the method can be applied to other scanners to derive similar dose regimens. Such dosing rules more closely correlate with weight than with height, girth, or body mass index.[24]

Nearly all PET imaging procedures are now performed using hybrid PET/CT scanners, in which the CT is used for attenuation correction and coregistration of functional PET images with anatomical CT images. Because a substantial portion of the total procedure radiation absorbed dose is delivered from the CT, it is important to also adjust the CT technique for pediatric patients. A protocol has been developed for PET/CT imaging in pediatric patients, which strives to optimize image quality while minimizing radiation exposure.[25]

Summary

In summary, pediatric doses of radiopharmaceuticals are generally adjusted by body weight or by BSA. Many imaging procedures in newborns and infants may require administration of a minimum dose of the radiopharmaceutical. Pediatric dose adjustments typically allow high-quality imaging with radiation absorbed doses at values similar to those received by adults.

BREAST MILK EXCRETION OF RADIOPHARMACEUTICALS

Performance of a nuclear medicine procedure in a breast-feeding woman, although an infrequent event, typically causes concern about excretion of the radiopharmaceutical in breast milk and ingestion by the nursing infant. In addition to ethical concerns relating to radiation exposure of the infant, regulatory concerns arise regarding compliance with rules and proper record keeping.

All women of lactating potential should be asked about breast-feeding prior to a nuclear medicine procedure.[26–28] If a woman is breast-feeding, careful consideration should be given to the medical necessity of the nuclear medicine procedure (i.e., could a nonnuclear procedure be performed instead?) and to its urgency (i.e., could it be electively delayed until the child is weaned?).[27,29,30] For nuclear medicine procedures that are performed on breast-feeding women, the activity administered should be the smallest amount possible to obtain a satisfactory result.[26,27,29] If a choice between different radiopharmaceuticals is feasible (e.g., ^{67}Ga-citrate versus ^{111}In-leukocytes for infection imaging), selection of the radiopharmaceutical that is less excreted in milk or gives a lower radiation dose to the infant is warranted.[26,31]

Excretion in Breast Milk

Lactation becomes fully established within the first week after giving birth. For the first few days, colostrum is secreted, a fluid high in protein derived from the mother's plasma protein. Over the next week, the protein content rapidly falls while the fat content increases to achieve mature milk. Mature milk is an emulsion composed of 95% to 97% water and 3% to 5% lipid; it includes about 7% dissolved carbohydrates (predominantly lactose), about 1% proteins (mostly α-lactalbumin with lesser amounts of immunoglobulins and casein), and mineral electrolytes.[32,33] The mammary epithelium serves as a semipermeable lipid membrane.

Drugs in the mother's blood may be excreted into breast milk, depending on many factors.[30,32] Accumulation in breast milk due to passive diffusion along a concentration gradient is likely, especially for drugs with molecular weight of 200 or less, but such excretion generally results in relatively low concentrations. Higher concentrations in milk generally occur with drugs that are actively secreted by ion or other transport pathways. On the basis of pKa/ionization, weak bases can be concentrated in milk because the pH is slightly lower than that of plasma. Lipid-soluble drugs can concentrate in milk fat, while certain other drugs can concentrate in milk via protein binding.

There is wide variation, ranging over several orders of magnitude, in the percent excretion of radiopharmaceuticals in breast milk (Table 16-3).[27,29,30,34–46] Reasons for this variability include biologic differences in milk production among women, different stages of lactation, different sampling times, different kinetic modeling or extrapolation, different uptake in mother's target organs (e.g., thyroid uptake of radioiodide), and different radiochemical purity.[26,29,30]

With few exceptions, very small fractions of administered doses of radiopharmaceuticals are excreted in milk. Only ^{99m}Tc-sodium pertechnetate, ^{131}I-sodium iodide, and ^{67}Ga-citrate may exhibit cumulative excretions >10%.[27,29,30,33] Pertechnetate and iodide ions are actively secreted into milk, as are other ions such as chloride and bicarbonate, with concentrations up to 30 times higher than that in plasma.[30] ^{67}Ga-citrate can achieve similar concentrations in milk, about 10 to 30 times higher than that in plasma.[47] ^{67}Ga, however, is probably secreted as gallate ion, $[Ga(OH)_4]^-$, via the PO_4^- pathway and, once in the milk, is highly bound to lactoferrin.[27,30,47] Lactoferrin has an affinity for gallium approximately 90 times greater than does plasma transferrin.[47]

Length of Time to Interrupt Breast-Feeding

It is generally agreed that a reasonable dose criterion for interrupting breast-feeding is 0.1 rem effective dose (or effective dose equivalent) to the infant.[27,29,30,34] For several radiopharmaceuticals, there is substantial variation in recommendations for interruption (Table 16-4).[27–30,34,35,37,38,40–46,48] The reasons for this variability, in addition to differences in excretion in milk as discussed above, are related to different assumptions used in the dose calculations, including chemical form, absorption and biodistribution in the infant, infant dosimetry models and dose conversion factors, whether or not external radiation exposure from close contact was considered, and radionuclidic purity.[29,30]

For many radiopharmaceuticals, especially those with low excretion fractions, interruption of breast-feeding is not necessary (Table 16-4). However, given that most excretion in milk

TABLE 16-3 Radiopharmaceutical Breast Milk Excretion Data[a]

Radiopharmaceutical	Fraction (%) of Administered Dose Excreted in Milk	Biologic Half-time Excretion in Milk	Administered Dose (mCi) Requiring Instructions	Administered Dose (mCi) Requiring Records
^{18}F-fludeoxyglucose (^{18}F-FDG)	<1	NL	NL	NL
^{32}P-sodium phosphate	NL	35 days	NL	NL
^{67}Ga-citrate	3.2–16	20–537 hr	0.04	0.2
^{99m}Tc-disofenin (^{99m}Tc-DISIDA)	0.1–0.28	9–10 hr	30	150
^{99m}Tc-exametazime (^{99m}Tc-HMPAO)	0.15	5 hr	NL	NL
^{99m}Tc-leukocytes	0.7–1.6	7–12 hr	4	15
^{99m}Tc-medronate (^{99m}Tc-MDP)	0.01–2.7	8–34 hr	30	150
^{99m}Tc-mertiatide (^{99m}Tc-MAG3)	0.7–1.6	7–12 hr	30	150
^{99m}Tc-albumin aggregated (^{99m}Tc-MAA)	0.2–6	4–54 hr	1.3	6.5
^{99m}Tc-oxidronate (^{99m}Tc-HDP)	0.02	8–34 hr	NL	NL
^{99m}Tc-pentetate (^{99m}Tc-DTPA)	0.01–7.9	6–30 hr	30	150
^{99m}Tc-pentetate (^{99m}Tc-DTPA) aerosol	NL	NL	30	150
^{99m}Tc-sodium pertechnetate	0.13–33	3–66 hr	3	15
^{99m}Tc-pyrophosphate (^{99m}Tc-PYP)	0.15–0.44	7–8 hr	25	120
^{99m}Tc-red blood cells, in vivo labeling	0.006–1.0	7–10 hr	10	50
^{99m}Tc-red blood cells, in vitro labeling	0.02–0.03	8–9 hr	30	150
^{99m}Tc-sestamibi	0.01–0.03	6–23 hr	30	150
^{99m}Tc-succimer (^{99m}Tc-DMSA)	0.03	354 hr	NL	NL
^{99m}Tc-sulfur colloid	0.16–1.5	8–35 hr	7	35
^{111}In-leukocytes	0.12–1.0	10–140 hr	0.2	1
^{111}In-pentetate (^{111}In-DTPA)	0.015–0.2	10–37 hr	NL	NL
^{111}In-pentetreotide	0.025	12 hr	NL	NL
^{111}In-platelets	0.019	NL	NL	NL
^{123}I-sodium iodide	2.6	10 hr	0.5	3
^{123}I-iobenguane (^{123}I-MIBG)	0.35	85 hr	2	10
^{125}I-albumin (^{125}I-HSA)	20	76 hr	NL	NL
^{131}I-sodium iodide	23–46	7–526 hr	0.0004	0.002
^{131}I-iobenguane (^{131}I-MIBG)	0.03	11 hr	NL	NL
^{201}Tl-thallous chloride	0.6	13–362 hr	1	5
^{133}Xe-gas	Trivial	NL	NL	NL
^{90}Y-microspheres	<0.0001	A few hr	NL	NL

[a] NL = not listed, no data reported.
Source: References 27, 29, 30, 34–46.

occurs in the first few hours after administration,[27,30] and to alleviate parents' concerns,[28] a conservative recommendation to discard the first milk produced (e.g., 4 hours) after radiopharmaceutical administration seems reasonable and appropriate. On the other hand, a few radiopharmaceuticals (e.g., ^{131}I-sodium iodide, ^{67}Ga-citrate) require such long periods of interruption that maintenance of lactation is difficult or impractical and thus complete cessation is advised. An interesting case is ^{123}I-sodium iodide; if radionuclidicly pure, interruption is recommended for only a brief time (e.g., 5–24 hours), whereas prolonged interruption (23–106 days) or complete cessation is recommended for products that contain ^{124}I or ^{125}I impurities.[27,30,48]

Because excretion fractions of radiopharmaceuticals in milk are variable between patients, and even between different times for the same patient, literature data may not be predictive for the individual case. Hence, for an individualized determination, the patient's breast milk can be periodically pumped, milk samples counted for radioactivity concentration, and calculations performed to establish the time when it would be safe to resume breast-feeding.[26–30] For calculation purposes, milk output (and infant ingestion) can be assumed to be 142 mL

TABLE 16-4 Recommended Times to Interrupt Breast-Feeding[a]

Radiopharmaceutical	34	30	28	29	27	Other
^{18}F-fludeoxyglucose (^{18}F-FDG)	NL	NL	NL	NL	NL	None[37,b]
^{32}P-sodium phosphate	NL	NL	NL	NL	C	
^{67}Ga-citrate	1–4 wk	C	C	1 mo	18 days	
^{99m}Tc-disofenin (^{99m}Tc-DISIDA)	None	None	None	None	NL	
^{99m}Tc-exametazime (^{99m}Tc-HMPAO)	NL	NL	NL	NL	NL	None[44]
^{99m}Tc-leukocytes	12–24 hr	48 hr	NL	NL	NL	
^{99m}Tc-medronate (^{99m}Tc-MDP)	None	None	None	NL	None	4 hr[35]
^{99m}Tc-mertiatide (^{99m}Tc-MAG3)	None	None	None	NL	NL	7 hr[40]
^{99m}Tc-albumin aggregated (^{99m}Tc-MAA)	12 hr	12 hr	12 hr	NL	6 hr	12 hr[35]
^{99m}Tc-oxidronate (^{99m}Tc-HDP)	NL	NL	None	NL	None	
^{99m}Tc-pentetate (^{99m}Tc-DTPA)	None	None	12 hr	6–32 hr	None	4 hr[35]
^{99m}Tc-pentetate (^{99m}Tc-DTPA) aerosol	None	None	NL	NL	NL	
^{99m}Tc-sodium pertechnetate	12–24 hr	4 hr	>24 hr	30–62 hr	36 hr	12 hr[35]
^{99m}Tc-pyrophosphate (^{99m}Tc-PYP)	None	None	12 hr	8 hr	NL	
^{99m}Tc-red blood cells, in vivo labeling	6 hr	12 hr	12 hr	NL	13 hr	4 hr[35]
^{99m}Tc-red blood cells, in vitro labeling	None	None	12 hr	None[b]	NL	
^{99m}Tc-sestamibi	None	None	None[b]	None[b]	NL	
^{99m}Tc-succimer (^{99m}Tc-DMSA)	NL	NL	None	NL	None	
^{99m}Tc-sulfur colloid	6 hr	None	None	None	NL	
^{111}In-leukocytes	1 wk	None	None	NL	None	
^{111}In-pentetate (^{111}In-DTPA)	NL	NL	NL	NL	NL	None[43]
^{111}In-pentetreotide	NL	NL	NL	NL	NL	None[38,b]
^{111}In-platelets	NL	NL	NL	NL	NL	None[38,b]
^{123}I-sodium iodide	None	C	>24 hr	NL	5 hr	23–106 days[48]
^{123}I-iobenguane (^{123}I-MIBG)	12–24 hr	48 hr	NL	NL	NL	24 hr[41]
^{125}I-iodinated albumin (^{125}I-HSA)	NL	NL	NL	NL	206 hr	
^{131}I-sodium iodide, diagnostic dose	C	C	C	6–80 days	72 days	
^{131}I-sodium iodide, cancer therapy dose	C	C	C	C	C	
^{131}I-iobenguane (^{131}I-MIBG)	NL	NL	NL	NL	NL	14 hr[42]
^{201}Tl-thallous chloride	2 wk	96 hr	C	NL	NL	48 hr[45]
^{133}Xe-gas	NL	None	NL	NL	NL	
^{90}Y-microspheres	NL	NL	NL	NL	NL	None[46]

[a] NL = not listed, no recommendation offered; C = cessation; None = interruption is not necessary, but an interruption of 4 hours may be recommended to assuage parent's concerns.[28]
[b] Close contact should be avoided or restricted to no more than 5 hours in the first 24 hours after radiopharmaceutical administration because of high external dose rate from the mother's chest region.
Source: References 27–30, 34, 35, 37, 38, 40–46, 48.

every 4 hours (850 mL/day)[27,29,30,35] and dose conversion factors (e.g., millirem per millicurie ingested by the infant) can be obtained from published values (Table 16-5).[27,29,30] The variation in dose conversion factor values is due to differences in assumptions for absorption and biodistribution in infants, use of different dosimetry models, and use of different organ weighting factors (i.e., to yield effective dose versus effective dose equivalent). An example of such a calculation is given in the case study below.

Regulatory Requirements

The Nuclear Regulatory Commission (NRC) has established instructions and record-keeping requirements relating to breast-feeding.[34] In 10 CFR 35.75(b), the licensee is required to provide the released individual with instructions, including written instructions, on actions recommended to maintain doses to other individuals as low as is reasonably achievable if the total effective dose equivalent to any other individual is

TABLE 16-5 Dose Conversion Factors (Radiation Absorbed Dose per Unit Activity Ingested by the Infant)[a]

Radioisotope	Newborn (mrem/mCi)[b]	4 kg Infant (mrem/mCi)[b]	4 kg Infant (mrem/mCi)[c]	1-Year-Old (mrem/mCi)[b]
^{18}F	NL	NL	3,150	NL
^{67}Ga	4,400	12,600	15,700	1,181
^{99m}Tc	520	1,040	1,260	229
^{111}In	20,000	21,100	25,200	8,100
^{123}I	10,000	8,510	14,000	7,000
^{131}I	NL	907,000	1,570,000	NL
^{201}Tl	13,000	NL	4,180	7,800

[a] NL = not listed.
[b] Effective dose
[c] Effective dose equivalent (values for effective dose differ from those for effective dose equivalent because of different tissue-weighting factors; see Table 6-2 and discussion in Chapter 6).
Source: References 27, 29, 30.

likely to exceed 0.1 rem. If the dose to a breast-feeding infant or child could exceed 0.1 rem with no interruption of breast-feeding, the instructions are also to include guidance on the interruption or discontinuation of breast-feeding and information on the consequences of failure to follow the guidance (e.g., ^{131}I ingested from breast milk may harm the infant's thyroid). In 10 CFR 35.75(d), the licensee is required to maintain a record, for 3 years after the date of release, that instructions were provided to a breast-feeding woman if the radiation dose to the infant or child from continued breast-feeding could result in a total effective dose equivalent exceeding 0.5 rem. Guidance on determining when instructions and records are required is included in NRC Regulatory Guide 8.39[34] (Table 16-3). For radiopharmaceuticals not listed in the regulatory guide, it is the licensee's responsibility to evaluate whether instructions or records (or both) are required. If information on the excretion of a radiopharmaceutical is not available, an acceptable approach is to assume that 50% of the administered activity is excreted in the milk and calculate the absorbed dose to the infant by using dose conversion factors for a newborn infant.[34]

Other Issues

Radiochemical Purity

Most of the ^{99m}Tc found in breast milk is in the chemical form of pertechnetate.[27,30,35] Similarly, most ^{123}I and ^{131}I found in breast milk is in the chemical form of iodide.[30,35] Hence, administration of a ^{99m}Tc-labeled radiopharmaceutical or a radioiodinated radiopharmaceutical with low radiochemical purity (i.e., a high fraction of free pertechnetate or free iodide impurity) will exhibit greater accumulation in milk.[27,29,30,35,49] For example, the unusually large fraction of ^{99m}Tc excreted in one woman's milk following injection of ^{99m}Tc-pentetate (^{99m}Tc-DTPA), 7.9% versus typical values of 0.05% to 0.24%, was likely due to a high percentage of free ^{99m}Tc pertechnetate in the administered preparation.[29] ^{99m}Tc-red blood cells labeled in vivo exhibit higher amounts of free ^{99m}Tc-pertechnetate in blood and excretion in milk compared with ^{99m}Tc-red blood cells labeled in vitro.[29,49] Different ^{99m}Tc-albumin aggregated (^{99m}Tc-MAA) products may have different initial labeling efficiencies and thus different amounts excreted in milk.[36] Therefore, radiochemical purity testing to verify a high labeling efficiency should be performed immediately before administration of a radiolabeled radiopharmaceutical to a breast-feeding woman.[27,29]

The chemical form of ^{111}In excreted in milk is unknown. However, given its chemical similarity to gallium, it may be bound to lactoferrin in milk. Highly protein-bound ^{111}In present in milk samples obtained shortly after injection is likely related to free ^{111}In residually present in ^{111}In-labeled radiopharmaceuticals such as ^{111}In-leukoyctes.[31]

Metabolism/Decomposition

Most ^{99m}Tc radiopharmaceuticals exhibit low fractional excretion in milk because of their rapid blood clearance or their chemical stability. For these radiopharmaceuticals, milk concentrations peak by about 3 to 4 hours and then decrease monoexponentially.[27,30] ^{99m}Tc-albumin aggregated (^{99m}Tc-MAA) and ^{99m}Tc-red blood cells, however, exhibit slow breakdown, so there is delayed excretion of ^{99m}Tc-pertechnetate in milk.[27,30,35] Peak concentration of activity in breast milk for ^{99m}Tc-MAA and for ^{99m}Tc-red blood cells may not occur until about 5 to 8 hours and 7 to 10 hours, respectively.[27,29,35] After peak, excretion in milk decreases exponentially (Figure 16-1). For ^{99m}Tc-MAA, some variation in the rate and extent of excretion in milk may be related to different rates of breakdown among different commercial products.[36]

^{131}I-sodium iodide exhibits a biexponential clearance and secretion into milk.[26,30,50] Initially, some fraction of free ^{131}I iodide is secreted into milk while some other fraction is taken up in the thyroid. Later, metabolism of radioiodinated thyroid

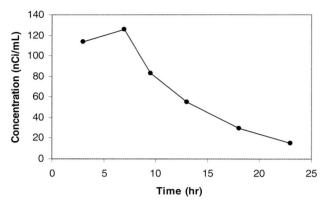

FIGURE 16-1 Concentration of ^{99m}Tc activity excreted in breast milk after an intravenous dose of ^{99m}Tc-MAA. See text for detailed discussion.

hormone provides a slow prolonged release of ^{131}I iodide, a portion of which is secreted into the milk.[30] Similarly, radioiodinated compounds (e.g., ^{125}I-albumin) will be slowly metabolized and liberate free radioiodide, which will be secreted into milk.[27,30]

Concurrent Drugs

The observation that ^{99m}Tc concentrations in breast milk were much lower in women undergoing ^{99m}Tc-sodium pertechnetate brain imaging than in women undergoing ^{99m}Tc-sodium pertechnetate thyroid imaging suggests that potassium perchlorate, which was routinely used to block ^{99m}Tc-sodium pertechnetate uptake in the choroid plexus during brain imaging, may have also blocked secretion of ^{99m}Tc-pertechnetate into breast milk.[51] Studies in lactating goats found that perchlorate reduced secretion of ^{99m}Tc-pertechnetate and ^{123}I-iodide in milk by 70% to 80% and 60% to 67%, respectively.[51] Whether or not the low values for excretion in milk of radioiodinated radiopharmaceuticals such as ^{131}I-iobenguane might be related to inhibition of radioiodide secretion into milk by concurrent thyroid-blocking agents (e.g., Lugol's solution or potassium iodide) remains unknown. Nonetheless, intentional administration of potassium perchlorate, Lugol's solution, or potassium iodide to a lactating woman in order to allow early resumption of breast-feeding is not recommended because of potential adverse effects on the infant from ingesting abnormally large amounts of these substances.

No Excretion Data Available

A situation occasionally occurs when a lactating woman receives a radiopharmaceutical for which there are no data available regarding excretion in breast milk. A conservative recommendation would be interruption of breast-feeding for 10 physical half-lives if the radionuclide is short lived or complete cessation of breast-feeding if the radionuclide is long lived.[30] A better strategy, however, may be to obtain serial milk samples and perform specific dose calculations.[27,30]

Close Contact

In certain situations a nursing infant may receive more radiation from external sources than from internal radioactivity. Such situations typically involve radiopharmaceuticals that exhibit very little excretion in milk and are persistently localized in the mother's chest region (e.g., heart or breast tissue). For example, ^{99m}Tc-red blood cells, especially when radiolabeled in vitro, yield minimal excretion in milk but produce relatively high external dose rates from cardiac blood pool activity. Similarly, ^{99m}Tc-sestamibi is minimally excreted in milk but accumulates in cardiac muscle, liver, and lactating breast tissue. ^{18}F-fludeoxyglucose (^{18}F-FDG) is also minimally excreted in milk but accumulates in metabolically active glandular breast tissue in response to suckling.[37] Because external exposure to a nursing infant from these radiopharmaceuticals exceeds their internal radiation dose from ingested milk, close contact should be avoided or restricted to short periods totaling no more than 5 hours during the first 24 hours after injection.[28,29,37] For ^{111}In-labeled radiopharmaceuticals (e.g., ^{111}In-pentetreotide and ^{111}In-platelets), close contact should be limited for similar reasons.[38]

Case Study

A lactating woman was referred to the nuclear medicine department for a perfusion lung scan. She agreed to interrupt breast-feeding for 24 hours and collect milk samples for radioactivity analysis. She was injected with 4.3 mCi ^{99m}Tc-MAA. Six milk samples were obtained over the next 23 hours. Aliquots of milk samples were counted in a well counter, and the counts were converted to nCi based on the efficiency of the counter (Table 16-6). When the concentrations (nCi/mL) were plotted against time (Figure 16-1), peak concentration was observed in the 7 hour sample, consistent with delayed excretion of ^{99m}Tc-pertechnetate in milk following breakdown of the ^{99m}Tc-MAA as described above. After peak, the excreted activity closely followed a monoexponential decline. An exponential fit of the data from 7 to 23 hours yielded a slope of $e^{-0.13x}$, which gives an effective half-life of $0.693/0.13 = 5.3$ hours.

The steps involved in calculating the radiation dose to the infant, and what the radiation dose would have been if breast-feeding had continued without interruption, are summarized in Table 16-6. It was assumed that breast-feeding would resume 4 hours after the last measured sample (i.e., resume at 27 hours after injection) and continue at 4 hour intervals thereafter (Table 16-6, first column). Using an effective half-life of 5.3 hours, ^{99m}Tc concentrations in milk were calculated for the projected breast-feeding times through 43 hours, after which the ^{99m}Tc concentration in milk would be negligible (Table 16-6, second column). Assuming that the infant ingests 142 mL of milk at each feeding, the activity ingested at each time point was calculated (Table 16-6, third column). The ingested activities were then summed from each breast-feeding time point forward to determine the total activity ingested by the infant if breast-feeding was resumed at that point in time (Table 16-6, fourth column). Finally, the sum of

TABLE 16-6 ⁹⁹ᵐTc Albumin Aggregated (⁹⁹ᵐTc-MAA) Excretion in Milk and Radiation Absorbed Dose to Infant

Time after Injection (hr)[a]	^{99m}Tc Concentration in Milk (nCi/mL)[a]	Activity in 142 mL Feeding (nCi)	Summed Activity if Breast-Feeding Resumed at This Time (nCi)	Radiation Absorbed Dose to Infant from This Summed Activity (mrem)[b]
3	114	16,188	62,764	79
7	126	17,892	45,576	59
9.5	83	11,786	28,684	36
13	55	7,810	16,898	21
18	29	4,118	9,088	11
23	15	2,130	4,970	6
27	*9*	1,278	2,840	4
31	*5*	710	1,562	2
35	*3*	426	852	1
39	*2*	284	426	0.5
43	*1*	142	142	0.2

[a] For times in italics (i.e., 27 hr and after), ⁹⁹ᵐTc concentrations in milk (shown in italics) are calculated values.
[b] Effective dose equivalent.

ingested activity at each breast-feeding time point was multiplied by an infant dose conversion factor, 1260 mrem/mCi (= 0.00126 mrem/nCi) taken from Table 16-5, to obtain the effective dose equivalent that would be received by the infant if breast-feeding was resumed at that time (Table 16-6, fifth column).

In this case study, breast-feeding was resumed at 27 hours after injection of ⁹⁹ᵐTc-MAA. As shown in Table 16-6, the effective dose equivalent received by the infant from this and subsequent feedings is calculated to be about 4 mrem. If breast-feeding had initially been continued without interruption, at the times that milk samples were obtained and then every 4 hours thereafter, the infant would have received about 79 mrem effective dose equivalent. Assuming that periods of close contact (e.g., during nursing) totaled 5 hours during the first 24 hours after radiopharmaceutical administration, the infant would have received an additional radiation absorbed dose from external exposure of about 6 mrem (1.48 mrem/mCi × 4.3 mCi),[52] which would yield a grand total effective dose equivalent of about 85 mrem. In retrospect, this particular patient could have continued breast-feeding without interruption and still have been in compliance with NRC release rules (i.e., <0.1 rem). Nonetheless, as discussed above, large variations in the fractional excretion of radiopharmaceuticals in milk warrant caution and justify the use of conservative interruption times unless shorter times can be established by individual sample counting and calculation.

Summary

Radiopharmaceuticals are variably excreted in breast milk. In order to minimize radiation exposure of a breast-feeding infant, the lowest activity dose possible to permit adequate performance of the procedure should be administered, high radiochemical purity should be verified before administration, and breast-feeding should be interrupted for a suitable period of time. Interruption times should be conservatively long, unless shorter times can be established by individual sample counting and calculation. Instructions to the patient and maintenance of records may also be required.

ADVERSE REACTIONS TO RADIOPHARMACEUTICALS

Adverse reactions to radiopharmaceuticals are extremely rare. Radioactive drugs are typically administered in only tracer doses, masses so small that they do not produce pharmacologic effects. Moreover, radiopharmaceuticals are generally administered as single doses, so there is lack of chronic drug exposure. Hence, dose-related pharmacologic side effects or toxic effects are typically not observed with radiopharmaceuticals.

Definitions

The nomenclature regarding adverse drug reactions is not always clear. Over the years, FDA and the World Health Organization have used a number of terms, including side effect, adverse drug experience, adverse effect, and adverse reaction. None of these terms, however, is fully appropriate for radiopharmaceuticals because each term includes reference to pharmacologic or therapeutic action in its definition. As mentioned above, radiopharmaceuticals typically do not exhibit pharmacologic effects, and treatment effects, if any, are produced by the radiation rather than by the drug component. Hence, the Pharmacopeia Committee of the Society of Nuclear Medicine (SNM) established a working definition for an adverse reaction to a radiopharmaceutical as follows:[53]

> The reaction is a noxious and unintended clinical manifestation (signs, symptoms, laboratory abnormalities) following the administration of a radiopharmaceutical; the reaction is not the result of an overdose (which is a misadministration); the reaction is not the result of injury caused by poor injection technique; the reaction is not caused by a vasovagal response (slow pulse and low blood pressure); the reaction is not due to deterministic effects of therapeutic radiation (e.g., myelosuppression); the definition excludes altered biodistribution which causes no signs, symptoms or laboratory abnormalities.

A number of different types of adverse reactions to radiopharmaceuticals are possible (Table 16-7). Although they are listed in this table, vasovagal reactions and deterministic radiation effects should be excluded according to the SNM definition.

Reporting Systems

Characterization of adverse reactions to radiopharmaceuticals is extremely difficult for a number of reasons. Many adverse reactions may go unnoticed because the patient leaves the nuclear medicine department before a reaction is recognized. Adverse reactions, even if witnessed, are infrequently reported for a variety of reasons, including misunderstanding or ignorance of a reporting mechanism, lack of time to submit the report, perception that reporting is unimportant, lack of interest if a similar adverse reaction has already been reported in the literature, lack of proof of causality, liability concerns, and health information privacy concerns. These issues, along with the fact that the frequency of radiopharmaceutical adverse reactions is too low to observe trends at single institutions, underscore the need for a large-scale reporting system.

A large-scale reporting system might be beneficial through helping to define the types, characteristics, and frequency of adverse reactions; identify trends; alert the profession to problems; and provide feedback to manufacturers. Reporting systems in the United States and abroad are described briefly in Table 16-8. Since the discontinuation of USP's Drug Problem Reporting Program, a reporting system specifically for radiopharmaceutical adverse reactions has not been operating in the United States. However, serious adverse reactions to radiopharmaceuticals can be reported to FDA through its MedWatch program. The MedWatch program is intended for the reporting of serious adverse events and product problems for all drugs; it defines serious adverse events to mean death; a life-threatening event; an event causing hospitalization, disability, or congenital anomaly; or a medical/surgical intervention required to prevent permanent impairment or damage. MedWatch is not intended for the reporting of other, nonserious reactions.

Statistics

Figure 16-2 shows numbers of radiopharmaceutical adverse reactions reported in the United States.[53–55] (Because only a fraction of adverse reactions are reported, the actual numbers of adverse reactions to radiopharmaceuticals are higher than indicated in Figure 16-2.) Several aspects of this graph are of particular interest. The peak in 1968 was primarily due to allergic reactions to ^{113m}In-iron hydroxide aggregates used for lung scanning. The peak in 1971 was primarily due to cases of aseptic meningitis from ^{131}I iodinated albumin (RISA) injected intrathecally for cisternography procedures and to cases of

TABLE 16-7 Types of Adverse Reactions to Radiopharmaceuticals

Type of Reaction	Description
Allergic/hypersensitivity reactions	Antibody-mediated, release of histamine, kinins, and other substances from mast cells and basophils
-Type I: Immediate/anaphylactic reaction	Urticaria, angioedema, bronchospasm, shock
-Type II: Cytotoxic reaction	Complement cascade, lysis of blood cells
-Type III: "Serum sickness"	Complement cascade, damage to capillary endothelium
-Type IV: Delayed hypersensitivity	Dermatitis
Anaphylactoid/idiosyncratic reactions	Stimulated release of histamine, kinins, and other substances but not antibody-mediated
Pyrogenic/aseptic meningitis	Due to bacterial endotoxin, especially if intrathecal injection
Mechanical	Blockage of blood vessels (e.g., capillaries) by injected particulates
Other: Unpleasant or metallic taste; warmth, flushing, headache, dizziness, nausea	Usually related to high blood concentration following bolus injection
Vasovagal reactions	Excessive vagal nerve activity: hypotension, bradycardia, syncope; can be produced by anxiety/fear, often associated with venipuncture
Deterministic radiation effects	Examples are myelosuppression, sialadenitis, gastrointestinal symptoms

TABLE 16-8	Reporting Systems for Radiopharmaceutical Adverse Reactions		
Location	Year	Description of Reporting System	
United States	1967	Society of Nuclear Medicine (SNM) established an "Adverse Reactions Registry."	
	1976	SNM, in collaboration with the United States Pharmacopeia (USP) and the Food and Drug Administration (FDA), replaced the previous system with the "SNM Drug Problem Reporting System." A single form was used for reporting adverse reactions and drug quality problems.	
	1986	USP, in cooperation with SNM, replaced the previous system with the "Drug Product Problem Reporting Program" for adverse reactions and drug quality problems. This program was discontinued in August 2000.	
	1989–ca2004	The SNM Pharmacopeia Committee prospectively collected monthly reports of all adverse reactions and number of radiopharmaceutical administrations from ~18 participating institutions in order to estimate prevalence.	
	2006	Dr. Edward Silberstein (University of Cincinnati) began conducting a multi-institutional prospective study on adverse reactions to radiopharmaceuticals and ancillary drugs.	
Europe	1979	The European Association of Nuclear Medicine (EANM) established a coordinated European reporting system for adverse reactions. Annual reports were published in the *European Journal of Nuclear Medicine* through 2001 and thereafter on the EANM website.	
	1999	The British Nuclear Medicine Society in association with the UK Radiopharmacy Group established an Internet reporting system for adverse reactions.	
Japan	1975	Japan Radioisotope Association began conducting annual nationwide surveys of adverse reactions.	

flushing from ^{99m}Tc-iron hydroxide aggregates used for lung scanning. The peak in 1976 was probably not an increase in the actual number of adverse reactions but rather an increase in reporting of reactions consequent to the highly publicized initiation of the SNM/USP/FDA reporting system. The steady decline after 1976 was likely due to actions taken in response to earlier reports, including limulus amebocyte lysate (LAL) testing for endotoxins, better manufacturing practices, better radiopharmacy practices, avoidance of radiopharmaceuticals with known problems, and replacement of problem radiopharmaceuticals with newer, safer ones.

Various estimates of the frequency of adverse reactions to radiopharmaceuticals are summarized in Table 16-9.[53–62] The frequency of adverse reactions appears to have stabilized at a very low rate of about 3/100,000 following a decline from historically higher values.

Severity

The vast majority of reactions to radiopharmaceuticals are mild and self-limiting and require no treatment. A much smaller fraction are moderate or intermediate reactions, which require

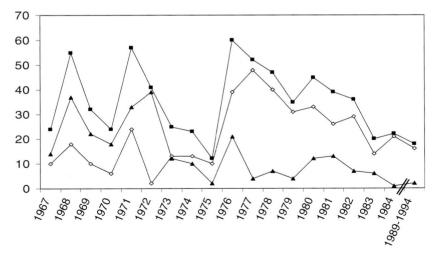

FIGURE 16-2 Numbers of radiopharmaceutical adverse reactions reported in the United States (■ = total number, ◊ = ^{99m}Tc products, ▲ = other products). Compiled from references 53–55.

TABLE 16-9 Estimated Frequency of Adverse Reactions to Radiopharmaceuticals		
Location	Time Period	Estimated Frequency (no. per 100,000 administrations)
United States	1967–1970	20–100
	1978	1–6
	1984	1–3
	1989–1994	2.3 (CI = 1.2–3.4)[a]
	1994–1997	0 (CI = 0–3.7) for PET radiopharmaceuticals
Europe	1977–1983	10–36
	1996	11 (CI = 3.3–19.2)
Japan	1977–1979	10.6
	1992–1993	1.1
	2004	1.3

[a] CI = confidence interval.
Source: References 53–62.

some treatment but are not life threatening. Only a very small percentage (≤3%) are severe, life-threatening reactions.[56] Deaths related to radiopharmaceutical adverse reactions are extremely rare but have been reported (Table 16-10).[53–56,63–69] Overall, the frequency of death related to radiopharmaceuticals is <1 in 10 million. The frequency of adverse reactions and death related to radiopharmaceuticals is 2 to 3 orders of magnitude lower than with other drugs used in hospitalized patients (Table 16-11).[70–73]

Specific Adverse Reactions

Specific adverse reactions reported with individual radiopharmaceuticals have been described elsewhere.[53–70,74] Listing them all is beyond the scope of this chapter, but most adverse reactions to most radiopharmaceuticals can be organized into a relatively short list. Table 16-12 lists adverse reactions typically reported with most radiopharmaceuticals.

Special Concerns

Several radiopharmaceuticals have unique issues that deserve special discussion. These special concerns are detailed below.

Radioiodine Radiopharmaceuticals

The administration of radioiodine radiopharmaceuticals to patients with a history of adverse reactions to iodinated x-ray contrast media or a claim to be "allergic to iodine" may cause concern. It is important to emphasize that adverse reactions to iodinated contrast media are not allergic reactions to iodine (as a chemical element). Rather, these adverse reactions are caused primarily by hyperosmolality of the contrast media and possibly also by direct chemotoxic effects, activation of the complement system, sequestration of calcium ions, or other mechanisms related to their chemical structures.[75,76] Also, the mass of iodine administered with radioiodine radiopharmaceuticals is much smaller than iodine masses associated with iodinated contrast media or dietary intake of iodine (Table 16-13).

TABLE 16-10 Deaths Reported from Radiopharmaceutical Adverse Reactions			
Location	Time Period	Reported Deaths	Radiopharmaceutical
United States	1965–1976	9	^{131}I-albumin aggregated
		3	^{99m}Tc-ferric hydroxide aggregates
		3	^{99m}Tc-albumin aggregated
	1976–1979	0	
	1982–1984	1	^{99m}Tc-medronate
	1987	1	^{99m}Tc-albumin colloid
	1989–1994	0	
	2004–2005	2	^{99m}Tc-fanolesomab
Canada	ca1994	1	^{99m}Tc-pentetate (intrathecal injection)
Europe	1960s	9	^{113m}In-ferric hydroxide aggregates
	1975–1992	1	^{99m}Tc-sulfur colloid
		1	^{99m}Tc-albumin aggregated (intracoronary injection)
		1	^{99m}Tc-pentetate (aerosol inhalation)
		2	^{99m}Tc-pentetate (intrathecal injection)
	1993–2002	0	
Australia/	1980	1	^{99m}Tc-tin colloid
New Zealand	ca1994	1	^{99m}Tc-pentetate (intrathecal injection)

Source: References 53–56 and 63–69.

TABLE 16-11 Frequency of Adverse Reactions and Death Related to Drugs

Drug Category	Adverse Reactions	Death
Radiopharmaceuticals	~3/100,000	<1/10,000,000
Iodine contrast media, ionic	3.8–12.7%	~1/75,000
Iodine contrast media, nonionic	0.6–3.1%	~1/75,000
Pharmacologic drugs[a]	0.6–6%	~1/1700

[a] In hospitalized patients.
Source: References 70–73.

Hence, based on differences in chemical structures, osmolality, and mass of iodine, the risk of adverse reactions to radioiodine radiopharmaceuticals in these patients is no greater than in the general patient population.

Colloids

Adverse reactions to colloid radiopharmaceuticals (e.g., ^{99m}Tc-sulfur colloid) are typically related to the stabilizing agents present in the formulation rather than to the colloid material itself. Gelatin, which is used in the currently marketed sulfur colloid product, has a much lower risk of adverse reactions than dextran and other stabilizing agents that were used in the past.[63]

Bone Agents

Skin rash, often with delayed onset, is reported as the most common adverse reaction to currently used radiopharmaceuticals. In rare cases, adverse reactions may be systemic (e.g., liver and kidney dysfunction).[77] ^{99m}Tc-pyrophosphate has been safely used in patients who previously manifested severe adverse reactions to ^{99m}Tc-diphosphonate agents.[78]

Lung Perfusion Agents

Deaths reported following administration of lung perfusion products generally involved patients with severe pulmonary hypertension and injection of a relatively large number of particles (up to 5 million) having a relatively wide range of particle size (10–100 μm).[67,79,80] Currently, safety is ensured in patients with pulmonary hypertension or right-to-left cardiac shunts by limiting ^{99m}Tc-MAA doses to no more than about 100,000 to 200,000 particles[81] and using modern ^{99m}Tc-MAA products in which most particles are 10 to 50 μm.

^{99m}Tc-Pentetate Administered into Cerebrospinal Fluid

Off-label uses of ^{99m}Tc-pentetate (^{99m}Tc-DTPA) include detection of cerebrospinal fluid (CSF) leaks, evaluation of ventriculoperitoneal shunt patency, and evaluation of intrathecal drug delivery with implantable pumps (see Chapter 19). When administered into the CSF, ^{99m}Tc-DTPA presents two special concerns: pyrogenic reactions and neurologic adverse reactions. Pyrogenic reactions can be avoided by ensuring that the bacterial endotoxin content is appropriate for intrathecal injection. This can be accomplished by performing LAL testing on the ^{99m}Tc-DTPA preparation at a sensitivity appropriate for intrathecal injection (≤14 endotoxin units [EU]/dose) or by limiting the injected volume to no more than 8% (i.e., the ratio of endotoxin content allowed for intrathecal injection to that allowed for intravenous injection, 14 EU/175 EU) of the vial contents.[82] Neurologic adverse reactions can be avoided by proper attention to the ^{99m}Tc-DTPA salt and mass. Because calcium and magnesium ions in the CSF can be sequestered by the free acid or sodium salt of ^{99m}Tc-DTPA, it is important that only a calcium trisodium ^{99m}Tc-DTPA formulation be used.[82,83] In addition, it is important to use only preservative-free ^{99m}Tc-DTPA products (i.e., no stabilizers or antioxidants) and to administer no more than 1 mg mass of ^{99m}Tc-DTPA.[82,83]

Murine Antibodies

Monoclonal antibodies of murine origin can elicit a human anti-mouse antibody (HAMA) response.[84] Subsequent administration of a murine monoclonal antibody may then result in hypersensitivity reactions. Also, although not an adverse

TABLE 16-12 Adverse Reactions Typically Reported for Most Radiopharmaceuticals

Organ System	Reaction
Cardiovascular	Hypotension, bradycardia, hypertension, tachycardia
Central nervous	Headache; faintness, dizziness, syncope
Gastrointestinal	Nausea, vomiting
Pulmonary	Respiratory reaction, dyspnea
Skin	Rash, pruritus, hives/urticaria
Systemic	Chills, fever; erythema, flushing, diaphoresis

TABLE 16-13 Comparison of Iodine Masses

Substance	Mass of Iodine
Iodinated x-ray contrast media	~300 mg/mL (up to 0.2 mg/mL I-)
RDA (dietary intake)	0.15 mg
^{131}I, 200 mCi treatment dose	0.002 mg
^{123}I, 400 μCi imaging dose	0.0000002 mg

reaction, HAMA may interfere with localization at intended target sites of a subsequently administered murine monoclonal antibody.[85]

Radiation Effects

Radiation effects are more properly categorized as dose-related treatment effects than as adverse reactions. Some examples of adverse effects due to radiation are as follows. Transient, short-term effects from ^{131}I-sodium iodide treatment include sialadenitis, nausea, vomiting, change in taste or smell, dry mouth, neck pain, dyspnea, dysphagia, and hyperthyroidism (thyroid storm).[86–89] Intermediate and long term-effects from ^{131}I-sodium iodide treatment include xerostomia, conjunctivitis, alopecia, myelosuppression, vocal cord paralysis, hypothyroidism, and hypoparathyroidism.[89–92] Radiopharmaceuticals used to treat metastatic bone pain may cause flare in bone pain and myelosuppression.[93] Radiolabeled antibodies to treat non-Hodgkin's lymphoma may cause hematologic toxicity.[94]

New Products

Adverse reactions observed in clinical trials do not necessarily predict adverse reactions in real-world practice. This is because rare adverse reactions may be seen only when the new product is used in a great number of patients; adverse reactions of other types, frequency, or severity may be seen when the product is used in other patient populations (e.g., different age group, presence of coexisting disease); and adverse reactions of other types, frequency, or severity may be seen with off-label uses of the product (e.g., different indication, dose, route). Therefore, postmarketing surveillance is important to update the adverse reaction profile of the product.

As a case example, consider [99mTc]-fanolesomab. Adverse reactions seen in clinical trials were generally infrequent and minor: flushing (2%); dyspnea (1%); and syncope, dizziness, hypotension, chest pressure, paresthesia, nausea, injection site burning/erythema, pain, and headache (each <1%). There were 4 severe reactions in 523 patients: hypotension, worsening of sepsis, chest pressure and decreased oxygen saturation, and pain. Two deaths were observed (severe hypotension, worsening of sepsis), both involving postsurgical patients, which were probably related to underlying disease.[95]

NeutroSpec, a kit marketed by Palatin for the preparation of [99mTc]-fanolesomab, was approved by FDA in July 2004. In the first 17 months after FDA approval, there were numerous reports of unpredictable, life-threatening adverse reactions involving shortness of breath and sudden drop in blood pressure within minutes after administration. Two patients died from cardiopulmonary failure and 15 others had severe reactions requiring treatment. Most but not all of these patients had preexisting cardiac or pulmonary conditions that may have placed them at higher risk. Forty-six additional patients experienced adverse reactions that were similar but less severe. As a result of these unanticipated severe adverse reactions, this product was withdrawn from the market in December 2005.[64]

Case Study

The following case study describes an unusual and intense allergic reaction to [99mTc]-medronate ([99mTc]-MDP) or [99mTc]-sulfur colloid, or both.

A 35 year old man was scheduled for nuclear medicine procedures to evaluate suspected infection of the left lower extremity several months after closed fracture of the proximal tibia with internal fixation. The following is a timeline of events.

10:00 am	Autologous blood was withdrawn for leukocyte labeling with ^{111}In-oxine.
10:20 am	10 mCi of [99mTc]-sulfur colloid was injected for bone marrow imaging.
11:25 am	After bone marrow imaging was complete, 20 mCi [99mTc]-MDP was injected for blood flow and blood pool phase imaging. Delayed bone imaging was scheduled for the next morning simultaneous with ^{111}In-leukocyte imaging.
11:30 am	A local weltlike skin reaction (redness and itching) was noted near the injection site and the patient was treated with oral diphenhydramine 25 mg.
1:00 pm	0.5 mCi autologous ^{111}In-leukocytes was injected.
8:00 pm	Patient self-medicated with more oral diphenhydramine because of increasing symptoms (large red blotches and severe itching).
10:10 pm	Patient called hospital triage center complaining of whole body itching and mild dyspnea. The patient was told to go to the emergency room (ER).
10:50 pm	ER physician administered epinephrine 0.3 mg intramuscularly for generalized erythematous urticarial rash over arms, neck, and trunk and throat tightness.
11:40 pm	Although the patient's rash resolved following epinephrine, it recurred after about 45 minutes. A second dose of epinephrine 0.3 mg was administered along with methylprednisolone 125 mg.
Next day	Patient was seen by an allergy specialist. Erythema was still present on lower abdomen, but rash and urticaria were gone.

The allergic reaction observed in this case is unusual in its intensity. Given the timing of the initial local reaction and the higher reported frequency of rash with [99mTc]-diphosphonates, the reaction was initially assumed to be due to [99mTc]-MDP. After additional consideration, however, the timing of the events could also have been consistent with reaction to sulfur colloid, and the greater severity of the allergic symptoms, especially hives, dyspnea, and throat tightness in contrast to just a simple skin rash, suggest that the reaction may more likely have been related to ^{99m}Tc-sulfur colloid. Skin rash and urticaria have also been reported as adverse reactions to ^{111}In-(oxine) leukocytes, but because the initial local rash appeared prior to injection of ^{111}In-leukocytes, it is unlikely that the later symptoms were related to ^{111}In-leukocytes. Without allergy skin testing or repeat injection with a single radiopharmaceutical, the causative agent remains unconfirmed.

Summary

In summary, adverse reactions to radiopharmaceuticals are exceedingly rare. The vast majority of adverse reactions that do occur are mild to moderate and require little if any medical attention. Nonetheless, life-threatening reactions to radiopharmaceuticals are possible, so emergency treatment drugs (e.g., epinephrine, pressor amines, corticosteroids, antihistamines) and advanced cardiopulmonary life support systems should be readily available where and when radiopharmaceuticals are administered. Ongoing collection and analysis of adverse reaction reports via large-scale reporting systems is important, especially as new radiopharmaceuticals are developed and marketed.

REFERENCES

1. Ponto JA, Swanson DP, Freitas JE. Clinical manifestations of radiopharmaceutical formulation problems. In: Hladik WB, Saha GB, Study KT, et al., eds. *Essentials of Nuclear Medicine Science*. Baltimore: Williams & Wilkins; 1987:268–89.
2. Ponto JA. Preparation and dispensing problems associated with technetium Tc-99m radiopharmaceuticals. In: Hladik WB, Newman J, eds. *Correspondence Continuing Education Courses for Nuclear Pharmacists and Nuclear Medicine Professionals*. Vol. 11, lesson 1. Albuquerque, NM: University of New Mexico Health Sciences Center College of Pharmacy; 2004.
3. Hung JC, Ponto JA, Hammes RJ. Radiopharmaceutical-related pitfalls and artifacts. *Semin Nucl Med*. 1996;26:208–55.
4. Hojelse C. Factors which affect the integrity of radiopharmaceuticals. In: Sampson CB, ed. *Textbook of Radiopharmacy—Theory and Practice, Third Edition*. Amsterdam: Gordon and Breach Science Publishers; 1999:187–94.
5. Hinkle GH, Loesch JA, Hill TL, et al. Indium-111-monoclonal antibodies in radioimmunoscintigraphy. *J Nucl Med Technol*. 1990;18:16–28.
6. Zimmer AM, Chinn PE, Morena RA, et al. Clinical experience with preparation of In-111 and Y-90 labeled Zevalin: radiochemical purity assessment. *Eur J Nucl Med*. 2001;28:1261(PS722).
7. Jacobson M, Danckwart HR, Mahoney DW. Radiolytic decomposition of 2-deoxy-2-[F-18]-fluoro-D-glucose. *J Nucl Med*. 2006;47(suppl):529P.
8. Webster EW, Alpert NM, Brownell GL. In: James AE, Wagner HN, Cooke RE, eds. *Pediatric Nuclear Medicine*. Philadelphia: WB Saunders; 1974:34–58.
9. Stabin MG, Kooij PPM, Bakker WH, et al. Radiation dosimetry for In-111-pentetreotide. *J Nucl Med*. 1997;38:1919–22.
10. Smith T, Gordon I. An update of radiopharmaceutical schedules in children. *Nucl Med Commun*. 1998;19:1023–36.
11. Treves ST, Davis RT, Fahey FH. Administered radiopharmaceutical doses in children: a survey of 13 pediatric hospitals in North America. *J Nucl Med*. 2008;49:1024–7.
12. Helton JD, Barron TL, Hung JC. Criteria for determining optimal pediatric dosage for a diagnostic nuclear medicine procedure. *J Nucl Med Technol*. 1996;24:35–8.
13. Shore RM, Hendee WR. Radiopharmaceutical dosage selection for pediatric nuclear medicine. *J Nucl Med*. 1986;27:287–98.
14. Kieffer CT, Suto PA. Management of the pediatric nuclear medicine patient (or children are not small adults). *J Nucl Med Technol*. 1983;11:13–7.
15. Piepsz A, Hahn K, Roca I, et al. A radiopharmaceuticals schedule for imaging in paediatrics. *Eur J Nucl Med*. 1990;17:127–9.
16. Gordon I. Issues surrounding preparation, information and handling the child and parent in nuclear medicine. *J Nucl Med*. 1998;39:490–4.
17. Solimando DA, Waddell JA. Verifying antineoplastic dosages calculated according to body surface area (BSA). *Hosp Pharm*. 2000;35:1036–41.
18. Shirkey HC, ed. *Pediatric Therapy*. 3rd ed. St Louis: CV Mosby;1968.
19. Heyman S. Toxicity and safety factors associated with lung perfusion studies with radiolabeled particles [letter]. *J Nucl Med*. 1979;20:1098–9.
20. Treves ST, Packard AB. Lungs. In: Treves ST, ed. *Pediatric Nuclear Medicine/PET*. 3rd ed. New York:Springer; 2007.
21. Davis MA, Taube RA. Toxicity and safety factors associated with lung perfusion studies with radiolabeled particles [reply]. *J Nucl Med*. 1979; 20:1099.
22. Levine G, Mazzetti C, Malhi B. A methodology for preparing pediatric doses of Tc-99m MAA for pulmonary perfusion studies. *J Nucl Med Technol*. 1980;8:94–6.
23. Neonatal deaths associated with use of benzyl alcohol—United States. *MMWR Wkly*. 1982;31:290–1.
24. Accorsi R, Karp JS, Surti S. Improved dose regimen in pediatric PET. *J Nucl Med*. 2010;51:293–300.
25. Alessio AM, Kinahan PE, Manchandra V, et al. Weight-based, low-dose pediatric wholebody PET/CT protocols. *J Nucl Med*. 2009;50:1570–8.
26. Romney BM, Nickoloff EL, Esser PD, et al. Radionuclide administration to nursing mothers: mathematically derived guidelines. *Radiology*. 1986;160:549–54.
27. Mountford PJ, Coakley AJ. A review of the secretion of radioactivity in human breast milk: data, quantitative analysis and recommendations. *Nucl Med Commun*. 1989;10:15–27.
28. Harding LK, Bossuyt A, Pellet S, et al. Recommendations for nuclear medicine physicians regarding breastfeeding mothers. *Eur J Nucl Med*. 1995;22:BP17.
29. Rubow S, Klopper J, Wasserman H, et al. The excretion of radiopharmaceuticals in human breast milk: additional data and dosimetry. *Eur J Nucl Med*. 1994;21:144–53.
30. Stabin MG, Breitz HB. Breast milk excretion of radiopharmaceuticals: mechanisms, findings, and radiation dosimetry. *J Nucl Med*. 2000;41:863–73.
31. Mountford PJ, Coakley AJ. Excretion of radioactivity in breast milk after an indium-111 leukocyte scan. *J Nucl Med*.1985;26:1096–7.
32. Logsdon BA. Drug use during lactation. *J Am Pharm Assoc*. 1997; NS37:407–18.
33. ICRP 95: Doses to infants from ingestion of radionuclides in mothers' milk. *Annals ICRP*. 2004;34(3–4):1–282.
34. *Regulatory guide 8.39. Release of Patients Administered Radioactive Materials*. Washington, DC: US Nuclear Regulatory Commission; April 1997.
35. Ahlgren L, Ivarsson S, Johansson L, et al. Excretion of radionuclides in human breast milk after the administration of radiopharmaceuticals [published correction appears in *J Nucl Med*. 1986;27:151]. *J Nucl Med*. 1985;26:1085–90.
36. Rose MR, Prescott MC, Lloyd JJ, et al. Radioactivity in breast milk following the administration of ^{99m}Tc MAA from different manufacturers, with and without prior administration of Technegas. *Eur J Nucl Med* 1997;24:89–90.
37. Hicks RJ, Binns D, Stabin MG. Pattern of uptake and excretion of ^{18}F-FDG in the lactating breast. *J Nucl Med*. 2001;42:1238–42.
38. Rubow SM, Ellmann A. Indium-111 scintigraphy and breastfeeding. *Eur J Nucl Med*. 2000;27:1057(PS185).
39. Hesselwood SR, Thornback JR, Brameld JM. Indium-111 in breast milk following administration of indium-111-labeled leukocytes. *J Nucl Med*. 1988;29:1301–2.

40. Evans JL, Mountford PJ, Herring AN, et al. Secretion of radioactivity in breast milk following administration of ^{99}Tc^m-MAG3. *Nucl Med Commun.* 1993;14:108–11.
41. Kettle AG, O'Doherty MJ, Blower PJ. Secretion of [^{123}I] iodide in breast milk following administration of [^{123}I] *meta*-iodobenzylguanidine. *Eur J Nucl Med.* 1994;21:181–2.
42. Wilkinson LE, Heggie JCP, Booth RJ. Secretion of [131]iodide in breast milk and infant dosimetry resulting from the administration of [131I]meta-iodobenzylguanidine. *Eur J Nucl Med.* 1995;22:1079–80.
43. Johnson TK, Stabin M. To the editor [letter]. *J Nucl Med.* 1995;36:1723–4.
44. Marshall DSC, Newberry NR, Ryan PJ. Measurement of the secretion of technetium-99m hexamethylpropyl amine oxine into breast milk. *Eur J Nucl Med.* 1996;23:1634–5.
45. Johnston RE, Mukherji SK, Perry JR, et al. Radiation dose from breastfeeding following administration of thallium-201. *J Nucl Med.* 1996;37:2079–82.
46. Gulec SA, Siegel JA. Posttherapy radiation safety considerations in radiomicrosphere treatment with ^{90}Y-microspheres. *J Nucl Med.* 2007;48:2080–6.
47. Weiner RE, Spencer RP. Quantification of gallium-67 citrate in breast milk. *Clin Nucl Med.* 1994;19:763–5.
48. Blue PW, Dydek GJ. Excretion of radioiodine in breast milk: reply. *J Nucl Med.* 1989;30:127–8.
49. Rose MR, Prescott MC, Herman KJ. Excretion of iodine-123-hippuran, technetium-99m-red blood cells, and technetium Tc-99m-macroaggregated albumin into breast milk. *J Nucl Med.* 1990;31:978–84.
50. Dydek GJ, Blue PW. Human breast milk excretion of iodine-131 following diagnostic and therapeutic administration to a lactating patient with Graves' disease. *J Nucl Med.* 1988;29:407–10.
51. Mountford PJ, Heap RB, Hamon M, et al. Suppression by perchlorate of technetium-99m and iodine-123 secretion in milk of lactating goats. *J Nucl Med.* 1987;28:1187–91.
52. Mountford PJ. Estimation of close contact doses to young infants from surface dose rates on radioactive adults. *Nucl Med Commun.* 1987;8:857–63.
53. Silberstein EB, Ryan J, and the Pharmacopeia Committee of the Society of Nuclear Medicine. Prevalence of adverse reactions in nuclear medicine [published correction appears in *J Nucl Med.* 1996;37:1064–7]. *J Nucl Med.* 1996;37:185–92.
54. Shani J, Atkins H, Wolf W. Adverse reactions to radiopharmaceuticals. *Semin Nucl Med.* 1976:6;305–28.
55. Cordova MA, Hladik WB, Rhodes BA, et al. Adverse reactions associated with radiopharmaceuticals. In: Hladik WB, Saha GB, Study KT, eds. *Essentials of Nuclear Medicine Science.* Baltimore: Williams & Wilkins; 1987:303–20.
56. Rhodes BA, Cordova MA. Adverse reactions to radiopharmaceuticals: incidence in 1978, and associated symptoms. Report of the adverse reactions committee of the Society of Nuclear Medicine. *J Nucl Med.* 1980;212:1107–10.
57. Atkins HL. Reported adverse reactions to radiopharmaceuticals remain low in 1984. *J Nucl Med.* 1986;27:327.
58. Silberstein EB and the Pharmacopeia Committee of the Society of Nuclear Medicine. Prevalence of adverse reactions to positron-emitting radiopharmaceuticals in nuclear medicine. *J Nucl Med.* 1998;39:2190–2.
59. Keeling DH, Sampson CB. Adverse reactions to radiopharmaceuticals. United Kingdom 1977–1983. *Br J Radiol.* 1984;57:1091–6.
60. Hesslewood SR, Keeling DH and the Radiopharmacy Committee of the European Association of Nuclear Medicine. Frequency of adverse reactions to radiopharmaceuticals in Europe. *Eur J Nucl Med.* 1997;24:1179–82.
61. Sasaki Y. Adverse reactions associated with radiopharmaceutical administrations—nationwide surveys in Japan. *Eur J Nucl Med.* 1995;22:746.
62. Kusakabe K, Okamura T, Kasagi K, et al. The 27th report on survey of the adverse reaction to radiopharmaceuticals (the 30th survey in 2004) [in Japanese]. *Kaku Igaku.* 2006;43(1):23–35.
63. Ponto JA. Adverse reactions to technetium-99m colloids. *J Nucl Med.* 1987;28:1781–2.
64. NeutroSpec withdrawn from market. *J Nucl Med.* 2006;47(2):36N.
65. Flanagan R. Intrathecal use of DTPA kits—some additional facts. *CARS [Canadian Association of Radiopharmaceutical Scientists] Newsletter.* 1995; no. 45:XIII–XIV.
66. Williams ES. Adverse reactions to radio-pharmaceuticals: a preliminary survey in the United Kingdom. *Br J Radiol.* 1974;47:54–9.
67. Keeling DH. Adverse reactions and untoward events associated with the use of radiopharmacy. In: Sampson CB, ed. *Textbook of Radiopharmacy: Theory and Practice.* 3rd ed. Langhorne, PA: Gordon and Breach Science Publishers. 1999;431–46.
68. European system for reporting adverse reactions to and defects in radiopharmaceuticals: annual report 2001. *Eur J Nucl Med.* 2003;30(10):87–94.
69. McQueen EG. New Zealand committee on adverse drug reactions: fifteenth annual report 1980. *NZ Med J.* 1981;93:194–8.
70. Silberstein EB. Adverse reactions to radiopharmaceuticals. In: Hladik WB, ed. *Correspondence Continuing Education Courses for Nuclear Pharmacists and Nuclear Medicine Professionals.* Vol V, no. 3. Albuquerque, NM: University of New Mexico; 1996.
71. Bates DW, Cullen DJ, Laird N, et al. Incidence of adverse drug events and potential adverse drug events: implications for prevention. *JAMA.* 1995;274:29–34.
72. Classen DC, Pestotnik SL, Evans RS, et al. Adverse drug events in hospitalized patients: excess length of stay, extra costs, and attributable mortality. *JAMA.* 1997;277:301–6.
73. Leape LL, Brennan TA, Laird N, et al. The nature of adverse events in hospitalized patients: results of the Harvard Medical Practice Study II. *N Engl J Med.* 1991;324:377–84.
74. Cordova MA, Hladik WB, Rhodes BA. Validation and characterization of adverse reactions to radiopharmaceuticals. *Noninvasive Med Imag.* 1984;1:17–24.
75. Thrall JH. Adverse reactions to contrast media: etiology, incidence, treatment, and prevention. In: Swanson DP, Chilton HM, Thrall JH. *Pharmaceuticals in Medical Imaging: Radiopaque Contrast Media, Radiopharmaceuticals, Enhancement Agents for Magnetic Imaging and Ultrasound.* New York: Macmillan; 1990;253–77.
76. Neff S. Why does a patient, who had a reaction to a radiology study that used iodine, not have a problem with nuclear medicine studies using iodine? *J Nucl Med Technol.* 1996;24:349.
77. Balan KK, Choudhary AK, Balan A, et al. Severe systemic reaction to ^{99m}Tc-methylene disphosphonate: a case report. *J Nucl Med Technol.* 2003;31:76–8.
78. Ramos-Gabatin A, Orzel JA, Maloney TR, et al. Severe systemic reaction to diphosphonate bone imaging agents: skin testing to predict allergic response and a safe alternative agent. *J Nucl Med.* 1986;27:1432–5.
79. Bolster AA, Murray T, Hilditch TE. Contraindications in the administration of ^{99}Tc^m-labelled macroaggregates or microspheres of albumin—any basis? *Nucl Med Commun.* 1994;15:188–91.
80. Barrington SF, O'Doherty MJ. Is perfusion lung scanning hazardous in pulmonary hypertension? *Nucl Med Commun.* 1995;16:125–7.
81. Parker JA, Coleman RE, Siegel BA, et al. Procedure guideline for lung scintigraphy. In: *Society of Nuclear Medicine Procedure Guidelines Manual 1999.* Reston, VA: Society of Nuclear Medicine; 1999:155–60.
82. Ponto JA. Special safety considerations in preparation of technetium Tc-99m DTPA for cerebrospinal fluid-related imaging procedures. *J Am Pharm Assoc.* 2008;48:413–6.
83. Verbruggen AM, de Roo MJK, Klopper JF. Technetium-99m diethylene triamine penta-acetic acid for intrathecal administration: are we playing with fire? *Eur J Nucl Med.* 1994;21:261–3.
84. Abdel-Nabi H, Harwood SJ, Collier BD, et al. Repeated administration of Oncoscint™ (In-111-CYT-103) in colorectal carcinoma patients. Interim report of safety, efficacy and HAMA development. *J Nucl Med.* 1993;34:213P.
85. Torres G, Berna L, Estorch M, et al. Preexisting human anti-murine antibodies and the effect of immune complexes on the outcome of immunoscintigraphy. *Clin Nucl Med.* 1993;18:477–81.
86. Hayek A. Thyroid storm following radioiodine for thyrotoxicosis. *J Pediatr.* 1978;93:978–80.

87. Van Nostrand D, Neutze J, Atkins F. Side effects of "rational dose" iodine-131 therapy for metastatic well-differentiated thyroid carcinoma. *J Nucl Med.* 1986;27:1519–27.
88. Khan S, Waxman A, Ramanna L, et al. Transient radiation effects following high dose I-131 therapy for differentiated thyroid cancer (DTC). *J Nucl Med.* 1994;35:15P.
89. Grewal RK, Larson SM, Pentlow CE, et al. Salivary gland side effects commonly develop several weeks after initial radioactive iodine ablation. *J Nucl Med.* 2009;50:1605–10.
90. Richards GE, Brewer ED, Conley SB, et al. Combined hypothyroidism and hypoparathyroidism in an infant after maternal ^{131}I administration. *J Pediatr.* 1981;99:141–3.
91. Lee TC, Harbert JC, Dejter SW, et al. Vocal cord paralysis following I-131 ablation of a postthyroidectomy remnant. *J Nucl Med.* 1985;26:49–50.
92. Alexander C, Bader JB, Schaefer A, et al. Intermediate and long-term side effects of high-dose radioiodine therapy for thyroid carcinoma. *J Nucl Med.* 1998;39:1551–4.
93. Serafini AN. Therapy of metastatic bone pain. *J Nucl Med.* 2001;42:895–906.
94. Juweid ME. Radioimmunotherapy of B-cell non-Hodgkin's lymphoma: from clinical trials to clinical practice. *J Nucl Med.* 2002;43:1507–29.
95. NeutroSpec™ [package insert]. St Louis: Mallinckrodt Inc; July 2004.

CHAPTER 17

Molecular Imaging and the Development of New Radiopharmaceuticals

Stephen M. Moerlein

The development of new radiopharmaceuticals is key to the future advancement of nuclear pharmacy and nuclear medicine. Although describing that complicated process in detail is beyond the scope of this chapter, an overview of the general radiopharmaceutical discovery process is presented for the purposes of giving readers an appreciation for the accomplishments of radiopharmaceutical scientists and identifying potential areas for involvement of nuclear pharmacists who want to participate in research.

Radiopharmaceutical development is a multidisciplinary effort that combines creativity with rigorous controls. The myriad research initiatives in nuclear medicine follow a common sequence in the creative process that generates new tracers for radiopharmaceutical use. This chapter will discuss that process and the considerable work that must be invested in a radiotracer before it can be brought to the clinical arena as a radioactive drug.

MOLECULAR IMAGING

An overarching consideration in the current design of new radiopharmaceuticals is the concept of molecular imaging, which arose from the intersection of techniques in molecular biology and in vivo imaging. Molecular imaging visualizes molecular events associated with disease. Because such molecular events precede anatomic changes, pathology can be identified at a preclinical or predisease stage, when medical interventions are expected to be most effective.

Molecular imaging is defined by the Society of Nuclear Medicine as the visualization, characterization, and measurement of biologic processes at the molecular and cellular level in humans and other living systems.[1] This definition specifically includes (in addition to radiotracer imaging) modalities such as magnetic resonance (MR) imaging, MR spectroscopy, optical imaging, and ultrasound. Molecular imaging probes are defined as agents used to visualize, characterize, and measure biologic processes in living systems.[1] These probes or biomarkers include both endogenous molecules and exogenous compounds and pertain to nuclear medicine imaging as well as alternative imaging methods.

Biomedical imaging techniques have in common the interaction of some form of energy with tissues of the body to enable noninvasive visualization of the tissues. These methods (Table 17-1) can be organized according to the type of energy employed. With the exception of optical imaging, all of the methods are in routine clinical use, and scaled-down instruments also exist for preclinical imaging of small animals. The imaging of small animal models of human disease is a particularly useful aspect of molecular imaging, as it promotes bench-to-bedside research development.

TABLE 17-1 Comparison of Imaging Techniques

	Energy	Image Type[a]	Advantages	Disadvantages
Nuclear	Radioactivity	M, F	High sensitivity Depth-independent	Radiation burden Moderate resolution
MRI	Magnetic	M, F, S	High resolution Depth-independent	Low sensitivity Magnetic flux
Optical	Light	M, F	No ionizing radiation High resolution	Poor depth penetration Moderate sensitivity
Ultrasound	Sound	F, S	No ionizing radiation Portable	Intravascular probes only
CT	X-ray	F, S	High resolution Depth-independent	Radiation burden

[a] M = molecular imaging; F = functional imaging; S = structural imaging.

The physics relevant to each technique affects the type of imaging possible, giving each modality particular strengths and weaknesses. Because the different techniques can be melded to make up for these weaknesses, they should be considered complementary rather than competitive. There is substantial effort to fuse different imaging platforms into hybrid systems to attain synergy beyond the individual imaging modes. A prime example of this approach is positron emission tomography–computed tomography (PET/CT), which has effectively supplanted standalone PET devices.

As indicated in Table 17-1, not all of the available imaging techniques are appropriate for molecular imaging. In particular, CT imaging and ultrasound have not found much utility for this purpose. CT imaging gives very high resolution structural images, and it is primarily used for the acquisition of anatomic information. However, data from molecular imaging methods can be fused with CT images to co-register the molecular or functional data with anatomic location. This is currently done with clinical PET/CT but in principle could be applied to other types of molecular imaging as well. A primary advantage of ultrasound is portability; bedside imaging can be attained in the absence of ionizing radiation. However, probes for ultrasound are confined to the vascular compartment, which is a severe limitation for molecular imaging.

The preeminent method for molecular imaging is nuclear scanning with PET or single-photon emission computed tomography (SPECT). This has major advantages over magnetic resonance imaging (MRI) or optical imaging. MRI results in high-resolution images but has relatively low sensitivity and requires large concentrations of biomarkers compared with radiopharmaceuticals. Optical imaging has the major limitation that penetration of light into tissue is poor, which precludes clinical imaging of deep tissues within human subjects. PET and SPECT remain virtually the only imaging techniques available for noninvasive measurement of biochemical receptor sites. Image data can be obtained in a depth-independent fashion, so noninvasive quantification of receptor sites is feasible. The remainder of this chapter discusses the development of molecular probes specifically for PET and SPECT imaging.

TARGET AND LEAD DEVELOPMENT

The general sequence for radiopharmaceutical development is summarized in Figure 17-1. This flow chart is not all-inclusive, but it shows the most important steps in the research

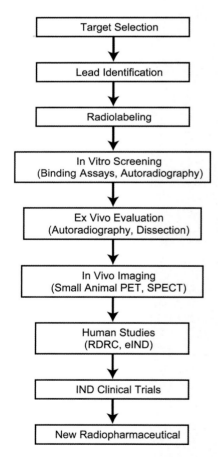

FIGURE 17-1 Sequence for development of new radiopharmaceuticals. RDRC = Radioactive Drug Research Committee program; eIND = exploratory investigational new drug program.

and development of a new radioactive drug. These key steps are the selection of an imaging target, identification of a biomarker for this target, development of a radiolabeling strategy, purification and reformulation of the radiotracer, in vitro screening experiments, in vivo validation, optimization for human application, and clinical application. Each of these topics is discussed in the following sections.

Selection of Imaging Target

The initial step in the development of a new biomarker, namely, selection of the imaging target, is probably the most critical to the success of radiopharmaceutical development. The target chosen for imaging must relate to a medically important process, so that biomarker development is driven by biomedical need. Without biomedical relevance, the effort expended in radiopharmaceutical development cannot be justified.

It is also important that imaging of the selected target fulfills an unmet biomedical need. That is, the impetus for development of the new imaging procedure should be to provide clinicians with substantial new information beyond what is already available with existing techniques. Thus, being able to image this new target will provide more than just an incremental improvement in health care benefit, and the research effort will be put to efficient use.

Although the selection of appropriate imaging targets may seem difficult, recent advances in genomic medicine have made possible a systematic approach to this problem. Mapping of the human genome was completed in 2001; this has made possible greater understanding of the intimate relationship between inheritance and disease.[2,3] For several common diseases, it has been noted that interactions between multiple genes and environmental factors can have protective or pathologic roles. Examples of such multifactorial diseases are cancer (breast and ovarian,[4–7] prostate,[7] lung,[8] colorectal,[9–10] and hematological[11]), cardiovascular disease,[12] Parkinson's disease,[13–15] Alzheimer's disease,[13,16–17] and infection.[18–20] In addition to genomic influences on the disease pathology itself, there are also genomic influences on drug efficacy,[21–22] which place certain subpopulations at risk for therapeutic failure. Future medical practice will undoubtedly incorporate gene-expression profiling to stratify patients with regard to disease risk and to individualize therapy for improved effectiveness.

The close relationship between genomes and disease derives from the fact that genomes, via mRNA, determine the proteins that constitute enzymes, receptors, and transporters used for all aspects of physiology. Aberrations in an individual's genomic composition can thus affect the quantity and nature of these key proteins. When essential proteins are altered, the chemical pathways in which they are involved are changed, leading to pathophysiology.

Because these key proteins are so closely related to disease and therapeutic response, they are a rational target for the development of diagnostic probes. Evaluation of biomarkers for these altered proteins thus serves as a logical starting point for molecular imaging and the development of new radiopharmaceuticals. To be most effective, the target selected for molecular imaging must not merely be associated with disease but must represent a control point for the initiation and progression of disease. When such targets are identified, the associated imaging procedures that are developed will have an impact on disease diagnosis as well as drug development.

Numerous targets are being actively investigated for molecular imaging. Some of these are listed in Table 17-2, which represents a synopsis of abstracts presented at recent scientific conferences concerned with radiopharmaceutical development: the 18th International Symposium on Radiopharmaceutical Sciences (Edmonton),[23] the World Molecular Imaging Congress (Montreal),[24] the Society of Nuclear Medicine Annual Meeting (Toronto),[25] and the Annual Congress of the European Association of Nuclear Medicine (Barcelona).[26] Virtually the entire gamut of pathophysiologic mechanisms is represented in the table, suggesting the widespread utility of radiotracer techniques for evaluating biologic processes.

TABLE 17-2 Targets for Biomarker Development

Cell death (apoptosis, autophagy, necrosis)
Neurotransmitter synthesis
Receptor expression (neurotransmitter, drug, peptide, antibody)
Signal transduction
Enzyme activity
Energy metabolism
Tissue inflammation and plaque
Stem cells and tissue regeneration
Gene expression
Tumor biology
Angiogenesis
Infection
Cellular immunity
p-Glycoprotein and transport systems
Nanotechnology

Source: References 23–26.

Identification of Lead Structures

Once the target for molecular imaging is selected, the next step in the development of a radiopharmaceutical is to identify a lead compound that has promise as a radiolabeled probe. This is very similar to the process that the pharmaceutical industry uses in new drug discovery, but the goal of radiopharmaceutical development is high image contrast rather than therapeutic efficacy (Table 17-3). The unique requirements for radiopharmaceuticals are discussed in more detail below. Although these differ in some respects from the characteristics of a new drug, both a molecular imaging probe and the corresponding therapeutic agent will bind to the same target site in a specific manner. For

TABLE 17-3 Comparison of Lead Development for Drugs and Radiopharmaceuticals

New Drug	New Radiopharmaceutical
High target affinity	High target affinity
High target specificity	High target specificity
Pharmacologic efficacy	NA
Acceptable side effects	NA
Oral bioavailability	NA
Inactive metabolites	Rapid elimination of radiometabolites
NA	Low nonspecific localization
NA	Acceptable dosimetry

NA = not applicable.

this reason, molecular probe development and drug discovery have much in common and can follow similar pathways from the laboratory to clinical use. Indeed, the two research endeavors can work in synergy; new radiopharmaceuticals can be based on the work of medicinal chemists, and molecular imaging can be used to enhance the effectiveness of drug development.

High-throughput screening (HTS) is used to identify chemical entities that interact with the target.[27] These automated assays are used to sieve through many new chemical entities (NCEs) to identify "hit" compounds, which are clustered on the basis of their potential for drug development.[28,29] This assessment includes evaluation of molecular weight, lipophilicity, and hydrogen bonding and other physicochemical characteristics that influence biologic behavior. Quantitative structure–activity relationships (QSARs)[30,31] are then applied to chemical libraries to identify lead compounds that have optimal characteristics for interaction with the target. The lead compounds serve as baseline structures for chemical modification to create molecular entities that are chemically similar yet different.[32] The goal of these slight chemical modifications is to retain or enhance affinity for the target while minimizing properties that negatively affect in vivo behavior. These slight chemical modifications change lead structures into druglike structures and, if successful, eventually into drugs.[33]

Optimization of drug properties can be accomplished with combinatorial chemistry and organic synthesis, and modifications can be either ligand-based or target structure–based.[34] The latter molecular modifications require structural knowledge of the target and often apply computer-aided design (CAD) tools.[35–37] Application of chemoinformatics makes the entire hit-to-lead process more efficient, rapid, and effective.[38–41]

RADIOLABELING STRATEGY
Choice of Radiolabel

In selecting a radiolabel, the biologic process under evaluation should be matched with the radioactive half-life of the label. This minimizes unnecessary radiation burden to the patient while ensuring that the radioactivity remains long enough for the noninvasive measurement to be completed.

Radionuclides for labeling lead compounds can be categorized as traditional and nontraditional (Table 17-4). The traditional radionuclides have the advantage of greater familiarity to the production staff and more widespread availability. Traditional single-photon emitters for the development of SPECT tracers include generator-available ^{99m}Tc and cyclotron-produced ^{111}In and ^{123}I. These nuclides emit photons of the appropriate energy for imaging in high abundance and have half-lives ranging from 6 hours to 2.8 days. All of the traditional single-photon emitters are commercially available and in clinical use.

The PET radionuclides of this category include ^{13}N, ^{11}C, and ^{18}F. These traditional positron emitters are already used in clinical radiopharmaceuticals and are produced in-house with medical cyclotrons. All of them decay almost entirely via positron emission, and half-lives range from 10 to 110 minutes. Note that ^{15}O is not listed, since its short half-life (2 minutes) precludes synthetic procedures of any complexity.

Nontraditional radionuclides are less prevalent in the nuclear medicine community but have distinct advantages for specific applications. Their predominant advantage is a long half-life, which permits distribution from specialized production sites, as well as longer image acquisition protocols. An exception to this generalization is ^{68}Ga, which has a convenient half-life of 67.7 minutes. ^{68}Ga is unique in that it is available from the decay of parent ^{68}Ge, and it has underutilized potential for use as a generator-based PET radiopharmaceutical label. ^{64}Cu, ^{76}Br, and ^{124}I are produced with medium-sized cyclotrons and have half-lives ranging from approximately 12 hours to more than 4 days. Although the yield of positrons from these nuclides is modest relative to the traditional positron-emitting nuclides, the long half-lives of these labels is an advantage for imaging slower biologic processes, for which longer intervals are needed to achieve adequate image contrast.

^{125}I is also listed as a nontraditional radionuclide for biomarker development. This nuclide is a reactor-produced isotope available with high specific activity from the decay of parent ^{125}Xe. ^{125}I has a long half-life of 59.4 days, which is convenient for nuclide distribution and laboratory procedures. The radioisotope decays via electron capture, with emission of 27 keV x-ray and 36 keV gamma photons. Although these photons are not useful for imaging, the widespread use of ^{125}I in biomedical research and its availability with high specific activity make it a useful surrogate nuclide for preliminary studies leading to eventual application with ^{123}I for SPECT or ^{124}I for PET.

Radiolabeling Chemistry

In addition to their differences in radioactive half-life, the radionuclides available for radiopharmaceutical development differ in their chemistry, since they are different elements.

TABLE 17-4 Radionuclides for Labeling Biomarkers

Radionuclide	Emission Energies (abundance [%])		Half-life
	Gamma[a]	Positron[b]	
Traditional			
^{99m}Tc	141 (89)		6.02 hr
^{111}In	171 (91)		2.80 days
	245 (94)		
^{123}I	159 (83)		13.2 hr
^{13}N		492 (100)	10.0 min
^{11}C		386 (100)	0.4 min
^{18}F		250 (97)	109.8 min
Nontraditional			
^{60}Cu		970 (93)	23.7 min
^{62}Cu		1314 (97)	9.7 min
^{64}Cu		278 (18)	12.7 hr
^{68}Ga		836 (88)	67.7 min
^{76}Br		1180 (55)	16.2 hr
^{86}Y		660 (32)	48 min
^{89}Zr		396 (23)	78.4 hr
^{94m}Tc		1072 (10.5)	52 min
^{124}I		820 (23)	4.2 days
^{125}I	27 (114),[c]		59.4 days
	31–32 (24),[c]		
	35 (7)		

[a] Photopeak energies in keV.
[b] Mean beta energies in keV.
[c] X-rays (keV).
Source: www.nndc.bnl.gov/chart/.

This affects the chemical reactivity of the different nuclides and hence the chemical mechanisms whereby they may be attached to the lead structures. For both single-photon and positron-emission imaging, there are available radionuclides that are metallic and nonmetallic.

Metallic elements belong to the s, p, d, and f blocks of the periodic table (groups 1–13). Attachment of metallic radionuclides to substrates is generally achieved via chelation, although formation of olefin complexes and aromatic complexes such as π- or σ-complexes and sandwich compounds are also available options.[42] Chelation involves the formation of ionic bonds between the metallic element and electron-donating atoms and necessitates the addition of a bulky chelating group to the lead compound structure. Of the radionuclides listed in Table 17-4, several are metallic and require the use of chelating groups for attachment of the radionuclide to the substrate. Specific chelates and labeling methods have been optimized for labeling with ^{99m}Tc,[43–49] ^{111}In,[50,51] ^{64}Cu,[52–54] and ^{68}Ga.[50,55]

Nonmetallic elements include carbon, nitrogen, and the halogens. Radionuclides of this category form covalent bonds and can be attached to lead compound structures using the vast armamentarium of organic chemistry reactions.[56] It is often possible to attach the radionuclide directly to the lead compound without the need for larger chemical synthons in radiopharmaceutical synthesis. Special chemical requirements have been optimized for radiolabeling compounds of high specific activity with the traditional PET radionuclides ^{11}C,[57–62] ^{13}N,[57,63] and ^{18}F.[57,58,64–66]

High specific-activity radiohalogenation methods have also been optimized for ^{123}I, ^{124}I, and ^{125}I,[67–69] as well as for ^{76}Br.[69–70] Radioiodination and radiobromination methods differ from those used for ^{18}F labeling, since electrophilic halogenation reactions can be used with the heavy halogens. In contrast, high specific-activity labeling with ^{18}F must focus on nucleophilic reaction pathways.

For both metallic and nonmetallic radionuclides, there are common requirements for radiopharmaceutical labeling despite differences in the details of the chemical reactions involved. The radiolabeling techniques should be rapid in comparison with the radionuclide half-life, give high radiochemical yield, and allow easy purification of the labeled product. If these common requirements are not attained, the targeted radiochemical has little chance of advancing to the level of a clinically used radiopharmaceutical.

Molecular Location of Radiolabel

Another key consideration in the preparation of a potential radiopharmaceutical is where the radionuclide will be attached to the lead structure. Most lead structures have specific structure–activity relationships (SARs) familiar to medicinal chemistry,[71] and interference with these through the attachment of the radiolabel may obviate the favorable characteristics of the lead compound. Although retention of the SAR for a lead compound does not guarantee that the radiolabeled analogue will progress to a bona fide radiopharmaceutical, disturbance of essential SARs by the radiolabeling process may ensure that the analogue does not retain the desired characteristics of the lead structure. This aspect of radiopharmaceutical development should be assessed in the choice of a labeling strategy.

It is sometimes possible to attach the radiolabel at a nonessential portion of the molecular structure. There may be substantial chemical liberty, allowing the introduction of bulky chelating groups, for example. However, when this is not possible, it is necessary to further examine the influence of the radiolabel on the lead structure, considering factors such as radiolabel size, polarity, hydrogen bonding, and lipophilicity. The size of the label is important because steric bulk may prevent the radiolabeled analogue from fitting the binding site for which the lead structure has affinity. An increase in polarity due to the radiolabeled moiety may cause problems by decreasing affinity for a hydrophobic site or increasing affinity for a competitive hydrophilic binding site. Likewise, a change in hydrogen bonding from that in the lead structure (introduction or elimination of bonding) may change binding to the original receptor site, as well as changing the conformation of the lead structure itself so that its interaction with the binding site is decreased. Finally, increasing the lipophilicity of the lead compound through the radiolabeling procedure is disadvantageous because nonspecific binding and protein binding may increase in concert, yielding a decrease in contrast for the image target.

A final caveat regarding SARs is that the labeling procedure should not alter the conformation of the lead compound itself. This is especially important for structurally sensitive compounds like proteins, with which the chemical reactions used for attachment of the radiolabel may alter the weak hydrogen bonds or van der Waals forces that maintain the tertiary structure of the protein. Such alterations can have dramatic effects on the biologic behavior of antibodies, antibody fragments, and other large molecular-weight materials.

It is essential that the nonradioactive form of the biomarker be synthesized early in the development sequence. This facilitates characterization of the molecular structure using classical analytical techniques, and it provides the reference standard for the radiolabeled tracer. The compound can also be used in binding assays, target-blocking studies, and (if the lead is successful) later toxicologic evaluation prior to approval for human use.

IN VITRO BINDING ASSAYS

Once an appropriate lead has been prepared, the next step is to evaluate the in vitro affinity of the lead compound for binding to the target. These binding studies can be accomplished experimentally using in vitro binding assays or autoradiography. Both methods measure biomarker–target binding, but they have different characteristics (Table 17-5). The basic premise behind both methods is the same: a biomarker and target bind reversibly to produce a biomarker–target complex plus unbound biomarker. These techniques were originally applied to the measurement of receptor-binding radioligands, but they also pertain in more generic fashion to other sorts of targets and probes.

Role of In Vitro Binding Assays

Homogenate-binding or cell-binding methods are less time-consuming than autoradiography and can be considered a preliminary test of the probe–target interaction prior to further evaluation as radiopharmaceuticals. Although these methods provide only a measure of binding to target sites within the sampled tissue, this binding is a necessary condition for further development of the lead structure. In vitro binding studies are therefore useful primarily as a screening tool to exclude structures that have no possibility of success as in vivo tracers. In vitro binding assays are commonly used in biomedical research for screening of radioligands, as well as to measure target concentrations in different tissues.[72–75]

In evaluating in vitro binding assay results, the greater is a biomarker's specific binding to the target, the more promise the biomarker has of becoming a radiopharmaceutical. This is because the in vitro bound-to-free (B:F) ratio is representative of the image contrast that can be attained, and the in vivo target-to-nontarget (T:NT) ratio is almost always lower. Biologic interactions of the biomarker, such as protein binding, metabolism, and perfusion, will decrease the in vivo binding; therefore, the in vivo T:NT ratio achieved will be below the B:F ratio measured under more ideal in vitro conditions. In vitro binding assays are thus a necessary, but not sufficient, condition for success. They are used to exclude unpromising lead structures and minimize wasted development effort.[76,77]

TABLE 17-5 Comparison of In Vitro Binding Assays and Autoradiography

In Vitro Assay	Autoradiography
Fast	Tedious
Wide variety of preparations	Solid tissue specimens only
Poor tissue discrimination	High-resolution binding
Whole cells, membranes, solubilized receptors	Whole body or tissue sections

Direct-Binding and Competitive-Binding Assays

In vitro binding assays can be performed in two different ways (Figure 17-2). Binding of the radiolabeled probe to the target can be measured directly, or the target-binding affinity of the nonradioactive form of the biomarker can be measured indirectly from its ability to displace a validated radioligand. The latter, indirect technique is known as a competitive-binding assay. Each of these approaches has advantages.

The direct assay has the advantage that target binding by the radiolabeled lead compound is measured directly, and binding information is more readily translated to subsequent in vivo evaluation measurements. Successful in vitro binding of the labeled biomarker to the target encourages further in vivo evaluation, because it demonstrates that the radiotracer can be prepared in sufficient purity for target recognition.

The competitive-binding method has the advantage that the radiolabeled probe is already validated, so there is no ambiguity as to the binding site. Also, it is usually easier to synthesize a lead compound in a nonradioactive form than to radiosynthesize the molecule in high specific activity. Thus, for those compounds showing inadequate affinity for the target, time is not wasted in preparing a high specific-activity radiochemical with no potential for further development. For those promising lead compounds that do progress to further levels of development, the nonradioactive version of the biomarker is needed in any event, because it is used as a standard for validation of chromatographic purification methods, as well as for target-blocking and toxicity studies in vivo.

Control Studies of Target-Binding Affinity

Most in vitro binding assays have focused on receptor sites, but in principle the same experimental procedure can be used for in vitro assay of other molecular imaging targets. Target-rich tissues are prepared for the assay in the form of isolated whole cells, cell membranes, or solubilized targets.[72] The target-bearing component is filtered, centrifuged, and the pellet washed with buffer. These steps are often performed cold to minimize denaturation of the target. The purified sample may then be stored frozen until the actual binding assay is performed.

The general procedure used for binding assays is illustrated in Figure 17-3. A homogenate is prepared from a sample of target-rich tissue. A buffered solution of the radiolabeled biomarker is added to the target-rich homogenate. After a defined incubation period, the target-bound activity is physically separated from free radioligand (still in solution). The ratio of radioactivity retained by the filter to the activity in the filtrate (the B:F ratio) is used as an index of target binding by the radioligand. In the direct assay method, the radioligand is the radiolabeled biomarker itself; in the indirect method, a validated radioligand is used instead.

There are many variations on this basic in vitro assay.[72] The target can be dispersed in solution and filtered away from the solution that contains unbound ligand so that bound-to-free radiopharmaceutical can be measured. Alternatively, the receptor-bearing material can be affixed to the walls of a container, such as a well. After the incubation period, the supernatant bearing free radioligand can be pipetted off, and radioactivity measured in the well and in the liquid layer to assess the B:F ratio. In addition to filtration or centrifugation, dialysis can be used to separate target-bound from free radioligand. Variables that must be optimized in performing an in vitro binding assay are given in Table 17-6.

For competitive-binding assays, the validated radioligand that is used is usually commercially available and labeled with either ^{3}H or ^{125}I. Tritium has the advantage of being an isotope of hydrogen (which is in every molecular structure and hence easy to label), and it has a long half-life (so radioligand does not have to be reordered frequently). Although labeling with ^{125}I usually results in a chemical analogue that needs further validation, these radiolabeled analogues have very high specific activity and are especially useful for targets that are expressed in low tissue densities.

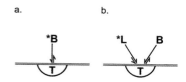

FIGURE 17-2 Direct and competitive-displacement (indirect) binding assays. T is the target; *B is radiolabeled biomarker; B is nonradioactive biomarker; *L is validated radioligand for T. (a) Direct binding; (b) Competitive-displacement (indirect) binding.

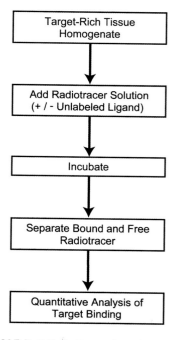

FIGURE 17-3 Procedure for in vitro binding assays.

TABLE 17-6 Variables of In Vitro Binding Assays

Radioligand
Incubation time
Buffer
Radioligand concentration
Receptor concentration
Concentration of nonradioactive drug
Specific binding

Specific and Nonspecific Target Binding

The total binding that is measured in these assays is the sum of target-specific binding (which has limited capacity) and nonspecific binding (which has high capacity) (Figure 17-4). The biologically interesting component is binding to the low-capacity target, which is saturable. By increasing the ligand concentration, the specific binding can be saturated, since at a finite ligand concentration all binding sites will be occupied.

Competitive-binding assays are used to measure the affinity of compounds for specific binding to the target. The general experimental procedure in Figure 17-3 is followed, except that during the incubation period a measured amount of unlabeled ligand with affinity for binding to the target is added to the solution. The percentage of radiolabeled ligand bound to the target is decreased in proportion to the affinity of the unlabeled compound for binding competition for the same target. Measurements of the affinity of new compounds can thus be ascertained in such competitive-binding assays.

Target-Binding Selectivity

More sophisticated experimental designs can be used to measure the selectivity of the biomarker for binding to the target site. Competitive binding can be used to measure the affinity for the target site, and ligands specific for nontarget sites can be used to estimate the specificity of the radioactive biomarker (Figure 17-5). If the potential radiopharmaceutical binds

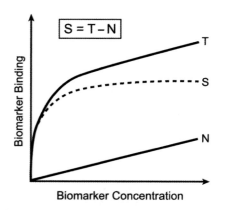

FIGURE 17-4 Total (T), specific (S), and nonspecific (N) binding as a function of biomarker concentration. Note that S = T − N.

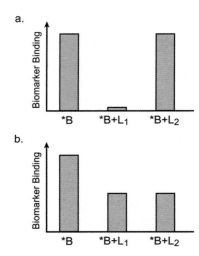

FIGURE 17-5 Binding selectivity of radiolabeled biomarker *B. L_1 is unlabeled ligand specific for T_1; L_2 is unlabeled ligand specific for T_2. (a) T_1-selective binding (*B binds to T_1 but not T_2); (b) non-T_1-selective binding (*B binds to both T_1 and T_2).

specifically to the target site, it will be competitively displaced by ligands specific for that site but not by ligands that are specific for other nontarget sites. Displacement of the binding of the potential radiopharmaceutical by ligands specific for nontarget sites means the binding by the radiolabeled biomarker is nonselective. This is generally a disadvantage, since it means that the target-binding signal obtained in imaging studies is diluted by additional binding to nontarget sites. Note that binding by a T_1-specific biomarker is completely blocked by target-specific ligand L_1 (Figure 17-5a) but only partially blocked by L_1 for a nonspecific biomarker (Figure 17-5b). Very specific biomarkers are the goal in radiopharmaceutical development, because they enhance the integrity of the image data representing the molecular target of interest.

Quantification of Biomarker–Target Binding

Several excellent review articles on the mathematical analysis of ligand–receptor binding assays are available.[73–75] These analyses can provide information on biomarker target recognition, the density of targets, and kinetic parameters of the interaction. This is very important information for radiopharmaceutical development and can be used for the development of new lead structures.

AUTORADIOGRAPHY

The main utility of in vitro binding assays is to identify biomarker target recognition and affinity. The additional information gathered by autoradiography is the regional distribution of the target within tissue. Autoradiography is a powerful experimental tool. It can be considered a type of invasive imaging in which the heterogeneity of biomarker binding is visualized. It is useful for measurement of the binding of biomarkers to neurotransmitter receptors, enzymes, and ion channels.[78,79] Tissue sections are placed onto film, which is exposed by particulate emissions from the radiolabeled biomarker. The basic experimental sequence for autoradiography is shown in Figure 17-6.[80,81]

Autoradiography acts as a bridge between in vitro and ex vivo testing, since the method can be applied in both situations. In vitro autoradiography refers to administration of the radiotracer directly to the tissue sections to determine heterogeneity of binding. Ex vivo autoradiography refers to administration of the radiotracer to the circulation prior to tissue sectioning.

There are several parallels in the experimental design of in vitro binding assays and autoradiographic studies. Both methods involve the common steps of tissue preparation, incubation of tissues with radioligand, removal of unbound ligand, and quantification of specific binding. The principles of bimolecular probe target binding with kinetic, saturation, and competition analysis apply to both in vitro binding assays and autoradiography. Detection and quantification of saturable, specific binding is a key aspect of both methods and can be accomplished by competitive-binding assays in which increasing concentrations of unlabeled ligand compete with a radiolabeled ligand for the limited target sites (Figure 17-5). Like in vitro binding assays, autoradiographs can be analyzed to quantify radioligand binding. Details of these analyses are given elsewhere.[73]

In Vitro Autoradiography

In this method, tissue samples are excised and cryogenically frozen so thin slices can be cut with use of a microtome. The tissue section is mounted onto a stationary backing and dipped into a solution containing the radioactive probe. After a predetermined incubation period, the solution is removed and the tissue slice is rinsed so only bound radioactivity remains. The tissue plate is placed onto film for exposure, which lasts days to weeks. After development of the film, the resulting image represents tissue binding by the radioactive probe, both target-specific and nonspecific. The images can be quantified by taping a scale of diluted standard onto the film before exposure and measuring the grain density for the tissue section and scale with a microdensitometer. The sections may also be stained for easier viewing.

As with in vitro binding assays, there can be direct or competitive-binding experimental designs with autoradiography to evaluate binding characteristics of a new compound. That is, the new compound can be prepared as a radioactive ligand and the effect of unlabeled validated ligands on its binding can be measured (direct binding), or a validated radioligand for the target can be used to assess the target-binding affinity of the new compound (competitive binding).

Ex Vivo Autoradiography

In ex vivo autoradiography, the radiopharmaceutical is injected into the intact animal and, at a specific time after injection, the animal is killed and the tissue sample prepared for autoradiographic imaging. This form of autoradiography is similar in concept to ex vivo biodistribution studies, since the distribution of radioactivity in the tissue section represents the in vivo distribution of the radiotracer at the time of death. As with in vitro autoradiography, ex vivo autoradiography studies can be performed in which competitive ligands are administered in challenge studies to examine the specificity of biomarker binding to the target.

Conventional Autoradiography

Images of tissue binding from autoradiography are largely dependent on the preparation of the tissue sample, since the section presented to the film is what will be reproduced in the autoradiograph. The orientation of these tissue sections is important, because they determine what information can be visualized. Ideally the sections contain the target as well as non–target-containing tissue so image contrast is generated with the radioactive probe. Tissue slices can be limited to a narrow field of view (such as a cross-section of the brain) or

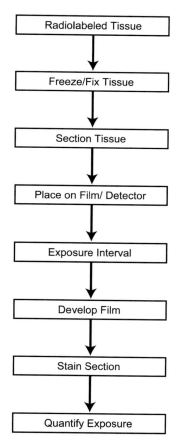

FIGURE 17-6 Experimental sequence for autoradiography.

may be representative of the entire body,[82] depending on the specific organs of interest to the investigator. Section thicknesses 5 to 100 μm are used, and standard radioactivity–density curves need to be developed to calibrate film response to exposure. These can be further validated by measurement of the tissue samples used for film exposure.

To be done successfully, autoradiography requires substantial investment to optimize empiric parameters (Table 17-7). Evaluation of the amount of radioactivity and exposure time is key to good image contrast. Section thickness is also important, since it influences the amount of radioactivity to which the film is exposed as well as the resolution of the final image. The solution to which the tissues are exposed must not disturb the essential biomarker–target binding interaction, and the incubation time must be adequate for binding equilibrium to occur. The type of radionuclide applied in the studies must be appropriate, since the film is sensitive to particulate emissions but is transparent to photons. Thus, nuclides with higher dose equivalent are detected more readily with autoradiography.

Classical receptor autoradiography can map receptors spatially with a resolution of about 100 μm.[83] This is much higher than the resolution achievable with noninvasive imaging. Resolution may be further enhanced from the regional level by application of light microscopy. Microscopic autoradiography is able to discriminate localization of radiolabeled biomarker at the subcellular level and allows binding heterogeneity to be compared with morphology.[84] An additional refinement is the use of hybrid image analysis to enhance the spatial mapping of receptors.[83] Such analysis might include in situ hybridization for comparison of the radioligand distribution with that of mRNA for the receptor protein. Experimental information obtained in this fashion sheds light on mechanisms of action, kinetic modeling, and toxicity of drugs, and it can be used in radiopharmaceutical design to better understand the biomarker–target interaction.

Indirect autoradiography techniques have also been developed to enhance autoradiography data.[80] Fluorography is a method in which the tissue sample is coated with a scintillator prior to exposure (Figure 17-7b). The rationale for this is that the scintillator gives a multiplier effect, so that each radioactive event is converted into hundreds of light photons in the scintillator. A variation on the indirect approach is the use of intensifying screens (Figure 17-7c). With intensifying screens, a scintillator is placed behind the film, so that ener-

 FIGURE 17-7 Enhancements to conventional autoradiography. (a) Standard autoradiography; (b) fluorography; (c) intensifier screen.

getic particles that pass through the emulsion undetected are converted to light that exposes the film from the reverse side. Indirect autoradiography increases the sensitivity of detection, but it is at the cost of decreased spatial resolution.

Applications with Radiopharmaceuticals

Special mention must be made concerning the application of autoradiography techniques with radiopharmaceuticals. Although autoradiography is a widely used technique in pharmacology and toxicology, most of its applications in biomedical research have used radioligands labeled with ^{3}H or ^{125}I. A major difference between these nuclides and the typical nuclear medicine nuclides in Table 17-4 is that they have weak particulate emissions that do not permeate tissue well. As shown in Table 17-8, the energies and ranges of particles emitted from nuclides such as ^{18}F, ^{99m}Tc, and ^{123}I are at least an order of magnitude higher than ^{3}H or ^{125}I. This means that tissue thickness affects autoradiographic exposure to a greater extent than for the lower-energy emissions. For this reason, autoradiography of radiopharmaceuticals requires particular attention to the uniformity of tissue sections.[85] The use of

TABLE 17-7 Parameters of Autoradiography

Section orientation
Section thickness
Radionuclide
Type of film emulsion
Radioactive dose
Exposure time
Backscatter from support material
Temperature
Conditions of development

TABLE 17-8 Selected Radionuclides for Autoradiography

Nuclide	Emission[a]	Energy[b]	Range[c]	Specific Activity[d]
^{3}H	β−	5.7	0.004	29
^{125}I	Au	24	0.012	2.2×10^3
^{18}F	β+	250	0.637	1.7×10^6
^{99m}Tc	IC	120	0.195	5.2×10^5
^{123}I	IC	127	0.214	2.4×10^5

[a] Au = Auger electron; IC = internal conversion electron.
[b] Average energy in keV.
[c] Average tissue range in mm.
[d] Maximum specific activity in Ci/mmol.
Source: Reference 85.

replicate slices and motorized cryostats decreases variability in autoradiograms of these higher-energy particle emitters.

Despite the challenges in making autoradiograms with the nuclides used for imaging in nuclear medicine, the technique has nonetheless been successfully employed. Pioneering experiments demonstrated that the autoradiographic method can be applied to biomarkers labeled with the positron emitters ^{18}F[86,87] and ^{11}C.[88,89] High-resolution images from autoradiography are much better than what is obtained by external detection, and the images can be quantified with use of densitometers. Autoradiography has proven useful not only for imaging low-energy betas and positrons but also for single-photon emitting nuclides, including ^{99m}Tc, ^{201}Tl, and ^{111}In.[90]

Some further advances in traditional autoradiography have been adapted to radiopharmaceutical research. Micro-autoradiography has demonstrated utility for evaluation of the microdosimetry of radiolabeled tracers in tumors, bone, other organs, and inflammation.[90] This method has also been used to determine the subcellular distribution of ^{99m}Tc-exametazime (^{99m}Tc-HMPAO) in leukocytes.[91] Fluorography has also been employed to improve autoradiography with ^{18}F.[92,93]

Electronic Autoradiography

From the radiopharmaceutical development perspective, there are two major disadvantages of autoradiography with film. First, the exposure times need to be optimized, and exposure can involve lengthy intervals. Second, the optical response to radiation is linear for only a limited portion of the film response curve. This makes quantitative autoradiography tedious and time-consuming. To circumvent this problem, an electronic digital radioimager has been developed that couples a multi-wire proportional chamber to a microchannel array detector.[94] Autoradiographic images can thereby be acquired without use of film. Such electronic autoradiography generates images much faster than traditional autoradiography, and the response is linear over a wide range of radioactivity. Electronic auto-radiography systems are available commercially and have proven useful for autoradiography of compounds labeled with the single-photon emitters ^{99m}Tc[94–96] and ^{111}In[95,96] as well as the positron emitters ^{11}C,[97] ^{18}F,[98] and ^{64}Cu.[97,98]

Disadvantages of such electronic autoradiography are lower spatial resolution and inability to do the hybrid imaging that is possible with traditional autoradiography. However, the disadvantages are outweighed by the advantage of rapid acquisition of image data (minutes instead of hours to days). This streamlined electronic audioradiographic methodology has found favor within the nuclear medicine research community for convenient measurement of biomarker distributions in a straightforward and timely manner.

Dual-Isotope Autoradiography

Another attribute of autoradiography that is valuable for radiopharmaceutical development is the option for dual-label autoradiography. Dual-label imaging promotes validation by comparing the distribution of a radiolabeled biomarker to a gold standard within the same animal. Interindividual differences thus do not have to be accounted for in comparing the binding of the two compounds. This method uses two sequential film exposures. The exposures will reproduce the localization of the radiolabeled biomarker (bearing a short-lived nuclide typical for use in nuclear medicine) and the distribution of a gold standard compound (labeled with a long-lived beta-emitting nuclide). The initial autoradiogram will represent predominantly the short-lived tracer's distribution, whereas the second exposure (performed after the first nuclide has decayed away) represents the localization of the gold standard labeled with ^{3}H, ^{14}C, or ^{125}I. The radiopharmaceutical development literature is replete with examples of dual-isotope autoradiography studies.[87,97,98–105]

Role of Autoradiography in Drug Development

As described above, autoradiography can play an important role in drug development in general[84,106] and radiopharmaceutical development in particular. Autoradiography can be used to measure the biodistribution of tracers for dosimetry estimation and to evaluate the effect of pharmacologic intervention on the binding of radiolabeled biomarkers to targets. The effect of coadministration of specific-binding ligands facilitates validation of the specificity of the radiolabeled probe. A major role of autoradiography in the radiopharmaceutical development sequence is preliminary evaluation of whether imaging studies with a potential biomarker have adequate power for detection of specific regional changes in the target.[107] This is a critical, go/no-go decision point in tracer development.

EX VIVO TISSUE BIODISTRIBUTION

Once the specificity of the radiopharmaceutical has been shown with in vitro testing, the next step in drug development is evaluation of tracer localization in vivo. This is an important evaluative step, because in vitro studies may demonstrate that the probe has affinity for the target, but bioavailability may prevent the tracer from coming into contact with the target to engage in binding.

A disadvantage of the in vitro binding studies described in the preceding section is that they offer no information on the tracer kinetics of the potential biomarker. Tracer kinetics is a key parameter of radiopharmaceutical development, because it influences delivery of the biomarker to the target site. In vitro binding studies will indicate whether a lead compound has affinity for the target, but they cannot ascertain whether the biomarker will distribute to the target after it is administered by injection.

Justification for Ex Vivo Studies

In vivo delivery of the tracer to the target site is affected by several factors, including plasma protein binding, metabolism, perfusion, and partitioning across membranes (Table 17-9).

TABLE 17-9 Factors Influencing Tracer Localization to the Target

Perfusion
Metabolism
Protein binding
Clearance
Nonspecific binding
Competitive displacement by endogenous compounds

These parameters come into play during in vivo studies in intact organisms but not in the simplified, more ideal conditions designed into in vitro binding assays. High protein binding and rapid metabolism with redistribution of radioactive metabolites will limit the amount of radiopharmaceutical available for interaction with the target.

Moreover, radiopharmaceuticals require high image contrast to be successful. Image contrast is affected by clearance of the tracer from nontarget sites, as well as by the redistribution of radiolabeled metabolites. Tracer kinetic characteristics like slow clearance or high nonspecific binding will lower image contrast for target binding by increasing the background activity. These parameters of tracer development cannot be assessed with in vitro binding studies, since they utilize excised tissue sections and do not examine the intact subject.

A further shortcoming of in vitro binding studies of lead structures is that they usually involve study of the lead compound in a fluid optimized to maximize binding to the target under in vitro conditions. The pH, isotonicity, and electrolytes used in media for binding studies do not necessarily correspond to a radiopharmaceutical vehicle that is designed to be physiologically compatible and to promote product stability. In vivo competition for the target binding site by endogenous ligands may displace the radiotracer from the binding site and decrease image contrast, but this cannot be duplicated in vitro.

All of these biologic parameters conspire to decrease the signal from radiopharmaceutical–target interaction and must be evaluated using ex vivo experiments in intact animals to assess the promise of the radiopharmaceutical candidate for successful in vivo imaging.

Control Studies of Biomarker Distribution

The most direct way to determine tracer kinetics is with ex vivo biodistribution studies. The method basically involves administering the tracer to rodents by intravenous injection at a peripheral site and euthanizing the animals at a predetermined time after injection. The animal is anesthetized for tracer administration but allowed to awaken afterwards. This minimizes potential iatrogenic influence of the anesthesia on the biodistribution of the radiotracer. Following death of the animal, the various organs are dissected out, blotted dry of blood, and weighed, and the radioactivity content is measured in a well-type scintillation counter. Large organs can be sectioned and the entire organ measured as the sum of the individual components, or a representative section may be used (for example, a single bone may be used instead of the entire skeleton). The scintillation counter may be manually operated or automated; automation has the advantage of faster throughput for analysis of several subjects.

Along with the tissue samples from the animal, a standard dilution of the injectate is also measured in the scintillation counter. For each tissue sample, the counts per minute (cpm) per gram of tissue can be calculated. The cpm can be compared with that in the standard sample of the injectate to calculate the percent injected dose per gram of tissue.

By performing such tests on multiple animals with differing euthanasia times, a tissue–activity curve can be constructed as a function of time for each organ sampled. Such tissue–activity curves give very good preliminary information about the tracer kinetics. The data provide a good assessment of the likelihood of a radiotracer being useful for noninvasive imaging of the target. Through evaluation of the accumulation of activity in the target-rich tissue, it is possible to ascertain the rate at which the biomarker localizes to the target. The retention of activity in the region of interest can also be evaluated from the rate at which the activity in the target tissue decreases after a peak accumulation is reached.

An index of image contrast can also be obtained by comparing the localization of activity within the target-rich tissue with that in the other organs and tissues that were sampled. This target-to-nontarget (T:NT) ratio should be as high as possible for optimal image contrast. Note that the T:NT ratio represents the accumulation of radioactivity from the biomarker and any radiometabolites that may have formed in vivo. The behavior of radiometabolites in vivo may dramatically change the T:NT ratio from that attributed to the nonmetabolized biomarker alone.

In addition to assessment of the tracer's kinetics, the tracer's biodistribution as a function of time can be used to estimate absorbed radiation dosimetry. These calculations are usually not performed in the preliminary evaluation of a lead compound, but the raw data used in these ex vivo studies can be archived for later use if the tracer goes on to be developed as a bona fide radiopharmaceutical to be used in human subjects.

Target-Blocking Studies

In addition to measuring tracer kinetics in the control case, ex vivo studies can be used with interventional measurements that validate the hypothesized mechanism of localization for the biomarker. That is, once control studies indicate that the biomarker selectively localizes within target-rich tissue with relatively low retention in nontarget tissues, target-blocking studies can be performed to demonstrate that the selective localization is indeed due to binding to the target. Unlabeled target-binding ligand administered by preinjection or co-injection will competitively bind to the target, making binding sites unavailable for the radiolabeled tracer. If localization of the tracer is due to binding to the target, selective uptake will be prevented in the blocking study. If selective localization is not prevented, it suggests that the localization in the tissue is not due to binding to the targeted site.

In a similar method, target specificity can also be measured by competitive ex vivo studies. As an example, consider the case in which two different targets (T_1 and T_2) are located in a given organ or tissue under study. If a biomarker selectively binds to only one of these, say T_1, then administration of unlabeled T_1-selective ligand will block localization in vivo, whereas administration of T_2-selective ligand will not. If there is partial blockade by administration of unlabeled T_1-selective or T_2-selective ligand, this indicates that the biomarker is binding to both targets T_1 and T_2 and is not selective for either one of these. Note that this description of ex vivo evaluation of biomarker selectivity is very similar to that for in vitro binding assays (Figure 17-5).

A further application of ex vivo studies is to evaluate the reversibility of biomarker binding to the target. For this application, timing of the study is critical. The radiolabeled biomarker is given to the test animal and allowed to accumulate in the target-rich tissue. The time for this is determined through control tests to evaluate the tracer kinetics. After optimum accumulation of tracer in the target-rich tissue, a second injection is given of unlabeled ligand with affinity for the same binding site. If the biomarker binds to the target in a reversible fashion, then the postadministration of competitive ligand will displace the radiotracer from its binding site. Displaceability is a valuable characteristic of a biomarker for measurement of drug potencies. For valid conclusions to be reached about the results, the design of such ex vivo displacement studies must account for optimization of the dose, pharmacokinetics, and pharmacologic effects of the unlabeled displacing ligand.[108]

Animal Care

Special mention needs to be made regarding animal care in these studies. Unlike in vitro studies, ex vivo (and in vivo, see below) studies involve experiments with intact animal subjects that must be treated humanely. Strict federal regulations have been enacted regarding the care and use of test animals, including housing and feeding conditions, transportation, and minimization of stress.[109] Institutional approval and oversight is required for all experiments involving live animals to ensure that all regulations are followed.

For all of the study types described above, the number of animals used should be the minimum needed to arrive at statistically valid data. This is for ethical reasons as well as to minimize the cost of radiopharmaceutical development. Guidelines have been developed for standardizing conditions and minimizing bias between experimental groups for interventional studies.[108] Groups of 5 to 12 animals are needed in most cases to achieve 80% power.[110,111]

SMALL-ANIMAL IMAGING

Although ex vivo studies generate substantial information for tracer development, they have a major drawback: for each individual subject, data are obtained for a single time point only, that at which the animal was euthanized. Assessment of tracer kinetics thus requires multiple subjects, with at least one animal killed to determine biodistribution data for each time point. Greater numbers are actually required, since multiple individual subjects are needed for each time point for the purpose of generating a mean and standard deviation with adequate power for statistical comparison between groups.[108,110,111]

These empirical shortcomings have been addressed by the development of small-animal scanners (microPET and microSPECT). These devices make possible noninvasive imaging of small research animals, usually rodents. Small-animal scanners are designed with a small field of view, and the closer placement of detector elements leads to improvements in spatial resolution and sensitivity compared with clinical scanners. The performance of PET and SPECT scanners is compared in Table 17-10.[112]

Small-Animal PET and SPECT

The complementary nature of small-animal PET and SPECT has been discussed elsewhere in detail.[113,114] Both microPET and microSPECT have improved spatial resolution compared with corresponding clinical scanners. Unlike human imaging, SPECT imaging and PET imaging of small animals have approximately the same image resolution. Pinhole technology vastly improves the resolution of microSPECT compared with clinical SPECT. The high resolution of microSPECT makes the method much more competitive with microPET for preclinical research applications. This is unlike the situation for clinical imaging, in which PET has superior image resolution. Despite the comparable spatial resolution of small-animal SPECT and PET, however, the sensitivity of SPECT remains low compared with PET. In some cases, this may have an impact on its relative utility for radiopharmaceutical development.

Small-animal PET is often more useful than SPECT for preclinical studies. The greater sensitivity of PET facilitates detection of very low levels of radioactivity. This is an advantage for small-animal scanning, in which pharmacologic and

TABLE 17-10 Comparative Performance of PET and SPECT Scanners

	Small Animal		Clinical	
	PET	SPECT	PET	SPECT
Sensitivity (%)	2–4	0.3	1–3	0.01–0.03
Spatial resolution (mm)	1.5	1.2	5	10

Source: Reference 112.

radiologic dose to the animal is a consideration (discussed below). PET is preferred for dynamic imaging of a biomarker, since tomographic data are acquired simultaneously for all angles around the field of view. SPECT is more suited for static imaging, as a complete rotation of detector elements around the subject is required for each image frame. Because of the accurate attenuation correction of PET, small-animal scanning can be quantitative, facilitating the development of tracer kinetic models for noninvasive estimation of various physiologic and pharmacologic descriptors of the target. This is more difficult with SPECT because of inaccurate attenuation correction. As in clinical imaging, PET tracers are labeled with relatively short-lived nuclides, which is convenient for repeat studies and mitigates radiation safety concerns.

Despite the distinct advantages of microPET, there are special attributes of small-animal SPECT that may have unique experimental utility in preclinical molecular imaging. As discussed above, SPECT tracers are labeled with nuclides of heavy elements that have half-lives longer than the nuclides typically used for PET. This means the observational window is generally wider with SPECT than PET, which is advantageous for imaging slower biologic processes. Whereas molecular imaging studies with PET typically occur over minutes to hours, SPECT study durations range from several hours to days. This experimental option is especially valuable for study of biomarker binding to targets when the nonspecifically bound tracer has a slow rate of clearance.

A further technical advantage that is unique to SPECT imaging is the option of performing dual-isotope imaging. Through discrimination of photopeak energies, simultaneous acquisition of SPECT data can be obtained for two separate single-photon emitting nuclides. This is not possible with PET, because all positron-emitting nuclides generate the same 511 keV annihilation photons. Dual-isotope SPECT imaging generates perfectly registered images for the two respective nuclides, since they are acquired concurrently in the same animal. This imaging method would be ideal for direct comparison of two different biomarkers, each labeled with a separate radionuclide.

Advantages of Small Animal Imaging

Regardless of whether PET or SPECT is employed, use of these methods in noninvasive small-animal scanning is invaluable in radiopharmaceutical development. The use of these imaging techniques fits into the spectrum of translational research as the interface between ex vivo studies in animals and large-scale clinical imaging of humans. In the development of biomarkers into radiopharmaceuticals, repeat studies can be performed in the same animal subject, which can decrease the statistical noise associated with interindividual differences when group comparisons are made.[115] Longitudinal studies can be performed, including imaging of the same subject before and after a pharmacologic intervention. In addition, molecular imaging procedures developed in small animals can often be directly transferred to imaging protocols for human subjects.

A particularly useful aspect of microPET and microSPECT is the fact that noninvasive imaging of biomarker localization in the mouse (as opposed to larger animal species) is now experimentally possible. The mouse has evolved into the preferred subject for biomarker development since sequencing of the mouse genome,[116] and manipulation of this genome has allowed the creation of models of human disease.[117] For a given target that is associated with a disease-related genomic alteration, imaging studies of the appropriate gene-knockout mouse are a straightforward means of validating the biomarker.[118] That is, gene-knockout mice and the respective wild type should be identical in all respects other than the absence of the target. Differences in PET or SPECT imaging of the two animals can be attributed solely to differences in binding of the biomarker to the target. Experimentally, this method gives much cleaner validation data than the alternative method of pharmacologically blocking the target, because the drugs used as blocking agents have side effects that may alter the physiologic equilibrium of the test subject and thereby introduce bias in the results.

Moreover, since animal death is not a prerequisite for data acquisition, experimental studies of expensive transgenic mice as models of human disease are feasible with these imaging methods. Small-animal imaging is the only practical way of performing radiopharmaceutical development experiments with transgenic mice. Because of the scarcity and expense of these animals, experimental designs in which the transgenic animal must be killed would be a waste of a precious resource. With noninvasive imaging in which the subjects survive, experimentation with these valuable test animals becomes cost-effective.

Limitations of Small Animal Imaging

It is important to be aware of intrinsic challenges in the imaging of small animals with PET or SPECT. The weight of an adult mouse is only 20 to 26 grams,[119] which is only 0.02% to 0.04% that of a 70 kg adult human. Also, the blood volume of a mouse is only 5.5 mL/100 grams, or 1.1 to 1.4 mL.[120] This means that the maximum injection volume for the biomarker should be about 0.1 mL. A further challenge is that because of the small mass of the mouse, it is not possible to perform microPET or microSPECT studies with a radiopharmaceutical dose (administered activity) only 0.04% that of a human study. To achieve adequate count statistics in the face of the modest scanner sensitivity shown in Table 17-10, it is necessary to give a dose similar to that used for clinical studies in humans.[121] This results in a much greater dose, in microcuries per gram, for mice than for humans.

The increased dose needed to achieve adequate images in mice creates difficulties in two ways. First, depending on the specific activity of the tracer, increased amounts of nonradioactive biomarker will be administered concomitantly with the higher doses of radiolabeled tracer. Depending on the pharmacologic and toxicologic profile of the biomarker, the physiologic equilibrium of the mouse under study may thereby be affected.[122,123] From this perspective, SPECT may be easier to use than PET, since the single-photon emitting nuclides typically have higher specific activity than positron emitters.

Second, in addition to any pharmacologic or toxicologic effect, the radiation burden to the mouse may be several orders of magnitude higher than in human subjects.[124] Any radiation damage resulting from such high doses may alter the normal physiology of the mouse. In this way, it is possible that radiation effects may introduce bias in longitudinal studies with mice.

A related caveat with regard to small-animal imaging is that rodents must be anesthetized during the imaging procedure.[121] The anesthetic used for this purpose may interfere pharmacologically with biomarker–target binding and thereby affect the imaging studies. In any event, anesthesia is an experimental parameter to be addressed in study design with small-animal scanning that is not an issue in the clinical imaging of human subjects.

Although challenges exist, small-animal scanning is an incredibly valuable research tool that can be considered the apex of the preclinical development of radiopharmaceuticals. The noninvasive techniques that are optimized for promising biomarkers often can be directly adapted to human imaging protocols, once species differences in physiology are addressed.[120]

DRUG DISCOVERY

Thus far, this chapter has outlined the development of tracers through the following pathway: identification of lead structures, radiosynthesis, and in vitro and ex vivo screening tests. For compounds that succeed through these stages as well as through in vivo validation by small-animal imaging, the next step to radiopharmaceutical development is preparation for human use.

However, even if a tracer is not further developed into a bona fide radiopharmaceutical for imaging human subjects, it may find widespread use as a method for preclinical evaluation of lead compounds with potential activity at a given target. Therefore, a brief discussion of the growing use of radiolabeled biomarkers in drug development by the pharmaceutical industry is in order.

Testing in knockout mice is a fast way to validate the localization of radiotracers to a specific target. Knockout mice have been used to study Alzheimer's disease, glycogen storage disease, and multidrug resistance (MDR).[125] Unlike pharmacologic validation studies, which are complicated by complex biochemical changes induced by the intervention drug, knockout mice produce a single clear biochemical alteration that is easier to interpret. That is, differences in the radiotracer binding in wild-type mice compared with target-knockout mice give a clear signal for the amount of binding to the target, since all other variables are the same. Biologic results are thus much easier to interpret, which streamlines drug development by making evaluation simpler and faster.

It has been suggested that small-animal scanning is most useful for deciding go/no-go late in the developmental process, when the number of potential drug candidates is less than 10.[125] In vivo scanning would allow determination of the pharmacokinetics or pharmacodynamics of lead compounds, together with validation of agent binding to the specific target. This would substantially shorten the drug approval process.

Molecular imaging is useful for selecting promising drug candidates for further rigorous clinical trials, as well as terminating candidates with less likelihood for success.[126] Such pivotal information saves time and expense in the drug development process, because it allows researchers to focus on those drug entities that have more probability of success, rather than devoting effort to unidentified unsuccessful agents. In this manner, the overall efficiency of drug development is improved. Several reviews of the use of molecular imaging in preclinical drug development have been published.[125–127]

Although this discussion has centered on the preclinical use of molecular imaging agents for drug development, the use of appropriately developed radiopharmaceuticals can likewise speed drug development through the application of clinical imaging studies in human subjects. In this regard, proof-of-concept is established when target binding measured in vivo is linked to a significant change in a clinical endpoint.[128]

CLINICAL RESEARCH WITH HUMAN SUBJECTS

For those lead structures that succeed through the many levels of preclinical scrutiny, a key step in further development is to gain approval for use in human subjects. The traditional, albeit expensive, way for new chemical entities (NCEs) to be approved for human use is through submission of an investigational new drug (IND) application to FDA.[129] A potential new drug can then be evaluated via Phases 1 through 4 clinical trials (Figure 17-8).

It is important to note that prior to the initiation of any human imaging studies, the preparation of the radiotracer must be upgraded from research grade to pharmaceutical grade. At this juncture, the biomarker makes the transition from a radiochemical to a radiopharmaceutical. That is, whereas preparation of the labeled biomarker for preclinical studies must meet high standards of purity, even higher requirements of product quality must be met for a tracer to be used in human subjects. Besides being pure in the radiochemical, radionuclidic, and chemical senses, the compound must also demonstrate sterility and apyrogenicity to be approved for human use. In addition to passing stringent product tests to meet all drug release specifications, a radiopharmaceutical must meet fastidious production controls. Several guidelines relevant to radiopharmaceutical production are available from official organizations in the United States and European Union.[130–139]

A difficulty in obtaining an IND approval is that extensive pharmacologic toxicity (pharm/tox) data are required prior to the initiation of human studies. Pharm/tox studies are very expensive, and the requirement to perform them would often prevent further development of potentially useful radiopharmaceuticals. There are two potential pathways for human imaging with new tracers that are exempt from

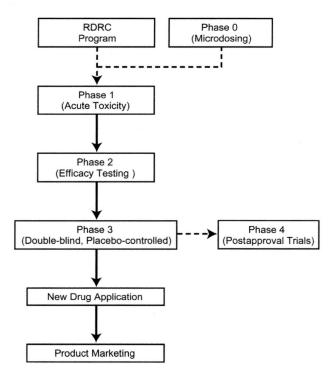

FIGURE 17-8 Regulatory aspects of radiopharmaceutical use in humans.

TABLE 17-11 Study Parameters Evaluated by Radioactive Drug Research Committee (RDRC)

1. Radiation dose—as low as practical
2. Pharmacologic toxicity—no clinically detectable pharmacologic effect
3. Qualification of the investigators is appropriate
4. Institution is licensed for radioactive materials
5. Selection and consent of subjects is appropriate
6. Radioactive drug quality meets appropriate standards
7. Rationale and scientific value of study protocol is valid
8. Potential adverse effects are reported immediately to RDRC and then to FDA
9. Concurrent approval by institutional review board

submission of the traditional IND application. These are the Radioactive Drug Research Committee (RDRC) program and the exploratory IND (eIND).

Radioactive Drug Research Committee

A regulatory pathway that is especially convenient for new imaging studies in humans involves evaluation and approval by the Radioactive Drug Research Committee (RDRC).[140] The RDRC acts as a local branch of FDA at research institutions; RDRC approval can be considered a bridge approval for scientific research prior to formal clinical trials associated with a traditional IND application. Historically, the RDRC has played an important role in radiopharmaceutical research with human subjects.[141] The RDRC reviews clinical studies of limited numbers of subjects (≤30) in research that is not intended for therapeutic or diagnostic purposes or to determine safety and effectiveness of the test substance. According to the Code of Federal Regulations in 21 CFR 361.1, study approval by the RDRC is strictly confined to scientific investigations regarding tracer localization, distribution, kinetics, or metabolism. Studies to determine safety and efficacy are specifically excluded from the RDRC domain. RDRC review of studies involves a broad spectrum of requirements addressing quality and safety (Table 17-11). Although studies approved by the RDRC are intended only to further scientific knowledge, the information gained from these investigations may have implications for subsequent therapeutic or diagnostic evaluation via the traditional IND.

A major limitation of the RDRC program is that it specifically precludes first-in-human studies. Research studies in human subjects may be allowed with already approved radiopharmaceuticals or when safety data are provided from other investigators who have previously used the radiotracer in human subjects. Typically, such data are acquired from clinical studies performed in countries other than the United States. RDRC approval would therefore not be an option for development of a novel biomarker that has never before been used in humans. Unfortunately, most cutting-edge biomarkers under investigation are unlikely to have been previously used in humans.

Exploratory IND

To promote drug development and shorten the time needed to bring new drugs to the market, FDA launched the Critical Path Initiative.[142] It was noted that many NCEs initiated with an IND application failed to develop into new drugs. To improve identification of poor candidates early in the developmental process, an exploratory IND (eIND) was designated for examination of test compounds in a Phase 0 clinical trial. Phase 0 trials are microdosing studies in which a single subtherapeutic dose of a test drug is administered to a small number of subjects (10–15) to ascertain preliminary pharmacokinetic or pharmacodynamic data.[143] Phase 0 trials have no therapeutic intent and provide no information on drug safety or efficacy. However, these clinical trials generate valuable human (rather than animal) data that improve the precision of go/no-go decisions regarding further development of an NCE via a traditional IND application. The initiation of the eIND is a paradigm shift in drug development because it eliminates the costs for extensive pharm/tox studies as well as for full-scale commercial manufacture of test substances. The pharmaceutical industry has cautiously initiated the use of Phase 0 studies for drug development.[144] Introduction of the eIND has led to increased application of biomarkers in drug development,[145,146] as well as the development of a Critical Path Institute (C-Path) for drug company cooperation with preclinical data sharing.[146,147]

From the radiopharmaceutical perspective, the eIND is a streamlined regulatory pathway for evaluation of first-in-human radiolabeled biomarkers. It therefore fills the gap left by the RDRC program concerning initial clinical development of novel radiopharmaceuticals. By definition, the very low mass of radiotracers should be subtherapeutic, making the eIND a valuable means for demonstration of proof-of-concept as well as for selection of optimal radiopharmaceutical candidates from a research portfolio.[148] Phase 0 studies can thus facilitate early identification of unpromising candidates so that further development can be focused on the best candidates for the standard Phase 1–4 clinical trials. The eIND has recently been used for commercial development of ^{18}F-labeled amyloid-imaging radiopharmaceuticals,[149] and it has been further suggested that radiopharmaceuticals be applied as surrogate markers for drug development in accordance with the Critical Path Initiative.[150]

Additional information on RDRC and IND regulations can be found in Chapter 18.

COMMERCIALIZATION OF RADIOPHARMACEUTICALS

Radiotracers deemed worthy of commercial investment require full evaluation via the traditional IND application; the exemptions described above for research studies do not apply. The IND procedure is identical to that used by FDA for conventional drug development and culminates in a new drug application (NDA).[151] Prior to the initiation of any clinical trial with an NCE, an IND application must be submitted to FDA for one or more phases of investigation.[129] The different types of clinical investigations for an IND leading to an NDA are shown in Figure 17-8. Note that a Phase 4 clinical trial is performed after marketing and is not required for initial approval of a new drug for commercialization.

IND applications contain details that are scrutinized by FDA prior to approval. The three broad areas described in the IND application are pharmacology and toxicology studies, manufacturing information, and clinical protocol and investigator information. Pharmacology and toxicology information includes data from preclinical studies in animals as well as any previous experience in human subjects from foreign sources. The manufacturing information is submitted to ensure that the drug can be produced and supplied in a consistent manner; it includes information regarding composition, manufacture, stability, and controls. IND protocols are evaluated to ensure that research subjects are not exposed to unnecessary risk and that the investigators are qualified to perform all professional responsibilities relating to the clinical trial. Clinical trials are allowed to enroll only subjects who provide informed consent, and the studies must also be reviewed and approved by the local institutional review board (IRB).

Phase 1 Trials

Phase 1 trials are the first time the drug is used in humans and are therefore designed to assess safety. Small numbers of healthy volunteers (20–100) are typically studied for several months. The focus of Phase 1 studies is to determine a safe dosage range and to identify any acute side effects. From the perspective of radiopharmaceutical development, tracer kinetics and radiation dosimetry might be examined during Phase 1.

Phase 2 Trials

In Phase 2 clinical trials, the drug is tested for the first time in patients with the targeted disease. Several hundred patients may be studied over a period of several months to years to demonstrate proof of concept that there is treatment benefit. In terms of radiopharmaceutical development, Phase 2 trials might involve imaging studies with the new tracer in diseased patients to demonstrate that target binding by the new radiopharmaceutical is different in patients than in healthy control subjects.

Phase 3 Trials

The goal of Phase 3 clinical trials is to demonstrate the utility of the drug in large populations of patients with the targeted disease. Typically, multicenter trials are performed over several years for this purpose. The test drug may be compared with placebo or existing treatments and, optimally, double-blind study designs are used. Clinical endpoints are used to validate treatment benefit. For radiopharmaceutical development, Phase 3 trials might involve the validation of surrogate endpoints that link imaging data to clinically significant endpoints.[150]

New Drug Application (NDA)

Upon successful completion of a Phase 3 clinical trial, the sponsor may submit an NDA to FDA. If the data are sufficient to demonstrate safety and efficacy, FDA will approve the NDA, which gives the sponsor the legal right to market the drug product.

CLINICAL RESEARCH INITIATIVES

The findings of research groups worldwide have led to several biomarkers with sufficient promise in preclinical studies to warrant further studies in human subjects. Tables 17-12[152–305] and 17-13[306–420] list examples of these promising biomarkers. These tables include biomarkers that have been used in clinical research, but the lists are not comprehensive; several additional biomarkers have been omitted in the interest of conciseness. The biomarkers that are listed have wide acceptance in the imaging community, as demonstrated by continued use for imaging of human subjects, often by several different

TABLE 17-12 Positron-Emitting Biomarkers Used for Clinical Research Studies with PET

Type of Biomarker	Molecular Target	References
Receptor Ligands		
^{11}C-raclopride	Dopamine D_2 receptors	152–161
^{11}C-NMSP	Dopamine D_2 and Serotonin 5-HT_{2A} receptors	162–166
^{11}C-NNC 112	Dopamine D_1 receptors	167–170
^{11}C-WAY 100635	Serotonin 5-HT_{1A} receptors	171–177
^{11}C-flumazenil (^{11}C-FMZ)	Benzodiazepine receptors	178–183
^{11}C-PK11195	Peripheral benzodiazepine receptors	184
^{11}C-carfentanil	Opioid receptors	185, 186
^{11}C-diprenorphine	Opioid receptors	187–193
^{18}F-NMB	Dopamine D_2 receptors	194
^{18}F-fallypride	Dopamine D_2 and D_3 receptors	195–198
^{18}F-FESP	Dopamine D_2 and serotonin 5-HT_2 receptors	199–203
^{18}F-MPPF	Serotonin 5-HT_{1A} receptors	204–208
^{18}F-altanserin	Serotonin 5-HT_{2A} receptors	209–218
^{18}F-CPFPX	Adenosine A_1 receptors	219, 220
^{18}F-FES	Estrogen receptors	221–225
^{64}Cu-TETA-octreotide	Somatostatin receptors	226
^{68}Ga-DOTA-NOC	Somatostatin receptors	227–229
Transporter Ligands		
^{11}C-DTBZ	Vesicular monoamine transporter (VMAT)	230–233
^{11}C-DASB	Serotonin transporter (SERT)	234–237
^{11}C-hydroxyephedrine	Norepinephrine transporter (NET)	238–243
^{11}C-verapamil	P-Glycoprotein multidrug transporter	244–246
^{18}F-FDG	Glucose transporter (and hexokinases)	247–252
^{18}F-FET	Amino acid transporter	253–256
Enzyme Substrates		
^{11}C-acetate	TCA cycle, fatty acid synthetase	257–260
^{11}C-MP4A	Acetylcholinesterase (AChE)	261–265
^{11}C-PMP	Acetylcholinesterase (AChE)	266–269
^{11}C-harmine (^{11}C-HAR)	Monoamine oxidase-A (MAO-A)	270, 271
^{11}C-metomidate (^{11}C-MTO)	11β-Hydroxylase	272, 273
^{18}F-fluorodopa (^{18}F-FDOPA)	Aromatic L-amino acid decarboxylase (AADC)	274–278
^{18}F-fluorodopamine (^{18}F-FDA)	Dopamine β-hydroxylase	279–283
^{18}F-fluorothymidine (^{18}F-FLT)	Thymidine kinase-1	284–289
^{124}I-FIAU	Herpes simplex virus thymidine kinase (HSV-TK)	290, 291
Antibodies		
^{89}Zr-U36	Cell membrane glycoprotein CD44 antigen	292
^{124}I-cG250	Carbonic anhydrase IX	293
Miscellaneous		
^{11}C-PIB ([^{11}C]6-OH-BTA)	Aggregates of β-amyloid peptide	294
^{18}F-BAY94-9172	Aggregates of β-amyloid peptide	295
^{18}F-fluoromisonidazole (^{18}F-FMISO)	Hypoxia	296–301
^{60}Cu-ATSM	Hypoxia	302–304
^{62}Cu-ATSM	Hypoxia	305

TABLE 17-13 Single Photon-Emitting Biomarkers Used for Clinical Research Studies with SPECT

Type of Biomarker	Molecular Target	References
Receptor Ligands		
^{99m}Tc-apcitide	Glycoprotein IIb/IIIa receptors	306–310
^{99m}Tc-TP3954	Vasoactive intestinal peptide (VIP) receptors	311, 312
^{99m}Tc-IL-2	Interleukin-2 (IL-2) receptors	313–315
^{99m}Tc-depreotide	Somatostatin receptors	316–321
^{99m}Tc-HYNIC-TOC	Somatostatin receptors	322–325
^{99m}Tc-HYNIC-annexin V	Phosphatidylserine	326–329
^{99m}Tc-NC100692	Integrin $\alpha_v\beta_3$	330–331
^{99m}Tc-DTPA-mannosyl-dextran	Macrophage mannose receptors (MR)	332–334
^{99m}Tc-GSA	Asialoglycoprotein	335–340
^{111}In-DTPA-folate	Folic acid receptors	341
^{111}In-DTPA-octreotide (pentetreotide)	Somatostatin receptor	342–344
^{111}In-biotin	Streptavidin/avidin	345–350
^{123}I-iomazenil (^{123}I-IMZ)	Central benzodiazepine (BZD) receptors	351–354
^{123}I-VEGF$_{165}$	Vascular endothelial growth factor (VEGF) receptors	355, 356
Transporter Ligands		
^{99m}Tc-TRODAT-1	Dopamine transporter (DAT)	357–364
^{99m}Tc-sestamibi	P-glycoprotein multidrug transporter	365–367
^{123}I-FP-CIT	Dopamine transporter (DAT)	368–370
^{123}I-ADAM	Serotonin transporter (SERT)	371–375
^{123}I-MIBG	Norepinephrine transporter (NET)	376–379
Enzyme Substrates		
^{99m}Tc-medronate	Aminoacyl-tRNA synthetase type 2	380–386
^{123}I-BMIPP	Enzymes of fatty acid metabolism	387–390
Antibodies		
^{99m}Tc-IMMU-4	Carcinoembryonic antigen (CEA)	391–393
^{99m}Tc-OKT3	CD3 antigen on human T-cells	394–396
^{99m}Tc-hR3	Epidermal growth factor (EGF) receptors	397, 398
^{99m}Tc-ior-egf/r3 MAb	Epidermal growth factor (EGF) receptors	399–401
^{99m}Tc-infliximab	Tumor necrosis factor-α	402–404
^{111}In-2B8	CD20 antigen on B-cell non-Hodgkin's lymphoma	405, 406
^{111}In-capromab pendetide	Prostate-specific membrane antigen (PSMA)	407–410
^{123}I-NP-4F(ab')$_2$MAb	Carcinoembryonic antigen (CEA)	411
^{123}I-T84.66 minibody	Carcinoembryonic antigen (CEA)	412
Miscellaneous		
^{99m}Tc-exametazime	Leukocytes (intracellular components)	413–418
^{99m}Tc-P483H	Leukocytes (unknown binding site)	419, 420

PET or SPECT centers. With further investigation of these radiotracers in clinical research, some of these investigational biomarkers may eventually become clinical radiopharmaceuticals for routine use in health care on a reimbursable basis. Review of the tracers in Tables 17-12 and 17-13 gives the reader an understanding of current research initiatives.

Table 17-12 lists positron-emitting biomarkers that are used with PET procedures, whereas Table 17-13 lists single-photon emitting biomarkers for application with SPECT procedures. Both tables are subdivided by type of molecular target (i.e., receptors, transporters, enzymes, antibodies, and miscellaneous) for the respective biomarkers, and the radionuclides used to label the biomarkers are listed. Note that virtually all of the radionuclides in Table 17-4 are represented in the human studies listed in Tables 17-12 and 17-13. For each biomarker, the radionuclide used is based on the chemical properties of the labeling substrate as well as the in vivo kinetics of the labeled compound.

Several of the biomarkers in Tables 17-12 and 17-13 are no longer strictly research tracers but are also used as radiopharmaceuticals for clinical nuclear medicine procedures. They are tabulated for two reasons. First, their in vivo localization characteristics make them suitable for specific molecular imaging applications, often in concert with other biomarkers that measure a different facet of physiology or pharmacologic mechanism. Second, these clinical radiopharmaceuticals were originally developed as research biomarkers, using the sequence of experiments discussed earlier in the chapter. The inclusion of these now clinically used radiopharmaceuticals emphasizes the pathway through which they were developed and underscores the potential for development of novel radiopharmaceuticals in the future using these same research methods.

As shown in the tables, a large number of biologic targets and associated biomarkers have been successfully applied in human imaging studies, demonstrating the great potential of molecular imaging in nuclear medicine. It is important to keep in mind that in addition to these biomarkers used in human studies, an even greater number of other biomarkers in the research pipeline are still in the preclinical stage of development. New and unique imaging methods for both clinical research and routine nuclear medicine procedures are expected to result from such ongoing research. The references cited in the tables provide details on these exciting developments taking place in molecular imaging.

SUMMARY

The development of new radiopharmaceuticals is a laborious process involving several sequential steps with rigorous requirements that must be met before advancing to the next level of evaluation. Molecular medicine offers a starting point in the quest for new radiopharmaceuticals, because it identifies aberrant proteins associated with disease processes. Screening numerous compounds for those with affinity for binding to these target proteins generates candidates for radiolabeling. Methods are developed for incorporation of PET and SPECT radionuclides into these molecular structures without destruction of target-binding affinity. Preclinical binding studies are performed with in vitro assays and autoradiography, and ex vivo studies are performed in rodents to assess the biologic behavior of the biomarker. Validation of the in vivo binding of tracers to the target is accomplished by small-animal scanning with PET or SPECT methods. The candidate biomarkers that survive this demanding battery of preclinical studies advance to clinical research in human subjects. Regulatory bodies exist to ensure the safety of subjects participating in clinical research with new tracers. Oversight of scientific studies in small numbers of subjects is attained through RDRCs, whereas development of a radiopharmaceutical for commercial use requires an IND application to FDA. INDs are evaluated through three phases of clinical trials and, if successful, the radiopharmaceutical is granted new drug status and can be marketed by drug companies for interstate commerce.

Although the development of new radiopharmaceuticals is very challenging, the impetus to develop new molecular imaging agents to assist in the detection, staging, and treatment of human disease motivates continued research.

REFERENCES

1. Mankoff DA. A definition of molecular imaging. *J Nucl Med.* 2007; 48:18N–21N.
2. Venter JC, Adams MD, Meyers EW, et al. The sequence of the human genome. *Science.* 2001;291:1304–51.
3. International Human Genome Sequencing Consortium. Initial sequencing and analysis of the human genome. *Nature.* 2001;409: 860–921.
4. Wooster R, Weber B. Breast and ovarian cancer. *N Engl J Med.* 2003;348:2339–47.
5. Turnbull C, Rahman N. Genetic predisposition to breast cancer: past, present and future. *Annu Rev Genomics Hum Genet.* 2008;9:321–45.
6. Ripperger T, Gadzicki D, Meindl A, et al. Breast cancer susceptibility: current knowledge and implication for genetic counseling. *Eur J Hum Genet.* 2009;17:722–31.
7. Febbo PG. Genomic approaches to outcome prediction in prostate cancer. *Cancer.* 2009;115(suppl 13): 3046–57.
8. Myerson M, Franklin WA, Keley MJ. Molecular classification and molecular genetics of human lung cancers. *Semin Oncol.* 2004; 31(suppl 1):4–19.
9. Lynch HT, de la Chapelle A. Hereditary colorectal cancer. *N Engl J Med.* 2003;48:919–32.
10. Markowitz SA, Bertagnollu MM. Molecular basis of colorectal cancer. *N Engl J Med.* 2009;361:49–60.
11. Staudt LM. Molecular diagnosis of the hematologic cancers. *N Engl J Med.* 2003;348:1777–85.
12. Nabel EG. Cardiovascular disease. *N Engl J Med.* 2003;349:60–72.
13. Nussbaum RL, Ellis CE. Alzheimer's disease and Parkinson's disease. *N Engl J Med.* 2003;348:1356–64.
14. Klein C, Schlossmacher MG. The genetics of Parkinson disease: implications for neurological care. *Nat Clin Pract Neurol.* 2006;2:136–46.
15. Mouradian MM. Recent advances in the genetics and pathogenesis of Parkinson's disease. *Neurology.* 2002;58:179–85.
16. Bertram L. Alzheimer's disease genetics: current status and future perspectives. *Int Rev Neurobiol.* 2009;84:167–84.
17. Brouwers N, Sleegers K, van Broeckhoven C. Molecular genetics of Alzheimer's disease: an update. *Ann Med.* 2008;40:562–83.
18. Van de Vosse E, Hoeve MA, Ottenhoff THM. Human genetics of intracellular infectious diseases: molecular and cellular immunity against mycobacteria and salmonellae. *Lancet Infect Dis.* 2004;4:739–49.
19. Small PM, Fujiwara PI. Management of tuberculosis in the United States. *N Engl J Med.* 2001;345:189–200.
20. Michael NL. Host genetics influences on HIV-1 pathogenesis. *Curr Opin Immunol.* 1999;11:466–74.
21. Evans WE, McLeod HL. Pharmacogenomics—drug disposition, drug targets and side effects. *N Engl J Med.* 2003;348:538–49.
22. Weinshilboum R. Inheritance and drug response. *N Engl J Med.* 2003;348:529–37.
23. Abstracts of the 18th International Symposium on Radiopharmaceutical Sciences. Edmonton, Canada. July 12–17, 2009. *J Lab Comp Radiopharm.* 2009;52(S1):S1–S539.
24. Abstracts of the 2009 World Molecular Imaging Congress. Montreal, Canada. September 23–26, 2009. *Mol Imaging Biol.* 2010;12(suppl 1): S3–461.
25. Abstracts of the Society of Nuclear Medicine 2009 Annual Meeting. Toronto, Canada. June 13–17, 2009. *J Nucl Med.* 2009;50(suppl 2): 1–1978.
26. Abstracts of the Annual Congress of the EANM 2009. Barcelona, Spain. October 10–14, 2009. *Eur J Nucl Med Mol Imag.* 2009;36 (suppl 2):S158–S539.

27. Manning HC, Lander A, McKinley E, et al. Accelerating the development of novel molecular imaging probes: a role for high-throughput screening. *J Nucl Med.* 2008;49:1401–4.
28. Lipinski CA, Lombardo F, Dominy BW, et al. Experimental and computational approaches to estimate solubility and permeability in drug discovery and development settings. *Adv Drug Disc Rev.* 2001;46:3–26.
29. Ghose AK, Viswanadhan VN, Wendoloski JJ. A knowledge-based approach in designing combinatorial or medicinal chemistry libraries for drug discovery. 1. A qualitative characterization of known drug databases. *J Comb Chem.* 1999;1:55–68.
30. Salum LB, Andricopulo AD. Fragment-based QSAR: perspectives in drug design. *Mol Divers.* 2009;13:277–85.
31. Dudek AZ, Arodz T, Gàlvez J. Computational methods in developing quantitative structure-activity relationships (QSARs): a review. *Comb Chem High Throughput Screen.* 2006;9:213–28.
32. Patani GA, LaVoie EJ. Bioisosterism: a rational approach in drug design. *Chem Rev.* 1996;96:3147–76.
33. Oprea TI, Davis AM, Teague SJ, et al. Is there a difference between leads and drugs? A historical perspective. *J Chem Inf Comput Sci.* 2001;41:1308–15.
34. Ikins S, Mestres J, Testa B. In silico pharmacology for drug discovery: methods for virtual ligand screening and profiling. *Br J Pharmacol.* 2007;152:9–20.
35. Klebe G. Recent developments in structure-based drug design. *J Mol Med.* 2000;78:269–81.
36. Whittle PJ, Blundell TL. Protein structure-based drug design. *Annu Rev Biophys Biomol Struct.* 1994;23:349–75.
37. Andricopoulo AD, Salum LB, Abraham DJ. Structure-based drug design strategies in medicinal chemistry. *Curr Top Med Chem.* 2009; 771–90.
38. Brown FK. Chemoinformatics: what it is and how does it impact drug discovery. *Annu Rep Med Chem.* 1998;33:375–84.
39. Howe TJ, Mahieu G, Marichal P, et al. Data reduction and representation in drug discovery. *Drug Disc Today.* 2007;12:45–53.
40. Oprea TI, Gottfries J, Sherbukhin V, et al. Chemical information management in drug discovery: optimizing the computational and combinatorial chemistry interfaces. *J Mol Graph Model.* 2000;18: 512–24.
41. Shaikh SA, Jain T, Sanhu G, et al. From drug target to leads—sketching a physicochemical pathway for lead molecule design in silico. *Curr Pharm Des.* 2007;13:3454–70.
42. Cotton FA, Wilkinson G, Murillo CA, et al. *Advanced Inorganic Chemistry.* 6th ed. New York: Wiley-Interscience; 1999.
43. Dewanjee MK. The chemistry of ^{99m}Tc-labeled radiopharmaceuticals. *Semin Nucl Med.*1990;20:5–27.
44. Steigman J, Richards P. Chemistry of technetium-99m. *Semin Nucl Med.* 1974;4 269–79.
45. Banerjee S, Pillai MR, Ramamoorthy N. Evolution of Tc-99m in diagnostic radiopharmaceuticals. *Semin Nucl Med.* 2001;31:260–77.
46. Mease RC, Lambert C. Newer methods of labeling diagnostic agents with Tc-99m. *Semin Nucl Med.* 2001;31:278–85.
47. Eckelman WC. Radiolabeling with technetium-99m to study high-capacity and low-capacity biochemical systems. *Eur J Nucl Med.* 1995;22:249–63.
48. Banerjee SR, Maresca KP, Francesconi L, et al. New directions in the coordination chemistry of ^{99m}Tc: a reflection on technetium core structures and a strategy for new chelate design. *Nucl Med Biol.* 2005;32:1–20.
49. Mahmood A, Jones AG. Technetium radiopharmaceuticals. In: Welch MJ, Redvanly CS, eds. *Handbook of Radiopharmaceuticals. Radiochemistry and Applications.* Hoboken, NJ: John Wiley and Sons; 2003:323–62.
50. Weiner RE, Thakur ML. Chemistry of gallium and indium radiopharmaceuticals. In: Welch MJ, Redvanly CS, eds. *Handbook of Radiopharmaceuticals. Radiochemistry and Applications.* Hoboken, NJ: John Wiley and Sons; 2003:363–99.
51. Welch MJ, Welch TJ. Solution chemistry of carrier-free indium. In: Subramanian G, Rhodes BA, Cooper JF, et al., eds. *Radiopharmaceuticals.* New York: Society of Nuclear Medicine; 1975:73–9.
52. Anderson CJ, Green MA, Fujibayashi Y. Chemistry of copper radionuclides and radiopharmaceutical products. In: Welch MJ, Redvanly CS, eds. *Handbook of Radiopharmaceuticals. Radiochemistry and Applications.* Hoboken, NJ: John Wiley and Sons; 2003: 401–22.
53. Wadas TJ, Wong EH, Weisman GR, et al. Copper chelation chemistry and its role in copper radiopharmaceuticals. *Curr Pharm Des.* 2007;13:3–16.
54. Shokeen M, Anderson CJ. Molecular imaging of cancer with copper-64 radiopharmaceuticals and positron emission tomography (PET). *Accts Chem Res.* 2009;42:832–41.
55. Fani M, André JP, Maecke HR. ^{68}Ga-PET: a powerful generator-based alternative to cyclotron-based PET radiopharmaceuticals. *Contrast Media Mol Imaging.* 2008;3:67–77.
56. Smith MB, March J. *March's Advanced Organic Chemistry.* 6th ed. Hoboken, NJ: John Wiley and Sons; 2007.
57. Miller PW, Long NJ, Vilar R, et al. Synthesis of ^{11}C, ^{18}F, ^{15}O, and ^{13}N radiolabels for positron emission tomography. *Angew Chem Int Ed.* 2008;47:8998–9033.
58. Ferrieri RA. Production and application of synthetic precursors labeled with carbon-11 and fluorine-18. In: Welch MJ, Redvanly CS, eds. *Handbook of Radiopharmaceuticals. Radiochemistry and Applications.* Hoboken, NJ: John Wiley and Sons; 2003:229–82.
59. Wuest F, Berndt M, Kniess T. Carbon-11 labeling chemistry based on [^{11}C]Methyl iodide. *Ernst Schering Res Found Workshop.* 2007;62: 183–213.
60. Långström B, Antoni G, Bjurling P, et al. Synthesis of compounds of interest for positron emission tomography with particular reference to synthetic strategies for ^{11}C labeling. *Acta Radiol Scand.* 1990;374: 147–51.
61. Allard M, Fouquet E, James D, et al. State of art in ^{11}C labeled radiotracers synthesis. *Curr Med Chem.* 2008;15:235–77.
62. Antoni G, Kihlberg T, Långström B. Aspects on the synthesis of ^{11}C-labelled compounds. In: Welch MJ, Redvanly CS, eds. *Handbook of Radiopharmaceuticals. Radiochemistry and Applications.* Hoboken, NJ: John Wiley and Sons; 2003:141–94.
63. Clark CJ, Aigbirhio FI. Chemistry of nitrogen-13 and oxygen-15. In: Welch MJ, Redvanly CS, eds. *Handbook of Radiopharmaceuticals. Radiochemistry and Applications.* Hoboken, NJ: John Wiley and Sons; 2003:119–40.
64. Snyder SE, Kilborn MR. Chemistry of fluorine-18 radiopharmaceuticals. In: Welch MJ, Redvanly CS, eds. *Handbook of Radiopharmaceuticals. Radiochemistry and Applications.* Hoboken, NJ: John Wiley and Sons; 2003:195–227.
65. Coenen HH. Fluorine-18 labeling methods: features and possibilities of basic reactions. *Ernst Schering Res Found Workshop.* 2007;62:15–50.
66. Wuest F. Fluorine-18 labeling of small molecules: the use of ^{18}F-labeled aryl fluorides derived from no-carrier-added [^{18}F]fluoride as labeling precursors. *Ernst Schering Res Found Workshop.* 2007;62: 51–78.
67. Finn R. Chemistry applied to iodine radioisotopes. In: Welch MJ, Redvanly CS, eds. *Handbook of Radiopharmaceuticals. Radiochemistry and Applications.* Hoboken, NJ: John Wiley and Sons; 2003:423–40.
68. Seevers RH, Counsell RE. Radioiodination techniques for small molecules. *Accts Chem Res.* 1982;82:575–90.
69. Coenen HH, Moerlein SM, Stöcklin G. No-carrier-added radiohalogenation methods with heavy halogens. *Radiochim Acta.* 1983;34: 47–68.
70. Rowland DJ, McCarthy TJ, Welch MJ. Radiobromine for imaging and therapy. In: Welch MJ, Redvanly CS, eds. *Handbook of Radiopharmaceuticals. Radiochemistry and Applications.* Hoboken, NJ: John Wiley and Sons; 2003:441–65.
71. Nogrady T, Weaver DF. *Medicinal Chemistry. A Molecular and Biochemical Approach.* New York: Oxford University Press; 2005.
72. Bylund BD, Toews ML. Radioligand binding methods: practical guide and tips. *Am J Physiol.* 1993;265:L421–9.
73. Keen M, MacDermot J. Analysis of receptors by radioligand binding. In: Wharton J, Polak JM, eds. *Receptor Autoradiography. Principles and Practice.* New York: Oxford University Press; 1993:23–55.
74. Martin IL, Davies M, Dunn SMJ. Characterization of receptors by radiolabelled ligand binding. In: Baker G, Dunn S, Holt A, eds.

Handbook of Neurochemistry and Molecular Neurobiology. 3rd ed. New York: Springer Science; 2007:258–74.
75. Haylett DG. Direct measurement of drug binding to receptors. In: Freman JC, Johansen T, eds. *Textbook of Receptor Pharmacology.* 2nd ed. New York: CRC Press; 2003:153–80.
76. Eckelman WC, Mathis CA. Targeting proteins in vivo: in vitro guidelines. *Nucl Med Biol.* 2006;33:161–4.
77. Kilbourn MR, Mathis CA. Discussion of targeting proteins in vivo: in vitro guidelines. *Nucl Med Biol.* 2006;33:449–51.
78. Wharton J, Polak JM. *Receptor Autoradiography. Principles and Practice.* New York: Oxford University Press; 1993.
79. Walsh DA, Wharton J. Autoradiography of enzymes, second messenger systems and ion channels. In: Davenport AP, ed. *Methods in Molecular Biology, vol 306: Receptor Binding.* 2nd ed. Totowa, NJ: Humana Press; 2005:139–54.
80. Harvey BW. Autoradiography and fluorography. In: Walker JM, Rapley R, eds. *Molecular Biomethods Handbook.* 2nd ed. Totowa, NJ: Humana Press; 2008;397–410.
81. Rogers AW. *Techniques of Autoradiography.* 3rd ed. New York: Elsevier North-Holland Biomedical Press; 1979.
82. Curtis CG, Cross SAM, McCulloch RJ, et al. *Whole Body Autoradiography.* New York: Academic Press; 1981.
83. Palacios JM, Mengold G, Vilaro MT, et al. Recent trends in receptor analysis techniques and instrumentation. *J Chem Neuroanatomy.* 1991;4:343–53.
84. Stumpf W. Drug localization and targeting with receptor microscopic autoradiography. *J Pharmacol Toxicol Meth.* 2005;51:25–40.
85. Frey KA, Albin RL. Receptor binding techniques. *Curr Protocols Neurosci.* 1997;1.4.1–1.4.14.
86. Yonekura Y, Brill AB, Som P, et al. Quantitative autoradiography with radiopharmaceuticals, part 1: digital film-analysis system by videodensitometry: concise communication. *J Nucl Med.* 1983;24:231–7.
87. Som P, Yonekura Y, Oster ZH, et al. Quantitative autoradiography with radiopharmaceuticals, part 2: applications in radiopharmaceutical research: concise communication. *J Nucl Med.* 1983;24:238–44.
88. Matsumura K, Bergström M, Onoe H, et al. In vitro positron emission tomography (PET): use of positron emission tracers in functional imaging in living brain slices. *Neurosci Res.* 1995;22:219–29.
89. Pike VW, Halldin C, McCarrib JA, et al. [Carbonyl-^{11}C]Desmethyl-WAY-100635 (DWAY) is a potent and selective radioligand for central 5-HT$_{1A}$ receptors in vitro and in vivo. *Eur J Nucl Med.* 1998;25:338–46.
90. Puncher MRB, Blower PJ. Radionuclide targeting and dosimetry at the microscopic level: the role of microautoradiography. *Eur J Nucl Med.* 1994;21:1347–65.
91. Puncher MRB, Blower PJ. Autoradiography and density separation of technetium-99m-examtazime (HMPAO) labeled leucocytes reveals selectivity for eosinophils. *Eur J Nucl Med.* 1994;21:1175–82.
92. Johnström P, Davenport AP. Imaging and characterization of radioligands for positron emission tomography using quantitative phosphor imaging autoradiography. In: Davenport AP, ed. *Methods in Molecular Biology, vol 306: Receptor Binding Techniques.* 2nd ed. Totowa, NJ: Humana Press; 2005:203–16.
93. Nikolaus S, Larisch R, Beu M, et al. Imaging system of striatal dopamine D$_2$ receptors with a PET system for small laboratory animals in comparison with storage phosphor autoradiography: a validation study with ^{18}F-(N-methyl)benperidol. *J Nucl Med.* 2001;42:1691–6.
94. Petegnief Y, Petiet A, Paker MC, et al. Quantitative autoradiography using a radioimager based on a multiwire proportional chamber. *Phys Med Biol.* 1998;43:3629–38.
95. Sarda-Mantel L, Michel J-B, Rouzet F, et al. ^{99m}Tc-Annexin V and ^{111}In-antimyosin antibody uptake in experimental myocardial infarction in rats. *Eur J Nucl Med Mol Imaging.* 2006;33:239–45.
96. Hwang J-J, Yen T-C, Hsieh B-T, et al. Application of macroautoradiography and instant image in radiopharmaceutical research. *J Radioanalyt Nucl Chem.* 1999;241:581–7.
97. Fujibayashi Y, Cutler CS, Anderson CJ, et al. Comparative studies of Cu-64-ATSM and C-11-acetate in an acute myocardial infarction model: ex vivo imaging of hypoxia in rats. *Nucl Med Biol.* 1999;26:117–21.
98. Lewis JS, Sharp TL, Laforest R, et al. Tumor uptake of copper-diacetyl-bis(N^4-methylthiosemicarbazone): effect of changes in tissue oxygenation. *J Nucl Med.* 2001: 42:655–61.
99. Olds J, Frey KA, Ehrenkaufer RL, et al. A sequential double-label autoradiographic method that quantifies altered rates of regional glucose metabolism. *Brain Res.* 1985;361:217–24.
100. Picchio M, Beck R, Haubner R, et al. Intratumoral spatial distribution of hypoxia and angiogenesis assessed by ^{18}F-FAZA and ^{125}I-Gluco-RGD autoradiography. *J Nucl Med.* 2008;49:597–605.
101. Salber D, Gunawan J, Langen K-J, et al. Comparison of ^{99m}Tc- and ^{18}F-Ubiquicidin autoradiography to anti-*Staphylococcus aureus* immunofluorescence in rat muscle abscesses. *J Nucl Med.* 2008;49:995–9.
102. Miyagawa T, Oku T, Sasajima T, et al. Assessment of treatment response by autoradiography with ^{14}C-aminocyclopentane carboxylic acid, ^{67}Ga-DPTA, and ^{18}F-FDG in a herpes simplex virus thymidine kinase/gancyclovir brain tumor model. *J Nucl Med.* 2003;44:1845–54.
103. Tokita N, Hasegawa S, Maruyama K, et al. ^{99m}Tc-HYNIC-annexin V imaging to evaluate inflammation and apoptosis in rats with autoimmune myocarditis. *Eur J Nucl Med.* 2003;30:232–8.
104. Iwasaki T, Suzuki T, Tateno M, et al. Dual-tracer autoradiography with thallium-201 and iodine-125-metaiodobenzylguanidine in experimental myocardial infarction of rat. *J Nucl Med.* 1996;37:680–4.
105. Weinstein H, Reinhardt CP, Leppo JA. Teboroxime, sestamibi and thallium-201 as markers of myocardial hypoperfusion: comparison by quantitative dual-isotope autoradiography in rabbits. *J Nucl Med.* 1993;34:1510–7.
106. Solon EG, Balani SK, Lee FW. Whole-body autoradiography in drug discovery. *Curr Drug Metab.* 2002;3:451–62.
107. Schmidt KC, Smith CB. Resolution, sensitivity and precision with autoradiography and small animal positron emission tomography: implications for functional brain imaging in animal research. *Nucl Med Biol.* 2005;32:719–25.
108. Eckelman WC, Kilbourn MR, Joyal JL, et al. Justifying the number of animals for each experiment. *Nucl Med Biol.* 2007;34:229–32.
109. National Institutes of Health, Office of Animal Care and Use. http://oacu.od.nih.gov/. Accessed September 22, 2010.
110. Eckelman WC. Further discussions on choosing the number of animals for each experiment. *Nucl Med Biol.* 2008;35:1–2.
111. Scheibe PO. Number of samples—hypothesis testing. *Nucl Med Biol.* 2008;35:3–9.
112. Jansen FP, Vanderheyden J-L. The future of SPECT in a time of PET. *Nucl Med Biol.* 2007;34:733–5.
113. Rowland DJ, Cherry SR. Small-animal preclinical nuclear medicine instrumentation and methodology. *Semin Nucl Med.* 2008;38:209–22.
114. Chatziionnou AF. Instrumentation for molecular imaging in preclinical research. Micro-PET and micro-SPECT. *Proc Am Thorac Soc.* 2005;2:533–6.
115. Scheibe PO, Vera DR, Eckelman WC. What is to be gained by imaging the same animal before and after treatment? *Nucl Med Biol.* 2005; 32:727–32.
116. Initial sequencing and comparative analysis of the mouse genome. *Nature.* 2002;420:520–62.
117. Malakoff D. The rise of the mouse, biomedicine's model mammal. *Science.* 2000;288:248–53.
118. Eckelman WC. The use of gene-manipulated mice in the validation of receptor binding radiotracer. *Nucl Med Biol.* 2003;30:851–60.
119. Mouse phenome database. The Jackson Laboratory, Bar Harbor, ME. http://phenome.jax.org/. Accessed September 22, 2010.
120. Bernstein SE. Physiological characteristics. In: Green EL, ed. *Biology of the Laboratory Mouse.* 2nd ed. New York: Dover; 1966: chapter 16. http://www.informatics.jax.org/greenbook/index.shtml. Accessed September 22, 2010.
121. Acton PD, Kung HF. Small animal imaging with high resolution single photon emission tomography. *Nucl Med Biol.* 2003;30:889–95.
122. Hume SP, Gunn RN, Jones T. Pharmacological constraints associated with positron emission tomographic scanning of small laboratory animals. *Eur J Nucl Med.* 1998;25:173–6.
123. Kung M-P, Kung HF. Mass effect of injected dose in small rodent imaging by SPECT and PET. *Nucl Med Biol.* 2005;673–8.

124. Funk T, Sun M, Hasegawa BH. Radiation dose estimate in small animal SPECT and PET. *Med Phys.* 2004;31:2680–6.
125. Eckelman WC. The use of PET and knockout mice in the drug discovery process. *Drug Disc Today.* 2003;8:404–10.
126. Willmann JK, van Bruggen N, Dinkelborg LM, et al. Molecular imaging in drug development. *Nat Rev Drug Disc.* 2008;7:591–607.
127. Rudlin M, Weissleder R. Molecular imaging in drug discovery and development. *Nat Rev Drug Disc.* 2003;2:123–31.
128. Hargreaves RJ. The role of molecular imaging in drug discovery and development. *Clin Pharmacol Ther.* 2008;83:349–53.
129. US Food and Drug Administration. Laws, regulations, policies and procedures for drug applications. www.fda.gov/Drugs/Development ApprovalProcess/ucm090410.htm. Accessed September 22, 2010.
130. *Nuclear Pharmacy Guidelines for the Compounding of Radiopharmaceuticals.* Washington, DC: American Pharmaceutical Association; 2001. www.pharmacist.com/AM/Template.cfm?Section=Nuclear_Pharmacy_Practice&CONTENTID=2460&TEMPLATE=/CM/ContentDisplay.cfm. Accessed September 25, 2010.
131. Radiopharmaceuticals for positron emission tomography—compounding (chapter 823). In: *The United States Pharmacopeia, 29th rev, and The National Formulary, 24rd ed.* Rockville, MD: United States Pharmacopeial Convention, Inc; 2006.
132. Automated radiochemical synthesis apparatus (chapter 1015). In: *The United States Pharmacopeia, 29th rev, and The National Formulary, 24rd ed.* Rockville, MD: United States Pharmacopeial Convention, Inc; 2006.
133. Code of Federal Regulations Title 21, Part 210. Current Good Manufacturing Practice in Manufacturing Processing, Packing, or Holding of Drugs. www.access.gpo.gov/nara/cfr/waisidx_03/21cfr210_03.html. Accessed September 22, 2010.
134. Code of Federal Regulations Title 21, Part 58. Good Laboratory Practice for Nonclinical Laboratory Studies. www.accessdata.fda.gov/scripts/cdrh/cfdocs/cfcfr/CFRSearch.cfm?CFRPart=58. Accessed September 22, 2010.
135. US Food and Drug Administration. Guidance: PET Drugs—Current Good Manufacturing Practice (CGMP). December 9, 2009. www.fda.gov/Drugs/DevelopmentApprovalProcess/Manufacturing/ucm085783.htm. Accessed September 22, 2010.
136. European Association of Nuclear Medicine Radiopharmacy Committee. *Guidelines on Current Good Radiopharmacy Practice (cGRPP) in the Preparation of Radiopharmaceuticals.* 2007. www.eanm.org/scientific_info/guidelines/gl_radioph_cgrpp.pdf. Accessed September 22, 2010.
137. Decristoforo C, Elsinga P, Faivre-Chauvet A, et al. The specific case of radiopharmaceuticals and GMP—activities of the Radiopharmacy Committee. *Eur J Nucl Med Mol Imaging.* 2008;35:1400–1.
138. Verbruggen A, Coenen HH, Deverre J-R, et al. Guideline to regulations for radiopharmaceuticals in early phase clinical trials in the EU. *Eur J Nucl Med Mol Imaging.* 2008;35:2144–51.
139. Elsinga P, Todde S, Penuelas I, et al. Guidance on current good radiopharmacy practice (cGRPP) for the small-scale preparation of radiopharmaceuticals. *Eur J Nucl Med Mol Imaging.* 2010;37:1049–62.
140. US Food and Drug Administration. Radioactive Drug Research Committee (RDRC) Program. www.fda.gov/AboutFDA/Centers Offices/CDER/ucm085831.htm. Accessed September 22, 2010.
141. Suleiman OH, Fejka R, Houn F, et al. The Radioactive Drug Research Committee: background and retrospective study of reported research data (1975–2004). *J Nucl Med.* 2006;47:1220–6.
142. US Food and Drug Administration. Innovation or Stagnation: Challenge and Opportunity on the Critical Path to New Medical Products. March 2004. www.fda.gov/ScienceResearch/SpecialTopics/CriticalPathInitiative/CriticalPathOpportunitiesReports/ucm077262.htm. Accessed September 22, 2010.
143. US Food and Drug Administration. Guidance for Industry, Investigators, and Reviewers. Exploratory IND Studies. January 2006. www.fda.gov/downloads/Drugs/GuidanceComplianceRegulatoryInformation/Guidances/UCM078933.pdf. Accessed September 22, 2010.
144. Karara AH, Edeki T, McLeod J, et al. PhRMA survey on the conduct of first-in-human clinical trials under exploratory investigational new drug applications. *J Clin Pharmacol.* 2010;50:380–91.
145. Karsdal MA, Henriksen K, Leeming DJ, et al. Biochemical markers and the FDA Critical Path: how biomarkers may contribute to the understanding of pathophysiology and provide unique and necessary tools for drug development. *Biomarkers.* 2009;14:181–202.
146. Woodcock J, Woosley R. The FDA Critical Path and its influence on new drug development. *Annu Rev Med.* 2008;59:1–12.
147. Woosley RL, Myers RT, Goodsaid F. The Critical Path Institute's approach to sharing and advancing regulatory science. *Clin Pharmacol Ther.* 2010;87:530–3.
148. Mills G. The exploratory IND. *J Nucl Med.* 2008;49:45N–47N.
149. Carpenter AP, Pontecorvo MJ, Hefti FF, et al. The use of the exploratory IND in the evaluation and development of ^{18}F-PET radiopharmaceuticals for amyloid imaging in the brain: a review of one company's experience. *Q J Nucl Med Mol Imaging.* 2009;53:387–93.
150. Richter WS. Imaging biomarkers as surrogate endpoints for drug development. *Eur J Nucl Med Mol Imaging.* 2006;33:S6–S10.
151. Lipsky MS, Sharp LK. From idea to market: the drug approval process. *J Am Board Fam Pract.* 2001;14:362–7.
152. Farde L, Ehrin E, Eriksson L, et al. Substituted benzamides as ligands for visualization of dopamine receptor binding in the human brain by positron emission tomography. *Proc Natl Acad Sci USA.* 1985;82:3863–7.
153. Nordström A-L, Farde L, Wiesel F-A, et al. Central D2-dopamine receptor occupancy in relation to antipsychotic drug effects: a double-blind PET study of schizophrenic patients. *Biol Psychiatry.* 1993;33:227–35.
154. Dewey SL, Smith GS, Logan J, et al. Effects of central cholinergic blockade on striatal dopamine release measured with positron emission tomography in normal human subjects. *Proc Natl Acad Sci USA.* 1993;90:11816–20.
155. Antonini A, Schwarz J, Oertel WH, et al. [^{11}C]Raclopride and positron emission tomography in previously untreated patients with Parkinson's disease: influence of L-dopa and lisuride therapy on striatal dopamine D2-receptors. *Neurology.* 1994;1325–9.
156. Lawrence AD, Weeks RA, Brooks DJ, et al. The relationship between striatal dopamine receptor binding and cognitive performance in Huntington's disease. *Brain.* 1998;121:1343–55.
157. Kapur S, Zipursky R, Jones C, et al. Relationship between dopamine D_2 occupancy, clinical response, and side effects: a double-blind PET study of first-episode schizophrenia. *Am J Psychiatry.* 2000;157:514–20.
158. Hilker R, Klein C, Ghaemi M, et al. Positron emission tomographic analysis of the nigrostriatal dopaminergic system in familial parkinsonism associated with mutations in the parkin gene. *Ann Neurol.* 2001;49:367–76.
159. Singer HS, Szymanski S, Giuliano J, et al. Elevated intrasynaptic dopamine release in Tourette's syndrome measured by PET. *Am J Psychiatry.* 2002;159:1329–36.
160. Goerendt IK, Messa C, Lawrence AD, et al. Dopamine release during sequential finger movements in health and Parkinson's disease: a PET study. *Brain.* 2003;126:312–25.
161. Wong DF, Kuwabara H, Schretlen DJ, et al. Increased occupancy of dopamine receptors in human striatum during cue-elicited cocaine craving. *Neuropsychopharmacology.* 2006;31:2716–27.
162. Gjedde A, Wong DF. Positron tomographic quantitation of neuroreceptors in human brain in vivo—with special reference to the D2 dopamine receptors in caudate nucleus. *Neurosurg Rev.* 1987;10:9–18.
163. Tune LE, Wong DF, Pearlson G, et al. Dopamine D2 receptor density estimates in schizophrenia: a positron emission tomography study with ^{11}C-N-methylspiperone. *Psychiatry Res.* 1993;49:219–37.
164. Brandt J, Folstein SE, Wong DF, et al. D2 receptors in Huntington's disease: positron emission tomography findings and clinical correlates. *J Neuropsychiatry Clin Neurosci.* 1990;2:20–7.
165. Nyberg S, Farde L, Eriksson L, et al. 5-HT2 and D2 dopamine receptor occupancy in the living human brain. A PET study with risperidone. *Psychopharmacology (Berl).* 1993;110:265–72.
166. Nyberg S, Eriksson B, Oxensterna G, et al. Suggested minimal effective dose of risperidone based on PET-measured D2 and 5-HT2A receptor occupancy in schizophrenic patients. *Am J Psychiatry.* 1999;156:869–75.

167. Halldin C, Foged C, Chou YH, et al. Carbon-11 NNC 112: a radioligand for PET examination of striatal and neocortical D1-dopamine receptors. *J Nucl Med.* 1998;39:2061–8.
168. Abi-Dargham A, Mawlawi O, Lombardo I, et al. Prefrontal dopamine D1 receptors and working memory in schizophrenia. *J Neurosci.* 2002;22:3708–19.
169. Narendran R, Frankle WG, Keefe R, et al. Altered prefrontal dopaminergic function in chronic recreational ketamine users. *Am J Psychiatry.* 2005;162:2352–9.
170. Slifstein M, Kegeles LS, Gonzales R, et al. [^{11}C]NNC 112 selectivity for dopamine D1 and serotonin 5-HT(2A) receptors: a PET study in healthy human subjects. *J Cereb Blood Flow Metab.* 2007;27:1733–41.
171. Pike VW, McCarron JA, Lammertsma AA, et al. First delineation of 5-HT$_{1A}$ receptors in human brain with PET and [^{11}C]WAY-100635. *Eur J Pharmacol.* 1995;283:R1–R3.
172. Bailer UF, Frank GK, Henry SE, et al. Altered brain serotonin 5-HT$_{1A}$ receptor binding after recovery from anorexia nervosa measured by positron emission tomography and [*carbonyl* ^{11}C]WAY-100635. *Arch Gen Psychiatry.* 2005;62:1032–41.
173. Bhagwagar Z, Rabiner EA, Sargent PA, et al. Persistent reduction in brain serotonin1A receptor binding in recovered depressed men measured by positron emission tomography with [^{11}C]WAY-100635. *Mol Psychiatry.* 2004;9:386–92.
174. Doder M, Rabiner EA, Turjanski N, et al. Tremor in Parkinson's disease and serotonergic dysfunction: an ^{11}C-WAY-100635 PET study. *Neurology.* 1003;60:601–5.
175. Cleare AJ, Messa C, Rabiner EA, et al. Brain 5-HT$_{1A}$ receptor binding in chronic fatigue syndrome measured using positron emission tomography and [^{11}C]WAY-100635. *Biol Psychiatry.* 2005;57:239–46.
176. Savic I, Lindstrom P, Gulyas B, et al. Limbic reductions of 5-HT$_{1A}$ receptor binding in human temporal lobe epilepsy. *Neurology.* 2004;62:1343–51.
177. Meltzer CC, Price JC, Mathis CA, et al. Serotonin 1A receptor binding and treatment response in late-life depression. *Neuropsychopharmacology.* 2004;29:2258–65.
178. Heiss WD, Kracht L, Grond M, et al. [^{11}C]Flumazenil/H$_2$O positron emission tomography predicts irreversible ischemic cortical damage in stroke patients receiving acute thrombolytic therapy. *Stroke.* 2000;31:366–9.
179. Heiss WD, Sobesky J, Smekal U, et al. Probability of cortical infarction predicted by flumazenil binding and diffusion-weighted imaging signal intensity: a comparative positron emission tomography/magnetic resonance imaging study in early ischemic stroke. *Stroke.* 2004;35:1892–8.
180. Koepp MJ, Richardson MP, Brooks DJ, et al. Cerebral benzodiazepine receptors in hippocampal sclerosis. An objective in vivo analysis. *Brain.* 1996;119:1677–87.
181. Sanabria-Bohorquez SM, de Volder AG, Arno P, et al. Decreased benzodiazepine receptor density in the cerebellum of early blind human subjects. *Brain.* 2001;888:203–11.
182. Holopainen IE, Metsahonkala EL, Kokkonen H, et al. Decreased binding of [^{11}C]flumazenil in Angelman syndrome patients with GABA(A) receptor β3 subunit deletions. *Ann Neurol.* 2001;49:110–3.
183. Asahina N, Shiga T, Egawa K, et al. [^{11}C]Flumazenil positron emission tomography analyses of brain gamma-aminobutyric acid type A receptors in Angelman syndrome. *J Pediatr.* 2008;152:546–9.
184. Charbonneau P, Syrota A, Crouzel C, et al. Peripheral-type benzodiazepine receptors in the living heart characterized by positron emission tomography. *Circulation.* 1986;73:476–83.
185. Frost JJ, Wagner HN, Dannals RF, et al. Imaging opiate receptors in the human brain by positron tomography. *J Comput Assist Tomogr.* 1985;9:231–6.
186. Mayberg HS, Sadzot B, Meltzer CC, et al. Quantification of mu and non-mu opiate receptors in temporal lobe epilepsy using positron emission tomography. *Ann Neurol.* 1991;30:3–11.
187. Sadzot B, Franck G. Non-invasive methods to study drug disposition: positron emission tomography. Detection and quantification of brain receptors in man. *Eur J Drug Metab Pharmacokinet.* 1990;15:135–42.
188. Jones AK, Cunningham VJ, Ha-Kawa S, et al. Changes in central opioid receptor binding in relation to inflammation and pain in patients with rheumatoid arthritis. *Br J Rheumatol.* 1994;33:909–16.
189. Jones AK, Liyi Q, Cunningham VJ, et al. Endogenous opiate response to pain in rheumatoid arthritis and cortical and subcortical response to pain in normal volunteers using positron emission tomography. *Int J Clin Pharmacol Res.* 1991;11:261–6.
190. Jones AK, Watabe H, Cunningham VJ, et al. Cerebral decreases in opioid receptor binding in patients with central neuropathic pain measured by [^{11}C]diprenorphine binding and PET. *Eur J Pain.* 2004;8:479–85.
191. Willoch F, Schindler F, Wester HJ, et al. Central poststroke pain and reduced opioid receptor binding within pain processing circuitries: a [^{11}C]diprenorphine PET study. *Pain.* 2004;108:213–20.
192. Weeks RA, Cunningham VJ, Piccini P, et al. ^{11}C-Diprenorphine binding in Huntington's disease: a comparison of region of interest analysis with statistical parametric mapping. *J Cereb Blood Flow Metab.* 1997;17:943–9.
193. Koepp MJ, Richardson MP, Brooks DJ, et al. Focal cortical release of endogenous opioids during reading-induced seizures. *Lancet.* 1998;352:952–5.
194. Karimi M, Moerlein SM, Videen TO, et al. Decreased striatal dopamine receptor binding in primary focal dystonia: a D2 or D3 effect? *Mov Disord.* In press.
195. Grunder G, Landvogt C, Vernalken I, et al. The striatal and extrastriatal D2/D3 receptor-binding profile of clozapine in patients with schizophrenia. *Neuropsychopharmacology.* 2006;31:1027–35.
196. Riccardi P, Li R, Ansari MS, et al. Amphetamine-induced displacement of [^{18}F]fallypride in striatum and extrastriatal regions in humans. *Neuropsychopharmacology.* 2006;31:1016–26.
197. Kessler RM, Ansari MS, Riccardi P, et al. Occupancy of striatal and extrastriatal dopamine D$_2$/D$_3$ receptors by olanzapine and haloperidol. *Neuropsychopharmacology.* 2005;30:2283–9.
198. Riccardi P, Baldwin R, Alomon R, et al. Estimation of baseline dopamine D$_2$ receptor occupancy in striatum and extrastriatal regions in human brain with positron emission tomography with [^{18}F]fallypride. *Biol Psychiatry.* 2008;63:241–4.
199. Barrio JR, Satyamurthy N, Huang SC, et al. 3-(2′-[^{18}F]fluoroethyl) spiperone: in vivo biochemical and kinetic characterization in rodents, nonhuman primates, and humans. *J Cereb Blood Flow Metab.* 1989;9:830–9.
200. Wienhard K, Coenen HH, Pawlik G, et al. PET studies of dopamine receptor distribution using [^{18}F]fluoroethylspiperone: findings in disorders related to the dopaminergic system. *J Neural Transm Gen Sect.* 1990;81:195–213.
201. Messa C, Colombo C, Moresco RM, et al. 5-HT$_{2A}$ Receptor binding is reduced in drug-naïve and unchanged in SSRI-responder depressed patients compared to healthy controls: a PET study. *Psychopharmacology (Berl).* 2003;167:72–8.
202. Moresco RM, Colombo C, Fazio F, et al. Effects of fluvoxamine treatment on the in vivo binding of [F-18]FESP in drug naïve depressed patients: a PET study. *Neuroimage.* 2000;12:452–65.
203. Lucignani G, Losa M, Moresco RM, et al. Differentiation of clinically non-functioning pituitary adenomas from meningiomas and craniopharyngiomas by positron emission tomography with [^{18}F]fluoroethyl-spiperone. *Eur J Nucl Med.* 1997;24:1149–55.
204. Passchier J, van Waarde A, Pieterman RM, et al. Quantitative imaging of 5-HT$_{1A}$ receptor binding in healthy volunteers with [^{18}F]p-MPPF. *Nucl Med Biol.* 2000;27:473–6.
205. Passchier J, van Waarde A, Pieterman RM, et al. In vivo delineation of 5-HT$_{1A}$ receptors in human brain with [^{18}F]MPPF. *J Nucl Med.* 2000;41:1830–5.
206. Udo de Haes JI, Bosker FJ, van Waarde A, et al. 5-HT$_{1A}$ Receptor imaging in the human brain: effect of tryptophan depletion and infusion on [^{18}F]MPPF binding. *Synapse.* 2002;46:108–15.
207. Merlet I, Ostrowsky K, Costes N, et al. 5-HT$_{1A}$ Receptor binding and intracerebral activity in temporal lobe epilepsy: an [^{18}F]MPPF-PET study. *Brain.* 2004;127:900–13.
208. Praschak-Rieder N, Hussey D, Wilson AA, et al. Tryptophan depletion and serotonin loss in selective serotonin reuptake inhibitor-treated depression: an [^{18}F]MPPF positron emission tomography study. *Biol Psychiatry.* 2004;56:587–91.

209. Biver F, Goldman S, Luxen A, et al. Multicompartmental study of fluorine-18 altanserin binding to brain 5HT$_2$ receptors in humans using positron emission tomography. *Eur J Nucl Med.* 1994;21:937–46.
210. Sadzot B, Lamaire C, Maquet P, et al. Serotonin 5HT$_2$ receptor imaging in the human brain using positron emission tomography and a new radioligand, [^{18}F]altanserin: results in young normal controls. *J Cereb Blood Flow Metab.* 1995;15:787–97.
211. Biver F, Wikler D, Lotstra F, et al. Serotonin 5-HT$_2$ receptor imaging in major depression: focal changes in orbito-insular cortex. *Br J Psychiatry.* 1997;171:444–8.
212. Mintun MA, Sheline YI, Moerlein SM, et al. Decreased hippocampal 5-HT$_{2A}$ receptor binding in major depressive disorder: in vivo measurement with [^{18}F]altanserin positron emission tomography. *Biol Psychiatry.* 2004;55:217–24.
213. Sheline YI, Mintun MA, Barch DM, et al. Decreased hippocampal 5-HT$_{2A}$ receptor binding in older depressed patients using [^{18}F]altanserin positron emission tomography. *Neuropsychopharmacology.* 2004;29: 2235–41.
214. Meltzer CC, Price JC, Mathis CA, et al. PET imaging of serotonin type 2A receptors in late-life neuropsychiatric disorders. *Am J Psychiatry.* 1999;156:1871–8.
215. Bailer UF, Price JC, Meltzer CC, et al. Altered 5-HT$_{2A}$ receptor binding after recovery from bulimia-type anorexia nervosa: relationships to harm avoidance and drive for thinness. *Neuropsychopharmacology.* 2004;29:1143–55.
216. Frank GK, Kaye WH, Meltzer CC, et al. Reduced 5-HT$_{2A}$ receptor binding after recovery from anorexia nervosa. *Biol Psychiatry.* 2002;52: 896–906.
217. Hurlemann R, Boy C, Meyer PT, et al. Decreased prefrontal 5-HT$_{2A}$ receptor binding in subjects at enhanced risk for schizophrenia. *Anat Embryol (Berlin).* 2005;210:519–23.
218. Adams KH, Hansen ES, Pinborg LH, et al. Patients with obsessive-compulsive disorder have increased 5-HT$_{2A}$ receptor binding in the caudate nuclei. *Int J Neuropsychopharmacol.* 2005;8:391–401.
219. Bauer A, Holschbach MH, Meyer PT, et al. In vivo imaging of adenosine A$_1$ receptors in the human brain with [^{18}F]CPFPX and positron emission tomography. *Neuroimage.* 2003;19:1760–9.
220. Elmenhorst D, Meyer PT, Winz OH, et al. Sleep deprivation increases A$_1$ adenosine receptor binding in the human brain: a positron emission tomography study. *J Neurosci.* 2007;27:2410–5.
221. Mintun MA, Welch MJ, Siegel BA, et al. Breast cancer: PET imaging of estrogen receptors. *Radiology.* 1988;169:45–8.
222. Dehdashti F, Mortimer JE, Siegel BA, et al. Positron tomographic assessment of estrogen receptors in breast cancer: comparison with FDG-PET and in vitro receptor assays. *J Nucl Med.* 1995;36:1766–74.
223. McGuire AH, Dehdashti F, Siegel BA, et al. Positron tomographic assessment of 16α-[^{18}F]fluoro-17β-estradiol uptake in metastatic breast carcinoma. *J Nucl Med.* 1991;32:1526–31.
224. Dehdashti F, Flanagan FL, Mortimer JE, et al. Positron emission tomographic assessment of "metabolic flare" to predict response of metastatic breast cancer to antiestrogen therapy. *Eur J Nucl Med.* 1999;26:51–6.
225. Peterson LM, Mankoff DA, Lawton T, et al. Quantitative imaging of estrogen receptor expression in breast cancer with PET and ^{18}F-fluoroestradiol. *J Nucl Med.* 2008;49:367–74.
226. Anderson CJ, Dehdashti F, Cutler PD, et al. ^{64}Cu-TETA-octreotide as a PET imaging agent for patients with neuroendocrine tumors. *J Nucl Med.* 2001;42:213–21.
227. Fanti S, Ambrosini V, Tomassetti P, et al. Evaluation of unusual neuroendocrine tumors by means of ^{68}Ga-DOTA-NOC PET. *Biomed Pharmacother.* 2008;62:667–71.
228. Ambrosini V, Tomassetti P, Castellucci P, et al. Comparison between ^{68}Ga-DOTA-NOC and ^{18}F-DOPA PET for the detection of gastro-entero-pancreatic and lung neuro-endocrine tumors. *Eur J Nucl Med Mol Imaging.* 2008;35:1431–8.
229. Ambrosini V, Castellucci P, Rubello D, et al. ^{68}Ga-DOTA-NOC: a new PET tracer for evaluating patients with bronchial carcinoid. *Nucl Med Commun.* 2009;30:281–6.
230. Koeppe RA, Frey KA, Vander Borght TM, et al. Kinetic evaluation of [^{11}C]dihydrotetrabenazine by dynamic PET: measurement of vesicular monoamine transporter. *J Cereb Blood Flow Metab.* 1996;16: 1288–99.
231. Bohnen NI, Koeppe RA, Meyer P, et al. Decreased striatal monoaminergic terminals in Huntington disease. *Neurology.* 2000;54:1753–9.
232. Frey KA, Koeppe RA, Kilbourn MR. Imaging the vesicular monoamine transporter. *Adv Neurol.* 2001;86:237–47.
233. Gilman S, Koeppe RA, Little R, et al. Striatal monoamine terminals in Lewy body dementia and Alzheimer's disease. *Ann Neurol.* 2004;55: 774–80.
234. Houle S, Ginovart N, Hussey D, et al. Imaging the serotonin transporter with positron emission tomography: initial imaging studies with [^{11}C]DAPP and [^{11}C]DASB. *Eur J Nucl Med.* 2000;27:1719–22.
235. Meyer JH. Imaging the serotonin receptor during major depressive disorder and antidepressant treatment. *J Psychiatry Neurosci.* 2007;32: 86–102.
236. Reimold M, Smolka MN, Zimmer A, et al. Reduced availability of serotonin transporters in obsessive-compulsive disorder correlates with symptom severity—a [^{11}C]DASB PET study. *J Neural Transm.* 2007; 114:1603–9.
237. Brown AK, George DT, Fujita M, et al. PET [^{11}C]DASB imaging of serotonin transporters in patients with alcoholism. *Alcohol Clin Exp Res.* 2007;31:28–32.
238. Schwaiger M, Kalff V, Rosenpire K, et al. Noninvasive evaluation of sympathetic nervous system in human heart by positron emission tomography. *Circulation.* 1990;82:457–64.
239. Shulkin BL, Wieland DM, Schwaiger M, et al. PET scanning with hydoxyephedrine: an approach to the localization of pheochromocytoma. *J Nucl Med.* 1992;33:1125–31.
240. Ziegler SI, Frey AW, Uberfuhr P, et al. Assessment of myocardial re-innervation in cardiac transplants by positron emission tomography: functional significance tested by heart rate variability. *Clin Sci (Lond).* 1996;91:126–8.
241. Shulkin BL, Wieland DM, Baro ME, et al. PET hydroxyephedrine imaging of neuroblastoma. *J Nucl Med.* 1996;37:16–21.
242. Pietila M, Malminiemi K, Ukkonen H, et al. Reduced myocardial retention is associated with poor prognosis in chronic heart failure. *Eur J Nucl Med.* 2001;28:373–6.
243. Trampal C, Engler H, Juhlin C, et al. Pheochromocytomas: detection with ^{11}C hydroxyephedrine PET. *Radiology.* 2004;230:423–8.
244. Hendrikse NH, de Vries EG, Franssen EJ, et al. In vivo measurement of [^{11}C]verapamil kinetics in human tissues. *Eur J Clin Pharmacol.* 2001;56:827–9.
245. Sasongko L, Link JM, Muzi M, et al. Imaging P-glycoprotein at the human blood-brain barrier with positron emission tomography. *Clin Pharmacol Ther.* 2005;77:503–14.
246. Kortekaas R, Leenders KL, van Oostrom JC, et al. Blood-brain barrier dysfunction in parkinsonian midbrain in vivo. *Ann Neurol.* 2005; 57:176–9.
247. Reivich M, Kuhl D, Wolf A, et al. The [^{18}F]fluorodeoxyglucose method for the measurement of local cerebral glucose utilization in man. *Circ Res.* 1979;44:127–37.
248. Huang SC, Phelps ME, Hoffman EJ, et al. Noninvasive determination of local cerebral metabolic rate of glucose in man. *Am J Physiol.* 1980; 238:E69–E82.
249. Heiss WD, Pawlik G, Herholz K, et al. Regional kinetic constants and cerebral metabolic rate for glucose in normal human volunteers determined by dynamic positron emission tomography of [^{18}F]-2-fluoro-2-deoxy-D-glucose. *J Cereb Blood Flow Metab.* 1984;4:212–23.
250. Newberg A, Alavi A, Reivich M. Determination of regional cerebral function with FDG-PET imaging in neuropsychiatric disorders. *Semin Nucl Med.* 2002;32:13–34.
251. Otsuka H, Graham M, Kubo A, et al. Clinical utility of FDG PET. *J Med Invest.* 2004;51:14–9.
252. Alavi A, Kung JW, Zhuang H. Implications of PET based molecular imaging on the current and future practice of medicine. *Semin Nucl Med.* 2004;34:56–69.
253. Weber WA, Wester HJ, Grosu AL, et al. O-(2-[^{18}F]Fluoroethyl)-L-tyrosine and L-[methyl-^{11}C]methionine uptake in brain tumors: initial results of a comparative study. *Eur J Nucl Med.* 2000;27:542–9.

254. Pauleit D, Floeth F, Tellmann L, et al. Comparison of O-(2-[18F]fluoroethyl)-L-tyrosine PET and 3-[123I]-iodo-α-methyl-L-tyrosine SPECT in brain tumors. *J Nucl Med.* 2004;45:374–81.
255. Pauleit D, Stoffels G, Schaden W, et al. PET with O-(2-[18F]fluoroethyl)-L-tyrosine in peripheral tumors: first clinical results. *J Nucl Med.* 2005;46:411–6.
256. Pauleit D, Floeth F, Hamacher K, et al. O-(2-[18F]fluoroethyl)-L-tyrosine PET combined with MRI improves the diagnostic assessment of cerebral gliomas. *Brain.* 2005;128(pt 3):678–87.
257. Walsh MN, Geltman EM, Borown MA, et al. Noninvasive estimation of regional myocardial oxygen consumption by positron emission tomography with carbon-11 acetate in patients with myocardial infarction. *J Nucl Med.* 1989;30:1798–808.
258. Gropler RJ, Siegel BA, Geltman EM. Myocardial uptake of carbon-11 acetate as an indirect estimate of regional myocardial blood flow. *J Nucl Med.* 1991;32:245–51.
259. Sciacca RR, Akinboboye O, Chou TL, et al. Measurement of myocardial blood flow with PET using 1-[11C]-acetate. *J Nucl Med.* 2001;42:63–70.
260. Oyama N, Miller TR, Dehdashti F, et al. [11C]-Acetate PET imaging of prostate cancer: detection of recurrent disease at PSA relapse. *J Nucl Med.* 2003;44:549–55.
261. Iyo M, Namba H, Fukushi K, et al. Measurement of acetylcholinesterase by positron emission tomography in the brains of healthy controls and patients with Alzheimer's disease. *Lancet.* 1997;349:1805–9.
262. Shinotoh H, Namba H, Fukushi K, et al. Progressive loss of cortical acetylcholinesterase activity in association with cognitive decline in Alzheimer disease: a positron emission tomography study. *Ann Neurol.* 2000;48:194–200.
263. Rinne JO, Kaasinen V, Jarvenpaa T, et al. Brain acetylcholinesterase activity in mild cognitive impairment and early Alzheimer's disease. *J Neurol Neurosurg Psychiatry.* 2003;74:113–5.
264. Herholz K, Weisenbach S, Zundorf G, et al. In vivo study of acetylcholine esterase in basal forebrain, amygdala, and cortex in mild to moderate Alzheimer disease. *Neuroimage.* 2004;21:136–43.
265. Hilker R, Thomas AV, Klein JC, et al. Dementia in Parkinson disease: functional imaging of cholinergic and dopaminergic pathways. *Neurology.* 2005;65:1716–22.
266. Kuhl DE, Koeppe RA, Minoshima S, et al. In vivo mapping of cerebral acetylcholinesterase activity in aging and Alzheimer's disease. *Neurology.* 1999;52:691–9.
267. Kuhl DE, Minoshima S, Frey KA, et al. Limited donepezil inhibition of acetylcholinesterase measured with positron emission tomography in living Alzheimer cerebral cortex. *Ann Neurol.* 2000;48:391–5.
268. Bohnen NI, Kaufer DI, Hendrickson R, et al. Cognitive correlates of alterations in acetylcholinesterase in Alzheimer's disease. *Neurosci Lett.* 2005;380:127–32.
269. Marshall GA, Shchelchkov E, Kaufer DI, et al. White matter hyperintensities and cortical acetylcholinesterase activity in parkinsonian dementia. *Acta Neurol Scand.* 2006;113:87–91.
270. Bergström M, Westerberg G, Nemeth G, et al. MAO-A Inhibition in brain after dosing with esuprone, moclobemide and placebo in healthy volunteers: in vivo studies with positron emission tomography. *Eur J Clin Pharmacol.* 1997;52:121–8.
271. Orlefors H, Sundin A, Fasth KJ, et al. Demonstration of high monoaminoxidase-A levels in neuroendocrine gastroenteropancreatic tumors in vitro and in vivo—tumor visualization using positron emission tomography with [11C]-harmine. *Nucl Med Biol.* 2003;30:669–79.
272. Bergström M, Juhlin C, Bonasera TA, et al. PET imaging of adrenal cortical tumors with the 11β-hydroxylase tracer [11C]-metomidate. *J Nucl Med.* 2000;41:275–82.
273. Khan TS, Sundin A, Juhlin C, et al. [11C]-Metomidate PET imaging of adrenocortical cancer. *Eur J Nucl Med Mol Imaging.* 2003;30:403–10.
274. Garnett E, Firnau G, Nahmias C. Dopamine visualized in the basal ganglia of living man. *Nature.* 1983;305:137–8.
275. Vingerhoets FJ, Snow BJ, Lee CS, et al. Longitudinal fluorodopa positron emission tomography studies of the evolution of idiopathic parkinsonism. *Ann Neurol.* 1994;36:759–64.
276. Kuwabara H, Cumming P, Yasuhara Y, et al. Regional striatal DOPA transport and decarboxylase activity in Parkinson's disease. *J Nucl Med.* 1995;36:1226–31.
277. Becherer A, Karanikas G, Szabo M, et al. Brain tumor imaging with PET: a comparison between [18F]fluorodopa and [11C]methionine. *Eur J Nucl Med Mol Imaging.* 2003;30:1561–7.
278. Becherer A, Szabo M, Karanikas G, et al. Imaging of advanced neuroendocrine tumors with [18F]-FDOPA PET. *J Nucl Med.* 2004;45:1161–7.
279. Goldstein DS, Holmes C, Stuhlmuller JE, et al. 6-[18F]fluorodopamine positron emission tomographic scanning in the assessment of cardiac sympathoneural function—studies in normal humans. *Clin Auton Res.* 1997;7:17–29.
280. Li ST, Dendi R, Holmes C, et al. Progressive loss of cardiac sympathetic innervation in Parkinson's disease. *Ann Neurol.* 2002;52:220–3.
281. Langer O, Halldin C. PET and SPET tracers for mapping the cardiac nervous system. *Eur J Nucl Med Mol Imaging.* 2002;29:416–34.
282. Ilias I, Shulkin B, Pacak K. New functional imaging modalities for chromaffin tumors, neuroblastomas and gangliomas. *Trends Endocrinol Metab.* 2005;16:66–72.
283. Goldstein DS, Eisenhofer G, Flynn JA, et al. Diagnosis and localization of pheochromocytomas. *Hypertension.* 2004;43:907–10.
284. Vesselle H, Grierson J, Muzi M, et al. In vivo validation of 3'-deoxy-3'-[18F]fluorothymidine as a proliferation imaging tracer in humans: correlation of [18F]FLT uptake by positron emission tomography with Ki-67 immunohistochemistry and flow cytometry in human lung tumors. *Clin Cancer Res.* 2002;8:3315–23.
285. Smyczek-Gargya B, Fersis N, Dittmann H, et al. PET with [18F]fluorothymidine for imaging of primary breast cancer: a pilot study. *Eur J Nucl Med Mol Imaging.* 2004;31:720–4.
286. Wagner M, Seitz U, Buck A, et al. 3'-[18F]fluoro-3'-deoxythymidine ([18F]FLT) as positron emission tomography tracer for mapping proliferation in a murine B-cell lymphoma model and in the human disease. *Cancer Res.* 2003;63:2681–7.
287. Buck AK, Halter G, Schirrmeister H, et al. Imaging proliferation in lung tumors with PET: 18F-FLT versus 18F-FDG. *J Nucl Med.* 2003;44:1426–31.
288. Cobben DC, Jager PL, Elsinga PH, et al. 3'-18F-fluoro-3'-deoxy-L-thymidine: a new tracer for staging metastatic melanoma? *J Nucl Med.* 2003;44:1927–32.
289. Dittmann H, Dohmen BM, Paulsen F, et al. [18F]FLT PET for diagnosis and staging of thoracic tumors. *Eur J Nucl Med Mol Imaging.* 2003;30:1407–12.
290. Jacobs A, Braunlich I, Graf R, et al. Quantitative kinetics of [124I]FIAU in cat and man. *J Nucl Med.* 2001;42:467–75.
291. Jacobs A, Voges J, Reszka R, et al. Positron-emission tomography of vector-mediated gene expression in gene therapy for gliomas. *Lancet.* 2001;358:727–9.
292. Borjesson PK, Jauw YW, Boellaard R, et al. Performance of immuno-positron emission tomography with zirconium-89-labeled chimeric monoclonal antibody U36 in the detection of lymph node metastases in head and neck cancer patients. *Clin Cancer Res.* 2006;12:2133–40.
293. Divgi CR, Pandit-Taskar N, Jungbluth AA, et al. Preoperative characterisation of clear-cell renal carcinoma using iodine-124-labelled antibody chimeric G250 (124I-cG250) and PET in patients with renal masses: a phase I trial. *Lancet Oncol.* 2007;8:304–10.
294. Klunk WE, Engler H, Nordberg A, et al. Imaging brain amyloid in Alzheimer's disease with Pittsbugh Compound-B. *Ann Neurol.* 2004;55:306–19.
295. Rowe CC, Ackerman U, Browne W, et al. Imaging of amyloid beta in Alzheimer's disease with 18F-BAY94-9172, a novel PET tracer: proof of mechanism. *Lancet Neurol.* 2008;7:129–35.
296. Rasey JS, Koh WJ, Evans ML, et al. Quantifying regional hypoxia in human tumors with positron emission tomography of [18F]fluoromisonidazole: a pretherapy study of 37 patients. *Int J Radiat Oncol Biol Phys.* 1996;36:417–28.
297. Rajendran JG, Wilson DC, Conrad EU, et al. [18F]FMISO and [18F]FDG PET imaging in soft tissue sarcomas: correlation of hypoxia, metabolism and VEGF expression. *Eur J Nucl Med Mol Imaging.* 2003;30:695–704.

298. Bruehlmeier M, Roelcke U, Schubiger PA, et al. Assessment of hypoxia and perfusion in human brain tumors using PET with ^{18}F-fluoromisonidazole and ^{15}O-H$_2$O. *J Nucl Med.* 2004;45:1851–9.
299. Markus R, Reutens DC, Kazui S, et al. Hypoxic tissue in ischaemic stroke: persistence and clinical consequences of spontaneous survival. *Brain.* 2004;127:1427–36.
300. Guadagno JV, Donnan GA, Markus R, et al. Imaging the ischaemic penumbra. *Curr Opin Neurol.* 2004;17:61–7.
301. Eschmann SM, Paulsen F, Reimold M. Prognostic impact of hypoxia imaging with ^{18}F-misonidazole PET in non-small cell lung cancer and head and neck cancer before radiotherapy. *J Nucl Med.* 2005;46:253–60.
302. Chao KS, Bosch WR, Mutic S, et al. A novel approach to overcome hypoxic tumor resistance: Cu-ATSM-guided intensity-modulated radiation therapy. *Int J Radiat Oncol Biol Phys.* 2001;49:1171–82.
303. Dehdashti F, Grigsby PW, Mintun MA, et al. Assessing tumor hypoxia in cervical cancer by positron emission tomography with ^{60}Cu-ATSM: relationship to therapeutic response—a preliminary report. *Int J Radiat Oncol Biol Phys.* 2003;55:1233–8.
304. Dehdashti F, Mintun MA, Lewis JS, et al. In vivo assessment of tumor hypoxia in lung cancer with ^{60}Cu-ATSM. *Eur J Nucl Med Mol Imaging.* 2003;30:844–50.
305. Takahashi N, Fujibayashi Y, Yonekura Y, et al. Evaluation of ^{62}Cu labeled diacetyl-bis(N^4-methylthiosemicarbazone) as a hypoxic tissue tracer in patients with lung cancer. *Ann Nucl Med.* 2000;14:323–8.
306. Muto P, Lastoria S, Varrella P, et al. Detecting deep venous thrombosis with technetium-99m-labeled synthetic peptide P280. *J Nucl Med.* 1995;36:1384–91.
307. Taillefer R, Therasse E, Turpin S, et al. Comparison of early and delayed scintigraphy with 99mTc-apcitide and correlation with contrast-enhanced venography in detection of acute deep vein thrombosis. *J Nucl Med.* 1999;40:2029–35.
308. Taillefer R, Edell S, Innes G, et al. Acute thromboscintigraphy with ^{99m}Tc-apcitide: results of the phase 3 multicenter clinical trial comparing ^{99m}Tc-apcitide scintigraphy with contrast venography for imaging acute DVT. Multicenter Trial Investigators. *J Nucl Med.* 2000;41:1214–23.
309. Bates SM, Lister-James J, Julian JA, et al. Imaging characteristics of a novel technetium Tc 99m-labeled platelet glycoprotein IIb/IIIa receptor antagonist in patients with acute deep vein thrombosis. *Arch Intern Med.* 2003;163:452–6.
310. Dunziger A, Hafner F, Schaffler G, et al. ^{99m}Tc-apcitide scintigraphy in patients with clinically suspected deep vein thrombosis and pulmonary embolism. *Eur J Nucl Med Mol Imaging.* 2008;35:2982–7.
311. Thakur ML, Marcus CS, Saeed S, et al. ^{99m}Tc-labeled vasoactive intestinal peptide analog for rapid localization of tumors in humans. *J Nucl Med.* 2000;41:107–10.
312. Rao PS, Thakur ML, Pallela V, et al. ^{99m}Tc Labeled VIP analog: evaluation for imaging colorectal cancer. *Nucl Med Biol.* 2001;28:445–50.
313. Annovazzi A, Biancone L, Caviglia R, et al. ^{99m}Tc-Interleukin-2 and ^{99m}Tc-HMPAO granulocyte scintigraphy in patients with inactive Crohn's disease. *Eur J Nucl Med Mol Imaging.* 2003;30:374–82.
314. Signore A, Annovazzi A, Barone R, et al. ^{99m}Tc-Interleukin-2 scintigraphy as a potential tool for evaluating tumor-infiltrating lymphocytes in melanoma lesions: a validation study. *J Nucl Med.* 2004;45:1647–52.
315. Annovazzi A, Bonanno E, Arca M, et al. ^{99m}Tc-Interleukin-2 scintigraphy for the in vivo imaging of vulnerable atherosclerotic plaques. *Eur J Nucl Med Mol Imaging.* 2006;33:117–26.
316. Mena E, Camacho V, Estorch M, et al. ^{99m}Tc-depreotide scintigraphy of bone lesions in patients with lung cancer. *Eur J Nucl Med Mol Imaging.* 2004;31:1399–404.
317. Martins T, Lino JS, Ramos S, et al. ^{99m}Tc-depreotide scintigraphy in the evaluation of indeterminate pulmonary lesions: clinical experience. *Cancer Biother Radiopharm.* 2004;19:253–9.
318. Van Den Bossche B, D'Haeninck E, De Winter F, et al. ^{99m}Tc depreotide scan compared with 99mTc-MDP bone scintigraphy for the detection of bone metastases and prediction of response to hormonal treatment in patients with breast cancer. *Nucl Med Commun.* 2004;25:787–92.
319. Cholewinski W, Kowalczyk JR, Stefaniak B, et al. Diagnosis and staging of children's lymphoma using the technetium-labelled somatostatin analogue, ^{99m}Tc-depreotide. *Eur J Nucl Med Mol Imaging.* 2004;31:820–4.
320. Boundas D, Karatzas N, Moralidis E, et al. Comparative evaluation of ^{99m}Tc-depreotide and ^{201}Tl chloride single photon emission tomography in the characterization of pulmonary lesions. *Nucl Med Commun.* 2007;28:533–40.
321. Papathanasiou ND, Rondogianni PE, Pianou NK, et al. ^{99m}Tc-depreotide in the evaluation of bone infection and inflammation. *Nucl Med Commun.* 2008;29:239–46.
322. Płachcińska A, Mikołajczak R, Maecke H, et al. Clinical usefulness of ^{99m}Tc-EDDA/HYNIC-TOC scintigraphy in oncological diagnostics: a preliminary communication. *Eur J Nucl Med Mol Imaging.* 2003;30:1402–6.
323. Płachcińska A, Mikołajczak R, Kozak J, et al. Differential diagnosis of solitary pulmonary nodules based on ^{99m}Tc-EDDA/HYNIC-TOC scintigraphy: the effect of tumour size on the optimal method of image assessment. *Eur J Nucl Med Mol Imaging.* 2006;33:1041–7.
324. Gonzalez-Vazquez A, Ferro-Flores G, Arteaga de Murphy C, et al. Biokinetics and dosimetry in patients of ^{99m}Tc-EDDA/HYNIC-Tyr3-octreotide prepared from lyophilized kits. *Appl Radiat Isot.* 2006;64:792–7.
325. Sun H, Jiang XF, Wang S, et al. ^{99m}Tc-HYNIC-TOC scintigraphy in evaluation of active Grave's ophthalmology (GO). *Endocrine.* 2007;31:305–10.
326. Vermeersch H, Loose D, Lahorte C, et al. ^{99m}Tc-HYNIC annexin-V imaging of primary head and neck carcinoma. *Nucl Med Commun.* 2004;25:259–63.
327. Vermeersch H, Ham H, Rottey S, et al. Intraobserver, interobserver, and day-to-day reproducibility of quantitative ^{99m}Tc-HYNIC annexin-V imaging in head and neck carcinoma. *Cancer Biother Radiopharm.* 2004;19:205–10.
328. Vermeersch H, Mervillie K, Lahorte C, et al. Relationship of ^{99m}Tc-HYNIC annexin V uptake to microvessel density, FasL and MMP-9 expression, and the number of tumor-infiltrating lymphocytes in head and neck carcinoma. *Eur J Nucl Med Mol Imaging.* 2004;72:333–9.
329. Kartachova M, Haas RL, Olmos RA. In vivo imaging of apoptosis by ^{99m}Tc-annexin V scintigraphy: visual analysis in relation to treatment response. *Radiother Oncol.* 2004;72:333–9.
330. Bach-Gansmo T, Danielsson R, Saracco A, et al. Integrin receptor imaging of breast cancer: a proof-of-concept study to evaluate ^{99m}Tc-NC100692. *J Nucl Med.* 2006;47:1434–9.
331. Bach-Gansmo T, Bogsrud TV, Skretting A. Integrin scintimammography using a dedicated breast imaging, solid-state gamma-camera and ^{99m}Tc-labelled NC100692. *Clin Physiol Funct Imaging.* 2008;28:235–9.
332. Ellner SJ, Hoh CK, Vera DR, et al. Dose-dependent biodistribution of [^{99m}Tc]DTPA-mannosyl-dextran for breast cancer sentinel lymph node mapping. *Nucl Med Biol.* 2003;30:805–10.
333. Wallace AM, Hoh CK, Vera DR, et al. Lymphoseek: a molecular radiopharmaceutical for sentinel node detection. *Ann Surg Oncol.* 2003;10:531–8.
334. Wallace AM, Hoh CK, Ellner SJ, et al. Lymphoseek: a molecular imaging agent for melanoma sentinel lymph node mapping. *Ann Surg Oncol.* 2007;14:913–21.
335. Virgolini I, Müller C, Angelberger P, et al. Quantification of human hepatic binding protein (HBP) via ^{99m}Tc-galactosyl-neoglycoalbumin (NGA) liver scintigraphy. *Wien Klin Wochenschr.* 1991;103:458–61.
336. Kurtaran A, Müller C, Novacek G, et al. Distinction between hepatic focal nodular hyperplasia and malignant liver lesions using technetium-99m-galactosyl-neoglycoalbumin. *J Nucl Med.* 1997;38:1912–5.
337. Kira T, Tomiguchi S, Takahashi M. Quantitative evaluation of the regional hepatic reserve by ^{99m}Tc-GSA dynamic SPECT before and after chemolipiodolization in patients with hepatocellular carcinoma. *Ann Nucl Med.* 1998;12:369–73.
338. Kira T, Tomiguchi S, Takahashi M, et al. Correlation of ^{99m}Tc-GSA hepatic scintigraphy with liver biopsies in patients with chronic active hepatitis type C. *Radiat Med.* 1999;17:125–30.

339. Sasaki N, Shiomi S, Iwata Y, et al. Clinical usefulness of scintigraphy with ^{99m}Tc-glactosyl-human serum albumin for prognosis of cirrhosis of the liver. *J Nucl Med.* 1999;40:1652–6.
340. Hirai I, Kimura W, Fuse A, et al. Evaluation of preoperative portal embolization for safe hepatectomy, with special reference to assessment of nonembolized lobe function with ^{99m}Tc-GSA SPECT scintigraphy. *Surgery.* 2003;133:495–506.
341. Siegel BA, Dehdasti F, Mutch DG, et al. Evaluation of ^{111}In-DTPA-folate as a receptor-targeted diagnostic agent for ovarian cancer: initial clinical results. *J Nucl Med.* 2003;44:700–7.
342. Krenning EP. Somatostain receptor scintigraphy with [^{111}In-DTPA-D-Phe1]- and [^{123}I-Tyr3]-octreotide: the Rotterdam experience with more than 1000 patients. *Eur J Nucl Med.* 1993;20:716–31.
343. Jamar F. Somatostatin receptor imaging with indium-111-pentetreotide in gastroenteropancreatic neuroendocrine tumors: safety, efficacy and impact on patient management. *J Nucl Med.* 1995;36:542–9.
344. Forssell-Aronsson E. Biodistribution data from 100 patients i.v. injected with ^{111}In-DTPA-D-Phe1-octreotide. *Acta Oncol.* 2004;43:436–42.
345. Chiesa R, Melissano G, Castellano R, et al. Avidin and ^{111}In-labeled biotin scan: a new radioisotopic method for localizing vascular graft infection. *Eur J Vasc Endovasc Surg.* 1995;10:405–14.
346. Samuel A, Paganelli G, Chiesa R, et al. Detection of prosthetic graft infection using avidin/indium-111 biotin scintigraphy. *J Nucl Med.* 1996;37:55–61.
347. Rusckowski M, Paganelli G, Hnatowich DJ, et al. Imaging osteomyelitis with streptavidin and indium-111 labeled biotin. *J Nucl Med.* 1996;37:1655–62.
348. Lazzeri E, Manca M, Molea N, et al. Clinical validation of the avidin/indium-111 biotin approach for imaging infection/inflammation in orthopedic patients. *Eur J Nucl Med.* 1999;26:606–14.
349. Lazzeri E, Pauwels EK, Erba PA, et al. Clinical feasibility of two-step streptavidin/^{111}In-biotin scintigraphy in patients with suspected vertebral osteomyelitis. *Eur J Nucl Med Mol Imaging.* 2004;31:1505–11.
350. Lazzeri E, Erba P, Perri M, et al. Scintigraphic imaging of vertebral osteomyelitis with ^{111}In-biotin. *Spine.* 2008;33:E198–204.
351. Zhogbi SS, Baldwin RM, Seibyl JP, et al. Pharmacokinetics of the SPECT benzodiazepine receptor radioligand [^{123}I]iomazenil in human and non-human primates. *Int J Rad Appl Instrum B.* 1992;19:881–8.
352. Sybirska E, Seibyl JP, Bremner JD, et al. [^{123}I]Iomazenil SPECT imaging demonstrates significant benzodiazepine receptor reserve in human and nonhuman primate brain. *Neuropharmacology.* 1933;32:671–80.
353. Dong Y, Fukuyama H, Nabatame H, et al. Assessment of benzodiazepine receptors using iodine-123-labeled iomazenil single-photon emission computed tomography in patients with ischemic cerebrovascular disease. A comparison with PET study. *Stroke.* 1997;28:1776–82.
354. Moriwaki H, Matsumoto M, Hashikawa K, et al. Iodine-123-iodoamphetamine SPECT in major cerebral artery occlusive disease. *J Nucl Med.* 1998;39:1348–53.
355. Li S, Peck-Radosavljevic M, Kienast O, et al. Imaging gastrointestinal tumors using vascular endothelial growth factor-165 (VEGF165) receptor scintigraphy. *Ann Oncol.* 2003;14:1274–7.
356. Li S, Peck-Radosavljevic M, Kienast O, et al. Iodine-123-vascular endothelial growth factor-165 (^{123}I-VEGF165). Biodistribution, safety and radiation dosimetry in patients with pancreatic carcinoma. *Q J Nucl Med Imaging.* 2004;48:198–206.
357. Dresel S, Krause J, Krause KH, et al. Attention deficit hyperactivity disorder: binding of [^{99m}Tc]TRODAT-1 to the dopamine transporter before and after methylphenidate treatment. *Eur J Nucl Med.* 2000;27:1518–24.
358. Brunswick DJ, Amsterdam JD, Mozley PD, et al. Greater availability of brain dopamine transporters in major depression shown by [^{99m}Tc]TRODAT-1 SPECT imaging. *Am J Psychiatry.* 2003;160:1836–41.
359. Gardiner SA, Morrison MF, Mozley PD, et al. Pilot study on the effect of estrogen replacement therapy on brain dopamine transporter availability in healthy, postmenopausal women. *Am J Geriatr Psychiatry.* 2004;12:621–30.
360. Weng YH, Yen TC, Chen MC, et al. Sensitivity and specificity of ^{99m}Tc-TRODAT-1 SPECT imaging in differentiating patients with idiopathic Parkinson's disease from healthy subjects. *J Nucl Med.* 2004;45:393–401.
361. Wang J, Jiang YP, Liu XD, et al. ^{99m}Tc-TRODAT-1 SPECT study in early Parkinson's disease and essential tremor. *Acta Neurol Scand.* 2005;112:280–5.
362. la Fougere C, Krause J, Krause KH, et al. Value of [^{99m}Tc]TRODAT-1 SPECT to predict clinical response to methylphenidate treatment in adults with attention deficit hyperactivity disorder. *Nucl Med Commun.* 2006;27:733–7.
363. Newberg A, Amsterdam J, Shults J. Dopamine transporter density may be associated with the depressed affect in healthy subjects. *Nucl Med Commun.* 2007;28:3–6.
364. Koch W, Pogarell O, Popperl G, et al. Extended studies of the striatal uptake of ^{99m}Tc-NC100697 in healthy volunteers. *J Nucl Med.* 2007;48:27–34.
365. Wackers FJ, Berman DS, Maddahi J, et al. Technetium-99m hexakis 2-methoxyisobutyl isonitrile: human biodistribution, dosimetry, safety and preliminary comparison to thallium-201 for myocardial perfusion imaging. *J Nucl Med.* 1989;30:301–11.
366. Del Vecchio S, Salvatore M. ^{99m}Tc-MIBI in the evaluation of breast cancer biology. *Eur J Nucl Med Mol Imaging.* 2004;31:S88–96.
367. Kawata K, Kanai M, Sasada T, et al. Usefulness of ^{99m}Tc-sestamibi scintigraphy in suggesting the therapeutic effect of chemotherapy against gastric cancer. *Clin Cancer Res.* 2004;10:3788–93.
368. Booij J, Habraken JB, Bermans P, et al. Imaging of dopamine transporters with iodine-123-FP-CIT SPECT in healthy controls and patients with Parkinson's disease. *J Nucl Med.* 1998;39:1879–84.
369. Benamer HT, Patterson J, Wyper DJ, et al. Correlation of Parkinson's disease severity and duration with ^{123}I-FP-CIT SPECT striatal uptake. *Move Disord.* 2000;15:692–8.
370. Covelli EM, Brunetti A, Di Lauro A, et al. Clinical impact of correlative [^{123}I]-FP-CIT brain imaging and neurological findings in suspect Parkinson's disease. *Radiol Med (Torino).* 2004;108:417–25.
371. Kauppinen TA, Bergström KA, Heikman P, et al. Biodistribution and radiation dosimetry of [^{123}I]ADAM in healthy human subjects: preliminary results. *Eur J Nucl Med Mol Imaging.* 2003;30:132–6.
372. Newberg AB, Plossl K, Mozley PD, et al. Biodistribution and imaging with ^{123}I-ADAM, a serotonin transporter imaging agent. *J Nucl Med.* 2004;45:834–41.
373. Newberg AB, Amsterdam JD, Wintering N. ^{123}I-ADAM binding to serotonin transporters in patients with major depression and healthy controls: a preliminary study. *J Nucl Med.* 2005;46:973–7.
374. Koskela AK, Keski-Rahkonen A, Sihvola E, et al. Serotonin transporter binding of [^{123}I]ADAM in bulimic women, their healthy twin sisters, and healthy women: a SPET study. *BMC Psychiatry.* 2007;7:19.
375. Schuh-Hofer S, Richter M, Geworski L, et al. Increased serotonin transporter availability in the brainstem of migrainers. *J Neurol.* 2007;254:789–96.
376. Dae MW, De Marco T, Botvinick EH, et al. Scintigraphic assessment of MIBG uptake in globally denervated human and canine hearts—implications for clinical studies. *J Nucl Med.* 1992;33:1444–50.
377. Kline RC, Swanson DP, Wieland DM, et al. Myocardial imaging in man with I-123 meta-iodobenzylguanidine. *J Nucl Med.* 1981;22:129–32.
378. Sisson JC, Shapiro B, Meyers L, et al. Metaiodobenzylguanidine to map scintigraphically the adrenergic nervous system in man. *J Nucl Med.* 1987;28:1625–36.
379. Moyes JS, Babich JW, Carter R, et al. Quantitative study of radioiodinated metaiodobenzylguanidine uptake in children with neuroblastoma: correlation with tumor histopathology. *J Nucl Med.* 1989;30:474–80.
380. Binnie D, Divoli A, McCready VR, et al. The potential use of ^{99m}Tc-MDP bone scans to plan high-activity ^{186}Re-HEDP targeted therapy of bony metastases from prostate cancer. *Cancer Biother Radiopharm.* 2005;20:189–94.
381. Wang K, Allen L, Fung E, et al. Bone scintigraphy in common tumors with osteolytic components. *Clin Nucl Med.* 2005;30:655–71.

382. Ali I, Johns W, Gupta SM. Visualization of hepatic metastases of medullary thyroid cardinoma on Tc-99m MDP bone scintigraphy. *Clin Nucl Med.* 2006;31:611–3.
383. Karacalioglu O, Ilgan S, Kuzhan O, et al. Disseminated metastatic disease of osteosarcoma of the femur in the abdomen: unusual metastatic pattern on Tc-99m MDP bone scan. *Ann Nucl Med.* 2006;20:437–40.
384. Lee EJ, Lee KH, Huh WS, et al. Incidence and radio-uptake patterns of femoral head avascular osteonecrosis at 1 year after renal transplantation: a prospective study with planar bone scintigraphy. *Nucl Med Commun.* 2006;27:919–24.
385. Chavdarova L, Piperkova E, Tsonevska A, et al. Bone scintigraphy in the monitoring of treatment effect of bisphosphonates in bone metastatic breast cancer. *J Buon.* 2006;11:499–504.
386. Namwongprom S, Nunez R, Kim EE, et al. Tc-99m bone scintigraphy and positron emission tomography (PET/CT) imaging in Erdheim-Chester disease. *Clin Nucl Med.* 2007;32:35–8.
387. Seki H, Toyama T, Higuchi K, et al. Prediction of functional improvement of ischemic myocardium with ^{123}I-BMIPP SPECT and ^{99m}Tc-tetrofosmin SPECT imaging: a study of patients with large acute myocardial infarction and receiving revascularization therapy. *Circ J.* 2005;69:311–9.
388. Kageyama H, Morita K, Katoh C, et al. Reduced ^{123}I-BMIPP uptake implies decreased myocardial flow reserve in patients with stable angina. *Eur J Nucl Med Mol Imaging.* 2006;33:6–12.
389. Nakae I, Mitsunami K, Matsuo S, et al. Creatinine depletion and altered fatty acid metabolism in diseased human hearts: clinical investigation using ^{1}H magnetic resonance spectroscopy and ^{123}I BMIPP myocardial scintigraphy. *Acta Radiol.* 2007;48:436–43.
390. Inoue A, Fujimoto S, Yamashina S, et al. Prediction of cardiac events in patients with dilated cardiomyopathy using ^{123}I-BMIPP and ^{201}Tl myocardial scintigraphy. *Ann Nucl Med.* 2007;21:399–404.
391. Behr T, Becker W, Hannappel E, et al. Targeting of liver metastases of colorectal cancer with IgF, F(ab′)2, and Fab′ anti-carcinoembryonic antigen antibodies labeled with ^{99m}Tc: the role of metabolism and kinetics. *Cancer Res.* 1995;55:5777s–85s.
392. Moffat FL, Pinsky SM, Hammershaimb L, et al. Clinical utility of external immunoscintigraphy with the IMMU-4 technetium-99m Fab′ antibody fragment in patients undergoing surgery for carcinoma of the colon and rectum: results of a pivotal, phase III trial. The Immunomedics Study Group. *J Clin Oncol.* 1996;14:2295–305.
393. Wegener WA, Patrelli N, Serafini A, et al. Safety and efficacy of arcitumomab imaging in colorectal cancer after repeated administration. *J Nucl Med.* 2000;41:1016–20.
394. Marcus C, Thakur ML, Huynh TV, et al. Imaging rheumatic joint disease with anti-T lymphocyte antibody OKT-3. *Nucl Med Commun.* 1994;15:824–30.
395. Martins FP, Souza SA, Gonçalves RT, et al. Preliminary results of [^{99m}Tc]OKT3 scintigraphy to evaluate acute rejection in renal transplants. *Transplant Proc.* 2004;36:2664–7.
396. Martins FP, Gutfilen B, de Souza SA, et al. Monitoring rheumatoid arthritis synovitis with ^{99m}Tc-anti-CD3. *Br J Radiol.* 2008;81:25–9.
397. Vallis KA, Reilly RM, Chen P, et al. A phase I study of ^{99m}Tc-hR3 (DiaCIM), a humanized immunoconjugate directed towards the epidermal growth factor receptor. *Nucl Med Commun.* 2002;23:1155–64.
398. Torres LA, Perera A, Batista JF, et al. Phase I/II clinical trial of the humanized ant-EGF-r monoclonal antibody h-R3 labeled with ^{99m}Tc in patients with tumour of epithelial origin. *Nucl Med Commun.* 2005;26:1049–57.
399. Iznaga-Escobar N, Torres LA, Morales A, et al. Technetium-99m-labeled anti-EGF-receptor antibody in patients with tumor of epithelial origin: I. Biodistribution and dosimetry for radioimmunotherapy. *J Nucl Med.* 1998;39:15–23.
400. Iznaga-Escobar N, Torres LA, Arocha LA, et al. Technetium-99m–epidermal growth factor receptor-antibody in patients with tumors of epithelial origin: Part II. Pharmacokinetics and clearances. *J Nucl Med.* 1998;39:1918–27.
401. Ramos-Suzarte M, Rodriguez N, Oliva JP, et al. ^{99m}Tc-labeled anti-human epidermal growth factor receptor antibody in patients with tumors of epithelial origin: Part III. Clinical trials safety and diagnostic efficacy. *J Nucl Med.* 1999;40:768–75.
402. Conti F, Priori R, Chimenti MS, et al. Successful treatment with intraarticular infliximab for resistant knee monarthritis in a patient with spondylarthropathy: a role for scintigraphy with ^{99m}Tc-infliximab. *Arthritis Rheum.* 2005;52:1224–6.
403. D'Alessandria C, Malviya G, Viscido A, et al. Use of a ^{99m}Tc labeled anti-TNFα monoclonal antibody in Crohn's disease: in vitro and in vivo studies. *Q J Nucl Med Mol Imaging.* 2007;51:334–42.
404. van der Laken CJ, Voskuyl AE, Roos JC, et al. Imaging and serum analysis of immune complex formation of radiolabelled infliximab and anti-infliximab in responders and non-responders to therapy for rheumatoid arthritis. *Ann Rheum Dis.* 2007;66:253–6.
405. Wiseman GA, White CA, Sparks RB, et al. Biodistribution and dosimetry results from a phase III prospectively randomized controlled trial of Zevalin radioimmunotherapy for low-grade, follicular, or transformed B-cell non-Hodgkin's lymphoma. *Crit Rev Oncol Hematol.* 2001;39:181–94.
406. Conti PS. Radioimmunotherapy with yttrium 90 ibritumomab tiuxetan (Zevalin): the role of the nuclear medicine physician. *Semin Nucl Med.* 2004;34:2–3.
407. Manyak MJ, Hinkle GH, Olsen JO, et al. Immunoscintigraphy with indium-111-capromab pendetide: evaluation before definitive therapy in patients with prostate cancer. *Urology.* 1999;54:1058–63.
408. Sodee DB, Malguria N, Faulhaber P, et al. Multicenter ProstaScint imaging findings in 2154 patients with prostate cancer. *Urology.* 2000;56:988–93.
409. Schettino CJ, Kramer EL, Noz ME, et al. Impact of fusion of indium-111 capromab pendetide volume data sets with those from MRI or CT in patients with recurrent prostate cancer. *Am J Roentgenol.* 2004;183:519–24.
410. Wong TZ, Turkington TG, Polascik TJ, et al. ProstaScint (capromab pendetide) imaging using hybrid gamma camera-CT technology. *Am J Roentgenol.* 2005;184:676–80.
411. Goldenberg DM, Wlodowski TJ, Sharkey RM, et al. Colorectal imaging with iodine-123-labeled CEA monoclonal antibody fragments. *J Nucl Med.* 1993;34:61–70.
412. Wong JY, Chu DZ, Williams LE, et al. Pilot trial evaluating an ^{123}I-labeled 80-kilodalton engineered anticarcinoembryonic antigen antibody fragment (cT84.66 minibody) in patients with colorectal cancer. *Clin Cancer Res.* 2004;10:5014–21.
413. Van Paesschen W. Ictal SPECT. *Epilepsia.* 2004;45:35–40.
414. Mesquita CT, Correa PL, Felix RC, et al. Autologous bone marrow mononuclear cells labeling with technetium-99m hexamethylpropylene amine oxime scintigraphy after intracoronary stem cell therapy in acute myocardial infarction. *J Nucl Cardiol.* 2005;12:610–2.
415. Filippi L, Schillaci O. Usefulness of hybrid SPECT/CT in ^{99m}Tc-HMPAO-labeled leukocyte scintigraphy for bone and joint infections. *J Nucl Med.* 2006;47:1908–13.
416. Uslu H, Varoglu E, Balik A, et al. Scintigraphic evaluation of acute pancreatitis patients with ^{99m}Tc-HMPAO-labeled leukocytes. *Nucl Med Commun.* 2007;28:289–95.
417. Nishimura T, Hashikawa K, Fukuyama H, et al. Decreased cerebral blood flow and prognosis of Alzheimer's disease: a multicenter HMPAO-SPECT study. *Ann Nucl Med.* 2007;21:15–23.
418. Cermik TF, Kaya M, Ugur-Altun B, et al. Regional cerebral blood flow abnormalities in patients with primary hyperparathyroidism. *Neuroradiology.* 2007;49:379–85.
419. Melendez-Alafort L, Rodriguez-Cortes J, Ferro-Flores G, et al. Biokinetics of ^{99m}Tc-UBI 29-41 in humans. *Nucl Med Biol.* 2004;31:373–9.
420. Akhtar MS, Qaisar A, Irfanullah J, et al. Antimicrobial peptide ^{99m}Tc-ubiquicidin 29-41 as human infection-imaging agent: clinical trial. *J Nucl Med.* 2005;46:567–73.

CHAPTER 18
Licensing and Regulatory Control

Neil A. Petry

Radiopharmaceuticals are radioactive prescription drug products that are internally administered and intended for use in the diagnosis, treatment, and mitigation of disease. These unique drug products are required for the practice of nuclear medicine, the clinical and laboratory medical specialty that utilizes the measured nuclear properties of radioactive and stable nuclides in diagnosis, therapy, and research and in evaluating metabolic, physiologic, and pathologic conditions of the body.[1] The term radiopharmaceutical is also applicable to nonradioactive reagent kits and radionuclide generators intended for use in the preparation of radioactive drugs and radioactive biologic drug products used in nuclear medicine practice. However, the definition of radiopharmaceuticals does not include substances that contain trace quantities of naturally occurring radionuclides or sealed radionuclide sources intended for brachytherapy.

Because radiopharmaceuticals have both a drug component and a radioactive or nuclear component, the two federal agencies with major responsibility for licensing and regulatory control of radioactive drug products are the Food and Drug Administration (FDA) and the Nuclear Regulatory Commission (NRC). Other federal agencies, including the Occupational Safety and Health Administration, the Environmental Protection Agency, and the Department of Transportation, primarily regulate the safe industrial production, handling, and transportation of radiopharmaceuticals, rather than issues related to their use in the course of medical or pharmacy practice.

Well-established legal precepts support the rights of states to regulate both the practice of medicine and the practice of pharmacy. Consequently, FDA authority over physicians and pharmacists and their use of drugs products, including radiopharmaceuticals, is attenuated at the state level. Exactly where federal authority and state authority intersect continues to be controversial. In contrast, NRC, through mutual agreement and with little controversy, most often delegates its authority to radiation control agencies within the various states.

This chapter focuses on issues related to the direct regulation of radiopharmaceuticals by FDA and NRC at the federal level. Before considering the specifics of federal regulation of radioactive drugs, the chapter presents an overview of the regulated products and practice environments, a brief introduction to existing nuclear pharmacy practice guidelines, and a historical perspective on drug regulations in general. The current regulatory framework should be viewed in the context of the evolution of nuclear medicine and nuclear pharmacy practice into vital components of today's high-quality health care systems.[2-7]

REGULATED PRODUCTS AND PRACTICE ENVIRONMENTS

Most of the radiopharmaceutical products used in nuclear medicine are procured from pharmaceutical companies that specialize in the manufacturing, marketing, and distribution of FDA-approved radioactive drug dosage forms. A majority of the radiopharmaceuticals used clinically are small-volume parenterals; however, oral solutions and capsules, aerosols, gases, and other unique dosage forms are also important. Some of these manufactured radiopharmaceutical products are provided as finished dosage forms ready to be dispensed for patient administration. However, most of the radiopharmaceutical products used in nuclear medicine must be prepared or compounded by a variety of methods just before administration to the patient, largely because of their short physical half-life (i.e., rapid radioactive decay of the radionuclide component) and in some cases because of limited radiochemical stability (i.e., dissociation of the radionuclide from the drug or molecular component).

Most frequently, radiopharmaceuticals are prepared or compounded by nuclear pharmacists practicing in community-based, centralized (i.e., commercial) nuclear pharmacies. Less frequently, they are prepared in hospital-based nuclear pharmacies or institutional nuclear medicine departments. In all practice settings, nuclear pharmacists strive to provide patient-specific unit doses of the highest quality. This typically requires a professional support staff that often includes nuclear pharmacy technicians with training and experience in nuclear medicine technology, pharmacy practice, or both.

Centralized nuclear pharmacies provide service to most hospitals and private clinics offering nuclear medicine services. Large medical institutions typically have an in-house nuclear pharmacy service to facilitate clinical research with both FDA-approved and investigational radiopharmaceutical products or preparations. Thus, nuclear pharmacy services are widely available in almost all practice settings and geographic areas. Occasionally, however, nuclear medicine technologists may perform basic pharmacy functions under the direct supervision of a qualified nuclear medicine physician. This arrangement is still permissible in any clinical practice setting, as it constitutes the practice of medicine. Usually, however, centralized unit dose radiopharmaceutical services are used because this may be more economical in terms of both personnel and overall cost. In addition, the recent development and publication of *United States Pharmacopeia (USP)* Chapter 797, Pharmaceutical Compounding—Sterile Preparations[8] and the need to comply with this enforceable practice standard at various levels of regulation may lead to greater use of centralized radiopharmaceutical services in some practice settings. Regardless of the model used for providing radiopharmaceutical services, it is important for hospital pharmacy directors to be aware of their oversight responsibility for the use of all drug products within their institutions, including those products and preparations used by the nuclear medicine service. Naturally, in settings that employ external services, this oversight responsibility extends to ensuring that the centralized radiopharmaceutical service is compliant with all existing regulations. A basic understanding of the associated licensing and regulatory controls, as summarized here, will be useful in a variety of practice settings.

NUCLEAR PHARMACY PRACTICE GUIDELINES

The American Pharmacists Association's *Nuclear Pharmacy Practice Guidelines* are well established.[9] They supplement competency-based pharmacy practice standards and identify those areas of responsibility that are unique to nuclear pharmacy. In addition to functioning as standards for nuclear pharmacy practitioners, the guidelines provide a practice-oriented foundation for the competency-based Board of Pharmaceutical Specialties (BPS) certification exam.[10] These standards may also be useful to nuclear medicine physicians and technologists involved in the clinical use of radiopharmaceuticals; however, the standards are not intended to directly govern the practices of these individuals. Many of these standards of practice are derived from current regulations governing the medical use of both drugs and radioactive materials; thus, there may be specific legal requirements for the activities described.

Nuclear pharmacy practice is a basic patient-oriented pharmaceutical service that embodies the scientific knowledge and professional judgment required to improve and promote health through assurance of the safe and efficacious use of radioactive drugs for diagnosis and therapy.[9] The most readily identifiable areas of responsibility, or domains, that constitute the practice of nuclear pharmacy are the procurement, compounding, quality control, dispensing, and distribution of radiopharmaceuticals and pharmaceutical drug products used in nuclear medicine. Additional domains of responsibility include health and safety, provision of information and consultation, monitoring patient outcome, and research and development.

Nuclear pharmacists work in a variety of practice settings; therefore, their responsibilities may vary significantly. For example, a nuclear pharmacist working in a centralized service facility dedicated primarily to the procurement, compounding, quality control, dispensing, and distribution of unit dose radiopharmaceuticals may have a set of regulatory responsibilities that differ substantially from those of the clinical nuclear pharmacist providing service in an institutional medical center where distribution may not be undertaken at all. On the other hand, the institution-based nuclear pharmacist invariably bears responsibility for regulatory issues associated with the compounding, use, and, in some cases, manufacturing of radiopharmaceuticals needed for clinical research and drug development projects. Regardless of practice setting, nuclear pharmacists must be knowledgeable about the many regulations governing each domain of responsibility as it applies to each unique practice setting. Furthermore, nuclear pharmacists may supervise a variety of individuals who assist in the provision of these specialized pharmacy services, so they are responsible for the activities of all such support personnel and for ensuring that they are appropriately trained and credentialed.

DRUG REGULATION: HISTORICAL PERSPECTIVE ON FDA

Current regulations on the medical use of drug products, including radiopharmaceuticals, can best be understood in the context of a series of federal regulatory actions over the past 100 years. A 2003 book by health and science reporter Philip J. Hilts[11] tells the fascinating story of FDA's maturation from its start in the Theodore Roosevelt administration through various crises and triumphs to the deregulatory climate of more recent years. With its many references and annotations, the book is a useful guide to understanding the inner workings of FDA and how the agency broadly regulates food and drug products in the United States. The FDA website also provides in-depth historical information.[12]

Until the early 1900s, a plethora of largely unregulated patent medicines of questionable value and safety were sold to the public. The Pure Food and Drugs Act of 1906 was signed into law as a result of unscrupulous manufacturing practices, adulterated foods and drugs, and unfounded claims of the therapeutic efficacy of patent medicines. The act prohibited interstate sales of misbranded and adulterated foods and drugs and paved the way for the establishment of FDA in 1931. The act did not, however, require premarket testing of drugs to determine their safety. When 107 people died after ingesting sulfanilamide elixir formulated with diethylene glycol, a substance now known to be toxic to both humans and animals, revisions in the 1906 act culminated in passage of the federal Food, Drug, and Cosmetic Act (FDCA) of 1938.

FDCA continues to be the basis of drug regulation in the United States today. Therefore, a working knowledge of this law is important for ensuring regulatory compliance of drug use in any clinical setting, including nuclear medicine practice. FDCA is both simple and complex.[13] It is simple in that it specifies only three illegal acts: adulteration, misbranding, and the placing of unapproved drug products into interstate commerce. At the same time, it is complex because many activities are included under the umbrella of these three illegal acts. When challenged to explain why something known to be illegal is in fact illegal under FDCA, stating that the activity is adulteration, misbranding, or the placing of an unapproved new drug into interstate commerce is a safe response. As passed in 1938, FDCA required premarket testing of new drugs for safety and presentation of safety data to FDA prior to approval of a new drug product for marketing. The act also eliminated the requirement for FDA to prove intent to defraud in drug misbranding cases.

In 1962, the Kefauver-Harris amendments to the 1938 FDCA increased federal control over methods of production and testing of drugs before their release for purchase by the public. The emotional impetus for these amendments came from another medical tragedy. An FDA medical officer, Frances Kelsey, blocked marketing approval of the drug thalidomide because of unexplained adverse effects; the subsequent discovery that thousands of deformed infants had been born to mothers in Europe and South America who had taken this supposedly safe sleeping pill during pregnancy caused Congress to vote for strong new drug controls in 1962. The most important change was the requirement that the efficacy of a new drug, as well as its safety, be established by "substantial evidence" before marketing approval. Among several new concepts contained in the amendment, the most important were investigational new drug (IND) procedures and procedures for the approval of a new drug application (NDA) prior to marketing of the product. Additional historical perspective and details regarding FDA's current regulation of radiopharmaceuticals can be found on the agency's website.[14]

RADIOISOTOPE REGULATION: HISTORICAL PERSPECTIVE ON NRC

The potential medical significance of radioisotopes was well recognized before World War II; nevertheless, the distribution of these radioactive materials was unregulated by the government.[15] Then, in 1942, the Manhattan Project was initiated by the United States Army to conduct atomic research for the purpose of providing new technology to facilitate an end to World War II. The postwar program for radioisotope distribution grew out of the Manhattan Project Isotopes Division, which during the war years had developed top technical expertise for producing and handling radioisotopes at the division's facilities in Oak Ridge, Tennessee. In 1946, the Manhattan Project formally publicized its groundbreaking program for distribution of radionuclides for peaceful research purposes. This new radioisotope distribution program for the first time placed nominal constraints on the procurement of radionuclides for research. Thus, radionuclides for medical research could not be ordered casually; an application for each proposed purchase had to be submitted, reviewed, and approved. A special subcommittee had to review and approve each application for human medical use.

Subsequently, with congressional passage of the Atomic Energy Act of 1946, the authority to continue radioisotope research was transferred from the Army to the United States Atomic Energy Commission (AEC). The Atomic Energy Act gave AEC exclusive governmental control over atomic research and the development of related technologies. The existing AEC radioisotope distribution review subcommittee was then renamed the Subcommittee on Human Applications. On June 28, 1946, the subcommittee held its first meeting, during which a system of local isotope committees was recommended. According to the recommended action plan, each local committee would include (1) a physician well versed in the physiology and pathology of the blood-forming organs, (2) a physician well versed in metabolism and metabolic disorders, and (3) a competent biophysicist, radiologist, or radiation physiologist qualified in the techniques of radioisotopes.

By October 1946, radioisotope distribution was well under way, with over 200 radioisotope requests received, reviewed, and approved. Of those approved, nearly 100 were for use in medical research in humans. In 1959, the subcommittee was absorbed into the Advisory Committee on Medical Uses of Isotopes (ACMUI), which still exists today.[16]

As the subcommittee gained experience in developing specific procedures for review and approval of a variety of radioisotope applications, it began to recognize that only a few of the requests for radioisotope use represented unusual cases; most applications were routine and did not require continuous review. The subcommittee delegated review of the routine radioisotope applications to the AEC Isotopes Division, and the Isotopes Division developed a procedure whereby an individual wishing to procure byproduct material had to file an application and receive an Authorization for Radioisotope Procurement prior to obtaining and using byproduct materials. This authorization functioned in much the same way as a license for a byproduct material does today. In 1951, an additional technical adjustment was made to the approval process with the introduction of "general authorizations" that delegated more authority to local radioisotope committees of approved research institutions. As a result, research institutions possessing general authorizations could for the first time obtain specified radioisotopes for approved purposes pursuant to filing a single application each year; they no longer needed to file a separate application for each radioisotope order.

During the 1950s a series of regulatory changes affected many administrative procedures governing radioisotope use. In 1954, licensing and regulation were added to AEC's authority. Concerns about radioisotope procedures were disseminated through various communications made public by the Isotopes Division. Throughout the 1960s and early 1970s, administrative procedures governing the licensing and regulation of radioisotope use continued to evolve. In 1975, as a result of congressional passage of the Energy Reorganization Act of 1974, a major change was implemented when AEC was split into the Energy Research and Development Administration, which later became the Department of Energy, and the Nuclear Regulatory Commission (NRC).

At several junctures during the development of procedures for licensing and regulating byproduct material intended for medical use, these government units published official circulars or guidance documents for use by the regulated community. A guidance document published in 1948 was only a few pages long and was simple to comprehend and follow. By 1957, that initial guidance was replaced by a 26 page AEC document entitled The Medical Use of Radioisotopes—Recommendations and Requirements. Then, in 1965, AEC published its Guide for the Preparation of Applications for the Medical Use of Radioisotopes, followed in 1980 by NRC Regulatory Guide 10.8, entitled Guide for the Preparation of Applications of Medical Programs.[17] Many additional changes in NRC regulations and licensing guidance have occurred since Regulatory Guide 10.8, including major revisions to Title 10 of the Code of Federal Regulations (CFR), Parts 20 and 35, and publication of NUREG-1556, Volume 9, entitled Consolidated Guidance About Material Licenses: Program-Specific Guidance About Medical Use Licenses.[18] Most important, NUREG-1556 provided valuable guidance on the recently revised 10 CFR 35 rule. Additional historical perspective and details regarding NRC and current regulation of radioisotopes can be found on the NRC website.[19]

REGULATORY CONTROL OF RADIOPHARMACEUTICALS

The regulation of radiopharmaceuticals has a complicated history. The 1938 FDCA applied to all drugs, including a unique and promising new class of radioactive materials with a variety of potential medical applications that would eventually become recognized as radiopharmaceutical drug products. Six years later, the Public Health Services Act of 1944 authorized the FDA Bureau of Biologics to regulate radioactive biologic drug products. However, because of their unique radioactive properties, radiopharmaceuticals were also under the control of AEC. During these formative years of nuclear medicine, radiopharmaceuticals were controlled chiefly by AEC. The 1954 Atomic Energy Act authorized AEC to license the possession, use, and transfer of byproduct material (i.e., radioisotopes produced in a nuclear reactor).

In 1963, after enactment of the Kefauver-Harris amendments, FDA began implementing IND procedures and requirements. These posed a substantial threat to the availability of radiopharmaceuticals and the emergence of nuclear medicine practice as a clinical specialty. FDA began to recognize both the potential clinical value of radiopharmaceuticals and the possibility that their safety and efficacy were not adequately controlled by the agency as required for all drugs by the 1962 FDCA amendments. Fortunately, existing AEC regulatory controls were deemed adequate to ensure radiation safety in the possession, use, and transfer of radioactive materials for medical use. However, FDA's overriding concern related to the medical safety and efficacy of radioactive drugs once administered to patients. FDA also had to come to terms with the fact that, as an agency, it was not prepared to be the sole regulator of radioactive drugs. The immediate problem was addressed in 1963 when FDA allowed a temporary exemption for radioactive new drugs and biologics from the IND regulations, provided these agents were distributed in complete compliance with existing AEC regulations.[20] The main purpose of the temporary exemption was to allow the continued availability of radioactive drugs manufactured from reactor-produced radionuclides until FDA and AEC could reach agreement on the establishment of effective regulations that would minimize unnecessary duplication of regulatory control. The exemption, however, did not include naturally occurring or accelerator-produced radionuclides. The temporary exemption was rescinded, in part, on November 3, 1971, when FDA actively entered the regulatory arena by publishing NDA requirements for radioactive drugs.[21] The new regulations, as outlined in 21 CFR 310.503, identified specific reactor-produced radioisotopes that for certain stated uses were no longer exempt from the new drug regulations because they were considered well-established drugs in nuclear medicine practice. Both FDA and AEC concluded that it was inappropriate for these radioactive drugs to be distributed under a claimed IND exemption when they were clearly intended for routine clinical use.[21] Therefore, manufacturers and distributors of these drug products were required to submit adequate evidence of safety and effectiveness for use as recommended in the product labeling.

During the regulatory transition period that followed, radiopharmaceutical manufacturers were allowed to distribute only those radioactive drug products for which FDA had approved an NDA or biologic product license or accepted an IND application. Accordingly, 52 NDAs for radiopharmaceuticals alone were submitted and approved by FDA between 1971 and 1975, compared with a total of 31 drug (pharmaceutical) NDA approvals between 1951 and 1970.[22] On July 25, 1975, FDA issued a final rule that totally revoked the 1963 IND exemption and placed radiopharmaceuticals completely under FDA regulatory authority, as are all other drug products.[23]

Another regulatory milestone was reached on January 19, 1975, when NRC and the Energy Research and Development Administration (which later became the Department of Energy) superseded AEC under the Energy Reorganization Act of 1974. Thus, with limited exception, NRC took responsibility for all licensing and regulatory functions originally assigned to AEC by the Atomic Energy Act of 1954 as amended. Although NRC's authority covered radioactive drug products containing reactor-produced byproduct materials, it did not include the regulation of such products that contain naturally occurring or accelerator-produced radioactive materials, because regulation of these two categories of radioactive materials was left to the states. States retained regulatory responsibility for these two categories of radioactive materials over the ensuing 30 years.

More recently, the Energy Policy Act of 2005 (EPAct), enacted August 8, 2005, significantly expanded the definition of byproduct material. This expansion of the definition in the outdated Atomic Energy Act of 1954 was intended to eliminate growing public safety concerns related to the long-standing regulatory exception, which had become a serious gap in regulatory control rather than a justified feature of the current regulatory structure. The new definition included any discrete source of radium 226, any material made radioactive by use of a particle accelerator, and any discrete source of naturally occurring radioactive material, other than source material, that the NRC, in consultation with other federal officials named in the EPAct, determines would pose a similar threat to the public health and safety or the common defense and security as a discrete source of radium 226, that are extracted or converted after extraction for use for a commercial, medical, or research activity. The EPAct placed these materials directly under NRC regulatory authority. Accordingly, on October 1, 2007, NRC amended its regulations to include jurisdiction over "discrete sources of radium 226, accelerator-produced radioactive materials, and discrete sources of naturally occurring radioactive material," which NRC began to refer to collectively as "naturally occurring and accelerator-produced radioactive material" (NARM).[24] Licensees, individuals, and other entities engaged in activities involving the newly defined byproduct material in agreement states (see below), non–agreement states, and U.S. territories are now affected by this regulation.

It is useful to note that the NRC maintains an Office of Federal and State Materials and Environmental Management Programs.[25] This office is primarily responsible for establishing and maintaining effective communication and working relationship between the NRC and states, local government, other federal agencies, and Native American tribal governments. It serves as the primary contact for policy matters between NRC and these external groups. In addition, the office keeps the external groups informed about NRC activities while keeping the agency apprised of these groups' activities as they may affect NRC, and it conveys to NRC management these groups' views on NRC policies, plans, and activities. This office also administers the agreement state program, which allows NRC to transfer its authority for control of the use of radioactive materials to individual states (agreement states). Currently, there are 36 agreement states, 13 non–agreement states, and 1 state that has filed a letter of intent with NRC to become an agreement state.[26] (The District of Columbia is treated as a non–agreement state.) It is important for nuclear pharmacists to understand how this regulatory scheme applies to a wide variety of medical and pharmacy practice settings. The NRC website provides up-to-date information and additional details regarding the interface between federal and state regulation of radioactive materials.[26]

REGULATORY AUTHORITY OF FDA AND NRC

After the termination of the 1963 exemption for radiopharmaceuticals from IND regulations, FDA stated that it would regulate the safety and efficacy of radioactive drugs with respect to patients. At the same time, NRC withdrew from regulating radioactive drug safety and efficacy and stated that it would regulate the radiation safety of workers and the general public.

In 1979 NRC developed and published the following three-part policy statement to guide the regulation of medical uses of radioisotopes (44 FR 8242, effective February 9, 1979):[27]

1. NRC will continue to regulate the medical uses of radioisotopes as necessary to provide for the radiation safety of workers and the general public.
2. NRC will regulate the radiation safety of patients where justified by the risk to patients and where voluntary standards or compliance with these standards is inadequate.

3. NRC will minimize intrusion into medical judgments affecting patients and into other areas traditionally considered to be part of the practice of medicine.

The policy states further in 44 FR 8242 that

> The NRC intends not to exercise regulatory control in those areas (regarding patients) where, upon careful examination, it determines that there are adequate regulations by other Federal or State agencies or well administered professional standards. The Commission recognizes that the FDA regulates the manufacture, interstate distribution, investigational, and research use of drugs, including radiopharmaceuticals, but does not have authority to restrict the routine use of drugs to the procedures (described in the product labeling) that the FDA has approved as safe and effective. The NRC sees itself as the only Federal Agency that is currently authorized to regulate the routine use of radioactive drugs from the standpoint of reducing the unnecessary radiation exposure of patients.

REGULATION OF POSITRON EMISSION TOMOGRAPHY DRUGS

Positron emission tomography (PET) drugs are by definition radiopharmaceuticals that have a radionuclide component consisting of a short-lived positron-emitting radioisotope. The most common positron emitters currently used in the production of PET drugs are ^{11}C, ^{13}N, ^{15}O, ^{18}F, and ^{82}Rb; however, there are a variety of short- and longer-lived positron emitters that may be used in clinical PET imaging in the near future. PET radiopharmaceuticals are discussed in detail in Chapter 12 of this book.

The development of a suitable regulatory framework for PET drugs and the associated imaging technology has been quite challenging, owing to the unique nature of PET as a medical imaging technology and the attempts of the federal government to apply regulatory policy devised for other, and often quite different, applications of medical technology. The struggle to regulate PET drugs and imaging technology and the emerging regulatory framework have been well documented. In 1998 Keppler and associates[28] published an excellent case study on PET regulation. The authors provide valuable theoretical perspective on why regulations exist, as well as analysis of why regulation of the maturing PET imaging industry has unfolded as it has. The insights provided in the article have supported and will continue to support efforts to expand clinical PET and introduce new imaging technologies into health care. The section of the article that summarizes the history of PET regulation is included here, with permission. (To facilitate identification of the regulatory documents cited in the original publication, the reference numbers have been edited to correlate with the reference list provided at the end of this chapter.)

Before 1975, the Food and Drug Administration (FDA) had delegated the regulation of the radiopharmaceutical industry to the Nuclear Regulatory Commission. When the FDA asserted its jurisdiction in 1975, it chose to regulate the industry substantially the same way as the traditional drug industry. Therefore, the FDA began evaluating new drugs—that is, radiopharmaceuticals—through investigational new drug applications. New drugs then were approved for use on the basis of sufficient information provided in new drug applications.[29] The FDA then would enforce current good manufacturing practices for production of the approved drug to ensure quality. Early on, however, it was clear that the short half-lives of many of the PET and other radiopharmaceuticals limited the traditional application of production and manufacturing regulations.

In the late 1970s, the FDA established a subcommittee to evaluate the special circumstances of these short-lived radiopharmaceuticals. Exemption from manufacturing regulations at the site of final use was considered for sites qualifying as a nuclear pharmacy or a medical facility under the provisions of the Durham-Humphrey amendment to the Food, Drug and Cosmetic Act. Thus, facilities preparing radiopharmaceuticals could operate under these provisions, eliminating the need to register with the FDA as a drug manufacturer for these activities. The report detailing the exemption requirements, titled "Nuclear Pharmacy Guideline: Criteria for Determining When to Register as a Drug Establishment" (Nuclear Pharmacy Guidelines), was adopted by FDA in 1984.[30] These guidelines not only covered production activities for traditional nuclear medicine isotopes but also described activities consistent with the preparation of PET isotopes.[30] The activities detailed affirmed the practice of medicine and pharmacy.

In the late 1980s, the concept of the "clinical" PET center developed because of the promise of clinical usefulness shown with early trials of ^{18}F-fluorodeoxyglucose (FDG). At about the same time, the health care industry was in the midst of transition. Hospitals were concerned about reductions in revenues because of Medicare Diagnostic Related Group payment schemes and managed care. Private sector insurance providers were faced with a rapidly rising cost base and were cutting reimbursement rates. The capital commitment ($5–$7.25 million) required to develop a clinical PET center made PET an unlikely venture for hospitals fearing the future contraction of the industry.[31] Furthermore, the revenues that could be proposed to offset the more than $2 million per year in operating expenses[31] were viewed as risky in part because its regulatory path and reimbursement potential remained unclear.[32]

Many of those attending to these early clinical activities proposed using the Nuclear Pharmacy Guidelines and the practice of medicine and pharmacy as the basis for operations at the new or planned clinical centers. Concern centered only on isotope approvals because PET scanners and the subsequently developed dual-use or coincidence imaging devices received FDA clearance for marketing as "PET devices," grandfathered in with changes in device regulations. The concept of the practice of medicine and pharmacy would allow a fast-track mechanism for clinical utilization of these new PET compounds. Many in the physician and pharmacist communities believed rationale existed for this approach, citing case law, the amendments to the Food, Drug and Cosmetic Act, as well as the Nuclear Pharmacy Guidelines and other FDA publications in support of the exemption for drugs not intended for interstate commerce.[32] But as Coleman et al. pointed out, the FDA contended then—and now—that the practice of medicine and pharmacy cannot be applied to unapproved drugs.[32] The physician and pharmacist communities' assertion that the practice of medicine and pharmacy should be the course of regulations did not convince many hospitals to invest in this new technology.

Industry representatives recognized that clarification of the regulatory, and thus the reimbursement, pathway for PET was essential to commercial growth. A dialogue was initiated with the FDA, ostensibly to seek clarification of their position. A less visible, but nonetheless plausible, concern of industry may have been that if the FDA did not regulate the end drug, then it might instead regulate the equipment used to produce the isotopes.

In 1989, members of the PET industry reached consensus with the FDA on a mechanism to regulate PET; the end users would develop new drug applications for all PET radiopharmaceuticals. At that time, the user community did not want the responsibility of organizing the data to obtain FDA approval, nor were prospective data on the clinical use of PET available to submit for evaluation. Moreover, the community feared that the costs of filing the new drug applications as well as bringing operations in line with manufacturing guidelines would be prohibitive, limiting the clinical proliferation of the technology. Despite appeals from the user community that an alternative approach be devised, the FDA began laying out its plan for regulation. The FDA had little rationale to change its course. Individual cyclotron sites would be expected to obtain new drug applications or abbreviated new drug applications (the generic equivalent) and register as drug manufacturing sites. In exchange for agreeing to their regulatory method, the FDA promised to consider retrospective data in their review of the drug master file, to review the data expeditiously, and to develop modifications of drug manufacturing guidelines so they would be better suited to PET. The FDA clearly stated their position: the practice of medicine and pharmacy would not be an acceptable method of practice; it could be used only until the initial new drug application was approved.

That the user community would be forced to comply with the FDA's plans was clear by 1991. It was believed that the Health Care Financing Administration would not act on a petition for PET reimbursement until FDG was approved, which would lead to a continued lack of Medicare reimbursement. Therefore, the Institute for Clinical PET, with funding from industry, led efforts to develop a single drug master file for FDG. With the continued support of industry and the diligent efforts of members of the community, a single PET site in Peoria, IL, filed a new drug application in 1992.

Neither the clinical drug master file nor the new drug application was reviewed expeditiously, as promised. Additional prospective data, reaffirming earlier conclusions, were required. Finally, the FDA approved the efficacy of FDG for a single application (epilepsy) and, in 1994, the new drug application from the Peoria PET facility. The FDA published a notice in the [*Federal Register*] in February 1995, which detailed the process that sites should follow in filing their own new drug applications or abbreviated new drug applications. An approved status for sites would be required or sites would face closure, a stance the FDA may have believed necessary because of the degree of opposition already expressed by the community.

Fluorine-18-fluorodeoxyglucose was now approved by the FDA, but Medicare reimbursement was still not forthcoming. In fact rubidium-82, previously approved by the FDA, at that time still had not been approved for coverage by the Health Care Financing Administration. The status of FDA approval for PET did not bring the reimbursement hoped for. Moreover, the technical requirements placed on the Peoria facility by compliance with the new drug application and the FDA's manufacturing requirements were stringent and required substantial additional operating monies to sustain. The community and industry had serious concerns whether, as now required by the FDA, PET sites could file abbreviated new drug applications and bring their operations into similar good manufacturing practices at a reasonable cost. Although the new drug application, sponsored by a consortium of industry, was now available to PET sites, no easy path existed for centers to comply with FDA oversight.

Since then, the business, physician, and pharmacist communities petitioned the FDA to modify their position. State boards of pharmacy have supported the practice of medicine and pharmacy approach, and most of the commonly used cyclotron-produced clinical compounds have been added to the *United States Pharmacopeia*. Individuals and professional societies, such as the Institute for Clinical PET, proposed alternative regulatory mechanisms to the FDA directly, methods not requiring hospital pharmacies to register as manufacturing facilities. A commercial radiopharmacy company filed a citizen's petition requesting that the FDA evaluate equipment approvals rather than manufacturing site approvals. This same company filed a lawsuit against the FDA for alleged rule-making violations and for not responding to the community's stated concerns. The initial ruling on this lawsuit was in favor of the FDA, stating that the FDA acted within its purview to create rules. Recently, the United States Court of Appeals overturned the earlier ruling, affirming that the FDA had indeed violated rule-making requirements.[33]

Efforts such as these had failed to modify the FDA's regulatory stance. Although indications were forthcoming from the FDA that some modifications to manufacturing process would be possible, they would not willingly retreat from their overall position. As anticipated in the December 1994 citizen's petition, the regulatory challenges have dramatically impeded the clinical practice of PET. Nearly 10 years after use of FDG became acceptable clinical practice in the minds of experienced physicians, FDG was still not broadly available.

Relief came through attempts to influence regulatory policy through legislative initiatives. Efforts focused on both circumventing one of the levers that the FDA was using to force compliance (Health Care Financing Administration approval for reimbursement of Medicare patients) and legislating changes in the mandate to the FDA with respect to PET. Reform of the process for the approval and oversight of the manufacturing of radiopharmaceuticals was accomplished legislatively through the recently passed FDA Modernization Act of 1997 ("FDA reform" act).[34] The legislation contains specific language requiring the FDA to adopt "appropriate" procedures for approval of PET new drug applications and abbreviated new drug applications and "appropriate" current good manufacturing practices. These procedures and practices are to be determined jointly by the FDA, industry, and the user community. In addition, the FDA is required to take "due account of any relevant differences" between commercial PET centers and not-for-profit PET facilities, with the hope of reducing the costs of coming into regulatory compliance at the hospital level. During the time period that these new processes are developed and for 2 years thereafter, neither new drug applications nor abbreviated new drug applications are required as the medical community brings itself into compliance with the new regulations.[34]

The legislation rescinded all of the recent rules published by the FDA on PET, including the notices entitled "Regulation of Positron Emission Tomography Radiopharmaceutical Drug Products: Guidance," published in the *Federal Register* on February 27, 1995;[35] "Draft Guideline on the Manufacture of Positron Emission Tomography Radiopharmaceutical Drug Products: Availability," published February 27, 1995;[36] and a final rule entitled "Current Good Manufacturing Practice for Finished Pharmaceuticals: Positron Emission Tomography" published in the *Federal Register* on April 22, 1997.[37] The rescission was published December 19, 1997, in the *Federal Register*.[38,39] Effectively the legislation reverts regulatory guidelines to those provided in the 1984 Nuclear Pharmacy Guidelines until new guidelines are established. In the interim, *United States Pharmacopeia* standards are to be met for drugs to be considered unadulterated.

Most importantly, a new Medicare approval process also resulted from the political efforts. Despite lack of FDA approval, the Health Care Financing Administration agreed that Medicare would begin to cover PET scans for characterization of solitary pulmonary nodules and initial staging of lung cancer as of January 1, 1998. It was agreed that a fast-track review of several other indications for PET, including evaluation of brain tumors, myocardial viability, colorectal cancer, head and neck cancer, melanoma, breast cancer, Hodgkin's lymphoma, and ovarian cancer, would be initiated over the next 18 months.

In late May 1998, although the prescribed regulatory process had not started in earnest, discussions between the PET community and the Health Care Financing Administration (HCFA) were under way to seek broader approval for use of ^{18}F-fludeoxyglucose (^{18}F-FDG, previously referred to as ^{18}F-fluorodeoxyglucose) in oncology, neurology, and cardiology applications. As a result of the persistent collaborative efforts between the PET community and HCFA, PET imaging reimbursement approvals expanded dramatically over the next 5 years.[40] (HCFA was renamed the Centers for Medicare and Medicaid Services [CMS] in 2000.) These initial reimbursement approvals combined with those implemented more recently via the National Oncologic PET Registry (NOPR) program[41] have, along with a variety of other factors, allowed clinical PET imaging to grow substantially over the last 12 year period. According to published market data from surveys of more than 2000 sites with fixed and mobile PET and PET/CT units, an estimated 456,200 clinical PET procedures were performed in the United States in 2002, up 79% from 255,300 in 2001.[42] From 2005 to 2008, the total number of PET and PET/CT patient imaging studies conducted in the United States increased 35%, from 1.13 million to 1.52 million.[43] However, the annual rate of growth in recent years has been more modest; the 2008 total represents only a 4% increase over the 2007 total of 1.48 million, compared with an average annual growth rate of 10.4% from 2005 to 2008. Even though the clinical use of PET and eventually PET/CT imaging in routine health care was expanding substantially during these developmental years, significant reimbursement and regulatory issues persisted.

As required by the FDA Modernization Act of 1997 (FDAMA),[34] representatives from the Institute of Clinical PET (ICP), United States Pharmacopoeia (USP), Society of Nuclear Medicine (SNM), and Council on Radionuclides and Radiopharmaceuticals (CORAR) have maintained a continuing collaborative dialogue with FDA in an attempt to define the future approval process and requirements for all radiopharmaceuticals. FDAMA gave FDA 2 years, and not more than 4 years, from November 1997 to establish new guidelines for the manufacture of PET drugs. Subsequently, new manufacturing guidelines were developed, with some input from the regulated stakeholders but largely within FDA, but not until many years after the implied November 2001 deadline. In fact, FDA did not finalize the new PET drug good manufacturing practice regulations until December 10, 2009.[44]

During this long transitional period, the PET community continued to prepare PET drugs according to applicable USP standards, in particular *USP* Chapter 823, Radiopharmaceuticals for Positron Emission Tomography— Compounding (*USP* 823), without being required to file NDAs or abbreviated new drug applications (ANDAs) with FDA. Now, however, since the new FDA guidelines have been published as a final rule in the *Federal Register,* the PET community has until December 12, 2012, to reach compliance with the new regulations.

Concerted efforts were made to develop standards for regulating PET drug production sites that take into account the genuine differences between commercial production centers and nonprofit hospitals compounding PET drugs for on-site use; however, FDA consistently resisted, and these important differences are essentially overlooked in the final rule. It is critical to recognize and emphasize that FDAMA also established a precedent-setting regulatory policy that all activities associated with the preparation or production of PET drugs will by definition be considered manufacturing, which essentially eliminates all compounding of PET imaging drugs. Clearly, there are still many regulatory issues to be resolved regarding the manufacture and clinical use of PET drugs, and the PET community must continue to monitor regulatory developments and prepare for compliance by the end of 2012. Most recently, FDA published a series of questions and answers regarding the new regulation on current good manufacturing practices (CGMPs) for PET drugs.[45,46] It is particularly important that nuclear pharmacists monitor new developments regarding the regulation of PET drugs.

Current information regarding the drug development process and the IND, NDA, and ANDA submission processes is available on the FDA website.[47–52]

ROLE OF FDA IN PUBLIC HEALTH PROTECTION

FDAMA (Public Law 105-115) affirmed FDA's public health protection role and defined the agency's mission as follows: (1) to promote the public health by promptly and efficiently reviewing clinical research and taking appropriate action on the marketing of regulated products in a timely manner; (2) with respect to such products, to protect the public health by ensuring that foods are safe, wholesome, sanitary, and properly labeled; human and veterinary drugs are safe and effective; there is reasonable assurance of the safety and effectiveness of devices intended for human use; cosmetics are safe and properly labeled; and public health and safety are protected from electronic product radiation; (3) to participate through appropriate processes with representatives of other countries to reduce the burden of regulation, harmonize regulatory requirements, and achieve appropriate reciprocal arrangements; and (4) as determined to be appropriate by the Secretary, carry out items (1) through (3) in consultation with experts in science, medicine, and public health, and in cooperation with consumers, users, manufacturers, importers, packers, distributors, and retailers of regulated products.

FDA's public health protection role as defined in FDAMA is extremely broad. With respect to the practices of medicine and pharmacy, the essence of the FDA mission is to ensure the safety and efficacy of marketed drugs and medical devices. The process by which this mission is achieved is authorized by Congress; formalized by codes, regulations, and guidelines; and interpreted and implemented by scientists, lawyers, biostatisticians, engineers, and project managers of varied backgrounds.[53] Thus, FDA plays a significant role in the development and approval of all radiopharmaceuticals in clinical use. FDA's Center for Drug Evaluation and Research (CDER) regulates formats for clinical trials and review of radiopharmaceuticals prior to their approval for marketing.

INVESTIGATIONAL NEW DRUG PROCEDURES FOR RADIOPHARMACEUTICALS

From the viewpoint of FDA, a "new drug" may be a new molecular (i.e., chemical) entity that requires proof of safety and efficacy for its intended clinical use (i.e., an investigational new drug) or a known entity that has been recently shown to be safe and efficacious for an intended clinical use (i.e., an approved new drug). A new drug may also be a new dosage form or new route of administration for an old drug (i.e., an approved drug that has been marketed or sold on a prescription basis for a significant period of time), or an old drug being used for a new clinical indication or purpose. Situations involving new forms, routes, or uses of approved drugs occur only occasionally in clinical nuclear medicine practice.

Before 1962, there was no requirement to notify FDA that drugs were being tested in humans. However, since the 1962 FDCA amendments, new drugs lacking NDA approval and intended for investigational use in human subjects may not enter into interstate commerce (commercial distribution) unless a responsible individual or a pharmaceutical company sponsors well-controlled, scientifically designed safety and efficacy studies under an FDA-accepted investigational new drug application (referred to simply as an IND). FDA authorization must be secured in advance of the interstate shipment and administration of the new drug to humans enrolled in the planned clinical studies.

The applicant, or "drug sponsor," is the person or entity that assumes responsibility for the marketing of a new drug, including responsibility for compliance with applicable provisions of the FDCA and all related regulations. The sponsor is typically an individual, partnership, corporation, government agency, manufacturer, or scientific institution. In many medical practice settings, including nuclear medicine, there is frequently no drug company interested in sponsoring and conducting important clinical investigations; therefore, the sponsor may be an institution-based physician, who is referred to as the "physician-sponsor" or "investigator-sponsor." As an alternative, the investigator-sponsor may be a clinical radiopharmacologist or nuclear pharmacist qualified by training and experience in the evaluation of new radioactive drug products.

PRECLINICAL RADIOPHARMACEUTICAL STUDIES

In the earliest phase of development of a radioactive drug product, data from animal studies and data on manufacturing and quality control are collected and summarized for eventual inclusion in an IND if the drug shows promise for use in humans. The data must clearly establish that the radioactive drug is reasonably safe for administration to human subjects during the proposed clinical trials. In addition, the actual procedures and methods used for generating and gathering the preliminary safety data must be described in detail. Preclinical studies are conducted in relevant animal models to assess the drug's relative safety rather than its efficacy, although some potentially relevant efficacy data may be obtained. Characterization and quantification of the radiochemical and radionuclidic purity of the radioactive drug are essential for the evaluation of radiation dosimetry; any trace radiocontaminants (including daughter radionuclides) and altered chemical forms that might significantly influence biodistribution and radiation absorbed dose estimates must be identified. Preclinical studies usually include studies of both biodistribution and toxicity in animals. Animal biodistribution studies are used to determine normal organ distribution patterns, assess translocation, and identify the routes and extent of radiopharmaceutical excretion. These data are essential for obtaining meaningful radiation absorbed dose estimates, the principal measure of radiopharmaceutical safety. Animal toxicity studies usually focus on the potential chemical toxicities of components other than the radionuclide, since only trace amounts of the radioactive element are typically present. Acute toxicity tests are usually required in at least two animal species to determine the acute LD_{50} (amount sufficient to kill 50% of a population of animals) and to demonstrate the lack of acute toxicity at doses several orders of magnitude higher on a dose-per-kilogram basis than those proposed for human studies using the same route of administration. Subacute toxicity testing over a 2 to 3 week period in two animal species, a rodent and a nonrodent, at several dosage levels is required to demonstrate adequate safety margins relative to equivalent maximum clinical dosage. Chronic toxicity studies are typically not required for radiopharmaceuticals, which are administered on a one-time basis to most patients.

Investigators may obtain the required data from their own experiments, but gathering the data can be quite challenging, costly, and time consuming. It may be advantageous to use data from the literature or other valid sources when possible, if the investigator can demonstrate that those data are applicable to the drug product under consideration (i.e., similar dosage form, same route of administration).

RADIOACTIVE DRUG RESEARCH COMMITTEE STUDIES

Limited use of radioactive drug products in human research subjects prior to FDA acceptance of an IND is allowed under specific conditions set forth in 21 CFR 361.1.[52] Radioactive drugs, as defined in 21 CFR 310.3(n), are generally recognized as safe and effective when administered to human research subjects, under the conditions set forth in 21 CFR 361.1(b), during the course of a research project intended to characterize the basic pharmacodynamic and pharmacokinetic properties of the radioactive drug product. Data regarding the metabolism (including kinetics, distribution, and localization) of the radioactive drug or regarding human physiology, pathophysiology, or biochemistry are extremely valuable. However, such studies must not be intended for immediate therapeutic, diagnostic, or similar purposes or intended to determine the safety and effectiveness of the radioactive drug in humans. In other words, an investigator must not have clinical intent regarding the medical care of subjects receiving the radioactive drug or carry out a clinical trial under this set of regulations. Certain basic research (e.g., studies to determine whether a drug localizes in a particular organ or fluid space and to describe the kinetics of that localization) may have eventual diagnostic or therapeutic implications; however, the initial studies are considered basic research within the meaning of this specific set of regulations.

Before these limited human studies are conducted, approval must be obtained from a local or contract radioactive drug research committee (RDRC). The RDRC must be approved by FDA in accordance with the regulations set forth in 21 CFR 361.1(c) that govern committee membership, function, reports, approvals, and monitoring responsibilities. Under the section describing reports that the RDRC must provide to FDA, there is an important federal control point regarding the number of research subjects that may be studied under these regulations. If at any time the RDRC approves a research proposal that involves exposure of more than 30 research subjects, or of any research subject under 18 years of age, the committee must immediately submit a special informational summary to FDA. The reporting of such RDRC approvals thus provides an opportunity for FDA to intervene if necessary. This may explain why most committees encourage protocols requesting fewer than 30 adult research subjects when appropriate and discourage the use of individuals under age 18 unless absolutely necessary. As a practical rule, the RDRC may choose to approve studies in only a few subjects (e.g., six) and require the investigator to report these results before additional studies are approved. In this way, the RDRC can ensure that the number of subjects is kept to the minimum needed to answer pertinent scientific questions. Standards set forth in 21 CFR 361.1(d) are used by the RDRC to determine if the pharmacologic dose and radiation dose are within the required limits and that the radiation exposure is justified by the quality of the proposed study and the importance of the information it seeks to obtain. The RDRC must also ensure that other requirements are in place regarding qualifications of the investigator, proper radioactive material licensure, selection and consent of research subjects, quality of radioactive drug product used, research protocol design, reporting of potential adverse reactions, and approval by an appropriate local or contract institutional review board (IRB).

Compliance with strict pharmacologic dose limits is compulsory in the conduct of human research under an RDRC approval. The amount of active ingredient or combination of active ingredients to be administered must be known not to cause any clinically detectable pharmacologic effect in humans. This fundamental regulatory principle implies that, in the absence of any known human pharmacology data, the study cannot be approved by the RDRC nor conducted by the investigator simply because there are no human data, and the active ingredient will be administered only in minuscule trace amounts. The lack of human data is most often the single reason why potential RDRC studies cannot be approved. In a few isolated cases, investigators have successfully undertaken IRB-approved subpharmacologic dose studies in humans using a nonradioactive form of the active ingredient to demonstrate the lack of clinically detectable pharmacologic effects. Once a subpharmacologic dose was determined and made available to the RDRC, it served as the basis for approval of studies that had otherwise satisfied all RDRC requirements. If such a pathway is approved by the RDRC and undertaken by the investigator, both must recognize that without proper operational controls and safeguards, the dose of active ingredient could theoretically exceed the established subpharmacologic threshold. For example, if short half-life positron-emitting ^{11}C is the radionuclide component of a very potent study drug, the actual mass of active ingredient in the administered human dose could vary dramatically with time after synthesis, resulting in the potential administration of a pharmacologic dose. In this case, it is critically important to require that the total amount of the active ingredient in any batch of the radioactive drug be substantially less than the established subpharmacologic threshold. Finally, when the same active ingredient without the radionuclide component is administered simultaneously under an IND or in accordance with the approved labeling for a therapeutic drug, the total amount of active ingredients including the radionuclide must be known not to exceed the dose limitations applicable to the separate administration of the active ingredients excluding the radionuclide.

The regulations governing RDRC studies also impose important limits on acceptable radiation dose. These limits are significantly more restrictive than those allowed under IND-approved studies, perhaps because normal volunteer populations are most frequently selected and enrolled in RDRC studies. The key here is that the amount of radioactive material to be administered must be such that human research subjects receive the minimum practical radiation dose necessary to perform the study without jeopardizing the benefits to be gained from each study. If the single-study or cumulative multiple-study radiation dose to an adult research subject in a given year is below the levels specified in these regulations, the study can be recognized as safe and thus approved by the RDRC. Accordingly, the whole body, active blood-forming organs, lens of the eye, and gonads may not receive more than 3 rem from a single dosage or an annual and total dose commitment of 5 rem. Other organs may not receive more than 5 rem from a single dosage or an annual and total dose commitment of 15 rem. Occasionally, investigators may wish to conduct studies in research subjects less than 18 years of age. This is possible; however, the radiation dose cannot exceed 10% of the specifications for adult subjects. All sources of radiation exposure associated with each study must be included in the determination of total radiation doses and dose commitments. Therefore, exposures from all radioactive components in the drug product must be included in this determination, whether they are essential or present as significant contaminants or impurities. Radiation doses from x-ray procedures that are part of the research protocol must be included. Finally, the numerical definitions of dose must be based on an absorbed-fraction method of radiation absorbed dose calculation. The RDRC is required to use either the Medical Internal Radiation Dose (MIRD) or the International Commission on Radiological Protection (ICRP) system for these calculations.

Investigators who decide to conduct studies under RDRC approval must realize there are valid reasons for ensuring that such studies are well designed, controlled, and managed, since the data could become critically important at some later date. According to 21 CFR 361.1(e), the results of any research conducted under an RDRC approval become part of any clinical evaluation of the radioactive drug product pursuant to 21 CFR 312, which specifies the requirements for submitting INDs. Therefore, if at any point the intent of an RDRC-approved study is altered in any way to begin use of the radioactive drug in a clinical trial, the RDRC must immediately call for the termination of the study and require the investigator to obtain an IND before additional human subjects are studied. A summary of all study results must be reported to the RDRC, and this report must be made part of the IND submitted to FDA.

According to 21 CFR 361.1(f), radioactive drugs prepared, packaged, distributed, and primarily intended for use in RDRC studies are exempt from misbranding (FDCA 502(f)(1)) and requirements for adequate directions for use (21 CFR 201.5, 201.100) if the packaging, label, and labeling are in compliance with federal, state, and local law regarding radioactive materials and if the label of the immediate container and shielded container, if any, either separate from or as part of any label and labeling required for radioactive materials by NRC or by state or local radiologic health authorities, bear certain specific information. Two important legal statements are required in the labeling: "Caution: Federal law prohibits dispensing without prescription" and "To be administered in compliance with the requirements of Federal regulations regarding radioactive drugs for research use (21 CFR 361.1)." There are numerous other requirements that may seem impractical but nevertheless must be satisfied to ensure full compliance. The published regulations provide details and list all labeling requirements for radioactive drugs used under RDRC approval.

EXPLORATORY IND STUDIES

Historically, the transition of preclinical radiotracers into candidate radiopharmaceutical products in clinical evaluation has always been a time-consuming, resource-intensive, and costly process, largely because of the regulatory requirements that govern this developmental transition. In some limited cases, in which applicable requirements have been satisfied, the RDRC study approach has been used successfully to make this transition. However, RDRC regulations strictly prohibit first-in-human trials and therefore eliminate this approach as a potential pathway for beginning to evaluate candidate products in human subjects. Thus, in a vast majority of cases, an IND is required for crossing this developmental threshold. The IND requirement has significantly challenged and often discouraged potential sponsors from developing many useful and potentially significant radiopharmaceutical drugs, largely because separate INDs were required for each candidate. Not surprisingly, the regulated industry and nuclear medicine community began to encourage FDA to address the need for new regulatory processes and mechanisms to facilitate the development of new radiopharmaceuticals.

In its March 2004 Critical Path Report, FDA stated that to reduce the time and resources expended on candidate products that are unlikely to succeed, new tools were needed to distinguish earlier in the process those candidates that hold promise from those that do not.[54] Thus, FDA recognized the need for innovative regulatory processes and mechanisms that would accelerate the process of candidate development, with particular focus on the transition from preclinical to clinical applications—the crucial decision point in the drug development process. In less than 2 years, responding to this critical need for new processes and mechanisms, FDA developed the Exploratory Investigational New Drug (eIND) option. In January 2006, FDA issued "Guidance for Industry, Investigators and Reviewers: Exploratory IND Studies," introducing the new mechanism for facilitating drug development.[55] The new guidance document describes "early phase 1 exploratory approaches that are consistent with regulatory requirements while maintaining needed human subject protection, but that involve fewer resources than is customary, enabling sponsors to move ahead more efficiently with the development of promising candidates." The eIND is designed to allow more flexibility in the early IND process and is strategically positioned to aid imaging development and to lower barriers at key points in the drug development process. An eIND study, as defined in the guidance document, is a clinical trial that is conducted in early phase 1, involves very limited human exposure, and has no therapeutic or diagnostic intent (e.g., screening studies, microdose studies). The newly developed eIND approach is ideal for conducting human proof-of-mechanism trials on novel radiopharmaceuticals. As an example, Avid Radiopharmaceuticals (Philadelphia, PA) recently used the eIND approach to efficiently conduct early clinical trials on a large number of novel radiopharmaceuticals, including four related novel ^{18}F-labeled PET amyloid imaging agents.[56] The eIND process was found to be a rapid and efficient mechanism for generating first-in-human efficacy data (amyloid binding), kinetics, and dosimetry, with significant advantages over other possible approaches (foreign trials, traditional corporate INDs, and physician-sponsored INDs).[56] This novel developmental approach led to the efficient identification of the styrylpyridine ^{18}F-AV-45, now known as ^{18}F-florbetapir, as a potential imaging agent for the early detection of Alzheimer's disease. Additional regulatory perspectives on the radiopharmaceutical development process, including the potential utility of the eIND, were recently provided by VanBrocklin.[57] The eIND process is described in detail on the FDA website.[48]

THE INVESTIGATIONAL NEW DRUG APPLICATION

The IND is an application that a drug sponsor must submit to FDA before beginning a clinical trial with a new drug in humans. It is not an application for marketing approval. Technically, it is a request for exemption from the federal statute that prohibits an unapproved drug product from being shipped in interstate commerce. Federal law requires that a drug be the subject of an approved marketing application before it is transported or distributed across state lines. Because a drug sponsor will almost certainly ship the investigational drug to several clinical sites in multiple states, an exemption must be obtained to satisfy this legal requirement. The IND is the means through which a sponsor obtains an exemption from FDA.

The main purpose of the IND is to provide detailed plans for well-controlled clinical drug studies in human subjects. It gives an overview of all that is currently known about the investigational drug product, including its structural formula, animal test results, human data, if any, and manufacturing information. Thus, an IND is typically quite lengthy and is labor intensive despite the use of innovative electronic processing systems. Most important, the IND serves as the basic documentation for FDA's acceptance of proposals to initiate clinical investigations in human subjects.

There are two categories of INDs, commercial and noncommercial. Pharmaceutical companies submit commercial INDs with the ultimate goal of obtaining marketing approval for a new drug product. Most INDs, however, are noncommercial; they are filed in support of conducting clinical research, with no intent to obtain marketing approval. Three types of INDs may be submitted to FDA for acceptance prior to conducting clinical investigations: the investigator IND, the emergency use IND, and the treatment IND.

A physician who wishes to both initiate and conduct a clinical investigation with a new drug product must submit an investigator IND, which may also be referred to as a physician-sponsored IND. Under this type of IND the physician accepts immediate responsibility for directing the preparation, dispensing, and administration of the investigational drug product. A physician might also wish to submit an investigator IND to conduct studies of an unapproved drug, or of an approved product for a new indication or in a new patient population.

A physician may obtain an emergency use IND when urgent medical conditions occur and patients may benefit from the use of an investigational drug product. With this type of IND, FDA authorizes use of an investigational drug in an emergency situation when time is insufficient for submission and review of an IND in accordance with 21CFR 312.23 or 312.24. The emergency use IND may also be useful for patients who do not meet the inclusion criteria of an existing clinical study protocol, or when there is no approved study protocol.

A treatment IND may be submitted for investigational drugs showing promise in clinical testing for serious or immediately life-threatening conditions during the period when the final clinical work is being conducted and FDA review is taking place. The treatment IND is thus a mechanism that allows promising investigational drugs to be used in "expanded-access" protocols—relatively unrestricted studies whose intent is both to learn more about the drugs, especially their safety, and to provide treatment for patients with immediately life-threatening or otherwise serious diseases for which there is no real alternative. These expanded-access protocols require researchers to formally investigate the drugs in well-controlled studies and to supply some evidence that the drugs are likely to be helpful. Of course, the drugs cannot expose patients to unreasonable risk. There are only a few therapeutic radiopharmaceuticals in development and, historically, treatment INDs for these important orphan therapy agents have been accepted by FDA.

The IND application must contain information in three broad areas: animal pharmacology and toxicology studies, manufacturing information, and protocols and investigator information. Preclinical animal pharmacology and toxicology data must be sufficient to permit assessment of whether the product is likely to be reasonably safe for initial human testing. Any previous experience with the drug product in humans must also be included. Manufacturing information pertaining to the composition, the manufacturing processes, stability, and the controls used for manufacturing the drug substance and the drug product must be included. This information is assessed to ensure that the manufacturer can adequately produce and supply consistent batches of the investigational drug product. Detailed clinical protocols for proposed studies must be provided to determine whether initial-phase studies will expose human subjects to unnecessary safety risks. The qualifications of the clinical investigators who oversee the administration of the investigational drug must also be included, for assessment of whether these individuals are qualified to fulfill their clinical trials duties. Finally, the investigator must provide a commitment to obtain informed consent from research subjects, to obtain initial review and maintain continuing review by an institutional review board (IRB), and to adhere to all existing IND regulations. Once the IND is submitted for FDA review, the sponsor must wait at least 30 calendar days before initiating any human studies. This allows FDA time to review the IND for safety in order to ensure that research subjects will not be subjected to unreasonable risks. Additional details and guidance regarding the IND application process can be found on the FDA website.[48]

THE IND PROCESS

The current general regulatory dicta for all IND submissions are found in CFR 21, Subparts 312.1 through 312.70.49. FDA's Center for Drug Evaluation and Research (CDER) has several on-line resources that summarize IND content, format, and classification and the IND review process.[48]

Given the nature of this set of regulations, securing an IND approval is a lengthy, labor-intensive process that is underappreciated by the novice who has never before accepted this responsibility. Novices may wonder where and how to initiate the intricate process of preparing and submitting an IND; two key FDA regulatory forms not only guide the process well but serve as both the starting and ending points. The process of securing an IND approval should in all cases revolve around the creation and assembly of the required documentation identified in Form FDA-1571, Investigational New Drug Application (IND) (21 CFR 312.23(a)(1)), and Form FDA-1572, Statement of Investigator (21 CFR 312.53(c)(1)). Both of these forms, with supporting documentation, must be part of the IND submission, and both are available from the FDA website.[58] Form FDA-1571 serves as the cover sheet for the entire IND submission document. Table 18-1 lists the information requested in the April 2006 version of this on-line electronic form. Form FDA-1572 serves as a means for the IND sponsor to document that investigators are appropriately qualified and sufficiently informed to begin participation in the clinical investigations. Table 18-2 lists the information requested in the May 2006 version of this on-line electronic form. Both forms warn that a willfully false statement is a criminal offense in accordance with U.S.C. Title 18, Section 1001.

Each of these forms states that the estimated "public reporting burden" for collecting the information is 100 hours, including the time for reviewing instructions, searching existing data sources, gathering and maintaining the data needed, and completing and reviewing the collected information.

TABLE 18-1 Information Requested by Form FDA-1571, Investigational New Drug Application

1. Name of sponsor
2. Date of submission
3. Address (number, street, city, state, and ZIP code)
4. Telephone number
5. Name(s) of drug, including all available names such as trade, generic, chemical, and code
6. IND number if previously assigned
7. Indication(s) covered by the submission
8. Phase(s) of clinical investigation to be conducted
9. List of numbers of all INDs (21 CFR Part 312), new drug or antibiotic applications (21 CFR Part 314), drug master files (21 CFR 314.420), and product license applications (21 CFR 601) referenced in the application
10. Consecutive or serial number assigned to the submission
11. Checklist indicating purpose of submission. If the submission includes a request for a treatment IND (21 CFR 312.35(b)), treatment protocol (21 CFR 312.35(a)), or charge request/notification (21 CFR 312.7(d)), a justification must be attached satisfying the elements specified in the cited CFR section
12. Checklist identifying items included in the initial application. Form FDA-1571 (21 CFR 312.23(a)(1)), the table of contents (21 CFR 312.23(a)(2)), and several other critical documentation sections are necessary to facilitate both the clinical review and the nonclinical review of the IND submission. The introductory statement (21 CFR 312.23(a)(3)), the general investigational plan (21 CFR 312.23(a)(3)), the investigator's brochure (21 CFR 312.23(a)(5)), and the protocol(s) (21 CFR 312.23(a)(6)) are the main subject of the clinical review. The key protocols for clinical review include study protocol(s) (21 CFR 312.23(a)(6)), investigator data (21 CFR 312.23(a)(6)(iii)(b)) or completed Form FDA-1572, facilities data (21 CFR 312.23(a)(6)(iii)(b)) or completed Form FDA-1572, and IRB data (21 CFR 312.23(a)(6)(iii)(b)) or completed Form FDA-1572. The nonclinical review focuses primarily on the chemistry, manufacturing, and control (CMC) data (21 CFR 312.23(a)(7)), the pharmacology and toxicology data (21 CFR 312.23(a)(8)), previous human experience data (21 CFR 312.23(a)(9)), and any additional information (21 CFR 312.23(a)(10)).
13. Statement identifying contract research organization (CRO) involvement or transfer of sponsor obligations to CRO. The attached statement should contain the name and address of the CRO, identification of the clinical study, and a list of the obligations transferred. If all obligations governing the conduct of the study have been transferred, a general statement of this transfer—in lieu of a list of the specific obligations transferred—may be submitted.
14. Name and title of person responsible for monitoring the conduct and progress of the clinical investigations
15. Name(s) and title(s) of person(s) responsible for review and evaluation of information relevant to the safety of the drug. The form lists the following sponsor commitments: agreement not to begin clinical investigations until 30 days after FDA's receipt of the IND unless otherwise notified by FDA that studies may begin; not to begin or continue clinical investigations covered by the IND if those studies are placed on clinical hold; that an institutional review board (IRB) that complies with the requirements set forth in 21 CFR Part 56 will be responsible for the initial and continuing review and approval of each of the studies in the proposed clinical investigation; and to conduct the investigation in accordance with all other applicable regulatory requirements
16. Name of sponsor or sponsor's authorized representative
17. Signature of sponsor or sponsor's authorized representative
18. Address (number, street, city, state, and ZIP code). If the person signing the application does not reside or have a place of business within the United States, the IND is required to contain the name and address of, and be countersigned by, an attorney, agent, or other authorized official who resides or maintains a place of business within the United States.
19. Telephone number
20. Date of signature

TABLE 18-2 Information Requested by Form FDA-1572, Statement of Investigator

1. Name and address of investigator
2. Education, training, and experience that qualifies the investigator as an expert in the clinical investigation of the drug for the use under investigation. A curriculum vitae or other statement of qualifications must be provided as an attachment.
3. Name and address of any medical school, hospital, or other research facility where the clinical investigation(s) will be conducted
4. Name and address of any clinical laboratory facilities to be used in the study
5. Name and address of the IRB responsible for review and approval of the study(ies)
6. Names of subinvestigators (e.g., research fellows, residents, associates) who will be assisting the investigator
7. Name and code number, if any, of the protocol(s) in the IND identifying the study(ies) to be conducted by the investigator
8. As attachments, the clinical protocol(s) for each planned phase of study. For Phase 1 investigations, a general outline of the planned investigation, including the estimated duration of the study and the maximum number of subjects that will be involved. For Phase 2 or 3 investigations, an outline of the study protocol, including an approximation of the number of subjects to be treated with the drug and the number to be employed as controls, if any; the clinical uses to be investigated; characteristics of subjects by age, sex, and condition; the kind of clinical observations and laboratory tests to be conducted; the estimated duration of the study; and copies or a description of case report forms to be used.
9. A commitment by the investigator agreeing to do the following: conduct the study(ies) in accordance with the relevant, current protocol(s) and make changes in a protocol only after notifying the sponsor, except when necessary to protect the safety, rights, or welfare of subjects; personally conduct or supervise the described investigation(s); inform any patients, or any persons used as controls, that the drugs are being used for investigational purposes and ensure that the requirements relating to obtaining informed consent in 21 CFR Part 50 and IRB review and approval in 21 CFR Part 56 are met; report to the sponsor adverse experiences that occur in the course of the investigation(s) in accordance with 21 CFR 312.64; read and understand the information in the investigator's brochure, including the potential risks and side effects of the drug; ensure that all associates, colleagues, and employees assisting in the conduct of the study(ies) are informed about their obligations in meeting the above commitments; maintain adequate records in accordance with 21 CFR 312.62 and make those records available for inspection in accordance with 21 CFR 312.68; ensure that an IRB that complies with the requirements in 21 CFR Part 56 will be responsible for the initial and continuing review and approval of the clinical investigation and that the investigator will promptly report to the IRB all changes in the research activity and all unanticipated problems involving risks to human subjects or others, and will not make any changes in the research without IRB approval, except where necessary to eliminate apparent immediate hazards to the human subjects; and comply with all requirements regarding the obligations of clinical investigators and all other pertinent requirements in 21 CFR Part 312.
10. Signature of investigator
11. Date of signature

However, the time required can easily surpass this estimate, and the novice sponsor or investigator should plan accordingly and consider seeking assistance from someone experienced in IND applications.

After the completed IND is submitted to CDER for review according to established policies and procedures[59] generally described in Figure 18-1, the sponsor must wait 30 calendar days before initiating any clinical trials. During this period, CDER must review the IND for safety to ensure that research subjects will not be subjected to unreasonable risk. If CDER concludes that the clinical trials present unreasonable risk to subjects, or if the data are insufficient to make such a determination, the IND will be placed on clinical hold and the reviewing division will contact the sponsor within the 30 day period. Customarily, drug review divisions do not contact the sponsor if no concerns arise about drug safety and the proposed clinical trials. Therefore, if the sponsor hears nothing from CDER by day 31 after submission of the IND, the study may proceed as described in the submitted IND protocol(s).

CLINICAL RADIOPHARMACEUTICAL STUDIES

Clinical research in human subjects must be conducted under an FDA-accepted IND generally consisting of three temporal phases. These developmental phases may be combined for a number of practical reasons and therefore may not be distinct. Phase 1 studies, known as clinical pharmacology studies, are carefully controlled and well documented, because they most often involve the initial administration of the radioactive drug product to a limited number of human subjects. Children and pregnant or lactating females must be

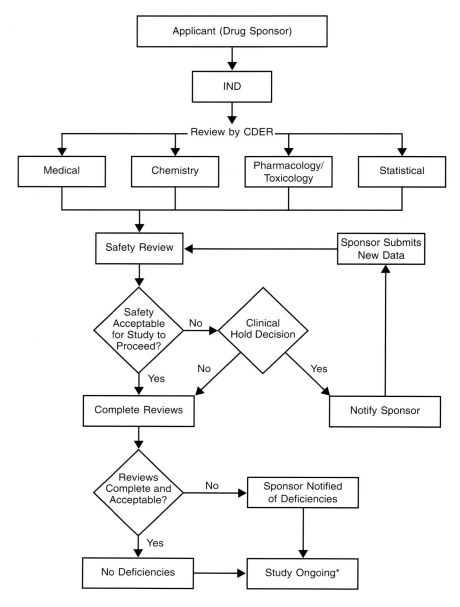

FIGURE 18-1 Overview of Food and Drug Administration's investigational new drug application (IND) review and acceptance process.

excluded from Phase 1 studies. For diagnostic radiopharmaceuticals, normal volunteers may be ideal; however, it may also be desirable and appropriate to enroll diseased subjects in the study. When therapeutic radiopharmaceuticals are being evaluated, only diseased patients should be enrolled. The goal of these initial human studies is to determine absorption, normal biodistribution patterns, organs receiving maximal concentrations, extent of metabolism, route of excretion and half-life, potential toxicity, preferred route of administration, optimum imaging and sampling times, and optimal and safe dosage range. A variety of clinical laboratory tests may be required for the assessment of safety. Radiation absorbed dose should be kept as low as practical for these studies; however, an adequate number of usable particles or photons should be available to ensure that statistically significant images or counting results are obtained with the instrumentation used. When imaging is performed, imaging times must be reasonable to prevent potential image degradation due to patient motion.

Phase 2 studies, known as clinical investigations, are designed to extend the safety evaluation of the radioactive drug product in a larger but controlled number of subjects for a specific disease state and to provide initial evidence of diagnostic or therapeutic efficacy. Phase 2 studies often require extensive laboratory testing, but somewhat less than is required for Phase 1.

Phase 3 studies, known as clinical trials, require the study of sufficient numbers of patients by two or more investigators to expand the evidence of the drug's safety and effectiveness and desirable dosage and to establish directions for use in the diagnosis or treatment of a specific disease. A risk-versus-benefit assessment is also made during this phase. Phase 3 studies typically require significantly less laboratory testing than is required in Phase 1 and 2 testing. The Phase 3 clinical trial protocol, with minor to moderate modification, often becomes the recommended clinical protocol once the radioactive drug product is approved by FDA for general marketing.

Upon completion of all required studies for each phase of development, the sponsor of the new drug (the pharmaceutical company) submits supporting data to FDA in the form of an NDA, which must be approved before the drug may be sold for routine clinical use. Phase 4 studies, also known as therapeutic use or postmarketing studies, may be required by FDA as a condition of approval, or they may be conducted on a voluntary basis by the drug sponsor wishing to get to market efficiently with a single approved indication and then seek approval for additional indications after marketing. Thus, Phase 4 studies relate to the original approved indication but go beyond the prior demonstration of the drug's safety, effectiveness, and dose range definition. FDA seldom requires Phase 4 studies of new diagnostic radiopharmaceuticals, and sponsors rarely conduct them on a voluntary basis because economic incentives are insufficient. However, it is reasonable and highly likely that Phase 4 studies with therapeutic radiopharmaceuticals, such as the recently approved radioimmunotherapy drug products for non-Hodgkin's lymphoma, will be undertaken either as a condition for approval or in an effort to optimize the drug's use and maximize the market potential of the original indication.

PHYSICIAN-SPONSORED IND APPLICATIONS

As stated previously, an IND is usually sponsored either by a pharmaceutical company that enlists a group of investigators to conduct the study or by a physician. Often the physician-sponsored IND is a necessary pathway for the investigator who wishes to conduct human studies with a drug product that otherwise has no true sponsor. Such so-called "orphan drugs" have medical applications but have potential utility in very limited numbers of patients with relatively rare conditions or diseases, making the drugs unappealing candidates for the costly and time-consuming IND–NDA approval process. Radiopharmaceuticals can easily fall into the orphan drug category, and this issue has been identified and addressed in the literature.[60] In 2002, it cost an estimated $802 million and took about 15 years to get a drug product from the laboratory to the marketplace.[61] If a "blockbuster" radioactive drug product could be brought to market rapidly for only 10% of this estimated cost, it would still be a challenge to attract a company to sponsor the drug's commercial development. For orphan drugs, an abbreviated form of IND submission is acceptable, and the sponsoring physician can deal directly with FDA. This enables both the investigator and FDA to accumulate data on safety and efficacy that can be shared with other physicians. If these studies can demonstrate potential utility in additional medical conditions or diseases, there is greater likelihood of a pharmaceutical company sponsor bringing the drug product to market. As a further economic incentive, and as specified in the Orphan Drug Act (Public Law 97-414), which amended the Food, Drug, and Cosmetic Act as of January 4, 1983, companies that sponsor orphan drugs and bring them to market may qualify for certain tax breaks and a guarantee of market exclusivity for up to 7 years after drug approval.[62] FDA approval of diagnostic iobenguane sulfate I 131 (^{131}I-MIBG) is an example of how this process can facilitate the commercial development of orphan radiopharmaceutical drug products.

The need to use a radiopharmaceutical without an approved NDA or to use an approved agent for an unapproved use or by a different route of administration occasionally arises in nuclear medicine practice. In such cases the question of the need to file a physician-sponsored IND also arises. Depending on the circumstances of use, composition of the radiopharmaceutical, and type of radioactive materials license, an IND may or may not be required. Such questions are difficult to resolve because of the numerous regulations and regulatory bodies (Figure 18-2) that govern the use of radiopharmaceuticals and the overlap and inconsistencies among regulations.[63] These issues are addressed in an article by Swanson and Lieto,[64] which provides specific examples that help clarify when and why a radiopharmaceutical IND may be required.

THE NDA PROCESS

Since 1938, the regulation and control of new drugs by FDA has been based on the NDA. Thus, for many decades every new drug in the United States has been subject to NDA approval prior to marketing. The NDA is the vehicle through which drug sponsors formally propose that FDA approve a new drug for human use and for sale and marketing in U.S. interstate commerce. The goal of the NDA is to provide enough well-organized information to permit FDA reviewers to conclude that (1) the product is safe and effective for its proposed use(s), and potential benefits outweigh the associated risks, (2) the proposed product labeling (i.e., the package insert) is suitable and contains sufficient information to promote safe use, and (3) the methods used in manufacturing the drug and the controls used to maintain the drug's quality are adequate to preserve the drug's identity, strength, quality, and purity. The documentation required in the NDA must convey the entire developmental history of the drug product. The data gathered in preclinical studies in animals and clinical trials in humans under the IND become a substantial part of the NDA. In addition, the NDA must identify the drug product ingredients, how the drug behaves in the body, and how it is manufactured, processed, and packaged. General requirements for an NDA submission are specified in 21 CFR, Subparts 314.1 through 314.170. To facilitate understanding of this complicated process, FDA's CDER has on-line resources that summarize NDA content, format, and classification and

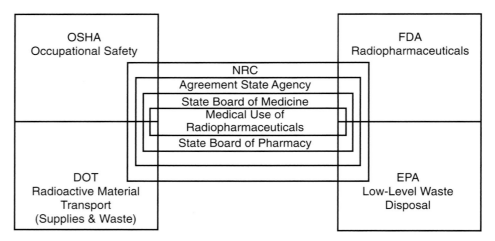

FIGURE 18-2 Regulatory environment for the medical use of radiopharmaceuticals, involving the federal Nuclear Regulatory Commission (NRC), Food and Drug Administration (FDA), Environmental Protection Agency (EPA), Occupational Safety and Health Administration (OSHA), and Department of Transportation (DOT).

the NDA review process.[50] Figure 18-3 provides a general overview of the NDA review process, and additional details can be found on the FDA website.[50]

Pharmacists and nuclear pharmacists are well qualified to participate in the complex but necessary U.S. drug approval process. Traditionally, however, most nuclear pharmacists practice in settings that neither require nor provide the occasion for their direct participation in IND–NDA preparation, approval, and management. Nevertheless, growing numbers of experienced nuclear pharmacists working for commercial companies, academic medical centers, and the federal government are playing key roles in a vital translational regulatory process that is being applied to PET drug products, a unique and important class of radiopharmaceuticals that is key to developing and expanding the potential of molecular imaging technologies. Therefore, nuclear pharmacists should consider participating in the preparation of FDA IND–NDA applications for a variety of PET drugs, especially those with current *USP* monographs (Table 18-3).

Since at least one ^{18}F-FDG NDA already exists, the potential next step for other ^{18}F-FDG manufacturing facilities will be to prepare, submit, and gain approval of site-specific abbreviated new drug applications (ANDAs), since each facility will be seeking approval to produce what FDA will view as a generic PET drug product.

ABBREVIATED NEW DRUG APPLICATIONS FOR PET DRUGS

An ANDA for submission to the CDER Office of Generic Drugs provides for the review and approval of a generic drug product.[51] After approval, an applicant may manufacture and market the generic product to provide the public with a safe, effective, and potentially lower-cost alternative. By definition,

a generic drug product is comparable to an innovator drug product in dosage form, strength, quality, administration route, performance characteristics, and intended clinical use. Generic drug applications are termed "abbreviated" because they are generally not required to include preclinical (animal) and clinical (human) data to establish safety and effectiveness. Instead, ANDA applicants must scientifically demonstrate that their product is bioequivalent (i.e., performs in the same manner as the innovator drug). Bioequivalence is most often demonstrated by measuring the time it takes the generic drug to reach the bloodstream in healthy volunteers. These studies provide data on the rate of absorption, or bioavailability, of the generic drug, which can then be compared with that of the innovator drug. To be considered bioequivalent, the generic version must deliver the same amount of active ingredients into the bloodstream in the same amount of time as the innovator drug. Since radiopharmaceuticals are different from other drug products, FDA has provided guidance documents regarding the content and format of NDAs and ANDAs for PET radiopharmaceutical drug products.[65]

IND–NDA–ANDA REVIEW PROCESS FOR RADIOPHARMACEUTICALS

A team of FDA medical officers at CDER initiates the extensive review process for each submitted radiopharmaceutical IND and NDA. In the near future, the FDA medical team will also be responsible for reviewing each submitted PET drug ANDA, most of which will be for ^{18}F-FDG. FDA medical officers have previously described the pathway for FDA review and approval of new radiopharmaceuticals.[53] The guidance, concepts, and principal elements identified by these

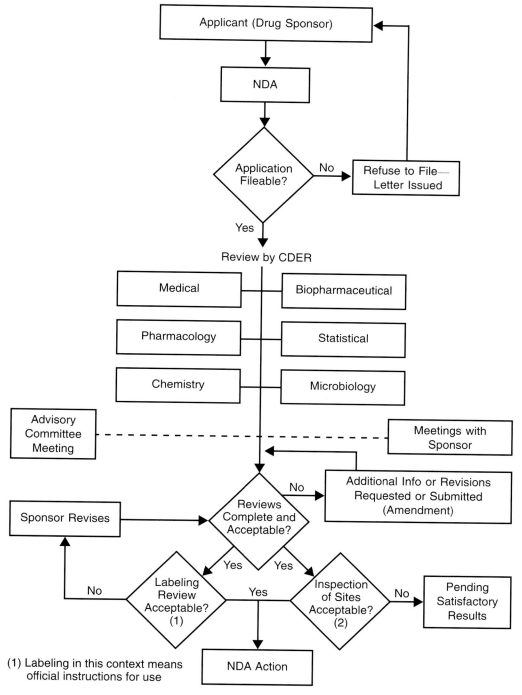

FIGURE 18-3 Overview of Food and Drug Administration's new drug application (NDA) review and approval process.

experts can be of great assistance to individuals charged with the task of gaining FDA clearance for the production, clinical evaluation, and eventual marketing of new or generic radiopharmaceutical products.

Radiopharmaceuticals, whether intended for diagnostic imaging procedures or therapeutic applications, differ significantly from other pharmaceuticals used for therapeutic interventions. The most obvious and important differences between these two types of drugs are the frequency of administration, pharmacologic response, pharmacokinetics, and pharmacodynamics. The FDA review team will pay particular attention to the unique aspects of radiopharmaceuticals.

TABLE 18-3 PET Radiopharmaceuticals with Monographs in *The United States Pharmacopeia*

Ammonia N 13 Injection
Carbon Monoxide C 11
Fludeoxyglucose F 18 Injection
Flumazenil C 11 Injection
Fluorodopa F 18 Injection
Mespiperone C 11 Injection
Methionine C 11 Injection
Raclopride C 11 Injection
Rubidium Chloride Rb 82 Injection
Sodium Acetate C 11 Injection
Sodium Fluoride F 18 Injection
Water O 15 Injection

The IND-NDA review process generally has two parts, the nonclinical review and the medical or clinical review. The nonclinical review includes assessment of pharmacology and toxicology, chemistry, and safety and efficacy biostatistics, whereas the clinical review focuses on overall study safety, rights of human subjects, and quality of the scientific evaluation to be conducted during Phases 1, 2, and 3 of the planned clinical trials.

Nonclinical Review

Pharmacology and Toxicology Review

Both pharmacologists and toxicologists serve on the review team to evaluate the results of animal testing and attempt to relate drug effects in animals to potential effects in humans. Title 21 CFR 312.23(8) and 314.50(2) require the inclusion of a section describing the pharmacologic effects and mechanism(s) of action of the drug in animals and information regarding the absorption, distribution, metabolism, and excretion of the drug, if known.[53] The regulations do not further describe the data to be presented, in contrast to the more detailed description of how to submit toxicologic data. A pharmacology summary report, without individual animal records or individual study results, usually satisfies this requirement. Diagnostic radiopharmaceuticals seldom elicit a detectable pharmacologic response, because only a trace mass of the radiopharmaceutical is typically required. For essentially the same reason, the toxicity of diagnostic agents due to the radionuclide moiety is seldom a significant concern; however, the potential toxicity of a chemical moiety or other chemical components must still be adequately addressed. Therefore, the customary biodistribution and excretion studies in separate animal species are principally designed to obtain essential data for the required calculation of radiation dosimetry estimates that establish relative safety. The ideal approach to estimating dosimetry is to determine the concentration of the radiopharmaceutical in all major organs, tissues, and blood at selected times subsequent to administration of the radioactive drug product to selected test animal species. From the cumulative animal data, the organs receiving the highest radiation absorbed dose can be identified, the blood-to-organ ratios can be calculated, and, for imaging purposes, target-to-nontarget concentrations can be readily established. The evaluation process for therapeutic radiopharmaceuticals is similar to that for conventional therapeutic drugs designed to yield pharmacologic effects. In addition to evaluation of the pharmacokinetics and pharmacodynamics of the therapeutic radiopharmaceutical, changes in vital signs, serum biochemistry, hematogram, and renal function after therapy must be documented. (A hematogram includes measurement of hemoglobin, hematocrit, mean corpuscular volume, and mean corpuscular hemoglobin.) All adverse reactions, regardless of degree of severity, must be evaluated, recorded, and reported. In the end, a complete, integrated summary of the pharmacologic and toxicologic effects of the drug must be provided for review.

Chemistry Review

Each FDA review division employs a team of chemists responsible for reviewing the chemistry, manufacturing, and control (CMC) sections of submitted INDs, NDAs, and ANDAs. The chemistry reviewers consider issues related to drug identity, manufacturing controls, and analysis. The reviewing chemist evaluates the manufacturing and processing procedures associated with each drug product to make certain the drug formulation is adequately reproducible and stable. If the drug were either unstable or not reproducible, the validity of any clinical trial would be significantly undermined because the investigators would not know precisely what chemical entities were actually administered to the human subjects; more important, the study could pose significant risks to the subjects.

At the beginning of the CMC section, the drug sponsor should state whether it believes the chemistry of either the drug substance or the drug product, or the manufacturing of either the drug substance or the drug product, presents any indicators of potential human risk.[59] If so, these indicators should be completely discussed, with steps proposed to monitor for the associated risks. In addition, sponsors should describe any chemistry and manufacturing differences between the drug product proposed for clinical use and the drug product used in the animal toxicology trials that formed the basis for the sponsor's conclusion that it was safe to continue with the proposed clinical study.[59] Likewise, the sponsor should describe how the known differences might affect the overall safety profile of the drug product as formulated for clinical trials. If there are no differences in the products, this should be stated.

Title 21 CFR 312.23(7) states that sufficient required information is to be submitted to ensure the proper identification, quality, purity, and strength of the investigational drug. The documentation that satisfies this regulatory requirement is commonly referred to as the CMC section of the IND application and is often a primary responsibility of a nuclear pharmacist. The CMC section should include all relevant information to show that the drug product can be prepared with sufficient purity and quality to be safe for administration to human subjects. The essential parts of a suitable CMC section are specified in the applicable regulations and have also been summarized as guidance for industry.[66] The four major elements of the CMC section describe the control processes related to the drug substance, drug product, quality controls, and container and closure system for the final drug product. The NDA will generally have the same documentation format as the IND CMC section; however, there are differences regarding the level of detail required. For example, the NDA requires additional emphasis on structure proofs and establishment of purity profiles and quality controls for raw materials, starting materials, intermediates, drug substance excipients, and the finished drug product. More extensive stability data must be provided, and the data must clearly support the assignment of a valid expiration date or time.

Biostatistics Review

Biostatisticians evaluate the potential statistical significance of data to be collected under an IND and compute the statistical relevance of data submitted in an NDA. Thus, the biostatistics reviewer takes full responsibility for the main tasks of evaluating the methods used to conduct clinical studies and to analyze the acquired data. The overall purpose of these statistical evaluations is to provide the medical officers with enhanced information about the power of the findings as they are extrapolated to the larger patient populations.

Title 21 CFR 314.126(a) indicates that the purpose of conducting clinical investigations of a drug is to distinguish the effect of a drug from other influences, such as spontaneous change in the course of disease, placebo effect, or biased observation. Data submitted to substantiate the safety and efficacy of a new radiopharmaceutical must undergo valid statistical analysis to help authenticate claims made and to confirm the degree to which a hypothesis has been proved or disproved.[67] The elements of the required statistical analysis are identified in 21 CFR 314.50(d)(5, 6). The requisite elements of a well-designed protocol have also been identified and explained in a guidance document on the format and content of the clinical and statistical section of an application.[67] More recently, regulatory experts have summarized these requisite elements for biostatistics evaluations as they apply to the review of radiopharmaceutical drug products.[53]

Clinical Review

Clinical reviewers, also called medical officers, are almost exclusively physicians; consequently, the term medical review may also be used. Nonphysician professionals may be used as medical officers to evaluate certain types of drug data. Clinical reviewers are held responsible for fully evaluating the clinical sections of submissions, such as the overall safety of the clinical protocols described in an IND and the clinical test results as submitted in an NDA. In most divisions, clinical reviewers take the lead role in IND or NDA review and are thus responsible for synthesizing the results of the animal toxicology, human pharmacology, and clinical reviews to formulate the overall basis for a recommended agency action on the application.[59]

Throughout the IND clinical review process, the reviewer evaluates each clinical trial protocol to determine (1) if the research subjects will be protected from unwarranted risks and (2) if the study design will provide data pertinent to the overall safety and effectiveness of the investigational drug product. Under federal regulations, proposed Phase 1 studies are evaluated almost exclusively for overall safety. Since the late 1980s, FDA reviewers have been instructed to provide drug sponsors with greater freedom during Phase 1, as long as the investigations do not expose research subjects to undue risks. In evaluating Phase 2 and Phase 3 investigations, however, clinical reviewers must also provide assurance that proposed studies are of sufficient scientific quality to be capable of yielding significant data that can support final marketing approval.

Title 21 CFR 312.11 describes the customary framework applied to the clinical review of each IND. FDA's primary objectives in reviewing an IND are, in all phases of the investigation, to ensure the safety and rights of subjects and, in Phases 2 and 3, to help ensure that the quality of the scientific evaluation of drugs is adequate to permit an evaluation of the drug's effectiveness and safety. In describing the design of the clinical trial investigational plan, the proposed indication for use of the new radiopharmaceutical should be clearly stated, along with the methods to be used for gathering the necessary data to support the proposed clinical use.

Phase 1

This opening phase of study includes the first introduction of the investigational new drug into human subjects. The purpose of this phase is to validate safety of the new drug or radiopharmaceutical product. These early studies must determine several key radiopharmaceutical characteristics, including blood clearance half-life, normal biologic distribution, critical organ(s) (i.e., those receiving maximum radiopharmaceutical concentration), routes of excretion, and optimal times for sampling and imaging. Additional requisites include descriptive information about the planned subject populations, dosage ranges, radiation absorbed dose ranges, clinical laboratory tests, vital signs, and radiopharmaceutical biodistribution.

Phases 2 and 3

Typically, the differences between Phase 2 and Phase 3 for diagnostic radiopharmaceuticals are minimal in comparison with the differences between these study phases for therapeutic radiopharmaceuticals or drugs. Diagnostic radiopharmaceutical safety evaluations require less detailed pharmacologic toxicity assessments and have a primary focus of establishing valid radiation absorbed dose estimates. Diagnostic efficacy can be established if use of the radiopharmaceutical contributes to making effective decisions about the presence, absence, or extent of disease. Although the exact nature of the disease or abnormality may not be specified, the clinical utility of the diagnostic radiopharmaceutical must be clearly demonstrated. Diagnostic value is a function of a radiopharmaceutical's biodistribution and the character of its emitted radiation. The extent to which radiopharmaceutical biodistribution is altered by disease, as defined by imaging studies, in vivo uptake studies, and in vitro tests, helps to validate efficacy of the agent. Each phase of the study may require fewer research subjects to satisfy the statistical determinants set for each study. Depending on the clarity of response, as few as 20 to 60 subjects may suffice for Phase 2 protocols and 200 to 300 for Phase 3 studies, in contrast to the much larger numbers of subjects often needed for therapeutic drug evaluations. In diagnostic radiopharmaceutical clinical trials the time required to collect the necessary pharmacokinetic, pharmacodynamic, and imaging data is often much shorter because of the nature of these studies and the usual ease of demonstrating presence or lack of efficacy. One of the most critical requirements for Phase 3 evaluations is the performance of studies under a common protocol by at least two separate and independent institutional sites and investigators.[53] The separate studies validate the ability to replicate the safety and efficacy determinants of the new agent with the particular dosage form and for the proposed indication.[68] The critical elements required by FDA for all clinical study designs have been outlined and the common specific elements for all phases of IND–NDA protocols have been summarized elsewhere.[53]

NDA SUBMISSIONS

Although nuclear pharmacists and other nuclear medicine professionals rarely become involved in the process of gaining FDA marketing approval for diagnostic and therapeutic radiopharmaceuticals, it is still useful for these individuals to know the NDA submission process as defined by drug law. Title 21 CFR 314.50 outlines the elements needed to meet the legal requirements for NDA approval to market a new radiopharmaceutical. As indicated previously, the submitted application collates and summarizes all of the nonclinical and clinical data derived to validate the safety and efficacy of the new radioactive drug product. The principal segments of the NDA, as identified in 21 CFR 314.50(c)(1-2ix), include summary sections devoted to CMC, nonclinical pharmacology and toxicology, human pharmacokinetics and bioavailability, and clinical data, including the results of statistical analysis of the clinical trials, plus a concluding discussion that presents the benefit and risk considerations related to use of the drug. The final distillation of information that must accompany the NDA during the review and approval process is the label or package insert. The package insert, as specified in 21 CFR 201.50, 201.56, and 201.57, provides a summary of the essential scientific information needed for the safe and effective use of the drug product. This labeling information must be both accurate and informative and cannot be promotional, misleading, or false. Information included in the package insert should be based on proven data that are presented in the NDA. The package insert must also have a prescribed format that includes (1) indication(s) and approved clinical uses, (2) dosage and administration, and (3) warnings and adverse reactions documented during the clinical trials. There are additional considerations regarding the evaluation, summarization, and presentation of completed studies in the NDA, along with several modifications that are appropriate for therapeutic radiopharmaceuticals.[53]

Every IND, NDA, or ANDA submission is a distinct document because of the unique physical and chemical properties associated with each diagnostic or therapeutic radiopharmaceutical product and the variation in the facilities used and personnel employed for their production. This chapter and its referenced sources will be useful to individuals who take responsibility for preparing and maintaining the various regulatory applications. To optimize the challenging process of gaining radiopharmaceutical drug product approvals, FDA, the nuclear medicine community, commercial sponsors, and investigators must focus on effective communication at many levels. When questions or potential problems arise at any stage in the drug development process, it is often best to contact FDA to obtain answers or identify a means of resolving problems in the marketing approval process. Recent experience has shown that improved communication can help decrease the time required to review and bring new radiopharmaceuticals into clinical trials and onto the market. The regulations, further guidance regarding IND procedures, relevant forms, and a variety of other resources are available online.[47–49,58,59]

NRC REGULATIONS APPLICABLE TO THE MEDICAL USE OF RADIOPHARMACEUTICALS

The stated mission of NRC is to regulate the civilian use of byproduct, source, and special nuclear materials to ensure adequate protection of public health and safety, to promote the common defense and security, and to protect the environment.[69] This regulatory mission covers three main areas: (1) reactors—commercial reactors for generating electric power and nonpower reactors used for research, testing, and training, (2) materials—uses of nuclear materials in medical, industrial, and academic settings and facilities that produce nuclear fuel, and (3) waste—transportation, storage, and disposal of nuclear materials and waste, and decommissioning of nuclear facilities from service. NRC thus has extremely broad legislative authority regarding radioactive materials. The following discussion, however, focuses on specific issues related to NRC regulation of the medical use of radiopharmaceuticals and the related implications for nuclear pharmacy and nuclear medicine practice.

All NRC regulations are published in the *Federal Register* (FR) and codified in 10 CFR Chapter 1. The specific NRC regulations pertaining to the medical use of byproduct materials are also identified in NRC Regulatory Guide 10.817 and listed here in Table 18-4. The medical use of radionuclides

TABLE 18-4 Nuclear Regulatory Commission Rules Pertaining to Medical Use of Byproduct Materials

10 CFR Part 19	Notice, Instructions, and Reports to Workers: Inspections and Investigations
10 CFR Part 20	Standards for Protection Against Radiation
10 CFR Part 21	Reporting of Defects and Noncompliance
10 CFR Part 30	Rules of General Applicability to Domestic Licensing of Byproduct Material
10 CFR Part 35	Medical Use of Byproduct Material
10 CFR Part 71	Packaging and Transportation of Radioactive Material
10 CFR Part 170	Fees for Facilities, Materials, Import and Export Licenses, and Other Regulatory Services Under the Atomic Energy Act of 1954, as Amended
10 CFR Part 171	Annual Fees for Reactor Licenses and Fuel Cycle Licenses and Materials Licenses, Including Holders of Certificates of Compliance, Registrations, and Quality Assurance Program Approvals and Government Agencies Licensed by the NRC

requires specific licensure, since it involves the intentional internal or external administration of byproduct material, or the radiation therefrom, to human beings. The regulations that most directly affect nuclear pharmacy and nuclear medicine practice are found in 10 CFR 35 (Medical Use of Byproduct Material), 10 CFR 20 (Standards for Protection Against Radiation), and 10 CFR 71 (Packaging and Transportation of Radioactive Materials). The other regulations listed in the table are equally important in that nuclear pharmacies and nuclear medicine clinics must operate in total compliance with all applicable sections of each part; however, those regulations are primarily administrative and do not directly affect routine practice activities. Important issues related to 10 CFR 20 and 10 CFR 71 are addressed in Chapters 6 and 7 of this book, so the following discussion will focus on sections of 10 CFR 35 in an effort to characterize the regulatory relationship between NRC and the practice of nuclear pharmacy and nuclear medicine. *Federal Register* notices affecting Title 10 of the Code of Federal Regulations and other relevant information can be found on the NRC website.[70] Regulatory changes in the past few years have focused mainly on personnel training requirements.

MEDICAL USE OF BYPRODUCT MATERIAL

Historically, the regulations contained in 10 CFR 35 have set forth requirements and provisions for the medical use of byproduct material, the issuance of specific licenses authorizing the medical use of this material, and the radiation safety of workers, the general public, patients, and human research subjects. The requirements and provisions of Part 35 are in addition to, and not a substitution for, others in 10 CFR. Unless specifically exempted, Parts 19, 20, 21, 30, 71, 170, and 171 of 10 CFR apply to applicants and licensees subject to Part 35. The nuclear pharmacy practice guidelines are based in part on these federal regulations, and the practice standards state that nuclear pharmacists have a professional responsibility to ensure compliance with NRC and state licenses under which the nuclear pharmacy operates and with federal, state, and institutional rules regulating radiation and radiopharmaceuticals.

A revision of 10 CFR 35 was published in the *Federal Register* (67 FR 20249) on April 24, 2002, and became effective on October 24, 2002.[71] This most recent revision was an effort to create a risk-informed and performance-based regulation that focuses on those medical procedures that pose the highest radiologic risk to workers, patients, and the public. Information is presented here about regulatory requirements pertinent to most nuclear pharmacy practice settings. (In the following discussion of those requirements, "this chapter" means 10 CFR Chapter 1.) For some practice settings, the regulatory requirements may not be adequately addressed here. The complete Part 35 is available on the NRC website.[71,72]

General Information

The stated purpose and scope of the revised regulations is to identify requirements and provisions for medical use of byproduct material and for issuance of specific licenses authorizing medical use of this material. These requirements and provisions provide for the radiation safety of workers, the general public, patients, and human research subjects. Part 35.2 defines key terms used in the regulations; the definitions most directly applicable to the practice of nuclear pharmacy and nuclear medicine follow.

An *authorized medical physicist* (AMP) is an individual who (1) meets the requirements in §§35.51(a) and 35.59; or (2) is identified as an AMP or teletherapy physicist on a specific medical use license issued by NRC or an agreement state, a medical use permit issued by an NRC master material licensee, a permit issued by an NRC or agreement state broad scope medical use licensee, or a permit issued by an NRC master material license broad scope medical use permittee.

An *authorized nuclear pharmacist* (ANP) is a pharmacist who (1) meets the requirements in §§35.55(a) and 35.59; or (2) is identified as an ANP on a specific license issued by NRC or an agreement state that authorizes medical use or the practice of nuclear pharmacy, a permit issued by an NRC master material licensee that authorizes medical use or the practice of nuclear pharmacy, a permit issued by an NRC or agreement state broad scope medical use licensee that authorizes medical use or the practice of nuclear pharmacy, or a permit issued by an NRC master material license broad scope medical use permittee that authorizes medical use or the practice of nuclear pharmacy; or (3) is identified as an ANP by a commercial nuclear pharmacy that has been authorized to identify ANPs; or (4) is designated as an ANP in accordance with §32.72(b)(4).

An *authorized user* (AU) is a physician, dentist, or podiatrist who (1) meets the requirements in §35.59 and 35.190(a), 35.290(a), 35.390(a), 35.392(a), 35.394(a), 35.490(a), 35.590(a), or 35.690(a); or (2) is identified as an AU on an NRC or agreement state license that authorizes the medical use of byproduct material, a permit issued by an NRC master material licensee that is authorized to permit the medical use of byproduct material, a permit issued by an NRC or agreement state specific licensee of broad scope that is authorized to permit the medical use of byproduct material, or a permit issued by an NRC master material license broad scope permittee that is authorized to permit the medical use of byproduct material.

A *dedicated check source* is a radioactive source that is used to ensure the constant operation of a radiation detection or measurement device over several months or years.

A *medical event* is an event that meets the criteria in §35.3045(a) or (b) and which may require reporting. Criteria in §35.3045(a) state that a licensee must report any event, except for an event that results from patient intervention, in which the administration of byproduct material or radiation from byproduct material results in (1) a dose that differs from the prescribed dose or dose that would have resulted from the prescribed dosage by more than 0.05 Sv (5 rem) effective dose equivalent, 0.5 Sv (50 rem) to an organ or tissue, or 0.5 Sv (50 rem) shallow dose equivalent to the skin, and the total dose delivered differs from the prescribed dose by 20% or more, the total dosage delivered differs from the prescribed dosage by 20% or more or falls outside the prescribed dosage range, or the fractionated dose delivered differs from the prescribed dose, for a single fraction, by 50% or more; (2) a dose that exceeds 0.05 Sv (5 rem) effective dose equivalent, 0.5 Sv (50 rem) to an organ or tissue, or 0.5 Sv (50 rem) shallow dose equivalent to the skin from any of the following: administration of a wrong radioactive drug containing byproduct material, administration of a radioactive drug containing byproduct material by the wrong route of administration, administration of a dose or dosage to the wrong individual or human research subject, administration of a dose or dosage delivered by the wrong mode of treatment, or a leaking sealed source; (3) a dose to the skin or an organ or tissue other than the treatment site that exceeds by 0.5 Sv (50 rem) to an organ or tissue and 50% or more of the dose expected from the administration defined in the written directive (excluding, for permanent implants, seeds that were implanted in the correct site but migrated outside the treatment site). Criteria in §35.3045(b) state that a licensee must also report any event resulting from intervention of a patient or human research subject in which the administration of byproduct material or radiation from byproduct material results or will result in unintended permanent functional damage to an organ or a physiologic system, as determined by a physician.

Medical use is the intentional internal or external administration of byproduct material or the radiation from byproduct material to patients or human research subjects under the supervision of an authorized user.

A *pharmacist* is an individual licensed by a state or territory of the United States, the District of Columbia, or the Commonwealth of Puerto Rico to practice pharmacy.

A *physician* is a medical doctor or doctor of osteopathy licensed by a state or territory of the United States, the District of Columbia, or the Commonwealth of Puerto Rico to prescribe drugs in the practice of medicine.

A *prescribed dosage* is the specified activity or range of activity of unsealed byproduct material as documented (1) in a written directive or (2) in accordance with the directions of the authorized user for procedures performed pursuant to §35.100 (use of unsealed byproduct material for uptake, dilution, and excretion studies for which a written directive is not required) and §35.200 (use of unsealed byproduct material for imaging and localization studies for which a written directive is not required). According to this definition, the term prescribed dosage is reserved for indicating the amount of radioactivity (e.g., microcuries or millicuries) to be administered. Thus, the term prescribed dosage is distinct from the term "prescribed dose," which refers to the radiation absorbed dose to be administered.

A *prescribed dose* is (1) for gamma stereotactic radiosurgery, the total dose as documented in the written directive; (2) for teletherapy, the total dose and dose per fraction as documented in the written directive; (3) for manual brachytherapy, either the total source strength and exposure time or the total dose, as documented in the written directive; and (4) for remote brachytherapy afterloaders, the total dose and dose per fraction as documented in the written directive. According to this definition, the term prescribed dose is reserved for indicating the radiation absorbed dose to be administered. Thus, the term prescribed dose is distinct from the term "prescribed dosage," which refers to the activity amount of radiopharmaceutical to be administered.

A *radiation safety officer* (RSO) is an individual who (1) meets the requirements in §§35.50(a) and 35.59; or (2) is identified as an RSO on a specific medical use license issued by NRC or an agreement state or on a medical use permit issued by an NRC master material licensee.

A *sealed source* is any byproduct material that is encased in a capsule designed to prevent leakage or escape of the byproduct material.

A *therapeutic dosage* is a dosage of unsealed byproduct material (e.g., a radiopharmaceutical) that is intended to deliver a radiation dose to a patient or human research subject for palliative or curative treatment.

A *therapeutic dose* is a radiation dose delivered from a source containing byproduct material to a patient or human research subject for palliative or curative treatment.

A *unit dosage* is a dosage prepared for medical use for administration as a single dosage to a patient or human research subject without any further manipulation of the dosage after it is initially prepared.

A *written directive* is an authorized user's written order for the administration of byproduct material or radiation from byproduct material to a specific patient or human research subject, as specified in §35.40.

Part 35.5 specifies requirements for *maintaining applicable records*. The records required by this part must be kept in a readily retrievable, legible form throughout the retention period specified by regulation. Records may be original or reproduced copies or a microform, provided that authorized personnel authenticate the copy or microform and that the microform is capable of producing a clear copy. Records may also be stored in electronic media with the capability for producing legible, accurate, and complete records. Records such as letters, drawings, and specifications must include all pertinent information such as stamps, initials, and signatures. The licensee must maintain adequate safeguards against tampering with and loss of all pertinent records.

Part 35.6 delineates provisions for the *protection of human research subjects* and therefore is an immensely important set of regulations for licensees involved, directly or indirectly, in human research involving radioactive drug products. Consequently, even the nuclear pharmacy licensee that participates only indirectly in such studies by providing radiopharmaceutical service must be aware of the regulatory requirements for protecting human subjects. A licensee may conduct research involving human subjects only if the byproduct materials to be used are specified in the current license and the uses are authorized by this license. If the research is conducted, funded, supported, or regulated by another federal agency that has implemented the federal policy for the protection of human subjects (Federal Policy), the licensee must, before conducting research (1) obtain review and approval of the research from an institutional review board (IRB), as defined and described in the Federal Policy; and (2) obtain informed consent, as defined and described in the Federal Policy, from the human research subject. If the research will not be conducted, funded, supported, or regulated by another federal agency that has implemented the Federal Policy, the licensee must, before conducting research, apply for and receive a specific amendment to its NRC medical use license. The amendment request must include a written commitment that the licensee will, before conducting research, (1) obtain review and approval of the research from an IRB, as defined and described in the Federal Policy; and (2) obtain informed consent, as defined and described in the Federal Policy, from the human research subject. Nothing in this section relieves licensees from complying with the other requirements in this part.

Part 35.11 specifies that a person *shall not manufacture, produce, acquire, receive, possess, use, or transfer* byproduct material for medical use except in accordance with a specific license issued by NRC or an agreement state that authorizes such activities. This licensure requirement is clearly reflected in the procurement practice guidelines that obligate nuclear pharmacists to act with proper authority (i.e., under a valid radioactive materials license) to purchase radioactive material. A specific license is not needed for an individual who (1) receives, possesses, uses, or transfers byproduct material in accordance with the regulations in this chapter under the supervision of an AU as provided in §35.27, unless prohibited by license condition; or (2) prepares unsealed byproduct material for medical use in accordance with the regulations in this chapter under the supervision of an ANP or AU as provided in §35.27, unless prohibited by license condition.

The requirements for *license application, amendment, or renewal* relative to nuclear medicine are specified in §35.12. The applicant or licensee's management must sign each license application. An application for a license for medical use of byproduct material as described in §35.100 (use of unsealed byproduct material for uptake, dilution, and excretion studies for which a written directive is not required), §35.200 (use of unsealed byproduct material for imaging and localization studies for which a written directive is not required), or §35.300 (use of unsealed byproduct material for which a written directive is required) must be made by filing an original and one copy of NRC Form 313, Application for Material License, that includes the facility diagram, equipment, and training and experience qualifications of the RSO, AU(s), ANP(s), and AMP(s) if any. Requests for a license amendment or renewal must be made by submitting an original and one copy of either NRC Form 313 or a letter requesting the amendment or renewal. The applicant or licensee must also provide any other information requested by NRC in its review of the application. An applicant satisfying the requirements specified in §33.13 governing specific domestic licenses of broad scope for byproduct material may apply for a Type A specific license of broad scope.

License amendments are addressed further in §35.13: A licensee must apply for and receive an approved license amendment (a) before it receives, prepares, or uses byproduct material for a type of use that is permitted under this part, but that is not authorized on the licensee's current license issued under this part; or (b) before it permits anyone to work as an AU, ANP, or AMP under the license. Important exceptions to §35.13(b) are allowed. For example, exception may be made for AUs who meet the training requirements specified in §§35.190(a) for uptake, dilution, and excretion studies, 35.290(a) for imaging and localization studies, 35.390(a) for use of unsealed byproduct material for which a written directive is required, 35.392(a) for the oral administration of ^{131}I-sodium iodide requiring a written directive in quantities not exceeding 1.22 GBq (33 mCi), and 35.394(a) for the oral administration of ^{131}I-sodium iodide requiring a written directive in quantities greater than 1.22 GBq (33 mCi). To qualify for the exception, the AU must be certified by a medical specialty board, the board certification process must adequately address all of the NRC training requirements, and the board certification must be recognized by NRC or an agreement state. An exception may also be made for ANPs who meet the training requirements specified in §35.55(a) or

35.980(a) and 35.59. Likewise, exceptions may be made for AMPs who meet the training requirements specified in §§35.51(a) or 35.961(a) or (b) and 35.59. Finally, exceptions may be made for an individual who is identified as an AU, an ANP, or an AMP on an NRC or agreement state license or other equivalent permit or license recognized by NRC that authorizes the use of byproduct material in medical use or in the practice of nuclear pharmacy; on a permit issued by NRC or an agreement state specific license of broad scope that is authorized to permit the use of byproduct material in medical use or in the practice of nuclear pharmacy; on a permit issued by an NRC master material licensee that is authorized to permit the use of byproduct material in medical use or in the practice of nuclear pharmacy; or by a commercial nuclear pharmacy that has been authorized to identify ANPs.

According to §35.13(c), the licensee must apply for and receive a license amendment before it changes RSOs, except as provided in §35.24(c). This exception allows the licensee to permit a qualified individual to serve as a temporary RSO and to perform the associated functions under the prescribed conditions for up to 60 days each year, provided the NRC is properly notified in accordance with §35.14(b). In accordance with 35.13(d), the licensee must apply for and receive a license amendment before it receives byproduct material in excess of the amount or in a different form, or receives a different radionuclide, than is authorized on the license. Further, in accordance with §35.13(e), a licensee must apply for and receive a license amendment before it adds to or changes the areas of use identified in the application or on the license, except for areas of use where byproduct material is used only in accordance with either §§35.100 (uptake, dilution, and excretion studies) or 35.200 (imaging and localization studies). Finally, §35.13(f) requires licensees to apply for and receive a license amendment prior to changing the address(es) of use identified in the application or on the radioactive materials license.

Licensees are obligated under §35.14 to provide several types of timely *notification to NRC*. Part 35.14(a) requires the licensee to provide a copy of the board certification, the NRC or agreement state license, the permit issued by an NRC master material licensee, the permit issued by an NRC or agreement state licensee of broad scope, or the permit issued by an NRC master material license broad scope permittee for each individual no later than 30 days after the date that the licensee permits the individual to work as an AU, ANP, or AMP under §35.13 (b)(1) through (b)(4). In addition, in accordance with §35.14(b), the licensee must notify NRC by letter no later than 30 days after (1) an AU, ANP, RSO, or AMP permanently discontinues performance of duties under the license or has a name change; (2) the licensee's mailing address changes; (3) the licensee's name changes but the name change does not constitute a transfer of control of the license as described in §30.34(b); or (4) the licensee has added to or changed the areas of use identified in the application or on the license where byproduct material is used in accordance with either §§35.100 (uptake, dilution, and excretion studies) or 35.200 (imaging and localization studies).

General Administrative Requirements

The general administrative requirements associated with properly managing a radioactive materials license in compliance with applicable regulations are quite significant and must be adequately addressed by the licensee. Part 35.24 identifies the authority and responsibilities of the required radiation protection program that each licensee must maintain. These radiation protection program requirements are in addition to those specified elsewhere under §20.1101.

According to §35.24(a), a licensee's management must approve in writing (1) requests for a license application, renewal, or amendment before submittal to NRC; (2) any individual before allowing that individual to work as an AU, ANP, or AMP; and (3) radiation protection program changes that do not require a license amendment and are permitted under §35.26 (radiation protection program changes).

Pursuant to §35.24(b), a licensee's management must appoint an RSO who agrees in writing to be responsible for implementing the radiation protection program. The licensee, through the RSO, must ensure that radiation safety activities are being performed in accordance with licensee-approved procedures and regulatory requirements. As mentioned earlier, §35.24(c) allows the licensee to permit a qualified individual to serve as a temporary RSO and to perform the associated functions under the prescribed conditions for up to 60 days each year, provided the NRC is properly notified. This arrangement can be extended under §35.24(d) to the simultaneous appointment of more than one temporary RSO if needed to ensure that the licensee satisfies the requirements for each of the different types of uses of byproduct material permitted by the license. The licensee must establish the authority, duties, and responsibilities of the RSO in writing as required by §35.24(e).

In accordance with §35.24(f), licensees authorized for two or more different types of uses of byproduct material under Subpart E (Unsealed Byproduct Material—Written Directive Required) must establish a *radiation safety committee* (RSC) to oversee all uses of byproduct material permitted by the license. The RSC must include an AU for each type of use permitted, the RSO, a representative of the nursing service, and a representative of management who is neither an AU nor an RSO. The RSC may include other members the licensee considers appropriate, such as an ANP. The RSO, according to §35.24(g), must be provided sufficient authority, organizational freedom, time, resources, and management prerogative to (1) identify radiation safety problems; (2) initiate, recommend, or provide corrective actions; (3) stop unsafe operations; and (4) verify implementation of corrective actions. Finally, §35.24(h) states that the licensee must retain a record of actions taken under (a), (b), and (e) of this section in accordance with §35.2024.

Part 35.27 addresses the important issue of *proper supervision* of individuals involved in the medical use of licensed materials. As stipulated in §35.27(a), a licensee that permits the receipt, possession, use, or transfer of byproduct material by an individual under the supervision of an AU, as allowed by §35.11(b)(1), must (1) in addition to the requirements in §19.12, instruct the supervised individual in the licensee's written radiation protection procedures, written directive procedures, regulations of this chapter, and license conditions with respect to the use of byproduct material; and (2) require the supervised individual to follow the instructions of the supervising AU, written radiation protection procedures established by the licensee, written directive procedures, regulations of this chapter, and license conditions with respect to the medical use of byproduct material. Likewise, according to §35.27(b), a licensee that permits the preparation of byproduct material for medical use by an individual under the supervision of an ANP or physician who is an AU, as allowed by §35.11(b)(2), must (1) in addition to the requirements in §19.12, instruct the supervised individual in the preparation of byproduct material for medical use, as appropriate to that individual's involvement with byproduct material; and (2) require the supervised individual to follow the instructions of the supervising AU or ANP regarding the preparation of byproduct material for medical use, written radiation protection procedures established by the licensee, the regulations of this chapter, and license conditions. §35.27(c) stipulates that the licensee that permits supervised activities under (a) and (b) is responsible for the acts and omissions of the supervised individual.

AUs must write orders for the administration of byproduct material or radiation from byproduct material to a specific patient or human research subject in the form of a *written directive*. In many ways these directives are analogous to the written prescription required in the customary practice of medicine and pharmacy. According to §35.40(a), a written directive must be dated and signed by an AU prior to the administration of ^{131}I-sodium iodide in amounts greater than 1.11 MBq (30 µCi), any therapeutic dosage of unsealed byproduct material, or any therapeutic dose of radiation from byproduct material. If, because of the emergent nature of the patient's condition, a delay for the purpose of providing a written directive would jeopardize the patient's health, an oral directive is acceptable. The information contained in the oral directive must be documented as soon as possible in writing in the patient's record. A written directive must be prepared within 48 hours of the oral directive. In addition, according to §35.40(b), the written directive for a radiopharmaceutical must contain the patient or human research subject's name and the following information: (1) for any administration of quantities greater than 1.11 MBq (30 µCi) of ^{131}I-sodium iodide, the dosage; and (2) for administration of a therapeutic dosage of unsealed byproduct material other than ^{131}I-sodium iodide, the radioactive drug, dosage, and route of administration. Revisions to existing written directives may be made under §35.40(c) if the revision is dated and signed by an AU before administration of the dosage of unsealed byproduct material. If, because of the patient's condition, a delay for the purpose of providing a written revision to an existing written directive would jeopardize the patient's health, an oral revision to an existing written directive is acceptable. The oral revision must be documented as soon as possible in the patient's record. The AU must sign a revised written directive within 48 hours of the oral revision. As is customary with written prescriptions, the licensee must retain a copy of the written directive in accordance with §35.2040.

§35.41(a) requires that the procedures for administrations requiring a written directive be well documented. The licensee must develop, implement, and maintain written procedures to provide high confidence that (1) the patient's or human research subject's identity is verified before each administration and (2) each administration is in accordance with the written directive. At a minimum, these required procedures must address the following items that are applicable to the licensee's use of byproduct material: (1) verifying the identity of the patient or human research subject; (2) verifying that the administration is in accordance with the treatment plan, if applicable, and the written directive; and (3) checking both manual and computer-generated dose calculations. Finally, as is customary in medical and pharmacy practice, the licensee must retain a copy of the procedures required under paragraph (a), in accordance with §35.2041.

RSO *training requirements,* as delineated in §35.50, are important to consider, since nuclear pharmacists, nuclear medicine physicians, and medical physicists may bear responsibility for these associated duties. Except as provided in §35.57, the licensee must require an individual fulfilling the responsibilities of the RSO as provided in §35.24 to be an individual who (a) is certified by a specialty board whose certification process has been formally recognized by the NRC or an agreement state and who meets the requirements specified in paragraphs (d) concerning a written and signed attestation of RSO preceptor and (e) concerning training in radiation safety, regulatory issues, and emergency procedures. Alternatively, individuals may qualify under (b)(1) of this section if they have completed a structured educational program consisting of both 200 hours of didactic training in the following areas: radiation physics and instrumentation, radiation protection, mathematics pertaining to the use and measurement of radioactivity, radiation biology, and radiation dosimetry; and 1 year of full-time radiation safety experience under the supervision of the individual identified as the RSO on an NRC or agreement state license or permit issued by an NRC master material licensee that authorizes similar types of uses of byproduct material involving the following: shipping, receiving, and performing related radiation surveys; using and performing checks for proper operation of instruments used to determine the activity of dosages, survey meters, and instruments used to measure radionuclides; securing and controlling byproduct material; using administrative controls to avoid mistakes in the administration of byproduct material; using procedures to prevent or minimize radioactive contamination and using proper decontamination procedures; using emergency procedures to control byproduct material; and disposing of byproduct material. Otherwise, medical physicists may qualify under (c)(1) of this section, and experienced authorized nuclear pharmacists, authorized users, or authorized medical physicists identified by the licensee may qualify under (c)(2).

The *AMP training requirements,* as delineated in §35.51, are important to consider. Except as provided in §35.57, the licensee must require the AMP to be an individual who is certified by a specialty board whose certification process has been formally recognized by the NRC or an agreement state and who meets all of the training and experience requirements in paragraphs (b)(2) concerning a written and signed attestation of the AMP preceptor and (c) concerning training for the type(s) of use for which authorization is sought. Alternatively, individuals may qualify under (b)(1) if they hold a master's or doctor's degree in physics, medical physics, other physical science, engineering, or applied mathematics and have completed 1 year of full-time training in medical physics and an additional year of full-time work experience under the supervision of an AMP for the type(s) of use for which the individual is seeking authorization. This training and work experience must be conducted in a clinical radiation facility.

The *ANP training requirements,* as delineated in §35.55, are of obvious importance to nuclear pharmacists and nuclear pharmacy operations in any practice setting. Except as provided in §35.57, the licensee must require the ANP to be a pharmacist who is certified as a nuclear pharmacist by a specialty board whose certification process includes the requirements in (b)(2) concerning a written and signed attestation of the ANP preceptor and whose certification has been recognized by NRC or an agreement state. Alternatively, individuals may qualify under (b)(1) if they have completed 700 hours in a structured educational program consisting of both didactic training in the following areas: radiation physics and instrumentation, radiation protection, mathematics pertaining to the use and measurement of radioactivity, chemistry of byproduct material for medical use, and radiation biology; and supervised practical experience in a nuclear pharmacy involving shipping, receiving, and performing related radiation surveys; using and performing checks for proper operation of instruments used to determine the activity of dosages, survey meters, and, if appropriate, instruments used to measure alpha- or beta-emitting radionuclides; calculating, assaying, and safely preparing dosages for patients or human research subjects; using administrative controls to avoid medical events in the administration of byproduct material; and using procedures to prevent or minimize radioactive contamination and using proper decontamination procedures; and if under (b)(2) they have obtained written certification, signed by a preceptor ANP, that the individual has satisfactorily completed the requirements in (b)(1) and have achieved a level of competency sufficient to function independently as an ANP.

§35.57 addresses *exceptions to the training requirements* for experienced RSOs, medical physicists, AUs, and nuclear pharmacists. According to §35.57(a), an individual identified as an RSO, medical physicist, or nuclear pharmacist on an NRC or agreement state license or a permit issued by an NRC or agreement state broad scope licensee or master material license permit or by a master material license permittee of broad scope before October 24, 2002, need not comply with the training requirements of §§35.50, 35.51, or 35.55, respectively. In addition, according to §35.57(b), physicians, dentists, or podiatrists identified as AUs for the medical use of byproduct material on a license issued by NRC or an agreement state, a permit issued by an NRC master material licensee, a permit issued by an NRC or agreement state broad scope licensee, or a permit issued by an NRC master material license broad scope permittee before October 24, 2002, who perform only those medical uses for which they were authorized on that date need not comply with the training requirements of Subparts D through H of this part.

The issue of the *recentness of professional training* often surfaces in regard to the medical use of byproduct materials. §35.59 states that the training and experience specified in Subparts B, D, E, F, G, and H must have been obtained within the 7 years preceding the date of application or the individual must have had related continuing education and experience since the required training and experience was completed.

General Technical Requirements

Because of the importance of safety in the medical use of radioactive materials, numerous technical requirements must be satisfied to remain in regulatory compliance. Part 35.60 addresses the *possession, use, and calibration of instruments* used to measure the activity of unsealed byproduct material. For direct measurements performed in accordance with §35.63, a licensee must possess and use instrumentation to measure the activity of unsealed byproduct material before it is administered to each patient or human research subject. In addition, the licensee must calibrate the required instrumentation in accordance with nationally recognized standards or the manufacturer's instructions. Finally, the licensee must retain a record of each required instrument calibration in accordance with §35.2060.

Radiation survey instruments are used for a number of purposes in complying with regulations for the use of radioactive materials. Part 35.61 deals with the calibration of survey instruments used to make radiation measurements. The licensee must calibrate the survey instruments used in order to show compliance with this part and 10 CFR Part 20 (Standards for Protection Against Radiation) before first use, annually, and after a repair that affects the calibration. The licensee must (1) calibrate all scales with readings up to 10 mSv (1000 mrem) per hour with a radiation source; (2) calibrate two separate readings on each scale or decade that will be used to show compliance; and (3) conspicuously note on the instrument the date of calibration. A licensee may not use survey instruments if the difference between the indicated exposure rate and the calculated exposure rate is more than 20%. The licensee must retain a record of each survey instrument calibration in accordance with §35.2061.

According to §35.63, licensees must make an accurate determination of the *dosages of unsealed byproduct material* for medical use. Thus, the licensee must determine and record the activity of each dosage before release for medical use. For a unit dosage, this determination must be made by (1) direct measurement of radioactivity; or (2) a decay correction, based on the activity or activity concentration determined by a manufacturer or preparer (e.g., nuclear pharmacy) licensed under §32.72 or equivalent agreement state requirements, or by an NRC or agreement state licensee for use in research in accordance with an RDRC-approved protocol or an IND protocol accepted by FDA, or by a PET drug producer licensed under §32.32(j) of this chapter or equivalent agreement state requirement. For other than unit dosages, this determination must be made by (1) direct measurement of radioactivity; (2) combination of measurement of radioactivity and mathematical calculations; or (3) combination of volumetric measurements and mathematical calculations, based on the measurement made by a manufacturer or preparer licensed under §32.72 of this chapter or equivalent agreement state requirements. It is important to note that unless otherwise directed by the AU, the licensee may not use a dosage if the dosage does not fall within the prescribed dosage range or if the dosage differs from the prescribed dosage by more than 20%. Once again, the licensee must retain a record of the dosage determination required by this section in accordance with §35.2063.

Part 35.65 provides authorization for *calibration, transmission, and reference sources* needed to ensure the safe medical use of byproduct materials. Any person authorized by §35.11 for medical use of byproduct material may receive, possess, and use any of the following byproduct materials for check, calibration, transmission, and reference use: (a) sealed sources, not exceeding 1.11 GBq (30 mCi) each, manufactured and distributed by a person licensed under §32.74 of this chapter or equivalent agreement state regulations; (b) sealed sources, not exceeding 1.11 GBq (30 mCi) each, redistributed by a licensee authorized to redistribute the sealed sources manufactured and distributed by a person licensed under §32.74 of this chapter or equivalent agreement state regulations, provided the redistributed sealed sources are in the original packaging and shielding and are accompanied by the manufacturer's approved instructions; (c) any byproduct material with a half-life not longer than 120 days in individual amounts not to exceed 0.56 GBq (15 mCi); (d) any byproduct material with a half-life longer than 120 days in individual amounts not to exceed the smaller of 7.4 MBq (200 µCi) or 1000 times the quantities in Appendix B of 10 CFR Part 30 (Rules of General Applicability to Domestic Licensing of Byproduct Material); or (e) ^{99m}Tc in amounts as needed.

Sealed sources are routinely used in nuclear pharmacy and nuclear medicine practice, and brachytherapy sources may on rare occasions be in the possession of a nuclear pharmacy licensee. According to §35.67, certain requirements concerning the possession of sealed sources and brachytherapy sources must be satisfied. A licensee in possession of any sealed source or brachytherapy source must follow the radiation safety and handling instructions supplied by the manufacturer. A licensee in possession of a sealed source must (1) test the source for leakage before its first use unless the licensee has a certificate from the supplier indicating that the source was tested within 6 months before transfer to the licensee; and (2) test the source for leakage at intervals not to exceed 6 months or at other intervals approved by NRC or an agreement state in the Sealed Source and Device Registry. To satisfy the leak test requirements, the licensee must measure leak test samples in a manner that will allow for detecting the presence of 185 Bq (0.005 µCi) of radioactive material in the sample. The licensee must retain leak test records in accordance with §35.2067(a) for a period of 3 years. Additional requirements come into play if the leak test reveals the presence of 185 Bq (0.005 µCi) or more of removable contamination. If this occurs, the licensee must (1) immediately withdraw the sealed source from use and store, dispose, or cause it to be repaired in accordance with the requirements in 10 CFR Parts 20 and 30; and (2) file a report within 5 days of the leak test in accordance with §35.3067. The licensee does not need to perform a leak test on (1) sources containing only byproduct material with a half-life of less than 30 days; (2) sources containing only byproduct material as a gas; (3) sources containing 3.7 MBq (100 µCi) or less of beta- or gamma-emitting material or 0.37 MBq (10 µCi) or less of alpha-emitting material; (4) seeds of ^{192}Ir encased in nylon ribbon; and (5) sources stored and not being used. However, the licensee must test each such source for leakage before any use or transfer unless it has been leak tested within 6 months before the date of use or transfer. Licensees in possession of sealed sources or brachytherapy sources, except for gamma stereotactic radiosurgery sources, must conduct a semiannual physical inventory of all such sources in their possession. Finally, licensees must retain each inventory record in accordance with §35.2067(b) for a period of 3 years.

The proper *labeling of radioactive materials* is key to their safe handling and appropriate use. According to §35.69, each syringe and vial that contains unsealed byproduct material must be labeled appropriately to adequately identify the radioactive drug. Each syringe shield and vial shield must also be labeled unless the label on the syringe or vial is visible when shielded. This issue is also dealt with in §32.72(a)(4)(ii); accordingly, the licensee must affix a label to each syringe, vial, or other container used to hold a radioactive drug that is to be transferred for commercial distribution. The label must include the radiation symbol and the words "Caution, Radioactive Material" or "Danger, Radioactive Material" and an identifier that ensures that the syringe, vial, or other container can be correlated with the information on the transport radiation shield label. Beyond this, nuclear pharmacy services label radioactive drugs with the standard elements of required prescription labeling.

Surveys of ambient radiation exposure rates are required according to §35.70, and this requirement is in addition to the surveys required by 10 CFR Part 20. The licensee must survey with a radiation detection survey instrument at the end of each day of use all areas where unsealed byproduct material requiring a written directive were prepared for use or administered. However, the licensee does not need to perform such surveys in areas where patients or human research subjects are confined when they cannot be released under §35.75. The licensee must retain a record of each survey in accordance with §35.2070 for a period of 3 years.

A revised §35.75 was published in the *Federal Register* (62 FR 4120) on January 29, 1997, and became effective on May 29, 1997. This revision delineates the *criteria for release* of individuals containing unsealed byproduct material or implants containing byproduct material. The licensee may authorize the release of any individual from its control who has been administered unsealed byproduct material or implants containing byproduct material if the total effective dose equivalent (TEDE) to any other individual from exposure to the released individual is not likely to exceed 5 mSv (0.5 rem). The required methods for calculating such doses to other individuals and tables of activities not likely to cause doses exceeding this imposed limit can be found in the current revision of NUREG-1556, Volume 9. If the requirements for release are satisfied, the licensee must provide the released individual, or the individual's parent or guardian, with instructions, including written instructions, on actions recommended to maintain radiation doses to other individuals as low as reasonably achievable if the TEDE to any other individual is likely to exceed 1 mSv (0.1 rem). If the TEDE to a nursing infant or child could exceed 1 mSv (0.1 rem) assuming there were no interruption of breast-feeding, the instructions must also include (1) guidance on the interruption or discontinuation of breast-feeding and (2) information on the potential consequences, if any, of failure to follow the guidance. To be fully compliant, the licensee must maintain a record of the basis for authorizing the release of an individual in accordance with §35.2075(a) and the instructions provided to a breast-feeding female in accordance with §35.2075(b). These records must be maintained for a period of 3 years after the date of release of each individual.

In some practice settings licensees may be involved in the provision of *regional mobile medical services*. A licensee involved in providing such services must satisfy the conditions set forth in §35.80. The licensee must (1) obtain a letter signed by the management of each client for which services are rendered that permits the use of byproduct material at the client's address and clearly delineates the authority and responsibility of the licensee and the client; (2) check instruments used to measure the activity of unsealed byproduct material for proper function before medical use at each client's address or on each day of use, whichever is more frequent; at a minimum, the check for proper function must include a constancy check; (3) check survey instruments for proper operation with a dedicated check source before use at each client's address; and (4) before leaving a client's address, survey all areas of use to ensure compliance with the requirements in 10 CFR Part 20. Mobile medical services may not have byproduct material delivered from a manufacturer or a distributor (e.g., a nuclear pharmacy) to the client unless the client has a license allowing possession of the byproduct material. Byproduct material delivered to the client must be received and handled in conformance with the client's license. In addition, the licensee providing such services must retain each signed management letter that permits the provision of service and must record the results of the required radiation area surveys in accordance with §35.2080(a) and (b), respectively.

Holding short-lived radioactive materials for *decay-in-storage* is a very attractive and cost-effective management practice for both nuclear pharmacies and nuclear medicine clinics. Part 35.92(a) identifies the requirements associated with this practice. A licensee may hold byproduct material with a physical half-life of less than 120 days for decay-in-storage before disposal without regard to its radioactivity if the licensee (1) monitors byproduct material at the surface before disposal and determines that any residual radioactivity cannot be distinguished from the background radiation level as determined with an appropriate radiation detection survey meter set on its most sensitive scale and with no interposed shielding; and (2) removes or obliterates all radiation labels, except for radiation labels on materials that are within containers and that will be managed as biomedical waste after their release by the licensee. Thus, according to (a)(2), radiation labels on biomedical waste (e.g., sharps containers or individual needles and syringes) do not have to be removed or obliterated where there is a biohazard associated with retrieving such material from their outer container. In many cases, biomedical waste containers are packaged in barrels and incinerated, but this may not be done in all cases. Licensees must ensure that released biomedical waste either is incinerated or contains no legible radioactive labels that could otherwise cause a potential incident upon discovery by the general public. Regardless of this new flexibility to support occupational safety, good practice still necessitates the obliteration of all radioactive labels possible by using safe measures that do not constitute a biohazard to personnel. To be fully compliant, the licensee must retain a record of each disposal permitted in accordance with §35.2092 for a period of 3 years.

Unsealed Byproduct Material—Written Directive Not Required

According to §35.100, unsealed byproduct material may be used for uptake, dilution, and excretion studies without a written directive under certain circumstances. Except for quantities that require a written directive under §35.40(b), a licensee may use any unsealed byproduct material prepared for medical use for uptake, dilution, or excretion studies that is (a) obtained from a manufacturer or preparer licensed under §32.72 or a PET drug producer licensed under §30.32(j) or equivalent agreement state requirements; or (b) excluding production of PET radionuclides, prepared by an ANP, a physician who is an AU and who meets the requirements specified in §§35.290 or 35.390 and 35.290(c)(1)(ii)(G), or an individual under the supervision as specified in §35.27 of an ANP or AU; or (c) obtained from and prepared by an NRC or agreement state licensee for use in research in accordance with an RDRC-approved protocol or an IND protocol accepted by FDA; or (d) prepared by the licensee for use in research in accordance with an RDRC-approved application or an IND protocol accepted by FDA.

As specified in §35.190, appropriate training is required for uptake, dilution, and excretion studies. Except as provided in §35.57, the licensee must require an AU of unsealed byproduct material for the uses authorized under §35.100 to be a physician who (a) is certified by a medical specialty board whose certification process has been recognized by NRC or an agreement state and who meets the requirements specified in paragraph (c)(2) concerning a written and signed attestation from a qualified AU preceptor; or (b) is an AU under §§35.290, 35.390 or equivalent agreement state requirements; or (c)(1) has completed 60 hours of training and experience in basic radionuclide handling techniques applicable to the medical use of unsealed byproduct material for uptake, dilution, and excretion studies. The training and experience must include a minimum of 8 hours of classroom and laboratory training in the following areas: radiation physics and instrumentation, radiation protection, mathematics pertaining to the use and measurement of radioactivity, chemistry of byproduct material for medical use, and radiation biology; and work experience, under the supervision of an AU who meets the requirements in §35.190, §35.290, or §35.390 or equivalent agreement state requirements, involving ordering, receiving, and unpacking radioactive materials safely and performing the related radiation surveys; performing quality control procedures on instruments used to determine the activity of dosages and performing checks for proper operation of survey meters; calculating, measuring, and safely preparing patient or human research subject dosages; using administrative controls to prevent a medical event involving the use of unsealed byproduct material; using procedures to contain spilled byproduct material safely and using proper decontamination procedures; and administering dosages of radioactive drugs to patients or human research subjects; and (2) has obtained written attestation, signed by a preceptor AU who meets the requirements in §§35.190, 35.290, or 35.390 or equivalent agreement state requirements, that the individual has satisfactorily completed the requirements in (c)(1) and has achieved a level of competency sufficient to function independently as an AU for the medical uses authorized under §35.100.

In accordance with §35.200, unsealed byproduct material may be used for imaging and localization studies without a written directive under certain conditions. Except for quantities that require a written directive under §35.40(b), a licensee may use any unsealed byproduct material prepared for medical use for imaging and localization studies that is (a) obtained from a manufacturer or preparer licensed under §32.72 or a PET drug producer licensed under §30.32(j) or equivalent agreement state requirements; or (b) excluding production of PET radionuclides, prepared by an ANP, a physician who is an AU and who meets the requirements specified in §§35.290 or 35.390 and 35.290(c)(1)(ii)(G); or an individual under the supervision as specified in §35.27 of an ANP or an AU; or (c) obtained from and prepared by an NRC or agreement state licensee for use in research in accordance with an RDRC-approved protocol or an IND protocol accepted by FDA; or (d) prepared by the licensee for use in research in accordance with an RDRC-approved application or an IND protocol accepted by FDA.

There is always a potential that radionuclidic impurities may contaminate medical radionuclide generator system eluates. When these generators are functioning properly, their eluates typically contain the desired radionuclide as the sole radioactive component. However, if a significant breakthrough of potential radionuclidic impurities were to occur, excessive amounts of these impurities in administered dosages would contribute unnecessary radiation dose to patients. Because of this potential safety concern, §35.204 identifies the permissible ^{99}Mo, ^{82}Sr, and ^{85}Sr concentrations for radionuclide generator eluates. According to the stated regulatory limit for ^{99}Mo–^{99m}Tc generator eluates, a licensee may not administer to humans a radiopharmaceutical that contains more than 0.15 kilobecquerel of ^{99}Mo per megabecquerel of ^{99m}Tc (equivalent to 0.15 microcurie of ^{99}Mo per millicurie of ^{99m}Tc). In addition, a licensee that uses ^{99}Mo–^{99m}Tc generators for preparing a ^{99m}Tc radiopharmaceutical must measure the ^{99}Mo concentration of the first eluate after receipt of a generator to demonstrate compliance with paragraph (a) of this section. Historically, licensees were required to check for ^{99}Mo breakthrough after each elution of the ^{99}Mo–^{99m}Tc generator; this is another example of increased regulatory flexibility. According to the stated regulatory limit for ^{82}Sr–^{82}Rb generator eluates, a licensee may not administer to humans a radiopharmaceutical that contains more than 0.02 kilobecquerel of ^{82}Sr per megabecquerel of ^{82}Rb chloride injection (equivalent to 0.02 microcurie of ^{82}Sr per millicurie of ^{82}Rb) or more than 0.2 kilobecquerel of ^{85}Sr per megabecquerel of ^{82}Rb chloride injection (equivalent to 0.2 microcurie of ^{85}Sr per millicurie of ^{82}Rb). In addition, a licensee that uses ^{82}Sr–^{82}Rb generators for preparing an ^{82}Rb radiopharmaceutical must measure the ^{82}Sr and ^{85}Sr concentrations of the first eluate before the first patient use of the day to demonstrate compliance with paragraph (a) of this section. Finally, if a licensee is required to measure the ^{99}Mo concentration or the ^{82}Sr and ^{85}Sr concentrations, the licensee must retain a record of each measurement in accordance with §35.2204 for a period of 3 years. The record must include, for each measured elution of ^{99m}Tc, the ratio of the measures expressed as kilobecquerels of ^{99}Mo per megabecquerel of ^{99m}Tc (equivalent to microcuries of ^{99}Mo per millicurie of ^{99m}Tc). Records for each measured elution of ^{82}Rb must include the ratio of the measures expressed as kilobecquerels of ^{82}Sr per megabecquerel of ^{82}Rb (equivalent to microcuries of ^{82}Sr per millicurie of ^{82}Rb) and kilobecquerels of ^{85}Sr per megabecquerel of ^{82}Rb (equivalent to microcuries of ^{85}Sr per millicurie of ^{82}Rb). Finally, each of these radionuclidic purity measurements records must also include the time and date of the measurement, and the name of the individual who made the measurement.

As spelled out in §35.290, AUs must be properly trained to take responsibility for imaging and localization studies. Except as provided in §35.57, the licensee must require an AU of unsealed byproduct material for the uses authorized under §35.200 to be a physician who (a) is certified by a medical specialty board whose certification process has been recognized by NRC or an agreement state and who meets all of the requirements in paragraph (c)(2) concerning written and signed attestation from a qualified AU preceptor; or (b) is an AU under §35.390 and meets the requirements in §35.290(c)(1)(ii)(G), or equivalent agreement state requirements; or (c)(1) has completed 700 hours of training and experience, including a minimum of 80 hours of classroom and laboratory training, in basic radionuclide handling techniques applicable to the medical use of unsealed byproduct material for imaging and localization studies; this training and experience must include, at a minimum, classroom and laboratory training in the following areas: radiation physics and instrumentation, radiation protection, mathematics pertaining to the use and measurement of radioactivity, chemistry of byproduct material for medical use, and radiation biology; and work experience, under the supervision of an AU who meets the requirements in §§35.57, 35.290 or 35.390 and 35.290(c)(1)(ii)(G), or equivalent agreement state requirements, involving ordering, receiving, and unpacking radioactive materials safely and performing the related radiation surveys; performing quality control procedures on instruments used to determine the activity of dosages and performing checks for proper operation of survey meters; calculating, measuring, and safely preparing patient or human research subject dosages; using administrative controls to prevent a medical event involving the use of unsealed byproduct material; using procedures to safely contain spilled radioactive material and using proper decontamination procedures; administering dosages of radioactive drugs to patients or human research subjects; and eluting generator systems appropriate for preparation of radioactive drugs for imaging and localization studies, measuring and testing the eluate for radionuclidic purity, and processing the eluate with reagent kits to prepare labeled radioactive drugs; and (2) has obtained written certification, signed by a preceptor AU who meets the requirements in §§35.57, 35.290 or 35.390, and 35.290(c)(1)(ii)(G), or equivalent agreement state requirements, that the individual has satisfactorily completed the requirements in (a)(1) or (c)(1) and has achieved a level of competency sufficient to function independently as an AU for the medical uses authorized under §§35.100 and 35.200.

Unsealed Byproduct Material—Written Directive Required

According to §35.300, a licensee may use any unsealed byproduct material prepared for medical use and for which a written directive is required that is (a) obtained from a manufacturer or preparer licensed under §32.72 or a PET drug producer licensed under §30.32(j) or equivalent agreement state requirements; or (b) excluding production of PET radionuclides, prepared by an ANP, a physician who is an AU and who meets the requirements specified in §§35.290 or 35.390, or an individual under the supervision of either as specified in §35.27; or (c) obtained from and prepared by an NRC or agreement state licensee for use in research in accordance with an IND protocol accepted by FDA; or (d) prepared by the licensee for use in research in accordance with an IND protocol accepted by FDA.

The provision of safety instruction is a key component of any required radiation safety program. As required in §35.310, the licensee must provide radiation safety instruction, initially and at least annually, for all personnel caring for patients or human research subjects receiving radiopharmaceutical therapy and hospitalized, for compliance with §35.75 of this chapter. To satisfy this requirement, the instruction must describe the licensee's procedures for (1) patient or human research subject control; (2) visitor control; (3) contamination control; (4) waste control; and (5) notification of the RSO, or his or her designee, and the AU if the patient or the human research subject has a medical emergency or dies. The licensee must retain a record of individuals receiving instruction in accordance with §35.2310 for a period of 3 years.

Proper safety precautions, as required under §35.315, must be used with each patient or human research subject who cannot be released under §35.75. Accordingly, the licensee must (1) quarter these individuals either in a private room with a private sanitary facility, or in a room with a private sanitary facility with another individual who also has received therapy with unsealed byproduct material and who also cannot be released under §35.75; (2) visibly post the therapy room with a "Radioactive Materials" sign; (3) note on the door or in the individual's chart where and how long visitors may stay in the therapy room; and (4) either monitor material and items removed from the therapy room to determine that their radioactivity cannot be distinguished from the natural background radiation level with a radiation detection survey instrument set on its most sensitive scale and with no interposed shielding, or handle the material and items as radioactive waste. The licensee must notify the RSO, or his or her designee, and the AU as soon as possible if a patient or human research subject has a medical emergency or dies.

Appropriate training, as stipulated in §35.390, is necessary for use of unsealed byproduct material for which a written directive is required. Except as provided in §35.57, the licensee must require an AU of unsealed byproduct material for the uses authorized under §35.300 to be a physician who (a) is certified by a medical specialty board whose certification process has been recognized by NRC or an agreement state and who meets the requirements in paragraphs (b)(1)(ii)(G) and (b)(2) concerning written and signed attestation from a qualified AU preceptor; or (b)(1) has completed 700 hours of training and experience, including 200 hours of classroom and laboratory training, in basic radionuclide handling techniques applicable to the medical use of unsealed byproduct material requiring a written directive; this training and experience must include (i) classroom and laboratory training in the following areas: radiation physics and instrumentation, radiation protection, mathematics pertaining to the use and measurement of radioactivity, chemistry of byproduct material for medical use, and radiation biology; and (ii) work experience, under the supervision of an AU who meets the requirements in §35.390(a), §35.390(b), or equivalent agreement state requirements; a supervising AU who meets the requirements in §35.390(b) must have experience in administering dosages in the same dosage category or categories (i.e., §35.390(b)(1)(ii)(G)) as the individual requesting AU status; the work experience must involve: ordering, receiving, and unpacking radioactive materials safely and performing the related radiation surveys; performing quality control procedures on instruments used to determine the activity of dosages, and performing checks for proper operation of survey meters; calculating, measuring, and safely preparing patient or human research subject dosages; using administrative controls to prevent a medical event involving the use of unsealed byproduct material; using procedures to contain spilled byproduct material safely and using proper decontamination procedures; and administering dosages of radioactive drugs to patients or human research subjects involving a minimum of three cases in each of the following categories for which the individual is requesting AU status: (1) oral administration of no more than 1.22 GBq (33 mCi) of ^{131}I-sodium iodide; (2) oral administration of more than 1.22 GBq (33 mCi) of ^{131}I-sodium iodide (experience with at least three cases in category (G)(2) also satisfies the requirement in category (G)(1)); (3) parenteral administration of any beta emitter or a photon-emitting radionuclide with a photon energy less than 150 keV; and/or (4) parenteral administration of any other radionuclide requiring a written directive; and (b)(2) has obtained written certification that the individual has satisfactorily completed the requirements in paragraphs (a)(1) and (b)(1)(ii)(G) or (b)(1) of this section, and has achieved a level of competency sufficient to function independently as an AU for the medical uses authorized under §35.300. A preceptor AU who meets the requirements in §§35.57, 35.390, or equivalent agreement state requirements must sign the written certification. The preceptor AU, who meets the requirements in §35.390(b), must have experience in administering dosages in the same dosage category or categories (i.e., §35.390(b)(1)(ii)(G)) as the individual requesting AU status.

Specific training is required, as delineated in §35.392, for the oral administration of ^{131}I-sodium iodide requiring a written directive in quantities less than or equal to 1.22 GBq (33 mCi). These training requirements are essentially equivalent to those specified in §35.394 for the oral administration of ^{131}I-sodium iodide requiring a written directive in quantities greater than 33 mCi except for specified dosage limits. The purpose of these sections is to identify and clarify the requirements for physicians (e.g., endocrinologists) who seek only limited authorization for oral administration of ^{131}I-sodium iodide in dosages less than or equal to or dosages greater than 33 mCi and do not seek authorization to prepare radioactive drugs using generators and reagent kits. In the §35.392 limited-authorization setting, except as provided in §35.57, the licensee must require an AU to be a physician who (a) is certified by a medical specialty board whose certification process includes the requirements in paragraph (c)(1) and (c)(2) of this section and whose certification has been recognized by the NRC or an agreement state and who meets the requirements of paragraph (c)(3) or (b) is an AU under §35.390 for uses listed in §35.390(b)(1)(ii)(G)(1) or (2), §35.394, or equivalent agreement state requirements; or (c)(1) has successfully completed 80 hours of classroom and laboratory training applicable to the medical use of ^{131}I-sodium iodide for procedures requiring a written directive; this training must include radiation physics and instrumentation, radiation protection, mathematics pertaining to the use and measurement of radioactivity; chemistry of byproduct material for medical use, and radiation biology; and (c)(2) has work experience, under the supervision of an AU who meets the requirements in §35.57, 35.390, 35.392, 35.394, or equivalent agreement state requirements; a supervising AU who meets the requirements in §35.390(b) must have experience in administering dosages as specified in §35.390(b)(1)(ii)(G)(1) or (2); the work experience must involve ordering, receiving, and unpacking radioactive materials safely and performing the related radiation surveys; calibrating instruments used to determine the activity of dosages and performing checks for proper operation for survey meters; calculating, measuring, and safely preparing patient or human research subject dosages; using administrative controls to prevent a medical event involving the use of byproduct material; using procedures to contain spilled byproduct material safely and using proper decontamination procedures; and administering dosages to patients or human research subjects, including at least three cases involving the oral administration of less than or equal to 33 millicuries of ^{131}I-sodium iodide; and (c)(3) has obtained written certification that the individual has satisfactorily completed the requirements in (c)(1) and (c)(2) and has achieved a level of competency sufficient to function independently as an AU for medical uses authorized under §35.300. A preceptor AU who meets the requirements in §35.57, 35.390, 35.392, 35.394, or equivalent agreement state requirements must sign the written certification. Finally, a preceptor AU who meets the requirement in §35.390(b) must also have experience in administering dosages as specified in §35.390(b)(1)(ii)(G)(1) or (2).

Training and Experience Requirements

Applicable training and experience requirements for RSOs, AUs, AMPs, and ANPs are now specified in general terms in the various subparts and sections of the 10 CFR Part 35 regulations as identified above. Specific training and experience requirements are primarily defined by various specialty boards whose certification process has been recognized by the NRC or an agreement state. These recognized specialty certification boards are now listed on the NRC website.[73] Details of the training and experience requirements for a given professional certification can be found on the NRC web page entitled "Specialty Board(s) Certification Recognized by NRC Under 10 CFR Part 35" and the website of the specific specialty board.

Records

According to §35.2024, records of authority and responsibilities for the radiation protection program must be maintained. The licensee must retain a record of actions taken by the licensee's management in accordance with §35.24(a) for 5 years. The record must include a summary of the actions taken and a signature of licensee management. In addition, the licensee must retain a copy of authority, duties, and responsibilities of the RSO as required by §35.24(e) and a signed copy of each RSO's agreement to be responsible for implementing the radiation safety program, as required by §35.24(b), for the duration of the license. The records must include the signature of the RSO and licensee management.

Records of radiation protection program changes must be maintained according to §35.2026. The licensee must retain a record of each radiation protection program change made in accordance with §35.26(a) for 5 years. The record must include a copy of both the old and new procedures, the effective date of the change, and the signature of the licensee management that reviewed and approved the change.

According to §35.2040, the licensee must retain records of written directives as required under §35.40 for a period of 3 years. In addition, according to §35.2041, the licensee must retain records for procedures for administrations requiring a written directive, as required by §35.41(a), for the duration of the license.

Records of calibrations of instruments used to measure the activity of unsealed byproduct material must be kept as required by §35.2060. The licensee must maintain a record of instrument calibrations, as required by §35.60, for 3 years. In addition, the records must include the model and serial number of the instrument, the date of the calibration, the results of the calibration, and the name of the individual who performed the calibration.

Records of radiation survey instrument calibrations are required under §35.2061. The licensee must maintain a record of the radiation survey instrument calibrations, required by §35.61, for 3 years. In addition, the record must include the model and serial number of the instrument, the date of the calibration, the results of the calibration, and the name of the individual who performed the calibration.

Records of dosages of unsealed byproduct material for medical use must be maintained according to §35.2063. The licensee must maintain a record of dosage determinations required by §35.63 for 3 years. The record must contain (1) the radiopharmaceutical, (2) the patient's or human research subject's name, or identification number if one has been assigned, (3) the prescribed dosage, the determined dosage, or a notation that the total activity is less than 1.1 MBq (30 µCi), (4) the date and time of the dosage determination, and (5) the name of the individual who determined the dosage.

According to §35.2067, records of leak tests and the inventory of sealed sources and brachytherapy sources must be maintained. The licensee must retain records of leak tests required by §35.67(b) for 3 years. The records must include the model number, and serial number if one has been assigned, of each source tested; the identity of each source by radionuclide and its estimated activity; the results of the test; the date of the test; and the name of the individual who performed the test. In addition, the licensee must retain records of the semiannual physical inventory of sealed sources and brachytherapy sources required by §35.67(g) for 3 years. The inventory records must contain the model number of each source, and serial number if one has been assigned; the identity of each source by radionuclide and its nominal activity; the location of each source; and the name of the individual who performed the inventory.

Records of surveys for ambient radiation exposure rate, as required under §35.2070, must be maintained. The licensee must retain a record of each survey required by §35.70 for 3 years. The record must include the date of the survey, the results of the survey, the instrument used to make the survey, and the name of the individual who performed the survey.

Records of the release of individuals containing unsealed byproduct material or implants containing byproduct material must be maintained according to §35.2075. The licensee must retain a record of the basis for authorizing the release of an individual, in accordance with §35.75, if total effective dose equivalent (TEDE) is calculated by (1) using the retained activity rather than the activity administered, (2) using an occupancy factor less than 0.25 at 1 m, (3) using the biologic or effective half-life, or (4) considering the shielding by tissue. In addition, the licensee must retain a record that the instructions required by §35.75(b) were provided to a breast-feeding female if the radiation dose to the infant or child from continued breast-feeding could result in a TEDE exceeding 5 mSv (0.5 rem). These records must be retained for 3 years after the date of release of the individual.

Mobile medical services must keep records according to §35.2080. The licensee must retain a copy of each letter that permits the use of byproduct material at a client's address, as required by §35.80(a)(1). Each letter must clearly delineate the authority and responsibility of the licensee and the client and must be retained for 3 years after the last service is provided. A licensee must retain the record of each survey required by §35.80(a)(4) for 3 years. The record must include the date of the survey, the results of the survey, the instrument used to make the survey, and the name of the individual who performed the survey.

Decay-in-storage records are also required according to §35.2092. The licensee must maintain records of the disposal of licensed materials, as required by §35.92, for 3 years. The record must include the date of the disposal, the survey instrument used, the background radiation level, the radiation level measured at the surface of each waste container, and the name of the individual who performed the survey.

According to §35.2204, records of ^{99}Mo, ^{82}Sr, and ^{85}Sr concentrations in radionuclide generator eluates must be maintained by the licensee as required by §35.204(b) for 3 years. For each measured elution of ^{99m}Tc, the record must include the ratio of the measures expressed as kilobecquerels of ^{99}Mo per megabecquerel of ^{99m}Tc (equivalent to microcuries of ^{99}Mo per millicurie of ^{99m}Tc), the time and date of the measurement, and the name of the individual who made the measurement. For each measured elution of ^{82}Rb, the record must include the ratio of the measures expressed as kilobecquerels of ^{82}Sr per megabecquerel of ^{82}Rb (equivalent to microcuries of ^{82}Sr per millicurie of ^{82}Rb), kilobecquerels of ^{85}Sr per megabecquerel of ^{82}Rb (equivalent to microcuries of ^{85}Sr per millicurie of ^{82}Rb), the time and date of the measurement, and the name of the individual who made the measurement.

Safety instruction records must be maintained according to §35.2310. The licensee must maintain a record of safety instructions required by §§35.310, 35.410, and 35.610 for 3 years. The record must include a list of topics covered, the date of instruction, the names of the attendees, and the names of the individuals who provided the instruction.

Reports

Medical Events

Medical events must be managed properly, as stipulated in §35.3045, to ensure regulatory compliance. The licensee must report any event, except for an event that results from patient intervention, in which the administration of byproduct material or radiation from byproduct material results in (1) a dose that differs from the prescribed dose or dose that would have resulted from the prescribed dosage by more than 0.05 Sv (5 rem) effective dose equivalent, 0.5 Sv (50 rem) to an organ or tissue, or 0.5 Sv (50 rem) shallow dose equivalent to the skin; and the total dose delivered differs from the prescribed dose by 20% or more, the total dosage delivered differs from the prescribed dosage by 20% or more or falls outside the prescribed dosage range, or the fractionated dose delivered differs from the prescribed dose, for a single fraction, by 50% or more; (2) a dose that exceeds 0.05 Sv (5 rem) effective dose equivalent, 0.5 Sv (50 rem) to an organ or tissue, or 0.5 Sv (50 rem) shallow dose equivalent to the skin from any of the following: administration of a wrong radioactive drug containing byproduct material, administration of a radioactive drug containing byproduct material by the wrong route of administration, administration of a dose or dosage to the wrong individual or human research subject, administration of a dose or dosage delivered by the wrong mode of treatment, or a leaking sealed source; (3) a dose to the skin or an organ or tissue other than the treatment site that exceeds by 0.5 Sv (50 rem) to an organ or tissue and 50% or more of the dose expected from the administration defined in the written directive (excluding, for permanent implants, seeds that were implanted in the correct site but migrated outside the treatment site).

A licensee must report any event resulting from intervention of a patient or human research subject in which the administration of byproduct material or radiation from byproduct material results or will result in unintended permanent functional damage to an organ or a physiologic system, as determined by a physician. The licensee must notify by telephone the NRC operations center no later than the next calendar day after discovery of the medical event. The licensee must submit a written report to the appropriate NRC regional office (listed in §30.6(a) of this chapter) within 15 days after discovery of the medical event. The written report must include the licensee's name; the name of the prescribing physician; a brief description of the event; why the event occurred; the effect, if any, on the individual(s) who received the administration; what actions, if any, have been taken or are planned to prevent recurrence; and certification that the licensee notified the individual (or the individual's responsible relative or guardian), and if not, why not. The report may not contain the individual's name or any other information that could lead to identification of the individual.

The licensee must provide notification of the event to the referring physician and also notify the individual who is the subject of the medical event no later than 24 hours after its discovery, unless the referring physician personally informs the licensee either that he or she will inform the individual or that, in his or her medical judgment, telling the individual would be harmful. The licensee is not required to notify the individual without first consulting the referring physician. If the referring physician or the affected individual cannot be reached within 24 hours, the licensee must notify the individual as soon as possible thereafter. The licensee may not delay any appropriate medical care for the individual, including any necessary remedial care as a result of the medical event, because of any delay in notification. To meet these requirements, notification of the individual who is the subject of the medical event may be made instead to that individual's responsible relative or guardian. If oral notification is used, the licensee must inform the individual, or appropriate responsible relative or guardian, that a written description of the event can be obtained from the licensee upon request. The licensee must provide such a written description if requested.

Aside from the notification requirement, nothing in this section affects any rights or duties of licensees and physicians in relation to each other, to individuals affected by the medical event, or to the individual's responsible relatives or guardians. A licensee must (1) annotate a copy of the report provided to NRC with the name of the individual who is the subject of the event and the Social Security number or other identification number, if one has been assigned, of the individual who is the subject of the event; and (2) provide a copy of the annotated report to the referring physician, if other than the licensee, no later than 15 days after discovery of the event.

Leaking Source

Occasionally, licensees discover leaking sealed sources in the course of conducting required leak tests. According to §35.3067, a licensee must file a report within 5 days if a leak test required by §35.67 reveals the presence of 185 Bq (0.005 µCi) or more of removable contamination. The report must be filed with the appropriate NRC regional office listed in §30.6 and by an appropriate method listed in §30.6(a) of this chapter. A copy of the written report must also be delivered to the Director, Office of Federal and State Materials and Environmental Management Programs. The written report must include the model number and serial number, if assigned, of the leaking source; the radionuclide and its estimated activity; the results of the test; the date of the test; and the action taken.

NRC Requirements and Nuclear Pharmacy Practice Standards

10 CFR §35.24 specifies the authority and responsibilities for the radiation protection program that each licensee's management must approve in writing. §35.24 also makes reference to additional requirements for radiation protection programs in §20.1101; one of those, under paragraph (b), requires the licensee, to the extent practical, to employ procedures and engineering controls based on sound radiation protection principles to achieve occupational doses and doses to members of the public that are as low as reasonably achievable (ALARA). The ALARA concept is embraced throughout the practice standards; nuclear pharmacists are obligated to work jointly with the RSO, health physicist, and nuclear medicine physician in developing radiation protection procedures that comply with this legal standard. Numerous radiation protection procedures, such as the use of time, distance, and shielding techniques, are prescribed in the practice standards, and many of these procedures are specifically required in regulatory guidance. For example, the NUREG-1556 guidance document[74] specifies, among many other requirements for commercial radiopharmacy licenses, that the licensee must require each individual who prepares (i.e., compounds) a radiopharmaceutical kit to use a syringe radiation shield and must also require individuals to use a syringe radiation shield when administering a radiopharmaceutical by injection unless the use of the shield is contraindicated for the given patient. This regulation also requires each syringe or syringe radiation shield that contains a syringe with a radiopharmaceutical to be conspicuously labeled so that the contents can be readily identified; however, this requirement is limited to providing the radiopharmaceutical name or its abbreviation, the clinical procedure to be performed, or the patient's name. In this regard, it is interesting to note that nuclear pharmacists are obligated, under both federal and state laws and pharmacy practice standards, to provide for each radiopharmaceutical dosage an appropriate prescription label containing all legally required information.[22] In addition to this required information, radiopharmaceutical prescription labels must include the amount of radioactivity, the calibration and expiration time, and a cautionary statement and symbol indicating that the dosage is radioactive, in order to further ensure that the dosage is used in a safe and efficacious manner.

In some instances, conflict may exist between NRC regulations and the standards of nuclear pharmacy practice. For example, 10 CFR 35.200 specifies that a licensee may use any byproduct material in a diagnostic radiopharmaceutical or any generator or any reagent kit for the preparation and diagnostic use of a radiopharmaceutical containing byproduct material that has an FDA-accepted IND or FDA-approved NDA. The regulation also specifies that a licensee must elute generators and prepare kits in accordance with the manufacturer's instructions. This latter requirement is seldom a problem, because nuclear pharmacists usually adhere to the manufacturer's instructions. However, as professional practitioners, nuclear pharmacists must reserve the right to exercise professional judgment as appropriate to the provision of quality nuclear pharmacy services and the overall safety of nuclear medicine procedures. Thus, according to a physician's prescription, patient needs, and individual experience with the prescribed radiopharmaceutical, nuclear pharmacists may modify the manufacturer's instructions or in some cases develop acceptable compounding procedures that are not described by the manufacturer. As a classic example, nuclear pharmacists may establish specialized procedures for preparing pediatric dosages of ^{99m}Tc-albumin aggregated for pulmonary perfusion studies, because the instructions for preparing the radiopharmaceutical for pediatric patients in a reasonable volume of administration or desired particle number are not provided in the product labeling. Obviously, nuclear pharmacists must consider the legal ramifications associated with this course of action or any other practice activities that may deviate from applicable NRC regulations.

REGULATORY OUTLOOK

NRC's recently amended regulations, discussed above, are one component of the agency's overall program for adjusting its regulatory scheme for the medical use of radioactive byproduct materials. NRC's goals are to focus regulations on those medical procedures that pose the highest risk to workers, patients, and the public and to structure the regulations to be risk-informed and performance-based, consistent with the agency's strategic plan.[69] In the future, all regulated stakeholders will be expected to focus their compliance improvement efforts on high-risk procedures and to continue to adjust their practices to this new risk-informed, performance-based approach to regulation.

FDA is implementing FDAMA, which was passed by Congress after 3 years of FDA scrutiny.[34] In this act, which is very broad and covers all of the agency's activities and programs, Congress recognized that protecting the public health is a responsibility shared by the entire health care community. The law directs the agency to carry out its mission in consultation and cooperation with all FDA stakeholders, including consumer and patient groups, the regulated industry, health care professionals, and FDA's regulatory counterparts abroad. FDA is currently working to cooperate with its stakeholders in the United States and abroad to continue protecting consumers and the public health in a new era of unprecedented technologic and scientific advances.

Progress in clarifying the FDA regulatory framework for PET, the lead imaging technology in nuclear medicine practice, has been much slower than mandated by FDAMA, but a variety of challenging new FDA regulations found in the new Part 212 are sure to be applied to the production of PET drugs in the very near future. All facilities engaged in the production of PET drugs, whether for commercial sale or on-site use, will be required to comply. PET industry stakeholders are moving toward compliance now that FDA has delineated the specific Part 212 requirements. A recent article provides an overview of FDA's current drug development and approval process, with special emphasis on radiopharmaceuticals, and attempts to clarify many regulatory issues and questions.[75] Implementation of the updated regulations will continue to be a major area of concern for PET stakeholders well into the future.

Nuclear pharmacists and other health care professionals must continually strive to increase their working knowledge of the numerous regulations on the possession and use of radioactive materials and radiopharmaceuticals intended for medical use. Federal regulations promulgated by NRC and FDA are most important; however, other federal, state, and local regulations must also be considered. Nuclear pharmacists, by virtue of their professional licensure and unique practice setting, bear major responsibility for ensuring that regulatory requirements are satisfied in the course of providing the services necessary for the safe and efficacious use of all radiopharmaceuticals and drugs in nuclear medicine practice. To fulfill this professional obligation, nuclear pharmacists must constantly monitor regulatory changes and modify the practice environment to maintain a comprehensive compliance program.

REFERENCES

1. Accreditation Council for Graduate Medical Education, Suite 2000, 515 North State Street, Chicago, IL 60610-4322, 2004.
2. Croll MN. Mileposts in nuclear medicine history. In: Henkin RE, Boles MA, Dillehay GL, et al., eds. *Nuclear Medicine*. St Louis: Mosby-Year Book; 1996:3–9.
3. Rhodes BA, Hladik WB, Norenberg JP. Clinical radiopharmacy: principles and practices. *Semin Nucl Med*. 1996;26:77–84.
4. Callahan RJ. The role of commercial nuclear pharmacy in the future practice of nuclear medicine. *Semin Nucl Med*. 1996;26:85–90.
5. Shaw SM, Ice RD. Nuclear pharmacy. I: Emergence of the specialty of nuclear pharmacy. *J Nucl Med Technol*. 2000;28:8–11.
6. Ponto JA, Hung JC. Nuclear pharmacy. II: Nuclear pharmacy practice today. *J Nucl Med Technol*. 2000;28:76–81.
7. McCready VR. Milestones in nuclear medicine. *Eur J Nucl Med*. 2000;27(1 suppl):S49–79.
8. United States Pharmacopeia. USP <797> Guidebook to Pharmaceutical Compounding—Sterile Preparations. www.usp.org/products/797Guidebook/. Accessed May 2, 2010.
9. American Pharmaceutical Association. *Nuclear Pharmacy Practice Guidelines*. Washington, DC: American Pharmaceutical Association; 1995.
10. Board of Pharmaceutical Specialties, 2215 Constitution Ave NW, Washington, DC, 20037-2985. www.bpsweb.org/. Accessed May 2, 2010.
11. Hilts PJ. *Protecting America's Health: The FDA, Business and One Hundred Years of Regulation*. New York: Knopf; 2003.
12. US Food and Drug Administration. History. www.fda.gov/AboutFDA/WhatWeDo/History/default.htm. Accessed May 2, 2010.
13. Burns K. *Pharmacist Legal Responsibilities Under Federal and Pennsylvania State Law*. Boca Raton, FL: RxLaw.org; 2010:47.
14. US Food and Drug Administration. History. www.fda.gov/default.htm. Accessed May 2, 2010.
15. Siegel JA. History and organization of regulation of nuclear medicine. In: *Nuclear Regulatory Commission Regulation of Nuclear Medicine: Guide for Diagnostic Nuclear Medicine*. Reston, VA: Society of Nuclear Medicine; 2001:7–8.
16. US Nuclear Regulatory Commission. Advisory Committee on Medical Use of Isotopes. www.nrc.gov/about-nrc/regulatory/advisory/acmui.html. Accessed May 2, 2010.
17. US Nuclear Regulatory Commission. *Regulatory Guide 10.8: Guide for the Preparation of Applications of Medical Programs*. Revision 2. August 1987. www.nrc.gov/reading-rm/doc-collections/reg-guides/general/rg/. Accessed May 4, 2010.
18. US Nuclear Regulatory Commission. *Consolidated Guidance About Material Licenses: Program-Specific Guidance About Medical Use Licenses*. NUREG-1556, vol 9. August 1998. www.nrc.gov/reading-rm/doc-collections/nuregs/staff/sr1556/. Accessed May 2, 2010.
19. US Nuclear Regulatory Commission. Our history. www.nrc.gov/about-nrc/history.html. Accessed May 2, 2010.
20. US Food and Drug Administration. Radioactive new drugs for investigational use. 28 *Federal Register* 183 (1963).
21. US Food and Drug Administration. Radioactive new drugs: new drug application requirements. 36 *Federal Register* 21026 (1971).
22. Halperin JA, Stringer SA, Eastep RD. Regulation of radiopharmaceuticals: analysis of radiopharmaceuticals approved for marketing in the United States 1951–1978. In: *Radiopharmaceuticals II: Proceedings of the Second International Symposium on Radiopharmaceuticals*. New York: Society of Nuclear Medicine; 1979:xxi.
23. US Food and Drug Administration. Radioactive new drugs and radioactive biologics: termination of exemptions. 40 *Federal Register* 31298 (1975).
24. US Nuclear Regulatory Commission. Naturally-occurring and accelerator-produced radioactive material (NARM) toolbox. http://nrc-stp.ornl.gov/narmtoolbox.html. Accessed May 2, 2010.
25. US Nuclear Regulatory Commission. Office of Federal and State Materials and Environmental Management Programs. www.nrc.gov/about-nrc/organization/fsmefuncdesc.html. Accessed May 2, 2010.
26. US Nuclear Regulatory Commission. State and tribal programs: agreement and non-agreement states. http://nrc-stp.ornl.gov/. Accessed May 2, 2010.
27. US Nuclear Regulatory Commission. Regulation of the medical uses of radioisotopes; statement of policy. www.nrc.gov/reading-rm/doc-collections/commission/secys/2000/secy2000-0113/2000-0113scy.html. Accessed May 2, 2010.
28. Keppler JS, Thornberg CF, Conti PS. Regulation of positron emission tomography: a case study. *AJR Am J Roentgenol*. 1998;171:1187–92.
29. Kessler DA. The regulation of investigational drugs. *N Engl J Med*. 1989;320:281–8.
30. US Food and Drug Administration. *Nuclear Pharmacy: Criteria for Determining When to Register as a Drug Establishment*. Rockville, MD: Division of Drug Labeling Compliance, Center for Drugs and Biologics, Food and Drug Administration; 1984.
31. Conti PS, Keppler JS, Halls JM. Positron emission tomography: a financial and operational analysis. *AJR Am J Roentgenol*. 1994;162:1279–86.
32. Coleman RE, Robbins MS, Siegel BA. The future of positron emission tomography in clinical medicine and the impact of drug regulation. *Semin Nucl Med*. 1992;22:193–201.
33. *Syncor v Shalala*, 96-5371 (DC Cir 1997).
34. US Food and Drug Administration Modernization Act (FDAMA). Bill S. 830, Pub L No. 105-115 §121, 11 Stat 2296 (1997). www.fda.gov/RegulatoryInformation/Legislation/FederalFoodDrugandCosmeticActFDCAct/SignificantAmendmentstotheFDCAct/FDAMA/default.htm. Accessed May 4, 2010.
35. Regulation of positron emission tomography radiopharmaceutical drug products: guidance. 60 *Federal Register* 10594 (1995).
36. Draft guideline on the manufacture of positron emission tomography radiopharmaceutical drug products: availability. 60 *Federal Register* 10593 (1995).
37. Current good manufacturing practice for finished pharmaceuticals: positron emission tomography. 62 *Federal Register* 19493 (1997).
38. Revocation of certain guidance documents on positron emission tomography drug products. 62 *Federal Register* 66636 (1997).
39. Revocation of regulation on positron emission tomography drug products. 62 *Federal Register* 66522 (1997).
40. Centers For Medicare & Medicaid Services. National coverage determinations (NCDs): positron emission tomography scans. October 1, 2002. www.cms.gov/transmittals/downloads/R156CIM.pdf. Accessed May 2, 2010.
41. The National Oncologic PET Registry (NOPR). What is the NOPR. www.cancerpetregistry.org/what.htm. Accessed May 2010.

42. IMV Medical Information Division of Des Plaines. Latest IMV PET census shows fast growth in PET procedures [news release]. August 4, 2003. www.imvinfo.com/index.asp?sec=abt&sub=def. Accessed October 26, 2010.
43. IMV Medical Information Division of Des Plaines. Latest IMV PET census shows slower growth in PET patient studies from 2007 to 2008 [news release]. February 17, 2009. www.imvinfo.com/index.asp?sec=abt&sub=def. Accessed October 26, 2010.
44. US Food and Drug Administration. Current good manufacturing practice for positron emission tomography drugs. 74 *Federal Register* 65409 (2009). http://edocket.access.gpo.gov/2009/pdf/E9-29285.pdf. Accessed May 2, 2010.
45. US Food and Drug Administration. Positron emission tomography (PET): questions and answers about CGMP regulations for PET drugs. www.fda.gov/Drugs/DevelopmentApprovalProcess/Manufacturing/ucm193476.htm#PETDRUGSANDUSERFEES. Accessed May 2, 2010.
46. US Food and Drug Administration. Positron emission tomography (PET): Additional questions and answers based on December 9, 2009 stakeholder call. www.fda.gov/Drugs/DevelopmentApprovalProcess/Manufacturing/ucm207021.htm. Accessed May 2, 2010.
47. US Department of Health and Human Services, Food and Drug Administration, Center for Drug Evaluation and Research. Development and approval process (drugs). www.fda.gov/Drugs/DevelopmentApprovalProcess/default.htm. Accessed May 2, 2010.
48. US Department of Health and Human Services, Food and Drug Administration, Center for Drug Evaluation and Research. Investigational New Drug (IND) Application. www.fda.gov/Drugs/DevelopmentApprovalProcess/HowDrugsareDevelopedandApproved/ApprovalApplications/InvestigationalNewDrugINDApplication/default.htm. Accessed May 2, 2010.
49. US Food and Drug Administration Code of Federal Regulations, Part 312, investigational new drug application. www.accessdata.fda.gov/scripts/cdrh/cfdocs/cfcfr/CFRSearch.cfm?CFRPart=312. Accessed May 2, 2010.
50. US Department of Health and Human Services, Food and Drug Administration, Center for Drug Evaluation and Research. New drug application (NDA) Introduction. www.fda.gov/Drugs/DevelopmentApprovalProcess/HowDrugsareDevelopedandApproved/ApprovalApplications/NewDrugApplicationNDA/default.htm. Accessed May 2, 2010.
51. US Department of Health and Human Services, Food and Drug Administration, Center for Drug Evaluation and Research. Abbreviated new drug application (ANDA): generics. www.fda.gov/Drugs/DevelopmentApprovalProcess/HowDrugsareDevelopedandApproved/ApprovalApplications/AbbreviatedNewDrugApplicationANDAGenerics/default.htm. Accessed May 2, 2010.
52. US Department of Health and Human Services, Food and Drug Administration, Center for Drug Evaluation and Research. Radioactive Drug Research Committee (RDRC) program. www.fda.gov/Drugs/ScienceResearch/ResearchAreas/Oncology/ucm093322.htm. Accessed May 2, 2010.
53. Woodbury DH, Leiberman L, Leutzinger E, et al. Pathway for FDA approval of new radiopharmaceuticals. In: Henkin RE, Boles MA, Dillehay GL, et al., eds. *Nuclear Medicine*. St Louis: Mosby-Year Book; 1996:350–6.
54. Mills G. The exploratory IND. *J Nucl Med*. 2008;49(6):45–47N.
55. US Department of Health and Human Services, Food and Drug Administration. Guidance for industry, investigators and reviewers: exploratory IND studies. www.fda.gov/downloads/Drugs/GuidanceComplianceRegulatoryInformation/Guidances/UCM078933.pdf. Accessed May 2, 2010.
56. Skovronsky D. Use of eINDs for evaluation of multiple related PET amyloid plaque imaging agents. *J Nucl Med*. 2008;49(6):47–48N.
57. VanBrocklin HF. Radiopharmaceuticals for drug development: United States regulatory perspective. *Curr Radiopharm*. 2008;1:2–6.
58. US Department of Health and Human Services, Food and Drug Administration. IND forms and instructions, information for sponsor-investigators submitting investigational new drug applications (INDs). www.fda.gov/Drugs/DevelopmentApprovalProcess/HowDrugsareDevelopedandApproved/ApprovalApplications/InvestigationalNewDrugINDApplication/ucm071073.htm. Accessed May 3, 2010.
59. US Department of Health and Human Services, Food and Drug Administration, Center for Drug Evaluation and Research. Manual of policies and procedures (MaPPs), 6030.1 IND process and review procedures (including clinical holds). www.fda.gov/Drugs/DevelopmentApprovalProcess/HowDrugsareDevelopedandApproved/ApprovalApplications/InvestigationalNewDrugINDApplication/default.htm. Accessed May 4, 2010.
60. Swanson DP. Radiopharmaceuticals as orphan drugs. *Semin Nucl Med*. 1996;36:91–5.
61. Koppal T. Inside the FDA: how CDER's Janet Woodcock helps companies achieve compliance, speed the drug review process and improve time to market for new drugs. *Drug Discovery Dev*. November 2002;5:32–8.
62. US Department of Health and Human Services, Food and Drug Administration. Office of Orphan Product Development (OOPD). www.fda.gov/AboutFDA/CentersOffices/OC/OfficeofScienceandHealthCoordination/OfficeofOrphanProductDevelopment/default.htm. Accessed May 3, 2010.
63. Petry NA. NRC and FDA regulations affecting nuclear pharmacy practice. *J Pharm Pract*. 1989;2:306–13.
64. Swanson DP, Lieto RP. The submission of IND applications for radiopharmaceutical research: when and why. *J Nucl Med*. 1984;25:714–9.
65. US Department of Health and Human Services, Food and Drug Administration, Center for Drug Evaluation and Research. Guidance for industry: PET drug applications—content and format for NDAs and ANDAs. March 2000. www.fda.gov/downloads/Drugs/GuidanceComplianceRegulatoryInformation/Guidances/UCM078738.pdf. Accessed May 3, 2010.
66. US Department of Health and Human Services, Food and Drug Administration. Guidance for Industry. INDs for Phase 2 and Phase 3 studies. Chemistry, manufacturing and controls information. http://www.fda.gov/downloads/Drugs/GuidanceComplianceRegulatoryInformation/Guidances/ucm070567.pdf. Accessed October 14, 2010.
67. US Department of Health and Human Services, Food and Drug Administration, Center for Drug Evaluation and Research. Guidelines for format and content of the clinical and statistical section of an application. www.fda.gov/downloads/Drugs/GuidanceComplianceRegulatoryInformation/Guidances/UCM071665.pdf. Accessed May 3, 2010.
68. Johnson JR, Temple R. Food and Drug Administration requirements for approval of new anticancer drugs. *Cancer Treat Rep*. 1985;69:1155–7.
69. US Nuclear Regulatory Commission. NRC mission. www.nrc.gov/about-nrc.html. Accessed May 2, 2010.
70. US Nuclear Regulatory Commission. Federal Register notices affecting Title 10 of the Code of Federal Regulations. www.nrc.gov/reading-rm/doc-collections/cfr/fr/. Accessed May 3, 2010.
71. US Nuclear Regulatory Commission. Part 35: medical use of byproduct material. April 24, 2002. www.nrc.gov/reading-rm/doc-collections/cfr/part035/. Accessed May 4, 2010.
72. US Nuclear Regulatory Commission. Medical use of byproduct material: *Federal Register* notice April 24, 2002. www.nrc.gov/reading-rm/doc-collections/cfr/fr/2002/20020424.html. Accessed May 3, 2010.
73. US Nuclear Regulatory Commission. Specialty board(s) certification recognized by NRC under 10 CFR Part 35. www.nrc.gov/materials/miau/med-use-toolkit/spec-board-cert.html. Accessed May 3, 2010.
74. US Nuclear Regulatory Commission. Consolidated guidance about materials licenses: program-specific guidance about commercial radiopharmacy licenses (NUREG-1556, vol 13, rev 1). www.nrc.gov/reading-rm/doc-collections/nuregs/staff/sr1556/v13/r1/. Accessed May 3, 2010.
75. Harapanhalli RS. Food and Drug Administration requirements for testing and approval of new radiopharmaceuticals. *Semin Nucl Med*. 2010;40:364–84.

CHAPTER 19

Brain

Nuclear medicine procedures for imaging the central nervous system (CNS) provide functional information for the evaluation of diseases that affect the brain and cerebrospinal fluid. Radiopharmaceuticals employed in these procedures can be divided into five main groups: (1) nondiffusible tracers, (2) diffusible tracers, (3) metabolism agents, (4) neuronal agents, and (5) cerebrospinal fluid (CSF) agents.

Nondiffusible tracers were the earliest group of agents used in brain imaging. They do not enter the normal brain because they are ionized and unable to diffuse through an intact capillary endothelium in the brain. However, when disease is present, nondiffusible tracers can egress from the blood through the disrupted capillary endothelium and localize in the affected area of the brain. For example, nondiffusible tracers have been used to detect vascular brain lesions, such as tumors, subdural hematomas, and arteriovenous malformations. The lesions on a brain scan appear as areas of increased accumulation of radioactivity, which contrast with normal brain tissue that is devoid of activity, showing up as "hot" lesions in a "cold" background. Included in the nondiffusible tracer group are ^{99m}Tc-sodium pertechnetate, ^{99m}Tc-pentetate (^{99m}Tc-DTPA), ^{99m}Tc-gluceptate (^{99m}Tc-GH), and the positron emission tomography (PET) imaging agent ^{82}Rb-chloride. Brain imaging procedures with these types of agents have essentially been replaced by procedures with other diagnostic imaging modalities that are more suitable for assessing anatomic brain pathology, such as computed tomography (CT) and magnetic resonance imaging (MRI).

Diffusible tracers were developed later in the course of radiopharmaceutical development. They are typically neutral lipophilic complexes capable of localizing in normal brain by passive diffusion across an intact capillary endothelium. The agents used in single-photon emission computed tomography (SPECT) brain imaging are ^{99m}Tc-exametazime (^{99m}Tc-HMPAO) and ^{99m}Tc-bicisate (^{99m}Tc-ECD). These complexes enter normal brain tissue proportional to regional cerebral blood flow. Brain lesions that are associated with increased blood flow, such as seizure foci, will appear hotter than normal brain, and lesions associated with diminished or absent blood flow, such as infarcts, will appear colder than normal brain. The development of diffusible tracers revitalized brain imaging in nuclear medicine, because they provide functional information that can complement the anatomic findings of CT or MRI.

Metabolism agents localize in brain lesions that have increased metabolic activity. The principal metabolic marker in PET imaging is ^{18}F-fludeoxyglucose (^{18}F-FDG). As a glucose analogue, ^{18}F-FDG localizes in regions that have increased glucose metabolism, such as seizure foci and active brain tumors, demonstrating increased uptake of ^{18}F activity relative to normal brain. ^{18}F-FDG is also used in evaluating dementias, which demonstrate a pattern of progressive cortical atrophy with diminished glucose metabolism over time.

Several radiopharmaceuticals have been investigated for imaging neurologic function in the brain. *Neuronal agents* labeled with ^{11}C, ^{18}F, ^{123}I, and ^{99m}Tc have been developed to measure functional status within the brain regarding neurotransmission, neuronal biochemistry, and receptor density. They hold promise for performing diagnostic evaluation, monitoring disease progression, and assessing treatment regimens. In addition, specific compounds are being developed for imaging dementias, particularly Alzheimer's disease.

CSF agents are radiopharmaceuticals that remain confined to the CSF space after intrathecal injection. These radiotracers move with the spinal fluid and can be used to evaluate conditions that alter CSF dynamics, such as hydrocephalus and spinal fluid leaks. The standard agent used in these studies is ^{111}In-pentetate (^{111}In-DTPA).

The ability to measure cerebral blood flow and biochemical processes in the brain with SPECT and PET has changed the role of brain imaging from anatomic visualization of brain disease to functional assessment. It complements the anatomic information provided by CT and MRI, enhancing the ability to diagnose brain disease.

PHYSIOLOGIC PRINCIPLES

Neural and Nonneural Capillaries

Neural capillaries in the brain differ from nonneural capillaries in other parts of the body. With some exceptions, the walls of *nonneural capillaries,* through which fluid and ion exchange occur, are composed of a unicellular layer of endothelial cells on the luminal (blood) side and a basement membrane on the abluminal (extracellular fluid) side (Figure 19-1). Intercellular clefts form a thin (6–7 nm) space between the endothelial cells, through which water and water-soluble ions can pass from the blood into the extracellular fluid.[1] Nonneural endothelial cells contain pinocytotic vesicles that serve as transporters of high molecular weight substances and small particles. Some capillary endothelial cells contain fenestrae (oval windows) through which large volumes of substances can pass by simple diffusion. These types of capillaries can be found in the kidney glomeruli, where plasma filtration and urine formation occur. Fenestrae are also found in the hepatic portal blood vessels, where they facilitate the flow of large amounts of nutrients from the blood to the liver parenchymal cells. The intercellular junction between endothelial cells is open, permitting paracellular diffusion of small molecules. In addition, the basement membrane is discontinuous to allow free flow of substances into the extracellular space. Nonneural capillary endothelial cells also contain proteins that facilitate transport of specific substances, such as glucose and amino acids, from the blood into the tissues.

By contrast, *neural capillaries* in the brain differ from nonneural capillaries in at least three important ways (Figure 19-1). First, the intercellular clefts or junctions are tightly apposed, preventing paracellular diffusion. The tight junctions consist of fibrils that connect adjacent cell plasma membranes and occur as complete bands or belts around connected cells.[2] The fibrils surround the apical margins of epithelial cells of the choroid plexus and also form between endothelial cells of blood vessels in the brain. They are also present between barrier cells at the arachnoid membrane (Figure 19-2).[2] A second difference in neural capillaries is that they have a basement membrane that is continuous, thereby further restricting simple diffusion processes. Third, neural capillary cells have a paucity of pinocytotic vesicles to transport high molecular weight substances.

These structural differences in neural capillaries present a barrier to simple diffusion of substances between the blood and the brain extracellular space; this is known as the blood–brain barrier (BBB). Only uncharged, lipophilic molecules can diffuse across the intact capillary endothelium in the brain. A few water-soluble substances that are also soluble in the lipid of cell membranes and can cross the BBB are oxygen, carbon dioxide, alcohol, and fatty acids.[1] One similarity between neural and nonneural capillaries is the presence of cell membrane transport proteins for glucose and amino acids.

Blood–Brain Barrier

Knowledge of the various compartments in the brain (Figure 19-2) is necessary for understanding drug and radiotracer localization in the brain. The BBB separates the two main compartments of the CNS (the brain and CSF) from the third compartment, the blood, which is supplied to the brain

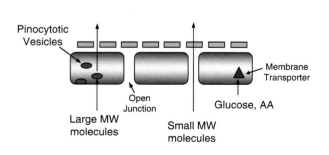
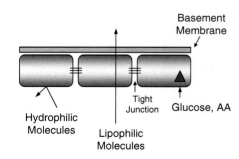

FIGURE 19-1 Transport processes through neural capillaries and nonneural capillaries from blood into extracellular fluid (ECF). See text for detailed explanation.

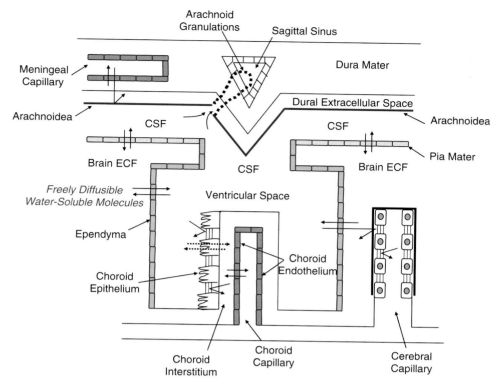

FIGURE 19-2 Sites of the blood–brain barrier in the brain are 3-fold: the blood–brain extracellular fluid (BECF) barrier, the blood–cerebrospinal fluid (CSF) barrier, and the brain–CSF barrier. See text for detailed explanation.

by the cerebral capillaries, the meningeal capillaries, and the capillaries of the choroid plexus.[3]

The *blood–brain extracellular fluid barrier* is found at the interface of the brain extracellular fluid (ECF) and the cerebral capillaries. Here, because of tight junctions, water-soluble molecules in the blood are immediately restricted from crossing the endothelium into the brain ECF. This barrier's selective permeability permits the passive diffusion of only uncharged, lipid-soluble molecules.

The *blood–CSF barrier* is found at the interface of the CSF and the capillaries of the choroid plexus. The cells of the choroid endothelium are fenestrated and freely permeable to water-soluble molecules, which can diffuse from the choroid capillary blood into the choroid interstitium and between the choroid epithelial cells up to the tight junctions.[3] However, the tight junctions between the cells of the choroid epithelium prevent substances from entering the CSF.

Another potential site of exchange between the CSF and blood is the arachnoid membrane (arachnoidea). However, though the meningeal capillaries within the dura are fenestrated, permitting free passage of water-soluble substances into the dural extracellular space, substances cannot penetrate into the CSF because of tight junctions in the outermost layers of the arachnoid membrane.[3]

The *CSF–brain barrier* is at the pia mater overlying the brain surface and at the ependyma lining the ventricular system.[3] At this interface no barrier appears to exist because the ependyma and pia allow rapid equilibration of water-soluble molecules between the brain ECF and the CSF. This distribution has been demonstrated by the injection of horseradish peroxidase (molecular weight [MW] 43,000) directly into the CSF; the molecule not only penetrates the ependyma and brain parenchyma but also permeates the basement membrane and the clefts between adjacent cerebral capillary endothelial cells up to the tight junctions.[2,4] Injection directly into the CSF is therefore an effective way to deliver drugs to the brain. The CSF itself and small substances in the CSF less than 1 μm in size can egress from the CSF space through the porous arachnoid granulations into the venous blood of the superior sagittal sinus.

Historically, the concept of the BBB developed from the observation that trypan blue dye, which binds to plasma protein, caused staining of tissues except the brain after intravenous injection into rabbits. The brain, however, could be stained when the dye was injected directly into the CSF.[2] The barrier, therefore, appears to protect the brain from various water-soluble substances present in the blood in order to maintain its homeostatic neuronal environment.[5] This barrier, however, can be disrupted by various pathologic conditions, which was the basis for brain imaging with hydrophilic radiotracers.

Cerebral Circulation

Blood is supplied to the brain by the carotid and vertebral arteries, as shown in Figures 19-3 and 19-4 (only the major

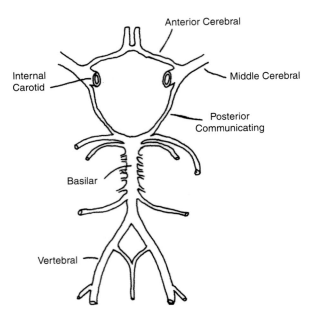

FIGURE 19-3 Arterial circulation at the base of the brain.

vessels are shown; the minor arteries have been omitted for clarity). The right and left vertebral arteries join to form the single basilar artery that supplies blood to the brain stem and the occipital cortex through the posterior cerebral arteries. The internal carotid arteries divide into the anterior and middle cerebral arteries and contribute communicating branches that anastomose with the posterior cerebrals. The circle of Willis at the base of the brain is formed by the interconnection of the two anterior cerebrals, the two posterior cerebrals, the two internal carotids, and the anterior and posterior communicating arteries.

The right and left anterior cerebral arteries run side by side in the longitudinal fissure along the medial surface of each hemisphere and end near the terminal branches of the posterior cerebral arteries. The right and left middle cerebral arteries arise as the largest branches of the internal carotid arteries. Each runs, at first, laterally in the sylvian fissure, then back and up, where its eight discrete branches distribute blood on the lateral surface of each hemisphere. The internal carotids and the anterior and middle cerebral arteries are the principal vessels visualized during the arterial phase of a brain blood flow study.

Blood drains from the brain through large venous sinuses. The superior sagittal sinus is the large venous channel that runs posterior from the nasal cavity over the top of the brain

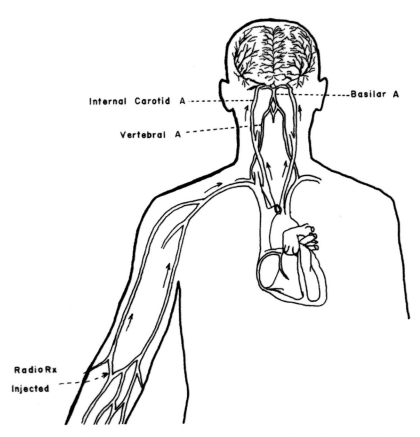

FIGURE 19-4 Cerebral delivery of radioactivity to the brain following intravenous injection of a radiopharmaceutical.

between the two hemispheres to the occipital region, ending in the confluence of sinuses (torcula). Other major sinuses that drain into the torcula include the straight, occipital, and inferior sagittal sinuses. From the torcula, blood drains bilaterally into the right and left transverse sinuses that eventually terminate as the internal jugular veins, which return blood to the heart. These are the primary vessels seen on the venous phase of a blood flow study.

Cerebrospinal Fluid

The entire cavity enclosing the brain and spinal cord has a volume of approximately 1650 mL.[1] The major structures are shown in Figure 19-5. The brain and spinal cord are bathed in CSF, which cushions the brain. In humans, the total volume of the CSF space is about 150 mL, 30 mL of which is in the spinal canal. The rate of CSF formation is about 30 to 35 mL per hour; CSF is chiefly produced by choroid plexus secretion.

Normally about 800 mL of CSF is produced each day. The composition of CSF is different from that of interstitial fluid in that CSF's concentration of sodium is 7% higher, glucose is 30% less, and potassium is 40% less. Protein concentration is extremely low (only about 0.025%) and is similar to that of brain interstitial fluid. Normal CSF pressure is 10 mm Hg (equivalent to 135 mm H_2O), ranging from 6 to 14 mm Hg.

The choroid plexus is a cauliflowerlike tuft of infolding capillaries covered by a thin coat of ependymal cells, which produce the CSF. These cells also actively transport foreign substances from the CSF into the blood. The plexus projects into the temporal horns of the lateral ventricles, the posterior portions of the third ventricle, and the roof of the fourth ventricle. Fluid normally passes from the lateral ventricles through the foramen of Monro into the third ventricle and through the aqueduct of Sylvius into the fourth ventricle. Fluid escapes from the fourth ventricle through the median foramen of Magendie and two lateral foramina of Luschka to enter the

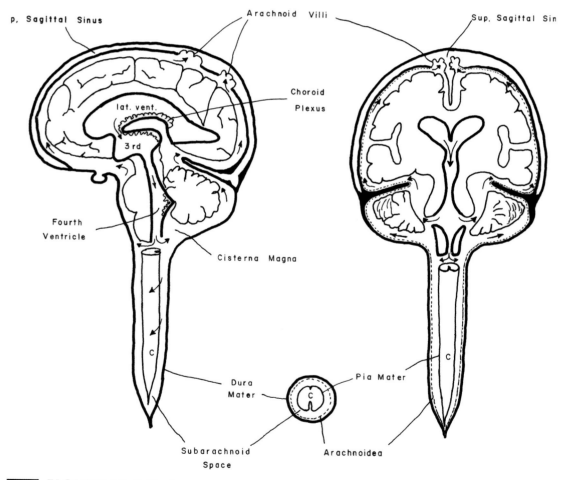

FIGURE 19-5 The brain and CSF space showing the site of CSF production (choroid plexus) in the lateral, third, and fourth ventricles. CSF flow proceeds out of the ventricles in a caudad direction around the spinal cord and cephalad over the cerebral hemispheres and is absorbed at the arachnoid villi into the superior sagittal sinus. The cord cross-section demonstrates the meninges and subarachnoid space.

cisterna magna. From the cisterna magna it flows in the subarachnoid space through the tentorial opening and out over the cerebrum, where it eventually passes through the arachnoid granulations into the venous sinuses of the brain. Flow through the one-way valves of the arachnoid membrane has minimal resistance; large molecules, such as proteins and other substances up to 1 μm in size, readily pass through the membrane into the sagittal sinus. As the CSF flows over the ventricular and pial surfaces of the brain, it sweeps away substances that diffuse from the brain into the CSF space.

A small quantity of CSF is formed by the blood vessels of the brain and spinal cord parenchyma and meninges. This fluid combines with that from the choroid plexus to flow through the subarachnoid space of higher levels. Most of the CSF bathing the spinal cord comes from this source.

A single layer of ependymal cells lines the walls of the ventricles and spinal cord (Figures 19-2 and 19-5). Tight junctions do not connect these epithelial cells, and the ependyma therefore is not a diffusion barrier to small solutes. Drugs placed into the ventricles readily pass into the brain extracellular space.[2] The apical microvilli of the choroid epithelial cells facing the CSF have tight junctions that prevent diffusion of large molecules but are "leaky" to small ions (e.g., Na^+ and K^+).[2]

BRAIN IMAGING AGENTS

Nondiffusible Tracers

Tracer Development

Brain imaging was initially conceived to localize intracranial tumors. George Moore,[6] a neurosurgeon at the University of Minnesota, first attempted to visualize tumors during brain surgery by using ultraviolet light to detect previously injected fluorescein, which concentrated in tumors. This was followed by the use of radiolabeled ^{131}I-diiodofluorescein and then ^{131}I-labeled human serum albumin (^{131}I-HSA).[7,8] Although ^{131}I-HSA demonstrated high tumor-to-brain ratios, blood clearance was slow.

In 1959 Blau and Bender[9] introduced ^{203}Hg-chlormerodrin for brain imaging. This shortened the time between dose administration and imaging because this agent was cleared quickly from the blood by the kidney. It also had better physical properties for imaging. This was followed soon by the introduction of ^{197}Hg-chlormerodrin.[10] Both of these agents had shortcomings. The ^{203}Hg label produced a high kidney radiation dose because of its beta radiation and long effective half-life, and the 77 keV photons of ^{197}Hg were easily attenuated in brain tissue.

In the 1960s, ^{99m}Tc was introduced. The ^{99m}Tc generator was developed at Brookhaven National Laboratory and refined for medical use by Richards.[11,12] Harper et al.[13] were the first to demonstrate in mice that ^{99m}Tc-pertechnetate activity localized in the stomach, salivary glands, and thyroid gland. The physical properties of ^{99m}Tc (short half-life, efficient photon detection, no particulate radiation) were more suited to the gamma camera, and large amounts of activity could be given with a smaller radiation dose. ^{99m}Tc-sodium pertechnetate soon became the agent of choice for brain scanning, requiring only prior administration of potassium perchlorate to retard its uptake in the choroid plexus. When the ^{82}Sr–^{82}Rb generator became available, the principal nondiffusible tracer for PET imaging of the brain was ^{82}Rb-chloride.

A desire to shorten the time interval between ^{99m}Tc-sodium pertechnetate administration and imaging, normally 3 to 4 hours, led to the use of technetium complexes, which had faster blood clearance.[14] When ^{99m}Tc-DTPA and ^{99m}Tc-GH were compared with ^{99m}Tc-sodium pertechnetate, ^{99m}Tc-GH was shown to have better accuracy at early and delayed times after injection and became the agent of choice for brain imaging.[15–18] Eventually, brain imaging with nondiffusible tracers diminished once CT became available, providing superior anatomic detail in brain images.

Although ^{99m}Tc-sodium pertechnetate is no longer used in brain imaging, its biologic properties are described in the following section. The properties of ^{99m}Tc-DTPA and ^{99m}Tc-GH are discussed in Chapter 24.

Sodium Pertechnetate Tc 99m Injection

The clinical use of ^{99m}Tc as the pertechnetate anion (^{99m}TcO$_4^-$) began in 1961 at the University of Chicago, where its biologic behavior was reported by Lathrop and Harper.[19] Sodium pertechnetate Tc 99m injection (^{99m}Tc-sodium pertechnetate) is usually administered by intravenous injection, but it may be administered orally. Of the total activity circulating in the bloodstream 1 hour after injection, an average of 30% is contained in the red cell fraction and 70% in the plasma.[14] Pertechnetate is freely diffusible into and out of red blood cells and can be removed by serial washing of cells in saline. Approximately 75% of plasma activity is protein bound, with one-third of this very loosely bound. The disappearance of activity in blood after intravenous injection is multiexponential and shows wide variation in individuals, with half-lives in the ranges of 1 to 2 minutes (50%–60%), 5 to 20 minutes (15%), and 100 to 300 minutes (20%–30%), without the coadministration of potassium perchlorate. These times are significantly prolonged if perchlorate is given.[19]

One hour after intravenous administration, ^{99m}Tc-sodium pertechnetate has the following organ distribution: 30% in gastric mucosa and juice, 2% in the thyroid gland, and 5% in salivary glands and saliva.[19] Similar to the iodide ion, pertechnetate is concentrated and secreted by the mucoid cells of the gastric glands but not by the peptic (chief) cells or oxyntic (parietal) cells (Figure 19-6).[14,20] In the thyroid gland, pertechnetate is not metabolized as is iodide, and its accumulation is limited to the ion-concentrating mechanism of thyroid epithelial cells (i.e., it is trapped but not organified). The striated epithelial cells of the salivary glands also have concentrating mechanisms for the group VII anions, including iodide and its analogues (such as pertechnetate, thiocyanate, and perchlorate).

Although ^{99m}Tc-sodium pertechnetate is similar to iodide in its tissue distribution and blood clearance, its biologic elimination is somewhat different. The renal clearance of pertechnetate (17 mL/minute) is about half that of iodide,[21] representing about 14% of inulin clearance. Thus, about 86% of pertechnetate is reabsorbed by the renal tubule, assuming that the filtra-

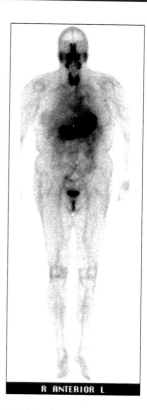

FIGURE 19-6 Total body image 2 hours after intravenous administration of 10 mCi (370 MBq) of ^{99m}Tc-sodium pertechnetate. Normal uptake of activity is seen in the salivary glands, thyroid gland, and stomach. There is also activity seen in the oral and nasopharyngeal regions and the urinary bladder.

tion fractions for pertechnetate and inulin are similar. This explains why plasma clearance is so slow after intravenous administration of ^{99m}Tc-sodium pertechnetate. Another important difference between pertechnetate and iodide occurs in the bowel. Although iodide is completely absorbed by the intestine, pertechnetate is partly bound to fecal material in the intestine and is excreted with a half-life dependent upon intestinal evacuation.[22] For these reasons, 72 hours after dosing only 50% of pertechnetate is eliminated from the body by urinary and fecal routes, whereas iodide excretion exclusive of thyroid activity is greater than 98% by this time.

The long-term retention of technetium in humans after ^{99m}Tc-sodium pertechnetate administration has been measured using the longer-lived isotopes ^{95m}Tc ($T_{1/2}$ = 60 days) and ^{96}Tc ($T_{1/2}$ = 43 days).[23] After oral or intravenous administration, urinary excretion of pertechnetate is rapid within the first 24 hours but drops dramatically on the second and third days, with less than 1% excreted per day thereafter. Fecal excretion begins more slowly but is the principal route of elimination 1 day after administration; it reaches a maximum by the fourth or fifth day and decreases thereafter. Cumulative urinary and fecal excretion is 30% in 1 day (27% urine, 3% fecal), 72% in 4 days (31% urine, 41% fecal), and 90% in 8 days (34% urine, 56% fecal).[24] Long-term retention studies estimate that 77% of the dose is eliminated with a biologic half-life of 1.6 days, 19% with a half-life of 3.7 days, and 4% with a half-life of 22 days.[23]

^{99m}Tc-sodium pertechnetate will concentrate in the choroid plexus of the brain.[24] It has been shown previously that radioiodide also concentrates in the choroid plexus by a process of active transport from spinal fluid into the blood and that this process can be retarded by perchlorate. Coben et al.[25] demonstrated that perchlorate increases the blood-to-CSF transport of iodide and decreases the reverse process of CSF-to-blood. This phenomenon suggests that the blood–CSF barrier is due to active transport from CSF to blood rather than to membrane hindrance of transport from blood to spinal fluid.[25] That is, iodide that readily diffuses from blood to spinal fluid is rapidly transported back to blood by an active process, and this creates the blood–CSF barrier. Harper et al.[26] visualized the CSF space with ^{99m}Tc-sodium pertechnetate by pretreating the patient with perchlorate to prevent active transport of pertechnetate from the CSF to blood. Oral administration of perchlorate prevents accumulation of pertechnetate in the choroid plexus and readily displaces what has already accumulated.[24] Perchlorate may be given orally at any time before or after the injection of ^{99m}Tc-sodium pertechnetate, provided the perchlorate administration precedes imaging by at least 60 minutes.[27] The usual oral dose of sodium or potassium perchlorate is 200 mg to 1000 mg. Up to 450 mg of sodium perchlorate has also been given intravenously.[28] Uptake of ^{99m}Tc-sodium pertechnetate in the thyroid and stomach is also retarded by perchlorate.

^{99m}Tc-sodium pertechnetate is excreted in human milk, and it is recommended that breast-feeding be suspended for at least 48 hours after radionuclide studies.[29,30]

^{99m}Tc-sodium pertechnetate also undergoes placental transfer, which is reduced by perchlorate; however, posttreatment with perchlorate does not release ^{99m}Tc from the fetus.[18] ^{99m}Tc-sodium pertechnetate, like all radiopharmaceuticals, is contraindicated during pregnancy.

Mechanisms of Localization

Nondiffusible tracers typically are charged hydrophilic compounds and therefore cannot permeate through the BBB. They are able to enter the brain, however, when this barrier is disrupted in some way as a result of the pathologic condition. Basic mechanisms for localization of these agents in brain tumors have been described by Tator.[31]

Vascularity

Many tumors are highly vascularized, and a large fraction of their radioactive content is due to their increased amount of blood. Examples include hemangioblastomas, vascular meningiomas, arteriovenous malformations, and certain malignant gliomas.

Interstitial Fluid

Almost all brain tumors contain more interstitial fluid than does normal brain tissue. Many substances have a tendency to accumulate in this space, especially small molecules such as ^{99m}Tc-sodium pertechnetate and its complexes.

Capillary Permeability

The endothelial cell junctions in many tumors widen, and large molecules can readily diffuse through the open pores into the brain. In addition, pinocytosis, not widely present in normal endothelial cells, is present to a greater extent in brain tumors. Front and Israel[32] have shown that meningiomas and glioblastomas that take up ^{99m}Tc-sodium pertechnetate and ^{57}Co-bleomycin also demonstrate, histologically, many pinocytotic vesicles and fenestrated endothelial cells, whereas tumors that exhibit no uptake of activity have normal endothelial architecture with tight junctions and few pinocytotic vesicles.

Tumor Cell Uptake

Mechanisms of uptake directly within neoplastic cells may involve substances that are substrates for energy requirements (glucose and phosphates) and for growth (amino acids). FDG localization is a good example. Also, tumors may contain some intracellular protein not present in normal cells that binds tracer. This has been noted with ^{67}Ga uptake in tumors outside the brain.

Diffusible Tracers

Tracer Development

In 1978, Oldendorf[33] called attention to the decreasing number of brain scans being performed in nuclear medicine and pointed out that the lipophilic agent ^{123}I-iodoantipyrine had shown brain uptake proportional to cerebral blood flow (CBF).[34] Although iodoantipyrine readily crossed the BBB, it also diffused back out and therefore had limited usefulness without a brain trapping mechanism. Nuclear medicine scientists began to direct their efforts toward the development of lipophilic ^{99m}Tc tracers that could cross the BBB so that CBF could be measured. To this end, Loberg et al.[35] prepared a series of ^{99m}Tc-labeled iminodiacetic acid analogues with varying octanol-to-water partition coefficients. They demonstrated that brain uptake was proportional to lipophilicity, but high plasma protein binding limited the clinical potential of these analogues. About the same time, Kung and Blau[36] developed two diamine compounds, di-β-(piperidinoethyl)-selenide (PIPSE) and di-β-(morpholinoethyl)-selenide (MOSE), labeled with ^{75}Se for brain localization. Their work was based on the pH-gradient hypothesis (see Chapter 9). Although PIPSE and MOSE never became useful brain imaging agents, this work introduced a mechanism that could be exploited for localizing radiotracers in the brain and led to the development of agents for measuring regional cerebral blood flow (rCBF). The importance of being able to measure rCBF lies in the fact that blood flow and metabolism in the brain are coupled and that cerebrovascular disease can produce local changes in rCBF.

There are three basic requirements for SPECT and PET radiotracers for assessing rCBF: (1) ability to cross the BBB, (2) brain retention long enough to acquire images, and (3) lack of redistribution in the brain.[37]

In the early 1980s, two radioiodinated amine compounds with high first-pass extraction and potential for measuring rCBF were introduced for brain imaging: N-isopropyl-p-^{123}I-iodoamphetamine (^{123}I-IMP or iofetamine)[38] and hydroxy ^{123}I-iodobenzyl propyl diamine (^{123}I-HIPDM)[39] (Figure 19-7). ^{123}I-IMP was approved by FDA in 1988, but it was eventually removed from the market because it redistributed in the brain 1 hour after dosing, leading to inaccurate imaging at later times.[40] The in vivo profiles of ^{123}I-IMP and ^{123}I-HIPDM were similar,[39] and neither of these compounds became routine brain imaging agents for several reasons, including uncertain brain retention mechanisms, inconvenience, and the cost of using ^{123}I.

Also in the early 1980s, a number of technetium coordination compounds were being developed for perfusion brain imaging. Significant among these were the technetium bis-aminethiol complexes (^{99m}Tc-BAT) by Burns et al.[41] and Kung et al.[42] and the technetium propyleneamine oxime (^{99m}Tc-PnAO) complex by Volkert et al.[43] BAT, also known as diaminedithiol (DADT), is an N_2S_2 ligand, and PnAO is a tetraamine (N_4) ligand. The first generation of these neutral lipophilic complexes demonstrated brain uptake, but retention in brain was poor and unsuitable for SPECT imaging.[44] Subsequently, several derivatives of these ligands were prepared with amine side chains as functional groups that could potentially improve brain retention. The derivatives, however, yielded diasterioisomers that had to be separated by high-performance liquid chromatography in order to isolate the isomer with highest brain uptake. Most of the ^{99m}Tc-BAT derivatives that exhibited increased retention in the brain had side chains containing carboxyester groups or pendent amine functionalities.[45] The most successful of these was the L,L-isomer of Tc(V)oxo-1,2-N,N'-ethylenedylbis-L-cysteine diethyl ester.[46,47] Otherwise known as ^{99m}Tc-bicisate (^{99m}Tc-ECD; Neurolite [Lantheus]), it is a neutral diester derivative that undergoes in vivo hydrolysis in the brain to yield an ionized metabolite that is retained.[47]

Although the ^{99m}Tc-PnAO complex showed rapid brain uptake after intravenous administration, its rapid washout pre-

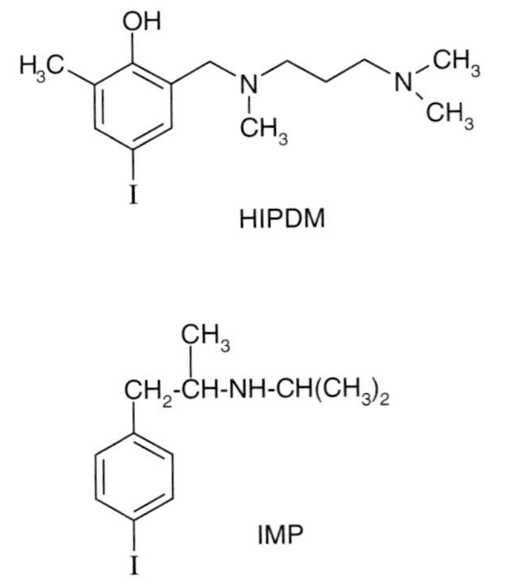

FIGURE 19-7 Chemical structures of radioiodinated amine-type first-generation diffusible tracers for brain imaging.

cluded its use for SPECT imaging. Subsequently, several derivatives of ^{99m}Tc-PnAO were synthesized with methyl groups on the amineoxime backbone, in the hope of finding an agent that remained fixed in the brain. One of these was the hexamethyl derivative, ^{99m}Tc-hexamethylpropyleneamine oxime (^{99m}Tc-HMPAO), otherwise known as ^{99m}Tc-exametazime (Ceretec [GE Healthcare]). This complex is neutral and lipophilic but unstable in aqueous solution. The instability was found to be a conversion from the primary lipophilic complex to a secondary hydrophilic complex and was mediated by reducing agents.[48] Studies in animals and humans demonstrated that the *meso*-isomer had greater in vitro stability but little brain retention, while the D,L-isomer had poor in vitro stability but high brain retention. It was then surmised that brain uptake was caused by the lipophilic complex and brain retention was caused by its intracellular conversion to the nondiffusible hydrophilic complex. The brain conversion was eventually shown to be mediated by the intracellular reducing agent glutathione.[48]

^{99m}Tc-exametazime was eventually approved by FDA in December 1988 and ^{99m}Tc-bicisate, in November 1994. Both of these agents have high first-pass extraction into the brain (^{99m}Tc-exametazime 70%[49] and ^{99m}Tc-bicisate 60%[50]). This property makes them useful markers for assessing rCBF in the evaluation of cerebrovascular disease, particularly stroke.

Technetium Tc 99m Exametazime Injection

Technetium Tc 99m exametazime injection (^{99m}Tc-HMPAO, ^{99m}Tc-exametazime) is a neutral lipophilic complex with MW of 384. Its octanol-to-water partition coefficient is 80 (i.e., log P_{oct} 1.9).[49] These properties permit ^{99m}Tc-HMPAO to passively diffuse through the capillary endothelium of the BBB. The compound exists in two diastereoisomeric forms, D,L- and *meso*-.[51] However, only the D,L-isomer demonstrates significant brain uptake (Figure 19-8).[52] The two isomeric forms are separated by fractional crystallization so that the kit contains only

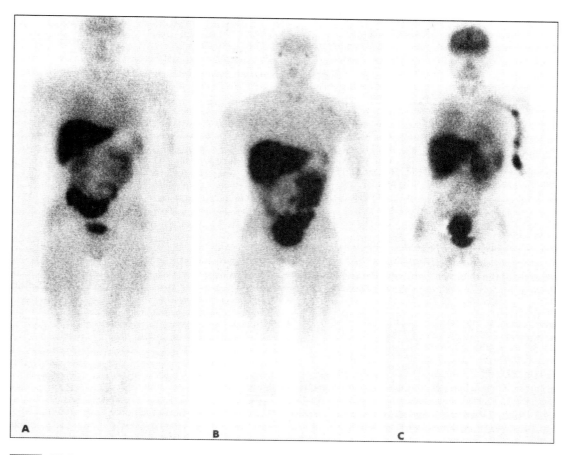

FIGURE 19-8 Anterior whole-body scans of ^{99m}Tc-HMPAO isomers 4 hours after injection. (A) Mixture of D,L- and *meso*- isomers, with uptake seen in brain, skeletal muscle, and lung. Excretion is hepatobiliary; kidneys, bladder, liver, and small intestine are all visible. (B) *Meso*-isomer, with distribution similar to the mixture but with lower lung uptake and obvious concentration of material in lacrimal glands. Brain uptake is only slightly higher than soft tissue uptake. (C) D,L-isomer, showing high uptake in brain. Uptake also is clearly seen in myocardium, subcutaneous fat of buttocks, and medial aspect of thighs. Retention of material in left brachiocephalic vein, into which material is injected, is a common feature of these materials. Activity is also seen in the urinary bladder, but with relatively small amounts in the intestine. (Reprinted with permission of the Society of Nuclear Medicine from reference 52).

the D,L-enantiomer. The ^{99m}Tc-D,L-HMPAO isomer (Figure 19-9) will be referred to henceforth as ^{99m}Tc-HMPAO.

After intravenous injection, a mean of 72% of the primary lipophilic complex is extracted during the first pass into the brain at resting CBF (59 mL/minute per 100 grams), but this extraction decreases at higher flow rates.[49] Once the primary complex has crossed the BBB, its fate is determined by a competition between rapid conversion to a nondiffusible hydrophilic complex and washout to the blood. These two routes of loss of diffusible complex are of approximately equal importance, so that only about 50% of the ^{99m}Tc activity entering the brain is retained.[53] When ^{99m}Tc-HMPAO is injected into an arm vein, only a fraction (approximately 15%) of the injected dose (ID) reaches the brain, and a mean of 4.1% (3.5%–7.0%) of the ID actually localizes there.[52] The amount trapped in the brain reaches equilibrium by 2 minutes and remains steady over the first 8 hours. Brain retention is due to a glutathione-mediated conversion of the lipophilic complex to a hydrophilic form that cannot diffuse back out of the brain.[48]

Lassen et al.[53] described a kinetic model in which ^{99m}Tc-HMPAO exists in the blood and brain compartments in an exchangeable lipophilic form and in a retained brain compartment as a nonexchangeable hydrophilic form. The model explains that the decline in extraction at high CBF is due to back diffusion of the exchangeable lipophilic complex from brain to blood. The amount that back diffuses was shown to be around 15%, and the back diffusion occurs within the first 2 minutes after injection.[49,52] When a correction is made for this back diffusion, the relationship between rCBF and ^{99m}Tc-HMPAO distribution is more linear.[53] One hundred percent extraction can never be achieved with ^{99m}Tc-HMPAO; a portion of the ID is already in the hydrophilic form because of in vitro conversion prior to injection and in vivo conversion once in the blood.[54,55] The radiochemical purity of lipophilic ^{99m}Tc-HMPAO, assessed by octanol extraction, has been shown to be approximately 90% in normal saline, 40% in plasma, and 20% in whole blood by 10 minutes after tracer addition.[55] Use of the stabilized kit to prepare ^{99m}Tc-HMPAO reduces the chance of in vitro conversion.

^{99m}Tc-HMPAO exhibits both hepatobiliary and urinary excretion. Twenty minutes after injection, liver uptake is 10% and urinary excretion is about 2.5%, increasing to 35% in 24 hours.[52] About 30% of the ID is in the gastrointestinal tract immediately after injection, and about one-half of this is excreted via the intestinal tract by 48 hours.[56] About 40% of the ID is excreted by the kidneys into urine over 48 hours.[56] Soft-tissue distribution is predominantly in skeletal muscle. Twelve percent of activity remains in the blood 1 hour after injection.[52] The biologic half-life in the brain is estimated to be 71 hours.[57,58]

^{99m}Tc-HMPAO, with or without stabilization, is indicated for detection of altered cerebral perfusion in stroke. The usual adult administered activity is 10 to 30 mCi (370–1110 MBq). The critical organ listed in the package insert is the lacrimal glands, with a radiation absorbed dose of 5.6 rad(cGy)/20 mCi; however, this figure has been challenged and reported to be significantly lower, at 1.02 rad(cGy)/20 mCi, and to occur in only 11% of patients.[59]

Technetium Tc 99m Bicisate Injection

Technetium Tc 99m bicisate injection (^{99m}Tc-ECD) is a complex of technetium with ethyl cysteinate dimer. It is a neutral lipophilic complex with MW of 436. It can have four possible isomeric forms (D,D-, L,L-, L,D-, or D,L-), depending on whether it was synthesized with L-cysteine, D-cysteine, or D,L-cysteine.[60] Both the L,L- and D,D-isomers demonstrate brain uptake, but only the L,L-isomer exhibits brain retention.[47] Brain retention is not only stereospecific but also species-specific in that ^{99m}Tc-L,L-ECD (Figure 19-9) localizes only in the brains of primates (monkeys and humans). Its octanol-to-water partition coefficient is 51, and its gray-to-white-matter ratio is about 4.5, demonstrating its potential usefulness in assessing rCBF.[47] The L,L-isomer will henceforth be referred to as ^{99m}Tc-ECD.

After intravenous injection, ^{99m}Tc-ECD demonstrates a high first-pass extraction into brain (47%[61], 60%[62]). Friberg et al.[63] measured brain dynamics and demonstrated that uptake and retention were triexponential, representing a vascular input spike, a back diffusion from brain to blood, and a very slow loss due to incomplete retention of hydrophilic metabolite. The distribution, however, does not change with time in the brain, and the loss appears to be the same from all regions. The retained fraction in the brain is 44%.[63] Walovitch et al.[60] demonstrated that ^{99m}Tc-ECD is rapidly metabolized in brain tissue, primarily in the cytosol, to a

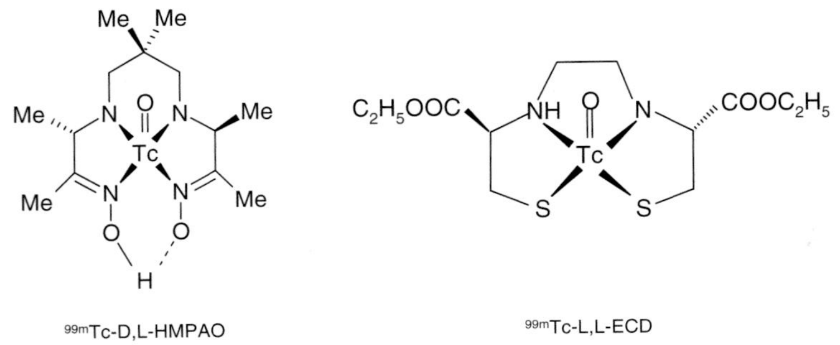

FIGURE 19-9 Chemical structures of ^{99m}Tc-D,L-HMPAO and ^{99m}Tc-L,L-ECD.

monoacid ester that is selectively trapped in primate brains but not in the brains of other species.

^{99m}Tc-ECD brain uptake correlates with CBF, but above 50 mL/minute per 100 grams it underestimates flow by as much as 20%.[64] Lassen and Sperling[65] compared the distribution of ^{99m}Tc-ECD and CBF measured with ^{133}Xe in patients with dementia, head trauma, epilepsy, brain tumor, and stroke. Good agreement was found in all cases except in subacute stroke patients, who failed to show reflow hyperemia in the infarct area. It was noted that this finding may be useful, particularly in subacute cases, when other SPECT methods present difficulties because of reflow masking the size and severity of the lesion.

In normal human subjects after intravenous administration, ^{99m}Tc-ECD demonstrates a maximum brain uptake of 6.5% in 5 minutes, slowly declining thereafter to 5.2% by 1 hour and 3.8% by 4 hours.[66,67] Figure 19-10 illustrates the total body distribution of ^{99m}Tc-ECD. Kinetic analysis demonstrates that 40% of brain activity washes out quickly ($T_{1/2} = 1.3$ hour), while 60% clears much more slowly ($T_{1/2} = 42.3$ hour).[66] Blood activity declines rapidly to 4.9% in 1 hour because of rapid plasma conversion to the monoethyl ester metabolite, which has high renal clearance (75% ID in urine within 6 hours).[68] Some of the tracer is excreted through the hepatobiliary system, with initial liver uptake of 17% at 5 minutes declining to 2.5% by 4 hours, with prominent gallbladder activity. Fecal excretion is 11% in 48 hours.[66]

The use of ^{99m}Tc-ECD in brain imaging is indicated as an adjunct to CT and MRI in localization of stroke in patients diagnosed with stroke. The usual adult administered activity for brain imaging is 10 to 30 mCi (370–1110 MBq) by intravenous injection. The critical organ is the urinary bladder wall, with a radiation absorbed dose of 0.11 rad(cGy)/mCi.[69]

Table 19-1 summarizes the properties of ^{99m}Tc-ECD and ^{99m}Tc-HMPAO. ^{99m}Tc-HMPAO demonstrates slow blood clearance of its metabolite, causing lower target-to-background ratios than ^{99m}Tc-ECD, which exhibits rapid removal of its plasma metabolite.[70] Both agents are stable in kit form, although ^{99m}Tc-ECD is stable for a longer time (4 hours for stabilized ^{99m}Tc-HMPAO versus 6 hours for ^{99m}Tc-ECD).

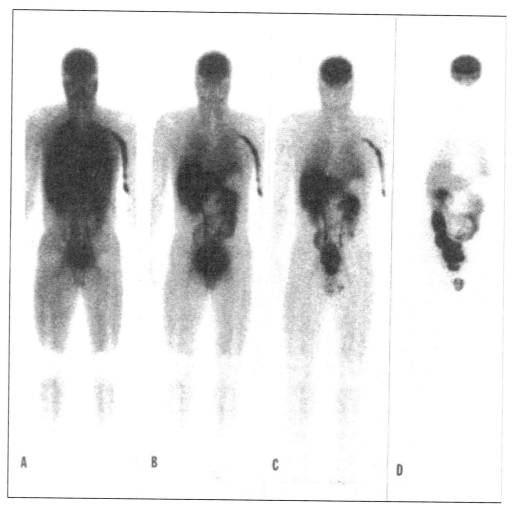

FIGURE 19-10 Whole-body distribution of ^{99m}Tc-L,L-ECD in a normal subject. Anterior views were obtained at (A) 5 minutes, (B) 1 hour, (C) 2 hours, and (D) 4 hours. (Reprinted with permission of the Society of Nuclear Medicine from reference 66).

TABLE 19-1 Properties of SPECT Blood Flow Markers for Brain Imaging

Property	^{99m}Tc-HMPAO	^{99m}Tc-ECD
Molecular charge/lipophilicity	Neutral/lipophilic	Neutral/lipophilic
Brain extraction efficiency	72%–80%	47%–60%
Maximum brain uptake	4.1% (20 min)	6.5% (5 min)
Brain washout	15% over 2 min $T_{1/2}$ = 72 hr (slow component)	20% over 1 hr $T_{1/2}$ = 42.3 hr (slow component)
Blood levels[a]	12% ID (1 hr)	5% ID (1 hr)
Excretion	Urine (40% in 48 hr) Hepatobiliary (30% immediate)	Urine (72% in 24 hr) Hepatobiliary (12% in 48 hr)
Critical organ	Lacrimal glands 5.16 rad/20 mCi	Urinary bladder wall 5.6 rad/20 mCi

[a] ID = injected dose.

Neither agent redistributes in the brain, and both are retained long enough for SPECT imaging. ^{99m}Tc-ECD has a higher brain uptake initially than ^{99m}Tc-HMPAO, but ^{99m}Tc-ECD washes out of the brain more rapidly than ^{99m}Tc-HMPAO. However, brain washout of each agent is slow and does not appear to affect imaging.

Metabolism Agents

Agent Development

After the development of CT, the usefulness of conventional planar brain imaging with the gamma camera declined significantly. Only those studies that provided dynamic or physiologic information were of interest, and these were few because of the limited arsenal of radiopharmaceuticals. The desire for physiologic information from brain imaging led by necessity to the development of compounds labeled with carbon, nitrogen, and oxygen isotopes. However, the only practical radionuclide choices of these elements are the short-lived positron emitters ^{11}C, ^{13}N, and ^{15}O, which require PET. Investigations were subsequently conducted with 2-deoxy-D-glucose labeled with ^{14}C for animal studies and ^{11}C and ^{18}F for clinical studies,[50,71–73] demonstrating that ^{18}F-2-deoxy-2-fluoro-D-glucose (^{18}F-fludeoxyglucose or ^{18}F-FDG) was an accurate marker of glucose utilization in the brain and that it was a useful tool for studying the brain's response to normal, pathologic, and interventional stimuli.[72] These investigations were able to demonstrate further that deoxyglucose crosses the BBB and is phosphorylated similarly to glucose; however, unlike glucose, deoxyglucose-6-phosphate is retained in tissue for an extended time, which facilitates imaging studies.

FDA originally approved ^{18}F-FDG in July 1994 at the Downstate Clinical PET Center for the identification of regions of abnormal glucose metabolism associated with epileptic seizure foci. Since then, ^{18}F-FDG has shown wide applicability in cancer diagnosis in many different organ systems, for identifying primary tumors and differentiating recurrent tumor from radiation necrosis.

Fludeoxyglucose F 18 Injection

Fludeoxyglucose F 18 injection (^{18}F-FDG) is a glucose analogue with a fluorine atom replacing a hydroxyl group at the C-2 position of D-glucose (Figure 19-11). This modification permits FDG to be a substrate for glucose transport and partial metabolism within the cell. Both glucose and FDG are taken up by facilitated diffusion and phosphorylated by hexokinase to glucose-6-phosphate and FDG-6-phosphate, respectively (Figure 19-12). However, FDG cannot be metabolized beyond this first step, whereas glucose is metabolized further to glycogen or enters the glycolytic pathway with ultimate conversion to carbon dioxide and water. Consequently,

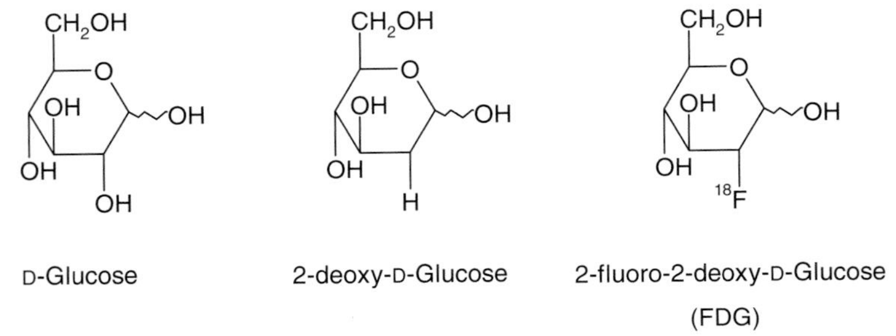

FIGURE 19-11 Chemical structures of D-glucose, 2-deoxy-D-glucose, and 2-fluoro-2-deoxy-D-glucose (FDG).

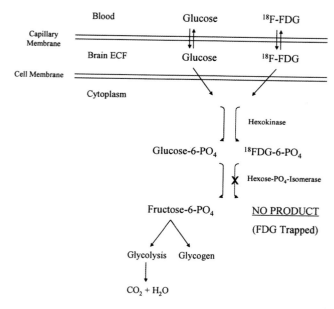

FIGURE 19-12 Metabolic scheme for glucose and FDG transport and metabolism in the brain.

^{18}F-FDG is trapped in tissue as FDG-6-phosphate, which permits PET imaging of its distribution. FDG also differs from glucose in that it does not undergo tubular reabsorption and is readily excreted in the urine.

Routine procedures for brain imaging require that the patient fast for at least 4 hours before the study; if the patient has an elevated blood glucose level, poor brain uptake of FDG will occur. The patient is injected with ^{18}F-FDG in a quiet room and should remain inactive for at least 45 minutes to 1 hour to minimize uptake of ^{18}F-FDG in muscle and other tissue. The waiting time is necessary to maximize brain uptake because the first-pass extraction of FDG is low. During this time ^{18}F-FDG activity will accumulate in the urine, and the patient should void just before entering the PET scanner to reduce radiation dose to the bladder and surrounding organs.

The use of ^{18}F-FDG for measuring regional cerebral glucose utilization (rCMRglc) was validated by Phelps et al.[73] This investigation demonstrated that total ^{18}F activity, as FDG and FDG-6-phosphate, in brain tissue increases slowly after intravenous injection of 5 to 10 mCi (185–370 MBq). Accumulation in gray and white matter reaches a plateau at about 90 minutes. The average concentration of ^{18}F in gray matter was 2.1 times that in white matter. Tissue concentration of total ^{18}F decreases after 120 minutes. The average tissue clearance half-time of FDG from brain was reported to be 9.1 hours. Within 1 hour after injection, the blood activity of ^{18}F-FDG falls to about 15% of its initial value.[50,73] For measurement of rCMRglc, the optimal time for data acquisition is between 60 and 120 minutes after injection, because during this time the rCMRglc remains constant.[74]

^{18}F-FDG has fairly wide clinical application as a metabolic marker and has been used to assess brain disorders such as dementias, Parkinson's disease, and epilepsy. It has also been shown to be effective in the diagnosis and staging of several types of cancer, including brain, lung, colorectal, breast, and prostate cancer. Because tumor growth requires glucose utilization, ^{18}F-FDG can readily identify primary and metastatic lesions. ^{18}F-FDG PET can determine if a previously resected or radiation-treated brain tumor has become viable.[75–77] A patient's deterioration in the months after treatment may be due to tumor regrowth or to radiation injury, and it is usually not possible to distinguish the difference by radiographic (CT) or clinical findings.[77] The differentiation is important because resection of necrotic tissue may halt deterioration and should be done at an early stage, whereas tumor recurrence requires early institution of chemotherapy. ^{18}F-FDG PET therefore provides functional information that improves diagnosis and facilitates patient care. At the time of diagnosis, ^{18}F-FDG PET provides information concerning the degree of malignancy and patient prognosis. After therapy, ^{18}F-FDG PET is able to assess persistence of tumor, determine degree of malignancy, monitor progression, differentiate recurrence from necrosis, and assess prognosis.[78]

The usual adult dosage of ^{18}F-FDG is 6 to 15 mCi (222–555 MBq). Brain imaging typically begins 45 to 60 minutes after ^{18}F-FDG injection. The principal route of elimination is urinary, with 20% of the dose excreted 2 hours after injection.[79] The critical organ is the urinary bladder wall, with a radiation absorbed dose of 7 rad(cGy)/10 mCi based on a 4.8 hour bladder voiding interval. The effective dose equivalent is 1.1 rem(cSv)/10 mCi.[80]

Thallous Chloride Tl 201 Injection

Thallous chloride Tl 201 injection (^{201}Tl-chloride) has been shown to be effective in localizing brain tumors with SPECT imaging. Its mechanism is believed to involve thallium's uptake by the Na-K ATPase pump in the membrane of viable tumor cells. Generally, imaging of brain tumors is begun within 5 minutes of administering 2 to 4 mCi (74–148 MBq) of ^{201}Tl-chloride by intravenous injection. In a comparative study of ^{201}Tl-chloride, ^{99m}Tc-gluceptate, and ^{67}Ga-citrate with pathologic correlation in patients with gliomas, ^{201}Tl-chloride identified viable tumor more accurately than the other agents and was minimally affected by concomitant corticosteroid therapy.[81] Thallium was also reported to be taken up in viable malignant tumors but not in areas of radiation necrosis.[82] Kim et al.[83] compared a region of interest (ROI) around a tumor with a contralateral ROI in normal brain. They found that a ratio of 3.5:1 of maximum counts per pixel in the tumor ROI to counts in the normal ROI was statistically significant for identifying tumor presence.

Neuronal Agents

Neurotransmission

Neuronal systems in the brain are complex and involve signal transmission between presynaptic and postsynaptic nerve terminals. A nerve cell is composed of a soma (cell body), an axon, and dendrites. The synaptic cleft between neurons is where interneuronal communication occurs via neurotransmitter–

receptor interaction. Neurotransmitters can be excitatory or inhibitory. Receptors for neurotransmitters in the synapse for signal transduction are located on the postsynaptic nerve fiber. Other key elements of neurons are vesicles that store neurotransmitter and membrane transporters for reuptake of neurotransmitter from the synaptic cleft, both located in the presynaptic nerve terminal (Figure 19-13). For example, specific transporters for reuptake of dopamine, serotonin, or norepinephrine from the synapse are located on the presynaptic membrane, whereas uptake of these monoamines from the presynaptic neuron cytoplasm into vesicles occurs by a common adenosine triphosphate (ATP)-dependent transporter, the type 2 vesicular monoamine transporter (VMAT2).

A sensory impulse to receptors on the soma or dendrites of a nerve cell initiates an action potential that travels down the axon of the nerve, causing depolarization of the membrane on the presynaptic axon terminal. This initiates an influx of calcium ions into the axon cytoplasm, causing a series of events that result in the release of neurotransmitter from storage vesicles into the synaptic cleft.[84] Once in the cleft the neurotransmitter interacts with receptors located on the postsynaptic nerve fiber, transmitting the signal to the nerve. Unused neurotransmitter is either metabolized or transported back into the presynaptic nerve fiber where it is stored in vesicles for future need and is protected from catabolism by monoamine oxidase (MAO) in the cytosol.

The principal neurotransmitters in the brain are (1) acetylcholine; (2) the monoamines norepinephrine, dopamine, serotonin (5-hydroxytryptamine or 5-HT), melatonin, and histamine; (3) the amino acids glutamate, gamma amino butyric acid (GABA), aspartate, and glycine; and (4) the purines ATP and guanosine triphosphate (GTP). The principal areas of the brain that have been investigated with neuroimaging agents are the dopamine system, the serotonin system, the cholinergic system, the γ-aminobutyric acid (GABA) system, and opioid receptors. Various aspects of neuronal function in these systems have been explored with radiotracers, including dopamine synthesis, receptor density, and transporter activity at the presynaptic and vesicular membranes.[85–87] A multitude of PET and SPECT radiotracers for imaging the brain have been investigated.[88,89]

Dopamine System Imaging Agents

The dopamine system has undergone extensive investigation with radiotracers. Major midbrain structures associated with this system—the substantia nigra, the striatum (caudate nucleus and putamen), nucleus accumbens, and subthalamic nucleus—reside in the basal ganglia of each hemisphere. Dopaminergic neurons play a principal role in reward, addiction, and movement.

A major disease associated with the dopamine system is Parkinson's disease (PD). A key pathophysiologic feature of PD is the loss of dopamine-containing nerve cells in the substantia nigra, most prominently in ventrolateral nigral projections to the putamen.[90] A wide variety of radiotracers have been investigated that measure different aspects of dopaminergic function, in particular, presynaptic dopamine synthesis, the dopamine transporter (DAT), and VMAT2 (Table 19-2).

^{18}F-6-fluorodopa (^{18}F-DOPA) is a fluorinated analogue of 3,4-dihydroxyphenylalanine (DOPA). It is a marker of presynaptic nerve terminal function because it traces dopamine synthesis. ^{18}F-DOPA readily crosses the BBB by the neutral amino acid transporter into the brain ECF. It is then actively transported into dopaminergic neurons similar to DOPA, where it is decarboxylated to ^{18}F-fluorodopamine by the aromatic amino acid decarboxylase (AADC) enzyme. ^{18}F-DOPA is therefore useful in assessing a deficiency in dopamine synthesis and storage in presynaptic nerve terminals. It has been investigated in PET imaging studies in the diagnostic work-up of PD to evaluate the effects of PD therapy with dopamine agonists, to monitor brain implants of human fetal mesencephalic cells to help restore dopaminergic function, and to help differentiate PD from other neurodegenerative diseases that involve the dopaminergic system.[90,91] PD typically begins with unilateral symptoms progressing to bilateral involvement. Studies have shown that at the time of symptom onset, there is a 30% reduction of ^{18}F-DOPA uptake in the putamen contralateral to the affected limbs. ^{18}F-DOPA uptake declines as PD progresses.

Imaging agents that target the DAT bind to a transport protein located on the presynaptic membrane, providing a measure of transporter density as an indirect measure of nerve terminal integrity.[87] Radiotracers developed for PET and SPECT imaging that target the DAT on presynaptic neurons include the tropane-based tracers ^{123}I-βCIT, ^{123}I-FP-CIT (^{123}I-ioflupane), ^{123}I-altropane, and ^{99m}Tc-TRODAT-1 (Table 19-2).

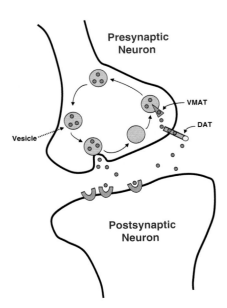

FIGURE 19-13 Presynaptic neuron with neurotransmitter (dopamine) in storage vesicles. Signal transmission releases neurotransmitter into the synaptic cleft, where it activates receptor sites on the postsynaptic neuron. Excess neurotransmitter is returned to the presynaptic neuronal cytoplasm via the dopamine transporter (DAT) and taken back up into storage vesicles via the vesicular monoamine transporter (VMAT).

TABLE 19-2 Radiopharmaceuticals for Evaluation of Presynaptic Neuronal Function in Parkinson's Disease

Radiotracer	Site of Action
^{18}F-FD, or ^{18}F-6-fluorodopa	Dopamine synthesis
^{123}I-βCIT, or ^{123}I-2β-carbomethyl-3β-(4-iodophenyl)tropane	Dopamine transporter
^{123}I-FP-CIT, or ^{123}I-ioflupane, or ^{123}I-N-ϖ-fluoropropyl-2β-carbomethyl-3β-(4-iodophenyl)nortropane	Dopamine transporter
^{11}C-DTBZ, or ^{11}C-dihydrotetrabenazine	Vesicular monoamine transporter 2
^{18}F-FP-DTBZ, or 9-[^{18}F]-fluoropropyl-9-O-desmethyldihydrotetrabenazine	
^{18}F-AV-133, or ^{18}F-(+)fluoropropyldihydrotetrabenazine	
^{99m}Tc-TRODAT-1, or ^{99m}Tc-[2-[[2-[[[3-(4-chlorophenyl)-8-methyl-8-azabicyclo[3.2.1]oct-2-yl]methyl](2-mercaptoetyl)amino]ethyl]amino] ethanethiolato(3-)-N2,N2'S2,S2']oxo-[1R-(exo-exo)]	Dopamine transporter
^{123}I-altropane, or ^{123}I-2β-carbomethoxy-3β-(4-fluorophenyl)-N-(1-iodoprop-1-en-3-yl)nortropane	Dopamine transporter

Source: References 85–91 and Okamura N, Villemagne VL, Drago J, et al. In vivo measurement of vesicular monoamine transporter type 2 density in Parkinson disease with ^{18}F-AV-133. J Nucl Med. 2010;51:223–8.

Another useful target for radiotracer studies in PD is the vesicular monoamine transporter (VMAT) located on the amine storage vesicles within the presynaptic nerve terminal. A type 2 transporter (VMAT2) has high concentration in dopaminergic neurons of the striatum. Radiotracers developed to target VMAT2 are ^{11}C-DTBZ, ^{18}F-FP-DTBZ, and ^{18}F-AV-133 (Table 19-2). A similar transporter, VMAT1, is found in the adrenal medulla.

A radiotracer that targets postsynaptic dopamine receptors is ^{11}C-raclopride. There are five dopamine receptors (D_1–D_5), but only D_1 and D_2 have been imaged in humans. D_1 is an excitatory receptor with much higher density in the brain than the D_2 inhibitory receptor. ^{11}C-raclopride is a postsynaptic receptor antagonist with high affinity for the D_2 receptor. It has been used to evaluate the function of the dopaminergic system because it can be displaced by dopamine. ^{11}C-raclopride is discussed in more detail in Chapter 12.

Despite the variety of radiotracers investigated with PD, no clear consensus recommends one agent over another for effectiveness in diagnosis and prognosis. A report that reviewed study outcomes conducted with four radiotracers—^{18}F-FDG, ^{18}F-fluorodopa (^{18}F-FD), ^{123}I-βCIT, and ^{11}C-DHTZ—in PD concluded that existing data do not support their use in determining diagnosis or prognosis of PD.[91] Although ^{18}F-FDG does not directly assess specific dopaminergic function, it does measure altered patterns of cerebral metabolism associated with PD. One confounding issue with radiotracer studies is that they may underestimate the degree of neuronal loss due to compensatory changes that can occur as PD progresses, such as upregulation of AADC and downregulation of DAT.[92] Although these agents have been able to distinguish normal individuals from groups of patients with known PD, there have been few studies with cases of diagnostic uncertainty in which imaging was able to predict a subsequent clinical diagnosis.[91]

The development of radioligands to target specific receptors must meet certain requirements. Radioligands must have stereochemical compatibility with the receptor to achieve selectivity (high ligand–receptor interaction), have high binding affinity, and have rapid permeation through the BBB to reach the receptor. In addition, any metabolites formed should be rapidly removed from the receptor region in the brain. Because of stereochemical requirements, most radioligands are labeled with ^{11}C and ^{18}F for PET imaging. SPECT agents typically are labeled with radioiodine, but some technetium-labeled agents are being developed.

Pathology in the brain can affect nerve cell function, and radiotracers are used to assess altered function. The distribution, density, and activity of receptors located on presynaptic nerve terminals and storage vesicles in vivo can be visualized and quantified by radioligands. PET is a more sensitive method for quantitation of receptor density compared with SPECT, which has lower detection efficiency. However, the longer-lived radiolabels of SPECT agents may have an advantage for procedures that require delayed imaging to reduce interference from nonspecific binding.

Dementia Imaging Agents

Among the different types of dementias, Alzheimer's disease (AD) has received much focus, particularly because of the cognitive detriment it causes in an aging population. Consequently, a great amount of effort has been placed on developing radiotracers to detect AD in the brain, with the desired outcome being early diagnosis so that treatment can be instituted to arrest or at least mitigate disease progression. Radiotracers that can specifically target tissue changes associated with AD in the brain would also be beneficial for assessing the effectiveness of drugs being developed to treat the disease and to differentiate AD from other causes of dementia.

The pathophysiologic characteristics in the brain most prominently associated with AD are the aggregation of beta amyloid peptide (Aβ) into plaques and the hyperphosphorylation of the tau protein into neurofibrillary tangles, both of which increase with the progression of AD. Early work with the amyloid-binding histological dye thioflavin-T led to the development of N-methyl-[^{11}C]2-(4'-methylaminophenyl)-6-hydroxybenzothiazole, otherwise known as Pittsburgh compound B or PIB, a radiotracer that targets Aβ.[93] Initial studies in human subjects with PIB demonstrated marked retention of PIB in areas of the cerebral cortex known to contain large amounts of amyloid deposits in AD patients and low retention in normal subjects. PIB has undergone extensive clinical investigation. One of its limitations is the short half-life (20 minutes) of the ^{11}C label. More recently, an ^{18}F-labeled compound, (E)-4-(2-(6-(2-(2-(2-^{18}F-fluoroethoxy)ethoxy)ethoxy)pyridin-3-yl)vinyl)-N-methyl benzenamine (^{18}F-AV-45 or florbetapir F 18), has been developed and is currently in clinical trials.[94] These recent studies in humans demonstrated that florbetapir F 18 could readily discriminate between AD patients and control subjects. Florbetapir F 18 is undergoing multicenter brain imaging trials.

CSF Agents

Agent Development

The technique of radioisotope cisternography was first described by DiChiro et al.,[95] who injected ^{131}I-radioiodinated serum albumin (RISA, or ^{131}I-HSA) into the lumbar subarachnoid space to evaluate CSF dynamics. ^{131}I-HSA was not ideal because the ^{131}I label restricted the administered activity to 100 μCi (3.7 MBq), and there were isolated reports of aseptic meningitis that was ascribed to the amount of albumin administered.[96–98]

The desire for a higher photon yield and lower radiation dose led to the development of other radiopharmaceuticals. ^{99m}Tc-albumin was introduced for CSF rhinorrhea studies in 1968, which allowed administration of 2 mCi (74 MBq) doses.[99] However, the use of ^{99m}Tc in cisternography was limited to short-term studies, such as for CSF leaks, because of its 6 hour half-life. Routine cisternography for hydrocephalus evaluation requires imaging periods ranging from 6 to 72 hours, which necessitates a radiotracer with a longer half-life.

In 1970, ^{169}Yb-DTPA was introduced for cisternography.[100] Its advantages were a highly stable complex and a 32 day half-life that permitted strict quality control before human use. In addition, it had a biologic half-life in the CSF compartment of about 10 hours, which was long enough for the study but short enough to keep the radiation dose low, except in patients with reduced renal clearance.

In 1971, ^{111}In was introduced for CSF studies as a transferrin complex and as a colloid.[101] The physical properties of ^{111}In (2.8 day half-life, no beta emission, and two photons) were well suited for cisternography. The colloid preparation was unsatisfactory because it collected in the basal cisterns. The transferrin complex was inconvenient because it required in-house labeling of the patient's own serum. Also, when compared with DTPA or EDTA complexes, the transferrin complex had a slower rate of clearance from the blood and the CSF space, which was attributed to its high molecular weight.[102] ^{111}In-DTPA had essentially the same biologic properties as ^{169}Yb-DTPA but significantly lower radiation absorbed dose to the spinal cord. Its half-life permitted commercial production and availability with a reasonable shelf life. All of these properties made ^{111}In-DTPA the agent of choice for cisternography, and FDA approved it for routine use in 1982.

For short-term studies such as the localization of CSF leaks (rhinorrhea and otorrhea), ^{99m}Tc-DTPA has been used, although this is not an approved indication in the product labeling.[103,104] ^{99m}Tc-pertechnetate cannot be used because, as noted earlier, it is actively transported from the CSF into blood.

Indium In 111 Pentetate Injection

The production and physical properties of indium In 111 pentetate injection (^{111}In-DTPA) have been well described and are discussed in Chapter 11.[105] After injection into the lumbar subarachnoid space, ^{111}In-DTPA moves slowly, with the natural flow of spinal fluid, away from the injection site toward the head (Figure 19-14). Leakage at the injection site can be minimized by administering the radiopharmaceutical in two to three times its volume of sterile 10% dextrose injection.[106] This also improves its rate of transport cephalad. In normal human subjects, the tracer migrates first to the basal cisterns. Activity appears there in about 1 hour, achieving peak concentration at 4 hours.[107] Tracer then flows over the cerebral convexities to the parasagittal region. Activity first appears in this region at 4 hours, reaching peak levels at about 14 to 17 hours. The activity in this region then falls, decreasing to half the peak values 10 to 14 hours later, having been absorbed into the blood through the arachnoid granulations. Activity does not normally enter the ventricular system; however, in certain types of hydrocephalus, activity may reflux into the ventricles.

Upon absorption into the blood, ^{111}In-DTPA follows a normal urinary route of excretion through glomerular filtra-

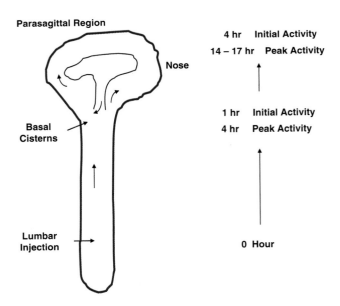

FIGURE 19-14 Right lateral view diagram of the brain and spinal canal illustrating the temporal movement of radiotracer from the lumbar injection site to the parasagittal region of the brain.

tion. About 65% of the ^{111}In-DTPA chelate is eliminated in the urine in 1 day, increasing to 85% in 3 days.[102] The systemic distribution of ^{111}In-DTPA is described in Chapter 24.

The usual adult intrathecal dosage of ^{111}In-DTPA is 500 µCi (18.5 MBq; maximum dosage). The critical organ is the spinal cord surface, with a radiation absorbed dose of 5 rad(cGy)/500 µCi.

Mechanisms of Drug Transport in the CSF

To remain in the CSF, a radiopharmaceutical must have certain properties, as outlined by Bell et al.[108] (Table 19-3). First, it must not be lipid soluble, or it will diffuse through the pia mater into the underlying nervous tissue. Also, CSF enzymes must not metabolize it. This may pose a problem for protein tracers, but the currently used DTPA chelates are not metabolized. Removal of substances from the CSF space occurs primarily through the arachnoid membrane in the sagittal sinus, which, because of its porosity, permits the egress of substances with a wide range of molecular weights. Substances can also diffuse through the pia mater and ependymal cells into the brain ECF. This diffusion is favored for lipophilic molecules, but it has also been shown to occur with water-soluble molecules, such as inulin and radiographic contrast material (metrizamide).[3,109,110]

Transport of tracer molecules in the spinal fluid occurs by bulk flow or diffusion. Smaller molecules favor diffusion; proteins favor bulk flow. Theoretically, a tracer that mobilizes by bulk transport in the CSF would be a better marker of CSF flow. This would favor the use of radiolabeled proteins, but there are inherent disadvantages to their use. Albumin has been associated with aseptic meningitis and its ^{131}I label confers a high radiation dose, and labeled transferrin has to be prepared extemporaneously from patient serum. These disadvantages have favored the use of radiolabeled chelates.

^{111}In-DTPA appears to mobilize by bulk flow and diffusion.[108,111] Egress of this chelate through the arachnoid granulations is facilitated by the porosity and by the difference in pressure between the dural venous blood (about 90 mm H$_2$O) and the mean CSF pressure (about 135 mm H$_2$O).[111] Agents such as ^{111}In-DTPA may also undergo transependymal diffusion, because nearly normal clearance has been shown to occur with ^{169}Yb-DTPA during spinal canal obstruction.[111] High molecular weight molecules, such as RISA, have been shown to clear the CSF space much more slowly than chelates, leading to potentially excessive radiation dose to the spinal cord.[112] Some agents are cleared rapidly from the CSF space by active transport through the choroid plexus epithelium. These agents include iodide, bromide, thiocyanate, phenol red, and phenolsulphonthalein.[113] Notably, ^{99m}Tc-pertechnetate clears by this mechanism, but its rate of clearance can be significantly reduced by oral perchlorate.[26]

Safety Considerations for Intrathecal Injections

Injection of foreign material into the spinal fluid, where it may come into intimate contact with the spinal nerves and the brain, deserves special attention because of potential adverse reactions. The integrity of nerve function is closely related to proper control of fluid pH, electrolyte balance, and osmolarity.[114] In addition, drug substances may have a direct effect on nerve function.[115,116] In particular, depletion of calcium ion readily causes tetany,[115] and low pH causes dilation of pial blood vessels.[115] Several cases of aseptic meningitis have been reported that were related to the chemical amounts of albumin administered during radioisotope cisternography.[96–98] Reactions believed to be associated with pyrogenic contamination in supposedly pyrogen-free injections have also been reported.[117] According to this report, anion exchange resins and buffers used in the manufacturing of ^{131}I-HSA and ^{111}In-DTPA were contaminated with pyrogens at levels below those detectable by the traditional rabbit pyrogen test. It was surmised that the reactions might have been due to the amount of endotoxin present in the radiotracer. Endotoxin is much more potent in producing a febrile response when administered by the intrathecal route than by the intravenous route. Limulus (bacterial endotoxin) testing is 5 to 10 times more sensitive than the rabbit test in detecting endotoxin.

All radiopharmaceuticals must now meet the USP bacterial endotoxins test by not exceeding a specified number of endotoxin units (EU) per milliliter in the administered dose. For radiopharmaceuticals intended to be administered intravenously, the limit is 175 EU/V, where V is the maximum dose in milliliters, at the expiration time. For intrathecal radiopharmaceuticals, such as ^{111}In-DTPA, the limit is lower (14 EU/V) because of the enhanced sensitivity of spinal nerve

TABLE 19-3 Properties of Radiopharmaceuticals for Cisternography, Past and Present

Agent	Decay Mode	Physical Half-life	Effective Half-life[a]	Photon Energy keV	%	Administered Activity (mCi)	rad/Administered Activity[b] (spinal cord)	Study Beyond 24 hr
^{131}I-HSA	β−	8 days	26 hr	364	83	0.1	7.1	Yes
^{169}Yb-DTPA	EC	32 days	12 hr	177	22	0.5	8.0	Yes
				198	35			
^{111}In-DTPA	EC	2.8 days	10 hr	171	91	0.5	1.9	Yes
				245	94			
^{99m}Tc-DTPA	IT	6 hr	5 hr	140	88	2.0	5.0	No

[a]Data from reference 2.
[b]Data from reference 108.

roots to endotoxin. In some instances, ^{99m}Tc-DTPA has been employed in off-label use for a CSF procedure, such as CSF leak studies or the evaluation of ventriculoperitoneal shunt patency. Workers should be aware that kits with DTPA as the free acid or trisodium salt have the potential to chelate Ca^{2+} and Mg^{2++} in the spinal fluid and may cause an adverse reaction in the patient. Therefore, if ^{99m}Tc-DTPA is used in a CSF procedure, a kit with the $CaNa_3DTPA$ is preferred. Since ^{99m}Tc-DTPA kits are permitted a higher endotoxin limit compared with ^{111}In-DTPA, it is advisable to prepare a DTPA kit with a high specific activity to limit the amount of DTPA and endotoxin administered. Endotoxin testing of the end product could also be performed before intrathecal administration so that the acceptable number of EUs per administered dose is not exceeded. ^{99m}Tc-DTPA has been used effectively and safely for intrathecal procedures when prepared and administered appropriately.[104,118]

Certain cisternography procedures require the injection of substantial amounts of fluid into the spinal canal.[104] For such procedures it is recommended that the fluid used have a composition similar to Elliott's B Solution (artificial CSF [Orphan Medical]), to avoid possible complications arising from changes in spinal fluid pH, electrolytes, and osmolarity. Table 19-4 gives the composition of artificial CSF.

NUCLEAR MEDICINE PROCEDURES

Nuclear medicine imaging of the CNS can be divided into two categories: imaging of the brain and imaging of the CSF. Although there continues to be an important role in nuclear medicine for brain imaging, most CNS imaging is now done by MRI or CT. These modalities offer good anatomic information. However, nuclear techniques are valuable when there is a question of abnormal regional blood flow in the brain or abnormal flow of CSF.

Normally, the BBB restricts many substances in the blood from entering the brain. Four main categories of brain imaging agents are used on the basis of this principle. One category, nondiffusible tracers, cannot cross the BBB. These radiopharmaceuticals can be used to evaluate blood flow to the brain and to determine if there is a focal abnormality or breakdown in the BBB. The second category, diffusible radiotracers, are more commonly used for brain imaging. They are typically lipophilic and readily cross the BBB to localize in brain tissue in proportion to blood flow. The third category is radiopharmaceuticals for evaluating metabolic activity in the brain, principally with PET imaging. A fourth category is radiotracers for assessing neurologic disease; it includes agents for measuring receptor density and binding affinity[119] and agents to diagnose dementia, such as Alzheimer's disease.[93,94]

BRAIN IMAGING

Brain Death

Rationale

Various events, such as head trauma, anoxia, and cerebrovascular accidents, can cause fluid accumulation in the confined space of the calvaria. The resulting increased intracranial pressure causes cessation of cerebral blood flow, leading to brain death, which is defined as irreversible cessation of all brain

TABLE 19-4 Artificial CSF (Elliott's B Solution)[a]

Formula		
1. NaCl	5.608 g	Dissolve ingredients 1 to 5 in 970 mL SWFI. Add dropwise (with glass pipette) the acid salt solution (~ 2 mL) with stirring to pH 7.4 and qs to 1000 mL. Filter through a 0.22 μm sterile membrane into vials or syringes for immediate use.
2. $Na_2CO_3 \cdot H_2O$	2.557 g	
3. KCl	0.285 g	
4. Na_2HPO_4	0.076 g	
5. Glucose	0.758 g	
6. Acid salt solution qs[a] to pH 7.4		
7. Sterile water for injection (SWFI) qs	1000 mL	
Acid salt solution		
$CaCl_2$	2.0 g	Dissolve salts in 20 mL of HCl with heating. Cool and qs to 25 mL.
$MgCl_2 \cdot 6 H_2O$	1.0 g	
HCl 12 M qs	25.0 mL	
Content of final preparation (mg/100 mL)		
Na^+	318.0	
Cl^-	450.0	
K^+	14.9	
Ca^{++}	5.5	
Mg^{++}	0.9	
P	1.7	
HCO_3^-	126.0	
Glucose	76.0	

[a] qs = Add a sufficient quantity.
Source: Reference 114.

and brain stem function.[120] Brain death is assessed clinically by standard neurologic tests, and a lack of blood flow to the brain can confirm the clinical diagnosis. A perfusion imaging study of the head is done to determine the presence or absence of cerebral blood flow in a patient who is suspected to be brain dead.

Procedure

The study typically consists of a radionuclide cerebral angiogram, which assesses blood flow in the vessels of the brain, and static images, which assess parenchymal uptake and distribution of tracer activity within the brain. Nondiffusible tracers, such as 15 to 20 mCi (555–740 MBq) of ^{99m}Tc-DTPA, or diffusible tracers, such as 10 to 30 mCi (370–1100 MBq) of ^{99m}Tc-HMPAO or ^{99m}Tc-ECD, can be used.[121] A bolus of activity is usually administered intravenously into an antecubital vein. Injection technique with a high-quality bolus injection is more important for nondiffusible tracers than for diffusible tracers, which rely more on assessment of parenchymal uptake on static images for the interpretation of brain death. During the angiogram portion of the study, a series of 1 to 3 second per frame anterior images of the head is obtained. Image acquisition is started just as the bolus of radiotracer is administered and is continued for 1 to 2 minutes. Static anterior and lateral blood pool images are generally obtained 20 minutes or more after completion of the blood flow portion of the study. It can be helpful to place an elastic band over the patient's head just above the orbits to minimize blood flow to the scalp vessels.

Interpretation

Normally, when the radiotracer bolus is injected into a peripheral arm vein, it travels to the right side of the heart, then to the lungs and back to the left side of the heart. It is visible in the carotid arteries shortly after this. As it enters the brain, radiotracer becomes apparent in the cerebral arteries and in the sagittal sinus (Figure 19-15). If a patient is suspected to be brain dead by clinical evaluation and there is no evidence of cerebral perfusion on the radionuclide angiogram, the diagnosis is certain. Interpretation of brain death with nondiffusible tracers relies on the dynamic flow study, because the tracer does not normally cross the BBB and no radiotracer uptake will be evident in the brain on the delayed images (Figure 19-16). Diffusible tracers, however, can assess both dynamic flow and

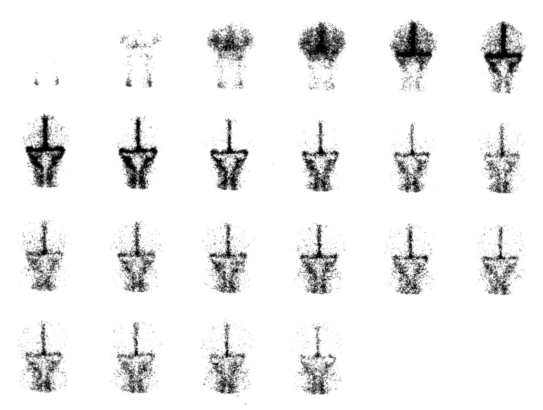

FIGURE 19-15 Normal posterior radionuclide cerebral angiogram (flow study). Images shown are made at 2 second intervals after intravenous injection of 20 mCi (740 MBq) of ^{99m}Tc-gluceptate (^{99m}Tc-GH). The arterial phase (first 2 to 6 seconds) is characterized by visualization of the two internal carotid, the two middle cerebral, and the paramedial posterior cerebral arteries. This is followed by a capillary "blush" phase leading quickly to the venous phase, which is recognized by the appearance of activity in the superior sagittal sinus (midline). Subsequent images show venous drainage through the lateral (transverse) venous sinuses and the jugular veins, which return blood to the heart.

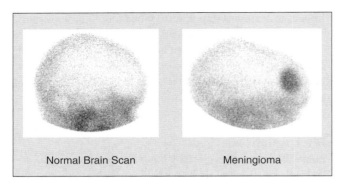

FIGURE 19-16 Brain scan (right lateral view) illustrating typical distribution pattern of a non-diffusible tracer, 99mTc-sodium pertechnetate. Normal brain, with intact BBB, is devoid of activity. Tumors, such as meningiomas, typically demonstrate increased uptake of activity.

parenchymal uptake in the brain because these agents diffuse across the BBB. Thus, activity will accumulate in the brain over time if brain perfusion is preserved (Figures 19-8 and 19-10).

When there is brain death, internal carotid artery blood flow ceases because of increased intracranial pressure or clotting. During the angiographic phase of the study, there is flow to both carotids, but the flow stops at this level (Figure 19-17). There is no blush of radiotracer activity in the cerebral artery territories or pooling of radiotracer in the sagittal sinus. After the angiogram phase, the blood pool images in brain death fail to demonstrate radiotracer activity in the sagittal or transverse sinuses. Since blood flow is blocked in the internal carotid arteries, blood is shunted to the external carotid arteries, where it may cause increased accumulation of activity in the nasopharynx. This is often called the "hot nose" sign (Figure 19-18).[122] On occasion, a patient with a clinical determination of brain death will exhibit some blood flow to the brain, and a diagnosis of brain death needs to carefully consider potential confounding factors in the clinical diagnosis and image interpretation (Figure 19-19).[120,123]

Epilepsy

Epilepsy is a disorder of the brain characterized by recurring excessive neuronal discharge resulting in repeated episodes of seizures. Seizures occur when there is an abnormal focus of neuronal discharge. Epilepsy affects approximately 0.5% to 1% of the population. For patients with refractory partial seizures that cannot be adequately controlled with medication, surgery is an important treatment option if the seizure focus can be located. Nuclear medicine imaging plays an important role in identifying the epileptogenic focus that requires resection.[124]

Rationale

During a seizure, there is an increase in blood flow in the region of the neuronal discharge associated with the seizure focus.[125] In the period between seizures, there is normal or decreased blood flow to the region of the seizure focus. Thus, radiopharmaceuticals that cross the BBB and are taken up in the cortex in proportion to blood flow are useful in identifying the seizure focus. SPECT imaging with diffusible perfusion tracers, such as 99mTc-HMPAO and 99mTc-ECD, is most commonly used to evaluate brain perfusion for epilepsy. SPECT imaging of the brain can be done during a seizure

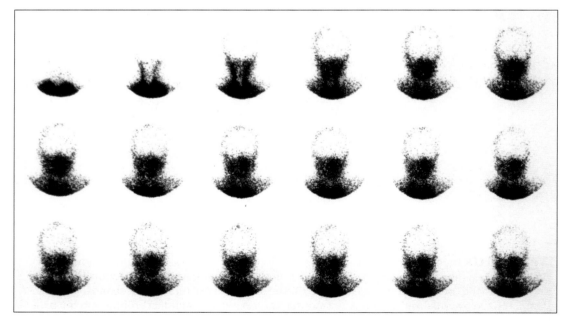

FIGURE 19-17 Brain death. Absence of cerebral perfusion after intravenous injection of 99mTc-gluceptate (99mTc-GH) is seen in this 11 month old child, a victim of smoke inhalation in a house fire.

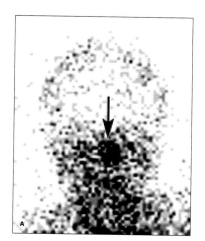
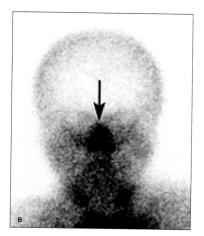

FIGURE 19-18 Anterior brain images obtained in a 40 year old woman with multiple traumatic injuries after a motor vehicle collision. Images were obtained following intravenous administration of 25 mCi (925 MBq) of ^{99m}Tc-HMPAO. (A) Early arterial phase image and (B) delayed phase image both demonstrate no flow to the cerebral hemispheres and abnormal increased nasopharyngeal activity (arrow). (Reprinted with permission of the Radiological Society of North America from reference 122.)

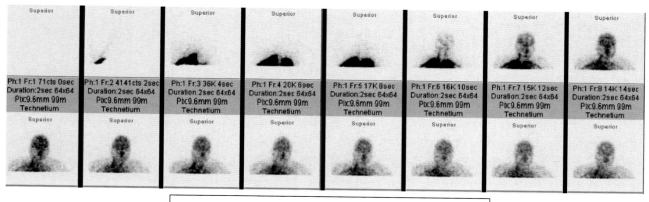
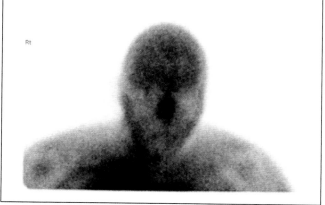

FIGURE 19-19 Anterior brain images obtained in a 22 year old man with a shotgun wound to the neck. Images were obtained following intravenous administration of 22 mCi (888 MBq) of ^{99m}Tc-HMPAO. Upper images show a normal brain flow study with evidence of flow to the carotid, anterior, and middle cerebral arteries. Static image shows accumulation of radiotracer in the cerebrum with a "hot nose" sign due to shunting of blood flow to the nasopharynx via the external carotid arteries. Although there is evidence of brain flow in this study, the patient was declared brain dead from all other clinical signs.

(ictal SPECT), between seizures (interictal SPECT), or after a seizure has occurred (postictal SPECT).[126]

Since blood flow can be normal between seizures, ictal SPECT is more sensitive for detecting the seizure focus. Because perfusion and metabolism are normally coupled in the brain, PET imaging with ^{18}F-FDG can also be used to localize the seizure focus. During a seizure there is increased glucose metabolism at the seizure focus.[127] Between seizures, the foci generally demonstrate reduced glucose metabolism. Because ^{18}F has a physical half-life of only 110 minutes, it is difficult to have a dose ready and immediately available for an ictal study. Also, FDG uptake occurs by a slower process of facilitated diffusion and continues to accumulate in the brain over 30 to 40 minutes, which is typically much longer than the duration of a complex partial seizure. It does not have the high first-pass extraction and retention that the ^{99m}Tc diffusible agents have. Thus, metabolic brain imaging with ^{18}F-FDG is usually performed during the interictal period. However, ^{18}F-FDG can identify an ictal focus if it is administered while a seizure is occurring (see Figure 27-42 in Chapter 27). Often an ictal SPECT study done with either ^{99m}Tc-HMPAO or ^{99m}Tc-ECD is compared with an interictal ^{18}F-FDG PET study.

Procedure

Prior to either ictal or interictal imaging studies, patients are monitored by electroencephalography. If an ictal study is desired, either stabilized ^{99m}Tc-HMPAO or ^{99m}Tc-ECD is prepared for use and is kept readily available. The administered activity is usually 15 to 30 mCi (555–1110 MBq). The patient is monitored for seizure activity. At the onset of a seizure, the dose is administered intravenously. To obtain an adequate ictal study, it is important that the dose be administered either during the seizure or within 30 seconds after completion of the seizure. SPECT imaging of the brain is usually performed 30 to 60 minutes later. Because these radiopharmaceuticals do not redistribute, imaging can be delayed as much as 4 hours after administration.

A meta-analysis of ictal, interictal, and postictal SPECT brain imaging with ^{99m}Tc-HMPAO relative to standard diagnostic evaluation and postsurgical outcome in patients with refractory partial seizures found that the sensitivities of SPECT imaging for localizing the seizure focus were 0.44 (interictal), 0.75 (postictal), and 0.97 (ictal). It was concluded that institutions using SPECT imaging in epilepsy should perform ictal, preferably, or postictal imaging in combination with interictal imaging.[126] Interictal ^{18}F-FDG-PET imaging in patients with temporal lobe seizures indicated that hypometabolism on PET imaging was associated with marked improvement of seizure control after surgery in 94% of the patients.[128] Thus, ^{18}F-FDG appears to be more sensitive in identifying a seizure focus during the interictal state, whereas ^{99m}Tc-HMPAO is more sensitive during the ictal state.

Interpretation

If an ictal SPECT study is performed with ^{99m}Tc-HMPAO, seizure foci are seen as areas of increased activity because of the increased perfusion at the seizure focus. During interictal studies, seizure foci demonstrate either areas of decreased radiotracer uptake or normal uptake. If ^{18}F-FDG is used for an interictal study, the seizure focus may be seen as an area of decreased uptake related to hypometabolism at the seizure focus (Figure 19-20).

Dementia

Dementia is a general mental deterioration due to organic or physiologic factors that is categorized by some degree of disorientation along with impairment in judgment, intellect, and memory. In vascular dementias, such as multi-infarct

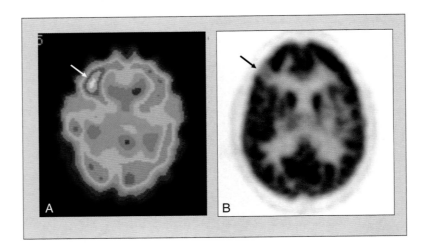

FIGURE 19-20 (A) ^{99m}Tc-HMPAO ictal brain SPECT scan showing a focal area of increased uptake in the right inferior frontal lobe in a patient with complex partial seizures. (B) ^{18}F-FDG interictal PET brain study in the same patient showing a focal area of hypometabolism in the inferior right frontal lobe that corresponds to the focus of increased uptake on the ictal scan.

dementia, cognitive decline is often abrupt and stepwise.[129] In Alzheimer's disease, mental deterioration is typically gradual. Although clinical evaluation along with anatomic imaging such as CT and MRI is important in screening for reversible causes of dementia, nuclear medicine techniques can often improve diagnostic accuracy.

Rationale

Several forms of dementia are treatable. For example, dementia may occur in normal-pressure hydrocephalus. Thus, it is important to try to determine the cause of new-onset dementia. Both PET and SPECT have been useful in helping to determine the cause of dementia by evaluating regional blood flow and metabolic abnormalities in the brain.

Interpretation

There is normally symmetric perfusion to the cerebral hemispheres in dementia. However, in Alzheimer's disease there is a classic pattern demonstrating decreased perfusion and metabolism of ^{18}F-FDG in the temporoparietal regions. As mental deterioration worsens, hypometabolism in the frontal lobes is also observed (Figure 19-21).[130,131] Multi-infarct dementia is caused by repeated infarcts in the brain. Brain imaging with either perfusion agents or metabolic agents typically demonstrates multiple asymmetric defects occurring in the brain. These defects can occur anywhere in the cortex.

Multicenter clinical trials with ^{11}C- and ^{18}F-labeled compounds that target beta amyloid plaque in the brain have shown promise in the detection of Alzheimer's disease.[93,94] Figures 19-22 and 19-23 demonstrate significant differences between AD patients and healthy control subjects in the uptake of florbetapir F 18.

PET Imaging for Tumor Recurrence

^{18}F-FDG is taken up into the brain tissues similarly to glucose. Normally, the gray matter, basal ganglia, and thalami show the greatest amount of uptake in the brain, with much less uptake in the white matter. The amount of uptake in the brain tissue is related to blood flow and metabolic activity.

Tumors often have increased metabolic activity compared with most tissues, which makes PET imaging very important in oncology. However, since the brain is highly metabolic, PET imaging is not always ideal for tumor imaging in the brain. An important exception to this is after a brain tumor has been surgically removed. Most often the patient has also had adjuvant radiation therapy to the tumor region. In this case, if the tumor has been removed, there should be only scar tissue in the surgical bed. Using CT or MRI, it is often difficult to determine whether the remaining tissue in the surgical bed is residual or recurrent tumor or scar tissue secondary to radiation necrosis. Scar tissue is not hypermetabolic like malignant tumor. In this scenario, if the brain is imaged with PET using ^{18}F-FDG, focal areas of hypermetabolism in the surgical bed suggest residual or recurrent tumor. Lack of significant radiotracer uptake in the surgical bed is more consistent with radiation necrosis (Figure 19-24).

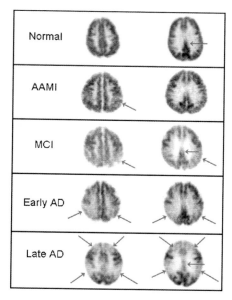

FIGURE 19-21 Changes in cortical metabolism typical for various degrees of impairment, from normal to late Alzheimer's disease (AD). ^{18}F-FDG-PET images of transaxial planes from five patients, shown at comparable axial levels. Top row: Normal pattern, provided for reference. Note how the posterior cingulate cortex (arrow) normally has activity that is visibly higher than in average cortex. Second row: Patient with age-associated memory impairment (AAMI) not meeting the criteria for mild cognitive impairment (MCI). Arrow denotes patchiness beginning to emerge in the inferior parietal cortex, and activity in the posterior cingulate cortex is also seen to be less robust than in a normal subject. Third row: MCI, with clear hypometabolism in the parietal, parietotemporal, and posterior cingulate cortex. Fourth row: Early AD, demonstrating posterior-predominant cortical hypometabolism. Fifth row: Late AD, with bilateral prefrontal, parietal, temporal, and posterior cingulate cortical regions markedly hypometabolic but with continued relative preservation of sensorimotor and visual cortex. At lower planes than shown here, metabolism in the basal ganglia, thalamus, cerebellum, and brainstem would also be seen to be relatively preserved at all stages. (Reprinted with permission of Elsevier from reference 131.)

Cerebrovascular Reserve

Cerebral ischemia can be caused by a variety of conditions, such as diabetes mellitus, carotid stenosis, transient ischemic attack (TIA), and arteriovenous malformations. Conditions that cause regional ischemia in the brain create an oxygen and nutrient deficiency. Consequently, significant amounts of carbon dioxide and hydrogen ion accumulate in the ischemic region, triggering an autoregulatory vasodilation to improve blood flow.[132] The ability of the brain to produce a maximal

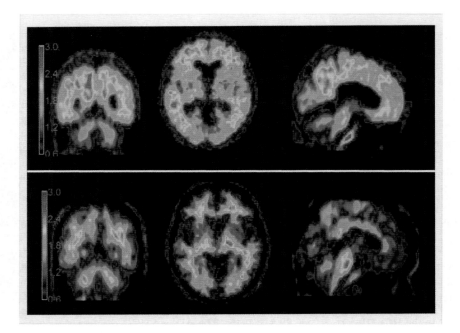

FIGURE 19-22 Average of two consecutive 5 minute PET brain images (obtained 50 to 60 minutes after injection of 10 mCi (370 MBq) of florbetapir F 18 in (top) 77 year old woman with mild Alzheimer's disease and Mini-Mental State Examination (MMSE) score of 24 and (bottom) 82 year old cognitively healthy man with MMSE of 30. Experimental conditions and imaging and computational parameters were identical for the two subjects. Counts are shown as a ratio to average of gray matter in the cerebellum for each subject (standard uptake value ratio). (Reprinted with permission of the Society of Nuclear Medicine from reference 94.)

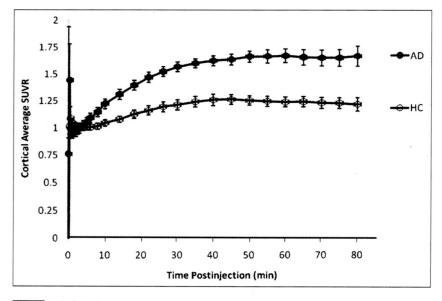

FIGURE 19-23 Mean cortical average standard uptake value ratio (SUVR) for subjects who were healthy cognitively (HC) and patients with Alzheimer's disease (AD). Cortical target-to-cerebellum SUVRs for both AD patients and HC subjects approached an asymptote at 50 minutes and remained essentially unchanged between 50 and 90 minutes after injection of florbetapir F 18. Subjects were scanned for approximately 90 minutes (horizontal axis represents beginning of each imaging time point). (Reprinted with permission of the Society of Nuclear Medicine from reference 94.)

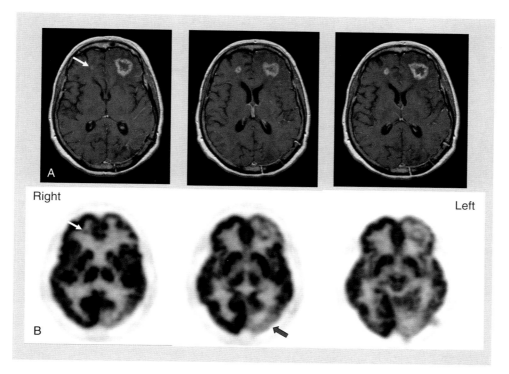

FIGURE 19-24 (A) MRI of the brain in a patient after surgery and radiation therapy for brain metastasis. There is some enhancement along the posterior aspect of the lesion in the left frontal lobe, which suggests recurrence. There is also evidence of a small metastasis in the right frontal lobe in the white matter (arrow). (B) ^{18}F-FDG PET brain study demonstrating two focal areas of hypermetabolism in the frontal lobes corresponding to the MRI findings and consistent with metastases. The focus on the left is adjacent to an area of decreased activity, likely a central area of necrosis. There is decreased metabolism in the left occipital region (red arrow) where the patient had surgery and radiation therapy, consistent with no evidence of recurrence in this region.

vasodilatory response via autoregulation is called the cerebrovascular reserve (CVR). If the autoregulatory response becomes exhausted and inadequate collateral circulation is present, patients may be unable to maintain adequate blood flow against any further decreases in perfusion pressure and may be at risk of a cerebrovascular accident. Regions with reduced CVR often exhibit normal blood flow under resting conditions but demonstrate reduced flow, compared with normally perfused areas, when challenged by a vasodilator.

CVR can be assessed in nuclear medicine by the acetazolamide stress test. Acetazolamide is a carbonic anhydrase inhibitor that is believed to induce hypercapnic vasodilation by elevating carbon dioxide levels in the blood. The stress study involves intravenous injection of 1 gram of acetazolamide over 2 minutes. After a 15 to 20 minute wait, 30 mCi (1110 MBq) of ^{99m}Tc-HMPAO is administered; imaging is begun in 15 minutes to assess regional blood flow in the brain. A separate baseline rest study with ^{99m}Tc-HMPAO is done for comparison with the stress study, usually on the day following the stress study. Regions of the brain with compromised CVR will be unable to produce a normal vasodilatory response with increased blood flow following the acetazolamide challenge, and uptake of radioactivity will be reduced compared with the baseline rest study (Figure 19-25).[133] The procedure provides objective evidence of reduced CVR in patients being considered for carotid endarterectomy to improve cerebral blood flow.[134] It has also been shown to be useful in identifying patients at risk for ischemic stroke during endarterectomy, who may need carotid shunting during the procedure.[135]

The acetazolamide stress test is indicated for evaluating CVR in TIA, completed stroke, or vascular anomalies (e.g., arteriovenous malformation) and for distinguishing vascular from neuronal causes of dementia.[136] The test is contraindicated in patients who are allergic to sulfa drugs or prone to migraine headaches and within 3 days of acute stroke.

Parkinson's Disease

PD is a neurodegenerative condition whose key pathophysiologic feature is the loss of dopamine-containing nerve cells.[90] Thus, an agent that traces dopamine in nerve cells can be useful for monitoring the progress of PD. Two agents have been shown to be valuable in this regard: PET imaging with ^{18}F-FD, which traces dopamine synthesis, and SPECT imaging with ^{123}I-βCIT, which targets the DAT located on the presynaptic membrane (Figure 19-26).[87]

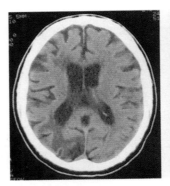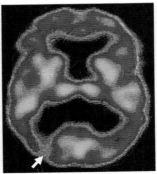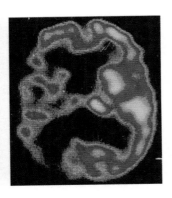

FIGURE 19-25 A patient with an old infarct (left) showed a corresponding small area of right parietal hypoperfusion (arrow) on routine ^{99m}Tc-HMPAO SPECT scan (center). The acetazolamide stress study (right) revealed a much larger area of reduced vascular reserve in the right middle cerebral artery territory, reflecting the area at risk for further vascular compromise that is not apparent on the baseline SPECT or CT study. (Reprinted with permission of Elsevier from reference 133.)

CSF IMAGING

Most of the CSF is formed by the choroid plexuses of the lateral, third, and fourth ventricles. The CSF produced in the lateral ventricles flows though the interventricular foramina into the third ventricle. The flow continues through the cerebral aqueduct into the fourth ventricle and from there into the subarachnoid space around the brain and spinal cord.[137] Most of the flow of CSF in the subarachnoid space is cephalad around the cerebral convexities toward the superior sagittal sinus. The main site of absorption of the CSF back into the venous system is through the arachnoid villi.

CSF imaging is often used to detect and evaluate normal-pressure hydrocephalus in a patient with clinical symptoms of dementia, gait disturbance, and urinary incontinence. In these patients, there is dilatation of the ventricles with normal CSF pressure. Imaging is also useful for detecting suspected CSF leaks and evaluating existing ventriculoperitoneal shunt function.

Normal-Pressure Hydrocephalus

Rationale

Hydrocephalus refers to enlargement of the ventricles caused by excessive accumulation of CSF. This can be due to overproduction of CSF by the choroid plexus, obstruction of flow to the arachnoid villi, or an abnormality in absorption. Enlargement of the ventricles can sometimes be a normal finding on MRI and CT imaging, secondary to age-related cerebral atrophy. The clinical symptoms associated with normal-pressure hydrocephalus can show improvement after placement of a shunt to divert CSF from the ventricles back to the venous system, such as a ventriculoperitoneal (VP) shunt.

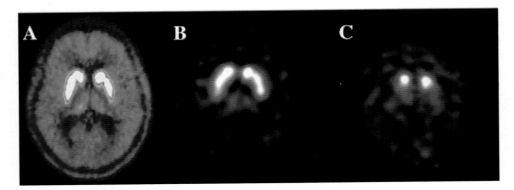

FIGURE 19-26 Presynaptic dopamine neuron imaging with ^{18}F-FD (A, healthy control) and the dopamine transporter agent ^{123}I-βCIT (B, healthy control; C, patient with Parkinson's disease). Despite different molecular targets, the presynaptic PET and SPECT agents demonstrate similar patterns of uptake in healthy and diseased states and are all useful for the visual evaluation of the integrity of the dopaminergic projections to striatum. (Reprinted with permission of Elsevier from reference 87.)

Evaluation of CSF flow is usually accomplished by administering a radiopharmaceutical intrathecally. The radiotracer has to be diffusible throughout the CSF space but remain in the CSF space until it can be absorbed with the CSF using the normal pathway through the arachnoid villi.

Procedure

A lumbar puncture is performed using a small-gauge spinal needle. Typically, 500 µCi (18.5 MBq) of ^{111}In-DTPA is administered intrathecally into the subarachnoid space. Initial anterior images of the head are obtained 6 hours after administration of the radiopharmaceutical. Sometimes, posterior images of the back can be obtained to evaluate whether the injection was successful. Anterior images of the head are obtained at 24 hours and 48 hours after injection. Sometimes images are also obtained at 72 hours, which is an advantage of ^{111}In, which has a 2.8 day half-life.

Interpretation

After the radiopharmaceutical has been successfully injected into the lumbar subarachnoid space, it begins to ascend through the spinal canal. In adult patients, activity can normally be seen accumulating in the basal cisterns by 2 to 4 hours. Activity can also be seen in the interhemispheric and sylvian fissures at this time. Normally, radiotracer is not seen entering the lateral ventricles at any time. Over the next 24 hours, radiotracer should ascend over the cerebral convexities to the sagittal sinus, and activity in the basal cisterns should begin to clear (Figure 19-27).

Patients with normal-pressure hydrocephalus demonstrate a different flow pattern. Early on, there is reflux of radiotracer into the lateral ventricles. This will persist on the delayed images. In addition, ascent over the cerebral convexities is usually markedly delayed (Figure 19-28).

CSF Leak

Rationale

The most common cause of CSF leaks is trauma. Most CSF leaks are located in the skull base between the region of the sphenoid sinus and temporal bone. CT imaging is most often used to evaluate a CSF leak. However, when this is nondiagnostic, nuclear imaging can be useful in helping to confirm and localize the leak site.

Procedure

To evaluate a possible CSF leak, pledgets are typically placed at the suspected leak site prior to intrathecal administration of

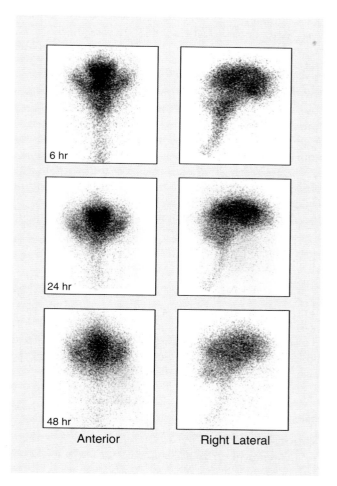

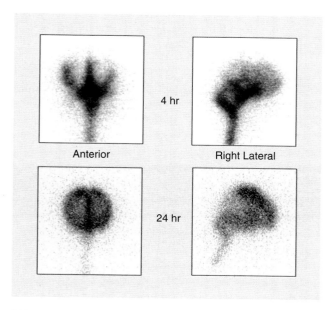

FIGURE 19-27 Normal CSF cisternogram. ^{111}In-DTPA study showing normal radiotracer accumulation in the basal cisterns and interhemispheric and sylvian fissures on the 4 hour images (trident appearance). Images at 24 hours demonstrate normal ascent of the radiotracer over the convexities to the superior sagittal sinus.

FIGURE 19-28 Normal-pressure hydrocephalus. The 6 hour images demonstrate ^{111}In-DTPA activity in the spinal subarachnoid space, basal cisterns, and lateral ventricles. At 24 hours the activity persists in the lateral ventricles, and in this patient there is very slow progression over the hemispheres. This pattern is essentially the same at 48 hours and indicates extraventricular obstruction and normal-pressure hydrocephalus.

the radiopharmaceutical. ^{111}In-DTPA is approved for intrathecal injection at a maximum dose of 0.5 mCi (18.5 MBq). ^{99m}Tc-DTPA has also been used off-label at a dose of 1 to 2 mCi (37–74 MBq) for CSF leak studies.[104] If the suspected leak is from the nose, pledgets are usually placed in locations near the sphenoethmoidal recess, the cribriform plate, and the middle meatus, bilaterally. If the suspected leak is from the ear, a pledget is placed in the ear canal. As in the evaluation of normal-pressure hydrocephalus, a lumbar puncture is then performed and the radiopharmaceutical is administered intrathecally. The patient is placed in the prone position or a position that most exacerbates the leak. Anterior, posterior, and lateral imaging of the head is performed, usually 2 to 4 hours later. A sample of the patient's blood is collected to measure activity in the plasma. The pledgets are also collected and measured for activity using a well counter. The pledget volume and plasma sample should be equal, about 0.5 mL each.[138] Pledget and plasma activity are usually measured as counts per minute per gram, and pledget-to-plasma activity ratios are calculated. The pledget-to-plasma activity ratio in 16 normal subjects was reported to not exceed 1.3 to 1. A patient with an occult leak had a ratio of 6.2 to 1.[138]

Interpretation

After the radiopharmaceutical has been successfully injected into the lumbar subarachnoid space, activity is usually seen in the basal cisterns by 2 to 4 hours. It is important to wait until the radiotracer accumulates in the suspected site of the leak before imaging the head and subsequently removing the pledgets. Normally, there should be no accumulation of radiotracer outside the cranial vault. Images positive for CSF leak demonstrate focal accumulations of radiotracer outside the cranium (Figure 19-29).

Radiotracer activity should not be seen in the systemic circulation until the radiotracer is absorbed into the venous system by the arachnoid villi. Thus, there should be no appreciable activity in the blood at 4 hours. Likewise, if there is no CSF leak, there should be no appreciable activity in the pledgets, and the pledget-to-plasma activity ratio should be 1 to 1. A pledget-to-plasma ratio greater than 1.5 to 1 is considered positive for CSF leak.[139] In the nose, the pledget with the highest ratio suggests the location of the leak (Table 19-5).

Shunt Evaluation

Rationale

VP and ventriculoatrial (VA) shunts are used to treat patients with obstructive hydrocephalus. If clinical symptoms begin to return or interval enlargement of the ventricles is seen on MRI or CT, the shunt may be obstructed. Anatomic studies such as plain film x-rays can determine if the shunt tubing is broken or kinked. If there is no evidence of this, nuclear medicine techniques can examine shunt function.

Procedure

To evaluate for shunt patency, radiotracer is injected into the shunt port using sterile technique (Figure 19-30). Typically,

FIGURE 19-29 CSF Leak in a patient with rhinorrhea after motor vehicle accident. Lateral image of the head taken 1 hour after intrathecal injection of 2 mCi (74 MBq) of ^{99m}Tc-DTPA at the lumbar spine level demonstrates activity in the nose consistent with a CSF leak. Pledget-to-plasma ratios (Table 19-5) demonstrate the main area of the leak to be in the left cribriform/middle meatus region.

1 to 2 mCi (37–74 MBq) of ^{99m}Tc-DTPA is used. There are three parts to a VP or VA shunt: (1) the shunt port, (2) the proximal limb from the port to the ventricle, and (3) the distal limb from the port to either the atrium or the peritoneal cavity. To evaluate the proximal limb of the shunt, radiotracer is injected while manual compression is maintained on the distal limb near the port. This should force the radiotracer into the proximal limb and into the ventricle. Dynamic images of the head can be obtained by using a transmission source behind the patient to verify radiotracer accumulation in the ventricle. After radiotracer is seen in the ventricle, pressure is

TABLE 19-5 Nasal Pledget-to-Plasma Activity Ratios in the Assessment of a Positive CSF Leak with ^{99m}Tc-DTPA (see Figure 19-29)[a]

Pledget Placement	Net Counts per Minute	Pledget-to-Plasma Ratio
Right cribriform	84	0.24
Right middle meatus	590	1.7
Right sphenoid	2,627	7.5
Left cribriform	126,169	360
Left middle meatus	111,795	319
Right sphenoid	53,881	154
Plasma	350	1

[a]0.5 mL of plasma, obtained at end of the procedure, is counted along with each nasal pledget suspended in 0.5 mL of saline.

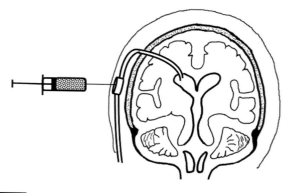

FIGURE 19-30 Diagram illustrating the injection port for radiotracer administration during ventriculoperitoneal shunt evaluation.

released from the distal limb and serial images are obtained to follow the flow of radiotracer through the distal limb.

Interpretation

With proper manual pressure on the distal limb, there should be prompt visualization of activity in the ventricle after injection. If the port is accessed properly and radiotracer fails to appear in the ventricle, this is evidence of a proximal limb obstruction.

Once proximal limb patency is observed, pressure is released from the distal limb. There should be prompt passage of radiotracer through the distal limb. If the patient has a VP shunt, activity should be seen spilling freely into the peritoneal cavity in a few minutes to an hour (Figures 19-31 and 19-32). There is evidence of obstruction if the radiotracer fails to advance through the shunt tubing or pools at the distal tip (Figure 19-33). If the patient has a VA shunt, the

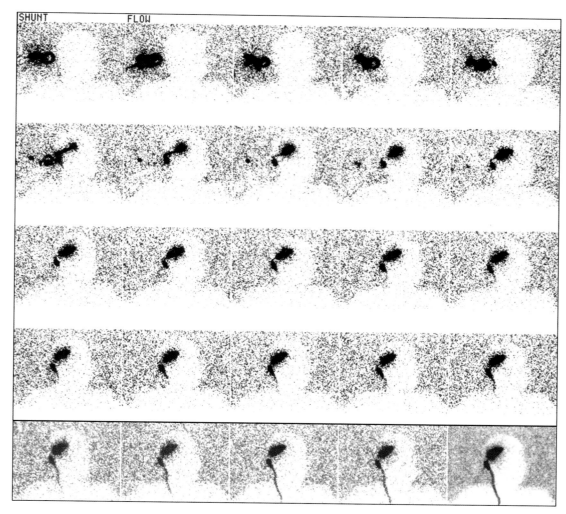

FIGURE 19-31 Ventriculoperitoneal (VP) shunt evaluation. ^{99m}Tc-DTPA study demonstrating patent VP shunt. With thumb pressure on the distal limb of the VP shunt, radiotracer is injected into the VP shunt port during the flow phase of the study. Once activity is seen in the lateral ventricle and obstruction of the proximal limb is ruled out, manual pressure is taken off the distal limb. Activity is seen to flow freely through the shunt toward the peritoneal cavity.

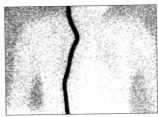 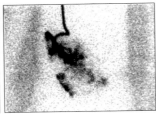

FIGURE 19-32 Patent ventriculoperitoneal shunt. Abdominal view of the same patient as Figure 19-31, demonstrating movement of radiotracer down the patent distal limb and spilling into the peritoneal cavity.

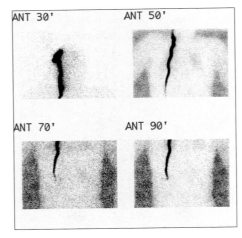

FIGURE 19-33 Obstructed ventriculoperitoneal shunt. ^{99m}Tc-DTPA study demonstrating delayed progression of radiotracer through the distal limb of the shunt. There is no evidence of tracer in the peritoneal cavity after 90 minutes.

radiotracer will reach the systemic circulation and will be seen in the kidneys.

REFERENCES

1. Guyton AC. *Textbook of Medical Physiology.* 6th ed. Philadelphia: WB Saunders; 1981:232–5.
2. Rapoport SI. *Blood-Brain Barrier in Physiology and Medicine.* New York: Raven Press; 1976:1–16.
3. Sage MR, Turski PA, Levin A. CNS imaging and the brain barriers. In: Neuwelt EA, ed. *Implications of the Blood-Brain Barrier and Its Manipulation. Vol 2: Clinical Aspects.* New York: Plenum; 1989:14–51.
4. Sage MR. Blood-brain barrier: phenomenon of increasing importance to the imaging clinician. *AJNR Am J Neuroradiol.* 1981;3:127–38.
5. Bradbury MWB. Why a blood-brain barrier? *Trends Neurosci.* 1979; 2:36–8.
6. Moore GE. Fluorescein as an agent in the differentiation of normal and malignant tissues. *Science.* 1947;106:130–1.
7. Moore GE. Use of radioactive diiodofluorescein in the diagnosis and localization of brain tumors. *Science.* 1948;107:569–71.
8. Chou SN, Aust JB, Moore GE, et al. Radioactive iodinated human serum albumin as a tracer agent for diagnosing and localizing intracranial lesions. *Proc Soc Exp Biol Med.* 1951;77:193–5.
9. Blau M, Bender MA. Radiomercury (Hg-203) labeled Neohydrin: A new agent for brain tumor localization. *J Nucl Med.* 1962;3:83–93.
10. Sodee DB. A new scanning isotope, mercury-197: a preliminary report. *J Nucl Med.* 1963;4:335–44.
11. Lindeman JF, Quinn JL III. The recent history of clinical procedures in nuclear medicine. In: Gottschalk A, Potchen EJ, eds. *Diagnostic Nuclear Imaging.* Baltimore: Williams & Wilkins; 1976:8–13.
12. Richards P. A survey of the production at the Brookhaven National Laboratory of radioisotopes for medical research. In: *5 Congresso Nucleare.* Vol 2. Rome: Comitato Nazionale Ricerche Nucleare; 1960:223–44.
13. Harper PV, Lathrop KA, Gottschalk A. Pharmacodynamics of some technetium-99m preparations. In: Andrews GA, Kniseley RM, Wagner HN Jr, eds. *Radioactive Pharmaceuticals.* Symposium 6. Washington, DC: US Atomic Energy Commission; 1966:335–58.
14. McAfee JG, Fuegar CF, Stern HS, et al. Tc-99m pertechnetate for brain scanning. *J Nucl Med.* 1964;5:811–27.
15. Eckelman W, Richards P. Instant Tc-99m DTPA. *J Nucl Med.* 1970;11:761.
16. Wolfstein RS, Tanasescu D, Sakimura IT, et al. Brain imaging with Tc-99m DTPA: a clinical comparison of early and delayed studies. *J Nucl Med.* 1974;15:1135–7.
17. Waxman A, Tanasescu DT, Siemsen J, et al. Tc-99m glucoheptonate as a brain scanning agent: critical comparison with pertechnetate. *J Nucl Med.* 1976;17:345–8.
18. Rollo DF, Cavalieri RR, Born M, et al. Comparative evaluation of Tc-99m GH, TcO4, and Tc-99m DTPA as brain imaging agents. *Radiology.* 1977;123:379–83.
19. Lathrop KA, Harper PV. Biologic behavior of Tc-99m from Tc-99m pertechnetate ion. *Prog Nucl Med.* 1972;1:145–62.
20. Brown-Grant K. Extrathyroidal iodide concentrating mechanisms. *Physiol Rev.* 1961;41:189–213.
21. Dayton DA, Maher FT, Elveback LR. Renal clearance of technetium (Tc-99m) as pertechnetate. *Mayo Clin Proc.* 1969;44:549–51.
22. Andros G, Harper PV, Lathrop KA, et al. Pertechnetate-99m localization in man with application to thyroid scanning and the study of thyroid physiology. *J Clin Endocrinol.* 1965;25:1067–76.
23. Beasley TM, Palmer HE, Nelp WB. Distribution and excretion of technetium in humans. *Health Phys.* 1966;12:1425–35.
24. Witcofski RL, Janeway R, Maynard CD, et al. Visualization of the choroid plexus on the technetium-99m brain scan: clinical significance and blocking by potassium perchlorate. *Arch Neurol.* 1967;16:286–9.
25. Coben LA, Loeffler JD, Elsasser JC. Spinal fluid iodide transport in the dog. *Am J Physiol.* 1964;206:1373–8.
26. Harper PV, Lathrop KA, Jiminez F, et al. Technetium-99m as a scanning agent. *Radiology.* 1965;85:101–9.
27. Alazraki N, Littenberg RL, Hurwitz S, et al. Differences in choroid plexus concentration of pertechnetate produced by varying time of perchlorate administration and brain imaging. *J Nucl Med.* 1974;15:884–6.
28. Scheu JD, Tetalman MR, Araujo O, et al. The efficacy of intravenous sodium perchlorate in choroid plexus blocking [abstract]. *J Nucl Med.* 1976;17:528.
29. Rumble WF, Aamodt RL, Jones AE, et al. Accidental ingestion of Tc-99m in breast milk by a 10 week-old child. *J Nucl Med.* 1978;19:913–5.
30. Vagenakis AG. Duration of radioactivity in the milk of a nursing mother after Tc-99m administration [abstract]. *J Nucl Med.* 1971;12:188.
31. Tator CH. Radiopharmaceuticals for tumor localization with special emphasis on brain tumors, In: Subramanian G, Rhodes BA, Cooper JF, et al., eds. *Radiopharmaceuticals.* New York: Society of Nuclear Medicine; 1975:474–81.
32. Front D, Israel O. Scintigraphy and ultrastructure of the blood-brain and blood-tissue barriers of human brain tumors: radionuclides as in vivo indicators of tumor permeability. In: Magistretti PL, ed. *Functional Radionuclide Imaging of the Brain.* Vol 5. New York: Raven Press; 1983:47–59.
33. Oldendorf WH. Need for new radiopharmaceuticals [letter]. *J Nucl Med.* 1978;19:1182.
34. Uzler JM, Bennett LR, Mena I, et al. Human CNS perfusion scanning with ^{123}I-iodoantipyrine. *Radiology.* 1975;115:197–200.

35. Loberg MD, Corder EH, Fields AT, et al. Membrane transport of Tc-99m-labeled radiopharmaceuticals. I: Brain uptake by passive transport. *J Nucl Med.* 1979;20:1181–8.
36. Kung HF, Blau M. Regional intracellular pH shift: a new mechanism for radiopharmaceutical uptake in brain and other tissues. *J Nucl Med.* 1980;21:147–52.
37. Kung HF. Brain radiopharmaceuticals, In: Fritzberg AR, ed. *Radiopharmaceuticals: Progress and Clinical Perspectives.* Vol II. Boca Raton, FL: CRC Press; 1986:21–39.
38. Winchell HS, Horst WD, Braun L, et al. N-isopropyl-[^{123}I] p-iodoamphetamine single-pass brain uptake and washout: binding to brain synaptosomes and localization in dog and monkey brain. *J Nucl Med.* 1980;21:947–52.
39. Kung HF, Tramposch KM, Blau M. A new brain perfusion agent: [I-123] HIPDM: N,N,N′-trimethyl-N′-[2-hydroxy-3-methyl-5-iodobenzyl]-1,3-propanediamine. *J Nucl Med.* 1983;24:66–72.
40. Creutzig H, Shober O, Gielow P, et al. Cerebral dynamics of N-isopropyl-(^{123}I) p-iodoamphetamine. *J Nucl Med.* 1986;27:178–83.
41. Burns HD, Manspeaker H, Miller R, et al. Preparation and biodistribution of neutral, lipid-soluble Tc-99m complexes of bis (2-mercaptoethyl) amine ligands [abstract]. *J Nucl Med.* 1979;20:654.
42. Kung HF, Molnar M, Billings J, et al. Synthesis and biodistribution of neutral lipid-soluble Tc-99m complexes which cross the blood brain barrier. *J Nucl Med.* 1984;25:326–32.
43. Volkert WA, Hoffman TJ, Seger RM, et al. ^{99m}Tc-Propylene amine oxime (^{99m}Tc-PnAO), a potential brain radiopharmaceutical. *Eur J Nucl Med.* 1984;9:511–6.
44. Kung HF. New technetium 99m-labeled brain perfusion imaging agents. *Semin Nucl Med.* 1990;20:150–8.
45. Volkert WA. Stereoreactivity of ^{99m}Tc-chelates at chemical and physiological levels. In: Nicolini M, Bandoli G, Mazzi U, eds. *Technetium and Rhenium in Chemistry and Nuclear Medicine 3.* Verona, Italy: Cortina International; 1990:343–52.
46. Cheesman EH, Blanchette MA, Ganey MV, et al. Technetium-99m ECD: ester-derivatized diamine-dithiol Tc complexes for imaging brain perfusion [abstract]. *J Nucl Med.* 1988;29:788.
47. Walovitch RC, Hill TC, Garrity ST, et al. Characterization of technetium-99m-L,L-ECD for brain perfusion imaging. 1: Pharmacology of technetium-99m ECD in nonhuman primates. *J Nucl Med.* 1989;30:1892–1901.
48. Neirinckx RD, Burke JF, Harrison RC, et al. The retention mechanism of technetium-99m-HM-PAO: intracellular reaction with glutathione. *J Cereb Blood Flow Metab.* 1988;8(suppl 1):S4–12.
49. Andersen AR, Friberg H, Knudsen GM, et al. Extraction of Tc-99m d,l-HM-PAO across the blood-brain barrier. *J Cereb Blood Flow Metab.* 1988;8(suppl 1):S44–51.
50. Gallagher BM, Ansari A, Atkins H, et al. Radiopharmaceuticals XXVII. 18F-labeled 2-deoxy-2-fluoro-D-glucose as a radiopharmaceutical for measuring regional myocardial glucose metabolism in vivo: tissue distribution and imaging studies in animals. *J Nucl Med.* 1977;18:990–6.
51. Neirinckx RD, Canning LR, Piper IM, et al. Technetium-99m d,l-HM-PAO: a new radiopharmaceutical for SPECT imaging of regional cerebral blood perfusion. *J Nucl Med.* 1987;28:191–202.
52. Sharp PF, Smith FW, Gemmell HG, et al. Technetium-99m HM-PAO sterioisomers as potential agents for imaging cerebral blood flow: human volunteer studies. *J Nucl Med.* 1986;27:171–7.
53. Lassen NA, Andersen AR, Friberg L, et al. The retention of [^{99m}Tc]-d,l-HM-PAO in the human brain after intracarotid bolus injection: a kinetic analysis. *J Cereb Blood Flow Metab.* 1988;8:S13–22.
54. Hung JC, Volkert WA, Holmes RA. Stabilization of technetium-99m-D,L-hexamethylpropyleneamine oxime (^{99m}Tc-D,L-HMPAO) using gentisic acid. *Nucl Med Biol.* 1989;16:675–80.
55. Andersen AR, Friberg H, Lassen NA, et al. Assessment of the arterial input curve for [^{99m}Tc]-d,l-HM-PAO by rapid octanol extraction. *J Cereb Blood Flow Metab.* 1988;8:S23–30.
56. Tc-99m Exametazime [package insert]. Arlington Heights, IL: MediPhysics Inc/Amersham Healthcare; 1996.
57. DeJong BM, VanRoyen EA. Uptake of SPECT radiopharmaceuticals in neocortical brain cultures. *Eur J Nucl Med.* 1989;15:16–20.
58. Nakamuira K, Tukatani Y, Kubo A, et al. The behavior of ^{99m}Tc-hexamethylpropyleneamineoxime (^{99m}Tc-HMPAO) in blood and brain. *Eur J Nucl Med.* 1989;15:100–7.
59. Villanueva-Meyer J, Thompson D, Mena I, et al. Lacrimal gland dosimetry for the brain imaging agent technetium-99m-HM-PAO. *J Nucl Med.* 1990;31:1237–9.
60. Walovitch RC, Cheesman EH, Maheu LJ, et al. Studies of the retention mechanism of the brain perfusion imaging agent ^{99m}Tc-bicisate (^{99m}Tc-ECD). *J Cereb Blood Flow Metab.* 1994;14(suppl 1):S4–11.
61. Knudsen GM, Andersen AR, Somnier FE, et al. Brain extraction and distribution of ^{99m}Tc-bicisate in humans and rats. *J Cereb Blood Flow Metab.* 1994;14(suppl 1):S12–18.
62. Ishizu K, Yonekura Y, Magata Y, et al. Extraction and retention of technetium-99m-ECD in human brain: dynamic SPECT and oxygen-15-water PET studies. *J Nucl Med.* 1996;37:1600–4.
63. Friberg L, Andersen AR, Lassen NA, et al. Retention of ^{99m}Tc-bicisate in the human brain after intracarotid injection. *J Cereb Blood Flow Metab.* 1994;14(suppl 1):S19–27.
64. Greenberg JH, Araki N, Karp A. Correlation between ^{99m}Tc-bicisate and regional CBF measured with iodo-[^{14}C] antipyrine in a primate focal ischemia model. *J Cereb Blood Flow Metab.* 1994;14(suppl 1):S36–43.
65. Lassen NA, Sperling B. ^{99m}Tc-bicisate reliably images CBF in chronic brain diseases but fails to show reflow hyperemia in subacute stroke: report of a multicenter trial of 105 cases comparing ^{133}Xe and ^{99m}Tc-bicisate (ECD, Neurolite) measured by SPECT on same day. *J Cereb Blood Flow Metab.* 1994;14(suppl 1):S44–8.
66. Vallabhajosula S, Zimmerman RE, Picard M, et al. Technetium-99m ECD: a new brain imaging agent: in vivo kinetics and biodistribution studies in normal human subjects. *J Nucl Med.* 1989;30:599–604.
67. Holman BL, Hellman RS, Goldsmith SJ, et al. Biodistribution, dosimetry, and clinical evaluation of technetium-99m ethyl cysteinate dimer in normal subjects and in patients with chronic cerebral infarction. *J Nucl Med.* 1989;30:1018–24.
68. Walovitch RC, Franceschi M, Picard M, et al. Metabolism of ^{99m}Tc-L,L-ethyl cysteinate dimer in healthy volunteers. *Neuropharmacology.* 1991;30:283–92.
69. Tc-99m Bicisate [package insert]. Billerica, MA: Dupont Pharmaceuticals; 1998.
70. Léveillé J, Demonceau G, Walovitch RC. Intrasubject comparison between technetium-99m-ECD and technetium-99m-HMPAO in healthy human subjects. *J Nucl Med.* 1992;33:480–4.
71. Kuhl D, Reivich M, Wolf A, et al. Determination of local cerebral glucose utilization by means of radionuclide computed tomography and (F-18) 2-fluoro-2-deoxy-D-glucose [abstract]. *J Nucl Med.* 1977;18:614.
72. Sokoloff L. Localization of functional activity in the central nervous system by measurement of glucose utilization with radioactive deoxyglucose. *J Cereb Blood Flow Metab.* 1981;1:7–36.
73. Phelps ME, Huang SC, Hoffman EJ, et al. Tomographic measurement of local cerebral glucose metabolic rate in humans with (F-18) 2-fluoro-2-deoxy-d-glucose: validation of method. *Ann Neurol.* 1979;6:371–88.
74. Lucignani G, Schmidt KC, Moresco RM, et al. Measurement of regional cerebral glucose utilization with fluorine-18-FDG and PET in heterogeneous tissues: theoretical considerations and practical procedure. *J Nucl Med.* 1993;34:360–9.
75. DiChiro G, Oldfield E, Wright DC, et al. Cerebral necrosis after radiotherapy and/or intra-arterial chemotherapy for brain tumors. *AJR Am J Roentgenol.* 1988;150:189–97.
76. Doyle WK, Budinger TF, Valk PE, et al. Differentiation of cerebral radiation necrosis from tumor recurrence by [^{18}F]FDG and ^{82}Rb positron emission tomography. *J Comput Assist Tomogr.* 1987;11:563–70.
77. Valk PE, Budinger FT, Levin VA, et al. PET of malignant cerebral tumors after interstitial brachytherapy: demonstration of metabolic activity and correlation with clinical outcome. *J Neurosurg.* 1988;69:830–8.

78. Coleman RE, Hoffman JM, Hanson MW, et al. Clinical application of PET for the evaluation of brain tumors. *J Nucl Med.* 1991;32:616–22.
79. Jones SC, Alavi A, Christman D, et al. The radiation dosimetry of 2-[F-18] Fluoro-2-deoxy-D-glucose in man. *J Nucl Med.* 1982;23:613–7.
80. Radiation Internal Dose Information Center. *Radiation Dose Estimates for F-18 FDG.* Oak Ridge, TN: Oak Ridge Institute for Science and Education; 1992.
81. Kaplan WD, Takvorian T, Morris JH, et al. Thallium-201 brain tumor imaging: a comparative study with pathologic correlation. *J Nucl Med.* 1987;28:47–52.
82. Yoshi Y, Satou M, Yamamoto T, et al. The role of thallium-201 single photon emission tomography in the investigation and characterization of brain tumors in man and their response to treatment. *Eur J Nucl Med.* 1993;20:39–45.
83. Kim KT, Black KL, Marciano D, et al. Thallium SPECT imaging of brain tumors: methods and results. *J Nucl Med.* 1990;32:965–9.
84. Westfall TC, Westfall DP. Neurotransmission: The autonomic and somatic motor nervous systems. In: Brunton LL, Lazo JS, Parker KL, eds. *Goodman & Gilman's The Pharmacological Basis of Therapeutics.* 11th ed. New York: McGraw-Hill Medical Publishing Division; 2006.
85. Heiss WD, Herholz K. Brain receptor imaging. *J Nucl Med.* 2006;47:302–12.
86. Meyer JH. Applying neuroimaging ligands to study major depressive disorder. *Semin Nucl Med.* 2008;38:287–304.
87. Seibyl JP. Single-photon emission computed tomography and positron emission tomography evaluations of patients with central motor disorders. *Semin Nucl Med.* 2008;38:274–86.
88. Kadir A, Nordberg A. Target-specific PET probes for neurodegenerative disorders related to dementia. *J Nucl Med.* 2010;51:1418–30.
89. Kung HF, Kung MP, Choi SR. Radiopharmaceuticals for single-photon emission computed tomography brain imaging. *Semin Nucl Med.* 2003;33:2–13.
90. Heiss W-D, Hilker R. The sensitivity of 18-fluorodopa positron emission tomography and magnetic resonance imaging in Parkinson's disease. *Eur J Neurol.* 2004;11:5–12.
91. Ravina B, Eidelberg D, Ahlskog JE, et al. The role of radiotracer imaging in Parkinson's disease. *Neurology.* 2005;64:208–15.
92. Lee CS, Samii A, Sossi V, et al. In vivo positron emission tomographic evidence for compensatory changes in presynaptic dopaminergic nerve terminals in Parkinson's disease. *Ann Neurol.* 2000;47:493–503.
93. Klunk WE, Engler H, Nordberg A, et al. Imaging brain amyloid in Alzheimer's disease with Pittsburgh Compound-B. *Ann Neurol.* 2004;55:306–19.
94. Wong DF, Rosenberg PB, Zhou Y, et al. In vivo imaging of amyloid deposition in Alzheimer disease using radioligand ^{18}F-AV-45 (Flobetapir F 18). *J Nucl Med.* 2010;51:913–20.
95. DiChiro G, Reames PM, Matthews WB Jr. RISA-ventriculography and RISA-cisternography. *Neurology.* 1964;14:185–91.
96. Detmer DE, Blocker HM. A case of aseptic meningitis secondary to intrathecal injection of I-131 human serum albumin. *Neurology.* 1965;15:642–3.
97. Nicol CP. A second case of aseptic meningitis after isotope cisternography using I-131 human serum albumin. *Neurology.* 1967;17:199–200.
98. Oldham RK, Staab EV. Aseptic meningitis after the intrathecal injection of radioiodinated serum albumin. *Radiology.* 1970;97:317–21.
99. Ashburn WL, Harbert JC, Briner WH, et al. Cerebrospinal fluid rhinorrhea studied with the gamma scintillation camera. *J Nucl Med.* 1968;9:523–9.
100. Wagner HN Jr, Hosain F, DeLand FH, et al. A new radiopharmaceutical for cisternography: chelated ytterbium 169. *Radiology.* 1970;95:121–5.
101. Matin P, Goodwin DA. Cerebrospinal fluid scanning with In-111. *J Nucl Med.* 1971;12:668–72.
102. Goodwin DA, Song CH, Finston R, et al. Preparation, physiology, and dosimetry of In-111 labeled radiopharmaceuticals for cisternography. *Radiology.* 1973;108:91–8.
103. Som P, Hosain F, Wagner HN Jr, et al. Cisternography with chelated complex of Tc-99m. *J Nucl Med.* 1972;13:551–3.
104. Curnes JT, Vincent LM, Kowalsky RJ, et al. CSF rhinorrhea: detection and localization using overpressure cisternography with Tc-99m DTPA. *Radiology.* 1985;154:795–9.
105. Welch MJ, Welch TJ. Solution chemistry of carrier-free indium. In: Subramanian G, Rhodes BA, Cooper JF, et al., eds. *Radiopharmaceuticals.* New York: Society of Nuclear Medicine; 1975:73–9.
106. Alazraki NP, Halpern SE, Ashburn WL, et al. Hyperbaric cisternography: experience in humans. *J Nucl Med.* 1973;14:226–9.
107. Partain CL, Alderson PO, Donovan RL, et al. Regional kinetics of indium-111 DTPA in CSF imaging of normal volunteers. In: Cloutier RJ, Coffey JL, Synder WS, et al., eds. *Radiopharmaceutical Dosimetry Symposium.* Rockville, MD: US Department of Health, Education, and Welfare, Public Health Service, Food and Drug Administration, Bureau of Radiological Health; 1976:404–9. HEW Publication (FDA) 76–8044.
108. Bell EG, Maher B, McAfee JH, et al. Radiopharmaceuticals for gamma cisternography, In: Subramanian G, Rhodes BA, Cooper JF, et al., eds. *Radiopharmaceuticals.* New York: Society of Nuclear Medicine; 1975: 399–410.
109. Drayer BP, Rosenbaum AE. Metrizamide brain penetrance. *Acta Radiol Suppl.* 1977;355:280–93.
110. Sage MR, Wilcox J, Evill CA, et al. Brain parenchyma penetration by intrathecal ionic and non-ionic contrast media. *AJNR Am J Neuroradiol.* 1982;3:481–3.
111. DeLand FH, Simmons GH. Spinal cord and cerebrospinal fluid. In: Cloutier RJ, Coffey JL, Synder WS, et al., eds. *Radiopharmaceutical Dosimetry Symposium.* Rockville, MD: US Department of Health, Education, and Welfare, Public Health Service, Food and Drug Administration, Bureau of Radiological Health; 1976:390–7. HEW Publication (FDA) 76–8044.
112. Harbert JC, McCullough D, Zeiger LS, et al. Spinal cord dosimetry in I-131 HSA cisternography. *J Nucl Med.* 1970;11:534–41.
113. McAfee JG, Subramanian G. Radioactive agents for imaging. In: Freeman LM, Johnson PM, eds. *Clinical Scintillation Imaging.* 2nd ed. New York: Grune & Stratton; 1975:13–114.
114. Elliott KAC, Jasper HH. Physiological salt solutions for brain surgery. *J Neurosurg.* 1949;6:140–52.
115. Mahaley MS Jr, Odom GL. Complication after intrathecal injection of fluorescein. *J Neurosurg.* 1966;25:298–9.
116. Huggins CB, Hastings AB. Effect of calcium and citrate injections into cerebrospinal fluid. *Proc Soc Exp Biol Med.* 1933;30:459–60.
117. Cooper JF, Harbert JC. Endotoxin as a cause of aseptic meningitis after radionuclide cisternography. *J Nucl Med.* 1975;16:809–13.
118. Verbruggen AM, de Roo MJK, Klopper JF. Technetium-99m diethylene triamine penta-acetic acid for intrathecal administration: are we playing with fire? *Eur J Nucl Med.* 1994;21:261–3.
119. Meegalla S, Plossl K, Kung MP, et al. ^{99m}Tc labeled tropanes as dopamine transporter imaging agents. *Bioconjug Chem.* 1996;7:421–9.
120. Zuckier LS, Kolano J. Radionuclide studies in the determination of brain death: criteria, concepts, and controversies. *Semin Nucl Med.* 2008;38:262–73.
121. Donohoe KJ, Frey KA, Gerbaudo VH, et al. Procedure guideline for brain death scintigraphy. *J Nucl Med.* 2003;44:846–51.
122. Huang AH. The hot nose sign. *Radiology.* 2005;235:216–7.
123. Brill DR, Schwartz JA, Baxter JA. Variant flow patterns in radionuclide cerebral imaging performed for brain death. *Clin Nuc Med.* 1985;10:346–52.
124. Goffin K, Dedeurwaerdere S, Van Laere K, et al. Neuronuclear assessment of patients with epilepsy. *Semin Nucl Med.* 2008;38:227–39.
125. Hougaard K, Oikawa T, Sveinsdottir E, et al. Regional cerebral blood flow in focal cortical epilepsy. *Arch Neurol.* 1979;33:527–35.
126. Devous MD Sr, Thisted RA, Morgan GF, et al. SPECT brain imaging in epilepsy: A meta-analysis. *J Nucl Med.* 1998;39:285–93.
127. Kuhl D, Engel J, Phelps M, et al. Epileptic patterns of local cerebral metabolism and perfusion in humans determined by emission computed tomography of ^{18}FDG and ^{13}NH$_3$. *Ann Neurol.* 1980;8:47–60.
128. Delbeke D, Lawrence SK, Abou-Khalil BW, et al. Postsurgical outcome of patients with uncontrolled complex partial seizures and temporal lobe hypometabolism on ^{18}FDG-positron emission tomography. *Invest Radiol.* 1996;31:261–5.

129. Butler CRE, Costa DC, Walker Z, et al. PET and SPECT imaging in the dementias. In: Murray IPC and Ell PJ, eds. *Nuclear Medicine in Clinical Diagnosis and Treatment.* London: Churchill Livingstone; 1998:713–28.
130. Faulstich ME. Brain imaging in dementia of the Alzheimer type. *Int J Neurosci.* 1991;57:39–49.
131. Silverman DHS, Mosconi L, Ercoli L, et al. Positron emission tomography scans obtained for the evaluation of cognitive dysfunction. *Semin Nucl Med.* 2008;38:251–61.
132. Frenkel A, Front D. Clinical investigation of cerebral blood flow. In: Magistretti PL, ed. *Functional Radionuclide Imaging of the Brain.* Vol 5. New York: Raven Press; 1983:31–45.
133. Leung DK, Van Heertum RL. Interventional nuclear brain imaging. *Semin Nucl Med.* 2009;39:195–203.
134. Cikrit DF, Burt RW, Dalsing MC, et al. Acetazolamide enhanced single photon emission computed tomography (SPECT) evaluation of cerebral perfusion before and after carotid endarterectomy. *J Vasc Surg.* 1992;15:747–54.
135. Kim JS, Moon DH, Kim GE, et al. Acetazolamide stress brain-perfusion SPECT predicts the need for carotid shunting during carotid endarterectomy. *J Nucl Med.* 2000;41:1836–41.
136. Juni JE, Waxman AD, Devous MD Sr, et al. Society of nuclear medicine procedure guideline for brain perfusion single photon computed tomography (SPECT) using Tc-99m radiopharmaceuticals. Version 3.0. *Practice Management Procedure Guidelines.* Reston VA: Society of Nuclear Medicine; 2009. (www.snm.org/guidelines). Accessed September 27, 2010.
137. Moore KL. *Clinically Oriented Anatomy.* Baltimore: Williams & Wilkins; 1980:941–6.
138. McKusick KA, Malmud LS, Kordela PA, et al. Radionuclide cisternography: normal values for nasal secretion of intrathecally injected ^{111}In-DTPA. *J Nucl Med.* 1973;14:933–4.
139. Ommaya A, O'Tuama L, Lorenzo A. Hydrocephalus, shunts and cerebrospinal fluid leaks. In: Wagner H, Szabo Z, Buchanan J, eds. *Principles of Nuclear Medicine.* Philadelphia: WB Saunders; 1995: 576–84.

CHAPTER 20
Thyroid

Nuclear medicine procedures of the thyroid gland include the radioactive iodine uptake (RAIU) test to assess thyroid gland function, radionuclide therapy for hyperthyroidism and thyroid cancer, and diagnostic imaging studies to detect disease within the thyroid gland and to scan the total body for metastases from thyroid cancer.

PHYSIOLOGIC PRINCIPLES

The thyroid gland is composed of a large number of follicles that contain epithelial cells filled with a substance called colloid.[1] The major constituent of colloid is thyroglobulin, which is the base for production and storage of thyroid hormone. Ingested iodide is actively transported from the blood by thyroid epithelial cells (iodide trapping), in which it is rapidly oxidized to iodine by peroxidase enzymes and hydrogen peroxide. The thyroid gland is capable of concentrating iodide to 40 times the plasma concentration under normal circumstances, and concentrations may increase 10-fold in the hyperthyroid state. Iodine then reacts with tyrosine residues in thyroglobulin to form thyroid hormone, which is stored in the colloid. The binding of one atom of iodine to tyrosine forms monoiodotyrosine (MIT), and addition of a second atom forms diiodotyrosine (DIT). Triiodothyronine (T_3) is formed by coupling one molecule each of MIT and DIT, and thyroxine (T_4) is formed by coupling two molecules of DIT (Figure 20-1).

The daily turnover of iodine in the body is shown in Figure 20-2. Iodine metabolism has a three-compartment kinetic model: compartment I, the extrathyroidal iodide pool (75 μg iodine); compartment II, the thyroid iodine pool (6000 μg iodine); and compartment III, the extrathyroidal iodine hormone pool (500 μg iodine).[2,3] The average daily intake of iodide is about 300 μg, but this value can vary widely according to geographic location and dietary habits. Iodide intake is balanced by a urinary loss of 285 μg and a fecal loss of 15 μg. About 75 μg of iodide is made into thyroid hormone and released into the blood. Of this amount, 15 μg is metabolized in the liver and excreted in the feces. The remaining hormone undergoes enzymatic deiodination in tissues, with recycling of 60 μg of iodide back into compartment I for reuse or renal excretion. During an average day, about 20% of the ingested iodide is organified into hormone and 80% is excreted.

A number of substances either promote or block the synthesis of thyroid hormone. Thyroid-stimulating hormone (TSH, or thyrotropin) released from the anterior pituitary gland controls several of these functions, including iodide trapping, the coupling reactions between MIT and DIT, and hormone release from the colloid into the blood. Excess plasma iodide suppresses thyroid gland uptake of iodide.

The drugs methimazole and propylthiouracil block the reaction of iodine with tyrosine and the coupling reactions that form thyroid hormone. Several halogen oxyanions and thiocyanate are capable of inhibiting the iodide-trapping mechanism and also cause discharge of iodide from the propylthiouracil-blocked gland. Of the halogenated oxyanions, perchlorate (ClO_4^-) is the most effective agent in discharging trapped iodide, being 10 times more potent than thiocyanate.[4]

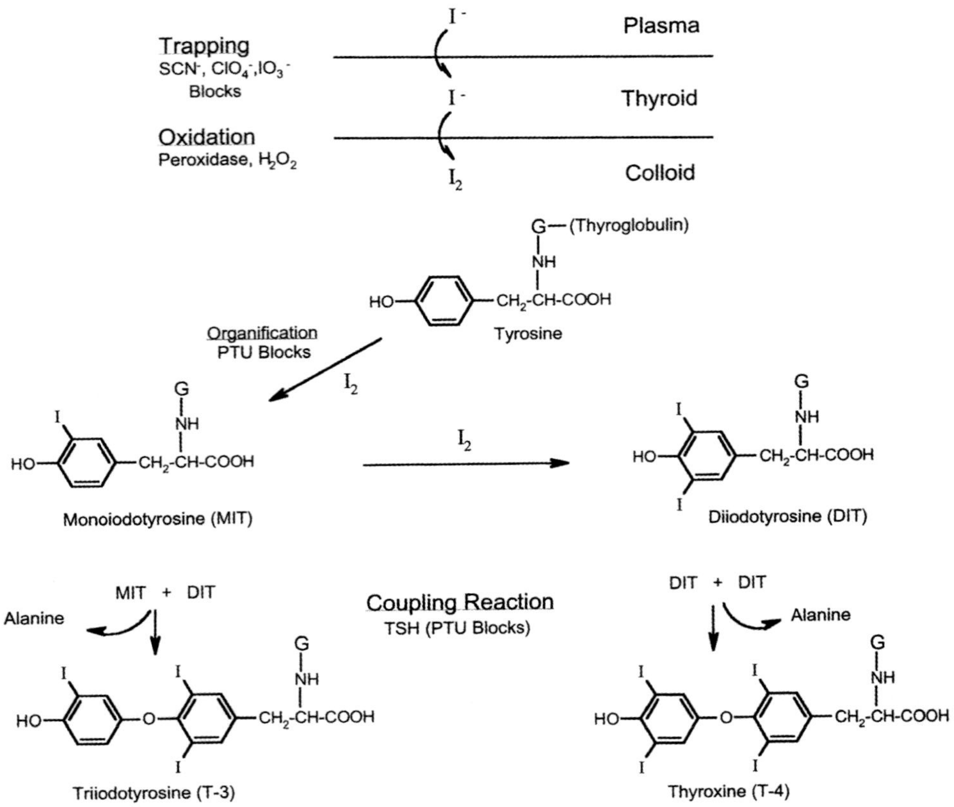

FIGURE 20-1 Schematic representation of iodine metabolism in the thyroid gland.

Thyroid gland function is controlled by the hypothalamic–pituitary axis through a feedback mechanism. When the levels of circulating thyroid hormone decrease, thyrotropin-releasing factor is secreted from the hypothalamus, causing release of TSH from the anterior pituitary, which stimulates the thyroid gland to produce more hormone. Excess circulating thyroid hormone reverses this process through negative feedback to the hypothalamus, which causes TSH levels to fall and thyroid hormone production to decrease. In the absence of disease, this process maintains thyroid gland homeostasis.

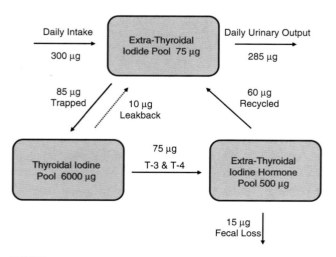

FIGURE 20-2 Daily turnover of iodine in the body.

HISTORICAL PERSPECTIVES

The first production of radioiodine for clinical application was at the Massachusetts Institute of Technology in 1937, when Robley Evans produced ^{128}I by neutron irradiation of ethyl iodide.[5] Intravenous injection of ^{128}I into rabbits clearly demonstrated, for the first time, the rapid accumulation of radioiodide by the thyroid gland.[6]

After the discovery of artificial radioactivity in 1934, cyclotrons were built at several research centers in the United States to produce radionuclides, mainly for medical research. Collaboration between institutions led to the rapid discovery of several iodine isotopes. Large quantities of ^{128}I were produced at Berkeley, but its 25 minute half-life was too short for metabolic studies. In 1938, ^{126}I ($T_{1/2}$ 13.3 days) was produced at Michigan by Tape and Cork,[7] while ^{131}I ($T_{1/2}$ 8.04 days) was made by Livingood and Seaborg[8] at Berkeley. Hamilton and Soley[9] reported on the first use of ^{131}I in human subjects with thyroid disease. Keston et al.[10] first reported on the uptake of radioiodine in thyroid metastasis in 1942. Probably the most significant application of ^{131}I, which heralded the value of radioisotopes in medicine, was use of the [131]I "atomic cocktail" in the treatment of metastatic thyroid cancer, reported by Seidlin et al.[11] in 1946. Before that, ^{131}I had been used to treat Graves' disease,[12] but a "cancer cure" had a much more dramatic impact than a treatment for hyperthyroidism. Strong public and monetary support for the fledgling discipline of nuclear medicine began with this experience.

RADIOPHARMACEUTICALS FOR THYROID STUDIES

Sodium Iodide I 123 and I 131 Capsules and Solution

[131]I is the standard radionuclide used for routine thyroid studies. It is relatively inexpensive to produce, and its 8 day half-life allows it to be available when needed in the nuclear medicine clinic. ^{131}I-sodium iodide is available from commercial suppliers in hard gelatin capsules and in aqueous solution for oral administration. Dosage forms are available in various activity amounts for diagnostic and therapeutic procedures from commercial suppliers and nuclear pharmacies (see Table 11-3, Chapter 11). ^{123}I-sodium iodide is also available in capsule form for diagnostic applications. Its cost is significantly higher than ^{131}I because it is produced in a cyclotron and has a 13.2 hour half-life, which necessitates daily delivery. It can be used for RAIU studies but is usually reserved for total body imaging.

Biologic Properties of Radioiodide

After oral administration of sodium radioiodide, the rate of gastrointestinal absorption is rapid, on the order of 5% per minute, and absorption is nearly complete within 1 to 2 hours.[13] The absorption rate may be delayed if food is present, and it is directly influenced by thyroid function, being increased in hyperthyroidism and decreased in hypothyroidism.

Iodide is cleared from the plasma primarily by the thyroid gland but also by the salivary glands, gastric mucosal cells, mammary glands, and kidneys. It is widely distributed in the body after administration. Most is excreted within the first 24 hours, and the remainder is localized mainly in the thyroid gland. Bodily distribution of radioiodide between 1 hour and 80 days is summarized in Figure 20-3.

Renal Excretion

Renal clearance of iodide is by glomerular filtration at a mean rate of 34 mL/minute, or 27% of the normal glomerular filtration rate. Thus about 73% of filtered iodide is reabsorbed by the tubule. Iodide is not bound in the kidney. Renal clearance of iodide is fairly constant over a wide range of plasma concentrations. After a dose of 10 μCi (370 kBq) of ^{131}I-sodium iodide with carrier doses of 0.001, 0.1, 1.0, and 10.0 mg of stable iodide, no difference in renal clearance of iodide occurs; however, a significant reduction of thyroid uptake occurs with the 1.0 and 10.0 mg doses, with uptakes similar to euthyroidism and the athyroid state, respectively.[14] In some instances it is necessary to administer stable iodide to protect the thyroid gland from unnecessary radiation exposure from

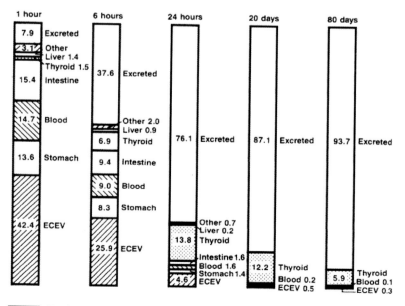

FIGURE 20-3 Estimated percentage of administered radioiodine in tissues of the body at various times after a single oral dose of radioiodide corrected for radioactive decay. The maximum thyroid uptake is assumed to be 15%. ECEV = extracellular extravascular space. (Reprinted with permission of the Society of Nuclear Medicine from reference 18.)

TABLE 20-1 FDA-Recommended Dose of Potassium Iodide for Thyroid Gland Protection

Age Group	Threshold Dose of Predicted Thyroid Exposure (rad or cGy)	Potassium Iodide Dose per Day (mg)
Adults over 40 yr	≥500	130[a]
Adults 18 to 40 yr	≥10	
Adolescents ≥70 kg	≥5	
Pregnant or lactating women	≥5	
3 yr to 18 yr	≥5	65
1 month to 3 yr	≥5	32
Birth to 1 month	≥5	16

[a] 130 mg of potassium iodide (99.4 mg of free iodide) is contained in each of the following: 1 potassium iodide 130 mg tablet, Lugol's solution 0.8 mL, saturated solution of potassium iodide (SSKI) 0.13 mL.
Source: Reference 15.

free radioiodide, such as may occur from radioiodide released by the metabolism of radioiodinated radiopharmaceuticals or from accidental ingestion of radioiodide.

Thyroid-blocking doses of potassium iodide recommended by the Food and Drug Administration, based on studies after the Chernobyl nuclear reactor disaster, are shown in Table 20-1.[15] Doses are based on subject weight and predicted thyroid exposure dose. A single dose protects the thyroid gland for 24 hours. The dose of potassium iodide should be administered with milk, fruit juice, or a large volume of water to minimize gastric irritation. These doses of potassium iodide can block 90% of radioiodine absorption if the first dose is given a few hours before or immediately after intake of radioiodine; the drug can still block 50% of radioiodine absorption if the first dose is administered within 4 hours after exposure.[16] Potassium iodide (KI) is available as potassium iodide tablets, 65 mg and 130 mg (equivalent to 50 mg and 100 mg iodine, respectively), as saturated solution of potassium iodide (SSKI), 1 gram KI per mL, or as Lugol's solution, 130 mg KI per 0.8 mL. Potential adverse effects of potassium iodide include gastrointestinal disturbances, allergic reactions, and minor rashes.[15]

The fraction of an administered dose of radioiodide excreted by the kidneys over 24 hours is inversely related to thyroid gland function.[17] Seventy-six percent of a dose of radioiodide is excreted in 24 hours in normal subjects with a thyroid uptake of 15% (Figure 20-3).[18]

Thyroid Gland Uptake

Measurement of thyroid gland function with radioiodide is predicated on an amount of radiotracer that is physiologic (i.e., will not alter the gland's normal function). The amount administered in thyroid studies easily meets this requirement; the average daily dietary intake of iodide is 300 µg, and a 10 µCi (370 kBq) diagnostic dose of ^{131}I-sodium iodide contains only 8×10^{-5} µg of iodine (one eighty-millionth of total body iodine).

The RAIU study is one of the oldest in vivo function studies performed in nuclear medicine. It is based on physiologic incorporation of radioiodide tracer into the thyroid gland followed by determination of the fraction of the dose taken up in the gland over a given time period. This study is described later in the chapter. For a successful test, it is important that no substance be present in the blood that will interfere with uptake of radioiodide by the thyroid gland, and patients must be questioned before the test to identify any interfering substances. Suppression of RAIU by stable iodide has been reported.[19] Table 20-2 lists common interfering substances.[20,21] The table does not list all of the radiographic contrast agents available, but all of them contain large amounts of iodine, and several weeks should elapse before the RAIU test is conducted in a patient who has received contrast material.

Salivary Glands

A significant amount of iodide is secreted by the salivary glands. Most of this is swallowed, but expectoration of saliva is a potential source of clothing contamination after therapeutic dosing of ^{131}I. This may lead to artifacts on whole-body images. Similar artifacts are more likely to result from urinary contamination; therefore, it is wise to provide clean hospital gowns before any imaging studies. The concentration of radioiodide in salivary glands can produce a metallic taste in the mouth within a few hours after administration of a therapeutic dose of ^{131}I. In addition, radiation sialadenitis has been reported to occur after large therapeutic doses of 150 mCi (5550 MBq), producing dry mouth and swelling and tenderness of the submaxillary glands.[22] Chewing gum or using lozenges such as lemon drops that promote salivation can shorten the residence time of ^{131}I in the salivary glands and may reduce the frequency of sialadenitis.

Gastric Glands

Radioiodide is highly concentrated in the gastric mucosa. Plasma clearance of iodide by the gastric glands is on the order of 25 mL/minute, and gastric juice-to-plasma ratios may be as high as 40 to 1.[23] This high concentration is inhibited by stable iodide and by perchlorate, suggesting an active transport process by the gastric cells. This gastric activity is associated with the mucoid cells rather than the parietal cells. The significance of the gastric concentrating mechanism is not known. Secretion of iodide into the gastric juice at high concentration increases the apparent iodide space of the body, but the iodide is normally rapidly reabsorbed after passing into the small intestine.[20]

Mammary Glands

Clearance of iodide by the mammary glands may achieve milk-to-plasma ratios up to 33 to 1.[24] Radioiodide and other radiopharmaceuticals are known to be excreted in human milk.[25,26] Precautions must be taken to prevent a nursing infant from ingesting contaminated milk from a mother who has received radioactive material. This is especially important in

TABLE 20-2 Drugs and Chemical Substances that Decrease 24-Hour Thyroid Uptake

Substance	Avg. Duration of Effect
Iodide-Containing Drugs	
SSKI, Lugol's Solution	1–4 weeks
Vitamin–mineral products	
Pima Syrup (Fleming)	
Isopropamide iodide	
Amiodarone, benziodarone	
Calcium iodide in Calcidrine syrup (Abbott)	
Hydriodic acid syrup	
Topical Iodide Products	
Iodochlorhydroxyquin–Clioquinol (Clay-Adams)	1–9 months
Iodine tincture	2 weeks
X-Ray Contrast Media	
Hypaque Meglumine, Hypaque Sodium (Winthrop)	1–2 weeks
Lipiodol, Ethiodol (Fougera)	1 year or more
Cholografin Meglumine (Bracco)	3 months
Telepaque (Nycomed)	2 months
Antithyroid Drugs	
Methimazole, propylthiouracil	2–8 days
Thyroid Medication	
Thyroid hormone, thyroxine, liothyronine	1–2 weeks
Other Drugs	
Phenylbutazone, sulfonamides	1 week
Adrenal and gonadal steroids, ACTH	8 days

Source: References 20 and 21.

regard to ^{131}I because of the potentially high radiation dose to the infant's thyroid gland. It is important for a nursing mother who receives radioiodide to know when she can resume breast-feeding her infant.[27] See Chapter 16 for more details on excretion of radiopharmaceuticals in human milk.

Placental Transport of Iodide

Several studies reviewed by Brown-Grant[23] indicate that placental transport of radioiodide occurs and high fetal-to-maternal thyroid ratios are achieved near term in many instances. Studies in humans have demonstrated that the fetal thyroid gland has the ability to accumulate ^{131}I iodide by the 12th to 14th week of gestation.[28,29] Consequently, the use of radioiodine and any radioactive material during pregnancy is contraindicated.

Radiation Dose from Radioiodines

The critical organ for radioiodine is the thyroid gland. The magnitude of the radiation dose to the gland and other organs and the total body depends on the radionuclide administered and the uptake by the gland. Estimates of radiation dose to the thyroid and the total body from ^{123}I, ^{124}I, ^{125}I, and ^{131}I are shown in Table 20-3. The ^{124}I and ^{125}I isotopes are included in the table because they may occur as radionuclidic contaminants in radioiodine preparations.

Sodium Pertechnetate Tc 99m Injection

After intravenous administration of sodium pertechnetate Tc 99m injection (^{99m}Tc-sodium pertechnetate), the pertechnetate

TABLE 20-3 Radiation Absorbed Dose from Radioiodines and Technetium

Target Organ	Absorbed Dose in rad(cGy)/mCi of Radioiodine Administered[a]				
	^{123}I	^{124}I	^{125}I	^{131}I	^{99m}Tc
Thyroid gland	13.0	890	790.0	1300.0	0.2
Total body	0.029	0.83	0.49	0.71	0.014

[a] Values assume a 25% maximum uptake by the thyroid.
Source: Reference 18.

TABLE 20-4 Radiopharmaceuticals for Thyroid Imaging

Radiopharmaceutical	Administered Activity (organ imaged)	Route	Time from Dose to Image
^{131}I-sodium iodide	50–100 µCi (thyroid) 2 mCi (total body)	Oral	24 hr
^{123}I-sodium iodide	200–400 µCi (thyroid) 2 mCi (total body)	Oral	24 hr
^{99m}Tc-sodium pertechnetate	2–10 mCi (thyroid)	IV	15–30 min

anion is trapped by the thyroid epithelial cells in a manner similar to iodide because of its similar ionic charge and volume. Its accumulation in the gland is limited to the trapping mechanism, and it is metabolized no further. The normal thyroid gland handles pertechnetate in the same manner that the propylthiouracil-blocked gland handles iodide; that is, pertechnetate is discharged from the gland by perchlorate.[30] Uptake of ^{99m}Tc-sodium pertechnetate by the thyroid gland is between 1% and 2% of the administered activity in euthyroid subjects but may be 10 times greater in thyrotoxicosis.[30] Multimillicurie amounts of ^{99m}Tc-sodium pertechnetate can be administered because the radiation dose to the gland is low compared with the dose from radioiodines (Table 20-4). ^{99m}Tc-sodium pertechnetate is administered intravenously for thyroid imaging at a dose of 2 to 10 mCi (74–370 MBq). Imaging is begun at the time of maximal uptake, which occurs in 15 to 30 minutes. The oral route may be used for imaging done at 1 hour. Chapter 19 contains more complete biologic information on ^{99m}Tc-sodium pertechnetate.

THYROID PATHOPHYSIOLOGY

Nuclear medicine procedures can be used in the diagnosis of several diseases of the thyroid gland, the most common being hyperthyroidism, thyroiditis, and thyroid nodules. In addition, hyperthyroidism and thyroid cancer can be treated with radioiodine.

Hyperthyroidism

Hyperthyroidism is characterized by hyperplastic thyroid tissue. The gland is increased in size 2 to 3 times and secretes excessive amounts of thyroid hormone, as much as 5 to 15 times normal.[1] Plasma TSH levels are far below normal or essentially zero because of the suppressive feedback effect on the anterior pituitary gland. A common cause of hyperthyroidism is Graves' disease, in which the gland is a diffusely enlarged goiter. This disease has an autoimmune origin; thyroid-stimulating immunoglobulins stimulate the TSH receptors on thyroid cells, causing hyperthyroidism. Another cause of hyperthyroidism is toxic nodular goiter, in which the gland is enlarged either with multiple hyperfunctioning nodules (Plummer's disease) or with a solitary nodule that is a hyperfunctioning adenoma. In each of these conditions, the hyperfunctioning tissues demonstrate increased accumulation of radioiodine or pertechnetate on a thyroid scan.

Thyroiditis

Thyroiditis is an inflammation of the thyroid gland. Subacute thyroiditis is a benign, self-limiting condition thought, but not proven, to be of viral origin.[31,32] Initially there is a hyperthyroid phase, lasting for weeks to several months. This phase is caused by inflammation-induced release of stored thyroid hormone from the gland. The RAIU is also reduced because the trapping mechanism is impaired. Hypothyroidism may develop, which can be treated with thyroid hormone. Upon resolution of the disease, most patients return to normal thyroid function.

Chronic thyroiditis, also known as lymphocytic or Hashimoto's thyroiditis, is an autoimmune inflammatory disease with evidence of elevated circulating antithyroid antibodies similar to those associated with Graves' disease.[32,33] The thyroid parenchymal cells are increasingly replaced by lymphocytes and plasma cells and eventually by fibrosis. Hypothyroidism is the end result. In the initial stages of the disease, TSH levels are elevated but thyroid hormone levels are low normal. The gland is diffusely enlarged, but the condition differs from Graves' disease in that patients in the latter stages of thyroiditis often exhibit hypothyroidism, and thyroid scans demonstrate reduced, nonhomogeneous uptake of activity.

Thyroid Nodules

A frequent indication for thyroid scanning is the presence of one or more palpable nodules in the thyroid gland. Nodules are usually benign, but they must be evaluated to rule out the presence of cancer. Nodules that accumulate radioiodine or pertechnetate and appear "hot" on the thyroid scan are called functioning nodules. Functioning nodules may or may not be under the control of TSH. A nodule that is not controlled by TSH is called an autonomous nodule. Sometimes, an autonomous nodule can achieve large size and produce enough thyroid hormone to supply the entire needs of the body. It may also produce excessive amounts of hormone, causing the patient to become thyrotoxic (hyperthyroid). Such a toxic nodule can suppress TSH release from the pituitary and cause suppression of the remaining normal thyroid tissue.[33] Functioning or hot nodules occur with low frequency (approximately 10%) and typically are benign. Most nodules are nonfunctioning (i.e., they do not take up radiotracer); however, some nonfunctioning nodules (approximately 6% to 10%) may be cancerous, so further studies are needed to establish a diagnosis.

NUCLEAR MEDICINE PROCEDURES

Two nuclear medicine procedures are commonly used to evaluate patients with suspected thyroid abnormalities: the RAIU test and the thyroid scan. In addition, ^{131}I therapy is used to treat patients with known thyroid gland disease. These procedures are well established and have been used in nuclear medicine for over 50 years.[34]

Radioactive Iodine Uptake

Rationale
RAIU is a functional study that measures iodine metabolism in the thyroid gland, because the thyroid gland both traps iodine and organifies it into thyroid hormone.[35] Determination of the fraction of a dose of radioactive iodine that accumulates in the thyroid gland at specific times after oral administration can be used to estimate thyroid function. This can be helpful in evaluating both hypothyroid and hyperthyroid conditions. In hyperthyroid conditions, the 4 and 24 hour thyroid uptake values are commonly used in determining iodine metabolism and turnover rates in the thyroid. These uptake measurements can assist in the diagnosis of hyperthyroidism and are useful in determining appropriate ^{131}I therapeutic dosages. RAIU along with a thyroid scan is also useful in differentiating causes of hyperthyroidism, such as Graves' disease, Plummer's disease (toxic multinodular goiter), and subacute thyroiditis.

Procedure
Patient preparation is important. Many substances can interfere with the uptake of iodine in the thyroid gland (Table 20-2). Patients who are taking these interfering substances should not be scheduled for RAIU testing until the effects of these substances have cleared. Administration of antithyroid drugs such as propylthiouracil and methimazole should be stopped at least 4 to 5 days before an RAIU test. Iodine-containing medications and food rich in iodine can also decrease radioiodine uptake in the thyroid gland. Prior imaging procedures such as computed tomography or angiography that use iodinated contrast agents can affect uptake of iodine in the thyroid for weeks. The patient should take nothing by mouth for 4 hours before administration of radioiodine to ensure adequate intestinal absorption. Food in the stomach can interfere with early iodine uptake.

Typically the patient is given an oral dose of 4 to 10 μCi (148–370 kBq) of ^{131}I-sodium iodide. Thyroid gland function is assessed by measuring the amount of administered activity, in counts per minute (cpm), of ^{131}I that is taken up into the patient's thyroid gland at 4 hours and 24 hours after administration of the capsule. Before the capsule is given to the patient, activity (cpm) of the capsule is determined in a neck phantom (a Plexiglas cylinder that simulates the geometry of the patient's neck). Room background cpm is also obtained and is subtracted from the neck phantom cpm to give the net cpm in the ^{131}I-sodium iodide capsule. All of these counts are obtained with a sodium iodide probe detector. After the 4 hour and 24 hour thyroid counts are obtained over the patient's neck, blood background cpm is obtained over the patient's thigh. The thigh cpm is subtracted from the patient's neck cpm to give net cpm in the thyroid gland. All counts are stored in the computer of the thyroid uptake probe and are automatically decay-corrected to the time of patient counting. The resulting RAIU is expressed as a percentage.

$$\text{RAIU} = \frac{\text{Neck (cpm)} - \text{Thigh (cpm)}}{\text{Administered dose (cpm)} - \text{Background (cpm)}} \times 100$$

Interpretation

The normal range for RAIU values is different for populations with different iodine intakes, and results should be interpreted in the context of other clinical information. In general, the normal range is considered to be 5% to 15% for 4 hour uptake and 10% to 35% for 24 hour uptake.[36]

In certain hyperthyroid individuals, such as those with thyrotoxicosis, the 4 hour uptake will be higher than the 24 hour uptake because the radioiodine incorporated into thyroid hormone will be largely discharged from the gland before the 24 hour measurement (Figure 20-4). For these

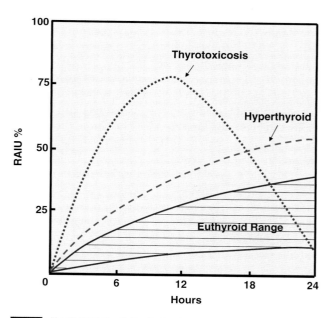

FIGURE 20-4 Radioactive iodine uptake (RAIU) over time under various thyroid conditions. (Adapted from Powers TA. Radioiodine thyroid uptake measurement. In: Sandler MP, Patton JA, Partain CL, eds. *Thyroid and Parathyroid Imaging.* Norwalk, CT: Appleton-Century-Crofts; 1986:181.)

individuals who demonstrate more rapid than normal turnover of iodine in the thyroid gland, it may be appropriate to use a higher ^{131}I dose in radioiodine therapy.

Thyroid Scan

Rationale

Thyroid scanning can be used to relate the structure of the gland to its function. Thyroid scans are often obtained to evaluate a single palpable nodule, multiple nodules, or an enlarging gland. The scan is useful in determining the functional nature of a palpable thyroid abnormality and is often useful in distinguishing benign from malignant disease.

In patients with hyperthyroidism, thyroid scans can be useful in distinguishing Graves' disease from toxic adenoma or toxic multinodular goiter. This information helps to determine the amount of ^{131}I-sodium iodide needed for appropriate thyroid therapy.

Scanning is useful in determining whether a thyroid nodule is functional. Although most thyroid nodules are benign, hypofunctioning or "cold" nodules are at increased risk for malignancy.[37]

Scanning is also useful in differentiating Graves' disease from subacute thyroiditis, postpartum thyroiditis, and factitious thyroiditis. Graves' disease causes prominent diffuse radiotracer uptake throughout the thyroid gland, whereas the other conditions demonstrate low radiotracer uptake and poor visualization of the thyroid gland.

Thyroid scans can also be used to locate ectopic thyroid tissue, such as in substernal goiter or lingual thyroid.

Procedure

Just as in RAIU testing, attention should be paid to patient preparation. The patient should not be pregnant or lactating, especially if radioiodine therapy is being considered. The patient should avoid interfering medications such as antithyroid drugs, iodine-rich foods such as kelp products, and iodine-containing medications such as amiodarone for an appropriate period of time. In many instances, the thyroid scan is obtained after the RAIU, and patient preparation has already been considered.

The patient is positioned supine with the neck extended. This is usually accomplished by placing a pillow or blanket under the patient's shoulders. Images of the neck are obtained using a gamma camera equipped with a pinhole collimator. The time from radiopharmaceutical dosing to imaging varies with the radiopharmaceutical used (Table 20-4).

Images are usually obtained 15 to 30 minutes after intravenous administration of [99mTc]-sodium pertechnetate. The usual administered dose is 2 to 10 mCi (74–370 MBq). Anterior, right anterior oblique, and left anterior oblique images typically are obtained. If there is a palpable nodule, imaging may be repeated with a lead marker or point source placed on the patient's skin overlying the palpable abnormality. This can be used to confirm that the palpable abnormality corresponds to a cold, hot, or warm nodule.

If ^{123}I-sodium iodide is administered orally for the thyroid scan, images are usually obtained 16 to 24 hours later. The administered dose is usually 200 to 400 µCi (7.4–14.8 kBq) [123]I.

Both ^{123}I-sodium iodide and [99mTc]-sodium pertechnetate are used for thyroid imaging. Both of these are trapped by the thyroid gland (i.e., transported into follicular cells of the thyroid). However, only the iodine is organified or synthesized into thyroid hormone. Both [99mTc]-sodium pertechnetate and ^{123}I-sodium iodide are adequate for anatomic imaging, but ^{123}I is more accurate for functional imaging. [131]I can also be used for imaging the thyroid, but it is not preferred because of its higher radiation dose to the gland, which is the result of a long half-life of 8.04 days and beta particle emission.

[123]I-sodium iodide is the imaging agent of choice because of its excellent imaging characteristics. It has a short half-life of approximately 13 hours, a gamma energy (159 keV) that is efficiently detected with the gamma camera, and absence of beta emissions. However, it is more expensive than [99mTc]-sodium pertechnetate and less readily available. Thus, in most institutions, [99mTc]-sodium pertechnetate is the imaging agent of choice for thyroid scanning.

Interpretation

The thyroid gland is a bilobed structure that normally demonstrates homogeneous radiotracer uptake throughout both lobes (Figure 20-5). The two lobes are joined, inferiorly and medially, by the thyroid isthmus. The isthmus often demonstrates less uptake of radioactivity than the remainder of the gland. Commonly, patients have a pyramidal lobe, which arises from the isthmus or the medial aspect of one of the thyroid lobes, extending medially and superiorly. In adults, the thyroid gland usually weighs between 15 and 25 grams. The right thyroid lobe is often larger than the left, extending more superiorly and inferiorly.

Palpable thyroid nodules are the most common indication for thyroid scanning. Imaging of the thyroid demonstrates the functional status of the palpable nodule and will sometimes identify other nodules. Nodules that correspond to focal areas of increased radiotracer accumulation in the thyroid gland are referred to as "hot" or hyperfunctioning nodules. Nodules that correspond to focal areas of absent radiotracer

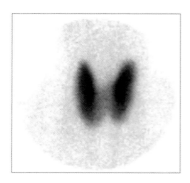

FIGURE 20-5 Normal thyroid gland in a 25 year old woman. Anterior pinhole image obtained approximately 30 minutes after administration of [99mTc]-sodium pertechnetate 10 mCi.

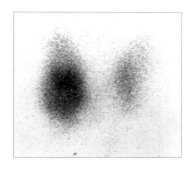

FIGURE 20-6 Large, toxic thyroid adenoma (hot nodule). Anterior image of the thyroid gland 30 minutes after intravenous administration of ^{99m}Tc-sodium pertechnetate 10 mCi demonstrates a hyperfunctioning nodule in the right lobe of the thyroid with suppression of the remaining gland.

accumulation in the thyroid gland are referred to as "cold" or hypofunctioning nodules. Nodules that demonstrate some activity or activity similar to the rest of the gland are often called indeterminate or "warm" nodules. The main role of thyroid scintigraphy is to determine which nodules should undergo fine-needle aspiration biopsy.

Hot nodules are almost always benign hyperfunctioning thyroid adenomas. Autonomous functioning nodules can produce enough thyroid hormone to block the secretion of TSH from the pituitary gland, causing suppression of the remaining normal thyroid tissue (Figure 20-6).

About 85% to 90% of palpable thyroid nodules are cold nodules. Cold nodules are most commonly benign colloid cysts or other benign lesions (Figure 20-7). However, some 6% to 10% of cold nodules are malignant.[38] Therefore, cold nodules demand further evaluation, such as by ultrasonography or fine-needle aspiration biopsy. Warm nodules can be cold nodules that are embedded in the gland with overlying normal thyroid tissue. Warm nodules are approached procedurally as if they were cold nodules.

Multinodular goiter is usually seen as an enlarged gland with heterogeneous radiotracer uptake throughout the gland with hot, cold, and warm nodules (Figure 20-8). Cold nodules in multinodular goiter are less likely to be cancerous. However, a dominant cold nodule in a multinodular goiter also warrants further investigation (Figure 20-9).

Patients with Graves' disease usually have some degree of thyromegaly along with suppressed TSH. Thyroid scan usually demonstrates prominent, homogeneous radiotracer uptake throughout the thyroid gland. Many times there is a prominent pyramidal lobe (Figure 20-10).

Thyroid scanning can also be useful in hypothyroid states. An example of this is documenting the lack of thyroid tissue in a newborn with elevated TSH levels (Figure 20-11).

Radioiodine Therapy

Rationale

Radioiodine therapy is an important option in the treatment of hyperthyroidism associated with Graves' disease, toxic thyroid adenoma, and toxic multinodular goiter or Plummer's disease. The options for treating hyperthyroidism include antithyroid medications, surgery, and ^{131}I-sodium iodide ablation therapy. Before radioiodine therapy is administered, it is important to confirm that the patient has hyperthyroidism both clinically and biochemically. The nature of the hyperthyroidism should also be determined. RAIU and thyroid scanning can help in differentiating thyroiditis from other conditions. Special considerations must be made if the patient is pregnant or is at high risk for thyroid storm.

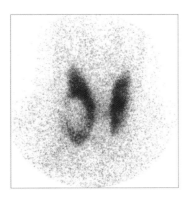

FIGURE 20-7 Cold nodule. Palpable right thyroid nodule in a 30 year old woman. Anterior pinhole image of the neck obtained 30 minutes after intravenous administration of ^{99m}Tc-sodium pertechnetate 10 mCi demonstrates a focal area of absent radiotracer accumulation in the right thyroid lobe consistent with a cold nodule.

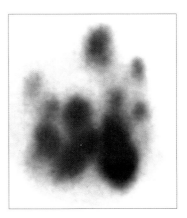

FIGURE 20-8 Multinodular goiter. Sixty-seven year old woman with hyperthyroidism and an enlarged, nodular thyroid by palpation. Anterior pinhole image of the neck obtained 30 minutes after intravenous administration of ^{99m}Tc-sodium pertechnetate 10 mCi demonstrates multiple focal areas of both increased and decreased radiotracer accumulation consistent with toxic multinodular goiter.

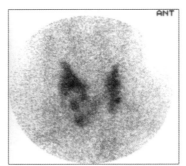
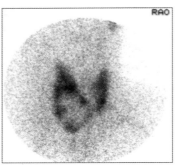

FIGURE 20-9 Thyroid scan demonstrating heterogeneous uptake in the thyroid gland with multiple focal areas of increased uptake in both lobes compatible with multinodular goiter. There is also a dominant focal area of decreased uptake involving the mid to lower right thyroid lobe, which correlated with a palpable nodule. Although most "cold" nodules in a multinodular goiter are benign, dominant cold nodules are usually investigated further with ultrasound or fine-needle aspiration biopsy. This cold nodule was found to be a Hürthle cell neoplasm.

Radioiodine therapy is also an option in nontoxic or euthyroid multinodular goiter in a patient with dysphagia who is not a good surgical candidate. ^{131}I therapy is also used as adjuvant treatment after surgery for papillary or follicular cell–type thyroid cancers. Controversies still exist regarding the administered dosage of ^{131}I-sodium iodide in each of these conditions.

Procedures

Graves' Disease

There are several approaches to selecting the dose of ^{131}I-sodium iodide for the treatment of Graves' disease. One of the most common methods of dose determination involves estimating the size of the thyroid gland and determining the 4 and 24 hour thyroid uptake values. The administered oral therapy dose of ^{131}I-sodium iodide is calculated by the following formula:

$$^{131}\text{I Dose (mCi)} = \frac{\text{Estimated gland weight (g)} \times \text{Number of } \mu\text{Ci desired per gram of tissue}}{\% \text{ uptake at 24 hours} \times 10}$$

There is wide variation in the recommended dose in activity per gram of tissue. Generally, this varies between 55 and 200 μCi (2.035 and 7.4 kBq) per gram of tissue.[39–41] The higher doses are used for patients with severe hyperthyroid symptoms or underlying cardiac problems, for whom it would be advantageous to induce clinical hypothyroidism as soon as possible.

The therapy dose is based in part on the RAIU result, with the assumption that a diagnostic radioiodide capsule used in the RAIU test has the same bioavailability as radioiodide solution used for therapy. One report has shown that differences in bioavailability are possible.[42] A potential consequence is that a falsely low RAIU result caused by capsules with lower bioavailability may result in potential overdose during therapy of hyperthyroidism patients treated with radioiodide solution. Using radioiodide solution for both diagnostic RAIU tests and therapy dosages would eliminate any potential discrepancies. (See Chapter 11 for more discussion on bioavailability of radioiodide from capsules.)

β-Adrenergic blockers such as atenolol are administered to help control hyperthyroid symptoms until the patient becomes euthyroid. A single ^{131}I dose of 10 mCi (370 MBq) will induce a euthyroid or hypothyroid state in 90% of patients, with a relapse rate of only 10% to 25%.[43]

Uninodular and Multinodular Goiter

Most hot or hyperfunctioning thyroid nodules are benign.[44] Toxic autonomously functioning thyroid adenomas are relatively radioresistant, and larger doses of radioiodine are used. Toxic multinodular goiters also require a larger dose of ^{131}I than

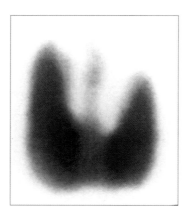

FIGURE 20-10 Graves' disease. Twenty-nine year old man with hyperthyroidism. Anterior pinhole image of the neck obtained 30 minutes after intravenous administration of ^{99m}Tc-sodium pertechnetate. Scan demonstrates prominent homogeneous uptake in an enlarged thyroid gland. Uptake is also seen in a prominent pyramidal lobe.

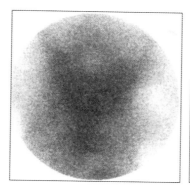
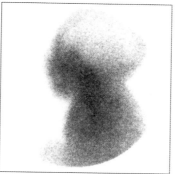

FIGURE 20-11 Agenesis of the thyroid. Twelve day old girl with elevated TSH. Anterior and left lateral pinhole images obtained after intravenous administration of ^{99m}Tc-sodium pertechnetate. There is no evidence of radiotracer accumulation in the thyroid gland.

is used for Graves' disease. Typically, the maximum outpatient dose that has been used to treat these conditions is around 30 mCi (1110 MBq) of ^{131}I-sodium iodide.

Well-Differentiated Thyroid Carcinoma

Thyroid carcinomas typically start out as cold nodules on thyroid scans. Well-differentiated thyroid carcinomas such as papillary, follicular, and mixed tumors will concentrate radioiodine. However, thyroid cancer does not concentrate iodine as well as normal thyroid tissue. Because of this, it is necessary to surgically remove as much normal thyroid tissue as possible before definitive ^{131}I therapy for thyroid cancer. When most of the normal thyroid tissue is absent, TSH will rise, which will increase the function of the thyroid cancer cells.

Therapy for thyroid cancer starts with near-total thyroidectomy, followed by remnant ablation with ^{131}I-sodium iodide to destroy any remaining thyroid tissue. The traditional procedure is to withhold thyroid hormone replacement therapy after surgery to allow the patient to become hypothyroid. With this approach, replacement therapy is withheld for 6 weeks to allow the endogenous TSH to rise, which promotes radioiodine uptake into thyroid tissue. When the TSH is high (>30 µIU/mL), a diagnostic ^{123}I total-body scan is done to visualize the remaining functional thyroid tissue. The administered dose is usually around 2 mCi (74 MBq) of ^{123}I-sodium iodide, and imaging is performed the next day.

Depending on the amount of residual thyroid tissue or tumor and the location and extent of disease, an appropriate dose of ^{131}I-sodium iodide is administered. Typically, doses range from 150 to 200 mCi (5550–7400 MBq). Scans are often obtained a few days after treatment to further evaluate the extent of disease (Figure 20-12).

Because hypothyroidism is associated with adverse effects such as fatigue, difficulty concentrating, short-term memory impairment, and depression, an alternative to the hypothyroid approach following thyroidectomy is to place the patient on thyroid replacement therapy to achieve a euthyroid state. Prior to diagnostic imaging to visualize thyroid remnants, TSH levels are elevated by administration of recombinant thyroid-stimulating hormone (rhTSH). Two doses of rhTSH, 0.9 mg each, are given by intramuscular injection 1 day apart, followed by administration of a 2 to 4 mCi (74–148 MBq) diagnostic dose of ^{123}I- or ^{131}I-sodium iodide 24 hours after the last dose of rhTSH. With this approach, the patient is spared the side effects associated with the hypothyroid state.

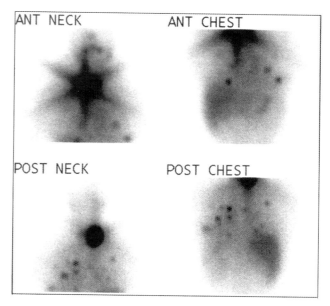

FIGURE 20-12 Metastatic thyroid cancer in a patient after thyroidectomy. Scan obtained 5 days after oral administration of ^{131}I-sodium iodide 175 mCi. The prominent focus of activity in the neck in the region of the thyroid bed is consistent with uptake in residual thyroid tissue or thyroid cancer. There is a starlike artifact associated with the prominent focus of activity in the neck. The high-energy photons of ^{131}I penetrate the collimator septa, causing this star artifact known as septal penetration. The multiple focal areas of accumulation in the lungs are consistent with metastatic thyroid cancer.

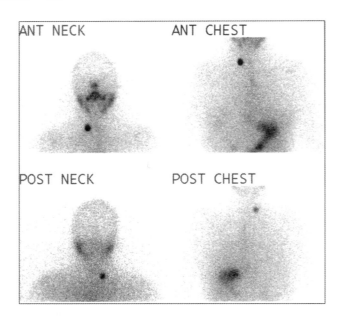

FIGURE 20-13 Thyroid cancer after therapy. ^{123}I scan of the same patient as in Figure 20-12, obtained 1 year after ^{131}I therapy. Scan shows only a single focus of residual cancer in the anterior neck. The patient subsequently underwent a second course of therapy with ^{131}I-sodium iodide.

Recent approval has been granted for rhTSH use in thyroid ablation with therapeutic doses of ^{131}I-sodium iodide. Compared with radioiodine treatment for patients in the hypothyroid state, this approach has been shown to have equivalent effectiveness but with fewer side effects and one-third less radiation dose to the blood.[45]

Follow-up ^{123}I total-body scans are usually done 1 year after ^{131}I therapy to assess for residual or recurrent thyroid cancer after the patient has been off thyroid hormone replacement for 6 weeks (Figure 20-13). Thyroglobulin (Tg) levels are also used to evaluate for recurrent disease.

Sometimes there is recurrent thyroid carcinoma, as evidenced by rising Tg, but the ^{123}I or ^{131}I whole-body scan is negative. In this subset of patients, positron emission tomography using ^{18}F-fludeoxyglucose has shown some value in identifying and localizing recurrent disease.[46,47]

SAFETY CONSIDERATIONS IN RADIOIODINE THERAPY

Patients who receive ^{131}I radioiodine therapy need to take precautions to minimize radiation exposure of others. In the United States, anyone who uses radioiodine must be licensed by the Nuclear Regulatory Commission (NRC). Until recently, patients treated with more than 30 mCi (1110 MBq) of ^{131}I needed to be hospitalized in a private room with a private toilet and monitored until the administered activity fell below 30 mCi (1110 MBq). However, this guideline is no longer in effect, and patients now are frequently released after having received much larger doses, even cancer therapy doses of 150 to 200 mCi (5550–7400 MBq) ^{131}I. In accordance with Title 10 of the Code of Federal Regulations, Part 35.75, NRC now permits a licensee to "authorize the release from its control of any individual who has been administered radiopharmaceuticals or permanent implants containing radioactive material if the total effective dose equivalent to any other individual from exposure to the released individual is not likely to exceed 5 millisievert (0.5 rem)."[48] The licensee must "provide the released individual with instructions, including written instructions, on actions recommended to maintain doses to other individuals as low as reasonably achievable if the total effective dose equivalent to any other individual is likely to exceed 1 millisievert (0.1 rem)."[48] These authorizations for release are based on patient-specific calculations. In most cases, NRC requires that records of the basis for authorizing patient release be maintained for 3 years.[48] A summary of release criteria, instructions, and records required for patients who receive ^{131}I is given in Table 20-5.

TABLE 20-5 Release Criteria, Instructions, and Records for Patients Treated with ^{131}I

Patient Group	Basis for Release	Criteria for Release	Instructions Needed?	Records Needed?
All patients, including patients who are breast-feeding	Administered activity Retained activity Measured dose rate Patient-specific calculations	If ≤33 mCi is given If ≤33 mCi is retained If ≤7 mr/hr @ 1 m If EDE[a] is ≤0.5 rem to any individual	Yes, if >7 mCi is given Yes, if >7 mCi is retained Yes, if >2 mr/hr @ 1 m Yes, if EDE is >0.1 rem	No Yes Yes Yes
Patients who are breast-feeding	All the above bases for release apply		Added instructions[b] required if >0.4 μCi given or if calculated dose to the child is >0.1 rem	Need record that instructions were provided if >2.0 μCi given or if calculated dose to the child is >0.5 rem

[a] Effective dose equivalent.
[b] NRC recommends discontinuance or interruption of breast-feeding.
Source: Reference 48.

REFERENCES

1. Guyton AC. *Textbook of Medical Physiology*. 6th ed. Philadelphia; WB Saunders; 1981:931–43.
2. Hoffer PB, Gottschalk A, Quinn J III. Thyroid in vivo studies. In: Gottschalk A, Potchen EJ, eds. *Diagnostic Nuclear Medicine*. Baltimore: Williams & Wilkins; 1976:255–77.
3. Rapoport B, DeGroot LJ. Current concepts of thyroid physiology. *Semin Nucl Med*. 1971;1:265–86.
4. Wyngaarden JB, Wright BM, Ways P. The effect of certain anions upon the accumulation and retention of iodide by the thyroid gland. *Endocrinology*. 1952;50:537–49.
5. Brucer M. The genesis of thyroid-iodine. In: *Vignettes in Nuclear Medicine*. No. 90. St Louis, MO: Mallinckrodt Inc; 1978.
6. Hertz S, Roberts A, Evans RD. Radioactive iodine as an indicator in the study of thyroid physiology. *Proc Soc Exp Biol Med*. 1938;38:510–3.
7. Tape GF, Cork JM. Induced radioactivity in tellurium [abstract]. *Phys Rev*. 1938;53:676.
8. Livingood JJ, Seaborg GT. Radioactive isotopes of iodine. *Phys Rev*. 1938;54:775–82.
9. Hamilton JG, Soley MH. Studies in iodine metabolism by use of a new radioactive isotope of iodine. *Am J Physiol*. 1939;127:557–8.
10. Keston AS, Ball RP, Frantz VK. Storage of radioactive iodine in a metastasis from thyroid carcinoma. *Science*. 1942;95:362–3.
11. Seidlin SM, Marinelli LD, Oshry E. Radioactive iodine therapy: effect of functioning metastasis of adenocarcinoma of the thyroid. *JAMA*. 1946;132:838–47.
12. Hertz S, Roberts A. Application of radioactive iodine in therapy of Graves' disease [abstract]. *J Clin Invest*. 1942;21:624.
13. Keating FR Jr, Albert A. The metabolism of iodine in man as disclosed with the use of radioiodine. *Recent Prog Horm Res*. 1949;4:429–81.
14. Childs DS Jr, Keating FR Jr, Rall JE, et al. The effect of varying quantities of inorganic iodide (carrier) on the urinary excretion and thyroidal accumulation of radioiodine in exophthalmic goiter. *J Clin Invest*. 1950;29:726–38.
15. US Department of Health and Human Services, Food and Drug Administration, Center for Drug Evaluation and Research. *Guidance: Potassium Iodide as a Thyroid Blocking Agent in Radiation Emergencies*. December 2001. www.fda.gov/downloads/Drugs/GuidanceComplianceRegulatoryInformation/Guidances/ucm080542.pdf. Accessed September 25, 2010.
16. US Food and Drug Administration. *Final Recommendations on Potassium Iodide for Radiation Emergencies*. Washington, DC: US Department of Health and Human Services, Food and Drug Administration; 1982. Bureau of Radiological Health Bulletin, vol 16, no. 7.
17. Skanse B. Radioactive iodine: its use in studying the urinary excretion of iodine by humans in various states of thyroid function. *Acta Med Scand*. 1948;131:251–68.
18. Loevinger R, Budinger TF, Watson EE. *MIRD Primer for Absorbed Dose Calculations*. New York: Society of Nuclear Medicine; 1988:53.
19. Sternthal E, Lipworth L, Stanley B, et al. Suppression of thyroid radioiodine uptake by various doses of stable iodine. *N Engl J Med*. 1980;303:1083–8.
20. Magalotti MF, Hummon IF, Hierschbiel E. The effect of disease and drugs on the twenty-four hour I-131 thyroid uptake. *AJR Am J Roentgenol*. 1959;81:47–60.
21. Hladik WB III, Nigg KK, Rhodes BA. Drug-induced changes in the biologic distribution of radiopharmaceuticals. *Semin Nucl Med*. 1982;12:184–218.
22. Sisson JC. Treatment of thyroid cancer. In: Wagner HN Jr, Szabo Z, Buchanan JW, eds. *Principles of Nuclear Medicine*. 2nd ed. Philadelphia: WB Saunders; 1995:629–37.
23. Brown-Grant K. Extrathyroidal iodide concentrating mechanisms. *Physiol Rev*. 1961;41:189–213.
24. Honour AJ, Myant NB, Rowlands EN. Secretion of radioiodide in digestive juices and milk in man. *Clin Sci*. 1952;11:447–62.
25. Romney BM, Nickoloff EL, Esser PD, et al. Radionuclide administration to nursing mothers: mathematically derived guidelines. *Radiology*. 1986;160:549–54.
26. Nurnberger C, Lipscomb A. Transmission of I-131 to infants through human maternal milk. *JAMA*. 1952;150:1398–1400.
27. Toohey RE, Stelson AT. *Guidelines for Breastfeeding Mothers in Nuclear Medicine*. Oak Ridge, TN: Radiation Internal Dose Information Center; 1996.
28. Chapman EM, Corner GW, Robinson D, et al. The collection of radioactive iodine by the human fetal thyroid. *J Clin Endocrinol*. 1948;8:717–20.
29. Hodges RE, Evans TC, Bradbury JT, et al. The accumulation of radioiodine by human fetal thyroids. *J Clin Endocrinol*. 1955;15:661–7.
30. Andros G, Harper PV, Lathrop LA, et al. Pertechnetate-99m localization in man with applications to thyroid scanning and the study of thyroid physiology. *J Clin Endocrinol*. 1965;25:1067–76.
31. Volpe R. Thyroiditis: current views of pathogenesis. *Med Clin North Am*. 1975;59:1163–75.
32. Sandler MP, Martin WH, Powers TA. Thyroid imaging. In: Sandler MP, Coleman RE, Wackers FJ, et al., eds. *Diagnostic Nuclear Medicine*. 3rd ed. Baltimore: Williams & Wilkins; 1996:911–42.
33. Palmer EL, Scott JA, Strauss HW. *Practical Nuclear Medicine*. Philadelphia: WB Saunders; 1992:311–41.
34. Chapman EM, Maloof F. The use of radioactive iodine in the diagnosis and treatment of hyperthyroidism; ten years' experience. *Medicine*. 1955;34:261–70.
35. Wolff J. Transport of iodide and other anions in the thyroid gland. *Physiol Rev*. January 1964;44:45–90.
36. Balon H, Dworkin H. In vivo nonimaging studies. In: Henkin RE, Boles MA, Dillehay GL, et al., eds. *Nuclear Medicine*. St Louis, MO: Mosby-Year Book; 1996:445–71.
37. Attie JN. The use of radioactive iodine in the evaluation of thyroid nodules. *Surgery*. 1959;47:611–4.
38. Freitas JE, Freitas AE. Thyroid and parathyroid imaging. *Semin Nucl Med*. 1994;24:234–45.
39. Kaplan MM, Meier DA, Dworkin HJ. Treatment of hyperthyroidism with radioactive iodine. *Endocrinol Metab Clin North Am*. March 1998;27:205–22.
40. Cooper DS. Treatment of thyrotoxicosis. In: Braverman LE, Utiger RD, eds. *Werner and Ingbar's The Thyroid: A Fundamental and Clinical Text*. 7th ed. Philadelphia: Lippincott-Raven; 1996:708–34.
41. Becker DV, Hurley JR. Radioiodine treatment of hyperthyroidism. In: Sandler MP, Coleman RE, Wackers FJ, et al., eds. *Diagnostic Nuclear Medicine*. New York: Williams & Wilkins; 1996:943–58.
42. Rini JN, Vallabhajosula S, Zanzanico P, et al. Thyroid uptake of liquid versus capsule I-131 tracers in hyperthyroid patients treated with liquid I-131. *Thyroid*. 1999;9:347–52.
43. Caruso DR, Mazzaferri EL. Intervention in Graves' disease: choosing among imperfect but effective treatment options. *Postgrad Med*. December 1992;92:117–34.
44. Ross DS. Evaluation of the thyroid nodule. *J Nucl Med*. 1991;32:2181–92.
45. Pacini F, Ladenson PW, Schlumberger M, et al. Radioiodine ablation of thyroid remnants after preparation with recombinant human thyrotropin in differentiated thyroid carcinoma: results of an international, randomized, controlled study. *J Clin Endocrinol Metab*. 2006;91:926–32.
46. Hooft L, Hoekstra OS, Deville W, et al. Diagnostic accuracy of ^{18}F-fluorodeoxyglucose positron emission tomography in the follow-up of papillary or follicular thyroid cancer. *J Clin Endocrinol Metab*. 2001;86:3779–86.
47. Wang W, Macapinlac H, Larson S, et al. [^{18}F]-2-fluoro-2-deoxy-D-glucose positron emission tomography localizes residual thyroid cancer in patients with negative diagnostic ^{131}I whole body scans and elevated serum thyroxine levels. *J Clin Endocrinol Metab*. 1999;84:2291–2302.
48. US Nuclear Regulatory Commission, Office of Nuclear Regulatory Research. *Release of Patients Administered Radioactive Materials*. Washington, DC: US Nuclear Regulatory Commission; April 1997. Regulatory Guide 8.39.

CHAPTER 21

Heart

Heart disease is the leading cause of death in the United States.[1] The principal factor contributing to cardiac death is coronary artery disease (CAD). The progression of CAD results in arterial stenosis that predisposes the heart muscle to myocardial ischemia, abnormal ventricular function, and eventual heart failure. Any situation that creates a workload on the heart creates a myocardial need for more oxygen. This induces an autoregulatory mechanism to increase blood flow to the myocardium to meet the oxygen need. A major component of autoregulation is adenosine-mediated vasodilation of coronary arteries. Normal coronary arteries can dilate sufficiently to increase blood flow; however, hardened, stenosed arteries respond inadequately because of loss of compliance, and ischemia results. When coronary arterial lumen diameter is reduced by 50% because of atherosclerotic plaque, perfusion abnormalities begin to appear when the heart is under stress but patients are usually asymptomatic.[2] When lumen diameter is reduced by 70%, clinical symptoms (angina) may occur during physical exertion. Under these conditions myocytes can have altered membrane potentials, and membrane transport of essential electrolytes and nutrients will be impaired. In advanced CAD stenosed arteries may not be able to provide adequate blood flow and oxygenation at resting conditions, and myocardial cells will begin to lose viability. Metabolic function in these regions will cease and myocardial infarction (MI) will occur. Abnormalities of electrical conductivity in the myocardium can also occur when myocardial ischemia results in ventricular arrhythmias. Infarction and arrhythmias are the leading causes of heart failure. Thus, it is vitally important that ischemic heart disease be identified in its early stages, since ischemic, viable myocardium can be restored to health through appropriate medical treatment and revascularization procedures.

Most clinical nuclear medicine studies today use single-photon emission computed tomography (SPECT) methods for data acquisition; however, positron emission tomography (PET) methods are increasing as PET centers become established. Most of these studies involve myocardial perfusion imaging (MPI) and radionuclide ventriculography, while the remainder involve metabolism studies.[2] MPI seeks to evaluate the extent and severity of perfusion abnormalities in the myocardium and their reversibility, an indirect measure of muscle viability. SPECT studies are performed with ^{201}Tl-chloride, ^{99m}Tc-sestamibi, and ^{99m}Tc-tetrofosmin and PET studies with ^{15}O-water, ^{13}N-ammonia, and ^{82}Rb-chloride (Table 21-1). Ventriculography studies evaluate heart function, assessing contraction abnormalities and ventricular ejection fraction and volumes. These studies can be accomplished with ^{99m}Tc-labeled red blood cells or ^{99m}Tc-sestamibi and ^{99m}Tc-tetrofosmin. Metabolism studies provide information about the metabolic status of the heart muscle and its viability. Oxidative metabolism can be assessed with ^{11}C-acetate; fatty acid metabolism with agents such as ^{11}C-palmitate, ^{123}I-iodophenylpentadecanoic acid (^{123}I-IPPA), and ^{123}I-iodofiltic acid; glucose metabolism with ^{18}F-fludeoxyglucose; and myocyte membrane function with ^{201}Tl-chloride. In routine nuclear medicine studies myocardial viability is assessed primarily with ^{201}Tl-chloride or ^{18}F-fludeoxyglucose.

TABLE 21-1 SPECT and PET Myocardial Imaging Agents

SPECT Agents	PET Agents
Perfusion Agents	**Perfusion Agents**
^{201}Tl-chloride	^{82}Rb-chloride
^{99m}Tc-sestamibi	^{15}O-Water
^{99m}Tc-tetrofosmin	^{13}N-ammonia
Blood Pool Agents	**Metabolism Agents**
^{99m}Tc-red blood cells	^{11}C-acetate
	^{11}C-palmitate
	^{18}F-fludeoxyglucose

PHYSIOLOGIC PRINCIPLES

Myocardial Blood Flow

Blood flow to the heart is supplied by a system of coronary arteries that originate from the aorta. They are the right coronary artery (RCA) and the left coronary artery mainstem, which bifurcates into the left circumflex artery (LCx) and the left anterior descending artery (LAD) (Figure 21-1). Blood flow within the heart varies, with the greatest flow per gram of heart muscle to the left ventricle and the least flow to the atria. The left ventricle receives about 80% of blood flow in the heart because of its greater mass compared with other heart chambers.[3] Also, because of its greater thickness, the left ventricle is the predominant region seen in cardiac imaging, while the right ventricle is faintly visualized.

Resting myocardial blood flow to most regions of the left ventricle ranges from 0.6 to 0.8 mL/minute per gram.[3] At rest, the oxygen extraction efficiency of the heart is quite high, about 70%, compared with 20% for skeletal muscle.[4] Consequently, increased coronary blood flow is more important than increased extraction efficiency in meeting the heart's need for oxygen. Coronary blood flow can increase in almost direct proportion to the metabolic consumption of oxygen by the heart. Under normal conditions and after appropriate stimuli, such as exercise or the administration of specific pharmacologic agents, blood flow can increase several fold, up to 3 to 4 mL/min/gram.[3] However, in the clinical setting with healthy volunteers and patients, maximal induced flows are typically in the range of 2 to 2.5 mL/min/gram.[5–7]

The difference between baseline flow and maximal flow is known as coronary flow reserve. This reserve is progressively diminished as vessels become stenosed and hardened by atherosclerosis. The stenosis may or may not cause significant impairment of blood flow to a particular region of the heart, because compensatory collateral circulation to the region may develop. Thus, a region perfused by a 90% stenosed coronary artery with a rich supply of collaterals may experience less ischemia than one with 60% stenosis and poor collaterals. However, over time, a chronic decrease in perfusion to regions of the myocardium can develop, causing a condition known as hibernating myocardium. Hibernating myocardium may

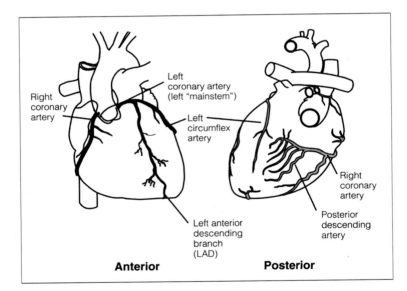

FIGURE 21-1 Anterior and posterior projections of the two main coronary arteries arising from the aorta. The right coronary artery supplies the lateral and posterior walls of the right ventricle and inferior wall of the left ventricle; the left anterior descending branch supplies the anterior wall of the right ventricle and the septum, apex, and anterior wall of the left ventricle; the left circumflex artery supplies the lateral and posterior wall of the left ventricle. (Reprinted with permission of Lantheus Medical Imaging Inc. from *Introduction to Nuclear Cardiology*, 3rd ed. North Billerica, MA: DuPont Pharma Radiopharmaceuticals; 1993:75.)

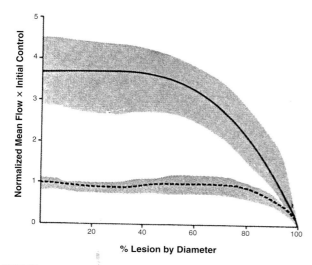

FIGURE 21-2 Relationship of coronary artery stenosis (x-coordinate) to alteration in coronary blood flow (y-coordinate) at resting flow (bottom curve) and at hyperemic flow (top curve). (Reprinted with permission of Elsevier from reference 8.)

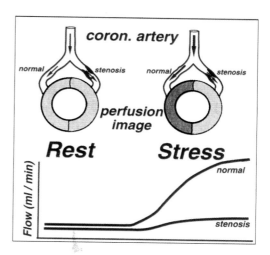

FIGURE 21-3 Schematic representation of the principle of rest–stress myocardial perfusion imaging. Top: Rest and stress diagrams of two branches of a coronary artery, one is normal (left) and one stenotic (right). Middle: Myocardial perfusion images of the territories supplied by the two branches. Bottom: Coronary blood flow in the branches at rest and during stress. At rest, myocardial blood flow is equal in both coronary artery branches. When a myocardial radiotracer is injected at rest, uptake is homogeneous (normal image). During stress, coronary blood flow increases 2.0 to 2.5 times in the normal branch but not to the same extent in the stenosed branch, resulting in heterogeneous distribution of blood flow. This heterogeneity of blood flow can be visualized with SPECT and PET perfusion imaging agents as an area with decreased uptake (abnormal image with a myocardial perfusion defect). (Adapted from reference 9 and used with permission of the Society of Nuclear Medicine.)

have adequate blood flow and function at rest but become severely compromised under stress. The major goal of MPI studies is to be able to identify these regions of hibernating myocardium and assess whether they contain viable muscle. If they do, methods for improving blood supply to these regions early on can be used to improve overall myocardial function and quality of life.

Studies in dogs have shown that resting coronary blood flow does not diminish until arterial stenosis exceeds 85%, whereas maximal coronary blood flow begins to decrease when stenosis exceeds 45% (Figure 21-2).[8] Thus, only the most severe coronary obstruction is likely to be detected by perfusion imaging under resting conditions. It follows, then, that assessment of regional myocardial perfusion under conditions of elevated coronary blood flow (cardiac stress) substantially increases the sensitivity for detecting CAD. Consequently, myocardial perfusion imaging studies to identify regions of subcritical coronary stenosis are conducted both at resting blood flow and at elevated flow under stress.

The principal aspect of perfusion imaging under stress is the creation of a disparity of flow between well-perfused normal myocardium and poorly perfused jeopardized myocardium. Under these circumstances, radiotracer uptake is increased in normally perfused regions of the heart and decreased in underperfused regions of the heart (Figure 21-3).[9] The current methods used to achieve maximal coronary dilatation and increased blood flow are intense exercise or administration of pharmacologic agents.

Exercise Stress

Exercise on a treadmill is the preferred method of inducing cardiac stress in patients who complain of chest pain on exertion, because it can reproduce the patient's real-life symptoms. It also provides information about the patient's exercise capacity, which is important for risk evaluation and patient management. Exercise is a natural method of stress that increases cardiac work and metabolic demand on the heart, causing coronary vasodilation. It can also provide additional diagnostic information such as the magnitude of blood pressure increase, an indicator of cardiac function; heart rate, an indicator of electrical conduction system function; electrocardiogram (ECG) abnormalities; and chest pain on exertion.[10]

In general, treadmill exercise is used with a modified Bruce protocol. In this method, speed and grade of the treadmill are increased in a stepwise manner to achieve the required workload on the heart.[11] Endpoints for achieving adequate levels of cardiac stress are 85% of an age-related maximal predicted heart rate determined by the formula 220 minus the patient's age in years and a pressure–rate product (systolic blood pressure times heart rate) of 25,000 or higher. Heart rate[12] and the pressure–rate product[13] have been shown to increase linearly with workload, and coronary blood flow is closely related to the pressure–rate product.[14] In general, exercise stress under this protocol increases blood flow to about 2 times the resting flow, which is adequate for diagnostic evaluation. Exercise stress provides important prognostic information. Experience has shown that patients with CAD have a better survival rate if they can exercise longer than stage IV of the Bruce protocol (at least 12 minutes), achieve a heart rate higher than 160 beats per minute (bpm) at peak exercise, and show no ST-segment depression on the exercise ECG.[9]

A maximal increase in coronary blood flow after exercise is sometimes difficult to achieve clinically, because some patients suspected of having heart disease may not have the physical ability to exercise at the intense level required to produce maximal coronary dilatation. Furthermore, exercise cannot be used in some patients because of claudication, cerebrovascular accidents, arthritis, amputation, severe anxiety, or current use of medications that limit heart rate, such as beta blockers and calcium channel blockers.

Pharmacologic Stress

An alternative to exercise stress is pharmacologic stress with agents such as dipyridamole, adenosine, regadenoson, or dobutamine (Figure 21-4).[15–20] Dipyridamole, adenosine, and regadenoson are coronary vasodilators. After administration of these agents, normal coronary vessels dilate maximally, but stenosed, noncompliant vessels fail to dilate sufficiently, creating a perfusion deficit. Hence, a heterogeneity of blood flow to the heart is created relative to the severity of CAD. Maximal coronary dilatation is consistently achieved with these agents compared with exercise.

Dobutamine is a predominant β_1-receptor agonist, increasing heart rate and myocardial contractility and systolic blood pressure.[20] Dobutamine increases myocardial oxygen demand, being akin to exercise. Normal coronary arteries dilate to increase perfusion in order to meet the demand, while stenotic arteries may not be able to increase flow to the same degree as normal vessels, creating a perfusion deficit as in exercise stress.

Adenosine

Adenosine is an endogenous nucleoside present in all cells of the body, including the myocardium. A major pathway of production is by enzymatic dephosphorylation of ATP (Figure 21-5).[18] Interstitial concentrations of adenosine rise in response to ischemia and increased metabolic oxygen requirements in the heart.[21] Adenosine readily diffuses into the extracellular space, where it causes vasodilation by activating the adenosine A_2 receptors on coronary endothelial cells and by increasing intracellular cAMP. Adenosine is rapidly transported back to the intracellular space, where it is metabolized to uric acid, phosphorylated to adenosine monophosphate, or conjugated to homocysteine to form S-adenosyl homocysteine.[18] Adenosine's plasma half-life is very short; estimates range from less than 2 seconds[22] to less than 10 seconds.[23]

Adenosine's stimulatory effect on A_2-receptors is blocked by methylxanthine compounds such as caffeine, theophylline, and aminophylline.[24] Thus, food or beverages containing caffeine and methylxanthine medications must be discontinued for at least 12 hours before adenosine administration. Because of potentiation, patients taking oral dipyridamole should discontinue this medication for 12 hours before adenosine stress testing.[16] Adenosine also activates other receptor subtypes, including A_1, A_{2B}, and A_3.[25,26]

Activation of A_1 receptors on myocardial cells causes a slowing of electrical conduction through the atrioventricular node, potentially causing first-, second-, and third-degree heart block. Thus, the use of adenosine (and dipyridamole) in patients with these conditions should be approached with cau-

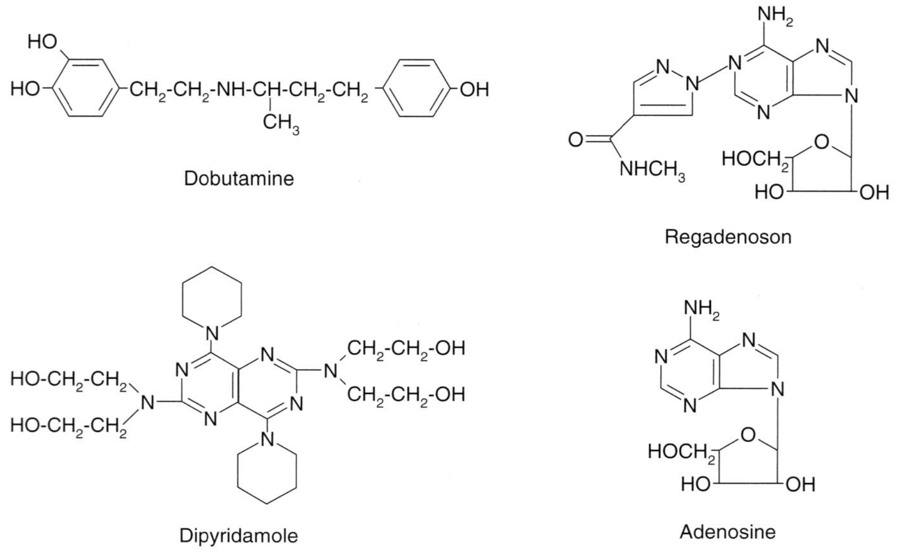

FIGURE 21-4 Chemical structures of the pharmacologic interventional agents dobutamine, regadenoson, dipyridamole, and adenosine.

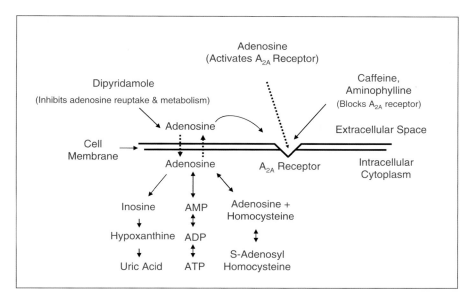

FIGURE 21-5 Mechanisms of adenosine production and action. Adenosine is produced by myocytes and other cells upon demand via dephosphorylation of ATP by 5′ nucleotidase. Adenosine causes vasodilation by activation of A_2 receptors on vascular endothelial cells. Receptor activation is blocked by methylxanthine compounds such as caffeine and aminophylline. Adenosine is rapidly reabsorbed into cells, where it is converted back to ATP by adenosine kinase or metabolized by adenosine deaminase to uric acid. Adenosine reuptake into the intracellular space and its metabolism are inhibited by dipyridamole.

tion and may be contraindicated.[27] Bronchospasm is mediated by the A_{2B} and A_3 receptors. Thus, both adenosine and dipyridamole are contraindicated in asthmatic patients.[28,29] Because of the potential side effects from stimulation of these receptor subtypes, specific A_{2A} receptor agonists have been developed and are discussed below.

For cardiac imaging, adenosine is administered at a standard dosage of 140 μg/kg per minute for 6 minutes to achieve maximal coronary vasodilation. An alternative 4-minute protocol can also be used. Radiotracer is injected halfway through the adenosine infusion, when vasodilation is maximal, and continued for an additional 2 or 3 minutes. Upon cessation of adenosine infusion, imaging can begin with ^{201}Tl-chloride within a few minutes, with ^{99m}Tc-tetrofosmin after waiting 45 minutes, or with ^{99m}Tc-sestamibi after waiting 60 minutes, to allow for hepatobiliary clearance of these agents.[30,31] Thallium imaging must commence and be completed before significant redistribution occurs, whereas the technetium agents are more firmly fixed in the heart and imaging can begin once background activity clears. With the standard dosage of adenosine, 84% of normal patients achieve an average coronary vasodilation equal to 4.4 times the resting coronary blood flow.[21] The average time from onset of infusion until maximal vasodilation is 84 ± 46 seconds. The time from cessation of infusion until coronary blood flow velocity returns to baseline levels is 145 ± 67 seconds. In normal patients the heart rate rises 24 ± 14 beats/minute, and mean arterial pressure falls 6 ± 7 mm Hg.[21]

Dipyridamole

Dipyridamole increases plasma adenosine levels indirectly by inhibiting adenosine's metabolism by adenosine deaminase and phosphodiesterase and by inhibiting adenosine reuptake into cells. The standard dosage of dipyridamole is 0.56 mg/kg administered over 4 minutes. Because its action is indirect, dipyridamole achieves a maximum dilator effect 6 to 8 minutes after the start of the infusion, indicated by a drop in blood pressure and a rise in heart rate.[7] It increases coronary blood flow about 4 times baseline, similar to adenosine. Radiotracer should be injected at the time of maximal effect, approximately 7 minutes from the start of the infusion. Imaging is begun at about 12 to 15 minutes from the start of infusion with ^{201}Tl-chloride and at later times with ^{99m}Tc-sestamibi or ^{99m}Tc-tetrofosmin. The biologic half-life of adenosine is prolonged after dipyridamole administration, and the patient should be monitored for side effects 10–20 minutes after a study.[15] Adverse effects, such as severe chest pain, bronchospasm, severe headache, significant drop in blood pressure, and nausea and vomiting, can be mitigated by intravenous administration of aminophylline 50–100 mg over 30–60 seconds in doses ranging from 50 to 250 mg.[28]

Splanchnic uptake of radiotracer (more so with ^{99m}Tc agents than with ^{201}Tl) is greater with vasodilator stress than with exercise stress. Walking on a treadmill (1.7 mph, 0 grade) during a 4- or 6-minute infusion of dipyridamole or adenosine is an effective way to reduce hepatic and gut tracer uptake and lower blood pressure.[30]

Adenosine A$_{2A}$ Receptor Agonists

A class of adenosine A$_{2A}$ receptor agonists has been investigated with the goal of identifying coronary vasodilators equivalent to adenosine but with fewer side effects. The major differences between adenosine and the A$_{2A}$ receptor agonists are selectivity for the A$_{2A}$ receptor, receptor affinity, and duration of effect of coronary artery dilation. Greater selectivity potentially maximizes coronary vasodilation, with less chance of adverse effects associated with the A$_1$ receptor (AV block) and the A$_{2B}$ and A$_3$ receptors (bronchospasm). The duration of coronary artery dilation is directly related to A$_{2A}$ receptor affinity. Affinity is lowest for adenosine, higher for regadenoson, and higher yet for binodenoson.[32] The higher affinity and longer duration of effect of the A$_{2A}$ receptor agonists permits a shorter duration of dose administration, as a bolus or short infusion. Thus, regadenoson is administered as a 0.4 mg bolus injection over 10 seconds followed by a 5 mL saline flush and achieves a 2.5-fold increase in coronary artery blood flow with a rapid onset (30 seconds) for a duration of 2.5 minutes or longer.[33] The radiotracer for MPI is injected within 10 to 20 seconds after the saline flush. Imaging is begun 45 to 60 minutes after ^{99m}Tc tracer injection.[34] The dosage of binodenoson found to be most concordant with adenosine for MPI is 1.5 µg/kg infused over 30 seconds[35] or by bolus infusion. Bolus infusion produced maximal coronary hyperemia by 4.5 ± 3.7 minutes that was sustained for 7.4 ± 6.8 minutes, producing modest changes in blood pressure and heart rate and no evidence of second- or third-degree AV block.[36] These parameters for dose administration and biologic effects fit well with a myocardial perfusion study. Regadenoson has received drug approval status for routine use. Binodenoson and another agent, apadenoson, are still under investigation.

In summary, compared with adenosine in reported trials, regadenoson and binodenoson exhibit concordant efficacy for myocardial perfusion imaging,[37] are easy to administer, and cause no AV block or bronchospasm. However, despite a higher affinity for A$_{2A}$ receptors, these agents are associated with the same subjective side effects (chest pain, dyspnea, and flushing) of the other receptor subtypes, similar to but at a slightly lower frequency than with adenosine. Other reports of regadenoson use in patients indicate a low frequency of first-degree (3%) and second-degree (0.1%) AV block.[34] In addition, in patients with moderate to severe chronic obstructive pulmonary disease (COPD), the rate of bronchoconstriction was 12% and 6% for regadenoson and placebo groups, respectively, but bronchoconstriction did not occur in patients with moderate asthma. The rate of dyspnea was 61% in COPD patients and 34% in patients with asthma.

Dobutamine

Principally a β$_1$-receptor agonist, dobutamine causes a combined increased inotropic and chronotropic effect on the myocardium, increasing contractility and oxygen demand and causing coronary vasodilation. A general imaging protocol is to administer the dosage slowly, ramping up in 10 µg/kg per minute steps every 3 minutes for 12 minutes to a maximal dose of 40 µg/kg per minute.[20] Radiotracer is injected 1 minute after the highest dose is begun (10 minutes from start of infusion), and dobutamine infusion is continued for 2 minutes more while tracer is localizing in the myocardium. If heart rate is less than 85% of the predicted rate after the maximal dose of dobutamine, intravenous atropine 0.2 mg to 1 mg may be administered.[38] Imaging with ^{201}Tl is begun at about 3 minutes from cessation of infusion, while imaging with ^{99m}Tc agents is begun in 60 minutes. The plasma half-life of dobutamine is 2 minutes.

Dobutamine is typically diluted with 0.9% sodium chloride injection for intravenous infusion and must be used within 24 hours. It should not be mixed in solutions containing sodium bicarbonate, sodium bisulfite, or ethanol.

Table 21-2 summarizes the infusion protocols for pharmacologic intervention during myocardial perfusion imaging.

Precautions, Adverse Effects, and Contraindications

Adenosine, dipyridamole, regadenoson, and dobutamine are potent pharmacologic agents that should be administered with caution and with full awareness of their adverse effects and potential toxicities. Blood pressure, heart rate, and a 12-lead ECG should be monitored continuously. Preferably, an electronic infusion pump should be used to administer these agents to provide precise control of the dosage.[39] Regadenoson is an exception to this; it is given by bolus injection. Contraindications for these agents are summarized in Table 21-3.[40] Contraindications for regadenoson are similar to those for adenosine.

The relative merits of adenosine and dipyridamole have been compared[19] and their adverse effects have been reported.[41-45] The most common adverse effects of dipyridamole and adenosine are chest pain, headache, dizziness, flushing, dyspnea, ST-T changes on ECG, and ventricular extrasystoles. Other effects occurring less frequently include nausea, hypotension, and tachycardia (Table 21-4). These effects are related to the plasma level of vasodilator and can be readily reversed—in the case of dipyridamole by giving intravenous aminophylline (50 to 100 mg over 30–60 seconds), and for adenosine by simply stopping the infusion, because of its short plasma half-life. Patients who receive aminophylline should be monitored for drug-induced ischemia before being released from the nuclear medicine department. These agents should be used with caution in patients with heart block and are contraindicated in asthma and COPD.

The distribution of cardiac output to various organs is different with adenosine and dipyridamole than with exercise stress.[3] The relative blood flow to abdominal viscera increases with these agents, whereas with exercise it decreases. Thus, perfusion imaging agents, particularly ^{99m}Tc-sestamibi and ^{99m}Tc-tetrofosmin, tend to accumulate in the liver and spleen area to a greater extent with adenosine and dipyridamole than with exercise. This visceral activity may interfere with assessing perfusion of the inferior wall of the left ventricle.[3] Walking after dipyridamole infusion or during adenosine infusion can help reduce gut-related activity. In addition, as noted above, adenosine and dipyridamole do not work effectively if patients

TABLE 21-2 Pharmacologic Stress Protocols in Cardiac Perfusion Imaging

Drug	Dosage	Infusion Protocol
Adenosine (standard)	140 µg/kg/minute	• 3 minute infusion • Inject radiotracer • 3 minute infusion
Adenosine (alternate)	140 µg/kg/minute	• 2 minute infusion • Inject radiotracer • 2 minute infusion
Dipyridamole	140 µg/kg/minute	• 4 minute infusion • Wait 3 to 5 minutes • Inject radiotracer
Adenosine or dipyridamole with exercise	140 µg/kg/minute	• Treadmill exercise (1.7 mph, 0 grade) during entire 4- or 6-minute infusion
Regadenoson	400 µg in 5 mL	• 10 second bolus infusion • 5 mL saline flush • Inject radiotracer
Dobutamine	Graded infusion from 10 to 40 µg/kg/minute	• 3 minutes each at 10, 20, 30 µg/kg/minute • 1 minute at 40 µg/kg/minute • Inject radiotracer • 2 minutes at 40 µg/kg/minute

TABLE 21-3 Contraindications for Use of Pharmacologic Agents in Stress Testing

Contraindications	Dipyridamole	Adenosine	Dobutamine
Unstable angina or resting ischemia	X	X	X
Poor LV function (EF <15%)	X	X	X
Hypertension (>200 mm Hg systolic)		X	X
Hypotension (<90 mm Hg systolic)	X	X	
Severe aortic stenosis			X
History of asthma	X	X	
Active bronchospastic disease	X	X	
History of tachyarrhythmias			X
Second-degree AV block		X	
Oral dipyridamole		X	
Xanthine compounds	X	X	
Atrial fibrillation with rapid ventricular response			X

LV = left ventricular; EF = ejection fraction; AV = atrioventricular.
Source: Reprinted with permission from reference 40.

TABLE 21-4 Percentages of Patients Experiencing Adverse Effects of Pharmacologic Agents Used in Stress Testing

Effect	Adenosine	Dipyridamole	Dobutamine
Flushing	36.5	3.4	14
Dyspnea	35.2	2.6	14
Chest pain	34.6	19.7	31
Gastrointestinal	14.6	5.6	—
Headache	14.2	12.2	14
Dizziness	8.5	11.8	4
Palpitation	—	—	29

Source: References 41, 42, and 45 for adenosine, dipyridamole, and dobutamine, respectively.

have taken caffeine or other methylxanthine substances; these substances are contraindicated for vasodilator studies.

Regadenoson has contraindications and adverse effect cautions similar to those for adenosine, but with reported lower frequency of side effects. It has a longer plasma half-life than adenosine, and adverse effects can be reversed by intravenous aminophylline 50–100 mg administered over 30–60 seconds.[34]

Dobutamine is not affected by caffeine or methylxanthine drugs and is an alternative agent if adenosine, dipyridamole, or regadenoson is contraindicated. The principal adverse effects of dobutamine are chest pain, palpitation, headache, flushing, and dyspnea (Table 21-4).[45] These effects should readily abate with cessation of the infusion, since dobutamine's half-life is short (2 minutes). However, a β_1-receptor-blocking agent such as metoprolol succinate should be available in case problems arise; it can be administered as an intravenous dose of 0.25 mg, in repeated doses if needed, to slow heart rate. Dobutamine is contraindicated in the same conditions that apply to exercise stress. Dobutamine stress does not produce as great an increase in coronary blood flow as does vasodilator stress, making it less than ideal; therefore, its use should be restricted to patients in whom vasodilators are contraindicated.

Extraction and Retention of Myocardial Tracers

The principal requirements for a useful myocardial perfusion tracer are high myocardial extraction efficiency proportional to blood flow over the range of flows seen clinically and heart retention long enough to conduct imaging studies. Myocardial extraction versus flow has been measured for several perfusion tracers (Figure 21-6).[46–49] Ideally, if a radiotracer's extraction fraction is 1.0 (100%) at any flow rate, regional differences in myocardial tracer concentration will be proportional to regional blood flow. Only labeled microspheres exhibit this property, having 100% first-pass extraction at any flow rate. Most diffusible tracers used clinically have extraction fractions less than 1.0 when blood flow is above resting flow. This occurs because radiotracer delivered to the heart that is not fixed in the myocardium back-diffuses and washes away. As blood flow increases, capillary recruitment occurs, increasing the capillary surface area for exchange between the blood and interstitial space and making more tracer available for extraction (i.e., an increase in the capillary permeability surface area product occurs). However, permeability, or the rate of tracer flux through the capillary membrane, is independent of surface area. Thus, at higher flows and increased capillary surface area, more tracer is made available for exchange and tracer uptake into the heart increases, but the fraction of total tracer extracted declines because the time available for tracer to diffuse across the capillary surface is shortened, causing some tracer to wash away. Thus, myocardial extraction becomes nonproportional to flow because radiotracer is diffusion limited at elevated flows.[47–51] Nonproportional extraction with flow is generally not a significant issue in myocardial perfusion imaging because absolute measure of blood flow is not the goal; relative blood flow is the goal. However, a tracer that exhibits low extraction at higher blood flow may have limited sensitivity in detecting mild ischemia, because the difference in count density between normal and ischemic myocardium may be difficult to assess (see discussion under ^{99m}Tc-tetrofosmin).

Once a radiotracer diffuses across the capillary membrane into the interstitial space and is transported across the sarcolemmal membrane into the myocyte intracellular space, it must remain fixed there for the period of time required for imaging; if it does not, a correction must be made for back-diffusion of tracer. For example, imaging with ^{201}Tl-chloride must be completed within 35 minutes after injection because it redistributes out of myocytes. One technetium radiopharmaceutical, ^{99m}Tc-teboroxime, required completion of imaging within 10 to 15 minutes of injection because of its rapid back-diffusion; this agent is no longer marketed because of the practical limitations imposed by its rapid washout from the heart. ^{99m}Tc-sestamibi and ^{99m}Tc-tetrofosmin are fixed in the myocardium, so imaging with these agents can be con-

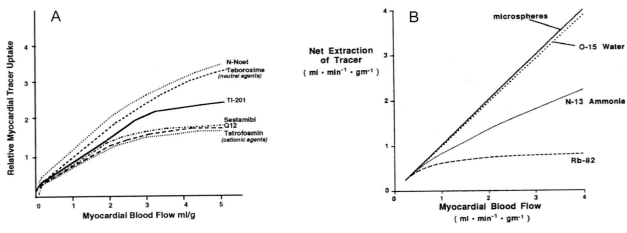

FIGURE 21-6 (A) Relationship between myocardial blood flow and uptake of SPECT perfusion tracers. (Reprinted with permission of Elsevier from reference 47). (B) Myocardial extraction of PET perfusion tracers versus coronary artery blood flow. (Reprinted with permission of W. B. Saunders from reference 49.)

ducted without regard to time and repeat imaging can be done if needed, which is one advantage of these agents.

Retention of radiotracer activity in heart muscle is a function of the tracer's unique biochemical properties and the physiologic integrity of heart muscle, which is affected by disease. For example, the mechanisms of localization of SPECT perfusion imaging agents differ. Thallium is taken up by active transport via the membrane pump, similar to potassium, but it is retained longer than potassium. Thallium is not bound to myocytes, however, and it eventually diffuses back out of these cells in proportion to blood flow.[52] When myocyte viability is compromised, thallium extraction and retention are diminished. ^{99m}Tc-sestamibi and ^{99m}Tc-tetrofosmin, as lipophilic monovalent cations, are taken up into myocytes by passive diffusion associated with intact negative sarcolemmal and mitochondrial membrane potentials. Retention is dependent on maintenance of these membrane potentials.[53]

RADIOPHARMACEUTICALS FOR HEART IMAGING

SPECT Perfusion Imaging Agents

MPI agents are also known as "cold spot" markers because they demonstrate decreased accumulation of activity in poorly perfused regions of myocardium. Table 21-5 summarizes the biochemical properties of SPECT MPI agents. Most myocardial perfusion studies are performed using SPECT procedures, which have proven value for measuring the extent and severity of perfusion abnormalities and have been useful in risk stratification and patient management.

The low photon energy limitations of ^{201}Tl led to the development of ^{99m}Tc-labeled complexes that would allow administration of larger amounts of activity with improved image quality and lower radiation dose. In addition, technetium agents permit gated imaging so that ventricular wall thickening and function can be assessed. The properties of ^{201}Tl and ^{99m}Tc are compared in Table 21-6. A significant advantage of the ^{99m}Tc-labeled agents is that gated studies can be done to evaluate ventricular function because these agents are fixed in the myocardium and large amounts of activity are administered, which is necessary for gated ventriculography studies.

In 1982, Jones et al.[54] prepared the hexakis(isonitrile) Tc(I) complex [^{99m}Tc(CNR)$_6$]$^+$, in which R could be one of several alkyl groups, the most successful being 2-methoxy isobutyl. It produced the stable Tc(I) complex ^{99m}Tc-sestamibi, which was easy to prepare and achieved myocardial uptake proportional to blood flow when compared with thallium.[55] Subsequently, ^{99m}Tc-tetrofosmin was developed. It is a phosphine complex of Tc(V) that exhibits biologic properties similar to those of ^{99m}Tc-sestamibi.[56,57] ^{99m}Tc-sestamibi and ^{99m}Tc-tetrofosmin were shown to be comparable to ^{201}Tl-chloride in MPI, but their heart activity was fixed and did not redistribute as did thallium. Thus, while a thallium stress–rest imaging protocol can be accomplished with one dose of thallium because of its redistribution in the myocardium, two separate doses of the technetium agents must be administered, one during stress and one during rest.

Thallous Chloride Tl 201 Injection

Thallous chloride Tl 201 injection (^{201}Tl-chloride) is available commercially as a prepared, ready-to-use injection. Thallium is a potassium analogue that localizes in the heart in proportion to myocardial blood flow. After intravenous administration of ^{201}Tl-chloride, thallium disappears rapidly from the blood, with only 5% to 8% remaining 5 minutes after injection.[58] Maximum myocardial uptake occurs in 10 to 30 min-

TABLE 21-5 Biochemical Properties of SPECT and PET Myocardial Perfusion Imaging Agents

Property	^{201}Tl-Chloride	^{99m}Tc-Sestamibi	^{99m}Tc-Tetrofosmin	^{82}Rb-Chloride	^{13}N-Ammonia
Molecular charge	+1 Cation Hydrophilic	+1 Cation Lipophilic	+1 Cation Lipophilic	+1 Cation Hydrophilic	Neutral Lipophilic
Uptake mechanism	Active transport Na-K ATPase	Passive diffusion	Passive diffusion	Active transport Na-K ATPase	Passive diffusion $NH_4^+ \rightarrow NH_3 + H^+$ (19 μsec equilibrium)
First-pass extraction (at resting blood flow)	~ 85% Diffusion limited >2.5 mL/min/gram	~ 66% Diffusion limited >2.0 mL/min/gram	~ 54% Diffusion limited > 1.7 mL/min/gram	59% Diffusion limited >1 mL/min/gram	82% Diffusion limited >2 mL/min/gram
Heart uptake (% injected dose at rest)	~ 4%	~ 1.2%	~ 1.0%		
Heart retention	Redistributes	Bound in mitochondria	Bound in mitochondria	Redistribution (not possible) (rapid physical half-life = 76 sec)	(^{13}NH$_3$ + Glutamic acid → ^{13}N-Glutamine) Biologic half-life = 1–2 hr (rapid physical half-life = 10 min)

TABLE 21-6 Comparison of ^{201}Tl and ^{99m}Tc for Myocardial Perfusion Imaging

Property	^{201}Tl	^{99m}Tc
Photon energy	69–82 keV Low energy Wider energy window More scatter More attenuation Low resolution	140 keV High energy Narrow energy window Less scatter Less attenuation High resolution
Half-life	73 hr	6 hr
Administered activity	Low dose (2–4 mCi) Long collection times Low count densities	High dose (10–30 mCi) Short collection times High count densities
Gated images	No	Yes Wall thickening Ventricular function
First-pass study	No	Yes
Effective dose	1.3 rem(cSv)/mCi (3.9 rem/3 mCi)	0.056 rem(cSv)/mCi (1.68 rem/30 mCi)

utes in the resting state and within 5 minutes after exercise stress. In humans, approximately 4% of the injected dose (ID) is localized in the heart. Thallium heart activity is sustained long enough after exercise stress that imaging can be performed; however, imaging should be completed by 35 minutes after injection, before thallium begins to redistribute from the heart into blood.[52] When ^{201}Tl-chloride is given by intracoronary injection in the presence of very low levels of plasma ^{201}Tl activity, the half-life of ^{201}Tl release from the heart has been found to be rapid (about 45–90 minutes).[52] The half-life from the heart after intravenous injection in clinical studies was found to be much slower (4.5 to 8 hours) because of equilibration of heart activity with ^{201}Tl activity in the blood and other organs.

Thallium's myocardial uptake measured in dogs at resting blood flow was found to be 0.039% of the ID per gram, rising to 0.061% after dipyridamole administration.[59] The fact that the increased uptake with dipyridamole was only 60% higher when blood flow increased to 3 to 4 times normal indicates a nonlinear extraction of ^{201}Tl with increased blood flow. The linear extraction of thallium with flow begins to fall off when blood flow is 2 to 2.5 times normal, which is about the level at which flow increases with exercise stress.[59] A similar nonlinear relationship was shown with adenosine treatment.[60]

Total body elimination of thallium is slow, with a biologic half-life of 10 days.[58] Only 5% is excreted in the urine by 24 hours, and insignificant fecal excretion occurs. These factors plus the long physical half-life of 73 hours are undesirable properties of ^{201}Tl and contribute to its high radiation absorbed dose. The critical organs are the thyroid (2.3 rad[cGy]/mCi) and the testes (3 rad[cGy]/mCi), and the effective dose is 1.3 rem(cSv)/mCi.

After transcapillary diffusion from blood into the interstitial space, thallium is taken up into myocytes by the membrane-bound Na-K ATPase pump. Factors cited for the similarity of action between K$^+$ and its biologic analogues are a monovalent cationic charge and similar hydrated ionic radii. Cation uptake by the membrane pump is known to be partially inhibited by ouabain and hypoxia.[60–62] While there is a close relationship between thallium uptake and blood flow, adequate tissue oxygenation to support cellular metabolism is also required for uptake.[63] Figure 21-7 illustrates sodium and potassium ion transport in myocardial cells. High intracellular potassium and extracellular sodium ion concentrations exist in normal cells, but this situation is reversed when cells are damaged.[64,65] A significant reduction in blood flow and an oxygen deficit impair the intracellular–extracellular exchange of ions and cause loss of lactic dehydrogenase from heart tissue.[66]

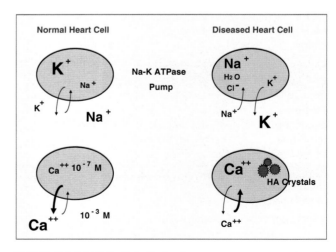

FIGURE 21-7 Sodium, potassium, and calcium ion transport between myocyte cytoplasm and interstitial fluid, demonstrating alterations in ion concentration in normal and necrotic cells.

Although uptake of thallium into the myocardium depends on adequate blood delivery of tracer to the myocardium, active transport across the myocyte sarcolemmal membrane is independent of flow and occurs only if myocytes have not experienced irreversible injury.[66] Thus, thallium is well suited as a marker of myocardial viability.

Technetium Tc 99m Sestamibi Injection

Technetium Tc 99m sestamibi injection (^{99m}Tc-sestamibi) is a monovalent cationic lipophilic complex composed of one atom of Tc(I) and six ligand molecules of 2-methoxy isobutyl isonitrile (MIBI). Its preparation from a radiopharmaceutical kit and its chemistry are discussed in Chapter 10, and its biochemical properties are compared with those of other perfusion imaging agents in Table 21-5.

After intravenous injection, ^{99m}Tc-sestamibi clears rapidly from the blood and remains relatively fixed in the myocardium, with a resting heart uptake of 1% of the ID. This amount is about one-fourth the uptake of ^{201}Tl, but the 10 to 30 mCi (370–1110 MBq) administered activity of ^{99m}Tc-sestamibi results in a much higher count density in the heart. Compared with ^{201}Tl, ^{99m}Tc-sestamibi's 140 keV photons have less tissue attenuation, are detected efficiently, and produce sharper, higher-contrast images. Table 21-6 compares the general properties of ^{201}Tl and ^{99m}Tc for MPI studies.

^{99m}Tc-sestamibi is extracted by the myocardium in proportion to blood flow up to 2.0 to 2.5 mL/minute per gram.[67,68] In canine myocardium at rest, its first-pass extraction is about 65%, decreasing above resting flow, in comparison with 82% extraction by ^{201}Tl.[67] In a rabbit perfused heart model, ^{201}Tl also shows higher extraction than ^{99m}Tc-sestamibi.[51] The higher extraction of thallium relative to ^{99m}Tc-sestamibi is due to higher transcapillary exchange (i.e., the plasma-to-interstitial space transport of thallium is greater). However, ^{99m}Tc-sestamibi has a higher parenchymal cell permeability and volume of distribution, contributing to slower cellular washout than with thallium.[51]

Myocardial uptake studies indicate that ^{99m}Tc-sestamibi enters the myocardium by passive diffusion because of its lipophilicity and is retained for a prolonged period of time. Uptake is related to the integrity of the plasma and mitochondrial membrane potentials in myocytes.[69] Subcellular distribution studies in rat heart demonstrate that approximately 90% of ^{99m}Tc-sestamibi is associated with mitochondria in an energy-dependent manner as a free cationic complex.[70] The attraction of ^{99m}Tc-sestamibi to mitochondria is promoted by a negative potential generated in the mitochondrial membrane, implying that ^{99m}Tc-sestamibi is a marker of both myocardial perfusion and viability.[70] Other studies show that the uptake of ^{99m}Tc-sestamibi by the heart is not affected by ouabain, and thus it is not extracted by the Na-K ATPase membrane pump.[61] Therefore, sestamibi is not a potassium analogue.

In humans, the mean heart retention of ^{99m}Tc-sestamibi is 1% ± 0.4% of the ID 60 minutes after intravenous injection at rest and 1.4% ± 0.3% after exercise.[55] The activity that is fixed in the myocardium demonstrates insignificant redistribution.[71–73] After intravenous injection, over 90% of ^{99m}Tc-sestamibi clears from the blood in less than 5 minutes. Blood levels immediately after injection with ^{99m}Tc-sestamibi are higher than those after injection with ^{201}Tl-thallous chloride, presumably because of lower extraction, whereas late blood levels are lower, presumably because of lack of redistribution.[55] ^{99m}Tc-sestamibi is excreted intact principally by the kidneys and the hepatobiliary system. By 24 hours, urinary excretion is 30% and 24% of the ID after rest and exercise studies, respectively. By 48 hours, fecal excretion is 37% at rest and 29% after exercise.[55] This difference in fecal excretion via the hepatobiliary system is due to reduced splanchnic blood flow and lower liver uptake during exercise. The highest activity is achieved in the gallbladder and liver, followed by the heart, spleen, and lungs. However, activity in the liver, lungs, and spleen decreases more rapidly with time than heart activity and more so with exercise studies than at rest. The biologic half-life from the heart is 6 hours and from the liver 0.5 hours after rest or exercise.[74] Thus, the heart-to-liver ratios at 30, 60, and 120 minutes are 0.5, 0.6, and 1.1, respectively, after rest injection and 1.4, 1.89, and 2.3 after exercise injection. Rest imaging thus is best begun 60 minutes or more after injection of ^{99m}Tc-sestamibi; however, some practitioners shorten the dose-to-image time to 15 to 30 minutes for SPECT exercise studies because of higher heart-to-liver ratios. Dose-to-image time should be 60 minutes if adenosine or dipyridamole stress is used, to allow more time for splanchnic activity to clear.[30]

The critical organ after rest injection is the upper large intestinal wall (3.7 rad[cGy]/20 mCi), and the effective dose is 1.11 rad(cGy)/20 mCi.[75]

Technetium Tc 99m Tetrofosmin Injection

Technetium Tc 99m tetrofosmin injection (^{99m}Tc-tetrofosmin) or [^{99m}Tc-(tetrofosmin)$_2$O$_2$]$^+$ is a monovalent cationic lipophilic complex of the Tc(V) dioxo core, O=Tc=O$^+$, and two bis[(2-ethoxyethyl)phosphino]ethane ligands.[56] Its preparation from a radiopharmaceutical kit and its chemistry are discussed in Chapter 10, and its biochemical properties are compared with other perfusion imaging agents in Table 21-5.

After intravenous injection in humans, ^{99m}Tc-tetrofosmin clears rapidly from the blood; by 10 minutes, less than 5% of the ID remains, with less remaining after exercise.[57] Uptake in skeletal muscle is the main reason for high background clearance after exercise. After rest injection, heart activity declines slowly over time; it is 1.2%, 1.0%, and 0.7% of the ID at 1, 2, and 4 hours, respectively. Values are slightly higher after exercise injection. Liver uptake at these same times is 2.1%, 0.9%, and 0.3% of the ID. Heart-to-liver ratios at 30 and 60 minutes after dosing, respectively, are 0.6 and 1.2 at rest and 1.2 and 3.1 after exercise, reflecting high and rapid hepatobiliary clearance. The rapid clearance of ^{99m}Tc-tetrofosmin from abdominal organs allows imaging to be performed 30 to 45 minutes after injection at rest and 15 minutes after exercise.[30,47]

^{99m}Tc-tetrofosmin's uptake in the heart is proportional to coronary blood flow up to 1.7 to 2.0 mL/minute per gram.[76] This extraction is somewhat less than with ^{99m}Tc-sestamibi and ^{201}Tl-chloride. All the SPECT agents underestimate flow at high flow rates and overestimate flow at very low flow rates, which is a function of the time tracer is available for extraction.[76] When compared with thallium extraction in the rabbit heart, tetrofosmin exhibits a myocardial extraction 60% that of thallium, but heart uptake of both tracers is correlated with blood flow.[77]

Approximately 66% of the injected activity is excreted in 48 hours, about 40% in urine and 26% in feces.[78] The critical organ is the gallbladder wall, with a radiation absorbed dose of 0.123 rad(cGy)/mCi (stress) and 0.180 rad(cGy)/mCi (rest).[78]

^{99m}Tc-tetrofosmin uptake in myocytes is by a metabolism-dependent process that does not involve cation channel transport. The most likely mechanism for this is potential-driven diffusion of the lipophilic cation across the sarcolemmal and mitochondrial membranes.[79] ^{99m}Tc-tetrofosmin appears to be associated more with the cytosol than within mitochondria, whereas ^{99m}Tc-sestamibi demonstrates a higher concentration within the mitochondria.[80] Quantitative analysis of ^{99m}Tc-tetrofosmin retention in the heart demonstrates that washout is very slow (4% per hour after exercise and 0.6% per hour after rest).[81] Ischemic-to-normal myocardium ratios of ^{99m}Tc-tetrofosmin range only from 0.75 to 0.72 from 5 minutes to 4 hours after injection, indicating insignificant redistribution of this agent over time. ^{99m}Tc-sestamibi, however, has been shown to exhibit a small degree of redistribution between 1 and 3 hours after dosing because of faster clearance rates from normal segments.[82] Ischemic-to-normal wall ratios with ^{99m}Tc-sestamibi were significantly higher at 3 hours (0.84) than at 1 hour (0.73), which may affect detection of CAD in cases where the ischemic defect is slight or mild. Indeed, ^{99m}Tc-tetrofosmin has been shown to underestimate perfusion abnormalities compared with ^{99m}Tc-sestamibi in mild-to-moderate CAD during dipyridamole stress SPECT.[83]

PET Perfusion Imaging Agents

PET perfusion studies offer the advantage of higher photon energy and greater tissue penetration, improving specificity in patients who have large body mass or breast tissue that may cause significant photon attenuation, leading to false positive diagnoses of nonreversible defects on resting studies. Also, the ability to see calcium deposits on the CT images provides additional prognostic information for the patient. Other advantages include increased sensitivity and spatial resolution (PET, 6 to 10 mm; SPECT, 10 to 15 mm), which translates into a higher degree of accuracy of PET versus SPECT in the detection of CAD with perfusion tracers.[84] Table 21-5 summarizes the biochemical properties of PET MPI agents.

Ammonia N 13 Injection

Ammonia N 13 injection (^{13}N-ammonia) has proven to be one of the most effective MPI tracers in PET. Its first-pass extraction is high (82%) because of the rapid diffusion of uncharged lipophilic ammonia across the capillary endothelium and myocyte membrane (Figure 21-6). However, back-diffusion of unfixed tracer occurs, so the amount retained decreases as coronary blood flow increases. At coronary blood flows of 1 and 3 mL/minute per gram, the average first-pass retention is 83% and 60%, respectively.[49] Once taken up into the myocyte, ammonia is rapidly fixed as ^{13}N-glutamine by the enzymatic conversion of glutamic acid by glutamine synthetase.

^{13}N-ammonia is cyclotron produced, and its 10 minute half-life restricts its use to facilities with an on-site cyclotron. The production of ^{13}N and other positron emitters and their chemistry are discussed in Chapter 12.

Rubidium Chloride Rb 82 Injection

Rubidium chloride Rb 82 injection (^{82}Rb-chloride) is a generator-produced nuclide with a half-life of 76 seconds. The availability of a generator in the nuclear medicine clinic gives ^{82}Rb-chloride a big advantage over ^{13}N-ammonia. ^{82}Rb is the daughter nuclide of ^{82}Sr, which has a 25 day half-life, giving the generator a useful life of 4 to 6 weeks. ^{82}Rb is eluted from the generator with 0.9% sodium chloride injection as rubidium chloride. Patients receive doses of ^{82}Rb directly from the generator by way of an automated delivery system. Repeat doses can be obtained from the generator every 10 minutes. After intravenous injection of ^{82}Rb-chloride, the rubidium cation, an analogue of potassium, is taken up across the sarcolemmal membrane via the Na-K ATPase pump. A substantial amount of nontransported tracer back-diffuses from the interstitial space and is washed away in increasing amounts nonlinearly as coronary blood flow increases.[49] The mean extraction fraction at 1 and 3 mL/minute per gram is 59% and 26%, respectively (Figure 21-6).[49,85] A disadvantage of ^{82}Rb is its high-energy positron, which has a long travel path before annihilation, decreasing intrinsic resolution.[85] Table 21-7 summarizes the physical properties of radionuclides used in PET heart imaging.[86]

Water O 15 Injection

Water O 15 injection (^{15}O-water) requires an on-site cyclotron for its production and use. ^{15}O ($T_{1/2}$ 2.04 minutes) as labeled water is a freely diffusible tracer whose first-pass extraction is 95%. It is independent of blood flow and metabolically inert and thus appears to be an ideal tracer for perfusion studies. However, ^{15}O-water also distributes into tissues adjacent to the heart (lung and heart blood pool), which complicates perfusion imaging by requiring the use of background subtraction techniques.[49]

^{15}O-water is administered as an intravenous bolus of 30 mCi (1110 MBq), with imaging beginning immediately. An

TABLE 21-7 Physical Properties of PET Radionuclides for Cardiac Imaging

Nuclide	$T_{1/2}$ (minutes)	E max β+ (MeV)	Root Mean Square Range in Water (mm)
^{18}F	109.8	0.64	0.23
^{13}N	9.97	1.19	0.58
^{15}O	2.04	1.73	1.0
^{82}Rb	1.27	3.38	2.6

alternative to injection of ^{15}O-water is inhalation of ^{15}O-labeled carbon dioxide for 3 to 4 minutes with continuous imaging. In vivo, the ^{15}O-carbon dioxide is converted to ^{15}O-water by carbonic anhydrase.[49] After the ^{15}O-water perfusion imaging study, and allowing for complete decay, the blood pool is labeled by a single-breath dose of ^{15}O-carbon monoxide, which labels red blood cells by forming ^{15}O-carboxyhemoglobin. Blood pool images are acquired and subtracted from heart perfusion images to assess net myocardial perfusion. The challenging requirements of ^{15}O-water limit its use for routine PET perfusion imaging.

Ventriculography Imaging Agents

Ventriculography studies are done with agents that produce a high photon flux either in the cardiac blood pool (e.g., ^{99m}Tc-labeled red blood cells) or within the myocardium (e.g., ^{99m}Tc-sestamibi and ^{99m}Tc-tetrofosmin for SPECT studies or ^{13}N-ammonia and ^{82}Rb-chloride for PET studies). This requirement eliminates the use of ^{201}Tl, whose high radiation absorbed dose limits the amount of administered activity to 2 to 4 mCi. ^{99m}Tc is an ideal nuclide for ventriculography studies because of its low radiation absorbed dose at high administered activities. ^{13}N-ammonia and ^{82}Rb-rubidium are also useful because of the high amounts of activity administered in PET MPI.

Technetium Tc 99m Red Blood Cells

The various methods of labeling red blood cells (RBCs) with technetium are discussed in Chapter 10. A popular labeling method for cardiac blood pool imaging is the *in vivo method*. With this method, a dose of "cold" stannous pyrophosphate (Sn-PPi) is injected intravenously and allowed to equilibrate for 20 to 30 minutes to "tin" the RBCs in vivo.[87] Subsequently, ^{99m}Tc-sodium pertechnetate is injected, whereupon the pertechnetate ion enters the RBCs and is reduced by the intracellular stannous ion and bound to intracellular hemoglobin. This provides a fixed blood pool of activity with minimal background interference from other organs around the heart. Although much of the administered stannous ion in Sn-PPi is widely distributed in the body, the amount associated with the RBCs produces a labeling efficiency of about 75%.[88,89]

An alternative method of labeling RBCs is the *modified in vivo method*. It has the advantage of higher labeling yields. As in the in vivo method, a dose of "cold" Sn-PPi is injected intravenously and allowed to equilibrate for 20 to 30 minutes. Subsequently, 10 mL of tinned RBCs is removed into a heparinized syringe containing 20 to 30 mCi of ^{99m}Tc-sodium pertechnetate and allowed to incubate for 10 minutes. This method produces a 90% or greater labeling yield.

^{99m}Tc-RBCs circulate with a long biologic half-life, providing good photon flux for imaging from the blood pool activity in the cardiac chambers. ^{99m}Tc-RBCs are useful for ventriculography studies not associated with perfusion imaging, such as evaluation of heart function in patients undergoing chemotherapy with drugs that are potentially cardiotoxic. The in vivo method of labeling RBCs is more suitable for first-pass ventriculography studies, which require a small-volume bolus injection of activity that is easy to accomplish with a high-concentration ^{99m}Tc-sodium pertechnetate dose. The modified in vivo method involves injection of at least a 10 mL volume of labeled cells and is more suitable for gated equilibrium studies that do not require a small bolus injection.

Imaging Agents for Myocardial Metabolism and Viability

Myocardial cells are considered to be nonviable when some basic aspect of cell behavior no longer functions.[90] When CAD is present, the progression of atherosclerosis causes arterial stenosis and a gradual loss of arterial compliance, resulting in a temporary reduction in blood flow (ischemia) to meet myocardial oxygen demand.[12] A principal symptom is chest pain upon exertion, believed to be caused by lactate accumulation. Other changes include electrical conduction abnormalities and loss of contractile function in the affected muscle. The loss of functional integrity in ischemic but viable myocardium can be reversed if blood flow is restored. However, sustained insufficient perfusion and oxygenation eventually causes loss of metabolic function and cellular death.

There are several unique metabolic pathways in the myocyte that can be investigated. (Figure 21-8). Imaging agents that probe these pathways can provide useful information regarding myocardial function and muscle viability.

Thallium and Rubidium

Two imaging agents that probe parenchymal membrane function are the cations ^{201}Tl$^+$ and ^{82}Rb$^+$, described earlier. As analogues of potassium, these radioactive cations are taken up into the myocyte via the sodium–potassium (Na-K) ATPase membrane pump, which maintains cell function and volume through high extracellular sodium ion and high intracellular potassium ion concentrations (Figure 21-7).[4] It is believed that the energy available for the pump is decreased during ischemia, so that sodium ions along with chloride ions and water accumulate within the cell and potassium ions leak out into the extracellular space. This process alters the sodium–potassium ion concentration ratio, causing a marked effect on membrane polarity and heart muscle function. This disparity in ion shift is the basis for using ^{201}Tl$^+$ and ^{82}Rb$^+$ for imaging. These agents participate in the ion transport process similarly to potassium; they are taken up and redistributed out of myocytes over time and therefore can be used to assess membrane integrity and cellular viability.[90]

Fludeoxyglucose F 18 Injection

Fludeoxyglucose F 18 Injection (^{18}F-FDG, FDG) contains ^{18}F ($T_{1/2}$ 110 minutes) labeled as 2-fluoro-2-deoxyglucose. FDG is the premier metabolic marker for glucose metabolism. This hydrophilic molecule is taken up into cells via facilitated diffusion, similar to glucose uptake, but its metabolism is different (Figure 21-8). The first step of glucose and FDG metabolism is similar. Hexokinase converts glucose and FDG into glucose-6-phosphate and FDG-6-phosphate, respectively.

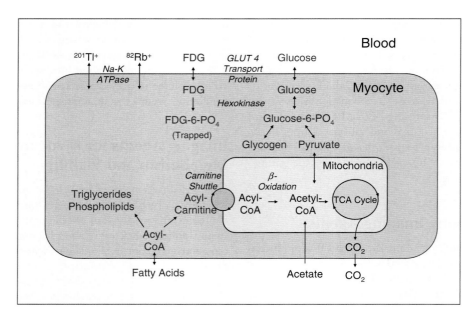

FIGURE 21-8 Metabolic processes occurring in the myocyte, illustrating the membrane uptake mechanisms for monovalent cations (^{201}Tl$^+$ and ^{82}Rb$^+$) via Na-K ATPase and FDG and glucose via the glucose transporter. While glucose is metabolized completely, FDG is trapped as FDG-6-phosphate following enzymatic conversion by hexokinase. Fatty acids, as the preferred metabolic substrate for energy, can be converted into triglycerides or phospholipid stores or shunted, via the carnitine shuttle, into mitochondria for beta oxidation. (Reprinted with permission of Elsevier from Schwaiger M, Hutchins GD. Evaluation of coronary artery disease with positron emission tomography. *Semin Nucl Med.* 1992;22:210–23.)

While glucose-6-phosphate participates further as a substrate for glycolysis or glycogen synthesis, FDG-6-phosphate is not a substrate for these pathways. Because it cannot readily diffuse out of the cell, FDG-6-phosphate becomes trapped and accumulates in cells that are in active metabolism. The phosphorylation of FDG is independent of blood flow, and therefore FDG is not a useful tracer of myocardial perfusion. The phosphorylation step is slow and requires about 45 minutes for maximum accumulation of the FDG dose into the heart before imaging can commence.

Under aerobic fasting conditions, the heart uses fatty acids for its energy needs. However, after a glucose load, such as after a carbohydrate meal, elevated plasma insulin levels cause an increase in myocardial glucose metabolism in preference to fatty acid metabolism. Glucose metabolism is also upregulated in ischemic heart muscle, because of a switch to anaerobic glycolysis. If the ischemic regions contain viable muscle, they will accumulate FDG. Areas of scarring will not accumulate FDG. This is the basis for using FDG as a metabolic marker of myocardial viability. It is the principal agent used to assess viability in PET imaging.

Carbon C 11 Palmitate

Carbon 11 palmitate (^{11}C-palmitate) is used to provide information about myocardial fatty acid metabolism (Figures 21-8 and 21-9). The free fatty acids are largely distributed in the myocardium in proportion to blood flow, whereupon they cross the parenchymal membrane, presumably by passive diffusion or possibly by facilitated transport.[39] Retention or trapping of ^{11}C-palmitate requires energy-dependent esterification to ^{11}C acyl-CoA, which can enter either of two routes. In one fraction it moves via the carnitine shuttle to the inner mitochondrial membrane, where β-oxidation cleaves two carbon fragments off the long carbon chain, directing acetyl CoA to the tricarboxylic acid (TCA) cycle for complete oxidation to carbon dioxide and water. The other fraction of acyl-CoA is further esterified and deposited as triglyceride and phospholipid stores.

After intravenous administration of 15 to 20 mCi (555–740 MBq) of ^{11}C-palmitate and image acquisition, stored images of the ventricles are used to identify regions of interest, from which time–activity curves are generated for analysis of uptake and clearance kinetics. The slopes of these curves can then be used to assess regional metabolism in the myocardium.[49]

Iodofiltic Acid I 123 Injection

Iodofiltic acid I 123 (β-methyl-*p*-[^{123}I]-iodophenyl-pentadecanoic acid, ^{123}I-BMIPP) is a methyl-branched analogue of iodophenylpentadecanoic acid (^{123}I-IPPA). ^{123}I-IPPA is a straight-chain fatty acid taken up into myocytes during aerobic metabolism. It is rapidly metabolized and eliminated following β-oxidation in mitochondria, which presents time constraints for SPECT imaging. ^{123}I-BMIPP is also taken up into myocytes but does not undergo β-oxidation because of the branched β-methyl group (Figure 21-9) and is trapped within

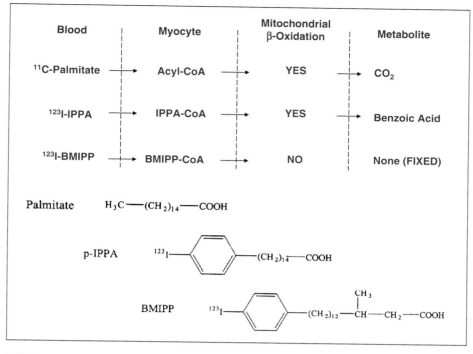

FIGURE 21-9 Metabolism of radiolabeled fatty acid analogues in the heart. The straight-chain analogues ^{11}C-palmitate and ^{123}I-IPPA undergo beta oxidation to metabolites that are readily eliminated, whereas the branched-chain ^{123}I-BMIPP analogue cannot undergo beta oxidation and is trapped in the myocardium.

the myocyte. This permits sufficient time to perform SPECT imaging to assess fatty acid metabolism. ^{123}I-BMIPP is being investigated for SPECT imaging in myocardial ischemia.[91]

Heart muscle derives its energy from the metabolism of fatty acids in the fasting aerobic state. Myocardial ischemia causes hypoxic conditions within the ischemic region, shifting energy production from fatty acid metabolism to glucose metabolism. This metabolic shift can be assessed with radiolabeled fatty acids, such as ^{123}I-BMIPP, as regions of decreased count density relative to normal muscle. It has been observed that ^{123}I-BMIPP is capable of identifying regions of myocardial ischemia for a long period of time (up to 30 hours) after an exercise-induced ischemic event and restoration of perfusion.[91] This phenomenon of delayed recovery of regional fatty acid metabolism following ischemia and reperfusion is known as metabolic stunning. It is analogous to the delayed recovery of regional function after reperfusion following an acute ischemic event, known as myocardial stunning. The potential importance of this observation is that ^{123}I-BMIPP imaging may extend the time window for identifying myocardial ischemia long after resolution of chest pain and restoration of resting myocardial blood flow.[91]

Sodium Acetate C 11 Injection

Sodium acetate C 11 injection (^{11}C-acetate) is avidly extracted by the myocardium and activated to acetyl-CoA, which is oxidized in the mitochondria to ^{11}C-carbon dioxide and water (Figure 21-8). ^{11}C-acetate permits evaluation of flux through the TCA cycle and overall myocardial oxygen consumption because of its close link to oxidative phosphorylation.[49] Myocardial oxidative capacity can be measured by analyzing uptake and clearance curves. These curves change with regional abnormalities in blood flow and metabolism. Thus, in ischemic regions, the initial uptake of ^{11}C-acetate decreases in proportion to myocardial blood flow, and clearance curves are reduced significantly.[49]

^{18}F-FDG and ^{11}C-palmitate are tracers that measure specific substrates in myocardial metabolism, whereas ^{11}C-acetate measures overall oxidative metabolism. ^{18}F-FDG measures the initial steps of exogenous glucose utilization, whereas ^{11}C-palmitate traces the entire pathway of fatty acid metabolism. In the fasted state, when fatty acid metabolism is highest in the myocardium, ^{18}F-FDG demonstrates very little uptake in the heart, but ^{11}C-palmitate uptake and clearance curves reflect rapid kinetics. The combination of ^{11}C-palmitate or ^{18}F-FDG with ^{11}C-acetate permits assessment of the contribution of fatty acid or glucose metabolism to overall oxidative metabolism or, in the ischemic condition, assessment of anaerobic glucose metabolism.[49]

NUCLEAR MEDICINE PROCEDURES

There have been many advances in MPI since it was first introduced by Zaret et al.[92] in 1973, using ^{43}K to evaluate for the presence or absence of CAD. Since that time, advances in hardware and the development of new cardiac perfusion agents have allowed for excellent gated cardiac imaging. The main nuclear medicine applications of cardiac imaging are MPI to assess the distribution of coronary blood flow to the myocardium and gated equilibrium radionuclide ventriculography (RNV) to assess left ventricular performance. First-pass techniques can also be used to evaluate right and left ventricular performance.

MPI is now well established as an effective tool in the evaluation and management of CAD. It is commonly used in patients with known or suspected CAD to assess for the presence of significant ischemia as well as to determine the location and severity of the lesions. MPI can also be used to assess myocardial viability. For patients with known MI, both the size and the severity of the infarct territory can be evaluated. Left ventricular ejection fraction (LVEF) and left ventricular volumes can also be determined with gated MPI. Event-free cardiac survival has been found to be directly proportional to the LVEF after MI.[93,94] Information obtained during MPI can be used to stratify risk and help determine prognosis.

Gated equilibrium RNV is useful in evaluating ventricular function in patients with congestive heart failure (CHF) and in patients undergoing chemotherapy with cardiotoxic drugs. Variables such as LVEF, regional wall motion, end systolic (ES) and end diastolic (ED) volumes, and peak filling rate can be estimated. If first-pass imaging is performed, right ventricular performance can also be evaluated.

Myocardial Perfusion Imaging

A variety of imaging protocols utilizing SPECT and PET radiotracers have evolved to assess myocardial perfusion.[30] Some procedures originally developed to assess myocardial perfusion only have been modified to give some assessment of myocardial viability as well. In addition, perfusion studies with ^{99m}Tc, ^{82}Rb, and ^{13}N can be gated to provide additional functional information during the perfusion study.

Rationale
MPI uses intravenously administered radiopharmaceuticals to assess coronary blood flow to the myocardium. MPI radiopharmaceuticals are taken up into the myocardium in proportion to blood flow and can be used to evaluate areas in the myocardium with reduced blood flow associated with ischemia or scarring. If there is significant coronary artery stenosis, there will be an area or areas of reduced radiotracer uptake in the territory supplied by the affected arteries. If the area of reduced uptake in the myocardium is worse during conditions of stress than at rest, the perfusion abnormality is most likely due to ischemia. If the area shows reduced uptake during both stress and rest conditions, the perfusion abnormality is more likely due to scarring and may need to be evaluated further to rule out potential artifacts.

Information gained during the MPI study can be used not only to identify significant CAD but also to give insight into the patient's prognosis, such as the probability of a hard cardiac event (i.e., MI or cardiac-related death). MPI can help to distinguish low-risk from high-risk cardiac patients, which is important in their clinical management.

Procedures
Myocardial perfusion imaging (MPI) studies are carried out to evaluate myocardial ischemia. Studies are acquired at rest and during stress, either exercise induced or pharmacologic. The imaging protocol depends on the type of radiopharmaceutical administered and whether the study is done on a single day or over 2 days.

A medical history should be obtained before the stress study. This should include the reason for the exam, symptoms that led to the exam, risk factors for CAD, cardiac history, respiratory history, medications, allergies, and results of any prior studies. A 12-lead ECG should be examined to evaluate for left bundle-branch block, significant arrhythmias, and acute ischemia. Patients should fast for a minimum of 4 hours before the stress exam. The contraindications to both exercise and pharmacologic stress should be reviewed before the study.[95]

If there are no contraindications to exercise stress, it is generally the preferred method. Exercise stress gives further information about the patient, such as the degree of exercise tolerance, time to maximal heart rate, and blood pressure response. If feasible, medications that would interfere with the heart-rate and blood-pressure response, such as β-blockers, should be stopped for an appropriate time before the test.

Often exercise is not an alternative for the patient. In these cases, pharmacologic stress with either a vasodilator such as adenosine or an inotropic/chronotropic agent such as dobutamine can be used. There are contraindications for each of these agents. The major contraindications for adenosine include second- or third-degree atrioventricular block, sick sinus syndrome, asthma, wheezing, pulmonary hypertension, and a systolic blood pressure of less than 90 mm Hg. Dobutamine should not be used if the patient has a ventricular tachyarrhythmia. Neither type of agent should be used just after an acute MI or with unstable angina.

An intravenous line is placed before the stress portion of the exam. If the patient is to undergo exercise stress on a treadmill, the intravenous access must be secure and easily accessible for radiopharmaceutical administration at peak stress.

Thallium MPI Protocols
Thallium has been used as a marker of myocardial perfusion and viability for many years, beginning in the mid 1970s.[96] The first thallium procedures were conducted with two separate injections: one for the exercise stress study and one for the rest study. The rest study was conducted 1 to 2 weeks after the stress study because of thallium's long effective half-life in the body. The thallium procedure changed to a 1 day protocol when it was observed that thallium redistributed in the heart over time. By 3 to 4 hours after an exercise study, normal myocardial regions demonstrated a decrease in thallium activity while ischemic regions showed an increase in activity.[97] Subsequently, the standard procedure became a 1 day thallium protocol (exercise study followed by a 3 to 4 hour redistribution study) to differentiate ischemia from infarction. This procedure was shown to provide an accurate assessment of myocardial viability when perfusion defects on the exercise study showed redistribution on delayed imaging.

Although this procedure had high sensitivity for predicting viability of reversible defects, its specificity for predicting nonviability of irreversible defects was low. This became apparent when the size of irreversible defects decreased in a number of patients who had a resting thallium study the next day.[98,99] This indicated that incomplete redistribution of thallium had occurred on the initial redistribution image, causing an overestimation of infarct size at that time. Other studies underscored the high rate of false-positive diagnosis with delayed thallium imaging. One study showed that 45% of fixed defects demonstrated normal perfusion after revascularization[100] and another showed that 58% of fixed ^{201}Tl defects demonstrated ^{18}F-FDG uptake, consistent with viability.[101]

Henceforth, late (8 to 24 hour) redistribution imaging was instituted to allow more time for thallium to redistribute.[102] Kiat et al.[103] demonstrated that 61% of irreversible thallium defects on 3 to 4 hour images reversed on late redistribution at 18 to 72 hours. Late redistribution imaging improved diagnosis, but study specificity was still low, as 37% of patients with apparent irreversible defects had perfusion restored after myocardial revascularization procedures. Thus, although late imaging improved diagnosis, it still overestimated the frequency and severity of myocardial fibrosis.

To improve specificity, a thallium reinjection technique was introduced with the idea that increased plasma levels of thallium during redistribution would improve the count density in underperfused areas. Reinjection of 1 mCi (37 MBq) of ^{201}Tl-chloride, either after the standard 3 to 4 hour redistribution study or after a late 24 hour redistribution study, facilitates uptake of thallium into viable regions of myocardium with apparently irreversible defects. With this technique, up to 49% of apparent irreversible defects on the 4 hour redistribution study[104] and 39% of such defects on the 24 hour redistribution study[105] showed improved or normal uptake after thallium reinjection. A number of studies have also shown the ability of thallium reinjection to predict improved ventricular function after revascularization.[104,106–108] Thus, the thallium stress–redistribution–reinjection protocol has become a standard procedure for evaluating myocardial perfusion and viability (Figure 21-10).

Generally, 2 to 4 mCi (74–148 MBq) of ^{201}Tl-chloride is administered intravenously at peak exercise or peak pharmacologic stress. The patient is maintained at peak stress for 1 to 2 more minutes. SPECT imaging is performed as soon as possible after stress. If patients undergo treadmill exercise, they are imaged as soon as deep breathing in response to exercise normalizes, usually 5 to 10 minutes after the end of stress, to prevent motion artifacts. The net washout rate of ^{201}Tl-chloride after exercise has a half-life of about 4 hours.[109] SPECT imaging of the heart is usually repeated 4 hours after the stress images for comparison.

Images of the heart are usually displayed in three planes as short-axis, vertical long-axis, and horizontal long-axis views (Figure 21-11), with the stress images on top and the corresponding rest images below. Regional defects seen on the stress images are compared with the same regions on the rest images. The stress defects are usually described as fixed, reversible, or partially reversible. If the defect appears fixed or only partially reversible, delayed imaging at 24 hours can be useful to further evaluate the extent of reversibility. An alternative is to

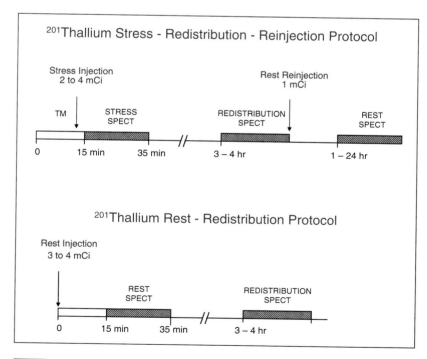

FIGURE 21-10 Thallium rest–redistribution–imaging protocol and stress–redistribution–reinjection imaging protocol. TM = treadmill.

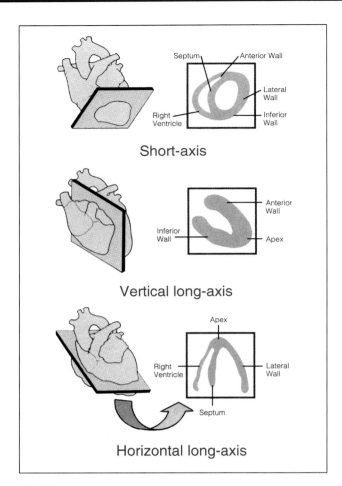

FIGURE 21-11 Cardiac tomography. Diagram of short-axis, vertical long-axis, and horizontal long-axis views of myocardial walls on SPECT and PET imaging. (Reprinted with permission of Lantheus Medical Imaging Inc. from *Introduction to Nuclear Cardiology*, 3rd ed. North Billerica, MA: DuPont Pharma Radiopharmaceuticals; 1993:261.)

reinject the patient with 1 mCi (37 Mbq) of ^{201}Tl-chloride immediately after obtaining the redistribution images to overcome poor count statistics. The reinjection images help to normalize perfusion defects, as does the 24-hour imaging.[104] Reinjection can also be done at 24 hours. Fixed defects seen on the 4 hour redistribution images that show normal or improved ^{201}Tl uptake after reinjection are consistent with ischemic and viable myocardium (Figure 21-12). ^{201}Tl MPI studies are usually not gated in standard practice because of the low photon energies associated with thallium and the long half-life, which limits the administered dose to 2 to 4 mCi (74–148 MBq).

Technetium MPI Protocols

Because of the short 6 hour half-life of ^{99m}Tc, a much larger dose of ^{99m}Tc agents than of ^{201}Tl-chloride can be administered. Also, there is less attenuation and scatter compared with ^{201}Tl because of the higher 140 keV photopeak of ^{99m}Tc. These properties result in a larger photon flux, which allows for gated imaging to evaluate regional ventricular wall motion. The higher administered dose also improves resolution of the images (Table 21-6). The most commonly used compounds are ^{99m}Tc-sestamibi and ^{99m}Tc-tetrofosmin. Like ^{201}Tl-chloride, these compounds are taken up in the myocardium in proportion to blood flow. Unlike ^{201}Tl-chloride, they passively diffuse across the cell membrane, bind to mitochondria, and exhibit limited redistribution. Because of the limited redistribution, ^{99m}Tc-agent protocols are different from ^{201}Tl-chloride protocols, requiring separate injection of stress and rest doses.

Several injection and imaging protocols can be used. One day protocols are most frequently used. However, if the patient is obese, 2 day protocols can be used to maximize the radiotracer dose both at rest and during stress. A common 1 day protocol is the rest–stress protocol (Figures 21-13 and 21-14). This protocol is useful if the goal is to induce ischemia in a patient suspected of having CAD. Exercise is the preferred mode of stress, but if the patient cannot perform adequate exercise, pharmacologic stress is used. Typically, 8 to 10 mCi (296–370 MBq) of a ^{99m}Tc agent is injected at rest, and the patient is imaged 1 hour later. One to four hours after the rest images, 25 to 30 mCi (925–1110 MBq) of the ^{99m}Tc agent is injected at peak stress. The activity in the heart during the stress study overwhelms the residual activity from the rest dose because of the increased blood flow to the heart and the large stress dose administered. Imaging is repeated 10 to 20 minutes later if the patient underwent exercise stress or 45 to 60 minutes later if the stress was pharmacologic. If a 2 day protocol is used, 25 to 30 mCi (925–1110 MBq) of the ^{99m}Tc agent is administered both at rest and during stress, and either the rest or the stress measurements can be done on the first day (Figure 21-15). There is no significant difference in diagnostic accuracy between 1 day and 2 day protocols.[31,110] With the 2 day protocol, doing the stress study on the first day has appeal for a patient whose pretest likelihood of CAD is low, because the rest study can be canceled if the stress study is normal.[31] It is also useful for patients with potential attenuation artifacts caused by large body mass (>250 pounds) or breast tissue.

An advantage of ^{99m}Tc protocols is the ability to evaluate myocardial perfusion and ventricular function with a single injection of tracer by means of ECG gating. This permits the simultaneous evaluation of stress perfusion with resting ventricular function for an estimate of myocardial viability. A stress perfusion defect with preserved regional wall motion and thickening at rest implies ischemia, whereas regional hypokinesis or akinesis and decreased wall thickening could be associated with either scar or severely ischemic or stunned myocardium.[111]

Dual-Isotope Protocols

Some nuclear medicine clinics use a dual-isotope protocol to decrease the overall study time.[31] In this protocol 3 to 4 mCi (111–148 MBq) of ^{201}Tl-chloride is administered at rest, and images are obtained about 15 minutes later (Figure 21-16). About 30 minutes after the rest images are completed, the patient is stressed with exercise or pharmacologically, and 25 to 30 mCi (925–1110 MBq) of a ^{99m}Tc agent is administered at peak stress. The stress images are obtained 30 to 60 min-

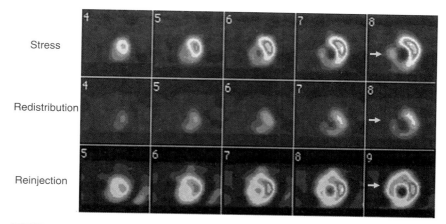

FIGURE 21-12 Adenosine–²⁰¹Tl stress–rest–reinjection myocardial perfusion study. Short-axis stress images (top row) and redistribution images obtained 4 hours later (second row) demonstrate perfusion defects (arrows) involving the septum and inferior walls of the heart, with poor washout over the 4 hour period. Images obtained 24 hours later just after reinjection with 1 mCi (37 MBq) ²⁰¹Tl at rest (bottom row) show improvement in the septum and inferior walls consistent with ischemia and viable myocardium in these regions.

utes later, depending on whether the patient underwent exercise or pharmacologic stress, respectively. A downside to this protocol is the higher radiation dose to the patient.

PET MPI Protocols

As PET technology continues to advance and the number of clinical PET/CT scanners increases, PET MPI is also increasing. Currently ⁸²Rb-chloride and ¹³N-ammonia are approved for PET MPI.

Rubidium. ⁸²Rb-chloride is available from the ⁸²Sr–⁸²Rb generator. ⁸²Rb is a positron-emitting cation that is taken up into the myocardial cells in relation to coronary artery blood flow, similarly to potassium and ²⁰¹Tl.[112] There are advantages to using this agent because of its very short half-life (76 seconds) and high photon energy (511 keV). The short half-life allows for a larger administered dose, faster protocols, and lower radiation exposure of the patient. The 511 keV photon energy offers better attenuation correction and higher sensitivity as the result of electronic collimation. ⁸²Rb-chloride is administered to the patient directly from the generator via an automated system of delivery that monitors the radioactive concentration of ⁸²Rb in the infusion line. The first portion of the generator eluate is dumped to waste, and when the radioactive concentration is high the dose is administered to the patient. A second dose of ⁸²Rb can be obtained from the generator 10 minutes after a previous dose. A typical ⁸²Rb-chloride protocol involves injecting 30 to 60 mCi (1110–2220 MBq) ⁸²Rb-chloride at rest, followed by gated imaging (Table 21-8).

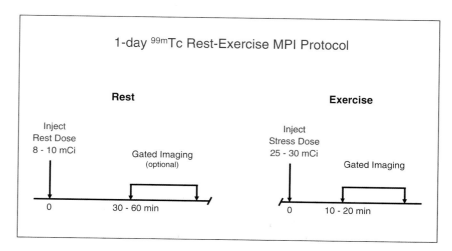

FIGURE 21-13 One-day ⁹⁹ᵐTc rest–exercise myocardial perfusion imaging protocol. (Adapted from reference 30.)

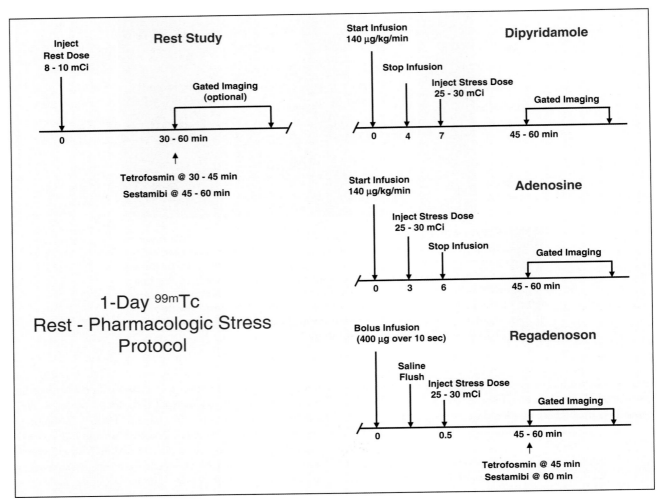

FIGURE 21-14 One-day 99mTc rest–pharmacologic stress myocardial perfusion imaging protocols for dipyridamole, adenosine, and regadenoson. (Adapted from reference 30.)

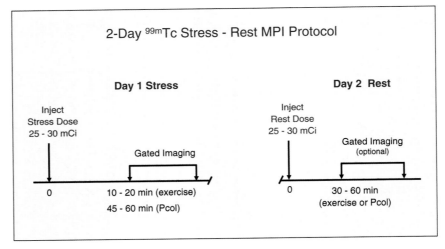

FIGURE 21-15 Two-day 99mTc stress–rest myocardial perfusion imaging protocol. (Adapted from reference 30.)

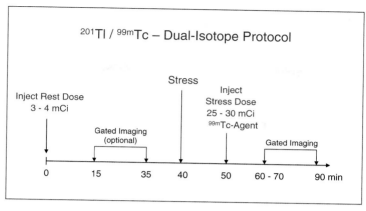

FIGURE 21-16 ^{201}Tl–^{99m}Tc dual-isotope imaging protocol. (Adapted from reference 30.)

Imaging begins at the end of ^{82}Rb infusion. Image processing occurs at 90 to 120 seconds after ^{82}Rb injection, depending on the patient's cardiac function, to allow for blood clearance of ^{82}Rb activity. Redistribution imaging as with ^{201}Tl is not possible with ^{82}Rb because of its short half-life. Thus, ^{82}Rb requires separate injections for rest and stress studies. After rest imaging is completed, the patient is immediately stressed with a pharmacologic agent. Table 21-8 illustrates the stress protocol with adenosine. ^{82}Rb-chloride 30 to 60 mCi is injected at peak stress. Repeat gated imaging begins immediately after the ^{82}Rb infusion while adenosine infusion continues for an additional 4 minutes. Image processing is similar to that in the rest study. The length of the entire protocol is often less than 1 hour.

Ammonia. The use of ^{13}N-ammonia requires an on-site cyclotron. For evaluating myocardial blood flow, 10 to 20 mCi of ^{13}N-ammonia (370–740 MBq) is administered intravenously, with imaging starting about 5 minutes later to allow clearance of excess tracer from the blood. The 10 minute half-life of ^{13}N requires about a 30 minute wait between rest and stress injections to allow for decay.[84] Because of expense, ^{13}N-ammonia perfusion studies are typically performed only in combination with an ^{18}F-FDG study for viability assessment.

Interpretation

Normally, when there is increased demand on the heart due to exercise or pharmacologic stress, blood flow increases in the coronary vessels and there is increased radiotracer uptake in the myocardium related to the increased blood flow. If there are no perfusion abnormalities during stress, there is homogeneous radiotracer uptake throughout the myocardium. In a normal study using a ^{99m}Tc agent or ^{82}Rb-chloride, no perfusion abnormalities are seen on either the stress or rest images (Figures 21-17 and 21-18). If ^{201}Tl-chloride is used, there are no perfusion defects seen in the stress images and there is appropriate washout from the myocardium 4 hours later on the redistribution images (Figure 21-19). If there is a persistent defect on the 4 hour redistribution image, ^{201}Tl-chloride is reinjected and the patient is reimaged at a later time to differentiate ischemia from infarction (Figure 21-12).

Myocardial regions with reduced radiopharmaceutical uptake on the stress portion of the myocardial perfusion study that normalize or partially normalize on the rest study represent regions of stress-induced ischemia associated with flow-limiting CAD (Figures 21-20 through 21-22). Myocardial regions with fixed perfusion defects on both the stress and rest portions of the exam can represent either scarring from prior MI (Figures 21-23 and 21-24) or possibly an attenuation artifact, usually caused by the diaphragm or breast. Gated SPECT allows further evaluation of fixed perfusion defects through examination of wall motion and myocardial thickening. Fixed perfusion defects with normal wall motion and thickening are likely related to attenuation artifacts. Fixed defects with abnormal focal wall motion abnormalities and decreased thickening are generally associated with scar.

Equilibrium Radionuclide Ventriculography

Equilibrium radionuclide ventriculography, also known as gated blood pool imaging or multigated acquisition (MUGA),

TABLE 21-8 Perfusion Imaging Protocol for ^{82}Rb-Chloride

Perform rest CT attenuation

Rest study
— Start ^{82}Rb generator
— Direct low-concentration ^{82}Rb to waste
— Infuse high-concentration ^{82}Rb (30–60 mCi) at 50 mL/min
— Acquire rest images

Stress study
— Adenosine 140 μg/kg for 2 minutes
— Start ^{82}Rb generator
— Direct low-concentration ^{82}Rb to waste
— Infuse high-concentration ^{82}Rb (30–60 mCi) at 3 minute mark
— Begin acquiring stress images
— Adenosine continued for 4 minutes

Perform stress CT attenuation

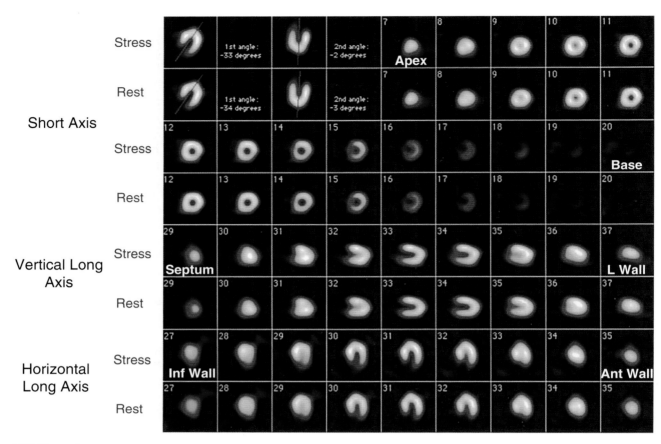

FIGURE 21-17 Normal ^{99m}Tc-sestamibi myocardial perfusion study. Images of the heart are shown in three planes. The top row represents short-axis views of the heart from apex on the left to mid-heart on the right at stress. Matching views on the rest portion of the study are shown in the next row. The third and fourth rows are continuation of the short-axis views at stress and rest from the mid-heart to the base or valve plane. The next two rows (frames 29 to 37) are the vertical long-axis views, stress on top and rest below. These views are from the septum on the left to the lateral wall on the right. The last two rows are the horizontal long-axis views, stress on top and rest below. These views run from the inferior wall on the left to the anterior wall on the right. There are no significant perfusion defects seen either during stress or at rest to suggest flow-limiting coronary artery disease.

is an imaging technique developed in the early 1970s.[113] In this technique the patient's red blood cells are labeled and images are collected as the activity of the labeled red cell blood pool passes through the heart. Image data are analyzed by computer.

Rationale

RNV is a noninvasive way of evaluating left ventricular performance. LVEF can be calculated from the data, and other parameters such as left ventricular ED and ES volumes and peak filling rate (a measure of left ventricular compliance) can be estimated. These values can be used to assess performance after an acute MI. They can also serve to determine baseline function as well as follow-up after treatment in cardiomyopathies and CHF. One of the most common indications is following the performance of the heart after administration of cardiotoxic drugs such as doxorubicin used in cancer chemotherapy. Usually, baseline RNV is performed before the first chemotherapy treatment. RNV is then performed just before each cycle of chemotherapy to determine if there has been a significant decrease in cardiac performance and if it is safe to administer another cycle.

Procedure

The patient's red blood cells are labeled by the in vivo or modified in vivo procedure described previously. The patient is usually positioned in the left anterior oblique (LAO) position for visualization of the heart. Sometimes a craniocaudal tilt is necessary to visualize the left ventricle and septum. If a craniocaudal tilt is applied, this is called a modified LAO position. Gated images of the heart are then obtained. Gating of the images is based on the R-R cardiac interval. The R-R interval is typically divided into 32 segments (previously, 16 segments as shown in Figure 21-25), and the computer begins data collection at the beginning of the R wave. Data are collected in these time bins over several cardiac cycles. The images can then be played back

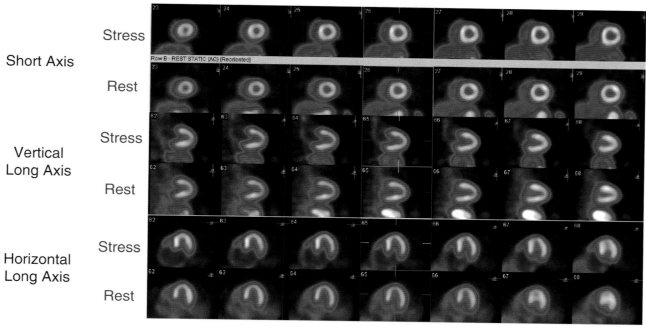

FIGURE 21-18 Normal ^{82}Rb-chloride rest–stress myocardial perfusion study in a 45 year old woman with chest pain and prior history of coronary artery disease and percutaneous transluminal coronary angioplasty. Images were obtained after intravenous injection of 30 mCi of ^{82}Rb-chloride following the protocol outlined in Table 21-8.

as a set of dynamic images to visualize cardiac contraction and evaluate for regional wall motion abnormalities. Regions of interest can be drawn around the left ventricle in diastole and in systole to determine LVEF using the formula below.

$$\text{LVEF} = \frac{(\text{ED counts}) - (\text{ES counts})}{(\text{ED counts}) - (\text{Bkg counts})}$$

Ventricular function can also be assessed during myocardial perfusion imaging studies that are gated. Assessment of LVEF provides valuable prognostic information regarding patient survival,[114] and assessment of wall motion and thickening can help determine whether nonreversible defects are caused by attenuation artifact.[115] Gated SPECT studies can be done with ^{99m}Tc-sestamibi or ^{99m}Tc-tetrofosmin and gated PET studies with ^{82}Rb-chloride or ^{13}N-ammonia.

Interpretation

The cardiac images are played back in a cinematic loop to visually evaluate regional left and right ventricular wall motion. The ventricular walls should contract and the septum should thicken during systole (Figure 21-26). Cardiac wall motion is referred to as normal, hypokinetic, akinetic, or dyskinetic. Hypokinesis refers to diminished wall motion compared with normal (Figure 21-27). Akinetic means lack of wall motion, and dyskinetic refers to paradoxical wall motion as might be seen with an aneurysm. Normally the septum moves less than the other walls, with the greatest motion usually seen in the anterior, lateral, and posterior walls. The septum should thicken during systole. Dyskinetic septal motion can be seen with scarring from an MI. It can also be seen secondary to prior coronary artery bypass surgery.

First-Pass Radionuclide Angiocardiography

First-pass radionuclide angiocardiography involves imaging a bolus of radiopharmaceutical as it passes through the heart. The technique allows for evaluation of both the right and left ventricular ejection fraction. Since a bolus of activity is followed through the heart, the acquisition can be performed in just a few seconds. The technique can be used either at rest or during stress.

Rationale

The first-pass technique can be used to evaluate both right and left heart performance. However, because the configuration of the right ventricle makes it difficult to evaluate with equilibrium techniques, first-pass is a practical technique for determining right ventricular ejection fraction. Right heart function may be of value in assessing cardiomyopathies and pulmonary-related cardiac disease.

Procedure

In the first-pass technique, a small-volume, high-concentration radiotracer bolus (typically 20 to 30 mCi [740–1110 MBq] of ^{99m}Tc-sodium pertechnetate) is administered rapidly into a

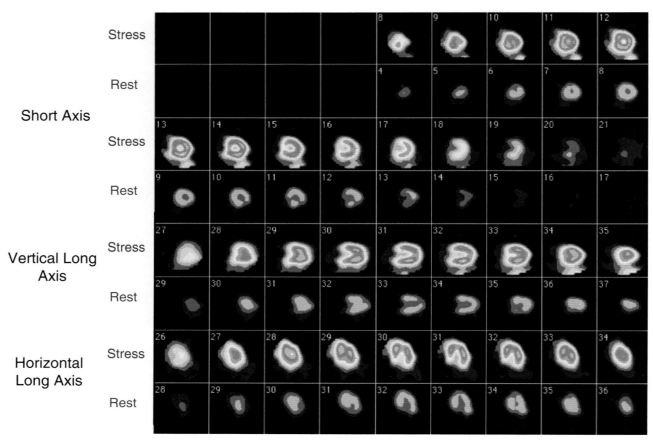

FIGURE 21-19 Normal ^{201}Tl-chloride myocardial perfusion study. Stress images are on top and rest images below. Images of the heart are shown in three planes. The top four rows are the short-axis views from apex on the left to base on the right. The next two rows are the vertical long-axis views, stress on top and rest below. The last two rows are the horizontal long-axis views, stress on top and rest below. On this study, there are no significant perfusion defects during stress, and there is good washout of the radiotracer on the redistribution images obtained 4 hours after stress.

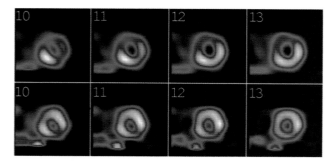

FIGURE 21-20 ^{99m}Tc-sestamibi study demonstrating myocardial ischemia. Short-axis views at stress (top row) and at rest (bottom row). Significant perfusion defects are seen on the stress portion of the study that are not present at rest, consistent with multivessel flow-limiting coronary artery disease.

vein near the heart. Some practitioners prefer to use the external jugular vein for this purpose; however, in most nuclear medicine clinics a large antecubital vein is used. The patient is imaged in the 30° right anterior oblique position or anterior position over several heart cycles. Rapidly acquired image frames are obtained to observe the bolus as it passes through the heart. Ejection fraction measurements for both the right and left ventricles can be obtained by measuring the change in activity over time in each of these regions.

Interpretation

When the radiotracer bolus is administered into a large antecubital vein such as the median basilic vein, serial images show progression of the bolus front from the basilic vein into the subclavian vein, then to the superior vena cava and into the right atrium. When the right atrium contracts, the bolus advances through the tricuspid valve into the right ventricle. Next, when the right ventricle contracts, radiotracer enters the pulmonary outflow tract and moves into the lungs, left atrium, left ventricle, and aorta. During the first couple of beats, the right ventricle can be isolated on the images (Figure 21-28).

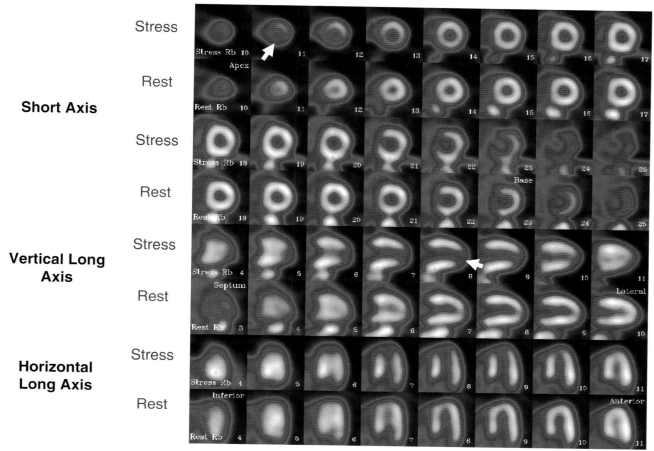

FIGURE 21-21 Abnormal ⁸²Rb-chloride rest–stress study in a 77 year old man with prior coronary artery bypass graft, evaluated prior to left upper lobectomy for lung cancer. Images were obtained after intravenous injection of 30 mCi of ⁸²Rb-chloride following the protocol outlined in Table 21-8. There is a moderate-sized reversible defect involving the apical and apical inferior walls (arrows) seen on the stress portion of the study that is not present at rest, consistent with myocardial ischemia. There was a moderate calcification of the left anterior descending artery seen on the CT attenuation scan (image not shown).

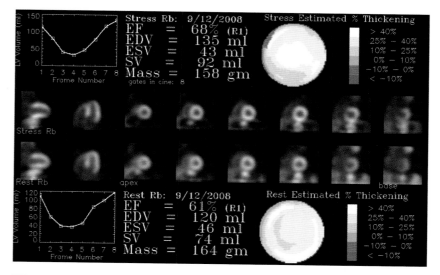

FIGURE 21-22 ECG-gated ventricular function study with ⁸²Rb-chloride obtained during the perfusion study in the same patient as Figure 21-21. There is normal global systolic function and wall thickening and a left ventricular ejection fraction of 68%.

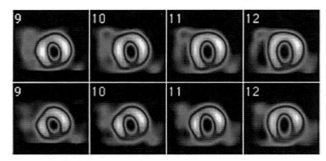

FIGURE 21-23 ^{99m}Tc-sestamibi study demonstrating a fixed defect in the inferior wall both during stress and at rest. There was hypokinesis in the inferior wall on the gated images as well as decreased thickening in the inferior wall during systole, most consistent with scarring from a prior myocardial infarction.

Regions of interest can be drawn around the right ventricle during diastole and systole and a right ventricular ejection fraction (RVEF) can be estimated using the same formula previously described for calculating LVEF.

Myocardial Viability

Rationale

The primary goal of viability assessment is to predict whether underperfused dysfunctional myocardium can be restored to normal function after revascularization. The four principal conditions that can cause left ventricular dysfunction are (1) transmural MI, which involves full-thickness myocardial necrosis; (2) nontransmural MI, in which necrosis is limited to the subendocardium or scattered throughout the myocardium; (3) myocardial stunning; and (4) hibernating myocardium.[116] Revascularization is not expected to be beneficial for the first two conditions. Myocardial stunning results from severe acute ischemia followed by reperfusion, both causing myocardial injury and dysfunction.[117] Myocardial hibernation results in dysfunction caused by chronic reduction in coronary blood flow. In stunning, reduced contraction is mismatched with increased perfusion; in hibernation, reduced contraction is matched by reduced perfusion.[117] In both situations, the dysfunction is reversible and will resolve after restoration of myocardial perfusion. Time is required in the case of stunning to restore function; improved blood flow is required in the case of hibernation.

Procedures and Interpretation

Nuclear medicine studies have been shown to be valuable in predicting viability of stunned and hibernating myocardium.[96] Myocardial viability can be assessed by SPECT procedures with ^{201}Tl-chloride or a ^{99m}Tc agent or by PET procedures with an ^{18}F-FDG metabolism study paired with a perfusion study with ^{82}Rb-chloride, ^{13}N-ammonia, or a ^{99m}Tc agent.

The terms positive predictive value (PPV) and negative predictive value (NPV) are often used in viability assessment. PPV predicts the percentage of reversible defects that will improve after revascularization, and NPV predicts the percent-

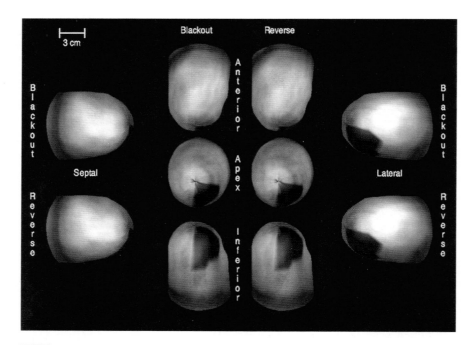

FIGURE 21-24 Three-dimensional blackout-reversibility map for the study shown in Figure 21-23, demonstrating a fixed perfusion defect (black area) in the inferior wall. Significant reversibility or ischemia would be seen as a white area in the blackout region on the reverse images but is not seen in this case.

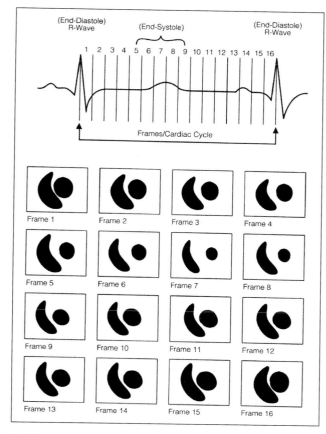

FIGURE 21-25 Gating mechanism to obtain serial single-frame images of left ventricle contraction through one cardiac cycle. See text for explanation. (Reprinted with permission of Lantheus Medical Imaging Inc. from *Introduction to Nuclear Cardiology*, 3rd ed. North Billerica, MA: DuPont Pharma Radiopharmaceuticals; 1993:226.)

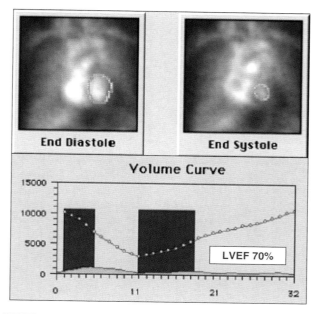

FIGURE 21-26 Normal multigated acquisition (MUGA) study. Left anterior oblique images of the heart. Regions of interest drawn around the left ventricle at end-diastole and end-systole gave a calculated left ventricular ejection fraction of 70%. There were no regional wall motion abnormalities on the gated images.

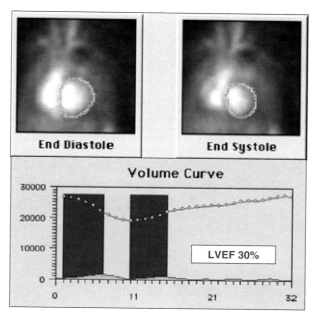

FIGURE 21-27 Abnormal MUGA study in a patient with congestive heart failure. Left anterior oblique images of the heart. Regions of interest drawn around the left ventricle at end-diastole and at end-systole gave a calculated left ventricular ejection fraction of only 30%. Although no focal wall motion abnormalities were noted, there was global left ventricular hypokinesis.

age of nonreversible defects that will not improve after revascularization. It is desirable that both these values be high.

Thallium Procedures

Two procedures are used to assess myocardial viability with thallium, rest–redistribution and stress–redistribution–reinjection (Figure 21-10). In patients with known CAD and left ventricular dysfunction, the ^{201}Tl rest–redistribution protocol has been shown to predict improved ventricular function after revascularization.[118] Several thallium viability studies, however, have been reviewed and summarized.[96,116,119–121] Several reports demonstrate high PPVs for reversible defects on rest–redistribution thallium imaging but low NPVs for nonreversible defects (underestimating viability), as evidenced by an undesirably high number of nonreversible defects that had improved function following revascularization. Subsequently, it was shown that when a quantitative method of reversibility assessment is used, the NPV of this procedure improves significantly.[122] Quantitative assessment is now recommended in the standard SPECT procedure.[30]

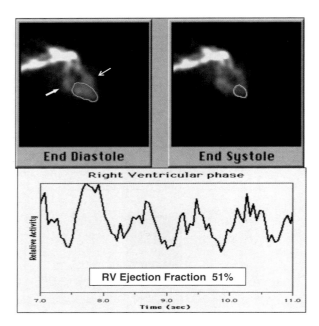

FIGURE 21-28 Normal first-pass study to evaluate right ventricular performance. Shortly after a 30 mCi (1110 MBq) bolus of ^{99m}Tc-sodium pertechnetate into a right arm vein, radiotracer can be seen in the superior vena cava, right atrium (left arrow), right ventricle (dotted region) and pulmonary outflow tract (right arrow). The right ventricle can be isolated because radiotracer has not yet advanced to the lungs or left ventricle. A time–activity curve for the right ventricle was used to determine right ventricular end-diastole and end-systole. Regions of interest drawn around the right ventricle during end-diastole and end-systole gave a normal estimated right ventricular ejection fraction of 51%.

In a rest–redistribution study, the thallium dose is injected into a patient at rest who has fasted for at least 4 hours. Imaging is begun in 15 minutes and proceeds for about 20 minutes. A redistribution image is acquired 3 to 4 hours later to observe any change in thallium activity in the heart from the initial rest image. A perfusion defect seen on the resting study that reverses on the 3 to 4 hour redistribution study indicates viable myocardium. A defect that does not reverse indicates myocardial scar.

An alternative procedure is the stress–redistribution–reinjection ^{201}Tl protocol, which is useful for assessing inducible ischemia and viability information in a patient who is suspected of having CAD. This protocol was discussed earlier under thallium perfusion imaging (Figure 21-10). A stress study is acquired followed by a 3 to 4 hour redistribution study. Defect reversibility on the redistribution study indicates ischemia. If the defect does not reverse, a booster dose of thallium is given and the patient is reimaged in 1 to 24 hours. Reversal of the defect following reinjection indicates viable hibernating myocardium.

After reinjection, a higher plasma concentration of thallium is made available to underperfused areas during the rest study, improving the probability of defect reversibility. The benefit of this technique has been shown in several studies.[104,107,123] After exercise–redistribution–reinjection thallium imaging, NPVs in studies conducted without ^{201}Tl reinjection were 43%, 53%, and 48%, but with reinjection these values improved to 100%, 75%, and 75%, respectively, demonstrating the benefit of thallium reinjection in improving diagnostic accuracy. Studies have compared thallium reinjection with FDG PET in viability assessment. Severe irreversible defects identified by thallium reinjection were found to have 88% concordance with FDG PET.[124] Two additional studies reported similar predictive accuracies between thallium reinjection and FDG PET, with PPVs of 80% to 87% and NPVs of 82% to 100%.[104,107] Thus, thallium SPECT procedures have been shown to be effective for assessing myocardial viability and predicting patients who are likely to benefit from bypass surgery.

Technetium Procedures

Most institutions now use a ^{99m}Tc agent for myocardial perfusion studies to achieve better quality images. There is debate, however, regarding viability assessment with ^{99m}Tc perfusion studies in patients with hibernating myocardium. ^{99m}Tc-sestamibi, as a perfusion marker, may be able to assess viability in stunned myocardium when perfusion is adequate, such as following reperfusion therapy for acute myocardial infarction, but it may not be able to adequately identify viability in hibernating myocardium in chronic CAD with sustained reduction in blood flow.[117] From a physiologic standpoint, ^{99m}Tc-sestamibi and ^{99m}Tc-tetrofosmin are potential viability agents, since their myocardial uptake is dependent on intact parenchymal cell and mitochondrial membrane electrochemical gradients.[69,70] Consequently, reversibility of stress defects on a rest study should indicate viability. In addition, ^{99m}Tc-sestamibi has been shown to exhibit a minor degree of redistribution. However, because these agents do not redistribute significantly, they lack thallium's advantage of sustained presence in blood over time and availability for uptake into viable myocardium. Thus, compared with thallium, the ^{99m}Tc agents have a higher probability of underestimating viability in underperfused regions of myocardium.[117,125–128] ^{99m}Tc-sestamibi has also been found to underestimate myocardial viability when compared with FDG PET in patients with CAD.[129–131]

A number of techniques have been proposed to improve viability assessment with ^{99m}Tc agents, including quantitative analysis of images, attenuation correction, gating perfusion studies to assess wall motion, and use of vasodilators to enhance perfusion at rest. When images are analyzed by a quantitative method, regional myocardial activities of both ^{201}Tl and ^{99m}Tc-sestamibi have been shown to differentiate viable from nonviable myocardium after resting injections, and both agents comparably predict reversibility of significant wall motion abnormalities after revascularizaion.[122] Gating perfusion studies permit differentiation of attenuation artifacts from scar and reduce the rate of incorrect interpretation in borderline normal and abnormal studies.[132] Besides gating, other techniques are used for identifying apparent defects caused by breast and diaphragmatic tissue attenuation. These include imaging patients prone instead of supine and using attenuation correction transmission maps.[10]

Another technique for reducing underestimation of viability at rest with ^{99m}Tc-sestamibi is nitrate enhancement. The rationale for this technique is to improve collateral blood flow to hypoperfused areas. The technique had been shown to be effective in previous studies with ^{201}Tl.[133,134] A typical method involves intravenous administration of 10 mg of isosorbide dinitrate in 100 mL of sterile isotonic saline over 20 minutes. ^{99m}Tc-sestamibi is injected after systolic blood pressure drops >20 mm Hg from baseline or reaches <90 mm Hg. If these criteria are not met, tracer is injected 15 minutes after the start of nitrate infusion, followed by a 2 minute additional infusion.[135] Alternatively, ^{99m}Tc radiotracer can be administered 5 to 10 minutes after 0.4 to 0.8 mg of sublingual nitroglycerin. Nitrate-enhanced images of ^{99m}Tc-sestamibi at rest have compared favorably with thallium rest–redistribution and reinjection studies in predicting, with high sensitivity, improvement in ventricular function after coronary revascularization.[136,137]

Clinical studies have shown that ^{99m}Tc-tetrofosmin accurately detects CAD when compared with ^{201}Tl-chloride[138–140] and ^{99m}Tc-sestamibi.[141] Viability data for ^{99m}Tc-tetrofosmin, however, are limited. A qualitative comparison for viability assessment in CAD after exercise–rest studies demonstrated that ^{99m}Tc-tetrofosmin underestimated myocardial viability compared with thallium reinjection, and the difference correlated with severity of the persistent defect on rest images.[142] However, one study reported that quantitative rest ^{99m}Tc-tetrofosmin imaging predicts functional recovery after revascularization comparable to that of rest–redistribution ^{201}Tl.[143] Another study demonstrated that uptake of ^{99m}Tc-tetrofosmin in dysfunctional myocardial regions at rest after nitrate enhancement correlated with metabolism assessed by FDG.[144]

PET Procedures

PET imaging, in general, offers advantages over SPECT that include better tissue penetration due to higher photon energy, quantitative assessment due to attenuation correction, independent perfusion and viability studies, and shorter imaging protocols. Newer SPECT imaging systems now have the capability for attenuation correction, but PET myocardial imaging has been considered the reference standard for myocardial viability assessment for several years. While SPECT methods have improved for measuring viability, comparative studies have shown higher accuracies with PET. The differences in reported accuracies between SPECT and PET imaging for assessing viability have been attributed to the different abilities of these modalities to detect ongoing physiologic processes. PET seems to have an advantage because of its better resolution, ability to correct for photon attenuation, and ability to assess metabolic function independent of perfusion.[145] A review of comparative studies reported high predictive accuracies with PET for assessing myocardial viability.[116]

A typical PET procedure for viability assessment is a perfusion–metabolism study performed at rest (Figure 21-29).[30] The perfusion study is performed first and can be done with ^{13}N-ammonia, ^{82}Rb-chloride, or ^{15}O-water. Information from a prior SPECT perfusion study can also be used but is not ideal. ^{13}N-ammonia is frequently used because its half-life allows longer imaging times and higher count density images compared with ^{82}Rb-chloride or ^{15}O-water.[116] The ^{82}Rb generator offers the advantage of availability at sites that do not have a cyclotron. ^{15}O-water perfusion studies require additional analytical methods and have limited routine application.

The metabolism phase of the study involves intravenous injection of ^{18}F-FDG after a loading dose of glucose. Glucose uptake into the myocardium is via the insulin-dependent GLUT4 transporter, and elevated serum glucose is necessary to facilitate FDG uptake into the myocardium. Subjects given 50 grams of glucose orally prior to an FDG study demonstrated heart-to-blood activity ratios 2.5 times higher than subjects who fasted for 12 hours.[146] Glucose loading can be done by intravenous infusion of glucose and insulin or as a high-carbohydrate meal prior to the study. Frequent monitoring of blood glucose is needed. Levels should rise, indicating

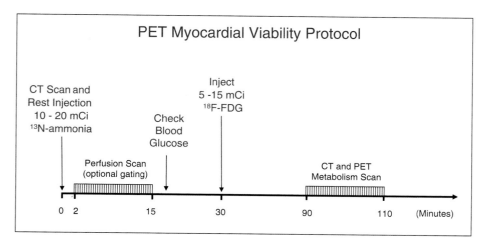

FIGURE 21-29 Standard PET ^{13}N-ammonia perfusion–^{18}F-FDG metabolism imaging protocol to assess myocardial viability. (Adapted from reference 30.)

glucose absorption, and then decline, indicating an insulin response, which facilitates FDG uptake into the myocardium. When serum glucose is in the correct range, 15 mCi of ^{18}F-FDG is injected followed by a saline flush. The patient should rest comfortably for 45 minutes for FDG uptake prior to imaging. Imaging should begin ideally between 60 and 90 minutes from the time of FDG injection to provide adequate heart uptake and reduction in background activity.[30]

The pattern of activity uptake between the perfusion and metabolism images is key to the diagnosis. Figure 21-30 illustrates a normal study. If the ^{13}N-ammonia and ^{18}F-FDG activity distributions are identical in the region of a perfusion defect (a match), the defect is likely caused by lack of both perfusion and active metabolism and is due to scar (Figure 21-31). If there is increased uptake of ^{18}F-FDG in a region seen as a perfusion deficit with ^{13}N-ammonia (a mismatch), the interpretation is that the region represents hibernating myocardium, which is active metabolically and therefore viable (Figure 21-32).

Early studies demonstrated the value of FDG PET in predicting functional recovery in small groups of patients with reduced left ventricular dysfunction undergoing coronary artery bypass surgery. Two studies suggested PPVs of functional recovery after surgery in regions of perfusion–metabolism mismatch of 78% to 85% and NPVs in regions with a match of 78% to 92%.[147,148] Larger studies have borne out these early reports, demonstrating that perfusion–metabolism mismatch is a good predictor of functional improvement after revascularization, with additional improvement of CHF symptoms, exercise capacity, and prognosis.[149–151]

Infarct Imaging

The early years of myocardial imaging in nuclear medicine involved the use of infarct-avid imaging agents to help identify whether a patient had had a myocardial infarction. Infarct-avid agents, otherwise known as "hot spot" markers, demonstrate an increased accumulation of radiotracer activity in regions of infarcted myocardium. Initially, infarct-avid imaging with ^{99m}Tc-pyrophosphate was widely used.[152–154] Infarct localization was based on disruption of the myocyte membrane, causing calcium ions to diffuse into the infarcted cells with deposition of intracellular hydroxyapatite crystals, which bound ^{99m}Tc-pyrophosphate. The procedure was an important contribution to the work-up of patients with an acute cardiac event, because at the time there was a lack of sensitive diagnostic tests for identifying infarction.

Several other radiolabeled agents were subsequently developed for infarct imaging based on the pathophysiologic changes that occur in the infarct region.[155–161] Some of these mechanisms have been well elucidated, while others are still being pursued in the continued development of new radiotracers for imaging infarction. ^{111}In-imciromab pentetate is a monoclonal antibody Fab fragment with specificity for myosin. Its localization is based on affinity for exposed myosin in

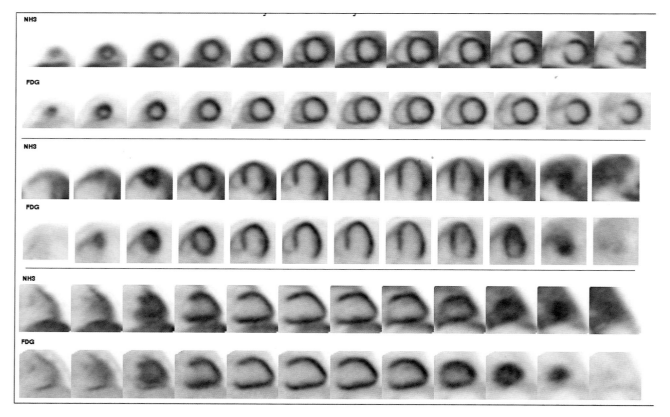

FIGURE 21-30 Normal PET ^{13}N-ammonia perfusion–^{18}F-FDG metabolism study demonstrating absence of any defect on the perfusion or metabolism images.

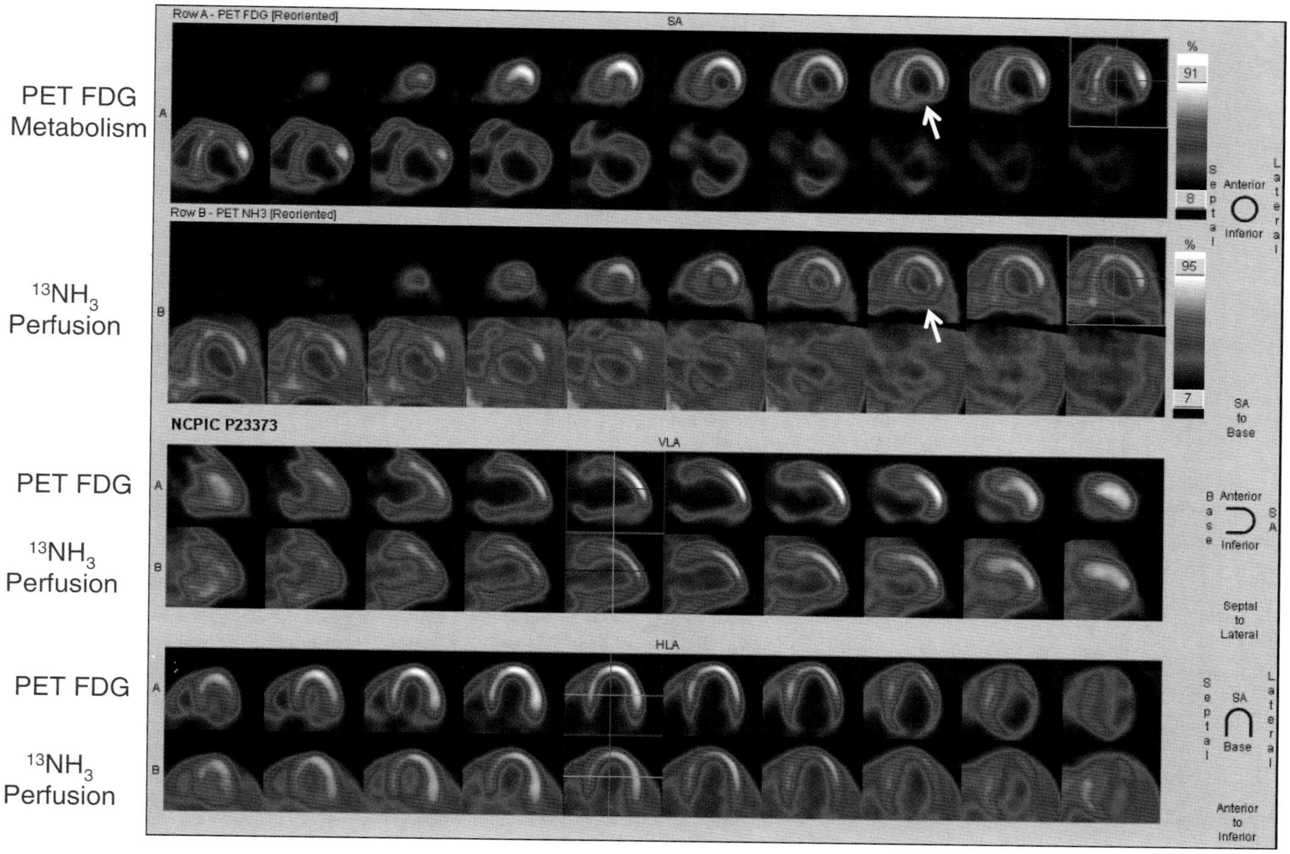

FIGURE 21-31 Abnormal PET ^{13}N-ammonia perfusion–^{18}F-FDG metabolism study performed at rest in a 54 year old male with ischemic cardiomyopathy and advanced three-vessel coronary artery disease. There is a matching perfusion–metabolism defect in the inferior wall (white arrows) most consistent with scarring from a prior myocardial infarction.

dying myocytes after disruption of the sarcolemmal membrane.[155,156] This radiopharmaceutical is no longer marketed. ^{99m}Tc-glucarate is a six-carbon sugar taken up into irreparably damaged myocytes, where it is associated with highly basic histones. It has been shown to localize in infarcts within 9 hours of infarct occurrence.[157,158] ^{99m}Tc-annexin V is a radiolabeled form of annexin V, an endogenous protein that has a high affinity for exposed phosphatidylserine on apoptotic cells.[159] Apoptotic cells have been identified in areas of severe ischemia and infarction. During natural apoptosis (programmed cell death) in the body, an enzymatic process is initiated that causes the phospholipid phosphatidylserine to become expressed on the outer membrane surface of dying cells, where it is able to bind the annexin V ligand. The purpose of this interaction is unknown, but the binding affinity between annexin V and phosphatidylserine is high, making ^{99m}Tc-annexin V a potentially good imaging agent.[160,161] Table 21-9 summarizes the infarct-avid radiotracers and their mechanisms of localization. Infarct-avid imaging has declined over the years with the development of sensitive serum enzyme assays (CK-MB, myoglobin, and troponin), which can provide the necessary information soon after the MI event, and advancement in methods for assessing myocardial perfusion and metabolism.

Other imaging modalities, such as contrast enhanced magnetic resonance imaging (ce-MRI), can directly demonstrate the extent of myocardial scarring. Cardiac MRI, compared with MPI, offers superior spatial resolution and is not subject to scatter and attenuation artifacts. The di-N-methylglucamine salt of gadolinium complexed with diethylenetriaminepentaacetic acid (Gd-DTPA) is a contrast material typically used in MRI imaging. Studies in dogs demonstrate that ce-MRI with Gd-DTPA produces hyperenhancement of infarcted myocardium but not viable myocardium.[162] The hyperenhancement is related to the increased concentration of Gd-DTPA in infarcted myocardium compared with normal myocardium, yielding high tissue contrast between the two regions.[163] The distinct difference in tissue contrast, coupled with high-resolution MRI images, permits accurate assessment of the extent of muscle necrosis and viability (Figure 21-33).[164] Although the evidence is not conclusive, it is believed that the increased localization of Gd-DTPA in myocardial infarcts is related to cellular necrosis and loss of myocyte membrane integrity, allowing Gd-DTPA to accumulate within the infarcted cells. These findings have led to the suggestion that ce-MRI in combination with cine-MRI, to assess wall motion, can be used to distinguish between acute MI (hyperenhanced with con-

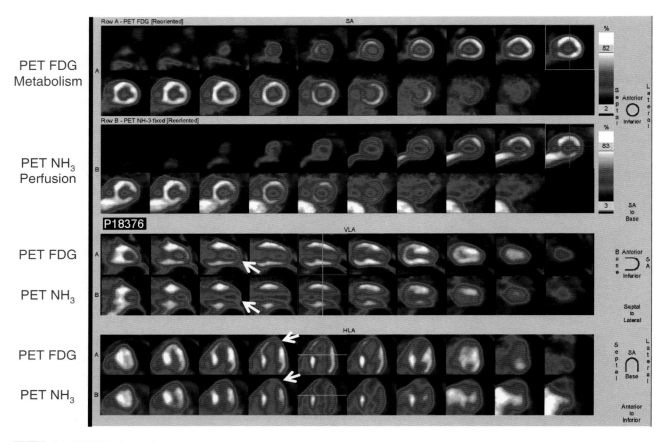

FIGURE 21-32 Abnormal PET ^{13}N-ammonia perfusion–^{18}F-FDG metabolism study performed at rest in a 69 year old male with diabetes, hypertension, hyperlipidemia, and CAD. A prior ^{99m}Tc adenosine stress myocardial perfusion study showed a large, fixed apical defect consistent with infarction. The left ventricular ejection fraction was 22%. There is a perfusion–metabolism mismatch in the large apical defect (white arrows) consistent with viable hibernating myocardium. Reperfusion in this patient will likely lead to functional improvement.

TABLE 21-9 Infarct-Avid Radiopharmaceuticals

Radiopharmaceutical	Mechanism of Localization
^{99m}Tc-pyrophosphate	Binds to hydroxyapatite crystals in mitochondria of infarcted myocytes. Most positive 24–48 hr after infarct. Had moderate to poor sensitivity and specificity.
^{111}In-imciromab pentetate	Antibody that binds to myosin exposed in infarcted muscle. Had high sensitivity and specificity.
^{99m}Tc-glucaric acid	Binds to histones in infarcted muscle. Can detect infarcts within 9 hr of occurrence.
^{99m}Tc-annexin V	Annexin is an endogenous protein that binds to phosphatidylserine expressed on outer membrane of apoptotic myocytes.

tractile dysfunction), injured but viable myocardium (not hyperenhanced but with contractile dysfunction), and normal myocardium (not hyperenhanced with normal function).[162] Kim et al. demonstrated that the transmural extent of infarction correlated with functional recovery following revascularization, reporting that patients with hypokinetic regions of subendocardial infarction less than 25% of wall thickness had up to 80% likelihood of improvement, whereas those with hypokinetic regions involving greater than 50% wall thickness had low probability of functional recovery after revascularization (Figure 21-34).[165] The detection of myocardial scar and viability by late gadolinium enhancement MRI has been endorsed by the American College of Cardiology.[166]

Cardiac Neuronal Imaging

Cardiac sympathetic innervation can be imaged by SPECT and PET with radiolabeled neurotransmitters that target presynaptic nerve terminals. Investigations in this area of imaging have been reviewed.[162] The principal SPECT agent for this purpose is m-iodobenzylguanidine labeled with ^{123}I (^{123}I-

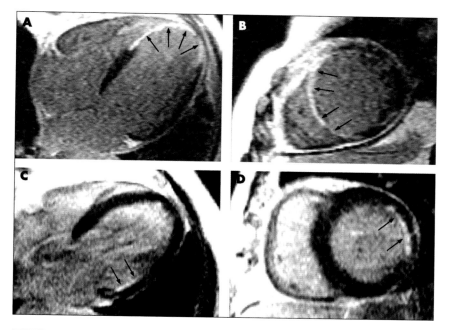

FIGURE 21-33 Transmural versus subendocardial myocardial infarction. Panels A and B (four-chamber and apical short-axis views, respectively) show a patient with transmural myocardial infarct in the apical left ventricle (LV) on late gadolinium enhancement (LGE) images. The arrows delineate the bright white signal in the apical LV reflecting transmural myocardial infarction with no viable myocardium. Revascularization would not result in improved function of the apical LV. Panels C and D (four-chamber and short-axis views, respectively) show a patient with a limited basal lateral wall subendocardial infarction. The arrows delineate an area of bright signal involving 25% to 50% of the myocardial wall (viable myocardial is nulled to appear black). Revascularization of an appropriate target lesion would be expected to result in improved function of the affected segments. (Reprinted with permission of BMJ Publishing Group Ltd. from reference 164.)

MIBG), an analogue of norepinephrine. For PET studies the principal agent is ^{11}C-labeled hydroxy ephedrine (^{11}C-HED). ^{123}I-MIBG localizes in adrenergic neurons of the heart principally by the presynaptic norepinephrine transporter uptake-1 pathway and is stored in the sympathetic nerve terminals. ^{11}C-HED uptake also occurs via the uptake-1 pathway; however, ^{11}C-HED seems to exhibit lower vesicular storage with higher rates of turnover.

Uptake of adrenergic imaging agents into cardiac sympathetic nerve tissue has been studied in a variety of heart diseases, for conditions that affect the heart, and for monitoring drug effects in heart disease treatment.[167] Various parameters are used to assess heart uptake of neuronal imaging agents, including size of the region of reduced uptake (defect size), washout rates, and the heart-to-mediastinum ratio (HMR). HMR is obtained by dividing the mean pixel counts in a selected region of the heart by the mean pixel counts in the upper region of the mediastinum. In general, defect size and washout rates increase and the HMR decreases in conditions that cause sympathetic denervation in the heart.

It has been shown that adrenergic neurons in the heart have higher sensitivity to oxygen deprivation than do myocytes. Thus, defects in adrenergic tissue may become evident earlier and be larger in size compared with perfusion defects in patients with ischemic heart disease. Indeed, ischemic regions assessed by SPECT perfusion–^{123}I-MIBG or PET perfusion–^{11}C-HED studies have demonstrated greater adrenergic defects compared with perfusion defects. For example, in CAD patients without prior MI, ^{123}I-MIBG innervation defects were shown to be present in the absence of perfusion defects.[168] In addition, patients with dilated cardiomyopathy demonstrated reduced uptake of ^{11}C-HED (larger defects) compared with ^{13}N-ammonia, suggesting loss of neurons or downregulation of the uptake-1 transporter.[169]

The heightened sensitivity to sympathetic stimulation of denervated but viable myocardium has been considered to be a significant factor contributing to cardiac arrhythmias in patients with ischemic heart disease and death in heart failure patients. Numerous studies have investigated the HMR for predicting cardiac death. Wakabayashi et al.[170] evaluated the

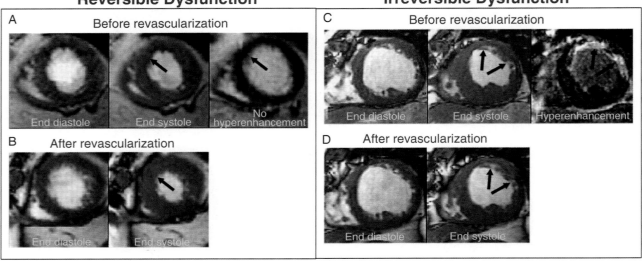

FIGURE 21-34 Representative cine images and contrast-enhanced images obtained by magnetic resonance imaging in a patient with reversible ventricular dysfunction (panels A and B) and another patient with irreversible ventricular dysfunction (panels C and D). The patient with reversible dysfunction had severe hypokinesia of the anteroseptal wall (arrows), and this area was not hyperenhanced before revascularization. The contractility of the wall improved after revascularization. The patient with irreversible dysfunction had akinesia of the anterolateral wall (arrows), and this area was hyperenhanced before revascularization. The contractility of the wall did not improve after revascularization. (Reprinted with permission from reference 165.)

long-term prognostic value of ^{123}I-MIBG for assessing cardiac death in patients with either ischemic cardiomyopathy (ISM) or idiopathic cardiomyopathy (ICM). HMRs were measured at 30 minutes (early) and 3 to 4 hours (late) after injection of 3 mCi (111 MBq) of ^{123}I-MIBG. Analysis showed that the late HMR was the most powerful predictor of cardiac death for both groups, having an identical threshold (1.82) for identifying patients at risk for cardiac death. This threshold was significantly lower than the HMRs for the ISM and ICM control groups, which were 2.30 and 2.38, respectively. ^{123}I-MIBG washout rates were 42% and 34%, respectively, for the ICM and ISM groups, significantly higher than the 20% rate for the control group.

A meta-analysis of 290 heart failure patients from 1993 to 2002 demonstrated that major cardiac events (MCE = cardiac death, cardiac transplant, potentially fatal arrhythmia) occurred in 67 patients (26%).[171] The mean HMR was 1.51 ± 0.30 for the MCE group and 1.97 ± 0.54 for the non-MCE group ($p < 0.001$). Two-year event-free survival using an optimum HMR threshold of 1.75 was 62% for HMR less than 1.75 and 95% for HMR greater than or equal to 1.75 ($p < 0.0001$).

A number of possible mechanisms have been proposed regarding ^{123}I-MIBG myocardial kinetics in heart failure.[170] The localization of ^{123}I-MIBG can be affected by sympathetic nerve damage due to severe ischemia or infarction and impairment of uptake-1 active transport due to ATP depletion. Increased washout rates may be caused by impairment of the norepinephrine storage and reuptake mechanism.

RISK STRATIFICATION

There is a large accumulation of experience with MPI. The sensitivity and specificity for detecting CAD are similar for SPECT procedures with ^{201}Tl, ^{99m}Tc-sestamibi, and ^{99m}Tc-tetrofosmin. For PET MPI studies with ^{82}Rb-chloride and ^{13}N-ammonia, the sensitivity for detecting CAD is similar to that of SPECT but the specificity is higher (Table 21-10).[172]

A principal benefit of diagnostic information acquired from cardiac imaging is improved prognosis and care of patients at risk. A wealth of useful information acquired from numerous studies has improved the ability to make recommendations regarding patient management and treatment and predictions regarding the risk of adverse events, such as cardiac death or nonfatal myocardial infarction. Diagnostic information from cardiac studies helps clinicians decide whether a patient should be treated medically or undergo coronary revascularization and whether a patient is at low, moderate, or high risk of a future cardiac event. Reports show that the risk of a hard event, such as cardiac death or nonfatal MI, is low (<1%) in the first year following a normal myocardial perfusion scan (Figure 21-35).[173,174] A major benefit of this finding is that such patients can be spared unnecessary referral for invasive evaluation. The 1 year risk, however, may be somewhat higher in subsets of patients with normal studies, depending on the patient's clinical characteristics and co-existing disease, such as diabetes mellitus. The risk for diabetic patients with a normal perfusion study is three times that of patients without diabetes mellitus.[175] Table 21-11 shows that risk increases with factors such

TABLE 21-10 Sensitivity and Specificity of Procedures for Detecting Coronary Artery Disease with ≥50% Stenosis

Procedure	% Sensitivity (Patients with CAD)	% Specificity (Patients without CAD)
SPECT		
— Exercise stress	87	73
— Vasodilator stress	89	75
PET		
— Exercise and vasodilator stress	89	86

TABLE 21-11 Hard Event Rate after a Normal Perfusion Scan

Patient Characteristics	Predicted Event Rate (%)	
	Year 1	Year 2
50 year old male, exercise stress, no history of CAD	0.1	0.1
80 year old male, adenosine stress, no history of CAD	1.5	1.6
50 year old male, history of CAD	0.9	1.5
80 year old male, history of CAD	1.4	2.4

Adapted from reference 174.

as age, time, and prior CAD and if pharmacologic stress is required instead of exercise.[175]

Risk in patients with an abnormal perfusion scan is based on the extent and severity of the perfusion defect. Assessment of perfusion defects is made by noting defect size (small, medium, or large), severity (mild, moderate, or severe), and type (reversible, fixed, or mixed). This can be done by qualitative, semiquantitative, and quantitative methods of assessment.[176] Hachamovitch and Berman[174] noted that patients with mildly abnormal perfusion scans are at intermediate risk for MI but at low risk for cardiac death and could benefit from aggressive medical therapy versus revascularization. The highest rates of adverse events occur in patients with moderate to severely abnormal perfusion defects, and these patients are most likely to benefit from revascularization. An analysis of cardiac death rate in patients over 2 years as a function of the extent of inducible ischemia with SPECT demonstrated that if ischemia involved 10% or less of the myocardium, medical treatment yielded better survival than revascularization. Conversely, if ischemic involvement was greater than 10% of the myocardium, revascularization produced better survival.[177] In certain situations, however, such as mild ischemia with significant uncontrolled chest pain, revascularization could be justified instead of medical therapy.

Other prognostic indicators of risk that should be considered include underlying patient characteristics such as history of CAD, advanced age, diabetes mellitus, and other heart conditions. In addition, nonperfusion markers noted during MPI are valuable indicators of risk. Increased lung uptake of ^{201}Tl, a marker of left ventricular dysfunction after exercise, was found to be the best predictor of a future cardiac event compared with angina, previous MI, and ST-segment depression.[178] Transient ischemic dilation (TID) has also been described as a marker for high-risk CAD. TID occurs when the left ventricular cavity appears significantly larger in poststress images than at rest; it is considered to represent critical stenosis in vessels supplying a large portion of the myocardium.[174,179,180]

Left ventricular (LV) function information can be vitally important for risk assessment and prognosis. LV dysfunction is not always caused by irreversible myocardial scar but may result from a loss of myocyte function due to ischemia. Uncorrected LV dysfunction that results in CHF predisposes patients to a high risk for cardiac death. Ejection fraction is an important tool for predicting survival in patients with CAD because patients with depressed LV function have a worsened prognosis. In the Coronary Artery Surgery Study (CASS) registry cohort of CAD patients treated medically, survival declined significantly when LV ejection fraction fell below 50% (Table 21-12).[114] LV function is readily assessed by gating myocardial perfusion studies and has been shown to provide added diagnostic and prognostic value to perfusion imaging.[115,132,181]

Among patients at highest risk of death from CAD are those with left main vessel and three-vessel disease (3VD), two groups in whom revascularization can improve survival.[182] Multivessel disease affects global perfusion, which

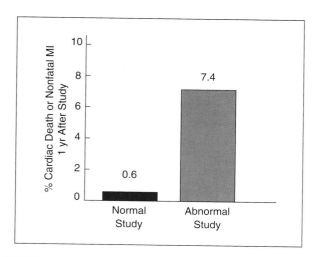

FIGURE 21-35 Rate of hard cardiac events (death or nonfatal myocardial infarction) in patients with normal and abnormal stress SPECT myocardial perfusion studies. (Adapted from reference 173.)

TABLE 21-12 Patient Prognosis Based on Ventricular Function Studies

LVEF (%)	12 Year Survival (%)	
	1 Vessel Disease	3 Vessel Disease
≥50	73	58
35–49	54	35
<35	21	10

Adapted from reference 114.

may result in uniform ischemia throughout the myocardium. A principal challenge in diagnosing multivessel disease with MPI, which relies on heterogeneity of blood flow in the heart, is that there may be no reference region of normal perfusion to compare with regions that have perfusion defects. In patients with angiographic 3VD, Lima et al.[183] found that only 46% of perfusion defects could be identified by perfusion studies alone, but that if gated functional assessment was included detection increased to 60%. Specificity in patients without 3VD was not significantly decreased by combining functional assessment with perfusion. Furthermore, in gated exercise or adenosine stress SPECT perfusion studies with ^{99m}Tc-sestamibi in patients with angiographically confirmed left main vessel disease, no significant perfusion defects could be identified in 13% of patients.[184] Thus, although gating improves the diagnostic accuracy of perfusion studies, MPI still has a significant deficiency for detecting multivessel disease.

REFERENCES

1. Centers for Disease Control and Prevention. Deaths: Final Data for 2002. National Vital Statistics Reports, vol. 53, no. 5 (October 2004). www.cdc.gov/nchs/data/nvsr/nvsr53/nvsr53_05acc.pdf. Accessed October 13, 2010.
2. Bianco JA, Wilson MA. Myocardial ischemia and viability. In: Wilson MA, ed. *Textbook of Nuclear Medicine*. Philadelphia: Lippincott-Raven; 1998: 33–66.
3. Marcus ML, Harrison DG. Physiologic basis for myocardial perfusion imaging. In: Marcus ML, Skorton DH, Schelbert HR, et al., eds. *Cardiac Imaging*. Philadelphia: WB Saunders; 1991.
4. Guyton AC. *Textbook of Medical Physiology*. Philadelphia: WB Saunders; 1981.
5. Holmberg S, Serzysko W, Varnauskas E. Coronary circulation during heavy exercise in control subjects and patients with coronary heart disease. *Acta Med Scand*. 1971;190:465–80.
6. Heiss HW, Barmeyer J, Wink K, et al. Studies on the regulation of myocardial blood flow in man, I: training effects on blood flow and metabolism of the healthy heart at rest and during standardized heavy exercise. *Basic Res Cardiol*. 1976;71:658–75.
7. Brown GB, Josephson MA, Petersen RB, et al. Intravenous dipyridamole combined with isometric handgrip for near maximal acute increase in coronary flow in patients with coronary artery disease. *Am J Cardiol*. 1981;48:1077–85.
8. Gould KL, Lipscomb K, Hamilton GW. Physiologic basis for assessing critical coronary stenosis. *Am J Cardiol*. 1974;33:87–94.
9. Wackers FJ. Exercise myocardial perfusion imaging. *J Nucl Med*. 1994;35:726–9.
10. Vitola JV, Delbeke D. Cardiovascular imaging. In: Habibian MR, Delbeke D, Martin WH, et al., eds. *Nuclear Medicine Imaging: A Teaching File*. 2nd ed. Philadelphia: Wolters Kluwer/Lippincott Williams & Wilkins; 2009.
11. Bruce RA. Multi-stage treadmill test of submaximal and maximal exercise. In: *Exercise Testing and Training of Apparently Healthy Individuals: A Handbook for Physicians*. New York: American Heart Association; 1972.
12. Wolthuis RA, Froelicher VF, Fischer J, et al. The response of healthy men to treadmill exercise. *Circulation*. 1977;55:153–7.
13. Kariv I, Kellerman JJ. Effects of exercise on blood pressure. *Mal Cardiovasc*. 1969;10:247–52.
14. Holmberg S, Serysko W, Varnauskas E. Coronary circulation during heavy exercise in control subjects and patients with coronary heart disease. *Acta Med Scand*. 1971;190:465–80.
15. Leppo JA. Dipyridamole myocardial perfusion imaging. *J Nucl Med*. 1994;35:730–3.
16. Iskandrian AS. Adenosine myocardial perfusion imaging. *J Nucl Med*. 1994;35:734–6.
17. Leppo JA. Dipyridamole-thallium imaging: the lazy man's stress test. *J Nucl Med*. 1989;30:281–7.
18. Verani MS, Mahmarian JJ. Myocardial perfusion scintigraphy during maximal coronary artery vasodilation with adenosine. *Am J Cardiol*. 1991;67:12D–17D.
19. Wackers FJ. Adenosine or dipyridamole: which is preferred for myocardial perfusion imaging? *J Am Coll Cardiol*. 1991;17:1295–6.
20. Verani MS. Dobutamine myocardial perfusion imaging. *J Nucl Med*. 1994; 5:737–9.
21. Wilson RF, Wyche K, Christensen BV, et al. Effects of adenosine on human coronary arterial circulation. *Circulation*. 1990;82;1595–1606.
22. Möser GH, Schrader J, Deussen A. Turnover of adenosine in plasma of human and dog blood. *Am J Physiol*. 1989;256:C799–806.
23. Klabunde RE. Dipyridamole inhibition of adenosine metabolism in human blood. *Eur J Pharmacol*. 1983;93:21–6.
24. Smits P, Aengevaeren WRM, Corstens FHM, et al. Caffeine reduces dipyridamole-induced myocardial ischemia. *J Nucl Med*. 1989;30: 1723–6.
25. Glover DK, Ruiz M, Takehana K, et al. Pharmacological stress myocardial perfusion imaging with the potent and selective A_{2A} adenosine receptor agonists ATL193 and ATL146e administered by either intravenous infusion or bolus injection. *Circulation*. 2001;104:1181–7.
26. Gao Z, Li Z, Baker SP, et al. Novel short-acting A_{2A} adenosine receptor agonists for coronary vasodilation: inverse relationship between affinity and duration of action of A_{2A} agonists. *J Pharmacol Exp Ther*. 2001;298:209–18.
27. Lee J, Heo J, Ogilby JS, et al. Atrioventricular block during adenosine thallium imaging. *Am Heart J*. 1992;123:1569–73.
28. Dipyridamole [product monograph]. Billerica, MA: Dupont Pharma; 1990.
29. Holgate ST, Mann JS, Cushley MJ. Adenosine as a bronchoconstrictor mediator in asthma and its antagonism by methylxanthines. *J Allergy Clin Immunol*. 1984;74:302–6.
30. American Society of Nuclear Cardiology. Imaging guidelines for nuclear cardiology procedures 2006. www.asnc.org/section_73.cfm. Accessed October 13, 2010.
31. Berman DS, Kiat HS, Van Train KF, et al. Myocardial perfusion imaging with technetium-99m-sestamibi: comparative analysis of available imaging protocols. *J Nucl Med*. 1994;35:681–8.
32. Cerqueira MD. Advances in pharmacologic agents in imaging: new A_{2A} receptor agonists. *Curr Cardiol Rep*. 2006;8:119–22.
33. Hendel RC, Bateman TM, Cerqueira MD, et al. Initial clinical experience with regadenoson, a novel selective A_{2A} agonist for pharmacologic stress single-photon emission computed tomography myocardial perfusion imaging. *J Am Coll Cardiol*. 2005;46:2069–75.
34. Lexiscan [prescribing information]. Deerfield, IL: Astellas Pharma US, Inc; 2008.
35. Udelson JE, Heller GV, Wackers FJ, et al. Randomized, controlled dose-ranging study of the selective adenosine A_{2A} receptor agonist binodenoson for pharmacological stress as an adjunct to myocardial perfusion imaging. *Circulation*. 2004;109:457–64.

36. Hodgson JM, Dib N, Kern MJ, et al. Coronary circulation responses to binodenoson, a selective adenosine A_{2A} receptor agonist. *Am J Cardiol.* 2007;99:1507–12.
37. Udelson JE. Lessons from the development of new adenosine A_{2A} receptor agonists. *J Am Coll Cardiol Img.* 2008;1:317–20.
38. McNeil AJ, Fioretti PM, El-Said ME-S, et al. Enhanced sensitivity for detection of coronary artery disease by addition of atropine to dobutamine stress echocardiography. *Am J Cardiol.* 1992;70:41–6.
39. Kowalsky RJ, Perry JR. Evaluation of intravenous infusion systems for adenosine. *J Nucl Med Technol.* 1992; 20: 215–9.
40. Blust JS, Boyce TM, Moore WH. Pharmacologic cardiac intervention: comparison of adenosine, dipyridamole, and dobutamine. *J Nucl Med Technol.* 1992;20:53–61.
41. Cerqueira MD, Verani MS, Schwaiger M, et al. Safety profile of adenosine stress perfusion imaging: results from the Adenoscan multicenter trial registry. *J Am Coll Cardiol.* 1994;23:384–9.
42. Ranhosky A, Kempthrone-Rawson J. The safety of intravenous dipyridamole thallium myocardial perfusion imaging. *Circulation.* 1990:81:1205–9.
43. Wackers FJ. Adenosine-thallium imaging: faster and better? *J Am Coll Cardiol.* 1990;16:1384–6.
44. Miller DD, Labovitz AJ. Dipyridamole and adenosine vasodilator stress for myocardial imaging: vive la difference! *J Am Coll Cardiol.* 1994;23:390–2.
45. Hays JT, Mahmarian JJ, Cochran AJ, et al. Dobutamine thallium-201 tomography for evaluating patients with suspected coronary artery disease unable to undergo exercise or vasodilator pharmacologic stress testing. *J Am Coll Cardiol.* 1993;21:1583–90.
46. Weich H, Strauss WH, D'Agostino R, et al. Determination of extraction fraction by a double-tracer method. *J Nucl Med.* 1977;18:226–30.
47. Jain D. Technetium-99m labeled myocardial perfusion imaging agents. *Semin Nucl Med.* 1999;29:221–36.
48. Leppo JA, Meerdink DJ. Comparative myocardial extraction of two technetium-labeled BATO derivatives (SQ30217, SQ32014) and thallium. *J Nucl Med.* 1990;31:67–74.
49. Schelbert HR. Principles of positron emission tomography. In: Marcus ML, Skorton DH, Schelbert HR, et al., eds. *Cardiac Imaging.* Philadelphia: WB Saunders; 1991;1140–68.
50. Berman DS, Kiat H, Leppo J, et al. Technetium-99m myocardial perfusion imaging agents. In: Marcus ML, Skorton DH, Schelbert HR, et al., eds. *Cardiac Imaging.* Philadelphia: WB Saunders; 1991: 1097–1109.
51. Leppo JA, Meerdink DJ. Comparison of the myocardial uptake of a technetium-labeled isonitrile analogue and thallium. *Circ Res.* 1989;65:632–9.
52. Pohost GM, Alpert NM, Ingwall JS, et al. Thallium redistribution: mechanisms and clinical utility. *Semin Nucl Med.* 1980;10:70–93.
53. Chiu ML, Kronauge JF, Piwnica-Worms D. Effect of mitochondrial and plasma membrane potentials on accumulation of hexakis(2-methoxyisobutylisonitrile) technetium(I) in cultured mouse fibroblasts. *J Nucl Med.* 1990;31:1646–53.
54. Jones AG, Abrams MJ, Davison A, et al. Biological studies of a new class of technetium complexes: the hexakis(alkylisonitrile)technetium (I) complexes. *Int J Nucl Med Biol.* 1984;11:225–34.
55. Wackers FJ, Berman DS, Maddahi J, et al. Technetium-99m hexakis 2-methoxyisobutyl isonitrile: human biodistribution, dosimetry, safety, and preliminary comparison to thallium-201 for myocardial perfusion imaging. *J Nucl Med.* 1989;30:301–11.
56. Kelly JD, Forster AM, Higley B, et al. Technetium-99m-tetrofosmin as a new radiopharmaceutical for myocardial perfusion imaging. *J Nucl Med.* 1993;34:222–7.
57. Higley B, Smith FW, Smith T, et al. Technetium-99m-1,2-bis[bis(2-ethoxyethyl)phosphino]ethane: human biodistribution, dosimetry and safety of a new myocardial perfusion imaging agent. *J Nucl Med.* 1993;34:30–8.
58. Atkins HL, Budinger TF, Lebowitz E, et al. Thallium-201 for medical use. Part 3: human distribution and physical imaging properties. *J Nucl Med.* 1977;18:133–40.
59. Hamilton GW, Narahara KA, Yee H, et al. Myocardial imaging with thallium-201: effect of cardiac drugs on myocardial images and absolute tissue distribution. *J Nucl Med.* 1978;19:10–6.
60. Weich HF, Strauss WH, Pitt B. The extraction of thallium-201 by the myocardium. *Circulation.* 1977;56:188–91.
61. Meerdink DJ, Leppo JA. Comparison of hypoxia and ouabain effects on the myocardial uptake kinetics of technetium-99m hexakis 2-methoxyisobutyl isonitrile and thallium-201. *J Nucl Med.* 1989; 30:1500–6.
62. Maublant JC, Gachon P, Moins N. Hexakis (2-methoxyisobutylisonitrile) technetium-99m and thallium-201 chloride: uptake and release in cultured myocardial cells. *J Nucl Med.* 1988;29:48–54.
63. Ritchie JL, Hamilton GW. Biologic properties of thallium. In: Ritchie JL, Hamilton GW, Wackers FJ, eds. *Thallium-201 Myocardial Imaging.* New York: Raven Press; 1978:9–28.
64. Buja LM, Burton KP, Hagler HK, et al. Quantitative x-ray microanalysis of the elemental composition of individual myocytes in hypoxic rabbit myocardium. *Circulation.* 1983; 68:872–82.
65. Jennings RB, Sommers HM, Kaltenbach JP, et al. Electrolyte alterations in acute myocardial ischemic injury. *Circ Res.* 1964;14:260–9.
66. Goldhaber SZ, Newell JB, Alpert NM, et al. Effects of ischemic-like insult on myocardial 201 thallium accumulation. *Circulation.* 1983; 67:778–86.
67. Mousa SA, Cooney JM, Williams SJ. Relationship between regional myocardial blood flow and the distribution of Tc-99m sestamibi in the presence of total coronary artery occlusion. *Am Heart J.* 1990; 119:842–7.
68. Heller GV. Tracer selection with different stress modalities based on tracer kinetics. *J Nucl Cardiol.* 1996;3:S15–S21.
69. Piwnica-Worms D, Kronauge JF, Chiu ML. Uptake and retention of hexakis (2-methoxyisobutyl isonitrile) technetium(I) in cultured chick myocardial cells. Mitochondrial and plasma membrane potential dependence. *Circulation.* 1990;5:1826–38.
70. Carvalho PA, Chiu ML, Kronauge JF, et al. Subcellular distribution and analysis of technetium-99m-MIBI in isolated perfused rat hearts. *J Nucl Med.* 1992;33:1516–21.
71. Okada RD, Glover D, Gaffney T, et al. Myocardial kinetics of technetium-99m-hexakis-2-methoxy-2-methylpropyl-isonitrile. *Circulation.* 1988; 77:491–8.
72. Glover DK, Okada RD. Myocardial kinetics of Tc-MIBI in canine myocardium after dipyridamole. *Circulation.* 1990;81:628–37.
73. Li QS, Solot G, Frank TL, et al. Myocardial redistribution of technetium 99m-hexakis-2-methoxy-2-methylpropyl-isonitrile (SESTAMIBI). *J Nucl Med.* 1990;31:1069–76.
74. Cardiolite [package insert]. Billerica, MA: DuPont Pharmaceuticals; 2000.
75. Radiation Dose Estimates for Radiopharmaceuticals. Oak Ridge, TN: Radiation Internal Dose Information Center; 1996.
76. Sinusas AJ, Shi O, Saltzberg MT, et al. Technetium-99m-tetrofosmin to assess myocardial blood flow: experimental validation in an intact canine model of ischemia. *J Nucl Med.*1994;35:664–71.
77. Dahlbert ST, Gilmore MP, Leppo JA. Effect of coronary blood flow on the "uptake" of tetrofosmin in the isolated rabbit heart [abstract]. *J Nucl Med.* 1992;33:846.
78. Myoview [package insert]. Arlington Heights, IL: Nycomed-Amersham; 1998.
79. Platts EA, North TL, Pickett RD, et al. Mechanism of uptake of technetium-tetrofosmin. I: uptake into isolated adult rat ventricular myocytes and subcellular localization. *J Nucl Cardiol.* 1995;2:317–26.
80. Arbab AS, Koizumi K, Toyama K, et al. Technetium-99m-tetrofosmin, technetium-99m-MIBI and thallium-201 uptake in rat myocardial cells. *J Nucl Med.* 1998;39:266–71.
81. Sridhara BS, Braat S, Rigo P, et al. Comparison of myocardial perfusion imaging with ^{99m}Tc-tetrofosmin versus ^{201}Tl in coronary artery disease. *Am J Cardiol.* 1993;72:1015–9.
82. Taillefer R, Primeau M, Costi P, et al. Technetium-99m-sestamibi myocardial perfusion imaging in detection of coronary artery disease: comparison between initial (1-hour) and delayed (3-hour) postexercise images. *J Nucl Med.* 1991;32:1961–5.

83. Soman P, Taillefer R, DePuey GE, et al. Enhanced detection of reversible perfusion defects by Tc-99m sestamibi compared to Tc-99m tetrofosmin during vasodilator stress SPECT imaging in mild-to-moderate coronary artery disease. *J Am Coll Cardiol*. 2001;37:458–62.
84. Schwaiger M. Myocardial perfusion imaging with PET. *J Nucl Med*. 1994;35:693–8.
85. Budinger TF, Yano Y, Derenzo SE, et al. Myocardial extraction of Rb-82 vs. flow determined by positron emission tomography [abstract]. *Circulation*. 1983;68:III–81.
86. Cherry SR, Sorenson JA, Phelps ME. *Physics in Nuclear Medicine*. 3rd ed. Philadelphia: Saunders; 2003.
87. Pavel DG, Zimmer AM, Patterson VN. In vivo labeling of red blood cells with Tc-99m: a new approach to blood pool visualization. *J Nucl Med*. 1977;18:305–8.
88. Dewanjee MK, Rao SA, Penniston GT. Mechanism of red blood cell labeling with ^{99m}Tc-pertechnetate. The role of the cation pump at RBC membrane on the distribution and binding of Sn^{2+} and ^{99m}Tc with the membrane proteins and hemoglobin. *J Labelled Comp Radiopharm*. 1982;19:1464–5.
89. Dewanjee MK, Anderson GS, Wahner HW. Pharmacodynamics of stannous-chelates administered with technetium-99m chelates for radionuclide imaging. In: Sodd VS, Hoogland DR, Allen DR, et al., eds. *Radiopharmaceuticals II*. New York: Society of Nuclear Medicine; 1979:421–34.
90. Gould KL. Myocardial viability: what does it mean and how do we measure it? *Circulation*. 1991;83:333–5.
91. Dilsizian V, Bateman TM, Bergmann SR, et al. Metabolic imaging with β-methyl-p-[^{123}I]-iodophenyl-pentadecanoic acid identifies ischemic memory after demand ischemia. *Circulation*. 2005;112:2169–74.
92. Zaret BL, Strauss HW, Martin WD, et al. Noninvasive regional myocardial perfusion with radioactive potassium. *N Engl J Med*. 1973;288:809–12.
93. Gaudron P, Eilles C, Kugler I, et al. Progressive left ventricular dysfunction and remodeling after myocardial infarction: potential mechanisms and early predictors. *Circulation*. 1993;87:755–63.
94. The Multicenter Postinfarction Research Group. Risk stratification and survival after myocardial infarction. *N Engl J Med*. 1983;309:331–6.
95. Strauss HW, Miller DM, Wittry MD, et al. Procedure Guideline for Myocardial Perfusion Imaging 3.3. In: *Procedure Guidelines Manual 2008*. Reston VA: Society of Nuclear Medicine; 2008. www.snm.org/guidelines. Accessed October 13, 2010.
96. Dilsizian V, Bonow RO. Current diagnostic techniques of assessing myocardial viability in patients with hibernating and stunned myocardium. *Circulation*. 1993;87:1–20.
97. Pohost GM, Zir LM, Moore RH, et al. Differentiation of transiently ischemic from infarcted myocardium by serial imaging after a single dose of thallium-201. *Circulation*. 1977;55:294–302.
98. Blood DK, McCarthy DM, Sciacca RR, et al. Comparison of single-dose and double-dose thallium-201 myocardial perfusion scintigraphy for the detection of coronary artery disease and prior myocardial infarction. *Circulation*. 1978;58:777–8.
99. Ritchie JL, Albro PC, Caldwell JH, et al. Thallium-201 myocardial imaging: a comparison of the redistribution and rest images. *J Nucl Med*. 1979;20:477–83.
100. Gibson RS, Watson DD, Taylor GJ, et al. Prospective assessment of regional myocardial perfusion before and after coronary revascularization surgery by quantitative thallium-201 scintigraphy. *J Am Coll Cardiol*. 1983;1:804–15.
101. Brunken R, Schwaiger M, Grover-McKay M, et al. Positron emission tomography detects tissue metabolic activity in myocardial segments with persistent thallium perfusion defects. *J Am Coll Cardiol*. 1987;10:557–67.
102. Gutman J, Berman DS, Freeman M, et al. Time to completed redistribution of thallium-201 in exercise myocardial scintigraphy: relationship to the degree of coronary artery stenosis. *Am Heart J*. 1983;106:989–95.
103. Kiat H, Berman DS, Maddahi J, et al. Late reversibility of tomographic myocardial thallium-201 defects: an accurate marker of myocardial viability. *J Am Coll Cardiol*. 1988;12:1456–63.
104. Dilsizian V, Rocco TP, Freedman NMT, et al. Enhanced detection of ischemic but viable myocardium by the reinjection of thallium after stress-redistribution imaging. *N Engl J Med*. 1990;323:141–6.
105. Kayden DS, Sigal S, Soufer R, et al. Thallium-201 for assessment of myocardial viability: quantitative comparison of 24-hour redistribution imaging with imaging after reinjection at rest. *J Am Coll Cardiol*. 1991;18:1480–6.
106. Bonow RO. Identification of viable myocardium. *Circulation*. 1996;94:2674–80.
107. Ohtani H, Tamaki N, Yonekura Y, et al. Value of thallium-201 reinjection after delayed SPECT imaging for predicting reversible ischemia after coronary artery bypass grafting. *Am J Cardiol*. 1990;66:394–9.
108. Kitsou AN, Srinivasan G, Quyyumi AA, et al. Stress-induced reversible and mild-to-moderate irreversible thallium defects: are they equally accurate for predicting recovery of regional left ventricular function after revascularization? *Circulation*. 1998;98:501–8.
109. Garcia E, Maddahi J, Berman D, et al. Space-time quantitation of ^{201}Tl myocardial scintigraphy. *J Nucl Med*. 1981;22:309–17.
110. Taillefer R, Laflamme L, Dupras G, et al. Myocardial perfusion imaging with Tc-99m-methoxy-isobutyl-isonitrile (MIBI): comparison of short and long time intervals between rest and stress injections: preliminary results. *Eur J Nucl Med*. 1988;13:515–22.
111. Leppo JA, DePuey GE, Johnson LL. A review of cardiac imaging with sestamibi and teboroxime. *J Nucl Med*. 1991;32:2012–22.
112. Bergman SR. Positron emission tomography of the heart. In: Gerson M, ed. *Cardiac Nuclear Medicine*. 3rd ed. New York: McGraw-Hill; 1996:299–335.
113. Strauss W, Zaret BL, Hinley RJ, et al. A scintiphotographic method for measuring left ventricular ejection fraction in a man without cardiac catheterization. *Am J Cardiol*. 1971;28:757.
114. Emond M, Mock MB, Davis KB, et al. Long-term survival of medically treated patients in the Coronary Artery Surgery Study (CASS) registry. *Circulation*. 1994;90:2645–57.
115. Taillefer R, DePuey GE, Udelson JE, et al. Comparative diagnostic accuracy of Tl-201 and Tc-99m sestamibi SPECT imaging (perfusion and ECG-gated SPECT) in detecting coronary artery disease in women. *J Am Coll Cardiol*. 1997;29:69–77.
116. Maddahi J, Schelbert J, Brunken R, et al. Role of thallium-201 and PET imaging in evaluation of myocardial viability and management of patients with coronary artery disease and left ventricular dysfunction. *J Nucl Med*. 1994;35:707–15.
117. Bonow RO, Dilsizian V. Thallium-201 and technetium-99m-sestamibi for assessing viable myocardium. *J Nucl Med*. 1992;33:815–8.
118. Iskandrian AS, Hakki A, Kane SA, et al. Rest and redistribution thallium-201 myocardial scintigraphy to predict improvement in left ventricular function after coronary arterial bypass grafting. *Am J Cardiol*. 1983;51:1312–6.
119. Mori T, Minamiji K, Kurongane H, et al. Rest-injected thallium-21 imaging for assessing viability of severe asynergic regions. *J Nucl Med*. 1991;23:1718–24.
120. Alfieri O, LaCanna G, Guibbini R, et al. Recovery of myocardial function. *Eur J Cardiothorac Surg*. 1993;7:325–30.
121. Ragosta M, Beller GA, Watson DD, et al. Quantitative planar rest-redistribution Tl-201 imaging in detection of myocardial viability and prediction of improvement in left ventricular function after coronary bypass surgery in patients with severely depressed left ventricular function. *Circulation*. 1993;87:1630–41.
122. Udelson JE, Coleman PS, Metherall J, et al. Predicting recovery of severe regional ventricular dysfunction: comparison of resting scintigraphy with ^{201}Tl and ^{99m}Tc-sestamibi. *Circulation*. 1994;89:2552–61.
123. Tamaki N, Ohtani H, Yamashita K, et al. Metabolic activity in the areas of new fill-in after thallium-201 reinjection: comparison with positron emission tomography using fluorine-18-deoxyglucose. *J Nucl Med*. 1991;32:673–8.
124. Bonow RO, Dilsizian V, Cuocolo A, et al. Myocardial viability in patients with chronic coronary artery disease and left ventricular dys-

function: thallium-201 reinjection versus ^{18}F-fluorodeoxyglucose. *Circulation.* 1991;83:26–37.

125. Cuocolo A, Pace L, Ricciardelli B, et al. Identification of viable myocardium in patients with chronic coronary artery disease: comparison of thallium-201 scintigraphy with reinjection and technetium-99m methoxyisobutyl isonitrile. *J Nucl Med.* 1992;33:505–11.

126. Maurea S, Cuocolo A, Nicolai E, et al. Improved detection of viable myocardium with thallium-201 reinjection in chronic coronary artery disease: comparison with technetium-99m-MIBI imaging. *J Nucl Med.* 1994;35:621–4.

127. Marzullo P, Sambuceti G, Parodi O. The role of sestamibi scintigraphy in the radioisotopic assessment of myocardial viability. *J Nucl Med.* 1992;33:1925–30.

128. Marcassa C, Galli M, Cuocolo A, et al. Rest-redistribution thallium-201 and rest technetium-99m-sestamibi SPECT in patients with stable coronary artery disease and ventricular dysfunction. *J Nucl Med.* 1997;38:419–24.

129. Altehoefer C, vom Dahl J, Biedermann M, et al. Significance of defect severity in technetium-99m-MIBI SPECT at rest to assess myocardial viability: comparison with fluorine-18-FDG PET. *J Nucl Med.* 1994;35:569–74.

130. Soufer R, Dey HM, Ng CK, et al. Comparison of MIBI single-photon emission computed tomography with positron emission tomography for estimating left ventricular myocardial viability. *Am J Cardiol.* 1995;75:1214–9.

131. Sawada S, Allman KC, Muzik O, et al. Positron emission tomography detects evidence of viability in rest technetium-99m MIBI defects. *J Am Coll Cardiol.* 1994;23:92–8.

132. Smanio PEP, Watson DD, Segalla DL, et al. Value of gating of technetium-99m sestamibi single-photon emission computed tomographic imaging. *J Am Coll Cardiol.* 1997;30:1687–92.

133. Aoki M, Sakai K, Koyanagi S, et al. Effect of nitroglycerin on coronary collateral function during exercise evaluated by quantitative analysis of thallium-201 single photon emission computed tomography. *Am Heart J.* 1991;121:1361–6.

134. He A-X, Darcourt J, Guigner A, et al. Nitrates improve detection of ischemic but viable myocardium by thallium-201 reinjection SPECT. *J Nucl Med.* 1993;34:1472–7.

135. Sciagra R, Bisi G, Santoro GM, et al. Comparison of baseline-nitrate technetium-99m sestamibi with rest-redistribution thallium-201 tomography in detecting viable hibernating myocardium and predicting postrevascularization recovery. *J Am Coll Cardiol.* 1997;30:384–91.

136. Batista JF, Pereztol O, Valdes JA, et al. Improved detection of myocardial perfusion reversibility by rest-nitroglycerin Tc-99m-MIBI: comparison with Tl-201 reinjection. *J Nucl Cardiol.* 1999;6:480–6.

137. Oudiz RJ, Smith DE, Pollack AJ, et al. Nitrate-enhanced thallium-201 single photon emission tomography imaging in hibernating myocardium. *Am Heart J.* 1999;138:206–9.

138. Zaret BL, Rigo P, Wackers FJ, et al. Myocardial perfusion imaging with ^{99m}Tc-tetrofosmin: comparison to ^{201}Tl imaging and coronary angiography in phase III multicenter trial. *Circulation.* 1995;91:313–9.

139. Takahashi N, Tamaki N, Tadamura E, et al. Combined assessment of regional perfusion and wall motion in patients with coronary artery disease with ^{99m}Tc-tetrofosmin. *J Nucl Cardiol.* 1994;1:29–38.

140. Rigo P, Leclercq B, Itti R, et al. Technetium-99m-tetrofosmin myocardial imaging: a comparison with ^{201}Tl and angiography. *J Nucl Med.* 1994;35:587–93.

141. Flamen P, Bossuyt A, Franken PR. Technetium-99m-tetrofosmin in dipyridamole-stress myocardial SPECT imaging: intraindividual comparison with technetium-99m-sestamibi. *J Nucl Med.* 1995;36:2009–15.

142. Matsunari I, Fujino S, Taki J, et al. Myocardial viability assessment with technetium-99m-tetrofosmin and thallium-201 reinjection in coronary artery disease. *J Nucl Med.* 1995;36:1961–7.

143. Matsunari I, Fujino S, Taki J, et al. Quantitative rest technetium-99m tetrofosmin imaging in predicting functional recovery after revascularization: comparison with rest-redistribution thallium-201. *J Am Coll Cardiol.* 1997;29:1226–33.

144. He W, Acampa W, Mainolfi C, et al. Tc-99m tetrofosmin tomography after nitrate administration in patients with ischemic left ventricular dysfunction: relation to metabolic imaging by PET. *J Nucl Cardiol.* 2003;10:599–606.

145. Bonow RO. Assessment of myocardial viability. In: Sandler MP, Coleman RE, Patton JA, et al., eds. *Diagnostic Nuclear Medicine.* 4th ed. Philadelphia: Lippincott Williams & Wilkins; 2003:319–32.

146. Berry JJ, Baker JA, Pieper KS, et al. The effect of metabolic milieu on cardiac PET imaging using fluorine-18-deoxyglucose and nitrogen-13-ammonia in normal volunteers. *J Nucl Med.* 1991;32:1518–25.

147. Tillisch JH, Brunker R, Marshall R, et al. Reversibility of cardiac wall-motion abnormalities predicted by positron tomography. *N Engl J Med.* 1986;314:884–8.

148. Tamaki N, Yonekura Y, Yamashita K, et al. Positron emission tomography using fluorine-18 deoxyglucose in evaluation of coronary artery bypass grafting. *Am J Cardiol.* 1989;64:860–5.

149. Bax JJ, Wijns W, Corner JH, et al. Accuracy of currently available techniques for prediction of functional recovery after revascularization in patients with left ventricular dysfunction due to chronic coronary artery disease: comparison of pooled data. *J Am Coll Cardiol.* 1997;30:1451–60.

150. Allman KC, Shaw IJ, Hachamovitch R, et al. Myocardial viability testing and impact of revascularization on prognosis in patients with coronary artery disease and left ventricular dysfunction: a meta-analysis. *J Am Coll Cardiol.* 2002;39:1151–8.

151. Bax JJ, Visser FC, Poldermans D, et al. Relationship between preoperative viability and postoperative improvement in LVEF and heart failure symptoms. *J Nucl Med.* 2001;42:79–86.

152. Bonte FJ, Parkey RW, Graham KD, et al. A new method for radionuclide imaging of acute myocardial infarction. *Radiology.* 1974;110:473–4.

153. Buja LM, Tofe AJ, Kulkarni PV, et al. Sites and mechanisms of localization of technetium 99m phosphorous radiopharmaceuticals in acute myocardial infarcts and other tissues. *J Clin Invest.* 1977;60:724–40.

154. Willerson JT, Parkey RN, Bonte FJ, et al. Pathophysiologic considerations and clinicopathological correlates of technetium-99m stannous pyrophosphate myocardial scintigraphy. *Semin Nucl Med.* 1980;10:54–69.

155. Volpini M, Giubbini R, Gei P, et al. Diagnosis of acute myocardial infarction by indium-111 antimyosin antibodies and correlation with the traditional techniques for the evaluation of extent and localization. *Am J Cardiol.* 1989;63:7–13.

156. Johnson LL, Seldin DW, Becker LC, et al. Antimyosin imaging in acute transmural myocardial infarctions: results of a multicenter clinical trial. *J Am Coll Cardiol.* 1989;13:27–35.

157. Khaw BA, Nakazawa A, O'Donnell SM, et al. Avidity of technetium-99m glucarate for the necrotic myocardium: in vivo and in vitro assessment. *J Nucl Cardiol.* 1997;4:283–90.

158. Mariani G, Villa G, Rossettin PF, et al. Detection of acute myocardial infarction by ^{99m}Tc-labeled D-glucaric acid imaging in patients with acute chest pain. *J Nucl Med.* 1999;40:1832–9.

159. Johnson LL. Myocardial hotspot imaging. In: Sandler MP, Colemen RE, Patton JA, et al., eds. *Diagnostic Nuclear Medicine.* 4th ed. Philadelphia: Lippincott Williams & Wilkins; 2003: 333–41.

160. Tait JF, Brown DS, Gibson DF, et al. Development and characterization of annexin V mutants with endogenous chelation sites for (99m) Tc. *Bioconjug Chem.* 2000;11;6:918–25.

161. Hofstra L, Liem IH, Dumont EA, et al. Visualization of cell death in vivo in patients with acute myocardial infarction. *Lancet.* 2000;356:209–12.

162. Kim RJ, Fieno DS, Parrish TB, et al. Relationship of MRI delayed contrast enhancement to irreversible injury, infarct age, and contractile function. *Circulation.* 1999;100:1992–2002.

163. Rehwald WG, Fieno DS, Chen EL, et al. Myocardial magnetic resonance imaging contrast agent concentrations after reversible and irreversible ischemic injury. *Circulation.* 2002;105:224–9.

164. Assomull RG, Pennell DJ, Prasad SK. Cardiovascular magnetic resonance in the evaluation of heart failure. *Heart.* 2007;93:985–92.

165. Kim RJ, Wu E, Rafael A, et al. The use of contrast-enhanced magnetic resonance imaging to identify reversible myocardial dysfunction. *N Engl J Med.* 2000;343:1445–53.
166. Hendel RC, Kramer CM, Patel MR, et al. ACCF/ACR/SCCT/SCMR/ASNC/NASCI/SCAI/SIR appropriateness criteria for cardiac computed tomography and cardiac magnetic resonance imaging: a report of the American College of Cardiology Foundation Quality Strategic Directions Committee Appropriateness Criteria Working Group, American College of Radiology, Society of Cardiovascular Computed Tomography, Society for Cardiovascular Magnetic Resonance, American Society of Nuclear Cardiology, North American Society for Cardiac Imaging, Society for Cardiovascular Angiography and Interventions, and Society of Interventional Radiology. *J Am Coll Cardiol.* 2006;48:1475–97.
167. Henneman MM, Bengal FM, van der Waal EE, et al. Cardiac neuronal imaging: application in the evaluation of cardiac disease. *J Nucl Cardiol.* 2008;15:442–55.
168. Hartikainen J, Mustonen J, Kuikka J, et al. Cardiac sympathetic denervation in patients with coronary artery disease without previous myocardial infarction. *Am J Cardiol.* 1997;80:273–7.
169. Ungerer M, Hartmann F, Karoglan M, et al. Regional in vivo and in vitro characterization of autonomic innervation in cardiomyopathic human heart. *Circulation.* 1998;97:174–80.
170. Wakabayashi T, Nakata T, Hashimoto A, et al. Assessment of underlying etiology and cardiac sympathetic innervation to identify patients at high risk of cardiac death. *J Nucl Med.* 2001;42:1757–67.
171. Agostini D, Verberne HJ, Burchert W, et al. I-123-*m*IBG myocardial imaging for assessment of risk for a major cardiac event in heart failure patients: insights from a retrospective European multicenter study. *Eur J Nucl Med Mol Img.* 2008;35:535–46.
172. American Society of Nuclear Cardiology. ACC/AHA/ASNC Guidelines for the Clinical Use of Cardiac Radionuclide Imaging. 2003. www.asnc.org/section_73.cfm.
173. Iskander S, Iskandrian AE. Risk assessment using single-photon emission computed tomographic technetium-99m sestamibi imaging. *J Am Coll Cardiol.* 1998;32:57–62.
174. Hachamovitch R, Berman DS. The use of nuclear cardiology in clinical decision making. *Semin Nucl Med.* 2005;35:62–72.
175. Hachamovitch R, Hayes S, Friedman JD, et al. Determinants of risk and its temporal variation in patients with normal stress myocardial perfusion scans: what is the warranty period of a normal scan? *J Am Coll Cardiol.* 2003;41:1329–40.
176. American Society of Nuclear Cardiology. Imaging guidelines for nuclear cardiology procedures. Myocardial perfusion and function: single photon emission computed tomography. 2007. www.asnc.org/section_73.cfm 2006. www.asnc.org/section_73.cfm. Accessed October 13, 2010.
177. Hachamovitch R, Hayes S, Friedman JD, et al. Comparison of the short-term survival benefit associated with revascularization compared with medical therapy in patients with no prior coronary artery disease undergoing stress myocardial perfusion single photon emission computed tomography. *Circulation.* 2003;107:2900–7.
178. Gill JB, Ruddy TD, Newell JB, et al. Prognostic importance of thallium uptake by the lungs during exercise in coronary artery disease. *N Engl J Med.* 1987;317:1486–9.
179. Weiss AT, Berman DS, Lew AS, et al. Transient ischemic dilation of the left ventricle on stress thallium-201 scintigraphy: a marker of severe and extensive coronary artery disease. *J Am Coll Cardiol.* 1987; 9:752–9.
180. Mazzanti M, Germano G, Kiat H, et al. Identification of severe and extensive coronary artery disease by automatic measurement of transient ischemic dilation of the left ventricle in dual-isotope myocardial perfusion SPECT. *J Am Coll Cardiol.* 1996;27:1612–20.
181. Sharir T, Germano G, Kavanagh PB, et al. Incremental prognostic value of post-stress left ventricular ejection fraction and volume by gated myocardial perfusion single photon emission computed tomography. *Circulation.* 1999;100:1035–42.
182. CASS Principal Investigators and Associates. Myocardial infarction and mortality in Coronary Artery Surgery Study (CASS) randomized trial. *N Engl J Med.* 1984;10:750–8.
183. Lima RSL, Watson DD, Goode AR, et al. Incremental value of combined perfusion and function over perfusion alone by gated SPECT myocardial perfusion imaging for detection of severe three-vessel coronary artery disease. *J Am Coll Cardiol.* 2003;42:64–70.
184. Berman DS, Kang X, Slomka PJ, et al. Underestimation of extent of ischemia by gated SPECT myocardial perfusion imaging in patients with left main coronary artery disease. *J Nucl Cardiol.* 2007;14:492–6.

CHAPTER 22

Lung

Radiopharmaceuticals for lung imaging can be divided into two main groups: (1) lung perfusion agents and (2) lung ventilation agents. Perfusion agents typically are radio-labeled particles that are temporarily trapped in the lung's arterioles and capillaries after intravenous injection and provide diagnostic information about regional blood flow to the lung. Ventilation agents are radioactive gases or radioaerosols that, after inhalation, demonstrate patency of the airways and alveolar system.

The diagnosis of pulmonary embolism (PE) is the principal reason for lung imaging. PE is not a disease itself but a complication of an underlying condition of coagulopathy, namely, deep-vein thrombosis (DVT). Because PE is potentially fatal and is treatable with anticoagulants, a quick and accurate diagnosis is essential to limit risk.

PHYSIOLOGIC PRINCIPLES

Lung Ventilation

The airways of the lung are divided into three functional zones: the *conducting zone*, consisting of the bronchi and bronchioles, delivers inspired air via the *transition zone* to the *respiratory zone*, which is made up of a complex of alveoli and capillaries where air and blood come into close contact to facilitate the exchange of oxygen and carbon dioxide (Figure 22-1).[1,2] The respiratory zone is sometimes called the gas exchange apparatus.

The airway conduction system has a branching pattern from the trachea to the alveolar sacs.[2] After inspiration, fresh air moves into the lungs by bulk flow as far as the respiratory bronchioles, where the transition zone begins, but from that point the movement of air into the alveolar system occurs by diffusion. Higher tidal volumes of air penetrate deeper but never reach the end of the alveolar system by bulk flow.[3]

Under normal conditions, airway resistance is minimal and air moves freely into and out of the airways and respiratory zone. In the presence of disease, airway resistance may increase significantly, affecting inspiration and expiration. In less severe cases of asthma or emphysema, for example, air may enter the bronchioles and alveoli readily because chest expansion and the inflow of air inflate these structures. Expiration, however, is more difficult because the weakened, diseased bronchioles collapse from the pressure of the thoracic cage against the lungs, causing air to be trapped in the alveoli. In advanced airway disease, air intake also may be obstructed because of inflamed bronchi and bronchioles and the presence of mucus plugs, often found in bronchial asthma. These alterations in air distribution can be observed with the use of radioactive gases and aerosols.

Lung Perfusion

The lungs are perfused by pulmonary arteries and veins, which distribute blood to and remove blood from the capillary beds where gas exchange occurs. The pulmonary artery has

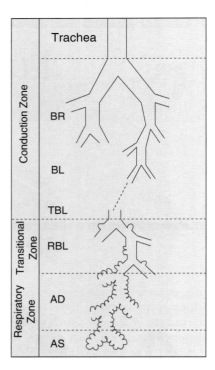

FIGURE 22-1 Branching pattern of the airways from the trachea to the alveolar sacs. BR = bronchi; BL = bronchioles; TBL = terminal bronchioles; RBL = respiratory bronchioles; AD = alveolar duct; AS = alveolar sac. (Adapted from reference 2.)

22 to 26 branches. At about the 24th branch, a short connector artery of 125 μm diameter divides at right angles to form two distribution arteries (Figure 22-2).[4] The smallest-diameter distribution arteries range in size from 60 to 100 μm, and from these vessels short precapillary arterioles of 25 μm diameter arise at right angles. These vessels then divide into alveolar capillary beds. The basic elements of the alveolar capillary bed are the capillary segments, which have the shape of short cylindrical tubes. The segments are modified at their ends to form wedges that allow each segment to join at either end with two adjacent segments. The average internal diameter of a capillary segment is 8 μm (range, 6 to 10 μm).[2]

It is clear from the structure of the alveolar capillary network that blood entering each precapillary arteriole has alternative routes for reaching the postcapillary venule. This capillary arrangement is important during lung scanning, because pulmonary hemodynamics are not appreciably affected by the temporary occlusion of a small percentage of capillary segments by radiolabeled particles unless there is advanced lung disease and pulmonary hypertension.

Normal blood flow in the lung is influenced by hydrostatic pressure. In the upright position, the mean pulmonary arterial pressure is 3 mm Hg at the apex of the lung, 13 mm Hg in the midzone, and 21 mm Hg at the base of the lung. This pressure difference alters blood distribution between the upper and lower parts of the lung. Studies conducted with the injection of xenon 133 gas dissolved in saline have shown that, in normal subjects, a change from the upright to the supine

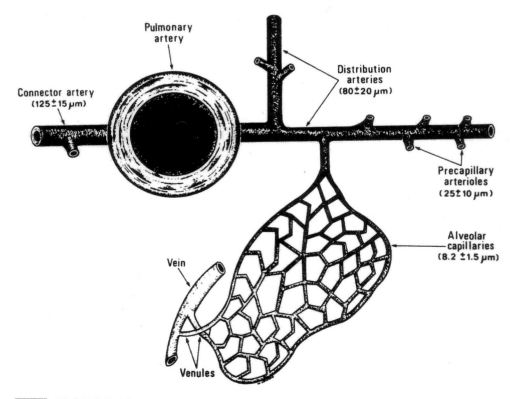

FIGURE 22-2 Schematic representation of pulmonary vasculature, emphasizing the alveolar capillary network and the anastomotic nature of the capillary bed. (Reprinted with permission of the Society of Nuclear Medicine from reference 4.)

position results in a doubling of blood perfusion to the upper lung zones at the cost of the lower zones, with the midzones recording no change.[5] This equalization in hydrostatic pressure is the reason that radiolabeled particles for lung scanning are injected while the patient is supine, which evens the distribution of radioactivity in the lung.

Blood flow to the lung is affected by various pathologic conditions. In the case of PE, the embolus not only mechanically impedes blood flow to the lung area distal to the blockage but also may cause local vasospasm, which further decreases blood flow to the region.[6] In the case of airway disease such as emphysema, destruction of many alveolar walls also causes destruction of their capillaries, resulting in increased vascular resistance to blood flow. In addition, a physiologic decrease in local blood perfusion occurs in emphysema because emphysematous alveoli exhibit poor gas exchange. The resulting alveolar hypoxia causes reflex vasoconstriction of local blood vessels, shunting blood to areas of the lung that have better aeration.[6] For these reasons, ventilatory disease may cause perfusion defects to appear on the lung scan.

LUNG PERFUSION AGENTS

Site-specific localization of radioactivity in the lungs began with the therapeutic application of gold; ^{198}Au was labeled to 50 μm carbon particles to irradiate tumor tissue.[7] The diagnostic utility of radioactive particles for lung scanning was initially investigated in 1963 by Taplin et al., who prepared ^{131}I-labeled human serum albumin macroaggregates (^{131}I-MAA). Their studies demonstrated that these particles could be trapped and eventually cleared from the lungs.[8-10] These investigators also performed the first toxicity studies to demonstrate the safety of injecting large numbers of particles intravenously. After these studies, which occurred from 1965 to 1970, ^{131}I-MAA became the agent of choice for imaging the presence of suspected pulmonary emboli.

The search for an imaging agent with properties more favorable than ^{131}I led to the development of particles labeled with radionuclides that have shorter half-lives. These included ^{113m}In-labeled ferric hydroxide and ^{99m}Tc-labeled ferrous hydroxide macroaggregates; however, toxic reactions associated with these preparations led to their disuse.[11,12]

In 1969, ^{99m}Tc-labeled human albumin microspheres (^{99m}Tc-HAM) were developed by Zolle et al.[13] and Rhodes et al.[14] ^{99m}Tc-HAM could be microsieved to control particle size and were biodegradable in vivo. A ^{99m}Tc-HAM kit using sodium thiosulfate as the tagging agent became available from the 3M Company.[14] Labeling efficiency was only 60% to 70%, however, and insonation was necessary to break up any potential microsphere aggregations. Labeling efficiency of the ^{99m}Tc-HAM kit improved when stannous chloride was used as the reducing agent. The kit was used for several years before the company stopped its manufacture.

Subramanian et al.[15] developed technetium Tc 99m albumin aggregated injection (^{99m}Tc-MAA) as an "instant kit" using stannous chloride as the tagging agent. Labeling yields were quantitative, and the kit became commercially available in the mid-1970s. It is still the agent of choice for perfusion lung imaging.

Technetium Tc 99m Albumin Aggregated

Kits for routine preparation of ^{99m}Tc-MAA are commercially available (see Tables 10-5 and 10-6 in Chapter 10). In general, stannous MAA kits are prepared by mixing sterile solutions of human serum albumin and stannous chloride in acetate buffer at pH 5. The resulting solution is heated at controlled temperature and mixed to form aggregated particles, which are in a size range of 10 to 90 μm in commercial kits. The sterile lyophilized kits are sealed under nitrogen. The number of particles in commercial kits varies, averaging about 5 million. Labeling of the MAA particles with ^{99m}Tc is accomplished by adding the required amount of ^{99m}Tc-sodium pertechnetate to a kit to achieve the desired particle concentration and specific activity. Dilution may be necessary to adjust the particle concentration for pediatric patients or patients with pulmonary hypertension. Chapter 10 includes a detailed discussion of ^{99m}Tc-MAA chemistry and kit preparation.

A number of factors are important in the preparation of ^{99m}Tc-MAA particles for perfusion lung imaging, including particle size, number, hardness, and chemical composition. These factors influence the biodistribution, metabolic fate, and potential toxicity of the lung perfusion agent.

Particle Size

Lung localization of ^{99m}Tc-MAA requires particles that are larger than the smallest capillaries, which have an internal diameter of 6 to 10 μm. Kits must contain particles no larger than 150 μm to reduce the possibility of a significant reduction in blood flow and elevation of pulmonary arterial pressure. Davis[4] investigated particle size and concluded that the capillary segments were the ideal vessels to block in the lung, because they are highly anastomotic and blockade of one or even several capillary segments will not substantially alter blood flow or pressure. The particle size recommended was 13.5 ± 1.5 μm. This size range, however, was found to be impractical to produce, and it would cause tracer to clear from the lung too rapidly to be useful for lung imaging. For these reasons, commercial MAA kits contain particles in the 10 to 90 μm range, with most particles between 20 and 60 μm (Figure 22-3).

Particle Number

Reports of acute toxicity from administering too many particles for lung scans led to investigations to determine the ideal number of particles for a satisfactory lung scan. Heck and Duley[16] reported that having too few particles (15,000 to 30,000) of ^{99m}Tc-albumin microspheres in the 15 to 30 μm range produced patchy-looking lung scans, particularly at the lung periphery. They determined that the minimum number of particles for a satisfactory lung scan was 60,000. Dworkin et al.[17] used 10 to 50 μm particles of ^{99m}Tc-MAA in dogs. They confirmed the work of Heck and

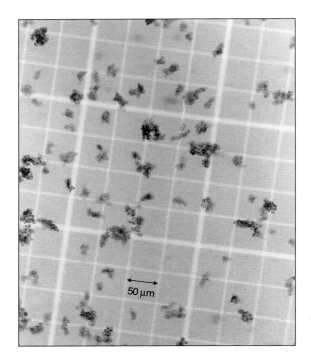

FIGURE 22-3 Photomicrograph of macroaggregated albumin (MAA) particles.

Duley, concluding that the minimum number of particles required for a satisfactory lung scan was 60 particles per gram of lung, which translates to 60,000 particles for an average weight of adult human lung tissue of 1000 grams. An upper limit of 250,000 particles for a lung scan was suggested because little is gained above this number and the chance of toxicity is increased. In practice, it is difficult to determine the exact number of particles administered because of the wide range of particle numbers per kit. If desired, an accurate particle number in a kit can be determined by using a microscope and hemacytometer, discussed in Chapter 14.

Particle Hardness and Composition

Ideally, particles that lodge in the pulmonary blood vessels should be biodegradable and should not produce local tissue reactions. Human serum albumin (HSA), a natural body constituent, meets these requirements when heated to the desired hardness for proper lung clearance. Generally, higher heating temperatures produce harder particles that take longer to clear from the lung. For example, Zolle et al.[13] demonstrated that albumin microspheres prepared at 118°C, 146°C, and 165°C had biologic half-lives in dog lungs of 2.4, 7.2, and 144 hours, respectively.

Early concerns about potential antigenic effects of denatured protein particles in the lung were dispelled when Iio and Wagner[18] found no evidence to prove that aggregated human albumin is antigenic to humans. This finding was corroborated by Taplin et al.[9] Human serum albumin became the preferred agent for making ^{99m}Tc-MAA. ^{99m}Tc-labeled iron hydroxide aggregates were also used for lung imaging, but their use was discontinued when fatalities were reported and there was an unresolved question of iron-induced reactions.[12]

Biologic Properties

After intravenous injection of ^{99m}Tc-MAA, more than 90% of the dose is extracted by the pulmonary arterial bed on the first pass through the lung. The mechanism of localization is physical entrapment of particles larger than the blood vessel diameter. The quantitative amount of particle activity (Q) distributed in the lung of a known concentration of radiotracer in the blood (C) is related to regional blood flow (F), i.e., Q = FC. This assumes that the radiotracer is not metabolized or eliminated prior to delivery to the lung. Consequently, based on this principle, the amount of radioactivity in a ^{99m}Tc-MAA dose delivered to any region of the lung distal to an obstruction will be diminished and will appear as a perfusion defect on the lung scan (Figure 22-4). Big emboli will occlude larger vessels and produce more extensive perfusion defects.

Biodegradation of MAA particles in the lung is slow enough to allow ample time for imaging. Using an in vivo cinemicroscopic technique, Taplin and MacDonald[19] demonstrated that the mechanism of MAA clearance from the lungs is due to particle fragmentation by blood cell bombardment and by continuous forward and backward movement within arterioles until aggregates are small enough to traverse the capillary lumen.

The rate of particle clearance from the lung is a function of particle size, distribution, and number; method of preparation (related to particle hardness); and the state of lung health. Smaller particles are expected to have faster clearance. Taplin and MacDonald[19] demonstrated in dogs that ^{131}I-MAA doses with particle sizes of 5 to 25 μm, 10 to 70 μm, and 10 to 150 μm had biologic half-lives of 30 minutes, 4 to 6 hours, and 18 to 24 hours, respectively. Particles that are too soft because of inadequate heating or are too small readily pass through the lungs and become localized in the liver, potentially interfering with the lung scan (Figure 22-5).

A patient's condition may also affect lung clearance of particles. Davis[20] determined that the average clearance half-life of ^{99m}Tc-iron hydroxide particles from the lung was 19 hours in normal subjects but was significantly slower in patients with various degrees of PE (105 hours) and chronic lung disease (222 hours). Busse et al.[21] demonstrated that the lung clearance rate of ^{131}I-MAA was slower in asthmatics and in patients receiving corticosteroid and immunosuppressive therapy, with viral pneumonia, and with chronic interstitial lung disease. Half-lives of ^{99m}Tc-MAA in the lung are listed in Table 10-6 in Chapter 10. An important point in this regard is that the rate of activity loss and the rate of particle loss from the lungs are not identical; a particle may partially break up and release a portion of its activity but still remain in the lung until it is small enough to pass through.[4]

Once particles are cleared from the lungs, they are phagocytosed by the reticuloendothelial system primarily in the liver (Figure 22-5). Taplin and MacDonald[19] demonstrated that ^{131}I-MAA particles undergo proteolytic digestion in Kupffer's cells, evidenced by the presence of ^{131}I-labeled tyrosine,

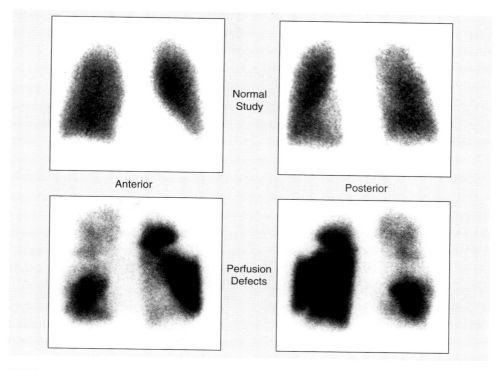

FIGURE 22-4 Perfusion lung scan with ^{99m}Tc-MAA. Normal scan and abnormal scan demonstrating multiple segmental defects due to pulmonary embolism.

peptides, and free radioiodide in the plasma and urine a few hours after injection. The biologic half-life in the liver was reported to be 9 to 10 hours. ^{99m}Tc-MAA is also metabolized in the liver, with urinary excretion of 30% to 75% of the dose in 24 hours, depending on the kit. These particles are also most likely digested by liver enzymes, with the release of ^{99m}Tc-labeled amino acids and pertechnetate, which are excreted in the urine. A fraction of the activity also remains in the liver, probably as insoluble ^{99m}Tc-TcO$_2$ colloid. The biologic fate of dual-labeled HSA colloid labeled with ^{14}C and ^{99m}Tc supports this mechanism.[22] Such an agent used for liver imaging demonstrated a biologic half-life of ^{14}C activity in the liver of 4 hours and of ^{99m}Tc activity of 11 hours.

Particle Toxicity

Safety and effectiveness considerations with particulate lung-scanning agents have been reviewed.[23] In the early days of lung scanning with ^{131}I-MAA, a typical lung scanning dose (LSD) contained 10^6 particles. Studies performed by Taplin and MacDonald in normal dogs demonstrated that a wide margin of safety exists when such doses are administered. These studies showed that the first sign of acute toxicity, observed as a rise in pulmonary arterial pressure (PAP), occurred at a dose of 20 mg/kg body weight (1000 times the average LSD). Such a large safety factor is possible because in the normal lung less than 1% of the arterioles and capillaries are blocked by such doses.[4]

Despite this wide margin of safety in normal subjects, several deaths have been reported after the administration of ^{131}I-MAA for lung scanning.[24–27] An evaluation of these cases revealed that the patients suffered from severe pulmonary hypertension and that their underlying diseases had caused narrowing and occlusion of pulmonary blood vessels. In each case, immediately after injection of the MAA dose, clinical deterioration occurred, manifested by respiratory distress, cyanosis, hypertension, and eventual death. Each of these reports discussed the need in such cases to decrease the

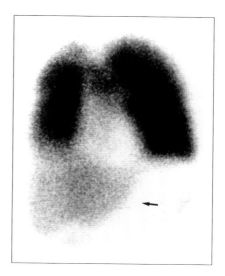

FIGURE 22-5 Anterior perfusion lung scan with ^{99m}Tc-MAA. Liver activity (arrow) is due to phagocytosis of particles that pass through the lung capillaries.

number of particles injected and to restrict their size to less than 50 μm, preferably to the 10 to 30 μm range.

The primary cause of cardiopulmonary toxicity associated with lung-scanning agents is the size and number of particles injected.[28,29] Large particles, when compared with the same size dose of small particles, are more effective in raising PAP. The inverse relationship between particle size and number in producing an acute toxic response is derived from the fact that larger particles block larger arterioles, of which there are fewer. Allen et al.[30] demonstrated that the safety factor for a LSD of 10^6 particles of ^{99m}Tc-MAA (30 to 50 μm) was 125, based on an elevation of PAP of 10% to 20%. This safety factor increases to 1250 if only 100,000 particles are administered, which emphasizes why it is desirable to administer only as many particles as are needed for a satisfactory scan (not more than 250,000). It should be pointed out that these safety factors are for normal patients. Patients with pulmonary hypertension should be administered a minimum number of particles (60,000) for safety reasons.

Pediatric patients also require special consideration regarding number of particles in the LSD. Heyman[31] noted that a significant increase in the number of alveoli and pulmonary arteries occurs during the first few years of life, reaching adult levels at about 8 years of age. The number is 10% to 30% of adult values during the first year of life and up to 50% of the adult number at 3 years. Heyman suggested limiting the number of particles to 50,000 in the newborn infant and 165,000 in children up to 1 year old. A technique has been described for preparing pediatric doses whereby excess stannous MAA particles from ^{99m}Tc-MAA kits are discarded and the number remaining are radiolabeled with ^{99m}Tc-sodium pertechnetate to achieve the desired concentration for pediatric lung doses.[32] An example of a dilution procedure for reducing particle concentration is shown in Table 22-1. Manufacturer package inserts for ^{99m}Tc-MAA kits contain a table of recommended dosages of particles for pediatric patients (Table 22-2).

Sufficient time must be allowed for MAA particles to be tagged with technetium during the labeling process to ensure high radiochemical purity. This is particularly important if a large fraction of particles is removed from a kit prior to the addition of ^{99m}Tc-sodium pertechnetate, such as when pediatric doses are prepared. As little as 5% to 10% free pertechnetate in a kit will be evident as thyroid uptake on lung scans (Figure 22-6).

An early concern with lung scanning was the potential threat of cerebral microembolization from particles that enter the systemic circulation either after degradation in the lung or through a right-to-left cardiac shunt. In this regard, Taplin et al.[9] reported that suspensions of albumin particles initially trapped in the lung are subsequently cleared and transposed to the liver and spleen. It was stated that if small particles were able to traverse the pulmonary capillaries they would also traverse cerebral vessels without significant danger of microembolization. In other studies, Taplin and MacDonald[19] estimated the margin of safety for particles that were not degraded into smaller sizes in the lungs but entered the systemic circulation directly through a right-to-left cardiac shunt. Studies were performed in monkeys receiving direct carotid arterial injections of MAA. After repeated injections of aggregates (10 to 100 μm), there was no evidence of histologic or observable lesions in brain tissue at administered doses less than 6 mg MAA per 100 grams of brain tissue. On the basis of these data, Taplin and MacDonald estimated that the margin of safety for a 1 mg MAA dose for a lung scan was greater than 2000. This safety margin was based on the assumption that 50% of the dose was shunted to the general circulation, of which 10% went to the head and 3% went to each hemisphere.[19]

LUNG VENTILATION AGENTS

The ventilation lung scan became an important diagnostic tool because it improved the specificity of the perfusion lung scan in the diagnosis of PE.[33] Taplin and Chopra[34] reviewed combined perfusion–ventilation lung scanning. Before the ventilation lung scan was established as a routine procedure, a diagnosis of PE was based on clinical suspicion and on the finding of one or more segmental or lobar perfusion defects on a normal chest radiograph. When it became evident that a normal chest radiograph could not exclude all nonembolic

TABLE 22-1 Dilution Procedure for Preparing Reduced Particle Concentration of ^{99m}Tc-MAA

1. Add 10 mL saline to 5 million particle MAA kit (500,000 particles/mL).
2. Remove 8.8 mL (4.4 million particles) and discard.
3. Add 30 mCi Na^{99m}TcO$_4$ to the remaining 600,000 particles in the kit.
4. Incubate for 15 minutes.
5. Add saline to make 5 mL final volume.
6. Final concentration: 3 to 6 mCi (60,000 to 120,000 particles)/0.5 to 1.0 mL.

TABLE 22-2 MAA Particles per Pediatric Dose (in thousands)[a]

Variable	Newborn	1 Year	5 Years	10 Years	15 Years
Weight (kg)	3.5	12.1	20.3	33.5	55.0
Activity (mCi)	0.2	0.4	0.6	1.0	1.7
Range of particles given	10–50	50–100	200–300	200–300	200–700

[a] The suggested average pediatric dose is based on 30 μCi/kg body weight, except for newborns. A minimum dose of 200 μCi of ^{99m}Tc-MAA should be used for the lung scan procedure.
Source: Pulmolite package insert, Pharmalucence, Bedford, MA.

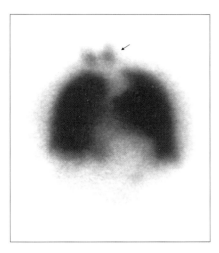

FIGURE 22-6 Perfusion lung scan demonstrating thyroid gland uptake (arrow) of free pertechnetate activity present as an excessive impurity in the ^{99m}Tc-MAA preparation.

causes of perfusion defects, particularly in cases involving chronic obstructive pulmonary disease (COPD), the ventilation lung scan (V) gained acceptance as a method for evaluating regional ventilation. It was shown to have diagnostic value in the early detection of obstructive pulmonary disease, for which chest radiographs are relatively insensitive. More important, it added specificity to lung perfusion scanning (Q) by demonstrating that perfusion defects of embolic origin (proven angiographically) were nearly always well ventilated, whereas those caused by parenchymal or obstructive airway disease were neither perfused nor ventilated (VQ matching). The finding of poorly perfused but well-ventilated regions (VQ mismatch) in radiographically normal lung was thus considered strong evidence for PE.

Xenon Xe 133 Gas

The most widely used agents for lung ventilation studies are the radioactive noble gases. Their use in lung ventilation imaging is based on their poor aqueous solubility in body fluids, which confines the inhaled activity to the lung air space, where regional distribution is proportional to airflow in patent airways. Several inert gases possess physical properties that are useful in ventilation imaging (Table 22-3). Although the properties of ^{133}Xe are not ideal, it is routinely used in clinical practice today because it is readily available. The other gases, namely ^{127}Xe, ^{135}Xe, and ^{81m}Kr, which have more favorable physical properties for imaging, are not routinely used because of higher cost and logistical or production limitations.[35–40]

^{133}Xe was first used for ventilation imaging in the mid-1960s and became more widely used clinically in the 1970s because it is relatively inexpensive to produce in large quantities in a nuclear reactor as a byproduct of uranium fission. It is supplied commercially in unit dose vials of 10 mCi and 20 mCi (370 and 740 MBq). It has also been available in bulk curie amounts for in-house packaging. ^{133}Xe, with a half-life of 5.24 days, can be stored in the nuclear medicine lab ready for use. It is typically replenished by weekly shipments from the supplier.

The primary disadvantages of ^{133}Xe are poor image quality due to attenuation of its 81 keV gamma ray and low photon abundance (38%). In addition, the low-energy photon requires that a ventilation scan be performed prior to a ^{99m}Tc-MAA perfusion lung scan, which would not be necessary if the perfusion scan was normal. Some institutions perform a postperfusion xenon ventilation study by increasing the administered activity of ^{133}Xe to 20 to 25 mCi (740–925 MBq) to override the technetium activity present in the lung from the ^{99m}Tc-MAA study. Alternatively, a ^{99m}Tc-DTPA aerosol scan can be done in lieu of ^{133}Xe.

Xenon Administration

A ventilation study with ^{133}Xe is performed with the patient attached to a closed ventilation system (Figure 22-7). The essential elements are a leak-resistant breathing circuit, a carbon dioxide absorber, a moisture absorber (essential for efficient charcoal trap function), an expansion device (spirometer or breathing bag), a bacterial filter (to protect the system), valves to control airflow and to admit oxygen on demand, and a device to trap and contain the xenon when a study is complete.

During a lung ventilation study xenon gas is admitted into the system near the patient's mouth during an inspired breath (wash-in phase). Xenon activity is flushed from its vial with a special device (Figure 22-8). A double needle punctures

TABLE 22-3 Radioactive Gases for Lung Ventilation Imaging

Gas	Production	Half-life	Gamma keV	Abundance %
^{133}Xe	^{235}U(n, f) ^{133}Xe $\xrightarrow{\beta^-}$ ^{133}Cs	5.24 days	81	38
^{135}Xe	^{235}U(n, f) ^{135}Xe $\xrightarrow{\beta^-}$ ^{135}Cs	9.1 hr	250	90
^{127}Xe	^{133}Cs(p, 2p5n) ^{127}Xe $\xrightarrow{EC}$ ^{127}I	36.4 days	172 203 375	26 68 17
^{81m}Kr	^{81}Rb $\xrightarrow{4.5hr}$ ^{81m}Kr $\xrightarrow{IT}$ ^{81}Kr	13.1 sec	190	68

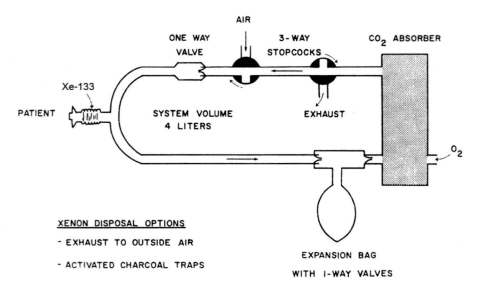

FIGURE 22-7 Diagram of a simplified closed-circuit breathing device for administering radioactive xenon gas for lung ventilation imaging. The charcoal trap, moisture absorber, bacterial filter, and motorized blowers are omitted for simplicity.

the vial stopper; air is pumped into the vial through one needle, and xenon activity exits through the other needle into the breathing system. As the patient rebreathes into the system, xenon activity equilibrates between the system and the lungs, achieving a maximum accumulation of lung activity (equilibrium phase). At this point in the study a valve is adjusted to permit room air to enter the system, and xenon expelled from the patient's lungs is directed into a lead-shielded charcoal trap (washout phase). The activated charcoal retains the ^{133}Xe while it decays.

Safety Control of Xenon

Federal and state regulations have been established to limit the concentration of radioactive material that can be expelled into the environment so that the total effective dose equivalent (TEDE) to the public does not exceed 0.1 rem per year and the dose rate from an external source to an unrestricted area does not exceed 2 mrem in any 1 hour. The regulations related to xenon release into the work environment and effluents into public places are described in Chapter 7, with illustrative calculations to ensure that release of ^{133}Xe does not exceed regulatory limits. Chapter 7 also presents an example of how to determine the evacuation time in an imaging room if an accidental spill of xenon occurs.

Biologic Distribution of Xenon

The amount of xenon that enters the body during a ventilation study is directly related to the lung air concentration, the duration of exposure, and xenon's solubility in tissue fluid. The poor solubility of xenon in water (12% at 25°C) is the reason for its slow absorption into the body. About 30 hours of rebreathing is required for xenon to reach an equilibrium concentration in other tissues.[41] Within the short time of 10 minutes required for a ventilation study, only about one-third of the administered activity of xenon enters the body, assuming a lung volume of 2.5 L and a ventilation system volume of 10 L. About two-thirds of this amount is in the lung and the remainder in other tissues (Table 22-4).

The amount of xenon distributed to various tissues and their subsequent rates of clearance are related to blood flow, tissue mass, and the tissue-to-blood partition coefficient. For example, the fat-to-blood coefficient is 7.9, whereas the skeletal muscle-to-blood coefficient is only 0.7.[42] Therefore, all things being equal, fatty tissue will concentrate far more xenon than muscle, and its rate of clearance will be slower. A greater fraction of the administered dose is distributed in muscle, however, because of its larger mass and greater blood flow.

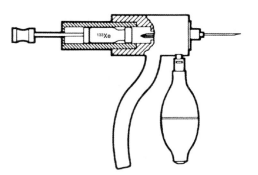

FIGURE 22-8 Hand-held device for administration of xenon gas. Forced air from the squeeze bulb transfers gas from the unit dose vial to the patient breathing system.

TABLE 22-4 Biologic Distribution of Xenon

Body Compartment	% Distribution in Body	Biologic Half-life
Lung (air)	68	22 sec
RBC (hemoglobin)	9	3 min
Muscle	11	0.4 hr
Fat	8	2.7 hr
Other tissue	4	7.6–17 hr

Source: Reference 41.

Activity in the liver is sometimes seen on xenon washout studies because of the organ's fatty content.

The clearance rate of xenon from the tissues varies considerably, being fastest from the lungs ($T_{1/2}$ = 22 seconds) and slowest from fat ($T_{1/2}$ = 8 to 17 hours).[41] Obese people will therefore retain xenon for longer periods of time than lean people.

Radioaerosols

Technetium Tc 99m Pentetate

Radioaerosols predated the use of radioactive gases for lung ventilation imaging but were not widely accepted because of the inability to generate aerosol droplets small enough to achieve deep penetration into the lung. Droplets larger than 3 to 5 μm in diameter cause hyperdeposition of aerosol in the trachea and major airways. Hence, studies in normal subjects without airway obstruction may lead to false-positive scans. A droplet size smaller than 2 μm is necessary for good distribution and minimal large-airway deposition.[43] Eventually, nebulization systems were developed that efficiently generated aerosol droplets 1 μm or less. Disposable devices are now commercially available for routine radioaerosol studies (UltraVent [Covidien]). Figure 22-9 illustrates the essential components of a radioaerosol device.

A typical radioaerosol procedure involves placing 30 to 50 mCi (1110–1850 MBq) of ^{99m}Tc-pentetate (^{99m}Tc-DTPA) in a 2 to 3 mL volume into the nebulizer. Aerosol droplets are generated by forcing air or oxygen through the nebulizer at 8 to 10 L/minute at 25 to 50 psi. The patient inhales the radioaerosol during normal breathing through the mouth with the nose clamped shut. Radioaerosol not inhaled by the patient or exhaled during the procedure is trapped in a particle-retentive filter. The amount of radioactivity deposited in the patient's lungs depends on the initial nebulizer concentration, the length of breathing time, and the patient's condition. Generally, with a normal subject and an initial ^{99m}Tc-DTPA concentration of 30 mCi (1110 MBq) in 3 mL, the lung deposition of activity is approximately 0.1 mCi (3.7 MBq) per minute of breathing time. A typical study requires between 5 and 10 minutes of breathing time to acquire sufficient activity to perform inhalation imaging. The efficiency of delivering activity from the nebulizer to the lungs is only about 2% to 5%, which is why such a large amount of activity is placed in the nebulizer. After inhalation of radioaerosol, the unit is removed and the patient is transported to the gamma camera for imaging.

The agent of choice for radioaerosol imaging is ^{99m}Tc-DTPA. Its rate of lung clearance into the blood is somewhat slower than that of ^{99m}Tc-sodium pertechnetate because of DTPA's larger molecular weight. The clearance rate across the pulmonary epithelium into the blood in normal subjects is 1.5% per minute for ^{99m}Tc-DTPA and 5.1% per minute for ^{99m}Tc-pertechnetate.[44] Clearance rates increase in smokers and patients with interstitial lung disease, in whom epithelial permeability is increased.[44,45]

Radioaerosols provide an alternative to xenon gas for ventilation imaging. Aerosols have the inherent advantage of ready availability of technetium and DTPA kits and do not require special facilities and equipment to control a radioactive gas. Because aerosol particles are fixed in the lungs, it is possible to record ventilation images in multiple projections to provide comparable orientation with perfusion images. Although the washout phase of the xenon study is a sensitive method for detecting localized obstructive airway disease, radioaerosol studies provide a useful alternative method if xenon is not available.

Because both the perfusion and aerosol ventilation agents employ a ^{99m}Tc label, a potential interference problem occurs during imaging. This can be overcome by using different amounts of activity for the two studies. A routine technique is to perform the aerosol study first by nebulizing 30 to 40 mCi (1110–1850 MBq) in 2 mL of ^{99m}Tc-DTPA for 3 to 5 minutes, depositing approximately 1.0 mCi in the lung. This is followed by a ^{99m}Tc-MAA perfusion study with 3 to 4 mCi (111–148 MBq). The MAA activity ideally should be 3 to 4 times larger than the aerosol activity. Another method is to perform the ^{99m}Tc-MAA perfusion scan first with 1 mCi (37 MBq), followed by the aerosol study with 60 to 80 mCi in 2 mL of ^{99m}Tc-DTPA, depositing 3 mCi (111 MBq) in the lung.[34,46] The rationale for a postperfusion aerosol study is that defects caused by embolism seen on the perfusion study will demonstrate fill-in on the aerosol study. Complete fill-in, however, requires that the activity of the ^{99m}Tc aerosol dose in the lung exceed that of the ^{99m}Tc-MAA dose by 2 to 3 times. A simple technique is to record the maximum count rate over the lungs from the ^{99m}Tc-MAA dose and then administer the aerosol with the patient supine until the lung count rate doubles or triples.

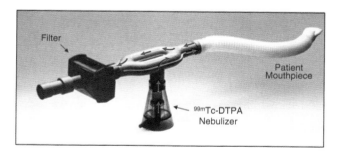

FIGURE 22-9 Disposable system for the generation of ^{99m}Tc-DTPA radioaerosol.

Technegas

A unique technetium aerosol called Technegas was developed in 1986.[47] Technegas particles are hexagonal platelets of metallic technetium encapsulated within a thin layer of graphite. The aerosol is produced by heating ^{99m}Tc-sodium pertechnetate solution to high temperature (2550°C for 15 seconds) in a graphite crucible. According to the following reaction, during the heating process pertechnetate is reduced to metallic technetium in a two-step reduction by carbon and encased in graphite a few nanometers thick, preventing its reoxidation:[48]

$$2\ NaTcO_4 + 3\ C \rightarrow CO_2 \uparrow + 2\ CO \uparrow + 2\ NaTcO_2$$
$$2\ NaTcO_2 \rightarrow 2\ NaO_2 \uparrow + 2\ Tc$$

The average particle is 30 to 60 nm wide and 5 nm thick, with 80% of the particles below 100 nm.[48] The particles are hydrophobic and not subject to increases in size in the airways during inhalation, enabling them to penetrate deeply into the lung.

Technegas is manufactured by an Australian firm (Vita Medical Limited) for use in that country, Europe, Africa, the Middle East, and South America. The product is not licensed in the United States.

Technetium Tc 99m Sulfur Colloid

^{99m}Tc-sulfur colloid has also been used for radioaerosol imaging, offering the advantage of prolonged retention in the lung airways, which may be especially advantageous in smokers or patients with other inflammatory lung disease who have a very rapid clearance of ^{99m}Tc-DTPA from the lungs.[49]

NUCLEAR MEDICINE PROCEDURES

Ventilation–Perfusion Lung Scan

Rationale

Functional imaging of the lungs in nuclear medicine is done to evaluate lung ventilation and perfusion. Lung ventilation is evaluated by imaging the lungs during inspiration of an inert gas such as xenon ^{133}Xe or a radiolabeled aerosol such as ^{99m}Tc-DTPA. Evaluation of perfusion involves intravenous administration of particles such as ^{99m}Tc-MAA and is based on capillary blockade. The most common indication for ventilation and perfusion lung imaging is evaluation of suspected acute PE. Other indications include evaluation of patients for lung transplantation (e.g., patients with cystic fibrosis), evaluation of patients with suspected chronic pulmonary emboli as a cause for pulmonary hypertension, preoperative evaluation of patients with COPD, and evaluation of differential lung function prior to surgical lobectomy or pneumonectomy.

At least 94,000 new cases of PE are diagnosed in the United States each year; 25% of these patients die within 7 days, even with the availability of heparin prophylaxis.[50] Treatment involves timely identification of PE, appropriate anticoagulant therapy, and possible interventional therapies such as vena cava filter placement.

Clinically, PE is suspected when a patient experiences sudden-onset shortness of breath and pleuritic chest pain. These symptoms are even more worrisome in a person who is at high risk for PE. Many factors can lead to increased risk for PE; they include a history of previous PE or DVT, use of oral contraceptives,[51] malignant neoplasms, surgery, trauma, immobilization, paralysis, long airplane flights, and certain blood factors.

VQ lung scans represent a safe, noninvasive means of evaluating patients with suspected PE. These studies are especially effective if the patient has a normal chest x-ray. In patients with normal chest x-rays, VQ scans have been shown to yield a definitive diagnosis of either PE or no PE in 83% of studies.[52] However, in many cases chest x-rays are abnormal, which can lead to an increase in intermediate-probability scans. In these cases, helical computed tomography (CT) and multichannel CT have proven useful in evaluating suspected PE.

Procedure

The first step is to examine a current chest x-ray. If the patient is experiencing acute changes in symptoms, the chest x-ray should be performed and evaluated just before the VQ scan. Otherwise, a chest x-ray performed within 24 hours of the VQ scan is generally acceptable. If there are significant abnormalities on the chest x-ray that would warrant further anatomic imaging, such as a suspected tumor, then CT may be the more appropriate test for PE as well as for other abnormalities on the film. However, if there are no significant abnormalities on the chest x-ray, ventilation–perfusion imaging is an effective test for PE.

Before the ventilation or perfusion study is performed, a clinical history should be obtained to evaluate the likelihood of PE. Also, certain patients require special considerations. The radiation dose should be minimized in pregnant women. The number of ^{99m}Tc-MAA particles should be reduced in pediatric patients and patients with severe pulmonary hypertension or a known right-to-left ventricular shunt. In addition, patient positioning should be considered during dose administration. With the patient in the upright position, the dependent or lower lung zones usually demonstrate better ventilation and perfusion. This is more obvious on perfusion images than on ventilation images.[53] To compensate for the effects of gravity, patients are usually injected with ^{99m}Tc-MAA particles in the supine position. For the ventilation study, the patient can be in either upright or supine position during administration of the gas or aerosol.

The ventilation study is usually performed before the perfusion study. If the perfusion study is done first, activity from ^{99m}Tc-MAA will add background activity to the ^{133}Xe or ^{99m}Tc-DTPA ventilation study. If the ventilation study is done first with ^{133}Xe, the lower photon energy of ^{133}Xe will not contribute to the background activity in the higher-energy window of ^{99m}Tc. If the ventilation study is performed with ^{99m}Tc-DTPA aerosol prior to the perfusion study, a smaller amount of activity, in the range of 0.5 to 1 mCi (18.5–37 MBq), should be used so that at least 3 to 4 times this amount of activity can be used in the subsequent ^{99m}Tc-MAA perfusion study. If postperfusion aerosol imaging is done, the ^{99m}Tc-MAA dose is reduced as described earlier. One advantage of postperfusion aerosol imaging is that the patient can be placed in the same positions as those used in the perfusion study for a more direct comparison. The patient is positioned either upright or supine in front of the camera. If ^{133}Xe is used, the ventilation study is usually done in three phases. The first phase is the first-breath or wash-in imaging phase. The second is the equilibrium phase, and the final part is the washout phase. The wash-in image is obtained when the patient first takes in a deep breath of the ^{133}Xe and holds it for several seconds. The next set of images (the equilibrium phase) occurs while the patient is breathing into a closed system containing the ^{133}Xe and some oxygen. After this, the ^{133}Xe gas is exhausted into a trap of activated charcoal, and images are obtained as the patient breathes fresh air or oxygen while the xenon clears from the lungs. Posterior or anterior and posterior images of the lungs are obtained during the wash-in, equilibrium, and washout phases of the study.

For the perfusion study, the patient is first instructed to cough and take a couple of deep breaths. The typical dosage of ^{99m}Tc-MAA in adults is 3 mCi (111 MBq), with a range of 1 to 5 mCi (37–185 MBq), and the number of particles injected is typically between 200,000 and 700,000.[54] The number of particles should be reduced in patients with known severe pulmonary hypertension or right-to-left ventricular shunt. The dose is administered intravenously while the patient is in the supine position. Since ^{99m}Tc-MAA particles may settle out in the syringe, the syringe contents should be thoroughly mixed before administration. In addition, blood drawn into the syringe during injection should not be allowed to stand too long, or clotting may occur. A delay may occur if there is some difficulty during the injection, and a new dose should be used if clotting is suspected. Injection of clotted particles will result in spurious information on the lung scan. After administration of ^{99m}Tc-MAA, images are obtained in the anterior, posterior, right anterior oblique, left anterior oblique, right posterior oblique, left posterior oblique, right lateral, and left lateral projections.

Interpretation

In a normal ventilation study, homogeneous radiotracer activity is seen in both lungs in the first-breath or wash-in image (Figure 22-10A). During the washout phase, activity usually clears quickly from both lungs within 2 to 3 minutes. Xenon is fat soluble and is often seen accumulating in the right upper quadrant during the washout phase in a patient with a fatty liver. Wash-in defects that normalize during the equilibrium phase and fail to clear normally in the washout phase (gas trapping) are often associated with obstructive pulmonary disease.

A normal perfusion study demonstrates homogeneous perfusion to both lungs (Figure 22-10B). A PE is typically seen as a wedge-shaped, pleural-based defect on perfusion images of the lungs secondary to obstruction of the associated segmental pulmonary arterial flow distal to the embolus. However, ventilation to this region is usually not affected. This results in a perfusion defect with no corresponding ventilation abnormality (a ventilation–perfusion mismatch).

Analysis of the perfusion portion of the study involves looking at the type of defects (classifying them as either segmental or nonsegmental), the number of defects, and the size of the defects. Segmental defects are associated with bronchopulmonary segmental territories. Subsegmental defects are associated with a portion or subset of a bronchopulmonary segmental territory. The size of a segmental perfusion defect is usually described as large, moderate, or small. A large segmental defect is defined as a perfusion defect that involves more than 75% of an anatomic lung segment. A moderate segmental defect involves 25% to 75% of a pulmonary segment, and a small segmental defect involves less than 25% of a segment. Nonsegmental defects are perfusion defects that do not correspond to any of the anatomic lung segments. Examples of nonsegmental perfusion defects include a large cardiac silhouette, tumor, pleural effusion, or elevation of a hemidiaphragm. Nonsegmental defects are not usually associated with pulmonary emboli. Once the perfusion defects have been categorized, they are compared with corresponding areas on the ventilation study. Matching ventilation–perfusion defects are usually not associated with PE.

A normal perfusion study is one in which there are no perfusion defects and pulmonary perfusion matches the shape of the lungs seen on the chest x-ray (Figure 22-10B). A normal perfusion study essentially rules out the diagnosis of PE, regardless of any abnormalities on the ventilation study. In perfusion scans that demonstrate abnormalities, a set of diagnostic criteria is used to determine the probability of PE. These criteria are based on a large study, the Prospective Investigation of Pulmonary Embolism Diagnosis (PIOPED),[55] and have been redefined on the basis of retrospective analysis of the data.[56] The modified PIOPED criteria define high probability (≥80%), intermediate probability (20%–79%), and low probability (≤20%) for PE on the basis of combined results from the ventilation and perfusion studies.

Using these criteria, if two or more large segmental perfusion defects are seen without a corresponding ventilation abnormality (ventilation–perfusion mismatch), the probability of PE is considered high. The probability is also considered high with any combination of moderate or large mismatches that are the equivalent of two large segmental defects. Two moderate defects are considered equivalent to one large defect. In the PIOPED study, the specificity of a high-probability ventilation–perfusion study for PE was 97% (Figure 22-11).

The probability of PE is low if the study shows only small perfusion defects, matched perfusion and ventilation defects with a normal chest x-ray, nonsegmental perfusion defects, or any perfusion defect with a substantially larger chest x-ray abnormality. Studies that are difficult to categorize as either high or low probability are considered to show an intermediate probability for PE. An example of this would be a study with one or two moderate mismatched segmental perfusion defects.

Matching ventilation and perfusion defects can be seen with a number of conditions, including COPD, congestive

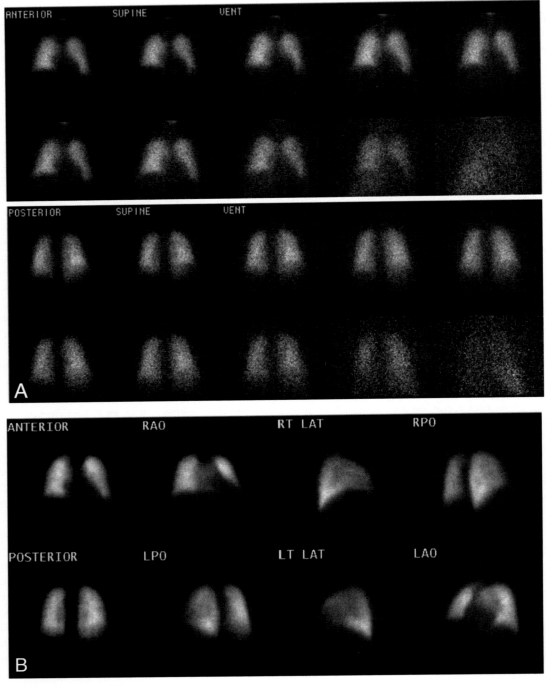

FIGURE 22-10 Normal ventilation/perfusion study. (A) Anterior and posterior ventilation images of the lungs were obtained after administration of 8.6 mCi ^{133}Xe gas by face mask. No significant ventilation defects are seen on either the wash-in or washout phase of the study. There is some evidence of ^{133}Xe accumulation in the liver, likely representing some degree of fatty infiltration. (B) Anterior, posterior, lateral, and oblique images of the lungs were obtained after intravenous administration of 3 mCi (111 MBq) of ^{99m}Tc-MAA. No perfusion defects are seen to suggest pulmonary emboli.

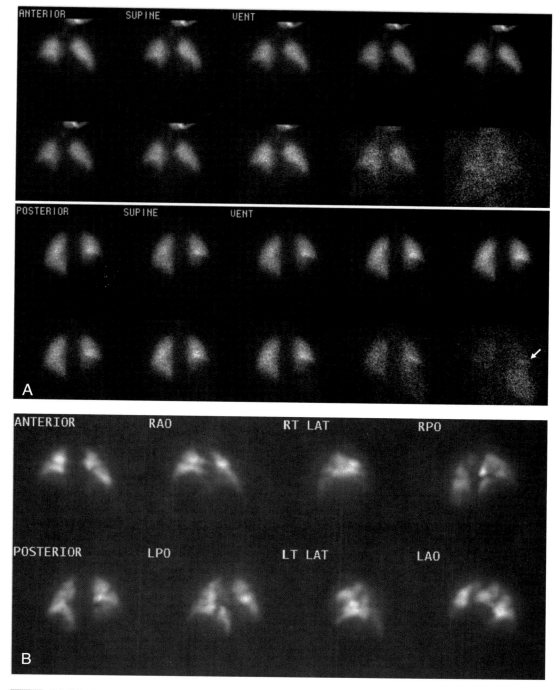

FIGURE 22-11 High probability for pulmonary emboli. (A) ¹³³Xe ventilation study with no significant ventilation defects seen on the initial first-breath or wash-in image. There is evidence of mild gas trapping at the right base (arrow) in the washout phase. Also notice ¹³³Xe uptake in the liver below the region of mild gas trapping. (B) A ⁹⁹ᵐTc-MAA perfusion study done immediately after the ventilation study demonstrated multiple wedge-shaped, pleural-based perfusion defects consistent with a high probability for pulmonary embolism.

heart failure, pleural effusion, lung tumors, bullous disease, and mucus plugs (Figure 22-12). Two simplified algorithms for distinguishing PE from chronic obstructive pulmonary disease (COPD) in patients after either a perfusion first–ventilation second or a ventilation first–perfusion second protocol are shown in Figures 22-13 and 22-14.

An update to PIOPED II was designed to redefine the categories for VQ scan diagnosis in PE, maintaining the criteria for high probability scans, revising the intermediate and low probabilities, and adding a very low probability criterion.[57] In addition, the VQ study was compared with contrast-enhanced spiral CT pulmonary angiography (CTA) in

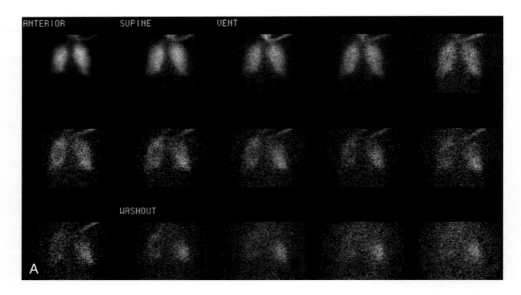
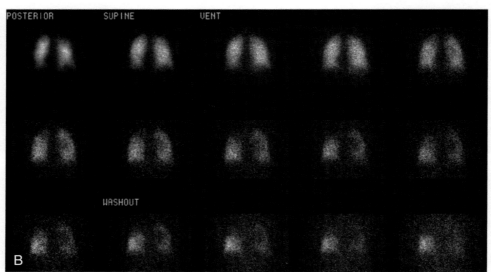
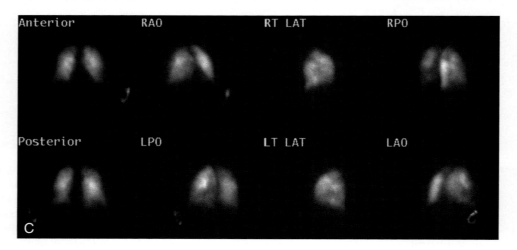

FIGURE 22-12 Obstructive pulmonary disease. (A) Anterior and (B) posterior images from a ^{133}Xe ventilation study demonstrating defects at both lung bases and the right lung apex on the first-breath image. These defects fill in during the equilibrium phase of the study and then demonstrate gas trapping during the washout phase. (C) On the ^{99m}Tc-MAA perfusion study, there is heterogeneous perfusion in both lungs with matching perfusion defects seen at the right lung apex and both bases. Matching ventilation and perfusion defects are most likely related to obstructive pulmonary disease and not to pulmonary emboli.

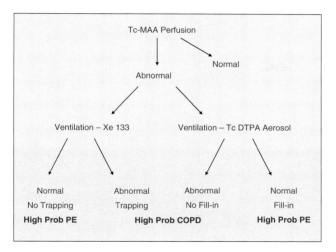

FIGURE 22-13 Perfusion–ventilation algorithm for VQ lung imaging.

diagnosing PE.[58] These studies favored CTA as a first choice in patients with a questionable chest x-ray. However, because CTA requires intravenous iodinated contrast, the VQ study is still the recommended procedure in patients with renal failure or allergy to iodinated contrast and in patients in whom the radiation dose is of concern, such as with breast exposure in young women. A key point in the diagnosis of PE is the requirement of a test with low positive predictive value (PPV) for the presence of PE to decide not to treat patients with anticoagulant therapy. In this regard the PPV of a very low probability VQ scan based on the revised PIOPED II criteria was found to be 8.2%, and when combined with a low probability objective clinical assessment it was 3.1%.[57] In women ≤40 years old, the PPV of a very low probability VQ scan with a low objective clinical assessment was 2%. Thus, the VQ scan is considered a first choice in patients with a normal chest x-ray, when CTA is contraindicated,

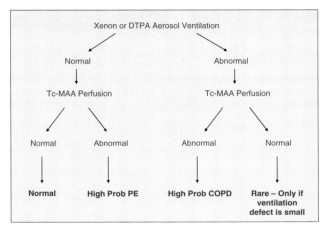

FIGURE 22-14 Ventilation–perfusion algorithm for VQ lung imaging.

and it is a valuable study for assessing patients on anticoagulant therapy.

Differential Lung Perfusion

Surgical resection of tumors is a consideration in patients with non–small-cell lung carcinoma that has not advanced to stage IIIb or stage IV. The surgeon's decision to perform surgery for lung cancer depends on the patient's ability to tolerate the procedure. Differential perfusion studies along with pulmonary function studies can help to determine if the patient will be able to tolerate lobectomy or pneumonectomy. Surgical resection of all or a portion of the lung can lead to significant patient disability if the forced expiratory volume at 1 second (FEV_1) is reduced to less than 0.8 L.[59] Differential lung perfusion along with pulmonary function measurement has been found to be an accurate predictor of postoperative lung function.[60]

Anterior and posterior images of the lungs are obtained after intravenous administration of 1 to 5 mCi (37–185 MBq) of ^{99m}Tc-MAA. Regions of interest are drawn around each lung in both the anterior and posterior images. Counts are obtained for the anterior and posterior regions, and geometric means are calculated to determine differential lung perfusion. If partial lobectomy is a consideration, the lungs can be further divided into upper, middle, and lower lung zones for a more complete evaluation of function. The differential perfusion values, along with pulmonary function test data, can be used to estimate postoperative lung function (Figure 22-15).

Right-to-Left Shunt Evaluation

Radionuclide imaging can be used to detect and quantify right-to-left shunting in congenital heart disease. Imaging can also be useful for follow-up evaluation of a right-to-left shunt to determine worsening or improvement and after corrective surgery. The most common method is to use ^{99m}Tc-MAA.[61] In a normal perfusion scan, intravenous administration of ^{99m}Tc-MAA results in the particles being trapped in the lungs by capillary blockade. Normally, about 95% of the ^{99m}Tc-MAA particles become trapped in the pulmonary capillary bed and the remainder escape into the systemic circulation via normal anatomic shunting. When there is a right-to-left shunt, some of the intravenous particles will bypass the lungs and increase the amount entering the systemic circulation (Figure 22-16). Once in the systemic circulation, they will become trapped in systemic capillary beds. Total-body scanning after administration of ^{99m}Tc-MAA permits determination of the percentage of particles trapped in the lungs versus those diverted into the systemic circulation:

$$\% \text{Right-to-left shunt} = \frac{(\text{Total body counts} - \text{Total lung counts}) \times 100}{\text{Total body counts}}$$

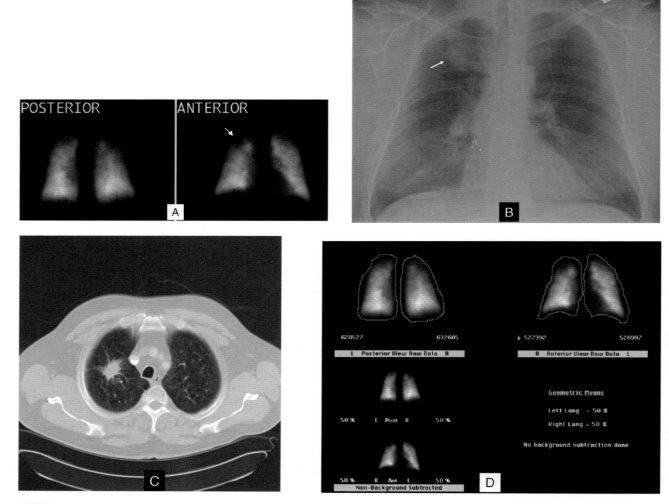

FIGURE 22-15 Differential lung perfusion study in a person with lung cancer. (A) Anterior and posterior images of the lungs were obtained after intravenous administration of 3 mCi (111 MBq) of ^{99m}Tc-MAA. There is a perfusion defect seen in the right upper lobe that corresponds to the patient's lung mass seen on a chest x-ray (B) and chest CT (C). Regions of interest were drawn around each lung in the anterior and posterior perfusion images, and geometric means were calculated to determine the differential lung perfusion (D), which was equal in this study.

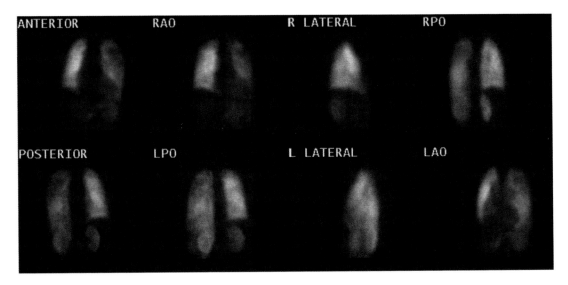

FIGURE 22-16 Right-to-left shunt in a patient with ventricular septal defect and left pulmonary atresia. ^{99m}Tc-MAA perfusion images demonstrate decreased perfusion to the left lung consistent with the patient's history of left pulmonary atresia. Radiotracer activity is seen outside the lungs in the kidneys, consistent with a right-to-left shunt.

REFERENCES

1. Weibel ER. Morphological basis of alveolar-capillary gas exchange. *Physiol Rev.* 1973;53:419–95.
2. Weibel ER. *Morphometry of the Human Lung.* New York: Academic Press; 1963:73–86.
3. Weibel ER. Morphological basis for V/Q distribution. In: Hutas I, Debreczeni LA, eds. *Advances in Physiological Sciences. Vol 10: Respiration.* Budapest: Pergamon Press; 1980:179–89.
4. Davis MA. Particulate radiopharmaceuticals for pulmonary studies. In: Subramanian G, Rhodes BA, Cooper JF, et al., eds. *Radiopharmaceuticals.* New York: Society of Nuclear Medicine; 1975:267–81.
5. Bryan AC, Bentivoglio LG, Beerel F, et al. Determination of regional variations in ventilation and perfusion in normal subjects using xenon 133. *J Appl Physiol.* 1964;19:395–402.
6. Guyton AC. *Textbook of Medical Physiology.* 6th ed. Philadelphia: WB Saunders; 1981:296.
7. Muller JH, Rossier PH. A new method for the treatment of cancer of the lungs by means of artificial radioactivity. *Acta Radiol.* 1951;35:449–68.
8. Haynie TP, Calhoon JD, Nasjleti CE, et al. Visualization of pulmonary artery occlusion by photoscanning. *JAMA.* 1963;185:306–8.
9. Taplin GV, Johnson DE, Dore EK, et al. Suspensions of radioalbumin aggregates for photoscanning the liver, spleen, and lung and other organs. *J Nucl Med.* 1964;5:259–75.
10. Lindeman JF, Quinn JL III. The recent history of clinical procedures in nuclear medicine. In: Gottschalk A, Potchen EJ, eds. *Diagnostic Nuclear Medicine.* Baltimore: Williams & Wilkins; 1976:8–13.
11. Stern HS, Goodwin DA, Wagner HN Jr, et al. In-113m—a short-lived isotope for lung scanning. *Nucleonics.* 1966;24:57–60.
12. Robinowitz M, Mathew J, Eckelman W, et al. Fatal reactions after Tc-99m ferrous hydroxide lung scans [abstract]. *J Nucl Med.* 1973;14:445.
13. Zolle I, Rhodes BA, Wagner HN Jr. Preparation of metabolizable radioactive human serum albumin microspheres for studies of the circulation. *Int J Appl Radiat Isot.* 1970;21:155–67.
14. Rhodes BA, Stern HS, Buchannan JA, et al. Lung scanning with Tc-99m microspheres. *Radiology.* 1971;99:613–21.
15. Subramanian G, Arnold RW, Thomas FD, et al. Evaluation of an instant Tc-99m labeled lung scanning agent [abstract]. *J Nucl Med.* 1972;13:790.
16. Heck LL, Duley JW. Statistical considerations in lung imaging with Tc-99m albumin particles. *Radiology.* 1974;113:675–9.
17. Dworkin HJ, Gutkowski RF, Porter W, et al. Effect of particle number on lung perfusion images. *J Nucl Med.* 1977;18:260–2.
18. Iio M, Wagner HN Jr. Studies of reticuloendothelial system (RES). I: Measurement of the phagocytic capacity of the RES in man and dog. *J Clin Invest.* 1963;42:417–26.
19. Taplin GV, MacDonald NS. Radiochemistry of macroaggregated albumin and newer lung scanning agents. *Semin Nucl Med.* 1971;1:132–52.
20. Davis MA. Long-term retention and biologic fate of Tc-99m iron hydroxide aggregates. In: *Radiopharmaceuticals and Labeled Compounds.* Vol 2. Vienna, Austria: International Atomic Energy Agency; 1973:43–63.
21. Busse W, Reed C, Tyson I, et al. Prolonged retention of radioactivity after perfusion lung scan in asthmatic patients. *J Nucl Med.* 1973;14:837–9.
22. Gallagher BM. Personal communication. North Billerica, MA: New England Nuclear Corp; 1983.
23. Kowalsky RJ. Safety and effectiveness considerations with particulate lung scanning agents. *J Nucl Med Technol.* 1982;10:223–7.
24. Dworkin HJ, Smith JR, Bull FE. A reaction after administration of macroaggregated albumin (MAA) for a lung scan. *AJR Am J Roentgenol.* 1966;98:427–33.
25. Vincent WR, Goldberg SJ, Desilets D. Fatality immediately after rapid infusion of macroaggregates of Tc-99 albumin (MAA) for lung scan. *Radiology.* 1968;91:1181–4.
26. Williams JO. Death after injection of lung scanning agent in a case of pulmonary hypertension. *Br J Radiol.* 1974;47:61–3.
27. Child JS, Wolfe JD, Tashkin D, et al. Fatal lung scan in a case of pulmonary hypertension due to obliterative pulmonary vascular disease. *Chest.* 1975;67:308–10.
28. Davis MA, Taube RA. Pulmonary perfusion imaging: acute toxicity and safety factors as a function of particle size. *J Nucl Med.* 1978;19:1209–13.
29. Allen DR, Ferens JM, Cheney FW, et al. Critical evaluation of acute cardiopulmonary toxicity of microspheres. *J Nucl Med.* 1978;19:1204–8.
30. Allen DR, Nelp WB, Hartnett DE, et al. Critical assessment of changes in the pulmonary circulation after injection of lung scanning agent (MAA). In: *Radiopharmaceuticals and Labeled Compounds.* Vol 2. Vienna, Austria: International Atomic Energy Agency; 1973:37–41.

31. Heyman S. Toxicity and safety factors associated with lung perfusion studies with radiolabeled particles [letter]. *J Nucl Med.* 1979;20:1098–9.
32. Davis MA, Taube RA. Toxicity and safety factors associated with lung perfusion studies with radiolabeled particles [reply]. *J Nucl Med.* 1979;20:1099.
33. Wagner HN Jr, Lopez-Majano V, Langan JK, et al. Radioactive xenon in the differential diagnosis of pulmonary embolism. *Radiology.* 1968;91:1168–74.
34. Taplin GV, Chopra SK. Lung perfusion-ventilation scintigraphy in obstructive airway disease and pulmonary embolism. *Radiol Clin North Am.* 1978;16:491–513.
35. Newhouse MT, Wright FJ, Ingham GK, et al. Use of scintillation camera and xenon-135 for study of topographic pulmonary function. *Respir Physiol.* 1968;4:141–53.
36. Hoffer PB, Harper PV, Beck RN, et al. Improved xenon images with Xe-127. *J Nucl Med.* 1973;14:172–4.
37. Atkins HL, Susskind H, Klopper JF, et al. A clinical comparison of Xe-127 and Xe-133 for ventilation studies. *J Nucl Med.* 1977;18:653–9.
38. Chilton HM, Cooper JF, Friedman BI. Xe-127 ventilation imaging immediately after Tc-99m perfusion studies. *Clin Nucl Med.* 1977;2:152–4.
39. Goddard BA, Ackery DM. Xenon-133, Xe-127, and Xe-125 for lung function investigations: a dosimetric comparison. *J Nucl Med.* 1975;16:780–6.
40. McCartney WH, Perry JR, Staab EV, et al. Comparison of Xe-127 and Xe-133 in ventilation-perfusion imaging in diagnosis of pulmonary embolus [abstract]. *J Nucl Med.* 1978;19:675.
41. Susskind H, Atkins HL, Cohn SH, et al. Whole-body retention of radioxenon. *J Nucl Med.* 1977;18:462–71.
42. Ponto RA, Loken MK. Radioactive gases: production, properties, handling and uses. In: Subramanian G, Rhodes BA, Cooper JF, et al., eds. *Radiopharmaceuticals.* New York: Society of Nuclear Medicine; 1975:296–304.
43. Hayes M, Taplin GV, Chopra SK, et al. Improved radioaerosol administration system for routine inhalation lung imaging. *Radiology.* 1979;131:256–8.
44. Rinderknecht J, Shapiro L, Krouthhammer M, et al. Accelerated clearance of small solutes from the lungs in interstitial lung disease. *Am Rev Respir Dis.* 1980;121:105–17.
45. Jones JG, Lawler P, Crawley JCW, et al. Increased alveolar epithelial permeability in cigarette smokers. *Lancet.* 1980;1(January 12):66–8.
46. Hayes M, Taplin GV. Lung imaging with radioaerosols for the assessment of airway disease. *Semin Nucl Med.* 1980;10:243–51.
47. Burch W, Sullivan P, McLaren C. Technegas: a new ventilation agent for lung scanning. *Nucl Med Commun.* 1986;7:865–71.
48. Senden TJ, Moock KH, Fitz Gerald J, et al. The physical and chemical nature of technegas. *J Nucl Med.* 1997;38:1327–33.
49. Ponto JA, Graham MM, Bricker JA. Alternative to technetium-99m pentetate for radioaerosol inhalation lung imaging. *J Am Pharm Assoc.* 2002;42:112–4.
50. Gray HW. The natural history of venous thromboembolism: impact on ventilation/perfusion scan reporting. *Semin Nucl Med.* 2002;32:159–72.
51. Koster T, Small RA, Rosendaal FR, et al. Oral contraceptives and venous thromboembolism: a quantitative discussion of the uncertainties. *J Intern Med.* 1995;238:31–7.
52. Gottschalk A. New criteria for ventilation-perfusion lung scan interpretation: a basis for optimal interaction with helical CT angiography. *Radiographics.* 2000;20:1206–10.
53. Elgazzar AH, Khadada M. Respiratory system. In: Elgazzar AH, ed. *The Pathophysiologic Basis of Nuclear Medicine.* New York: Springer-Verlag; 2001:189–99.
54. Parker JA, Coleman RE, Siegel BA, et al. Procedure guideline for lung scintigraphy. In: *Procedure Guidelines Manual 2001–2002.* Reston, VA: Society of Nuclear Medicine; 2001:181–6.
55. PIOPED Investigators. Value of the ventilation/perfusion scan in acute pulmonary embolism: results of the Prospective Investigation of Pulmonary Embolism Diagnosis (PIOPED). *JAMA.* 1990;263:2753–9.
56. Gottschalk A, Sostman HD, Coleman RE, et al. Ventilation-perfusion scintigraphy in the PIOPED study. II: Evaluation of the scintigraphic criteria and interpretations. *J Nucl Med.* 1993;34:1119–26.
57. Gottschalk A, Stein PD, Dirk Sostman H, et al. Very low probability interpretation of V/Q lung scans in combination with low probability objective clinical assessment reliably excludes pulmonary embolism: data from PIOPED II. *J Nucl Med.* 2007;48:1411–5.
58. Stein PD, Fowler SE, Goodman LR, et al. Multidetector computed tomography for acute pulmonary embolism. *N Engl J Med.* 2006;354:2317–27.
59. Williams CD, Brenowitz JB. "Prohibitive" lung function and major surgical procedures. *Am J Surg.* 1976;132:763–6.
60. Wernly JA, DeMeester TR, Kirchner PT, et al. Clinical value of quantitative ventilation-perfusion lung scans in the surgical management of bronchogenic carcinoma. *J Thorac Cardiovasc Surg.* 1980;80:535–43.
61. Gates GF, Orme HW, Dore EK. Measurement of cardiac shunting with technetium-labeled albumin aggregates. *J Nucl Med.* 1971;12:746–9.

CHAPTER 23

Liver, Spleen, and Gastrointestinal System

Radiopharmaceuticals for liver, spleen, and gastrointestinal (GI) studies were developed to assess changes in normal biologic processes due to disease. The liver and spleen are a major part of the reticuloendothelial system (RES); they contain cells that participate in the body's defense mechanism. Macrophages in the liver (Kupffer's cells) sequester foreign particles from the sinusoidal blood as it flows through the liver. Reticulum cells in the spleen and the splenic sinusoidal tissue play a role in the removal of foreign particles and effete red blood cells from the blood. The liver and spleen were two of the first organs to be studied in nuclear medicine. Early studies used radiocolloids labeled with ^{198}Au and ^{113m}In to target these organs; they were based on phagocytosis by macrophages present in the organs. Altered activity distribution patterns were evidence of disease. Soon after ^{99m}Tc-sodium pertechnetate became available in nuclear medicine, ^{99m}Tc-sulfur colloid (^{99m}Tc-SC) was developed for liver and spleen imaging. This radiopharmaceutical produced better-quality images at lower radiation dose. Although they are still useful on occasion, such studies eventually diminished in nuclear medicine, being replaced by more accurate anatomic imaging modalities such as computed tomography and magnetic resonance imaging. ^{99m}Tc-SC, however, has become a key radiotracer for other applications in nuclear medicine. For example, procedures have been developed using ^{99m}Tc-SC to evaluate gastroesophageal reflux and gastric emptying.

Spleen-specific imaging studies were made possible by the development of techniques for preparing radiolabeled denatured red blood cells that could localize in splenic tissue. Procedures have also been developed to identify GI bleeding sites after intravenous injection of ^{99m}Tc-SC or ^{99m}Tc-red blood cells, which can extravasate into a bleeding site. The liver also contains functional cells (hepatocytes) that play a significant role in removing xenobiotics and degraded endogenous substances from the blood. Thus, a group of radiotracers labeled with ^{99m}Tc, known as iminodiacetic acid analogues, have been developed that trace biochemical transport properties in hepatocytes and bile flow in the hepatobiliary system, permitting an evaluation of liver function in vivo in various disease states.

LIVER IMAGING AGENTS

Physiologic Anatomy

A diagram of the liver and hepatobiliary system is shown in Figure 23-1. The two principal types of liver cells are the hepatocytes and the sinusoidal cells. The hepatocytes, also called polygonal cells because of their shape, account for 85% of the cells in the liver and are responsible for its major metabolic functions. The sinusoidal cells constitute the remaining 15% of liver cells.

The functional units of the liver are the lobules, which number between 50,000 and 100,000.[1] Each lobule consists of a segment of the central vein surrounded by a number of sinusoids in an arrangement similar to spokes around the hub of a wheel (Figure 23-2). The sinusoids transport blood from the portal vein to the lobule. Blood is then directed to the central vein and ultimately to the hepatic vein.

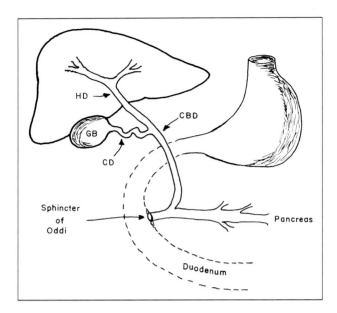

FIGURE 23-1 Liver and hepatobiliary system. Hepatic duct (HD), gallbladder (GB), cystic duct (CD), common bile duct (CBD).

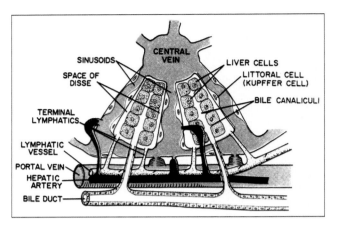

FIGURE 23-2 Basic structure of a liver lobule showing the hepatic cellular plates, the blood vessels, the bile-collecting system, and the lymph flow system composed of the space of Disse and the interlobular lymphatics. (Reprinted with permission of Elsevier Science from reference 1.)

Two main types of cells make up the sinusoids: endothelial cells and Kupffer's cells. Endothelial cells form the main structure, the sinusoidal conduit, which transports blood through the lobule. Kupffer's cells, which are dispersed along the sinusoid, are macrophages that remove foreign substances from the blood as it passes through the liver. Surrounding the sinusoids are the hepatocytes. Between the sinusoid and the hepatocytes is the space of Disse. This space ranges from 0.25 μm to 2.0 μm in depth and contains a reinforcing network of collagen fibers. Features of liver microanatomy are depicted in Figure 23-3 and have been described by Elias and Serrick.[2]

Sinusoidal Cells

The sinusoidal endothelial cells are flat, irregularly shaped cells characterized by numerous pores (fenestrae) that appear throughout their cytoplasm. The pores are approximately

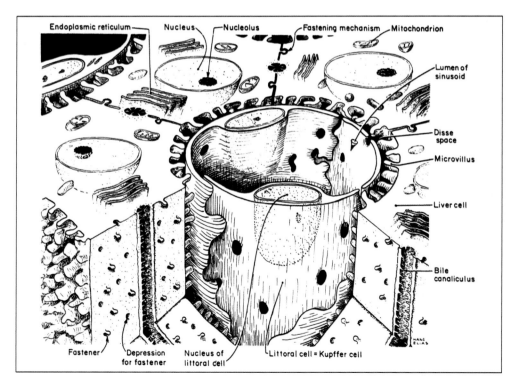

FIGURE 23-3 Ultrastructure of the mammalian liver. (Reprinted with author permission from Elias H, Pauly JE. *Human Microanatomy*. Chicago: DaVinci Publishers; 1960:135. [Publication is out of print.])

0.1 μm in diameter. Molecular substances and small particles can leave the sinusoidal blood through the pores to enter the space of Disse, where they have access to the underlying hepatocytes. Larger particles can enter the space through the 0.02 to 1.0 μm wide space between the endothelial cells. In addition to this sievelike function, endothelial cells contain pinocytotic vesicles that can sequester particles less than 0.1 μm in size. This process has been observed with antimony sulfide particles.

Kupffer's cells vary in shape but are basically stellate. They lie on or embedded in the endothelial lining and can also lie at least partly in the space of Disse, where their microvilli intermingle with the microvilli of the hepatocytes. They also tend to accumulate near the branches of the portal vein. Their variability in shape and position suggests that Kupffer's cells are mobile. Kupffer's cells have bulky cytoplasm that is rich in lysosomes, indicating their ability to digest various substances. Their membrane is covered with a 70 nm thick, fuzzy coat of proteinlike material that contains pinocytotic structures capable of trapping particles less than 0.1 μm in size. In addition, Kupffer's cells are capable of phagocytosis, having pseudopodia that engulf particles larger than 0.1 μm.

Some general properties of phagocytosis by Kupffer's cells have been observed. Microscopic studies have shown in several instances, although not with all types of substances, that intravenously injected particles become coated with an opsonin, which may be a plasma protein or an antibody that renders foreign particles susceptible to phagocytosis. The coating frequently causes the particles to adhere to each other and to the walls of the liver sinusoids. Thus, because of aggregation, particle size in vivo can be much larger than the preinjection size. A coated particle that adheres to the fuzzy coat of a Kupffer's cell induces the phagocytic process. Once particles are inside the cell, lysosomes digest metabolizable particles and dispose of them by reutilization or excretion. Indigestible particles may be stored in lysosomes or distributed over their daughter cells or to other organs by the migration of loaded Kupffer's cells. Sinusoidal cell function has been described in detail by Wisse.[3,4]

Hepatocytes

Hepatocytes are polygon shaped, with eight or more surfaces, and about 30 μm in diameter. The cells are arranged in plates one cell thick that form an irregular wall surrounding the sinusoids. The hepatocyte has three physiologic membrane surfaces: a surface that contacts neighboring hepatocytes, a grooved surface that delimits bile canaliculi, and a sinusoidal surface in contact with the blood that projects numerous microvilli into the space of Disse.[5]

Hepatocytes are responsible for excretory and metabolic functions in the liver. Soluble substances, such as xenobiotics and products of metabolism, are removed from the sinusoidal blood at specific receptor sites on the sinusoidal membrane surface capable of transporting organic anions, organic cations, neutral compounds, and conjugated bile salts (Figure 23-4).[6] Specific molecular transporters have been identified for handling these substances.[7]

The liver is also an exocrine gland, secreting bile. Bile has two principal functions: (1) to excrete waste products, such as bilirubin, and (2) to secrete bile salts into the intestinal tract to

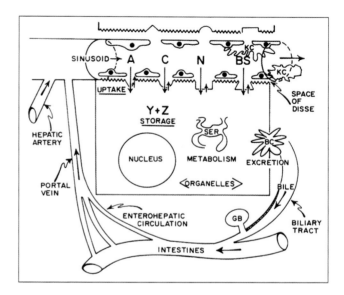

FIGURE 23-4 Schematic representation of the hepatocyte. Substrates in the blood diffuse through pores in the endothelial lining of the sinusoids and bind to the hepatocyte at one of four membrane-bound carriers: anionic (A), cationic (C), nonionic (N), and bile salt (BS). Within the hepatocyte a substrate may be stored at specific binding sites such as Y and Z proteins, and it also may undergo metabolic conversion at other sites, including the smooth endoplasmic reticulum (SER). Biliary excretion occurs at a biliary canaliculus (BC). Subsequently, the substrate in the bile may be stored and concentrated in the gallbladder (GB) or excreted into the intestine. Some biliary components are reabsorbed from the intestine into the portal vein and reextracted by the hepatocyte (enterohepatic circulation). The sinusoids are lined by Kupffer's cells (KC), which are a part of the reticuloendothelial system. (Reprinted with permission of the Society of Nuclear Medicine from reference 6.)

aid digestion. Bile is produced continuously by the hepatocytes. The hepatobiliary system (Figures 23-1 and 23-2) consists of the biliary canaliculi, the ductal system (hepatic, cystic, and common bile ducts), the gallbladder, and the sphincter of Oddi.[8] Bile formed in the hepatocytes drains into the canaliculi, collecting in the right and left hepatic ducts. These two ducts form the common hepatic duct, which becomes the common bile duct distally. The cystic duct of the gallbladder branches from this ductal pathway. The common bile duct is the segment between the cystic duct and the ampulla of Vater, where the common bile duct and the pancreatic duct join.

The muscular sphincter of Oddi in the ampulla maintains a high tone during fasting, directing bile into the gallbladder. This sphincter also regulates bile delivery to the duodenum as it relaxes in response to peristalsis and endogenous cholecystokinin (CCK) released when food (gastric chyme) enters the duodenum. It also prevents reflux of intestinal contents into the bile and pancreatic ducts. Substances that relax the sphincter include CCK, anticholinergics, β-adrenergics, calcium-channel blockers, and glucagon.

Agents that increase sphincter tone include opiates, gastrin, cholinergics, α-adrenergics, and motilin.[8]

The gallbladder wall has smooth muscle, which relaxes to allow filling and contracts to empty the gallbladder. The epithelial cells of the gallbladder absorb sodium, bicarbonate, chloride, and water, to concentrate the bile. This mechanism has a high capacity; the gallbladder's nominal volume is only 40 to 70 mL, while the volume of bile delivered to it from the liver is 800 to 1000 mL per day. Bile enters the gallbladder when common duct pressure is greater than intraluminal pressure in the gallbladder. The gallbladder does not fill continuously; filling is interrupted periodically by partial emptying of concentrated bile and aspiration of diluted bile. In the fasting state, about 25% of gallbladder contents is released every 2 hours in response to late phase 2 of the migratory myoelectric complex.[9] In response to gastric chyme released from the stomach into the duodenum, the gallbladder empties more than 75% of its contents into the duodenum. The gallbladder refills with bile in about 3 to 4 hours.[9]

Radiocolloids

Several radioactive agents have been used to study the liver. Investigators in the 1950s used colloidal ^{32}P-chromic phosphate to study liver blood flow.[10] Human serum albumin (HSA) labeled with ^{131}I and colloidal gold (^{198}Au) were used to perform the first liver scans with the newly developed rectilinear scanner.[11,12] An interesting approach to liver scanning was the administration of ^{99}Mo as the molybdate anion that localized in hepatocytes by incorporation into xanthine oxidase.[13] Scans were obtained in 24 hours by detection of accumulated ^{99m}Tc activity in the liver. In 1956 Benacerraf et al.[14] developed colloidal particles of ^{131}I-HSA to study liver phagocytosis, but this agent was rapidly metabolized and cleared from the liver and thus was not satisfactory for slow-moving rectilinear scanners.

In 1963, ^{99m}Tc-SC was developed by Richards.[15] Tagging ^{99m}Tc to sulfur particles was accomplished by air oxidation of hydrogen sulfide gas bubbling through an acidified solution of ^{99m}Tc-sodium pertechnetate and gelatin. The preparation was sterilized by filtration through a 0.22 μm membrane. The particle size was estimated to be 0.05 to 0.15 μm.[16] Soon thereafter, sulfur colloid preparation was greatly simplified by use of a sodium thiosulfate kit developed by Stern et al.[17]

Other radiocolloids were developed for diagnostic studies, but none have supplanted the clinical utility of ^{99m}Tc-SC. ^{99m}Tc-antimony sulfide colloid prepared from preformed particles in the 8 to 12 nm range became more useful for studying the lymphatic system.[18] ^{113m}In-hydroxide colloid was useful in laboratories with ^{113}Sn–^{113m}In generators.[19] ^{99m}Tc-phytate (inositol hexaphosphate), which forms an insoluble calcium chelate in the blood, was investigated as a means of altering biologic localization among liver, spleen, and bone marrow.[20] ^{99m}Tc-stannous albumin colloid kit (Microlite) was developed to reduce radiation dose to the liver because of its hepatic metabolism, but it never achieved widespread use. In Europe a kit for preparing ^{99m}Tc-albumin colloid (Nanocoll) is routinely used for lymphoscintigraphy.

^{99m}Tc-SC remains the agent of choice for RES imaging because of its ready availability, ease of preparation, and long history of use. Its method of preparation, chemistry, and kit formulation are described in Chapter 10. Table 23-1 summarizes the properties of several radiocolloids developed for liver imaging.

Biologic Properties of ^{99m}Tc-Sulfur Colloid

After intravenous administration in humans, ^{99m}Tc-SC clears rapidly from the vascular space with a half-life of 2 to 3 minutes, localizing in the liver, spleen, and bone marrow.[16] The normal adult dose is 5 mCi (185 MBq). Liver imaging can begin in 10 to 15 minutes, when about 97% of the dose is cleared from the blood. However, in patients with severely diseased livers, imaging may be delayed because of slow blood clearance. By 92 hours, 4% of the injected activity is excreted in the urine and 3% in the feces.[16] The remaining activity is retained in the body with an effective half-life of 6 hours.

Several factors have been shown to influence the blood clearance and distribution of radiocolloids.[21] The most important of these are organ blood flow, disease state, particle size and dose, and serum factors.

Blood Flow

Extraction of radiotracer by an organ is directly related to its blood flow. The liver receives about 30% of the cardiac output, compared with only about 5% for the spleen.[22] This high blood flow, in part, can explain why in a healthy individual,

TABLE 23-1 Radiocolloids Developed for Liver Imaging

Agent	Particle Size (nm)	Half-life Physical	Half-life Biologic	Decay Mode	Gamma (keV)	Administered Activity (mCi)	Liver Dose rad(cGy)/mCi
^{198}Au-colloidal gold	5–50	2.8 days	Very long	β−	411	0.3	40
^{99m}Tc-SC	100–1000	6 hr	Very long	IT	140	5.0	0.34
^{131}I-HSA colloid	10–20	8 days	60 min	β−	364	0.2	0.8
^{113m}In-hydroxide colloid	10–20	1.7 hr	30 days	IT	393	2.0	0.5
^{99m}Tc-HSA colloid	200–1000 (80%) <200 (15%)	6 hr	11 hr	IT	140	5.0	0.34

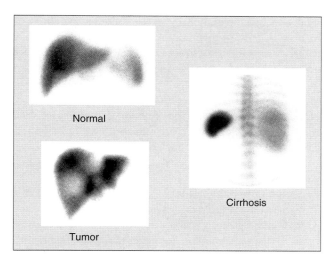

FIGURE 23-5 99mTc-sulfur colloid liver scans. Anterior view of a normal liver and spleen, anterior liver scan with a tumor mass in the right lobe, and posterior liver–spleen scan in a patient with hepatocellular dysfunction secondary to alcoholic cirrhosis, demonstrating decreased uptake of radiotracer in the liver with compensatory increased uptake in the spleen and bone marrow.

about 85% of a ^{99m}Tc-SC dose localizes in the liver, 4% to 8% in the spleen, and the remainder in the bone marrow (Figure 23-5).[23] Significant global liver disease may restrict blood flow through the liver, diminishing delivery and localization of radiocolloid. In addition, space-occupying lesions, such as tumors and cysts, will exclude radiocolloid from these sites, where they appear as regions of absent radioactivity.

Disease State

Liver disease may destroy a significant number of cells, diminishing the amount of radiocolloid deposited in the liver. For example, in severe cirrhosis, normal tissue is replaced by scar, and altered biodistribution occurs with shunting of radiocolloid to spleen and bone marrow (Figure 23-5).

Using metabolizable microaggregates of radioiodinated albumin, Wagner and Iio[24] demonstrated that bacterial infections may cause increased RES functional capacity in humans due to high levels of endotoxin, which stimulates overall RES activity in the body. This may lead to uptake of radiocolloid in organs that do not normally demonstrate uptake. For example, increased lung uptake of ^{99m}Tc-SC has been demonstrated in endotoxin-treated animals compared with nontreated controls.[25]

Particle Size

^{99m}Tc-SC prepared by the original hydrogen sulfide method produces an essentially monodispersed colloidal preparation, with 90% of the particles being 0.09 ± 0.01 μm.[26] ^{99m}Tc-SC prepared by the thiosulfate kit method yields a more diverse size distribution. Using Nuclepore (Whatman) filtration analysis, Davis et al.[27] have shown size distribution of thiosulfate-generated particles as follows: less than 0.1 μm, 15%; less than 0.4 μm, 70%; 0.1 to 1.0 μm, 80%; and greater than 1.0 μm, 5%.

In general, larger particles are cleared faster from the blood and have greater deposition in the liver and spleen and less in the bone marrow. Atkins et al.[28] compared small particles of ^{99m}Tc-SC produced by the hydrogen sulfide method with particles of ^{99m}Tc-SC 10 times larger produced by the thiosulfate method. Their studies demonstrated that as the dose of smaller particles is increased the percentage of uptake in the liver decreases, uptake in bone marrow increases, and spleen uptake is unaffected.

Blood Clearance and Opsonization

When colloidal particles of ^{198}Au are administered to rats in small numbers ($<1 \times 10^{13}$ particles/kg body weight), the rate of blood clearance is constant with a half-life of 2.5 minutes,[29] similar to ^{99m}Tc-SC.[16] With these trace amounts of particles the rate of liver uptake is related to blood flow rather than RES capacity. With radiocolloid doses greater than 1×10^{13} particles/kg, the rate of blood disappearance is slowed with increasing dose, suggesting RES depression. This phenomenon has also been observed in humans. It has been suggested that the depression is due to saturation of the reticuloendothelial cell capacity, but experimental studies reveal that it may also be due to depletion of a specific serum opsonin pool for that particular colloid.[21] Supporting the role of opsonins is the fact that RES depression in animals and humans appears to be particle specific; the injection of one type of colloid induces a state of RES depression relative to the subsequent blood clearance of that particular colloid, whereas the clearance of a dissimilar colloid is less affected.[24] It can be inferred that the localization of particles by phagocytosis may be controlled by both antibodylike opsonins in the blood and specific macrophages that recognize a specific opsonin–particle complex. Recognition may also be influenced by particle charge or other chemical properties of the particle surface.

Localization and Metabolic Fate

A common belief is that colloidal particles are localized primarily in Kupffer's cells. Although this is true, it may not explain the process entirely. Reports indicate that other mechanisms of localization may be involved. Using autoradiography, Chaudhuri et al.[30] demonstrated in mice that although colloidal ^{198}Au was primarily engulfed by Kupffer's cells, ^{99m}Tc-SC maintained a generalized distribution throughout the liver with no apparent concentration in Kupffer's cells. This may be due to differences in particle size of these two colloids or to their opsonized chemical properties. According to Brucer,[31] phagocytosis of small colloids (<0.1 μm) is a primary function of Kupffer's cells, whereas larger particles (0.1 to 1.0 μm), which leave the sinusoidal blood through the slits between endothelial cells, may become trapped in the network of hepatocellular microvilli and collagen fibrils in the space of Disse. Other studies demonstrate that a size distribution of sinusoidal cells exists with a preferential localization of small colloids (0.005 to 0.05 μm) such as ^{198}Au in the smaller cells, which are in the majority, whereas larger colloids (0.8 to 1.5 μm), such as ^{99m}Tc-SC, localize primarily in the larger sinusoidal cells, which make up only 15% to 25% of all cells.[32]

Particles trapped in the space of Disse are presumed to drain into portal and hepatic lymph nodes, a process requiring weeks to months.[31] Although no experimental evidence supports this, it seems reasonable at least for the smaller colloids, which are readily transported in lymph. Similarly, no firm evidence is available to demonstrate the fate of particles engulfed by phagocytes. Unless specific enzyme systems are present to transform these inert sulfur particles into chemically excretable forms (e.g., SO_4^{2-}), these particles are probably stored in the cytoplasm of Kupffer's cells. Because these cells are eventually replaced by new cells, the particles may simply be transferred from cell to cell with each succeeding generation. One study demonstrated that colloidal particles in the liver can enter the systemic circulation and migrate to the lungs, where they can pass into the airways and be expelled in the saliva or swallowed and excreted in the feces.[33]

Adverse Reactions and Toxicity

When a substance is retained for an indefinite period of time in the body, toxicity is a concern. However, chronic and acute toxicity is of little concern with diagnostic radiocolloids because they are usually administered only once or twice to the same patient and in extremely small amounts. In addition, when the RES is challenged with high doses of colloids, recovery from RES depression is rapid and complete because of the system's regenerative capacity.[24] Colloidal indium hydroxide has been shown to produce hepatocyte toxicity in mice, but only at doses 10,000 times larger than typical doses used in nuclear medicine studies.[34]

One circumstance in which adverse effects may be serious is the potential misadministration of a beta-emitting ^{32}P radiopharmaceutical for therapeutic use. A serious situation may occur if the wrong chemical form of ^{32}P is used. The treatment of peritoneal effusions typically involves an intracavitary injection of insoluble ^{32}P-chromic phosphate colloid. If, by error, the soluble ^{32}P-sodium phosphate salt is administered, severe bone marrow depression may result.[35] The sodium salt is readily absorbed into the bloodstream and translocated to the bone marrow, where it can cause severe bone marrow depression.

No toxic reactions with ^{99m}Tc-SC have been observed in mice given intravenous doses 1000 times the usual adult dose.[36] However, pyrogenic or allergic reactions have been reported with the use of ^{99m}Tc-SC; these were attributed to stabilizers used in the formulations.[35] Such reactions are rare.

Radiation Dose

Radiation dose estimates for ^{99m}Tc-SC are shown in Table 23-2.[37] The doses listed are for normal individuals and those with advanced parenchymal disease. As with all radiopharmaceuticals, the radiation dose to organs may be altered by disease. In the case of severe liver disease, radiocolloid is shifted to other organs and their radiation dose is increased while the dose to the liver is decreased. For example, when diffuse parenchymal disease (cirrhosis) is present, the radiation dose to the liver may be halved while the dose to the spleen is doubled.[38]

Hepatobiliary Agents

In 1955, an interest in hepatic reticuloendothelial function led George Taplin to investigate the excretion of rose bengal dye into the biliary system. It was presumed at the time that excretion of the dye was through Kupffer's cells. However, further investigation with ^{131}I-labeled rose bengal led to the discovery that the dye was excreted by hepatocytes and was not absorbed by the bowel. This information eventually led Taplin and his colleagues[39] to introduce ^{131}I-rose bengal as the radiopharmaceutical for studying hepatobiliary excretion. It remained in use for nearly 20 years.

In the mid 1970s, a serendipitous event occurred as Loberg et al.[40] were attempting to develop a technetium-labeled heart imaging agent based on structure–distribution

TABLE 23-2 Radiation Dose Estimates for ^{99m}Tc-Sulfur Colloid

Organ	Dose after IV Administration (rad[cGy]/5 mCi)		Dose after Oral Administration (rad[cGy]/mCi)
	Normal Liver	Advanced Parenchymal Disease	
Liver	16.9	8.1	–
Spleen	10.6	21.3	–
Bone marrow	1.4	3.9	–
Testes	0.055	0.16	0.005
Ovaries	0.281	0.600	0.096
Total body	0.94	0.88	0.018
Stomach wall	–	–	0.140
Small Intestine	–	–	0.260
Upper large intestine	–	–	0.480
Lower large intestine	–	–	0.330

Source: Reference 37.

relationships. They chose to work with a molecule similar in structure to lidocaine. The first two agents studied were ^{99m}Tc-methyl-iminodiacetic acid (^{99m}Tc-MIDA) and ^{99m}Tc-2,6-di-methylacetanilido-iminodiacetic acid (^{99m}Tc-HIDA). Biodistribution studies in animals demonstrated that ^{99m}Tc-MIDA had rapid excretion into the urine, whereas ^{99m}Tc-HIDA was excreted primarily into bile.[40] This excretory pattern was not completely unpredictable; MIDA is quite hydrophilic, and the HIDA compound has lipophilic properties because of its substituted ring system. The high liver extraction prompted the acronym HIDA, for hepatobiliary-IDA. Since no good technetium-labeled agent for hepatobiliary imaging was available at the time, the course of development shifted away from heart agents to hepatobiliary agents.

Several ^{99m}Tc-IDA compounds were produced by alkyl substitution into the aromatic ring; they include the dimethyl (lidofenin), diethyl (etifenin), paraisopropyl (iprofenin), and parabutyl (butilfenin) analogues, which have differing hepatocyte transit times.[41] The diisopropyl analogue 2,6-diisopropylacetanilido-iminodiacetic acid (DISIDA, or disofenin) and the trimethylbromo analogue 2,4,6-trimethyl,5-bromoacetanilido-iminodiacetic acid (BRIDA, or mebrofenin) proved to have the best pharmacokinetic properties.[42,43] Three of these compounds were eventually marketed in the United States: ^{99m}Tc-lidofenin, ^{99m}Tc-disofenin, and ^{99m}Tc-mebrofenin. The latter two are the only agents currently on the market. The labeling chemistry of these analogues and their chemical structures are described in Chapter 10.

Physicochemical Properties of ^{99m}Tc-IDA Analogues

An interesting finding from the original work with ^{99m}Tc-HIDA was the discovery that the final complex formed with technetium was a dimer, with two molecules of the chelating agent (HIDA) reacting with one atom of ^{99m}Tc (Figure 23-6). The complex was technetium essential; without the technetium atom the ^{14}C-labeled HIDA ligand itself had less than 1% hepatobiliary excretion, whereas ^{99m}Tc-HIDA had greater than 70% excretion.[44] The dimeric structure of the complex also confers stability. In vivo/in vitro ligand exchange experiments between ^{99m}Tc-HIDA and EDTA demonstrated that at physiologic pH, technetium release from HIDA is very slow. Thus, while ^{99m}Tc-HIDA does not possess the thermodynamic stability of ^{99m}Tc-EDTA, it is kinetically inert and quite stable in vivo.[45] This fact has been borne out in dogs injected with ^{99m}Tc-HIDA in procedures in which the contents of the urinary bladder and gallbladder were obtained and reinjected.[40] The results showed an excretory pattern similar to that of the original compound, suggesting that ^{99m}Tc-HIDA is excreted in its original radiochemical form, minimally dissociated or metabolized. Other ^{99m}Tc-IDA analogues should behave similarly.

The hepatobiliary excretion of a substance has been shown to be related to several physicochemical properties.[46,47] These include (1) a molecular weight between 300 and 1000; (2) the presence of a strong polar group, usually ionized at plasma pH and typically anionic; (3) the presence of nonpolar groups, usually as aromatic rings; (4) a lipophilic character enhanced by ring substitution; and (5) binding to plasma albumin, which may promote transfer into the hepatocyte and limit urinary excretion. The ^{99m}Tc-IDA complexes appear to satisfy these requirements. In a review of structure–distribution relationships for hepatobiliary agents, Nunn and Loberg[48] described how the alkyl substitution pattern on the phenyl ring modifies receptor binding of the complex to plasma protein, the hepatocyte membrane, and the intracellular hepatocyte proteins.

Biodistribution of ^{99m}Tc-IDA Analogues

After intravenous injection of either ^{99m}Tc-disofenin or ^{99m}Tc-mebrofenin, liver uptake in normal individuals is evident within 5 minutes, peaking at around 10 minutes. Gallbladder visualization is generally evident within 10 to 15 minutes, peaking in 30 to 60 minutes, with intestinal activity also evident at this time. The radiotracer is actively extracted from blood via the anionic pathway on the hepatocyte membrane and is excreted unconjugated into the bile. Specific proteins are responsible for hepatobiliary uptake and transport.[7,49] Investigational studies demonstrate that the organic anion-transporting polypeptides OATP1B1 and OATP1B3 are involved in sinusoidal uptake of ^{99m}Tc-mebrofenin from blood and the multidrug resistance-associated proteins MRP2 and MRP3 are involved in its canalicular transport into bile (Figure 23-7).[50]

The ideal hepatobiliary agent should have high hepatocyte extraction and transit and low renal excretion and should be an effective competitor for bilirubin excretion.[51] During

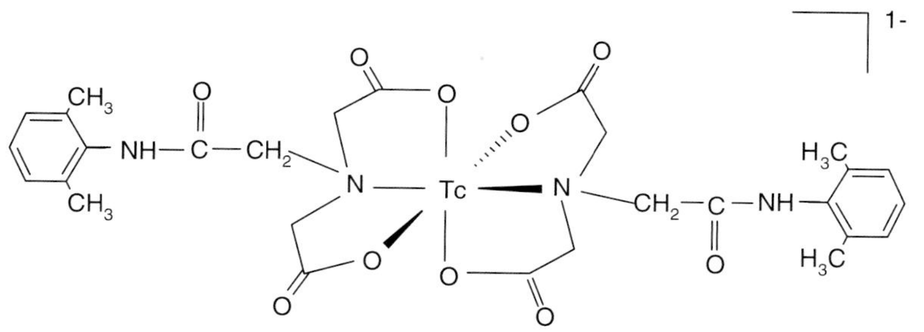

FIGURE 23-6 Chemical structure of ^{99m}Tc-2,6-dimethylacetanilidominodiacetic acid (^{99m}Tc-lidofenin), the first ^{99m}Tc iminodiacetic acid analogue for hepatobiliary imaging.

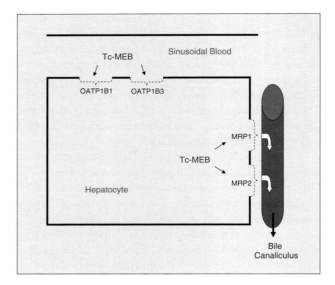

FIGURE 23-7 Hepatobiliary transport of ^{99m}Tc-mebrofenin (Tc-MEB). Hepatocyte uptake from sinusoidal blood via OATP1B1 and OATP1B3 transporters on the basolateral membrane and efflux into bile via MRP1 and MRP2 transporters on the canalicular membrane.

TABLE 23-3 Radiation Dose Estimates for ^{99m}Tc-Disofenin and ^{99m}Tc-Mebrofenin

	Radiation Dose Estimate (rad[cGy]/5 mCi)	
Organ	^{99m}Tc-Disofenin	^{99m}Tc-Mebrofenin
Upper large intestine wall	1.75	1.24
Lower large intestine wall	1.25	0.99
Urinary bladder wall	0.45	1.21
Small intestine	0.9	0.8
Gallbladder wall	0.7	0.63
Liver	0.15	0.40
Ovaries	0.39	0.32
Testes	0.03	0.06
Total body	0.07	0.08

Source: References 54 and 55.

bilirubinemia, liver uptake of IDA analogues is reduced and kidney excretion increased. Some institutions increase the administered activity of the IDA analogue when serum bilirubin is high. A typical dosing scheme is less than 2 mg/dL bilirubin, 5 mCi (185 MBq); 2 mg/dL to 10 mg/dL bilirubin, 7.5 mCi (277.5 MBq); and more than 10 mg/dL bilirubin, 10 mCi (370 MBq).

Lidofenin is effectively extracted up to 5 mg/dL of plasma bilirubin concentration, whereas disofenin and mebrofenin are effective at higher levels. Mebrofenin extraction is less affected by bilirubin than is disofenin. In studies of isolated rat hepatocytes, disofenin uptake decreased 64% at 10 mg/dL bilirubin, whereas mebrofenin uptake decreased only 29%.[52] The latter two agents exhibit similar hepatocyte extraction efficiency (disofenin 60%, mebrofenin 66%) and parenchymal transit time ($T_{1/2}$ liver clearance: disofenin, 19 minutes; mebrofenin, 17 minutes), but mebrofenin exhibits less renal excretion than disofenin.[53] At 3 hours postinjection, 7% of disofenin and 1% of mebrofenin appears in the urine of patients with normal bilirubin levels, rising to about 30% for disofenin and 6% for mebrofenin at 24 mg/dL bilirubin.[53] Thus, mebrofenin demonstrates a greater hepatic specificity than disofenin and should be more effective in patients with reduced hepatocellular function.

Radiation Dose

Radiation dose estimates for ^{99m}Tc-disofenin and ^{99m}Tc-mebrofenin are listed in Table 23-3.[54,55] In jaundiced patients these dose estimates change because of altered tracer kinetics.

Pharmacologic Interventions

Several pharmacologic agents have become quite useful for augmenting hepatobiliary studies in nuclear medicine. The principal agents are CCK, morphine sulfate, and phenobarbital. Each of these substances has a role in particular clinical situations.

Cholecystokinin

CCK is an endogenous 33 amino acid polypeptide hormone released from the intestinal mucosa in response to the presence of fats and protein substances in the intestinal contents. After its release, CCK is absorbed into the blood and transported to the gallbladder, causing it to contract. CCK also stimulates the pancreas to release its digestive enzymes and relaxes the sphincter of Oddi so that bile and enzymes can enter the duodenum to participate in the digestive process.

Sincalide is the synthetic C-terminal octapeptide of CCK. It is marketed as a sterile lyophilized powder (5 μg/vial) under the trade name Kinevac (Bracco). After reconstitution with sterile water for injection to a concentration of 1 μg/mL, sincalide injection is stable for 24 hours at room temperature. It has pharmacologic actions similar to those of CCK and a plasma half-life of about 2.5 minutes.

There are two principal reasons for using sincalide in hepatobiliary imaging.[56] One reason is to empty the gallbladder of its contents in situations that might lead to a false-positive study (nonvisualization of the gallbladder despite a patent cystic duct). A full gallbladder interferes with radiotracer uptake because of biliary stasis and sludge. Patients suspected to have acute cholecystitis who have a full gallbladder because of fasting 24 hours or more, are receiving total parenteral nutrition, or are anorexic should have their gallbladder emptied before the study. To facilitate gallbladder uptake of radiotracer after sincalide-stimulated emptying of the gallbladder, a waiting period of 30 to 60 minutes should elapse before injection of the ^{99m}Tc-IDA tracer to allow the gallbladder to relax.

The second reason for sincalide cholescintigraphy is to measure gallbladder ejection fraction (GBEF), which is useful in the diagnostic work-up of patients with chronic acalculous cholecystitis (cholecystitis without stones).[56,57] The rationale for sincalide cholescintigraphy is based on the hypothesis that a partially obstructed, functionally impaired gallbladder will respond differently to exogenous sincalide than a normal gallbladder.[56] The GBEF provides an objective test to preoperatively confirm a clinician's impression of acalculous disease.

The dosing of sincalide for GBEF studies is critical.[57] The dose for these studies (0.02 μg/kg) was originally recommended to be given over 1 to 3 minutes; this was modified to 30 minutes[57] and more recently to 60 minutes.[58] The longer dosing time is reported to provide more reliable gallbladder contraction with less variability among normals and fewer adverse effects. A shorter infusion time (1 to 3 minutes) has the potential of causing gallbladder neck spasm with inadequate gallbladder contraction and a falsely low GBEF. In addition, abdominal cramping, nausea, and occasional vomiting may occur when the infusion rate is too rapid.

Besides sincalide, other approaches for stimulating gallbladder contraction have been used. These generally involve oral administration of a fatty substance to stimulate endogenous release of CCK by the small intestine. This method has greater variability than sincalide in normal subjects, but it may be useful when sincalide is not available.

Morphine Sulfate

In a typical hepatobiliary study, 5 mCi (185 MBq) of ^{99m}Tc-labeled disofenin or mebrofenin is administered to a patient who has had no oral intake for 4 hours. Routine images are obtained every 10 minutes for 1 hour. If the gallbladder is visualized within this time, the study is normal. If the gallbladder has not been visualized by this time, delayed imaging may be conducted for up to 4 hours to distinguish chronic cholecystitis from acute cholecystitis. The time delay is necessary to allow radiotracer to penetrate the viscous sludge present in the cystic duct and gallbladder that occurs in 20% of patients with chronic cholecystitis.[59] To shorten the time from 4 hours to about 1.5 hours, morphine sulfate (0.04 mg/kg in 10 mL normal saline infused intravenously over 3 minutes) can be administered, provided radiotracer is seen within the small bowel. Morphine enhances sphincter of Oddi tone, causing increased intraluminal pressure in the common bile duct, facilitating the flow of tracer into the gallbladder.[56] If the gallbladder is visualized during the delayed study, cystic duct obstruction and acute cholecystitis can be ruled out. If it is not visualized within 30 minutes after morphine administration, acute cholecystitis is deemed present. The specificity of delayed hepatobiliary imaging is 93% to 96%.[60] Before morphine is administered, it is important to view images to ensure that sufficient radiotracer is present within the biliary radicals; otherwise a false-positive study may result. If there is insufficient activity, 1.5 mCi (55.5 MBq) of additional tracer should be given before administering the morphine.

Phenobarbital

Phenobarbital is an enzyme inducer that promotes the conjugation and excretion of bilirubin. It is a useful adjunct in hepatobiliary imaging for differentiating neonatal hepatitis from biliary atresia. Neonatal hepatitis manifests itself at 1 to 4 weeks of age. It may be caused by hepatocytes that are unable to conjugate bilirubin as a result of infection, toxins, or unknown causes.[56] Atresia is an obstructive condition caused by sclerosing cholangitis or by the absence of intrahepatic or extrahepatic bile ducts. Neonatal hepatitis requires medical management and is self-limiting, whereas biliary atresia requires surgical intervention. Hence, the differential diagnosis is critical.

Prior to the ^{99m}Tc-IDA hepatobiliary study, the patient is pretreated with phenobarbital 5 mg/kg per day orally in two divided doses for 5 consecutive days.[56] After injection of ^{99m}Tc-disofenin or ^{99m}Tc-mebrofenin at 5 mCi (185 MBq)/1.7 m^2, images are obtained every 10 minutes for 1 hour. Delayed images may be required for up to 24 hours. If biliary excretion of the radiotracer occurs within that time frame, biliary atresia can be excluded, and the patient is treated for neonatal hepatitis. If no excretion occurs, biliary atresia is the likely cause.

SPLEEN IMAGING AGENTS

Physiologic Anatomy

The spleen consists of two main compartments: the white pulp, composed of small lymphocytes and plasma cells, and the red pulp, a swampy mass of vascular spaces (sinuses) separated by a supporting structure (cords) that contains phagocytic cells of the RES. The spleen has two main functions: antibody production in the white pulp and particle filtration in the red pulp. There is no significant storage of red or white blood cells in the human spleen as there is in certain animal species; however, about 30% of the body's platelets are sequestered by slow transit in the spleen and can be immediately released into the circulation when needed.[61]

Blood entering the spleen through the splenic arterioles empties first into the white pulp, where plasma skimming occurs to remove soluble antigens. Passing through the white pulp, blood enters the red pulp through an open meshwork of large oval cells at the junction of the white and red pulp, termed the marginal zone. This cellular maze appears to serve as an initial filter that removes abnormal cells and allows normal cells to proceed unhindered. Upon entering the red pulp proper, blood cells may enter either the sinuses or the cords. Blood that enters the sinuses directly passes out easily from the spleen through the efferent veins. Blood cells that enter the cords, however, must pass through a fenestrated, screen-like basement membrane separating the cords from the sinus before gaining access to the venous drainage system. Thus, they are delayed to varying degrees in their transit. Because these fenestrae are only 3 μm in diameter, the blood cells must squeeze through and are deformed in their passage. It is at this point that abnormal, misshapen, and chemically altered cells are detained or destroyed. The macrophages that line the cord side of the sinus basement membrane can then phagocytize the cellular debris. The spleen's physiologic function has been described in detail by Williams et al.[62]

Imaging Agents

Two basic types of radioactive agents are used to image the spleen: (1) radiocolloids, which are localized by splenic phagocytes; and (2) denatured radiolabeled red blood cells (RBCs). ^{99m}Tc-SC is used to image the spleen because of convenience, but it lacks specificity because the liver sequesters most of the radiopharmaceutical. Heat-denatured radiolabeled RBCs are more spleen-specific, but their use requires collection of a blood sample from the patient, radiolabeling the RBCs, and heat-denaturing the RBCs prior to reinjection.

Heat-Denatured ^{99m}Tc-Red Blood Cells

Techniques for labeling and heat-denaturing RBCs were developed by Smith and Richards[63] at Brookhaven National Laboratory. The method involved a stannous citrate kit to "tin" the RBCs prior to labeling with pertechnetate. The ^{99m}Tc-labeled red blood cells (^{99m}Tc-RBCs) were then mixed with saline and heated for 15 minutes at 49°C to 50°C. The unique feature of this method was that excess tin, not associated with the cells, was removed before the addition of pertechnetate. This maximized the labeling yield at more than 97%. A commercial kit now on the market (UltraTag-RBC [Covidien]) is based on the Brookhaven kit. It has been modified to allow labeling to occur in whole blood, obviating the need for a centrifugation step to remove excess stannous ion from the tinned RBCs prior to pertechnetate addition. The UltraTag kit is not approved for spleen imaging, but the method could easily be adapted for this purpose with a heat-treating step after the RBCs are labeled. An alternative method is use of the stannous pyrophosphate in vivo or in vitro RBC-labeling technique (see Chapter 10), modified by adding a heating step once the cells are labeled with technetium.

^{99m}Tc-labeled heat-denatured RBCs have been evaluated for splenic sequestration in humans.[64] Blood clearance of the cells is rapid ($T_{1/2}$, 6.3 minutes), and splenic uptake reaches a plateau by 30 minutes. Spleen uptake varies, with a low value of 42% of the administered activity and the mean at 72%. Two hours after dosing, 5% of the administered activity is excreted in the urine. The reliability of splenic uptake with this method has been attributed to two factors: (1) The small volume of cells (4 mL) during the heating step provides more uniform heating so that adequate cell preparation is achieved in a short time, and (2) the small volume of cells administered reduces the possibility of overloading the spleen's sequestering ability, compared with a larger volume of cells.[64]

Mechanism of Localization

Heating RBCs changes their shape from tough biconcave disks to spherocytes with knobby projections and a fragile cell membrane.[65] When these cells squeeze through the 3 μm pores in the cords of the red pulp they are lysed, releasing their radioactive contents within the spleen. Splenic removal of RBCs is a more selective process than removal by the liver and other RES tissue. Insufficient heating of cells produces incomplete denaturation and decreased sequestration by the spleen, whereas overheating causes localization in the liver due to cell lysis in the bloodstream. Heating must be controlled carefully to produce a spleen-specific agent. Figure 23-8 illustrates a normal spleen scan and the localization of an accessory spleen using ^{99m}Tc-labeled heat-denatured RBCs.

GASTROINTESTINAL IMAGING AGENTS

Several useful procedures have been developed to evaluate the GI system with radiopharmaceuticals. Routine procedures have been developed to evaluate gastroesophageal reflux and

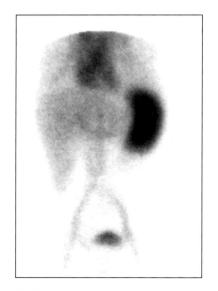
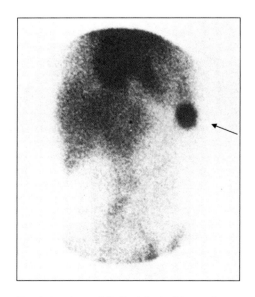

FIGURE 23-8 Spleen scans after injection of ^{99m}Tc-labeled heat denatured red blood cells. Normal anterior spleen scan (left), and spleen scan in the left anterior oblique projection (right) identifying an accessory spleen in a patient who underwent a splenectomy 10 years earlier.

gastric emptying, detect a GI bleeding site, and localize a Meckel's diverticulum. GI protein loss can also be assessed.

Gastroesophageal Reflux

^{99m}Tc-SC is typically used in gastroesophageal reflux studies because it is insoluble in gastric juices. Significant amounts of soluble ^{99m}Tc-sodium pertechnetate are undesirable because its systemic absorption may contribute to increased background activity in the esophageal region. ^{99m}Tc-SC is usually given as an acidified liquid meal, which tends to aggravate reflux because an acid load in the stomach appears to delay gastric emptying and lower the resting pressure gradient of the lower esophageal sphincter.[66]

Gastric Emptying

Radionuclide gastric emptying studies measure the rate of removal of radiolabeled liquids and solids from the stomach. Such studies provide a noninvasive method of evaluating gastric physiology. Under normal conditions, liquids clear in an exponential manner and at a faster rate than solids. Although gastric emptying studies are relatively simple to perform, accurate quantitation requires attention to several aspects of the study:

1. Radionuclide markers must have high labeling efficiency and remain stable in vivo during the study.
2. Meal size and composition should be standardized.
3. A standard patient position and posture should be maintained for imaging times.
4. Correction techniques should be applied when needed to compensate for radionuclide decay, multiple radionuclide interference, geometry changes, septal penetration, and scatter from high-energy gamma rays.

The ideal tracer should not be absorbed through nor bound to the gastric mucosa, should have no effect on gastric emptying, and should mix evenly with ingested food. Liquid markers and solid markers have been used in gastric emptying studies. Liquid marker radiopharmaceuticals are miscible with aqueous liquids and will trace the movement of liquids from the stomach. They must be nonabsorbable and stable and not localize in the stomach. Liquid markers that have been used include ^{111}In-DTPA, ^{99m}Tc-DTPA, and ^{99m}Tc-SC. ^{99m}Tc-sodium pertechnetate cannot be used because it localizes in the gastric mucosa. Solid markers are typically ^{99m}Tc-SC incorporated into a scrambled egg.[67]

To obtain meaningful results, a standard institutional protocol must be established for performing gastric emptying studies. The size of the meal and its composition will affect emptying rates. Liquids empty from the stomach monoexponentially and at a faster rate than solids, which tend to empty at a constant rate, being restricted by the pylorus.[68,69] The emptying rate is also influenced by the type of food and the amount. In general, large meals and high-calorie and fatty meals empty more slowly. A standard procedure that has been used is to incorporate 1 mCi of ^{99m}Tc-SC into a scrambled egg, which is eaten, followed by 4 ounces of water. An institution should use a standard meal for its gastric emptying studies. A consensus-recommended method for gastric emptying studies is a solid meal using ^{99m}Tc-SC incorporated into 4 ounces of scrambled egg white eaten with two slices of white bread toast, 30 grams of jam or jelly, and 120 mL of water.[67]

GI Bleeding

Successful management of acute GI bleeding depends upon accurate localization of the bleeding site. Most GI bleeding is intermittent. This bleeding pattern presents a problem for invasive diagnostic methods such as angiography, which requires active bleeding during the procedure to accurately localize the site. The significant morbidity associated with this procedure led to the development of scintigraphic methods using radiopharmaceuticals.

Two radiopharmaceuticals have been found to be useful for localizing GI bleeding sites. The first agent was ^{99m}Tc-SC, and the current agent is ^{99m}Tc-RBCs. Each agent has unique properties related to its use for detecting GI bleeding (Table 23-4). An ideal agent for detection of GI bleeding would extravasate into the bleeding site and be cleared from the blood and excreted rapidly. This would provide a high target-to-background ratio to facilitate localization of the site. Neither ^{99m}Tc-SC nor ^{99m}Tc-RBCs meets these stringent requirements. ^{99m}Tc-SC clears rapidly from the blood after intravenous injection and therefore requires active bleeding at the time of injection. In addition, its localization in the liver and spleen may obscure bleeding sites in these areas of the abdomen. The advantage of ^{99m}Tc-RBCs is that the radiopharmaceutical's prolonged circulation time provides a reservoir of radioactive blood when bleeding occurs and is thus ideal for intermittent bleeding. This is a double-edged sword, because the prolonged blood clearance creates high tissue background. However, this has not appeared to interfere significantly with the detection of bleeding sites. The use of both these agents has been reviewed.[70-72]

^{99m}Tc-RBCs are now the agent of choice for identifying GI bleeding. A comparison study of ^{99m}Tc-SC and ^{99m}Tc-RBCs was conducted in a group of 100 patients. Within this group, 41 had active bleeding and 38 of these cases were accurately identified by ^{99m}Tc-RBCs. Only five cases were identified by ^{99m}Tc-SC, and in no instance did ^{99m}Tc-SC demonstrate

TABLE 23-4 Radiopharmaceuticals for Detection of Gastrointestinal Bleeding

^{99m}Tc-Sulfur Colloid	^{99m}Tc-Red Blood Cells
Fast blood clearance	Slow blood clearance
High target-to-background ratio	Low target-to-background ratio
Requires active bleeding	Best for intermittent bleeding

active hemorrhage that was not subsequently identified by ^{99m}Tc-RBCs.[72]

A key requirement for using ^{99m}Tc-RBCs is a high tagging efficiency, because any significant free pertechnetate will localize in the gastric mucosa, where it may interfere with diagnostic interpretation. A kit for the in vitro preparation of ^{99m}Tc-RBCs (UltraTag-RBC [Covidien]) permits labeling in whole blood and routinely produces yields in excess of 95%. The method for labeling RBCs with this kit is described in Chapter 10.

Meckel's Diverticulum

A Meckel's diverticulum is a vestigial remnant of the omphalomesenteric duct located on the ileum about 50 to 80 cm from the ileocecal valve.[73] About half of Meckel's diverticula have gastric mucosa. Meckel's diverticulum is found most commonly in children, although it has been diagnosed at every age from birth onward. The most common symptom is rectal bleeding that results from peptic ulceration of the bowel by acid secreted from the gastric mucosa in the diverticulum.[74] The frequent presence of gastric mucosa in a Meckel's diverticulum led to the use of ^{99m}Tc-sodium pertechnetate for scintigraphic studies because of its inherent biologic localization in gastric mucosa. Scintigraphic evaluation is recommended only when a patient is not actively bleeding. Active bleeding is best evaluated with a ^{99m}Tc-RBC study.[73] The use of ^{99m}Tc-sodium pertechnetate in this application has shown a sensitivity of 85% and a specificity of 95% in cases of surgically proven Meckel's diverticula with ectopic mucosa.[75,76] The procedure does not directly detect diverticula that do not contain gastric mucosa.

GI Protein Loss

Under normal circumstances, approximately 100 grams of protein enters the GI tract daily as a result of digestive processes involving pancreatic and biliary secretions; however, this protein is largely reabsorbed after intestinal catabolism. In addition, total body catabolism of albumin accounts for a small amount of the normal intestinal loss of protein. Protein-losing enteropathy (PLE) is characterized by excessive loss of protein, primarily albumin, in the stool. It may be caused by diseases that produce inflammation and ulceration of the GI tract with increased mucosal permeability and leakage of serum albumin into the intestinal tract. The clinical consequence of PLE is hypoalbuminemia caused by an albumin loss in excess of normal synthesis in the body.

Nuclear medicine procedures can assist in the diagnosis of PLE. An in vitro test for protein loss involves recovering stool containing radiolabeled albumin eliminated into the GI tract.[77] Ideally, the radiolabel should remain firmly bound to the protein and not be excreted in the urine, which can contaminate stool samples. The labeled protein should enter the intestinal tract only through abnormal leakage sites and should not be reabsorbed.

^{131}I-HSA was one of the first agents used to study PLE. However, it had the disadvantage of being catabolized in the gut, with reabsorption of the radioiodine. Studies using ^{51}Cr-labeled albumin or ^{51}Cr-chromic chloride demonstrated that these agents could be recovered almost completely from the stool after oral administration, indicating no reabsorption of the radiolabel.[78,79] Consequently, ^{51}Cr-labeled products became the agents of choice for in vitro assessment of PLE. A comparison study demonstrated that ^{51}Cr-albumin had less urinary excretion (13%) than ^{51}Cr-chromic chloride (60%) 3 days after dosing;[78] however, ^{51}Cr-albumin is no longer on the market. ^{51}Cr-chromic chloride is available as the radiochemical and can be compounded into a sterile injection by a nuclear pharmacist. ^{51}Cr-chromic chloride adheres tenaciously to glass, which should be avoided in its preparation; loss can be minimized by using plastic or siliconized glass containers and maintaining the solution at or below pH 4. When given intravenously, ^{51}Cr-chromic chloride labels plasma albumin in vivo and is effective for a PLE study.

In Vitro Procedure for PLE

A dose of 0.25 to 0.5 µCi (9.25–18.5 kBq) ^{51}Cr-chromic chloride per pound of body weight is administered intravenously. All stools passed during the next 4 days are collected by the patient. Patients with some conditions (e.g., Crohn's disease) may need multiple containers because of the large volume of liquid stool. Empty metal paint cans containing 10 mL phenol preservative are satisfactory for collecting stool specimens. Each sample is adjusted to a standard weight with water and assayed for radioactivity with a scintillation counter. A ^{51}Cr standard, identical to the patient dose administered, and a background can are prepared and counted along with the stool specimens. The percentage of the dose excreted in the cumulative stool samples is calculated. Normal subjects excrete 1% or less of the administered dose in 4 days.[77,78] Patients with PLE have been shown to excrete 2% to 40% of the dose. Care must be taken to instruct patients not to contaminate stool samples with urine.

Scintigraphic Imaging for PLE

PLE can also be assessed by scintigraphic imaging with ^{99m}Tc-labeled human serum albumin (^{99m}Tc-HSA).[80] Unfortunately, ^{99m}Tc-HSA is no longer marketed in the United States, but it can be compounded extemporaneously (see Chapter 10).

Following intravenous injection of 20 mCi (740 MBq) of ^{99m}Tc-HSA, gamma camera images are acquired for up to 24 hours.[80] In the presence of PLE ^{99m}Tc-HSA exudes into the gut, where accumulation of activity is evident on abdominal images (Figure 23-9).

NUCLEAR MEDICINE PROCEDURES

Magnetic resonance imaging (MRI), computed tomography (CT), and ultrasound are commonly used to evaluate the anatomy of the liver, hepatobiliary system, and spleen.

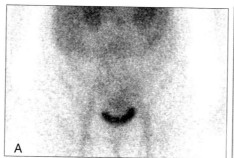
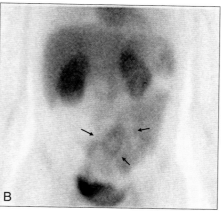

FIGURE 23-9 Protein-losing enteropathy. (A) Anterior abdominal image 24 hours after intravenous injection of 21 mCi (777 MBq) of ^{99m}Tc-HSA in a 37 year old woman with a prior history of protein-losing enteropathy, colectomy, and juvenile polyposis in the stomach and distal small bowel. Activity is evident in the liver, kidney, and bladder. Activity superior to the bladder most likely represents radiotracer within the uterus. There is no evidence of radiotracer accumulation in the small bowel to suggest protein-losing enteropathy up to 24 hours. (B) Anterior abdominal image 4 hours after intravenous injection of 20 mCi (740 MBq) of ^{99m}Tc-HSA in a 16 year old male with very low levels of serum albumin. Accumulation of small bowel activity (arrows) suggests gastrointestinal protein loss of ^{99m}Tc-HSA. There is no evidence of gastric uptake caused by any free ^{99m}Tc-pertechnetate impurity in the ^{99m}Tc-HSA, which could have given false-positive results.

However, imaging with radionuclides can offer more information about the physiology and function of these organs. Several radiopharmaceuticals with different mechanisms of uptake can be used to image these organ systems.

Liver–Spleen Imaging

Rationale

Liver–spleen imaging uses radiolabeled colloids that are phagocytized by RES cells in the liver, spleen, and bone marrow. Most of the particles (80% to 90%) are taken up in the liver, about 4% to 8% in the spleen, and the remainder in the bone marrow (Figure 23-10).[23] This permits functional imaging of the RES and can be useful for evaluation of hepatocellular disease such as cirrhosis, hepatomegaly, splenomegaly, and certain hepatic or splenic lesions seen in an anatomic imaging study such as CT. Liver–spleen nuclear imaging studies can also be useful to localize splenic tissue.

In patients with hepatic metastatic disease, hepatic artery infusion pumps are surgically placed to deliver strong chemotherapy directly to the liver. Proper placement is essential so that the chemotherapy drugs are delivered only to the liver and not to other organs. Imaging evaluation of pump function is accomplished with ^{99m}Tc-albumin aggregated (^{99m}Tc-MAA). Once injected via the pump, these small particles are trapped in the liver by capillary blockade. If radiotracer accumulation is seen outside the liver, the position of the pump catheter needs to be adjusted or collateral circulation to these other areas needs to be identified and ligated. ^{99m}Tc-SC liver–spleen imaging can be performed for comparison if there is a question about the results of the ^{99m}Tc-MAA study. ^{99m}Tc-SC liver–spleen imaging is especially important in patients who have had liver surgery prior to pump placement. In these patients, the liver may have an abnormal contour that can be difficult to evaluate (Figure 23-11).

Blood-pool studies with ^{99m}Tc-RBCs can be used to evaluate suspected cavernous hemangiomas in the liver. Because there is sluggish blood flow in hemangiomas, these are typically seen as focal defects on the early flow images that demonstrate increased uptake on delayed images.

Splenic imaging is useful for evaluating abdominal masses thought to be residual functional splenic tissue in patients who have had surgical splenectomy for thrombocytopenia, for confirming suspected splenosis in patients who have undergone splenectomy after trauma, and for diagnosing congenital abnormalities such as asplenia or polysplenia in children. Splenic imaging can be accomplished with colloid agents as part of the liver–spleen scan. More specific imaging of the spleen can be accomplished with autologous heat-denatured ^{99m}Tc-RBCs; this is useful in determining the presence of functioning splenic tissue. Heat-denatured ^{99m}Tc-RBCs are sequestered by the functioning splenic tissue, which allows for its identification with imaging (Figure 23-12). The red blood cells are not sequestered by the liver, so the liver is not imaged as in a colloid liver–spleen scan. This can be helpful in differentiating spleen from adjacent structures, especially in patients with a liver that has a large left lobe adjacent to the spleen or in a patient with questionable situs inversus.

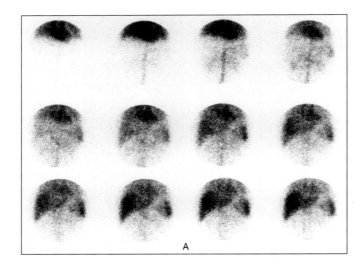
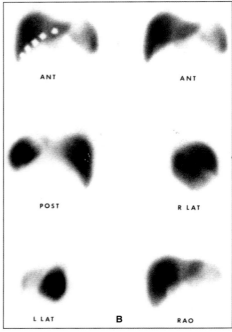

FIGURE 23-10 (A) Normal anterior liver blood flow study in a patient after intravenous injection of 5 mCi (185 MBq) of ^{99m}Tc-sulfur colloid. Sequential images taken at 5 second intervals demonstrate the normal distribution of activity in the heart, abdominal aorta, liver, and spleen. (B) Normal static liver–spleen scan in the same patient obtained 15 minutes after the liver flow study. Four standard views (anterior, right lateral, posterior, and left lateral) and a right anterior oblique view are shown.

Procedure

Liver–spleen imaging requires no specific patient preparation. The adult patient receives an intravenous injection of 4 to 6 mCi (148–222 MBq) of ^{99m}Tc-SC. A large field-of-view camera with a low-energy all-purpose (LEAP) or low-energy high-resolution (LEHR) collimator is used for imaging. If blood flow imaging of the liver is desired, the patient is injected under the camera and 1 or 2 second frames can be obtained over the first minute to evaluate variations in blood flow to the liver (Figure 23-10A). Usually, static images are obtained 15 to 20 minutes after administration of the radiotracer. Images of the abdomen are usually obtained in the anterior, posterior, lateral, and oblique positions (Figure 23-10B).[81] Delayed images of the liver are obtained 1 to 2 hours later. Single-photon emission computed tomography (SPECT) imaging can sometimes be helpful if suspected lesions are small or there are several lesions.

Hepatobiliary Imaging

Rationale

Hepatobiliary scintigraphy is accomplished with radiopharmaceuticals that are taken up by hepatocytes and excreted into the bile, where they advance through the hepatic ducts and common bile duct into the small intestine. Some of the radiotracer also enters the cystic duct and accumulates in the gallbladder. Hepatobiliary function can be estimated by the amount and timing of liver uptake and clearance. Patency of the hepatobiliary system can be evaluated by following the flow of radiotracer through the biliary tree and intestines. Radionuclide imaging of the hepatobiliary system is commonly used to evaluate acute cholecystitis, GBEF, common bile duct obstruction, bile leak, and congenital anomalies such as biliary atresia.

The most common indication for hepatobiliary scintigraphy is evaluation of suspected acute cholecystitis. Most cases of acute cholecystitis are caused by obstruction of the cystic duct by a gallstone. It is important to obtain a pertinent patient history before performing this exam. The history should include details of any prior GI surgeries (such as cholecystectomy), when the last meal was eaten, whether the patient is taking opioid medications for pain, if the patient has had recent ultrasonography, and if the patient has had recent laboratory tests to determine liver enzyme levels.

The liver makes bile continuously. Normally, about two-thirds of the bile flows through the common bile duct into the duodenum. Approximately one-third flows up the cystic duct into the gallbladder, where it is stored.[82] The gallbladder contracts and empties its contents into the duodenum in response to endogenous CCK, which is produced by duodenal mucosal cells in response to fats and proteins in the intestinal contents. If the patient has recently eaten a meal, circulating endogenous CCK will prevent the gallbladder from filling. The patient should fast for 4 hours before beginning this exam. However, if the patient has not eaten for more than 24 hours or has been receiving total parenteral nutrition (TPN), the gallbladder may be full of thick, viscous bile. In this case the gallbladder may not accumulate radiotracer. Both

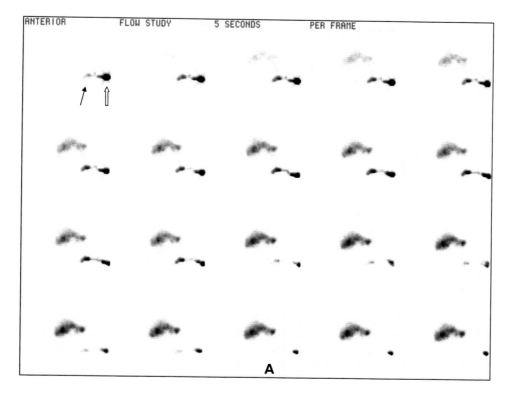

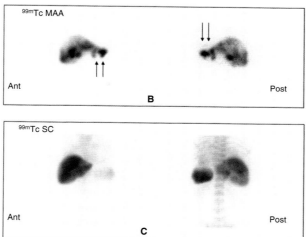

FIGURE 23-11 Hepatic artery infusion pump study. The patient has had a left hepatic lobectomy and placement of a hepatic artery infusion pump because of metastatic involvement by colon cancer. (A) Anterior 5-second-per-frame images from a ^{99m}Tc-MAA study. On the initial image, radiotracer is seen in the injection syringe (block arrow) and in the hepatic artery infusion pump (solid arrow). As the dose is injected, uptake is seen in the liver. (B) Anterior and posterior images of the liver from the ^{99m}Tc-MAA study show patchy uptake in the liver with apparent uptake in the left lobe (double arrows). (C) Anterior and posterior images from a ^{99m}Tc-SC study done the next day demonstrate absence of a left lobe, indicating that the ^{99m}Tc-MAA activity seen the previous day in this region is outside the liver parenchyma in bowel.

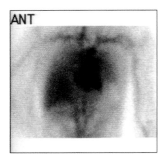
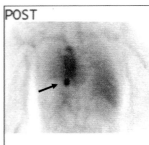
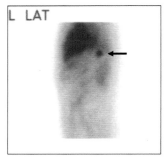

FIGURE 23-12 ^{99m}Tc-labeled heat-denatured RBC study in a patient with a history of prior splenectomy. Anterior, posterior, and left lateral views of the abdomen show a small focal area of ^{99m}Tc-RBC accumulation posteriorly in the left upper quadrant consistent with an accessory spleen.

of these conditions can result in false-positive studies (absence of gallbladder uptake of radiotracer activity).

If the patient is currently taking opioid drugs, this may interfere with the exam. The administration of opioids such as morphine sulfate results in constriction of the sphincter of Oddi, which increases the intraluminal pressure in the common bile duct and can prevent bile from entering the duodenum. Although the gallbladder may be visualized, failure of the radiotracer to enter the small bowel may be interpreted as obstruction of the common bile duct. Therefore, the study should be delayed for 4 hours after opioid administration.

If the patient's liver enzymes are elevated, the hepatocytes may not be able to concentrate enough radiotracer for visualization of the gallbladder. Higher dosage of the radiopharmaceutical may be needed if the patient has marked hyperbilirubinemia.

Procedure

The patient should fast for about 4 hours before administration of the radiotracer. If the patient has not eaten for more than 24 hours or is receiving TPN, the patient may be pretreated with sincalide, a synthetic C-terminal octapeptide of CCK. The usual pretreatment dose of sincalide is 0.02 μg/kg. This is preferably given as a slow drip over 60 minutes to lessen the possibility of adverse effects and to have less variability in the gallbladder ejection fraction measurement.[58] Sincalide should not be administered as a bolus because it can cause gallbladder spasm and abdominal discomfort. Sincalide should also not be used if the patient is currently receiving morphine, which acts to constrict the sphincter of Oddi. If pretreatment with sincalide is considered appropriate, the hepatobiliary radiotracer is usually administered 30 to 60 minutes after the sincalide infusion is over.

Hepatobiliary imaging is performed with ^{99m}Tc-IDA compounds. The two radiopharmaceuticals currently in use in the United States are ^{99m}Tc-disofenin and ^{99m}Tc-mebrofenin. ^{99m}Tc-mebrofenin has a higher hepatic extraction and lower renal excretion than ^{99m}Tc-disofenin and may be more useful in patients with high bilirubin levels.

Typically, 1.5 to 5 mCi (55.5–185 MBq) of a ^{99m}Tc-labeled IDA compound (^{99m}Tc-disofenin or ^{99m}Tc-mebrofenin) is administered to an adult. When these compounds are introduced into the blood, they are bound to albumin, which minimizes clearance from the kidneys. They are actively taken up into the hepatocytes by a carrier-mediated mechanism similar to that of bilirubin uptake. Once in the hepatocytes, they are excreted unchanged into the bile canaliculi by carrier-mediated transport mechanisms.[49,50] Images are obtained using a large field-of-view gamma camera with a LEAP or LEHR collimator. Anterior spot images of the abdomen can be obtained every 10 minutes, or continuous 1-minute-per-frame images of the abdomen can be obtained until radiotracer is seen in both the proximal small bowel and the gallbladder.[83] In most normal studies, the gallbladder and small bowel are visualized within 60 minutes (Figure 23-13). If there is questionable activity in the region of the gallbladder fossa, which may be either gallbladder accumulation or pooling in the proximal duodenum, a right lateral view can be helpful. The gallbladder is usually an anterior structure. If the focal accumulation is still questionable and the patient can have some water by mouth, this will clear the activity if it is in the duodenum.

If the gallbladder is not visualized after 60 minutes, delayed imaging should be performed after 4 hours. If there is activity in the proximal bowel, an alternative to 4 hour delayed imaging is augmentation with morphine sulfate. Morphine sulfate administration constricts the sphincter of Oddi, which increases the intraluminal pressure in the common bile duct. This forces radiotracer into the cystic duct and into the gallbladder if there is not an obstruction. Usually, a morphine dose of 0.04 mg/kg is given intravenously and imaging is continued for another 30 minutes. If there is very little radiotracer activity in the liver and a decision is made to administer morphine sulfate, a "booster dose" of ^{99m}Tc-disofenin or ^{99m}Tc-mebrofenin should be given 10 to 20 minutes before morphine administration. The booster dose is usually 1 to 1.5 mCi (37–55.5 MBq).

In some patients with a history of right upper quadrant pain, no evidence of gallstones on ultrasound, and a normal-appearing hepatobiliary scan, it can be useful to evaluate the GBEF. A poor ejection fraction can be seen in conditions such as chronic acalculous cholecystitis, biliary dyskinesia, cystic duct syndrome, and sphincter of Oddi dysfunction. Sincalide administration is used to assess GBEF after the gallbladder is maximally filled with the radiotracer and most of the radiotracer has cleared from the liver. Typically, sincalide

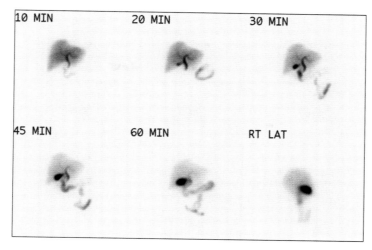

FIGURE 23-13 Normal hepatobiliary study after intravenous injection of 5 mCi of ^{99m}Tc-disofenin. On the initial image taken at 10 minutes, radiotracer accumulation is seen in the liver, biliary tree, common bile duct, and duodenum. Radiotracer is seen in the gallbladder on the 20 minute image. During the 60 minute study there is prompt clearance of radiotracer from the liver with increasing accumulation in the gallbladder and advancement through the bowel.

0.02 µg/kg is administered over 30 to 60 minutes. Anterior 1-minute-per-frame images of the gallbladder are then obtained over a 20 to 30 minute interval. Regions of interest are drawn around the gallbladder and the adjacent liver to determine background (Figure 23-14). GBEF is calculated as

$$\text{GBEF \%} = \frac{(\text{Net maximum GB counts} - \text{Net minimum GB counts}) \times 100}{\text{Net maximum GB counts}}$$

A 60 minute sincalide infusion of 0.02 µg/kg yields a GBEF lower limit of normal of 38%.[58] Below this limit is considered abnormal. Because sincalide has a plasma half-life of only about 2.5 minutes, it can be used both for pretreatment in patients who have fasted longer than 24 hours and for GBEF. However, sincalide should not be used in patients who required morphine stimulation for visualization of the gallbladder. In these patients, the sphincter of Oddi is contracted because of morphine administration. Sincalide will cause the gallbladder to contract against a closed sphincter.

Interpretation

If blood flow images of the abdomen are obtained, the liver is normally seen a few seconds after the kidneys and spleen. This is because most of the blood flow to the liver comes from the portal circulation. In a normal study, there is prompt, homogeneous uptake of the radiotracer in both lobes of the liver. After only about 5 minutes there should be good clearance from the blood pool and visualization of the liver parenchyma. Delay in clearance of the blood pool is seen with hepatic insufficiency. After about 10 minutes there should be evidence of radiotracer accumulation in the biliary tree (biliary excretion). Next there should be radiotracer in the common bile duct, followed by accumulation in the gallbladder and the proximal bowel. Normally, radiotracer activity should be seen in the gallbladder and bowel within 1 hour (Figure 23-13).

Acute Cholecystitis

The most common cause of acute cholecystitis is a stone in the cystic duct that causes an obstruction. In the proper clinical setting (i.e., the patient has fasted for 4 hours, has been pretreated with sincalide if he or she has fasted for more than 24 hours, and has not had a cholecystectomy), nonvisualization of the gallbladder by 4 hours is consistent with acute cholecystitis. If morphine augmentation is used, nonvisualization 30 minutes after morphine administration is considered consistent with acute cholecystitis. In some cases of acute cholecystitis, there is a rim of increased activity in the gallbladder fossa along the inferior hepatic edge with nonvisualization of the gallbladder. This is referred to as the rim sign and is secondary to inflammation and hyperemia around the gallbladder (Figure 23-15).

Chronic Cholecystitis

Most patients with chronic cholecystitis have normal hepatobiliary scans. Some patients may show delayed gallbladder filling several hours after radiotracer injection or delayed appearance of activity in the bowel. The time of gallbladder filling in chronic cholecystitis can be shortened by use of morphine sulfate, which will increase the intraluminal pressure in the common bile duct to facilitate gallbladder filling (Figure 23-16). Many people with symptoms associated with either calculous or acalculous chronic gallbladder disease have reduced GBEF.

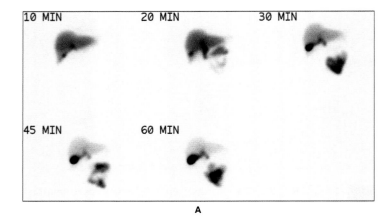
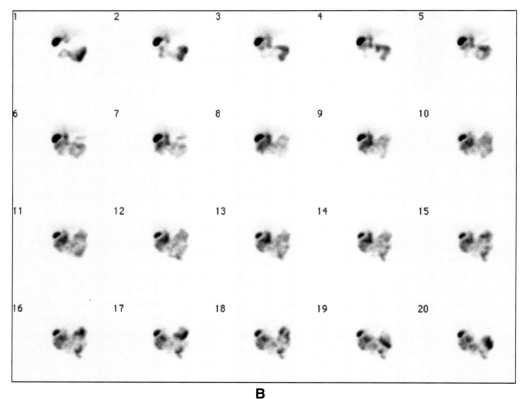

FIGURE 23-14 Normal hepatobiliary study after intravenous injection of 5 mCi (185 MBq) of ^{99m}Tc-disofenin. (A) On the initial image taken at 10 minutes, radiotracer accumulation is seen in the liver, biliary tree, and gallbladder. Radiotracer is seen in the proximal bowel on the 20 minute image. During the 60 minute study there is good clearance of radiotracer from the liver, with increasing accumulation in the gallbladder and advancement through the bowel. (B) After this, serial 1-minute-per-frame images of the abdomen were obtained during intravenous administration of cholecystokinin (CCK). There is normal gallbladder emptying during the 20 minute CCK administration.

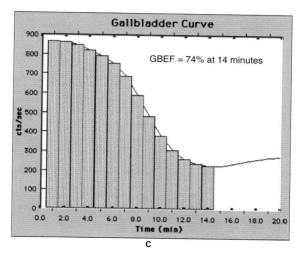

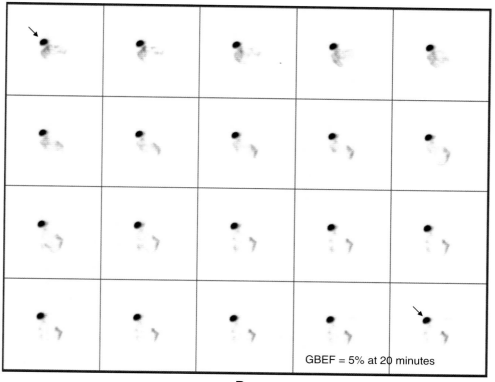

FIGURE 23-14 (Continued). Normal hepatobiliary study after intravenous injection of 5 mCi (185 MBq) of ^{99m}Tc-disofenin. (C) A gallbladder time–activity curve shows prompt emptying of the gallbladder, with 74% emptying at 14 minutes. (D) Another patient with an abnormal gallbladder ejection fraction (GBEF). During the 20 minute CCK administration there is very little change in the appearance of the gallbladder (arrows). GBEF was only 5%.

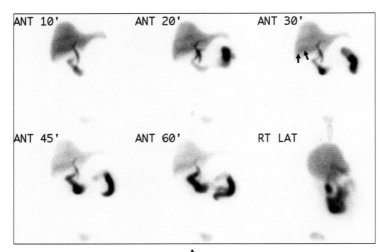
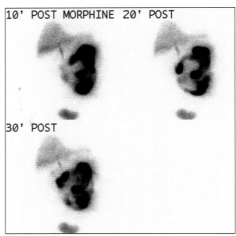

FIGURE 23-15 Acute cholecystitis. (A) Serial anterior images of the abdomen were obtained for 60 minutes after intravenous administration of 5 mCi (185 MBq) of ^{99m}Tc-disofenin. On the initial image at 10 minutes, there is radiotracer in the liver, biliary tree, common bile duct, and proximal bowel. Radiotracer is not seen in the gallbladder at 60 minutes. A rim of parenchymal activity is seen in the area of the gallbladder fossa on these images, which is often associated with acute cholecystitis (arrows). It is evident in about 25% of acute cholecystitis patients who are at an advanced stage of the disease. (B) Activity was not seen in the gallbladder even after augmentation with morphine to constrict the sphincter of Oddi.

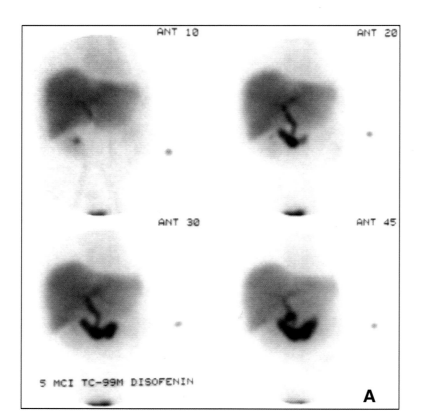
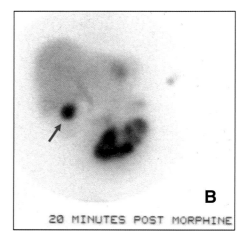

FIGURE 23-16 Chronic cholecystitis. (A) Serial anterior images of the abdomen were obtained for 45 minutes after intravenous administration of 5 mCi (185 MBq) of ^{99m}Tc-disofenin. A right lateral view was obtained at 60 minutes (image not shown). Normal activity is seen in the liver at 10 minutes and bowel activity at 20 minutes after injection. No activity was seen in the gallbladder at 60 minutes. (B) Morphine sulfate 3.38 mg was given intravenously and additional images acquired for 20 minutes, revealing accumulation of activity in the gallbladder consistent with chronic cholecystitis.

In these patients GBEF after sincalide administration can be clinically useful.

Common Duct Obstruction

If there is a high-grade common duct obstruction, radiotracer will not be able to enter the bowel. Early on, there is good visualization of the liver parenchyma but no evidence of accumulation in the biliary tree or bowel (Figures 23-17 and 23-18 A and B). Kidney activity may also be evident. If the obstruction is prolonged, there will eventually be decreased hepatic function. Partial common duct obstruction will cause a delayed or prolonged biliary-to-bowel transit time. Nonvisualization of activity in the bowel after 1 hour suggests partial common duct obstruction. However, delayed visualization of radiotracer in the bowel can be normal in some people (20% to 25%). In partial obstruction, there is often persistent accumulation of radiotracer in the common duct proximal to the obstruction.

Bile Gastritis

During the hepatobiliary scan, radiotracer can reflux from the duodenum into the stomach (bile reflux). This can sometimes be a cause of the patient's symptoms (Figure 23-19).

Biliary Atresia

Hepatobiliary imaging can be useful in a jaundiced infant with suspected biliary atresia. The main differential diagnosis in a neonate with prolonged hepatic dysfunction and hyperbilirubinemia is neonatal hepatitis versus biliary atresia. Biliary atresia requires prompt surgical intervention before irreversible damage to the liver occurs. If biliary atresia is suspected, the infant is usually pretreated with oral phenobarbital 2.5 mg/kg twice daily for 5 to 7 days prior to the hepatobiliary scan to stimulate liver excretion. After administration of the radiopharmaceutical, imaging is performed until definite bowel activity is seen. Demonstrating radiotracer in the extrahepatic biliary tree and bowel rules out the diagnosis of biliary atresia (Figure 23-20). If definite bowel activity is not seen after several hours, a delayed 24 hour image can be helpful. If there is no evidence of bowel activity after 24 hours, the findings likely represent biliary atresia (Figure 23-21).

Bile Leak

Hepatobiliary imaging is useful in evaluating patients with suspected bile leak after either trauma (hepatic laceration) or surgery (such as laparoscopic cholecystectomy, liver transplant, or hepatectomy). If fluid collection in the abdomen is detected by another imaging modality such as ultrasound or CT, hepatobiliary imaging can be used to evaluate for clinically significant leaks, as well as obstruction that may require further surgery (Figure 23-22). Small leaks often are clinically insignificant and can be watched.

Gastrointestinal Studies

Several other nuclear medicine studies involve the GI tract. Some of the more common studies involve evaluation of GI

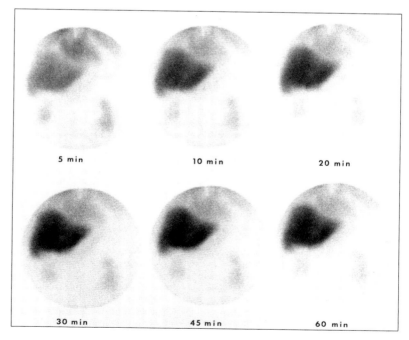

FIGURE 23-17 Biliary obstruction. Common bile duct obstruction decreases clearance of ^{99m}Tc-labeled IDA analogue tracer from the blood and liver. Kidney activity is evident as well, because of decreased hepatobiliary clearance of tracer from the blood.

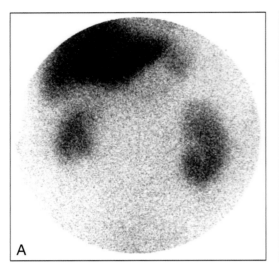
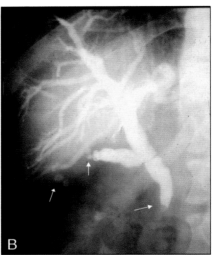

FIGURE 23-18 Five hour delayed image (A, left) of the same patient shown in Figure 23-17, demonstrating lack of bowel excretion and prominent kidney activity. A 24 hour image (not shown) was similar to the 5 hour image, consistent with complete obstruction. Transhepatic cholangiogram in this patient (B, right) demonstrates stone obstruction in the common bile duct and cystic duct (arrows). Two stones are also evident in the gallbladder.

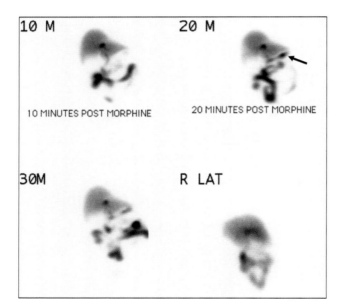

FIGURE 23-19 Acute cholecystitis with reflux gastritis. ^{99m}Tc-disofenin study with morphine augmentation. The gallbladder was not visualized during the first 60 minutes of the study, so morphine sulfate was administered. There was still no radiotracer accumulation in the gallbladder 30 minutes after morphine in this patient with acute cholecystitis. The linear area of radiotracer accumulation just inferior to the left lobe of the liver is bile reflux of the radiotracer into the stomach.

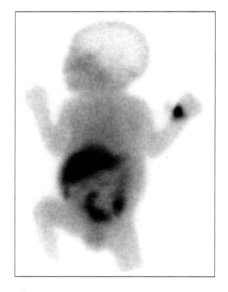

FIGURE 23-20 Neonatal hepatitis. Normal anterior hepatobiliary study in a child 2 hours after ^{99m}Tc-disofenin following 5 days of pretreatment with phenobarbital. Excretion of radiotracer into the intestinal tract confirms the presence of neonatal hepatitis and rules out biliary atresia.

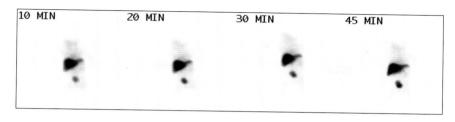

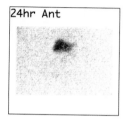

FIGURE 23-21 Biliary atresia. Anterior images of the body taken out to 24 hours after administration of 500 µCi (18.5 MBq) of ^{99m}Tc-disofenin. On the initial image taken at 10 minutes after injection of the radiopharmaceutical, there is uptake in the liver. Activity is also seen in the bladder and (faintly) in the kidneys. There was no evidence of radiotracer accumulation in the bowel or gallbladder after 24 hours in this 8 week old patient.

bleeding, suspected Meckel's diverticulum, gastroesophageal reflux, and gastric emptying.

GI Bleeding

Effective management of patients suspected of having active GI bleeding often depends on identifying the bleeding site. Upper GI bleeding, which originates from the GI tract proximal to the ligament of Treitz, is successfully evaluated with endoscopy. The lower GI tract is much more difficult to evaluate with endoscopy. Many patients bleed only intermittently, making it difficult to identify the bleeding site with angiography. Nuclear imaging using ^{99m}Tc-RBCs is the method of choice for evaluation of active lower GI bleeding. Labeled RBCs remain in the intravascular space and can be imaged for up to 24 hours. This increases the chances of detecting an intermittent hemorrhage site.

Procedure

A sample of the patient's blood is removed for radiolabeling using an in vitro technique. Dosage is typically 20 to 30 mCi (740–1100 MBq) of ^{99m}Tc-labeled autologous RBCs. With the patient supine, dynamic 60-second-per-frame anterior images of the abdomen and pelvis are obtained for 60 to 90 minutes using a large field-of-view camera with a LEAP parallel hole collimator. The images can then be played back in cine mode for evaluation. Sometimes, if there is no evidence of bleeding during the 90 minute study, a delayed image at 16 to 24 hours can be useful. Activity seen within loops of bowel on the delayed images indicates that there has been active GI bleeding and may suggest the severity of the bleeding. However, delayed imaging may not be able to demonstrate the intermittent bleeding site. Activity seen in the bowel may be downstream from the bleeding site because of peristalsis.[73]

Interpretation

Normally, ^{99m}Tc-labeled RBCs remain in the circulation. After administration, they come into equilibrium in the intravascular space. In a normal RBC study, activity can be seen in the heart, great vessels, liver, and spleen. It is normal to see some activity in the urine. Active GI bleeding is identified as an area of radiotracer accumulation within the bowel that increases and moves with time (Figures 23-23 to 23-25).

Meckel's Diverticulum

A Meckel's diverticulum is a blind-ending tube a few centimeters long arising from the antimesenteric border of the terminal ileum about 50 to 80 cm from the ileocecal valve. It is a vestigial remnant of the omphalomesenteric duct that is present in approximately 2% of the population. About half of these can contain ectopic gastric mucosa or pancreatic cells that can lead to ulceration from acid secretion and can become clinically apparent because of rectal bleeding. Most of the cases of rectal bleeding resulting from a Meckel's diverticulum occur before the age of 2 years, and almost all of the diverticula responsible for rectal bleeding contain ectopic gastric mucosa.[84] ^{99m}Tc-sodium pertechnetate is the agent of choice for imaging Meckel's diverticula containing ectopic gastric mucosa because it concentrates in the mucous cells of the gastric mucosa.

Procedure

With the patient supine, dynamic 60-second-per-frame anterior images of the abdomen and pelvis are obtained for 60 to 90 minutes using a large field-of-view camera with a LEAP parallel hole collimator. In children, about 200 to 300 µCi/kg (7.4–11.1 MBq/kg) of ^{99m}Tc-sodium pertechnetate is administered intravenously. In adults, the dose is usually 10 to 15 mCi

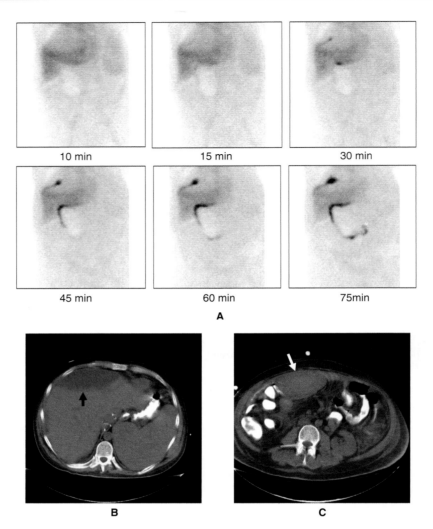

FIGURE 23-22 Bile leak. (A) Serial anterior images of the abdomen were obtained for 75 minutes after intravenous administration of 5 mCi (185 MBq) of ^{99m}Tc-disofenin in this patient after liver transplant. The images show a large photopenic defect at the dome of the liver. On the 30 minute image a linear accumulation of radiotracer is seen in the middle of the defect, which grows during the study and is consistent with an active bile leak. There is another large photopenic area below the liver. Between 45 and 75 minutes there was radiotracer accumulation in the proximal small bowel, which borders this photopenic area. (B) An axial CT image showing a low-density fluid collection anterior to the liver (black arrow). (C) Another axial CT image inferior to the liver shows a large fluid collection in the abdomen (white arrow).

(370–555 MBq).[82] Pretreatment with various pharmaceuticals can increase the sensitivity of a Meckel's scan. Histamine$_2$ (H$_2$)-blockers such as cimetidine block secretion from the gastric mucosa cells, resulting in increased uptake. In children, oral cimetidine can be given for 2 days prior to the study at a dose of 20 mg/kg per day. In adults the dose is 300 mg four times daily for 2 days before the study. Pretreatment with pentagastrin is an alternative to H$_2$-blockers. Pentagastrin stimulates gastric secretion and increases gastric mucosa uptake.[73] Glucagon has also been used as an interventional agent. Glucagon is a smooth-muscle relaxant in the intestine, which causes a decrease in peristalsis, allowing pooling of ^{99m}Tc-pertechnetate at the site of secretion. Pharmacologic interventions are not widely used.

Interpretation

Initial anterior images of the abdomen and pelvis show activity in the intravascular space, including the heart, great vessels, liver, spleen, and kidneys. After a few minutes, physiologic activity appears in the gastric mucosa of the stomach. Over time the stomach activity becomes intense as the background activity in the heart, liver, and spleen decreases. Radiotracer uptake in a Meckel's diverticulum containing ectopic gastric mucosa should appear at about the same time activity appears

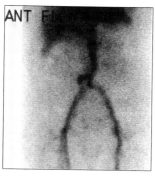 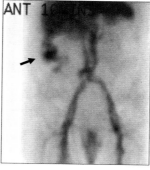

FIGURE 23-23 Abnormal GI bleeding study. Anterior images of the abdomen and pelvis from a ^{99m}Tc-RBC study in a patient with a history of passing bright red blood per rectum. (A) An early blood flow image shows activity in the heart, great vessels, and liver. (B) By 10 minutes, an abnormal focal area of accumulation is seen in the ascending colon near the hepatic flexure (arrow), demonstrating the active GI bleeding site.

in the stomach. This usually is identified as a focal area of radiotracer accumulation in the right lower quadrant or mid abdominal region (Figure 23-26). However, it can appear anywhere in the abdomen. On lateral views of the abdomen, this focus is anterior.

Gastroesophageal Reflux

Nuclear medicine imaging can be used to evaluate the presence and degree of gastroesophageal reflux, reflux of stomach contents into the esophagus manifested as a burning sensation in the chest referred to as heartburn. If this becomes severe and frequent, it is referred to as gastroesophageal reflux disease (GERD). The three different mechanisms thought to be responsible for esophageal reflux are transient relaxation of the lower esophageal sphincter, transient increase in intra-abdominal pressure, and low resting pressure of the lower esophageal sphincter.[85] GERD can lead to respiratory complications resulting from aspiration of gastric bacteria.[86]

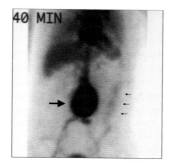

FIGURE 23-24 Abnormal GI bleeding study demonstrating a large abdominal aortic aneurysm (large arrow). During the study some radiotracer accumulation was also seen in the descending colon (small arrows), consistent with active GI bleeding. Normal activity is seen in the heart, liver, and spleen.

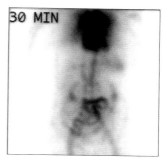

FIGURE 23-25 GI bleeding study showing active small bowel bleeding site. Anterior image taken 30 minutes after intravenous administration of ^{99m}Tc-RBCs. Activity can be seen more centrally in loops of small bowel.

The radionuclide gastroesophageal reflux study is a sensitive, noninvasive test that is able to detect reflux in 90% of symptomatic patients.[66,87] In addition, the study permits quantitation of the extent of reflux into the esophagus and can be extended to detect pulmonary aspiration of gastric contents.[88,89]

The procedure is well tolerated by children. Two hundred microcuries (7.4 MBq) of ^{99m}Tc-SC is administered orally in 30 mL of apple juice followed by additional juice or formula until the patient is sated. Abdominal pressure measurements are performed using a blood pressure cuff in small children to demonstrate reflux. The value of the radionuclide gastric reflux study in children was demonstrated in a study that showed an 80% rate of detection in subjects with gastroesophageal reflux previously documented by other acid reflux methods.[90]

Gastroesophageal reflux is a common problem in infants and children but also occurs in adults. It frequently leads to respiratory complications, presumably resulting from aspiration of gastric contents.[88] The reflux study can be extended to detect pulmonary aspiration as a consequence of gastroesophageal reflux and has been shown to be more sensitive than conventional nonradionuclide procedures.[89] The dose of ^{99m}Tc-SC given as a liquid meal at bedtime with lung scanning the following day can identify lung aspiration of refluxed gastric contents.

Adult patients should fast for 4 to 6 hours before the reflux study. The radiopharmaceutical most often used is ^{99m}Tc-SC. Typically 0.3 mCi (11.1 MBq) ^{99m}Tc-SC is added to 300 mL of acidic orange juice, which is a combination of 150 mL of orange juice and 150 mL of 0.1 N hydrochloric acid. Prior to administration of the radiotracer, the patient is fitted with an abdominal binder that can be inflated to increase pressure around the upper abdomen. The patient drinks the radiolabeled orange juice and then is positioned supine under the gamma camera. The field of view should include the entire esophagus and as much of the stomach as possible. If there is activity in the esophagus on the initial image, it may be residual swallowed activity. A small amount of water can be given to the patient to clear this activity. Typically, dynamic anterior 30-second-per-frame images are obtained during the study. If there is no evidence of activity in the esophagus, pressure is increased in the abdominal binder in 20 mm Hg increments up to 100 mm Hg. Serial 30-second-per-frame images are obtained at each increment.[91]

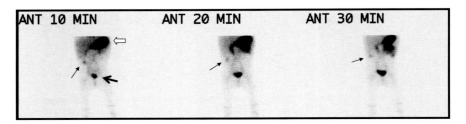

FIGURE 23-26 Meckel's diverticulum. Anterior images of the abdomen and pelvis of a 14 month old girl with intermittent GI bleeding, obtained at 10, 20, and 30 minutes after intravenous administration of 1 mCi (37 MBq) of ^{99m}Tc-sodium pertechnetate. Radiotracer accumulation is seen in the stomach (white arrow) and bladder (black arrow). There is also a small focal area of radiotracer accumulation in the right abdomen (small arrows), which was found to be a Meckel's diverticulum containing ectopic gastric mucosa.

Normally, there should be no activity above the stomach during the study. Activity visualized in the esophagus is considered abnormal and is consistent with esophageal reflux (Figure 23-27). The degree of reflux can be calculated using the following formula:

$$R = \frac{E_t - E_b}{G_0} \times 100$$

where:

R = the percent of gastroesophageal reflux
E_t = the esophageal counts at time t when the reflux is maximal
E_b = the esophageal background counts at the beginning of the study
G_0 = the stomach counts at the beginning of the study

The upper limit of normal is considered to be 3% reflux. Over 4% is considered abnormal.[87] Attention should also be paid to the lungs. If aspiration is present during the study, there will be radiotracer accumulation in the lungs.

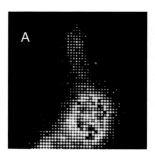 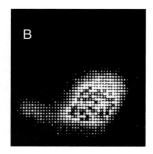

FIGURE 23-27 Gastroesophageal scintigraphy displayed on an oscilloscope of a data processor. (A) This study demonstrates reflux above the stomach into the esophagus. (B) An asymptomatic normal volunteer demonstrating gastric activity and filling of the pylorus but no reflux cephalad. (Reprinted with permission of Elsevier Science from reference 66.)

Gastric Emptying

The stomach can be divided into four regions: the fundus, body, antrum, and pylorus. The fundus and proximal body of the stomach predominantly serve as a storage reservoir for food. The distal stomach and antrum demonstrate peristaltic contractions that break down the food into small particles and mix them with gastric secretions. The pyloric sphincter acts as a sieve, allowing particles smaller than 1 mm to enter the duodenum with each peristaltic contraction.[92] Liquids empty from the stomach in an exponential fashion. However, solids demonstrate a lag phase while the food is being broken down into smaller particles. The lag phase is then followed by a more linear emptying phase.

Gastric emptying disorders fall into two different categories: delayed emptying and abnormally fast emptying (dumping). Delayed gastric emptying symptoms can include nausea, vomiting, abdominal discomfort, bloating, and early satiety. There are many potential causes for delayed gastric emptying, including endocrine disorders such as diabetes mellitus, mechanical obstruction, postsurgical effects, certain drugs, and idiopathic factors. Nuclear medicine gastric emptying studies are commonly ordered to evaluate for diabetic gastroparesis. Dumping can be associated with gastric surgery such as gastrectomy, antrectomy, and pyloroplasty that allows the passage of larger particles into the small bowel. In these patients, a dumping syndrome can occur after eating, with symptoms of dizziness, weakness, nausea, vomiting, sweating, and palpitations.

Scintigraphic evaluation of gastric emptying can be done with solids, liquids, or a combination of the two. In practice, gastric emptying studies are almost always done with a solid meal because the solid phase is more sensitive than the liquid phase for detecting delayed gastric emptying. The patient should fast for 8 hours before the study. Often patients are instructed to fast after midnight and the study is done in the morning. Typically, 1.0 mCi (37MBq) ^{99m}Tc-SC is mixed with an egg, which is subsequently cooked and served scrambled. This is served either alone or as an egg sandwich.[67] For infants 0.5 mCi (18.5 MBq) of ^{99m}Tc-SC can be added to milk or formula. The patient should eat the radiolabeled food quickly, and imaging should start immediately after completion of the meal. The patient is positioned so that anterior images of the

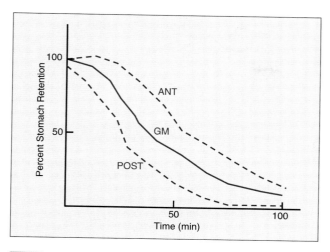

FIGURE 23-28 Gastric emptying curve. Stomach activity after a radioactive meal (1 mCi [37 MBq] of ^{99m}Tc-SC with scrambled egg), acquired by simultaneous recording of anterior and posterior counts over time. The geometric mean curve (GM) corrects for differences in depth attenuation of counts acquired from the anterior and posterior projections.

abdomen that include the distal esophagus, stomach, and small bowel can be obtained.[91] The patient can be standing or supine during the study. If geometric means of the stomach counts are to be calculated, a dual-headed camera can be used to obtain anterior and posterior images of the stomach. The geometric mean $\left(\sqrt{\text{anterior counts} \times \text{posterior counts}}\right)$ corrects for attenuation differences between the anterior and posterior counts (Figure 23-28). Alternatively, the left anterior oblique (LAO) viewing angle can be used to acquire images. In this projection, the stomach contents move roughly parallel to the head of the gamma camera, minimizing the effect of attenuation. This method has the advantages of simple acquisition and no need for mathematical correction; however, it is not as accurate as the geometric mean method.[93] Dynamic 1-minute-per-frame images can be obtained for at least 90 minutes. Alternatively, static images can be obtained every 15 minutes. Regions of interest are drawn around the stomach in each of the images, and a gastric time–activity curve is obtained to calculate percentage of gastric emptying.

The rate of gastric emptying depends on many variables and on the type of meal, patient position, and protocol used. Ideally, normal curves should be established in the institution performing the study. In general, for a solid gastric emptying study, there should be at least 50% emptying by 90 minutes (Figures 23-29 through 23-31).[84]

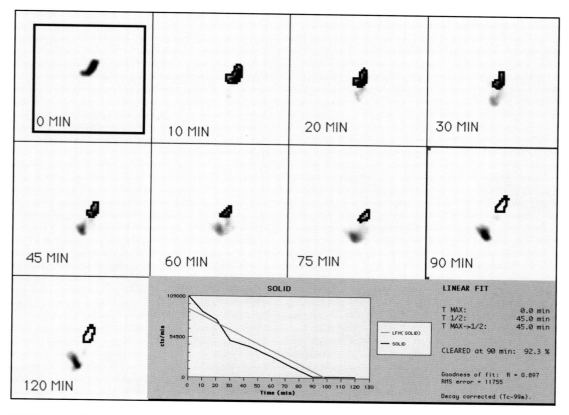

FIGURE 23-29 Normal gastric emptying study in a 2 month old girl. Serial anterior images of the abdomen were obtained for 120 minutes after oral administration of 0.5 mCi (18.5 MBq) of ^{99m}Tc-SC in 50 mL of formula. In the initial image, labeled 0 min, all of the radiolabeled formula is seen in the stomach. Regions of interest are drawn around the stomach in the subsequent images to obtain a gastric time–activity curve (inset). There was 92.3% gastric emptying by 90 minutes.

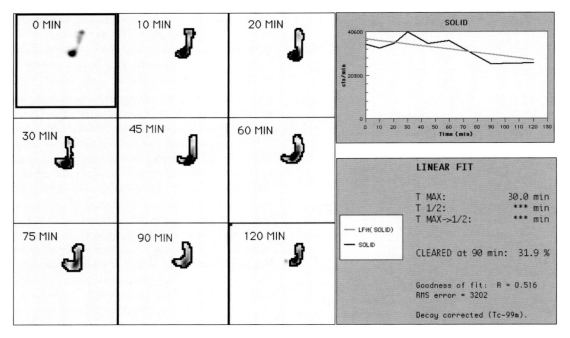

FIGURE 23-30 Delayed gastric emptying in a 50 year old patient. Serial anterior images of the abdomen were obtained out to 120 minutes after the patient ingested a ^{99m}Tc-SC-labeled scrambled egg. On the initial image, labeled 0 min, all of the radiolabeled egg is in the stomach. Regions of interest drawn around the stomach in the subsequent images show delayed gastric emptying, with only 31.9% gastric emptying by 90 minutes.

Dietary changes, including the reduction of extra fat and bulk and frequent small meals, can be useful in the treatment of both diabetic and nondiabetic gastroparesis. Prokinetic drugs can be used to improve gastric emptying. Metoclopramide has been used for this purpose.[94] Erythromycin can also be used to increase gastric emptying in both diabetic and nondiabetic gastroparesis (Figure 23-32).[95,96] Follow-up gastric emptying studies can be performed after initiation of medical therapy to assess the effectiveness of treatment.

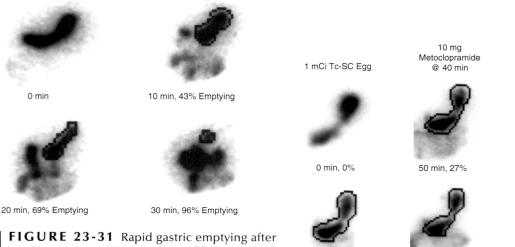

FIGURE 23-31 Rapid gastric emptying after 1 mCi (37 MBq) of ^{99m}Tc-SC scrambled egg meal. Regions of interest drawn around the stomach at various times represent activity remaining in the stomach, which is subtracted from the baseline stomach activity at time zero (100%) to give the percent emptying.

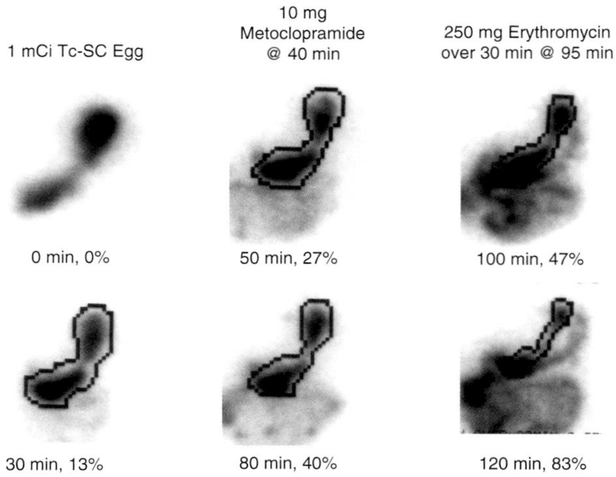

FIGURE 23-32 Delayed gastric emptying with pharmacologic intervention. Images at various time points represent percent gastric emptying.

REFERENCES

1. Guyton AC. *Circulatory Physiology II: Dynamics and Control of the Body Fluids*. Philadelphia: WB Saunders; 1975.
2. Elias H, Sherrick JC. *Morphology of the Liver*. New York: Academic Press; 1969.
3. Wisse E. Ultrastructure and function of Kupffer cells and other sinusoidal cells. In: Wisse E, Knook DL, eds. *Kupffer Cells and Other Liver Sinusoidal Cells*. Amsterdam: Elsevier North Holland; 1977:33–60.
4. Wisse E, DeZanger RB. On the morphology and other aspects of Kupffer cell function: observations and speculations concerning pinocytosis and phagocytosis. In: Liehr H, Grun M, eds. *The Reticuloendothelial System and the Pathogenesis of Liver Disease*. Amsterdam: Elsevier North Holland; 1979:3–9.
5. Rouiller CH. *The Liver: Morphology, Biochemistry, Physiology*. Vol 1. New York: Academic Press; 1963.
6. Loberg MD, Porter DW, Ryan JW. Review and current status of hepatobiliary imaging agents. In: Sorenson JA, ed. *Radiopharmaceuticals II: Proceedings of the Second International Symposium on Radiopharmaceuticals*. New York: Society of Nuclear Medicine; 1979:519–43.
7. Ho RH, Kim RB. Transporters and drug therapy: implications for drug disposition and disease. *Clin Pharmacol Ther*. 2005;78:260–77.
8. Shaffer EA. Gallbladder motility in health and disease. In: Feldman M, LaRusso NF, eds. *Gastroenterology and Hepatology*. Vol 6: Gallbladder and Bile Ducts. Philadelphia: Current Medicine; 1997:5.1–33.
9. Marzio L, Neri M, Capone F, et al. Gallbladder contraction and its relationship to interdigestive duodenal motor activity in normal human subjects. *Dig Dis Sci*. 1988;33:540–4.
10. Dobson EL, Jones HB. The behavior of intravenously injected particulate material: its rate of disappearance from the blood stream as a measure of liver blood flow. *Acta Med Scand*. 1952;144(suppl 273):1–71.
11. Stirrett LA, Yuhl ET, Libby RL. The hepatic radioactivity survey. *Radiology*. 1953;61:930–3.
12. Stirrett LA, Yuhl ET, Cassen B. Clinical applications of hepatic radioactivity surveys. *Am J Gastroenterol*. 1954;21:310–7.
13. Sorensen LB, Archambault M. Visualization of the liver by scanning with Mo-99 (molybdate) as tracer. *J Lab Clin Med*. 1963;62:330–40.
14. Benacerraf B, Biozzi G, Halpern B, et al. A study of phagocytic activity of the reticuloendothelial system toward heat denatured human serum albumin tagged with I-131. *Res Bull*. 1956;2:19–29.
15. Richards P. A survey of the production at Brookhaven National Laboratory of radioisotopes for medical research. In: *V. Congresso Nucleare*. Vol 2. Rome: Comitato Nazionale Ricerche Nucleari; 1960: 223–44.
16. Harper PV, Lathrop KA, Richards P. Tc-99m as a radiocolloid [abstract]. *J Nucl Med*. 1964;5:382.
17. Stern HS, McAfee JG, Subramanian G. Preparation, distribution and utilization of technetium Tc-99m sulfur colloid. *J Nucl Med*. 1966;7:665–75.
18. Ege GN, Warbick A. Lymphoscintigraphy: a comparison of Tc-99m antimony sulfide colloid and Tc-99m stannous phytate. *Br J Radiol*. 1979;52:124–9.
19. Goodwin DA, Stern HS, Wagner HN Jr, et al. In-113m colloid: a new radiopharmaceutical for liver scanning. *Nucleonics*. 1966;24:65–9.
20. Subramanian G, McAfee JG, Mehta A, et al. Tc-99m stannous phytate: a new in-vivo colloid for imaging the reticuloendothelial system [abstract]. *J Nucl Med*. 1973;14:459.
21. Saba TM. Physiology and physiopathology of the reticuloendothelial system. *Arch Intern Med*. 1970;126:1031–52.
22. Guyton AC. *Textbook of Medical Physiology*. 6th ed. Philadelphia: WB Saunders; 1981.
23. Nelp WB. An evaluation of colloids for RES function studies. In: Subramanian G, Rhodes B, Cooper JF, et al., eds. *Radiopharmaceuticals*. New York: Society of Nuclear Medicine; 1975:349–56.
24. Wagner HN, Iio M. Studies of the reticuloendothelial system (RES). III: blockade of the RES in man. *J Clin Invest*. 1964;43:1525–32.
25. Quinones JD. Localization of technetium-sulfur colloid after RES stimulation [abstract]. *J Nucl Med*. 1973;14:443–4.
26. Srivastava SC, Richards P. Technetium-labeled compounds. In: Rayudu GUS, ed. *Radiotracers for Medical Applications*. Boca Raton, FL: CRC Press; 1981:90–2.
27. Davis MA, Jones AG, Trindade H. A rapid and accurate method for sizing radiocolloids. *J Nucl Med*. 1974;15:923–8.
28. Atkins HL, Hauser W, Richards P. Factors affecting distribution of Tc-99m sulfur colloid. *J Nucl Med*. 1969;10:319–20.
29. Cohen Y, Ingrand J, Caro RA. Kinetics of the disappearance of gelatin protected radiogold colloids from the blood stream. *Int J Appl Radiat Isot*. 1968;19:703–5.
30. Chaudhuri TK, Evand TC, Chaudhuri TK. Autoradiographic studies of distribution in the liver of Au-198 and Tc-99m sulfur colloids. *Radiology*. 1973;109:633–7.
31. Brucer M. *Liver Scans, Clearances, and Perfusions*. New York: Krieger; 1977.
32. Bissell DM, Hammaker L, Schmid R. Liver sinusoidal cells: identification of a subpopulation for erythrocyte catabolism. *J Cell Biol*. 1972;54:107–19.
33. Easton TW. The role of macrophage movements in the transport and elimination of intravenous thorium dioxide in mice. *Am J Anat*. 1952;90:1–33.
34. Castronovo FP, Wagner HN. Comparative toxicity and pharmacodynamics of ionic indium chloride and hydrated indium oxide. *J Nucl Med*. 1973;14:677–82.
35. Atkins HL. Adverse reactions. In: Rhodes BA, ed. *Quality Control in Nuclear Medicine*. St Louis: Mosby; 1977:263–7.
36. Haney TA, Ascanio I, Gigliotti JA, et al. Physical and biological properties of a Tc-99m sulfur colloid preparation containing disodium edetate. *J Nucl Med*. 1971;12:64–8.
37. Sulfur colloid kit [package insert]. Bedford, MA: Pharmalucence; 2008.
38. MIRD report no. 3: summary of current radiation dose estimates to humans with various liver conditions from Tc-99m sulfur colloid. *J Nucl Med*. 1975;16:108A.
39. Taplin GV, Meredith OM, Kade H. The radioactive (I-131 tagged) rose bengal uptake: excretion test for liver function using external gamma-ray scintillation counting techniques. *J Lab Clin Med*. 1955;45:665–75.
40. Loberg MD, Cooper M, Harvey E, et al. Development of new radiopharmaceuticals based on N-substitution of iminodiacetic acid. *J Nucl Med*. 1976;17:633–8.
41. Wistow BW, Subramanian G, Van Heertum RL, et al. An evaluation of Tc-99m labeled hepatobiliary agents. *J Nucl Med*. 1977;18:455–61.
42. Wistow BW, Subramanian G, Gagne GM, et al. Experimental and clinical trials of new Tc-99m-labeled hepatobiliary agents. *Radiology*. 1978;128:793–4.
43. Nunn AD, Loberg MD, Conley RA, et al. The development of a new cholescintigraphic agent, Tc-SQ 26,962, using a structure-distribution relationship approach. *J Nucl Med*. 1981;22:P51.
44. Ryan J, Cooper M, Loberg M, et al. Technetium-99m-labeled N-(2,6-dimethylphenyl carbamoylmethyl) iminodiacetic acid (Tc-99m HIDA): a new radiopharmaceutical for hepatobiliary imaging studies. *J Nucl Med*. 1977;18:997–1004.
45. Loberg MD, Fields AT. Stability of Tc-99m labeled N-substituted iminodiacetic acids: ligand exchange reaction between Tc-99m HIDA and EDTA. *Int J Appl Radiat Isot*. 1977;28:687–92.
46. Firnau G. Why do Tc-99m chelates work for cholescintigraphy? *Eur J Nucl Med*. 1976;1:137–9.
47. Hirom PC. The physicochemical factors required for the biliary excretion of organic cations and anions. *Biochem Soc Trans*. 1974;2:327–9.
48. Nunn AD, Loberg MD. Hepatobiliary agents. In: Spencer RP, ed. *Radiopharmaceuticals: Structure-Activity Relationships*. New York: Grune & Stratton; 1981:539–48.
49. Annaert P, Swift B, Lee JK, et al. Drug transport in the liver. In: You G, Morris ME, eds. *Drug Transporters: Molecular Characterization and Role in Drug Disposition*. Hoboken, NJ: Wiley; 2007:359–500.
50. Ghibellini G, Leslie EM, Pollack GM, et al. Use of Tc-99m mebrofenin as a clinical probe to assess altered hepatobiliary transport: integration of in vitro, pharmacokinetic modeling, and simulation studies. *Pharm Res*. 2008;8:1851–60.

51. Fritzberg AR, Klingensmith WC III. Quest for the perfect hepatobiliary radiopharmaceutical. *J Nucl Med.* 1982;23:543–6.
52. Krishnamurthy S, Krishnamurthy T. Quantitative assessment of hepatobiliary diseases with Tc-99m-IDA scintigraphy. In: Freeman LM, Weissmann HS, eds. *Nuclear Medicine Annual 1988.* New York: Raven Press; 1988:309–30.
53. Klingensmith WC III, Fritzberg AR, Spitzer VM, et al. Work in progress: clinical evaluation of Tc-99m-trimethylbromo-IDA and Tc-99m-diisopropyl-IDA for hepatobiliary imaging. *Radiology.* 1983; 146:181–4.
54. Tc-99m disofenin [package insert]. Bedford, MA: Pharmalucence; 2008.
55. Tc-99m mebrofenin [package insert]. Princeton, NJ: Bracco Diagnostics Inc, 2000.
56. Fink-Bennett, D. Augmented cholescintigraphy: its role in detecting acute and chronic disorders of the hepatobiliary tree. *Semin Nucl Med.* 1991;21:128–39.
57. Ziessman HA. Cholecystokinin cholescintigraphy: victim of its own success? *J Nucl Med.* 1999;40:2038–42.
58. Ziessman HA, Tulchinsky M, Lavely WC, et al. Sincalide-stimulated cholescintigraphy: a multicenter investigation to determine optimal infusion methodology and gallbladder ejection fraction normal values. *J Nucl Med.* 2010;51:277–81.
59. Weissman HS, Sugarman LA, Freeman LM. The clinical role of technetium-99m iminodiacetic acid cholescintigraphy. In: Freeman LM, Weissman HS, eds. *Nuclear Medicine Annual 1981.* New York: Raven Press; 1981:35–90.
60. Fink-Bennett D. Hepatobiliary imaging. In: Sandler MP, Patton JA, Coleman RE, et al., eds. *Diagnostic Nuclear Medicine.* 3rd ed. Baltimore: Williams & Wilkins; 1997:759–72.
61. Erslev AJ, Gabuzda TG. *Pathophysiology of Blood.* 2nd ed. Philadelphia: WB Saunders; 1979:15–18.
62. Williams WJ, Beutler E, Erslev AJ, et al. *Hematology.* New York: McGraw-Hill; 1972:13–14.
63. Smith TD, Richards P. A simple kit for the preparation of Tc-99m labeled red blood cells. *J Nucl Med.* 1976;17:126–32.
64. Atkins HL, Goldman AG, Fairchild RG, et al. Splenic sequestration of Tc-99m labeled, heat treated red blood cells. *Radiology.* 1980; 136:501–3.
65. Wagner HJ Jr, Razzak MA, Gaertner RA, et al. Removal of erythrocytes from the circulation. *Arch Intern Med.* 1962;110:90–7.
66. Malmud LS, Fisher RS. Radionuclide studies of esophageal transit and gastroesophageal reflux. *Semin Nucl Med.* 1982;12:104–15.
67. Donohoe KJ, Maurer AH, Ziessman HA, et al. Adult solid-meal gastric-emptying study 3.0. In: *Procedure Guidelines Manual.* Reston, VA: Society of Nuclear Medicine; 2009. www.snm.org/guidelines. Accessed October 16, 2010.
68. Christian PE, Datz FL, Sorenson JA, et al. Technical factors in gastric emptying studies: teaching editorial. *J Nucl Med.* 1983;24:264–8.
69. Malmud LS, Fisher RS, Knight LC, et al. Scintigraphic evaluation of gastric emptying. *Semin Nucl Med.* 1982;12:116–25.
70. Alavi A. Detection of gastrointestinal bleeding with Tc-99m sulfur colloid. *Semin Nucl Med.* 1982;12:126–38.
71. Winzelberg GG, McKusick KA, Frolich JW, et al. Detection of gastrointestinal bleeding with Tc-99m labeled red blood cells. *Semin Nucl Med.* 1982;12:139–46.
72. Bunker S, Lull R, Tanasescu D, et al. Scintigraphy of gastrointestinal hemorrhage: superiority of Tc-99m red blood cells over Tc-99m sulfur colloid. *Am J Roentgenol.* 1984;143:543–8.
73. Ford PV, Bartold SP, Fink-Bennett DM, et al. Gastrintestinal bleeding and Meckel's diverticulum scintigraphy 1.0. In: *Procedure Guidelines Manual.* Reston, VA: Society of Nuclear Medicine; 1999. www.snm.org/guidelines. Accessed October 16, 2010.
74. Kilpatrick ZM. Scanning in diagnosis of Meckel's diverticulum. *Hosp Pract.* 1974;9:131–8.
75. Sfakianakis GN, Conway JJ. Detection of ectopic gastric mucosa in Meckel's diverticulum and in other aberrations by scintigraphy. I: Pathophysiology and 10 year experience. *J Nucl Med.* 1981;22:647–54.
76. Sfakianakis GN, Conway JJ. Detection of ectopic gastric mucosa in Meckel's diverticulum and in other aberrations by scintigraphy. II: Indications and methods—a 10 year experience. *J Nucl Med.* 1981;22: 732–8.
77. Stanley MM, Cerniak G. Protein-losing enteropathy. In: Rothfeld D, ed. *Nuclear Medicine In Vitro.* Philadelphia: JB Lippincott; 1974:341.
78. Mabry CC, Greenlaw RH, DeVore WD. Measurements of gastrointestinal loss of plasma albumin: a clinical and laboratory evaluation of chromium-51 labeled albumin. *J Nucl Med.* 1965;6:93–108.
79. Rubini ME, Sheehy TW, Johnson CR. Exudative enteropathy. 1: A comparative study of Cr-51 chloride and I-131 PVP. *J Lab Clin Med.* 1961;58:892–901.
80. Chiu NT, Lee BF, Hwang SJ, et al. Protein-losing enteropathy: diagnosis with ^{99m}Tc-labeled human serum albumin scintigraphy. *Radiology.* 2001;219:86–90.
81. Royal HD, Brown ML, Drum DE, et al. Hepatic and splenic imaging 3.0. In: *Procedure Guidelines Manual.* Reston, VA: Society of Nuclear Medicine; 2003:53–7. www.snm.org/guidelines. Accessed October 16, 2010.
82. Mettler FA, Guiberteau MJ. Gastrointestinal tract. In: *Essentials of Nuclear Medicine Imaging.* 4th ed. Philadelphia: WB Saunders, 1988:237–84.
83. Tulchinsky M, Ciak BW, Delbeke D, et al. Hepatobiliary scintigraphy 4.0. In: *Procedure Guidelines Manual.* Reston, VA: Society of Nuclear Medicine; 2010:1–20. www.snm.org/guidelines. Accessed October 16, 2010.
84. Hassan F, Naddaf S, Al-Enizi E, et al. Gastrointestinal tract. In: Elgazzar AH, ed. *The Pathophysiologic Basis of Nuclear Medicine.* New York: Springer-Verlag; 2001:284–93.
85. Dodds WJ, Dent J, Hogan WJ, et al. Mechanisms of esophageal reflux in patients with reflux esophagitis. *N Engl J Med.* 1982;307:1547–52.
86. DuMoulin GC, Paterson DG, Hedley-White J, et al. Aspiration of gastric bacteria in antacid-treated patients: a frequent cause of postoperative colonization of the airway. *Lancet.* 1982;1:242–5.
87. Fisher RS, Malmud LS, Roberts GS, et al. Gastroesophageal (GE) scintiscanning to detect and quantitate GE reflux. *Gastroenterology.* 1976;70:301–8.
88. Chernow B, Johnson LF, Janowitz WR, et al. Pulmonary aspiration as a consequence of gastroesophageal reflux. *Dig Dis Sci.* 1979;24:839–44.
89. Heyman S, Kirkpatrick JA, Winton HS, et al. An improved radionuclide method for the diagnosis of gastroesophageal reflux and aspiration in children (milk scan). *Radiology.* 1979;131:479–2.
90. Rudd TG, Christie DL. Demonstration of gastroesophageal reflux in children by radionuclide gastroesophagography. *Radiology.* 1979;131: 483–6.
91. Donohoe KJ, Maurer AH, Ziessman HA, et al. Procedure guideline for gastric emptying and motility 1.0. In: *Procedure Guidelines Manual 2001–2002.* Reston, VA: Society of Nuclear Medicine; 2001;37–40.
92. Meyer JH, Ohashi H, Jehn D, et al. Size of liver particles emptied from the human stomach. *Gastroenterology.* 1981;80:1489–96.
93. Fahey FH, Ziessman HA, Collen MJ, et al. Left anterior oblique projection and peak-to-scatter ratio for attenuation compensation of gastric emptying studies. *J Nucl Med.* 1989;30:233–9.
94. McCallum RW, Brown RI. Diabetic and nondiabetic gastroparesis. *Curr Treat Options Gastroenterol.* December 1998;1:1–7.
95. Urbain JLC, Vantrappen G, Janssens J, et al. Intravenous erythromycin dramatically accelerates gastric emptying in gastroparesis diabeticorum and normals and abolishes the emptying discrimination between solids and liquids. *J Nucl Med.* 1990;31:1490–3.
96. Urbain JLC, Vehemans MCM, Malmud LS. Esophageal transit, gastroesophageal reflux, and gastric emptying. In: Sandler MP, Patton JA, Coleman RD, et al., eds. *Diagnostic Nuclear Medicine.* 3rd ed. Baltimore: Williams & Wilkins; 1997:733–47.

CHAPTER 24
Kidney

The kidney is the primary organ for removing metabolic waste products, drugs, and their metabolites from the blood. Radiopharmaceuticals have been designed to exploit renal excretion mechanisms to provide noninvasive methods of assessing kidney function and morphologic alterations caused by disease. A unique feature of renal nuclear medicine studies is the ability to assess both individual kidney function and overall renal function. This is particularly useful for diagnosing diseases that may involve only one kidney, such as renovascular hypertension and urinary obstruction. Scintigraphic studies have been developed to assess glomerular and tubular function, to detect parenchymal abnormalities, and to measure relative function between the left and right kidneys. Genitourinary problems, including ureteral obstruction, residual urine after voiding, and vesicoureteral reflux, can also be evaluated. Pharmacologic agents are also used to enhance the diagnostic power of renal studies. Although structural abnormalities of the kidney are best evaluated by imaging modalities such as computed tomography (CT) or magnetic resonance imaging (MRI), renal nuclear medicine studies play a key role is assessing functional abnormalities.

PHYSIOLOGIC PRINCIPLES

A bisected kidney reveals two distinct regions: a pale inner region, the medulla, and a dark outer region, the cortex. The medulla is divided into several conical masses, the renal pyramids. The apex of each pyramid is oriented toward the renal pelvis. The cortex forms a cap over the base of each pyramid and descends downward between pyramids to form the renal columns of Bertin.

The functional unit of the kidney is the nephron, illustrated in Figure 24-1. It consists of the glomerulus, proximal convoluted tubule, loop of Henle, distal convoluted tubule, and a collecting duct that empties into the renal pelvis.[1] Tubular urine flows through the nephron in that order. There are approximately 1.5 million nephrons in each kidney. About 85% of the nephrons reside in the kidney cortex, with their glomeruli located in the outer two-thirds; about 15% of the nephrons are juxtamedullary, with glomeruli in the corticomedullary junction; and no glomeruli are found in the kidney medulla. In cortical nephrons, the loops of Henle penetrate only a short distance into the medulla and have thin descending and thick ascending limbs; in juxtamedullary nephrons, the loops of Henle are long, penetrate deep into the medulla, and return to the cortex.

Approximately 20% to 25% (about 1200 mL) of the cardiac output flows through the kidneys. About 20% of this amount is filtered by the glomeruli, producing a filtrate that is essentially identical to plasma minus the protein. Blood enters the nephron through the afferent arteriole, which branches off the interlobular artery, subsequently flows through the glomerular capillaries, exits the glomerulus through the efferent arteriole, and enters the peritubular capillaries that bathe the tubules. Blood then leaves the nephron through the cortical

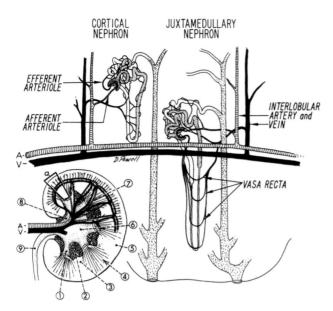

FIGURE 24-1 Diagram of the sagittal surface of a bisected kidney (lower left). Numbers 1 through 9 indicate the following: 1, minor calyx; 2, fat in sinus; 3, renal column of Bertin; 4, medullary ray; 5, cortex; 6, pelvis; 7, interlobar artery; 8, major calyx; and 9, ureter. The letter A indicates the renal artery, and the letter V indicates the renal vein. Insert (a) from the upper pole is enlarged to illustrate the relationships between the juxtamedullary and the cortical nephrons and the renal vasculature. (Reprinted with permission of Elsevier Science from reference 1.)

venule, which flows into the interlobular vein. About 85% of renal blood perfuses the cortex and bypasses the medulla, 10% perfuses the corticomedullary junction, 2% perfuses the medulla, and 3% flows through arteriovenous shunts that bypass the cortical glomerular and peritubular capillary systems and does not contribute to glomerular filtration or tubular secretion and reabsorption.

RENAL EXCRETION MECHANISMS

Renal elimination of drug substances from the blood and into urine involves transport processes that typically occur across biologic membranes: passive diffusion, facilitated diffusion, and active transport. These processes are described in Chapter 9. Renal excretion mechanisms have been discussed in depth.[2] In general, substances are removed from the blood by glomerular filtration and tubular secretion. Reabsorption processes also occur in the tubule, where essential substances in the glomerular filtrate, such as glucose, electrolytes, and water, are reabsorbed into the blood.

A substance must satisfy two basic criteria to enter the glomerular filtrate: It must be non–protein bound and it must have a small molecular size. The glomerular capillaries are about 100-fold more permeable to water and salt than capillaries in general; however, substances with molecular weights greater than 50,000 are almost completely excluded from the glomerular filtrate. Non–protein-bound drugs with molecular weights less than 5000 are freely filterable.

Any drug that enters the glomerular filtrate will be excreted in the urine if it is not reabsorbed into the blood. In general, un-ionized, lipid-soluble drugs are freely diffusible, whereas ionized, water-soluble species remain in the urine unless they are reabsorbed by a carrier-mediated transport process.[3] Therefore, substances used to assess the glomerular filtration rate (GFR) should possess strong polar properties that favor urinary excretion, as opposed to nonpolar properties that favor passive tubular reabsorption.

The degree of ionization of a drug depends on its pKa and the urinary pH, in accordance with the Henderson-Hasselbalch equation. For drugs that are weak organic acids, this equation is expressed as follows:

$$pH = pKa + \log \frac{\text{ionized drug}}{\text{un-ionized drug}}$$

Consider, for example, a weak-acid drug with a pKa of 6. At a urinary pH of 5 it is 90% un-ionized, favoring passive tubular reabsorption, whereas at a urinary pH of 7 it is 90% ionized, favoring urinary excretion. Since most drugs are either weak acids or weak bases, their renal elimination is affected by the urine's pH.

Active tubular secretion is known to occur by at least two independent pathways involving either organic anions or cations. These pathways are utilized by endogenous compounds such as choline, creatinine, epinephrine, and dopamine and by drugs such as atropine, cimetidine, and morphine.[4,5] Of importance to nuclear medicine is the active tubular secretion of organic anions such as ^{131}I-o-iodohippurate (^{131}I-OIH) and ^{99m}Tc-mertiatide (^{99m}Tc-MAG3). Several compounds that are secreted by the anionic pathway fit a general structural requirement proposed by Despopoulos:[6] possession of a primary anionic binding site such as a carboxylic acid (CO_2^-) or a sulfonic acid (SO_3^-) group and a secondary site provided by a carboxyl oxygen or hydroxyl group.[7] A few compounds that meet these requirements include p-aminohippurate (PAH), hydroxybenzoates, fatty acids, penicillin, probenecid, salicylates, and the renal radiopharmaceuticals ^{131}I-OIH and ^{99m}Tc-MAG3.

Glomerular filtration of a drug requires that it be non–protein bound and freely distributed in plasma water, but elimination of a drug by active tubular secretion is less affected by protein binding, provided the binding is reversible.[4] If a drug exhibits reversible protein binding, all of the drug will be available for tubular transport because as free drug is removed from plasma by the tubular cells, bound drug dissociates rapidly into plasma water and becomes immediately available for elimination.

Renal Clearance

Extraction of a drug from blood by the kidney can be quantitated by measuring renal clearance of the drug, defined as the minimal volume of plasma required to supply the amount of drug excreted in the urine in a given period of time.[8] The clearance concept is applicable to all substances excreted in the urine and is expressed by the formula

$$Cl_x = \frac{U_x \cdot V}{P_x} \qquad (24\text{-}1)$$

where:

U_x = urine concentration of drug
P_x = plasma concentration of drug
V = urine volume per unit time
Cl_x = renal clearance in milliliters per unit time

Consider, for example, the clearance of Na⁺. Given that V = 1 mL/minute, U_{Na} = 280 mEq/L, and P_{Na} = 140 mEq/L, the clearance of sodium (Cl_{Na}) is 2 mL/minute. In other words, an amount of sodium ion equal to that contained in 2 mL of plasma (0.28 mEq) is excreted into the urine each minute. Another way to describe clearance, therefore, is that it is the volume of plasma that has its amount of drug (Na⁺ in this case) completely removed per unit time. The clearance volume is virtual, however, not real, and one should not infer that all of a drug (Na⁺ in this case) is removed from each 2 mL of plasma passing through the kidney. On the contrary, only some of the sodium is removed from a much larger volume of plasma perfusing the kidney.

Drug clearance occurs by glomerular filtration, tubular secretion, or a combination of these processes. Reabsorptive processes work against clearance. If a drug is known to be cleared by only one particular process, such as glomerular filtration, then the drug can be used to measure that function. Normal GFR is about 125 mL/minute per 1.73 m² in humans and represents about 20% of the renal plasma flow (RPF). The mean RPF is about 650 mL/minute. A small molecular weight substance that is not bound to plasma protein can enter the glomerular filtrate. If it is not metabolized, reabsorbed, or secreted by the tubule, it will remain in the urine with a clearance of 125 mL/minute. The polysaccharide inulin best satisfies these requirements and is the standard substance used to measure GFR. Thus, during steady-state intravenous infusion of inulin, when GFR is normal the amount of inulin contained in 125 mL of plasma will appear in the urine each minute as a result of glomerular filtration. A disease process that affects glomerular function will change the inulin clearance. Hence, inulin clearance can be used to assess kidney function in a variety of conditions.

Although the clearance of a drug does not delineate a specific mechanism involved in its renal excretion, some idea of the processes involved can be gained. For example, if a freely filterable drug has a clearance less than the GFR, then it must undergo some degree of tubular reabsorption. Glucose is such a substance, being completely reabsorbed by the tubule. Its clearance in the normal individual, therefore, is zero. On the other hand, if a substance has a clearance greater than the GFR, it must undergo tubular secretion as well. PAH is an example of such a substance because its renal clearance is nearly equal to the RPF.

Extraction Efficiency

Renal clearance provides information only about the amount of drug removed from the blood that appears in the urine. It is not applicable to substances that are stored, synthesized, or metabolized by the kidney.[9] On the other hand, the extraction efficiency or extraction ratio (ER) is a measure of the amounts of drug eliminated into the urine and retained by the kidney. ER is defined by Smith[10] as the fraction of a substance removed from plasma during one circulation through the kidney. It is expressed as follows:

$$ER = \frac{A - V}{A} \qquad (24\text{-}2)$$

where:

A = the renal arterial concentration of a substance
V = the renal venous concentration of a substance

Consider, for example, a drug that has an ER of 1.0. It will be completely removed in a single pass through the kidney. If none of the drug appears in the urine, then all of it is retained by the kidney. A radiopharmaceutical with this property would be an ideal renal imaging agent because all of the injected activity would end up being bound in the kidney. No radiopharmaceutical comes close to having these properties, but ⁹⁹ᵐTc-succimer (⁹⁹ᵐTc-DMSA) has a significant amount of the injected dose retained in the kidney. On the other hand, if all of the drug appears in the urine, none is retained in the kidney and its clearance is equal to the RPF (650 mL/minute). A radiopharmaceutical with this property would be an ideal agent for measuring renal plasma flow and tubular function, because most of the activity entering the urine would occur by tubular secretion. Two radiopharmaceuticals that closely approximate these properties are ¹³¹I-OIH and ⁹⁹ᵐTc-MAG3.

If a drug has an ER of 0.2 and none of it is bound, secreted, or reabsorbed, it will have a clearance of 125 mL/minute, making it a good GFR agent. ¹²⁵I-iothalamate and ⁹⁹ᵐTc-pentetate (⁹⁹ᵐTc-DTPA) are radiopharmaceuticals that have essentially these properties, and they both are used to quantitate GFR. Figure 24-2 and Table 24-1 summarize these concepts.

In renal physiology, the clearance of PAH (Cl_{PAH}) is used to estimate the renal plasma flow. However, its clearance is not equal to the renal plasma flow, because its ER (E_{PAH}) is about 0.92. Therefore, the clearance of PAH measures only the effective renal plasma flow (ERPF).[10] The ERPF of PAH in humans averages 600 mL/minute. RPF can be determined from the clearance and extraction ratio of PAH by the following relationship:

$$RPF = \frac{Cl_{PAH}}{E_{PAH}} = \frac{600 \, mL/min}{0.92} = 650 \, mL/min$$

Renal blood flow (RBF) can be determined from RPF and the hematocrit by the relationship RPF/(1 − hematocrit).[10]

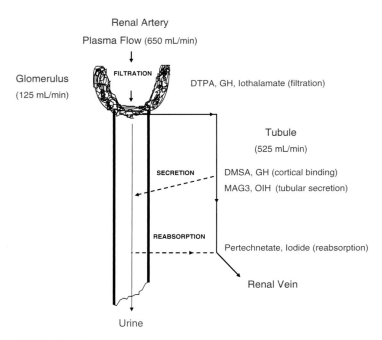

FIGURE 24-2 Schematic of the nephron, demonstrating the sites of radiopharmaceutical processing in the kidney. Normally, 20% (125 mL/min) of the renal plasma flow undergoes filtration at the glomerulus, carrying non–protein-bound small molecular weight substances into the tubular urine, from which they can undergo tubular reabsorption (pertechnetate and iodide) or no reabsorption (DTPA, gluceptate [GH], and iothalamate). Approximately 80% (525 mL/min) of the renal plasma flow bypasses the glomerulus and perfuses the renal tubules, from which substances can undergo transport into the tubular urine (OIH and MAG3) or be extracted and partially bound in the tubule (DMSA and GH).

Hence, an average individual with a hematocrit of 45% would have RBF of approximately 1200 mL/minute.

The reasons for incomplete extraction of PAH have been examined. Wesson[11] lists several possible causes: (1) parenchymal bypass, in which a small amount of blood that enters the renal vein does not pass through the tubules; (2) red blood cell transport, in which drug that enters red blood cells does not readily diffuse out during blood transit through the cortical tubules; (3) incomplete cortical extraction due to incomplete drug dissociation from plasma protein; and (4) unavailability of drug from blood that passes through the medulla.

OIH also exhibits renal extraction of less than 100% for reasons similar to those given for PAH. Because of free iodide present in OIH, which undergoes tubular reabsorption, clearance of OIH is less than clearance of PAH. OIH is no longer used in renal studies.

Drug Elimination

The renal clearance tells us the amount of drug that is excreted in the urine per unit time, but it tells us nothing about the fraction of total drug that is eliminated by the kidney. For this we need to know the drug's volume of distribution (V_d).[4] Consider, for example, a hypothetical class of drugs that are eliminated solely by renal excretion. After an intravenous bolus injection of a dose (I_0), the rate of change in plasma concentration with respect to time (dC/dt) is given by the equation

TABLE 24-1 Ideal Properties of Renal Radiopharmaceuticals

Extraction Ratio	Clearance (mL/min)	Significance
1.0	0	All bound in kidney (good renal anatomic agent)
1.0	650	None bound in kidney (good renal tubular agent)
0.2	125	None bound, secreted, or reabsorbed (good glomerular filtration agent)

$$\frac{dC}{dt} = -k \cdot C \qquad (24\text{-}3)$$

where k is the elimination rate constant determined by the ratio of clearance to apparent volume of distribution:

$$k = \frac{\text{Cl}_R}{V_d} \qquad (24\text{-}4)$$

The plasma concentration of drug immediately after injection (C_0) is

$$C_0 = \frac{I_0}{V_d} \qquad (24\text{-}5)$$

Integration of Equation 24-3 yields

$$C = C_0 e^{-kt} \quad \text{or} \quad C = \frac{I_0}{V_d} e^{-kt} \qquad (24\text{-}6)$$

This relationship between plasma concentration and time can be linearized by taking the natural logarithm of both sides of the equation:

$$\ln C = \ln C_0 - k \cdot t \qquad (24\text{-}7)$$

The clearance of a drug (Cl_R) is the product of renal extraction efficiency (ER) and renal plasma flow (Q_p):

$$\text{Cl}_R = Q_p \cdot \text{ER} \qquad (24\text{-}8)$$

This set of equations forms the basis for determining a drug's elimination rate and the influence that clearance and volume of distribution have on the rate.

Consider, for example, two hypothetical drugs, A and B, that are excreted entirely in the urine. V_d is smaller for drug A than for drug B, but the extraction efficiencies of the two drugs are identical (i.e., renal clearance is identical). If equimolar doses of the two drugs are administered, Equation 24-5 predicts that the initial concentration of A will exceed that of B. On the other hand, Equations 24-4 and 24-7 predict that the plasma concentration of A will decline more rapidly than that of B. The result of this situation is shown on the left in Figure 24-3. Now, consider a case in which the extraction efficiency (and renal clearance) of A is decreased (e.g., because of high plasma protein binding) such that the ratio of extraction efficiencies (B to A) is equal to the ratio of distribution volumes (B to A). Equations 24-4 and 24-7 then predict that the plasma concentration of A will decline at the same rate as the plasma concentration of B. The result of this situation is shown on the right in Figure 24-3. Thus, it can be shown that the rate of decline in plasma concentration (slope of k) is determined by both the volume of distribution and the clearance.

These concepts, particularly the second hypothetical situation described, explain why the elimination rates of ^{131}I-OIH and ^{99m}Tc-MAG3 are similar even though their biologic properties are different. A comparison of these agents is made later in the chapter.

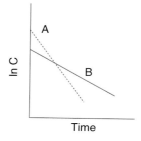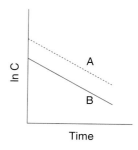

FIGURE 24-3 Plasma elimination rates for two hypothetical drugs, A and B, eliminated entirely by renal excretion and administered in equimolar doses. Left: Drugs A and B have identical renal extraction efficiencies (identical clearances) but drug A has a smaller volume of distribution than B. The elimination rate of drug A is higher because its initial plasma concentration is higher because of a smaller volume of distribution. Right: Drug A has a lower extraction efficiency than drug B (i.e., lower clearance) and the ratio of extraction efficiencies (B to A) is equal to the ratio of the distribution volumes (B to A). Drug A is now eliminated at the same rate as drug B. Even though the extraction of drug A is less than drug B, its plasma concentration is higher (because of a smaller distribution volume), which compensates for its lower clearance.

Often, elimination rates are discussed in terms of a drug's half-life. Substituting $0.693/T_{1/2}$ for k and rearranging Equation 24-4 yields

$$T_{1/2} = \frac{0.693 \cdot V_d}{\text{Cl}} \qquad (24\text{-}9)$$

This relationship demonstrates that as the volume of distribution of a drug increases, its half-life of elimination increases proportionately (i.e., its excretion rate declines). Goldstein et al.[4] have shown that, theoretically, if a drug were distributed only in plasma water ($V_d = 3$ L) and completely cleared per pass through the kidney (ER = 1.0, Cl = 650 mL/minute), its plasma elimination half-life would be 3 minutes. If its distribution were in extracellular fluid ($V_d = 12$ L) or the total body water ($V_d = 41$ L), its half-life would be 13 minutes or 44 minutes, respectively. Hence, any radiotracer distributed only in plasma water and having a high renal extraction will be cleared most rapidly from the bloodstream.

KIDNEY IMAGING AGENTS
Development

A large number of radiopharmaceuticals have been investigated for renal imaging studies. The chemical properties of agents routinely used in renal scintigraphy are described in Chapter 10. Several review articles have compared renal imaging agents.[5,7,12–14]

The radioisotope renogram was introduced by Taplin et al.[15] in 1956 in response to the need for a method of evaluating unilateral kidney disease in a noninvasive manner. The original method used a two-channel detection system, each kidney having its own scintillation detector connected to a rate meter and recorder. After intravenous injection of an agent cleared primarily by tubular secretion, time–activity curves were recorded over each kidney. Trials with different radioactive compounds and failures to radiolabel PAH with a gamma emitter eventually led to the development of ^{131}I-OIH (hippuran) in 1960 by Tubis et al.[16] OIH is structurally similar to PAH but can be labeled with radioiodine. It is eliminated primarily by tubular secretion, and for over 30 years it was the agent of choice for assessing renal function. Because OIH was not bound in the kidney, it could not be used for imaging kidney morphology.

The lack of an agent that would remain fixed in the kidney for imaging space-occupying lesions led to the development of ^{203}Hg- and ^{197}Hg-labeled chlormerodrin. These mercury compounds bind to sulfhydryl groups in tubular cell proteins, permitting visualization of kidney morphology.[17] However, the high radiation dose from ^{203}Hg and the poor resolution scans from ^{197}Hg prompted the development of a ^{99m}Tc-labeled agent.

In 1966, Harper et al.[18] developed a ^{99m}Tc-iron ascorbate complex for renal imaging, which permitted the ability to administer large amounts of activity with improved sensitivity and resolution and low radiation dose. This was followed by ^{99m}Tc complexes with diethylenetriaminepentaacetic acid (DTPA) labeled with either iron ascorbate or stannous ion as the reducing agent.[19,20] The iron-reduced complex did not behave biologically as a true chelate, and stannous-reduced ^{99m}Tc-pentetate (^{99m}Tc-DTPA) became the agent of choice for GFR studies.[20]

In the early 1970s ^{99m}Tc-glucepate (^{99m}Tc-GH) and ^{99m}Tc-DMSA were developed. Both agents bind to the renal cortex, making them good choices for imaging the renal parenchyma. ^{99m}Tc-DMSA reaches high concentrations in the renal cortex but is cleared slowly from the blood into urine because of high plasma protein binding, making it a poor choice for evaluating the pelvocaliceal collecting system.[21] Because of lower protein binding, ^{99m}Tc-GH demonstrates the collecting system well soon after injection, similar to ^{99m}Tc-DTPA, whereas later images show renal parenchyma alone, similar to ^{99m}Tc-DMSA.[13] The fraction of ^{99m}Tc-GH bound in kidney (~12%) is lower than the fraction of ^{99m}Tc-DMSA (~40%).

In 1965 Sigman et al.[22] used ^{131}I-iothalamate to measure GFR. This agent was subsequently shown to have renal clearance identical to that of inulin.[23] Iothalamate labeled with ^{125}I has remained the radioactive agent of choice for quantitative assessment of GFR over the years, although ^{99m}Tc-DTPA is also used for this purpose.

^{111}In-DTPA and ^{99m}Tc-DTPA have similar biologic properties in humans but are not identical.[24] The ^{111}In complex has a slightly faster total-body clearance. The slightly greater retention of ^{99m}Tc-DTPA in tissues probably represents ^{99m}Tc not chelated with DTPA. ^{99m}Tc-DTPA also slightly underestimates GFR (by a few percentage points) because of a small amount of protein binding, so corrections are required for its use in quantitative GFR measurements.

The first technetium complex designed to replace ^{131}I-OIH was ^{99m}Tc-N,N'-bis(mercaptoacetyl) ethylenediamine (^{99m}Tc-DADS), but it proved to be unsatisfactory, particularly in patients with increased creatinine levels.[25–27] The addition of a carboxylate group to the ethylenediamine backbone of ^{99m}Tc-DADS improved renal excretion but created an asymmetric carbon atom resulting in two chelate ring stereoisomers, ^{99m}Tc-CO_2-DADS-A and ^{99m}Tc-CO_2-DADS-B.[6] The A isomer had renal excretion properties similar to those of ^{131}I-OIH, but the B isomer did not. The requirement for high-performance liquid chromatographic (HPLC) purification to separate the A and B isomers precluded development of an easily usable kit, which limited the potential usefulness of ^{99m}Tc-CO_2-DADS in the nuclear medicine clinic.

The development work with ^{99m}Tc-DADS demonstrated that a carboxylate group was necessary for high specificity of the renal tubular transport system. To prevent the formation of stereoisomers from substituents placed on the N_2S_2 backbone of DADS, the ligand was changed to N_3S or triamide monomercaptide. Changing the core donor ligand and placement of the carboxyl group on the third amido nitrogen produced a radiochemically pure product without an asymmetric carbon.[28] The simplest ligand having the necessary groups for renal excretion was mercaptoacetyltriglycine (MAG3). The absence of stereoisomers permitted a kit formulation that could be easily used in the nuclear medicine clinic. Chapter 10 describes the chemistry of ^{99m}Tc-MAG3 in detail.

Animal and clinical studies have shown that the renal excretion profile of ^{99m}Tc-MAG3 is essentially identical to that of ^{131}I-OIH, and ^{99m}Tc-MAG3 has replaced ^{131}I-OIH as the renal function agent of choice in routine nuclear medicine studies.

Biologic Properties

Radiopharmaceuticals for renal studies belong to two principal groups: (1) renal clearance agents, for assessing GFR and tubular function; and (2) renal imaging agents, for assessing renal morphology and relative function (Table 24-2). The agents most commonly used are ^{99m}Tc-MAG3, ^{99m}Tc-DMSA, ^{99m}Tc-DTPA, and ^{125}I-iothalamate. ^{131}I-OIH, no longer used, is included for comparison.

Technetium Tc 99m Pentetate Injection

Technetium Tc 99m pentetate injection (^{99m}Tc-DTPA) is a complex of Tc(IV) with diethylenetriaminepentaacetic acid (DTPA). Its low plasma protein binding (around 5%) and insignificant red blood cell binding produces rapid renal excretion, with 50% of the dose in the urine 2 hours after injection and 96% excreted by 24 hours.[13,24] The complex is quite stable and is excreted unchanged in the urine. It is

TABLE 24-2 Renal Radiopharmaceuticals

Renal Clearance Agents
1. Glomerular filtration agents
   a. ^{125}I-iothalamate
   b. ^{99m}Tc-pentetate (^{99m}Tc-DTPA)
2. Tubular secretion agents
   a. ^{131}I-o-iodohippurate (^{131}I-OIH)
   b. ^{99m}Tc-mertiatide (^{99m}Tc-MAG3)

Renal Imaging Agents
1. ^{99m}Tc-gluceptate (^{99m}Tc-GH)
2. ^{99m}Tc-succimer (^{99m}Tc-DMSA)

excreted by glomerular filtration, is not secreted or reabsorbed by the kidney tubule, and has no appreciable binding to the renal parenchyma. Its renal distribution is restricted to the renal vascular space initially and the tubular urine thereafter. After an intravenous dose of 10 mCi (370 Mbq), ^{99m}Tc-DTPA is useful in evaluating gross blood flow to the kidneys and visualizing obstructions to urine flow in the collecting system and ureters. Its routine use for these applications, however, has diminished with the availability of ^{99m}Tc-MAG3.

The renal clearance of ^{99m}Tc-DTPA is approximately 20% of renal plasma flow (i.e., 125 mL/minute), making it useful for the quantitative assessment of GFR.[29–34] Methods that measure GFR by plasma clearance of ^{99m}Tc-DTPA are reliable when GFR is greater than 30 mL/minute.[33] The method used should employ a correction for plasma protein binding of ^{99m}Tc-DTPA.[30,31] While most methods are relatively simple in principle, they require careful attention to technique. In addition, it is important to report GFR values that are normalized to body surface area of 1.73 m².[34]

Technetium Tc 99m Succimer Injection

Technetium Tc-99m succimer injection (^{99m}Tc-DMSA) is a complex of Tc(III) with 2,3-dimercaptosuccinic acid (DMSA). Several ^{99m}Tc-DMSA complexes can be formed, depending on the labeling conditions.[35,36] Complex I is formed at a pH of 2.5, with a DMSA:Sn ratio of 3 and the absence of oxygen. It reverts to complex II within 10 minutes of incubation. Complex II localizes in the renal cortex, making it useful for morphologic imaging. If ^{99m}Tc-DMSA is prepared at a higher pH, renal cortical concentration decreases.[37] Unstabilized preparations of ^{99m}Tc-DMSA are useful for 30 minutes, while the product prepared from a stabilized kit is useful for 4 hours. Chapter 10 provides details on chemistry and preparation.

After intravenous injection in humans, ^{99m}Tc-DMSA becomes loosely bound to plasma protein (75% at 1 hour after injection, increasing to 90% by 24 hours), with little or no diffusion into red blood cells.[13,38] Renal excretion is slow, with only 16% of the dose in urine 2 hours after injection. The tracer accumulates slowly in the renal cortex, where it becomes fixed, primarily in the cells of the proximal convoluted tubule.[21,39] Kidney micropuncture studies demonstrate that its mechanism of localization is via extraction from peritubular blood and fixation within the cortical cells.[40,41] Evidence suggests that this occurs by an active process that is not inhibited by probenecid; thus, ^{99m}Tc-DMSA apparently localizes by a transport system different from that of other tubular agents (e.g., ^{131}I-OIH and ^{99m}Tc-MAG3).[41,42] Although a small amount of non–protein-bound ^{99m}Tc-DMSA is filtered, this fraction is not believed to be reabsorbed.[40] Subcellular localization within cortical cells has been reported to be primarily within cytosol proteins[43] and microsomes.[44] In rats with acidic urine[42] and patients with proximal tubular acidosis,[45] renal uptake of ^{99m}Tc-DMSA declines significantly.

A maximum of about 40% of an injected dose of 5 mCi (185 MBq) is eventually bound in the two kidneys 6 hours after injection.[13] Since the uptake half-life for renal accumulation of ^{99m}Tc-DMSA is 1 hour, imaging is best performed 4 to 5 hours after injection, but it can be done as early as 2 hours. The high degree of protein binding limits the amount of activity that is filtered, precluding visualization of the collecting system. The low urinary excretion and high cortical binding of ^{99m}Tc-DMSA make it an excellent agent for detecting focal abnormalities in the renal cortex, and it is useful for assessing relative function between right and left kidney. Normal distribution in the kidney demonstrates high uptake of activity within the cortical regions. Extension of the relatively active cortical columns of Bertin into the colder regions of the medulla gives the kidneys a unique pattern of hot/cold activity distribution that must be appreciated in reading DMSA scans. Lesions such as tumors and cysts appear as cold areas within an otherwise hot kidney. An alternative route of elimination for ^{99m}Tc-DMSA is the hepatobiliary system, which can be visualized in renal failure patients studied with this agent (Figure 24-4).

Technetium Tc 99m Gluceptate Injection

Technetium Tc-99m gluceptate injection (^{99m}Tc-GH) is a complex of Tc(V) and the complexing agent glucoheptonic acid. Its chemical structure and method of preparation are discussed in Chapter 10.

After intravenous injection, ^{99m}Tc-GH distributes into the extracellular space. It is loosely bound to plasma protein (50% to 75% 1 to 6 hours after injection), with no significant binding to red blood cells.[13] Renal excretion is rapid,

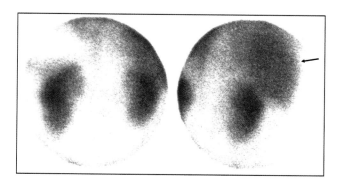

FIGURE 24-4 Renal scan with ^{99m}Tc-succimer (^{99m}Tc-DMSA), demonstrating liver uptake (arrow) secondary to renal failure.

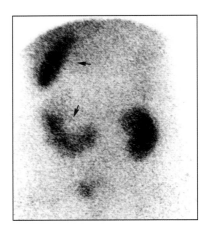

FIGURE 24-5 Hepatobiliary excretion (arrows) of ^{99m}Tc-gluceptate (^{99m}Tc-GH) in a renal transplant patient.

with 50% of the dose in urine 2 hours after injection and 71% excreted by 24 hours. Its renal mechanism is glomerular filtration and tubular secretion.[46] Soon after injection, the renal distribution of ^{99m}Tc-GH resembles that of ^{99m}Tc-DTPA, being limited to the vascular space initially, followed by a significant amount of activity accumulating in the collecting system. In later images, ^{99m}Tc-GH differs from ^{99m}Tc-DTPA in that about 12% of the injected dose is retained in the cortex because of tubular binding.[13] Thus, ^{99m}Tc-GH has the versatility of visualizing the collecting system early after injection, similar to ^{99m}Tc-DTPA, and the renal parenchyma at later times, similar to ^{99m}Tc-DMSA. Compared with ^{99m}Tc-DMSA, however, a smaller fraction of the injected dose of ^{99m}Tc-GH localizes in the kidney, and thus its 10 to 15 mCi (370–555 MBq) dosage is proportionately larger. Clinically, therefore, ^{99m}Tc-GH is useful for evaluating renal perfusion, obstructive uropathy, relative kidney function, and renal masses. An alternative route of elimination for ^{99m}Tc-GH is via the hepatobiliary system, but this is not typically seen except in patients with severe renal insufficiency (Figure 24-5). ^{99m}Tc-GH is not widely used in the United States.

Technetium Tc 99m Mertiatide Injection

Technetium Tc-99m mertiatide injection (^{99m}Tc-MAG3) is a complex of Tc(V) and mercaptoacetyltriglycine.[28,47] Its chemical structure and a detailed method of preparation are discussed in Chapter 10. The primary advantage of ^{99m}Tc-MAG3 is that it exhibits the rapid excretion of ^{131}I-OIH but has the more desirable imaging properties of ^{99m}Tc.

After intravenous injection, ^{99m}Tc-MAG3 clears quickly from the blood pool in a biphasic pattern. The first component clears with a $T_{1/2}$ of 3.2 minutes, and the second component with a $T_{1/2}$ of 16.9 minutes.[48] Although ^{99m}Tc-MAG3 is highly protein bound in plasma (~90%), restricting its filtration fraction, the binding is reversible, permitting rapid renal excretion via tubular secretion.[14,48,49] Plasma concentrations of ^{99m}Tc-MAG3 in humans are around 1.5 times higher than those of ^{131}I-OIH, but its clearance is 55% to 65% that of ^{131}I-OIH. However, both agents have similar rates of loss over time, with about 70% of the dose excreted in the urine 30 minutes after injection and greater than 90% by 180 minutes.[14,49] Although ^{99m}Tc-MAG3's clearance is less than that of ^{131}I-OIH, because of a lower extraction efficiency, its plasma concentration is higher because of a smaller volume of distribution.[49] The lower clearance is balanced by the smaller volume of distribution (Equation 24-4), giving ^{99m}Tc-MAG3 an elimination rate equal to that of ^{131}I-OIH.[48,49]

The renogram curves for ^{99m}Tc-MAG3 and ^{131}I-OIH are similar, with a mean time to peak of 3 to 5 minutes for both agents.[50] The tracer is not bound by the kidney, nor does it have significant red blood cell uptake. Properties of ^{131}I-OIH and ^{99m}Tc-MAG3 are listed in Table 24-3, and the properties of all renal imaging radiopharmaceuticals are summarized in Table 24-4.

Because of its mechanism of excretion, ^{99m}Tc-MAG3 is an excellent agent for visualizing the renal collecting system, evaluating urinary obstruction, and assessing renal tubular function. Studies have shown that, in general, ^{99m}Tc-MAG3 clearance is proportional to that of ^{131}I-OIH but that some disproportionate differences occur clinically.[14,49] Thus, it has been suggested that clearances of the two agents should be reported directly and not be related to each other by a conver-

TABLE 24-3 Properties of ^{99m}Tc-MAG3 and ^{131}I-OIH

Property	^{99m}Tc-MAG3	^{131}I-OIH	^{99m}Tc-MAG3/^{131}I-OIH Ratio	Ref
Plasma protein binding (%)	87.5 ± 2.6	66.2 ± 6.9	1.32	48
	90.1 ± 2.8	70.7 ± 5.0	1.27	49
RBC uptake (%)	5.1 ± 3.3	15.3 ± 4.1	0.33	49
Tubular extraction coefficient[a]	0.55	0.83	0.66	49
Plasma clearance, Cl (mL/min)	420 ± 120	600 ± 100	0.70	48
	265 ± 98	412 ± 169	0.64	49
Volume of distribution, V (mL)	5210	7030	0.74	48
	3380	5540	0.62	49
Elimination rate, k (min^{-1})	0.0806	0.0853	0.94	48
	0.0784	0.0744	1.05	49

[a] Fraction of drug extracted as a function of its clearance and protein binding.

TABLE 24-4 Biologic Properties of Renal Imaging Radiopharmaceuticals

Property	99mTc-DTPA	99mTc-GH	99mTc-DMSA	99mTc-MAG3	131I-OIH
Glomerular filtration	Yes	Yes	Not significant[a]	Not significant[a]	Yes
Tubular transport	No	Yes (cortical binding)	Yes (cortical binding)	Yes (secretion)	Yes (secretion)
Tubular reabsorption	No	No	No	No	No
Collecting system	Yes	Yes	No	Yes	Yes
Cortical binding	No	Yes (~12% ID[b])	Yes (~40% ID[b])	No	No
Dosage	10 mCi (blood flow) 3 mCi (renogram)	10 mCi (blood flow) 15 mCi (static)	5 mCi (static)	5 mCi (blood flow) 1–3 mCi (renogram)	75 µCi (1 kidney) 200 µCi (2 kidneys) (renogram)
Critical organ	Bladder wall	Bladder wall	Kidney cortex	Bladder wall	Bladder wall
rad(cGy)/mCi (4.8 hr void)	0.27	0.28	0.85	0.48	5.71

[a] 85% to 90% plasma protein binding.
[b] Injected dose.

sion factor when these agents are used to quantitate ERPF.[14] Although the renogram curves of the two agents are similar, superior image quality and anatomic detail are achieved with 99mTc-MAG3 because of its 99mTc label. In the past, renal perfusion in kidney transplants was evaluated with a combination of 99mTc-DTPA for RBF and 131I-OIH for tubular function. The use of 99mTc-MAG3 allows both determinations to be made with one agent, simplifying the procedure and improving overall image quality. The relatively high renal extraction of 99mTc-MAG3 (around 50%) compared with 99mTc-DTPA (around 20%) provides superior images in patients with impaired renal function.

Hepatobiliary excretion has been reported to occur occasionally with 99mTc-MAG3, and it increases in patients with impaired renal function. Investigational studies suggest that this occurrence can be associated with radiopharmaceutical impurities that can vary in a radiolabeled kit, depending on labeling conditions.[51] Kits prepared in a 10 mL final volume produced less impurities than kits prepared in a 5 mL volume; this suggests a simple way to improve radiochemical purity.

Iothalamate Sodium I 125 Injection

Iothalamate sodium I 125 injection (125I-iothalamate; Glofil-125 [Iso-Tex Diagnostics]) has been used for many years as a diagnostic agent for measuring GFR. Its renal clearance closely approximates that of inulin, and 125I-iothalamate is cleared by glomerular filtration without tubular reabsorption or secretion. 125I-iothalamate is not a renal imaging agent, because 125I photons (27 to 35 keV) are almost entirely absorbed by the tissues.

Iodohippurate Sodium I 131 Injection

After intravenous administration, iodohippurate sodium I 131 injection (131I-OIH) rapidly disappears from the blood by diffusion into the extracellular space and through renal elimination. It is not metabolized in the body and is excreted unchanged in the urine.[52] It is not bound to renal tubular cells.[9] In the normal hydrated subject, 70% of a single dose of 131I-OIH is excreted in the urine 30 minutes after injection.[53]

Renal excretion of 131I-OIH is primarily by active tubular secretion and a fraction by glomerular filtration, but the exact amounts excreted by each process are not well defined. Approximately 70% of 131I-OIH is protein bound in plasma.[49] The overall renal extraction efficiency is less than 100%, varying between 65% and 85%.[9] The wide range of reported values depends on the conditions of measurement and several intrinsic factors, such as binding to blood components, the presence of radioiodide or other impurities in the injected dose, and reabsorption by the tubule.[9,10,49,54,55]

131I-OIH has been used to measure ERPF by several different methods.[33,34,53] The procedure demands attention to technique. Technical errors, such as inaccuracy in dose preparation, incomplete injection of the dose, inattention to the plasma sampling time, and inaccurate sample dilutions, must be avoided.[56]

A simpler study for routine assessment of renal function is the renogram. This procedure has been performed for over 30 years with 131I-OIH, but it is now done primarily with 99mTc-MAG3.

NUCLEAR MEDICINE PROCEDURES

Anatomic imaging using ultrasound, CT, and MRI has an important role in evaluations of the genitourinary system. However, renal scintigraphy continues to play an important role in the evaluation of renal perfusion, renal function, and, in certain cases, anatomic abnormalities. Imaging with radionuclides can provide a combination of both anatomic and physiologic information about the kidneys. In most cases, it is the functional information that makes nuclear imaging unique and important.

Rationale

Renal scintigraphy is performed to provide information about renal perfusion, renal parenchymal function, or function of the collecting system. Hydronephrosis is one of the most common indications for scintigraphic evaluation of the kidneys. Hydronephrosis or obstructive uropathy is dilatation of the pelvis and calices of one or both kidneys, usually identified on an anatomic study. Hydronephrosis can be acute, such as in passage of a calculus with impaction in the ureter, or it may be chronic or congenital. Other common indications for renal and genitourinary scintigraphy include evaluation of acute or chronic renal failure, renal function after trauma, renovascular disease, acute pyelonephritis, cortical scarring in patients with vesicoureteral reflux, differential renal function, postoperative perfusion, RBF, renal agenesis, congenital abnormalities, vesicoureteral reflux, and acute testicular pain, as well as evaluation of potential kidney donors and of kidney transplants.

Procedures

The choice of procedure for evaluation of the kidneys and genitourinary system depends on the indication. The type of study and radiopharmaceutical are chosen to specifically answer questions about the renal or urologic problem being investigated.

Perfusion Studies

Renal perfusion studies are obtained to evaluate blood flow to the kidneys. The patient is placed in the supine position with the gamma camera facing the patient's lower back. A large field-of-view camera equipped with a low-energy, high-resolution parallel hole collimator is used so that the kidneys and bladder can be visualized. The agent of choice for renal blood flow is ^{99m}Tc-MAG3. The radiotracer is administered as a bolus of 3 to 5 mCi (111–185 MBq) into an antecubital vein. One- to five-second-per-frame images are obtained for the first minute.

Renography

Renography refers to a recording of the amount of radioactivity in the kidney or kidneys over time after administration of a radiopharmaceutical that measures renal clearance. A renogram graphically displays the uptake and clearance of the radiotracer in the kidneys in the form of a renal time–activity curve (Figure 24-6). Time–activity curves are calculated by obtaining multiple serial images of the kidneys over 20 to 40 minutes. Regions of interest are drawn around the kidneys, the renal cortices, and the renal pelves in each of the images to calculate the amount of activity in each of these structures at different time points over the length of the study. Renograms are used to evaluate suspected renal obstruction, renal artery stenosis, acute tubular necrosis, and transplant rejection.[57–59]

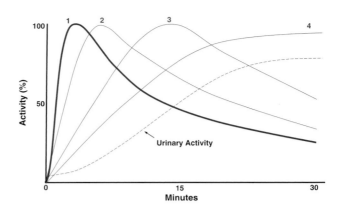

FIGURE 24-6 Normal renogram curve (1) with associated urinary activity and curves depicting progressive kidney deterioration (2, 3, and 4).

Before any renogram study, the patient should be well hydrated. Poor hydration can result in delayed radiotracer uptake and clearance, suggesting poor renal function. Typically, 500 mL of water 15 to 30 minutes before the study is recommended. The patient should void before the start of the study. Normally, the patient is placed in the supine position and posterior images are obtained for native kidneys. If a transplanted kidney is being evaluated, anterior images are obtained, because a transplanted kidney is usually located anteriorly in the iliac fossa.

A perfusion study is typically obtained for the first minute after intravenous injection of 3 mCi (111 MBq) ^{99m}Tc-MAG3. After this, static 1- to 5-minute-per-frame images of the abdomen and pelvis are obtained for the next 20 to 40 minutes to evaluate both uptake and clearance of the radiotracer in the kidneys.

Diuresis Renography

Once hydronephrosis has been established by another imaging modality such as ultrasound, nuclear medicine evaluation of the dilated upper urinary tract is usually accomplished with diuresis renography after intravenous administration of 3 mCi (111 MBq) ^{99m}Tc-MAG3. During the renogram, if there is poor clearance from one or both of the collecting systems after 20 minutes, a diuretic can be given. Furosemide (1 mg/kg in infants, 0.5 mg/kg in children age 1 year to 16 years, and 40 mg in adults) is administered intravenously to stimulate tubular function and increase urine flow. The renogram study is continued for another 20 minutes to determine if the poor clearance initially observed is caused by either a functional or mechanical obstruction. If the obstruction is functional, there is typically a good response with washout of activity from the obstructed kidney. If the obstruction is mechanical, there is a poor response with little or no washout of activity (Figure 24-7).

Captopril Renography

Normally, systemic blood pressure is controlled by the renin–angiotensin system. When there is a drop in systemic blood pressure, there is an associated drop in GFR in both kidneys. As a response to the drop in GFR, there is release of

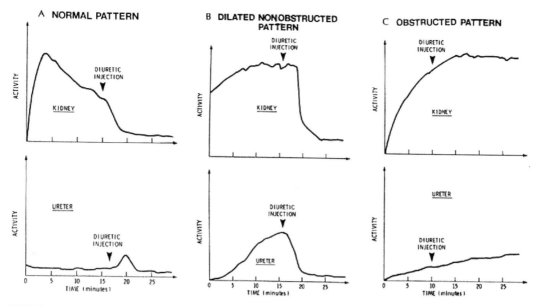

FIGURE 24-7 Representative furosemide renograms showing time–activity curves in a normal patient (A) and in patients with dilated nonobstruction (B) and obstruction (C) of the urinary collecting system. (Reprinted with permission of Elsevier Science from Thrall JH, Koff SA, Keyes JW Jr. Diuretic radionuclide renography and scintigraphy in the differential diagnosis of hydroureteronephrosis. *Semin Nucl Med.* 1981;11:89–104.)

renin from the juxtaglomerular cells in the kidneys. Renin then converts circulating angiotensinogen, which is made in the liver, into angiotensin I. Angiotensin I is then converted to the potent vasoconstrictor angiotensin II by angiotensin-converting enzyme (ACE), predominantly found in vascular endothelium in the lungs. Angiotensin II increases GFR by constricting the efferent glomerular arterioles in the kidneys. The resultant increase in GFR leads to an inhibition of renin release in the juxtaglomerular apparatus. Angiotensin II increases the systemic blood pressure not only by vasoconstriction but also by increasing salt and water reabsorption by the renal tubular cells (Figure 24-8).[60–62]

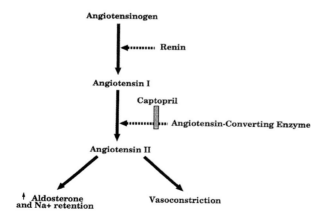

FIGURE 24-8 Schematic of the renin–angiotensin–aldosterone system showing the cascade of events leading to vasoconstriction. (Reprinted with permission of Elsevier Science from reference 60.)

Most patients with hypertension have what is referred to as essential hypertension—hypertension of unknown origin that is treated with lifelong medical management. However, a small subset of patients have hypertension secondary to stenosis of one of the renal arteries.[62] Many of these patients have new-onset hypertension that is difficult to control with medications. If the degree of renal artery stenosis is great enough to significantly decrease GFR, the renin–angiotensin system is activated, resulting in increased systemic blood pressure. An elevation in blood pressure that is caused by renal artery stenosis is referred to as renovascular hypertension (RVH). Renal artery stenosis is often caused by either atherosclerosis or fibromuscular dysplasia. If the patient's hypertension is secondary to renal artery stenosis, revascularization procedures such as balloon angioplasty and stent placement can cure or improve the hypertension.

Radionuclide evaluation of RVH is accomplished with captopril renography. The effect of captopril on split renal function in patients with unilateral renal artery stenosis was described by Wenting et al.[63] Captopril is an ACE inhibitor that works by blocking the conversion of angiotensin I to the potent vasoconstrictor angiotensin II. If significant renal artery stenosis is the cause of the patient's hypertension, inhibiting the effect of angiotensin II on the efferent arterioles in the kidneys will decrease GFR in the affected kidney (Figure 24-9). The drop in GFR will result in delayed uptake of the radiopharmaceutical as well as increased cortical retention in the affected kidney.

Captopril renography can be done as either a 1 day or a 2 day study. The study usually consists of two renograms, one with captopril and a baseline study without captopril. If the patient or ordering physician agrees to a 2 day study, the

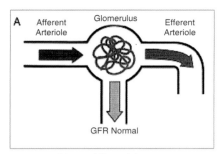
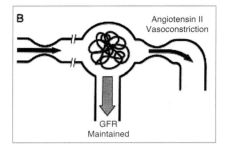
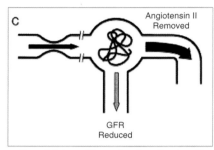

FIGURE 24-9 Schematic of captopril's effect on renal function. (A) normal renal artery and normal GFR; (B) renal artery stenosis, causing angiotensin II-mediated constriction of the efferent arteriole, with GFR maintained; (C) renal artery stenosis plus captopril, demonstrating relaxation of efferent arteriole due to ACE inhibition. (Reprinted with permission of Elsevier Science from reference 60.)

captopril renogram is done first. If the captopril renogram is normal, there is little chance that the patient has RVH, and there is no need to obtain a baseline study. However, if the captopril study is abnormal, a baseline study with the patient off ACE inhibitors is needed for comparison. For a 1 day study, a baseline renogram is usually performed first with a small amount of radiotracer. Typically, 1 mCi (37 MBq) of ^{99m}Tc-MAG3 is used. ^{99m}Tc-DTPA can also be used. After the baseline renogram, there should be a delay of about 1 hour before beginning the captopril study. The delay allows time to clear out possible interfering residual activity in the kidneys from the baseline study. The patient is then given 25 to 50 mg of captopril by mouth. The captopril is crushed and dissolved in water before administration. Patients should be well hydrated for the study. However, except for fluids, patients should not eat for 4 hours before the study, to enhance absorption of the captopril. The patient is then monitored for 1 hour while the blood level of captopril comes to peak, because significant hypotension can be associated with captopril administration. The renogram is then repeated with a larger amount of activity, usually 3 mCi (111 MBq) of ^{99m}Tc-MAG3. The higher dose is used to compensate for any residual activity from the baseline study. An alternative ACE inhibitor to oral captopril is intravenous enalaprilat, which is dosed at 40 μg/kg and administered over 3 to 5 min with a maximum dose of 2.5 mg. The radiopharmaceutical is then administered 15 minutes after the enalaprilat.

The renal time–activity curves for the studies done before and after captopril administration are then compared. With ^{99m}Tc-MAG3, RVH is associated with parenchymal retention of activity. Typical renogram curve patterns associated with parenchymal retention of activity after administration of captopril are shown in Figure 24-6.[62,64] For example, a diagnosis of high probability for unilateral RVH can be made if the ^{99m}Tc-MAG3 renogram pattern before captopril changes from pattern 1 to 3 or from pattern 2 to 4 after captopril. If ^{99m}Tc-DTPA is used for the renogram, RVH is associated with a reduction in relative uptake greater than 10% after ACE inhibition.[65]

Cortical Scintigraphy

Acute pyelonephritis in children usually results from vesicoureteral reflux of infected urine. Renal cortical scintigraphy using tubular fixation agents such as ^{99m}Tc-DMSA or ^{99m}Tc-GH has been shown to be a sensitive technique for diagnosis of acute pyelonephritis and assessment of renal scarring. After intravenous administration of either agent, accumulation in the renal parenchyma is related to the amount of functioning renal tissue. In acute pyelonephritis there is decreased radiotracer accumulation in the affected renal parenchyma as a result of both ischemia and tubular cell dysfunction.[66] If acute pyelonephritis is detected early and treated with appropriate antibiotics, the infection can be healed without scar formation. Repeated infections can lead to renal scarring and eventually to renal failure. Cortical scintigraphy can also be used to evaluate the relative functioning of each kidney.

There is no specific patient preparation for administration of the radiotracer. Typically, ^{99m}Tc-DMSA is used. The dose in children is 0.04 to 0.05 mCi/kg (1.5–1.9 MBq/kg) administered intravenously.[67] An adult dose is 5 mCi (185 MBq). Anterior and posterior images of the kidneys are usually obtained approximately 4 hours later. However, delayed imaging can be obtained up to 24 hours. Oblique images or single-photon emission computed tomography (SPECT) images of the kidneys can also be obtained to help evaluate for cortical defects.

Radionuclide Cystography

Radionuclide cystography is generally accepted as a sensitive test for vesicoureteral reflux. Urinary tract infection (UTI) with vesicoureteral reflux can lead to renal infections that result in scarring and hypertension. Although vesicoureteral reflux can be evaluated with conventional radiographic techniques such as the voiding cystourethrogram, radionuclide cystography results in significantly less gonadal radiation dose. Radionuclide cystography is considered the technique of choice for evaluating UTI in young girls. Radionuclide cystography can also be used for follow-up after surgical intervention or medical management of vesicoureteral reflux.

The most common method for evaluating vesicoureteral reflux involves catheterization of the bladder and instillation of 1 mCi (37 MBq) of ^{99m}Tc-sodium pertechnetate through the catheter into the bladder. The catheter is then attached to a saline drip and the bladder is passively filled to capacity. For the filling phase of the study, the patient is supine with the camera under the patient. Serial posterior 5-second-per-frame images of the abdomen and pelvis are obtained during the filling, maximal fill, and voiding phases of the study. In an older, more cooperative child, the voiding phase is usually accomplished with the child sitting on a bedpan with the camera positioned behind the patient. An image is obtained after voiding. Regions of interest can be drawn around the bladder and the ureters to obtain an estimate of the relative amount of vesicoureteral reflux.

Scrotal Scintigraphy

Although ultrasound is more commonly used to evaluate acute scrotal pain, scrotal scintigraphy can be used as an alternative method. Scrotal scintigraphy can be useful in differentiating acute pain from inflammation (e.g., in epididymitis) from testicular torsion, which is a surgical emergency.

With the patient in the supine position, the testicles must be positioned properly for anterior imaging. Usually, a lead shield is placed behind the testicles to shield underlying activity from the thighs. This can be an appropriately sized piece of lead wrapped in a towel. The penis is taped to the abdomen so that it does not overlie the testicles, and the testicles are positioned on the shield so that side-to-side comparison can be made. After positioning, 2-seconds-per-frame blood flow images are obtained for the first minute after administration of the radiopharmaceutical. In an adult, the dose is usually 10 mCi (370 MBq) of ^{99m}Tc-sodium pertechnetate. Static blood pool images are obtained after the flow phase of the study.

Interpretation

To evaluate renal perfusion, the radiotracer is given as a bolus, usually into an antecubital vein. With the patient in the supine position, posterior 1- to 5-second-per-frame images of the abdomen are obtained for the first minute. Imaging is started just as the bolus of radiotracer is administered. The bolus of activity advances through the subclavian vein into the superior vena cava and then into the right heart. After this, radiotracer is seen in the lungs, and then in the left heart and the aorta. Blood flow to the kidneys should be seen approximately 1 second after the bolus of radiotracer in the abdominal aorta passes the renal arteries. Normally, there should be prompt, symmetric blood flow to both kidneys seen at approximately the same time the abdominal aorta is visualized (Figure 24-10).

There are three distinct phases to the normal renogram. Initially, after administration of the radiotracer there is uptake in the kidneys related to blood flow. Radiotracer is seen in the kidneys within about 1 second after the injected radiotracer in the abdominal aorta passes the renal arteries. This phase occurs over the first minute and is referred to as the vascular phase. The next 2 to 4 minutes are referred to as the parenchymal phase, in which radiotracer is concentrated in the renal parenchyma. Maximal parenchymal activity is usually seen between 3 and 5 minutes after administration of the radiopharmaceutical. The time to peak is the duration between administration of the radiopharmaceutical and the maximum renal cortical activity; it is a measure of renal function. At the

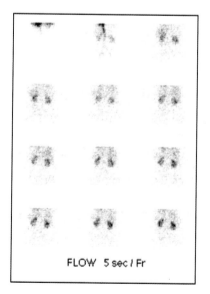

FIGURE 24-10 Normal ^{99m}Tc-MAG3 renal blood flow study. Prompt, symmetric activity is seen in both kidneys. Activity in the kidneys is seen at the same time as in the abdominal aorta.

end of the parenchymal phase, radiotracer accumulation begins to be seen in the collecting systems. The next phase is the excretion phase. During the excretion phase, there should be continued clearance from the renal parenchyma. If there is no obstruction, there also should be continued clearance from the collecting systems (Figure 24-11).

A captopril (or ACE inhibitor) renogram is performed to evaluate for RVH. A positive study is one that demonstrates a significant change in the baseline renogram after administration of an ACE inhibitor such as captopril or enalaprilat. In patients with normal renal function, radiotracer retention in one kidney after administration of captopril is consistent with a high probability for RVH (Figure 24-12). Many of these patients can be cured or significantly improved through balloon angioplasty or stent placement.

When there is a question of obstructive uropathy, a diuretic renogram is performed. When patients present with a dilated collecting system, it is important to determine if the collecting system is obstructed. Hydronephrosis can be the result of mechanical obstruction, which can lead to loss of renal function, or it can be secondary to functional abnormality such as muscle atony. In either the mechanically obstructed kidney or the functionally dilated but non-obstructed kidney, there is a delayed time to peak for the radiotracer as well as little to no clearance prior to administration of a diuretic. In the obstructed kidney, administration of the diuretic has little to no effect on the renal time–activity curve. However, in the case of nonobstructed hydronephrosis, there is usually prompt response to the administration of the diuretic (Figure 24-13). After administration of the diuretic, the kidney usually clears half the radioactivity in less than 10 minutes when there is no significant obstruction. If the time to half-maximum radiotracer is greater than 20 minutes after diuretic administration, the kidney is obstructed. A time to

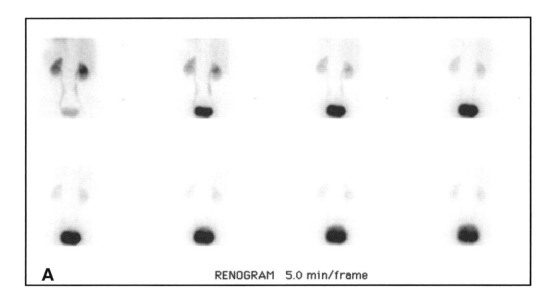

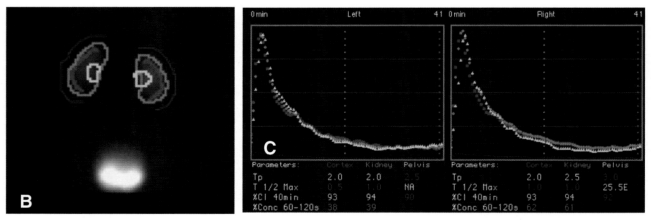

FIGURE 24-11 Normal ^{99m}Tc-MAG3 renogram. (A) Static posterior 5-minute-per-frame images of the kidneys were obtained for 40 minutes. There is good clearance in both kidneys. (B) Regions of interest were drawn around the renal cortex, kidney, and pelvis of each kidney to measure activity over time in these regions, as well as background activity. (C) Renal time–activity curves demonstrate the maximum parenchymal phase activity at approximately 2 minutes. There is good clearance from both kidneys during the study.

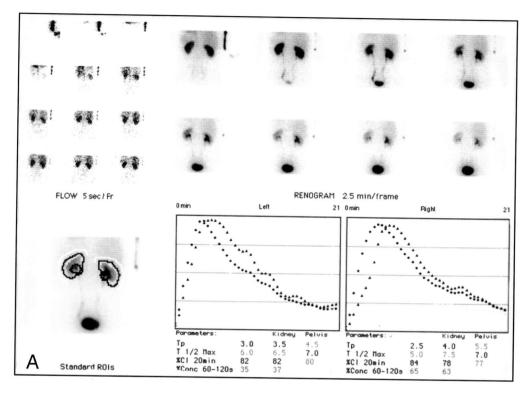
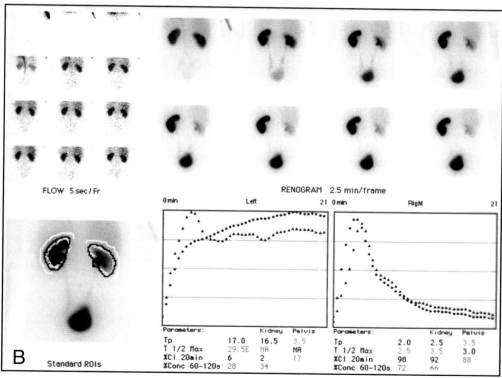

FIGURE 24-12 Captopril renogram with 99mTc-MAG3. (A) Pre-captopril renogram with 1 mCi (37 MBq) of 99mTc-MAG3 shows normal renal time–activity curves in both kidneys. (B) Post-captopril study demonstrates a significant increase in the time to peak and delayed clearance in the left kidney consistent with renal artery stenosis.

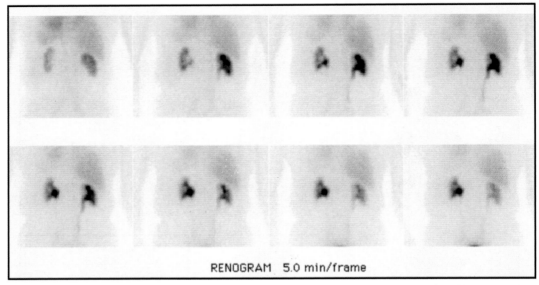
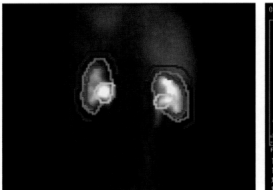
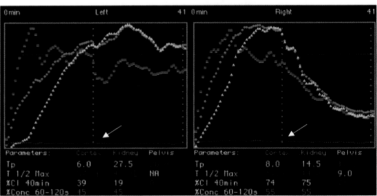

FIGURE 24-13 Diuretic renogram with ^{99m}Tc-MAG3 in a patient with bilateral hydronephrosis. There is a delayed time to peak in both kidneys and no significant clearance from either kidney 20 minutes into the study. A diuretic was administered at 20 minutes (arrows). There is prompt response to the diuretic in the right kidney, with 75% clearance at 40 minutes, and little to no response in the left kidney in this patient with partial obstruction in the left kidney and nonobstructed hydronephrosis in the right kidney.

half-maximum between 10 and 20 minutes is indeterminate for partial obstruction.

Kidney transplants are placed in the anterior iliac fossa. Renograms are sometimes obtained shortly after surgery to determine whether the transplanted kidney is functioning. A complication that is often present soon after transplantation is ischemic damage to the donor kidney. This presents as poor renal function and poor urine output, but with good renal perfusion. This is referred to as acute tubular necrosis and usually resolves over the first 3 weeks. On the renogram it is seen as a delayed time to peak and poor clearance (Figure 24-14). After the first week and most commonly during the first 3 months after transplantation, acute rejection is a concern. On the renogram, acute rejection is seen as delayed renal uptake and reduced excretion with worsening renal perfusion on the flow images.

In a normal renal cortical scan, there is homogeneous radiotracer accumulation in the renal parenchyma of both kidneys. Both kidneys should demonstrate reniform shape. Normal differential renal function can vary from 50% in each kidney to 44% and 56% (Figure 24-15). Defects sometimes can be associated with nonfunctioning renal tissue. Figure 24-16 demonstrates a cortical defect associated with a renal cell carcinoma. Infection such as acute pyelonephritis can be seen as a single cortical defect in one kidney or multiple cortical defects involving one or both kidneys. Scarring is usually associated with volume loss. Scarring may appear as a wedge-shaped cortical defect, a flattening of the renal contour, or a concave defect (Figure 24-17).

Although not routinely used for renal imaging, positron emission tomography with ^{18}F-fludeoxyglucose (^{18}F-FDG) has been used to image renal masses and to assess for distant metastatic involvement. Since ^{18}F-FDG is normally excreted by the kidneys, a diuretic such as furosemide can be used to clear activity from the kidneys prior to imaging (Figure 24-18).

Renal cortical scintigraphy can also be useful in evaluating renal function in congenital renal abnormalities. Figure 24-19 demonstrates lack of function in a multicystic dysplastic kidney.

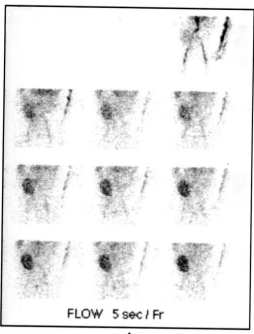

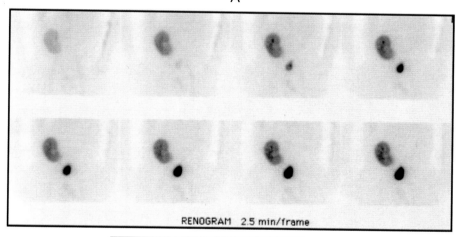

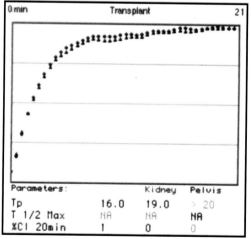

FIGURE 24-14 Renal transplant evaluation using ⁹⁹ᵐTc-MAG3. (A) Anterior 5-second-per-frame blood flow images of the transplant kidney in the right iliac fossa. There is prompt blood flow to the transplant kidney. Activity is seen in the kidney at the same time as in the iliac vessels. (B) Static anterior 2.5-minute-per-frame images over the next 20 minutes and the associated renal time–activity curve demonstrate a delayed time to peak in the kidney and poor clearance, although there is some excretion into the bladder. The patient's transplant surgery was only a few days earlier. These findings are most consistent with acute tubular necrosis.

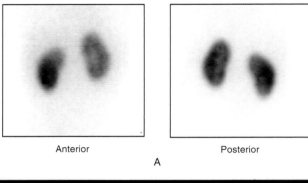
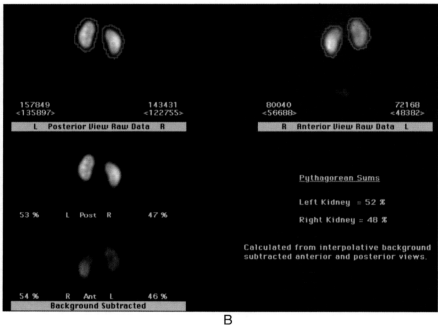

FIGURE 24-15 Normal renal cortical scan using ^{99m}Tc-DMSA. (A) There is relatively homogeneous radiotracer accumulation in both kidneys. There are no cortical defects to suggest infection or scarring. (B) Regions of interest were drawn around each kidney in both the anterior and posterior projections, and geometric means were calculated to determine differential renal function. The differential function was calculated to be 52% on the left and 48% on the right, which is considered normal.

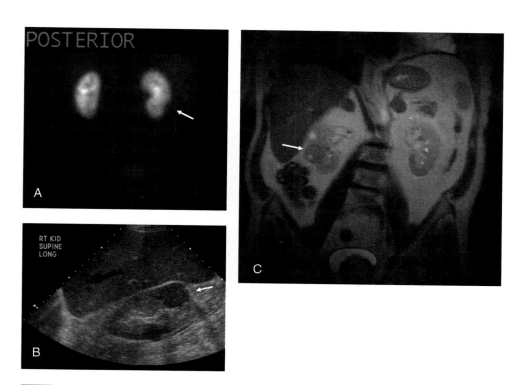

FIGURE 24-16 Renal cell carcinoma. (A) Posterior renal cortical scan with 99mTc-DMSA demonstrating a concave cortical defect in the lateral lower pole of the right kidney (arrow). (B) Renal ultrasound long-axis image of the right kidney showing a corresponding hypoechoic mass in the right kidney (arrow). (C) Coronal magnetic resonance image also demonstrating the mass in the lateral lower pole of the right kidney (arrow).

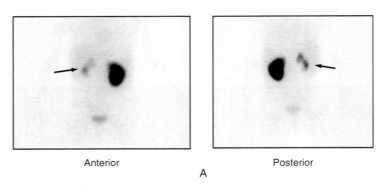
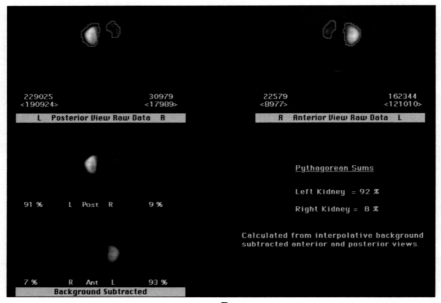

FIGURE 24-17 Renal cortical scan of a 5 year old female with history of urinary tract infections, obtained 6 hours after intravenous administration of 2 mCi (7.4 MBq) of ^{99m}Tc-DMSA. (A) Anterior and posterior images show a small scarred right kidney (black arrows) with a normal-appearing left kidney. (B) The differential renal function was calculated to be only 8% in the right kidney.

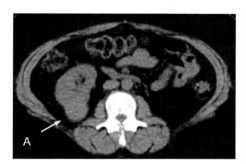
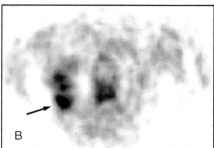
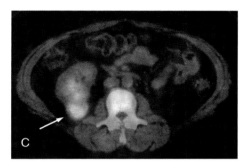

FIGURE 24-18 Renal sarcoma. (A) Axial noncontrast computed tomography (CT) image through the right kidney shows an exophytic soft tissue mass in the posterior aspect of the right kidney (white arrow). (B) A corresponding axial positron emission tomography (PET) image using ^{18}F-fludeoxyglucose shows a focal area of hypermetabolism in the same region (black arrow). (C) Axial PET/CT fusion image demonstrates that the focal area of hypermetabolism corresponds to the exophytic mass (white arrow). The mass was found to be a renal sarcoma.

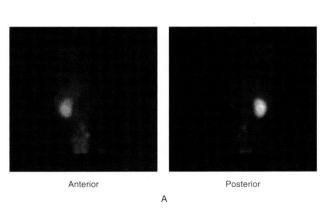
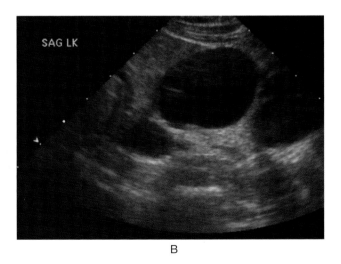

FIGURE 24-19 Multicystic dysplastic kidney in a 5 month old child. (A) ^{99m}Tc-DMSA renal cortical scan shows normal-appearing right kidney with no evidence of a functioning left kidney. (B) Sagittal ultrasound image of the left kidney shows multiple cysts in this patient with multicystic dysplastic kidney.

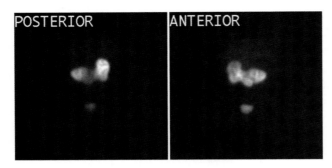

FIGURE 24-20 Horseshoe kidney. Anterior and posterior images from a renal cortical 99mTc-DMSA study demonstrating a union of the lower poles of the kidneys in this patient with a horseshoe kidney.

Figure 24-20 shows a horseshoe kidney joined at the lower poles.

A normal radionuclide cystogram will show radiotracer activity in a normally shaped bladder only during the filling, full bladder, and voiding stages of the exam. On the postvoid image, there should be no significant residual activity in the bladder. In the presence of vesicoureteral reflux, there will be evidence of radiotracer accumulation in one or both of the ureters during the exam. If radiotracer refluxes only into the distal portion of the ureter, this is considered mild or grade 1 reflux. There is moderate or grade 2 reflux if radiotracer is seen to the level of the renal pelvis of what appears to be a nondilated renal collecting system (Figure 24-21). If there is

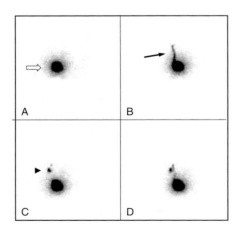

FIGURE 24-21 Vesicoureteral reflux. After catheterization of the bladder, 1 mCi (37 MBq) of 99mTc-sodium pertechnetate was administered into the bladder. (A) Radiotracer is seen in the bladder (open arrow). (B) On a subsequent image there is reflux of radiotracer into the left ureter on this posterior image of the abdomen and pelvis (black arrow). (C) On a later image there is accumulation in the left renal pelvis (black arrowhead). (D) Radiotracer is seen in the left renal pelvis and ureter. There is no evidence of reflux into the right ureter.

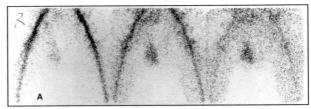

Flow Study 5 sec/Image

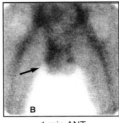
1 min ANT

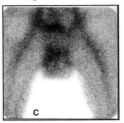
15 min ANT

FIGURE 24-22 Scrotal imaging with 99mTc-sodium pertechnetate. (A) Blood flow, (B) blood pool, and (C) 15 minute delayed anterior images of the pelvis demonstrate increased blood flow and blood pool to the superior and lateral right scrotum in this patient with acute-onset right scrotal pain. The pattern is most consistent with acute epididymitis.

reflux into a dilated renal collecting system, the reflux is severe or grade 3.

Scrotal imaging is performed to evaluate acute-onset scrotal pain. In the normal scan, there is symmetric, diffuse, mild perfusion to both testicles as well as symmetric blood pooling on the delayed images. Acute epididymitis is seen as increased blood flow and pooling in the region of the epididymis (Figure 24-22). Early in testicular torsion there may be decreased blood pooling in the torsed testicle. After several hours, there can be increased blood flow and pooling to the scrotum around the torsed testicle. The testicle appears as a photopenic defect surrounded by increased radiotracer (Figure 24-23).

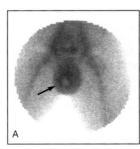
5 min ANT

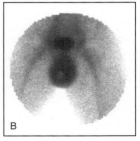
15 min ANT

FIGURE 24-23 Delayed testicular torsion. Anterior images of the scrotum taken at (A) 5 and (B) 15 minutes after intravenous administration of 99mTc-sodium pertechnetate. Images show increased scrotal activity on the right with a photopenic central area corresponding to the ischemic right testicle (arrow). The patient's right scrotal pain began the day before the study.

Renal Function Studies

A quantitative assessment of renal function can be quite useful because so many factors affect kidney function, notably disease processes and the effects of nephrotoxic drugs. Renal function measurements are also used to adjust drug therapy, particularly chemotherapy regimens in oncology patients. Renal clearance of radiotracers has been used for many years to obtain a measure of renal function through the assessment of GFR and tubular secretion. The routine methods have been reviewed.[68]

GFR measurement is the most familiar to clinicians and is the clearance measurement most widely sought. Simplified plasma sampling methods for GFR measurement have been developed for ^{125}I-iothalamate[53] and validated for ^{99m}Tc-DTPA with one or two plasma samples taken at 1 to 4 hours.[30–32] Camera-based methods are also used.[68] Simplified plasma sample methods have also been devised for measurement of ^{99m}Tc-MAG3 clearance to assess renal tubular function.[33,34,69] However, the reproducibility of ^{99m}Tc-MAG3 clearance methods has been questioned.[70] Chapter 26 contains additional discussion of GFR measurement.

Recently, a new radiopharmaceutical, ^{99m}Tc(CO)$_3$-nitrilotriacetic acid [^{99m}Tc(CO)$_3$-NTA] has been investigated.[71] ^{99m}Tc(CO)$_3$-NTA, prepared using the Isolink kit (Covidien), was shown to have a plasma clearance and urinary excretion almost identical to ^{131}I-OIH. ^{99m}Tc(CO)$_3$-NTA exhibits less protein binding and red blood cell uptake than ^{131}I-OIH, which may make it a useful agent for assessing renal tubular function and effective renal plasma flow.

REFERENCES

1. Brenner BM, Rector FC Jr. *The Kidney*. Philadelphia: Saunders; 1981.
2. Widmaier EP, Raff H, Strang KT. *Vander, Sherman, and Luciano's Human Physiology, The Mechanisms of Body Function*. 9th ed. Boston: McGraw Hill; 2004:513–62.
3. Gonzales FJ, Tukey RH. Drug metabolism. In: Brunton LL, Lazo JS, Parker KL, eds. *Goodman & Gilman's The Pharmacological Basis of Therapeutics*. 11th ed. New York: McGraw-Hill Medical Publishing Division; 2006.
4. Goldstein A, Aronow L, Kalman SM. *Principles of Drug Action*. New York: Harper & Row; 1969.
5. Fritzberg AR. Advances in renal radiopharmaceuticals. In: Fritzberg AR, ed. *Radiopharmaceuticals: Progress and Clinical Perspectives*. Vol 1. Boca Raton, FL: CRC Press; 1986:61–87.
6. Despopoulos A. A definition of substrate specificity in renal transport of organic anions. *J Theor Biol*. 1965;8:163–92.
7. Fritzberg AR. Current status of renal radiopharmaceuticals. *J Nucl Med Technol*. 1984;12:177–89.
8. Gottschalk CW, Lassiter WE. Mechanisms of urine formation. In: Mountcastle VB, ed. *Medical Physiology*. Vol 2. 14th ed. St Louis: Mosby; 1980.
9. McAfee JG, Grossman ZD, Gagne G, et al. Comparison of renal extraction efficiencies for radioactive agents in the normal dog. *J Nucl Med*. 1981;22:333–8.
10. Smith HB. *The Kidney: Structure and Function in Health and Disease*. New York: Oxford University Press; 1951:154.
11. Wesson LG. *Physiology of the Human Kidney*. New York: Grune & Stratton; 1969.
12. Chervu RL, Blaufox MD. Renal radiopharmaceuticals: an update. *Semin Nucl Med*. 1982;12:224–45.
13. Arnold RW, Subramanian G, McAfee JG, et al. Comparison of Tc-99m complexes for renal imaging. *J Nucl Med*. 1975;16:357–67.
14. Eshima D, Fritzberg AR, Taylor A Jr. Tc-99m renal tubular function agents: current status. *Semin Nucl Med*. 1990;20:28–40.
15. Taplin GV, Meredith OM Jr, Kade H, et al. The radioisotope renogram. *J Lab Clin Med*. 1956;48:886–901.
16. Tubis M, Posnick E, Nordyke RA. Preparation and use of I-131 labeled sodium iodohippurate in kidney function tests. *Proc Soc Exp Biol Med*. 1960;103:497–8.
17. McAfee JG, Wagner HN Jr. Visualization of renal parenchyma: scintiscanning with Hg-203 Neohydrin. *Radiology*. 1960;75:820–1.
18. Harper PV, Lathrop KA, Hinn GM, et al. Technetium-99m iron complex. In: Andrews GA, Kniseley RM, Wagner HN Jr, eds. *Radioactive Pharmaceuticals*. USAEC Symposium Series 9. Oak Ridge, TN: US Atomic Energy Commission; 1966:347–51.
19. Eckelman W, Richards P. Instant Tc-99m DTPA. *J Nucl Med*. 1970;11:761.
20. Atkins HL, Cardinale KG, Eckelman WC, et al. Evaluation of Tc-99m DTPA prepared by three different methods. *Radiology*. 1971;98:674–7.
21. Lin TH, Khentigen A, Winchell HS. A Tc-99m chelate substitute for organoradiomercurial renal agents. *J Nucl Med*. 1974;15:34–5.
22. Sigman EM, Elwood CM, Knox F. The measurement of glomerular filtration rate in man with sodium iothalamate I-131 (Conray). *J Nucl Med*. 1965;7:60–8.
23. Elwood CM, Sigman EM. The measurement of glomerular filtration and effective renal plasma flow in man by iothalamate I-131 and iodpyracet I-131. *Circulation*. 1967;36:441–8.
24. McAfee JG, Gagne G, Atkins HL, et al. Biological distribution and excretion of DTPA labeled with Tc-99m and In-111. *J Nucl Med*. 1979;20:1273–8.
25. Jones AG, Davison A, LaTegola MR, et al. Chemical and in vivo studies of the anion oxo[N,N'-ethylenebis(2-mercaptoacetimido)]technetate(V). *J Nucl Med*. 1982;23:801–9.
26. Fritzberg AR, Whitney WP, Kuni CC, et al. Biodistribution and renal excretion of ^{99m}Tc-N,N'-bis(mercaptoacetyl) ethylenediamine: effect of renal tubular transport inhibitors. *Int J Nucl Med Biol*. 1982;9:79–82.
27. Klingensmith WC III, Gerhold JP, Fritzberg AR, et al. Clinical comparison of Tc-99m N,N'-bis(mercaptoacetyl) ethylenediamine and ^{131}I-orthoiodohippurate for evaluation of renal tubular function. *J Nucl Med*. 1982;23:377–80.
28. Fritzberg AR, Kasina S, Eshima D, et al. Synthesis and biological evaluation of technetium-99m MAG3 as a hippuran replacement. *J Nucl Med*. 27;1986:111–6.
29. Gates GF. Split renal function testing using Tc-99m DTPA: a rapid technique for determining differential glomerular filtration. *Clin Nucl Med*. 1982;8:400–7.
30. Russell CD, Bischoff PG, Kontzen FN, et al. Measurement of glomerular filtration rate: single injection plasma clearance method without urine collection. *J Nucl Med*. 1985;26:1243–7.
31. Rowell KL, Kontzen FN, Stutzman ME, et al. Technical aspects of a new technique for estimating glomerular filtration rate using technetium-99m-DTPA. *J Nucl Med Technol*. 1986;14:196–8.
32. Ham HR, Piepsz A. Estimation of glomerular filtration rate in infants and in children using a single-plasma sample method. *J Nucl Med*. 1991;32:1294–7.
33. Blaufox MD, Aurell M, Bubeck B, et al. Report of the radionuclides in nephrourology committee on renal clearance. *J Nucl Med*. 1996;37:1883–90.
34. Bubeck B. Renal clearance determination with one blood sample: improved accuracy and universal applicability by a new calculation principle. *Semin Nucl Med*. 1993;23:73–86.
35. Ikeda I, Inoue O, Kurata K. Chemical and biological studies on Tc-99m DMS-II: effect of Sn(II) on the formation of various Tc-DMS complexes. *Int J Appl Radiat Isot*. 1976;27:681–88.
36. Ikeda I, Inoue O, Kurata K. Preparation of various Tc-99m dimercaptosuccinate complexes and their evaluation as radiotracers. *J Nucl Med*. 1977;18:1222–9.
37. Kubiatowicz KO, Bolles TF, Nova JC, et al. Localization of low molecular weight Tc-99m labeled dimercaptodicarboxylic acids in kidney tissue. *J Pharm Sci*. 1979;68:621–3.

38. Dewanjee MK. Binding of diagnostic radiopharmaceuticals to human serum albumin by sequential and equilibrium dialysis. *J Nucl Med.* 1982;23:753–4.
39. Hosokawa S, Kawamura J, Yoshida O, et al. Basic studies on intrarenal localization of renal scanning agent ^{99m}Tc-DMSA. *Acta Urol Jap.* 1978;24:61–5.
40. Suur-Müller R, Gutsche HU. Tubular reabsorption of technetium-99m-DMSA. *J Nucl Med.* 1995;36:1654–8.
41. Goldraich NP, Alvarenga AR, Goldraich IH, et al. Renal accumulation of ^{99m}Tc-DMSA in the artificially perfused isolated rat kidney. *J Urol.* 1985;134:1282–6.
42. Yee CA, Lee HB, Blaufox MD. Tc-99m DMSA renal uptake: influence of biochemical and physiologic factors. *J Nucl Med.* 1981;22:1054–8.
43. Razumeni NV, Petrovi J. Biochemical studies of the renal radiopharmaceutical compound dimercaptosuccinate. II: Subcellular localization of ^{99m}Tc-DMS complex in the rat kidney in vivo. *Eur J Nucl Med.* 1982;7:304–7.
44. Moretti JL, Rapin JR, Saccavini JC, et al. 2,3-Dimercaptosuccinic acid chelates 2: renal localization. *Int J Nucl Med Biol.* 1984;11:275–9.
45. Van Luyk WHJ, Ensing GJ, Piers DA. Low renal uptake of ^{99m}Tc-DMSA in patients with proximal tubular dysfunction. *Eur J Med.* 1983;8:404–5.
46. Lee HB, Blaufox MD. Mechanism of renal concentration of technetium-99m glucoheptonate. *J Nucl Med.* 1985;26:1308–13.
47. Nosco DL, Wolfangel RG, Bushman MJ, et al. Technetium-99m MAG3: labeling conditions and quality control. *J Nucl Med Technol.* 1993;21:69–74.
48. Taylor A Jr, Eshima D, Fritzberg AR, et al. Comparison of iodine-131 OIH and technetium MAG3 renal imaging in volunteers. *J Nucl Med.* 1986;27:795–803.
49. Bubeck B, Brandau W, Weber E, et al. Pharmacokinetics of technetium-99m-MAG3 in humans. *J Nucl Med.* 1990;31:1285–93.
50. Jafri RA, Britton KE, Nimmon CC, et al. Technetium-99m MAG3, a comparison with iodine-123 and iodine-131 orthoiodohippurate, in patients with renal disorders. *J Nucl Med.* 1988;29:147–58.
51. Sattuck LA, Eshima D, Taylor AT Jr, et al. Evaluation of the hepatobiliary excretion of technetium-99m-MAG3 and reconstitution factors affecting radiochemical purity. *J Nucl Med.* 1994;35:349–55.
52. Burbank MK, Tauxe WN, Maher FT, et al. Evaluation of radioiodinated Hippuran for the estimation of renal plasma flow. *Proc Staff Mayo Clin.* 1961;36:372–86.
53. Dubovsky EV, Russell CD. Quantitation of renal function with glomerular and tubular agents. *Semin Nucl Med.* 1982;12:308–29.
54. Maher FT, Tauxe WN. Renal clearance in man of pharmaceuticals containing radioactive iodine: influence of plasma binding. *JAMA.* 1969;207:97–104.
55. Magnusson G. Influence of varying amounts of I-131 in radiohippuran on the radioactivity measurements. *Acta Med Scand.* 1962;(suppl 378):111–5.
56. Kontzen FN, Tobin M, Dubovsky EV, et al. Comprehensive renal function studies: technical aspects. *J Nucl Med Technol.* 1977;5:81–4.
57. Eshima D, Taylor A Jr. Technetium-99m (99mTc) mercaptoacetyltriglycine: update on the new ^{99m}Tc renal tubular function agent. *Semin Nucl Med.* 1992;22:61–73.
58. O'Reilly P, Aurell M, Britton K, et al. Consensus on diuresis renography for investigating the dilated upper urinary tract. *J Nucl Med.* 1996;37:1872–6.
59. Blaufox MD. Procedures of choice in renal nuclear medicine. *J Nucl Med.* 1991;32:1301–9.
60. Nally JV, Black HR. State-of-the-art review: captopril renography—pathophysiological considerations and clinical observations. *Semin Nucl Med* 1992;22:85–97.
61. Sarkar SD, Singhal PG. Basis of renal scintigraphy. In: Elgazzar AH, ed. *The Pathophysiologic Basis of Nuclear Medicine.* New York: Springer-Verlag; 2001:154–68.
62. Taylor A, Nally J, Aurell M, et al. Consensus report on ACE inhibitor renography for detecting renovascular hypertension. *J Nucl Med.* 1996;37:1876–82.
63. Wenting GJ, Tan-Tjiong HL, Derkx FHM, et al. Split renal function after captopril in unilateral renal artery stenosis. *BMJ.* 1984;288:886–90.
64. Fommei E, Ghione S, Hilson AJW, et al. Captopril radionuclide test in renovascular hypertension: a European multicenter study. *Eur J Nucl Med.* 1993;20:625–44.
65. Taylor AT, Blaufox MD, Dubovsky EV, et al. Diagnosis of renovascular hypertension 3.0. In: *Practice Management Procedure Guidelines.* Reston, VA: Society of Nuclear Medicine; 2003:97–104.
66. Majd M, Rushton HG. Renal cortical scintigraphy in the diagnosis of acute pyelonephritis. *Semin Nucl Med.* 1992;22:98–111.
67. Mandell GA, Eggli DF, Gilday DL, et al. Renal cortical scintigraphy in children 3.0. In: *Practice Management Procedure Guidelines.* Reston, VA: Society of Nuclear Medicine; 2003:195–8.
68. Taylor A. Radionuclide renography: a personal approach. *Semin Nucl Med.* 1999;29:102–27.
69. Russell CD, Taylor AT, Dubovsky EV. Measurement of renal function with technetium-99m-MAG3 in children and adults. *J Nucl Med.* 1996;37:588–93.
70. Piepsz A, Tondeur M, Kinthaert J, et al. Reproducibility of technetium-99m mercaptoacetylglycine clearance. *Eur J Nucl Med.* 1996;23:195–8.
71. Taylor AT, Lipowska M, Marzilli LG. ^{99m}Tc(CO)$_3$(NTA): a ^{99m}Tc renal tracer with pharmacokinetic properties comparable to those of ^{131}I-OIH in healthy volunteers. *J Nucl Med.* 2010;51:391–6.

CHAPTER 25
Bone

The major deficiency of a plain film x-ray exam for evaluating bone disease is that greater than 50% demineralization must have occurred for a lesion in bone to be detected.[1] Therefore, disease must be significantly advanced before it can be found on x-ray. The major advantage of the bone scan in nuclear medicine is its sensitivity for detecting early-stage disease associated with sites of bone remodeling. Although the bone scan has high diagnostic sensitivity for detecting bone lesions, high specificity requires correlation of the bone scan findings with additional patient information, because sites of bone remodeling can be caused by a wide variety of conditions. There are several possible reasons for performing a bone scan. Principal indications include evaluation of primary and metastatic disease in bone, bone infection, and benign skeletal disease.

BONE FORMATION

Bone is composed of minute crystals of hydroxyapatite (HA) associated with collagen fibers. The crystals are continually being produced and reabsorbed in the bone remodeling process. The surface area of bone mineral is quite large because of the small size of bone crystals. The composition of bone mineral is mainly calcium, phosphate, and hydroxyl ions. The presence of these ions on the large adsorptive surface of bone creates a chemically reactive site for many radionuclidic substances that have an affinity for bone. This provides a mechanism for studying bone physiology and for performing diagnostic imaging procedures to detect disease associated with bone.

Living bone consists of a variety of tissues, as shown in Figure 25-1.[2] The outer layer of bone, which imparts its shape and strength, is cortical bone. Internal to the cortical bone is spongy cancellous or trabecular bone. Cancellous bone contains the marrow, which is composed of fat and hematopoietic elements. The articulating surfaces of bone are covered with a layer of cartilage. Tendons, ligaments, and muscle attachments are inserted into cortical bone. Blood vessels penetrate the cortex and permeate the cancellous bone. A fibrous and cellular envelope covers the bone tissue surfaces. This envelope, demarcated by the periosteum externally and endosteum internally, contains the osteocytes, which are pluripotent in bone remodeling.

Fresh compact bone is composed of 9% water, 11% organic matrix, and 69% inorganic salts.[2] The organic matrix consists of a noncollagenous or ground substance (10%) and collagen fibers (90%). The noncollagenous matrix consists of multiple substances, including mucopolysaccharides, glycoproteins, phosphoproteins, and phospholipids. It serves as mineralization nucleator and inhibitor and as a "glue" that occupies the space between collagen fibers. Collagen gives bone its tensile strength and provides nucleation centers for the deposition of inorganic salts. These salts, composed essentially of calcium phosphate and hydroxyl ions, give bone its compressional strength. The principal inorganic salts found in bone are amorphous calcium phosphate (ACP) and hydroxyapatite (HA). Bone composition is summarized in Figure 25-2.

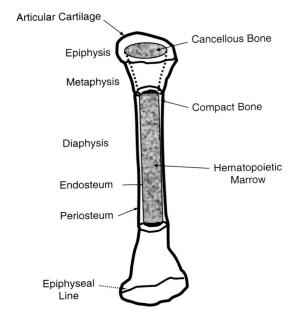

FIGURE 25-1 Principal tissues in bone.

or new bone, is not crystalline but is a mixture of hydrated calcium phosphates of varying Ca/P molar ratios, consisting mainly of calcium monohydrogen phosphate (CaHPO$_4$, Ca/P = 1.0), octacalcium phosphate [Ca$_4$H(PO$_4$)$_3$, Ca/P = 1.33], and tricalcium phosphate [Ca$_3$(PO$_4$)$_2$, Ca/P = 1.5].[3] By a process of substitution and addition of atoms, or resorption (through osteoclastic activity) and reprecipitation, these ACP salts are converted into the well-crystallized HA [Ca$_{10}$(OH)$_2$(PO$_4$)$_6$, Ca/P = 1.66].

HA crystals are a very stable end product of bone mineralization and do not readily redissolve. In addition, the crystals formed are microcrystalline in nature; under physiologic conditions HA almost never forms crystals that are larger than a few hundred angstroms in length, breadth, or thickness. Consequently, the HA system has a large surface area (200 m^2/gram) that provides an enormous opportunity for adsorption and surface exchange of a variety of ions. Prominent examples include Sr^{2+}, Pb^{2+}, Ra^{2+}, and Mg^{2+}, which exchange with Ca^{2+}; F$^-$, which exchanges with OH$^-$; and CO$_3^{2-}$, citrate, phosphate esters, diphosphonates, and pyrophosphate, which exchange with phosphate. It is precisely for this reason that radionuclide species of these substances have been used successfully to study and image the skeleton in humans and animals.

Mature bone is composed of minute crystals of HA associated with collagen fibers. The crystals are continually being produced and resorbed by osteocytes in the bone remodeling process. Osteocytes can function as osteoblasts or osteoclasts. Osteoclasts are instrumental in bone resorption. Osteoblasts are bone-forming cells that lie directly on bone surfaces; they are responsible for synthesizing the organic matrix, called osteoid, that occupies the space between osteoblasts and the underlying calcified bone. Soon after collagen fibers are formed by osteoblasts, ACP precipitates on their surfaces at periodic intervals to form minute nidi that rapidly multiply and grow over days and weeks to form HA crystals. The ACP,

DEVELOPMENT OF BONE-IMAGING AGENTS

The use of self-luminous paint on watch dials began after the discovery that alpha particles emitted from radium produced luminescence when striking a zinc sulfide surface. This discovery led to the use of radium paint in the production of self-luminescent watch dials and other products in the 1920s. Radium paint was composed of radium chloride and zinc sul-

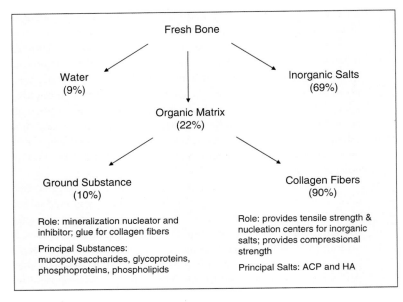

FIGURE 25-2 Chemical composition of bone.

fide with a binder and minute amounts of radioactive radium (^{226}Ra). It was common practice for the women who painted the watch dials to use their tongues and lips to put a fine tip on the paintbrush, causing them to ingest small amounts of radium. Being a bone seeker with a long half-life, radium accumulated in the jawbones of the dial painters. The first hint that radioactive material localized in bone came a few years later, when a dial painter developed jaw pain and loss of teeth, so-called "radium jaw."[4] Subsequent investigations revealed that radium was the cause of bone disease in the watch dial painters.

Many radionuclides have been used to study bone, but few of the earlier ones had desirable physical properties for bone imaging. The earliest use of bone-seeking radionuclides was for the treatment of bone disease and metastatic lesions. The radionuclides included primarily beta emitters such as ^{32}P, ^{45}Ca, ^{47}Ca, and ^{89}Sr. In 1949, ^{72}Ga was one of the first radionuclides used to detect bone metastases, but it was unacceptable because of its high 2.5 MeV gamma energy.[5] Investigations were switched to ^{68}Ga and then to ^{67}Ga in the early 1950s, but the serendipitous finding by Edwards and Hayes[6] in 1969 that carrier-free gallium localized in soft-tissue tumors all but terminated gallium's potential for becoming a bone-imaging agent. Furthermore, bone localization of radiotracer gallium required the coadministration of stable carrier gallium, which was thought to be potentially toxic.

Around 1961, clinical bone scanning had its true beginning when Charkes and Sklaroff[7] at the Einstein Medical Center in Philadelphia began to use ^{85}Sr bone scans to locate metastases for the rational application of radiotherapy. ^{85}Sr was selected because its metabolism simulated that of calcium and its gamma emission could be measured by external detection. In addition, bone metastasis could be detected with ^{85}Sr before it was evident on x-ray exam.[8] ^{85}Sr was available as the nitrate or chloride salt and was administered intravenously. Its long biologic and physical half-lives limited the adult dose to 100 μCi (3.7 MBq), which resulted in prolonged scanning times and poor counting statistics. Although the rate of bone uptake was rapid, the slow excretion required at least a 2 day delay before scanning to improve the bone-to-background ratio. In 1964, Meckelnburg[9] introduced ^{87m}Sr-citrate for bone imaging. The short 2.8 hour half-life and decay by isomeric transition lowered the radiation dose significantly below that of ^{85}Sr; this favored the use of ^{87m}Sr in children. However, its short half-life required bone scans to be performed before adequate excretion occurred, which increased the potential for false-positive and false-negative interpretation.

In 1962, Blau et al.[10] introduced ^{18}F as sodium fluoride for bone scanning. Its principal advantage was high bone-to-background ratios because of its rapid blood clearance after intravenous injection. However, there were significant logistical problems delivering ^{18}F to nuclear medicine facilities, and the introduction of ^{99m}Tc-labeled phosphate compounds for bone imaging brought ^{18}F use to a virtual halt.

^{99m}Tc-tripolyphosphate (^{99m}Tc-STPP) was developed by Subramanian and McAfee[11] on the basis that ^{32}P-labeled polyphosphate localized in the mineral phase of bone.[12] A few months later, Subramanian's group,[13] a Harvard group,[14] and several other groups investigating the properties of polyphosphate discovered that the desirable bone-localizing property of the original long-chain polyphosphates was due to the presence of pyrophosphate (PPi), either as an impurity or as a degradation product. The formal introduction of ^{99m}Tc-pyrophosphate (^{99m}Tc-PPi) came in 1972 by Perez and coworkers[15] in Paris, France. ^{99m}Tc-PPi had high chemical purity, labeling yields with ^{99m}Tc-sodium pertechnetate were high (>90%), and its blood clearance was more rapid than that of polyphosphate.

Also in 1972, a different class of ^{99m}Tc-labeled phosphorus compounds was introduced for bone imaging. These were the diphosphonates, which are organophosphorus compounds characterized by a phosphorus-to-carbon (P–C–P) bond, in contrast to the phosphorus-to-oxygen (P–O–P) bond found in polyphosphates and pyrophosphates. The diphosphonates had previously been shown to inhibit dissolution of bone and crystal growth of HA in certain bone diseases. Other studies indicated that the phosphorus-to-carbon bond was more stable in vivo, because it was not broken down by phosphatases as were the phosphates.

Three groups reported almost simultaneously on ^{99m}Tc-labeled ethane-1-hydroxy-1-diphosphonate (^{99m}Tc-etidronate, or ^{99m}Tc-EHDP) for bone imaging.[16–18] The advantages claimed for ^{99m}Tc-EHDP compared with ^{99m}Tc-PPi were a slightly greater bone concentration (50% to 55% of injected dose [ID] for ^{99m}Tc-EHDP versus 45% to 50% of ID for ^{99m}Tc-PPi), improved in vivo stability (although this was implied), and faster blood clearance, which was its primary advantage. The blood clearance, however, was still not as rapid as that of ^{18}F.

In 1975 Subramanian et al.[19] introduced ^{99m}Tc-labeled methylene diphosphonate (^{99m}Tc-medronate, or ^{99m}Tc-MDP) for bone imaging and compared it with ^{99m}Tc-EHDP, ^{99m}Tc-PPi, and ^{99m}Tc-labeled polyphosphate. Their findings in humans indicated that diphosphonates are cleared more rapidly from the blood than pyrophosphates or polyphosphates. ^{99m}Tc-labeled polyphosphates were found to be slowest; ^{99m}Tc-MDP was found to be fastest and equivalent to ^{18}F-fluoride clearance. The slower clearance of polyphosphates and pyrophosphates was attributed in part to their higher plasma protein binding and diffusion into red blood cells (RBCs). The fraction associated with RBCs 24 hours after injection was 22% for ^{99m}Tc-labeled polyphosphate, 60% for ^{99m}Tc-PPi, and negligible amounts for ^{99m}Tc-EHDP and ^{99m}Tc-MDP. Each agent produced excellent bone images, but high-quality images could, as a rule, be obtained 2 hours after injection with ^{99m}Tc-MDP, whereas 3 to 4 hours was often required for ^{99m}Tc-EHDP and 4 hours for pyrophosphates and polyphosphates.

In addition to achieving higher blood clearance for bone agents, higher bone affinity can also increase the bone-to-soft-tissue ratio. When one of the methylene hydrogen atoms in MDP is replaced by a hydroxyl group, hydroxymethylene diphosphonate (HMDP) is formed. The technetium complex is known as ^{99m}Tc-oxidronate (^{99m}Tc-HDP). The presence of

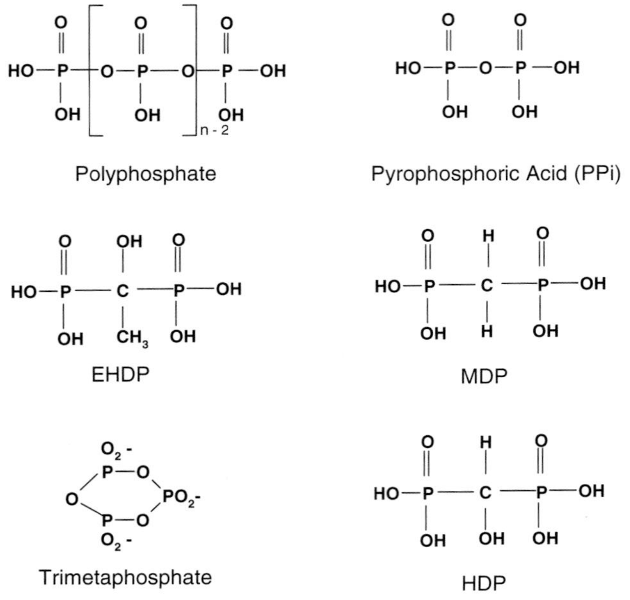

FIGURE 25-3 Various ligands used in preparing ^{99m}Tc bone agents.

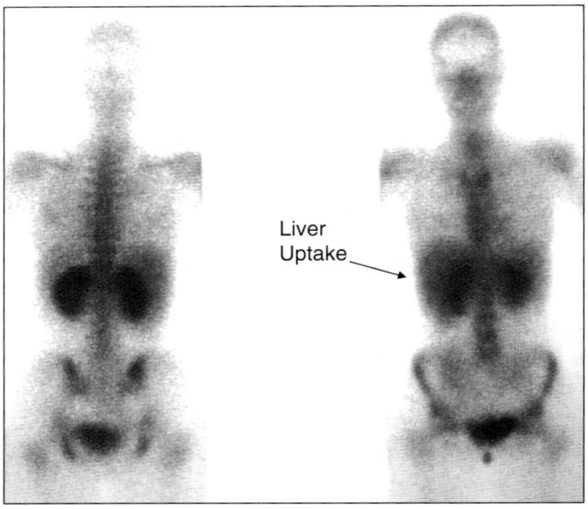

FIGURE 25-4 ^{99m}Tc-pyrophosphate (^{99m}Tc-PPi) bone scan demonstrating liver uptake due to hydrolyzed–reduced technetium impurity in the administered dose.

the hydroxyl group on the carbon atom appears to increase the bone uptake of diphosphonates because it provides the opportunity for tridentate binding to calcium on the growing surface of HA crystals.[20] Experimental work in animals demonstrated that ^{99m}Tc-HDP has a higher binding affinity for apatite crystals than do ^{99m}Tc-EHDP and ^{99m}Tc-MDP.[21]

Also in 1975, a ^{99m}Tc-trimetaphosphate complex was introduced for bone imaging.[22] This cyclic phosphate was reported to have bone uptake equivalent to that for ^{99m}Tc-EHDP and tripolyphosphate. Optimum bone uptake, with minimal liver uptake, was observed when 25 to 50 mg trimetaphosphate per milligram of stannous chloride was used.

The chemical structures of phosphate and diphosphonate ligands for preparing ^{99m}Tc bone agents are shown in Figure 25-3. The ^{99m}Tc kits currently available on the market contain MDP, HDP, or PPi. See Chapter 10 for kit composition and preparation.

^{99m}Tc BONE AGENTS

The ^{99m}Tc bone complexes are prepared by simply adding the required amount of pertechnetate to a sterile lyophilized kit containing a mixture of stannous ion and the appropriate ligand. Radiochemical purity is checked by instant thin-layer chromatography for the presence of free pertechnetate and hydrolyzed–reduced technetium colloid. Formation of the colloidal impurity is favored if the ratio of phosphate to tin is too low and the pH of the reaction mixture is too high.[22] Liver uptake on clinical images has been observed when the colloid impurity is significant (Figure 25-4). Excess pertechnetate impurity is evident on the bone scan as uptake in the thyroid gland and stomach (Figure 25-5).

Technetium bone complexes can degrade over time, producing free pertechnetate.[23] This can be minimized by increasing the amount of stannous ion in kits. However,

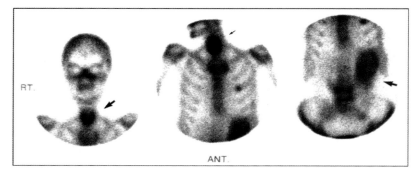

FIGURE 25-5 Bone scan demonstrating undesirable thyroid activity (left and middle images) and stomach activity (right image) from excess pertechnetate impurity in the injected bone agent.

increased amounts of Sn(II) in kits were shown to interfere with the activity distribution in pertechnetate brain scans for up to 2 weeks after the administration of ^{99m}Tc bone agents.[24] Bone kits developed with low amounts of tin were subject to stability problems and the resultant generation of free pertechnetate impurity over time because of the oxidative degradation of the ^{99m}Tc complex. The problems associated with low-tin kits were significantly reduced by the addition of ascorbic acid or gentisic acid as antioxidants to diphosphonate bone kits (see Chapter 10 Figure 10-6).[25,26] However, only gentisic acid can be used in pyrophosphate kits, because the presence of ascorbic acid causes the formation of a ^{99m}Tc-ascorbate complex that results in renal images.[27]

The antioxidants in kits are free-radical scavengers that stabilize by retarding free-radical reactions with the ^{99m}Tc bone complex. Radiolytically produced alkoxy (RO·) and peroxy (RO·$_2$) radicals can be stabilized by ascorbate or gentisate through the transfer of a hydrogen atom from the antioxidant molecule to the free radical, which yields a resonance-stabilized and nonreactive molecule, RO$_2$H. This reaction is shown in Figure 10-5, Chapter 10. Elimination of the intermediate radical by ascorbate (or gentisate) is believed to provide in vitro stability by inhibiting the slow oxidation of TcO$_2$ to TcO$_4^-$.[25] The mechanism of bone complex degradation to yield pertechnetate has been suggested to occur by initial dissociation of Tc(IV) from the stable chelate and hydrolysis into TcO$_2$ with subsequent oxidation to TcO$_4^-$, rather than by a direct interaction of the chelate with oxygen.[28]

Other kit formulation factors important to the clinical performance of technetium bone complexes include proper ligand-to-tin ratio and the amount of bone agent injected. Tofe and Francis[29] found that the optimum EHDP-to-stannous chloride weight ratio was between 5:1 and 50:1 based on binding affinity to HA in vitro. Ratios of 12:1 and 50:1 gave the same biodistribution in animals. Subramanian et al.[19] obtained optimal skeletal localization with MDP-to-tin ratios of 10:1 at doses between 0.01 and 0.5 mg MDP per kilogram of body weight, with no significant difference in distribution. Bevan et al.[30] found that liver uptake of ^{99m}Tc activity occurred when the Na$_2$HMDP load was above 0.1 mg per kilogram of body weight in rats, guinea pigs, and dogs. These findings were taken into consideration when HMDP kits were formulated for human use. Ponto[31] has compiled an excellent review of how formulation factors of ^{99m}Tc bone agents and other technetium compounds affect their biodistribution.

Analysis of technetium bone complexes indicates that they are not a single well-defined chemical species but are probably mixtures of short- and long-chain polymers.[32] The polymeric structure of [Tc(OH)MDP$^-$]$_n$ is characterized by technetium atoms bridged by MDP and OH ligands, and the MDP ligands are in turn bridged by technetium atoms, as shown in Figure 25-6. High-performance liquid chromatographic (HPLC) analysis of preparations with varying concentrations of technetium has shown that polymeric complexes, evidenced by multiple HPLC peaks, are formed when millimolar concentrations of technetium are present, whereas preparations made with no-carrier-added technetium produce only one peak.[33] It has been suggested that multiple complex formation may affect the biodistribution and elimination of ^{99m}Tc-diphosphonate complexes.

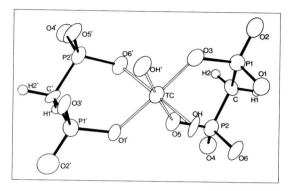

FIGURE 25-6 Perspective view of a portion of the [Tc(OH)(MDP)$^-$]$_n$ infinite polymer (MDP represents methylene diphosphonate in an unknown protonation state). Hydrogen atoms bonded to oxygen, lithium counterions, and waters of hydration are omitted for clarity. (Reprinted with permission of the Society of Nuclear Medicine from reference 32.)

Biologic Properties

The biodistribution of ^{99m}Tc bone agents in humans and animals has been compared.[19] After intravenous administration these agents are rapidly distributed in the body, and by 3 hours most of the activity is located in the skeleton, urine, and blood. The whole-blood activity as a fraction of the ID at this time is 8% for ^{99m}Tc-PPi, 5% for ^{99m}Tc-EHDP, and 3% for ^{99m}Tc-MDP. For ^{99m}Tc-PPi the blood activity is primarily found in plasma protein (43%) and RBCs (30%). The blood activity of the diphosphonates is primarily associated with protein, with no significant binding to RBCs. Consequently, the blood clearance rates for ^{99m}Tc-EHDP and ^{99m}Tc-MDP are significantly faster than those of ^{99m}Tc-PPi, with ^{99m}Tc-MDP being the most rapid (Figure 25-7). The 3 hour urine accumulation in humans is 43% for ^{99m}Tc-PPi, 56% for ^{99m}Tc-EHDP, and 59% for ^{99m}Tc-MDP. The absolute average bone concentration of the diphosphonates in rabbits has been shown to be 1.6 times higher than that of ^{99m}Tc-PPi. No statistically significant difference has been found between the bone concentration of ^{99m}Tc-EHDP and ^{99m}Tc-MDP.[19] In humans, approximately 45% to 55% of the injected activity of ^{99m}Tc-EHDP and ^{99m}Tc-MDP localizes in bone within 3 hours.[14]

In a comparison between diphosphonates, the blood clearance of ^{99m}Tc-HDP was similar to ^{99m}Tc-MDP up to 3 hours but faster thereafter.[30] In vitro and in vivo bone uptake experiments have demonstrated that ^{99m}Tc-HDP has significantly higher bone affinity than ^{99m}Tc-EHDP

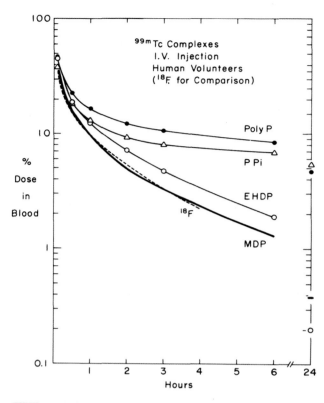

FIGURE 25-7 Blood clearance of 99mTc-MDP in humans compared with three other 99mTc complexes and 18F-sodium fluoride (corrected for physical decay). (Reprinted with permission of the Society of Nuclear Medicine from reference 19.)

and 99mTc-MDP.[34] Indirect whole-body retention measurements in humans have demonstrated that 99mTc-HDP has a 20% higher skeletal uptake than 99mTc-MDP.[35] Despite these differences, clinical comparisons of 99mTc-HDP, 99mTc-EHDP, and 99mTc-MDP have shown no significant difference in lesion detection.[36–38] Earlier comparisons between 99mTc-MDP and 99mTc-PPi showed higher-quality bone scans with 99mTc-MDP but no obvious difference in diagnostic sensitivity.[39]

In summary, several 99mTc bone agents have been developed since 1971. The first was 99mTc-STPP, then 99mTc-polyphosphate, followed by 99mTc-PPi, 99mTc-etidronate (99mTc-EHDP), 99mTc-medronate (99mTc-MDP), and finally 99mTc-oxidronate (99mTc-HDP). In general, each successive agent yielded improved scan quality because of higher bone-to-soft-tissue ratio. Although there appears to be no significant difference in lesion detection with the last four agents, 99mTc-MDP and 99mTc-HDP are preferred because of their higher-quality, cleaner-looking bone images. 99mTc-EHDP is no longer marketed.

Localization Mechanisms

Bone localization of the 99mTc complexes requires adequate blood flow to bone and the absence of any diffusional barrier during transit from blood to the bone surface.[40] The chemical composition of the bone surface and the structural properties of the bone agent itself are important factors in the binding of 99mTc complexes to bone.

The anionic 99mTc-labeled phosphate and phosphonate complexes are believed to interact with bone by binding to Ca^{2+} ions in bone crystals. The process is referred to as chemisorption. Localization occurs primarily in the mineral phase of bone, with insignificant binding to the organic matrix.[21] In vitro and in vivo experiments have shown that 99mTc-diphosphonates bind to ACP to a higher degree than to crystalline HA.[21,34] The hypothesis for this is that ACP contains newly formed apatite crystallites that have a crystal-growing face with a chemical configuration of calcium ions best suited for binding to oxygen atoms in the diphosphonate ligand.[21] As bone matures, this configuration is altered so that less uptake of bone agent occurs in mature bone. The difference in binding between ACP and HA is the basis for detecting bone lesions, because areas of increased osteogenic activity contain higher concentrations of ACP relative to HA. It is important to note that the bone scan does not identify specific disease processes in bone but only shows the effect of bone involvement associated with a variety of pathologic causes (e.g., trauma, infection, or cancer). However, characteristic patterns of activity distribution do occur in certain conditions such as Paget's disease.

The mechanism of localization of 99mTc-diphosphonate complexes in bone is not entirely known but appears to depend on the chemical structure of the ligand in the complex. The ligands EHDP, MDP, and HDP are geminal diphosphonates, which means that both phosphonate moieties are attached to the same carbon atom. This places the oxygen atoms on the phosphonate groups at a distance and position conducive to binding with calcium in bone.[20] The type of substitution at the two remaining sites on the central carbon atom can also influence the agent's binding to bone. Binding to calcium in bone is highest when one of these substituents is hydroxyl and the other is hydrogen, as in HDP.[21] This is illustrated in Figures 25-8 and 25-9. It is proposed that the triangular face of the HDP ligand allows optimal tridentate binding to calcium through both geminal diphosphonate oxygen atoms and the hydroxyl group. Binding of EHDP to calcium, which also has three oxygen-binding groups, is less likely because of steric hindrance by the methyl group. MDP has lower binding affinity because it can undergo only bidentate binding. An additional factor that may contribute to binding affinity is the relative solubilities of the complex's calcium salts formed when they react at the bone surface. In this regard the decreasing order of solubility has been determined to be MDP > EHDP > HMDP ≥ PPi.[21]

Dosimetry

Estimates of radiation dose to selected organs from 99mTc bone agents are listed in Table 25-1.[41] The dose to the skeleton is an average adult dose that assumes uniform skeletal

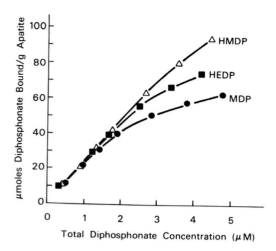

FIGURE 25-8 Adsorption of ^{14}C-labeled MDP, HEDP, and HMDP (HDP) on crystalline hydroxyapatite. (Reprinted with permission of the Society of Nuclear Medicine from reference 21.)

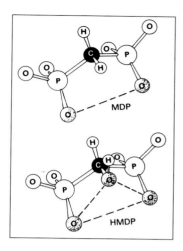

FIGURE 25-9 Comparison of HMDP (HDP) and MDP structures, indicating the potential of tridentate binding in hydroxylated molecules, such as HMDP. (Reprinted with permission of the Society of Nuclear Medicine from reference 30.)

distribution; however, bone distribution often is not uniform, with higher concentrations in the vertebrae and ribs than in long bones.[19] In addition, bone activity has been shown to be two to three times higher in the epiphyseal region than in the diaphysis of growing long bones. This distribution is evident particularly in pediatric patients; the growth plate has a much higher concentration than in adults. The radiation dose to the pediatric growth plate has been estimated to be as much as 6 to 8 times higher than the dose to the adult skeleton.[42] The critical organ is bone, and the next highest dose is to the urinary bladder wall, reflecting the high urinary excretion of the injected dose. The bladder dose is typically based on either a 2.4 hour or 4.8 hour voiding interval. The dose can be reduced if the patient voids more frequently; frequent voiding should be recommended to the patient at the time of dose administration. Dosimetry estimates reported in the literature and product package inserts differ somewhat because of variations in the biokinetic models used for calculating absorbed dose.

Adverse Effects

No significant toxic effects have been reported for ^{99m}Tc bone agents with the normally administered activity. The usual dosage range for ^{99m}Tc-PPi is 0.02 to 0.2 mg PPi per kilogram of body weight; for ^{99m}Tc-diphosphonate it is one-fifth to one-half that amount. A first level of potential toxic effect is related to the complexing of bone agents with serum calcium. With PPi, tetany in rats is produced by 22 mg/kg and electrocardiographic changes indicative of hypocalcemia are seen at 12 mg/kg intravenous doses.[43] If these doses are extrapolated to humans, the safety factor between a diagnostic dose and the lowest dose for minimally detectable hypocalcemia as noted by electrocardiogram is 55.[19]

Few adverse reactions to the administration of ^{99m}Tc bone complexes have been noted. The frequency of reactions to ^{99m}Tc-MDP reported in 1984 was 0.5 per 100,000.[44] Reported adverse effects of ^{99m}Tc-MDP include skin rash and transient symptoms of headache, dizziness, nausea, myalgia, and fever.[45]

TABLE 25-1 Radiation Dose Estimates for Bone Imaging Agents

	Adult Radiation Absorbed Dose		
	(rad[cGy] per 20 mCi)		(rad[cGy] per 10 mCi)
Tissue	^{99m}Tc-HDP	^{99m}Tc-MDP	^{18}F-NaF
Bone surfaces	3.80	2.60	2.20
Urinary bladder wall	1.60	2.40	9.10
Red marrow	0.56	0.40	1.00
Kidneys	0.44	0.64	0.70
Ovaries	0.26	0.24	0.40
Testes	0.17	0.17	0.30
Effective dose equivalent (rem)	0.46	0.44	1.00

Source: Reference 41 (dosimetry tables).

Altered Biodistribution

Altered biodistribution of bone-imaging agents labeled with ^{99m}Tc has been associated with drugs and treatment regimens.[46,47] Intense renal uptake of bone agent has been reported in children treated with vincristine, doxorubicin, and cyclophosphamide, either alone or in combination. The mechanism of renal uptake is unknown. Long-term steroid therapy induces bone mineral depletion and has been shown to cause a generalized decrease in skeletal uptake of bone-imaging agents. Bilateral breast uptake of ^{99m}Tc-PPi has been reported in a man treated with diethylstilbestrol for prostate cancer. Several reports have demonstrated abnormal distribution and localization of ^{99m}Tc bone agents because of iron. Localized activity has occurred at sites where intramuscular iron–dextran has been injected; this may be due to localized hyperemia or complexation of the bone agent with iron. Plasma iron overload has been associated with decreased skeletal uptake of bone agents in several cases. Technetium diphosphonates have shown splenic uptake in sickle cell disease, which may be related to increased iron concentration. Increased liver uptake of ^{99m}Tc-diphosphonates has been shown to occur with increased levels of plasma aluminum; this has also been documented in controlled animal experiments. A so-called sickle sign, an area of diffuse activity around the calvaria, has been observed on bone scans in 56% of breast cancer patients receiving intensive cytotoxic therapy.

Other maldistributions have been described and more detail provided by Hladik et al.[46] and Lentle et al.[47] Although it is difficult to determine the exact causes of these maldistributions, it is important to know that they can occur for many reasons that may not be associated with the particular disease being evaluated by the bone scan.

^{18}F-SODIUM FLUORIDE

^{18}F-sodium fluoride was first introduced for bone imaging in 1962 by Blau et al.;[10] however, distribution issues and suboptimal imaging with the gamma camera in comparison with ^{99m}Tc bone agents led to its disuse at that time. More recently, ^{18}F-sodium fluoride has been reintroduced because of its favorable biologic properties and the widespread availability of positron emission tomography (PET) scanners.[48] In addition, ^{18}F-sodium fluoride is readily available in high specific activity from cyclotron facilities that produce fludeoxyglucose.

The ^{18}F fluoride anion (F$^-$) is rapidly cleared from the plasma following intravenous injection (Figure 25-7) and localizes in bone by isomorphous exchange of fluoride with hydroxyl (OH$^-$) in HA. A factor that makes ^{18}F-sodium fluoride a useful bone-imaging agent is its high bone-to-background ratio because of low protein binding (around 5%) and lack of RBC binding at physiologic pH.[49] This allows imaging to commence soon after dosing. The average time from dose administration to imaging is 15 to 30 minutes with a total image acquisition time of 15 to 30 minutes.[48] The typical adult dosage is 10 mCi (370 MBq). This activity produces an effective dose of 1.0 rem (10 mSv), a bladder wall dose (critical organ) of 9.1 rad(cGy), and a bone surface dose of 2.2 rad(cGy) (Table 25-1).[41]

^{18}F-sodium fluoride bone scintigraphy has been useful for evaluating oncologic and benign disease of the skeleton (Figure 25-10). During a typical procedure an adult dose of 10 mCi (370 MBq) of ^{18}F-sodium fluoride, with a range of 8 to 15 mCi (296–555 mBq), is administered intravenously. Image acquisition begins 15 minutes after dose administration and is completed by 30 to 60 minutes. Studies comparing ^{18}F-sodium fluoride and ^{99m}Tc-MDP bone scans indicate that ^{18}F-sodium fluoride is more accurate in identifying metastatic lesions in patients with breast, prostate, thyroid, and lung cancer.[50–53] ^{18}F-sodium fluoride PET can detect more lesions than ^{99m}Tc-MDP planar or single-photon emission computed tomography (SPECT) imaging (Figure 25-11). In benign disease, ^{18}F-sodium fluoride was shown to be useful in the evaluation of back pain in children and adolescents and in predicting bone viability after trauma.[48]

NUCLEAR MEDICINE PROCEDURES

Rationale

Bone scanning continues to be a mainstay of general nuclear medicine imaging. The most common indication for a whole-body bone scan is in the work-up of metastatic disease. However, there are many other indications for bone imaging, including the evaluation of trauma or occult fractures, osteomyelitis, bone pain, bone tumors, bone infarcts, avascular necrosis, arthritic disease, reflex sympathetic dystrophy, certain metabolic disorders, viability of bone grafts, suspected loosening or infection of an orthopedic prosthesis, and heterotopic bone formation.

Bone scans have a very high sensitivity for detecting metastasis to bone. Bone lesions can usually be detected much earlier than with plain film x-rays, because bone scans demonstrate focal areas of increased osseous remodeling that often precede the structural changes seen on plain films. The degree of radiopharmaceutical uptake in bone depends on several factors. The most important are thought to be blood flow and extraction efficiency.[54] An increase in either blood flow or extraction efficiency leads to increased radiotracer uptake in bone. There is decreased radiotracer accumulation in areas of decreased blood flow, such as a bone infarct. Decreased focal accumulation of radiotracer is sometimes seen with metal prostheses and in the occasional site where tumor has replaced most of the bone.

Procedure

Before administration of the radiopharmaceutical, it is important to understand the reason for the exam and to obtain a relevant patient history. If the problem is localized or there is a question of infection, a three-phase bone scan may be useful in evaluating blood flow and blood pooling in the region in

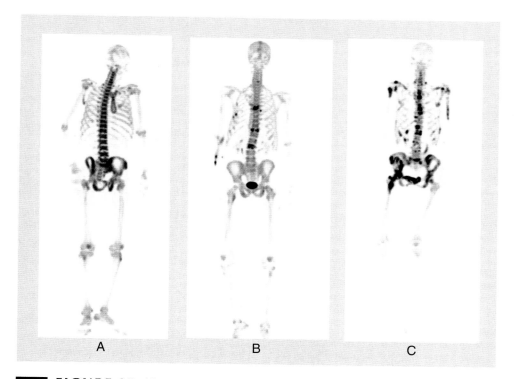

FIGURE 25-10 ¹⁸F-sodium fluoride PET maximum intensity projection (MIP) bone scans. (A) normal study, (B) abnormal study in a 47 year old woman with a history of non–small cell lung cancer, and (C) abnormal study in a 57 year old man with a history of small cell lung cancer.

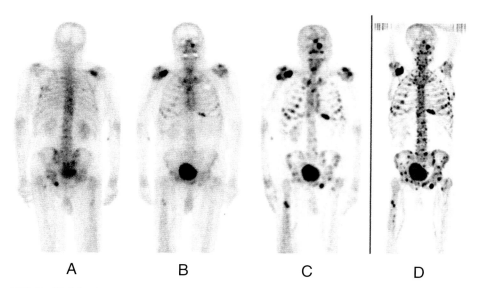

FIGURE 25-11 Posterior (A) and anterior (B) ⁹⁹ᵐTc-MDP planar scintigraphy, ⁹⁹ᵐTc-MDP multiple-field-of-view SPECT (C), and ¹⁸F-fluoride PET (D) of an 82 year old patient with numerous bone metastases. As in this patient, more lesions are typically detected by SPECT than by planar imaging, and ¹⁸F-sodium fluoride PET detects more lesions than does SPECT. (Reprinted with permission of the Society of Nuclear Medicine from reference 48.)

addition to standard delayed imaging to detect the presence of osseous remodeling. This requires that the patient be positioned and imaged with the gamma camera during intravenous administration of the radiopharmaceutical. Also, the site of radiopharmaceutical injection should be away from the site of suspected injury or pathology. Often there is partial extravasation of the radiotracer in the soft tissue at the site of administration, which can make this area difficult to evaluate.

Although bone scans are very sensitive for osseous remodeling, they are not very specific. A patient history can help increase the specificity of the exam. The patient should be asked about any history that might affect the bone scan results. This could include prior fractures, surgeries that have involved bone, recent trauma, cancer history, known bony lesions or metastatic disease to bone, prosthetic implants, radiation therapy, and osteomyelitis. If there is a question of a fracture in a younger person, localized spot imaging of the suspected area is usually performed. Whole-body bone scans are performed if there is a suspicion or history of cancer. SPECT with ^{99m}Tc-labeled diphosphonates or PET with ^{18}F-sodium fluoride can be used to improve detection or rule out bony lesions in patients with localized pain, such as in the lower back.

Once the history is obtained, the reason for the exam is understood, and the appropriate imaging procedure has been selected, the adult patient is usually injected with 20 to 30 mCi (740–1110 MBq) of ^{99m}Tc-MDP or ^{99m}Tc-HDP intravenously. In children, the dose is determined by weight, usually 250 to 300 µCi/kg (9.25–11.1 MBq/kg) with a minimum of 1 to 2.5 mCi (37–92.5 MBq).[54] Unless there are contraindications, patients should be well hydrated after injection. If blood flow imaging is to be performed, the region of interest is positioned on the gamma camera prior to injection. One- to two-second-per-frame images of the region are obtained for 30 to 60 seconds during administration of the radiotracer. After this, blood-pool images of the region are obtained within 10 minutes after administration of the radiotracer. Standard imaging is usually done 2 to 4 hours after injection. Diphosphonates are rapidly cleared from the blood and accumulate in the skeleton and in the urine. Two to four hours after administration, approximately 50% of the diphosphonate radiotracer should be adsorbed onto the mineral phase of bone; most of the rest is cleared by the kidneys. After administration of the radiopharmaceutical and before imaging, the patient is encouraged to drink fluids. Unbound diphosphonates are rapidly cleared by the kidneys. Having the patient drink fluids and void just prior to imaging helps reduce soft-tissue retention and minimize radiation dose to the bladder.

Interpretation

The normal whole-body bone scan varies with the age of the patient. In the child, there is normally intense, symmetric radiotracer uptake in the growth plate regions of the long bones (Figure 25-12). These become less intense in the older child and teenager (Figure 25-13). Once growth plate fusion is complete in the adult, the symmetric increased epiphyseal uptake is no longer present (Figure 25-14).

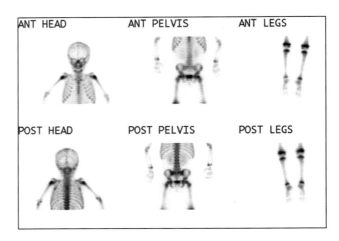

FIGURE 25-12 Normal bone scan in a 2 year old girl. Note the symmetric, increased uptake in the epiphyseal and metaphyseal growth plate regions, which is normal in this age group.

In the adult, focal areas of increased uptake in the cervical spine that have a lateral and posterior location typically represent degenerative change. Areas of more prominent skeletal uptake are normally seen in the acromioclavicular joints, sternoclavicular joints, sternum, and iliac crests. There is increased activity in the nasopharyngeal region in the skull. In general, the distribution of the radiopharmaceutical should be symmetric from side to side in the skeleton. Focal areas of increased activity represent areas of increased osseous remodeling or repair. Asymmetric focal areas of either increased or decreased

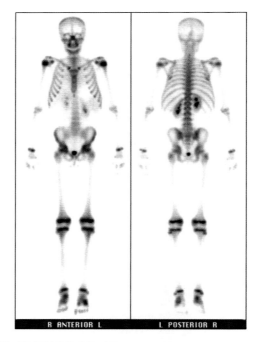

FIGURE 25-13 Normal bone scan in a 13 year old boy. There continues to be increased physiologic uptake in the growth plates, which is normal.

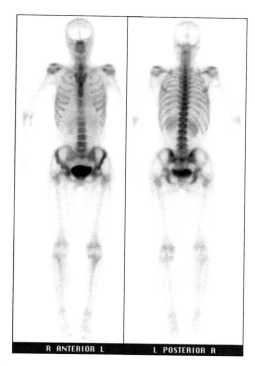

FIGURE 25-14 Normal bone scan in an adult. The epiphyseal growth centers have fused and no longer demonstrate increased uptake. The activity in the fused growth centers is now similar to that in the adjacent bone.

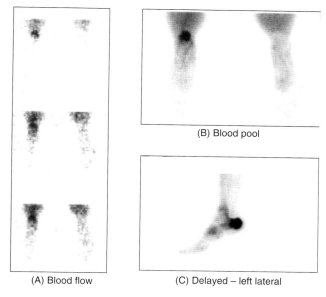

(A) Blood flow (B) Blood pool (C) Delayed – left lateral

FIGURE 25-15 Stress fracture in a patient with pain in the left foot after trauma. Three-phase bone scan of the feet demonstrates increased blood flow (A) and increased blood pool (B) in the posterior left foot. Delayed images (C) obtained 3 hours after administration of ^{99m}Tc-HDP demonstrate increased osseous remodeling in this patient with a calcaneal fracture. An x-ray obtained earlier was read as negative.

radiotracer uptake are abnormal and should be analyzed in the context of the patient's history.

Most fractures can be detected as focal areas of increased radiotracer uptake within 24 hours after injury.[55] Early on, the increased radiotracer uptake in bone is secondary to hyperemia. In the first few weeks after injury, there is generalized increased activity in the region of the fracture. The fracture becomes more prominent and focal over the next 2 to 3 months. After this, diphosphonate uptake at the fracture site gradually decreases to the same intensity as in the normal surrounding bone. Most fractures return to normal on the bone scan within 3 years.

Many traumatic fractures, such as displaced fractures, are easily seen on plain x-rays. However, some fractures can be subtle or show no abnormality on plain film x-ray. Stress fractures or fatigue fractures are the result of repeated trauma and are common in athletes. These fractures begin as microfractures that are painful but show no abnormality on plain film. It usually takes 7 to 14 days to demonstrate an anatomic abnormality on x-rays after a stress fracture. If there is continued trauma, the stress fracture can progress to a complete fracture. On a bone scan, a stress fracture is typically seen as a focal area of increased osseous remodeling at the fracture site. On a three-phase bone scan there is also increased blood flow and blood pooling to the area (Figure 25-15).

Shin splints, a painful periosteal reaction to stress along muscle insertions, are common in runners. Unlike stress fractures, shin splints do not progress to fracture. Shin splints are usually seen as increased radiotracer uptake along the cortical shaft of the tibia. They frequently involve more than one-third of the length of the tibial shaft. On a three-phase bone scan, they do not demonstrate increased blood flow and blood pooling of radiotracer. Shin splints and tibial stress fractures commonly coexist in the same patient, and it is important to differentiate between them to determine the proper therapy (Figure 25-16). If a delay

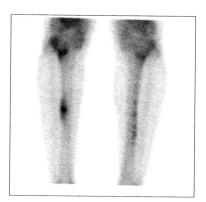

FIGURE 25-16 Anterior image of the tibias in a jogger with bilateral tibial pain. The mild, patchy increased cortical uptake seen along the left tibia represents shin splints. The more focal area of increased osseous remodeling in the right tibia represents a stress fracture.

in diagnosis is acceptable, plain film imaging after 10 to 14 days can be done. However, if a delay in diagnosis is not acceptable, such as in a marathon runner in training, three-phase bone scanning is a sensitive means of evaluating for stress fracture.[56]

Osteomyelitis, infection of the bone, can be related to a penetrating injury such as a puncture wound or to spread of infection from the adjacent soft tissues or via the blood. Early diagnosis is important for successful treatment. Osteomyelitis is more difficult to treat than soft-tissue infection (cellulitis). Usually, cellulitis responds to short-term treatment with antibiotics, whereas osteomyelitis requires long-term antibiotic therapy and possibly surgery. Thus, distinguishing between the two is important. Bone scanning will demonstrate increased osseous remodeling secondary to acute osteomyelitis before there are anatomic changes that can be seen on plain film x-rays. Improvements in bone scanning along with antibiotics and surgical techniques have helped reduce the morbidity associated with osteomyelitis.[57] In general, osteomyelitis will demonstrate increased blood flow, increased blood pool, and prominent delayed uptake in the bone within 48 to 72 hours of the onset of infection in a three-phase bone scan (Figure 25-17).[58] Evidence of bone destruction on x-rays is not appreciated for 10 to 14 days.[59] Cellulitis will demonstrate increased blood flow and blood pool but will not demonstrate prominent uptake in the bone on the delayed imaging. The interpretation of the three-phase bone scan is summarized in Table 25-2.

One of the most common indications for a whole-body bone scan is evaluation of metastatic involvement in bone for cancer staging. Also, serial bone scintigraphy in patients with known metastasis to bone can be helpful in planning radiation treatment and monitoring response. Bone scintigraphy has a high sensitivity for detecting metastatic disease to bone and evaluating the extent of involvement. Bone scanning is particularly useful in patients with nonosseous cancers that have a high tendency to metastasize to bone, such as malignancies of the prostate, breast, lung, kidney, and thyroid gland (Figure 25-18). Bone metastases most frequently involve the axial skeleton (skull, spine, and thoracic girdle). When there is extensive involvement of the skeleton, metastases to the appendicular skeleton, the pelvis, and upper and lower extremities may also be seen. When there is diffuse involvement of the skeleton in widespread metastatic disease, the lesions can become confluent, resulting in a deceptively normal-appearing bone scan called a "superscan." In such a scan there is little to no renal uptake, a high bone-to-background soft tissue ratio, and greater uptake in the axial skeleton than in the appendicular skeleton (Figure 25-19). Superscans can also be seen with metabolic conditions such as hyperparathyroidism.

Hypertrophic pulmonary osteoarthropathy is a nonneoplastic process often seen with lung cancer. Intense, often patchy, pericortical uptake is seen in the long bones; this is sometimes referred to as the "tramtrack sign" (Figure 25-20). These skeletal manifestations often regress after excision of the pulmonary lesion.[60]

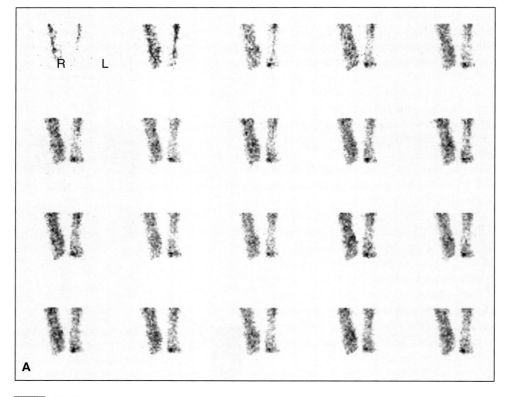

FIGURE 25-17 A patient with an ulcer on the right heel. Anterior (A).

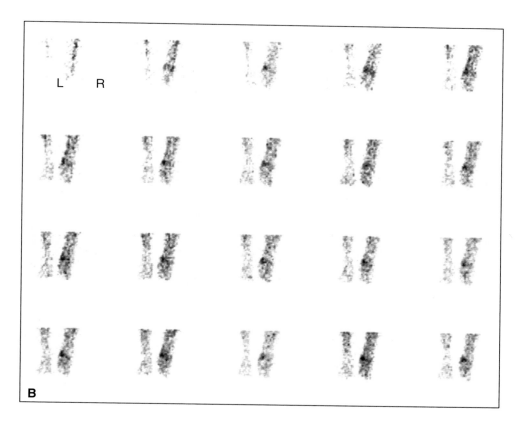
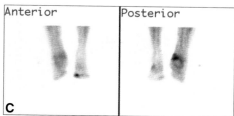
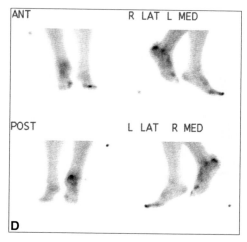

FIGURE 25-17 (Continued). A patient with an ulcer on the right heel. Posterior (B) blood flow images demonstrate increased blood flow to the right heel. There is also increased blood flow to the left great toe. Increased blood pooling (C) is also seen in these regions. On the delayed images (D) of the feet, there is increased osseous remodeling in these areas as well. Increased blood flow, blood pool, and osseous remodeling in the calcaneus was consistent with osteomyelitis. There was no associated ulceration on the left great toe. The similar pattern in the toe was likely due to trauma.

TABLE 25-2 Activity Distribution in a Three-Phase Bone Scan

Disease	Activity Accumulation		
	Phase 1: Blood Flow Image	Phase 2: Blood Pool Image	Phase 3: Delayed Image
Cellulitis	↑	↑	Normal
Osteomyelitis	↑	↑	↑
Noninflammatory[a]	Normal	Normal	↑

[a] For example, degenerative joint disease, chronic inactive osteomyelitis.

Primary bone tumors can often display prominent radiotracer uptake on bone scans (Figure 25-21). Computed tomography and magnetic resonance imaging are most useful for evaluating the extent of the tumor margin in bone and extension into the soft tissues. However, bone scintigraphy can be useful in determining whether there are distant metastases (Figure 25-22).

Bone scanning can also be useful in evaluating metabolic bone disease.[61] A bone scan can determine the extent of the disease and help evaluate response to therapy. Sometimes the bone scan can even be diagnostic. Paget's disease (osteitis deformans) is a disease of unknown etiology that begins with active resorption of bone. This is followed by a mixed phase of both resorption and bone formation and then a sclerotic phase of bone formation. It is relatively common in temperate climates, with a prevalence of 3% to 4% in people over 55 years of age.[62] The disease has a classic appearance on a bone scan. There is often very prominent uptake of the radiotracer in bone, involving the pelvis, segments of long bones, and the skull (Figure 25-23).

Gouty arthritis is a recurrent arthritis of the peripheral joints that occurs as a result of monosodium urate crystal deposition in joints and tendons. Hyperuricemia is most commonly associated with decreased renal clearance of urate. The initial symptom is acute monoarticular or polyarticular pain that is often nocturnal. Over time, insoluble crystals precipitate into joints, which can result in permanent joint deformities (Figure 25-24).

Avascular necrosis of bone is an uncommon condition characterized by occlusion of the nutrient artery to bone. The most common sites include the femoral head, distal femoral condyles, and humeral head, which are vulnerable because of

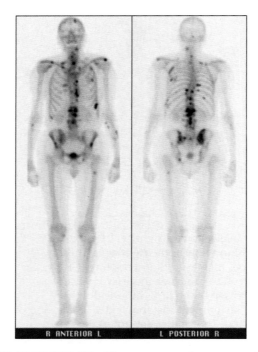

FIGURE 25-18 Anterior and posterior whole-body images demonstrating multiple abnormal focal areas of osseous remodeling in a patient with prostate cancer and metastatic disease to bone.

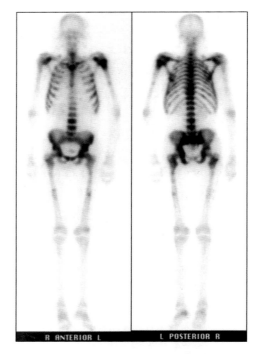

FIGURE 25-19 "Superscan" in a patient with prostate carcinoma and widespread metastatic disease to bone. The skeleton is diffusely involved with metastatic disease. There is a high bone-to-soft tissue background ratio, and the kidneys are barely visualized.

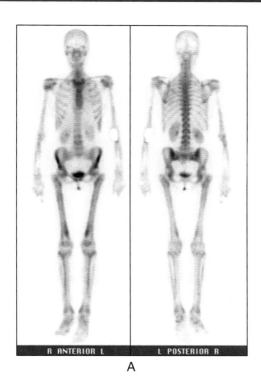
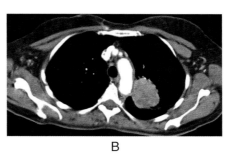

FIGURE 25-20 Hypertrophic pulmonary osteoarthropathy in a patient with lung cancer. (A) There is diffusely increased osseous remodeling along the cortical shafts of the femurs, which is secondary to lung cancer in this case. Note the focal area of increased soft tissue uptake in the left chest on the posterior whole-body image between the fifth and sixth ribs, representing uptake in the patient's lung mass. (B) CT image demonstrating left apical lung mass.

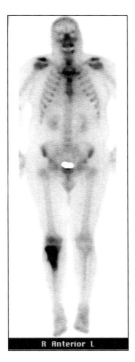

FIGURE 25-21 Adult male with knee pain. There is prominent, abnormal, increased osseous remodeling in the proximal one-third of the right tibia in this patient with chondrosarcoma.

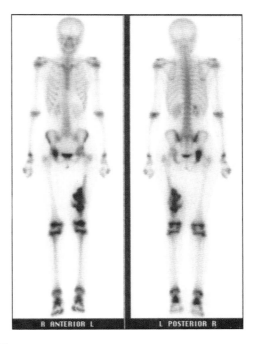

FIGURE 25-22 There is a large, expansile-appearing focal area of increased osseous remodeling involving the mid to distal left femur in this 14 year old patient with osteosarcoma. There is also metastatic involvement of the right acetabulum extending into the ischium. Note the physiologic increased uptake in the growth plates.

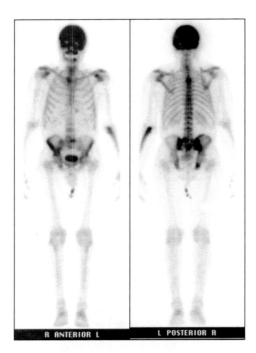

FIGURE 25-23 Anterior and posterior whole-body bone scan demonstrating prominent uptake in the skull, pelvis, and left ulna in a patient with Paget's disease.

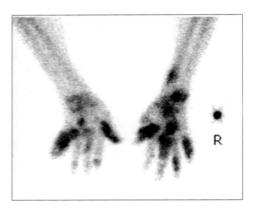

FIGURE 25-24 Bone scan of the hands demonstrating multiple abnormal focal areas of osseous remodeling in a patient with gout.

poor collateral blood supply. In the early stages of avascular necrosis, decreased blood activity and osseous uptake at the site appear as a cold spot on the bone scan (Figure 25-25). In later stages of the disease, collateral circulation causes increased blood activity with osseous remodeling and evidence of increased uptake at the site.

As seen in the preceding examples, an area of increased osseous remodeling or a focal hot spot on a bone scan typically catches a clinician's attention. Bone scintigraphy offers information not only about increased osseous remodeling but about normal, decreased, or absent osseous remodeling. Areas of decreased osseous remodeling can be seen in regions where a patient received prior radiation therapy. Areas of absent uptake in bone can be associated with bone infarctions, avascular necrosis, or metal prostheses, or may occur where tumor has replaced the bone. Diffusely increased soft tissue uptake in the body can be seen with poor renal function. More localized soft tissue uptake can be seen in certain tumors or in areas with poor venous return, such as areas of edema in an extremity. Metabolic information about osseous remodeling obtained through bone scanning continues to have an important place in clinical decision-making.

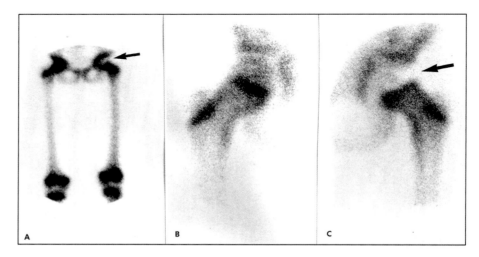

FIGURE 25-25 Avascular necrosis. Anterior bone scan on a young child illustrating the pelvis and thighs (A), right hip (B), and left hip (C). Note the absence of tracer in the left femoral head because of arterial occlusion to bone. Also evident is the increased concentration of activity normally seen in the epiphyseal growth plates of a child.

REFERENCES

1. Edelstyn GA, Gillespie PJ, Grebell FS. The radiological demonstration of osseous metastasis. Experimental observation. *Clin Radiol.* 1967:18:158–62.
2. Trifflitt JT. The organic matrix of bone tissue. In: Urist MR, ed. *Fundamental and Clinical Bone Physiology.* Philadelphia: JB Lippincott; 1980:45–82.
3. Neuman WF. Bone material and calcification mechanisms. In: Urist MR, ed. *Fundamental and Clinical Bone Physiology.* Philadelphia: JB Lippincott; 1980:83–107.
4. The Radium Girls. www.runet.edu/~wkovarik/envhist/radium.html. Accessed November 6, 2010.
5. Dudley HC, Maddox GE, LaRue HC. Studies of the metabolism of gallium. *J Pharmacol Exp Ther.* 1949;96:135–8.
6. Edwards CL, Hayes RL. Tumor scanning with Ga-67 citrate. *J Nucl Med.* 1969;10:103–5.
7. Charkes ND, Sklaroff DM. Early diagnosis of metastatic bone cancer by photoscanning with strontium-85. *J Nucl Med.* 1964;5:168–79.
8. Gynning I, Langeland P, Lindberg S, et al. Localization with Sr-85 of spinal metastasis in mammary cancer and changes in uptake after hormone and roentgen therapy: a preliminary report. *Acta Radiol.* 1961;55:119–28.
9. Meckelnburg RI. Clinical value of generator produced 87m-strontium. *J Nucl Med.* 1964;5:929–35.
10. Blau M, Nagler W, Bender WA. Fluorine-18: a new isotope for bone scanning. *J Nucl Med.* 1962;3:332–4.
11. Subramanian G, McAfee JG. A new complex of Tc-99m for skeletal imaging. *Radiology.* 1971;99:192–6.
12. Fels IG, Kaplan E, Greco J. Incorporation in vivo of P-32 from condensed phosphates. *Proc Soc Exp Biol Med.* 1959;100:53–5.
13. Subramanian G, McAfee JG, O'Mara RE, et al. Tc-99m polyphosphate PP_{46}: a new radiopharmaceutical for skeletal imaging. *J Nucl Med.* 1971;12:399–400.
14. Davis MA, Jones AG. Comparison of Tc-99m labeled phosphate and phosphonate agents for skeletal imaging. *Semin Nucl Med.* 1976;6:19–31.
15. Perez R, Cohen Y, Henry R, et al. A new radiopharmaceutical for Tc-99m bone scanning [abstract]. *J Nucl Med.* 1972;13:788.
16. Yano Y, McRae J, Van Dyke DC, et al. Technetium-99m labeled stannous ethane-1-hydroxy-1, 1-diphosphonate: a new bone scanning agent. *J Nucl Med.* 1973;14:73–8.
17. Castronovo FP, Callahan RJ. New bone scanning agent: Tc-99m-labeled 1-hydroxyethylidene-1, 1-disodium phosphonate. *J Nucl Med.* 1972;13:823–7.
18. Subramanian G, McAfee JG, Blair RJ, et al. Tc-99m EHDP: a potential radiopharmaceutical for skeletal imaging. *J Nucl Med.* 1972;13:947–50.
19. Subramanian G, McAfee JG, Blair RJ, et al. Technetium-99m methylene diphosphonate: a superior agent for skeletal imaging: comparison with other technetium complexes. *J Nucl Med.* 1975;16:744–55.
20. Barnett BL, Strickland LC. The structure of disodium dihydrogen-1-hydroxyethylidene diphosphonate tetrahydrate: a bone growth regulator. *Acta Crystallogr B.* 1979;B35:1212–4.
21. Francis MD, Tofe AJ, Benedict JJ, et al. Imaging the skeletal system. In: Sorenson JA, ed. *Radiopharmaceuticals II.* New York: Society of Nuclear Medicine; 1979:603–14.
22. Nelson MF, Melton RF, Van Wazer JR. Sodium trimetaphosphate as a bone-imaging agent. I: Animal studies. *J Nucl Med.* 1975;16:1043–8.
23. Kowalsky RJ, Dalton DR. Technical problems associated with the production of technetium Tc-99m tin (II) pyrophosphate kits. *Am J Hosp Pharm.* 1981;38:1722–6.
24. Chandler WM, Shuck LD. Abnormal technetium-99m pertechnetate imaging following stannous pyrophosphate bone imaging. *J Nucl Med.* 1975;16:518–9.
25. Tofe AJ, Francis MD. In vitro stabilization of low-tin bone-imaging agent (Tc-99m-Sn-HEDP) by ascorbic acid. *J Nucl Med.* 1976;17:820–5.
26. Tofe AJ, Bevan JA, Fawzi MD, et al. Gentisic acid: a new stabilizer for low tin skeletal imaging agents: concise communication. *J Nucl Med.* 1980;21:366–70.
27. Ballinger J, Der M, Bowen B. Stabilization of Tc-99m pyrophosphate injection with gentisic acid. *Eur J Nucl Med.* 1981;6:153–4.
28. Hambright P, McRae J, Valk PE, et al. Chemistry of technetium radiopharmaceuticals: exploration of the tissue distribution and oxidation state consequences of technetium (IV) in Tc-Sn-gluconate and Tc-Sn-EHDP using carrier ^{99}Tc. *J Nucl Med.* 1975;16:478–82.
29. Tofe AJ, Francis MD. Optimization of the ratio of stannous tin:ethane-1-hydroxy-1-diphosphonate for bone scanning with Tc-99m pertechnetate. *J Nucl Med.* 1974;15:69–74.
30. Bevan J, Tofe AJ, Benedict JJ, et al. Tc-99m HMDP (hydroxymethylene diphosphonate): a radiopharmaceutical for skeletal and acute myocardial infarct imaging. 1: Synthesis and distribution in animals. *J Nucl Med.* 1980;21:961–6.
31. Ponto JA. Preparation and dispensing problems associated with technetium Tc-99m radiopharmaceuticals. In: Hladik W B III, ed. *Continuing Education Course for Nuclear Pharmacists and Nuclear Medicine Professionals.* Vol 11, no. 1. Albuquerque, NM: University of New Mexico College of Pharmacy; 2004.
32. Deutsch E. Inorganic radiopharmaceuticals. In: Sorenson JA, ed. *Radiopharmaceuticals II.* New York: Society of Nuclear Medicine; 1979:129–46.
33. Pinkerton TC, Ferguson DL, Deutsch E, et al. In vivo distributions of some component fractions of $Tc(NaBH_4)$-HEDP mixtures separated by anion-exchange high performance liquid chromatography. *Int J Appl Radiat Isot.* 1982;33:907–15.
34. Francis MD, Ferguson DL, Tofe AJ, et al. Comparative evaluation of three diphosphonates: in vitro adsorption (C-14 labeled) and in vivo osteogenic uptake (Tc-99m complexed). *J Nucl Med.* 1980;21:1185–9.
35. Fogelman I, Pearson DW, Bessent RG, et al. A comparison of skeletal uptake of three diphosphonates by whole body retention: concise communication. *J Nucl Med.* 1981;22:880–3.
36. Silberstein EB. A radiopharmaceutical and clinical comparison of Tc-99m-Sn-hydroxy-methylene diphosphonate with Tc-99m-Sn-hydroxy-ethylidene diphosphonate. *Radiology.* 1980;136:747–51.
37. Littlefield JL, Rudd TG. Tc-99m hydroxy-methylene diphosphonate (HMDP) versus Tc-99m methylene diphosphonate (MDP): biological and clinical comparison [abstract]. *Clin Nucl Med.* 1980;5:S28.
38. Rosenthall L, Arzoumanian A, Damtew B, et al. A crossover study comparing Tc-99m labeled HMDP and MDP in patients. *Clin Nucl Med.* 1981;6:353–5.
39. Rudd TG, Allen DR, Hartnett DE. Tc-99m methylene diphosphonate versus Tc-99m pyrophosphate: biologic and clinical comparison. *J Nucl Med.* 1977;18:872–6.
40. Fogelman I, Citrin DL, McKillop JH. A clinical comparison of Tc-99m HEDP and Tc-99m MDP in the detection of bone metastasis. *J Nucl Med.* 1979;20:98–101.
41. Stabin MG, Stubbs JB, Toohey RE. *Radiation Dose Estimates for Radiopharmaceuticals.* Oak Ridge, TN: Radiation Internal Dose Information Center, Oak Ridge Institute for Science and Education (ORISE); 1996.
42. Thomas SR, Gelfand MJ, Kerades JG, et al. Dose to the metaphyseal growth complexes in children undergoing Tc-99m EHDP bone scans. *Radiology.* 1978;126:193–5.
43. Stevenson JS, Eckelman WC, Sobocinski PZ, et al. The toxicity of Sn-pyrophosphate: clinical manifestations prior to acute LD-50. *J Nucl Med.* 1974;15:252–6.
44. Atkins HL. Reported adverse reactions to radiopharmaceuticals remain low in 1984. *J Nucl Med.* 1986;27:327.
45. Hoving J, Roosjen HN, Brouwers JRBJ. Adverse reactions to technetium-99m methylene diphosphonate. *J Nucl Med.* 1988;29:1302–3.
46. Hladik WB III, Nigg KK, Rhodes BA. Drug induced changes in the biologic distribution of radiopharmaceuticals. *Semin Nucl Med.* 1982;12:184–218.
47. Lentle BC, Scott JR, Noujaim AA, et al. Iatrogenic alterations in radionuclide biodistributions. *Semin Nucl Med.* 1979;9:131–43.
48. Grant FD, Fahey FH, Packard AB, et al. Skeletal PET with ^{18}F-fluoride: applying new technology to an old tracer. *J Nucl Med.* 2008;49:68–78.

49. Jones AG, Francis MD, Davis MA. Bone scanning: radionuclidic reaction mechanisms. *Semin Nucl Med.* 1976;6:3–18.
50. Schirrmeister H, Guhlmann A, Elsner K, et al. Sensitivity in detecting osseous lesions depends on anatomic localization: planar bone scintigraphy versus ^{18}F PET. *J Nucl Med.* 1999;40:1623–9.
51. Schirrmeister H, Guhlmann A, Kotzerke J, et al. Early detection and accurate description of extent of metastatic bone disease in breast cancer with fluoride ion and positron emission tomography. *J Clin Oncol.* 1999;17:2381–9.
52. Schirrmeister H, Glatting G, Hetzel J, et al. Prospective evaluation of the clinical value of planar bone scans, SPECT, and ^{18}F-labeled NaF PET in newly diagnosed lung cancer. *J Nucl Med.* 2001;42:1800–04.
53. Hetzel M, Arslandemir C, Konig HH, et al. F-18 NaF PET for detection of bone metastases in lung cancer: accuracy, cost-effectiveness, and impact on patient management. *J Bone Miner Res.* 2003;18:2206–14.
54. Donohue KJ, Brown ML, Collier BD, et al. Procedure guideline for bone scintigraphy 3.0. In: *Procedure Guidelines Manual.* Reston, VA: Society of Nuclear Medicine; 2003.
55. Matin P. Appearance of bone scans following fractures: including immediate and long-term studies. *J Nucl Med.* 1979;20:1227–31.
56. Matin P. Basic principles of nuclear medicine techniques for detection and evaluation of trauma and sports medicine injuries. *Semin Nucl Med.* 1988;18:90–112.
57. Waldvogel FA, Vasey H. Osteomyelitis: the past decade. *N Engl J Med.* 1980;303:360–70.
58. Van der Wall H. Assessment of infection. In: Murray IPC, Ell PJ, eds. *Nuclear Medicine in Clinical Diagnosis and Treatment.* 2nd ed. Edinburgh: Churchill Livingstone; 1998:1185–1207.
59. Handmaker H, Leonards R. The bone scan in inflammatory osseous disease. *Semin Nucl Med.* 1976;6:95–105.
60. Rosenthall L, Kirsh J. Observations of radionuclide imaging in hypertrophic pulmonary osteoarthropathy. *Radiology.* 1976;120:359–62.
61. Hain SF, Fogelman I. Nuclear medicine studies in metabolic bone disease. *Semin Musculoskelet Radiol.* 2002;6:323–30.
62. Elgazzar AH, Shehab D, Malki A, et al. Musculoskeletal system. In: *The Physiologic Basis of Nuclear Medicine.* New York: Springer-Verlag; 2001:88–126.

CHAPTER 26

Total-Body SPECT Procedures

The previous chapters have focused on radiopharmaceuticals for evaluation of the major organ systems in the body. This chapter discusses radiopharmaceuticals employed in total-body soft-tissue single-photon emission computed tomography (SPECT) imaging, with principal focus on infection and tumor imaging and miscellaneous nuclear medicine imaging procedures.

INFECTION IMAGING

The Inflammatory Process

An inflammatory response may be caused by a noxious stimulus, such as infection, heat, chemical insult, or trauma. It may also be caused by a disease process, such as rheumatic or allergic conditions. The inflammatory reaction is characterized by a vascular response and occurs in two phases: a fluid phase and a cellular phase.[1,2] The fluid phase begins immediately after the initial injury, when chemical mediators such as histamine, bradykinin, and serotonin are liberated by the damaged tissue. These substances cause localized hyperemia and increased capillary permeability, which allows proteins and fluids to leak into the affected tissue, causing edema. The presence of fibrinogen in the fluid leads to the formation of clots that wall off the injured area, delaying the spread of bacteria and toxic products. Other substances are also instrumental, such as C-reactive protein, which is released from the liver and functions as a nonspecific opsonin to enhance phagocytosis of bacterial debris and toxins from an infectious process. Cytokines, such as the interleukins released from activated macrophages and other cells, participate in local and systemic responses to inflammation.[1]

The cellular phase of the inflammatory reaction has three stages.[2] The first involves phagocytic activity of local macrophages. The second is characterized by neutrophilia, with a large increase in circulating neutrophils in response to chemical mediators released in the inflamed tissues. The number of neutrophils can increase from 15,000 to 25,000 per microliter within a few hours. Neutrophils migrate to the inflamed area and localize by diapedesis through the porous capillary walls. They are attracted by bacterial toxins and cellular products (chemotaxis), whereupon they become attached and exert their phagocytic function. The third stage of the cellular response is slower but longer. It involves migration of lysosome-rich monocytes to the injured area over a period of 8 to 12 hours. Monocytes also exert a phagocytic function.

The adult human has approximately 7000 white blood cells (WBCs) per cubic millimeter (microliter) of blood. The WBC component of blood is composed of polymorphonuclear neutrophils (62%), eosinophils (2.3%), and basophils (0.4%), collectively called "polys" or granulocytes because of their granular appearance; monocytes (5.3%); and lymphocytes (30%). Once released from the bone marrow, granulocytes have a normal life span of 6 to

8 hours in circulation and 2 to 3 days in tissues. Circulation time is shortened during serious infection. Monocytes have a short life span in blood but may exist in tissues as macrophages for months. Lymphocytes have a prolonged life span of several months. Neutrophils and monocytes are the main types of WBCs involved in an inflammatory response to invading microorganisms.

Infection stimulates an acute inflammatory process in the body, and a variety of radiolabeled agents have been developed to localize active sites of infection by exploiting the various processes involved in inflammation. The most widely used agents for infection imaging are ^{67}Ga-citrate and ^{111}In- and ^{99m}Tc-labeled WBCs. Other agents include radiolabeled monoclonal antibodies and immunoglobulins, colloids, peptides, interleukins, antibiotics, and ^{18}F-fludeoxyglucose (^{18}F-FDG, or FDG).

Gallium Citrate Ga 67 Injection

Gallium citrate Ga 67 injection (^{67}Ga-citrate) was originally investigated as a bone imaging agent but was quickly redirected to tumor imaging when, serendipitously, it was discovered that carrier-free ^{67}Ga-citrate localized in the lymph nodes of a subject with Hodgkin's disease.[3] Soon thereafter, it was discovered that ^{67}Ga also localized in inflammatory processes.[4,5] Despite this lack of specificity for tumor imaging, ^{67}Ga-citrate has been used for over 30 years to image certain tumors and inflammatory processes.[6] Its role in nuclear medicine for infection and tumor imaging has diminished somewhat with the availability of FDG positron emission tomography (PET) imaging.

After intravenous injection of ^{67}Ga-citrate, the complex dissociates in blood and ^{67}Ga becomes protein bound to plasma transferrin.[7,8] The plasma clearance of ^{67}Ga activity is biphasic, with half-lives of 30 hours and 25 days for 17% and 85% of the injected activity, respectively.[9] Excretion from the body is slow, with 15% to 25% appearing in the urine within 24 hours. By 7 days, 35% has been excreted in urine and feces combined, and 65% remains in the body.[10] At 24, 48, and 72 hours after injection, about 20%, 10%, and 5%, respectively, still remains in the blood.[11]

Gallium normally localizes primarily in the liver, spleen, gastrointestinal (GI) tract, kidney, skeleton, and bone marrow (Figure 26-1),[9,12] and uptake occasionally occurs in the lacrimal and salivary glands and in the breasts during menarche, pregnancy, and lactation.[13–15] The high concentration of lactoferrin in many normal tissues is believed to be responsible for gallium uptake in these tissues. Gallium's nonspecific uptake in the GI tract is the major disadvantage of this radiopharmaceutical for imaging the abdominal region. Difficulties encountered with planar imaging in distinguishing bowel content of ^{67}Ga from abnormal foci in the abdomen or pelvis have been alleviated by SPECT imaging, which can more precisely define ^{67}Ga in bowel lumen or nearby organs (e.g., pancreas, kidney, and peritoneum) (Figure 26-2).

Although the uptake of ^{67}Ga in infectious processes and tumors has been investigated extensively, its precise mecha-

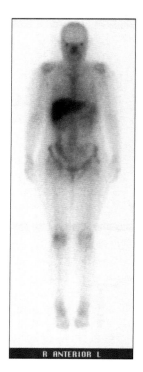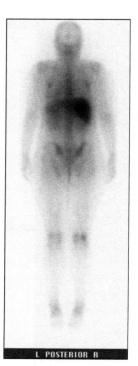

FIGURE 26-1 Normal total-body ^{67}Ga scan. Anterior and posterior whole-body images were obtained 48 hours after intravenous administration of 8 mCi (296 MBq) of ^{67}Ga-citrate. In the head, normal radiotracer accumulation is seen in the lacrimal glands and nasopharynx. Physiologic uptake is seen in the skeleton. In the abdomen, the most prominent uptake is seen in the liver, with a lesser amount in the spleen. Activity is often present in the colon.

nism is not clear. It is believed that at a site of infection or inflammation ^{67}Ga-transferrin diffuses through loose endothelial junctions into the extracellular space, which has high amounts of lactoferrin released by leukocytes. ^{67}Ga dissociates from transferrin and binds to lactoferrin, which has a higher binding affinity for gallium.[16,17] In addition, gallium can also bind to bacterial siderophores and macrophage ferritin, which bind gallium more firmly than transferrin and lactoferrin.[18,19]

Gallium is taken up in viable tumor cells and not in necrotic or scarred regions of tumor. However, the exact mechanism of gallium uptake in tumors remains unclear. One theory for tumor uptake is that Ga-transferrin in blood is taken up by tumor cell receptors and intracellular gallium dissociates and binds to lactoferrin or to a specific tumor protein or is localized in tumor cell organelles.[20–23]

The standard adult dosage is 3 to 5 mCi (111–185 MBq) for infection imaging and 8 to 10 mCi (296–370 MBq) for tumor imaging. Imaging for infection may be done at 4 hours and 24 hours if prompt information is required, with delayed imaging at 48 to 72 hours to provide better quality images because of less background activity. Tumor imaging is typically done 48 to 72 hours after dose administration. The criti-

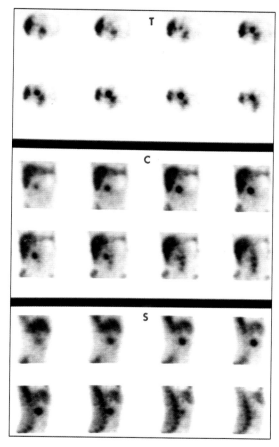

FIGURE 26-2 ^{67}Ga infection scan. Gallium SPECT transaxial (T), coronal (C), and sagittal (S) images demonstrating an abscess in the head of the pancreas. SPECT 3-D definition allowed differentiation of the pancreatic focus from bowel activity.

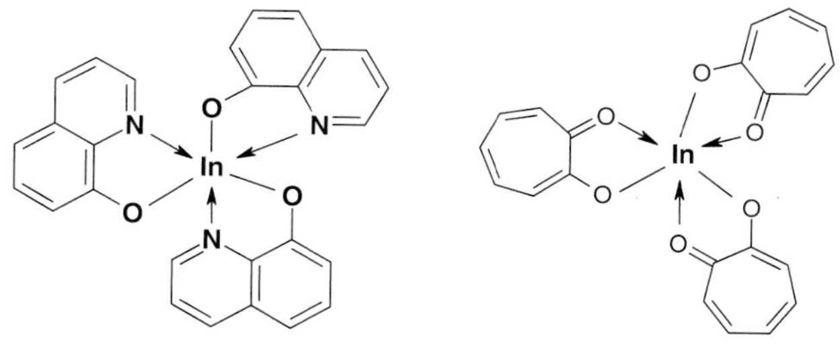

^{111}In-Oxine ^{111}In-Tropolone

FIGURE 26-3 Chemical structures of ^{111}In-oxine and ^{111}In-tropolone.

Subsequently, the lipophilic complex ^{99m}Tc-HMPAO was employed to label WBCs. A method for labeling WBCs with indium and technetium is illustrated in Figure 10-27 in Chapter 10 and discussed in Chapters 10 and 11.

Indium In 111 White Blood Cells

The maximum adult administered activity of ^{111}In-WBCs is 0.5 mCi (18.5 MBq), with a recommended range of 0.3 to 0.5 mCi (11.1–18.5 MBq). Labeled cells should be administered within 1 to 2 hours after labeling—no longer than 3 hours—because cells lose viability. The plasma half-life is about 7 hours.[28] Imaging is usually performed 18 to 24 hours after injection but may be done at 1 to 4 hours if information is urgently needed. Imaging at 1 to 4 hours, however, is associated with lower sensitivity for abscess detection (33%) than 24 hour images (95%).[29] Beyond 24 hours, radioactivity is normally present in the liver, spleen, and bone marrow, but body background is generally low (Figure 26-4). Early images after injection show transient lung activity due to margination of WBCs, which is mostly cleared by 4 hours. The intense spleen activity is due to labeled platelets and WBCs damaged during preparation. This distribution is the reverse of ^{67}Ga-citrate scans, in which liver activity is more intense than spleen activity. Areas of infection appear as focal regions of increased uptake of activity similar to that seen with ^{67}Ga-citrate (Figure 26-5). Because of the appearance of early lung activity with ^{111}In-WBCs, some physicians prefer ^{67}Ga-citrate when imaging chest inflammatory conditions such as sarcoidosis, diffuse pneumonia, and interstitial inflammatory reactions.[29] Suspected infections in the abdomen and pelvis, however, are best imaged with ^{111}In-WBCs because of the lack of activity in the bowel.

The critical organ with ^{111}In-WBCs is the spleen, with a radiation absorbed dose of 20 rad(cGy)/0.5 mCi. This includes the dose due to ^{111}In (99.75%) and ^{114m}In/^{114}In (0.25% impurities) at the time of expiration of the ^{111}In-oxine.

Technetium Tc 99m White Blood Cells

The usual adult administered activity of ^{99m}Tc-WBCs is 10 mCi (370 MBq), with a suggested range of 7 to 25 mCi (259–925 MBq). The plasma half-life is approximately 4 hours.[30]

cal organ is the lower large intestine, with a radiation absorbed dose of 0.9 rad(cGy)/mCi.

Radiolabeled White Blood Cells

The nonspecific localization of ^{67}Ga-citrate during infection imaging stimulated an investigation into methods for labeling WBCs with soluble and particulate agents.[24,25] Particulates were ineffective because of loosely attached radioactive particles on leukocyte surfaces after radiolabeling. Only nonpolar, lipid-soluble radionuclide chelates labeled cells effectively. The most efficient agents were ^{111}In chelates of oxine, tropolone, and acetylacetone, which form neutral, lipid-soluble 3:1 chelates with indium.[26,27] The oxine ligand was preferred over acetylacetone because the latter tended to elute from WBCs and had a higher degree of erythrocyte (red blood cell, or RBC) labeling.[24] Both ^{111}In-oxine and ^{111}In-tropolone have been shown to be effective labeling agents for WBCs and platelets; however, the only complex available commercially for routine use is ^{111}In-oxine. The chemical structures of these two complexes are shown in Figure 26-3.

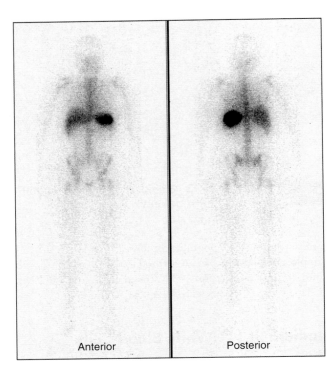

FIGURE 26-4 Normal ^{111}In-WBC scan. Anterior and posterior whole-body images were obtained 24 hours after intravenous administration of 500 μCi (18.5 MBq) of ^{111}In-oxine-labeled WBCs. Radiotracer accumulation is seen in the liver, spleen, and bone marrow.

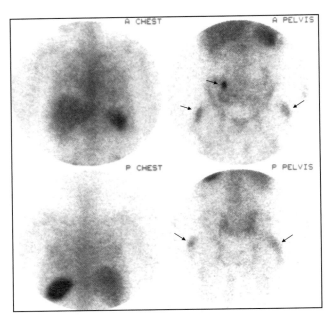

FIGURE 26-5 ^{111}In-WBC scan demonstrating infection. Anterior and posterior spot images of the chest, abdomen, and pelvis were obtained 24 hours after intravenous administration of 500 μCi (18.5 MBq) of ^{111}In-oxine-labeled WBCs. There are several abnormal focal areas of radiotracer accumulation; the most prominent are in the right pelvis and in the soft tissues lateral to both hips (arrows). There is also abnormal diffuse uptake in the lungs.

The biologic distribution of ^{99m}Tc-WBCs differs slightly from that of ^{111}In-WBCs, demonstrating some biliary excretion with bowel activity and activity in the renal collecting system and bladder.[30] These accumulations are attributed to release of secondary hydrophilic complexes of ^{99m}Tc-HMPAO.[30] Occasionally, gallbladder activity is seen. The nonspecific bowel activity does not appear until at least 2 hours after injection and increases beyond 4 hours after injection. It presents a problem in imaging inflammatory bowel disease (IBD) and abdominal infection. Typically, in imaging soft-tissue sepsis, sequential imaging at 1 to 4 hours is necessary to delineate abscess from nonspecific bowel activity. Infections localized away from the abdomen are more easily identified (Figure 26-6). Imaging can be done up to 24 hours after dosing, but significant bowel activity may interfere at this time and result in false-positive diagnosis. As with ^{111}In-WBCs, there is significant liver and spleen activity with ^{99m}Tc-WBCs. ^{99m}Tc-WBCs are the agent of choice for IBD and acute sepsis, in which an early diagnosis is mandatory.[30] Diagnostic accuracy in these conditions is considered equivalent to that of ^{111}In-WBCs.[30]

The critical organ with ^{99m}Tc-WBCs is the spleen, with a radiation absorbed dose of 13.9 rad(cGy)/25 mCi (925 MBq) maximum administered activity. Table 26-1 compares the properties of ^{67}Ga-citrate, ^{99m}Tc-WBCs, and ^{111}In-WBCs for infection imaging.

Miscellaneous Agents for Infection Imaging

A variety of agents for infection imaging have been developed with the goal of achieving infection-specific targeting, separate from noninfectious inflammatory processes, to improve diagnostic specificity. However, only radiolabeled WBCs have been found to be specific for imaging sites of acute infection.[31]

A serendipitous finding that radiolabeled immunoglobulin localized in sites of inflammation led to the development and investigation of ^{111}In- and ^{99m}Tc-labeled IgG for localizing sites of infection.[32,33] ^{111}In was bound to IgG via a DTPA-linked chelation site, and ^{99m}Tc was chelated to IgG via a hydrazinonicotinamide (HYNIC) chelation group. This was an exciting finding because it would greatly simplify infection imaging by eliminating the need for ex vivo labeling of leukocytes. Subsequent studies, however, revealed that radiolabeled IgG lacked specificity, localizing in regions of sterile inflammation as well as sites of bacterial and fungal infection. Additional investigations to confirm involvement of the Fc region of IgG with Fc receptors on inflammatory cells as a mechanism of localization revealed that no Fc binding was involved. The investigators concluded that radiolabeled IgG localized by increased influx into the expanded extracellular space characteristic of inflammatory foci.[34]

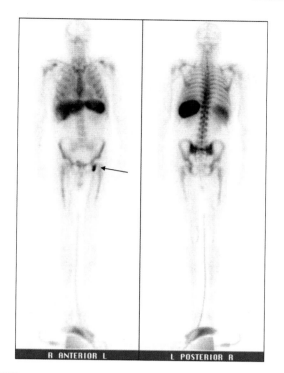

FIGURE 26-6 Femoral graft infection. Anterior and posterior whole-body images were obtained approximately 4 hours after intravenous administration of 10 mCi (370 MBq) ^{99m}Tc-HMPAO-labeled autologous WBCs. On the anterior image there is an abnormal focal accumulation of radiotracer in the left groin region that extends from the surface of the skin to the patient's femoral graft, consistent with infection at the graft insertion site (arrow).

The known specificity of radiolabeled WBCs for sites of infection led to the development of two technetium-labeled antigranulocyte monoclonal antibodies to label leukocytes for infection imaging. ^{99m}Tc-sulesomab (^{99m}Tc-LeukoScan [Immunomedics GmbH, Darmstadt]) reacts with nonspecific cross-reactive antigen (NCA-90) on the surface of granulocytes.[35] ^{99m}Tc-sulesomab is prepared from a kit. This product is not approved for routine use in the United States; however, clinical investigation has shown it to be effective for diagnosis of acute appendicitis in children.[36] LeukoScan is approved in Europe for the diagnosis of infection. ^{99m}Tc-fanolesomab (^{99m}Tc-NeutroSpec [Mallinckrodt, Hazelwood, MO]), formerly known as ^{99m}Tc-LeuTech, is an anti-CD15 immunoglobulin M monoclonal antibody that binds to the CD-15 epitope on activated neutrophils in vivo. A major advantage of this product was in vivo labeling of neutrophils after intravenous administration. NeutroSpec was approved by FDA for routine use in the diagnosis of appendicitis, but it was removed from the market for safety reasons because two deaths were associated with its use.

The idea of targeting microorganisms present in infections stimulated the development of a radiolabeled antibiotic, ^{99m}Tc-ciprofloxacin (Infecton [DraxImage, Quebec, Canada]), to image sites of infection.[37] Ciprofloxacin targets and inhibits the enzyme DNA-gyrase, which is necessary for bacterial replication. The radiolabeled product is prepared by mixing 2 mg of ciprofloxacin with 0.5 mg of stannous tartrate and 10 mCi of ^{99m}Tc-sodium pertechnetate. Although this agent has shown fairly high sensitivity in localizing infections, its specificity for excluding sites of inflammation without infection is marginal to low.[38,39] In line with these findings is the in vitro and in vivo evaluation of ^{18}F-ciprofloxacin, which demonstrated that this agent was not retained by bacterial

TABLE 26-1 Comparison of ^{67}Ga-Citrate and ^{99m}Tc- and ^{111}In-WBCs for Infection Imaging

Property	^{99m}Tc-WBCs	^{111}In-WBCs	^{67}Ga-Citrate
Availability	Same day as request	Next day after request	Same day as request
Administered activity	10 mCi	0.5 mCi	4–6 mCi
Target-to-background ratio	Fair	Good	Poor
Collimation	Low energy	Medium energy	Medium energy
Labeling	In ≤ 20% plasma	Remove plasma	None
EDE[a]	0.74 rem/10 mCi	1.2 rem/0.5 mCi	1.23 rem/3 mCi
Plasma $T_{1/2}$	~ 4 hr (^{99m}Tc label elutes)	~ 7 hr	30 hr (17%) 25 days (83%)
Imaging time	1 to 4 hr	18 to 24 hr	24 to 72 hr
Normal activity distribution	Liver, spleen, bone marrow, bowel, kidney, gallbladder	Liver, spleen, bone marrow	Liver, spleen, bone marrow, bowel, kidney, skeleton
General applications	Acute infection, inflammatory bowel disease, osteomyelitis	Acute and chronic infection, inflammatory bowel disease, osteomyelitis, FUO[b]	Chronic, nonbacterial, and spine infections, neutropenia, FUO[b]

[a] Effective dose equivalent data from Radiation Internal Dose Information Center, Oak Ridge Institute For Science and Education, Oak Ridge, TN 37831-0117.
[b] Fever of unknown origin.

cells in culture and had similar washout rates in infected and noninfected tissue in patients with confirmed sites of infection.[40] The investigators concluded that ^{18}F-ciprofloxacin localization is governed by increased blood flow and vascular permeability at the site of infection and that ^{18}F-ciprofloxacin is not suited for use as a bacteria-specific infection imaging agent.

The inflammatory process in the body is modulated by cytokines, such as interleukins and interferons, which are signaling molecules that transmit information between cells. Interleukins are a large group of proteins produced mainly by T cells and in some cases by mononuclear phagocytes. The interleukins participate in the inflammatory response, facilitating communication between leukocytes. Each type of interleukin modulates cell function by interacting with receptors on particular groups of cells, directing the cells to divide and differentiate. Radiolabeled interleukins have been developed as diagnostic markers for investigating inflammatory diseases. For example, Crohn's disease causes activation of mononuclear cells expressing interleukin-2 (IL-2) receptors. ^{123}I-labeled IL-2 has been shown to accumulate in the gut wall of Crohn's disease patients by binding to receptors of mononuclear cells infiltrating the gut wall.[41] Such methods would be useful for evaluating the presence of disease and monitoring the effects of therapy. ^{99m}Tc-labeled interleukin-8 (^{99m}Tc-IL-8) has been prepared by using an IL-8 HYNIC conjugate.[42] ^{99m}Tc-IL-8 was investigated for localizing infections in patients and was found to be well tolerated and to detect various infections 4 hours after intravenous injection. If labeled interleukins are proven over time to be effective for detecting sites of infection, they will provide a significant advantage over current methods that require the handling of patient blood and the use of ex vivo labeling methods.

Clinical experience with FDG PET in diagnosing infection has been reviewed.[43] Several applications that have shown benefit from FDG PET are as follows:

1. Chronic osteomyelitis, particularly involving the axial skeleton where labeled WBC sensitivity is low. FDG PET will likely replace ^{67}Ga-citrate for vertebral osteomyelitis;
2. Complicated diabetic foot infections, in which FDG can distinguish neuropathic osteoarthropathy from osteomyelitis;
3. Painful arthroplasty, to differentiate mechanical loosening of a prosthesis from an infected implant; and
4. Fever of unknown origin (FUO), particularly because FDG accumulates nonspecifically in infections, malignancies, and aseptic inflammatory diseases, which are the major causes of FUO.

In general, because of its rapid plasma clearance, FDG is most sensitive in the evaluation of soft-tissue infectious sites that are well perfused. It is probably the agent of choice for FUO and suspected abdominal sites of infection because it clears more rapidly than ^{67}Ga-citrate, which is bound to plasma transferrin and has interfering bowel activity.[31] However, even though the plasma clearance of ^{111}In- or ^{99m}Tc-labeled WBCs is slow, these agents are more specific for acute soft-tissue infections.

Clinical Considerations in Infection Imaging

It can be difficult to diagnose and localize a site of occult infection, particularly in patients who cannot cooperate or have no localizing signs. If detected early, most infections can be cured with proper treatment; delayed diagnosis is associated with higher mortality. If localizing signs are present, anatomic imaging with ultrasonography or computed tomography (CT) is often used first. However, these modalities cannot usually differentiate between infection, such as an abscess, and a noninfectious process, such as a sterile fluid collection or inflammation. Indications for nuclear medicine imaging in infection include FUO, suspected occult abscess, postoperative infection, suspected infection of a vascular graft, osteomyelitis, disk space infection, and suspected infections in immunocompromised patients. One advantage of a nuclear medicine study is that the whole body can be imaged. This is particularly important when there are no localizing signs.

Currently, the three most common agents used to localize foci of infection are ^{67}Ga-citrate, ^{111}In-oxine-labeled autologous WBCs, and ^{99m}Tc-HMPAO-labeled autologous WBCs. PET with ^{18}F-FDG has shown promise for evaluation of infection and inflammation. The appropriate choice of radiopharmaceutical for imaging infection depends on the clinical setting. Several factors should be considered in deciding which radiopharmaceutical to use, including whether the infection is acute or chronic, the suspected anatomic location, history of recent surgery, chemotherapy, blood transfusions, severe leukopenia, and existing malignancy.

Although ^{67}Ga-citrate is sensitive for detecting foci of infection, it is less specific than labeled WBCs, because it is also taken up in several tumors such as lymphoma and lung cancer and by sterile inflammation such as in a healing wound. Whole-body imaging is usually done 24 to 48 hours after administration. Accumulation in the colon can be significant, which may interfere with evaluation of the abdomen. Thus, laxatives are usually administered the day before imaging to reduce activity in the bowel. If the patient is receiving chemotherapy, blood transfusions, or iron therapy, ^{67}Ga-citrate can sometimes demonstrate an altered biodistribution that resembles a bone scan (Figure 26-7). This pattern is similar to the effect produced by iron saturation of transferrin,[44] and the pattern of altered distribution is likely caused by the decreased ability of gallium to bind to transferrin in the plasma. The finding of elevated serum iron levels after irradiation of animals strongly suggests this mechanism.[45] This can also be seen if the patient has received gadolinium for a magnetic resonance imaging (MRI) study just before ^{67}Ga-citrate administration. Whole-body x-irradiation can also alter the normal distribution of carrier-free ^{67}Ga-citrate, producing increased excretion and bone deposition together with decreased soft-tissue uptake.[46] A study comparing imaging with ^{67}Ga-citrate and with ^{111}In-WBCs in the same patients for diagnosis of occult

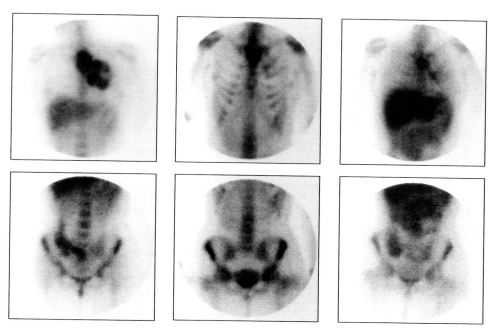

FIGURE 26-7 Lymphoma. (Left) Initial anterior spot images of the chest and abdomen were obtained 48 hours after intravenous administration of 10 mCi (370 MBq) of ^{67}Ga-citrate to a patient with lymphoma. The images demonstrate prominent abnormal focal accumulation of the radiotracer in the chest in the patient's tumor. (Center) A follow-up ^{67}Ga-citrate scan after treatment, demonstrating altered biodistribution resembling a diphosphonate bone scan. The patient had received chemotherapy and blood transfusions before the follow-up scan. With this altered distribution, neoplastic lesions can be missed. (Right) Another follow-up study in this patient at a later date demonstrated residual tumor in the chest.

sepsis showed that false-negatives with ^{67}Ga tended to occur in patients with infections of less than 1 week's duration, whereas false-negatives with ^{111}In-WBCs were more likely to occur in infections more than 2 weeks old.[47] Thus, ^{67}Ga-citrate may be a better choice for chronic infections. ^{67}Ga-citrate is preferred over labeled autologous WBCs for disk space infection and imaging of the spine.[48] ^{67}Ga-citrate is also more useful for evaluation of pulmonary disorders, such as evaluating response to therapy in AIDS patients with *Pneumocystis carinii* pneumonia,[49,50] drug-induced pulmonary toxicity in patients receiving amiodarone,[51] and granulomatous inflammatory processes such as sarcoidosis.[52] ^{67}Ga-citrate may also be appropriate in patients with suspected tumor or autoimmune disease and in evaluating FUO, a condition characterized by fever (38.3°C) of greater than 3 weeks' duration with no cause identified during 1 week of hospitalization.[29,30] However, a high percentage of FUOs are due to abdominal infections, and intestinal and renal excretion of gallium makes interpretation difficult in such cases.[53] Because of the lower specificity of gallium in this situation, a leukocyte study 24 hours after administration of 500 µCi of ^{111}In-WBC can be done first in chronic infection. If the study result is negative, a 24 to 48 hour gallium study can be performed after administration of a much larger dose (5 to 10 mCi) of ^{67}Ga-citrate.[53] However, FDG PET will likely replace gallium in the future as the preferred procedure for FUO.[54]

Radiolabeling WBCs is a somewhat tedious procedure, compared with a ^{67}Ga-citrate injection.[55] However, labeled WBCs are more specific for infection than ^{67}Ga-citrate. With ^{111}In-WBCs, physiologic uptake is seen in the liver, spleen, and bone marrow. There is normally no uptake in the urinary tract or the bowel. This can be useful for evaluating IBD or suspected urinary tract infection. Early imaging can be obtained 4 hours after administration. However, delayed imaging at 24 hours should be obtained if the early images are negative. Also, ^{111}In-WBCs have a higher target-to-background ratio than ^{67}Ga-citrate in acute pyogenic infections.[56]

Imaging with ^{99m}Tc-WBCs can be performed 2 to 4 hours after administration, offering an earlier diagnosis than with ^{67}Ga-citrate or ^{111}In-WBCs. ^{99m}Tc-WBCs are the agent of choice if an early diagnosis is urgently needed. Because of the higher photon flux of the ^{99m}Tc dose, better anatomic localization is sometimes possible. In contrast to ^{111}In-WBCs, there is physiologic bowel activity and urinary tract activity with ^{99m}Tc-WBCs that can complicate imaging of the abdomen.

Comparing ^{111}In-WBCs and ^{99m}Tc-WBCs, Peters[57] preferred ^{99m}Tc-WBCs when an early diagnosis is essential. However, in more chronic processes such as FUO or an infected hip prosthesis, where the turnover of granulocytes is slower, ^{111}In 24 hour imaging is preferred.

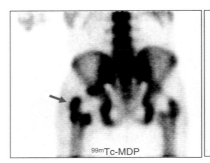 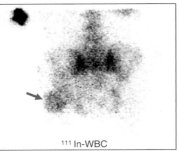

FIGURE 26-8 Osteomyelitis. Patient with pain at the site of repeat hip replacement performed after unresolved osteomyelitis several months prior to this study. Posterior images show increased uptake of ^{99m}Tc-MDP in the left acetabular region indicative of bone inflammation. Uptake of ^{111}In-labeled white blood cells in the same region confirms the presence of bone infection. (Images courtesy of Kara Duncan Weatherman, Purdue University, West Lafayette, IN.)

PET imaging with ^{18}F-FDG has proven to be clinically useful in the diagnosis and management of several malignancies. ^{18}F-FDG has also been shown to accumulate in a variety of infectious and inflammatory processes.[55,58] As PET imaging becomes more widely used in clinical medicine, it will undoubtedly prove to be a valuable tool in the diagnosis of infection and evaluation of response to therapy.[59] FDG has an advantage in FUO, which can be caused by infection, inflammation, or malignancy and can occur in patients with a nonfunctional immune system.[54,60]

In the bones and joints, there is little accumulation of ^{111}In-WBCs in noninflammatory regions of increased bone turnover, such as those caused by osteoarthritis. This is an advantage over ^{67}Ga-citrate in patients with complicated orthopedic problems. In suspected osteomyelitis, where a bone scan will show increased uptake due to bone remodeling, ^{111}In- or ^{99m}Tc-WBCs have been used to confirm or exclude a diagnosis of infection, in conjunction with a three-phase bone scan (Figure 26-8).

Table 26-2 summarizes the mechanisms of localization of infection imaging agents.

TUMOR IMAGING

Tumor imaging outside specific organs began in the 1960s with the development of ^{67}Ga-citrate, which was originally approved for lymphoma imaging and has been a valuable tumor-imaging agent for decades. Tumor imaging, for both benign and malignant tumors, has increased substantially since the introduction of ^{67}Ga-citrate. This can be attributed in part to the application of radiopharmaceuticals approved for other uses, such as ^{201}Tl-chloride and ^{99m}Tc-sestamibi, that were discovered to have avidity for tumors. The growth of tumor imaging has also benefited from the development of additional agents, such as ^{123}I- and ^{131}I-labeled *meta*-iodobenzylguanidine (MIBG), ^{111}In- and ^{99m}Tc-labeled antibodies and peptides, and a number of PET imaging agents, most notably ^{18}F-FDG.

The principal objectives of tumor imaging are staging disease to assess the extent of tumor involvement, evaluating response to therapy, and differentiating tumor recurrence from necrosis.[61] Traditional anatomic imaging modalities such as CT, MRI, and ultrasonography are primary methods employed in tumor detection. One limitation of these modal-

TABLE 26-2 Mechanisms of Localization of Infection Imaging Agents

Imaging Agent	Mechanism of Localization
^{67}Ga-citrate	^{67}Ga dissociates from ^{67}Ga-transferrin and binds to iron-containing proteins (lactoferrin and bacterial siderophores) at the infection site
^{111}In- and ^{99m}Tc-WBCs	Chemotaxis-induced migration of injected WBCs to the infection site
^{111}In- and ^{99m}Tc-IgG	Increased blood flow and endothelial membrane permeability at the infection site; no specific binding
^{99m}Tc-monoclonal antibodies	Binding to specific antigenic epitopes on WBCs that migrate to the infection site
^{123}I- and ^{99m}Tc-interleukins	Binding to specific interleukin receptors on WBCs at the site of infection
^{99m}Tc- and ^{18}F-ciprofloxacin	Increased blood flow and endothelial membrane permeability at the infection site; no specific binding
^{18}F-FDG	Increased glucose utilization by inflammatory cells

ities is their inability to distinguish between benign and malignant tumors or between pretherapeutic and posttherapeutic anatomic alterations such as scarring, inflammation, or necrosis and neoplastic processes.[62] Tumor imaging with radiopharmaceuticals having an avidity for viable tumor tissue can provide this information. For many years ^{67}Ga has been useful in this regard for planar and SPECT imaging of tumors, but now ^{18}F-FDG has risen to the premier spot for such studies. The proven diagnostic effectiveness of ^{18}F-FDG PET in detecting tumor viability led to the development of the hybrid single-unit CT/PET scanner that permits fusing high spatial-resolution CT images with functional information provided by PET. Fusion imaging has superior diagnostic power because it permits coregistration of anatomic and functional images during a single patient visit. The value of CT/PET for tumor imaging and other applications is discussed in Chapter 27, whereas the focus of this section is SPECT tumor-imaging agents.

Tumor Imaging with Gallium Citrate

Gallium citrate Ga 67 injection (^{67}Ga-citrate) was approved originally for the detection and staging of Hodgkin's and non-Hodgkin's lymphoma, and it still has value for these applications. It has been useful for identifying suspected recurrence of lymphoma and monitoring the effects of therapeutic regimens.

For tumor imaging, the standard dosage is 8 to 10 mCi (296–370 MBq), an amount necessary for good sensitivity with SPECT imaging. The three principal photopeaks, 93 keV (38%), 185 keV (24%), and 300 keV (16%), should be used for imaging to improve counting statistics. Because of gallium's slow excretion, imaging is typically done 48 to 72 hours or more after injection to improve the tumor-to-background ratio. Image interpretation is more difficult in regions of normal gallium uptake (liver, spleen, bone, bone marrow, and bowel). Delayed imaging of the abdomen is often necessary to distinguish tumor (fixed activity) from normal bowel accumulation that will clear over time. Laxatives can be used to facilitate bowel elimination.

The normal distribution of ^{67}Ga-citrate is shown in Figure 26-1. ^{67}Ga-citrate has been useful in evaluating the effectiveness of therapies in gallium-avid tumors. In these studies gallium is usually administered 3 weeks after the last treatment and 48 hours before the next treatment, because treatments are known to alter gallium's biodistribution. Response to therapy can be followed, because gallium uptake diminishes or disappears after successful treatment (Figure 26-9). Similarly, the appearance of new foci of gallium uptake is a sensitive predictor of disease recurrence.

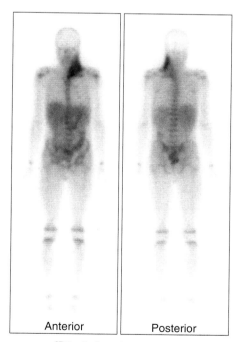
^{67}Ga Before Chemotherapy

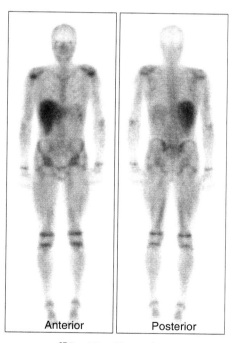
^{67}Ga After Chemotherapy

FIGURE 26-9 Hodgkin's lymphoma. (Left) Anterior and posterior whole-body images of a 14 year old boy with Hodgkin's lymphoma of the neck, obtained 48 hours after administration of 8 mCi (296 MBq) of ^{67}Ga-citrate. There is a prominent area of abnormal tracer uptake in the left neck that extends into the superior mediastinum. (Right) A follow-up ^{67}Ga-citrate whole-body scan obtained 2 months later demonstrates interval resolution of Hodgkin's lymphoma after chemotherapy.

The sensitivity of SPECT imaging using ^{67}Ga-citrate for tumor detection is high in Hodgkin's disease (>90%), non-Hodgkin's lymphoma (85%), hepatocellular carcinoma (90%), and soft-tissue sarcomas (93%). Other tumors that have shown gallium avidity, but with lower sensitivities for detection, include malignant melanomas (82%), lung tumors (85%), head and neck tumors (75%), and abdominal and pelvic tumors (55%).[63]

Tumor Imaging with Radiolabeled Peptides

The development of agents that target peptide receptors on tumors is a growing area in tumor imaging. Somatostatin-receptor imaging agents are an early example in this category. The widespread presence of somatostatin receptors in the body has allowed the imaging of many types of tumors that possess these receptors. The tissues containing somatostatin receptors are part of the neuroendocrine system, which is reviewed briefly in the following section.

Neuroendocrine System

The neuroendocrine system includes a diverse group of tissues whose cells have common features in their cytochemistry and ultrastructure.[64] Endocrine system cells have the ability to take up amine precursor substances such as dihydroxyphenyl-alanine (L-DOPA) and 5-hydroxytryptophan (5-HTP) and decarboxylate them, producing the corresponding biogenic amines dopamine and 5-hydroxytryptamine. This ability led to the early designation of endocrine system cells as amine precursor uptake and decarboxylation (APUD) cells. APUD cells have high levels of particular enzymes and the capacity to produce peptide hormones. They contain a prominent rough endoplasmic reticulum and Golgi complex and are rich in free ribosomes, but their most characteristic feature is the presence of round secretory granules that store amine substances.[64]

The "classic" APUD system included all peptide-producing cells of the stomach, duodenum, intestine, pancreatic islets, adrenal medulla, extra-adrenal paraganglia, anterior pituitary, parafollicular thyroid cells, and melanoblasts.[65] Observations over the years have changed and expanded the APUD cell concept. The same amines and peptides found in classic APUD cells are also present in the central nervous system, in peripheral nerves, and in more widely dispersed endocrine cells in organs of the GI and respiratory tracts. Studies have shown that some of these cells are derived during embryogenesis from the neural crest and some from the endoderm. The term neuroendocrine, therefore, has slowly replaced APUD, to reflect the close association between the neural and endocrine systems.

Somatostatin-Receptor Imaging Agents

Somatostatin, a 14-amino-acid peptide, is found principally in the hypothalamus, but it also occurs in the GI system, other parts of the brain, the spinal cord, peripheral nerves, placenta, retina, thymus, and adrenal medulla.[66] Physiologically, somatostatin plays a key role in the modulation of neuroendocrine cells. In the pituitary gland it acts as a neurohormone, inhibiting the release of growth hormone, adrenocorticotropic hormone (ACTH), and thyroid-stimulating hormone (TSH). In the nervous system it is a neurotransmitter or neuromodulator. Outside the brain it can act directly on cells (paracrine function) or as a true hormone. For example, circulating in the blood, it can act on the pancreas or GI tract to inhibit the release of intestinal peptides such as insulin, gastrin, and glucagon.

Somatostatin receptors (SSTRs) have been shown to be present on a variety of human tumors of neuroendocrine tissue, such as pituitary tumors, endocrine pancreatic tumors, carcinoids, paragangliomas, small cell lung cancers, medullary thyroid carcinomas, and pheochromocytomas, as well as meningiomas, astrocytomas, neuroblastomas, and some breast cancers.[67,68] Somatostatin is known to inhibit the secretion of hormones and growth factors that regulate tumor growth, inhibit angiogenesis, modulate immunologic activity, and exhibit direct antimitotic effects via somatostatin receptors on tumor cells.

Because of these actions, somatostatin should be a useful agent in the treatment of neuroendocrine tumors. From a practical point of view, however, the short plasma half-life (about 2 minutes) of somatostatin necessitated the development of a longer-acting derivative that could be used as a therapeutic agent.[66,69] This was eventually accomplished by chemical modification of the molecule to eliminate the amino acids responsible for somatostatin's rapid metabolism and retain the cyclic amino acid configuration required for receptor binding (Figure 26-10). The result was a synthetic 8-amino-acid sequenced peptide known as octreotide, which is currently marketed by Novartis as Sandostatin.

SSTRs are overexpressed in many tumors, which permits them to be imaged with radiolabeled somatostatin analogues. SSTRs exist in different subtypes ($SSTR_1$–$SSTR_5$), which differ in their interaction with somatostatin. Subtype 2 ($SSTR_2$) is most often expressed on the membrane surface of various tissues and tumors.[70] Several radiolabeled peptides have been developed to interact with $SSTR_2$ for tumor imaging, the most notable being ^{111}In-pentetreotide.

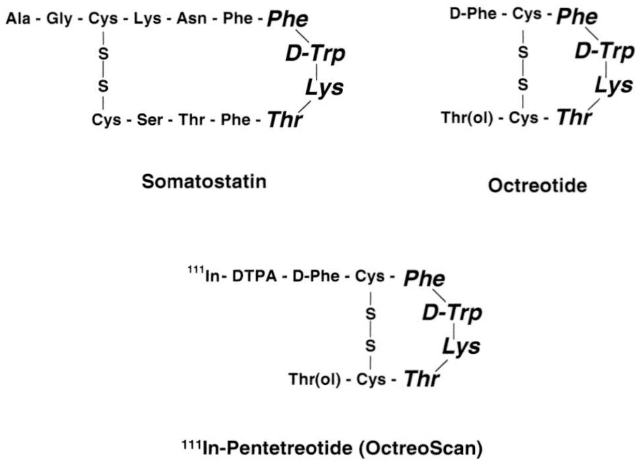

Ala - Gly - Cys - Lys - Asn - Phe - **Phe**
                  | |
                  S   **D-Trp**
                  | |
                  S   **Lys**
                  | |
Cys - Ser - Thr - Phe - **Thr**

Somatostatin

D-Phe - Cys - **Phe**
         | |
         S   **D-Trp**
         | |
         S   **Lys**
         | |
Thr(ol) - Cys - **Thr**

Octreotide

^{111}In - DTPA - D-Phe - Cys - **Phe**
                            | |
                            S   **D-Trp**
                            | |
                            S   **Lys**
                            | |
                  Thr(ol) - Cys - **Thr**

^{111}In-Pentetreotide (OctreoScan)

FIGURE 26-10 Chemical structures of somatostatin, octreotide, and ^{111}In-pentetreotide.

Indium In 111 Pentetreotide Injection

The high density of somatostatin receptors on primary and metastatic tumor cells permits the detection of these cells with radionuclide markers of octreotide analogues. One such agent is indium In 111 pentetreotide injection (^{111}In-pentetreotide, or OctreoScan [Covidien]), which has specific affinity for $SSTR_2$ on tumors. Positive scans of such tumors are useful in optimizing decisions about primary treatment and have been shown to predict good response to long-term therapy with somatostatin analogues (octreotide) for symptoms related to hormonal hypersecretion.

^{111}In-pentetreotide is prepared from a kit that contains a lyophilized mixture of pentetreotide and appropriate adjuvants for labeling. The kit also contains a separate vial of ^{111}In-chloride injection, which is added to the lyophilized pentetreotide at the time of use. The preparation and chemical properties of ^{111}In-pentetreotide are discussed in Chapter 11.

^{111}In-pentetreotide is indicated for diagnostic localization of primary and metastatic neuroendocrine tumors bearing somatostatin receptors.[69] After intravenous injection, ^{111}In-pentetreotide distributes rapidly into the tissues, with only about one-third of the injected dose remaining in the blood pool 10 minutes after injection. It is excreted primarily by the kidney. About 50% is eliminated in the urine by 6 hours and 90% by 48 hours. Minor amounts (<2%) undergo hepatobiliary excretion. The usual adult dosage for SPECT imaging is 6 mCi (222 MBq). Organs receiving the highest radiation absorbed dose are the spleen (14.8 rad), kidneys (10.8 rad), and urinary bladder (6 rad) per 6 mCi (222 MBq) dose.

The following precautions should be noted.[69] ^{111}In-pentetreotide should not be injected into intravenous lines used for parenteral nutrition, because the sugars present may cause a glycosyl–^{111}In-pentetreotide conjugate to form. Patients receiving octreotide therapy should suspend this medication for 72 hours before the administration of ^{111}In-pentetreotide. The patient should maintain good hydration (two 8 oz glasses of water prior to injection) to increase renal elimination and thus reduce background activity. Bowel cleansing is recommended before imaging to promote elimination of normal bowel accumulation of activity. Suggested bowel preparation agents are Dulcolax, magnesium citrate, and Metamucil.[71,72] Standard octreotide therapy may cause hypoglycemia in patients with insulinoma; therefore, as a precaution, an intravenous infusion of glucose should be given just before and during administration of ^{111}In-pentetreotide to patients with insulinoma.

Imaging is performed in 24 hours and may be repeated at 48 hours if needed to rule out suspected bowel accumulation of activity localized in the abdominal region. Uptake of ^{111}In-pentetreotide will be evident in tumors rich with somatostatin receptors. Uptake is also expected to occur in the normal pituitary gland, thyroid gland, liver, spleen, kidneys, and urinary bladder (Figure 26-11).[68] Many tumors have a high number of somatostatin receptors and will bind ^{111}In-pentetreotide (Figures 26-12 and 26-13). Krenning and colleagues[67,68] have comprehensively discussed ^{111}In-pentetreotide scintigraphy. Table 26-3 gives a partial list of tumors with their somatostatin receptor density, the detection rate of these tumors by ^{111}In-pentetreotide scintigraphy, and the correlation of ^{111}In-pentetreotide scintigraphy with other diagnostic methods. Fairly good agreement exists between the density of receptors present in tumors measured by positive ^{111}In-pentetreotide uptake and their detection by CT, MRI, ultrasonography, angiography, surgery, or biopsy. ^{111}In-pentetreotide may detect small tumors and metastatic lesions not found by CT, MRI, or

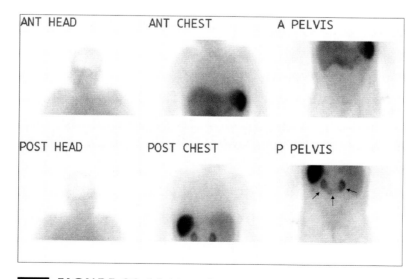

FIGURE 26-11 Normal ^{111}In-pentetreotide scan. Anterior and posterior spot images of the body were obtained 24 hours after intravenous administration of 5 mCi (185 MBq) of ^{111}In-pentetreotide. Physiologic radiotracer accumulation is seen in the liver, spleen, and kidneys. Note the curvilinear uptake in the kidneys in the posterior image of the pelvis in this patient with a horseshoe kidney (arrows).

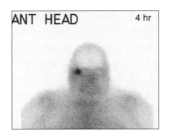 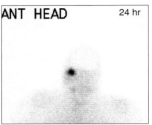

FIGURE 26-12 Abnormal ^{111}In-pentetreotide scan demonstrating a somatostatin receptor-positive tumor in the head. Anterior and posterior spot images of the head obtained at 4 and 24 hours after intravenous administration of 5 mCi (185 MBq) of ^{111}In-pentetreotide demonstrate an abnormal focal accumulation in the region of the right ear that corresponded to a right postauricular mass seen on a CT study.

ultrasonography. In a multicenter trial, 28% of tumors detected by ^{111}In-pentetreotide were missed by other methods.[71]

Imaging with ^{111}In-pentetreotide is helpful in the selection of a patient's treatment. For example, surgical treatment is more favorable when there is a solitary localized tumor. Widespread metastatic lesions are better treated with drugs. Posttreatment scans with ^{111}In-pentetreotide can also be used to monitor the effectiveness of treatment of receptor-positive tumors. Clinicians can better predict the response to octreotide therapy if tumors are receptor positive as shown by positive ^{111}In-pentetreotide uptake. Therapy and expense can be spared in patients who have negative scans, indicative of receptor-negative tumors.

Adrenal Medullary Imaging Agents

The body has two adrenal glands, each weighing about 5 grams, normally situated on the upper pole of each kidney.

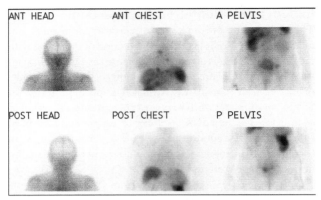
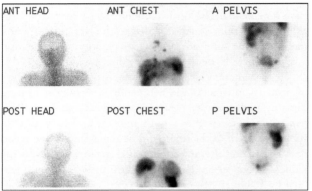

FIGURE 26-13 Abnormal ^{111}In-pentetreotide scan with multiple areas of increased uptake in a patient with metastatic carcinoid tumor. Anterior and posterior spot images of the head, chest, abdomen, and pelvis were obtained at 4 hours (top) and 24 hours (bottom) after intravenous administration of 5 mCi (185 MBq) of ^{111}In-pentetreotide. The images demonstrate multiple focal areas of abnormal accumulation in the chest, abdomen, and pelvis consistent with metastatic carcinoid.

TABLE 26-3 ^{111}In-Pentetreotide (OctreoScan) Correlation in Tumor Detection

Tumor Type	In Vitro Receptor Prevalence (%)[a]	Scintigraphy Detection Rate (%)[a]	Correlation with Other Methods (%)[b]
Carcinoid	55/62 (88)	74/78 (95)	190/237 (80)
Gastrinoma	6/6 (100)	13/14 (93)	40/42 (95)
Insulinoma	18/27 (67)	13/28 (46)	4/13 (31)
Medullary thyroid carcinoma	10/26 (38)	24/35 (69)	12/22 (54)
Paraganglioma	11/12 (92)	42/42 (100)	6/7 (86)
Pituitary adenoma	45/46 (98)	21/28 (75)	24/30 (80)
Small cell lung carcinoma	4/7 (57)	38/38 (100)	2/2 (100)
Pheochromocytoma	38/52 (73)	13/15 (87)	9/9 (100)

[a] Data from reference 68.
[b] Data from reference 69.

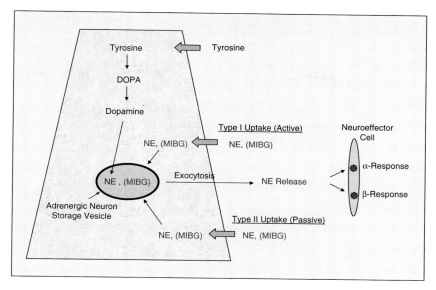

FIGURE 26-14 Mechanism of localization of *m*-iodobenzylguanidine ([123]I- or ^{131}I-MIBG) in adrenergic neurons and adrenomedullary cells.

Each adrenal gland has two zones. The adrenal cortex (outer zone) is responsible for producing mineralocorticoids, the primary one being aldosterone, and glucocorticoids, primarily cortisol. The cortex also produces androgenic hormones. The adrenal medulla (inner zone) produces epinephrine and norepinephrine, the pressor amines associated with stimulation of the sympathetic nervous system.

The role of the adrenal medulla in the synthesis and storage of catecholamines was key to the development of adrenal medullary imaging agents. Dopamine is the immediate precursor to norepinephrine, which is synthesized in the chromaffin cells of the adrenal medulla (Figure 26-14).

Unsuccessful early attempts to directly label dopamine with a gamma emitter[73] led to the development of radioiodinated benzylguanidines because of their potent antiadrenergic activity.[74] The ability to halogenate the aromatic ring in benzylguanidine led Wieland et al.[75] to investigate the *ortho-*, *meta-*, and *para-*iodinated derivatives. Further studies demonstrated that the *meta*-iodinated isomer *meta*-iodobenzylguanidine (MIBG) (Figure 26-15) was superior for imaging by virtue of its resistance to in vivo deiodination and its lower concentration in liver compared with the other isomers.[76]

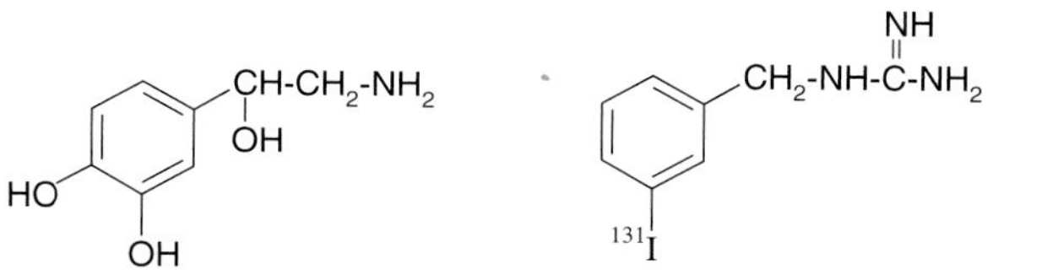

FIGURE 26-15 Chemical structures of norepinephrine and [131]I-*meta*-iodobenzylguanidine (^{131}I-MIBG).

MIBG is now commercially available as iobenguane I 123 injection (^{123}I-iobenguane) or iobenguane I 131 injection (^{131}I-iobenguane). These agents are radiolabeled by several methods.[77,78] A solid-phase isotope exchange method is described in Chapter 11. ^{123}I-iobenguane (AdreView [GE Healthcare]) is available as a 5 mL solution, 2 mCi(74 MBq)/mL and ^{131}I-iobenguane [Pharmalucence] is supplied as a frozen solution in a concentration of 2.3 mCi(85.1 MBq)/mL. [131]I- and ^{123}I-labeled MIBG have been shown to be effective in localizing neuroendocrine tumors derived from the APUD system. MIBG has approved indications for scintigraphic localization of pheochromocytomas and neuroblastomas.[79,80] ^{131}I-MIBG has also been shown to be useful in carcinoid tumors and medullary carcinoma of the thyroid.[81,82]

Tumors that take up ^{131}I-MIBG are believed to do so by a specific uptake mechanism, with storage in intracellular granules (Figure 26-14). Experimental evidence indicates that ^{131}I-MIBG is taken up into sympathetic neuroeffector cells similarly to norepinephrine by a specific catecholamine type I active uptake mechanism and is stored in adrenergic storage vesicles.[83] Reserpine, a drug known to inhibit the uptake of norepinephrine by chromaffin granules and to deplete stores of catecholamines in the adrenal medulla, has been shown to deplete 90% of MIBG stores in the adrenal medulla in dogs previously administered ^{131}I-MIBG.[76] ^{131}I-MIBG's uptake mechanism is similar to that of norepinephrine, but unlike norepinephrine it does not interact with postsynaptic α- and β-adrenergic receptors.

A number of drugs that participate in the same mechanisms of uptake or depletion of epinephrine as MIBG have been shown to reduce the uptake of MIBG in neuroendocrine tumors. Drugs that are known to interfere with MIBG scintigraphy or are expected to reduce MIBG uptake because of their known pharmacologic actions are listed in Table 26-4.[84] The

TABLE 26-4 Drugs Known or Expected to Reduce MIBG Uptake in Sympathetic Neuroeffector Cells

Group 1 (should not be taken for 6 weeks prior to MIBG administration)	Group 2 (should not be taken for 2 weeks prior to MIBG administration)
Tricyclic Antidepressants Amitriptyline, amoxapine, desipramine, doxepin, imipramine, maprotiline, nortriptyline, protriptyline, trazodone, trimipramine	**Tranquilizers** Acetophenazine, chlorpromazine, chlorprothixene, droperidol, fluphenazine, haloperidol, mesoridazine, perphenazine, pimozide, prochlorperazine, promazine, thioridazine, thiothixene, trifluoperazine, triflupromazine
	Sympathomimetics Albuterol, amphetamine, benzphetamine, cocaine, dextroamphetamine, diethylpropion, dobutamine, dopamine, fenfluramine, isoetharine, isoproterenol, mazindol, metaproterenol, metaraminol, methamphetamine, methylphenidate, phendimetrazine, phenmetrazine, phentermine, phenylephrine, phenylpropanolamine, pseudoephedrine, terbutaline
	Antihypertensive/Cardiovascular Drugs Bretylium, diltiazem, guanethidine, labetalol, nifedipine, reserpine, verapamil

Source: Reference 84.

length of time a patient should refrain from using the drug before MIBG administration is also given.

After intravenous injection, MIBG is found to localize normally in tissues with extensive sympathetic innervation (salivary glands, nasopharynx, and heart) and tissues involved in metabolism and excretion (liver, spleen, and urinary bladder) (Figure 26-16).[83,85] Normal adrenal medulla is seen only occasionally, because of its small size and its depth in the body. Most of the MIBG is excreted unchanged in the urine: 55% in 24 hours and 90% in 4 days.[86] The stability of ^{131}I- and ^{123}I-MIBG in vivo is attributed to the guanidine side chain. A small amount of in vivo metabolism does occur, however, with release of 2% to 6% free iodide. This requires administration of potassium iodide to protect the thyroid gland.

The usual intravenous adult administered activity of ^{131}I-MIBG is 0.5 mCi (18.5 MBq). In patients over 65 kg the dose is 0.3 mCi/m^2 (11.1 MBq/m^2) up to a maximum of 1 mCi (37 MBq). Children's doses of ^{131}I-MIBG are based on 0.3 mCi/m^2 (11.1 MBq/m^2) to a maximum dosage of 0.5 mCi. Dosing of ^{123}I-MIBG is based on 0.14 mCi/kg (5.18 MBq/kg), with a suggested maximum dose of 10 mCi (370 MBq) in adults.[87,88] The pediatric dose recommendation is scaled to the adult dose on the basis of patient weight.[89] A daily thyroid-blocking dose of potassium iodide should be administered 1 day before dosing and 5 to 7 days after dosing with

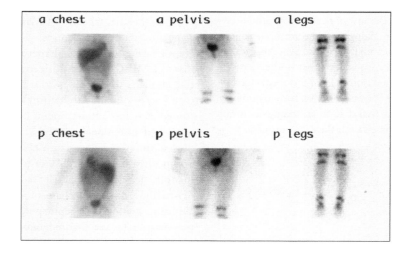

FIGURE 26-16 Normal MIBG scan. Anterior and posterior spot images of the chest, abdomen, and pelvis were obtained 24 hours after intravenous administration of 2.5 mCi (92.5 MBq) of ^{123}I-MIBG. Normal uptake is seen in the liver, bladder, and growth plates.

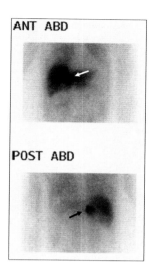

FIGURE 26-17 MIBG scan. Pheochromocytoma. Anterior and posterior images of the abdomen taken 24 hours after intravenous administration of 10 mCi (370 MBq) of ^{123}I-MIBG in this patient with a history of pheochromocytoma. There is an abnormal focal area of accumulation in the region of the right adrenal gland (arrow) consistent with the patient's history of pheochromocytoma.

^{131}I-MIBG.[90] A single blocking dose of potassium iodide should be administered 1 hour before ^{123}I-MIBG administration.[89] The daily FDA-recommended thyroid-protective doses of potassium iodide are listed in Table 20-1 (Chapter 20).

The critical organs for ^{131}I-MIBG in the adult are the urinary bladder wall and the liver, each with a radiation absorbed dose of 3 rad(cGy)/mCi.[90]

Figures 26-17 through 26-20 illustrate the value of MIBG in the diagnosis of pheochromocytoma, neuroblastoma, and paraganglioma.

Parathyroid Gland Imaging Agents

Normally, there are four parathyroid glands in the human body, one located behind each upper and lower pole of the thyroid. Ectopic locations of parathyroid glands can occur, particularly in the region of the thymus gland and mediastinum.

The parathyroid glands secrete parathormone (PTH), which regulates calcium and phosphorus metabolism in the body. PTH promotes calcium and phosphate resorption from bone, increased renal tubular reabsorption of calcium, and a diminished rate of phosphate reabsorption. Hypersecretion of PTH may be caused by a parathyroid adenoma (primary hyperparathyroidism) or through an indirect feedback mechanism initiated by hypocalcemia due to renal failure. The latter process results in parathyroid hyperplasia (secondary hyperparathyroidism) that can progress, leading to autonomous activity of one or more of the parathyroid glands (tertiary

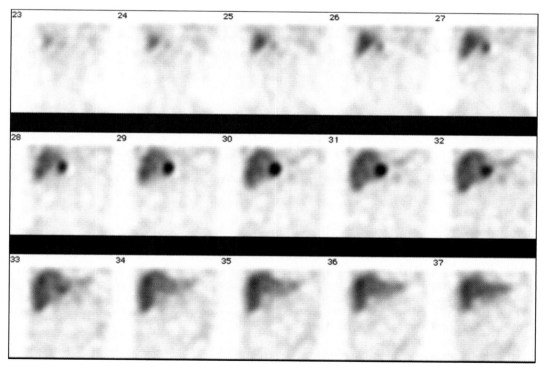

FIGURE 26-18 ^{123}I-MIBG SPECT. Pheochromocytoma. Coronal SPECT images through the abdomen of the patient in Figure 26-17 demonstrate a prominent abnormal focus in the region of the right adrenal gland.

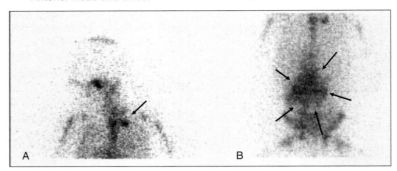

FIGURE 26-19 MIBG scan. Neuroblastoma. Anterior spot images of the head, chest, and abdomen taken 48 hours after intravenous administration of 2.2 mCi (81.4 MBq) of ^{123}I-MIBG. (A) Small focal area of abnormal radiotracer accumulation is seen near the manubrium (arrow). (B) There is a large focal area of abnormal accumulation in the abdomen and pelvis (arrows) representing uptake in neuroblastoma.

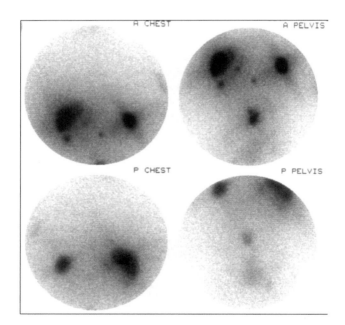

FIGURE 26-20 MIBG scan. Paraganglioma. Anterior and posterior spot images of the chest, abdomen, and pelvis taken 24 hours after intravenous administration of 475 µCi (17.58 MBq) of ^{131}I-MIBG in a patient who was treated for paraganglioma with metastases to liver and bone. There are abnormal focal areas of radiotracer uptake in the liver and left upper quadrant. Multiple smaller focal areas of abnormal uptake are identified in the abdomen. There is also a focus of radiotracer uptake in the midline lower abdomen that is more anteriorly located. These are all consistent with metastatic involvement. Radiotracer uptake was also noted in the right femur and left humerus (images not shown).

hyperparathyroidism). Serum calcium levels routinely monitored in blood screens are an important diagnostic parameter in identifying parathyroid disease.

There is no clear way to differentiate a hyperplastic gland from an adenoma. In general, an enlarged gland is classified as an adenoma if the remaining three glands are found to be normal. If all four parathyroids are enlarged, a diagnosis of hyperplasia is made.[91] About 85% of hyperparathyroid cases are caused by a solitary parathyroid adenoma that can be cured by surgery. The success rate for resection of parathyroid adenomas is 95%, which would appear to obviate the need for a presurgical scan of the parathyroids. However, scintigraphy is helpful in directing surgical exploration to the site of the adenoma, particularly if it is ectopic, and scanning is also useful in localizing suspected residual adenoma after a prior parathyroidectomy.

No radiopharmaceutical specifically targets the parathyroid glands, and so special imaging techniques have been devised to localize parathyroid adenomas. After an initial report that ^{201}Tl-chloride was taken up in parathyroid glands,[92] purportedly because of the increased cellular density and vascularity of parathyroid adenomas, a dual-isotope method was described to localize adenomas.[93] This technique used two agents, ^{201}Tl-chloride and ^{99m}Tc-sodium pertechnetate. The original method involved administration of pertechnetate first, followed by thallium. Others have reversed this order, acquiring the lower-energy thallium scan first to facilitate separation of the two photon energies by gamma spectrometry. Because thallium localizes in both parathyroid and thyroid tissue but pertechnetate localizes only in thyroid tissue, subtraction of the thyroid image isolates the parathyroid adenoma. If ^{99m}Tc-sodium pertechnetate is administered first, the 140 keV pertechnetate thyroid image is stored as a matrix in the computer. A downscatter image from ^{99m}Tc is also stored in the 75 keV ^{201}Tl window. The ^{99m}Tc 140 keV thyroid and 75 keV downscatter images are typically of 5 min-

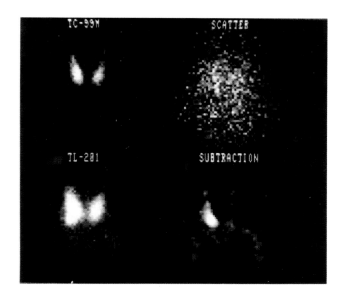

FIGURE 26-21 ^{201}Tl/^{99m}Tc-pertechnetate parathyroid imaging using a subtraction technique. The technetium thyroid image was subtracted from the thallium thyroid/parathyroid image to demonstrate the location of the parathyroid adenoma.

tracted serially from the thyroid/parathyroid image until the enlarged or hyperplastic parathyroid tissue becomes apparent in the image (Figure 26-21). Usually 2 mCi (74 MBq) of each tracer is used for this procedure.

In 1989, Coakley et al.[94] introduced ^{99m}Tc-sestamibi for imaging parathyroid adenoma. A subsequent report on preoperative localization of adenoma with this agent demonstrated its equivalency to ^{201}Tl-chloride.[95] Because ^{99m}Tc-sestamibi also localizes in both thyroid and parathyroid tissue, the method involved acquisition of ^{123}I thyroid images, which were then subtracted from ^{201}Tl-chloride or ^{99m}Tc-sestamibi thyroid/parathyroid images to isolate the parathyroid adenoma. Sestamibi was found to wash out of parathyroid tissue more slowly than thallium. The slower washout of sestamibi was attributed to its avidity for the increased number of mitochondria in parathyroid adenomas. In addition, ^{99m}Tc-sestamibi has an advantage over ^{201}Tl for better SPECT imaging and for radioguided surgery using an intraoperative probe detector.

To simplify the method of parathyroid adenoma scintigraphy, Taillefer et al.[96] described a double-phase differential washout technique. The method is based on the time-dependent variation in washout rates of ^{99m}Tc-sestamibi from the thyroid gland and parathyroid adenomas. The imaging technique consists of administration of 20 to 25 mCi (740–925 MBq) of ^{99m}Tc-sestamibi with images made at 15 minutes (early phase) and 2 hours (late phase). With this method, 90% (19 of 21) of parathyroid adenomas are localized on the basis of more rapid washout of sestamibi from thyroid tissue compared with slower washout from parathyroid adenoma. The parathyroid adenoma-to-thyroid activity ratio was reported to be 1.24 in early-phase imaging, increasing to 1.46 in late-phase imaging, where the adenoma was clearly seen (Figure 26-22).

utes' duration each. ^{201}Tl-chloride is injected, and the thyroid/parathyroid images are obtained in the 75 keV window. These images are acquired as an accumulation of 1 minute images for 25 minutes so that part of them can be discarded if patient motion occurs. During processing, the technetium downscatter image matrix is first subtracted from the thallium cumulative image. The ^{99m}Tc 140 keV thyroid image is then sub-

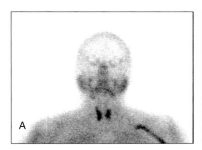
Anterior neck – 15 min

Anterior neck – 2 hr

FIGURE 26-22 ^{99m}Tc-sestamibi parathyroid scan demonstrating a parathyroid adenoma. Anterior spot images of the neck were obtained at 15 minutes and 2 hours after intravenous administration of 20 mCi (740 MBq) of ^{99m}Tc-sestamibi to this patient with hyperparathyroidism. (A) On the 15 minute image radiotracer accumulation is seen in both lobes of the thyroid gland. (B) On the 2 hour image there has been clearance of radiotracer from the thyroid gland. However, an abnormal focal area of residual radiotracer accumulation is seen in the region of the lower pole of the right lobe of the thyroid gland, consistent with a right inferior parathyroid adenoma.

LYMPHOSCINTIGRAPHY

The typical interstitial location of solid tumors predisposes their spread systemically via the lymphatic system. This concept was introduced in the 18th century by Le Dran, who described the progression of breast cancer to regional lymph nodes via the lymphatics.[97,98] Le Dran postulated that cancer metastasized systemically by this mechanism, and his work provided an early foundation from which the sentinel node concept has developed over the past 20 years.

The Sentinel Node Concept

The sentinel node concept postulates that the first lymph node, or sentinel node (SN), that receives lymph drainage from a tumor bed will contain cancer cells if the primary tumor has spread via the lymphatics. If this is true, then finding a SN that is cancer free is strong evidence that the tumor has not spread. The SN concept presupposes an orderly spread of cancer within the lymphatic chain beginning with the sentinel lymph node. The possibility exists that more than one lymph node bed may drain a tumor site, and thus more than one SN may be identified. Therefore, the mapping techniques that are used must be able to identify all lymph node basins draining a tumor site.

Morton and Chan[98] have described the application of the SN concept to malignant melanoma, tracing the concept from its early beginnings to the current methods of SN identification: lymphoscintigraphy, vital blue dye, and gamma probe counting during surgery. The rationale for the current diagnostic approach is that of those patients with intermediate-thickness primary melanoma who undergo lymph node dissection, only about 20% are expected to have metastasis in regional lymph nodes. Thus, about 80% of patients will undergo an unnecessary procedure that is costly and fraught with morbidity, including risk of acute wound problems, chronic lymphedema, and nerve injury. The finding of cancer-free SNs can spare many patients the expense and problems associated with lymph node dissection.

The SN concept was brought to the forefront of cancer management around the same time by Cabanas[99] and Holmes et al.[100] Cabanas, in the management of patients with penile cancer, introduced the term sentinel node and applied the SN concept to 43 penile carcinoma patients, finding that 31 patients with SNs negative for tumor had a 5 year survival rate of 90%.[101] SN identification was accomplished by a manual surgical technique. Holmes et al.[100] used lymphoscintigraphy in applying the SN concept to map lymph node basins and introduced the concept of selective lymphadenectomy in primary cutaneous melanomas. On the basis of their experience with radiopharmaceutical mapping, vital blue dye was introduced as a visual marker to facilitate localization of the lymphatic basin during surgery.[102] With the dye technique, they were able to identify SNs in 194 (82%) of 237 lymphatic basins. Of those nodes that were positive for tumor, only two SNs were negative when a nonsentinel node in the same basin was found to be positive (<1% false-negative rate), demonstrating the high accuracy rate of this procedure.

In 1993, Alex et al.[103] introduced the use of ^{99m}Tc-sulfur colloid (^{99m}Tc-SC) for melanoma lymphoscintigraphy. They initially compared lymphoscintigraphy and the use of a gamma scintillation probe with vital blue dye to identify sentinel lymph nodes in cats, demonstrating equal sensitivities of the two techniques. These findings were subsequently corroborated in patients with malignant melanoma.[104]

Melanoma Lymphoscintigraphy

Lymphoscintigraphy has been shown to improve the accuracy of staging patients with malignant melanoma.[105–107] The generally accepted staging for malignant melanoma is summarized in Table 26-5. The Breslow system is used to assign lesions within and between stages I and II.[105] T1 tumors are ≤0.75 mm thickness, T2 tumors 0.76–1.5 mm, T3 tumors 1.51–4.0 mm, and T4 tumors >4 mm thickness. T1 and T2 tumors constitute stage I disease. Approximately 85% of patients with malignant melanoma present with stage I or II disease.[105] Because of its significant diagnostic and prognostic information, lymphoscintigrapy is typically done in patients without either clinically apparent metastases or early-intermediate stage melanoma (Breslow thickness 0.76–4 mm).[108] Identification of nodal involvement in these patients is key to staging and treatment, because positive nodal involvement accurately predicts spread of disease via the lymphatics. Conversely, negative SN involvement removes the necessity for extensive nodal dissection and the associated morbidity. The value of vital blue dye combined with lymphoscintigraphy is that it allows accurate identification of not only the SNs but also the true nodal basins that drain from the melanoma lesions. This improves the localization of positive nodes in beds that would not have been predicted to receive lymphatic drainage from the primary melanoma on the basis of conventional estimates by surgeons.[109] The significant advantage from a pathology standpoint is that there is a higher probability of discovering cancer cells in a few thoroughly examined high-risk SNs than in less thorough examination of several nodes from an entire nodal basin.[97,105]

Variations in methods are reported in the literature, but a current recommended technique for localizing SNs in melanoma involves preoperative lymphoscintigraphy 1 to 4 hours before surgery with filtered ^{99m}Tc-SC injected intra-

TABLE 26-5 Staging Criteria for Malignant Melanoma

Melanoma Stage	Staging Criteria
Stage I	Thin (<1.5 mm) primary tumor
Stage II	Thicker (>1.5 mm) primary tumor
Stage III	Regional spread of disease to skin more than 5 cm from primary tumor or to regional lymph nodes
Stage IV	Evidence of distant metastases

Source: Reference 105.

dermally with a dosage of 0.5 to 0.8 mCi (18.5–29.6 MBq) at four to eight injection sites around the melanoma.[107] Gamma-camera imaging is performed to document the drainage pattern from the primary tumor area through the dermal lymphatics to the regional lymph nodes. The skin overlying the SN is then marked. The SN is usually identified within 30 minutes after injection. At the time of surgery, 1 to 2 mL of vital blue dye (patent blue or isosulfan blue) is injected around the primary tumor. A surgical incision with skin flaps is made in the lymph node basin to allow visual identification of the blue-stained lymphatic channel from the edge of the wound to the SN. Blue dye SN identification is corroborated by intraoperative gamma-probe counting. A node-to-background ratio of 2 or greater is used for positive identification. Blue-stained SNs are excised for histochemical assessment of metastasis. The rate of SN identification was found to be 99.1% when isosulfan blue dye was combined with radiocolloid lymphoscintigraphy, compared with 95.2% when isosulfan blue dye alone was used.[107]

The SN concept has been validated for melanoma, and it is in the process of being validated in breast and penile cancer, cancer of the vulva, and head and neck cancer.[110]

Breast Lymphoscintigraphy

The success that has been achieved in identifying SNs with melanoma has generated great interest in applying this concept to the staging of breast cancer. Alazraki et al.[110] summarized the findings of several studies using lymphoscintigraphy and other techniques. Although no standard procedure or preferred method for breast lymphoscintigraphy has been defined, the general techniques used have been described.

Both unfiltered and filtered ^{99m}Tc-SC has been used with success; however, unfiltered colloid requires larger administered activities (1 mCi [37 MBq]) to achieve adequate penetration within the lymphatics. This is attributed to the poor migration of larger particles in the unfiltered preparation. Filtered ^{99m}Tc-SC can be given in smaller amounts of 0.4 to 0.6 mCi (14.8–22.2 MBq) because a greater fraction of the particles injected are able to migrate through the lymphatic channels. The radiocolloid is typically injected either intradermally or interstitially around the tumor site (peritumoral injection).[110] One technique involves the injection of 450 µCi (16.65 MBq) in 6 mL, divided into six 1 mL aliquots injected in separate sites around the tumor. Another technique involves administering a total of 600 µCi (22.2 MBq): four 100 µCi (3.7 MBq) peritumoral injections and one 200 µCi (7.4 MBq) intradermal injection.

Imaging the area surrounding the injection site has been shown to be effective in locating SNs, providing a visual guide for the surgeon. Early sequential imaging reveals the pattern of radiocolloid progress in the lymphatic channels and helps to distinguish the first nodes to receive radioactivity, the SNs, from secondary nodes in the lymphatic chain that appear at later imaging times. Comparison of the sensitivity of various techniques for finding SNs in breast cancer (blue dye alone; intraoperative probe alone; blue dye plus probe; imaging plus probe; and blue dye, imaging, and probe combined) has shown that lymphoscintigraphic imaging combined with blue dye and the intraoperative probe is the most successful.[110]

Radiopharmaceuticals for Lymphoscintigraphy

Early diagnostic attempts to evaluate the lymphatic system were made by contrast lymphography after injection of radiographic contrast material (ethiodized oil) into cannulated lymphatic vessels. Later attempts, using ^{198}Au-colloidal gold by interstitial injection,[111] demonstrated the simplicity of radionuclide lymphoscintigraphy, with minimal complications and excellent correlation with contrast lymphography.[112,113] Because the radiocolloid did not require lymphatic vessel cannulation, the technique provided a means of evaluating lymphatic drainage in previously inaccessible areas.

Much of the early information on the ideal properties of radiocolloids for lymphoscintigraphy was gained from studies of internal mammary lymphoscintigraphy by Ege et al.[114] These investigations demonstrated that a uniform dispersion of small particles (<100 nm) was necessary for colloid to translocate from the interstitial injection site to the lymphatic channels and nodes. Large particles (500–2000 nm) remained trapped at the injection site. Although larger particles migrate through the lymphatics after intralymphatic injection, this precludes the advantage of interstitial administration.

A number of radiocolloid preparations have been investigated for use in lymphoscintigraphy. Early investigations demonstrated that ^{198}Au-colloidal gold had ideal particle size (5–50 nm), but its high radiation dose was unsatisfactory for routine diagnostic use. Important ^{99m}Tc agents that were evaluated included sulfur colloid, stannous phytate, and antimony sulfide. ^{99m}Tc-SC produced by the thiosulfate kit method was found to be unsatisfactory because of inadequate migration from the injection site as a result of the relatively large particle size range (100–1000 nm).[115] ^{99m}Tc-SC produced by the hydrogen sulfide method has smaller particles (<100 nm) with satisfactory lymph node scans, but the method of preparation is cumbersome.[116] A kit for the preparation of ^{99m}Tc-antimony sulfide produces particles in an ideal size range (3–30 nm), and this agent was shown to be satisfactory for mammary lymphoscintigraphy.[117,118] ^{99m}Tc-stannous phytate forms an in vivo colloid about the same size as ^{99m}Tc-antimony sulfide but was found to be inferior to it in clinical studies.[118] However, the ^{99m}Tc-antimony sulfide kit never achieved approval for routine use in the United States and is no longer available.

The resurgence in lymphoscintigraphy and the SN concept focused attention again on the preparation of satisfactory ^{99m}Tc-labeled agents for this application. Bergqvist et al.[119] revisited the particle size requirements for interstitial lymphoscintigraphic agents, reconfirming that a particle size less than 100 nm is necessary for adequate migration and uptake into lymph nodes after interstitial injection. They further noted that ^{99m}Tc-HSA, which is not a particulate agent, shows less retention within lymph nodes and that its rapid transit in the lymphatic channels may lead to missed detection of SNs. A ^{99m}Tc-albumin nanocolloid kit is available in Europe and has been found to provide good results in identifying SNs. Approximately 95% of the colloidal albumin particles in this kit are 5–80 nm in size.[120,121]

Filtered Sulfur Colloid

The relatively high proportion of large particles in traditionally prepared ^{99m}Tc-SC led to the use of filtration techniques to remove larger particles.[122–124] Dragotakes et al.[122] demonstrated, by laser light scattering analysis, that filtration of a standard preparation of ^{99m}Tc-SC through a 0.1 μm membrane filter yielded particles with a bimodal size distribution; the particles predominantly were 10 ± 2 nm in size, but a small portion (<0.1%) were in the 89 to 173 nm size range. Hung et al.[123] found that ^{99m}Tc-SC filtered through a 0.1 μm membrane filter produced about 90% of particles between 15 and 50 nm as determined by Nuclepore (Whatman) polycarbonate filter analysis, with a mean size of 38 ± 3.3 nm by electron microscopy. Laser light-scatter analysis showed the particles to be bimodal with mean particle sizes of 7.5 nm (minor peak) and 53.9 nm (major peak). The differences in particle size distribution reported for 0.1 μm membrane-filtered SC by these investigators may be due to differences in kit composition during the initial preparation of ^{99m}Tc-SC.

Eshima et al.[125] have shown that smaller colloidal particles are generated when ^{99m}Tc-SC is prepared using increased amounts of carrier ^{99}Tc from long-ingrowth generator eluates and a heating time that is shortened from 5 minutes to 3 minutes. The optimal preparation conditions were reported to be addition of 150 mCi (5550 MBq) of ^{99m}Tc-sodium pertechnetate from a 72 hour ingrowth generator in 3 mL to a lyophilized sulfur colloid kit (CIS-US). After addition of 1.5 mL 0.148 M hydrochloric acid, the vial is placed immediately into a boiling water bath for 1.5 minutes, removed and agitated, reheated for an additional 1.5 minutes, removed, and then cooled for 2 minutes at room temperature before addition of 1.5 mL of buffer. Nuclepore polycarbonate filter analysis demonstrated a shift to smaller particle sizes by this method of preparation. Table 26-6 shows the particle size distribution of ^{99m}Tc-SC prepared by the standard method and by the modified heating method described by Eshima et al.[125] ^{99m}Tc-SC prepared by this modified heating method followed by filtration through a 0.22 μm membrane filter was shown to be more effective for visualizing lymphatic channels and identifying sentinel lymph nodes than ^{99m}Tc-SC filtered through a 5.0 μm membrane filter.[124] It is worth noting that 0.22 μm membrane filtration after a standard heating method (5 minute boil) in the preparation of ^{99m}Tc-SC also results in satisfactory lymphoscintigraphy. The key point for success in lymphoscintigraphy is removal of larger particles that have a slowed migration rate from the interstitial space to the lymphatic channels. Figures 26-23 and 26-24 are lymphoscintigrams identifying sentinel lymph nodes after ^{99m}Tc-SC administration and imaging with a gamma camera.

TABLE 26-6 Particle Size Distribution of ^{99m}Tc-Sulfur Colloid

Particle Size (nm)	% of Total
Standard Preparation Method (5 minute heating)	
<100	15 to 20
100 to 600	70 to 80
700 to 5,000	2 to 4
>5,000	0.5 to 1.5
Modified Preparation Method (3 minute heating)	
<30	47
30 to 50	0
50 to 80	1
80 to 200	5
200 to 400	21
400 to 800	16
800 to 2,000	5
2,000 to 5,000	1
5,000 to 10,000	0
>10,000	5

^{99m}Tc-DTPA–Mannosyl–Dextran

^{99m}Tc-DTPA–Mannosyl–Dextran (^{99m}Tc-Lymphoseek) is a molecular imaging agent for SN detection in diagnostic imaging and during intraoperative mapping of the lymphatic system.[126,127] The basic molecule consists of a dextran backbone to which are bound mannose and DTPA units. The DTPA unit binds reduced ^{99m}Tc, and the mannose unit serves as a substrate for the macrophage mannose receptor (MR), a 165-kDa membrane glycoprotein located on macrophages within the lymphatic system. The mean diameter of ^{99m}Tc-Lymphoseek is 7 nm. ^{99m}Tc-Lymphoseek is prepared by adding ^{99m}Tc-sodium pertechnetate to a previously compounded kit containing a mixture of DTPA–mannose–dextran, stannous chloride, and ascorbic acid. After a 30 minute incubation, the product is buffered and sterile filtered prior to injection. When ^{99m}Tc-Lymphoseek was compared with filtered ^{99m}Tc-SC after intradermal injection, ^{99m}Tc-Lymphoseek demonstrated more rapid clearance from the injection site and lymph node uptake equivalent to ^{99m}Tc-SC but with a longer duration of retention in lymph nodes. Clinical investigations in human subjects have shown this agent to have promise for SN imaging in patients with breast cancer and melanoma.[128,129]

^{99m}Tc-Liposomes

Liposomes labeled with ^{99m}Tc-exametazime have been investigated for use in lymphoscintigraphy.[130,131] An extension of this technique has been to incorporate a blue dye within the liposomes to permit both radiometric and visual identification of the lymphatic channels and nodes with an intraoperative probe.[132]

DACRYOSCINTIGRAPHY

Dacryoscintigraphy is a useful method for assessing nasolacrimal drainage of tears.[133,134] Under normal conditions, the lacrimal glands release tear fluid to maintain a thin protective

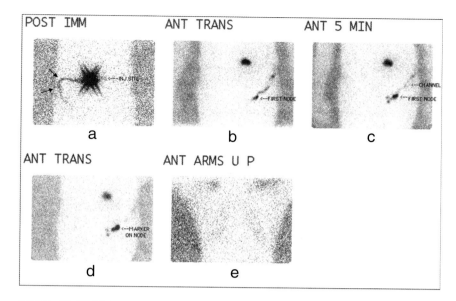

FIGURE 26-23 Lymphoscintigraphy for localization of the sentinel lymph node. (a) Posterior transmission image taken shortly after four intradermal injections of 250 μCi (9.25 MBq) of ^{99m}Tc-sulfur colloid were made around a melanoma lesion on the patient's lower left back. There is already radiotracer uptake in a draining lymph channel (arrows). (b) An anterior image of the pelvis shows the lymph channel leading to a focal accumulation in the left groin consistent with accumulation in the sentinel lymph node. (c) Shortly after this, two other smaller foci are seen in the left groin. (d) A ^{99m}Tc point source was used to locate the position on the patient's skin overlying the sentinel node, and a felt marker pen was used to mark the location on the patient's skin. (e) A transmission scan of the chest demonstrates no migration to the axillary regions.

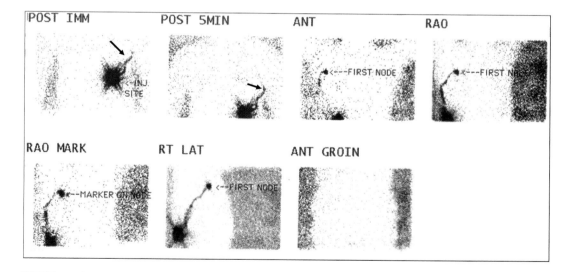

FIGURE 26-24 Lymphoscintigraphy in another patient with a melanoma lesion on the right mid back. Posterior transmission images taken immediately and 5 minutes after four intradermal injections of 250 μCi (9.25 MBq) of ^{99m}Tc-SC were made around the melanoma lesion. In this patient, the draining lymph channel courses toward the right axilla (arrows). Anterior, right anterior oblique (RAO), and right lateral (RT LAT) images clearly show the lymph channel and a focal accumulation in the right axilla consistent with accumulation of radiotracer in the sentinel node. A ^{99m}Tc point source was used to locate the position on the patient's skin overlying the node, and a felt marker pen was used to mark the patient's skin prior to surgery. An anterior transmission image of the pelvis showed no migration of the radiotracer to either groin region.

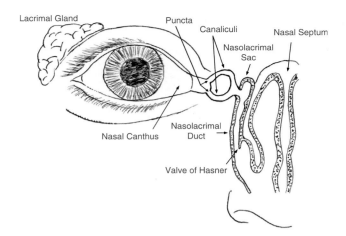

FIGURE 26-25 Diagram illustrating the principal structures in the nasolacrimal system.

film over the cornea through the blinking reflex (Figure 26-25). Excess tear fluid collects in the conjunctival sac and is cleared through the lacrimal puncta, located one each in the upper and lower eyelid in the region of the medial canthus. Tear fluid entering the puncta drains through the upper and lower canaliculi into the nasolacrimal sac and down the nasolacrimal duct through the valve of Hasner, emptying into the nasal cavity.

Epiphora (excess tear flow over the cheek) can be caused by obstruction of tear drainage in the nasolacrimal system. Obstruction of tear drainage can have many causes, such as trauma, inflammation, degenerative changes, and diseases of the nasal sinus. The traditional technique of dacryoscintigraphy involves topical application to the eye in the region of the nasal canthus of 50 to 100 μCi (1.85–3.7 MBq) of ^{99m}Tc-sodium pertechnetate solution in a volume of 10 to 30 μL.[134] Normal human lacrimal fluid volume is around 7 μL, but up to 30 μL can be added without overflow, provided the subject does not blink.[135] Blinking facilitates movement of tear fluid through the nasolacrimal system, and a smaller drop size would preclude overflow of activity from the conjunctival sac during blinking.[136] Ideally, the volume should be 10 to 15 μL, applied with a sterile 1 mL syringe to which a flexible intravenous catheter tip is affixed to avoid the risk of inadvertent injury to the eye.[137]

The asymptomatic eye can be examined first to familiarize the patient with the procedure and to provide a normal control for comparison. A specially constructed pinhole collimator with a 1 mm aperture can be used, which gives the best balance between resolution and sensitivity. After placement of a drop of ^{99m}Tc-sodium pertechnetate in the eye, serial scintigrams are obtained every 30 seconds for 5 minutes. Normal transit time from the conjunctiva to the nasolacrimal sac is less than 1.5 minutes.[133] Longer times indicate delay and provide a sensitive means of detecting obstruction to drainage. Partial obstruction can be demonstrated with a negative Valsalva's maneuver (pinching the nostrils while attempting to draw in air). Dacryoscintigrams demonstrating normal and abnormal drainage patterns are shown in Figure 26-26.

The radiation dose to the lens of a normally draining eye is estimated to be 0.014 to 0.021 rad (cGy) per 100 to 150 μCi (3.7–5.55 MBq), whereas in total obstruction the worst case would be 0.4 to 0.6 rad (cGy).[138] The radiation absorbed dose from this procedure is quite safe, since the threshold dose for initiating cataract formation in the eye is 200 rad (2 Gy).[139] Application of normal saline after a study facilitates clearance of radioactivity from the nasolacrimal system and lowers the radiation dose.

An alternative agent to ^{99m}Tc-sodium pertechnetate for dacryoscintigraphy is ^{99m}Tc-SC. The principal advantage of ^{99m}Tc-SC is a lower radiation dose to the thyroid gland because of its insolubility and therefore lack of absorption from the GI tract as occurs with ^{99m}Tc-sodium pertechnetate.[137]

BONE MARROW IMAGING

In adults, active bone marrow normally resides in the axial skeleton, primarily in the vertebral bodies, pelvis, sternum, ribs, and scapulae, and to a variable extent in the skull.[140] Its distribution in the appendicular skeleton is limited to the proximal one-third of the femurs and humeri.

Blood is supplied to the bone marrow through nutrient arteries that run longitudinally in the central portion of the marrow cavity and send out lateral branches that terminate in capillary beds within bone or at the periphery of the marrow space.[141] The arteriolar capillary blood is drained by postcapillary venules that reenter the marrow cavity and coalesce to form large venous sinuses in which the blood flow is back toward the center of the cavity to the central vein.[141] The erythropoietic marrow is in the form of cords that lie between the venous sinuses. The wall of the venous sinus, which consists of a lining cell, basement membrane, and adventitial cells, is fenestrated, requiring blood cells leaving the erythropoietic tissue to squeeze through pores to enter the venous circulation.[142] The adventitial cells, being phagocytic, remove foreign particles from the blood as it passes into the sinuses. They are responsible for trapping radiocolloids in the bone marrow.

Radiopharmaceuticals for imaging studies target cells in the erythropoietic, reticuloendothelial, or granulopoietic marrow.[140,143,144] The ideal radiotracer for the erythropoietic marrow is one that participates in RBC production. From a physiologic perspective, radionuclides of iron are ideal because erythropoiesis is responsible for 80% to 90% of plasma iron turnover in the body.[140] However, radionuclides of iron have potential imaging limitations. ^{52}Fe is a positron emitter with an 8.2 hour half-life and has been used for quantitative assessment of erythropoiesis in bone marrow expansion, but it requires a PET camera.[145] Furthermore, ^{52}Fe has the practical limitations of expense and availability because it is produced in a cyclotron. ^{59}Fe, with a half-life of 45 days, produces high-energy gamma rays of 1.1 MeV and 1.3 MeV that are unsatisfactory for imaging. Its use has been limited to ferrokinetic studies of erythropoietic bone marrow activity. Indium has chemical properties similar to those of iron and, after intravenous injection, ^{111}In-chloride labels plasma transferrin similarly to iron. However, indium's biologic properties are sig-

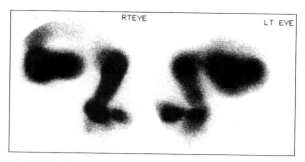
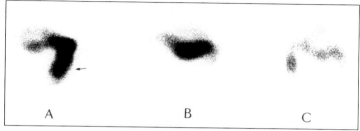

FIGURE 26-26 Dacryoscintigrams obtained with a pinhole collimator and scintillation camera after instillation of 100 µCi (3.7 MBq) of ^{99m}Tc-sodium pertechnetate into the lateral canthus of each eye. Upper panel: Normal dacryoscintigram demonstrating normal tear drainage from the right and left eyes. Lower panel: (A) The right eye exhibits normal drainage into the lacrimal duct (arrow); (B) the left eye demonstrates a lack of drainage under normal conditions, but drainage into the lacrimal duct is induced (C) after a negative Valsalva maneuver, which indicates a stenotic condition rather than complete obstruction.

nificantly different from those of iron. Studies in humans have demonstrated that the plasma clearance half-life of ^{111}In-transferrin (6.1 hours) is much slower than the plasma half-life of ^{59}Fe-transferrin (1 to 2 hours), and liver uptake of ^{111}In-transferrin is greater.[146] Animal studies demonstrate much less uptake of indium into erythrocytes compared with iron.[146] Other studies have shown that the distribution of indium is more like that of radiocolloids than that of iron.[140] Thus, although ^{111}In-chloride has been used for bone marrow imaging studies, it is not a true tracer of erythrogenesis.

These limitations of iron and indium nuclides led to the use of radiocolloids for bone marrow scintigraphy. ^{99m}Tc-SC is widely used, although in Europe ^{99m}Tc-human serum albumin nanocolloid is approved for use in bone marrow scintigraphy. A typical intravenous dose of ^{99m}Tc-SC for bone marrow imaging is 10 to 12 mCi, with imaging commencing in 20 to 30 minutes. This dose is larger than the typical liver-scanning dose but is necessary to provide an excess of particles and activity for visualizing the marrow. The particle size range of standard ^{99m}Tc-SC preparations is 0.1 to 1.0 µm. However, Atkins et al.[147] demonstrated that greater localization of ^{99m}Tc-SC in bone marrow occurs if smaller particles (around 0.1 µm) are used and if large doses are administered. Apparently, smaller particles are less likely to be trapped in the liver than larger particles, and larger doses decrease the efficiency of the liver and spleen in removing particles, so that extra particles localize in the marrow. With regard to this premise, filtering ^{99m}Tc-SC should improve bone marrow localization; this technique is used at some institutions. However, one study comparing filtered with unfiltered ^{99m}Tc-SC in rabbits reported no significant difference in bone marrow localization.[148]

Radiocolloids are a practical choice for studying the bone marrow because in normal individuals the erythropoietic marrow and reticuloendothelial marrow distributions are similar. An exception to this coincident marrow distribution pattern has been shown to occur mainly in diseases that cause ineffective erythropoiesis, in which the reticuloendothelial marrow expands but the erythropoietic marrow does not. This disparity has been demonstrated by comparing the activity distribution pattern between ^{52}Fe in erythropoietic marrow and radiocolloids in the reticuloendothelial marrow.[149] A similar disparity has been shown to occur after chemotherapy or irradiation of bone marrow; erythropoietic cells are depleted with little effect on reticuloendothelial cells.[140,150] A practical limitation of radiocolloids is that they are not useful for the evaluation of bone marrow in the lower thoracic and upper lumbar spine because of interfering radiocolloid accumulation in the liver and spleen. This type of interference does not occur with the use of iron radionuclides, which have insignificant activity uptake in liver and spleen. Despite these caveats, ^{99m}Tc-SC is convenient and quite useful for bone marrow imaging in many clinical situations. One example is myelofibrosis, in

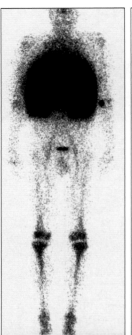

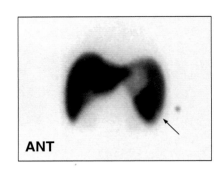
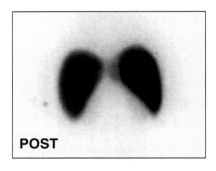

FIGURE 26-27 Bone marrow scan in a 55 year old man with pancytopenia and splenomegaly being evaluated for myelofibrosis. Anterior and posterior total body images (left panels) obtained 30 minutes after intravenous injection of 10 mCi (370MBq) of ^{99m}Tc-SC show absence of normal red marrow activity within the axial and proximal appendicular skeleton. Instead, there is abnormally increased radiotracer uptake within the distal femora and proximal tibias, reflecting red marrow expansion. Anterior and posterior liver–spleen scans (right panels) demonstrate splenomegaly with increased radiocolloid uptake, likely reflecting extramedullary hematopoiesis. (Photos courtesy of Dr. Amy Maszkiewicz, University of North Carolina Hospitals.)

which axial marrow is replaced by fibrous tissue, causing compensatory hyperplasia of marrow with peripheral expansion.[151] The spleen is also enlarged in this condition, compensating for the loss of normal erythropoietic tissue (Figure 26-27). Another example in which bone marrow imaging can be helpful is as an adjunctive study to ^{111}In-WBC imaging for osteomyelitis, to evaluate altered bone marrow uptake versus infection.[152]

Imaging the granulopoietic marrow became possible with the introduction of an antigranulocyte antibody labeled with ^{99m}Tc. The murine monoclonal antibody is an IgG$_1$ isotype anticarcinoembryonic antigen that cross-reacts with NCA-95 (a nonspecific cross-reacting antigen) that is present on the cellular membrane of human granulocytes. This antibody is designated in the literature as NCAA or BW 250/183.[143,153] The radiolabeled antibody is prepared from a kit by premixing ^{99m}Tc-sodium pertechnetate with a transfer ligand and stannous chloride and transferring this mixture to the lyophilized antibody. A gentle mixing process is used in order to limit the amount of liver uptake of the labeled antibody.[154] For clinical use in bone marrow imaging, 8 mCi (300 MBq) of the technetium-labeled antibody is administered intravenously and total body images are obtained in 3 to 4 hours. The use of this antibody for bone marrow imaging in several pathologic conditions has been described.[143,144]

REFERENCES

1. Widmaier EP, Raff H, Strang KT. *Vander, Sherman, and Luciano's Human Physiology, The Mechanisms of Body Function*. 9th ed. Boston: McGraw Hill; 2004:699–721.
2. Spector WG, Willoughby DA. The inflammatory response. *Bact Rev*. 1963;27:117–54.
3. Edwards CL, Hayes RL. Tumor scanning with Ga-67 citrate. *J Nucl Med*. 1969;10:103–5.
4. Bell EG, O'Mara RE, Henry CA, et al. Non-neoplastic localization of Ga-67 citrate. *J Nucl Med*. 1971;12:338–9.
5. Lavender JP, Lowe J, Barker JR. Gallium-67 citrate scanning in neoplastic and inflammatory lesions. *Br J Radiol*. 1971;44:361–6.
6. Staab EV, McCartney WH. Role of gallium-67 in inflammatory disease. *Semin Nucl Med*. 1978;8:219–34.
7. Larson SM, Allen DR, Rasey JS, et al. Kinetics of binding of carrier-free Ga-67 to human transferrin. *J Nucl Med*. 1978;19:1245–9.
8. Watson EE, Cloutier RJ, Gibbs WD. Whole-body retention of Ga-67 citrate. *J Nucl Med*. 1973;14:840–2.
9. Hnatowich DJ, Kulprathipanja S, Beh R. The importance of pH and citrate concentration on the in vitro and in vivo behaviour of radiogallium. *Int J Appl Radiat Isot*. 1977;28:925–31.

10. Nelson B, Hayes RL, Edwards CL, et al. Distribution of gallium in human tissues after intravenous administration. *J Nucl Med.* 1972;13:92–100.
11. Larson SM. Mechanisms of localization of gallium-67 in tumors. *Semin Nucl Med.* 1978;8:193–203.
12. Larson SM, Hoffer PB. Normal patterns of localization. In: Hoffer PB, Bekerman C, Henkin RE, eds. *Gallium-67 Imaging.* New York: Wiley; 1978:23.
13. Mishkin FS, Maynard WP. Lacrimal gland accumulation of Ga-67. *J Nucl Med.* 1975;15:630–1.
14. Larson SM, Schall GL. Gallium-67 concentration in human breast milk [abstract]. *JAMA.* 1971;218:257.
15. Bekerman C, Hoffer PB. Salivary gland uptake of Ga-67 citrate following radiation therapy. *J Nucl Med.* 1976;17:685–7.
16. Weiner RE, Hoffer PB, Thakur ML. Lactoferrin: its role as a Ga-57 binding protein from polymorphonuclear leukocytes. *J Nucl Med.* 1981;22:32–7.
17. Hoffer PB. Mechanisms of localization. In: Hoffer PB, Bekerman C, Henkin RE, eds. *Gallium-67 Imaging.* New York: Wiley; 1978:3.
18. Weiner RE, Schreiber GJ, Hoffer PB, et al. Compounds which mediate gallium-67 transfer from lactoferrin to ferritin. *J Nucl Med.* 1985;26:908–16.
19. Tzen KY, Oster Z, Wagner HN Jr, et al. Role of iron binding proteins and enhanced capillary permeability on the accumulation of gallium-67. *J Nucl Med.* 1980;21:31–5.
20. Vallabhajosula S, Harwig JF, Wolf W. The mechanism of tumor localization of ^{67}Ga-citrate: role of transferrin binding and the effect of tumor pH. *Int J Nucl Med Biol.* 1981;8:363–70.
21. Hayes RL. The tissue distribution of gallium radionuclides. *J Nucl Med.* 1977;18:740–2.
22. Hayes RL. Chemistry and radiochemistry of metal-ion nuclides commonly employed in radiopharmaceuticals. In: Heindel ND, Burns HD, Honda T, et al., eds. *The Chemistry of Radiopharmaceuticals.* New York: Masson; 1977:155.
23. Brown DH, Byrd BL, Carlton JE, et al. A quantitative study of the subcellular localization of Ga-67. *Cancer Res.* 1976;36:956–63.
24. McAfee JG, Thakur ML. Survey of radioactive agents for in vitro labeling of phagocytic leukocytes. I: Soluble agents. *J Nucl Med.* 1976;17:480–7.
25. McAfee JG, Thakur ML. Survey of radioactive agents for in vitro labeling of phagocytic leukocytes. II: Particles. *J Nucl Med.* 1976;17:488–92.
26. Sinn H, Silvester DJ. Simplified cell labeling with indium-111 acetylacetone. *Br J Radiol.* 1979;52:758–9.
27. Danpure HJ, Osman S, Brady F. The labeling of blood cells in plasma with In-111 tropolonate. *Br J Radiol.* 1982;55:247–9.
28. Goodwin DA, Doherty PW, McDougall IR. Clinical use of indium-111 labeled white cells: an analysis of 312 cases. In: Thakur ML, Gottschalk A, eds. *Indium-111 Labeled Neutrophils, Platelets, and Lymphocytes.* New York: Trivirum; 1981:131.
29. Datz FL. Indium-111-labeled leukocytes for the detection of infection: current status. *Semin Nucl Med.* 1994;24:92–109.
30. Peters AM. The utility of ^{99m}Tc-HMPAO-leukocytes for imaging infection. *Semin Nucl Med.* 1994;24:110–27.
31. Goldsmith SJ, Vallabhajosula S. Clinicaly proven radiopharmaceuticals for infection imaging: mechanisms and applications. *Semin Nucl Med.* 2009;39:2–10.
32. Fischman AJ, Rubin RH, Khaw BA, et al. Detection of acute inflammation with ^{111}In-labeled non-specific polyclonal IgG. *Semin Nucl Med.* 1988;18:335–44.
33. Rubin RH, Fischman AJ, Callahan RJ, et al. In-111-labeled non-specific immunoglobulin scanning in the detection of focal infection. *N Engl J Med.* 1989;321:935–40.
34. Fischman AJ, Fucello AJ, Pellegrino-Gensey J, et al. Effect of carbohydrate modification on the localization of human polyclonal IgG at focal sites of bacterial infection. *J Nucl Med.* 1992;33:1378–82.
35. Hansen HJ, Goldenberg DM, Newman ES, et al. Characterization of second-generation monoclonal antibodies against carcinoembryonic antigen. *Cancer.* 1993;71:3478–85.
36. Passalaqua AM, Klein RL, Wegener WA, et al. Diagnosing suspected acute nonclassic appendicitis with sulesomab, a radiolabeled antigranulocyte antibody imaging agent. *J Pediatr Surg.* 2004;39:1338–44.
37. Solanki KK, Bomanji J, Siraj Q, et al. Tc-99m "infecton": a new class of radiopharmaceutical for imaging infection [abstract]. *J Nucl Med.* 1993;34:11P.
38. DeWinter F, Gemmel F, Van Laere K, et al. ^{99m}Tc-ciprofloxacin planar and tomographic imaging for the diagnosis of infection in the postoperative spine: experience in 48 patients. *Eur J Nucl Med Mol I.* 2004;31:233–9.
39. Sarda L, Cremieux AC, Lebellec Y, et al. Inability of ^{99m}Tc-ciprofloxacin scintigraphy to discriminate between septic and sterile osteoarticular diseases. *J Nucl Med.* 2003;44:920–6.
40. Langer O, Brunner M, Zeitlinger M, et al. In vitro and in vivo evaluation of [18F]ciprofloxacin for the imaging of bacterial infections with PET. *Eur J Nucl Med Mol I.* 2005;32:143–50.
41. Signore A, Chianelli M, Annovazzi A, et al. ^{123}I-interleukin-2 scintigraphy for in vivo assessment of intestinal mononuclear cell infiltration in Crohn's disease. *J Nucl Med.* 2000;41:242–9.
42. Bleeker-Rovers CP, Rennen JHHM, Boerman OC, et al. ^{99m}Tc-labeled interleukin-8 for the scintigraphic detection of infection and inflammation. *J Nucl Med.* 2007;48:337–43.
43. Basu S, Alavi A. FDG-PET takes lead role in suspected or proven infection. Functional modality offers several practical advantages over structural imaging techniques or conventional planar scintigraphic methods. *Diagn Imaging.* 2007;23:59–66.
44. Hayes RL, Rafter JJ, Byrd BL, et al. Studies of the in vivo entry of Ga-67 into normal and malignant tissue. *J Nucl Med.* 1981;22:325–32.
45. Bradley WP, Alderson PO, Eckelman WC, et al. Decreased tumor uptake of gallium-67 in animals after whole-body irradiation. *J Nucl Med.* 1978;19:204–9.
46. Fletcher JW, Herbig FK, Donati RM. Ga-67 citrate distribution following whole-body irradiation or chemotherapy. *Radiology.* 1975;117:709–12.
47. Sfakianakis GN, Al-Sheikh W, Heal A, et al. Comparisons of scintigraphy with In-111 leukocytes and Ga-67 in the diagnosis of occult sepsis. *J Nucl Med.* 1982;23:618–26.
48. Seabold JE, Palestro CJ, Brown ML, et al. Procedure guideline for gallium scintigraphy in inflammation 2.0. In: *Procedure Guidelines Manual 2001–2002.* Reston, VA: Society of Nuclear Medicine; 2001:87–91.
49. Woolfenden JM, Carrasquillo JA, Larson SM, et al. Acquired immunodeficiency syndrome: Ga-67 citrate imaging. *Radiology.* 1987;162:383–7.
50. Palestro CJ. The current role of gallium imaging in infection. *Semin Nucl Med.* 1994;24:128–41.
51. Moinuddin M, Rackett J. Gallium scintigraphy in the detection of amiodarone lung toxicity. *Am J Radiol.* 1986;147:607–9.
52. Line BR, Hunninghake GW, Koegh BA, et al. Gallium-67 scanning to stage the alveolitis in sarcoidosis: correlation with clinical studies, pulmonary function studies and bronchoalveolar lavage. *Am Rev Resp Dis.* 1981;123:440–6.
53. Goldsmith SJ, Palestro CJ, Vallabhajosula S. Infectious disease. In: Wagner HN, Szabo Z, Buchanan JW, eds. *Principles of Nuclear Medicine.* Philadelphia: WB Saunders; 1995:728–45.
54. Bleeker-Rovers CP, van der Meer JWM, Oyen WJG. Fever of unknown origin. *Semin Nucl Med.* 2009;39:81–7.
55. Thakur ML, Coleman RE, Welch MJ. Indium-111-labeled leukocytes for the localization of abscesses: preparation, analysis, tissue distribution, and comparison with gallium-67 citrate in dogs. *J Lab Clin Med.* 1977;82:217–28.
56. McAfee JG, Subramanian G, Gagne G. Technique of leukocyte harvesting and labeling: problems and perspectives. *Semin Nucl Med.* 1984;14:83–106.
57. Peters AM. Imaging inflammation: current role of labeled autologous leukocytes. *J Nucl Med.* 1992;33:65–7.
58. Zhuang H, Alavi A. 18-fluorodeoxyglucose positron emission tomographic imaging in the detection and monitoring of infection and inflammation. *Semin Nucl Med.* 2002;32:47–59.

59. Chacko TK, Zhuang H, Nakhoda KZ, et al. Applications of fluorodeoxyglucose positron emission tomography in the diagnosis of infection. *Nucl Med Commun.* 2003;24:615–24.
60. Petruzzi N, Shantly N, Thakur M. Recent trends in soft-tissue infection imaging. *Semin Nucl Med.* 2009;39:115–23.
61. Abdel-Dayem HM, Omar W. Tumor imaging: scintigraphic and pathophysiologic correlation. In: Elgazzar AH, ed. *The Pathophysiologic Basis of Nuclear Medicine.* New York: Springer-Verlag; 2001:169–88.
62. Czernin J, Phelps ME. Positron emission tomography scanning: current and future applications. *Annu Rev Med.* 2002;53:89–112.
63. Thrall JH, Ziessman HA. *Nuclear Medicine: The Requisites.* 2nd ed. St Louis: Mosby; 2001:193–227.
64. DeGroot LJ, ed. *Endocrinology.* Vol 2. 3rd ed. Philadelphia: WB Saunders; 1995:1278.
65. Sherwood LM. Paraneoplastic endocrine disorders (ectopic hormone syndromes). In: DeGroot LJ, ed. *Endocrinology.* Vol 3. 3rd ed. Philadelphia: WB Saunders; 1995:2754–802.
66. DeGroot LJ, ed. *Endocrinology.* Vol 1. 2nd ed. Philadelphia: WB Saunders; 1989:250.
67. Krenning EP, Kwekkeboom DJ, Bakker WH, et al. Somatostatin receptor scintigraphy with [^{111}In-DTPA-D-Phe^1]- and [^{123}I-Tyr^3]-octreotide: the Rotterdam experience with more than 1000 patients. *Eur J Nucl Med.* 1993;20:716–31.
68. Krenning EP, Kwekkeboom DJ, Pauwels S, et al. Somatostatin receptor scintigraphy. In: Freeman LM, ed. *Nuclear Medicine Annual.* New York: Raven Press; 1995:1–50.
69. Somatostatin Receptor Imaging for Neuroendocrine Tumors [product monograph]. St Louis: Mallinckrodt Inc; 1994.
70. Bohuslavizki KH. Somatostatin receptor imaging; current status and future perspectives. *J Nucl Med.* 2001;42:1057–8.
71. Olsen JO, Pozderac RV, Hinkle G, et al. Somatostatin receptor imaging of neuroendocrine tumors with indium-111 pentetreotide (OctreoScan). *Semin Nucl Med.* 1995;25:251–61.
72. Balon HR, Goldsmith SJ, Siegel BA, et al. Procedure guideline for somatostatin receptor scintigraphy with ^{111}In-pentetreotide. *J Nucl Med.* 2001;42:1134–8.
73. Ice RD, Wieland DM, Beierwaltes WH, et al. Concentration of dopamine analogs in the adrenal medulla. *J Nucl Med.* 1975;16:1147–51.
74. Short JH, Darby TD. Sympathetic nervous system blocking agents. III: Derivations of benzylguanidine. *J Med Chem.* 1967;10:833–40.
75. Wieland DM, Wu JJ, Brown LE, et al. Radiolabeled adrenergic neuron-blocking agents: adrenomedullary imaging with I-131 iodobenzylguanidine. *J Nucl Med.* 1980;21:349–53.
76. Wieland DM, Brown LE, Tobes MC, et al. Imaging the primate adrenal medulla with I-123 and I-131 meta-iodo-benzylguanidine [concise communication]. *J Nucl Med.* 1981;22:358–64.
77. Wafelman AR, Konings MCP, Hoefnagel CA, et al. Synthesis, radiolabelling and stability of radioiodinated *m*-iodobenzylguanidine, a review. *Appl Radiat Isot.* 1994;45:997–1007.
78. Donovan AC, Valliant JF. A convenient solution-phase method for the preparation of meta-iodobenzylguanidine in high effective specific activity. *Nucl Med Biol.* 2008;35:741–6.
79. Shapiro B, Copp JE, Sisson JC, et al. Iodine-131 meta-iodobenzylguanidine for the locating of suspected pheochromocytoma: experience in 400 cases. *J Nucl Med.* 1985;26:576–85.
80. Geatti O, Shapiro B, Sisson JC, et al. 131-I-metaiodobenzylguanidine (131-I-MIBG) scintigraphy for the location of neuroblastoma: preliminary experience in 10 cases. *J Nucl Med.* 1984;26:736–62.
81. Fischer M, Kamanabroo D, Sonderkamp H, et al. Scintigraphic imaging of carcinoid tumours with 131-I-metaiodobenzylguanidine [letter]. *Lancet.* 1984;2:165.
82. Sonew T, Funkunaga M, Otsuka N, et al. Metastatic medullary thyroid cancer: localization with iodine-131-metaiodobenzylguanidine. *J Nucl Med.* 1985;26:604–8.
83. McEwan AJ, Shapiro B, Sisson JC, et al. Radioiodobenzylguanidine for the scintigraphic location and therapy of adrenergic tumors. *Semin Nucl Med.* 1985;15:132–53.
84. Petry NA, Shapiro B. Radiopharmaceuticals for endocrine imaging: adrenomedullary imaging. In: Swanson DP, Chilton HM, Thrall JH, eds. *Pharmaceuticals in Medical Imaging.* New York: Macmillan; 1990:343–93.
85. Nakajo M, Shapiro B, Copp J, et al. The normal and abnormal distribution of the adrenomedullary imaging agent m-[I-131]-iodobenzylguanidine (I-131-MIBG) in man: evaluation by scintigraphy. *J Nucl Med.* 1983;24:672–82.
86. Mangner TJ, Tobes MC, Wieland DM, et al. Metabolism of meta-I-131-iodobenzylguanidine in patients with metastatic pheochromocytoma: concise communication. *J Nucl Med.* 1986;27:37–44.
87. Gelfand MJ, Elgazzar AH, Kriss VM, et al. Iodine-123-MIBG SPECT versus planar imaging in children with neural crest tumors. *J Nucl Med.* 1994;35:1753–7.
88. Shapiro B, Gross MD. Radiochemistry, biochemistry, and kinetics of ^{131}I-metaiodobenzylguanidine (MIBG) and ^{123}I-MIBG: clinical implications of the use of ^{123}I-MIBG. *Med Pediatr Oncol.* 1987;15:170–7.
89. AdreView [package insert]. Arlington Heights, IL: GE Healthcare; 2008.
90. Iobenguane Sulfate I-131 [package insert]. Bedford, MA: Pharmalucence; 2008.
91. Gerald D, Marx SJ, Spiegel AM. Parathyroid hormone, calcitonin, and the calciferols. In: Wilson JD, Foster DW, eds. *Williams Textbook of Endocrinology.* 8th ed. Philadelphia: WB Saunders; 1992:1398–1476.
92. Fukunaga M, Morita R, Yonekura Y, et al. Accumulation of ^{201}Tl-chloride in a parathyroid adenoma. *Clin Nucl Med.* 1979;4:229–30.
93. Ferlin G, Borsato N, Camerani M, et al. New perspectives in localizing enlarged parathyroids by technetium-thallium subtraction scan. *J Nucl Med.* 1983;24:438–41.
94. Coakley AJ, Kettle AG, Wells CP, et al. ^{99m}Tc-sestamibi: a new agent for parathyroid imaging. *Nucl Med Commun.* 1989;10:791–4.
95. O'Doherty MJ, Kettle AG, Wells P, et al. Parathyroid imaging with technetium-99m-sestamibi: preoperative localization and tissue uptake studies. *J Nucl Med.* 1992;33:313–8.
96. Taillefer R, Boucher Y, Potvin C, et al. Detection and localization of parathyroid adenomas in patients with hyperparathyroidism using a single radionuclide imaging procedure with technetium-99m-sestamibi (double-phase study). *J Nucl Med.* 1992;33:1801–7.
97. Kardinal CG, Yarbro JW. A conceptual history of cancer. *Semin Oncol.* 1979;6:396–408.
98. Morton DL, Chan AD. The concept of sentinel node localization: how it started. *Semin Nucl Med.* 2000;30:4–10.
99. Cabanas RM. An approach for the treatment of penile carcinoma. *Cancer.* 1977;39:456–66.
100. Holmes EC, Moseley HS, Morton DL, et al. A rational approach to the surgical management of melanoma. *Ann Surg.* 1977;186:481–90.
101. Cabanas RM. Anatomy and biopsy of sentinel lymph nodes. *Urol Clin North Am.* 1992;19:267–76.
102. Morton DL, Wen DR, Wong JH, et al. Technical details of intraoperative lymphatic mapping for early stage melanoma. *Arch Surg.* 1992;127:392–9.
103. Alex JC, Krag DN. Gamma probe guided localization of lymph nodes. *Surg Oncol.* 1993;2:137–43.
104. Alex JC, Weaver DL, Fairbank JT, et al. Gamma probe guided lymph node localization in malignant melanoma. *Surg Oncol.* 1993;2:303–8.
105. Berman CG, Choi J, Hersh MR, et al. Melanoma lymphoscintigraphy and lymphatic mapping. *Semin Nucl Med.* 2000;30:49–55.
106. Morton DL, Chan AD. Current status of intraoperative lymphatic mapping and sentinel lymphandenectomy for melanoma: is it standard of care? *J Am Coll Surg.* 1999;189:214–23.
107. Morton DL, Thompson JF, Essner R, et al. Validation of the accuracy in a multicenter trial of intraoperative lymphatic mapping and sentinel lymphadenectomy for early-stage melanoma. *Ann Surg.* 1999;230:453–65.
108. Alazraki N, Glass EC, Castronovo F, et al. Society of Nuclear Medicine Procedure Guideline for Lymphoscintigraphy and the Use of Intraoperative Gamma Probe for Sentinal Lymph Node Localization in Melanoma of Intermediate Thickness 1.0. In: *Procedure Guidelines Manual 2002.* Reston VA: Society of Nuclear Medicine; 2002. www.snm.org/guidelines. Accessed November 11, 2010.

109. Alazraki NP, Eshima D, Eshima LA, et al. Lymphoscintigraphy, the sentinel node concept, and the intraoperative gamma probe in melanoma, breast cancer, and other potential cancers. *Semin Nucl Med.* 1997; 27:55–67.
110. Alazraki NP, Styblo T, Grant SF, et al. Sentinel node staging of early breast cancer using lymphoscintigraphy and the intraoperative gamma-detecting probe. *Semin Nucl Med.* 2000;30:56–64.
111. Sherman AL, Ter-Pogossian M. Lymph node concentration of radioactive colloidal gold following interstitial injection. *Cancer.* 1953;6: 1238–40.
112. Kazem I, Antionaides J, Brady LW, et al. Clinical evaluation of lymph node scanning utilizing colloidal gold 198. *Radiology.* 1968;90:905–11.
113. Kazem I, Nedwich A, Mortel R, et al. Comparative histological changes in the normal lymph node following ethiodol lymphography and colloidal gold-198 lymph scanning. *Clin Radiol.* 1971;22:382–8.
114. Ege GN. Internal mammary lymphoscintigraphy: the rationale, technique, interpretation and clinical application: a review. *Radiology.* 1976;118:101–7.
115. Warbick A, Ege GN, Henkelman RM, et al. An evaluation of radiocolloid size technique. *J Nucl Med.* 1977;18:827–34.
116. Dunson GL, Thrall JH, Stevenson JS, et al. Tc-99m minicolloid for radionuclide lymphoscintigraphy. *Radiology.* 1973;109:387–92.
117. Ege GN, Warbick A. Lymphoscintigraphy: a comparison of Tc-99m antimony sulfide colloid and Tc-99m stannous phytate. *Br J Radiol.* 1979;52:124–9.
118. Kaplan WD, Davis MA, Rose CM. A comparison of two technetium-99m-labeled radiopharmaceuticals for lymphoscintigraphy [concise communication]. *J Nucl Med.* 1979;20:933–7.
119. Bergqvist L, Strand S-E, Perssom BRR. Particle sizing and biokinetics of interstitial lymphoscintigraphic agents. *Semin Nucl Med.* 1983;12:9–19.
120. Nanocoll [package insert]. Amersham, England: GE Healthcare; 2003.
121. Wilhelm AJ, Mijnhout GS, Franssen EJ. Radiopharmaceuticals in sentinel lymph node dissection: an overview. *Eur J Nucl Med.* 1999; 26(suppl):S36–42.
122. Dragotakes SC, Callahan RJ, LaPointe LC, et al. Particle size characterization of a filtered ^{99m}Tc-sulfur colloid preparation for lymphoscintigraphy [abstract]. *J Nucl Med.* 1995;36:80P.
123. Hung JC, Wiseman GA, Wahner HW, et al. Filtered technetium-99m-sulfur colloid evaluated for lymphoscintigraphy. *J Nucl Med.* 1995;36:1895–1901.
124. Goldfarb LR, Alazraki NP, Eshima D, et al. Lymphoscintigraphic identification of sentinel lymph nodes: clinical evaluation of 0.22-μm filtration of Tc-99m sulfur colloid. *Radiology.* 1998;208:505–9.
125. Eshima D, Eshima LA, Gotti NM, et al. Technetium-99m-sulfur colloid for lymphoscintigraphy: effects of preparation parameters. *J Nucl Med.* 1996;37:1575–8.
126. Vera DR, Wallace AM, Hoh CK, et al. A synthetic macromolecule for sentinel node detection: ^{99m}Tc-DTPA-mannosyl-dextran. *J Nucl Med.* 2001;42:951–9.
127. Hoh CK, Wallace AM, Vera DR. Preclinical studies of [^{99m}Tc]-DTPA-mannosyl-dextran. *Nuc Med Biol.* 2003;30:457–63.
128. Ellner SJ, Hoh CK, Vera DR, et al. Dose-dependent biodistribution of [^{99m}Tc]DTPA-mannosyl-dextran for breast cancer sentinel lymph node mapping. *Nucl Med Biol.* 2003;30:805–10.
129. Wallace AM, Hoh CK, Ellner SJ, et al. Lymphoseek: a molecular imaging agent for melanoma sentinel lymph node mapping. *Ann Surg Oncol.* 2007;14:913–21.
130. Phillips WT, Rudolph AS, Goins B, et al. A simple method for producing a technetium-99m-labeled liposome which is stable in vivo. *Nucl Med Biol.* 1992;19:539–47.
131. Phillips WT, Andrews T, Liu HL, et al. Evaluation of [^{99m}Tc] liposomes as lymphoscintigraphic agents: comparison with [^{99m}Tc] sulfur colloid and [^{99m}Tc] human serum albumin. *Nucl Med Biol.* 2001;28:435–44.
132. Glut EM, Hinkle GH, Guo W, et al. Kit formulation for the preparation of radioactive blue liposomes for sentinel node lymphoscintigraphy. *J Pharm Sci.* 2002;91:1724–32.
133. Brown M, El Gammal TAM, Luxenberg MN, et al. The value, limitations, and applications of nuclear dacryocystography. *Semin Nucl Med.* 1981;11:250–7.
134. Brizel HE, Sheils WC, Brown M. The effects of radiotherapy on the nasolacrimal system as evaluated by dacryoscintigraphy. *Radiology.* 1975;116:373–81.
135. Mishima S, Gasset A, Klyce SD, et al. Determination of tear volume and tear flow. *Invest Ophthalmol.* 1966;5:264–76.
136. Daubert J, Nik N, Chandeyssoun PA, et al. Tear flow analysis through upper and lower systems. *Ophthal Plast Reconstr Surg.* 1990;6:193–6.
137. Ponto JA. Dispensing and administration of radiopharmaceuticals for dacryoscintigraphy. *J Am Pharm Assoc.* 2006;46:749–50.
138. Robertson JS, Brown ML, Colvard DM. Radiation absorbed dose to the lens in dacryoscintigraphy with pertechnetate. *Radiology.* 1979;113: 747–50.
139. Merriam GR, Szechter A, Focht EF. The effects of ionizing radiations on the eye. *Front Radiat Ther Oncol.* 1972;6:346–85.
140. Datz FL, Taylor A Jr. The clinical use of radionuclide bone marrow imaging. *Semin Nucl Med.* 1985;15:239–59.
141. Weiss L, Chen LT. The organization of hematopoietic cords and vascular sinuses in bone marrow. *Blood Cells.* 1975;1:617–38.
142. Weiss L. The structure of bone marrow. *J Morphol.* 1965;117:467–537.
143. Reske SN. Recent advances in bone marrow scanning. *Eur J Nucl Med.* 1991;18:203–21.
144. Yum M, Kim C, Sam J, et al. Bone marrow scintigraphy. In: Sandler MP, Coleman RE, Patton JA, et al., eds. *Diagnostic Nuclear Medicine.* 4th ed. Philadelphia: Lippincott Williams & Wilkins; 2003:565–87.
145. Ferrant A, Rodhain J, Leners N, et al. Quantitative assessment of erythropoiesis in bone marrow expansion areas using ^{52}Fe. *Br J Haematol.* 1986;62:247–55.
146. McIntyre PA. Agents for bone marrow imaging: an evaluation. In: Subramanian G, Rhodes B, Cooper JF, et al., eds. *Radiopharmaceuticals.* New York: Society of Nuclear Medicine; 1975:343–8.
147. Atkins HL, Hauser W, Richards P. Factors affecting distribution of ^{99m}Tc-sulfur colloid. *J Nucl Med.* 1969;10:319–20.
148. Callahan RJ, Achong DM, Wilkinson RA, et al. Can filtration improve the bone marrow imaging properties of Tc-99m sulfur colloid? [abstract] *J Nucl Med.* 1996;37:240P.
149. McIntyre PA. Newer developments in nuclear medicine applicable to hematology. In: Brown EB, ed. *Progress in Hematology.* Vol 10. Orlando, FL: Grune and Stratton; 1977:361–409.
150. Nelp WB, Gohil MN, Larson SM, et al. Long-term effects of local irradiation of the marrow on erythron and red cell function. *Blood.* 1970;36:617–22.
151. Dibos PE, Judisch JM, Spaulding MB. Scanning the reticuloendothelial system (RES) in hematological diseases. *Johns Hopkins Med J.* 1972;130:68–81.
152. Seabold JE, Nepola JV, Marsh JL, et al. Postoperative bone marrow alterations: potential pitfalls in the diagnosis of osteomyelitis with In-111-labeled leukocyte scintigraphy. *Radiology.* 1991;180:741–7.
153. Duncker CM, Carrió I, Berná L, et al. Radioimmune imaging of bone marrow in patients with suspected bone metastases from primary breast cancer. *J Nucl Med.* 1990;31:1450–5.
154. Reske S, Buell U. Reduced technetium-99m labeled NCA-95/CEA-antibody uptake in the liver due to gentle antibody reconstitution: technical note. *Eur J Nucl Med.* 1990;17:38–41.

CHAPTER 27

Clinical PET Procedures

Positron emission tomography (PET) is a functional imaging modality that can be performed with a variety of positron-emitting radiopharmaceuticals for diagnostic studies. However, only three agents are currently approved for reimbursement by the Centers for Medicare and Medicaid Services (CMS) for clinical procedures: ^{18}F-fludeoxyglucose (^{18}F-FDG, or FDG) for oncology and neurology, and ^{18}F-FDG, ^{13}N-ammonia, and ^{82}Rb-chloride for cardiology. FDG PET is now principally used for diagnosis, staging, and assessing response to therapy in oncology, but it also has significant value beyond oncology. In the brain, PET has been useful not only in grading primary tumor and evaluating for recurrent tumor but also in evaluating epilepsy and dementia. In the heart, PET can be used to evaluate function, for example, by assessing ischemia and viability.

Anatomic imaging modalities such as computed tomography (CT) and magnetic resonance imaging (MRI) have been very useful in staging and follow-up of patients with cancer. However, it has become clear that functional imaging can significantly complement anatomic imaging, resulting in increasing diagnostic confidence. It has been shown that hybrid imaging systems that combine PET and CT (PET/CT) are more accurate than either system alone or images from each modality viewed side by side. In addition, hybrid imaging results in more accurate localization of metabolic abnormalities and allows detection of lesions that are not associated with an anatomic abnormality, such as distant metastases. The combined technology also makes it easier to differentiate abnormal radiotracer uptake from physiologic accumulation in normal tissues. This chapter discusses procedures and protocols for clinical PET/CT imaging and presents examples of current clinical indications for PET/CT imaging.

PATIENT PREPARATION FOR ONCOLOGIC FDG PET/CT IMAGING

Before scheduling an FDG PET/CT exam, it is important to understand the patient's oncologic history and the reason for the evaluation. The evaluation may be for initial staging of disease or to provide diagnostic information after surgery, chemotherapy, radiation therapy, or some combination of these. After surgery, inflammation around the surgical site will demonstrate increased FDG uptake. Thus, sufficient healing time is necessary before the patient can be accurately evaluated for residual tumor near the operative site. Six weeks is usually sufficient, depending on whether there were postsurgical complications (e.g., abscess, infection). However, if the surgeon is not interested in imaging the operative site but in evaluating a distant abnormality (e.g., a lung nodule in a patient who had recent surgery for rectal cancer), this can be done without delay. An extensive inflammatory response can be associated with radiation therapy and will require sufficient time to resolve. Inflammation after radiation is variable with respect to the intensity and duration of FDG uptake, and the inflammation can last for months or years. The appropriate interval for follow-up imaging after radiation therapy is generally thought to be 2 to 3 months. Chemotherapy can also complicate evaluation for residual viable tumor or distant metastases. If the patient is receiving monthly chemotherapy, the exam should be scheduled at the end of the monthly cycle, just before the next round of chemotherapy.

PET imaging, unlike anatomic imaging modalities such as CT, requires careful attention to patient preparation because of its physiologic nature. Patients are usually contacted the day before the exam regarding diet, fasting, and diabetes medications. Because FDG is a glucose analogue and is competitively inhibited from uptake in tumor cells by existing glucose in the bloodstream, careful control of blood glucose is required prior to FDG administration. It is helpful if the patient can be on a low carbohydrate, low sugar diet the day before the scan. Patients are asked to have nothing to eat or drink other than water for 6 hours prior to the study. It is generally agreed that PET/CT should not be performed on patients with blood glucose concentrations above 200 mg/dL; with higher blood glucose levels, the sensitivity of the exam will decrease because of decreased uptake of FDG in the tumor cells and an overall increase in radiotracer background activity (Figure 27-1). Patients with diabetes need to be managed appropriately to maintain their blood glucose concentration below this level. Diabetes medications should not be taken the morning of the scan; insulin will force glucose and FDG into the muscles, increasing background activity. The patient should be well hydrated with water prior to and on the day of the scan. Adequate hydration reduces radiotracer background activity and facilitates urinary clearance, thus reducing the patient's radiation exposure. To minimize muscle uptake, strenuous exercise should be avoided the day before the scan (Figure 27-2).

Blood glucose is determined with a glucometer after the patient arrives for the exam. If the blood glucose is too high, the exam must be rescheduled. If the patient has been compliant with fasting and the blood glucose is acceptable, an intravenous line is inserted for administration of the radiopharmaceutical. The dose of FDG is based on the patient's weight and the time of administration. Because of the short 110 minute half-life of FDG, it may not be possible to use the dose for another patient if the initial study is canceled; this often results in loss of the radiopharmaceutical dose. The time from injection of the FDG dose until PET image acquisition begins is known as the FDG uptake time. It is generally thought that FDG uptake times of 60 to 90 minutes are sufficient for acceptable tumor-to-background ratios. During the FDG uptake period, the patient is asked to remain still to minimize muscle uptake. If the patient is cold, shivering, or nervous, prominent physiologic uptake of radiotracer can be seen in brown fat and muscle, which can make the study more difficult to interpret.[1] Uptake in brown fat occurs predominantly in the neck and shoulder regions but can also occur in the mediastinum, axillae, and upper abdomen (Figure 27-3). Focal areas of physiologic uptake in brown fat can be mistaken for tumor uptake or metastatic lymph nodes, especially by an untrained interpreter. However, with combined PET/CT technology, it is usually easy to identify this uptake as being in fat. Having the patient wear comfortable, warm clothing and using warming blankets can help reduce or prevent physiologic accumulation in brown fat. Beta-blocking medications (e.g., propranolol) can be useful in reducing brown fat uptake.[2] Alternatively, premedication with a short-acting benzodiazepine, such as lorazepam, can be useful, particularly in an anxious or claustrophobic patient, because of its anxiolytic, sedative, and muscle relaxant properties.

Removable metallic objects that will be in the field of view, such as jewelry, coins, metallic buttons, or belt buckles, should be removed before imaging. These items can absorb or block x-ray photons and may produce streak artifacts on the CT images, referred to as beam-hardening artifacts, which can make it difficult to evaluate adjacent structures and can lead to associated artifacts on the PET images as well.

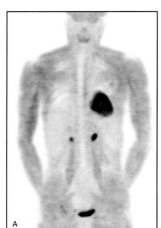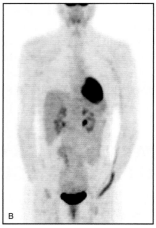

FIGURE 27-1 Anterior FDG PET maximum-intensity projection images of the same patient 7 days apart. The image on the left (A) shows diffuse muscle uptake. After the scan, the patient admitted eating only 45 minutes before arriving for the study. The image on the right (B), done after a 6 hour fast, shows physiologic uptake in the base of the brain, liver, spleen, kidneys, and bladder. Uptake in the left ventricle of the heart is variable and is prominent even on this 6 hour fasting study.

PET/CT ACQUISITION

Depending on the clinical indication, PET/CT studies can be performed on just a limited area of the body, such as the heart, or on a more extended area. Typical "whole-body" oncology studies are often done from the level of the forehead to the mid thighs, commonly referred to as "eyes to thighs." Scans of malignancies such as melanoma that can often involve extremities are done from the top of the head through the toes, for a "total-body scan."

When the patient is being positioned on the PET/CT scanner, it is important that he or she understand the need to limit motion during the study so that there is no mismatch

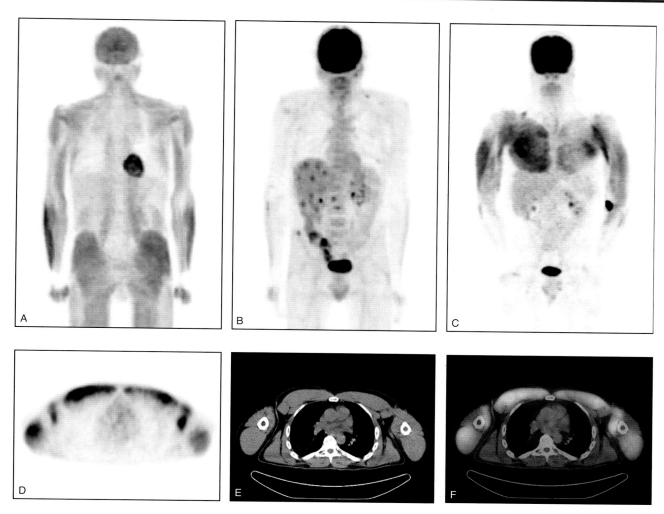

FIGURE 27-2 Anterior FDG PET maximum-intensity projection (MIP) images (A and B) of the same patient 10 days apart. Image A, done 1 day after the patient's training for a marathon, demonstrates diffuse increased muscle uptake. Increased muscle uptake can lower the sensitivity for detecting metastatic disease. Image B, done 10 days later after proper patient preparation and no recent exercise, shows metastases not apparent on the earlier study. Anterior FDG PET MIP image of a different patient (C) who worked out on a rowing machine a few hours before the exam; axial (D) FDG PET, (E) CT, and (F) FDG PET/CT fusion images through this patient's chest show diffuse increased uptake in muscle.

between the CT and PET acquisitions. Restraint devices can be used to minimize patient motion during the sequential CT and PET images.

The CT scan is performed first, followed by PET imaging. The CT scan is important for anatomic localization. Typically, it takes only about 30 seconds to perform an eyes-to-thighs study. The data from the CT scan are also used to create an attenuation-correction matrix, which will be used to correct the emission data obtained during PET imaging. PET imaging is accomplished in increments or body segments. These increments vary in size depending on the scanner but are often around 15 cm and are referred to as bed positions. PET imaging is usually performed in the caudocranial direction, imaging the pelvis and upper thighs first (first bed position). This minimizes artifacts from intense physiologic radiotracer accumulation in the urinary bladder and minimizes misregistration effects in the pelvis secondary to the change in the size of the bladder between the CT and PET image acquisitions as the bladder fills with urine. PET imaging is obtained in several contiguous bed positions. PET image quality can be optimized by adjusting the time per bed position according to patient size and weight and the administered dose. Typically, thinner patients or larger doses of radiopharmaceutical result in shorter times for each bed position; heavier patients or smaller doses require more time for each bed position. Bed position times usually are between 2 and 5 minutes per bed. The CT scan is obtained first, which allows reconstruction of the attenuation-corrected PET images to begin right after the first bed position is completed, while the next bed position image is being acquired. This is much more efficient than reconstructing the entire set of attenuation-corrected PET images after the study is completed.

The non–attenuation-corrected PET images should also be processed. Metallic objects such as dental metal, chemotherapy

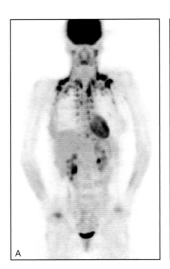
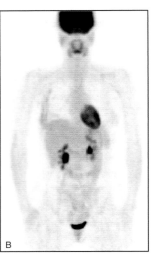
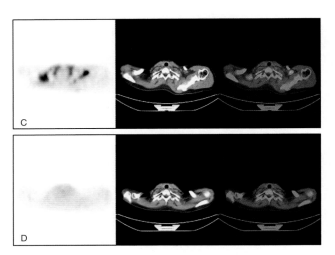

FIGURE 27-3 Anterior FDG PET maximum-intensity projection images in a breast cancer patient 3 weeks after a right mastectomy. The image on the left (A) shows extensive uptake in brown fat. A repeat study done 2 days later (B), after the patient was premedicated with lorazepam 1 mg and offered warming blankets, shows resolution of the brown fat uptake. There is only mild diffuse uptake in the lateral right chest representing mild residual inflammation from the mastectomy. FDG PET, CT, and FDG PET/CT fusion axial images (C) through the base of the neck from the first study demonstrate that the focal areas of radiotracer uptake correspond to fat on the noncontrast CT. Similar axial images from the second study (D) show resolution of the brown fat uptake.

ports, hip replacements, and other orthopedic devices are usually seen as photopenic defects on dedicated PET imaging. On PET/CT scanners, where attenuation correction is CT based, these areas often demonstrate falsely elevated (artifactual) FDG uptake. In these cases, it can be important to inspect both the CT attenuation-corrected and the non–attenuation-corrected PET images (Figure 27-4).

Once processing is completed, the study is typically sent to a work station that has software capable of displaying both CT and PET images along with PET/CT fusion images and 3-D maximum-intensity projections (MIPs). MIPs are essentially 3-D compilations of the entire PET study that can be rotated. Typical layouts display axial CT, axial PET, and colorized axial PET fused with the CT and MIP images. Coronal CT, coronal PET, sagittal CT, and sagittal PET, as well as fused coronal and sagittal images, can also be displayed.

THE NORMAL FDG PET/CT SCAN

Most organs and other body structures take up FDG to some degree. Many of these can demonstrate variable physiologic (normal) uptake. To be able to differentiate between this and abnormal or pathologic accumulation, it is important to understand the variations in normal physiologic uptake. On a normal FDG PET/CT exam, the brain, the urinary tract, and often the heart are the most prominent sites of FDG accumulation (Figure 27-5). In the neck, there is usually moderate symmetric uptake in tonsillar tissue. The thyroid gland can demonstrate variable uptake. In the chest,

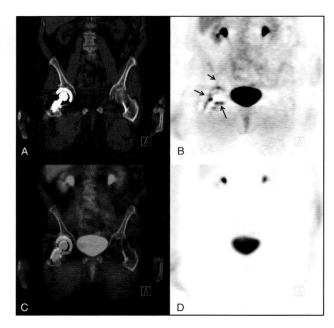

FIGURE 27-4 Coronal images through the pelvis in a patient with a right hip replacement. CT image (A) shows beam-hardening artifacts around the metal hip replacement. Attenuation-corrected FDG PET image (B) shows artifactual uptake around the prosthesis (arrows). FDG PET/CT fusion image (C). Non–attenuation-corrected FDG PET image (D) eliminates the artifacts.

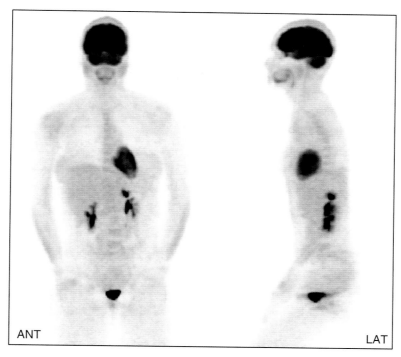

FIGURE 27-5 Normal FDG PET/CT scan. Anterior and lateral maximum-intensity projection images of a normal patient. Note that the most prominent uptake is in the brain, left ventricle of the heart (which can have variable uptake), kidneys, and bladder.

moderate blood pool activity in the great vessels and heart chambers can be seen, bordered by very low radiotracer accumulation in the surrounding lungs. The liver and spleen normally demonstrate uniform moderate uptake, slightly greater than the uptake in the blood pool. Bone marrow uptake is also typically slightly greater than uptake in the blood pool, but uptake is less than in the liver and spleen. Uptake in the stomach and bowel is variable. Uptake in the collecting systems of normally functioning kidneys and in the bladder is typically intense because, unlike glucose, FDG is not reabsorbed by the renal tubules.

FDG Uptake in the Brain

Glucose is the primary energy substrate for the brain. Thus, under normal physiologic circumstances, uptake in the brain of a living, conscious person is intense. There is strong uptake in the cerebral cortex (gray matter), caudate nucleus, and thalamus. Less intense uptake is seen in the white matter of the brain (Figure 27-6). The cerebral ventricles, which contain cerebrospinal fluid, demonstrate little to no appreciable uptake. Normally, the ventricles are not seen on PET images unless they are enlarged. Enlarged intra-axial or extra-axial fluid collections, which do not accumulate FDG, will be apparent as areas of absent radiotracer accumulation, or photopenic areas, next to normal brain uptake (Figure 27-7). Uptake

in the brain can be variable, depending on external stimuli the patient experiences during the first 30 minutes of the FDG uptake period. For example, if the patient is actively reading or experiencing visual stimuli, the primary visual cortex in the occipital lobes will show increased uptake. This can be minimized by placing the patient in a quiet, dark room with eyes closed during the uptake period. Brain perfusion studies with ^{15}O water have a similar appearance to studies with FDG.

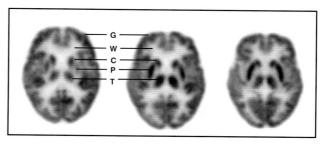

FIGURE 27-6 Three contiguous FDG PET axial sections through a normal brain. The most prominent FDG uptake is seen in the head of the caudate nucleus (C), putamen (P), and thalamus (T). There is also intense uptake in the gray matter (G), with little uptake in the white matter (W).

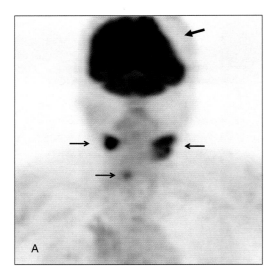
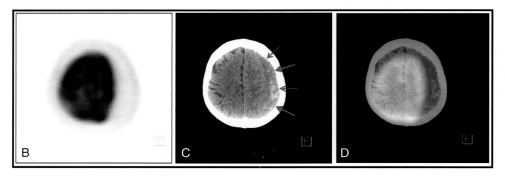

FIGURE 27-7 Anterior FDG PET maximum-intensity projection image (A) of a patient with recurrent laryngeal cancer with metastases in the neck (thin arrows). Note flattening of the normal brain contour (thick arrow). Axial images through the brain of this patient demonstrate flattening of the normal brain contour on the patient's left side on the FDG PET image (B). The corresponding noncontrast CT image (C) shows patchy areas of high attenuation within a large crescent-shaped abnormality in the brain, displacing the normal brain inward (arrows), consistent with acute hemorrhage within a chronic subdural hematoma. FDG PET/CT fusion image (D).

FDG Uptake in the Neck

Activity is often seen in the vocal cord region if the patient is talking during the FDG uptake period (Figure 27-8). This can be minimized if the patient is instructed to remain quiet, at least during the early uptake period; this is especially important for evaluation of head and neck cancer. Variable uptake can be seen in the tonsils and in the salivary glands. This uptake is more prominent in children and typically decreases with age. Normal salivary and tonsillar uptake should be symmetric (Figure 27-9). Mild physiologic thyroid uptake can sometimes be seen. Usually, however, no significant radiotracer accumulation is seen in the thyroid gland. Diffuse prominent uptake in the thyroid can be seen with Graves' disease or thyroiditis (Figure 27-10). Prominent focal uptake in the thyroid gland can be seen with either benign (thyroid adenoma) or malignant (papillary thyroid cancer) processes.

FDG Uptake in the Heart

FDG uptake in the heart is highly variable. The primary energy substrate in the heart depends on the substrate that is most available for metabolism. When the patient is in a fasting state, the heart relies on fatty acid metabolism instead of glucose metabolism for energy. In the fasting state, most individuals will have little FDG radiotracer accumulation in the heart. However, this is quite variable. In patients who have had nothing to eat for 4 to 6 hours, it can range from no cardiac uptake to diffuse, intense cardiac uptake. After eating, blood glucose and insulin levels increase. This increases GLUT4 transporter expression at the myocardial cell surface and in fat and skeletal muscle, which mediates glucose uptake into the cell.[3] The heart switches to glucose for energy, and accumulation of FDG in the heart increases. Therefore, for heart studies with FDG, the patient is given a glucose load before

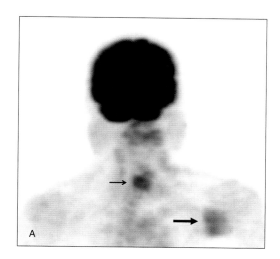
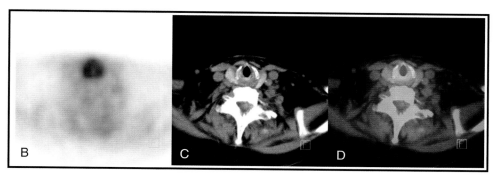
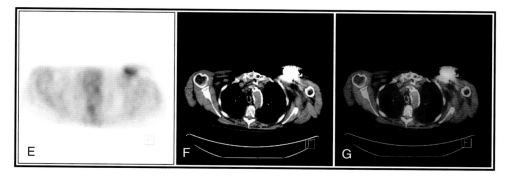

FIGURE 27-8 Anterior FDG PET maximum-intensity projection image (A) showing uptake in the midline base of the neck (small arrow) and in the left upper chest region (large arrow). Axial images through the focal area of radiotracer uptake in the midline neck (FDG PET image, B; CT image, C; and FDG PET/CT fusion image, D) correspond to the vocal cord region and represent physiologic uptake secondary to talking during the FDG uptake period. Axial images through the focal area of radiotracer uptake in the left upper chest (FDG PET image, E; CT image, F; and FDG PET/CT fusion image, G) correspond to a cardiac pacemaker implanted in the subcutaneous tissues of the chest. Attenuation artifacts from the metal in the pacemaker result in artifactual uptake in this region.

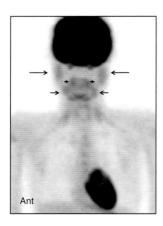
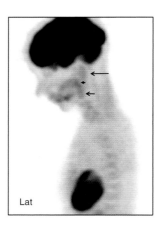

FIGURE 27-9 FDG PET study demonstrating normal tonsil and salivary gland uptake of FDG. Parotid glands (long arrows), submandibular glands (shorter arrows), tonsils (arrowheads).

administration of FDG and subsequent scanning. For oncology or infection imaging, the patient is given nothing by mouth except water for 4 to 6 hours before scanning to minimize uptake in the heart, skeletal muscle, and fat.

FDG Uptake in the Kidneys and Bladder

Although FDG is a glucose analogue, it differs from glucose in that it does not undergo tubular reabsorption in the kidneys and is excreted in the urine. Clearance of FDG from the blood pool occurs predominantly through renal excretion, so it is normal to see radiotracer accumulation in the kidneys, ureters, and bladder. Because of peristaltic activity, radiotracer accumulation is not usually seen along the entire course of the ureters unless there is at least a partial obstruction in the distal ureter (Figure 27-11). The ureters do often demonstrate focal areas of accumulation or accumulation in short segments, which is normal. This can sometimes be problematic in evaluation of para-aortic or iliac adenopathy adjacent to the ureters. However, the course of the ureters can often be followed on CT, which minimizes potential problems in image interpretation. Prominent radiotracer accumulation in the bladder can obscure abnormalities in the pelvis. This can often be avoided if the patient is instructed to void just before being placed on the scanner and if imaging is performed in a caudocranial direction with the pelvis first. Sometimes physiologic activity can be seen in the urethra. This is often apparent in male patients who have undergone a transurethral resection of the prostate for benign prostatic hypertrophy. When indeterminate or equivocal foci of FDG activity are seen in the kidney or along the course of

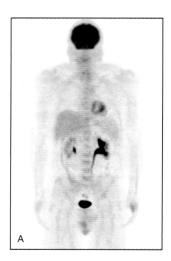

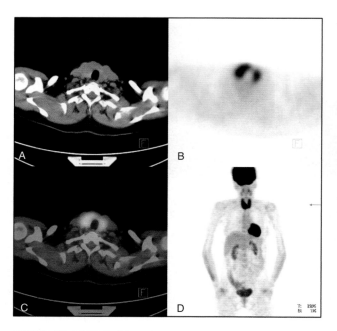

FIGURE 27-10 Patient with thyroiditis. Axial (A) noncontrast CT, (B) FDG PET, and (C) FDG PET/CT fusion image of the neck at the level of the arrow in the FDG PET maximum-intensity projection image (D), showing prominent radiotracer accumulation in a large thyroid gland.

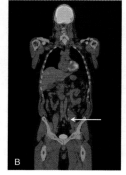
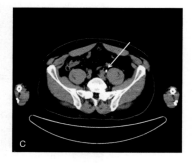

FIGURE 27-11 Anterior FDG PET maximum-intensity projection image (A) shows prominent radiotracer accumulation in the left renal collecting system with columning of radiotracer in the mid and proximal ureter. Coronal FDG PET/CT fusion image through the left ureter (B) shows radiotracer extending to a small calcification in the mid ureter (arrow). Axial noncontrast CT image (C) through this level shows the obstructing calcification in the left ureter (arrow).

the urinary tract, diuretics such as furosemide or delayed post-void imaging can be useful (Figure 27-12).[4]

FDG Uptake in the Digestive System

FDG uptake in the esophagus, stomach, and small and large bowel is variable. Normally, the esophagus does not demonstrate significant radiotracer accumulation. However, it can on occasion show mild low-level uptake along its length. It is common to see a small focus of mild-to-moderate uptake at the gastroesophageal junction, which may be related to activity of the esophageal sphincter. More diffuse moderate esophageal uptake can be seen with esophagitis. More prominent focal uptake or uptake associated with esophageal wall thickening on CT is suggestive of cancer. Normal stomach uptake can be variable. Often, mild-to-moderate radiotracer accumulation is seen diffusely in the stomach.[5] FDG uptake in the small bowel is usually less than uptake in the colon and is more typical in the low pelvis. There can be prominent diffuse or patchy uptake in the colon. This is usually more prominent in the right colon. Uptake in large segments of colon is often physiologic, although increased uptake can be seen in inflammatory bowel conditions such as Crohn's disease and ulcerative colitis. Focal areas of increased uptake in the colon can be physiologic or can be seen with diverticulitis, adenomatous polyps, and cancer. The CT portion of the PET/CT study is often helpful in differentiating among these conditions.

FDG Uptake in the Musculoskeletal System

Skeletal muscle at rest uses fatty acid metabolism for energy. However, when muscles are contracting they switch to glucose metabolism. Thus, skeletal muscle actively contracting during the FDG uptake period will show increased radiotracer accumulation (Figure 27-13). To prevent this, patients are usually placed in a comfortable chair and asked to rest peacefully during the uptake period. In patients with diabetes mellitus, insulin injected prior to FDG administration will increase glucose uptake in muscle, resulting in diffuse skeletal muscle uptake that can lower the sensitivity for detecting metastatic disease. Patients are also asked to refrain from strenuous exercise for 1 to 2 days before the study.

CLINICAL USE OF FDG PET IN ONCOLOGY

It has been known for some time that rapidly dividing tumor cells have increased metabolic demands requiring adenosine triphosphate (ATP) generated by glucose metabolism or glycolosis.[6] Glucose uptake in cancer cells is increased because of increased production of glucose transporters in the cell membrane.[7] FDG, as a glucose analogue, is taken up into cells similarly to glucose and is then phosphorylated by hexokinase to FDG-6-phosphate. FDG-6-phosphate is not a substrate for the next enzyme in the glycolytic pathway and remains trapped in the cell. The accumulation of the

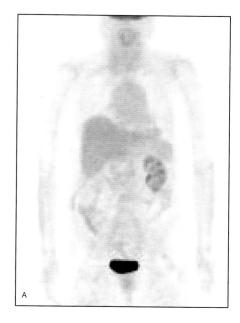

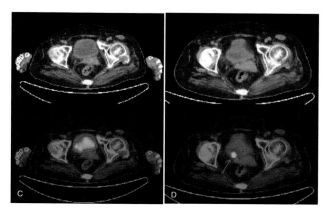

FIGURE 27-12 Anterior FDG PET maximum-intensity projection image of a patient with a history of bladder cancer (A). There is intense physiologic radiotracer accumulation in the bladder. Note that the patient only has one kidney; the right kidney is absent. The patient was asked to void and was given more water to drink, and a delayed single bed position image of the bladder was obtained (B), revealing a recurrent bladder cancer (arrow). Axial CT (upper panel) and FDG PET/CT fusion (lower panel) images of the bladder on the pre-void (C) and delayed post-void (D) parts of the study, showing recurrent tumor in the posterior aspect of the bladder on the right (red arrow).

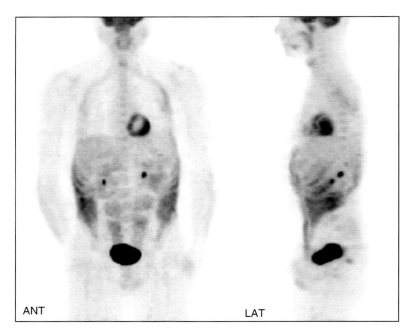

FIGURE 27-13 Anterior and lateral FDG PET maximum-intensity projection images showing diffuse abdominal wall muscle uptake in a patient who was coughing during the FDG uptake period.

positron emitter ^{18}F-FDG-6-phosphate in the cell allows for PET imaging.

Clinical use of FDG PET in oncology has grown rapidly in the past decade, becoming an important imaging modality for initial staging, monitoring of therapy, restaging, and assessment of cancer recurrence. The rise in clinical use of PET began in 1998 after a CMS decision to cover solitary pulmonary nodule imaging. After FDG PET imaging was found to be more accurate for staging cancer than other conventional anatomical imaging modalities such as CT, coverage was approved for PET imaging of other cancers. Coverage has now expanded to include diagnosis, staging, and monitoring of therapy for many cancers. Early diagnostic PET imaging used dedicated PET scanners, which produced functional (metabolic) images of FDG accumulation in the body. Dedicated PET findings were often difficult to understand because of a combination of relatively limited spatial resolution and large variations in physiologic uptake in normal tissues (e.g., bowel). In addition, false-positive uptake could be seen with infection or inflammation. Without careful anatomic correlation, PET findings were very difficult to interpret. Often a patient would be sent for a PET scan because of a finding on a previous CT scan. The PET images would have to be carefully reviewed along with the CT to put the PET findings in context with normal and abnormal morphology. Other complicating issues were the time between anatomic (CT) and functional (PET) imaging and differences in patient positioning between these studies.

The growth of clinical PET imaging was spurred on by advances in both hardware and software. In the early 1990s, David Townsend, Ron Nutt, and coworkers proposed to combine PET with CT in a single device.[8] This hybrid device would generate co-registered PET and CT images to combine both anatomy and function in a single imaging procedure. The first PET/CT scanners became commercially available in 2001. Since that time, more than 2500 PET/CT scanners have been delivered worldwide and PET/CT scanners have essentially replaced dedicated (standalone) PET scanners. Combined functional–anatomic imaging with PET and CT greatly increased the confidence of findings and was rapidly adopted for diagnosis, staging, and monitoring of therapy in a large variety of cancers.

Head and Neck Cancer

Head and neck cancers are cancers of the nasopharynx, oropharynx, oral cavity, hypopharynx, and larynx. A large majority of these are squamous cell carcinomas. It was estimated that more than 48,000 head and neck cancers would be diagnosed in the United States in 2009, including 35,720 cancers of the oral cavity and pharynx and 12,290 laryngeal cancers.[9] Most patients with head and neck cancer are middle-aged or elderly, and they often have a history of chronic tobacco exposure and excessive alcohol consumption.

The anatomy of the neck is complex, and there are multiple benign and physiologic causes for uptake in the neck on PET (e.g., uptake in the vocal cord region due to talking during the FDG uptake period). Combined PET and CT can help distinguish benign causes of uptake from malignancy. PET/CT has proven to be of value for staging and restaging head and neck cancer.[10–12] CT and MRI have most often been used to plan surgical treatment and radiotherapy in head and neck cancer patients. CT and MRI rely on nodal size and contrast enhancement patterns to identify sites of disease, evaluate the extent of disease, and detect lymph node metastases. In general, nodes greater than 10 mm in diameter are suspicious for malignancy. PET/CT combines both anatomic and physiologic criteria, which can potentially provide more accurate staging. For example, a normal size lymph node on CT would be interpreted as normal. However, if there is abnormal radiotracer accumulation in this lymph node on PET, the lymph node would be suspicious for metastatic involvement.

As discussed previously, metallic objects, such as metal in dental prosthetic devices, can cause attenuation artifacts on PET/CT, resulting in artifactual focal increased FDG uptake. If a dental prosthesis is removable, it should be removed prior to imaging, as should any ear or neck jewelry that may result in beam-hardening artifacts. When there are artifacts from nonremovable dental metal (fillings, caps, implants), it can be important to inspect both the CT attenuation-corrected and the non–attenuation-corrected PET images.

After injection of FDG, it is important for the patient to rest quietly during the 60 to 90 minute uptake period. Head motion during the uptake period can result in excessive muscle uptake. Reclining chairs can be used so that patients do not have to actively support their head during the uptake period. Talking during the uptake period can result in excessive uptake in the vocal cord region. Also, brown fat activity can be prevalent in the neck. It can be helpful to use warming blankets during the uptake period. Premedication with an anxiolytic agent, such as lorazepam or diazepam, can help reduce muscle uptake in an anxious patient.[13] Patients are typically imaged from eyes to thighs, with their arms down next to their sides to reduce streak artifacts in the head and neck on CT. Immobilization and patient comfort are especially important in head and neck imaging to prevent artifacts from motion and misregistration between the CT and PET parts of the study. Head immobilizers, rolled towels, and pillows can be used to optimize patient positioning and prevent motion during the study.

FDG PET/CT is used for initial staging, identification of unknown primary tumor, assessment of response to therapy or restaging after therapy, evaluation for locoregional recurrence, and detection of distant metastases (Figure 27-14). Radiation oncologists can use PET/CT data to guide assignments of radiation ports. An evaluation of early and advanced stage primary head and neck squamous cell carcinoma with FDG PET/CT showed a 31% change in clinical management based on the PET/CT findings.[14] When head and neck cancer is diagnosed, accurate staging is important to determine the best therapeutic option, which can include surgery, chemotherapy, radiation therapy, or a combination of these treatments. PET/CT can be useful in planning the extent of surgery and the radiation therapy. Surgery and radiation therapy can result in inflammation, edema, hyperemia, and scarring or fibrosis, which can be difficult to differentiate from recurrent tumor on CT or MRI. Scar tissue might look like residual tumor on anatomic imaging but, in contrast to active tumor, it would not demonstrate increased FDG uptake on PET (Figure 27-15). The timing of PET/CT imaging after therapy is important, particularly after radiation therapy. Inflammation induced by radiation therapy can result in increased FDG radiotracer

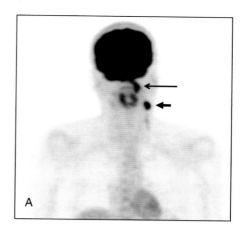
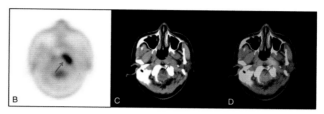
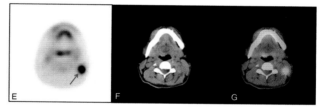

FIGURE 27-14 Anterior FDG PET maximum-intensity projection image (A) showing abnormal focal areas of hypermetabolism in the left nasopharynx and left neck. The patient had a poorly differentiated squamous cell carcinoma of the nasopharynx (long arrow) with metastases to cervical lymph nodes (short arrow) in the left neck. Axial (B) FDG PET, (C) CT, and (D) FDG PET/CT fusion images through the level of the long arrow in (A), showing the tumor in the nasopharynx (red arrows). Axial (E) FDG PET, (F) CT, and (G) FDG PET/CT fusion images through the level of the short arrow in A, showing a metastatic left cervical lymph node (blue arrows).

uptake. PET/CT has been shown to be more accurate when performed 2 to 3 months after radiation therapy.[15-18]

Thyroid Cancer

Thyroid carcinomas are classified as either differentiated or undifferentiated. The two main differentiated thyroid cancers are papillary and follicular; these account for about 90% of thyroid cancers. These cancers have the ability to concentrate iodine and are usually imaged with ^{123}I or ^{131}I. Hurthle cell is a differentiated thyroid cancer that does not concentrate iodine very well. Also, the poorly differentiated forms of thyroid cancer, medullary and anaplastic, typically do not concentrate

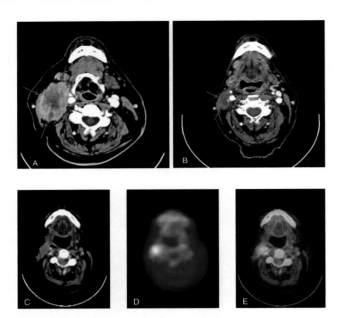

FIGURE 27-15 Axial CT image (A) through the neck, showing an unresectable squamous cell carcinoma in the right neck (red arrows). (B) 4 months later, after chemotherapy and radiation therapy, the mass has significantly decreased in size and could represent just residual scar tissue. Posttreatment axial CT image (C) with corresponding FDG PET (D) and FDG PET/CT fusion (E) images, showing abnormal hypermetabolism in the residual soft tissue consistent with active tumor.

iodine very well. For these cancers, imaging with ^{123}I or ^{131}I may be of little value. However, PET/CT with FDG can be useful for imaging both well differentiated and poorly differentiated thyroid cancers.

Typically, patients with papillary or follicular thyroid cancers undergo thyroidectomy. After thyroidectomy, ^{123}I or ^{131}I scans are done to evaluate for residual thyroid tissue in the neck and for local or distant metastatic disease. Residual thyroid tissue and metastatic disease identified on the thyroid scan is treated with ^{131}I therapy. Repeat diagnostic ^{123}I or ^{131}I whole-body imaging and measurements of serum thyroglobulin levels after administration of thyroid-stimulating hormone are obtained at regular intervals to evaluate for recurrent disease. Recurrent iodine-avid disease can be retreated with ^{131}I. However, deciding whether to retreat with ^{131}I becomes problematic when there is elevated serum thyroglobulin, signaling recurrent disease, but no evidence of iodine-avid disease on the whole-body ^{131}I scan. After therapy, radioiodine-avid differentiated thyroid cancers can dedifferentiate. Although dedifferentiated thyroid cancer cells can still produce thyroglobulin, they lose their ability to concentrate iodine, and retreatment with ^{131}I may not be of value in these patients. It has been shown that the more differentiated thyroid cancers are better imaged with ^{123}I or ^{131}I, while dedifferentiated cancers are better imaged with FDG PET/CT.[19] Therefore, PET/CT with FDG is particularly useful for patients with increasing serum thyroglobulin levels but negative radioiodine whole-body scans. If the site or sites of metastatic thyroid cancer are identified, localized, and amenable to surgery, a surgical cure may be possible (Figure 27-16).

Brain

In general, PET has a relatively low sensitivity for tumors and metastases in the brain. The normal intense uptake in the gray matter and basal ganglia can mask uptake in small primary tumors or metastases. MRI is considered more useful for initial evaluation of brain tumors or metastatic involvement. However, FDG PET or PET/CT can be useful in determining the aggressiveness or grading of tumors, which can be used to estimate prognosis.[20] High-grade gliomas contain regions of high increased FDG uptake and have a worse prognosis than low-grade gliomas with lower FDG uptake.[21] Other uses for PET in evaluating brain tumors include localizing the optimum biopsy site, defining target volumes for radiotherapy, assessing response to therapy, and evaluating for recurrence.[22]

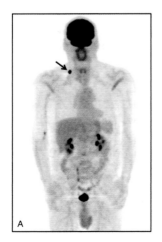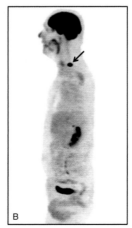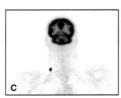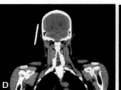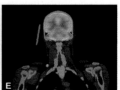

FIGURE 27-16 Patient with a history of thyroid cancer, increasing thyroglobulin, and a negative ^{131}I whole-body scan. Anterior (A) and lateral (B) maximum-intensity projection (MIP) images from an FDG PET/CT study show a focal area of hypermetabolism in the right base of the neck (arrows). Coronal (C) FDG PET, (D) CT, and (E) FDG PET/CT fusion images through the lesion seen on the MIP images show abnormal hypermetabolism in a nodule in the right base of the neck, which was found to be recurrent thyroid carcinoma.

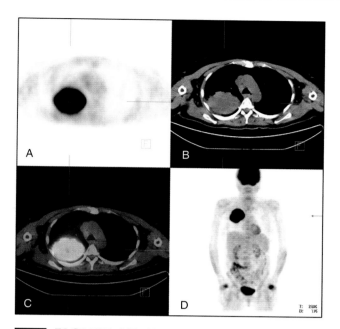
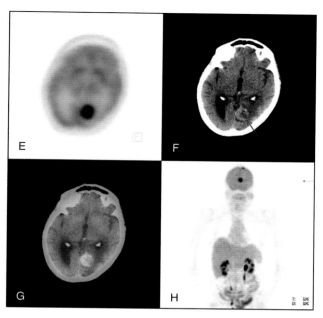

FIGURE 27-17 Patient with a large right upper lobe non–small cell lung cancer (adenocarcinoma). FDG PET maximum-intensity projection (MIP) image (D) shows a large area of hypermetabolism in the right upper lung. Axial (A) FDG PET, (B) noncontrast CT, and (C) FDG PET/CT fusion images at the level of the arrow in D show prominent radiotracer uptake in a large mass in the posterior right upper lobe. The large right upper lobe adenocarcinoma was resected and the patient received chemotherapy. A follow-up FDG PET/CT 1 year later showed metastatic lesions in the brain. FDG PET MIP image (H) shows two abnormal focal areas of hypermetabolism in the brain. Axial (E) FDG PET, (F) noncontrast CT, and (G) FDG PET/CT fusion images at the level of the arrow in H show prominent radiotracer uptake in a lesion in the left occipital lobe (red arrow).

After surgery, PET can be useful in differentiating scarring or gliosis from recurrent tumor. In evaluation for metastatic disease to the brain, MRI is considered the imaging modality of choice. However, it is not uncommon to identify brain metastases on PET/CT studies (Figure 27-17).

Outside oncology, PET/CT imaging of the brain can also be useful to evaluate seizures (epilepsy) and dementia, which will be discussed later in this chapter. For any of the brain imaging studies, the patient is asked to fast for 4 to 6 hours, except for water, prior to intravenous administration of 10 mCi (370 MBq) of FDG. After administration of FDG, patients are instructed to rest in a dimly lit room with their eyes closed. Imaging is begun 30 to 60 minutes after administration.

Lung Cancer and the Solitary Pulmonary Nodule

According to the American Cancer Society, lung cancer is the leading cause of cancer death for both men and women in the United States and smoking tobacco accounts for 8 out of 10 lung cancer cases. Lung cancer is classified histologically as non–small cell lung cancer (NSCLC) or small cell lung cancer (SCLC). NSCLCs include adenocarcinoma, bronchioloalveolar cancer (BAC), large cell carcinoma, and squamous cell carcinoma. NSCLC accounts for approximately 80% of primary lung cancers. SCLC or oat cell carcinoma is less common and is often metastatic at presentation.

The solitary pulmonary nodule, defined as a single pulmonary nodule less than 3 cm in diameter, is a significant clinical problem. It is often detected incidentally on a chest x-ray or CT scan. The decision to perform a PET/CT exam on a patient with a solitary pulmonary nodule depends on several factors. If the patient is young (less than 35 years old), with no risk factors and no suspicious nodule features on CT (e.g., spiculations), a solitary pulmonary nodule will most likely be related to granulomatous disease or other benign etiology. The uptake of FDG in granulomatous disease, infections, and fungal disease can be a source of false-positive findings in evaluation for malignancy on PET/CT. Follow-up imaging or fine-needle aspiration biopsy may be more appropriate in such a patient. However, in an older patient with risk factors such as smoking and a higher pretest probability for lung cancer, PET/CT may be very helpful not only in characterizing the nodule (Figures 27-18 and 27-19) but also in planning therapy and staging the disease.[23] If there are distant focal areas of increased radiotracer uptake, this may redirect the biopsy site not only for confirming the diagnosis but for changing the stage of the disease (Figure 27-20).

PET/CT is the preferred noninvasive method for staging NSCLC.[24] With the exception of BAC and carcinoid tumors, most primary lung cancers are FDG avid. Although PET/CT is routinely used to stage and to restage NSCLC, it has not traditionally been used for primary staging of SCLC, which is considered to be metastatic at the time of diagnosis. However, PET/CT can be useful for monitoring therapy in SCLC (Figure 27-21). For thoracic imaging of lung cancer, patients are positioned in the scanner with arms above the

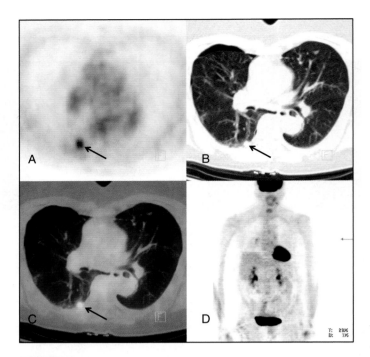

FIGURE 27-18 Solitary pulmonary nodule. Axial (A) FDG PET, (B) CT, and (C) FDG PET/CT fusion and (D) FDG PET maximum-intensity projection images demonstrate abnormal hypermetabolism in a solitary lung nodule in the posterior right lower lobe. The nodule had been recently biopsied and was found to be an adenocarcinoma. Note the small pneumothorax (red arrows) on the CT image (B) from the recent biopsy.

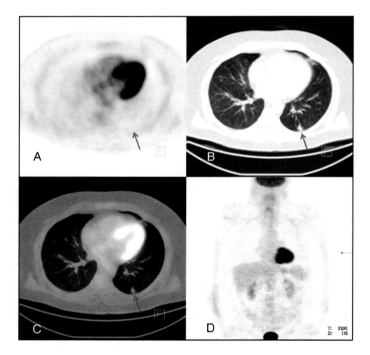

FIGURE 27-19 Solitary pulmonary nodule was found in this patient on a prior CT study. No abnormal uptake was seen in the nodule or elsewhere on the study to suggest malignancy or metastatic disease. Axial (A) FDG PET, (B) CT, and (C) FDG PET/CT fusion images show no significant uptake in the nodule (red arrows). The FDG PET maximum-intensity projection image (D) shows physiologic radiotracer accumulation in the body.

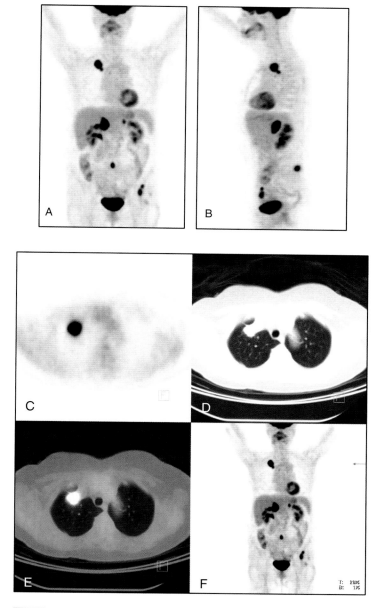

FIGURE 27-20 Lung cancer. Anterior (A) and lateral (B) FDG PET maximum-intensity projection (MIP) images of a female with history of a right upper lobe lung mass and a right adrenal lesion seen on a CT study. The PET images show prominent abnormal uptake in the right upper lobe mass, as well as several other abnormal focal areas of increased uptake most consistent with metastatic disease. Axial (C) FDG PET, (D) CT, and (E) FDG PET/CT fusion images at the level of the arrow on the anterior MIP image (F) showing prominent abnormal uptake in the right upper lobe cancer.

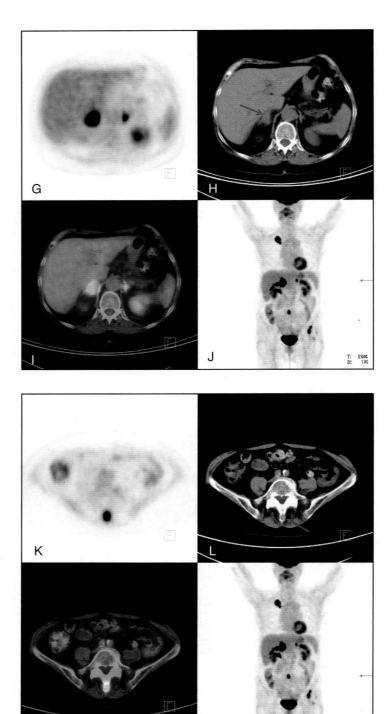

FIGURE 27-20 (Continued). Lung cancer. Axial (G) FDG PET, (H) CT, and (I) FDG PET/CT fusion images at the level of the arrow on the anterior MIP image (J) showing abnormal uptake in bilateral adrenal metastases (red arrows). Axial (K) FDG PET, (L) CT, and (M) FDG PET/CT fusion images at the level of the arrow on the anterior MIP image (N) showing abnormal uptake in a lytic bone metastasis involving the spinous process of L5 (red arrow).

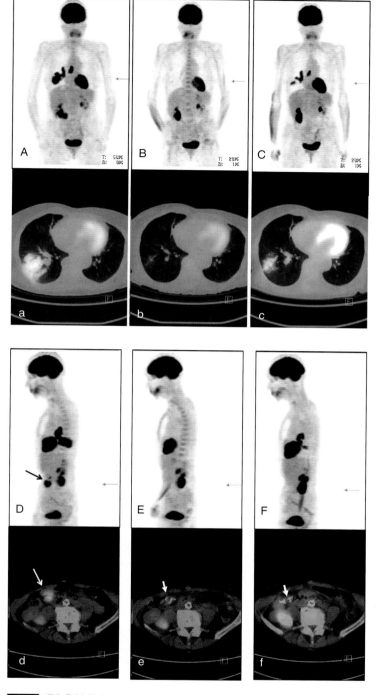

FIGURE 27-21 Three time points in a patient with small cell lung carcinoma. The initial staging anterior FDG PET maximum-intensity projection (MIP) image (A) shows extensive disease in the right chest with corresponding axial FDG PET/CT fusion image (a) at the level of the arrow. Follow-up study done 4 months later shows good response to chemotherapy (B and b). Follow-up study done 4 months later shows return of the patient's disease (C and c). Lateral MIP images of the same three time points (D–F). On the lateral image of the initial staging study (D), there is an abnormal focal area of hypermetabolism in the abdomen anterior to the kidney (black arrow), which was not well seen on the anterior MIP image. This was an unknown colon cancer. The corresponding axial FDG PET/CT fusion image (d) shows the lesion in the right anterior abdomen (white arrow). The colon cancer was surgically removed. The two follow-up studies show surgical sutures (short white arrows) in this area on the axial fusion images (e and f).

head if possible. This reduces beam-hardening artifacts from the CT scan in the chest, improving PET image quality.

Esophageal Cancer

The American Cancer Society estimated that there would be 16,470 new esophageal cancer cases in the United States in 2009 and 14,530 deaths due to this cancer. The two common histologic types of esophageal cancer are squamous cell and adenocarcinoma. Worldwide, most esophageal cancers are of the squamous cell type. However, the frequency of adenocarcinoma of the distal esophagus has been increasing in the Western world.[25] Usually patients present clinically with dysphagia and the primary tumor is identified with an upper endoscopy. The current role of PET/CT is in staging to evaluate for regional lymph node and distant metastatic involvement (Figure 27-22), as well as in restaging after therapy for recurrence (Figure 27-23). PET/CT has a high sensitivity for detecting recurrent tumor after curative resection of esophageal cancer.[26]

Lymphoma

Lymphomas are broadly classified into Hodgkin's lymphomas or Hodgkin's disease (HD) and non-Hodgkin's lymphoma (NHL). The lymphomas can be further classified into histologic subtypes. There are four histologic subtypes of HD: nodular sclerosis, lymphocyte predominance, mixed cellularity, and lymphocyte-depletion subtype. Hodgkin's lymphomas typically occur above the diaphragm and spread to contiguous lymph node chains (Figure 27-24). However, Hodgkin's lymphoma can involve areas below the diaphragm. NHLs are a more diverse group of lymphoproliferative disorders in which diffuse lymphatic and extranodal involvement is more common (Figure 27-25). There are many different histologic subtypes and variants of NHL, with diffuse large B-cell lymphoma being the most common.[27] The NHLs can be divided into slow-growing, aggressive, and fast-growing types.[28] Mesenteric involvement is common in NHL but uncommon in Hodgkin's lymphoma. Both Hodgkin's and non-Hodgkin's lymphomas accumulate FDG. However, some of the low-grade, slow-growing NHLs, like the small lymphocytic lymphomas, do not accumulate FDG very well. A study of 766 lymphoma patients with a histopathologic diagnosis indicated that with the exception of extranodal marginal zone lymphoma and small lymphocytic lymphoma, most lymphoma subtypes have high ^{18}F-FDG avidity.[29]

The treatment and prognosis of lymphoma depend on histologic features and the anatomic extent of disease or staging. Staging of both HD and NHL is determined by the Ann Arbor classification, which defines stage I as disease limited to a single lymph node or lymph node group. Stage II is defined as disease in two or more noncontiguous lymph node groups or spleen on the same side as the diaphragm. Stage III is defined as disease in two or more lymph node groups or spleen on both sides of the diaphragm. Stage IV is defined as disease in extranodal sites, often lung, liver, bone, or bone marrow, and less often other organs or tissues.

Gallium Ga 67 citrate was widely used to evaluate lymphoma before being replaced by FDG PET. The sensitivity of gallium for lymphoma depended on nodal size and location. Physiologic accumulation in the liver and colon interfered with evaluation of the abdomen. Also, lesions less than

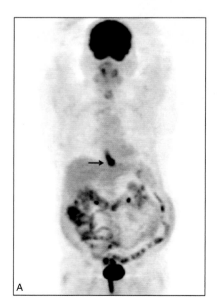
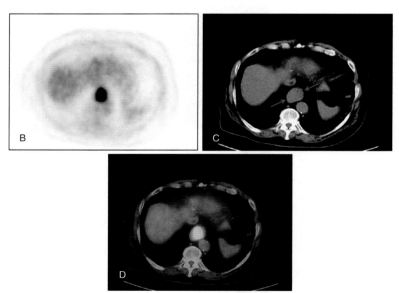

FIGURE 27-22 Anterior FDG PET maximum-intensity projection (MIP) image of a patient with biopsy-proven distal esophageal cancer (A) shows abnormal increased uptake in the region of the distal esophagus (arrow). Axial (B) FDG PET, (C) CT, and (D) FDG PET/CT fusion images at the level of the arrow in the MIP image show that the abnormal hypermetabolism corresponds to circumferential wall thickening in the distal esophagus (red arrows).

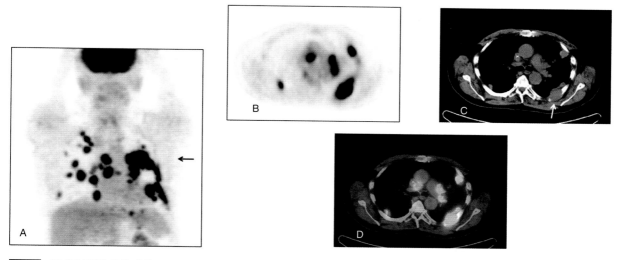

FIGURE 27-23 (A) Anterior FDG PET maximum-intensity projection image of a patient with advanced metastatic esophageal carcinoma. Axial (B) FDG PET, (C) CT, and (D) FDG PET/CT fusion images at the level of the black arrow in A demonstrate metastatic lesions in the lungs, mediastinum, and right hilum. Note the large lesion in the left posterior chest that has eroded through the rib (white arrow on image C).

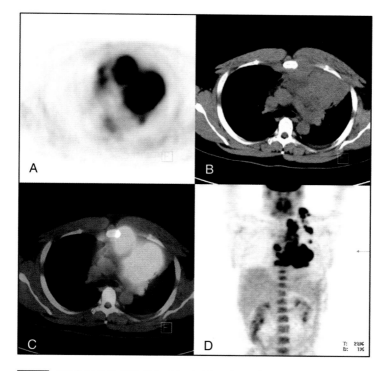

FIGURE 27-24 Hodgkin's lymphoma. Axial (A) FDG PET, (B) CT, and (C) FDG PET/CT fusion and (D) FDG PET maximum-intensity projection (MIP) images of a male with Hodgkin's disease. The axial images correspond to the level of the arrow on the MIP image (D), showing prominent uptake in a large mediastinal mass.

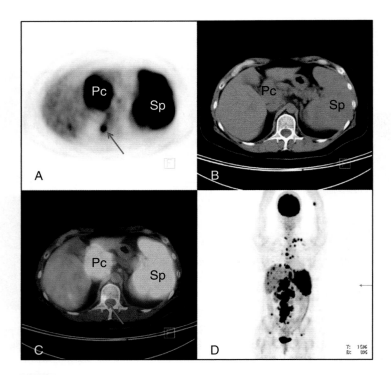

FIGURE 27-25 Non-Hodgkin's lymphoma (NHL). Axial (A) FDG PET, (B) CT, and (C) FDG PET/CT fusion and (D) FDG PET maximum-intensity projection (MIP) images of a female with stage IV NHL. The axial images correspond to the level of the arrow on the MIP image, showing extensive involvement in the spleen (Sp) and in a large portacaval mass (Pc). There is also a bone lesion at this level, seen only on the PET and fusion images (red arrows).

1.0 cm were often not detected. FDG PET/CT has become the preferred initial staging tool for lymphoma and is useful for both nodal and extranodal staging of lymphoma.[30] A significant proportion of lymphoma patients are cured by primary chemotherapy treatment. Patients not achieving complete remission, or those who relapse, have a second chance of cure with high-dose chemotherapy and hematopoietic stem cell transplantation.[31] Accurate assessment of remission after a planned course of treatment is essential for the purposes of avoiding unnecessary further treatment and the associated morbidity for those who have achieved a complete response (Figure 27-26) and identifying those who have not achieved a complete response and will need further therapy (Figure 27-27).

Breast Cancer

Breast cancer is the most common malignancy in women. In the United States, it was estimated that there would be 192,370 new cases of breast cancer and 40,170 breast cancer–related deaths in 2009.[9] FDG PET imaging for breast cancer has been approved by CMS for initial staging (except for lymph node status), monitoring response to therapy, and restaging patients with suspected recurrence. Whole-body PET/CT has minimal value for patients with early stage disease (stages 0 and 1), in whom the likelihood of metastatic disease is very low.[32] However, in patients with local invasive (clinical stage II or III) disease, detection of distant metastases may result in a change of therapy. PET/CT has been proven useful for staging patients with advanced disease, detecting recurrent disease, and monitoring therapy.

The ability of FDG PET/CT to detect primary breast cancers depends on the size of the tumor. Whole-body PET/CT scanners typically have a resolution of 4–6 mm, limiting the sensitivity of whole-body PET/CT for detection of small breast lesions. Also, PET/CT cannot identify micrometastases in axillary lymph nodes. Lymphoscintigraphy with ^{99m}Tc-sulfur colloid is still the best method for identifying the sentinel draining lymph node in patients who have nonpalpable lymph nodes and clinically suspected axillary lymph node involvement. Dedicated high-resolution positron emission mammography (PEM) scanners, which have a spatial resolution of 1–2 mm, may prove useful for evaluating small lesions in the breast (Figure 27-28). These scanners have been shown to be useful in evaluating dense breasts and breasts with implants, which can be difficult to evaluate with conventional mammography or MRI. PEM can also be useful in evaluating equivocal MRI findings.[33] Thus, there may be a role for PET in differentiating benign from malignant breast lesions. However, some low-grade, slow-growing breast carcinomas, such as tubular

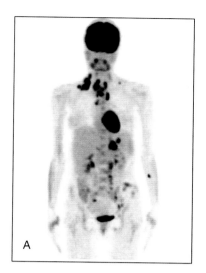
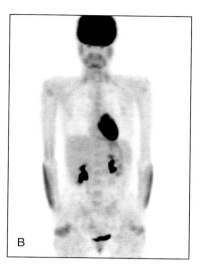

FIGURE 27-26 Hodgkin's lymphoma. Anterior FDG PET maximum-intensity projection images of a patient with Hodgkin's disease before (A) and after chemotherapy (B), showing good response.

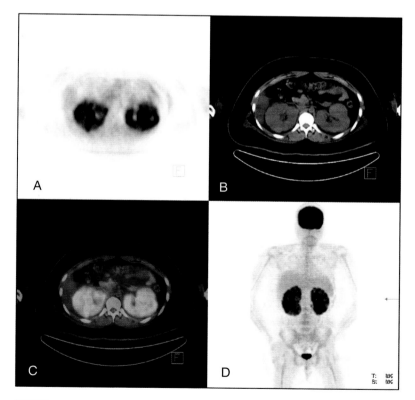

FIGURE 27-27 Patient with recurrent lymphoblastic lymphoma of the kidneys. Images show prominent abnormal radiotracer uptake in bilateral enlarged kidneys that are involved in lymphoma. Axial (A) FDG PET, (B) CT, and (C) FDG PET/CT fusion images through the kidneys. (D) FDG PET maximum-intensity projection image shows prominent uptake in the large kidneys. The axial images are at the level of the arrow in image D.

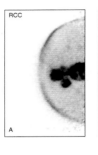
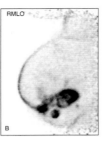
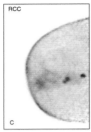
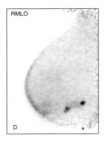

FIGURE 27-28 Craniocaudal (CC) and modified lateral oblique (MLO) FDG positron emission mammography (PEM) images of the right breast in a woman with breast cancer. The initial images (A and B) show prominent abnormal focal areas of hypermetabolism corresponding to cancer in the breast. Follow-up FDG PEM imaging of the right breast (C and D) after neoadjuvant chemotherapy shows much improvement. However, there continue to be two small focal areas of residual cancer.

and lobular carcinomas, are less glucose avid and are more likely to give a false-negative result on PET scans.

Breast cancer typically spreads to ipsilateral axillary lymph nodes (Figure 27-29). Intravenous administration of FDG should be in the arm opposite the known or suspected breast lesion. If this is not possible, a foot vein should be used. The reason for this is that if there is partial extravasation of the radiopharmaceutical at the site of administration, the dose can enter the lymph channels and result in uptake in sentinel draining lymph nodes in the axilla, which can result in a false-positive finding. Distant metastatic disease often involves bone, lung, liver, adrenal glands, and brain (Figure 27-30). Studies have shown that FDG PET is superior to ^{99m}Tc-MDP for detecting osteolytic bone metastases but inferior for detecting osteoblastic bone metastases.[34] However, PET with ^{18}F-sodium fluoride (Na^{18}F) is a highly sensitive imaging modality for the detection of both benign and malignant bone lesions. Although its uptake mechanism corresponds to osteoblastic activity, it is also sensitive for detecting lytic metastases by identifying their accompanying osteoblastic changes.[35]

Colorectal Cancer

Colorectal carcinoma is the third most common cancer in both men and women. In the United States over the past decade, the frequency and death rates for all cancers combined declined significantly for both men and women and for most racial and ethnic populations. The decreases were driven largely by declines in both frequency and death rates for the three most common cancers in men (lung, prostate, and colorectal carcinoma) and for two of the three leading cancers in women (breast and colorectal carcinoma).[36] However, it was estimated that there would be 146,970 new colorectal cancer cases in the United States in 2009 and 49,920 deaths related to colorectal cancer.[9] The diagnosis of colorectal carcinoma is usually made by colonoscopy with biopsy. The procedure may be a screening colonoscopy, which is advocated for patients over 50 years old, or it may be a colonoscopy performed in response to a positive fecal occult blood test or blood in the stool. Colorectal cancer is sometimes diagnosed emergently in response to a bowel obstruction, which is usually evaluated with CT. Cancers in small tumors and polyps can be missed because of high levels of FDG that can normally occur in the colon. Although PET/CT does not have a definite role in screening for colorectal cancer, it is not uncommon to identify an unsuspected colon cancer during an FDG PET/CT study for an unrelated reason (Figures 27-31 and 27-21D).[37]

FDG PET/CT has been found to be useful in the management of colorectal carcinoma. This was one of the first cancers for which FDG PET was approved for reimbursement by Medicare, in 1999, to detect recurrence in patients with rising blood levels of the tumor marker carcinoembryonic antigen. FDG PET is superior to CT for the initial staging of colon cancer. Local recurrence of rectal cancer occurs in approximately one-third of patients undergoing curative resection. However, early detection with CT can be difficult because of scarring related to surgery; without metabolic imaging, it is difficult to distinguish scar from residual or recurrent tumor. Early diagnosis of relapsing malignancies is important for planning therapeutic strategies that may cure the disease or prolong disease-free survival and improve the patient's quality of life.[38]

In approximately one-third of patients with newly diagnosed colon cancer, liver metastases will occur. Although PET/CT and contrast-enhanced CT have similar sensitivities for detection of hepatic metastases, PET/CT is superior for diagnosis of recurrent disease at or near the site of previous hepatic surgery, for recurrence at the site of the primary tumor, and for extrahepatic dissemination.[39]

Melanoma

The frequency of malignant melanoma has been increasing in both men and women.[36] Prognosis depends on the stage of the disease at the time of diagnosis. Staging depends on the thickness of the primary lesion and whether there are involved lymph nodes or spread to distant sites. Thickness of the primary tumor is the most important prognostic feature and is graded using either Breslow's classification (thickness of the tumor in millimeters) or Clark's classification

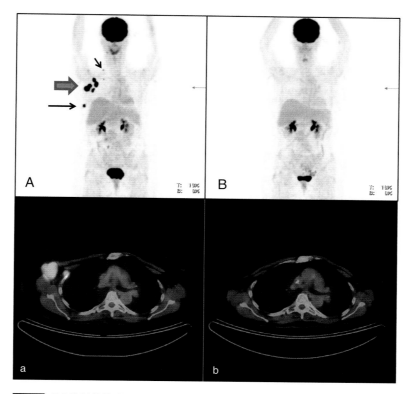

FIGURE 27-29 Woman with right breast cancer. Anterior FDG PET maximum-intensity projection image of the initial staging PET/CT (A) demonstrated prominent uptake in a right breast lesion (long arrow) with metastatic disease to axillary (blue arrow) and supraclavicular (small arrow) nodes. Corresponding axial FDG PET/CT fusion image (a) at the level of the pink arrow shows prominent uptake in a large right axillary node. A follow-up study showed excellent response to neoadjuvant chemotherapy (B and b).

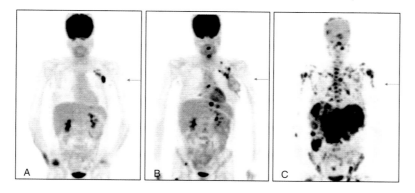

FIGURE 27-30 Three time points of the same patient showing progression of breast cancer. (A) Anterior initial staging FDG PET maximum-intensity projection showing metastatic lesions in the left axilla. (B) Follow-up study showing several new metastatic lesions. (C) Later study demonstrating widespread metastatic disease.

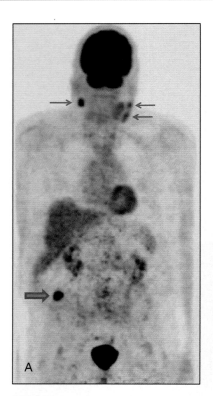
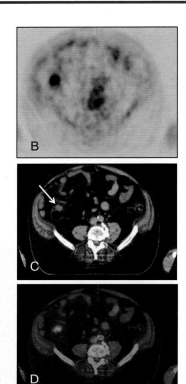

FIGURE 27-31 Unsuspected colon cancer in a head and neck cancer patient. Anterior FDG PET maximum-intensity projection image (A) of a head and neck cancer patient shows disease bilaterally in the neck (red arrows). However, there was also an abnormal focal area of hypermetabolism in the abdomen (blue arrow), which turned out to be an unsuspected colon cancer. Axial (B) FDG PET, (C) CT, and (D) FDG PET/CT fusion images through the focus in the abdomen show a small solid lesion in the colon (white arrow in C).

(anatomic level of skin invasion). Usually thin lesions, less than 1 mm, are excised with wide margins (1 cm) for surgical cure. For intermediate thickness lesions, 1 to 4 mm, lymphoscintigraphy with ^{99m}Tc-sulfur colloid is performed to evaluate whether sentinel draining lymph node(s) are involved. In this scenario, PET is not particularly useful because it cannot detect micrometastases.

PET/CT scanning is currently approved by CMS for diagnosis, staging, and restaging of malignant melanoma. However, PET is not approved for evaluation of regional lymph nodes, for the reason stated above. FDG PET/CT is most useful in staging patients who are at high risk of having advanced disease or have clinical findings suspicious for metastatic disease. FDG PET/CT is also useful for restaging patients with suspected recurrence or with known solitary recurrence being considered for surgical resection.

PET/CT imaging for melanoma is most often performed from the top of the head through the toes. The reason for this is that malignant melanoma can spread anywhere. Malignant melanoma is known for its propensity to metastasize to unusual sites such as the small bowel, myocardium, and leptomeninges. Because of its high tumor-to-background ratio, FDG PET can highlight metastases at unusual sites that are easily missed with conventional imaging modalities.[40]

Pancreatic and Hepatobiliary Malignancies

Patients with pancreatic cancer continue to have poor long-term outcomes. Surgery currently is the only treatment that offers a possible cure. Some 15% to 20% of patients have resectable disease at the time of diagnosis, but only around 20% of these patients survive to 5 years. For locally advanced, unresectable, and metastatic disease, treatment is palliative.[41] Determining resectability of the tumor is thus the main goal of imaging. Thin-cut CT has traditionally been the standard imaging procedure for the evaluation of pancreatic cancer, but FDG PET/CT has been shown to be useful in diagnosis and staging of pancreatic cancer (Figure 27-32). In a study of 54 patients with pancreatic masses, FDG PET/CT showed a sensitivity of 100%, specificity of 91%, and accuracy of 97%

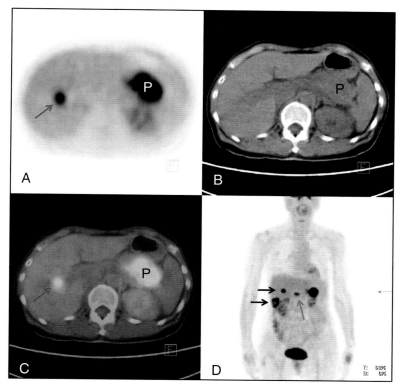

FIGURE 27-32 Pancreatic cancer. Axial (A) FDG PET, (B) CT, and (C) FDG PET/CT fusion and (D) FDG PET maximum-intensity projection (MIP) images. There is a large hypermetabolic lesion in the tail of the pancreas (P) in this patient. The MIP image (D) also shows a smaller lesion in the pancreatic head (green arrow) and two metastatic lesions in the liver (black arrows). The axial images were obtained at the level of the pink arrow on image D, showing the lesion in the tail of the pancreas (P) and a liver metastasis (red arrow), which is not well seen on the noncontrast CT.

in the detection of pancreatic lesions.[42] The investigators concluded that FDG PET/CT is a very useful tool for differentiating benign and malignant pancreatic masses and assessing tumor extension. Contrast-enhanced FDG PET/CT is also a feasible imaging protocol for staging and assessing resectability of pancreatic cancer.[43] In another study of 59 patients with pancreatic cancer deemed resectable, the clinical management of 16% of the patients was changed after PET/CT imaging detected additional distant metastases.[44]

The two most common primary liver neoplasms are hepatocellular carcinoma (HCC) and cholangiocarcinoma. FDG PET is not particularly useful for detecting primary HCC, having a sensitivity of 50% to 70%. The low sensitivity for detection of HCC is related to the relatively high physiologic uptake of radiotracer in the liver along with relatively low to moderate uptake in well-differentiated HCC tumors (Figure 27-33). PET/CT is more useful in detecting and staging moderately to poorly differentiated tumors and detecting extrahepatic spread. Results were evaluated in 67 patients with unresectable HCC who received nonoperative therapy (transcatheter chemo-embolization, transcatheter arterial chemotherapy, radiofrequency ablation, or systemic chemotherapy) and underwent FDG PET within 1 month after therapy.[45] Patients with posttherapy lesions that demonstrated FDG uptake less than or equal to uptake in the normal liver parenchyma had significantly longer survival than patients with lesion uptake higher than in adjacent liver parenchyma. The investigators concluded that posttherapeutic PET performed within 1 month after nonoperative therapy can be a good predictor of overall survival in patients with unresectable HCC.

Cholangiocarcinoma is a rare cancer of the bile ducts. Surgical resection of the tumor offers the best chance for long-term survival. Tumors near the hepatic hilum that demonstrate extensive biliary or vascular involvement are often not resectable. In a study of 61 patients with cancer of the biliary tract, PET/CT and contrast-enhanced CT had comparable accuracy for detecting primary intrahepatic and extrahepatic cholangiocarcinomas.[46] All distant metastases were detected by PET/CT, but only one-fourth were detected by contrast-enhanced CT. The PET/CT findings resulted in

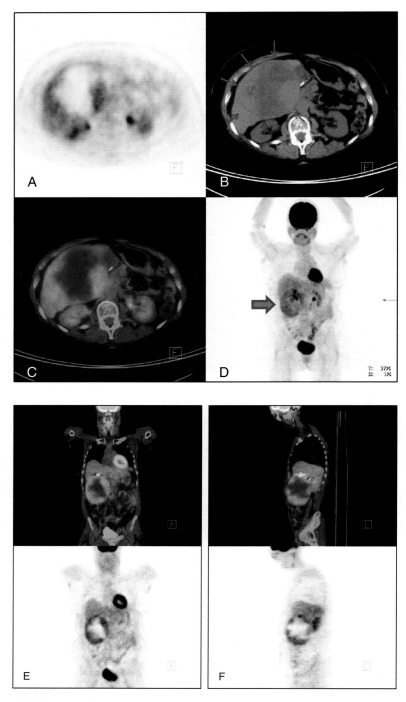

FIGURE 27-33 Hepatocellular carcinoma. Axial (A) FDG PET, (B) CT, and (C) FDG PET/CT fusion and (D) FDG PET maximum-intensity projection images of a patient with a large hepatocellular carcinoma. The mass involves the inferior right lobe of the liver (D) (large blue arrow), and appears as a heterogeneous soft-tissue mass on the CT (red arrows). Also shown are coronal FDG PET/CT fusion (E, upper panel) and FDG PET (E, lower panel) images, and (F) sagittal images through the liver mass. There are metallic surgical clips at the superior aspect of the mass. The mass demonstrates peripheral uptake with a large low-attenuation photopenic central area. The photopenic central area represents necrotic tumor.

a change in clinical management of 17% of the patients whose tumors had been deemed resectable after standard work-up. The investigators concluded that PET/CT is particularly valuable in detecting unsuspected distant metastases not diagnosed by standard imaging.

Cervical and Ovarian Cancers

An estimated 11,270 new cases of cervical cancer were predicted to occur in the United States in 2009, with 4,070 cervical cancer-related deaths.[9] When cervical cancer is localized, it can be treated effectively with surgery (Figure 27-34). However, if there is locally advanced disease, treatment includes a combination of surgery, radiation therapy, and chemotherapy. Cervical cancer spreads by local extension and by lymphatic spread to pelvic, para-aortic, and distant lymph nodes. The detection of nodal involvement or distant metastases affects clinical decision making.[47] In one study, the status of para-aortic lymph nodes, as determined by surgical staging, was the most significant prognostic factor.[48] FDG PET imaging has been found to be more accurate than CT and MRI for detection of metastatic lymph nodes.[49] Increasingly, FDG PET/CT has been integrated into radiation therapy treatment planning. PET/CT can influence the treatment goal of radiation therapy through better staging, and it can optimize the treatment plan to increase the radiation dose to the tumor while decreasing the dose to adjacent normal structures.[50]

Ovarian cancers have a wide range of histologic tumor types, including epithelial, germ cell, and sex cord stromal tumors. Almost 90% of ovarian cancers are epithelial in origin. The prognosis for patients with early localized ovarian cancer is very good. However, symptoms related to ovarian cancer often do not present until late in the disease process. Until it is advanced, ovarian cancer can be difficult to detect on physical exam. Also, it often metastasizes as small peritoneal implants, which can be hard to detect on CT. Thus, most patients present with stage III or IV disease. An estimated 21,550 new cases of ovarian cancer in the United States were predicted for 2009, with 14,600 ovarian cancer-related deaths.[9] Most ovarian cancers demonstrate a common pattern of spread, initially spreading to local organs, then diffusing or seeding into the peritoneal cavity (Figure 27-35). Ultrasound, CT, and tumor markers are used for initial evaluation. However, staging is often determined at surgery.

Physiologic increased FDG uptake in the ovaries is commonly seen in women of reproductive age, which limits the value of PET in the early diagnosis of ovarian cancer.[51] FDG PET is most useful in detecting and restaging recurrent ovarian cancer when there is increasing serum CA-125 tumor marker and equivocal CT or MRI.[52,53]

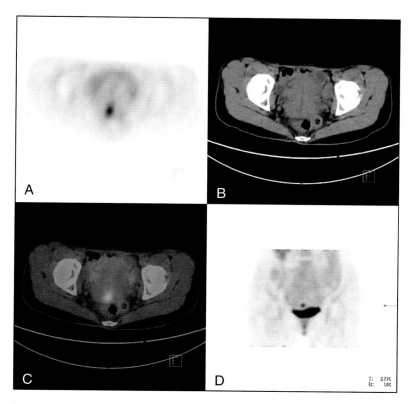

FIGURE 27-34 Small isolated cervical cancer. Axial (A) FDG PET, (B) CT, and (C) FDG PET/CT fusion and anterior (D) FDG PET maximum-intensity projection images of the pelvis demonstrate a small focal area of hypermetabolism in the uterine cervix.

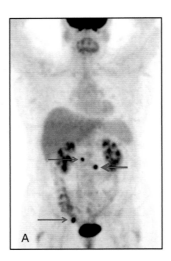

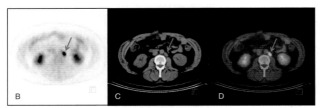

FIGURE 27-35 Recurrent ovarian carcinoma. Anterior FDG PET maximum-intensity projection (MIP) image of a woman who had undergone a hysterectomy, bilateral oophorectomy, and chemotherapy for ovarian cancer (A). There are three focal areas of hypermetabolism (red arrows) demonstrating recurrent disease. Axial (B) FDG PET, (C) CT, and (D) FDG PET/CT fusion images at the level of the middle (thick) arrow on the MIP image (A), showing that the focus corresponds to a left periaortic node (green arrows).

ASSESSING TUMOR RESPONSE TO THERAPY

When a patient is diagnosed with cancer, the oncologist has different treatment options depending on the individual's circumstances. In addition to surgical resection and radiation therapy, there often are several chemotherapeutic regimens available. The oncologist chooses the chemotherapy regimen that he or she believes will have the greatest benefit for the patient. However, in the United States, most drugs used in treatment plans are effective in fewer than 60% of patients.[54] The response to a drug in one patient may be much different in another patient with the same cancer. Early assessment of therapeutic response, particularly in aggressive cancers, is essential. In the past the patient was assessed, after completion of several cycles of chemotherapy, by anatomic imaging modalities such as CT. In some instances, increasing size of the tumor, new metastatic disease, or both showed that the chemotherapy had not resulted in a good outcome. Tumor shrinkage was used as the criterion for response to therapy. However, even if a tumor decreases in size, it can be difficult or impossible to distinguish between residual scarring or sterile tumor and residual viable tumor (Figure 27-15). The additional metabolic information offered by FDG PET/CT can distinguish between residual sterile tumor and viable tumor and can be used to assess whether the tumor has responded to even one cycle of chemotherapy (Figure 27-36). ^{18}F-FDG is gaining widespread acceptance for demonstrating early response or lack of response to therapy. The interest in using FDG PET/CT to monitor tumor response to therapy has been stimulated by the growing number of treatment regimens available (Figure 27-37). Several chemotherapy regimens can be combined with epidermal growth factor receptor (EGFR) or vascular endothelial growth factor (VEGF) antibodies in a large number of clinically used drug combinations. Effective use of these various drug regimens requires that nonresponding tumors be identified early.[55]

BONE SCINTIGRAPHY WITH ^{18}F-SODIUM FLUORIDE

^{18}F-sodium fluoride (Na^{18}F) is a positron-emitting bone agent that reflects blood flow and bone remodeling. It was one of the first bone scanning agents developed in the early 1960s;[56] however, the high photon energy of ^{18}F (511 keV) had to be imaged with rectilinear scanners with thick NaI crystals. Gamma cameras could image Na^{18}F, but the images were poor because gamma cameras performed best at lower photon energies. Also, the 110 minute half-life meant that a facility had to be located close to a cyclotron in order to obtain this radiopharmaceutical for clinical use. The initial ^{99m}Tc-phosphate compounds for bone imaging were developed in 1971 by Subramanian and McAfee.[57] ^{99m}Tc-diphosphonate radiopharmaceuticals were more suited for imaging with Anger cameras and were easily obtained with the use of a ^{99}Mo–^{99m}Tc generator. After this, Na^{18}F bone scanning was essentially abandoned.

At the time of this writing, there is a shortage of reactor-produced ^{99}Mo, resulting in a shortage of ^{99}Mo–^{99m}Tc generators. This shortage, coupled with the fact that there are now many available PET/CT scanners capable of excellent Na^{18}F bone scans, is leading to a resurgence in the use of this radiopharmaceutical. Aqueous ^{18}F-fluoride is produced in multicurie batches by proton irradiation of ^{18}O-water in the cyclotron. These high batch yields of Na^{18}F are then used as the precursor for production of PET radiopharmaceuticals such as ^{18}F-FDG, so Na^{18}F is readily available.

The uptake mechanism of Na^{18}F is similar to the uptake mechanism of ^{99m}Tc-MDP. After intravenous administration, Na^{18}F diffuses through the capillaries into bone extracellular fluid. Fluoride ions exchange with hydroxyl groups in hydroxyapatite crystal bone to form fluoroapatite, which is deposited mainly at the bone surface.[58] Although Na^{18}F, similarly to ^{99m}Tc-MDP, depends on blood flow and osteoblastic activity, the superior pharmacokinetic properties of Na^{18}F and the improved spatial resolution of PET make Na^{18}F a more sensitive modality for detecting both lytic and blastic

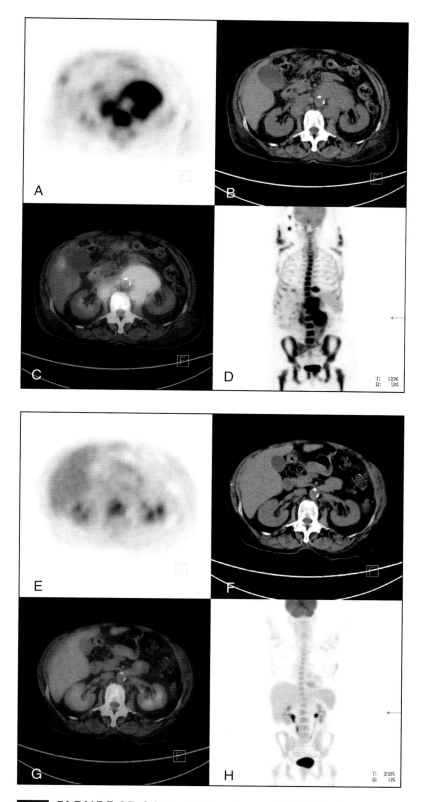

FIGURE 27-36 Burkitt's lymphoma. Axial (A) FDG PET, (B) CT, and (C) FDG PET/CT fusion and (D) FDG PET maximum-intensity projection (MIP) images. These initial staging images show prominent uptake in a bulky retroperitoneal adenopathy and throughout the bone marrow, which had extensive lymphoma involvement. The axial images are at the level of the arrow on the MIP image. When the patient returned for restaging 45 days later, after only two cycles of chemotherapy, imaging findings were consistent with a complete response to therapy. Axial (E) FDG PET, (F) CT, and (G) FDG PET/CT fusion and (H) FDG PET MIP images show complete resolution of the bulky retroperitoneal adenopathy seen on the previous study.

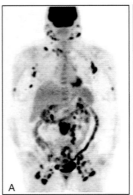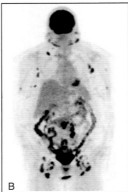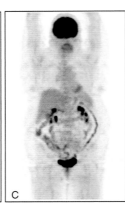

FIGURE 27-37 Follicular lymphoma. (A) Initial staging anterior FDG PET maximum-intensity projection image of a female with extensive lymphoma involving nodes in the neck, chest, abdomen, and pelvis. She began therapy with rituximab, a chimeric monoclonal antibody against the CD20 antigen expressed on the surface of B cells. (B) 4 months later, after four cycles of chemotherapy, there was only minimal response. It was decided to switch to a different chemotherapy regimen. (C) 4 months after treatment with bendamustine, a nitrogen mustard, there was no evidence of active lymphoma.

bone lesions.[59] Na^{18}F has several desirable characteristics compared with ^{99m}Tc bone agents. There is minimal binding to serum proteins, rapid bone uptake, and rapid clearance from the blood, which results in a high bone-to-background ratio in a relatively short time (Figure 27-38). Imaging can begin 30 to 60 minutes after administration. The overall bone uptake of Na^{18}F is estimated to be 2-fold higher than uptake of ^{99m}Tc-MDP. Another advantage for the patient is the shorter half-life.

Nononcologic Bone Scintigraphy with Na^{18}F

Although Na^{18}F is highly sensitive for bone remodeling, it is not specific. Increased Na^{18}F can also be seen in benign bone pathologies such as fibrous dysplasia, fractures, shin splints, osteomyelitis, and arthritis. The added anatomic information obtained with PET/CT can often help differentiate benign from malignant bone disease, thus improving specificity. In our clinic, Na^{18}F imaging has been useful in the evaluation of orthopedic problems, particularly back pain (Figure 27-39).

Patient Preparation for Na^{18}F Imaging

No special patient preparation is needed prior to a Na^{18}F PET/CT study. The injected dose in an adult is usually between 8 and 12 mCi (296–444 MBq). Scanning is typically performed 45 to 60 minutes after administration of the radiopharmaceutical. A low-dose CT scan is performed first, covering the area of interest. This is followed immediately by the PET scan without changing the patient position. Bed positions are usually acquired at 2 to 5 minutes per bed depending on the administered dose and the size of the patient.

NONONCOLOGIC IMAGING WITH PET/CT

Cardiac

As discussed in Chapter 21, PET myocardial studies have advantages over traditional SPECT imaging. One is the ability to provide attenuation-corrected images, which decrease the occurrence of attenuation artifacts and increase specificity.[60] Other advantages include increased sensitivity and increased spatial resolution.

^{82}Rb-chloride and ^{13}N-ammonia are the PET perfusion agents used clinically to diagnose coronary artery disease. The clinical use of PET for myocardial perfusion studies is increasing. However, cardiac perfusion studies generally are performed only at high-volume cardiac imaging facilities or at institutions with a nearby cyclotron. The high-volume cardiac centers typically use ^{82}Rb because, like ^{99m}Tc, it is generator produced. The half-life of the ^{82}Sr in the ^{82}Sr–^{82}Rb generator system is 25 days, so a new generator must be purchased every month. Because the generators are expensive, there must be a sufficient patient case load to justify the cost. Some centers share the generator with nearby cardiac PET imaging facilities to lower expenses. The alternative and preferred PET cardiac perfusion agent is ^{13}N-ammonia, which is cyclotron produced. Its 10 minute half-life restricts its use to facilities with an onsite or nearby cyclotron. Metabolic cardiac imaging with FDG can be combined with ^{13}N-ammonia perfusion imaging to diagnose viable hibernating myocardium.

^{13}N-ammonia has a high first-pass extraction (82%) because of the rapid diffusion of uncharged lipophilic ammonia across the capillary endothelium and sarcolemma of the

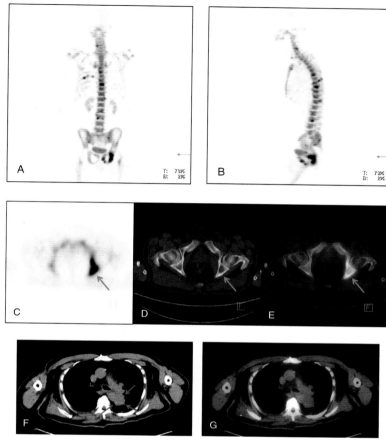

FIGURE 27-38 Anterior (A) and lateral (B) maximum-intensity projection (MIP) images from a Na^{18}F PET/CT bone scan in a patient who complained of back pain. The images demonstrate multiple focal areas of increased osseous remodeling suspicious for metastatic disease to bone. Axial (C) FDG PET, (D) CT, and (E) FDG PET/CT fusion images through the hips at the level of the pink arrow on the MIP images show a metastatic bone lesion in the left acetabulum (blue arrows). The noncontrast CT portion of the study shows a lung lesion in the medial left lower lobe (F). Although the lung lesion does not take up the bone agent on the fusion image (G), it was identified on the noncontrast CT study (red arrows) and was later determined to be lung cancer, the source of the bone metastases.

myocyte. Once taken up into the myocyte, ammonia is rapidly fixed as ^{13}N-glutamine by enzymatic conversion of glutamic acid by glutamine synthetase, resulting in high retention of radiotracer in the myocyte. Typically, 10 to 20 mCi (370–740 MBq) is administered intravenously, with imaging performed about 5 minutes later to allow adequate clearance of pulmonary background activity. To evaluate for coronary artery disease, a second ^{13}N-ammonia perfusion study can be obtained after pharmacologic stress, a method similar to rest–stress ^{99m}Tc myocardial perfusion studies with sestamibi or tetrofosmin. The second study is usually performed after a 50 minute interval to allow the initial ^{13}N-ammonia dose to decay to background levels. In evaluation for myocardial viability, the second study can be performed with FDG (Figures 27-40 and 27-41). Under fasting conditions, the heart uses fatty acids for energy. However, ischemic areas of the heart rely on glucose metabolism for energy. As noted in the previous discussion of normal heart uptake, the primary energy substrate in the heart depends on the substrate that is most available for metabolism. When the patient is in a fasting state, the heart relies on fatty acid metabolism instead of glucose metabolism for energy. After eating, blood glucose and insulin levels increase. This increases GLUT4 transporter expression at the myocardial cell surface as well as in fat and skeletal muscle, which mediates glucose uptake into the cell. Thus, high blood glucose and insulin concentrations promote FDG uptake in the heart. In preparation for the FDG metabolic imaging study, patients are fed a high-carbohydrate meal or offered an oral glucose load about 1 hour before administration of FDG.

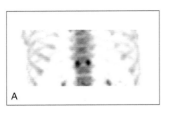
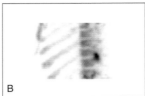
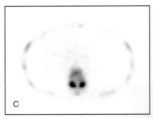
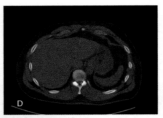
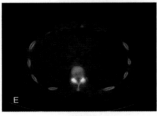

FIGURE 27-39 Anterior (A) and lateral (B) Na^{18}F PET maximum-intensity projection (MIP) images from the level of mid T7 to the base of T12 in a patient with low back pain. Focal areas of increased osseous remodeling are identified bilaterally at the T10–T11 level. Axial (C) PET, (D) CT, and (E) PET/CT fusion images show focal areas of increased osseous remodeling at the T10–T11 posterior facet joints secondary to trauma.

^{82}Rb is a potassium analogue with a physical half-life of only 76 seconds. ^{82}Rb is taken into the myocyte by active transport via the Na-K ATPase pump. Its extraction is less than that of ^{13}N-ammonia, being 50% to 60% at rest. Typically, 40 to 60 mCi (1480–2220 MBq) is administered intravenously over 20 to 30 seconds. Imaging is usually begun 2 minutes after administration to permit clearance of blood pool activity. The rapid decay of ^{82}Rb permits multiple studies to be performed about every 10 minutes. A disadvantage of ^{82}Rb is its high-energy positron (maximum energy 3.38 MeV), which travels farther from its site of origin before annihilation than lower energy positrons, decreasing intrinsic resolution.

Epilepsy

A seizure is an abrupt alteration in brain electrical activity. Many metabolic and anatomic abnormalities can result in seizure activity. The causes of seizures include drugs, alcohol, neoplasms, infections, and epilepsy. Epilepsy is a neurologic disorder characterized by chronic, recurrent, unprovoked seizure activity. Epilepsy affects an estimated 3% of the population during their lifetime.[61] Epilepsy can be idiopathic, or it can result from a neurologic injury. It can also be related to systemic medical disease. Epilepsy can be classified as either partial (focal) seizures or generalized seizures. Partial or focal seizures start in a focal area of the brain and either remain focal or secondarily spread or generalize throughout the brain. Most patients with epilepsy can be effectively managed with medications. However, many patients have seizure activity that is refractory to medical therapy. Many of these patients suffer from temporal lobe epilepsy. If the focal area

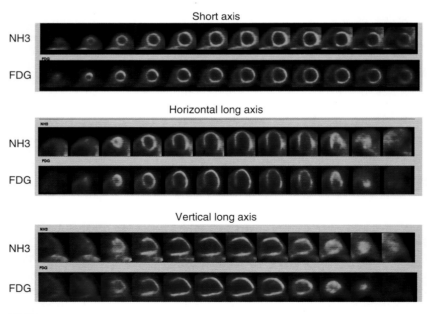

FIGURE 27-40 Normal cardiac ^{13}N-ammonia perfusion (NH3) and ^{18}F-FDG viability study demonstrating matching short axis, horizontal long axis, and vertical long axis views from ^{13}N-ammonia perfusion and ^{18}F-FDG metabolic studies. There are no perfusion defects, and there is good FDG uptake throughout the left ventricle of the heart.

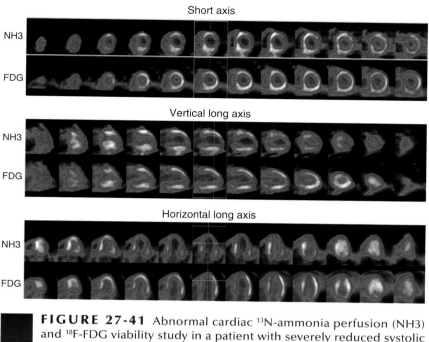

FIGURE 27-41 Abnormal cardiac ¹³N-ammonia perfusion (NH3) and ¹⁸F-FDG viability study in a patient with severely reduced systolic function. The perfusion study shows apical and lateral wall perfusion defects. There is improved regional uptake of FDG in the apex and lateral wall (perfusion–metabolism mismatch) consistent with viability in these areas. Findings suggest that reperfusion or revascularization will lead to functional improvement in these areas.

of abnormal activity (epileptogenic focus) can be identified, surgical resection of the epileptogenic cortex in these patients can be curative.

FDG PET/CT can be helpful in evaluating patients with partial epilepsy refractory to medical therapy. During seizure activity (ictal period), there is increased perfusion and glucose metabolism in the region of the seizure focus. Between seizures (interictal period), seizure foci demonstrate decreased glucose metabolism compared with other areas in the brain. Interictal brain FDG PET scanning can provide useful information for localizing the epileptogenic focus. Classically, the brain region with the most focally profound hypometabolism is considered to contain the epileptogenic focus.[62] FDG PET imaging to localize an epileptogenic focus is performed during the interictal period, when the seizure focus typically demonstrates reduced uptake. Because of the short half-life of FDG, ictal studies of epilepsy are rare. These occur only when a patient has an unexpected seizure during administration or shortly after administration of the radiopharmaceutical. In this case, the seizure focus will demonstrate increased radiotracer uptake (hypermetabolic focus) (Figure 27-42). Because of these opposite findings, it is important to know the status of the patient during the FDG uptake period.

Some seizures do not result in obvious physical manifestations and can be referred to as subclinical seizures. EEG monitoring is important during FDG administration and throughout the uptake period to determine whether subclinical seizure activity was present. Subclinical seizure activity during the uptake period could result in increased activity at the seizure focus, making the uptake in the contralateral brain seem hypometabolic. If the study was presumed to be interictal, the relative decreased uptake in the contralateral area of the brain may be falsely identified as the seizure focus.

Dementias

Dementia can be defined as a loss of cognitive ability beyond what is expected from normal aging. As life expectancy increases, dementia is having an increasing impact on the health care system. Dementia affects approximately 24 million people in the world, with about 5 million new cases occurring annually.[63] There are many causes of dementia, not all of which are related to the brain. Some causes, such as thyroid hormone deficiency and vitamin B_{12} deficiency, can be easily treated. The most common age-associated memory disorders include Alzheimer's disease (AD), multi-infarct or vascular dementia, Lewy body disease (LBD), frontotemporal dementia (FTD) or Pick's disease, Parkinson's disease, corticobasal degeneration, and primary progressive aphasia.[64] Identifying the specific cause of dementia can be a challenge. Since most of these age-associated dementias are related to abnormalities in the central nervous system, they cannot be readily diagnosed by routine laboratory testing. Different neurodegenerative disorders have different pathology and prognoses, and their appropriate treatments differ. As more disease-specific treatments are becoming

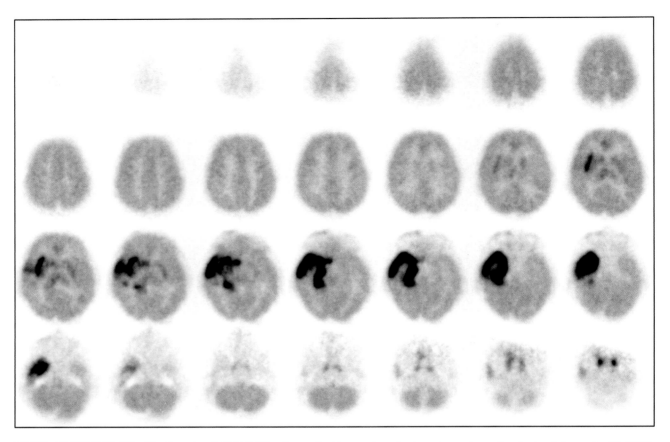

FIGURE 27-42 Rare ictal FDG PET brain study. The patient had a seizure just after being injected with FDG. Contiguous axial PET images of the brain show prominent uptake in the right temporal lobe, consistent with the seizure focus. If this had been an interictal study, the right temporal lobe would have demonstrated decreased uptake (relative hypometabolism) compared with the left temporal lobe.

available, it is becoming increasingly important to be able to identify the underlying cause of the dementia.

Clinically, most PET or PET/CT studies to evaluate dementia are performed using ^{18}F-FDG. Patterns of FDG uptake in the brain are somewhat comparable to the uptake patterns seen with SPECT brain perfusion radiotracers, but with increased spatial resolution and accuracy. Certain dementias demonstrate characteristic patterns of hypometabolism that can be detected on FDG PET (Figure 27-43). These characteristic distributions can aid in determining the correct diagnosis. The dementia best studied with FDG PET is AD. AD causes a pattern of glucose hypometabolism predominantly involving the posterior temporoparietal association cortex and posterior cingulate cortex, with sparing of the primary sensorimotor and visual (occipital) cortex as well as the basal ganglia and thalamus.[65] FTD causes glucose hypometabolism predominantly in the frontal lobes, anterior temporal lobes, and anterior cingulate cortex.[66] Distinguishing AD from FTD relies on clinical history and examination. However, FDG PET shows different patterns of hypometabolism in these disorders that can aid physicians in making the sometimes difficult clinical distinction between AD and FTD.[67]

Dementia associated with Parkinson's disease can show a similar pattern of hypometabolism to AD, but often with hypometabolism in the visual cortex, which is usually spared in AD. Dementia with Lewy bodies also demonstrates a pattern similar to AD, but with less sparing of the visual cortex. Huntington's disease can demonstrate hypometabolism in the caudate and putamen with cortical glucose metabolism relatively intact. Multi-infarct dementia classically demonstrates scattered focal areas of hypometabolism involving cortical and subcortical structures as well as the cerebellum.[68]

Infection and Inflammation

Infection refers to the detrimental invasion of a host by a foreign living microbe. Infection stimulates an inflammatory response by the host. Depending on the duration of the inflammatory response, inflammation can be categorized as acute (an early response to infection lasting days) or chronic (prolonged response lasting weeks or years). In acute inflammation, the inflammatory process is characterized by an accumulation of leukocytes, particularly polymorphonuclear leukocytes

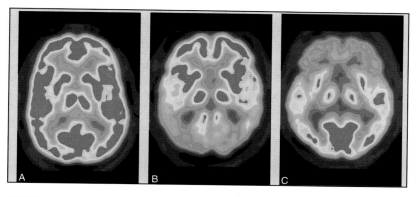

FIGURE 27-43 Axial FDG PET images of the brain in three different patients showing the differences between (A) normal, (B) Alzheimer's dementia, and (C) frontal lobe dementia (Pick's disease).

and macrophages, which ingest foreign particles and bacteria by phagocytosis. This usually leads to healing, with either restoration of normal tissue or development of a scar. Sometimes, instead of healing, the inflammation may become chronic. In the setting of chronic inflammation, there is a reduction in the number of polymorphonuclear leukocytes and increased infiltration of macrophages, lymphocytes, plasma cells, and fibroblasts.[69]

Activated inflammatory cells demonstrate increased glucose metabolism. In the activated state, inflammatory cells such as neutrophils and macrophages express high concentrations of glucose transporters, which facilitate the movement of FDG through the cell membrane.[70] For this reason, inflammation will result in increased FDG accumulation, and FDG PET/CT can be used to image sites of inflammation and infection. FDG has been shown to accumulate in a variety of infections and inflammatory processes.[71] The clinical settings in which FDG PET has been shown to be useful for imaging infection include chronic osteomyelitis, complicated diabetic foot infections, painful lower limb prostheses, fever of unknown origin, acquired immunodeficiency syndrome (AIDS), vascular graft infection, and fistula evaluation.[70]

prostate cancer, mucinous carcinomas, and certain other low-grade malignancies.

It is also important to understand that ^{18}F-FDG is not specific for malignancy. As discussed, there is increased FDG uptake at sites of inflammation or infection. If identifying the inflammatory site is the goal, as in studies to find the source of fever of unknown origin, this can be a good thing. However, in oncologic imaging, infections can make the study difficult to interpret. For example, pneumonia, abscess, or active granulomatous infections could lead to false-positive findings in a study being done to evaluate a pulmonary mass for cancer. Recent surgery, chemotherapy, or radiation therapy can also make the study more difficult to interpret. Furthermore, there are many benign tumors (e.g., fibrous dysplasia, colonic adenomatous polyp, fibrocystic breast disease) and conditions (e.g., thyroiditis, arthritis, recent bone fracture) that can accumulate ^{18}F-FDG.

It is important to understand the wide variations in normal physiologic uptake that can be seen. In some cases, physiologic uptake can mask an adjacent malignancy. Examples of this could include bladder cancers, renal cell carcinomas, a small lesion in the chest adjacent to prominent normal left ventricular uptake, or small lesions in the brain.

FALSE-POSITIVE AND FALSE-NEGATIVE STUDIES

Clearly, there are many factors that can influence a PET/CT imaging study. In oncology, lesions can be missed because of inadequate patient preparation (e.g., hyperglycemia often seen in patients with diabetes) or technical problems during the study (e.g., patient motion). If the lesion is too small, it may be below the resolution of the PET scanner (e.g., a 4 mm lung nodule). Also, there are some malignant tumors that often do not demonstrate significant FDG accumulation, such as BAC and carcinoid in the lung, hepatocellular carcinoma,

THE FUTURE

The growth of the PET/CT industry serves as evidence of its diagnostic value, predominantly in oncology but also in cardiovascular and neurologic disease. In a few years, PET/CT has gone from a method used predominantly in research to an accepted standard of care in many aspects of clinical oncology. As stated earlier in this chapter, the sharp rise of clinical PET began after favorable decisions by CMS in 1998 to reimburse for solitary pulmonary nodule evaluation. There are now thousands of PET/CT machines and enough cyclotrons to supply short-lived positron-emitting radiopharmaceuticals to almost anywhere in the country. This enlarging market will similarly continue to spark innovation in the equipment industry for higher resolution, faster scan times, and better software for analysis.

Most PET/CT scanners in use have a spatial resolution of 5 to 7 mm. Improvements in crystal design, hardware for coincidence timing, and signal processing have allowed high-resolution whole-body scanners to achieve a resolution of 2 mm. This will not only increase the detectability of smaller lesions but will also be valuable for radiation therapy planning. Integrated PET/MRI scanners may have a significant impact on the staging and treatment of head and neck cancers.[72,73] Higher-resolution scanners and PET/MRI may be very useful in radiation therapy planning. Stereotactic body radiotherapy is a relatively new technique that enables delivery of high doses of radiation to malignancies throughout the body with a greater degree of precision than conventional radiation modalities.[74]

Radiopharmaceutical industries are working on new molecular imaging agents to target processes such as cellular proliferation, angiogenesis, and hypoxia. As more specific, tailored radiopharmaceuticals become available, this will allow researchers to better evaluate new drugs and therapies. As cyclotrons become more readily available, radiotracers with short-lived ^{11}C (20 minute half-life) become more feasible.

REFERENCES

1. Yeung HW, Grewal RK, Gonen M, et al. Patterns of ^{18}F-FDG uptake in adipose tissue and muscle: a potential source of false-positives for PET. *J Nucl Med.* 2003;44:1789–96.
2. Soderlund V, Larsson SA, Jacobson H. Reduction in FDG uptake in brown adipose tissue in clinical patients by a single dose of propranolol. *Eur J Nucl Med Mol Imaging.* 2007;34:1018–22.
3. Stephens JM, Pilch PF. The metabolic regulation and vesicular transport of GLUT4, the major insulin-responsive glucose transporter. *Endocr Rev.* 1995;16:529–46.
4. Kamel EM, Jichlinski P, Prior JO, et al. Forced diuresis improves the diagnostic accuracy of 18-F-FDG PET in abdominopelvic malignancies. *J Nucl Med.* 2006;47:1803–7.
5. Salaun PY, Grewal RK, Dodamane I, et al. An analysis of the 18-F-FDG uptake pattern in the stomach. *J Nucl Med.* 2005;46:48–51.
6. Warburg O. On the origin of cancer cells. *Science.* 1956;123:309–14.
7. Brown RS, Leung JY, Kin PV, et al. Glucose transporters and FDG uptake in untreated primary human non-small cell lung cancer. *J Nucl Med.* 1999;40:556–65.
8. Townsend DW: Combined positron emission tomography/computed tomography: the historical perspective. *Semin Ultrasound CT MR.* 2008;29:232–5.
9. Jemal A, Siegel R, Ward E, et al. Cancer statistics, 2009. *CA Cancer J Clin.* 2009;59:225–49.
10. Hogaard L, Specht L. PET/CT in head and neck cancer. *Eur J Nucl Med Mol Imaging.* 2007;34:1329–33.
11. Ong SC, Schoder H, Lee NY, et al. Clinical utility of 18F-FDGPET/CT in assessing the neck after concurrent chemoradiotherapy for locoregional advanced head and neck cancer. *J Nucl Med.* 2008;49:532–40.
12. Schoder H, Carlson DL, Kraus DH, et al. 18F-FDG PET/CT for detecting nodal metastases in patients with oral cancer staged N0 by clinical examination and CT/MRI. *J Nucl Med.* 2006;47:755–62.
13. Barrington SF, Maisey MN. Skeletal muscle uptake of fluorine-18 FDG: effect of oral diazepam. *J Nucl Med.* 1996;37:1127–9.
14. Ha PK, Hdeib A, Goldenberg D, et al. The role of positron computed tomography and computed tomography fusion in the management of early-stage and advanced-stage primary head and neck squamous cell carcinoma. *Arch Otolaryngol Head Neck Surg.* 2006;132:12–16.
15. Ryan WR, Fee WE Jr, Le QT, et al. Positron emission tomography for surveillance of head and neck cancer. *Laryngoscope.* 2005;115:645–50.
16. Porceddu SV, Jarmolowski E, Hicks RJ, et al. Utility of positron emission tomography for the detection of disease in residual neck nodes after (chemo)-radiotherapy in head and neck cancer. *Head Neck.* 2005;27:175–81.
17. Quon A, Fischbein NJ, McDougall IR, et al. Clinical role of 18F-FDG PET/CT in the management of squamous cell carcinoma of the head and neck and thyroid carcinoma. *J Nucl Med.* 2007;48:58S–67S.
18. Schoder H, Fury M, Lee N, et al. PET monitoring of therapy response in head and neck squamous cell carcinoma. *J Nucl Med.* 2009;50:74S–88S.
19. Dietlein M, Scheidhauer K, Voth E, et al. Fluorine-18 fluorodeoxyglucose positron emission tomography and iodine-131 whole-body scintigraphy in the follow-up of differentiated thyroid cancer. *Eur J Nucl Med.* 1997;24:1342–8.
20. DiChiro G. Positron emission tomography using [18F] fluorodeoxyglucose in brain tumors: a powerful diagnostic and prognostic tool. *Invest Radiol.* 1987;22:360–71.
21. Padma MV, Said S, Jacobs M, et al. Prediction of pathology and survival by FDG PET in gliomas. *J Neurooncol.* 2003;64:227–37.
22. Spence AM, Mankoff DA, Muzi M. The role of PET in the management of brain tumors. *Appl Radiol.* 2007;36(6):8–21.
23. Fletcher JW, Kymes SM, Gould M, et al. A comparison of the diagnostic accuracy of 18F-FDG PET and CT in characterization of solitary pulmonary nodules. *J Nucl Med.* 2008;49:179–85.
24. Kligerman S, Digumarthy S. Staging of non-small cell lung cancer using integrated PET/CT. *AJR Am J Roentgenol.* 2009;193:1203–11.
25. Holmes RS, Vaughn TL. Epidemiology and pathogenesis of esophageal cancer. *Semin Radiat Oncol.* 2007;17:2–9.
26. Flamen P, Lerut A, Van Cutsem E, et al. Utility of positron emission tomography for the staging of patients with potentially operable esophageal carcinoma. *J Clin Oncol.* 2000;18:3202–10.
27. Armitage JO, Weisenburger DD. New approach to classifying non-Hodgkin's lymphomas: clinical features of the major histologic subtypes—Non-Hodgkin's Lymphoma Classification Project. *J Clin Oncol.* 1998;16:2780–95.
28. Jhanwar YS, Straus DJ. The role of PET in lymphoma. *J Nucl Med.* 2006; 47:1326–34.
29. Weiler-Sagie M, Bushelev O, Epelbaum R, et al. 18F-FDG avidity in lymphoma readdressed: a study of 766 patients. *J Nucl Med.* 2010;51:25–30.
30. Hicks RJ, Mac Manus MP, Seymour JF. Initial staging of lymphoma with positron emission tomography and computed tomography. *Semin Nucl Med.* 2005;35:165–75.
31. Mikhaeel NG. Use of FDG-PET to monitor response to chemotherapy and radiotherapy in patients with lymphomas. *Eur J Nucl Med Mol Imaging.* 2006;33:S22–S26.
32. Brunetti JC. PET and PET/CT imaging of breast cancer. *Appl Radiol.* 2009;38(9):9–16.
33. Schilling K, Conti P, Adler L, et al. The role of positron emission mammography in breast cancer imaging and management. *Appl Radiol.* 2008;37(4):26–36.
34. Abe K, Sasaki M, Kuabara Y, et al. Comparison of 18FDG-PET with 99m Tc-HMDP scintigraphy for the detection of bone metastases in patients with breast cancer. *Ann Nucl Med.* 2005;19:573–9.
35. Even-Sapir E, Mishani E, Flusser G, et al. 18F-fluoride positron emission tomography and positron emission tomography/computed tomography. *Semin Nucl Med.* 2007;37:462–9.
36. Edwards BK, Ward E, Kohler BA, et al. Cancer, 1975–2006, featuring colorectal cancer trends and impact of interventions (risk factors, screening, and treatment) to reduce future rates. *Cancer.* 2009;116:544–73.

37. Agress H, Cooper BZ. Detection of clinically unexpected and premalignant tumors with whole-body FDG PET: histopathologic comparison. *Radiology*. 2004;230:417–22.
38. Israel O, Kuten A. Early detection of cancer recurrence: 18F-FDG PET/CT can make a difference in diagnosis and patient care. *J Nucl Med*. 2007;48:28S–35S.
39. Selzer M, Hany TF, Wildbrett P, et al. Does the novel PET/CT imaging modality impact on the treatment of patients with metastatic colorectal cancer of the liver? *Ann Surg*. 2004;240:1027–34.
40. Schoder H, Larson SM, Yeung HWD. PET/CT in oncology: integration into clinical management of lymphoma, melanoma, and gastrointestinal malignancies. *J Nucl Med*. 2004;45:72S–81S.
41. Li D, Xie K, Wolff R, et al. Pancreatic cancer. *Lancet*. 2004;363:1049–57.
42. Maldonato A, Gonzalez F, Tamames S, et al. The role of 18F-FDG PET/CT in the evaluation of pancreatic lesions. *J Nucl Med*. 2007;48(suppl 2):26P.
43. Strobel K, Heinrich S, Bhure U, et al. Contrast enhanced 18F-FDG PET/CT: 1-stop-shop imaging for assessing the resectability of pancreatic cancer. *J Nucl Med*. 2008;49:1408–13.
44. Heinrich S, Goerres GW, Schafer M, et al. Positron emission tomography/computed tomography influences on the management of resectable pancreatic cancer and its cost-effectiveness. *Ann Surg*. 2005;242:235–43.
45. Higashi T, Hatano E, Ikai I, et al. FDG PET as a predictor in the early post-therapeutic evaluation for unresectable hepatocellular carcinoma. *Eur J Nucl Med Mol Imaging*. 2010;37:468–82.
46. Petrowsky H, Wildbrett P, Husarik DB, et al. Impact of integrated positron emission tomography and computed tomography on staging and management of gallbladder cancer and cholangiocarcinoma. *J Hepatol*. 2006;45:43–50.
47. Delbeke D, Schoder H, Martin WH, et al. Hybrid imaging (SPECT/CT and PET/CT): improving therapeutic decisions. *Semin Nucl Med*. 2009;39:308–40.
48. Stehman FB, Bundy BN, DiSaia PJ, et al. Carcinoma of the cervix treated with radiation therapy. I. A multi-variate analysis of prognostic variables in the Gynecologic Oncology Group. *Cancer*. 1991;67:2776–85.
49. Grigsby PW, Siegel BA, Dehdashti F. Lymph node staging by positron emission tomography in patients with carcinoma of the cervix. *J Clin Oncol*. 2001;19:3745–9.
50. Kuo PH, Chen Z, Weidhass JB. FDG-PET/CT for planning of radiation therapy. *Appl Radiol*. 2008;37(8):10–23.
51. Short S, Hoskin P, Wong W. Ovulation and increased FDG uptake on PET: potential for a false-positive result. *Clin Nucl Med*. 2005;30:707.
52. Schwarz JK, Grigsby PW, Dehdashti F, et al. The role of 18F-FDG PET in assessing therapy response in cancer of the cervix and ovaries. *J Nucl Med*. 2009;50(suppl):64S–73S.
53. Basu S, Li G, Alavi A. PET and PET/CT imaging of gynecological malignancies: present role and future promise. *Expert Rev Anticancer Ther*. 2009;9:75–96.
54. Aspinall MG, Hamermesh RG. Realizing the promise of personalized medicine. *Harv Bus Rev*. 2007;85:108–17.
55. Weber, WA. Assessing tumor response to therapy. *J Nucl Med*. 2009;50(suppl):1S–10S.
56. Blau M, Nagler W, Bender MA. A new isotope for bone scanning. *J Nucl Med*. 1962;3:332–4.
57. Subramanian G, McAfee JG. A new complex of Tc-99m for skeletal imaging. *Radiology*. 1971;99:192–6.
58. Even-Sapir E. Imaging of malignant bone involvement by morphologic, scintigraphic, and hybrid modalities. *J Nucl Med*. 2005;46:1356–67.
59. Even-Sapir E, Metser U, Flusser G, et al. Assessment of malignant skeletal disease: initial experience with 18F-fluoride PET/CT and comparison between 18F-fluoride PET and 18F-fluoride PET/CT. *J Nucl Med*. 2004;45:272–8.
60. Schwaiger M. Myocardial perfusion imaging with PET. *J Nucl Med*. 1994;35:693–8.
61. Hauser WA, Kurland LT. The epidemiology of epilepsy in Rochester, Minnesota, 1935 through 1967. *Epilepsia*. 1975;16:1–66.
62. Goffin K, Dedeurwaerdere S, Van Laere K, et al. Neuronuclear assessment of patients with epilepsy. *Semin Nucl Med*. 2008;38:227–39.
63. Ferri CP, Prince M, Brayne C, et al. Global prevalence of dementia: a Delphi consensus study. *Lancet*. 2005;366:2112–7.
64. Cohen RM. The application of positron-emitting molecular imaging tracers in Alzheimer's disease. *Mol Imaging Biol*. 2007;9:204–16.
65. Minoshima S, Giordani BJ, Berent S, et al. Metabolic reduction in the posterior cingulated cortex in very early Alzheimer's disease. *Ann Neurol*. 1997;42:85–94.
66. Ishii K, Sakamoto S, Sasaki M, et al. Cerebral glucose metabolism in patients with frontotemporal dementia. *J Nucl Med*. 1998;39:1875–8.
67. Foster NL, Heidebrink JL, Clark CM, et al. FDG-PET improves accuracy in distinguishing frontotemporal dementia and Alzheimer's disease. *Brain*. 2007;130:2616–35.
68. Silverman DHS, Mosconi L, Ercoli L, et al. Positron emission tomography scans obtained for the evaluation of cognitive dysfunction. *Semin Nucl Med*. 2008;38:251–61.
69. Goldsmith SJ, Vallabhajosula S. Clinically proven radiophamaceuticals for infection imaging: mechanisms and applications. *Semin Nucl Med*. 2009;39:2–10.
70. Basu S, Chryssikos T, Moghadam-Kia S, et al. Positron emission tomography as a diagnostic tool in infection: present role and future possibilities. *Semin Nucl Med*. 2009;39:36–51.
71. Zhuang H, Alavi A. 18-Fluorodeoxyglucose positron emission imaging in the detection and monitoring of infection and inflammation. *Semin Nucl Med*. 2002;32:47–59.
72. Judenhofer MS, Wehrl HF, Newport DF, et al. Simultaneous PET-MRI: a new approach for functional and morphological imaging. *Nat Med*. 2008;14:459–65.
73. Troost EGC, Schinagl DAX, Bussink J, et al. Innovations in radiotherapy planning of head and neck cancers: role of PET. *J Nucl Med*. 2010;51:66–76.
74. Rajagopalan MS, Heron DE. Role of PET/CT imaging in stereotactic body radiotherapy. *Future Oncol*. 2010;6:305–17.

CHAPTER 28

Antibodies and Peptides

The development of radiolabeled antibodies and peptides for tumor imaging and therapy evolved from research that identified the expression of tumor-specific antigenic markers and the overexpression of peptide receptors on tumor cells. The development of clinically useful radiolabeled antibodies has been a long and arduous process. The specificity of the antigen–antibody interaction offered hope that radiolabeled antibodies would become "magic bullets" in the diagnosis and treatment of disease. However, the results of early clinical studies targeting tumors in vivo were disappointing in comparison with the results of in vitro studies of antibodies and tumor cells. In vivo specificity for tumors, assessed by tumor-to-nontumor ratios, was hampered by nonspecific uptake in normal tissues. Persistent efforts by scientists and clinicians, however, have led to improved antibody development, radiolabeling technologies, and administration procedures that have decreased the rate of adverse reactions in humans and improved target specificity. For example, humanized antibodies have reduced the immunogenicity of murine antibodies, and pretargeting strategies with antibody administration have been developed to improve target-to-background ratios and minimize toxicities, particularly bone marrow suppression.

The hope of developing sensitive and specific radioactive methods for detecting cancer early in its progress via radioimmunodiagnosis (RID) has been one of the reasons for developing radiolabeled antibodies. Early identification and treatment of cancer can markedly reduce morbidity and mortality. Another important reason is the ability to deliver lethal doses of ionizing radiation to tumors via radioimmunotherapy (RIT). Both diagnostic and therapeutic antibodies have been developed; however, more success has been gained with therapeutic antibodies. Greater success has been achieved in the treatment of hematologic malignancies than of solid tumors, primarily because of antibodies' limited access to cells within a solid tumor.[1] The most successful application of RIT has been in the treatment of non-Hodgkin's lymphoma with ^{90}Y-ibritumomab tiuxetan and ^{131}I-tositumomab.

The overexpression of peptide receptors on tumor cells has attracted great interest in the development of radiolabeled peptides for diagnostic imaging and therapy in nuclear medicine.[2] Whereas antibodies and their fragments have relatively large molecular weights (50 kDa to 150 kDa), peptides usually contain fewer than 100 amino acids and have molecular weights of 10 kDa or less. The limited effectiveness of antibodies has been attributed, in part, to their lack of access to tumor cells, particularly in solid tumors, and to the heterogeneous distribution of tumor-associated antigens on the tumor surface. By contrast, peptide receptors are readily accessible on the external surface of cell membranes, and peptide–receptor affinity can be significantly greater than antigen–antibody affinity. Also, peptides are relatively easy to synthesize and modify to optimize receptor binding, and they can tolerate harsher chemical conditions for modification or radiolabeling. Being small molecules, peptides exhibit rapid blood clearance and are less likely to be immunogenic.

A disadvantage of peptides is their susceptibility to enzymatic degradation by plasma proteases and peptidases. However, several chemical modifications of amino acids can be made to improve stability. An example is substitution of D-tryptophan for its L-enantiomer in ^{111}In-pentetreotide, an indium chelation peptide for targeting somatostatin receptors in neuroendocrine tumors. Synthetic amino acids can also be used to improve in vivo stability and receptor-binding affinity. An example is ^{99m}Tc-apcitide, designed to mimic the peptide sequence –Arg-Gly-Asp (RGD) for targeting the GP IIb/IIIa receptor on activated platelets. The synthetic amino acid Apc (S-aminopropyl-L-cysteine) replaces arginine in the peptide sequence to improve receptor selectivity. Another confounding problem is the potential loss of receptor-binding affinity when the peptide is conjugated to a bifunctional chelating agent (BFCA) and labeled with a radionuclide. Small peptides with only four to six amino acid residues are particularly vulnerable in this regard. Fortunately, modifications can sometimes be made to mitigate this problem as well.

Antibodies and peptides play a significant role in diagnosis and therapy as standalone radiolabeled entities and as targeting moieties conjugated with BFCAs in the design of new radiopharmaceuticals. This topic is further addressed in Chapter 10.

THE IMMUNE SYSTEM

The human immune system originates from lymphocytic stem cells present in bone marrow at birth. These stem cells give rise to precursor cells that ultimately develop into mature T-lymphocytes and B-lymphocytes. Precursor T cells, or pre-T cells, undergo differentiation in the thymus gland, where they acquire the characteristics of T-lymphocytes. Precursor B cells are differentiated in the bone marrow, where they acquire the mature characteristics of B-lymphocytes. In birds, where this process was first studied, lymphocyte differentiation occurs in the bursa of Fabricius, which is the reason for the designation B cell. T-lymphocytes confer cellular immunity, whereas B-lymphocytes confer humoral immunity (Figure 28-1).

Mature cells leave the primary lymphoid tissue of the thymus gland and bone marrow to take up residence in the secondary lymphoid tissues of the body, composed mainly of the spleen, lymph nodes, and mucous-associated lymphoid tissues. In these places they wait, like armed warriors, to be activated by specific antigen invaders. Once activated, a sensitized T-lymphocyte can recognize and destroy specific antigens or cells directly, whereas activated B-lymphocytes are sensitized to become plasma cells that produce antibodies against specific antigens.

Antibody Structure

Enzyme digestion studies of the gamma globulin fraction of serum helped to determine the basic structure of the antibody, shown in Figure 28-2.[3,4] The antibody has a Y-shaped configuration consisting of two heavy-chain polypeptides (450 amino acid residues per chain) and two light-chain polypeptides (220 amino acid residues per chain) linked together by disulfide bonds and noncovalent forces (H-bonding, van der Waals, ionic, and hydrophobic interactions). Thus, a whole antibody consists of approximately 1340 amino acids. The disulfide bonds between the heavy chains occur in an area called the hinge region.

The antibody has two main functional regions. The fragment antigen-binding region (Fab), referred to as the amino terminus, contains the antigen recognition site (paratope), which binds the antibody with a specific antigenic determinant (epitope) on the antigen (e.g., tumor cell). Epitopes occur on the surface of the cell membrane of an invader cell. The other functional component of the antibody is the fragment crystallizable (Fc) region, sometimes referred to as the carboxyl terminus. The Fc region is responsible for linking the antibody to other molecules involved in the immunologic

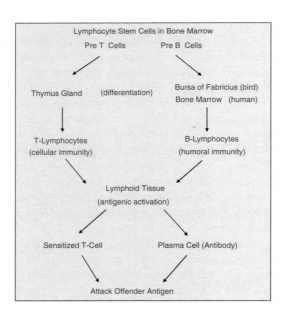

FIGURE 28-1 Schematic diagram of immune system development.

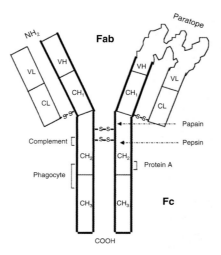

FIGURE 28-2 IgG antibody structure and key components. (Fab) antigen-binding region; (Fc) fragment crystallizable; (VL) variable light chain; (VH) variable heavy chain; (CL) constant light chain; (CH) constant heavy chain. See text for description of enzyme effects and Fc region functionalities.

response. For example, antibody interaction with a tumor cell can initiate effects such as complement-dependent cytotoxicity (CDC), binding to mononuclear cells with phagocytosis, or binding to Fc receptors on natural killer (NK) cells to effect destruction by antibody-dependent cell-mediated cytotoxicity (ADCC). The Fc region also contains oligosaccharides that reside just below the hinge region. Carbohydrate modifications in this region can be made to attach linker groups for binding radionuclides.

In the original work to determine antibody structure, pepsin digestion of whole antibody cleaved it below the hinge region, producing one bivalent F(ab')$_2$ fragment and two Fc fragments. When papain digestion was used, two monovalent Fab fragments were produced, each carrying a paratope at the end farthest from the cleavage site, demonstrating that there were two paratopes for each antibody molecule. A single Fc fragment, connected by disulfide bonds, was also produced in papain digestion because the cleavage occurred above the hinge region. This process is summarized in Figure 28-3.

Antibody Classification

There are five major immunoglobulin (Ig) classes in the blood: IgG, IgA, IgD, IgE, and IgM. The properties of these classes are summarized in Table 28-1. IgG, the most abundant class, has the highest serum concentration. The four-chain antibody structure described above is basic for all immunoglobulin classes, but whereas the light chains are similar between classes, composed of either kappa (κ) or lambda

TABLE 28-1 Antibody Classifications

Class (Isotype)	Serum Conc (mg/100 mL)	Molecular Weight (× 1000)	Chains Light	Chains Heavy
IgG	800–1600	150–180	2 λ or κ	2 γ
IgA	150–400	160	2 λ or κ	2 α
IgM	50–200	900	10 λ or κ	10 μ
IgD	0.3–40	180	2 λ or κ	2 δ
IgE	0.03	200	2 λ or κ	2 ε

(λ) chains, the heavy chains are specific for each class. The far right column of the table indicates that the IgG class contains two gamma (γ) heavy chains, the IgA class two alpha (α) heavy chains, and so forth. IgM is a composite of five basic antibody units and has five times the number of heavy and light chains of other antibody classes.

Enzymatic studies of immunoglobulins revealed that some fragments contained regions along the polypeptide chains where amino acid sequences either were relatively constant or varied between different antibody molecules. These regions are referred to as the constant and variable domains within the antibody molecule. IgG has one variable domain and one constant domain on the light and heavy chains in the Fab fragment and two constant domains in the Fc fragment of the heavy chains (Figure 28-2). The variable domains on the heavy and light chains extend from the amino terminus to approximately residue 120, and the constant domains extend from residues 121 to 220 on the light chain and from 121 to 450 on the heavy chain. By comparison, IgE and IgM each have one variable and four constant domains (three are in the Fc fragment).

Each variable domain contains amino acid positions with exceptional degrees of variability, termed the hypervariable regions or complementarity-determinant regions (CDRs) because these sites complement specific sites on the antigen. Sequence analysis has shown that CDRs occur along heavy and light chains at about residue 30, between residues 50 and 60, and between residues 90 and 100, with relatively constant "framework" regions between them.[5] The heavy chain has an additional CDR between residues 80 and 90. Each CDR consists of a peculiar amino acid sequence that determines the antibody's uniqueness. In the tertiary structure of the antibody, the heavy and light chains loop together in such a way that three of these hypervariable regions are exposed at the tip of each Fab fragment, forming the antigen-binding site (paratope). The CDR between residues 80 and 90 does not seem to participate in paratope formation.

Structural variation, particularly within the heavy chains, determines the immunoglobulin class (e.g., IgG versus IgA). Within each class are subclasses (isotypes) determined by specific amino acid sequence differences in the heavy chains and particularly by the number of disulfide bonds present in the hinge region. For example, human IgG contains four subclasses: IgG1, IgG2, IgG3, and IgG4 (Table 28-2). Subclasses

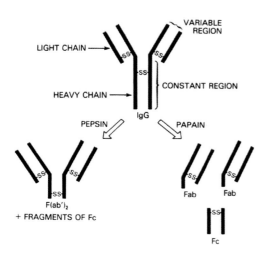

FIGURE 28-3 IgG molecules consist of two heavy and two light protein chains held together by disulfide bonds. Two variable regions can bind to specific antigenic sites, and the constant region interacts with the host immune system. Enzymatic digestion with pepsin removes part of the constant region to produce an F(ab')$_2$ fragment, whereas papain splits the molecule into an Fc fragment and two Fab fragments. (Reprinted with permission of the Society of Nuclear Medicine from reference 9.)

TABLE 28-2 Antibody Subclassification

Class	Subclass Human	Subclass Mouse
IgG	IgG1, IgG2, IgG3, IgG4	IgG1, IgG2a, IgG2b, IgG3
IgA	IgA1, IgA2	None
IgM	None	None
IgD	None	None
IgE	IgE1, IgE2	None

of an immunoglobulin differ in their ability to fix complement and to interact with effector cells such as macrophages or mast cells in the immunologic process. The subclass of an antibody available for clinical use is often listed in the product labeling.

The subclass or isotypic markers are constant within a given species but have distinct differences between species. For example, mouse IgG subclasses are IgG1, IgG2a, IgG2b, and IgG3. There are also allotypic determinants (genetic markers) that identify immunologic differences between IgG molecules within a given species. For example, blood from two individuals with different allotypic markers may be incompatible. Allotypic determinants are regions of variation found in the constant heavy domains, CH_1 and CH_2. In addition, the blood of individuals having identical allotypic markers may exhibit incompatibility because of differences in idiotypic markers. Idiotypes are specific markers found in the variable regions of the antibody and are likely to be associated with the hypervariable regions.[5]

Antibody Development

Antibodies against human tumor cell lines can easily be produced in the spleens of animals injected with tumor cells. Tumor cells as antigens, however, contain several different antigenic determinants (epitopes) on their cell membranes. Upon antigenic challenge with tumor cells, splenic B-lymphocytes are stimulated to develop antibodies against all epitopes expressed on the tumor cell. This results in a diverse group of antibodies (polyclonal antibodies) that have varying degrees of affinity (strength of binding) to the tumor cell. A key point in the development of antibodies for diagnosis and therapy is identification and isolation of the B-lymphocyte that produces the particular antibody (monoclonal antibody) that has the highest affinity for the tumor cell. Monoclonal antibodies are antibodies that possess identical specificities and recognize the same epitope on the antigen.[6] The advantage of monoclonal antibodies is that they react more predictably and exclusively with tumor antigen and have less interaction with nonspecific sites in the body.

A major problem discovered in early studies was that antigenically stimulated lymphocytes could not be grown in cell culture. The ability to grow these cells was essential for identifying desirable clones of B-lymphocytes that produce a specific monoclonal antibody. In 1975, Kohler and Milstein[7] made a discovery that solved this problem. They developed a method of producing monoclonal antibodies in vitro by fusing splenic lymphocytes from immunized mice with mouse myeloma cells. The resulting product was called a hybridoma. The myeloma cell provided immortality needed for the hybrid cell to reproduce itself, and the lymphocyte enabled the hybridoma to produce the desired monoclonal antibody.[8] Hybridomas can be grown in tissue culture, enabling isolation of the desired monoclonal antibody in buffer. Hybridomas can also be grown in mouse peritoneum, whereupon antibodies are harvested from ascites fluid. Commercial production of antibodies is carried out in mice and through the use of hybridoma technology (Figures 28-4 and 28-5).[9]

Although the antibody a hybridoma clone secretes is genetically derived from a single cell, it is not a monoclonal antibody in the immunologic sense because each cell of the clone has some chromosomes from the myeloma cell parent and some from the spleen cell parent.[9] As a result, the hybridoma antibody is frequently a mix of heavy and light chains of both parent cells. Because the aim is to find a clone with only heavy and light chains of the specifically immune spleen cell, hundreds of fusions and reclonings may be required before a single hybrid that secretes the desired antibody is formed (Figure 28-5). Once the desired clone is found, it can be frozen for long-term storage. At any time thereafter a sample of the clone can be injected into animals of the same strain as those that provided the original cells for fusing.[8] The injected hybridomas will grow and secrete the specific monoclonal antibody in high concentration in the animal's serum. Alternatively, a clone can be grown in mass culture in vitro and the antibody harvested from the medium.

Antibody Modification

Antibodies can be modified to remove undesirable characteristics and to change their kinetic properties. Hybridomas produce whole antibodies with the Fab and Fc regions intact. Radiolabeled antibodies for clinical use are of the IgG type and may be whole antibodies with a molecular weight of about 150 kDa, F(ab')$_2$ fragments of 100 kDa, and Fab fragments of 50 kDa. Intact, whole antibodies have properties distinctly different from antibody fragments. Compared with antibody fragments, whole antibodies have slower blood clearance because of their large molecular weight. This results in a slower but higher amount of tumor uptake with longer duration but lower tumor-to-background ratios.[10] Whole antibodies are more suited for longer-lived radionuclides such as ^{131}I and ^{111}In. By contrast, antibody fragments have faster blood clearance but more rapid, lower tumor uptake of shorter duration and higher tumor-to-background ratios. They are more suited for shorter-lived radionuclides such as ^{99m}Tc or ^{123}I and have a shorter dose-to-imaging time frame. Whole antibodies of murine origin have a higher rate of human antimouse antibody (HAMA) response because they retain the Fc portion. The HAMA response is an undesirable

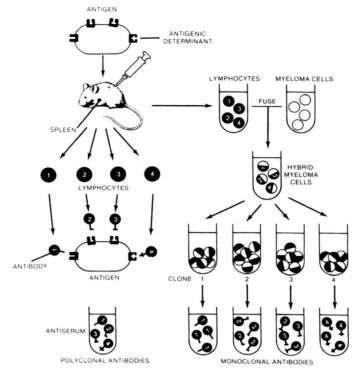

FIGURE 28-4 Injection of antigen into a mouse or other higher animal elicits a heterogeneous antibody response because of the stimulation of several B-lymphocytes by various determinants on the antigen, which results in polyclonal antibodies in the serum (left). If sensitized lymphocytes are removed from the spleen of an immunized animal and induced to fuse with myeloma cells, individual hybrid cells can be cloned, each producing monoclonal antibodies to a single antigenic determinant (right). (Reprinted with permission of the Society of Nuclear Medicine from reference 9.)

immunologic reaction that results from human antibodies being produced against the injected mouse antibody as the foreign antigen. This response is more probable after multiple exposures to the antibody. Because the Fc portion is removed in antibody fragments, they have less HAMA response and reduced binding to Fc receptors in the liver and uptake of metabolized antibody by macrophages, but they still can elicit an immunogenic response because they contain amino acid sequences unique to the mouse.[6]

Antibodies can be genetically modified by recombinant DNA and gene transfer methods to replace murine components with human components.[11] A *chimeric* antibody contains a murine variable domain and a human constant domain; a *humanized* antibody goes one step further and has only a murine CDR, with the remainder being human. While these modifications reduce the chance of HAMA response, human antichimeric antibodies (HACA) or human antihuman antibodies (HAHA) can still be produced because the CDR is of murine origin. Two other modifications can be made to shorten the time between antibody injection and imaging: single-chain antigen-binding proteins that consist of a variable heavy chain and a light chain linked together by a disulfide bond, and molecular recognition units that consist of the CDR alone.

In addition to reducing HAMA response, another advantage of chimeric and humanized antibodies is that they possess a much more potent immune effector function and are therefore more capable of inducing CDC and ADCC than are their murine parent antibodies, making them potentially more effective antitumor agents.[12]

Antigen–Antibody Interaction

The bond between antigen and antibody is noncovalent and may include hydrogen bonds, electrostatic bonds, van der Waals forces, and hydrophobic bonds. A whole antibody, having two antigen-binding sites (paratopes), is a bivalent ligand. Each of these sites can interact with a single antigenic epitope. Antibody *affinity* is a measure of the strength of attraction between a single paratope on the antibody and its

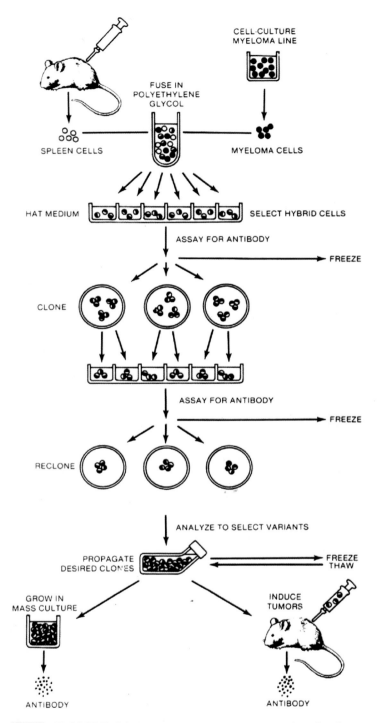

FIGURE 28-5 Fusion of antigen-stimulated spleen cells with myeloma cells in polyethylene glycol results in hybrid cells that can be cloned in hypoxanthine–aminopterin–thymidine (HAT) medium. Those clones that generate immunoglobulin are further propagated, and those producing antibodies to the desired antigen are selected to find the variant that produces antibody with the desired specificity and binding properties. Hybridomas can be maintained in mass culture or in mouse ascites fluid, and clones at any stage of development can be frozen for later use. (Reprinted with permission of the Society of Nuclear Medicine from reference 9.)

antigenic epitope.[9,13] It is a thermodynamic function of bond strength determined by equilibrium dialysis with known amounts of antigen and antibody. Affinity constants can be determined from a Scatchard plot of bound-to-free antigen ratios on the y-coordinate against the bound-antigen concentration on the x-coordinate. Affinity constants of useful monoclonal antibodies range from 10^8 to 10^{10} L/mole.[9] Antibody *avidity* is a measure of the integrity of the antigen–antibody complex once formed. It is a measure of binding strength between antigens with multiple epitopes and antibodies with more than one paratope. Thus, whole (intact) antibodies and F(ab')$_2$ fragments, being bivalent, have potential for greater avidity compared with monovalent Fab fragments.

The antigen-binding site may be altered when an antibody is radiolabeled, adversely affecting its affinity for the antigenic epitope. Antibody immunoreactivity can be tested by incubating the antibody with an excess of tumor cells (antigen) that express a specific epitope and measuring the fraction of labeled antibody bound to the cells.[14] Raji cells, which bear the antigenic determinants for lymphoma, have been used to assess the immunoreactivity of the LYM-1 antibody used to treat non-Hodgkin's lymphoma.

Antibody Nomenclature

A variety of terms are used to describe antibodies and their antigenic markers (Table 28-3). The name of an antibody reveals some of its characteristics. The generic name given to an antibody (e.g., ibritumomab or rituximab) is derived as follows.[15,16] The suffix *mab* is used for monoclonal antibodies and antibody fragments. Preceding the suffix is one or more letters identifying the antibody source (e.g., *o* = mouse, *u* = human, *xi* = chimeric). Preceding the antibody source designation is a letter code identifying the disease state subclass (e.g., *vir* = viral, *bac* = bacterial, *lim* = immunomodulator, *cir* = cardiovascular) or area of tumor involvement (e.g., *col* = colon, *mel* = melanoma, *mar* = mammary, *tum* or *tu* = miscellaneous tumor). Preceding the disease code is a unique prefix selected by the antibody manufacturer. If the antibody is radiolabeled, the antibody name is preceded by the name of the radionuclide, its elemental symbol, and its mass number. For example, Bexxar (Corixa and GlaxoSmithKline) is written as iodine I 131 tositumomab. If an antibody is a conjugate, the conjugate name follows its generic name. For example, Zevalin (Spectrum Pharmaceuticals) labeled with ^{90}Y or ^{111}In is a conjugate of the monoclonal antibody ibritumomab and the bifunctional chelating agent tiuxetan. The full names, thus, are yttrium Y 90 ibritumomab tiuxetan and indium In 111 ibritumomab tiuxetan. In addition to the generic name of an unconjugated antibody, there is a number/letter code designation for the antibody (e.g., B1 for tositumomab and 2B8 for ibritumomab).

Specific surface molecules on the lymphocyte cell membrane that are recognized by groups of monoclonal antibodies are called *clusters of differentiation* (CD). The CD designation is followed by an arbitrarily assigned number. These CD antigens are expressed on the surface of malignant B-lymphocytes in tumors and normal B-lymphocytes in the blood, spleen, lymph nodes, bone marrow, and some of their precursors in the marrow, depending on the antigen.[12] For example, the CD20 antigen, targeted by unconjugated rituximab, ^{111}In- or ^{90}Y-ibritumomab, and ^{131}I-tositumomab, is found on all cells, including pre-B cells of the bone marrow, but not on the stem cells.

ANTIBODY LABELING

A variety of radionuclides, mainly radiohalogens and radiometals, are used to label antibodies.[17–21] At present, the antibodies used in nuclear medicine practice are labeled with ^{131}I,

TABLE 28-3 Antibody Nomenclature

Disease[a]	Trade Name	Generic Name	Unconjugated Monoclonal Antibody Designation	Antigenic Marker
NHL	Zevalin	^{90}Y- or ^{111}In-ibritumomab tiuxetan	2B8	CD20
NHL	Bexxar	^{131}I-tositumomab	B1	CD20
NHL	Rituxan	Rituximab	C2B8	CD20
CR cancer	CEA-Scan	^{99m}Tc-arcitumomab	IMMU-4	CEA
CR/OV cancer	OncoScint CR/OV	^{111}In-satumomab pendetide	B72.3	TAG72
Prostate cancer	ProstaScint	^{111}In-capromab pendetide	7E11-C5.3	PSA
NHL		^{131}I-LYM-1	LYM-1	HLA-DR
NHL	Lymphocide	^{90}Y-epratuzumab	hLL2	CD22
Infection	LeukoScan	^{99m}Tc-sulesomab	IMMU-MN3	NCA-90
Infection	LeuTech	^{99m}Tc-LeuTech	Anti-SSEA-1	CD15

[a]NHL = non-Hodgkin's lymphoma; CR = colorectal; OV = ovarian.

TABLE 28-4 Properties of Selected Radiolabels for Diagnostic and Therapeutic Antibodies

Radionuclide	Half-life	Decay Mode	Particles Max MeV	Photons keV	Abundance (%)
^{67}Cu	61.8 hr	β–	0.56	93	16
				185	49
^{90}Y	64 hr	β–	2.28	—	—
^{99m}Tc	6.01 hr	IT	—	140.5	88
^{123}I	13.1 hr	EC	—	159	84
^{125}I	59.5 days	EC	—	27	115
				35	6.7
^{131}I	8.04 days	β–	0.61	364	82
				637	6.5
^{177}Lu	160 hr	β–	0.50	113	6.2
				208	10
^{186}Re	89 hr	β–	1.07	137	9
^{188}Re	17 hr	β–	2.12	155	16
^{211}At	7.2 hr	Alpha	5.87	—	—

^{99m}Tc, ^{111}In, and ^{90}Y, although other nuclides are being investigated. A list of radionuclides for RID and RIT is given in Table 28-4.

Radionuclide Considerations

Radiolabeled antibodies are used for both diagnosis and therapy. Currently, therapeutic applications are receiving much attention. Radionuclides for RID must necessarily emit gamma radiation, while those for RIT emit particulate radiation, typically beta particles. The development of alpha-particle emitters in RIT is under investigation. The properties of particulate radiations emitted from therapy radionuclides for antibodies are discussed in Chapter 29.

In general, the half-life of a radionuclide should be matched with the antibody's in vivo kinetics. Whole or intact antibodies, which have prolonged circulation times, are best labeled with longer half-lived nuclides, such as ^{131}I, ^{111}In, and ^{90}Y. Antibody fragments, which clear more quickly from the bloodstream, are preferably labeled with shorter-lived nuclides, such as ^{99m}Tc and ^{123}I. Ideally, a therapeutic radionuclide should also emit photons to permit imaging of antibody distribution in vivo. This greatly facilitates quantitative distribution studies for estimating radiation dosimetry. In some instances (e.g., with ibritumomab), the antibody is labeled with a beta emitter (^{90}Y) for therapy, but it can also be labeled with a gamma emitter (^{111}In) to facilitate dosimetry calculations.

Ideally, the radionuclide should not dissociate from the antibody in vivo, because this would result in an undesirable radiation dose to normal tissues and a reduced dose to the tumor. In general, the radioiodines are prone to enzymatic deiodination with uptake in the thyroid gland and stomach, whereas metallic radionuclides released from antibodies are taken up into bone. Significant effort has been directed toward the development of radiolabeling methods that mitigate these problems.

Radioiodine Labeling

Antibodies and proteins have been labeled with a variety of radiohalogens, including ^{123}I, ^{125}I, ^{131}I, ^{77}Br, ^{18}F, and ^{211}At.[17] Most antibody labeling, however, is done with the radioiodines. Both direct and indirect methods of labeling are used.

With direct iodination (via electrophilic substitution), radioiodide (I–) is oxidized to I+, which binds to hydroxylbenzyl moieties of tyrosyl residues in the antibody polypeptide chains (Figure 28-6). The direct method has the advantage of relatively easy iodine chemistry (see Chapter 11). Common oxidants for iodination are chloramine-T and the milder Iodo-Gen (Pierce), which is claimed to be less damaging to the antibody (Figure 28-7).[6] N-chlorosuccinimide (NCS) has been used as an oxidant with indirect methods of iodination. Iodo-Gen is preferred for labeling F(ab')$_2$ fragments because

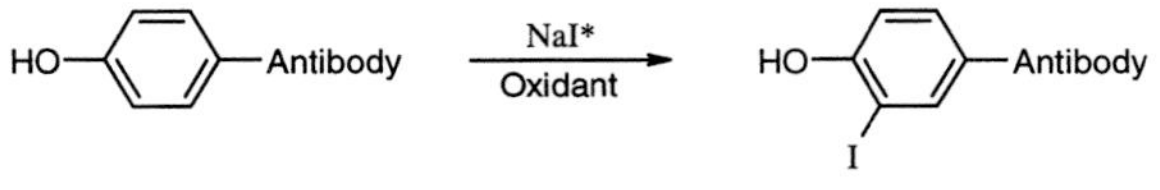

FIGURE 28-6 Electrophilic substitution of radioiodine into tyrosyl residues of an antibody.

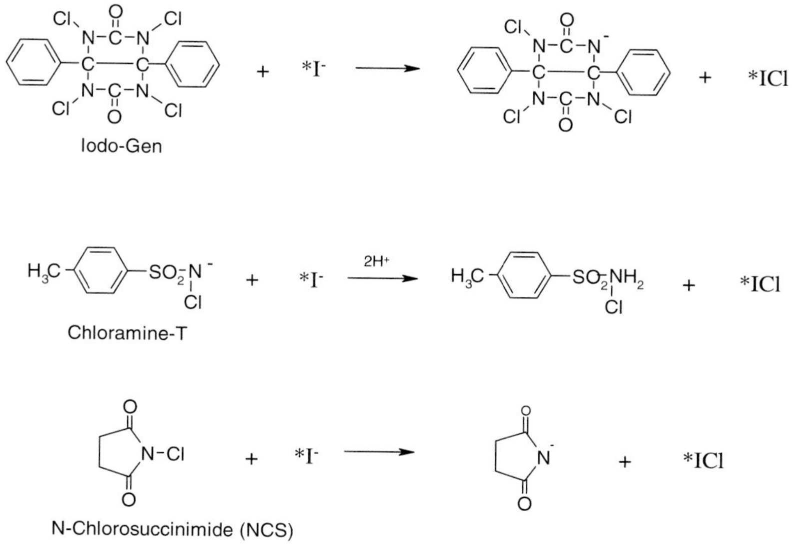

FIGURE 28-7 Oxidants used in the radioiodination of proteins and antibodies.

this method does not use sodium metabisulfite, which has the capacity to reduce F(ab′)$_2$ fragments to Fab.[20]

Indirect methods of labeling antibodies involve attaching the radiohalogen to an intermediate compound that conjugates with the antibody. The principal functional groups on antibodies involved in conjugation reactions are amines, sulfhydryls, and oxidized sugars. However, most conjugations are via acylation reactions between ε-amino groups in lysine and an activated N-hydroxysuccinimide ester (NHS). The classic agent of this type is the Bolton-Hunter reagent (N-succinimidyl 3-iodo-4-hydroxyphenylpropionate) (Figure 28-8). The main advantage offered by this agent is its ability to label proteins that do not contain tyrosyl residues or are particularly sensitive to the oxidizing and reducing conditions of direct iodination. Labeling yields, however, are lower with the Bolton-Hunter reagent because competing aqueous hydrolysis reactions of the active ester group occur during the radiolabeling. A significant problem with antibodies labeled directly or with the Bolton-Hunter reagent is that the radioiodine label is positioned *ortho* to the hydroxyl group in the aromatic ring. This makes it prone to in vivo deiodination by deiodinase enzymes, similar to the metabolism of thyroid hormone.

Other active ester intermediates for indirect iodination have been developed.[17-20] The prototypical method for iodination of an antibody by these intermediates occurs in two steps. The initial step involves synthesis of N-succinimidyl 3- or 4-^{131}I-iodobenzoate (SIB) from the corresponding N-succinimidyl (tri-n-butylstannyl)benzoate precursor via iododestannylation. This is followed by conjugation of SIB to lysine ε-amino groups on the antibody (Figure 28-9).[19,22] This method results in placement of the radioiodine atom in an aromatic ring without a hydroxyl group. Its position within the ring is *meta* or *para*, depending on the precursor used. Although it is more difficult to radiohalogenate nonphenolic aryl ring benzoates, they are more stable to dehalogenation in vivo. When compared with the Bolton-Hunter reagent or directly labeled antibody, the N-succinimidyl 3-iodobenzoate ester was shown to produce radiolabeled antibody most stable toward dehalogenation in vivo. Distribution studies in mice demonstrated that thyroid uptake of activity (a monitor of dehalogenation) of the iodobenzoyl conjugate was about one-half that of the Bolton-Hunter–labeled antibody and only 7% that of antibody iodinated directly with Iodo-Gen.[23] Additional reports demonstrated enhanced tumor uptake, therapeutic efficacy, and in vivo stability against dehalogenation with iodobenzoyl antibody conjugates.[24,25]

Many other methods can be used to radiolabel antibodies, as reviewed by Wilbur.[17]

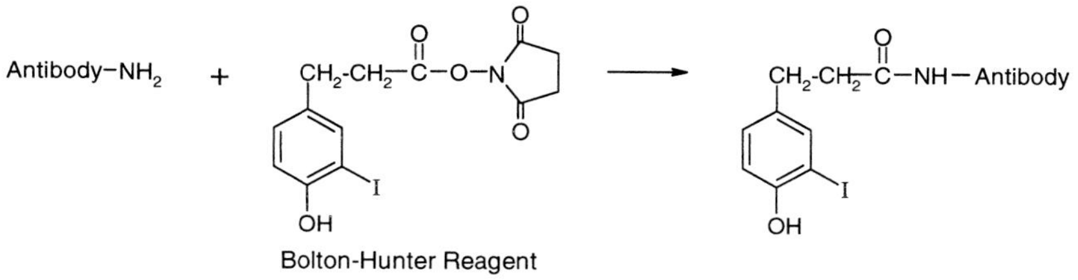

FIGURE 28-8 The radioiodinated Bolton-Hunter reagent is useful for indirect radiolabeling of antibodies that lack tyrosyl residues.

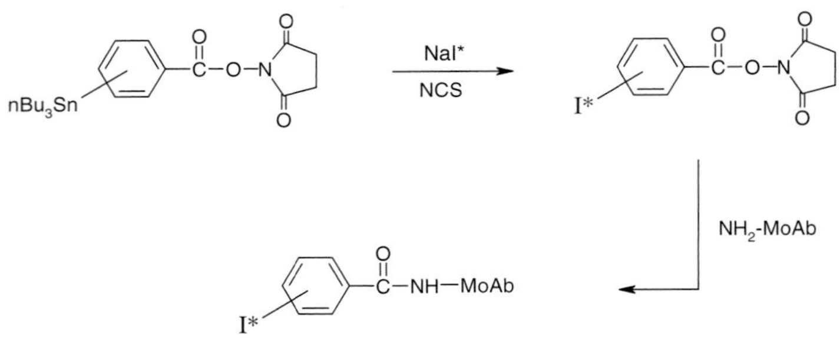

FIGURE 28-9 Synthesis of N-succinimidyl 3- or 4-^{131}I-iodobenzoate (SIB) from the corresponding N-succinimidyl (tri-n-butylstannyl)benzoate precursor via iododestannylation followed by conjugation of SIB to lysine ε-amino groups on the antibody.

Radiometal Labeling

The labeling of antibodies with radiometals advanced significantly with the development of BFCAs. A number of methods have been developed to label proteins and antibodies with radiometals.[25–35] For chelation with ^{111}In and ^{90}Y particularly, derivatives of ethylenediamine tetraacetic acid (EDTA), diethylenetriamine pentaacetic acid (DTPA), and 1,4,7,10-tetraazacyclododecane-1,4,7,10-tetraacetic acid (DOTA) have been explored. Antibodies currently approved for routine use in nuclear medicine use DTPA as the chelating ligand. They are labeled by the postlabeling approach, adding radionuclide to kits that contain the antibody–BFCA conjugate.

Two principal approaches have been employed to link BFCAs to antibodies. One approach is to conjugate the BFCA with the ε-amino group of lysine within the antibody. The disadvantage of this approach is that lysine is present throughout the antibody molecule so that site-specific conjugation is not possible. Depending on the labeling conditions, this approach may result in reduced antibody specificity if conjugation occurs in proximity to the antigen-binding region.

Another approach involves conjugation of the BFCA to oxidized oligosaccharide moieties within the antibody. Virtually all immunoglobulins contain carbohydrates that are linked to the constant regions of the heavy chains in the Fc region of the antibody. Consequently, conjugation and labeling of BFCAs in this region should result in less likelihood of interference with the antigen-binding site.

Initial antibody conjugation methods used DTPA dianhydride to covalently link DTPA to the antibody via lysine acylation (Figure 28-10). Labeling antibodies conjugated in this manner with ^{90}Y, for example, demonstrated that ^{90}Y

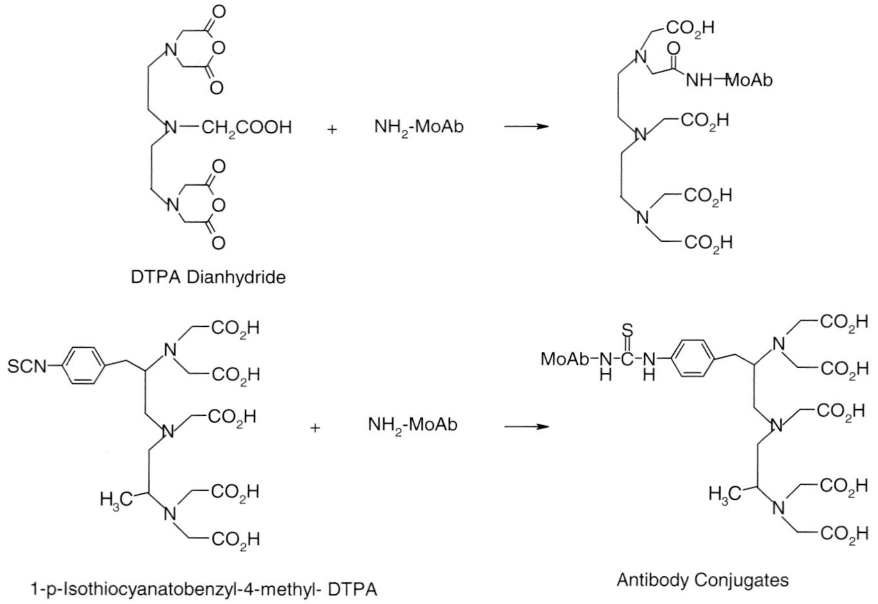

FIGURE 28-10 Antibody conjugation methods involving the use of DTPA dianhydride to covalently attach DTPA to the antibody via lysine acylation (top reaction) and by way of the DTPA derivative 1-p-isothiocyanatobenzyl-4-methyl-DTPA (bottom reaction), which forms a stable thiourea bond with the antibody. (MoAb is monoclonal antibody.)

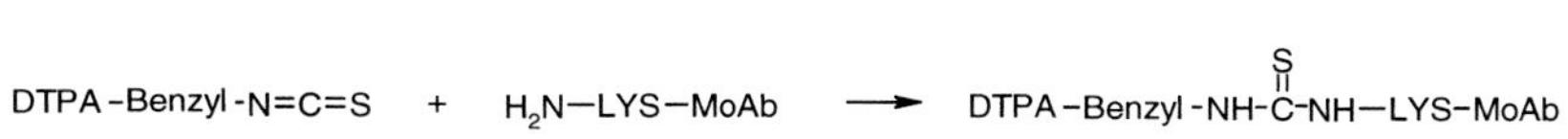

$$\text{DTPA-Benzyl-N=C=S} + \text{H}_2\text{N—LYS—MoAb} \longrightarrow \text{DTPA-Benzyl-NH-}\overset{\overset{\text{S}}{\|}}{\text{C}}\text{-NH—LYS—MoAb}$$

FIGURE 28-11 Covalent conjugation of a monoclonal antibody with 1-p-isothiocyanatobenzyl-DTPA via a stable thiourea bond, which permits subsequent labeling of the antibody via radionuclide metal chelation with DTPA.

slowly leached from the antibody in vivo and localized in bone.[36,37] It was thought that the loss of one DTPA carboxyl group to acylation rendered it unavailable for chelation and made the radiometal bond vulnerable.

Another conjugation approach employs DTPA derivatives. With this technique, a benzyl moiety containing an antibody-coupling group is covalently bound to a methylene carbon in the DTPA backbone (Figure 28-10).[27,30,31] An example is shown in Figure 28-11. This technique of conjugation frees all five DTPA carboxyl groups for chelation with the radiometal, significantly increasing the chelate's thermodynamic stability. Additional functional groups can be attached at other sites on the backbone to sterically hinder the release of the radiometal, improving stability further. An example is 1-p-isothiocyanatobenzyl-4-methyl-DTPA (1B4M-DTPA) (Figure 28-10). This BFCA, designated MX-DTPA, is conjugated with ibritumomab for labeling with ^{111}In or ^{90}Y, forming a stable 1:1 octadentate chelate with these metals.[31,32,38]

Another ligand that has been explored to improve chelation stability is DOTA. DOTA is a macrocyclic chelating agent known to produce exceedingly inert chelates with lanthanides. This significantly reduces radiometal dissociation and localization in bone. A major drawback to the clinical practicality of DOTA, however, is its slow reaction kinetics during chelation. This poses a problem during antibody labeling because of the detrimental radiolytic effects of hard beta emitters, such as ^{90}Y, on antibody immunoreactivity. Thus, new derivatives of DOTA have been sought to speed up its chelation kinetics.[21]

Oligosaccharide conjugation technology in antibody labeling involves the oxidation of sugar moieties in the Fc region to aldehydes, typically with sodium periodate (NaIO$_4$). The aldehyde group can selectively react with compounds containing amines, hydrazines, hydrazides, and semicarbazides.[39] This permits the selective attachment of BFCAs to the Fc region of the antibody, where radiolabeling will occur away from the antigen-binding site. A typical conjugation reaction is shown in Figure 28-12. The BFCA in ^{111}In-satumomab pendetide (OncoScint, Cytogen) and ^{111}In-capromab pendetide (ProstaScint, Cytogen) is glycyl-tyrosyl-(N-ε-DTPA)-lysine or GYK-DTPA. It is the linker molecule attached to oxidized sugars in the Fc region that chelates ^{111}In to the antibody.

Bifunctional chelation technology makes antibody labeling amenable to kit preparation; only the buffered radiometal must be added to the antibody conjugate in the kit. The labeling yield in commercially available products is typically greater than 90%. Radiolabeling via bifunctional chelation is discussed in Chapter 10.

Several approaches have been used to label antibodies with ^{99m}Tc.[33-35] A direct method relies on the reduction of disulfide bridges within the antibody to generate endogenous sulfhydryl groups. These groups are attachment sites for reduced technetium (Figure 28-13). An indirect method of labeling antibodies involves first conjugation of DTPA to the antibody and then addition of reduced technetium to the conjugate. This method is similar to the postlabeling approach described above for ^{111}In and ^{90}Y labeling. A third approach to labeling antibodies with technetium is to employ a prelabeled ligand (prelabeling approach).[35] In this method, dithionite-reduced technetium is complexed to an N$_2$S$_2$ ligand functionalized with a carboxylate group. This is then activated with an ester group, through which it is bound efficiently to the antibody via an acylation reaction with lysine amine residues (Figure 28-14). This labeling approach obviates nonspecific binding of technetium to the antibody, which has been found to occur with direct labeling by the postlabeling approach. Although this method produces a stable antibody label without nonspecific binding, it is somewhat cumbersome and less adaptable to simple kit formulation.

RADIOLABELED ANTIBODIES

Six radiolabeled antibodies for diagnosis and two for therapy have received FDA approval for routine use in nuclear medicine. Of these eight antibodies, only three remain on the market. The six diagnostic antibodies are ^{99m}Tc-arcitumomab

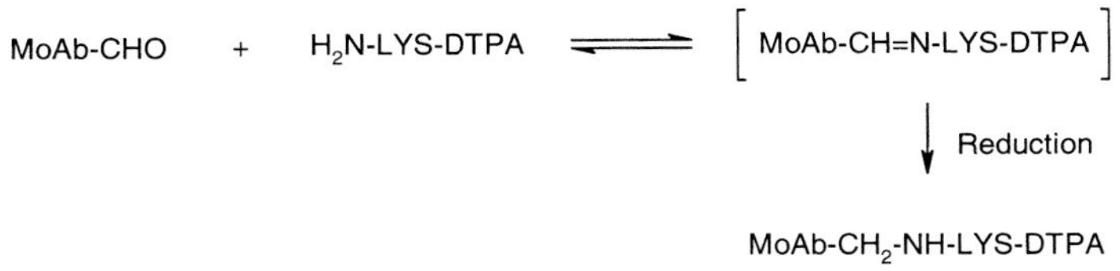

$$\text{MoAb-CHO} + \text{H}_2\text{N-LYS-DTPA} \rightleftharpoons [\text{MoAb-CH=N-LYS-DTPA}]$$
$$\downarrow \text{Reduction}$$
$$\text{MoAb-CH}_2\text{-NH-LYS-DTPA}$$

FIGURE 28-12 Conjugation reaction between a lysine amino derivative of DTPA and an antibody aldehyde group forming an imine or Schiff base with the antibody, which is then reduced to the stable secondary amine.

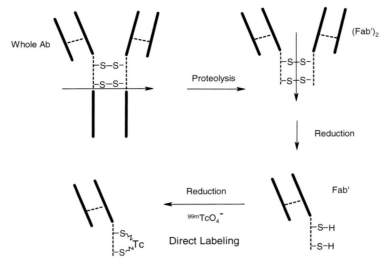

FIGURE 28-13 Generation of antibody fragments, reduction of their disulfide bonds, and direct labeling of fragments with technetium.

(CEA-Scan), a murine IgG Fab fragment of IMMU-4 monoclonal antibody that reacts with carcinoembryonic antigen (CEA), indicated for the detection of recurrent or metastatic colorectal cancer; ^{99m}Tc-nofetumomab merpentan (Verluma), a murine IgG Fab fragment of NR-LU-10 monoclonal antibody that reacts with a 40 kDa glycoprotein expressed on a variety of tumors, indicated for the detection of small cell lung cancer; ^{99m}Tc-fanolesomab (NeutroSpec), an intact murine anti-CD15 IgM monoclonal antibody that binds to the CD-15 epitope on activated neutrophils in vivo, indicated for the detection of appendicitis; ^{111}In-satumomab pendetide (OncoScint CR/OV), an intact murine IgG monoclonal antibody that reacts with a high molecular weight, tumor-associated glycoprotein (TAG-72) expressed on a variety of adenocarcinomas, indicated for colorectal and ovarian cancers; ^{111}In-imciromab pentetate (Myoscint), an antimyosin murine IgG Fab fragment of R11D10 monoclonal antibody, indicated for the detection of myocardial infarction; and ^{111}In-capromab pendetide (ProstaScint), an intact murine IgG monoclonal antibody that reacts with prostate-specific-antigen, indicated for imaging metastasis in patients with known prostate cancer. Only ProstaScint remains on the market. NeutroSpec was removed from the market for safety reasons because of two deaths associated with its use. The other diagnostic antibodies were removed from the market for economic reasons. The therapeutic antibodies approved for use are ^{90}Y-ibritumomab

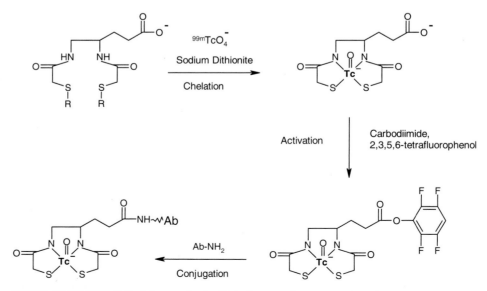

FIGURE 28-14 Method of labeling antibodies with technetium using the prelabeling approach (i.e., technetium chelation, activation, and conjugation to antibody).

tiuxetan (Zevalin) and ^{131}I-tositumomab (Bexxar) for the treatment of non-Hodgkin's lymphoma (NHL). These two antibodies and several new therapeutic antibodies under investigation are discussed in more detail in Chapter 29.

Indium In 111 Capromab Pendetide

[111]In-capromab pendetide (ProstaScint) is an intact antibody 7E11-C5.3 of the IgG1 subclass. It reacts with prostate-specific antigen, a glycoprotein expressed by the prostate epithelium. The antibody is labeled with ^{111}In via GYK-DTPA linker technology, as shown in Figure 28-15.

The kit is composed of the antibody in 1 mL of phosphate-buffered saline (PBS), which is stored in the refrigerator before use, a vial of acetate buffer, and a Millex-GV (Millipore) filter. ^{111}In-indium chloride is available from GE Healthcare and Covidien. Labeling is accomplished in a three-step process. First, 0.1 mL of acetate buffer is added to the indium chloride. This step produces the intermediate species, indium acetate, which keeps indium soluble at the pH necessary for antibody labeling. This is followed by addition of 6 to 7 mCi (222–259 MBq) of indium acetate to the antibody and incubation for 30 minutes at room temperature to effect complexation of ^{111}In to the antibody. At this point the remaining 1.9 mL of acetate buffer is added to the mixture to complete the preparation. To check radiochemical purity equal parts of ^{111}In-capromab pendetide and 0.05 M DTPA solution are mixed and allowed to stand for 1 minute. A drop of this mixture is then applied to an instant thin-layer chromatography with silica gel (ITLC-SG) strip and developed in normal saline.[40] The labeled antibody remains at the origin while the free indium, which is bound to DTPA, migrates to the solvent front. Radiochemical purity must be 90% or higher. The labeled antibody is drawn up through the Millex filter before use, and the product is stable for 8 hours at room temperature.

[111]In-capromab pendetide is indicated for use in patients with biopsy-proven and clinically localized prostate cancer who are at high risk for metastasis, to help clinicians decide on a course of therapy. ^{111}In-capromab pendetide is also indicated for use in postprostatectomy patients with a high clinical suspicion of occult recurrent or residual prostate cancer (Figures 28-16 and 28-17). The usual administered dosage is 5 mCi (185 MBq) given intravenously over 5 minutes.

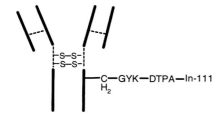

FIGURE 28-15 ^{111}In is chelated to an antibody via a GYK-DTPA linker that is covalently bound to a carbohydrate moiety in the Fc portion of the antibody.

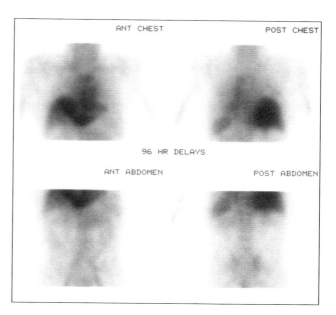

FIGURE 28-16 Normal scan in a 79 year old man with prostate cancer, obtained 96 hours after injection of 5 mCi (185 MBq) of ^{111}In-capromab pendetide. Increased activity is seen in the region of the prostate gland, consistent with the patient's known prostate cancer, but there is no evidence of metastasis in the central abdominal region.

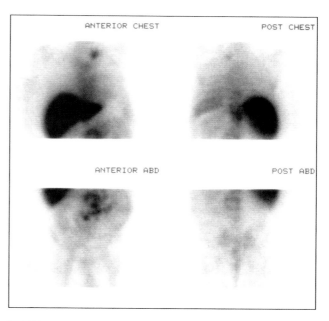

FIGURE 28-17 Prostate metastases in a 68 year old man with prostate cancer and elevated prostate-specific antigen level. Images obtained 30 minutes after injection of 4.7 mCi (174 MBq) of ^{111}In-capromab pendetide demonstrate evidence of metastatic lesions, seen as a focus of increased tracer accumulation in the left side of the neck and multiple foci of tracer accumulation in the central abdominal region.

Whole-body planar imaging is done of the pelvis, abdomen, and thorax between 72 and 120 hours after dosing. The patient should be prepared with a cathartic the night before and a cleansing enema and bladder void 1 hour before imaging. Single-photon emission computed tomography (SPECT) imaging is done, with a blood pool image acquired 30 minutes after dosing, followed by pelvis and abdomen images at 72 to 120 hours after dosing. Alternatively, simultaneous blood pool imaging can be performed using ^{99m}Tc-red blood cells at the time of antibody imaging, which offers the advantages of more precise image registration for interpretation and is more time efficient.[41] Lesions are seen as increased focal accumulations of activity (Figure 28-17). Some normal activity is seen in the liver, spleen, bone marrow, and genitalia. The critical organ is the liver, sustaining a radiation absorbed dose of 18.5 rad/5 mCi (185 MBq) administered activity.

Technetium Tc 99m Sulesomab

^{99m}Tc-sulesomab (LeukoScan [Immunomedics GmbH, Darmstadt, Germany]) is a murine Fab fragment of IMMU-MN3 antigranulocyte monoclonal antibody. It reacts with a 90 kDa glycoprotein, nonspecific cross-reactive antigen (NCA-90), on the surface of granulocytes.[42] LeukoScan cross-reacts with CEA and may interact with CEA-producing tumors.

^{99m}Tc-sulesomab is prepared from a kit containing 0.31 mg of the antibody sulesomab, 0.22 mg stannous chloride, and other adjuvants, at pH 5 to 7. The kit is stored at 2°C to 8°C prior to labeling. The lyophilized powder is reconstituted with 0.5 mL of sterile saline, followed by the addition of 30 to 40 mCi (1110–1480 MBq) ^{99m}Tc-sodium pertechnetate in a volume of 1.0 mL. The product is ready for use after a 5 minute incubation. The labeled antibody should be stored at room temperature and used within 4 hours after preparation. Its radiochemical purity must be 90% or higher, as determined by ITLC-SG in acetone. The labeled antibody remains at the origin and the free pertechnetate travels to the solvent front.

^{99m}Tc-sulesomab is indicated for the diagnostic localization of infection and inflammation in bone in patients with suspected osteomyelitis, including patients with diabetic foot ulcers. A typical administered dose is 20 to 30 mCi (740–1110 MBq) containing 0.25 mg of antibody. Imaging is performed 1 to 8 hours after injection as desired; there is no difference in detection of the presence or absence of osteomyelitis between 1 to 2 hour and 5 to 8 hour imaging times. Several studies have shown the value of ^{99m}Tc-sulesomab for diagnosing infection, compared with ^{111}In-oxine-labeled autologous leukocytes. Becker et al.[43] reported the sensitivity, specificity, and accuracy, respectively, of ^{99m}Tc-sulesomab as 90%, 84.6%, and 87.9% compared with 83.9%, 76.5%, and 81.3% for autologous leukocytes, whereas Hakki et al.[44] reported 93%, 89%, and 90% for ^{99m}Tc-sulesomab and 85%, 75%, and 79% for autologous leukocytes. Thus, ^{99m}Tc-sulesomab appears to be a better agent than ^{111}In-oxine leukocytes, requires less preparation time, and does not raise the safety concerns of labeling autologous leukocytes.

There have been no reports of positive HAMA response with a single dose of ^{99m}Tc-sulesomab. The critical organ is the kidney, with a radiation absorbed dose of 4.15 rad/25 mCi (925 MBq). The effective dose equivalent is 0.95 rem.

This product is not approved for routine use in the United States; however, it has undergone clinical investigation and has demonstrated effectiveness for diagnosing acute appendicitis in children.[45] LeukoScan is approved in Europe for the diagnosis of infection.

Antibodies under Investigation

The use of radiolabeled antibodies for RID and RIT is being explored in many clinical investigations. A list describing these studies can be accessed at http://clinicaltrial.gov/ct2/results?term=Radiolabeled+monoclonal+antibodies. A concise overview of the early development of radiolabeled antibodies has been written by Goldenberg.[46]

RADIOLABELED PEPTIDES

In the realm of molecular imaging, radiolabeled peptides have drawn great interest because of the diverse role that peptide hormones play in normal bodily functions and disease-related conditions. The discovery that neuropeptides are overexpressed on tumors stimulated interest in the development of radiolabeled peptides that have potential for diagnostic imaging and radiation therapy of cancer. Peptides are advantageous because they can be synthesized to meet specific receptor requirements and coordinated with a diagnostic or therapeutic radionuclide. Radiolabeled peptide tracers may have several components, and their synthesis must retain the integrity of the receptor-binding region. Possible constructs are (1) peptide–radionuclide; (2) peptide–BFCA–radionuclide; and (3) peptide–spacer–BFCA–radionuclide. In the first construct a radionuclide-binding region is incorporated within the peptide structure but away from the receptor-binding region. In the second construct a site on the peptide distant from the receptor-binding region is covalently linked to a BFCA, which coordinates the radionuclide. The third construct is similar to the second, except that a spacer molecule is inserted in the structure to distance the receptor-binding region from the radionuclide-binding region. Different BFCAs have been used to produce stable chelates with radionuclide metals, most notably DTPA, but newer agents include DOTA; 2,2′,2″-(1,4,7-triazacyclononane-1,4,7-triyl)triacetic acid (NOTA); triethylenetetramine (TETA); and 4,11-bis(carboxymethyl)-1,4,8,11-tetraazabicyclo[6.6.2]hexadecane (CB-TE2A) (Figure 28-18).[2] The BFCA is activated with a functional group to facilitate conjugation to the peptide, typically via an amide, thioether, or thiourea bond. The radionuclide label is incorporated into the construct by either the prelabeling method or postlabeling method. See Figures 10-13 and 10-14, Chapter 10 for examples of conjugation and labeling methods. The DOTA, NOTA, and TETA ligands have been shown to form stable chelates with ^{68}Ga and ^{111}In, but copper chelates (^{64}Cu)

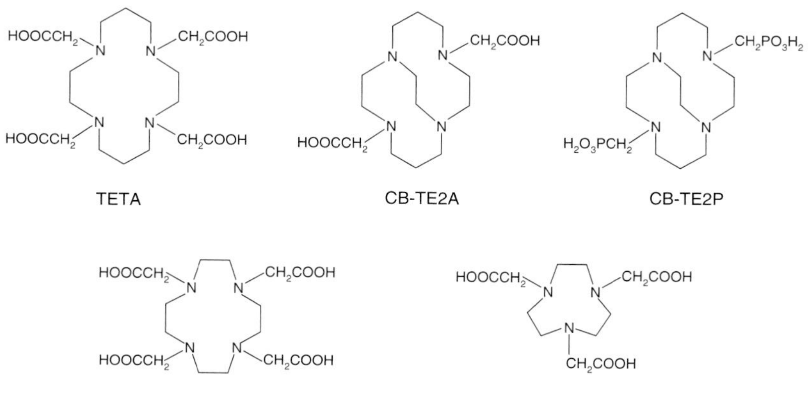

FIGURE 28-18 Chemical structures of triethylenetetramine (TETA); 4,11-bis(carboxymethyl)-1,4,8,11-tetraazabicyclo[6.6.2]hexadecane (CB-TE2A); 4,11-bis(phosphonatomethyl)-1,4,8,11-tetraazabicyclo[6.6.2]hexadecane (CB-TE2P); 1,4,7,10-tetraazacyclodecane-1,4,7,10-tetraacetic acid (DOTA); and 2,2',2''-(1,4,7-triazacyclononane-1,4,7-triyl)triacetic acid (NOTA).

with these ligands are kinetically unstable. However, the development of a ^{64}Cu somatostatin analogue, tyrosine-3-octreotate (Y3-TATE), demonstrated that a cross-bridged macrocyclic chelator, CB-TE2A, formed more stable chelates with copper than did TETA, as evidenced by higher tumor uptake and lower blood levels over time.[47] A disadvantage of the CB-TE2A ligand is its requirement for high-temperature labeling. However, a bisphosphonate analogue of CB-TE2A (CB-TE2P) permits room temperature labeling conditions and improved biodistribution.[48]

The important properties of radiolabeled peptides are favorable pharmacokinetics and high tumor affinity with a radiolabel that is kinetically inert, which is particularly important with therapeutic radionuclides that may target bone if the peptide is metabolized in vivo.[49,50] One disadvantage of peptides for cancer therapy is their rapid clearance from the bloodstream because of their small size, which reduces the radiation dose delivered to a tumor. A technique for extending peptide circulation time to increase tumor uptake involves linking polyethylene glycol to the peptide.[51]

Receptors that have attracted interest for radiolabeled peptide development include the somatostatin receptor, the arginine–glycine–aspartic acid (RGD) receptors (integrin $\alpha_v\beta_3$ and glycoprotein [GP] IIb/IIIa), the gastrin-releasing peptide receptor, and the melanocyte-stimulating hormone receptor.[2] Radiolabeled peptide analogues that target these receptors have been developed. Most studies have been in animal and tumor cell models.

Somatostatin Analogues

The development of radiolabeled peptides gained interest with the discovery that certain neuroendocrine tumors overexpressed the membrane-associated somatostatin receptor.[52,53] Somatostatin receptors (SSTRs) have been demonstrated on a variety of human tumors and exist in different subtypes (SSTR$_1$–SSTR$_5$), which differ in their interaction with somatostatin. Subtype 2 (SSTR$_2$) is most often expressed on the membrane surface of various tissues and tumors.[54] Somatostatin is a cyclopeptide with 14 or 28 amino acids that modulates neuroendocrine functions in the body. Octreotide is an octapeptide analogue of somatostatin that has been modified to make it more stable to enzymatic metabolism than natural somatostatin.

Various radiolabeled analogues of octreotide have been created to target SSTR$_2$ for tumor imaging, the most notable example being an indium-labeled DTPA-conjugate of octreotide, ^{111}In-pentetreotide.[55,56] This radiopharmaceutical is approved for routine use and is discussed in more detail in Chapter 26. Several octreotide analogues for PET imaging labeled with ^{68}Ga and ^{64}Cu have been investigated in animals and the findings summarized.[2,57] A kit-prepared ^{99m}Tc-labeled peptide to target somatostatin receptors, ^{99m}Tc-depreotide (NeoTect [Diatide]), was developed for imaging lung tumors. This peptide contained a somatostatin receptor-binding region and a ^{99m}Tc-binding region (Figure 28-19). This agent was approved for routine use but is no longer on the market.

RGD Analogues

When solid tumors begin to form they have no blood supply and cannot grow beyond a 2 to 3 mm size without a supply of oxygen and nutrients.[58] However, tumor cells induce angiogenesis, causing upregulation of various factors, such as vascular endothelial growth factor, integrins, and matrix metalloproteinases, to promote blood vessel formation.[59] The increased expression of these substances in tumor vasculature makes them ideal targets for radiotracers that can detect early biologic changes in tumors before advanced structural changes occur.

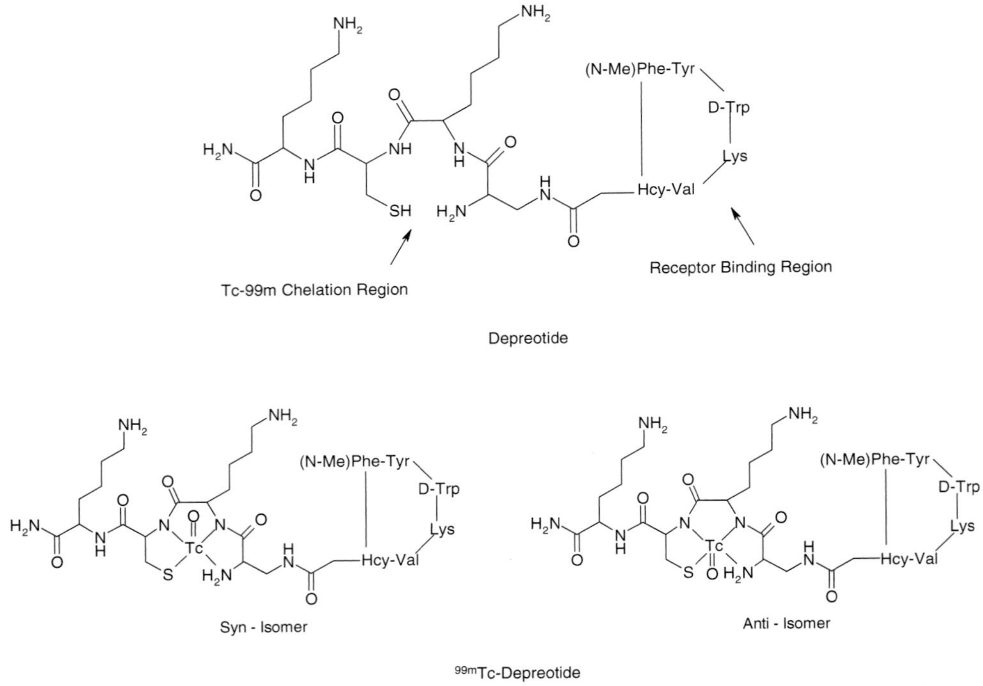

FIGURE 28-19 Chemical structures of the depreotide peptide and the *syn* and *anti* isomers of ^{99m}Tc-depreotide.

Integrins are a diverse family of transmembrane glycoproteins consisting of two noncovalently associated receptor subunits, α and β. These subunits form heterodimeric receptors on activated endothelial cells, facilitating their adhesion to the extracellular matrix (ECM) in the formation of new blood vessels.[60] Integrins facilitate a wide range of cell–cell and cell–ECM interactions.[61] A principal integrin expressed on activated endothelial cells is the vitronectin receptor, $α_vβ_3$, which binds to RGD-containing components of the ECM. Integrin $α_vβ_3$ is absent on quiescent blood vessels but is highly expressed on activated endothelial cells and modulates cell migration and survival during angiogenesis.[62]

Because integrins express an RGD binding site, radiolabeled RGD-containing peptides have been developed to investigate tumors via this mechanism; however, only a few have been tested in humans.[61] In SPECT imaging, ^{99m}Tc-NC100692 has been evaluated for imaging breast cancer. This agent was able to detect 19 of 22 malignant lesions (86%) no smaller than 7 mm; however, this success rate was noted by the investigators to be no better than that reported for ^{99m}Tc-sestamibi or ^{99m}Tc-tetrofosmin in breast imaging.[63] A series of ^{18}F-labeled RGD peptides have been investigated in humans for detection of cancer. The sugar-derivatized monomeric peptide ^{18}F-galacto-RDG demonstrated visualization of $α_vβ_3$ expression in melanoma and musculoskeletal sarcomas with high contrast.[64] Studies comparing ^{18}F-FDG with ^{18}F-galacto-RDG in patients with primary and metastatic cancer showed that although ^{18}F-FDG was more sensitive for tumor staging, the two agents do not correlate closely in malignant lesions, suggesting that these agents provide different information.[65] As a monomeric RGD peptide tracer, ^{18}F-galacto-RDG had relatively low integrin-binding affinity and only moderate standardized uptake values.[66] On the premise that integrin $α_vβ_3$ and RGD peptide interaction involves multivalent binding with clustering of integrins, a series of multimeric RGD peptides were developed to enhance tumor uptake over that achieved with the monomeric tracer. Monomeric dimeric, tetrameric, and octameric ^{18}F-labeled RGD peptides were compared to assess tumor uptake. Initial studies in animals demonstrated that multimeric RDG peptides significantly increase uptake in tumors.[61] Other PET analogues of RDG have been investigated, including ^{64}Cu conjugated with CB-TE2A linked to RGD and ^{68}Ga conjugated with DOTA-linked RGD. These conjugates and others have been reviewed.[57]

An RGD peptide that targets the GP IIb/IIIa receptor on activated platelets is ^{99m}Tc-apcitide (AcuTect [Diatide]). It was developed for detection of acute venous thrombosis (see Figures 10-17 and 10-18, Chapter 10). This peptide chelate prepared from a kit has a receptor-binding region (ApcGD) that mimics RDG. The synthetic amino acid Apc (S-aminopropyl-l-lysine) replaces arginine in the receptor-binding sequence and confers additional selectivity for the platelet receptor.[67] This radiopharmaceutical received approval for routine use but is no longer on the market.

Gastrin-Releasing Peptide Analogues

Gastrin is a peptide hormone released by G cells in the stomach, duodenum, and pancreas in response to physiologic factors and activation of the gastrin-releasing peptide receptor

(GRPr) by gastrin-releasing peptide (GRP), a 27 amino acid peptide. A principal function of gastrin is to stimulate release of hydrochloric acid by parietal cells in the stomach. Inhibition of gastrin release is modulated, in part, by the action of somatostatin on G cells. Gastrin-releasing peptide regulates numerous functions of the gastrointestinal and central nervous systems, including release of gastrointestinal hormones, smooth muscle cell contraction, and epithelial cell proliferation and is a potent mitogen for neoplastic tissues.

The GRPr is overexpressed on a variety of tumor types, including gastrointestinal, prostate, breast, and pancreas tumors and small cell lung cancer (SCLC), making it an attractive target for radiolabeled peptides with an affinity for the receptor.[68] Bombesin is an amphibian homologue of GRP with high affinity for the GRPr. Consequently, a variety of radiolabeled analogues of bombesin have been prepared and investigated for targeting tumors.[69] Both SPECT and PET bombesin analogues linked with HYNIC, DTPA, and DOTA have been labeled with ^{111}In, ^{64}Cu, ^{68}Ga, ^{99m}Tc, and ^{177}Lu, and studies with them have been reviewed.[2,69] Investigations have been limited to animal and in vitro studies.

Melanocyte-Stimulating Hormone Receptor Analogues

Of the group of melanocyte-stimulating hormones in the body, alpha melanocyte-stimulating hormone (αMSH) is most important in stimulating the production and release of melanin from melanocytes in the skin and hair. αMSH is a 13 amino acid peptide hormone produced by the pituitary gland. The αMSH receptor, known as melanocortin 1 receptor (MC1R), is primarily located on the surface of melanocytes. Its production is encoded by the MCR1 gene. MC1R has been found to be present in melanomas and has stimulated the development of radiolabeled peptide analogues to target this receptor.[70] Preliminary investigations with several DOTA-conjugated MSH analogues labeled with ^{111}In, ^{64}Cu, and ^{86}Y have been conducted in animal tumor models and summarized.[2,57]

REFERENCES

1. Sharkey RM, Goldenberg DM. Perspectives on cancer therapy with radiolabeled monoclonal antibodies. *J Nucl Med.* 2005;46:115S–127S.
2. Lee S, Xie J, Chen X. Peptides and peptide hormones for molecular imaging and disease diagnosis. *Chem Rev.* 2010;110:3087–111.
3. McCullough KC, Spier RE. *Monoclonal Antibodies in Biotechnology: Theoretical and Practical Aspects.* London: Cambridge University Press; 1990.
4. Steward MW. *Antibodies: Their Structure and Function.* New York: Chapman and Hall; 1984.
5. vanOss CJ, vanRegenmortal MHV, eds. *Immunochemistry.* New York: Marcel Dekker; 1994:12,31.
6. Lyster DM, Alcorn LN. Antibody imaging and therapy in human cancer. In: Fritzberg AR, ed. *Radiopharmaceuticals: Progress and Clinical Perspectives.* Vol 1. Boca Raton, FL: CRC Press;1986:41–59.
7. Kohler G, Milstein C. Continuous culture of fused cells secreting antibody of predefined specificity. *Nature.* 1975;256:495–7.
8. Milstein C. Monoclonal antibodies. *Sci Am.* 1980;243:66–74.
9. Keenan AM, Harbert JC, Larson SM. Monoclonal antibodies in nuclear medicine. *J Nucl Med.* 1985;26:531–7.
10. Goldenberg DM. Targeted therapy of cancer with radiolabeled antibodies. *J Nucl Med.* 2002;43:693–713.
11. Mayforth RD, Quintans J. Designer and catalytic antibodies. *N Engl J Med.* 1990;323:173–8.
12. Juweid ME. Radioimmunotherapy of B-cell non-Hodgkin's lymphoma: from clinical trial to clinical practice. *J Nucl Med.* 2002;43:1507–29.
13. Janeway CA Jr, Travers P. *Immunobiology.* 3rd ed. New York: Garland Publishing; 1997:2:19–2:21.
14. Lindmo T, Boven E, Cuttitta F, et al. Determination of the immunoreactive fraction of radiolabeled monoclonal antibodies by linear extrapolation to binding at infinite antigen excess. *J Immunol Meth.* 1984;72:77–89.
15. Augustine SC, Norenberg JP. The treatment of low grade B-cell non-Hodgkin's lymphomas with radiopharmaceuticals. *Correspondence Continuing Education Courses for Nuclear Pharmacists and Nuclear Medicine Professionals.* Albuquerque, NM: University of New Mexico Health Sciences Center College of Pharmacy; May 2002.
16. USAN Council. List no. 351, monoclonal antibodies. *Clin Pharmacol Ther.* 1993;54:114–6.
17. Wilbur DS. Radiohalogenation of proteins: an overview of radionuclides, labeling methods, and reagents for conjugate labeling. *Bioconjug Chem.* 1992;3:433–70.
18. Fritzberg AR, Meares CF. Metallic radionuclides for radioimmunotherapy. In: Abrams PG, Fritzberg AR, eds. *Radioimmunotherapy of Cancer.* New York: Marcel Dekker; 2000:57–79.
19. Zalutsky MR. Radiohalogens for radioimmunotherapy. In: Abrams PG, Fritzberg AR, eds. *Radioimmunotherapy of Cancer.* New York: Marcel Dekker; 2000:81–106.
20. Griffiths GL. Radiochemistry of therapeutic radionuclides. In: Goldenberg DM, ed. *Cancer Therapy with Radiolabeled Antibodies.* Boca Raton, FL: CRC Press; 1995:47–61.
21. Gansow OA, Wu C. Advanced methods for radiolabeling monoclonal antibodies with therapeutic radionuclides. In: Goldenberg DM, ed. *Cancer Therapy with Radiolabeled Antibodies.* Boca Raton, FL: CRC Press; 1995:63–76.
22. Wilbur DS, Hadley SW, Grant LM, et al. Radioiodinated iodobenzyl conjugates of a monoclonal antibody Fab fragment: in vivo comparisons with chloramine-T labeled Fab. *Bioconjug Chem.* 1991;2:111–6.
23. Vaidyanathan G, Zalutsky MR. Protein radiohalogenation: observations on the design of N-succinimidyl ester acylation agents. *Bioconjug Chem.* 1990;1:269–73.
24. Zalutsky MR, Noska MA, Colapinto EV, et al. Enhanced tumor localization and in vivo stability of a monoclonal antibody radioiodinated using N-succinimidyl 3-(tri-n-butylstannyl) benzoate. *Cancer Res.* 1989;49:5543–9.
25. Schuster JM, Garg PK, Bigner DD, et al. Improved therapeutic efficacy of a monoclonal antibody radioiodinated using N-succinimidyl 3-(tri-n-butylstannyl) benzoate. *Cancer Res.* 1991;51:4164–9.
26. Eckelman WC, Paik CH, Steigman J. Three approaches to radiolabeling antibodies with ^{99m}Tc. *Nucl Med Biol.* 1989;16:171–6.
27. Westerberg DA, Carney PL, Rogers PE, et al. Synthesis of novel bifunctional chelators and their use in preparing monoclonal antibody conjugates for tumor targeting. *J Med Chem.* 1989;32:236–43.
28. Meares CF, McCall MJ, Reardan DT, et al. Conjugation of antibodies with bifunctional chelating agents: isothiocyanate and bromoacetamide reagents, methods of analysis, and subsequent addition of metal ions. *Anal Biochem.* 1984;142:68–78.
29. Meares CF. Chelating agents for the binding of metal ions to antibodies. *Nucl Med Biol.* 1986;13:311–8.
30. Brechbiel MW, Gansow OA, Atcher RW, et al. Synthesis of 1-(p-isothiocyanatobenzyl) derivatives of DTPA and EDTA: antibody labeling and tumor imaging studies. *Inorg Chem.* 1986;25:2272–7.
31. Cummins CH, Rutter EW Jr, Fordyce WA. A convenient synthesis of bifunctional chelating agents based on diethylenetriaminepentaacetic acid and their coordination chemistry with yttrium (III). *Bioconjug Chem.* 1991;2:180–6.

32. Chinn PC, Leonard JE, Rosenberg J, et al. Preclinical evaluation of ^{90}Y-labeled anti-CD 20 monoclonal antibody for treatment of non-Hodgkin's lymphoma. *Int J Oncol*. 1999;15:1017–25.
33. Hnatowich DJ, Mardirossian G, Rusckowski M, et al. Directly and indirectly technetium-99m-labeled antibodies: a comparison of in vitro and animal in vivo properties. *J Nucl Med*. 1993;34:109–19.
34. Eckelman WC, Paik CH, Steigman J. Three approaches to radiolabeling antibodies with ^{99m}Tc. *Nucl Med Biol*. 1989;16:171–6.
35. Fritzberg AR, Abrahams PG, Beaumier PL, et al. Specific and stable labeling of antibodies with technetium-99m with a diamide dithiolate chelating agent. *Proc Natl Acad Sci USA*. 1989;85:4025–9.
36. Washburn L, Sun T, Lee Y-CC, et al. Comparison of five bifunctional chelate techniques for ^{90}Y-labeled monoclonal antibody CO17-1A. *Nucl Med Biol*. 1991;18:313–21.
37. Harrison A, Walker C, Parker D, et al. The in vivo release of ^{90}Y from cyclic and acyclic ligand-antibody conjugates. *Nucl Med Biol*. 1991;18:469–76.
38. Maecke HR, Riesen A, Ritter W. The molecular structure of indium-DTPA. *J Nucl Med*. 1989;30:1235–9.
39. Rodwell JD, Alvarez VL, Lee C, et al. Site-specific covalent modification of monoclonal antibodies: in vitro and in vivo evaluations. *Proc Natl Acad Sci USA*. 1986;83:2632–6.
40. Prostascint [package insert]. Princeton, NJ: Cytogen Corporation; 1997.
41. Kelty NL, Holder LE, Khan SH. Dual-isotope protocol for indium-111 capromab pendetide monoclonal antibody imaging. *J Nucl Med Technol*. 1998;26:174–7.
42. Hansen HJ, Goldenberg DM, Newman ES, et al. Characterization of second-generation monoclonal antibodies against carcinoembryonic antigen. *Cancer*. 1993;71:3478–85.
43. Becker W, Palestro CJ, Winship J, et al. Rapid imaging of infections with a monoclonal antibody fragment (LeukoScan). *Clin Orthoped*. 1996;329:263–72.
44. Hakki S, Harwood SJ, Morrissey MA, et al. Comparative study of monoclonal antibody scan in diagnosing orthopaedic infection. *Clin Orthoped*. 1997;335:275–85.
45. Passalaqua AM, Klein RL, Wegener WA, et al. Diagnosing suspected acute nonclassic appendicitis with sulesomab, a radiolabeled antigranulocyte antibody imaging agent. *J Pediatr Surg*. 2004;39:1338–44.
46. Goldenberg MD. Radiolabeled antibodies. *Sci Amer*. 1994;1(Mar/Apr):64–73.
47. Sprague JE, Peng Y, Sun X, et al. Preparation and biological evaluation of copper-64–labeled Tyr^3-octreotate using a cross-fridged macrocyclic chelator *Clin Cancer Res*. 2004;10:8674–82.
48. Stigers DJ, Ferdani R, Weisman GR, et al. A new phosphonate pendant-armed cross-bridged tetraamine chelator accelerates copper (II) binding for radiopharmaceutical applications. *Dalton Trans*. 2010;39:1699–1701.
49. Anderson CJ, Welch MJ. Radiometal-labeled agents (non-technetium) for diagnostic imaging. *Chem Rev*. 1999;99:2219–34.
50. Giblin MF, Veerendra B, Smith CJ. Radiometallation of receptor-specific peptides for diagnosis and treatment of human cancer. *In Vivo*. 2005;19:9–30.
51. Rogers BE, Manna DD, Safavy A. In vitro and in vivo evaluation of a ^{64}Cu-labeled polyethylene glycol-bombesin conjugate. *Cancer Biother Radiopharm*. 2004;19:25–34.
52. Reubi JC. Neuropeptide receptors in health and disease: the molecular basis for *in vivo* imaging. *J Nucl Med*. 1995;36:1825–35.
53. Bakker WH, Bruns AC, Breeman WAP, et al. [^{111}In-DTPA-D-Phe^1]-octreotide, a potential radiopharmaceutical for imaging of somatostatin receptor-positive tumors: synthesis, radiolabeling and in vitro validation. *Life Sci*. 1991;49:1583–91.
54. Bohuslavizki KH. Somatostatin receptor imaging: current status and future perspectives. *J Nucl Med*. 2001;42:1057–8.
55. Krenning EP, Kwekkeboom DJ, Bakker WH, et al. Somatostatin receptor scintigraphy with [^{111}In-DTPA-D-Phe^1]- and [^{123}I-Tyr^3]-octreotide: the Rotterdam experience with more than 1000 patients. *Eur J Nucl Med*. 1993;20:716–31.
56. Krenning EP, Kwekkeboom DJ, Pauwels S, et al. Somatostatin receptor scintigraphy. In: Freeman LM, ed. *Nuclear Medicine Annual*. New York: Raven Press; 1995:1–50.
57. Wadas T, Wong E, Weisman G, et al. Coordinating radiometals of copper, gallium, indium, yttrium, and zirconium for PET and SPECT imaging of disease. *Chem Rev*. 2010;110:2858–902.
58. Folkman J. Tumor angiogenesis: therapeutic implications. *N Engl J Med*. 1971;285:1182–6.
59. Rundhaug JE. Matrix metalloproteinases, angiogenesis, and cancer. *Clin Cancer Res*. 2003;9:551–4.
60. Hood JD, Cheresh DA. Role of integrins in cell invasion and migration. *Nat Rev Cancer*. 2002;2:91–100.
61. Niu G, Chen X. PET imaging of angiogenesis. In: Mach RH, ed. *New PET Radiotracers*. Philadelphia: Saunders; 2009:17–38.
62. Brooks PC, Clark RA, Cheresh DA. Requirements of vascular integrin alpha v beta 3 for angiogenesis. *Science*. 1994;264:569–71.
63. Bach-Gansmo T, Danielsson R, Saracco A, et al. Integrin receptor imaging of breast cancer: a proof-of-concept study to evaluate ^{99m}Tc-NC100692. *J Nucl Med*. 2006;47:1434–9.
64. Beer AJ, Haubner R, Goebel M, et al. Biodistribution and pharmacokinetics of the αvβ3-selective tracer 18F-galacto-RGD in cancer patients. *J Nucl Med*. 2005;46:1333–41.
65. Beer AJ, Lorenzen S, Metz S, et al. Comparison of integrin $α_vβ_3$ expression and glucose metabolism in primary and metastatic lesions in cancer patients: a PET study using ^{18}F-galacto-RGD and ^{18}F-FDG. *J Nucl Med*. 2008;49:22–9.
66. Zhang X, Xiong Z, Wu Y, et al. Quantitative PET imaging of tumor integrin $α_vβ_3$ expression with ^{18}F-FRGD2. *J Nucl Med*. 2006;47:113–21.
67. Taillefer R, Edell S, Innes G, et al. Acute thromboscintigraphy with Tc-99m-apcitide: results of the phase-3 multicenter clinical trial comparing Tc-99m apcitide scintigraphy with contrast venography for imaging acute DVT. *J Nucl Med*. 2000;41:1214–23.
68. Reubi JC. Peptide receptors as molecular targets for cancer diagnosis and therapy. *Endocr Rev*. 2003;24:389–427.
69. Smith CJ, Volkert WA, Hoffman TJ. Radiolabeled peptide conjugates for targeting of the bombesin receptor superfamily subtypes. *Nuc Med Biol*. 2005;32:733–40.
70. Tatro JB, Atkins M, Mier JW. Melanotropin receptors demonstrated in situ in human melanoma. *J Clin Invest*. 1990;85:1825–32.

CHAPTER 29
Therapeutic Applications of Radioactive Agents

A variety of radioactive agents and methods have been developed for radiotherapy in nuclear medicine. Approved clinical applications include radioimmunotherapy (RIT) of non-Hodgkin's lymphoma (NHL) with radiolabeled antibodies, treatment of hyperthyroidism and thyroid cancer with radioiodine, pain palliation of metastatic bone cancer with beta-emitting bone-localizing agents, treatment of inflamed joints (radiation synovectomy) and malignant effusions with in situ administration of radiolabeled colloids, treatment of polycythemia vera with radioactive phosphorus, and selective internal radiation therapy of liver cancer with radiolabeled microspheres. This chapter covers all these applications except thyroid therapy, which was discussed in Chapter 20.

RADIOIMMUNOTHERAPY

Advances in immunologic technology led to the development of monoclonal antibodies, which are now being used to destroy malignant tissue. The rationale for antibody therapy of cancer is based on the natural mechanisms of antibody immune effector functions, namely, lysis of tumor cells by complement-mediated cytotoxicity (CMC) and antibody-dependent cell-mediated cytotoxicity (ADCC) or induction of apoptosis (programmed cell death).

Initial clinical experience using unmodified antibodies to treat cancer was disappointing, but specific problems were identified and new approaches in antibody modification and use were developed. Significant problems included difficulties in antibody delivery to tumors and penetration into bulky tumors, the inability of murine antibodies to adequately exert cytotoxic effects through human immune effector functions, the development of immunogenic responses to mouse antibody (human antimouse antibody, or HAMA), and the development of antigen-negative tumor cells that render the tumor refractory to further treatment.[1]

Genetic engineering technology led to the creation of chimeric and humanized antibodies; this improved therapy significantly, because murine antibodies interacted poorly with complement and human effector cells in vivo. A good example of successful genetic engineering is the development of the chimeric monoclonal antibody rituximab (Rituxan [Genentech]), which targets the CD20 antigenic marker on malignant B-lymphocytes. This antibody contains murine variable regions and human constant regions. Studies have demonstrated this antibody's ability to lyse B cells by CMC and ADCC. In addition, its chimeric structure significantly reduces its immunogenicity compared with its murine parent.[2] Rituximab has been shown to be effective in the treatment of non-Hodgkin's lymphoma (NHL).[3,4] It produces objective tumor responses in approximately 50% of patients with relapsed or refractory low-grade or follicular NHL. It was the first antibody approved by the Food and Drug Administration (FDA) for treating cancer.

The desire for more effective therapy with antibodies led to the development of radiolabeled monoclonal antibodies. Studies indicated that NHL can be treated effectively by RIT for several reasons, including the inherent radiosensitivity of lymphocytes, the vascular accessibility of these malignancies, and the large number of target antigens on the surface of lymphocytes. For example, the B-cell antigens CD19, CD20, and CD22 are all expressed in high density on NHL cells. Approximately 99% of B-cell lymphomas express CD19, 95% express CD20, and 70% express CD22.[1]

RIT has the added advantage of causing toxic damage to tumor cells by radiolytic effects as well as by CMC and ADCC functions or apoptosis. Cytotoxic effects can occur even if a patient's immune system is depressed because of chemotherapy. Another advantage of RIT is the "crossfire" effect, whereby high-energy beta particles emitted from antibodies can reach tumor cells not interacting directly with antibody (Figure 29-1). This is especially important with poorly accessible bulky tumors or those with antigen-negative nontargeted cells.

Antibodies, particularly for treating NHL, have been labeled either directly with radioiodine (^{131}I) or with radiometal conjugates containing radionuclides such as ^{90}Y, ^{67}Cu, and ^{186}Re. The radiolabeled antibodies that have shown effectiveness and received approval for routine treatment of NHL are ^{131}I-tositumomab (Bexxar [Corixa and GlaxoSmithKline])

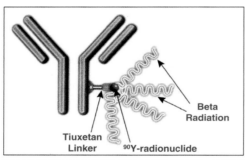

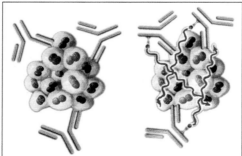

FIGURE 29-1 Upper panel illustrates ibritumomab labeled with ^{90}Y-yttrium via a tiuxetan linker molecule. Lower panel illustrates unlabeled antibody bound to tumor cells (naked antibody) and radiolabeled antibody bound to tumor cells, demonstrating the "crossfire" effect of beta radiation penetrating into tumor cells not bound by antibody. (Reprinted with permission of the Society of Nuclear Medicine from Fink-Bennett DM, Thomas K. ^{90}Y-ibritumomab tiuxetan in the treatment of relapsed or refractory B-cell non-Hodgkin's lymphoma. *J Nucl Med Technol.* 2003; 31:61–8.)

and ^{90}Y-ibritumomab tiuxetan (Zevalin [Cell Therapeutics and Spectrum Pharmaceuticals]), which target the CD20 antigen on B-lymphocytes. Investigational studies for NHL continue to be carried out for ^{131}I- and ^{67}Cu-labeled LYM-1, which target the human lymphocyte antigen (HLA-DR); and ^{131}I-, ^{186}Re-, and ^{90}Y-labeled hLL2IgG (epratuzumab), which target the CD22 antigen. Several other antibodies have been studied in clinical trials for radiation treatment of other types of cancer, as listed at http://clinicaltrial.gov/ct2/results?term=Radiolabeled+monoclonal+antibodies. Among these are ^{90}Y-hPAM4 for pancreatic cancer; ^{90}Y-Hu3S193, ^{90}Y-CC49, and ^{90}Y-MN-14 for ovarian cancer; ^{90}Y-B3, ^{90}Y-BrE3, ^{90}Y-m170, and ^{90}Y-MN-14 for breast cancer; ^{131}I-CC49-deltaCH2 for colorectal cancer; ^{131}I-BC8 for leukemia; ^{131}I-81C6, ^{131}I-TNT-1/B, and ^{131}I-8H9 for brain and central nervous system cancers; ^{131}I-A33 for colorectal cancer; ^{177}Lu-J591 for prostate cancer; and ^{177}Lu-CYT-500 for metastatic prostate cancer. Several other antibodies will likely be added to this list over time.

Radionuclide and Antibody Considerations

The objective of RIT is to deliver a lethal radiation dose to tumor cells. This is a function of the antibody's pharmacokinetic and immunologic properties (its blood clearance, tumor penetration, and tumor cell affinity) and the radionuclide's radiation properties (half-life, types of emission, and amount of radiation dose absorbed in critical centers within the tumor cell). In general, it takes 2 to 3 days for antibodies to penetrate solid tumors; therefore, a blood clearance shorter than this time would reduce tumor uptake.[5] Chimeric and humanized antibodies offer the advantage of a reduced HAMA response, which increases antibody circulation time because the HAMA response produces antibody–HAMA complexes that are quickly removed from the circulation by the reticuloendothelial system.[6]

The half-life of the antibody radiolabel is important in two respects. First, it must be long enough that most of the radionuclide has not decayed before peak uptake of antibody occurs in tumor. Second, it must be long enough to match the residence time of the antibody bound to tumor. A half-life of 3 to 8 days is considered optimal.[5] Whole antibodies and F(ab')$_2$ fragments have the potential for bivalent binding to antigens and therefore have a higher binding avidity than Fab fragments, particularly in tumor cells with high antigen density. Once antibodies bind to the surface of a tumor cell, they may dissociate from the cell or be internalized and catabolized. It generally takes 2 to 3 days for antibody bound to tumor cell surface to be gradually catabolized. Antibodies that are rapidly internalized after interaction with surface antigen are catabolized more quickly. These antibodies are more effective if the radionuclide label remains fixed within the tumor cell after lysosomal catabolism. Retention is more likely to occur with radiometal labels and less likely with conventionally iodinated antibodies.

Radionuclide decay properties are important and should match the intended application and target site within the tumor cell. In general, beta particles have the most useful application in RIT because of their variable effective ranges, which are related to their beta particle energy. Long-range high-energy beta particles are advantageous in tumors with inhomogeneous antibody uptake because cells not reached by antibody can be irradiated. For RIT, the radionuclide properties should be compatible with the size and location of the tumor. In general, large solid tumors and those with surface receptors that do not internalize antibody (e.g., CD20 antigens) require more energetic, penetrating radiation, while metastatic lesions, composed of only a few hundred cells, can be treated with short-range radiations. Tumors that internalize antibodies (e.g., CD22 antigens) can be more effectively killed by radionuclides that emit short-range radiation, such as Auger electrons (e.g., ^{125}I) and alpha particles (e.g., ^{211}At and ^{212}Bi). Low-energy Auger electrons have high linear energy transfer and possess the advantage of depositing their energy close to the cell nucleus. They also have the potential of sparing normal tissue surrounding the tumor. Larger tumors and those that bind antibodies on surface receptors are more effectively treated with more energetic beta emitters, such as ^{90}Y, ^{131}I, ^{186}Re, and ^{188}Re. Thus, ^{90}Y would be more effective in large, bulky tumors. A cluster of tumor cells may range from several thousand cells (~0.1 mm) to 10^5 cells (~1 mm). Thus, for 0.2 mm and 1.0 mm diameter tumors, the fraction of particle energy absorbed in each would be 17% and 54%, respectively, for a medium-energy beta emitter such as ^{131}I (β_{max} 0.61 MeV) but only 1.5% and 10%, respectively, for a high-energy beta emitter such as ^{90}Y (β_{max} 2.28 MeV) and 50% and 90%, respectively, for an alpha emitter such as ^{111}At.[7] Humm[7] has classified beta-emitting radionuclides as low-range (mean range, <200 μm), medium-range (mean range, 200 μm to <1 mm), and high range (mean range, >1 mm). See Chapter 28, Table 28-4 for a list of radionuclides for therapeutic antibodies.

Alpha particle emitters (^{211}At) and Auger electron emitters (^{125}I) are particularly effective over a short range. However, to be effective, antibodies labeled with these nuclides must reach all tumor cells and deposit nuclide close to sensitive targets within the cell. In view of these requirements, investigations with ^{211}At-labeled monoclonal antibodies have been directed toward malignancies such as neoplastic meningitis that have free-floating cells that are readily accessible and have a geometry more compatible with short-range alpha emitters.[8]

Advantages of ^{131}I are its availability and relatively low cost, its ease of labeling via chloramine-T or Iodo-Gen (Pierce), its emission of beta particles for therapy, and its emission of gamma rays for imaging, which is useful for making dosimetry estimates. Its principal disadvantage is the release of iodotyrosine and loss of label from the tumor during antibody catabolism. This results in reduced radiation dose to the tumor and causes a thyroidal radiation burden from the radioiodide released after deiodination of iodotyrosine. Radionuclide metal conjugates such as ^{90}Y are retained longer by tumor cells, because the radiolabel is residualized in the cell after lysosomal catabolism. This enhances tumor absorbed dose. Although ^{90}Y does not emit gamma radiation, antibody conjugates can be labeled with a ^{111}In surrogate for dosimetry purposes. This has worked well with ^{90}Y-ibritumomab tiuxetan.

Radioimmunotherapy Methods

The goal of any radiation therapy procedure is to optimize the therapeutic index, that is, to maximize the radiation dose to the target tissue and minimize dose to normal tissue. To accomplish this, radiation dosimetry estimates must be made. Issues involved in making dosimetry estimates have been reviewed by DeNardo et al.[9] Dosimetry studies are usually conducted in Phase 1 and Phase 2 investigations. Patients who are to be treated with a therapeutic dose of radiolabeled antibody are given a tracer dose of the same antibody, and the radiation absorbed dose to various organs is calculated by standard pharmacokinetic and Medical Internal Radiation Dose (MIRD) techniques. The cumulated activity ($\tilde{A}$) and the volume or mass of tumor and normal tissue needed for the "S" value calculation are the essential pieces of information required for dose estimates. Gamma-camera imaging provides a measure of tissue activity over time, while computed tomography (CT) or magnetic resonance imaging (MRI) is used to define tumor volumes. If the therapeutic radionuclide is a pure beta emitter, a photon-emitting surrogate with similar pharmacokinetic properties and physical half-life is labeled to the antibody for dosimetry estimates. This approach works well with ^{90}Y- and ^{111}In-labeled ibritumomab tiuxetan, which have similar biologic properties and decay half-lives. ^{99m}Tc is the gamma-emitting surrogate for ^{186}Re-labeled antibodies. A half-life match between the surrogate tracer nuclide and the therapy nuclide would be ideal, because a surrogate tracer having a shorter half-life than the therapy nuclide might miss important late pharmacokinetic data, resulting in unreliable time–activity curves and dosimetry estimates.

Two different approaches have been used to deliver a cytotoxic dose of radiation to tumors: (1) the nonmyeloablative or low-dose approach and (2) the myeloablative or high-dose approach.[6,9] Nonmyeloablative RIT is designed to spare bone marrow from toxicity. Consequently, dose escalation studies are required to determine the maximum tolerated dose (MTD). Dose-limiting toxicity in RIT usually occurs 2 to 3 weeks after therapy, with a nadir at about 4 to 8 weeks and full recovery usually within 12 weeks after therapy.[6] Myelotoxicity is typically manifested by thrombocytopenia and neutropenia.[9] The myelotoxicity response depends on patient-specific conditions and whether prior chemotherapy or immunotherapy has been received, in which case patients are more vulnerable to toxicity. In myeloablative RIT, the radiation dose is almost certain to result in bone marrow ablation and requires autologous stem cell transplantation (ASCT). The treatment dose is designed to deliver not more than the MTD to the dose-limiting normal tissues, those being the lung, kidney, liver, and gastrointestinal (GI) tract.[6,9]

Dosing Methods

Two dosing methods have been used in RIT: the radiation dose method and the radionuclide dose method.[6,9] In the radiation dose method, the amount of administered activity for the therapeutic dose is based on a prescribed radiation dose to the critical dose-limiting organ (red marrow or total body) in the case of nonmyeloablative RIT or the critical dose-limiting second organ (lung, kidney, liver, GI tract) in the case of myeloablative RIT.[6] In the radionuclide dose method, the amount of radioactivity administered is based on body weight or body surface area (i.e., mCi per kg or m^2).

Nonmyeloablative therapy in NHL has been conducted using both the radiation dose method and the radionuclide dose method.[6] In the radiation dose method, the amount of radioactivity to be administered is calculated by dividing the prescribed radiation dose (cGy) to the critical organ (red marrow or total body) by the radiation dose per unit activity (cGy/mCi) to the critical organ, estimated by a pretherapy tracer dose. The total-body dose is used as a marrow dose surrogate for ^{131}I-monoclonal antibodies because of its penetrating gamma radiation component, which contributes significantly to the red marrow dose.[10] This is the method used for dosing ^{131}I-tositumomab. The MTD for ^{131}I-tositumomab in nonmyeloablative therapy of NHL is 75 rad(cGy) to the total body. This was determined from a dose-escalation study beginning with a 25 rad(cGy) total-body dose, increasing in increments of 10 rad(cGy).[11]

With the radionuclide dose method in nonmyeloablative therapy, the amount of radioactivity administered is based on the maximum allowed activity per kilogram of body weight. This method does not require a tracer dose to calculate the administered activity, but tracer studies are typically done before therapy as a safety measure to ensure normal biodistribution to organs. This method has been used with ^{90}Y-ibritumomab tiuxetan. The MTD for ^{90}Y-ibritumomab tiuxetan in nonmyeloablative therapy of NHL is 0.4 mCi/kg (14.8 MBq/kg) in patients with platelet counts 150,000/μL or higher and 0.3 mCi/kg (11.1 MBq/kg) with platelet counts 100,000 to 149,000/μL. The maximum allowed dose per patient treatment is 32 mCi (1184 MBq) regardless of body weight, based on critical threshold doses of 300 rad(cGy) to the marrow and 2000 rad(cGy) to any secondary organ (lungs, liver, or kidneys). The MTD was determined from a dose escalation study beginning at a dose of 0.2 mCi/kg (7.4 MBq/kg) with increasing increments of 0.1 mCi/kg (3.7 MBq/kg).[12] The total-body dose method that is used for ^{131}I-tositumomab cannot be used with ^{90}Y-ibritumomab tiuxetan because the red marrow dose does not correlate well with total-body dose. The reason is the residualizing of ^{90}Y in the body and its lack of gamma emissions.[6]

Myeloablative therapy has also been used in NHL patients. The amount of administered activity in these patients is determined by the radiation dose method. In this approach, a Phase 1 dose-escalation study was conducted beginning at 1000 rad(cGy) to the lungs as the critical organ, with incremental increases to 3075 rad(cGy).[13] The MTD was determined to be 2725 rad(cGy). Patients also received ASCT in this treatment regimen because of the resulting bone marrow ablation. The results of myoablative and nonmyeloablative treatment methods in RIT are described below in the discussion of individual antibodies.

THERAPEUTIC ANTIBODIES

^{90}Y- and ^{111}In-Ibritumomab Tiuxetan (Zevalin)

Ibritumomab tiuxetan (Zevalin [Spectrum Pharmaceuticals, Irvine, CA]), which was approved for use in 2002, is an antibody conjugate of ibritumomab, an intact IgG1 kappa murine monoclonal antibody, and tiuxetan (MX-DTPA), an isothiocyanatobenzyl-derivatized DTPA linker, covalently bound to lysine and arginine amino groups in the antibody by a stable thiourea bond (see Figure 28-10, Chapter 28). ^{90}Y or ^{111}In is labeled to the antibody by chelation with the DTPA ligand. The radiolabeled antibody targets the CD20 antigen, a transmembrane phosphoprotein expressed on normal B-lymphocytes and malignant B-lymphocytes of NHL.

The therapeutic regimen of ibritumomab tiuxetan is supplied as two separate identical kits, one for preparing a single dose of ^{111}In-labeled diagnostic antibody and the other for a single dose of ^{90}Y-labeled therapeutic antibody. The kits should be stored in the refrigerator prior to use.

Antibody labeling follows specific steps that differ between ^{111}In and ^{90}Y in the amounts of activity and antibody used and the length of incubation. The labeling steps for each radionuclide are summarized in Table 29-1. In general, shaking and foaming of the antibody during labeling must be avoided to prevent denaturation of the antibody. ^{90}Y-chloride for labeling is supplied directly from MDS Nordion when the antibody kit is ordered from Spectrum Pharmaceuticals. Sterile ^{111}In-chloride must be obtained separately.

Radiochemical purity of the labeled antibody is assessed by instant thin-layer chromatography with silica gel in normal saline. The unbound nuclide travels to the solvent front and labeled antibody remains at the origin. Its purity must be 95% or higher. Labeled antibody is stored at 2°C to 8°C. ^{90}Y-ibritumomab tiuxetan should be administered within 8 hours and ^{111}In-ibritumomab tiuxetan within 12 hours after radiolabeling.

Developmental studies have shown that the immunoreactivity of ^{111}In-ibritumomab tiuxetan incubated for 15 minutes during radiolabeling is essentially conserved, while that of ^{90}Y-labeled antibody is about 60%.[14] The incubation time for ^{90}Y-ibritumomab tiuxetan is therefore limited to 5 minutes to minimize radiolytic degradation of antibody by the high-energy beta emission of ^{90}Y and to preserve immunoreactivity at around 75%. The formulation buffer added at the end of radiolabeling contains human serum albumin (HSA), which stabilizes the labeled antibody against radiolytic damage.[14] HSA was found to be effective in preserving antibody structure and immunoreactivity. The formulation buffer also contains DTPA to ensure that the small amount of ^{90}Y or ^{111}In not labeled to the antibody will be chelated and eliminated by renal excretion.

Ibritumomab tiuxetan is indicated for the treatment of patients with relapsed or refractory low-grade, follicular B-cell NHL or patients with previously untreated follicular NHL who achieve a partial or complete response to first-line chemother-

TABLE 29-1 Radiolabeling of Ibritumomab Tiuxetan with ^{111}In and ^{90}Y	
^{111}In Labeling Steps	**^{90}Y Labeling Steps**
1. Add required volume of 50 mM sodium acetate buffer to the reaction vial. (Buffer volume = 1.2 times the volume of 5.5 mCi ^{111}InCl$_3$.) Coat entire surface of reaction vial with buffer.	1. Add required volume of 50 mM sodium acetate buffer to the reaction vial. (Buffer volume = 1.2 times the volume of 40 mCi ^{90}YCl$_3$.) Coat entire surface of reaction vial with buffer.
2. Add 5.5 mCi ^{111}InCl$_3$ to the reaction vial and mix the solutions.	2. Add 40 mCi ^{90}YCl$_3$ to the reaction vial and mix the solutions.
3. Add 1.0 mL of the antibody to the reaction vial and roll/mix solutions gently. Do not cause foaming of the antibody.	3. Add 1.3 mL of the antibody to the reaction vial and roll/mix solutions gently. Do not cause foaming of the antibody.
4. Incubate at room temperature for 30 minutes.	4. Incubate at room temperature for 5 minutes.
5. Add the formulation buffer (FB) to quench the labeling. (FB volume = 10 mL minus the volumes of ^{111}InCl$_3$ + acetate buffer + antibody.)	5. Immediately add the formulation buffer to quench the labeling. (FB volume = 10 mL minus the volumes of ^{90}YCl$_3$ + acetate buffer + antibody.)

apy.[15] The first of these indications was included in the original approval in 2002. The latter indication was approved in 2009 and expands the population of patients who can receive Zevalin treatment.

Dosing Regimen for ^{90}Y-Ibritumomab Tiuxetan

The ibritumomab tiuxetan dosage schedule proceeds in two steps, each requiring a predose of rituximab prior to administration of either ^{111}In-ibritumomab tiuxetan or ^{90}Y-ibritumomab tiuxetan. The rituximab predose is given initially to block accessible CD20 sites in the peripheral circulation and prevent indiscriminate uptake of the radiolabeled antibody in the reticuloendothelial system. The pretreatment facilitates optimum biodistribution of the radiolabeled antibody to tumor sites.

Rituximab was the first monoclonal antibody approved by FDA for the treatment of cancer. It is a genetically engineered chimeric monoclonal antibody specific for the CD20 antigen on B-lymphocytes. It contains murine light- and heavy-chain variable region sequences and human constant region sequences. Ibritumomab is the murine antibody parent of rituximab. Rituximab alone in standard immunotherapy produces objective tumor responses. However, its effectiveness is significantly less than that of ^{90}Y-ibritumomab tiuxetan. In a prospective randomized Phase 3 trial comparing ^{90}Y-ibritumomab tiuxetan with rituximab in relapsed or refractory low-grade, follicular, or transformed NHL, the overall response rates for ^{90}Y-ibritumomab tiuxetan and rituximab were 80% and 56%, respectively.[16] The complete remission rate was 30% for ^{90}Y-ibritumomab tiuxetan and only 16% for rituximab.

Figure 29-2 illustrates a typical ibritumomab tiuxetan administration and diagnostic imaging schedule. The first step of the therapeutic regimen involves intravenous infusion of rituximab 250 mg/m^2. It is infused initially at a rate of 50 mg/hour, escalating in 50 mg/hour increments every 30 minutes to a maximum rate of 400 mg/hour, provided no hypersensitivity reactions occur. The major adverse effects are fever and chills, which can be controlled by reducing the infusion rate. Infusion reactions can be minimized by pretreating the patient with acetaminophen and diphenhydramine according to package insert instructions. Rituximab should be infused within 4 hours before the administration of either ^{111}In-ibritumomab tiuxetan or ^{90}Y-ibritumomab tiuxetan. For the diagnostic imaging phase of the dosing regimen rituximab infusion is followed by 5 mCi (185 MBq) of ^{111}In-ibritumomab tiuxetan, given intravenously over 10 minutes. Gamma-camera

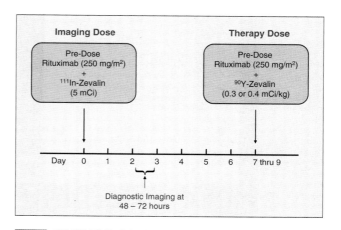

FIGURE 29-2 Dose administration and diagnostic imaging schedule for ^{111}In- and ^{90}Y-ibritumomab tiuxetan (Zevalin). A single whole-body image is made after ^{111}In-labeled antibody administration on day 2 or 3 to assess biodistribution.

imaging is performed 48 to 72 hours after dosing to confirm normal and abnormal biodistribution of antibody prior to ^{90}Y-ibritumomab tiuxetan administration (Figures 29-3, 29-4, and 29-5).[15,17,18] The distribution of ^{111}In-ibritumomab tiuxetan predicts the behavior of ^{90}Y-ibritumomab tiuxetan and ensures that normal tissue dose limits will not be exceeded, which would preclude administration of ^{90}Y-ibritumomab tiuxetan. Expected biodistribution includes (1) faintly visible activity in the blood pool areas of heart, abdomen, neck, and extremities, (2) moderately high to high uptake in normal liver and spleen, (3) moderately low or very low uptake in normal kidneys, urinary bladder, and normal uninvolved bowel, (4) non-fixed areas within the bowel lumen that change position with time; delayed imaging beyond the 48 to 72 hour image may be necessary to confirm GI clearance, and (5) focal fixed areas of uptake in the bowel wall (localization to lymphoid aggregates in bowel wall).[15] Altered biodistribution is evident as (1) intense localization of radiotracer in the liver, spleen, and bone marrow indicative of reticuloendothelial uptake, and (2) increased uptake in normal organs (not involved by tumor) such as (a) diffuse lung uptake more intense than liver uptake, (b) uptake in the kidneys more intense than in the liver on posterior view, (c) fixed areas of uptake in normal bowel greater than uptake in the liver, and (d) prominent bone marrow uptake seen as clear visualization of the long bones and ribs. Underlying factors that may cause increased bone marrow uptake should be considered, such as lymphoma, recent administration of hematopoietic growth factor, and HAMA or human antichimeric antibody (HACA) response.[15] Altered biodistribution has been found to occur in only 1.3% of patients evaluated in the Zevalin imaging registry.[18]

The second phase of the dosing regimen occurs in 7 to 9 days, with a second infusion of rituximab 250 mg/m^2 followed by 0.4 mCi/kg (14.8 MBq/kg) or 0.3 mCi/kg (11.1 MBq/kg) of ^{90}Y-ibritumomab tiuxetan given intravenously over 10 minutes. The infusion line should be flushed with 10 mL of saline after each infusion to ensure complete antibody administration.

The principal adverse effects of ^{90}Y-ibritumomab tiuxetan are hematologic: thrombocytopenia and neutropenia. These effects occur less frequently when platelet counts are greater than 150,000 cells/μL. The dose of ^{90}Y-ibritumomab tiuxetan should be reduced to 0.3 mCi/kg (11.1 MBq/kg) in patients with a baseline platelet count between 100,000 and 149,000 cells/μL. No patient with a platelet count less than 100,000 should be treated. The maximum allowable dose of ^{90}Y-ibritumomab tiuxetan is 32.0 mCi (1184 MBq) regardless of the patient's weight. The dose limit is to ensure that no

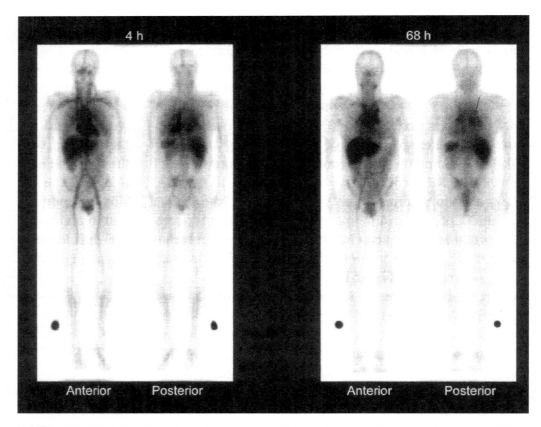

FIGURE 29-3 Whole-body 4- and 68-hour images after administration of ^{111}In-ibritumomab tiuxetan, demonstrating evident blood pool activity on the 4 hour image. Normal moderately high to high uptake is seen in the liver and spleen and moderately low to very low uptake in kidneys, urinary bladder, and bowel on 4- and 68-hour images. Tumor is evident in mediastinal lymph nodes (arrow). (Images reproduced with permission of Spectrum Pharmaceuticals, Inc.)

Therapeutic Applications of Radioactive Agents   679

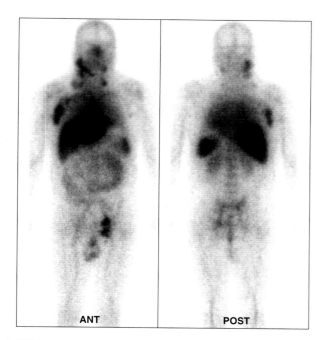

FIGURE 29-4 Fifty-seven-year-old man with low grade non-Hodgkin's lymphoma. ^{111}In-ibritumomab tiuxetan scan 48 hours after radiopharmaceutical administration showing multiple focal areas of accumulation in the neck, right axilla, and left inguinal region. The patient went on to have ^{90}Y-ibritumomab tiuxetan therapy 1 week later.

patient receives radiation absorbed doses to any normal organ greater than 2000 rad(cGy) or to the red marrow greater than 300 rad(cGy).[17] The median absorbed radiation doses per 32 mCi to selected organs are spleen, 1113 rad(cGy); liver, 568 rad(cGy); lung, 237 rad(cGy); kidney, 12 rad(cGy); red marrow, 154 rad(cGy); and total body, 59 rad(cGy).[15] The

median effective half-life of ^{90}Y-ibritumomab tiuxetan in the blood in clinical trials was 27 hours (range, 14 to 44 hours).[14] The frequency of HAMA response is 2.4% and of HACA response is 2%.[15]

From a therapeutic standpoint, ^{90}Y has a wider range of effectiveness than ^{131}I because of its higher beta energy. For example, the effective path lengths (χ_{90}) of ^{90}Y and ^{131}I are about 5 mm and 1 mm, respectively, meaning that 90% of their beta-particle energies are absorbed within a sphere of 5 mm or 1 mm radius, respectively.[17] A 5 mm path length corresponds to about 100 to 200 cell diameters. Thus, ^{90}Y betas emitted from antibody bound on the surface of tumor cells can deliver radiation dose to non–antibody-bound tumor cells located deep within the tumor (the crossfire effect). This consideration is important in bulky or poorly vascularized tumors not accessible to antibody. On the other hand, use of ^{90}Y in small tumors may result in higher radiation dose and possible damage to surrounding normal tissues, the so-called "bystander effect."

Radiation Safety of ^{90}Y-Ibritumomab Tiuxetan

Caution should be exercised in handling ^{90}Y during radiolabeling. The exposure rate per millicurie from the mouth of an open vial of ^{90}Y is 32 R/hour.[15] Its average beta particle range in air is about 3.7 m, unshielded. Although the beta particle energy is quite high, it can be completely absorbed by 1 cm of low-Z material such as Plexiglas or Lucite (see Table 6-8, Chapter 6). Syringe shields should be constructed of these materials to limit bremsstrahlung production. A thin layer of lead may be used over the plastic shield to absorb any bremsstrahlung. Geiger-Müller survey meters are extremely efficient for detecting low-energy photons but could give erroneously high readings unless they have been calibrated for ^{90}Y bremsstrahlung.[19]

^{90}Y exposure of family members and the public from patients treated with ^{90}Y-ibritumomab tiuxetan is very low. The specific bremsstrahlung dose constant for ^{90}Y in soft tissue from

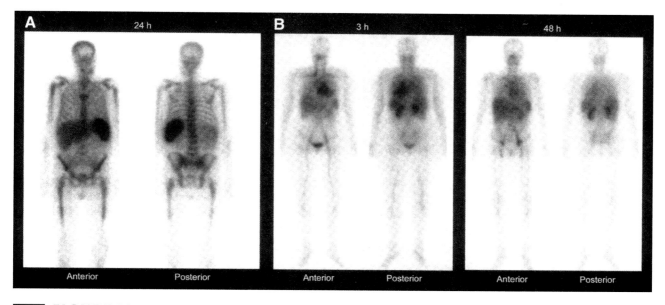

FIGURE 29-5 Abnormal whole-body biodistribution of ^{111}In-ibritumomab tiuxetan obtained at 24 hours after administration (A), showing prominent bone marrow and splenic uptake, and at 3 and 48 hours (B), showing greater than expected kidney uptake. (Images reproduced with permission of Spectrum Pharmaceuticals, Inc.)

a 70 kg patient is 0.00564 R-cm²/mCi-hr (1.52 μGy-cm²/MBq-hr).[19] Based on the Nuclear Regulatory Commission (NRC) threshold for exposure of members of the public (0.5 rem), excluding self-absorption of bremsstrahlung, a patient would need to be dosed above 38,500 mCi (1.42 × 10⁶ MBq) of ⁹⁰Y to exceed this limit.[19] Since the maximum dose of ⁹⁰Y-ibritumomab tiuxetan is 32 mCi (1184 MBq), therapy with this antibody is safe and requires no restrictions to prevent exposure of members of the public. The only recommendation for family members is to avoid contamination from body excretions, mainly urine, although this is not likely to be a safety issue because urinary excretion is quite low, about 7.3% per week. The recommended instructions for patients released after treatment with ⁹⁰Y-ibritumomab tiuxetan are as follows:[19] For 3 days after treatment, clean up any contaminated body fluid and urine and flush it down the toilet or place it in a plastic bag in household trash, and wash hands thoroughly after using the toilet. For 1 week after treatment, use condoms during sexual relations.

¹³¹I-Tositumomab (Bexxar)

Tositumomab and its radioiodinated derivative ¹³¹I-tositumomab (Bexxar [Corixa and GlaxoSmithKline]) is a murine IgG2a lambda monoclonal antibody directed against a specific epitope within the extracellular domain of the CD20 antigen, a transmembrane phosphoprotein, expressed on normal B-lymphocytes and malignant B-lymphocytes of NHL. Its mechanism of action includes CMC, ADCC, and induction of apoptosis.[20] Bexxar is indicated for the treatment of patients with CD20 antigen-expressing relapsed or refractory, low grade, follicular, or transformed NHL, including patients with rituximab-refractory NHL. It is not indicated for the initial treatment of patients with CD20-positive NHL.[20]

¹³¹I-tositumomab is supplied as both a dosimetric and a therapeutic package.[20] Each package contains unlabeled tositumomab in two single-use 225 mg vials for patient administration prior to infusion of the radiolabeled antibody and one single-use 35 mg vial for preparation of ¹³¹I-tositumomab. The unlabeled antibody is obtained separately from the radiolabeled antibody. The radiolabeled antibody is supplied frozen by MDS Nordion (Ottawa, Ontario, Canada). The nominal protein and activity concentrations at calibration for the dosimetric ¹³¹I-tositumomab are 0.1 mg/mL and 0.61 mCi/mL, respectively, and for the therapeutic ¹³¹I-tositumomab are 1.1 mg/mL and 5.6 mCi/mL. Prior to patient administration of the dosimetric or therapeutic dose, the frozen ¹³¹I-tositumomab must be thawed and adjusted to a total of 35 mg of protein with unlabeled tositumomab and to a total volume of 30 mL with 0.9% sodium chloride injection. Thawed dosimetric and therapeutic doses of ¹³¹I-tositumomab are stable for 8 hours at 2°C to 8°C or at room temperature. Diluted ¹³¹I-tositumomab solutions for infusion must not be frozen and should be kept at 2°C to 8°C prior to administration.

Dosing Regimen for ¹³¹I-Tositumomab

The dosing schedule for ¹³¹I-tositumomab involves administration of a 5 mCi dosimetric dose, gamma camera imaging over 7 days to evaluate biodistribution, and administration of a therapeutic dose for radiation treatment of the NHL. The thyroid gland must be protected from ¹³¹I released from metabolized antibody in the body. This is accomplished with a 130 mg daily oral dose of potassium iodide, which is begun 24 hours before the dosimetric dose and continued until 2 weeks after the therapeutic dose of ¹³¹I-tositumomab.

For the dosimetric study the patient receives a 50 mL IV infusion of tositumomab (450 mg) in 0.9% sodium chloride injection over 60 minutes followed by a separate 30 mL IV infusion of 5 mCi of ¹³¹I-tositumomab (35 mg) in 0.9% sodium chloride injection over 20 minutes. Gamma camera counts are obtained at three imaging time points: (1) on day 0, within an hour of the end of the infusion, (2) on day 2, 3, or 4, immediately after patient voiding, and (3) on day 6 or day 7 immediately after patient voiding, to assess biodistribution and total body residence time of ¹³¹I activity (Figure 29-6). If biodistribution is acceptable, the patient will receive a therapeutic dose of ¹³¹I-tositumomab on day 7 up to day 14.

Before the therapeutic dose of ¹³¹I-tositumomab is administered, a calculation is made to determine the amount of activity needed to deliver the required radiation dose. If the patient's platelet count is ≥150,000 cells/mm³ a total body dose of 75 cGy is chosen; if the platelet count is 100,000 to <150,000 cells/mm³ a dose of 65 cGy is chosen. For the therapeutic study the patient receives a 50 mL IV infusion of tositumomab (450 mg) in 0.9% sodium chloride injection over 60 minutes followed by a separate 30 mL IV infusion of the therapeutic dose of ¹³¹I-tositumomab (35 mg) in 0.9% sodium chloride injection over 20 minutes.

The amount of activity (mCi) for the patient's therapeutic dose is calculated from the following formula:

$$^{131}\text{I Activity (mCi)} = \frac{\text{Activity hours (mCi-hr)}}{\text{Residence time (hr)}} \times \frac{\text{Desired total body dose (cGy)}}{75 \text{ cGy}}$$

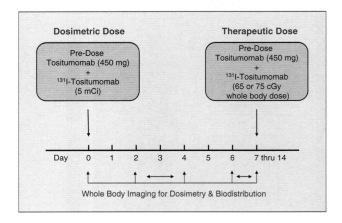

FIGURE 29-6 Dose administration and diagnostic imaging schedule for ¹³¹I-tositumomab. Imaging to assess biodistribution and residence time of the antibody is conducted 1 hour after dosimetric dose administration; on day 2, 3, or 4; and on day 6 or 7.

Activity hours in the formula is determined from a table of activity hours versus patient effective mass. A patient's effective mass is derived from a separate table of effective mass per patient height adjusted for patient sex. Residence time of ^{131}I activity in the patient's body is determined from a graphical plot of the percent of injected activity remaining in the body at each time point associated with the gamma camera imaging measurements made after the dosimetric dose. A line is constructed through the time points to intersect a horizontal line at 37% injected activity (Figure 29-7). A therapeutic dose should not be administered if the patient's residence time plot intersects the horizontal 37% line at less than 50 hours or more than 150 hours residence time, or if unexpected biodistribution is observed on gamma camera images. Biodistribution is assessed from the imaging data. Altered biodistribution is present if (1) on the first imaging time point there is (a) absence of blood pool activity, or (b) diffuse, intense tracer uptake in liver, spleen, or both, or (c) uptake suggestive of urinary obstruction, or (d) diffuse lung uptake greater than that of blood pool on the first day; or (2) on the second and third imaging time points there is uptake suggestive of urinary obstruction and diffuse lung uptake greater than blood pool.

Radiation Safety of ^{131}I-Tositumomab

Radiation safety guidelines for ^{131}I-tositumomab are similar to those used for any ^{131}I treatment procedure regarding release criteria, instructions, and records. The main issue with regard to ^{131}I therapy is the time a patient should observe precautionary measures so as not to exceed exposure limits for members of the public (500 mrem). The key issues for consideration are the amount of activity administered and the residence time of ^{131}I in the body, both of which are known for ^{131}I-tositumomab.

^{131}I-LYM-1

LYM-1 (Oncolym [Peregrine Inc., Tustin, CA]) is an IgG2a murine monoclonal antibody that selectively binds to an epitope on the β subunit of human leukocyte antigen HLA-DR that is present on malignant B-lymphocytes but less so on normal lymphocytes.[21,22] LYM-1 is produced by immunizing mice with human Burkitt's lymphoma cells. The antigen–antibody complex is not internalized, nor is the antigen shed into the blood of NHL patients.[22] This antibody has not reached the approval stage for routine use.

LYM-1 has been labeled with ^{131}I by standard chloramine-T iodination and with the ^{67}Cu-labeled bifunctional chelating agent 2-iminothiolane bromoacetamido benzyl TETA (^{67}Cu-2IT-BAT) to yield the radiolabeled antibody conjugate ^{67}Cu-2IT-BAT-LYM-1 (Figure 29-8). TETA (1,4,8,11-tetraazacyclotetradecane-1,4,8,11-tetraacetic acid) is a macrocyclic chelator that was specifically designed to produce a stable chelate with Cu(II).[23,24] The TETA chelator contains the iminothiolane functionality to link it to the antibody. ^{67}Cu is a potentially effective radionuclide for RIT. Its 2.58 day half-life is similar to the residence time of a typical antibody on the tumor.[23] As the TETA chelate, ^{67}Cu is quite stable (about 1% loss per day in in vitro studies),[23] and there is no biologic mechanism for its uptake in bone marrow to increase myelosuppression.[25] It emits beta particles of moderate energy with a mean range of 0.27 mm, and it has a useful photon energy (185 keV) for imaging and dosimetry estimation.

^{90}Y-Epratuzumab

^{90}Y-epratuzumab (^{90}Y-hLL2; Lymphocide [Immunomedics]) is the humanized monoclonal antibody of the murine IgG2a antibody (mLL2). LL2 reacts with the CD22 antigen on the surface of NHL B cells. This antibody does not react with normal peripheral blood cells but does react with germinal B cells of normal lymph nodes and the white pulp of the spleen.[26] It has been labeled with ^{131}I, ^{111}In, ^{90}Y, and ^{186}Re. This antibody has not reached the approval stage for routine use.

LL2, in contrast to anti-CD20 antibodies, becomes internalized within the cell after its CD22 interaction.[27] Subsequently, the CD22 antigen is resynthesized and re-expressed on the cell so that additional antibody can be localized and internalized. The radiation dose to a tumor cell of

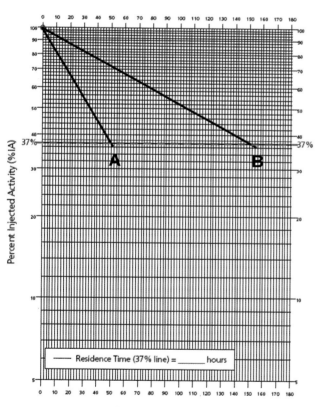

FIGURE 29-7 Percent injected activity remaining in the body over time after injection of a dosimetric dose of ^{131}I-tositumomab (residence time plot). A patient's residence time plot must intersect the 37% line between the intersect points of line A (50 hours) and line B (150 hours) before a therapeutic dose of ^{131}I-tositumomab can be administered.

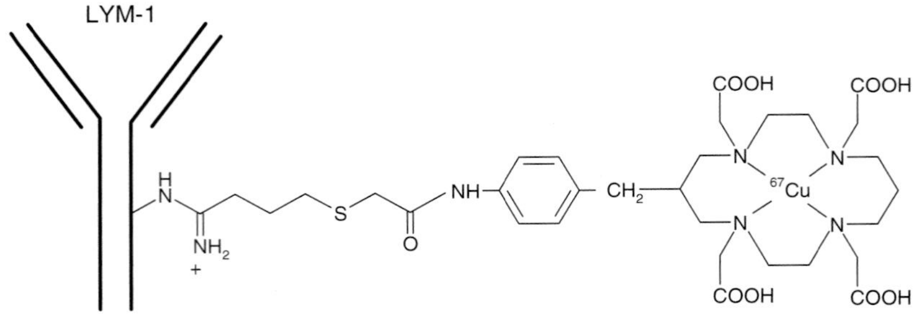

FIGURE 29-8 Chemical structure of ^{67}Cu-2-iminothiolane bromoacetamido benzyl TETA (^{67}Cu-2IT-BAT-LYM-1).

an internalized antibody depends on whether it is metabolized and excreted or is deposited (residualized) within the tumor cell. Radiation dose may be reduced if the rate of antibody metabolism and excretion is much higher than the radionuclide's decay rate. But if the radiolabel is residualized, its longer residence time will produce higher tumor-to-background ratios and radiation dose to the tumor.[28] Iodinated antibodies conventionally labeled with chloramine-T or Iodo-Gen release iodotyrosine as a principal metabolite, which readily diffuses out of the tumor cell[29,30] and undergoes deiodination by tyrosine deiodinase.[31] However, antibodies labeled with a residualizing radioiodine conjugate, such as dilactitol iodotyramine, have a longer residence time in tumors.[32,33] Studies have also shown that ^{111}In- and ^{90}Y-labeled antibodies have longer residence times in tumors than conventionally iodinated antibodies because of trapping of the radiolabeled metabolites within tumor cell lysosomes after catabolism.[32,29] Juweid et al.[34] compared epratuzumab (hLL2) radiolabeled with ^{131}I (by Iodo-Gen) and ^{111}In and ^{90}Y (via the benzyl-DTPA-hLL2 conjugate) in patients with relapsed refractory NHL. Radiation dose estimates to tumors larger than 3 cm was 2.4 ± 1.9 rad(cGy) for ^{131}I-hLL2 versus 21.5 ± 10.0 rad(cGy) for ^{90}Y-hLL2.

RADIOIMMUNOTHERAPY IN NON-HODGKIN'S LYMPHOMA

Most lymphomas of B-lymphocytic origin are incurable. Past experience has shown that standard chemotherapy, such as CHOP (cyclophosphamide, doxorubicin, vincristine, and prednisone), can improve prognosis and result in complete remission in about 50% of patients and cure in about 30% with advanced stages of intermediate and high-grade NHL.[35,36] Unfortunately, NHL eventually develops resistance to chemotherapy. Immunotherapy of NHL with rituximab, the first monoclonal antibody approved for human use, has shown treatment response equivalent to chemotherapy but with less systemic toxicity, and RIT has demonstrated response rates better than immunotherapy. However, RIT bears the risk of radiation-induced bone marrow suppression, which limits the amount of radioactivity that can be administered.

Rituximab is a chimeric antibody with mouse variable and human constant regions. This modification makes it a more effective immunotherapy agent for NHL than the murine only antibody in fixing complement and mediating ADCC.[2] Although rituximab kills both normal and NHL B cells, it spares stem cells because they lack the CD20 antigen, and the stem cells are able to regenerate normal B cells. The same is true in RIT with radiolabeled anti-CD20 antibodies, provided that radiation-induced bone marrow toxicity is limited to safe levels.

There are two major types of NHL, follicular and diffuse. About half of NHL patients have the follicular type, which is responsive to anti-CD20 immunotherapy with rituximab and RIT with ^{90}Y-ibritumomab tiuextan (Zevalin) and ^{131}I-tositumomab (Bexxar). Another monoclonal antibody for treating NHL is LYM-1 (Oncolym). More than 80% of B-cell NHL patients have malignancies that react with LYM-1.[22] Although unlabeled LYM-1 is not effective as immunotherapy for NHL in vivo,[37,38] the radiolabeled conjugate ^{131}I-LYM-1 is effective for RIT of NHL.

There are a number of approaches in the treatment of NHL that incorporate RIT into the treatment plan with the goal of improving outcome. The methods include myeloablative or high-dose treatment with ASCT and nonmyeloablative approaches, such as combining RIT with chemotherapy, RIT as consolidation after chemotherapy to target minimal residual disease, first-line RIT treatment, fractionated RIT, pretargeting RIT, and use of different antibodies.[22,39] Early nonmyeloablative studies demonstrated the effectiveness of RIT with ^{131}I-LYM-1 antibody.[40,41] A few years later, two pivotal studies revealed that RIT was more effective than treatment with unlabeled antibody or chemotherapy. Patients treated with ^{90}Y-ibritumomab tiuxetan had much higher response rates (80%) compared with rituximab (56%),[16] and in patients treated with ^{131}I-tositumomab 65% had complete or partial remission compared with only 28% in those treated with chemotherapy alone.[42] In addition, it was shown that retreatment of patients who previously responded to therapy with ^{131}I-tositumomab was relatively safe and effective and could produce a durable response (progression-free survival greater than 12 months) in some patients.[43] In addition to its therapeutic effectiveness, RIT has been associated with less morbidity than standard chemotherapy, with no therapy-associated mucositis, hair loss, or persistent nausea and vomiting.[22]

In early Phase 1 and 2 clinical studies in chemotherapy refractory or relapsed patients with NHL, Kaminski et al.[44] determined the total-body maximum tolerated dose (MTD) for ^{131}I-tositumomab to be 75 cGy for patients without ASCT and 45 cGy for patients after ASCT. Of the 59 patients treated, there were 42 (71%) responders, and 20 (34%) had a complete remission (CR). Median progression-free survival in all 42 responders was 12 months; for those with CR it was 20.3 months. Seven patients remained in CR for 3 to 7.5 years. Other studies have shown positive results with RIT as a first-line treatment. Kaminski et al.[45] evaluated the effectiveness of nonmyeloablative RIT alone in 76 patients with stage III or IV follicular lymphoma previously untreated with chemotherapy who received a single course of treatment with ^{131}I-tositumomab delivering a total body dose of 75 cGy. An overall response rate was achieved in 95% of patients treated, and 77% of the patients experienced complete remission, remaining disease free at 5 years. However, HAMA was observed in 63% of patients, in keeping with their being chemotherapy naïve. Also, the 5 year rate of progression-free survival was lower in patients whose HAMA levels were five times higher than the lowest levels within the first 7 weeks. Press et al.[46,47] treated patients with myeloablative doses of ^{131}I-tositumomab combined with ASCT, obtaining response rates as high as 86% and CR in 79%. Of these patients, 39% had recurrence-free survival for 5 to 10 years without additional therapy. The encouraging survival results in these studies support the use of RIT early in the course of treatment of NHL and with high-dose therapy.

RIT with LYM-1 and hLL2 monoclonal antibodies has been reviewed.[22,47–49] Investigational studies with radiolabeled analogues of these antibodies indicate that ^{67}Cu- and ^{90}Y-labeled analogues of LYM-1 achieve efficacy in treating NHL similar to that of ^{131}I-LYM-1. Response rates for anti-CD22 ^{90}Y-hLL2 were similar to those for anti-CD20 and HLA-DR monoclonal antibodies, with ^{90}Y-hLL2 being more efficacious than the ^{131}I- or ^{186}Re-labeled analogues of hLL2.

Initial approval for the routine use of RIT in NHL was restricted to patients whose disease was refractory to traditional forms of therapy. However, in 2009 the original indications for ^{90}Y-ibritumomab tiuxetan were amended to include patients with previously untreated follicular NHL who have achieved a partial or complete response to first-line chemotherapy.

PRETARGETED RADIOIMMUNOTHERAPY

Despite the success achieved in the treatment of NHL and other tumors, tumor targeting is fraught with problems. The amount of radiolabeled antibody taken up by tumors is typically only 0.001% to 0.1% of the injected dose because of heterogeneous antigen density, the development of antigen-negative tumor cells, nonuniform vascularity, and other factors such as circulating antigen that prevents injected antibody from reaching peripheral sites. All of these factors lead to low tumor-to-background ratios.

To overcome some of the causes of low tumor uptake of radiolabeled antibody, pretargeting methods have been developed. These techniques involve the administration of "cold" antibody derivatized to carry a secondary receptor that can bind a subsequently injected radiolabeled ligand. To enhance uptake of the radiolabeled ligand at tumor sites, a clearance technique is used to remove excess residual antibody in the circulation before the radiolabeled ligand is administered. These methods have been reviewed.[50–52]

A pretargeting modality that has been studied widely is the avidin–biotin system. Avidins are small oligomeric proteins made up of four identical subunits, each having a specific binding site for biotin. The affinity of biotin for avidins is very high, with the complex having a dissociation constant of 10^{-15} M. Avidins are isolated either from hen egg white (avidin) or from *Streptomyces avidinii* (streptavidin). Biotin is a 244 Da molecule that can be activated with a variety of functional groups to covalently attach it to proteins or chelating agents that bind radionuclides. One approach is to biotinylate an antibody by covalently attaching a biotin molecule to it. This conjugate can then be used to pretarget the tumor, followed by administration of radiolabeled streptavidin to deliver radioactivity to the tumor. Alternatively, the streptavidin can be covalently linked to the antibody as the pretargeting agent, followed by administration of radiolabeled biotin. In this case the biotin has been previously functionalized with a chelating agent to bind a radionuclide. The latter approach is being investigated with a DOTA–biotin conjugate labeled with ^{90}Y.[53–56]

Two approaches have been used in pretargeted therapy: the two-step method and the three-step method.[50] In the two-step method, biotinylated monoclonal antibody (or streptavidin monoclonal antibody) is administered first, followed in 1 to 2 days by administration of radiolabeled streptavidin (or radiolabeled biotin). This approach has appeal when tumor is confined to a small space, such as the peritoneal cavity. When disease is widespread and systemic administration is required, the three-step method is preferred. In this situation, a biotinylated monoclonal antibody is administered first to target the tumor. This is followed 1 day later by the injection of avidin, which removes excess circulating biotinylated monoclonal antibody, and streptavidin, which targets the tumor cells. After this step, radiolabeled biotin is injected, which targets the tumor-bound streptavidin, thus delivering the radioactivity to the tumor.

Another approach is the affinity enhancement system, which has an engineered antibody with two binding sites, each with an affinity for different antigens, one for tumor antigen and one for a radionuclide ligand administered after the antibody targets the tumor.[48,53] Goldenberg et al.[53] have comprehensively discussed pretargeting methods and reviewed FDA-approved antibodies and radionuclides for RIT.

RADIOTHERAPY OF BONE PAIN

Bone metastases eventually develop in 50% of patients with breast or prostate cancer[57] and are a major detriment to patient survival and quality of life. Skeletal metastases can cause severe morbidity, including bone pain, bone fracture, spinal cord compression, and problems resulting from bone marrow involvement. Radiopharmaceutical treatment of skeletal metastases has been shown to ameliorate bone pain 40% to 80% of the time.[58] Pain relief may last for a few weeks to several months. The goal of treatment is to preserve function and improve quality of life for the patient by relieving pain, reducing the amounts of narcotics needed for pain relief, and improving ambulation. Several modalities have been used to treat bone pain, including analgesics, hormone therapy, chemotherapy, surgery, radiopharmaceuticals, and external beam radiation.[59] External beam radiation has been effective in reducing pain from metastases, but its application is limited to treating localized sites in the body. More generalized external radiation treatments have debilitating adverse effects such as nausea and vomiting. Because metastases are often widespread, systemic radionuclide therapy is a better choice, providing more general and long-lasting relief with minimal adverse effects.

To be effective in reducing bone pain, a radiopharmaceutical must have certain characteristics.[60] First, it must have a high affinity for reactive bone; second, it must emit beta or electron radiation; third, the radiation must have sufficient energy to reach the cells responsible for the pain; and fourth, the half-life must be long enough to deliver damaging or lethal radiation to the cells.

Table 29-2 lists several radiopharmaceuticals that are approved for routine use or have been investigated. In general, most of these agents have an intrinsic affinity for mature bone (hydroxyapatite), with a particular affinity for regions of osteoblastic activity where amorphous calcium phosphate is being deposited. Thus, their sites of uptake in bone can be predicted by a standard bone scan with ^{99m}Tc-medronate (^{99m}Tc-MDP) or ^{99m}Tc-oxidronate (^{99m}Tc-HDP). ^{89}Sr undergoes biologic incorporation into bone similarly to calcium, and ^{186}Re and ^{153}Sm phosphonate complexes localize by chemisorption. ^{32}P as sodium phosphate is incorporated into hydroxyapatite as phosphate. Only ^{32}P-sodium phosphate, ^{89}Sr-chloride, and ^{153}Sm-lexidronam are approved by FDA for routine use.

These radiopharmaceuticals have widely varying half-lives and mean particle energies in the range of 0.13 to 0.70 MeV, but the response rate for all of these agents, measured by pain reduction, is 40% to 80% over a wide range of administered activities.[58] No dose–response effect has been demonstrated to occur with these agents.[60] The mechanism of action of any of these agents in relieving bone pain is not known.

Sodium Phosphate P 32 Solution

Sodium phosphate P 32 solution (^{32}P-sodium phosphate) is a clear, colorless sterile solution at pH 5.0 to 6.0 suitable for oral or intravenous administration. A commercial product that was available for many years had a radioactive concentration of 0.67 mCi/mL (24.8 MBq/mL) in 5 mCi (185 MBq) vials at the time of calibration. ^{32}P-sodium phosphate was the earliest agent used for treating painful bone metastases.[61] A review of ^{32}P-sodium phosphate for this indication showed effective pain response ranging from 60% to 90%.[62] One concern with ^{32}P as the phosphate is its involvement in many metabolic processes in the body, particularly in the hematopoietic system. Its principal toxicity has been bone marrow depression caused by its high-energy beta particle. When compared with ^{89}Sr-chloride for the treatment of painful bone metastases, ^{32}P-sodium phosphate was equally effective in reducing pain (90%) but caused a higher level of myelosuppression.[63] Pain reduction may be seen in 5 to 14 days after administration of ^{32}P-sodium phosphate.

Strontium Chloride Sr 89 Injection

Strontium chloride Sr 89 injection (^{89}Sr-chloride; Metastron [GE Healthcare]) was approved by FDA in June 1993 for the treatment of bone pain in patients with skeletal metastases. It is supplied as a sterile injection of ^{89}Sr-chloride in water at a concentration of 1 mCi/mL (37 MBq/mL) with a total of 4 mCi (148 MBq) in a 10 mL vial. Its expiration date is 28 days after calibration. The dosage range is 40 to 60 µCi/kg (1.48–2.22 MBq/kg), with an average dose of 4 mCi (148 MBq) given intravenously. The dose is administered over 1 to 2 minutes to minimize any "flushing" reaction. Administration can be repeated, but an interval of not less than 90 days is recommended. A single injection of 4 mCi (148 MBq) relieves pain

TABLE 29-2 Radiopharmaceuticals for Painful Bone Metastases

Radiopharmaceutical	Half-life (days)	Beta Max (MeV)	Beta Mean (MeV)	Mean Range (mm)	Gamma E (MeV) (abundance)
^{32}P-sodium phosphate	14.3	1.71	0.7	3.0	None
^{89}Sr-chloride	50.5	1.46	0.58	2.4	0.910 (0.009%)
^{186}Re-etidronate (^{186}Re-HEDP)	3.8	1.07	0.35	1.1	0.137 (9%)
^{153}Sm-lexidronam (^{153}Sm-EDTMP)	1.9	0.81	0.23	0.6	0.103 (28%)
^{117m}Sn-pentetate (^{117m}Sn-DTPA)	13.6	0.13[a] 0.15[a]	—	0.2 0.3	0.159 (86%)

[a] Conversion electrons.

in 65% to 80% of patients, with complete relief of pain in 20%. The time to response is 1 to 3 weeks, and the median duration of benefit is 4 to 6 months. Retreatment is possible at approximately 3 month intervals on up to eight occasions. ^{89}Sr-chloride is generally not recommended for patients with an expected survival of less than 3 months and patients with disseminated intravascular coagulation.[59]

Because ^{89}Sr suppresses bone marrow, at initial treatment the patient's platelet count should be greater than 60,000 and white blood cell count greater than 2400. A complete blood cell and platelet count should be obtained at least every other week to monitor the hematologic effects. The usual hematologic response is a 20% to 30% decrease in platelet count; the lowest point occurs at 5 to 6 weeks, with complete recovery by 12 weeks. The use of ^{89}Sr-chloride should be avoided in patients who have received previous chemotherapy with bone marrow suppression, because of the additive toxicity that may ensue.

Serafini[59] reviewed several studies performed with [89]Sr. In general, these studies indicated that pain relief was quite good, ranging from 60% to 85%, and that the degree of relief was independent of administered activity, which ranged from 1 to 10 mCi (37–370 MBq). Thus, a dose–response effect was not seen. Mild to moderate myelotoxicity (thrombocytopenia) was reported in most trials, with more severe toxicity occurring with higher doses.

The mechanism of ^{89}Sr uptake into bone is similar to that for calcium and occurs at sites of active osteogenesis. Increased uptake of longer duration occurs in metastases, compared with normal bone.[55] About 70% of the dose is retained in the skeleton, ranging from up to 88% retention with extensive metastases to 11% with few metastases. Of the fraction excreted, two-thirds is urinary and one-third is fecal.

Patients who are to receive ^{89}Sr-chloride should be advised that their pain may worsen in the 2 to 3 days after dosing. This so-called flare response, which occurs in 10% to 20% of patients, is transient and readily treated with nonaspirin analgesics. The type of analgesic is important because thrombocytopenia is exacerbated with anticlotting drugs like aspirin. Patients should be told that the onset of pain relief may take 1 to 3 weeks and that ^{89}Sr-chloride does not cause nausea, vomiting, or hair loss, which they might expect from chemotherapy or other forms of radiation treatment.

The critical organ is the bone surface, with a radiation dose of 63 rad(cGy)/mCi, equivalent to a 250 rad(cGy) dose from 4 mCi (148 MBq).

[89]Sr is considered to be a pure beta emitter, despite its 910 keV gamma ray of very low abundance (0.009%). Therefore, its assay in a dose calibrator relies on the measurement of bremsstrahlung, which can vary considerably depending on the assay geometry. Consequently, a method has been developed for the radioassay of ^{89}Sr dosages in the radionuclide dose calibrator.[64] The method is as follows: First, calculate the activity in the vial at the time and date of dispensing based on the label calibration data. Second, assay the vial in the dose calibrator, adjusting the calibration dial to read the calculated activity. Because of the insensitivity of the dose calibrator to beta emitters, an empiric calibration factor must be selected, typically a setting around 600, using the adjustable potentiometer dial, and a correction factor of 100 applied to the reading to get the correct value. Third, calculate the volume needed to obtain the patient dosage, and withdraw this amount into a syringe. Next, reassay the vial. The difference between the two readings is the activity in the syringe. Now assay the syringe and adjust the calibration dial to read the activity in the syringe. After injecting the patient, reassay the syringe for any residual activity. The difference between the two syringe readings is the activity injected into the patient.

Samarium Sm 153 Lexidronam Injection

The physical and chemical properties of this radiopharmaceutical are described in Chapter 11. Briefly, samarium Sm 153 lexidronam injection ([153]Sm-lexidronam; Quadramet [Berlex]) is available as a frozen solution at a concentration of 50 mCi/mL (1850 MBq/mL). It expires 48 hours after the calibration time or within 8 hours of thawing. It is indicated for the relief of pain in patients with confirmed osteoblastic metastatic bone disease.

[153]Sm is a beta and gamma emitter, and its 103 keV gamma ray permits bone imaging if desired. The biologic localization in lesions and normal bone of [153]Sm-lexidronam is similar to that of ^{99m}Tc-medronate.[65] Its half-life is 46.3 hours. It has been suggested that this short half-life results in a high dose rate over a short period of time, which should provide a rapid onset of pain relief and a limited amount of bone marrow suppression, but this has not been proven. After intravenous administration, [153]Sm-lexidronam localizes in bone metastases similarly to ^{99m}Tc-diphosphonate bone imaging agents.[66] By 4 to 6 hours after dosing, urinary excretion is essentially complete and uptake in bone is related to the extent of metastases present, not the dose administered.[67,68] Total bone uptake averages 65% of the injected dose. The critical organ is bone surface, with a radiation absorbed dose of 25 rad(cGy)/mCi.[68]

[153]Sm-lexidronam suppresses bone marrow in a manner similar to ^{89}Sr-chloride. Platelet counts reach a nadir in 3 to 5 weeks with recovery by 8 weeks; with ^{89}Sr-chloride the nadir occurs in 5 to 6 weeks with recovery in 12 weeks.[67] The standard dosage is 1 mCi/kg (37 MBq/kg) body weight. At this dosage, relief of bone pain occurs 1 week after dosing in 35% of patients, increasing to 70% by the fourth week.[67] By the 16th week, 39% of patients still note effective pain relief. After dosing at 1 mCi/kg (37 MBq/kg), the rate of flare response is low (7%, compared with 5.6% in untreated control subjects).[68] Retreatment with additional doses of [153]Sm-lexidronam has been shown to be safe and efficacious.[69]

Dose calibrator calibration factors must be established with [153]Sm-lexidronam because of the geometry dependence of activity measurements with beta emitters, despite the gamma emission of [153]Sm. This is typically done before the first dose is administered, by assay of a 15 mCi (555 MBq) calibration standard supplied by the manufacturer. Typically, calibration factors are determined by assaying a known activity of [153]Sm in its glass vial and in 3 mL and 5 mL plastic syringes used for patient dosing.

Rhenium Re 186 Etidronate Injection

The similarities of rhenium and technetium chemistry, the ability of both elements to label bone-seeking diphosphonates, and the beta–gamma emissions of ^{186}Re, which permit radiation therapy and gamma-camera imaging, prompted the development of ^{186}Re-etidronate (^{186}Re-HEDP) for bone pain palliation. The short 89 hour half-life is thought to permit relatively high dose rates for treatment of metastases and allow for repeated treatments. Dosimetry studies have demonstrated a high tumor-to-marrow dose ratio (mean, 34:1).[70] Kits have been developed for its preparation.[71,72] The labeled product is stabilized with ascorbic acid to prevent oxidation to perrhenate (^{86}ReO$_4^-$), which does not localize in bone.[73] This radiopharmaceutical is approved for use in Europe but not in the United States.

Tin Sn 117m Pentetate (DTPA) Injection

^{117m}Sn decays by isomeric transition, emitting a 159 keV gamma ray (86%) and conversion electrons of 127 to 129 keV and 152 keV (114%). Preliminary dosimetry calculations for ^{117m}Sn-DTPA in bone and bone marrow estimate a favorable bone-to-bone-marrow radiation dose ratio of 9 to 1, primarily because of its low-energy conversion electron radiation.[74] These estimates indicate that ^{117m}Sn might be effective in the palliative treatment of metastatic bone pain without significant bone marrow toxicity. The lower kinetic stability of the ^{117m}Sn-DTPA complex compared with its thermodynamic stability is believed to cause transchelation of the ^{117m}Sn from DTPA to the bone surface, where it is residualized as an oxide.[73] When compared with ^{32}P-sodium phosphate in a murine model, the low-energy conversion electrons of ^{117m}Sn-DTPA demonstrated an 8-fold therapeutic advantage over the high-energy beta particles of ^{32}P.[75] Investigational studies have been conducted with this agent, but it has not received approval for use.[76]

Radium Ra 223 Chloride Injection

^{223}Ra-chloride injection (Alpharadin [Algeta ASA, Oslo, Norway]) is being evaluated in a Phase 3 clinical investigation for the treatment of bone metastases in patients with hormone-refractory prostate cancer (HRPC). This double-blind study is designed to test standard-of-care treatment plus ^{223}Ra-chloride versus standard-of-care treatment plus placebo to evaluate the effect of ^{223}Ra on overall survival. Approximately 750 patients with histologically or cytologically confirmed adenocarcinoma of the prostate will be studied worldwide. The dosage of ^{223}Ra-chloride will be 1.35 µCi(50 kBq)/kg body weight administered 6 times at intervals of 4 weeks.

There are several ways to treat patients with prostate cancer, commonly involving surgery, drugs, or radiation.

The growth of prostate cancer cells is stimulated by male hormones, particularly testosterone and its metabolite dihydrotestosterone. Hormone therapy is often used in treating patients with prostate cancer to reduce the amount of male hormone available to cancer cells. Treatment may include use of a luteinizing-hormone releasing hormone agonist to prevent the testicles from producing testosterone or use of androgen-receptor blockers. However, patients ultimately become unresponsive to these agents and are referred to as having androgen-independent or hormone-refractory prostate cancer. HRPC has a poor prognosis, with a median survival of only about 1 to 2 years.

Radionuclide therapy has been shown to be effective in relieving pain associated with skeletal metastases. However, the standard agents used (^{89}Sr and ^{153}Sm) are beta emitters whose long range in tissue can cause bone marrow toxicity. The potential for marrow toxicity limits the dosage of beta-emitting radionuclides to palliative treatment only. ^{223}Ra is an alpha-emitting radionuclide that localizes in bone following intravenous injection.[77–79] The rationale for its use in treating bone metastases is lower radiation dose to bone marrow because of the limited range of alpha particles in tissue (<100 µm) and the high localized radiation dose due to the high LET alpha radiation. Thus, ^{223}Ra has the potential to improve survival outcome and quality of life in the treatment of bone metastases.

^{223}Ra decays with a physical half-life of 11.4 days to ^{219}Rn, emitting alpha radiation of 5.6 MeV (25%) and 5.7 MeV (52%) and principal gamma radiation of 0.154 MeV (5.7%) and 0.269 MeV (13.9%). Gamma emission permits imaging if necessary. ^{223}Ra-chloride injection (Alpharadin) is supplied as a clear, colorless ready-to-use sterile solution of ^{223}RaCl$_2$. It has a pH of 6–8 and a radioactive concentration of 0.027 mCi (1 MBq)/mL in a volume of 6 mL in a 20 mL vial with a 14 day precalibration. The product expires 28 days from its production date.

POLYCYTHEMIA VERA

Polycythemia vera is a rare disease characterized by elevated red blood cell mass, total blood volume, and peripheral leukocyte and platelet counts.[80] Hematocrits may be 50% to 60% or more, white blood cell counts greater than 12,000/µL, and platelets in excess of 400,000/µL. The high viscosity of blood causes sluggish blood flow through organs. If untreated, thrombosis and hemorrhage that can lead to death may result.[81] The condition has been treated with phlebotomy, chemotherapy (chlorambucil, hydroxyurea), and ^{32}P-sodium phosphate. The treatment of choice is hydroxyurea, with ^{32}P-sodium phosphate therapy as a second choice if hydroxyurea therapy fails or patients cannot comply with the requirements of hydroxyurea therapy.[82]

The dose of ^{32}P-sodium phosphate recommended by the Polycythemia Vera Study Group is 2 to 3 mCi/m^2 (7.4–11.1 MBq/m^2) given intravenously after phlebotomy to achieve a hematocrit level of 42% to 47%.[83] The dose should not exceed 5 mCi (185 MBq). This is usually sufficient to produce a remission. If remission does not occur in 3 months, the dose should be increased by 25%. Another dose increase by 25%, but not exceeding 7 mCi (259 MBq), may be tried as a third dose after a period of another 3 months. Retreatment is usually restricted for 6 months thereafter.

When soluble radiophosphate enters the miscible body phosphate pool, it is concentrated by rapidly proliferating tissue. Its use in polycythemia vera and other bone marrow diseases is based on the fact that blood cell precursors in the bone marrow divide and proliferate rapidly in health and to an even greater extent in these diseases. The radiophosphate selectively concentrates in the mitotically active cells of the bone marrow and in trabecular and cortical bone. The radiation dose to the bone marrow has been estimated to be 24 rad(cGy)/mCi, divided among marrow, 13 rad(cGy); trabecular bone, 10 rad(cGy); and cortical bone, 1 rad(cGy).[84] An increased rate of neoplasia (leukemia) occurs with bone marrow doses greater than 100 rad(cGy).[85] This is likely to occur with typical therapy; however, median survival has been reported to be better with ^{32}P-sodium phosphate (12 years) than with phlebotomy (8 years).[86]

EFFUSION THERAPY

Peritoneal and pleural effusions are complications of malignant disease. Intracavitary instillation of insoluble ^{32}P-chromic phosphate suspension into the peritoneal cavity or pleural cavity has been used for many years as palliative treatment of painful symptoms caused by accumulation of fluid in these cavities. The insoluble colloidal particles of chromic phosphate are engulfed by floating macrophages and eventually by fixed tissue macrophages lining the wall of the serous cavity. Beta radiation from ^{32}P causes fibrosis of the mesothelium and small blood vessels, which leads to reduced fluid production.[87] The dosage range for intraperitoneal instillation is 10 to 20 mCi (370–740 MBq), and for intrapleural instillation it is 6 to 12 mCi (222–444 MBq). Typically intracavitary instillation of ^{99m}Tc-sulfur colloid and gamma camera imaging is done prior to instillation of ^{32}P-chromic phosphate to ensure satisfactory cavitary distribution. Cavitary fluid is removed and ^{32}P-chromic phosphate is administered, dispersed in 30 to 50 mL of sterile saline.

In August 2008 Covidien published an adverse reaction warning to health care practitioners, noting radiation injury (necrosis and fibrosis) to the small bowel, cecum, and bladder after administration of ^{32}P-chromic phosphate into the peritoneal cavity. The manufacturer eventually removed this product from the market.

RADIATION SYNOVECTOMY

Diseases such as rheumatoid arthritis and hemarthrosis (hemorrhagic bleeding into the joint) in hemophilia cause synovial inflammation in the joints. The inflammatory process involves increased secretion of synovial fluid and the release of enzymes and cell factors that result in cartilage degradation and eventual joint destruction.[88] The failure of drug therapy and the disadvantages associated with surgical synovectomy (hospitalization, physical therapy, and cost) led to the use of radiation synovectomy. Radiation synovectomy involves the injection of a nondiffusible radioactive agent (radiocolloids) into the involved joint space. Local irradiation of the inflamed synovial lining decreases fluid production and the pain associated with intra-articular pressure. There is also destruction of the cells responsible for releasing enzymes and cytokines that mediate joint destruction.

Radiation synovectomy has been used effectively for the treatment of rheumatoid arthritis and hemarthrosis, with reported success rates of 40% to 90%.[89] The European Association of Nuclear Medicine describes ideal properties of an agent for radiation synovectomy: (1) colloid particle size that does not leak from the joint space, ideally in the range of 2 to 10 µm, (2) stable binding of radionuclide to the particle, and (3) homogeneous distribution in the joint space without causing an inflammatory response.[86,87] In addition, the colloid should ideally remain in the joint during complete decay of the radionuclide. Several radiocolloids have been used for treatment. Colloidal gold ^{198}Au was used initially, but its use was discontinued because its small particle size (5–50 nm) resulted in diffusion of particles out of the joint space. Other agents include ^{32}P-chromic phosphate, ^{90}Y-silicate/citrate, ^{186}Re-sulfide, and ^{169}Er-citrate.[89] No agent has achieved approval in the United States for radiation synovectomy, although ^{32}P-chromic phosphate has been used off-label effectively for many years.[89] ^{32}P-chromic phosphate has a favorable beta particle range in tissue of about 8 mm and a half-life of 14.3 days, which permits administration of small amounts of radioactivity (1 to 2 mCi [37–74 MBq]), to bring about a slow destruction of tissue without inducing an acute inflammatory response.[88] The ^{32}P-chromic phosphate particles are phagocytized by synovial macrophages, helping to retain the particles within the joint space. However, the commercial product sold as Phosphocol P-32 was removed from the market by the manufacturer (Covidien) following its warning in August 2008 of increased risk for leukemia, which occurred in two children with hemophilia who received intra-articular injections of Phosphocol P-32 for hemarthroses. Despite this warning, physicians who choose to perform radiation synovectomy using ^{32}P-chromic phosphate may obtain it from an alternative source (AnazaoHealth, Tampa, FL). In Europe, ^{90}Y-silicate/citrate, ^{186}Re-sulfide, and ^{169}Er-citrate are recommended for use.[90] At present, these agents can only be used investigationally in the United States.

Diffusion of particles from the joint space into other tissues is a concern in radiation synovectomy. Particles that are very small are more prone to diffuse out of the joint space and be transported into lymphatic vessels.[89] Leakage rates have been investigated. When a joint is immobilized for 48 hours after injection, leakage rates of ^{90}Y-citrate (1–2.5 µm), ^{90}Y-silicate (1–2.5 µm), and ^{186}Re-sulfide (0.05–0.5 µm) of 3.2%, 2.3%, and 2.5%, respectively, have been reported. Joint immobilization after injection significantly reduces the chance of

TABLE 29-3 Radiocolloids for Radiation Synovectomy

Radiocolloid	Half-life (days)	Mean Beta Energy (MeV)	Mean Range in Tissue (mm)	Clinical Application	Dosage (mCi)
^{90}Y-citrate ^{90}Y-silicate	2.7	0.94	3.6	Knee joint	5–6
^{186}Re-sulfur colloid	3.7	0.349	1.1	Hip	2–5
				Shoulder	2–5
				Elbow	2
				Wrist	1–2
				Ankle	2
				Subtalar joints	1–2
^{169}Er-citrate	9.4	0.099	0.3	Metacarpophalangeal joint	0.5–1
				Metatarsophalangeal joint	0.8–1
				Interphalangeal joints	0.3–0.5

Source: Reference 82.

particle leakage to nontarget tissues.[91] The properties of recommended radiopharmaceuticals and their applications are summarized in Table 29-3.

BRACHYTHERAPY OF LIVER CANCER

Liver cancer is of primary origin or metastatic. Typical treatment options for liver cancer include surgical resection, systemic chemotherapy, hepatic artery embolization with or without chemotherapy, cryotherapy, radiofrequency ablation, and radiation therapy. During the past two decades selective internal radiation therapy (SIRT) with ^{90}Y-labeled microspheres has been investigated for the treatment of primary and metastatic liver disease. This was encouraged by the development of implantable ports and pumps within the liver and the production, in the 1980s, of radionuclide labels such as ^{90}Y that do not leach from microspheres.[92] SIRT offers the potential advantages of localizing radiation dose to the tumor while sparing healthy liver tissue, reducing chemotherapy-associated morbidity, and increasing survival. ^{90}Y is a pure beta emitter with a physical half-life of 64.2 hours. Its average beta energy is 0.9367 MeV with a mean range in tissue of 2.5 mm and a maximum range of 10 mm.

Hepatocellular Carcinoma

Hepatocellular carcinoma (HCC) is a primary malignant tumor of liver parenchymal cells that is preceded by cirrhosis secondary to other causes, particularly hepatitis B (HBV) and hepatitis C (HCV) infection and alcoholic liver disease.[93] HCC is staged by Okuda's method, which considers the presence or absence of the following characteristics: bilirubin >51 μmol/L, serum albumin <30 g/L, the presence of ascites, and >50% tumor replacement of the liver.[94] Stage I patients have none of these characteristics, stage II patients have one or two characteristics, and stage III patients have more than two characteristics. The median survival of untreated patients with stage I and stage II disease is 8.3 months and 2 months, respectively. When surgical resection is possible, patients at stages I and II have median survival after resection of 25.6 and 12.2 months, respectively, compared with patients treated medically, whose respective survival is only 9.4 and 3.5 months.[94] The prognosis for cure of HCC is poor; thus, the goal of treatment is palliation.

Most patients are not good candidates for surgical resection of HCC because of three interrelated factors: large tumor size, poor liver function, and bilobar or metastatic disease.[95] The traditional treatment of unresectable HCC has been hepatic artery infusion chemotherapy or transhepatic artery chemoembolization (TACE). Each of these methods is associated with underlying liver toxicity because of the already compromised condition of the liver.[96] To mitigate adverse events associated with treatment, a variety of treatment options for unresectable HCC with radioactive agents have been investigated.[97] An attractive approach has been hepatic arterial embolization with ^{90}Y-labeled microspheres, which offers the potential advantage of targeting radiation dose to the tumor while sparing surrounding liver tissue.[98] Treatment with ^{90}Y-labeled glass microspheres (TheraSpheres [MDS Nordion, Ottawa, Ontario, Canada]) has been shown to provide effective palliation of unresectable HCC.[95,96,99] Patient survival has been shown to increase with radiolabeled microsphere treatment. When compared with historical survival of untreated Okuda stage I and II patients of 8.3 months and 2.0 months, respectively,[94] Carr et al.[96] reported median survival for stage I and II treated patients of 21.6 months and 10.6 months, respectively, while Salem et al.[99] reported similar survivals of 23 months and 11 months, respectively. Potential complications with ^{90}Y-microsphere therapy include treatment-associated cholecystitis, elevation of liver enzymes and bilirubin, and gastroduodenal ulceration and pneumonitis due to excessive extrahepatic leakage of microspheres. Overall, however, this treatment modality has been found to improve quality of life for patients with primary and metastatic liver cancer.[100–102]

Another radiation treatment approach for HCC has been adjuvant radionuclide therapy at the time of surgical resection. The procedure involves a single postoperative hepatic arterial injection of ^{131}I-lipiodol, a radiolabeled oil that is mostly embolized within the tumor. A limited number of patients have undergone this treatment. A progress report on a 5 year case–control study of this method indicated no significant difference between treatment and control groups of patients but a trend toward better survival after surgical resection, which pointed to the need for a larger study with longer follow-up.[103]

Hepatic Metastases

The principal cause of death from advanced colorectal cancer is liver metastases. In contrast to many other cancers, metastasis from colorectal cancer is frequently confined to the liver[104] and is amenable to locoregional therapy with microembolization. Hepatic artery embolization with radiolabeled microspheres in not new, having been used in the past for the treatment of liver cancer.[105–107]

Several studies for treating hepatic metastases with ^{90}Y brachytherapy have been conducted with ^{90}Y-resin microspheres (SIR-Spheres [Sirtex Medical, Lane Cove, Australia]) in Australia and New Zealand, either alone or in combination with chemotherapy. Stubbs et al.,[108] using ^{90}Y-SIR-Spheres as a first treatment in patients with metastases confined to the liver, reported that the median carcinoembryonic antigen (CEA) level 3 months after treatment declined to 18% of values before treatment and only 5 of 80 patients followed by CT exam showed disease progression. Morbidity associated with this study was seen in 7 of 87 patients (8%) who received microspheres via surgical portacath and 4 of 13 patients (31%) who received microspheres via femoral artery catheter. Gray et al.,[109] using SIR-Spheres in combination with chemotherapy, demonstrated that the median time for disease progression in the liver was significantly longer for patients who received hepatic artery infusion of floxuridine chemotherapy plus one treatment of ^{90}Y-microspheres compared with floxuridine infusion alone. The 1, 2, 3, and 5 year survival for patients receiving the ^{90}Y-SIR-Sphere treatment was 72%, 39%, 17%, and 3.5%, compared with 68%, 29%, 6.5%, and 0% for chemotherapy alone. There was no significant difference between treatment groups, with no loss of quality of life due to the addition of microsphere therapy to chemotherapy. Although SIR-Sphere therapy has been used primarily for palliation and salvage therapy in chemotherapy-refractory patients, there is also evidence of potential cure of colorectal metastasis in patients with small isolated lesions.[110]

A multi-institutional study in the United States was conducted in 208 patients with unresectable colorectal liver metastases refractory to oxaliplatin and irinotecan chemotherapy.[111] ^{90}Y-SIR-Spheres were administered in one, two, or three courses of treatment without concurrent chemotherapy. The ^{90}Y-SIR-Sphere dose ranged from 10.8 to 78.4 mCi (0.4–2.9 GBq) with a median dose of 45.9 mCi (1.7 GBq). Patients who had no measurable response by 6 weeks of treatment, assessed by CEA blood levels, CT, and FDG PET, had short survival (4.5 months), compared with responders' median survival of 10.5 months. The investigators concluded that microsphere therapy produced an encouraging median survival, with acceptable toxicity, and a significant objective response rate, suggesting further investigation. Additional studies have been instituted to investigate ^{90}Y-SIR-Sphere therapy combined with chemotherapy.[112]

[90]Y-Microspheres

Two radionuclide microsphere devices are approved for brachytherapy in the United States. ^{90}Y-labeled glass microspheres (TheraSpheres [MDS Nordion, Ottawa, Ontario, Canada]) are indicated for the treatment of unresectable HCC, a primary liver tumor, and ^{90}Y-labeled resin microspheres (SIR-Spheres [Sirtex Medical, Lane Cove, Australia]) are indicated for treating unresectable liver metastasis from primary colorectal cancer.

TheraSpheres are supplied in 0.6 mL of sterile, pyrogen-free water in a 1.0 mL vee-bottom vial.[113] Mean sphere diameter ranges from 20 μm to 30 μm with between 22,000 and 73,000 microspheres in each milligram. Six dose sizes are available ranging from 81 mCi (3 GBq) to 540 mCi (20 GBq). The 3 GBq vial contains 1.2 million spheres.

SIR-Spheres are supplied in a vial that contains 40 million to 80 million resin microspheres in 5 mL of water for injection labeled with 3 GBq of ^{90}Y activity at the time of calibration.[114] Microsphere size is 32.5 μm ± 2.5 μm, with less than 10% smaller than 30 μm and greater than 35 μm.

Both TheraSpheres and SIR-Spheres are infused into the liver via the hepatic artery. The microspheres localize in the tumor by embolization of small arterial vessels. Beta radiation emitted from ^{90}Y, having an average range of 2.5 mm in tissue, delivers a localized tumoricidal dose of radiation to the tumor. Most normal liver tissue is spared radiation exposure because the majority of blood flow to the liver is via the portal circulation, which is bypassed by this technique.

Because of arteriovenous shunts in tumors, there is the possibility of extrahepatic arterial shunting of microspheres to the lungs and GI tract that can cause adverse effects in these tissues.[115,116] Prior to administration of ^{90}Y-microspheres, 2 to 4 mCi (74–148 MBq) of ^{99m}Tc-MAA is administered via the hepatic arterial catheter, followed by gamma camera imaging to determine the extent of shunting to the lung and confirm absence of gastric and duodenal flow.[113] The dose fraction shunted to the lung is assessed from ^{99m}Tc activity localized in the liver and lung according to the following formula:

$$\% \text{ Lung shunting} = \frac{\text{Lung counts}}{\text{Liver counts} + \text{Lung counts}}$$

For TheraSpheres, the fraction shunted to the lungs multiplied by the ^{90}Y-microsphere activity injected cannot exceed 16.5 mCi (0.61 GBq), equivalent to an upper limit lung dose of 30 Gy.[113] For Sir-Spheres the ^{90}Y-microsphere dosage must be reduced by 20% if lung shunting is 10% to 15% and reduced by 40% if shunting is 15% to 20%. No SirSphere dosage should be given if shunting exceeds 20%.[114]

Accuracy of shunt assessment is related to the diagnostic method used. In a recent study in 68 patients, the sensitivity and specificity, respectively, for assessing extrahepatic shunting was 25% and 87% by planar imaging, 56% and 87% by SPECT, and 100% and 94% by SPECT/CT.[117]

Dose Calculation

For ^{90}Y-Theraspheres the recommended dose to the liver is between 80 Gy and 150 Gy (8,000 rad to 15,000 rad) and is determined by the following formula:[113]

$$\text{Activity required (GBq)} = \frac{[\text{Desired dose (Gy)}][\text{Liver mass (kg)}]}{50}$$

Liver mass is determined from volume measurements made by CT and the conversion factor 1.03 g/cm^3. The actual dose delivered to the liver after injection is determined by the following formula:

$$\text{Dose (Gy)} = \frac{50[\text{Injected activity (GBq)}][1-F]}{\text{Liver mass (kg)}}$$

where F is the fraction of injected activity localizing in the lungs measured by ^{99m}Tc-MAA scintigraphy. The actual activity injected into the liver is determined by multiplying the vial activity measured in a dose calibrator by the fraction of this activity delivered during the procedure. The fraction delivered is determined by measuring the ^{90}Y dose vial before injection and after injection with a calibrated ionization chamber in a defined geometry. The upper limit of injected activity shunted to the lungs is $F \times$ Activity injected = 16.5 mCi (0.61 GBq).

The risk of liver toxicity in patients with unresectable HCC has been shown to increase with increased pretreatment bilirubin levels and single ^{90}Y-microsphere liver dose up to 150 Gy.[101] Dose-induced toxicities were resolved with no evidence of radiation-induced liver disease. It was concluded that doses up to 150 Gy after single administration and up to 268 Gy after repeated administration were well tolerated.

For ^{90}Y-SIR-Spheres the patient dosage (GBq) is based on the extent of tumor involvement in the liver.[114] For tumor involvement >50%, the ^{90}Y dose is 3 GBq; between 25% and 50%, the dose is 2.5 GBq; and for <25%, the dose is 2 GBq. Also, an appropriate reduction of the dosage is made depending on the amount of shunting to the lung, as stated above.

Dose calibrator measurement of ^{90}Y-microspheres requires careful attention. A method must be established to correction for beta particle and bremsstrahlung attenuation to ensure an accurate dose calibrator measurement.[118]

TARGETED THERAPY WITH ALPHA PARTICLE EMITTERS

The efficient killing of tumors by radiation emitted from alpha particles localized within tumor cells has been of great interest for many years.[79,119–126] The high LET of alpha particles results in great amounts of energy being released in small volumes of the cell (see Figures 4-2 and 4-5, Chapter 4), creating double-strand chromosome breaks in DNA that have low probability of repair. Consequently, it is possible to achieve a level of tumor cell killing with intracellular deposition of alpha particles that is not achievable with beta particle emitters delivered to tumors. Despite this ideal concept for tumor treatment with radiation, the major challenge for the alpha particle approach is the same as that for any radiotracer that targets tumors: specificity for tumor while sparing normal tissue. Although no product has achieved approval for use in humans, novel dosage forms have been developed for targeted delivery of alpha emitters to tumors, and several clinical trials are being conducted to test feasibility (Table 29-4). The radiobiology and dosimetry of alpha particle emitters for targeted radioimmunotherapy was recently reviewed.[127]

NEW RADIOTHERAPEUTIC AGENTS

New radiolabeled molecular entities are being developed for the systemic treatment of a variety of diseases. Molecular Insight Pharmaceuticals (http://www.molecularinsight.com) has devel-

TABLE 29-4 Clinical Trial Studies with Alpha Particle Emitters

Radionuclide	Targeting Molecule	Cancer Target	Reference
^{211}At	Antitenascin IgG	Glioblastoma multiforme	120
	MX35 F(ab')$_2$	Ovarian carcinoma	121
^{213}Bi	Anti-CD33 IgG	Myelogenous leukemia (acute and chronic)	122, 123
	Antineurokinin receptor peptide	Glioblastoma	124
	Anti-CD20 IgG (rituximab)	Relapsed or refractory non-Hodgkin's lymphoma	125
	9.2.27 IgG	Melanoma	126
^{223}Ra	RaCl$_2$	Skeletal metastases of breast and prostate cancer	127, 128
^{225}Ac	Anti-CD33 IgG	Acute myelogenous leukemia	123

oped several agents that are in different stages of investigation. They include ^{131}I-iobenguane (Azedra), ^{131}I-ioflubenzamide (Solazed), and ^{90}Y-edotreotide (Onalta).

^{131}I-iobenguane (Azedra) is a neuroendocrine tumor-targeting agent being developed for the treatment of pheochromocytoma and paragangliomas in adult patients and neuroblastoma in children. The method for producing Azedra is designed to reduce the amount of nonradioactive iobenguane in the final product. The resulting high specific activity product has the potential benefits of higher tumor uptake and fewer side effects such as hypertension and nausea at the time of administration. Azedra is in Phase 2 trials in the United States and in commercial production in other countries.

^{131}I-ioflubenzamide (Solazed) is designed to treat malignant metastatic melanoma. This compound targets melanin, the pigment present in high concentration in approximately 50% of metastatic melanomas. Solazed is in early Phase 1 clinical trials in the United States.

^{90}Y-edotreotide (Onalta), also known as ^{90}Y-SMT487, ^{90}Y-DOTATOC, and ^{90}Y-DOTA-tyr^3-octreotide, is a radiolabeled analogue of octreotide, a peptide that binds to somatostatin receptors. This agent has been designed to target somatostatin-receptor positive tumors and is intended for the treatment of pancreatic neuroendocrine tumors and metastatic carcinoid tumors. Onalta has reached Phase 2 clinical trials in the United States and approval for Phase 3 trials in Europe. Metastatic carcinoid tumors, which are neuroendocrine tumors of the GI tract and lung bronchus, are incurable malignancies with debilitating symptoms. Palliation of symptoms is traditionally achieved with octreotide therapy; however, in some patients treatment becomes refractory to octreotide. Improved symptom treatment in such patients has been reported after administration of three cycles of 120 mCi (4.4 GBq) each of ^{90}Y-edotreotide given every 6 weeks.[128] The treatment regimen includes concurrent administration of intravenous amino acids to reduce radiation dose to the kidneys. However, the amino acid infusion itself can cause GI side effects, which are reversible upon cessation of the infusion.

REFERENCES

1. Krieger MS, Weiden PL, Breitz HB, et al. Radioimmunotherapy in the treatment of non-Hodgkin's lymphoma. In: Abrams PG, Fritzberg AR, eds. *Radioimmunotherapy of Cancer*. New York: Marcel Dekker; 2000:245–63.
2. Reff ME, Carner K, Chambers KS, et al. Depletion of B cells in vivo by a chimeric mouse human monoclonal antibody to CD20. *Blood*. 1994;83:435–45.
3. Maloney DG, Grillo-Lopez AJ, White CA, et al. IDEC-C2B8 (Rituximab) anti-CD20 monoclonal antibody therapy in patients with relapsed low-grade non-Hodgkin's lymphoma. *Blood*. 1997;90: 2188–95.
4. McLaughlin P, Grillo-Lopez AJ, Link BK, et al. Rituximab chimeric anti-CD20 monoclonal antibody therapy for relapsed indolent lymphoma: half of patients respond to a four-dose treatment program. *J Clin Oncol*. 1998;16:2825–33.
5. Mattes MJ. Pharmacokinetics of antibodies and their radiolabels. In: Goldenberg DM, ed. *Cancer Therapy with Radiolabeled Antibodies*. Boca Raton, FL: CRC Press; 1995:89–99.
6. Juweid ME. Radioimmunotherapy of B-cell non-Hodgkin's lymphoma: from clinical trial to clinical practice. *J Nucl Med*. 2002;43: 1507–29.
7. Humm JL. Dosimetric aspects of radiolabeled antibodies for tumor therapy. *J Nucl Med*. 1986;27;1490–97.
8. Zalutsky MR. Radiohalogens for radioimmunotherapy. In: Abrams PG, Fritzberg AR, eds. *Radioimmunotherapy of Cancer*. New York: Marcel Dekker; 2000:81–106.
9. DeNardo GL, Juweid ME, White CA, et al. Role of radiation dosimetry in radioimmunotherapy planning and treatment dosing. *Crit Rev Oncol Hematol*. 2001;39:203–18.
10. Juweid M, Dunn R, Zhang CH, et al. The contribution of whole-body radiation to red marrow dose in patients receiving radioimmunotherapy with ^{131}I-labeled anti-carcinoembryonic antigen monoclonal antibodies [abstract]. *Cancer Biother Radiopharm*. 1998;13:319.
11. Kaminski MS, Zasadny KR, Francis IR, et al. Iodine-131-anti-B1 radioimmunotherapy for B-cell lymphoma. *J Clin Oncol*. 1996;14: 1974–81.
12. Witzig T, White CA, Wiseman G, et al. Phase I/II trial of IDEC-Y2B8 radioimmunotherapy for treatment of relapsed or refractory CD20+ B-cell non-Hodgkin's lymphoma. *J Clin Oncol*. 1999;17:3793–803.
13. Press OW, Eary JF, Appelbaum FR, et al. Radiolabeled-antibody therapy of B-cell lymphoma with autologous bone marrow support. *N Engl J Med*. 1993;329:1219–24.
14. Chinn PC, Leonard JE, Rosenberg J, et al. Preclinical evaluation of ^{90}Y-labeled anti-CD 20 monoclonal antibody for treatment of non-Hodgkin's lymphoma. *Int J Oncol*. 1999;15:1017–25.
15. Zevalin [package insert]. Irvine, CA: Spectrum Pharmaceuticals Inc.; 2009.
16. Witzig TE, Gordon LI, Cabanillas F, et al. Randomized controlled trial of yttrium-90-labeled ibritumomab tiuxetan radioimmunotherapy versus rituximab immunotherapy for patients with relapsed or refractory low-grade, follicular, or transformed B-cell non-Hodgkin's lymphoma. *J Clin Oncol*. 2002;20:2453–63.
17. Wiseman GA, White CA, Stabin M, et al. Phase I/II ^{90}Y-Zevalin (yttrium-90 ibritumomab tiuxetan, IDEC-Y2B8) radioimmunotherapy dosimetry results in relapsed or refractory non-Hodgkin's lymphoma. *Eur J Nucl Med*. 2000;27:766–77.
18. Conti PS, White C, Pieslor P, et al. The role of imaging with ^{111}In-ibritumomab tiuxetan in the ibritumomab tiuxetan (Zevalin) regimen: results from a Zevalin imaging registry. *J Nucl Med* 2005; 46:1812–8.
19. Wagner HN Jr, Wiseman GA, Marcus CS, et al. Administration guidelines for radioimmunotherapy of non-Hodgkin's lymphoma with ^{90}Y-labeled anti-CD20 monoclonal antibody. *J Nucl Med*. 2002; 43:267–72.
20. Bexxar [package insert]. Seattle, WA: Corixa and Glaxo Smith Kline; 2003.
21. Rose LM, Gunasekera AH, DeNardo SJ, et al. Lymphoma-selective antibody Lym-1 recognizes a discontinuous epitope on the light chain of HLA-DR10. *Cancer Immunol Immunother*. 1996;43:26–30
22. DeNardo GL. Treatment of non-Hodgkin's lymphoma (NHL) with radiolabeled antibodies (mAbs). *Semin Nucl Med*. 2005;35:202–11.
23. Fritzberg AR, Meares CF. Metallic radionuclides for radioimmunotherapy. In: Abrams PG, Fritzberg AR, eds. *Radioimmunotherapy of Cancer*. New York: Marcel Dekker; 2000:57–79.
24. Griffiths GL. Radiochemistry of therapeutic radionuclides. In: Goldenberg DM, ed. *Cancer Therapy with Radiolabeled Antibodies*. Boca Raton, FL: CRC Press; 1995:47–61.
25. Linder MC, Hazegh-Azam M. Copper biochemistry and molecular biology. *Am J Clin Nutr*. 1996;63:797s-811s.
26. Juweid M, Sharkey RM, Markowitz A, et al. Treatment of non-Hodgkin's lymphoma with radiolabeled murine, chimeric, or humanized LL2, an *anti*-CD22 monoclonal antibody. *Cancer Res*. 1995;55: 5899–907.
27. Shih LB, Lu HH, Xuan H, et al. Internalization and intracellular processing of an anti-B-cell lymphoma monoclonal antibody, LL2. *Int J Cancer*. 1994;56:538–45.

28. Sharkey RM, Behr TM, Mattes MJ, et al. Advantage of residualizing radiolabels for an internalizing antibody against the B-cell lymphoma antigen, CD22. *Cancer Immunol Immunother.* 1997;44:179–88.
29. Naruki Y, Carrasquillo JA, Reynolds JC, et al. Differential cellular catabolism of ^{111}In, ^{90}Y, and ^{125}I radiolabeled T101 anti-CD5 monoclonal antibody. *Nucl Med Biol.* 1990;17:201–7.
30. Geissler F, Anderson SK, Venkatesan P, et al. Intracellular catabolism of radioiodinated anti-μ antibodies by malignant B-cells. *Cancer Res.* 1992;52:2907–15.
31. Dumas P, Maziere B, Autissoer N, et al. Specificite de l'iodotyrosine desiodase des microsomes thyroidiens et hepatiques. *Biochim Biophys Acta.* 1973;293:36–47.
32. Shih LB, Thorpe SR, Griffiths GL, et al. The processing and fate of antibodies and their radiolabels bound to the surface of tumor cells in vitro: a comparison of nine radiolabels. *J Nucl Med.* 1994;35:899–908.
33. Stein R, Goldenberg DM, Thorpe SR, et al. Effects of radiolabeling monoclonal antibodies with a residualizing iodine radiolabel on the accretion of radioisotope in tumors. *Cancer Res.* 1995;55:3132–9.
34. Juweid ME, Stadtmauer E, Hajjar G, et al. Pharmacokinetics, dosimetry, and initial therapeutic results with ^{131}I- and ^{111}In/^{90}Y-labeled humanized LL2 anti-CD22 monoclonal antibody in patients with relapsed, refractory non-Hodgkin's lymphoma. *Clin Cancer Res.* 1999;5(suppl):3292–3303s.
35. Fisher RI, Gaynor ER, Dahlberg S, et al. Comparison of a standard regimen (CHOP) with three intensive chemotherapy regimens for advanced non-Hodgkin's lymphoma. *N Engl J Med.* 1993;328:1002–6.
36. Longo DL, Duffey PL. Management of aggressive histology lymphoma: an approach based on data from the National Cancer Institute. *Hematol/Oncol Ann.* 1993;1:19–28.
37. Hu E, Epstein AL, Naeve GS, et al. A phase 1a clinical trial of Lym-1 monoclonal antibody serotherapy in patients with refractory B cell malignancies. *Hematol Oncol.* 1989;7:155–66.
38. Hu P, Hornick JL, Glasky MS, et al. A chimeric Lym-1/interleukin 2 fusion protein for increasing tumor vascular permeability and enhancing antibody uptake. *Cancer Res.* 1996;56:4998–5004
39. Bodet-Milin C, Chérel M, Faivre-Chauvet A, et al. Radioimmunotherapy of B-cell non-Hodgkin's lymphoma. *Cancer Ther.* 2008;6:247–56.
40. DeNardo SJ, DeNardo GL, O'Grady LF, et al. Treatment of a patient with B cell lymphoma by I-131 Lym-1 monoclonal antibodies. *Int J Biol Markers.* 1987;2:49–53.
41. DeNardo SJ, DeNardo GL, O'Grady LF, et al. Pilot studies of radioimmunotherapy of B cell lymphoma and leukemia using I-131 Lym-1 monoclonal antibody. *Antibody Immunoconj Radiopharm.* 1988;1:17–33.
42. Kaminski MS, Zelenetz AD, Press OW, et al. Pivotal study of iodine I-131 tositumomab for chemotherapy-refractory low-grade or transformed low-grade B-cell non-Hodgkin's lymphomas. *J Clin Oncol.* 2001;19:3918–28.
43. Kaminski MS, Radford JA, Gregory SA, et al. Re-treatment with I-131 tositumomab in patients with non-Hodgkin's lymphoma who had previously responded to I-131 tositumomab. *J Clin Oncol.* 2005;23:7985–93.
44. Kaminski MS, Estes J, Zasadny KR, et al. Radioimmunotherapy with iodine ^{131}I-tositumomab for relapsed or refractory B-cell non-Hodgkin's lymphoma: updated results and long-term follow-up of the University of Michigan experience. *Blood.* 2000;96:1259–66.
45. Kaminski MS, Tuck M, Estes J, et al. I-131 tositumomab therapy as initial treatment for follicular lymphoma. *N Engl J Med.* 2005;352:441–9.
46. Press OW, Eary JF, Appelbaum FR, et al. Treatment of relapsed B cell lymphomas with high-dose radioimmunotherapy and bone marrow transplantation. In: Goldenberg DM, ed. *Cancer Therapy with Radiolabeled Antibodies.* Boca Raton, Florida: CRC Press; 1995:229–38.
47. Press OW, Eary JF, Appelbaum FR, et al. Phase II trial of ^{131}I-B1 (anti-CD20) antibody therapy with autologous stem cell transplantation for relapsed B cell lymphomas. *Lancet.* 1995;346:336–40.
48. Goldenberg DM. Targeted therapy of cancer with radiolabeled antibodies. *J Nucl Med.* 2002;43:693–713.
49. DeNardo GL, O'Donnell RT, Oldham RK, et al. A revolution in the treatment of non-Hodgkin's lymphoma. *Cancer Biother Radiopharm.* 1998;13:213–23.
50. Chinol M, Grana C, Gennari R, et al. Pretargeted radioimmunotherapy of cancer. In: Abrams PG, Fritzberg AR, eds. *Radioimmunotherapy of Cancer.* New York: Marcel Dekker; 2000:169–93.
51. Theodore LJ, Fritzberg AR, Schultz JE, et al. Evolution of a pretarget radioimmunotherapeutic regimen. In: Abrams PG, Fritzberg AR, eds. *Radioimmunotherapy of Cancer.* New York: Marcel Dekker; 2000:195–221.
52. Paganelli G, Magnani P, Siccardi AG, et al. Clinical application of the avidin-biotin system for tumor targeting. In: Goldenberg DM, ed. *Cancer Therapy with Radiolabeled Antibodies.* Boca Raton, FL: CRC Press; 1995:239–54.
53. Goldenberg DM, Sharkey RM, Paganelli G, et al. Antibody pretargeting advances cancer radioimmunodetection and radioimmunotherapy. *J Clin Oncol.* 2006;24:823–34.
54. Meredith R, Buchsbaum D. Pretargeted radioimmunotherapy. *Int J Rad Oncol Biol Phys.* 2006;66:S57–S59.
55. Paganelli G, Grana C, Chinol M, et al. Antibody guided three-step therapy for high grade glioma with yttrium-90 biotin. *Eur J Nucl Med.* 1999;26:348–57.
56. Breitz HB, Weiden PL, Beaumier PL, et al. Clinical optimization of pretargeted radioimmunotherapy (PRIT™) with antibody–streptavidin conjugate and ^{90}Y-DOTA-biotin. *J Nucl Med.* 2000;41:131–40.
57. Robinson RG, Preston DF, Spicer JA, et al. Radionuclide therapy of intractable bone pain: emphasis on strontium-89. *Semin Nucl Med.* 1992;22:28–32.
58. Silberstein EB. Systemic radiopharmaceutical therapy of painful osteoblastic metastases. *Semin Radiat Oncol.* 2000;10:240–9.
59. Serafini AN. Therapy of metastatic bone pain. *J Nucl Med.* 2001;42:895–906.
60. Silberstein EB. Dosage and response in radiopharmaceutical therapy of painful osseous metastases. *J Nucl Med.* 1996;7:249–52.
61. Friedell HL, Storaasli JP. The use of radioactive phosphorus in the treatment of carcinoma of the breast with widespread metastases to bone. *AJR Am J Roentgenol.* 1950;64:559–75.
62. Silberstein EB. The treatment of painful osseous metastases with phosphorus-32-labeled phosphates. *Semin Oncol.* 1993;20(suppl 2):10–21.
63. Nair N. Relative efficacy of ^{32}P and ^{89}Sr in palliation in skeletal metastases. *J Nucl Med.* 1999;40:256–61.
64. Herold TJ, Gross GP, Hung JC. A technique for measurement of strontium-89 in a dose calibrator. *J Nucl Med Technol.* 1994;22:110.
65. Eary JF, Collins C, Stabin M, et al. Samarium-153-EDTMP biodistribution and dosimetry estimation. *J Nucl Med.* 1993;34:1031–36.
66. Goeckeler WF, Troutner DE, Volkert WA, et al. ^{153}Sm radiotherapeutic bone agents. *Nucl Med Biol.* 1986;13:479–82.
67. Resche I, Chatal J-F, Pecking A, et al. A dose-controlled study of ^{153}Sm-ethylenediaminetetramethylenephosphonate (EDTMP) in the treatment of patients with painful bone metastases. *Eur J Cancer.* 1997;33:1583–91.
68. Quadramet [package insert]. Richmond, CA: Berlex; 1999.
69. Menda Y, Bushnell DL, Williams RD. Efficacy and safety of repeated samarium-153 lexidronam treatment in a patient with prostate cancer and metastatic bone pain. *Clin Nucl Med.* 2000;25:698–700.
70. Maxon HR, Deutsch EA, Thomas SR, et al. Re-186 (sn) HEDP for treatment of multiple metastatic foci in bone: human biodistribution and dosimetric studies. *Radiology.* 1988;166:501–7.
71. Maxon HR III, Thomas SR, Hertzberg VS, et al. Rhenium-186 hydroxyethylidene diphosphonate for the treatment of painful osseous metastases. *Semin Nucl Med.* 1992;22:33–40.
72. de Klerk JMH, van Dijk A, van het Schip AD, et al. Pharmacokinetics of rhenium-186 after administration of rhenium-186-HEDP to patients with bone metastases. *J Nucl Med.* 1992;33:646–51.

73. Volkert WA, Hoffman TJ. Therapeutic radiopharmaceuticals. *Chem Rev.* 1999;99:2269–92.
74. Atkins HL, Mausner LF, Srivastava SC, et al. Tin-117m(4+)-DTPA for palliation of pain from osseous metastases: a pilot study. *J Nucl Med.* 1995;36:725–9.
75. Bishayee A, Rao DV, Srivastava SC, et al. Marrow-sparing effects of ^{117m}Sn(4+) diethylenetriaminepentaacetic acid for radionuclide therapy of bone cancer. *J Nucl Med.* 2000;41:2043–50.
76. Srivastava SC, Atkins HL, Krishnamurthy GT, et al. Treatment of metastatic bone pain with tin-117m stannic diethylenetriaminepentaacetic acid: a phase I/II clinical study. *Clin Cancer Res.* 1998;4:61–8.
77. Henriksen G, Breistøl K, Bruland ØS, et al. Significant antitumor effect from bone seeking, alpha-particle emitting ^{223}Ra demonstrated in an experimental skeletal metastases model. *Cancer Res.* 2002;62:3120–5.
78. Henriksen G, Fisher DR, Roeske JC, et al. Targeting of osseous sites with α-emitting ^{223}Ra: comparison with the β-emitter ^{89}Sr in mice. *J Nucl Med.* 2003;44: 252–9.
79. Nilsson S, Franzen L, Parker C, et al. Bone-targeted radium-223 in symptomatic, hormone-refractory prostate cancer: a randomised, multicentre, placebo-controlled Phase II study. *Lancet Oncol.* 2007;8:587–94.
80. Hoffman R, Wasserman LR. Natural history and management of polycythemia vera. *Adv Intern Med.* 1979;24:255–85.
81. Chievitz E, Thiede T. Complications and causes of death in polycythemia vera. *Acta Med Scand.* 1962;172:513–23.
82. Shastri KA, Logue GL. Polycythemia vera. In: Rakel RE, ed. *Conn's Current Therapy.* Philadelphia: WB Saunders; 1993.
83. Wasserman LR. The treatment of polycythemia vera. *Semin Hematol.* 1976;13:57–78.
84. Spiers FW, Beddoe AH, King SD. The absorbed dose to bone marrow in the treatment of polycythemia by P-32. *Br J Radiol.* 1976;49:133–40.
85. Silberstein EB. Radionuclide therapy of hematologic disorders. *Semin Nucl Med.* 1979;9:100–7.
86. Osgood EE. Polycythemia vera: age relationships and survival. *Blood.* 1965;26:243–56.
87. Hazra TA, Howells R. Use of beta emitters for intracavitary therapy. In: Spencer RP, ed. *Therapy in Nuclear Medicine.* New York: Grune & Stratton; 1978:307–12.
88. Shortkroff S, Sledge CB. Radiation synovectomy. In: Wagner HN Jr, Szabo Z, Buchanan JW, eds. *Principles of Nuclear Medicine.* Philadelphia: WB Saunders; 1995:1021–8.
89. Schneider P, Farahati J, Reiners C. Radiosynovectomy in rheumatology, orthopedics, and hemophilia. *J Nucl Med.* 2005;46:48S–54S.
90. Clunie G, Fischer M. EANM procedure guidelines for radiosynovectomy. *Eur J Nucl Med.* 2003;30(3):BP12–BP16.
91. Gratz S, Göbel D, Behr TM, et al. Correlation between radiation dose, synovial thickness, and efficacy of radiosynoviorthesis. *J Rheumatology.* 1999;26:1242–9.
92. Gray BN, Burton MA, Kelleher DK, et al. Selective internal radiation (SIR) therapy for treatment of liver metastases: measurement of response rate. *J Surg Oncol.* 1989;42:192–6.
93. Monto A, Wright TL. The epidemiology and prevention of hepatocellular carcinoma. *Semin Oncol.* 2001;28:441–9.
94. Okuda K, Ohtsuki T, Obata H, et al. Natural history of hepatocellular carcinoma and prognosis in relation to treatment. *Cancer.* 1985;56:918–28.
95. Dancey JE, Shephard FA, Paul K, et al. Treatment of nonresectable hepatocellular carcinoma with intrahepatic ^{90}Y-microspheres. *J Nucl Med.* 2000;41:1673–81.
96. Carr B. Hepatic arterial ^{90}Yttrium glass microspheres (Therasphere) for unresectable hepatocellular carcinoma: interim safety and survival data on 65 patients. *Liver Transplantation.* 2004;10 (Suppl 1):S107–10.
97. Lambert B, Van de Wiele C. Treatment of hepatocellular carcinoma by means of radiopharmaceuticals. *Eur J Nucl Med Mol I.* 2005;32:980–9.
98. Kennedy C, Nutting C, Coldwell D, et al. Pathologic response and microdosimetry of ^{90}Y microspheres in man: review of four explanted whole livers. *Int J Radiat Oncol Biol Phys.* 2004;60:1552–63.
99. Salem R, Thurston KG, Carr BI, et al. Yttrium-90 microspheres: radiation therapy for unresectable liver cancer. *J Vasc Interv Radiol.* 2002;13:S223–S229.
100. Ibrahim SM, Nikolaidis P, Miller FH, et al. Radiologic findings following Y-90 radioembolization for primary liver malignancies. *Abdom Imaging.* 2009;34:566–81.
101. Goin JE, Salem R, Carr BI, et al. Treatment of unresectable hepatocellular carcinoma with intrahepatic ^{90}Y-microspheres: factors associated with liver toxicities. *J Vasc Interv Radiol.* 2005;16(2):205–13.
102. Salem R, Thurston K. Radioembolization with ^{90}Yttrium microspheres: a state-of-the-art brachytherapy treatment for primary and secondary liver malignancies part 1: Technical and methodologic considerations. *J Vasc Interv Radiol.* 2006;17:1251–78.
103. Boucher E, Bouguen G, Garin E, et al. Adjuvant intraarterial injection of ^{131}I-labeled lipiodol after resection of hepatocellular carcinoma: progress report of a case-control study with a 5-year minimal follow-up. *J Nucl Med.* 2008;49:362–6.
104. Weiss L, Grundmann E, Torhorst J, et al. Haematogenous metastatic patterns in colonic carcinoma: an analysis of 1541 necropsies. *J Pathol.* 1986;150:195–203.
105. Ariel IM. Treatment of inoperable primary pancreatic and liver cancer by the intra-arterial administration of radioactive isotopes (Y-90 radiating microspheres). *Ann Surg.* 1965;162:267–78.
106. Ariel IM, Pack GT. Treatment of inoperable cancer of the liver by intra-arterial radioactive isotopes and chemotherapy. *Cancer.* 1967;20:793–804.
107. Mantravadi RV, Spigos DG, Tan WS, et al. Intraarterial yttrium-90 in the treatment of hepatic malignancy. *Radiology.* 1982;142:783–6.
108. Stubbs R, O'Brien I, Correia M. Selective internal radiation therapy with ^{90}Y-microspheres for colorectal liver metastases: single-centre experience with 100 patients. *ANZ J Surg.* 2006;76:696–703.
109. Gray B, Van Hazel G, Hope M, et al. Randomised trial of SIR-Spheres plus chemotherapy vs. chemotherapy alone for treating patients with liver metastases from primary large bowel cancer. *Ann Oncol.* 2001;12:1711–20.
110. Garrean S, Muhs A, Bui JT, et al. Complete eradication of hepatic metastasis from colorectal cancer by Yttrium-90 SIRT. *World J Gastroenterol.* 2007;13:3016–9.
111. Kennedy AS, Coldwell D, Nutting C, et al. Resin ^{90}Y-microsphere brachytherapy for unresectable colorectal liver metastases: modern USA experience. *Int J Radiat Oncol Biol Phys.* 2006;65:412–25.
112. Kuebler JP. Radioembolization of liver metastases in patients with colorectal cancer: a nonsurgical treatment with combined modality potential. *J Clin Oncol.* 2009;27:4041–2.
113. Therasphere [package insert]. Ottawa, ON, Canada: MDS Nordion; 2007.
114. SIR-Spheres [package insert]. Lane Cove NSW, Australia: Sirtex Medical Ltd; 2006.
115. Yip D, Allen R, Ashton C, et al. Radiation-induced ulceration of the stomach secondary to hepatic embolization with radioactive yttrium microspheres in the treatment of metastatic colon cancer. *J Gastroenterol Hepatol.* 2004;19:347–9.
116. Leung TW, Lau WY, Ho SK, et al. Radiation pneumonitis after selective internal radiation treatment with intraarterial ^{90}Yttrium-microspheres for inoperable hepatic tumors. *Int J Radiat Oncol Biol Phys.* 1995;33:919–24.
117. Hamami ME, Poeppel TC, Müller S, et al. SPECT/CT with ^{99m}Tc-MAA in radioembolization with ^{90}Y microspheres in patients with hepatocellular cancer. *J Nucl Med.* 2009;50:688–92.
118. Moore S, Park MA, Limpa-Amara N, et al. Measurement of Y-90 resin microsphere activity using dose calibrators. *J Nucl Med.* 2007;48(suppl):74P.
119. Zalutsky ME, Reardon DA, Akabani G, et al. Clinical experience with alpha-particle emitting ^{211}At: treatment of recurrent brain tumor

119. patients with ^{211}At-labeled chimeric antitenascin monoclonal antibody 81C6. *J Nucl Med.* 2008;49:30–8.
120. Hultborn R, Andersson H, Back T, et al. Pharmacokinetics and dosimetry of ^{211}At-MX35 F(AB')(2) in therapy of ovarian cancer: preliminary results from an ongoing phase I study [abstract]. *Cancer Biother Radiopharm.* 2006;21:395.
121. Jurcic JG, Larson SM, Sgouros G, et al. Targeted alpha particle immunotherapy for myeloid leukemia. *Blood.* 2002;100:1233–9.
122. Jurcic JG, McDevitt MR, Pandit-Taskar N, et al. Alpha-particle immunotherapy for acute myeloid leukemia (AML) with bismuth-213 and actinium-225 [abstract]. *Cancer Biother Radiopharm.* 2006;21:396.
123. Kneifel S, Cordier D, Good S, et al. Local targeting of malignant gliomas by the diffusible peptidic vector 1,4,7,10-tetraazacyclododecane-1-glutaric acid-4,7,10-triacetic acid-substance P. *Clin Cancer Res.* 2006;12:3843–50.
124. Heeger S, Moldenhauer G, Egerer G, et al. Alpha-radioimmunotherapy of B-lineage non-Hodgkin's lymphoma using ^{213}Bi-labelled anti-CD19- and anti-CD20-CHX-A″-DTPA conjugates [abstract]. *Abstr Pap Am Chem Soc.* 2003;225:U261.
125. Allen BJ, Raja C, Rizvi S, et al. Intralesional targeted alpha therapy for metastatic melanoma. *Cancer Biol Ther.* 2005;4:1318–24.
126. Bruland OS, Nilsson S, Fisher DR, et al. High-linear energy transfer irradiation targeted to skeletal metastases by the alpha-emitter Ra-223: adjuvant or alternative to conventional modalities? *Clin Cancer Res.* 2006;12(suppl):6250S–57S.
127. Sgouros G, Roeske JC, McDevitt MR, et al. MIRD pamphlet No. 22 (abridged): Radiobiology and dosimetry of α-particle emitters for targeted radionuclide therapy. *J Nucl Med.* 2010;51:311–28.
128. Bushnell DL Jr, O'Dorisio TM, O'Dorisio S, et al. ^{90}Y-edotreotide for metastatic carcinoid refractory to octreotide. *J Clin Oncol.* 2010;28:1652–9.

CHAPTER 30

In Vivo Function Studies

In vivo function studies are nonimaging procedures that use radiotracers to measure various physiologic functions in the body. The term *tracer* is particularly applicable to these studies. Simply defined, a tracer is a species that follows or outlines something else, the "tracee."[1] In nuclear medicine, a patient's bodily functions are assessed by measuring a radiotracer's absorption, dilution, concentration, or excretion. An effective radiotracer is handled by the body system of interest in the same way as its tracee. In general, this means that the tracer should be chemically identical to the tracee or should have characteristics similar to those of the tracee and should not alter the process being studied. Tracers commonly used in biomedical studies include radiographic contrast material, colored dyes, nonradioactive (stable) isotopes, and radioactive isotopes.

In vivo function studies using radioisotopic tracers have been used over the years to assess a variety of bodily functions. Examples include studies of radioactive iodine uptake (RAIU), blood volume, red blood cell survival, platelet survival, vitamin B_{12} absorption and metabolism, and glomerular filtration rate as a measure of renal function. In vivo function studies used in the diagnostic work-up of a patient are discussed in this chapter.

BLOOD VOLUME MEASUREMENT

Blood consists of a fluid fraction (plasma) and the cellular elements: red cells (erythrocytes, or RBCs), white cells (leukocytes, or WBCs), and platelets (thrombocytes). Each microliter (mm^3) of adult human blood contains an average of 5×10^6 RBCs, 7×10^3 WBCs, and 3×10^5 platelets. Most of the cellular volume is therefore composed of RBCs. When anticoagulated whole blood is centrifuged, it is separated into a volume of packed cells and supernatant plasma. The volume of RBCs, expressed as a percentage of the whole blood sample, is the hematocrit. It averages 45% in adult males and 42% percent in females.

Theoretically, a patient's blood volume can be determined from either the hematocrit or the plasmacrit (percentage of the blood volume occupied by plasma) and a measurement of the RBC volume or plasma volume, as follows:

$$\text{Whole blood volume} = \frac{\text{Red cell volume}}{\text{Hematocrit}} \text{ or } \frac{\text{Plasma volume}}{\text{Plasmacrit}} \quad (30\text{-}1)$$

Some error is associated with such a measurement, however, because of small differences between the large-vessel hematocrit (LVH) and the mean whole-body hematocrit (WBH). The WBH, which is the average distribution of RBCs in blood throughout the body, is calculated from independent measurements of the RBC volume and the plasma volume as follows:[2]

$$\text{Mean whole body hematocrit} = \frac{\text{Red cell volume}}{\text{Red cell volume} + \text{Plasma volume}} \quad (30\text{-}2)$$

The mean WBH is generally lower than the LVH. The mean WBH-to-LVH ratio is 0.915, with a range of 0.89 to 0.94.[2]

The LVH is determined by obtaining a sample of blood from an arm vein and centrifuging it in a capillary tube 75 mm long and 1.2 to 1.4 mm in diameter. A standard microhematocrit centrifuge spins the tube at 13,000 rpm for 4 to 5 minutes. The LVH is the ratio of the height of the packed RBCs to the height of the RBCs plus plasma. The LVH is somewhat exaggerated because of the amount of plasma trapped between the cells. The amount of trapped plasma varies with centrifugal force, spinning time, viscosity of the blood, and volume of RBCs present.[2] In general, a standard microhematocrit reading should be multiplied by 0.96 to correct for trapped plasma.[2] Thus, if the height of the RBC fraction in a microhematocrit tube is 24 mm and the height of the RBC and plasma fractions is 60 mm, the hematocrit is 40%. Correcting the microhematocrit for trapped plasma and LVH yields an approximate WBH, as follows:

$$0.40\,(\text{Hct\%}) \times 0.96 = 38.4\%\,(\text{LVH}) \times 0.915 = 34.94\%\,(\text{WBH})$$

A more accurate measurement of blood volume is made by measuring RBC volume and plasma volume independently and then adding them together to arrive at the whole blood volume.

Isotope Dilution Analysis

Measurements of blood volume and plasma volume are based on the principle of isotope dilution analysis. Following this principle, a radioactive tracer of known volume (V_1) and concentration (C_1) is added to an unknown volume (V_2). The tracer is allowed to equilibrate with the system, and a sample is then removed for analysis to determine the new tracer concentration (C_2) (Figure 30-1). The unknown volume is calculated from the following relationship:

$$V_2 = \frac{C_1 \cdot V_1}{C_2} \qquad (30\text{-}3)$$

An accurate determination of blood volume is predicated on the requirements that (1) the tracer does not degrade in or significantly leak from the compartment during the time of measurement and (2) the volume of tracer does not significantly change the volume of the compartment being measured. For routine nuclear medicine procedures, the second requirement is not a problem; however, the first requirement may be a concern in some circumstances, and corrections must be applied if necessary.

RBC Volume

RBC volume is routinely measured by labeling a sample of autologous RBCs with ^{51}Cr. An accurate volume of ^{51}Cr-labeled RBCs of known concentration (cpm/mL) is injected intravenously into an arm vein and allowed to reach equilibrium in the circulation, typically in 15 to 30 minutes. At this time a sample of blood is removed from the opposite arm. Samples are counted in a scintillation well counter, corrected for background count, and expressed as net counts per minute per milliliter. From Equation 30-3, the RBC volume is calculated as follows:

$$\text{Red cell volume}\,(\text{mL}) = \frac{\text{Injected }^{51}\text{Cr-RBC cpm}}{\text{Removed }^{51}\text{Cr-RBC cpm/mL}}$$

Even though the ^{51}Cr activity elutes from RBCs (approximately 1% per day), this small amount will not affect the results of the study, which is completed in 1 to 2 hours.

The following example illustrates a routine procedure for labeling RBCs with ^{51}Cr. Forty milliliters of whole blood from a patient's arm vein is drawn into a syringe containing 8 mL of anticoagulant citrate dextrose (ACD) solution. From this mixture, 14 mL is used to prepare background RBC and plasma samples. The remaining 34 mL is added to a sterile vented serum vial, followed by 150 µCi (5.55 MBq)

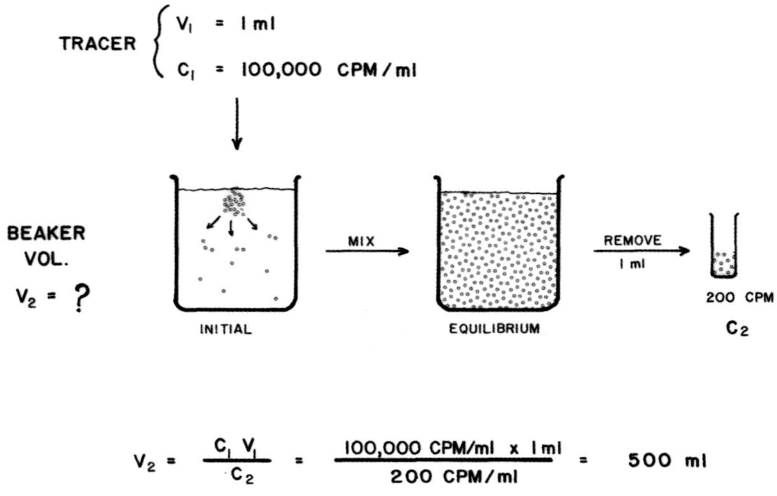

FIGURE 30-1 Determination of the unknown volume of a compartment (beaker) using the principle of isotope dilution analysis.

of ^{51}Cr-sodium chromate. This mixture is allowed to incubate for 20 to 30 minutes, with gentle mixing every 5 minutes. During the labeling reaction, the chromate anion (^{51}CrO$_4^{2-}$) diffuses into the RBCs and is reduced intracellularly to chromic ion (^{51}Cr^{3+}), which becomes bound to hemoglobin.[3-5] The maximum amount of chromium that labels RBCs is less than 0.5 µg/mL of RBCs and is nontoxic to the cells. Labeling efficiency is 85% to 90%. At the end of incubation, the blood can be centrifuged to separate the labeled cells from the unlabeled plasma activity or, alternatively, ascorbic acid (100 mg) can be added to the blood–chromate mixture and incubated for 5 minutes. The latter method is preferred by some investigators because it spares the cells from centrifuge and manipulation trauma. The ascorbic acid reduces the unlabeled chromate ion to chromic ion, preventing the in vivo labeling of RBCs when the labeled blood mixture is reinjected into the patient. When the ascorbic acid technique is used, however, plasma activity must be subtracted from whole blood measurements in the final analysis because chromic ion labels plasma protein. The labeling method is shown in Figure 30-2.

At completion of the labeling procedure, 1 mL of tagged blood is diluted to 100 mL with water. This 1:100 dilution is the ^{51}Cr-RBC standard, which is used to determine the total activity injected into the patient. After this, 20 mL of labeled blood containing about 80 µCi (2.96 MBq) of ^{51}Cr activity is injected intravenously into the patient. After a 30 minute equilibration time, a 15 mL sample of blood is removed from the opposite arm into a heparinized syringe. This represents the equilibrium concentration of the injected dose uniformly distributed in the unknown volume of blood. Two milliliters of the following samples is counted in a scintillation counter: standard whole blood (1:100 dilution), standard plasma (1:25 dilution), 30 minute whole blood, 30 minute plasma, whole blood background, and plasma background. Counts are adjusted for background, and dilution factors are applied where required. Corrections are made for large vein-to-whole body hematocrit and for trapped plasma (see Table 30-1). The RBC volume is then calculated. Blood volume measurement is explained in detail in standard textbooks.[2,6,7]

Plasma Volume

Plasma volume is measured with ^{125}I-human serum albumin (^{125}I-HSA). Typically, 5 to 10 µCi (185 to 370 kBq) of ^{125}I-HSA is injected intravenously and allowed to reach equilibrium in the body (in about 15 minutes). A blood sample is then removed from the opposite arm and the plasma activity is determined by sample counting in a scintillation counter. The plasma volume is then calculated from the total activity injected and the concentration of activity per milliliter of plasma at the time of injection (C_0), using Equation 30-3:

$$\text{Plasma volume(mL)} = \frac{\text{Injected cpm}\,^{125}\text{I-HSA}}{C_0 \text{ sample cpm/mL}}$$

Because ^{125}I-HSA slowly leaks out of the plasma compartment, plasma samples analyzed at different times after injection yield increasingly larger plasma volumes; in other words, the volume of ^{125}I-HSA distribution increases over time because the concentration in plasma is decreasing. Therefore, the initial equilibrium concentration (C_0) must be determined from the following equations, which are derived from a graphical

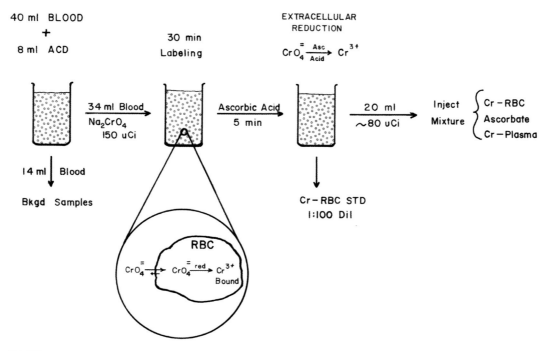

FIGURE 30-2 Method of labeling red blood cells with ^{51}Cr-sodium chromate for determining RBC volume by isotope dilution analysis.

TABLE 30-1 ^{51}Cr Red Blood Cell Volume Analysis Worksheet

Patient name: NS
Unit no:
Date:
Height: 59 in × 2.54 cm/in = 149.86 cm
Weight: 139 lb ÷ 2.2 lb/kg = 63.18 kg
Sex: Female
Age: 75
Dose ^{51}Cr-sodium chromate injected: 80 µCi

Sample volumes:
Counting time: 1 min
Hematocrits
Whole blood (WB) adm: 20 mL
Dilutions: Whole blood 1:100
Standard whole blood (std Hct) 0.41
Counting vol: 2 mL
Plasma 1:25
30 min patient vein (30 min Hct$_v$) 0.50

	(A) Gross Sample Cts (2 mL volume)	(B) Net Sample Cts (A − Relevant Bkg)	(C) Dilution Factor	(D) Net Corrected Cts (B × C)	
Std WB	6716	6684	100	668,400	(WBs)
Std plasma	2657	2625	25	65,625	(Pls)
30 min Pt WB	4143	4111	—	—	(WBp)
30 min Pt plasma	156	127	—	—	(Plp)
WB bkg	32	—	—	—	—
Plasma bkg	29	—	—	—	—

Corrected std Hct^a (Hct$_s$) = Std Hct 0.41 × 0.88 = 0.36
Corrected std plasmacrit (Pct$_s$) = (1.0 − Hct$_s$) = 0.64
Corrected 30 min Pt Hct (Hct30) = 30 min Hct$_v$ 0.50 × 0.88 = 0.44
Corrected 30 min Pt plasmacrit (Pct30) = (1.0 − Hct30) = 0.56

Red blood cell volume calculation

$$\frac{[WB_s - (Pl_s \times Pct_s)] \times (mL\ WB\ Adm) \times (Hct30)}{WB_p - (Pl_p \times Pct30)}$$

$$\frac{[668,400 - (65,625 \times 0.64)] \times 20\ mL \times 0.44}{4111 - (127 \times 0.56)} = 1364\ mL\ RBC\ volume$$

Parameter	Predicted Formula^b	Value	Measured
WB vol (mL)	Females: 24.8 H (cm)$^{0.725}$ W (kg)$^{0.425}$ − 1954 Males: 23.6 H (cm)$^{0.725}$ W (kg)$^{0.425}$ − 1229	3504	2665
RBC vol (mL)	Predicted WB vol × 0.4	1402	1364
RBC mass (mL/kg)	—	21.0	21.6
Plasma vol (mL)	Predicted WB vol × 0.6	2103	1301
Plasma mass (mL/kg)	—	31.5	20.6

^{a} The hematocrit correction factor (0.88) is derived from the product of the correction for whole-body hematocrit (0.92) and trapped plasma (0.96). Source: Early PJ, Sodee DB. *Principles & Practice of Nuclear Medicine*. Baltimore: Mosby; 1984:832.

^{b} Source: Reference 8.

plot of the natural logarithm of plasma activity versus time (Figure 30-3):

$$\ln C_0 = \frac{\ln C_2 t_1 - \ln C_1 t_2}{t_1 - t_2}$$

and $C_0 = e^{\ln C_0}$

where C_1 is the plasma cpm of the first sample time point (15 minutes) and C_2 is the plasma cpm at the second sample time point (30 minutes).

The following example illustrates the procedure for measuring plasma volume. An intravenous line is established and an infusion set attached, and a 5 mL sample of blood is obtained for background baseline measurement. Then 5 to 10 µCi (185 to 370 kBq) of ^{125}I-HSA is injected intravenously into the opposite arm and the exact time is recorded. From the initial arm, 5 mL samples of blood are removed into heparinized syringes at 15 and 30 minutes after the time of injection. These samples and the background sample are centrifuged, and the net activity (cpm/mL) in the plasma samples is determined in

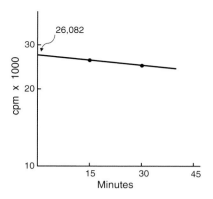

FIGURE 30-3 Graphical method of determining the zero-time plasma concentration of ^{125}I-human serum albumin in the measurement of plasma volume by isotope dilution analysis. Extrapolation of the line drawn through the 15 and 30 minute time points intersects the y-axis at the C_0 plasma activity (26,082 cpm).

a scintillation counter. These timed sample activity concentrations are plotted on semilog graph paper, and the resulting straight line is extrapolated to zero time to obtain the plasma concentration (C_0) of the injected dose at the time of injection (Figure 30-3).

Alternatively, the C_0 plasma concentration can be calculated from Equation 30-4 below. Its derivation is shown in Figure 30-4.

$$(C_0) = \frac{(C_1)^2}{C_2} \tag{30-4}$$

This equation is valid only if the second plasma sample is taken at a time that is exactly twice the sampling time of the first plasma sample (e.g., t_1 of 10 minutes and t_2 of 20 minutes, or t_1 of 15 minutes and t_2 of 30 minutes). This facilitates the determination of C_0.

The total activity of ^{125}I-HSA in the injected dose is determined by counting a standard dilution (1:4000) of the injected dose prepared from the same lot of ^{125}I-HSA. The values and calculations for a typical patient study are shown in Table 30-2.

An alternative to the two methods discussed above for measuring whole blood volume is an automated system available from Daxor Corporation (http://daxor.com/) that uses ^{131}I-HSA. The method requires five blood samples from the patient after intravenous injection of ^{131}I-HSA. The samples are processed in an automated sample counter to provide a blood volume measurement within 90 minutes.

Combined RBC–Plasma Volume Measurement

For convenience, ^{51}Cr-RBC volume and ^{125}I-HSA plasma volume studies are typically performed together. Using this approach, ^{125}I-HSA is injected first, and samples of blood for plasma volume measurement are obtained while the RBCs are

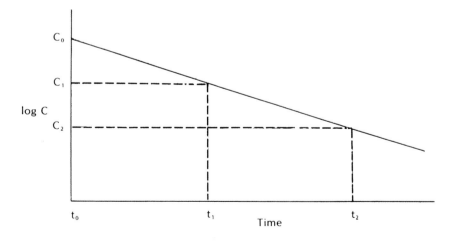

FIGURE 30-4 Derivation of the simplified equation for determining C_0 in plasma volume analysis. Assumptions: first-order exponential rate of loss of ^{125}I-HSA from plasma compartment, and t_2 equals $2 \times t_1$.

TABLE 30-2 ^{125}I-HSA Plasma Volume Analysis Worksheet

Patient name: NS	Unit no:
Height: 59 in × 2.54 cm/in = 149.86 cm	Weight: 139 lb ÷ 2.2 lb/kg = 66.18 kg
Sex: F	Age: 75
Dose (µCi ^{125}I-HSA): 7.0	Date: 6-3-02
Std dilution factor (DF): 1:4000	
Counting time: 2 min	Counting vol: 2 mL
15 min plasma gross counts: 25,285	Net counts (gross cts − bkg 1): 25,250
30 min plasma gross counts: 24,480	Net counts (gross cts − bkg 1): 24,445
Plasma background counts (bkg 1): 35	
Room background counts (bkg 2): 35	
Standard counts gross: 8520	Net counts (gross cts − bkg 2): 8485

$$C_0 \text{ counts} = \frac{(\text{Net } C_{15})^2}{\text{Net } C_{30}} = \frac{(25{,}250)^2}{24{,}445} = 26{,}082$$

$$\text{Plasma vol} = \frac{\text{Net Std Cts} \times \text{DF}}{C_0 \text{ counts}} = \frac{(8485) \times (4000)}{26{,}082} = 1301 \text{ mL}$$

being labeled with ^{51}Cr-sodium chromate. After the 15 minute and 30 minute samples are obtained for the plasma volume study, the tagged ^{51}Cr-RBCs are injected. A sample of blood is obtained 30 minutes later for the RBC volume determination. Although the blood sample for the RBC measurement will contain both ^{125}I and ^{51}Cr, the ^{125}I counts (photon E = 27 to 35 keV) are easily discriminated from the ^{51}Cr counting window (photon E = 320 keV) of the scintillation spectrometer. An example of a combined study in the same patient is shown in Tables 30-1 and 30-2.

Predicted normal values for whole blood volume are either determined from standard tables or calculated by a formula based on body surface area.[8] The predicted normal RBC volume or mass and plasma volume or mass are then calculated (see Table 30-1). The normal average values (mL/kg) for total blood volume (mass), RBC mass, and plasma mass for males and females are as follows:[9]

	Females	Males
Total blood mass	66	72
RBC mass	24	28
Plasma mass	42	44

RBC SURVIVAL

The ability to measure the survival of RBCs in the circulation is helpful for several reasons: (1) It provides information about the compatibility and suitability of donor blood used in transfusions, (2) it is helpful in validating whole blood collection and storage methods, and (3) in patients with hemolytic anemia, RBC survival studies may provide insight into the rate and mechanism of hemolysis.

A shortened half-life of disappearance of labeled RBCs from the circulation supports the diagnosis of intravascular hemolysis or hypersplenism. Because the spleen is the normal site for destruction of senescent RBCs, a splenic sequestration study may also yield useful information. The normal spleen-to-liver ratio of activity from ^{51}Cr-RBCs is 1 to 1. In cases of hypersplenism, it is greater than 2 to 1. Increased liver activity is indicative of intravascular hemolysis, because hemoglobin is metabolically processed by the liver.

The application of radiotracer techniques has proven to be an important advance in determining RBC survival. With this method, the patient's own blood, as well as donor blood, can be labeled with a radionuclide (e.g., ^{51}Cr or ^{99m}Tc), and survival of the cells in the circulation can be monitored by radioactive counting of blood samples obtained periodically over time.[10,11]

Cohort and Random Labeling of RBCs

Radionuclide methods of labeling RBCs can be divided into two groups: cohort labeling and random labeling.[12] In cohort labeling, the radiotracer is incorporated into cells as they are newly formed, and the survival of this group or cohort of cells of similar age is studied by monitoring their passage into and removal from the circulation. Although cohort labeling is the ideal method for studying RBC survival, it requires that labeling be accomplished within a short window of time, that the label remain within the cell throughout its life span, and that the label not be reused after destruction of the cell. Because no radiolabel meets all of these requirements, cohort labeling is not routinely used.

In random labeling, cells are labeled in such a way that the age distribution of the labeled sample reflects the age distribution of the parent population. Survival is studied by monitoring the disappearance of the labeled sample from the circulation. ^{51}Cr-sodium chromate is routinely used for random labeling of RBCs. The advantages of this agent are that the labeling is simple and convenient, the label is not reused, and external gamma counting can be performed with in vitro samples and, if necessary, in vivo detection to localize areas of RBC sequestration in the body. The latter is important for assessing the relative roles of different sites of RBC destruction in known cases of hemolytic anemia. One disadvantage of ^{51}Cr-sodium chromate is that the ^{51}Cr label elutes from the cells in vivo at an exponential rate of about 1% per day.[13] This elution requires a correction to obtain clinically satisfactory approximations of the mean RBC life span.

RBC Survival Study

The RBC survival method is useful for measuring the life span of RBCs in patients with known or suspected hemolytic anemia and for validating new procedures for collecting and storing whole blood. Survival is determined by monitoring the disappearance of ^{51}Cr-RBCs from the circulation. The half-life of ^{51}Cr-RBCs in normal subjects measured by this method is about 30 days.[6] This value is about twice the normal rate of removal by RBC destruction alone of 1% per day for senescent cells. The increased rate with ^{51}Cr-RBCs is due to an additional 1% loss per day from elution of the ^{51}Cr label. Consequently, ^{51}Cr-RBC survival data bear no simple relationship to the mean cell life span, which is the parameter required in clinical practice. If, however, as recommended by the International Committee for Standardization in Hematology, the ^{51}Cr-RBC survival data are corrected for ^{51}Cr elution, and the patient is in a steady state of RBC production and destruction during the course of study, then the corrected half-life value is a measure of RBC destruction only. Furthermore, multiplying this corrected half-life value by 1.443 gives the mean life of RBC survival.

A summary of the procedure follows; greater detail can be found elsewhere. Autologous RBCs are labeled with ^{51}Cr as described previously for RBC volume determination. This method conforms to the recommended methods established for labeling RBCs with ^{51}Cr. Labeled blood is then injected into the patient. On day 1 after injection, a sample of blood is withdrawn and the net count rate in the RBCs is determined. This procedure is repeated three times per week for 2 weeks. The percentage of the day 1 sample RBC activity that is present in each subsequent sample is determined as follows:

$$\% \text{ Survival of labeled red cells} = \frac{\text{Net cpm of sampled red cells} \times 100}{\text{Net cpm of red cells on day 1}}$$

These data are then corrected for ^{51}Cr decay and elution and plotted on semilog graph paper. The mean RBC life span is calculated as the reciprocal of the slope of the plotted line, or 1.443 times the half-life of disappearance over time (Figure 30-5). The normal mean RBC life span determined by this method is 115 days.

GLOMERULAR FILTRATION RATE

The glomerular filtration rate (GFR) is a useful measure of renal function in a variety of clinical situations that may affect the kidneys, such as in oncology patients receiving chemotherapy or patients undergoing renal transplantation. Nuclear

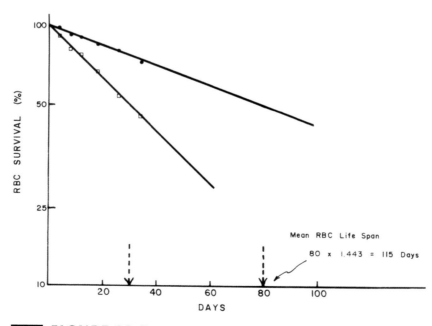

FIGURE 30-5 Determination of mean RBC life span from the in vivo survival of ^{51}Cr-labeled RBCs. □ = survival related to RBC destruction and ^{51}Cr elution; ● = survival related to RBC destruction only, corrected for ^{51}Cr elution.

medicine procedures have been developed using radiopharmaceuticals that are cleared from the blood by glomerular filtration. Typically, either ^{99m}Tc-DTPA or ^{125}I-iothalamate is used. GFR can be determined by measuring either the amount of intravenously administered radiopharmaceutical that is eliminated in the urine over time (urinary excretion method) or the rate of the radiopharmaceutical's disappearance from the blood over time (plasma clearance method). The plasma clearance method has become the most convenient and reliable means of measuring GFR. A typical procedure involves intravenous administration of ^{99m}Tc-DTPA or ^{125}I-iothalamate followed by collection of blood samples at prescribed time periods. The blood samples are centrifuged and plasma samples counted. Various methods have been validated with each of these radiopharmaceuticals, involving either multiple blood samples or one or two blood samples. The rate of disappearance of these agents from the blood is directly related to glomerular function.

When radiotracer is injected and blood samples are obtained at various time points and their respective plasma activities (cpm/mL) are plotted, a biexponential plasma disappearance curve is obtained (Figure 30-6). Subtraction of the slow component from the whole curve yields the fast component. From these curves the elimination rate constants of the fast (λ_β) and slow (λ_α) components and their respective y-intercepts, B and A, can be determined. The biexponential disappearance curve satisfies the equation $x = Ae^{-\lambda_\alpha t} + Be^{-\lambda_\beta t}$, from which the following clearance formula is derived:

$$\text{GFR} = \frac{D(\lambda_\alpha \cdot \lambda_\beta)}{A\lambda_\beta + B\lambda_\alpha} \quad (30\text{-}5)$$

Equation 30-5 can be rewritten in terms of half-life and the GFR normalized to a standard body surface area (BSA) of 1.73 m^2 as follows:

$$\text{GFR}\left(\text{mL/min}/1.73\text{ m}^2\right) = \frac{D(0.693)}{(AT_{1/2\alpha} + BT_{1/2\beta})} \times \frac{1.73\text{m}^2}{\text{Patient BSA}(\text{m}^2)} \quad (30\text{-}6)$$

Each patient's BSA is computed using the DuBois[14] formula shown below, where W is weight in kilograms and H is height in centimeters.

$$\text{BSA}(\text{m}^2) = (W^{0.425}) \times (H^{0.725}) \times 0.007174$$

In Equations 30-5 and 30-6, D is the dose in cpm injected into the patient, A and B are y-axis intercepts in cpm/mL obtained from a plot of plasma cpm versus postinjection time (Figure 30-6), λ_α and λ_β are slopes in min^{-1}, and $T_{1/2\alpha}$ and $T_{1/2\beta}$ are the half-lives, in minutes, of the two plasma sample curves fitted to a two-compartment pharmacokinetic model.[15]

Simplified plasma sampling methods for GFR measurement have been developed for ^{125}I-iothalamate,[15] and methods have been validated for ^{99m}Tc-DTPA with one or two plasma samples taken at 1 to 4 hours.[16–18] Procedures for using ^{125}I-iothalamate and ^{99m}Tc-DTPA to determine GFR are described in the following sections.

^{125}I-Iothalamate Procedure

^{125}I-iothalamate is commercially available from IsoTex Diagnostics (Friendswood, TX) as a 250 to 300 µCi/mL multidose vial. Different methods have been developed for measuring the GFR with ^{125}I-iothalamate. In the method used at our institution the patient dose for a GFR procedure is 25 µCi (925 kBq) administered intravenously. A standard is prepared by diluting 0.1 mL of ^{125}I-iothalamate to 500 mL. One milliliter of this standard is counted and a dilution correction factor is applied to the standard cpm to determine the dose in cpm injected into the patient. The patient is injected with a 25 µCi (925 kBq) dose. Initial blood samples are obtained at 5, 10, and 15 minutes after injection; later samples are obtained at 180, 210, and 240 minutes to allow for equilibration of ^{125}I-iothalamate in the vascular space. The patient samples and standard are then counted in a scintillation well counter and the data entered into a spreadsheet program. The program fits the data to the two-compartment model described above from which the A and B y-axis intercepts and slopes

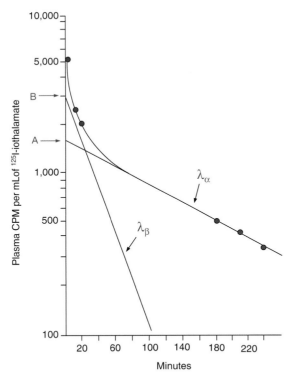

FIGURE 30-6 Biexponential plasma disappearance curve, obtained from multiple blood samples, of an agent used to measure glomerular filtration rate. The slow late component (λ_α) is subtracted from the curve to derive the early fast component (λ_β), from which the intercepts A and B are determined. The curve satisfies equation $x = Ae^{-\lambda_\alpha t} + Be^{-\lambda_\beta t}$ from which the GFR is determined.

are determined and GFR, corrected for BSA, is calculated using the formula in Equation 30-6.

^{99m}Tc-DTPA Procedure

Another plasma clearance procedure for measuring GFR uses ^{99m}Tc-DTPA and requires only two blood samples.[17,19] The patient receives a 3 mCi (111 MBq) intravenous injection of ^{99m}Tc-DTPA, and blood samples are obtained at 60 and 180 minutes after injection. A standard is prepared by making a 1:10,000 dilution of ^{99m}Tc-DTPA identical to the patient dose, and three 100 μL counting standards from this dilution are prepared in counting tubes. The 60 and 180 minute blood samples are centrifuged, and duplicate plasma samples obtained from each are transferred to ultrafiltration devices (Amicon Centrifree Micropartition System [Amicon Corporation, Danvers, MA]) and centrifuged to obtain an ultrafiltrate. Duplicate 100 μL ultrafiltrate samples are transferred to counting tubes and counted in a scintillation well counter along with the counting standards. The patient's BSA is determined and the GFR in mL/min/1.73 m^2 is calculated using the formula below, where D is the injected dose of ^{99m}Tc-DTPA in cpm and P_1 and P_2 are the mean cpm of the two ultrafiltrate samples at the T_1 and T_2 sample times in minutes, respectively.[17]

$$GFR = \left[D \ln(P_1/P_2)/(T_2 - T_1) \exp\{(T_1 \ln P_2 - T_2 \ln P_1)/(T_2 - T_1)\}\right]^{0.979}$$
$$\times \left[1.73 \, m^2 / Pt\, BSA\,(m^2)\right]$$

MISCELLANEOUS IN VIVO PROCEDURES

Other in vivo procedures performed in nuclear medicine include the thyroid uptake study, which was discussed in Chapter 20, the GI protein loss study discussed in Chapter 23, and the platelet survival study discussed in Chapter 11. In the recent past, a urinary excretion test for assessing malabsorption of vitamin B$_{12}$ from the GI tract (Schilling test) was routinely performed using radioactive vitamin B$_{12}$.[20,21] However, the use of this study has waned over the years, leading to a lack of availability of the radiopharmaceutical (^{57}Co-cyanocobalamin) involved.

The ^{14}C-urea breath test is an in vivo function study developed after *Helicobacter pylori* infection was discovered to be a principal cause of peptic ulcer disease and chronic gastritis.[22] This test is based on detection of the enzyme urease produced by *H. pylori*. Urease is normally not present in the stomach. However, in the presence of an *H. pylori* infection, orally administered ^{14}C-urea is hydrolyzed by urease in the stomach into ammonia and ^{14}CO$_2$. The ^{14}CO$_2$ is absorbed into the circulation and exhaled from the lungs, and the exhaled ^{14}CO$_2$ can be captured by specialized breathing equipment. The ^{14}C-urea breath test is a recommended procedure for determining the success of treatment to eradicate *H. pylori* infection in patients with duodenal ulcers and gastric ulcer disease.[23]

REFERENCES

1. Welch TJC, Potchen EJ, Welch MJ. *Fundamentals of the Tracer Method.* Philadelphia: WB Saunders; 1972.
2. Albert SN. *Blood Volume and Extracellular Fluid Volume.* 2nd ed. Springfield IL: Charles C Thomas; 1971.
3. Gray SJ, Sterling K. The tagging of red cells and plasma proteins with radioactive chromium. *J Clin Invest.* 1950;29:1604–13.
4. Person HA. The binding of Cr-51 to hemoglobin. I: In vitro studies. *Blood.* 1963;22:218–30.
5. Ebaugh FG, Samuels AJ, Dobrowlski P, et al. The site of the CrO$_4$-hemoglobin bond as determined by starch electrophoresis and chromatography [abstract]. *Fed Proc.* 1961;20:70.
6. Albert SN, Sodee DB. In vivo and in vitro studies of hemopoietic production and components. In: Sodee DB, Early PJ, eds. *Mosby's Manual of Nuclear Medicine Procedures.* 3rd ed. St Louis: Mosby; 1981:269.
7. Arnold JE. Blood volume. In: Gottschalk A, Potchen EJ, eds. *Diagnostic Nuclear Medicine.* Baltimore: Williams & Wilkins; 1976:130–8.
8. Hidalgo JU, Nadler SB, Bloch T. The use of the electronic digital computer to determine best fit of blood volume formulas. *J Nucl Med.* 1962;3:94–9.
9. Early PJ, Razzak MA, Sodee DB. *Textbook of Nuclear Medicine Technology.* 3rd ed. St Louis: Mosby; 1979:351.
10. Bandarenko N, Rose M, Kowalsky RJ, et al. In vivo and in vitro characteristics of double units of RBCs collected by apheresis with a single in-line WBC-reduction filter. *Transfusion.* 2001;41:1373–7.
11. Bandarenko N, Cancelas J, Snyder EL, et al. Successful in vivo recovery and extended storage of additive solution (AS)-5 red blood cells after deglycerolization and resuspension in AS-3 for 15 days with an automated closed system. *Transfusion.* 2007;47:680–6.
12. Bentley SA. Red cell survival studies reinterpreted. *Clin Haematol.* 1977;6:601–23.
13. Ebaugh FG Jr, Emerson CP, Ross JF. The use of radioactive chromium-51 as an erythrocyte tagging agent for the determination of red cell survival in vivo. *J Clin Invest.* 1953;32:1260–76.
14. DuBois EF. *Basal Metabolism in Health and Disease.* 3rd Ed. Philadelphia: Lea & Febiger;1936:223.
15. Dubovsky EV, Russell CD. Quantitation of renal function with glomerular and tubular agents. *Semin Nucl Med.* 1982;12:308–29.
16. Russell CD, Bischoff PG, Kontzen FN, et al. Measurement of glomerular filtration rate: single injection plasma clearance method without urine collection. *J Nucl Med.* 1985;26:1243–7.
17. Rowell KL, Kontzen FN, Stutzman ME, et al. Technical aspects of a new technique for estimating glomerular filtration rate using technetium-99m-DTPA. *J Nucl Med Technol.* 1986;14:196–8.
18. Ham HR, Piepsz A. Estimation of glomerular filtration rate in infants and in children using a single-plasma sample method. *J Nucl Med.* 1991;32:1294–7.
19. Blaufox MD, Aurell M, Bubeck B, et al. Report of the radionuclides in nephrourology committee on renal clearance. *J Nucl Med.* 1996; 37:1883–90.
20. Schilling RF. Intrinsic factor studies. II. The effect of gastric juice on the urinary excretion of radioactivity after the oral administration of radioactive vitamin B-12. *J Lab Clin Med.* 1953;42:860–6.
21. Katz JH, Dimase J, Donaldson RM Jr. Simultaneous administration of gastric juice bound and free radioactive cyanocobalamin. Rapid procedure for differentiating between intrinsic factor deficiency and other causes of vitamin B-12 malabsorption. *J Lab Clin Med.* 1963; 61:266–9.
22. Desroches JJ, Lahaie RG, Picard M, et al. Methodological validation and clinical usefulness of carbon-14-urea breath test for documentation of presence and eradication of *Helicobacter pylori* infection. *J Nucl Med.* 1997;38:1141–5.
23. Balon HR, Roff E, Freitas JE, et al. C-14 urea breath test 3.0 In: *Procedure Guidelines Manual.* Reston, VA: Society of Nuclear Medicine; 2001:37–9. www.snm.org/guidelines.

Index

Page numbers followed by t or f refer to tables or figures, respectively.

A

$\tilde{A}$ (cumulative activity), 92
A(n, γ)A*→B reaction, 32
abbreviated NDA (ANDA), 305, 391–394
    for PET drugs, 391
    review process for, 391–392
A(n,p)B reaction, 32–33
absolute configuration, 167
absolute risk model, 99
absorbed dose (D), 88, 91f
absorbed dose per unit cumulated activity (S), 92, 93t–94t
^{225}Ac. *See* actinium Ac 225
Academy of Molecular Imaging (AMI), 239, 254
accelerators, 33–34, 242–244
    charged-particle, 33–34, 33f, 242–244, 243t
    cyclotrons, 1, 33–34, 33f
        for gallium, indium, and thallium production, 215
        for iodine I 123 production, 209
        for PET radionuclide production, 241, 242–244, 243f, 244t
        radioiodine production by, 446
        targetry for, 243–244
        for xenon Xe 133 production, 220
    linear, 33–34, 33f
acceptance criteria, PET, 308–309
accuracy
    definition of, 79
    of dose calibrators, 285, 299t, 300
ACD. *See* anticoagulant citrate dextrose
ACE inhibitor, in renography, 556–558, 558f, 560, 561f
acentric chromosome, 131, 132f
acetazolamide stress test, 435, 436f
acetylcholine, 424
ACMUI. *See* Advisory Committee on Medical Uses of Isotopes
ACP. *See* amorphous calcium phosphate
actinium Ac 225 (^{225}Ac), 690t
active transport, 164f, 165–166, 548
activity, 25–28
    specific, 161–162, 162t, 293
    vs. time, 28, 28f
    units of, 25–28, 161
AcuTect. *See* technetium Tc 99m apcitide
acute effects, 138–139
acute incapacitation syndrome, 138, 139
acute radiation syndrome, 138
acute toxicity tests, 382
Adams, Edgar, 8
ADCC. *See* antibody-dependent cell-mediated cytotoxicity
additive risk model, 99
adenoma
    parathyroid, 604–605, 605f
    pituitary, 600t
adenosine
    chemical structure of, 462f
    for myocardial perfusion imaging, 462–463, 464–466, 474, 478f
    precautions, adverse effects, contraindications, 464–466, 465t, 474
    production of, 462, 463f
adenosine A_2 receptor agonists, for myocardial perfusion imaging, 463
adenosine triphosphate (ATP), 424
administration, routes of, 4–5, 4t
administrative requirements, 398–400
adrenal cortex, 601
adrenal glands, 600–603
adrenal medulla
    anatomy of, 601
    function of, 601
    tumors of, imaging of, 600–603, 603f, 604f
adverse reactions, 333–339
    case study of, 338–339
    definition of, 333–334
    fatal, 336, 336f, 337t
    to new products, 338
    reporting systems for, 334, 335t
    severity of, 335–336
    special concerns about, 336–338
    statistics on, 334–335, 335f, 336t
    types of, 334, 334t
Advisory Committee on Medical Uses of Isotopes (ACMUI), 376
AEC. *See* Atomic Energy Commission
afferent arteriole, 547
affinity, antibody, 659–661
affinity enhancement system, 683
age at exposure, and cancer risk, 101, 101t, 102t
age-based dose adjustment, 326
AIDS, infection imaging in, 595, 651
air, radiation absorption in, 95, 95t

airborne contamination, control of, 315–316
air filters, 261–262
ALARA. *See* as low as reasonably achievable
albumin
    carrier technetium and, 42–43, 43*f*
    chromium Cr 51 labeled albumin
        (^{51}Cr-labeled albumin), 528
    iodinated I 125 albumin (^{125}I-albumin
        or ^{125}I-HSA), 208, 214
        diagnostic application of, 158
        labeling with, 161
        pediatric dosing of, 327
        for plasma volume measurement,
            697–699, 699*f*, 700*t*
        radiochemical purity of, 286*t*
        for red blood cell–plasma measure-
            ment, combined, 699–700
    iodinated I 131 albumin (^{131}I-HSA), 214
        adverse reactions to, 334–335
        for CSF imaging, 426, 427*t*
        for gastrointestinal imaging, 528
        radiochemical purity of, 286*t*
    iodinated I 131 albumin aggregated
        (^{131}I-MAA), 286*t*
    iodinated I 131 albumin colloid
        (^{131}I-HSA colloid), 520, 520*t*
    technetium Tc 99m aggregated albumin
        (^{99m}Tc-MAA), 188–189
        adverse reactions to, 337,
            503–504
        appearance and color of, 294, 294*t*
        biologic properties of, 188, 502
        for differential lung perfusion
            studies, 513, 514*f*
        excessive technetium Tc 99 in, 320
        excretion in breast milk, 331–333,
            332*f*, 333*t*
        indications for, 189
        for liver imaging, 529, 531*f*
        localization in lung, 502, 503*f*
        for lung perfusion studies, 189,
            337, 501–504, 503*f*
        lung scanning dose of, 503–504
        particles of, 188, 189*f*
            cerebral microembolization
                of, 504
            clearance from lung, 502,
                503*f*
            hardness and composition
                of, 502
            number of, 294, 294*t*, 295*f*,
                501–502, 504
            reduced concentration of,
                504, 504*t*
            sedimentation of, 324
            size of, 294, 294*t*, 295*f*, 501,
                502*f*, 504
            toxicity of, 503–504
        pediatric dosing of, 325*t*, 327,
            504, 504*t*
        preparation of, 188, 188*t*, 501
        radioaerosols and, 507
        radiochemical purity of, 287*t*,
            288*t*
        for right-to-left shunt evaluation,
            513, 515*f*
        specific activity of, 162
        thyroid uptake of, 504, 505*f*
        for ventilation–perfusion lung
            scan, 508–509, 510*f*, 511*f*,
            512*f*, 513*f*
    technetium Tc 99m albumin
        (^{99m}Tc-HSA), 188
        for CSF imaging, 426
        for gastrointestinal imaging, 528,
            529*f*
        for lymphoscintigraphy, 607–608
        radiochemical purity of, 287*t*
    technetium Tc 99m albumin colloid,
        520, 520*t*
aldosterone, 601
ALI. *See* annual limit on intake
alkali metals, 148
alkaline-earth elements, 223
Allen, David, 8, 9*t*
allergic reactions, 334*t*
all-or-none effect, 98. *See also* stochastic
    effects
allotypic determinants, 658
alpha particle(s), 17, 47
    interactions with matter, 47–48, 48*f*, 49*f*
    path and range of, 48
    properties of, 48*t*
    size of, *vs.* electron, 48
    specific ionization for, 50, 51*f*
alpha particle decay, 24, 25*f*
alpha particle emitters, targeted therapy with,
    690, 690*t*
aluminum ions
    impurities of, 38–39, 293, 323
    interactions of, problems associated
        with, 323
    testing for in ^{99m}Tc generator, 38–39, 39*f*
Alzheimer's disease
    imaging agents for, 425–426
    PET/CT of, 629, 649–650, 651*f*
    reimbursable PET procedures in, 240
    SPECT of, 432–433, 433*f*, 434*f*
ambient radiation exposure rates, survey of,
    401, 406
American Chemical Society, 283
*American Journal of Pharmaceutical
    Education*, 10
American National Standards Institute
    (ANSI), 299–300
American Pharmaceutical Association
    (APhA) (currently American Pharmacists
    Association), 10–11, 266, 374
AMI. *See* Academy of Molecular Imaging
amine precursor uptake and decarboxylation
    (APUD) cells, 598
amino acids, Fischer configuration of, 167,
    167*f*
ammonia N 13 (^{13}N-ammonia)
    for heart imaging
        perfusion studies, 459, 460*t*, 466*f*,
            467*t*, 470, 479, 646–648, 648*f*,
            649*f*
        PET/CT, 646–648, 648*f*, 649*f*
        ventriculography, 471
        viability studies, 487–488, 488*f*,
            647–648, 648*f*, 649*f*
    impurities in, 293
    for PET, 234–235, 234*t*, 239,
        247–248, 617
        blood flow, 234–235
        reimbursable procedures,
            239–240, 239*t*
    radiochemical purity of, 288*t*
    radiosynthesis of, 247–248, 247*f*
amorphous calcium phosphate (ACP),
    571–572, 576
AMP. *See* authorized medical physicist
AMU. *See* atomic mass unit
analytic reagent (AR) grade, 283
anaphase, 130, 130*f*
anaphylactic reactions, 334*t*
ANDA. *See* abbreviated NDA
anemia, 138
angiocardiography, first-pass radionuclide,
    481–484
    interpretation of, 482–484, 486*f*
    procedure for, 481–482
    rationale for, 481
angiography, computed tomography pul-
    monary, of pulmonary embolism,
    511–513
angiotensin, 556–557, 557*f*
animal studies
    animal care in, 355
    ex vitro autoradiography, 351
    ex vivo tissue biodistribution, 353–355
    knockout mice, 357
    preclinical, 382
    of radiation effects in pregnancy, 143
    of risk assessment, 99
    small-animal imaging, 355–357
    toxicity, 382
annihilation radiation, 21, 21*f*, 230, 230*f*
annual dose limits, 111, 111*t*
annual effective dose equivalent, in U.S.,
    112, 112*t*
annual limit on intake (ALI), 115, 116–117,
    117*t*
anorexia, 138
ANP. *See* authorized nuclear pharmacist
ANSI. *See* American National Standards
    Institute
anterior descending artery, left, 460, 460*f*
anteroom, 259
antibiotics, radiolabeled
    for infection imaging, 593–594
    localization mechanisms of, 596*t*
antibody(ies), 655–668
    affinity of, 659–661
    antigen interaction with, 659–661
    avidity of, 661
    chimeric, 659, 673, 674
    classification of, 657–658, 657*t*
    development of, 658, 659*f*, 660*f*
    humanized, 659, 673, 674
    modification of, 658–659
    nomenclature for, 661, 661*t*
    radiolabeled, 160, 218, 224, 655,
        665–671
        adverse reactions to, 337–338,
            658–659, 673
        diagnostic, 655, 665–666
        FDA approval of, 665, 673
        under investigation, 668
        limited effectiveness of, 655
        as magic bullet, 655
        removed from market, 665, 666
        therapeutic, 655, 665, 666–667,
            673–674. *See also specific
            agents*
        for tumor imaging, 596

radiolabeled, monoclonal
    for bone marrow imaging, 612
    for infection imaging, 593
    localization mechanisms of, 596t
radiolabeling process for, 661–665, 674, 674f
    bifunctional chelates for, 186, 187f, 664–665, 664f, 665f
    radioiodine, 662–663, 662f, 663f, 664f, 674
    radiometal, 664–665, 664f, 665f, 674
    radionuclide considerations in, 662, 662t
structure of, 656–657, 656f, 657f
subclasses of, 657–658, 658t
in technetium-nonessential compounds, 185–186
antibody-dependent cell-mediated cytotoxicity (ADCC), 657, 659, 673–674
anticoagulant citrate dextrose (ACD), 696
anti configuration, 168, 168f
antigen–antibody interaction, 659–661
antimony sulfide colloid technetium Tc 99m. See technetium Tc 99m antimony sulfide colloid
antineutrino, 19
antioxidant(s), 176
antioxidant stabilizers
    in bone agent preparation, 575
    for technetium kit, 176–177, 177f, 177t
antiseptics, 314–315
APhA. See American Pharmaceutical Association
apoptosis, 673–674
appearance, of radiopharmaceuticals, 294, 294t
APUD cells. See amine precursor uptake and decarboxylation cells
AR. See analytic reagent grade
Archambault, George, 8
area monitors, 264, 264f
arginine–glycine–aspartic acid (RGD) analogues, 669–670
arginine–glycine–aspartic acid (RGD) receptors, 669
argon, in ionization detector, 59
arrhythmias, 459
arthritis, gouty, 584, 586f
arthroplasty, painful, 594
artifacts, in PET/CT, 618, 619–620, 620f, 627
artificial CSF, 428, 428t
AS. See authentic substance grade
ASCT. See autologous stem cell transplantation
aseptic meningitis, 334t
aseptic techniques, 266, 266t, 295, 306–307, 313–317
as low as reasonably achievable (ALARA), 110, 258, 408
aspartate, 424
astatine At 211 (^{211}At)
    clinical trials with, 690t
    labeling with, 662–663, 662t
^{211}At. See astatine At 211
atherosclerosis, 459, 460–461, 471
atom(s), 13

Bohr model of, 14f
central, 153
neutral, 13, 148
nuclide, 13–14
atom bomb. See Japanese survivors
Atomic Energy Act of 1946, 375
Atomic Energy Act of 1958, 376
Atomic Energy Commission (AEC), 8, 110, 375–377
atomic mass unit (AMU), 16
atomic number, 148
atomic orbital, 149, 152f
atomic weight, 147–148
ATP. See adenosine triphosphate
attenuation, in PET, 232, 232t
attenuation test, of dose calibrators, 301–302, 301f, 302f
AU. See authorized user
^{198}Au. See gold Au 198
Auger effect, 15, 15f
authentic substance (AS) grade, 283
Authorization for Radioisotope Procurement, 376
authorized medical physicist (AMP), 395, 400, 405
authorized nuclear pharmacist (ANP), 265, 395–396, 400, 405
authorized recipient, transfer to, 125
authorized user (AU), 396
autoclaving, 296
autologous stem cell transplantation (ASCT), 675
automated modular systems, for radionuclide production, 250–251
autoradiography, 350–353
    applications with radiopharmaceuticals, 352–353, 352t
    classical receptor, 352
    conventional, 351–352
    dual-isotope, 353
    electronic, 353
    enhancements in, 352, 352f
    experimental sequence for, 351f
    ex vitro autoradiography, 351
    indirect, 352
    intensifying screen for, 352, 352f
    in vitro, 351
    in vitro binding assay vs., 348, 349t, 350–351
    microscopic, 352
    parameters of, 352, 352t
    role in drug development, 353
autosomal-dominant disorders, 103t, 104
A value, of nuclides, 13
avascular necrosis of bone, 584–586, 586f
avidin–biotin system, 683
avidity, antibody, 661
Avogadro's number, 148, 161
Azedra. See iobenguane iodine I 131

B

^{137}Ba. See barium Ba 137
background radiation, 112
bacterial endotoxin testing (BET), 295, 296–299, 297t, 308, 315, 335
badges, monitoring, 114–116, 117f
Baker, William, 8, 9t
barium Ba 137 (^{137}Ba)

decay of, 23
pulse-height energy spectrum of, 69f, 70
as reference for dose calibrators, 300
base(s), DNA, 129–130
base level, in scintillation counter, 66
basilar artery, 414, 414f
basophils, 589–590
BATO. See boronic acid adducts of technetium oxime
BBB. See blood–brain barrier
BCNP. See board certified in nuclear pharmacy
BECF. See blood–brain extracellular fluid barrier
becquerel (Bq), 4, 25–28, 87, 161
Becquerel, Henri, 25, 127
Beightol, Robert, 8, 9t
BEIR. See Biologic Effects of Ionizing Radiation committee
benzyl alcohol, radiopharmaceuticals containing, pediatric dosing of, 327
Bertin, renal columns of, 547, 548f
beryllium, electronic configuration of, 152t
beta-minus (negatron) decay, 18, 19, 20f
beta particles, 17, 47
    emission of, 2
    in dose calibrator, 61
    in negatron decay, 19, 20f, 20t
    interactions with matter, 49–50, 49f, 50f
    liquid scintillation counter for, 73–74, 74f
    negatively charged. See negatron(s)
    positively charged. See positron(s)
    properties of, 48t
    protection from, 92–95, 113
    range in matter, 95, 95t
    therapeutic applications of, 160
Bexxar. See tositumomab I 131
beyond-use-date (BUD), 313, 315
BFCAs. See bifunctional chelating agents
BGO. See bismuth germanate
^{213}Bi. See bismuth Bi 213
bicisate technetium Tc 99m (^{99m}Tc-ECD). See technetium Tc 99m bicisate
bifunctional chelating agents (BFCAs), 179, 181
    for antibodies, 186, 187f, 664–665, 664f, 665f
    for peptides, 186, 187f, 656, 668–669
    for technetium-nonessential compounds, 185–186, 186f
bile, 519–520, 519f, 530–532
bile ducts, 518f, 519
bile gastritis, 537, 538f
bile leak, 530, 537, 540f
biliary atresia, 530, 537, 539f
biliary obstruction, 530, 537, 537f, 538f
binding assays, in vitro, 348–350
    autoradiography vs., 348, 349t
    competitive binding, 349, 349f
    control studies of target-binding affinity in, 349
    direct, 349, 349f
    disadvantages of, 353–354
    procedure for, 349, 349f
    quantification of biomarker–target binding in, 350

specific and nonspecific target binding in, 350, 350f
target-binding selectivity in, 350, 350f
variables in, 349, 350t
binding energy
electron, 14–15, 15f
nuclear, 16–17, 17f, 31
bioassays, 116–117
biologic effect(s), 87, 127–145
acute, 138–139
carcinogenic, 98, 99–102, 139–141
gastrointestinal, 135
hereditary (genetic), 98, 99, 102–104, 131–132, 141–142
measurement of, 88
overview of, 127–128, 128f
biologic effectiveness, relative, 88
Biologic Effects of Ionizing Radiation (BEIR) committee, 109
BEIR V, 99, 101, 103
BEIR VII, 100, 109
biologic safety cabinets (BSCs), 258f, 259, 261–262, 315
biomarker distribution, control studies of, 354
biomarker–target binding, quantification of, 350
biostatistics review, 393
bismuth Bi 213 (^{213}Bi), 690t
bismuth germanate (BGO), 78, 231
bisulfites, for radioiodination, 157
bladder, FDG uptake in, 624–625, 625f
bladder cancer, PET/CT of, 625f
bleeding, 138
gastrointestinal, 517, 527–528, 527t, 539, 541f
blood–brain barrier (BBB), 3, 164–165, 165f, 412–413, 412f, 413f
blood–brain extracellular fluid barrier (BECF), 413, 413f
blood–CSF barrier, 413, 413f
blood flow
cerebral, 413–415, 414f. See also cerebral blood flow studies
hepatic, 517, 519f, 530, 530f
myocardial, 459, 460–467. See also myocardial perfusion imaging
PET of, 234–235, 234t
renal, 547–548
spleen, 525
blood volume studies, 3, 158, 161, 234–236, 695–701
B-lymphocytes, 656, 656f
BMIPP iodine I 123. See iodofiltic acid iodine I 123
BNL. See Brookhaven National Laboratory
board certified in nuclear pharmacy (BCNP), 10–11, 11f
Board of Pharmaceutical Specialties (BPS), 10–11, 266, 374
body surface area (BSA), and dose, 326, 327t
Bolton-Hunter reagent, 211, 212–213, 212f, 663, 663f
bombesin, 671
bonds, chemical, 148–154
bone
composition of, 571–572, 572f
formation of, 571–572

principal tissues of, 571, 572f
tissue-weighting factor for, 104t
bone cancer
imaging of, 584, 585f
radiation-induced, 140
treatment of pain in, 4, 223–224, 225
bone imaging, 571–586
activity distribution in, 584t
in adults, normal, 580–581, 581f
agents for, 194, 195, 223, 573–578
adverse reactions to, 337, 577
altered biodistribution of, 578
development of, 572–574
dosimetry of, 576–577, 577t
impurities in, 574, 574f
localization mechanisms of, 576, 577f
preparation of, 574–575, 574f, 575f
of avascular necrosis, 584–586, 586f
of bone metastases, 582, 584f
in children, 580, 580f
of fractures, 581, 581f
of gouty arthritis, 584, 586f
of hypertrophic pulmonary osteoarthropathy, 582, 585f
of metabolic bone disease, 584, 586f
nuclear medicine procedures for, 578–586
interpretation of, 580–586
rationale for, 578
of osteomyelitis, 582, 582f–583f
PET, 577f, 578, 579f, 580
physiologic principles of, 571–572
plain x-ray, deficiencies of, 571
of primary bone tumors, 584, 585f
of remodeling process, 586, 646, 648f
of shin splints, 581–582, 581f
sodium fluoride F 18 scintigraphy for, 573, 576f, 578, 579f, 580, 644–646, 647f
specificity of, 580
bone marrow, 571, 572f, 610
imaging of, 610–612, 612f
suppression of, 138
tissue-weighting factor for, 104t
bone pain, radiotherapy for, 578, 582, 584f, 673, 683–686, 684t
Born, Gordon, 9, 9t
boron, electronic configuration of, 152t
boronic acid adducts of technetium oxime (BATO), 182
bound-to-free (B:F) ratio, 348
Bowen, Barry M., 9t
Bowman Gray School of Medicine, 8
BPS. See Board of Pharmaceutical Specialties
Bq. See becquerel
^{76}Br. See bromine Br 76
^{77}Br. See bromine Br 77
brachytherapy, 208
for liver cancer, 688–690
for liver metastases, 689–690
bradykinin, 589
Bragg curve, 51f
brain
circulation of, 413–415, 414f
compartments of, 412–413, 413f
FDG uptake in, 620, 621, 621f, 622f
neural and nonneural capillaries in, 412, 412f

brain cancer
PET/CT of, 628–629
reimbursable PET procedures in, 239t, 380
tumor recurrence in, imaging of, 433, 435f
brain death, imaging in, 428–430, 430f, 431f
brain imaging, 411–440
agents for, 416–428
capillary permeability and, 418
development of, 416, 418–419, 422
diffusible, 411, 418–422
diffusion of, 164–165, 165f
interstitial fluid and, 417–418
mechanisms of localization, 417
metabolism, 411, 422–423
neuronal, 411, 423–426
nondiffusible, 411, 416–418
tumor cell uptake and, 418
vascularity and, 417
in brain death, 428–430, 430f, 431f
of cerebrovascular reserve, 433–435, 436f
in dementia, 240, 425–426, 629, 649–650, 651f
in epilepsy, 431–432, 432f
in Parkinson's disease, 424–425, 425t, 435, 436f, 649–650
PET/CT, 629, 649–650, 651f
physiologic principles in, 412–416
static, 3, 3f
in tumor recurrence, 433, 435f
brain metastases, PET/CT of, 628–629, 629f
braking radiation (bremsstrahlung), 49f, 50, 50f, 61, 222
breast, tissue-weighting factor for, 104t
breast cancer
lymphoscintigraphy in, 606, 607, 636
PET/CT of, 620f, 636–638, 639f
positron emission mammography of, 636, 638f
radiation-induced, 140
reimbursable PET procedures in, 239t, 240–241, 380
breast-feeding
length of time to interrupt, 328–330, 330t
regulatory requirements on, 330–331
breast milk
dose conversion factors for, 329–330, 331t
metabolism/decomposition in, 331–332, 332f
radiopharmaceutical excretion in, 328–333, 417, 448–449
agents no data available on, 332
case study of, 332–333, 333t
concurrent drugs and, 332
breath test, carbon C 14 urea, 703
bremsstrahlung, 49f, 50, 50f, 61, 222
Breslow, Kenneth, 9t
BRIDA, ^{99m}Tc. See technetium Tc 99m mebrofenin
Briner, William, 8, 9t
bromine Br 76 (^{76}Br), as nontraditional radionuclide, 346, 347t
bromine Br 77 (^{77}Br), labeling with, 662–663
bronchi, 499, 500f

bronchioalveolar carcinoma, 629
bronchioles, 499, 500f
Brookhaven National Laboratory (BNL), 171
brown fat uptake, of FDG, 618, 620f, 627
Bruce protocol, modified, 461
BSA. *See* body surface area
BSCs. *See* biologic safety cabinets
bubble-point test, 315
BUD. *See* beyond-use-date
buffer zone, in nuclear pharmacy, 259
Bureau of Biologics, FDA, 376
burial, 125
Burkitt's lymphoma, 645f
byproduct material
   medical use of
      general administrative requirements for, 398–400
      general information for, 395–398
      general technical requirements for, 400
      NRC regulations on, 394–408, 395t
      records required for, 406–407
      report requirements for, 407
   sealed, 396, 401, 406
   unsealed
      dosages of, 400–401, 406
      regulatory requirements on, 400–401, 402–405
      written directive not required, 402–404
      written directive required, 404–405
bystander effect, 679

C

^{11}C. *See* carbon C 11
^{14}C. *See* carbon C 14
CA. *See* carrier added
^{45}Ca. *See* calcium Ca 45
^{47}Ca. *See* calcium Ca 47
CACIs. *See* compounding aseptic containment isolators
CAD. *See* computer-aided design; coronary artery disease
Cahn-Ingold-Prelog (CIP) configuration, 167–168, 168f
calcium, parathyroid hormone and, 603–604
calcium Ca 45 (^{45}Ca), for bone imaging, 573
calcium Ca 47 (^{47}Ca), for bone imaging, 573
calcium phosphate, amorphous, 571–572, 576
calculations
   dose, 91–92
   radioactivity, 28–30
calibration, dose. *See* dose calibrators
calibration, radioactive, 37–38, 38f
calibration, survey instrument, 304–305, 400
calibration time, 30
California Highway Patrol, 270
Callahan, Ronald, 9t, 10
cancellous bone, 571, 572f
cancer. *See also specific types*
   PET/CT in, 617–620
      clinical use of, 625–643
      patient preparation for, 617–618
      whole-body, 618
   PET procedures in, reimbursable, 239t, 240–241, 380, 617, 626
   radiation-induced, 99–102, 139–141
      age at exposure and site of exposure and, 101, 101t, 102t
      continuous exposure in adulthood and, 101, 101t
      continuous lifetime exposure and, 101, 101t
      dose–response relationship in, 100, 100f, 140–141, 141f
      exposure in pregnancy in, 143–144
      instantaneous exposure and, 101, 101t
      mechanisms of carcinogenesis, 140–141
      oncogenes and, 132, 141
      probability coefficients for, 102, 102t, 103t
      radiation risk *vs.* other risks, 104–105
      risk estimates for, 99, 100–102, 104–105, 128
      risk models for, 99
      as stochastic effect, 98, 128, 139
      suppressor genes and, 141
      tissue-weighting factor and, 104, 104t
      types of, 139–140
   radiotherapy for
      dose fractionation in, 137–138
      four Rs of, 137–138
      iodine I 125 in, 208
      multiple-dose, 137
      oxygen effects on, 136–137
      single-dose, 137
   treatment for, assessing response to, 628f, 644, 645f, 646f
capillaries, brain
   neural and nonneural, 412, 412f
   permeability of, 418
capromab pendetide (^{111}In-capromab pendetide), 218, 666, 667–668
   indication for, 666
   mixing order for, 321
   for prostate imaging, 218, 666, 667–668, 668f
   radiochemical purity of, 286t, 288t
   radiolabeling of, 665
captopril renography, 556–558, 558f, 560, 561f
capture reactions, neutron, 51–52
carbohydrates, Fischer configuration of, 167, 167f
carbon
   chemical bonds of, 149–153
   chemistry of, 225
   electronic configuration of, 152t, 225
   isotopes of, 18, 18t, 225
   on-site production of, 34
carbon C 11 (^{11}C), 5t, 26t, 225
   autoradiography of, 353
   for brain imaging, 411, 422
   decay of, 21, 21f
   half-life of, 207
   labeling precursors for, 244t
   labeling with, 160, 161t, 346–347
   for PET, 207, 230t, 378
   production of, 244t
   radionuclidic purity of, 276t
   as traditional radionuclide, 346, 347t
carbon C 11 carbon monoxide. *See* carbon monoxide C 11
carbon C 11 flumazenil. *See* flumazenil C 11
carbon C 11 hydroxy ephedrine (^{11}C-HED), for cardiac neuronal imaging, 491
carbon C 11 methyl iodide (^{11}C-methyl iodide), 244t, 245, 245f
carbon C 11 palmitate (^{11}C-palmitate)
   for heart imaging, 459, 460t, 472, 473, 473f
   substrate specificity of, 169
carbon C 11 raclopride. *See* raclopride C 11
carbon C 11 sodium acetate. *See* sodium acetate C 11
carbon C 14 (^{14}C), 5t, 26t, 225
   applications of, 207, 225
   autoradiography of, 353
   decay of, 19, 20f, 225, 225f
   labeling with, 161t
   production of, 225
carbon C 14 urea (^{14}C-urea) breath test, 703
carbon C 14 urea (^{14}C-urea) capsules, 225
carbon monoxide C 11 (^{11}C-carbon monoxide)
   for blood volume measurement, 235–236
   for PET, 234t, 235–236
   radiochemical purity of, 286t
carbon monoxide O 15 (^{15}O-carbon monoxide), for PET, 235–236
carcinogenic effects. *See* cancer, radiation-induced
carcinoid tumors
   imaging of, 600f, 600t
   treatment of, 691
cardiac neuronal imaging, 490–492
Cardinal Health, 9
cardiology. *See* heart imaging
cardiovascular parameters. *See also* heart imaging
   PET, 234–236, 234t
carotid artery, 413–414, 414f
carrier added (CA), 162, 178
carrier-free (CF), 162, 177
carrier iodine, 293
carrier technetium, 42–43, 43f, 177–178
CASS. *See* Coronary Artery Surgery Study
cataracts, radiation-induced, 128, 144–145, 145f
caution signs, 119, 119f
CB-TE2A (4,11-bis(carboxymethyl)-1,4,8,11-tetraazabicyclo[6.6.2]hexadecane), 668–669, 669f
CCK. *See* cholecystokinin
CDC. *See* complement-dependent cytotoxicity
CDE. *See* committed dose equivalent
CDRs. *See* complementarity-determinant regions
CDs. *See* clusters of differentiation
CEA-Scan. *See* technetium Tc 99m arcitumomab
CED. *See* coverage with evidence development
CEDE. *See* effective dose equivalent, committed

cell(s)
    hypoxic, and cancer treatment, 136–137
    radiosensitivity of, 135, 136t
cell cycle, 129–131, 130f
    and radiosensitivity, 137, 137f
cell death, programmed, 673–674
cell killing, definition of, 132
cell (plasma) membrane
    structure of, 163, 163f, 164f
    transport across, 163–166
cell survival curves, 132–135, 133f
    dose fractionation and, 134, 134f, 138
    linear energy transfer and, 133, 134f
    linear–quadratic model of, 134f, 135, 140
cellulitis, 582, 584t
Center for Drug Evaluation and Research (CDER), 381, 385, 389–390
Centers for Medicare and Medicaid Services (CMS), 239–241, 239t, 253, 380–381, 617, 626
centigray (cGy), 88
central atom, 153
centralized nuclear pharmacy, 8–9, 257, 374
central nervous system (CNS), imaging of, 411, 428. See also brain imaging; cerebrospinal fluid imaging
centrifuge, 265
cerebral arteries, 414, 414f
cerebral blood flow, 413–415, 414f
cerebral blood flow studies, 164–165
    diffusible tracers for, 418–422
    xenon Xe 133, 220
cerebrospinal fluid, 415–416
    artificial, 428, 428t
    drug transport in, mechanisms of, 427
    flow of, 415–416
    structures involved in, 415, 415f
    volume of, 415
cerebrospinal fluid imaging, 158, 217–218, 411, 436–440
    agents for, 411–412, 426–428, 427t
        indium In 111 pentetate injection, 158, 217–218, 412, 426–427, 426f, 427t
        technetium Tc 99m pentetate, 158, 195, 337, 427t, 428
    of CSF leaks, 412, 436, 437–438, 438f, 438t
    of hydrocephalus, 412, 436–437, 437f
    intrathecal administration for, safety considerations in, 427–428
    for shunt evaluation, 436, 438–440, 439f, 440f
cerebrospinal fluid leaks, 412, 436, 437–438, 438f, 438t
cerebrovascular reserve, imaging of, 433–435, 436f
cerebrovascular syndrome, 138, 139
certificate of analysis, PET, 307
certification, in nuclear pharmacy, 9–11, 10t, 11f, 266, 266t, 374, 405
certification, of pharmacy technicians, 265
cervical cancer
    PET/CT of, 643, 643f
    reimbursable PET procedures in, 239t, 240–241

cesium Cs 137 (^{137}Cs), 27t
    decay of, 23, 23f
    energy resolution of, 71, 72f
    pulse-height energy spectrum of, 67–70
    as reference for dose calibrators, 300
    shipping guidelines for, 122t
CF. See carrier-free
CFR. See Code of Federal Regulations
CGMPs. See current good manufacturing practices
cGy. See centigray
Chadwick, James, 51
Chandler, Robert, 8, 9t
characteristic x-ray peak, 70
charged-particle accelerators, 33–34, 33f, 242–244, 243t
charged-particle interactions, 47–51
chelates/chelating agents, 153–155, 157, 347
    ^{99m}Tc, common problems associated with, 320–321
    ancillary, for technetium kit, 175
    for antibody labeling, 664–665, 664f, 665f
    bifunctional, 179, 181
        for antibodies, 186, 187f, 664–665, 664f, 665f
        for peptides, 186, 187f, 656, 668–669
    for technetium-nonessential compounds, 185–186, 186f
    single amino acid, 182, 182f
chemical bonds, 148–154
chemical identity. See radiochemical identity
chemical purity, 293. See also radiochemical purity
    in ^{99m}Tc generator, 38–39, 39f
chemical quenching, 73–74, 75f
chemistry, manufacturing, and control (CMC), 392–393
chemistry, of radiopharmaceuticals, general topics of, 147–169
chemistry review, of new drugs/radiopharmaceuticals, 392–393
chemotherapy, assessing response to, 628f, 644, 645f, 646f
Chernobyl reactor accident, 127–128, 140
chest x-ray, for ventilation–perfusion scan, 508
CHF. See congestive heart failure
children
    bone imaging in, 580, 580f
    dosing for, 324–328
        age-based adjustment, 326
        body surface area and, 326, 327t
        height-based adjustment, 326–327
        minimum dose, 325–326
        PET/CT, 327–328
        radiation absorbed dose, 325, 325t
        radiopharmaceuticals containing benzyl alcohol, 327
        technetium Tc 99m aggregated albumin, 325t, 327
        weight-based adjustment, 326
Chilton, Henry M., 9t
chimeric antibody, 659, 673, 674
chiral molecules, 166–168
chi-square test, 84–85, 85t

chloramine-T, 211, 212, 662, 663f
cholangiocarcinoma, PET/CT of, 641–643
cholecystitis
    acute, 530, 533, 536f
    chronic, 533–537, 536f
cholecystokinin (CCK), 530–532
    for hepatobiliary imaging, 524–525, 534f
    sphincter of Oddi and, 519
choroid plexus, 415–416, 415f, 436
Christian, John, 8, 9, 9t
chromatid(s), 130, 131
    dicentric, 131, 132f
chromatogram, 277f, 278
chromatography, 275–284, 288t–291t
    gas, 275–278, 277f
    high-pressure liquid, 167, 278–279, 278f
        for PET radionuclide preparation, 249, 251–252
        for radiochemical identification, 275
    in nuclear pharmacy, 265
    paper, 275, 279–283
    streaking in, 282–283, 283f
    technical precautions in, 282–283, 283t
    thin-layer, 275, 279–283, 292f
chromic phosphate P 32 suspension (^{32}P-chromic phosphate), 222–223
    adverse reactions to, 522
    appearance and color of, 294, 294t
    for effusion therapy, 687
    intracavitary administration of, 222
    for liver imaging, 520, 522
    off-label uses of, 222–223
    production of, 222
    radiation absorbed dose of, 223, 223t
    for radiation synovectomy, 222–223, 687–688, 688t
    radiochemical purity of, 286t
chromium
    electronic configuration of, 221
    oxidation states of, 221
chromium Cr 51 (^{51}Cr), 5t, 26t
    applications of, 158–159, 207
    decay of, 22, 22f, 158–159, 221, 221f
    labeling with, 160
    production of, 221
    pulse-height energy spectrum of, 70f
    for red blood cell–plasma measurement, combined, 699–700
    for red blood cell survival measurement, 700–701, 701f
    for red blood cell volume measurement, 696–697, 697f, 698t
    shipping guidelines for, 121t, 122t
chromium Cr 51 chloride (^{51}CrCl$_3$), 221
chromium Cr 51 chromic chloride (^{51}Cr-chromic chloride), for gastrointestinal imaging, 528
chromium Cr 51-labeled albumin (^{51}Cr-labeled albumin), for gastrointestinal imaging, 528
chromium Cr 51 sodium chromate. See sodium chromate Cr 51
chromosomal abnormalities, 103t, 104, 131–132, 132f, 141. See also hereditary effects

chronic obstructive pulmonary disease (COPD), 505, 508, 509–513, 512f, 513f
chronic toxicity tests, 382
Ci. *See* curie
cimetidine, for Meckel's diverticulum imaging, 540
CIP. *See* Cahn-Ingold-Prelog configuration
ciprofloxacin, radiolabeled
    for infection imaging, 593–594
    localization mechanisms of, 596t
circle of Willis, 414
circulation
    cerebral, 413–415, 414f. *See also* cerebral blood flow studies
    hepatic, 517, 519f, 530, 530f
    myocardial, 460–461. *See also* myocardial perfusion imaging
    pulmonary, 499–501, 500f. *See also* lung perfusion studies
    renal, 547–548
    spleen, 525
circumflex artery, left, 460, 460f
cisterna magna, 415f, 416
cisternography, 158, 217–218, 411, 436–440
    agents for, 411–412, 426–428, 427t
        indium In 111 pentetate injection, 158, 217–218, 412, 426–427, 426f, 427t
        technetium Tc 99m pentetate, 158, 195, 337, 427t, 428
    of CSF leaks, 412, 436, 437–438, 438f, 438t
    of hydrocephalus, 412, 436–437, 437f
    intrathecal administration for, safety considerations in, 427–428
    for shunt evaluation, 436, 438–440, 439f, 440f
clearance, renal, 548–549, 550t
clearance agents, renal, 552, 553t
clearance time, for radioactive gases, 118
clinical research
    with human subjects, 357–359, 387–389
    initiatives on promising biomarkers, 359–362, 360t, 361t
clinical review, of new drugs/radiopharmaceuticals, 392, 393–394
clinical reviewers, 393
clinical trials, 357, 358f, 359, 387–389, 392, 393–394
clusters of differentiation (CDs), 661
CMC. *See* chemistry, manufacturing, and control
CMS. *See* Centers for Medicare and Medicaid Services
CNS. *See* central nervous system
^{57}Co. *See* cobalt Co 57
^{58}Co. *See* cobalt Co 58
^{60}Co. *See* cobalt Co 60
cobalt Co 57 (^{57}Co), 5t, 26t
    decay of, 22
    labeling with, 160
    pulse-height energy spectrum of, 71f
    as reference for dose calibrators, 300
    shipping guidelines for, 122t
cobalt Co 57 cyanocobalamin (^{57}Co-cyanocobalamin), 703
cobalt Co 58 (^{58}Co), 26t

cobalt Co 60 (^{60}Co)
    neoplastic transformation by, 140, 141f
    relative biologic effectiveness of, 88
Code of Federal Regulations (CFR), 376
    on clinical review, 393
    on medical use of byproduct material, 394–408, 395t
    on NDA approval, 394
    on Radioactive Drug Research Committee, 382–384
    on statistical analysis, 393
    Title 10, 110
        on area monitoring, 118
        on bioassays, 116–117
        on breast-feeding, 330–331
        on dose calibrators, 299–300
        on dose determination, 285–293
        on dose limits, 111
        on dose terminology, 90–91
        on exposure during pregnancy, 117–118
        on exposure to radiation from patients, 96, 456
        on generator testing, 38
        on inspections, 119–120
        on as low as reasonable achievable, 110
        on minimum detectable activity, 83
        on notices, instructions, and reports, 119
        on nuclear pharmacist training, 10
        on receipt of radioactive material, 120
        on shipment of radioactive material, 120–125
    Title 10, Part 19, 109
    Title 10, Part 20, 109
    Title 10, Part 35, 109, 285–293, 299–300
coefficient of variation (CV), 80–81, 81t
coincidence detection, in PET, 77–78, 77f, 230–231, 230f, 231f
coincidence lines, in PET, 77, 77f
coincidence loss, 82
coincidence sum peak, 70
cold spots, 2–3
cold thyroid nodules, 451, 453, 453f
Cole, Clyde N., 9t
collimator, in gamma camera, 76
colloids. *See also specific colloids*
    adverse reactions to, 337
    for liver imaging, 2–3, 520–522, 520t, 521f, 529–530
    for lymphoscintigraphy, 607–608
    for radiation synovectomy, 687
    for spleen imaging, 526, 529–530
colon, FDG uptake in, 625
color, of radiopharmaceuticals, 294, 294t
colorectal cancer
    liver metastases from, brachytherapy for, 689–690
    PET/CT of, 633f, 638, 640f
    radiolabeled antibodies for diagnosis of, 666
    reimbursable PET procedures in, 239t, 240–241, 380
colorimetric spot test, 38–39, 39f
color quenching, 73–74, 75f

columns
    in gas chromatography, 277
    in high-pressure liquid chromatography, 279
commercialization, of radiopharmaceuticals, 359
committed dose equivalent (CDE, $H_{T,50}$), 90–91
committed effective dose equivalent (CEDE, $H_{E,50}$), 91
common bile duct, 518f, 519
common bile duct obstruction, 530, 537, 537f, 538f
compact bone, 571–572, 572f
competitive-binding assays, 349, 349f
complaint handling, 309
complementarity-determinant regions (CDRs), 657
complement-dependent cytotoxicity (CDC), 657, 659, 673–674
component concentration, 322
compounded sterile products (CSPs), 313–317
compounding, 267–268, 267f, 268f
compounding area, 259–260
compounding aseptic containment isolators (CACIs), 259, 315
Compton edge, 69
Compton scatter, 53, 54–55, 55f, 55t, 56f, 57t
    in direct effects, 128, 128f
    in dose calibrators, 61
    in pulse-height spectrometry, 68–70
computed tomography (CT), 344, 344t
    average effective dose in, 105t–106t
    of head and neck cancer, 626–627
    hepatobiliary and spleen, 528–529
    of infection, 594
    PET with. *See* PET/CT
    of pulmonary embolism, 508
    renal, 547, 555
    SPECT with (SPECT/CT), 74
    of tumors, 598
computed tomography pulmonary angiography (CTA), of pulmonary embolism, 511–513
computer-aided design (CAD), 346
computer system, in nuclear pharmacy, 259–260, 259f
conducting zone, of lungs, 499, 500f
configuration
    electronic, 149, 152t
    isomer, 167–168, 167f, 168f
confinement, emergency, 125–126
congenital abnormalities, 103t, 104
congestive heart failure (CHF), 474, 480
Congress of the European Association of Nuclear Medicine, 345
constancy
    of dose calibrators, 299t, 300
    of survey instruments, 305
constitutional isomers, 166
container configuration, for dose calibrators, 303, 303t
container walls, interaction with, 324
contamination, microbial
    control/prevention of, 295–296, 313–317
    risk levels of, 313–314

contrast-enhanced MRI, of myocardial infarction, 489–490, 491f, 492f
contrast media, adverse reactions to, 336, 337t
control standard endotoxin (CSE), 297
conversion electron, 23
Cooper, James, 8, 9t, 10, 11
coordinate chemical bonds, 153
coordinating groups, 153
coordinating ligands, for technetium kits, 174–175, 175f, 179, 180f
coordination compounds, 153–154, 154f, 155f, 161
COPD. *See* chronic obstructive pulmonary disease
copper Cu 60 (^{60}Cu), as nontraditional radionuclide, 347t
copper Cu 62 (^{62}Cu), as nontraditional radionuclide, 347t
copper Cu 64 (^{64}Cu), 26t
  autoradiography of, 353
  as nontraditional radionuclide, 346, 347t
copper Cu 67 (^{67}Cu), 26t, 662t, 674
copper filter
  in dose calibrators, 62–63, 62f, 63f, 64t
  for film badges, 115
CORAR. *See* Council on Radionuclides and Radiopharmaceuticals
coregistration, 233
cores, technetium, 179–184, 180f
  [O = Tc = O]$^+$, 183, 183f
  [Tc]$^+$, 181, 181f
  [Tc]$^{3+}$, 182, 183
  [Tc(CO)$_3$]$^+$, 181–182, 181f
  [Tc]$^{4+}$, 182, 183
  [Tc-HYNIC], 184, 184f
  [Tc = O]$^{3+}$, 180f, 183
  [Tc ≡ N]$^{2+}$, 183f, 184
coronal plane, 3, 3f
coronary artery(ies), 471
  left, 460, 460f
  right, 460, 460f
  stenoses of, 460–461, 461t
  vasodilation of, 459. *See also* myocardial perfusion imaging
coronary artery disease (CAD)
  imaging techniques for, 459, 473–474. *See also* heart imaging
  risk stratification for, 492–494, 493f, 493t, 494t
Coronary Artery Surgery Study (CASS), 493
coronary flow reserve, 460–461
cortex, renal, 547, 548f
cortical bone, 571, 572f
cortical nephrons, 547, 548f
cortical scintigraphy, renal, 558–559, 562–568
  of horseshoe kidney, 568, 568f
  of multicystic dysplastic kidney, 562, 567f
  normal, 562, 564f
  of renal cell carcinoma, 562, 565f
  of renal sarcoma, 567f
  of urinary tract infection, 562, 566f
cortisol, 601
cosmic radiation, 53t, 112
Council on Radionuclides and Radiopharmaceuticals (CORAR), 381

counting
  chi-square test of, 84–85, 85t
  efficiency of, 81–82
  errors in, 78–79, 80–81, 81t
  instrumentation for, 71–78
  statistics in, 78–80, 80f, 80t
counts per minute (cpm), 79, 80t, 81
covalent bonds, 149–153
coverage with evidence development (CED), 239t, 240–241
C-PATH. *See* Critical Path Initiative
cpm. *See* counts per minute
^{51}Cr. *See* chromium Cr 51
C-reactive protein, 589
criteria for release, 402
critical organ, 91
Critical Path Initiative (C-PATH), 358–359, 384
critical sites
  definition of, 314
  protection of, 314–315
Crohn's disease, 594, 625
^{137}Cs. *See* cesium Cs 137
CSF. *See* cerebrospinal fluid
CSF–brain barrier, 413, 413f
CSPs. *See* compounded sterile products
CT. *See* computed tomography
CTA. *See* computed tomography pulmonary angiography
^{60}Cu. *See* copper Cu 60
^{62}Cu. *See* copper Cu 62
^{64}Cu. *See* copper Cu 64
^{67}Cu. *See* copper Cu 67
curie (Ci), 25–28, 87, 90, 161
Curie, Marie, 24
Curie, Pierre, 24, 127
current good manufacturing practices (CGMPs), 253, 274, 299, 305–306, 308–309, 381
current/voltage response curve, 59–60, 60f
Cutie Pie ionization chamber, 264f
CV. *See* coefficient of variation
cyanocobalamin cobalt Co 57. *See* cobalt Co 57 cyanocobalamin
cyclopentadienyl core complex, 181f, 182
cyclotrons, 1, 33–34, 33f
  for gallium, indium, and thallium production, 215
  for iodine I 123 production, 209
  for PET radionuclide production, 241, 242–244, 243f, 244t
  radioiodine production by, 446
  targetry for, 243–244
  for xenon Xe 133 production, 220
cystic duct, 518f, 519
cystography, radionuclide, 559, 568, 568f
cytokines, 589, 594

D

D. *See* absorbed dose
DAC. *See* derived air concentration
DADS technetium Tc 99m (^{99m}Tc-DADS). *See* technetium Tc 99m-N,N′-bis(mercaptoacetyl) ethylenediamine
daily regulatory tasks, 270, 270f
Dally, Clarence Madison, 140
Danlos, Henri, 1
DAT. *See* dopamine transporter

daughter nuclide, 19
D configuration, 167–168, 167f, 168f
DDE. *See* deep-dose equivalent
DDREF. *See* dose and dose rate effectiveness factor
dead cells, 136–137
dead time, 82–83, 232, 232t
death
  as adverse reaction, 336, 336f, 337t
  brain, imaging in, 428–430
decay, radioactive, 13, 17–30
  alpha particle, 24, 25f
  calculations of, 28–30
  diagnostic radiopharmaceuticals, 158–159
  dual-process, 21
  electron capture, 22–23, 22f, 158–159, 229
  negatron, 18, 19, 20f, 22–23, 51
  positron, 18, 21–23, 21f, 22f, 229–230, 231f
  rate of (activity), 25–28
  in storage, 125, 402, 406
decay constant, 25
decays or disintegrations per second (dps), 25–28
decay tables, 29–30
decay test, of dose calibrators, 301, 301f
dedicated check source, 396
deep-dose equivalent (DDE, Hd), 91
deep-vein thrombosis (DVT), 499
Dees, 33–34, 33f, 242
delayed gastric emptying, 542, 544f
deletion, chromosomal, 132, 132f
delivery area, 267
delivery personnel, 266
delta ray, 48, 49, 49f, 50f
dementia
  imaging agents for, 425–426
  PET/CT of, 629, 649–650, 651f
  reimbursable PET procedures in, 240
  SPECT of, 432–433, 433f, 434f
denatured ^{99m}Tc red blood cells, for spleen imaging, 526, 526f, 529, 532f
deoxyribonucleic acid. *See* DNA
Department of Energy (DOE), 376, 377
Department of Transportation (DOT), 109, 120–125, 253, 260, 270, 271, 373, 391f
depreotide technetium Tc 99m. *See* technetium Tc 99m depreotide
derived air concentration (DAC), 118
Desantis, Nunzio, 9, 9t
detection, radiation, 59–85
  counting, 78–85
    chi-square test of, 84–85, 85t
    efficiency in, 81–82
    errors in, 78–79, 80–81, 81t
    instrumentation for, 71–78
    statistics in, 78–80, 80f, 80t
  in imaging systems, 74–78
  instrumentation for, 71–85, 262–265
  ion collection methods of, 59–63
  maximum detectable activity in, 82–83, 84f
  minimum detectable activity in, 83–84
  in nuclear pharmacy, 262–265
  resolving time in, 82–83
  scintillation methods of, 63–71
detection efficiencies, 159, 159t

determinate errors, 78–79
deterministic effects, 98–99, 99f, 110, 127–128, 128f, 334t
    acute, 138
    threshold dose for, 98, 128
detriment, 104
developmental stage, 142–144
development of new radiopharmaceuticals. *See* radiopharmaceutical development
diabetic foot infections, 594, 651
diabetic gastroparesis, 544
diabetic medication, and PET/CT, 618
diagnostic radiopharmaceuticals, 158–160. *See also specific radiopharmaceuticals*
    chemistry of, 159–160
    decay mode of, 158–159
    detection efficiencies of, 159, 159t
    energy of, 158–159
    half-lives of, 159
diarrhea, 138–139
diastereomers, 166, 166f
dicentric chromatid, 131, 132f
dicentric chromosome, 131, 132f
diet, and PET/CT, 618, 618f
diethylenetriamine pentaacetic acid (DTPA), 664–665, 664f, 665f, 668–669, 669f
diethylenetriamine pentaacetic acid indium In 111. *See* indium In 111 pentetate
diethylenetriamine pentaacetic acid technetium Tc 99m. *See* technetium Tc 99m pentetate
diethylenetriamine pentaacetic acid ytterbium Yb 169 pentetate. *See* ytterbium Yb 169 pentetate
differential lung perfusion, 513, 514f
diffusible tracers
    for brain imaging, 411, 418–422
        chemical structure of, 418, 418f
        development of, 418–419
    for heart imaging, 466–467, 466f
diffusion
    facilitated, 165–166, 548
    passive, 163–165, 163f, 548
digestive system, FDG uptake in, 625
diiodotyrosine (DIT), 445
diphosphonates, technetium 99m, 573–578. *See also specific diphosphonates*
dipyridamole
    chemical structure of, 462f
    for myocardial perfusion imaging, 462, 463, 464–466, 478f
    precautions, adverse effects, contraindications, 464–466, 465t
direct binding assays, 349, 349f
direct effects, of radiation, 128–129, 128f
direct model, of risk, 102–103
Dirksen, J. William, 8, 9t
discovery, drug, 357. *See also* radiopharmaceutical development
DISIDA, ^{99m}Tc. *See* technetium Tc 99m disofenin
disinfection, 313–315
disintegration, 25. *See also* decay, radioactive
disintegrations per minute (dpm), 81
disofenin technetium Tc 99m (^{99m}Tc-DISIDA). *See* technetium Tc 99m disofenin
dispensing, 269–270, 269f
    delays before, problems associated with, 323–324

dispensing-syringe shield, 261, 261f
disposal, of radioactive waste, 125, 260, 260f
distal convoluted tubules, 547
distance
    from materials in transport, 123, 123t
    from radiation source, 96–97, 97t, 114
distribution, 270, 309. *See also* transport of radioactive material
DIT. *See* diiodotyrosine
diuresis renography, 556, 557f, 560–562, 562f
DMSA, ^{99m}Tc. *See* technetium Tc 99m succimer
DNA
    composition of, 129
    damage to, 99, 127, 141–142
    double-strand break in, 131, 131f, 135, 135f
    free radical reaction and, 129
    radiation effects on, 131–132
    single-strand break in, 131, 131f
    structure of, 129, 130f
    synthesis of, 129–131
dobutamine
    chemical structure of, 462f
    for myocardial perfusion imaging, 462, 463, 464–466, 474
    precautions, adverse effects, contraindications, 464–466, 465t, 474
DOE. *See* Department of Energy
dopamine
    adrenal synthesis of, 601
    imaging of, 237–238, 424–425, 435, 436f
dopamine transporter (DAT), 181–182, 424–425, 424f
dosage, prescribed, 396
dosage, radioactivity, 4
dosage, therapeutic, 396
dosage, unit, 397
dosage review, 270
dose
    average effective, in radiologic procedures, 105t–106t
    calculation of, 91–92
    conversion of measures, 91t
    determination of, 285–293
    effective, 88–90, 89t, 90t, 91f, 104
        average, in radiologic procedures, 105t–106t
    equivalent, 88–89, 90t, 91f, 104
    occupational, 111
    pediatric, 324–328
    prescribed, 396
    radiation, 4
    system used in protection, 91f
    terminology for, 88–91, 91t
    therapeutic, 396
dose and dose rate effectiveness factor (DDREF), 100
dose calibrators, 60–63
    accuracy of, 285, 299t, 300
    adjustment of measurements in, 61–63, 62f, 63f, 64t
    components of, 60–61, 60f
    constancy of, 299t, 300
    container configuration for, 303, 303f
    copper filter in, 62–63, 62f, 63f, 64t
    geometry of, 299t, 300, 302–304

    for iodine I 123 measurement, 62–63, 62f, 63f, 64t, 209–210
    linearity of, 299t, 300–302
        attenuation method for testing, 301–302, 301f, 302f
        decay method for testing, 301, 301f
        report on, 302t
    in nuclear pharmacy, 262–263, 285
    primary standards for, 61
    quality control for, 263, 285–293, 299–304
    for radiochemical purity measurements, 284
    secondary standards for, 61
    sensitivity of, 61, 61f
    syringe volume in, 63, 64t
    volume variation in, 303–304
dose equivalent (H), 89, 104
    committed, 90–91
    effective, 89, 104
        average annual, in U.S., 112, 112t
        committed, 91
        total, 91, 96, 402, 406, 506
dose fractionation
    and cell survival curve, 134, 134f, 138
    in tumor treatment, 137–138
dose limits
    annual, 111, 111t
    occupational, 111
    volunteer subjects, 111–112
dose preparation, radiation exposure in, 112–113
dose rate, and radiosensitivity, 133, 134f
dose rate effectiveness factor (DREF), 100
dose–response relationship, 140–141, 141f
    and cancer risk, 100, 100f
dosimeter(s)
    optically stimulated luminescent, 116
    thermoluminescent, 116
dosimetry, radiation, 90, 91–92
    for bone imaging agents, 576–577, 577t
    for personnel, 114–116, 116t, 117f
    for radioimmunotherapy, 675
DOT. *See* Department of Transportation
DOTA (1,4,7,10-tetraazacyclododecane-1,4,7,10-tetraacetic acid), 664, 668–669, 669f
double carbon bonds, 152–153, 152f
double-strand break, in DNA, 131, 131f, 135, 135f
doubling dose model, 102–103
dpm. *See* disintegrations per minute
dps. *See* decays or disintegrations per second
DREF. *See* dose rate effectiveness factor
drug development, *vs.* radiopharmaceutical development, 345–346, 346t
drug discovery, 357. *See also* radiopharmaceutical development
drug elimination, renal, 550–551
drug information, provision of, 270–271, 271t
drug order provision, 266–270
Drug Problem Reporting Program, 334
drug stability, common problems associated with, 323–324
dry system, of ^{99m}Tc generator, 35–38, 37f
DTPA, in radiolabeling, 664–665, 664f, 665f, 668–669, 669f

DTPA indium In 111. *See* indium In 111 pentetate
DTPA technetium Tc 99m. *See* technetium Tc 99m pentetate
DTPA tin Sn 117m. *See* tin Sn 117m pentetate
DTPA ytterbium Yb 169 pentetate. *See* ytterbium Yb 169 pentetate
$D_{T,R}$. *See* mean absorbed dose
dual-isotope autoradiography, 353
dual-isotope protocols, for myocardial perfusion imaging, 476–477, 479f
dual-process decay, 21
dumping (fast gastric emptying), 542, 544f
Durham-Humphrey amendment, 378
DVT. *See* deep-vein thrombosis
dynamic imaging studies, 2

E

ECD, ^{99m}Tc. *See* technetium Tc 99m bicisate
Edison, Thomas, 140
EDTA. *See* ethylenediaminetetraacetic acid
education programs, 9–10
effective dose, 88–90, 89t, 90t, 91f, 104
  average, in radiologic procedures, 105t–106t
effective dose equivalent (H_E), 89, 104
  average annual, in U.S., 112, 112t
  committed, 91
  total, 91, 96, 402, 406, 506
effective half-life, 159
effective rate of loss (R_{eff}), 159
effective renal plasma flow (ERPF), 549, 555
effluent
  monitoring of, 118
  release of, 125
effusions, malignant, treatment of, 673, 687
EHDP, technetium Tc 99m. *See* technetium Tc 99m etidronate
eIND. *See* exploratory IND
Einstein, Albert, 16
Einstein Medical Center, 573
elastic scatter, 52
electromagnetic radiation, 17, 47, 158. *See also* gamma rays; x-ray(s)
  interactions with matter, 52–56
  properties of, 52–53, 53t
electron(s), 13, 147–148
  Auger, 15, 15f
  conversion, 23
  energy levels of, 14–15, 15f, 148–149, 152f, 152t
  sharing of (covalent bonds), 149–153
  transfer of (electrovalent bonds), 149
  valence, 148
electron capture (EC), 22–23, 22f, 158–159, 229
electronegativity, 149–153, 153t
electronic autoradiography, 353
electronic collimation, 77, 230, 231
electronic configurations, 149, 152t
electron volt (eV), 16
electrophilic fluorine F 18, 244t, 246
electrophilic substitution, 210, 210f, 662, 662f
electrophoresis, 275
electrostatic attractive force, 47
electrostatic repulsive force, 16, 47

electrovalent bonds, 149
element(s)
  atomic number of, 148
  atomic weight of, 147–148
  chemical properties of, 148
  electronic configurations of, 149, 152t
  periodic table of, 14, 14f, 147–148, 150t–151t, 160, 160t
elimination, drug, in kidneys, 550–551
Elliott's B Solution, 428, 428t
embryo, radiation effects on, 117–118, 142–144
$E = mc^2$, 16
emergency procedures, 125–126
enantiomer(s), 166–168, 166f
enantiomeric purity, 293
endocytosis, 166
endothelial cells, hepatic, 518–519, 518f
endotoxins, testing for, 295, 296–299, 308, 315, 335
end-window Geiger-Müller detectors, 63, 65f, 263, 263f
energy
  recoil, 24
  thermal, 31–32
  transition, 19
energy-compensated probes, 262–263, 264
energy levels
  electron, 14–15, 15f, 148–149, 152f, 152t
  nuclear, 15–16, 16f
energy–mass relationship, 16
Energy Policy Act of 2005 (EPAct), 377
Energy Reorganization Act of 1974, 110, 377
Energy Research and Development Administration, 376, 377
energy resolution, 71, 72f
engineering control, primary, 313, 315–316
Environmental Protection Agency (EPA), 253, 270, 373, 391f
enzymes
  activity of, PET of, 237–238
  as transporters, 163, 165–166
eosinophils, 589–590
EPA. *See* Environmental Protection Agency
EPAct. *See* Energy Policy Act of 2005
epididymitis, acute, 568
epilation, 128
epilepsy
  PET/CT of, 629, 648–649, 650f
  SPECT of, 430–432, 432f
epinephrine, 601
epiphora, 610
epratuzumab yttrium Y 90. *See* yttrium Y 90 epratuzumab
equilibrium radionuclide ventriculography (RNV), 474, 479–480
  interpretation of, 481, 485f
  procedure for, 480–481, 485f
  rationale for, 480
equivalent dose (H_T), 88–89, 90t, 91f, 104
ER. *See* extraction efficiency or ratio
erbium Er 169 citrate (^{169}Er-citrate), for radiation synovectomy, 687–688, 688t
ERPF. *See* effective renal plasma flow
errors, counting, 78–79, 80–81, 81t
erythrocytes. *See* red blood cell(s)
erythromycin, for gastroparesis, 544
erythropoiesis, imaging of, 610–612, 612f

esophageal cancer
  PET/CT of, 634, 634f, 635f
  reimbursable PET procedures in, 239t, 240–241
esophagus, FDG uptake in, 625
ethylene, carbon bonds in, 152–153
ethylenediaminetetraacetic acid (EDTA), 153–154, 155
  in radiolabeling, 664
  in technetium Tc 99m sulfur colloid, 199, 200f
etidronate rhenium Re 186 (^{186}Re-HEDP). *See* rhenium Re 186 etidronate
etidronate technetium Tc 99m (^{99m}Tc-EHDP). *See* technetium Tc 99m etidronate
European Association of Nuclear Medicine, 292, 294, 345, 687
eV. *See* electron volt
Evans, Robley, 446
exametazime technetium Tc 99m (^{99m}Tc-HMPAO). *See* technetium Tc 99m exametazime
examination, certification, 10–11
excepted packages, 121–122
excessive technetium Tc 99, 320
excessive technetium Tc 99m, 320
excitation interactions
  alpha particle, 48, 48f
  beta particle, 49–50, 50f
excretion
  breast milk, 328–333, 417, 448–449
    agents no data available on, 332
    case study of, 332–333, 333t
    concurrent drugs and, 332
  hepatobiliary
    of technetium Tc 99m gluceptate, 554, 554f
    of technetium Tc 99m mertiatide, 555
  renal, 548–551
    of radioiodides, 447–448
exercise, strenuous, and PET/CT, 618, 619f
exercise stress, for myocardial perfusion imaging, 461–462, 474
exhaust hood, 260
expanded-access protocols, 385
experience requirements, 405
expiration date, of radiopharmaceuticals, 275, 277f
exploratory IND (eIND), 358–359, 384
exponential cell survival curve, 133, 133f
exposure, radiation. *See also* radiation safety
  ambient, survey of rates, 401, 406
  annual, 111, 111t
  distance and, 96–97, 97t, 114
  internal, sources of, 95–96
  as low as reasonably achievable, 110
  occupational, 96, 109, 111, 112–114
  from patients undergoing treatment, 96, 456
  population at risk from, 109
  in pregnancy, 117–118
  protection from, 92–97
  shielding from, 97, 98f, 98t, 114
  sources of, 112–114
  time of, 96, 96t, 114
  volunteer subject, 111–112

extraction efficiency or ratio (ER), 549–550, 550t, 551
extremity badges, 114–116, 117f
ex vitro autoradiography, 351
ex vivo tissue biodistribution, 353–355
    animal care in, 355
    control studies of biomarker distribution in, 354
    factors affecting, 353–354, 354t
    target-blocking studies in, 354–355
eyes, radiation exposure and, 128, 144–145, 145f
"eyes to thighs" PET/CT, 618, 627

F

^{18}F. See fluorine F 18
Fab. See fragment antigen-binding region
facial (fac) isomer, 168, 168f
facilitated diffusion, 165–166, 548
false-negative studies, PET/CT, 651
false-positive studies, PET/CT, 651
Farkas, Raymond, 8, 9t
fasting, for PET/CT, 618, 618f
fast neutrons, 51
fatty acid metabolism, cardiac, 459, 472–473
Fc. See fragment crystallizable region
FCC. See Food Chemicals Codes grade
FDA. See Food and Drug Administration
FDAMA. See Food and Drug Administration Modernization Act of 1997
FDG. See fludeoxyglucose F 18
Federal Register, 395, 402
femoral graft infection, 593f
fetal stage, 142–143
fetus, radiation effects on, 117–118, 142–144
fever of unknown origin (FUO), 594, 595, 651
film badges, 115–116, 117f
finger(s)
    monitoring equipment for, 114–116, 117f
    syringes and exposure of, 112–113, 112f, 113f
finger badges, 115–116, 117f
first-generation radiopharmaceuticals, 168–169
first-pass radionuclide angiocardiography, 481–484
    interpretation of, 482–484, 486f
    procedure for, 481–482
    rationale for, 481
Fischer configurations, 167, 167f
flame ionization detectors, 277
floor plan, nuclear pharmacy, 258, 258f
florbetapir fluorine F 18. See fluorine F 18 florbetapir
fludeoxyglucose F 18 (^{18}F-FDG)
    for brain imaging, 411, 422–423, 422f
        in dementia, 433, 433f
        in tumor recurrence, 433, 435f
    close contact and exposure to, 332
    formulation of, 251
    gas chromatography of, 277f
    for heart imaging, 459, 460t, 471–473, 472f, 484, 487–489, 489f, 490f
    impurities in, 293, 320
    for infection imaging, 594, 596
    labeling with, 161
    localization mechanisms of, 596t
    metabolic trapping of, 236, 236f, 422–423
    on-site preparation of, 34
    on-site production of, 34
    pediatric dosing of, 327
    for PET, 234t, 236, 248, 251, 617. See also positron emission tomography
        glucose utilization, 234t, 236
        reimbursable procedures, 239–241, 239t
    for PET/CT, 617–652. See also PET/CT
    radiochemical purity of, 283, 286t, 288t
    radiosynthesis of, 248, 248f
    regulation of, 378–381
    for renal imaging, 562, 567f
    residual solvents in, 293–294
    substrate specificity of, 169
    for tumor imaging, 598
    uptake of, 620–625
        in bladder, 624–625, 625f
        in brain, 620, 621, 621f, 622f
        in brown fat, 618, 620f, 627
        in digestive system, 625
        in heart, 622–624
        in kidneys, 624–625, 624f
        in musculoskeletal system, 625, 626f
        in neck, 622, 623f, 624f
        time for, 618
fluid phase, of inflammatory reaction, 589
flumazenil C 11 (^{11}C-flumazenil, ^{11}C-FMZ)
    formulation of, 251
    for neuroreceptor binding assessment, 238, 238f
    for PET, 234t, 237–238, 251
    radiochemical purity of, 286t
    radiosynthesis of, 249
fluorine, electronic configuration of, 152t
fluorine F 18 (^{18}F), 5t, 26t
    for autoradiography, 352–353, 352t
    for brain imaging, 411
    decay of, 21, 21f
    electrophilic, 244t, 246
    half-life of, 275
    labeling precursors for, 244t, 245–246
    labeling with, 346–347, 662–663
    for PET, 207, 230t, 378, 617
    production of, 207, 244, 244t
    pulse-height energy spectrum of, 71f
    radionuclidic purity of, 276t
    reactive, 244t, 245–246, 246f
    regulation of, 378–381
    shipping guidelines for, 121t, 122t
    as traditional radionuclide, 346, 347t
fluorine F 18 ciprofloxacin (^{18}F-ciprofloxacin), 593–594, 596t
fluorine F 18 florbetapir (^{18}F-florbetapir)
    for dementia imaging, 433, 433f
    FDA approval of, 384
fluorine F 18 fludeoxyglucose (^{18}F-FDG)
    for brain imaging, 411, 422–423, 422f
        in dementia, 433, 433f
        in tumor recurrence, 433, 435f
    close contact and exposure to, 332
    formulation of, 251
    gas chromatography of, 277f
    for heart imaging, 459, 460t, 471–473, 472f, 484, 487–489, 489f, 490f
    impurities in, 293, 320
    for infection imaging, 594, 596
    labeling with, 161
    localization mechanisms of, 596t
    metabolic trapping of, 236, 236f, 422–423
    on-site preparation of, 34
    on-site production of, 34
    pediatric dosing of, 327
    for PET, 234t, 236, 248, 251, 617. See also positron emission tomography
        glucose utilization, 234t, 236
        reimbursable procedures, 239–241, 239t
    for PET/CT, 617–652. See also PET/CT
    radiochemical purity of, 283, 286t, 288t
    radiosynthesis of, 248, 248f
    regulation of, 378–381
    for renal imaging, 562, 567f
    residual solvents in, 293–294
    substrate specificity of, 169
    for tumor imaging, 598
    uptake of, 620–625
        in bladder, 624–625, 625f
        in brain, 620, 621, 621f, 622f
        in brown fat, 618, 620f, 627
        in digestive system, 625
        in heart, 622–624
        in kidneys, 624–625, 624f
        in musculoskeletal system, 625, 626f
        in neck, 622, 623f, 624f
        time for, 618
fluorine F 18 fluorodopa. See fluorodopa F 18
fluorine F 18 galacto RGD, 669
fluorine F 18 labeled RGD peptides, 669
fluorine F 18 sodium fluoride. See sodium fluoride F 18
fluorodopa F 18 (^{18}F-FD)
    high-pressure liquid chromatography of, 278f
    impurities in, 320
    for neurotransmission imaging, 237–238, 424–425
    off-label uses of, 237t, 238
    for PET, 234t, 237–238, 248–249
    radiochemical purity of, 286t
    radiosynthesis of, 248–249, 249f
fluorography, 352, 352f
Food, Drug, and Cosmetic Act of 1938, 273, 375, 378
Food and Drug Administration (FDA), 373, 391f
    ^{82}Sr–^{82}Rb generator approved by, 44
    adverse reactions defined by, 333
    compounding regulations of, 268
    historical perspective on, 375
    new drug/radiopharmaceutical process in, 357–359, 358f, 375, 376–377, 381–394
    package insert approval by, 274
    PET regulations of, 253, 274, 305–306, 309, 315, 378–381

public health protection role of, 381
radiolabeled antibodies approved by, 665, 673
radiopharmaceutical definition of, 147
regulatory authority of, 377–378
regulatory outlook for, 408–409
reporting system of, 334
research regulation by, 111
safety regulations of, 7
Food and Drug Administration Modernization Act of 1997 (FDAMA), 253, 305, 381, 408
Food Chemicals Codes (FCC) grade, 283
foot monitors, 264, 264f
foramen of Magendie, 415
foramen of Monro, 415
foramina of Luschka, 415–416
foreign (nonisotopic) labeling, 160, 161
formulation problems, 319–324
four Rs, of radiobiology, 137–138
fractures, 581, 581f
fragment antigen-binding (Fab) region, 656, 656f
fragment crystallizable (Fc) region, 656–657, 656f
free radical reactions, 128–129, 176–177
free technetium Tc 99m, 285
freezing, 260, 265
frequency, of electromagnetic radiation, 52–53, 53t
frontotemporal dementia (FTD)
    PET of, 240
    PET/CT of, 649–650, 651f
Fu, Monty, 9, 9t
full width at half maximum (FWHM), 71
functioning (hot) thyroid nodules, 450–451, 452–453, 453f, 454–455
function studies, in vivo, 3, 158, 695–703
FUO. *See* fever of unknown origin
furifosmin technetium Tc 99m. *See* technetium Tc 99m furifosmin
furosemide, for renal imaging, 556, 557f, 562
FWHM. *See* full width at half maximum

G

G1 phase, 130, 137
G2 phase, 130, 137, 137f
^{67}Ga. *See* gallium Ga 67
^{68}Ga. *See* gallium Ga 68
^{72}Ga. *See* gallium Ga 72
GABA. *See* gamma aminobutyric acid
gadolinium-enhanced MRI, of myocardial infarction, 489–490, 491f, 492f
gadolinium silicate (GSO), 78, 231
gallbladder, 519–520
    emptying of, for imaging, 524
    imaging of. *See* hepatobiliary imaging
gallbladder ejection fraction (GBEF), 524–525, 530, 532–533, 535f
gallium
    chemistry of, 215–217
    electronic configuration of, 215
    physical properties of, 216t
    production of, 215–216
gallium Ga 67 (^{67}Ga), 5t, 26t
    applications of, 207
    for bone imaging, 573
    decay of, 22, 216, 216f
    for infection imaging, 593t
    labeling with, 160
    as nontraditional radionuclide, 346, 347t
    normal total-body scan with, 590, 590f
    pediatric dosing of, 325t, 327
    physical properties of, 216t
    production of, 215–216
    pulse-height energy spectrum of, 70f
    radionuclidic purity of, 276t
    shipping guidelines for, 121t, 122t
gallium citrate Ga 67 (^{67}Ga-citrate), 217
    altered biodistribution of, 594, 595f
    indications for, 217
    for infection imaging, 590–591, 590f, 593t, 594–595
    localization mechanisms of, 596t
    normal distribution of, 597
    preparation of, 215
    radiochemical purity of, 286t
    for tumor imaging, 595f, 596, 597–598, 597f, 634
gallium Ga 68 (^{68}Ga), 26t
    ^{68}Ge–^{68}Ga generator, 45, 242, 242t
    for bone imaging, 573
    decay of, 21, 22f
    for PET, 230t
gallium Ga 72 (^{72}Ga), for bone imaging, 573
gamma aminobutyric acid (GABA), 237, 238, 424
gamma camera, 2, 2f, 158
    in rectilinear scanner, 76
    in SPECT, 76, 76f
gamma globulin. *See* immunoglobulin G
gamma rays, 17, 47
    counting samples of, 71–85
    emission of, 2
        in alpha particle decay, 24
        in dose calibrator, 61
        in electron capture, 22
        in isomeric transition, 23
        in negatron decay, 19
        in nuclear de-excitation, 15–16, 16f
        in nuclear reaction, 31
    interactions with matter, 52–56
    properties of, 52–53
    protection from, 92–97
gamma spectrometry, 274–275
gas chromatography, 275–278, 277f
    chromatogram in, 277f, 278
    columns in, 277
    detectors in, 277
    mobile phase gases in, 277
    sample injection port in, 277
gaseous targets, of cyclotrons, 243–244
gases, radioactive. *See also specific agents*
    clearance times for, 118
    inhalation of, 95–96
gas exchange apparatus, 499, 500f
gastric emptying, 517, 527, 542–544
    delayed, 542, 544f
    fast, 542, 544f
    normal, 542, 543f
gastric glands, radioiodide concentration in, 448
gastrin, 669–670
gastrinoma, indium In 111 pentetreotide imaging of, 600t
gastrin-releasing peptide analogues, 669–670
gastroesophageal reflux, 517, 527, 541–542, 542f
gastrointestinal bleeding, 517, 527–528, 527t, 539, 541f
gastrointestinal effects, 135, 138–139
gastrointestinal imaging, 517, 537–544
    agents for, 517, 526–528
    of gastric emptying, 527
    of gastroesophageal reflux, 517, 527
    of gastrointestinal bleeding, 517, 527–528, 527t, 539, 541f
    of Meckel's diverticulum, 528, 539–541, 542f
    of protein loss, 528, 529f
gastrointestinal syndrome, 138–139
gated blood pool imaging. *See* equilibrium radionuclide ventriculography
Gaussian distribution, 79–80, 80f
GAW. *See* gram atomic weight
GBEF. *See* gallbladder ejection fraction
^{68}Ge. *See* germanium Ge 68
GE Healthcare, 9
Geiger-Müller (GM) alarm rate meter, 264, 264f
Geiger-Müller (GM) survey meters (counters), 60, 63, 65f, 88, 90, 304–305, 304f
    calibration of, 304–305
    in nuclear pharmacy, 262–264
    probes for, 263, 263f
    for radioactive packages, 120, 120t
Geiger region, 59–60, 60f
gel-clot limit test, 298, 298t
gene amplification, 141
general public, radiation exposure of, 109, 111
generators. *See* radionuclide generators
genetic effects, 131–132, 141–142. *See also* DNA; hereditary effects
geometry
    of dose calibrators, 299t, 300, 302–304
    of ion chamber well, 303, 303f
GERD. *See* gastroesophageal reflux
germanium Ge 68 (^{68}Ge), as line source in PET, 232
germanium Ge 68 (^{68}Ge)–gallium 68 (^{68}Ga) generator, 45, 242, 242t
germ cells, radiation damage to, 102, 141
gestational stages, 142–143
GFR. *See* glomerular filtration rate
GH, ^{99m}Tc. *See* technetium Tc 99m gluceptate
Glofil-125. *See* iothalamate sodium I 125
glomerular capillaries, 547–548
glomerular filtration, 548–549
glomerular filtration rate (GFR), 548
    in captopril renography, 556–558, 558f
    measurement of, 569, 701–703, 702f
        diagnostic radiopharmaceuticals for, 158
        as in vivo function study, 3, 208
        iothalamate sodium I 125 for, 3, 208, 215, 552, 569, 702–703
        technetium Tc 99m pentetate for, 158, 196, 552, 553, 569, 702, 703
    normal, 549
glomerulus, 547
glove box, 260
gloves, use of, 113, 314

glucagon, for Meckel's diverticulum imaging, 540
glucarate technetium Tc 99m. *See* technetium Tc 99m glucarate
gluceptate technetium Tc 99m (^{99m}Tc-GH). *See* technetium Tc 99m gluceptate
glucocorticoids, 601
glucoheptonate technetium Tc 99m. *See* technetium Tc 99m glucoheptonate
glucose, blood levels of, and PET/CT, 618
glucose utilization, 234t, 235. *See also* fluorine F 18 fludeoxyglucose; PET/CT
    in brain imaging, 422–423, 422f
    in heart imaging, 459, 471–472, 487–488
glutamate, 424
glycine, 424
GM. *See* Geiger-Müller
Gnau, Thomas, 8, 9t
goiter, 450
    imaging of, 453, 453f, 454f
    radioiodine therapy for, 453, 454–455
gold Au 198 (^{198}Au)
    for liver imaging, 517, 520, 521
    for lung perfusion studies, 198
    for lymphoscintigraphy, 607
    for radiation synovectomy, 687
    for spleen imaging, 517
gold Au 198 colloid, for liver imaging, 520, 520t, 521
gonads
    radiation effects on, 142, 142t
    radiosensitivity of, 135
    tissue-weighting factor for, 104t
goodness of fit (chi-square test), 84–85, 85t
gouty arthritis, 584, 586f
gram atomic weight (GAW), 148, 161
granulocytes, 589–590
Graves' disease, 450
    FDG uptake in, 622
    imaging of, 453, 454f
    radioiodine therapy for, 453, 454
gray (Gy), 4, 87, 88, 90
groups, of elements, 148
GSO. *See* gadolinium silicate
GTP. *See* guanosine triphosphate
guanosine triphosphate (GTP), 424
Gutkowski, Robert F., 9t
Gy. *See* gray

H

H. *See* dose equivalent
HA. *See* hydroxyapatite
HACA. *See* human antichimeric antibodies
HAHA. *See* human anti-human antibodies
hair loss, 128
half-life, 28, 28f, 147, 159–160
    effective, 159
    mean life and, 29
    radionuclide identification by, 274–275
half-value layer (HVL), 97, 98f, 98t
HAMA. *See* human anti-mouse antibody response
Hamilton, Donald R., 9t
Hamilton, Joseph, 1
hand(s)
    hygiene for, 314
    monitoring equipment for, 114–116, 117f
    syringes and exposure of, 112–113, 112f, 113f
handling techniques, 266, 266t
Harper, Paul, 8
Hashimoto's thyroiditis, 450
HCC. *See* hepatocellular carcinoma
HCFA. *See* Centers for Medicare and Medicaid Services
Hd. *See* deep-dose equivalent
HDP, ^{99m}Tc. *See* technetium Tc 99m oxidronate
H_E. *See* effective dose equivalent
$H_{E,50}$. *See* effective dose equivalent, committed
head and neck cancer
    PET/CT of, 626–627, 627f, 628f
    reimbursable PET procedures in, 239t, 240–241, 380
Health Care Financing Administration (HCFA). *See* Centers for Medicare and Medicaid Services
health care personnel
    consultation with, during inspections, 120
    dose limits for, 111, 111t
    dosimetry report for, 116t
    exposure to radiation, 96, 109, 111, 112–114
    monitoring for, 114–118
        devices for, 115–116, 117f
        investigational levels for, 115, 115t
    notices, instructions, and reports to, 119
    in nuclear pharmacy, 265–266
    pregnant, 117–118
Health Insurance Portability and Accountability Act of 1996, 253
heart
    blood flow in, 460–461. *See also* myocardial perfusion imaging
    FDG uptake in, 620, 622–624
heart imaging, 459–494
    exercise stress for, 461–462, 461f, 474
    extraction and retention of tracers in, 466–467, 466f
    metabolism and viability agents for, 471–473
    neuronal, 490–492
    nuclear medicine procedures for, 473–492
    perfusion. *See* myocardial perfusion imaging
    PET, 234–237, 234t, 459, 487–488, 617
        agents for, 459, 460t, 470–471, 471t
        reimbursable procedures, 239–240, 239t, 380, 617
    PET/CT, 646–648, 648f, 649f
    pharmacologic stress for, 462–466, 465t, 474
    physiologic principles of, 460–467
    risk stratification in, 492–494, 493f, 493t, 494t
    SPECT, 459, 460t, 467–470
    viability. *See* myocardial viability imaging
heart-to-mediastinum ratio (HMR), 491–492
heat-denatured ^{99m}Tc red blood cells, for spleen imaging, 526, 526f, 529, 532f
heating, improper, 321–322
heavy water, in nuclear reactor, 31, 32f
Hebner, Mark, 9, 9t
HED carbon C 11. *See* carbon C 11 hydroxy ephedrine
HEDP rhenium Re 186. *See* rhenium Re 186 etidronate
height-based dosage adjustment, 326–327
HeLa cells, 133, 141
Helicobacter pylori infection, 225, 703
helium
    electronic configuration of, 149, 152t
    in ionization detector, 59
hemacytometer, 265
hemangioma, hepatic, 529
hematocrit, 695–696
hematopoietic syndrome, 138
Henderson-Hasselbach equation, 164, 548
Henle, loops of, 547
HEPA. *See* high-efficiency particulate air filter
hepatic artery infusion pumps, 529, 531f
hepatic duct, 518f, 519
hepatic imaging. *See* liver imaging
hepatic metastases, brachytherapy for, 689–690
hepatic vein, 517
hepatitis, neonatal, 537, 538f
hepatobiliary excretion
    of technetium Tc 99m gluceptate, 554, 554f
    of technetium Tc 99m mertiatide, 555
hepatobiliary imaging, 178–179, 191, 530–532
    of acute cholecystitis, 530, 533, 536f
    agents for, 517, 522–525
    of bile gastritis, 537, 538f
    of bile leak, 530, 537, 540f
    of biliary atresia, 530, 537, 539f
    of biliary obstruction, 530, 537, 537f, 538f
    of chronic cholecystitis, 533–537, 536f
    indications for, 530
    interpretation of, 533–537
    of neonatal hepatitis, 537, 538f
    pharmacologic interventions for, 524–525
    physiologic principles in, 517–520, 518f
    procedure for, 532–533
    rationale for, 530–532
    technetium Tc 99m IDA analogues for, 193, 517, 523–524
hepatobiliary malignancies, PET/CT of, 640–643
hepatocellular carcinoma (HCC)
    brachytherapy for, 688–690
    PET/CT of, 641, 642f
hepatocytes, 517, 519–520, 519f
hereditary effects, 141–142
    risk estimates for, 99, 103–104, 103t
    risk models for, 102–104
    stochastic, 128
    as stochastic effect, 98
Hetzel, Kenneth R., 9t
Hevesy, George, 1
^{203}Hg. *See* mercury 203
^{197}Hg-chlormerodrin. *See* mercury Hg 197 chlormerodrin

^{203}Hg-chlormerodrin. *See* mercury Hg 203 chlormerodrin
hibernating myocardium, 460–461, 484
HIDA, ^{99m}Tc. *See* technetium Tc 99m lidofenin
high-efficiency particulate air (HEPA) filter, 261–262, 315
high-performance liquid chromatography. *See* high-pressure liquid chromatography
high-pressure liquid chromatography (HPLC), 167, 278–279
    analytic, 279
    detectors and recorders in, 279
    drawbacks in, 279
    normal-phase, 279
    for PET radionuclide preparation, 249, 251–252
    preparative, 279
    pump and damper in, 278–279
    for radiochemical identification, 275, 278–279
    reverse-phase, 279
    solvent reservoirs in, 278
high-throughput screening (HTS), 346
Hiroshima. *See* Japanese survivors
histamine, 424, 589
histamine H_2 blockers, for Meckel's diverticulum imaging, 540
hLL2 yttrium Y 90. *See* yttrium Y 90 epratuzumab
HMDP. *See* hydroxymethylene diphosphonate
HMPAO, ^{99m}Tc. *See* technetium Tc 99m exametazime
HMR. *See* heart-to-mediastinum ratio
Hodgkin's lymphoma
    gallium Ga 67 citrate imaging of, 597–598, 597f, 634
    PET/CT of, 634–636, 635f, 637f
hoods, 260, 261–262, 262f
Hoogland, Dennis R., 9t
horizontal-flow hood, 262, 262f
hospital workers. *See* health care personnel
hot spots, 2–3, 488
hot thyroid nodules, 450–451, 452–453, 453f, 454–455
HPLC. *See* high-pressure liquid chromatography
Hs. *See* shallow-dose equivalent
HSA. *See* human serum albumin
H_T. *See* equivalent dose
$H_{T,50}$. *See* committed dose equivalent
HTS. *See* high-throughput screening
human antichimeric antibodies (HACA), 659
human anti-human antibodies (HAHA), 659
human anti-mouse antibody (HAMA) response, 337–338, 658–659, 673, 674
humanized antibody, 659, 673, 674
human serum albumin (HSA). *See also* albumin
    carrier technetium and, 42–43, 43f
    iodinated I 125 albumin (^{125}I-albumin or ^{125}I-HSA), 208, 214
        diagnostic application of, 158
        labeling with, 161
        pediatric dosing of, 327
        for plasma volume measurement, 697–699, 699f, 700t
        radiochemical purity of, 286t
        for red blood cell–plasma measurement, combined, 699–700
    iodinated I 131 albumin (^{131}I-HSA), 214
        adverse reactions to, 334–335
        for CSF imaging, 426, 427t
        for gastrointestinal imaging, 528
        radiochemical purity of, 286t
    iodinated I 131 albumin colloid (^{131}I-HSA colloid), 520, 520t
    technetium Tc 99m albumin (^{99m}Tc-HSA), 188
        for CSF imaging, 426
        for gastrointestinal imaging, 528, 529f
        for lymphoscintigraphy, 607–608
        radiochemical purity of, 287t
human subjects
    clinical research with, 357–359
    dose limits for, 111–112
    protection of, 397
human use, preparation for, 357–362, 382–384
Hurwitz, David, 9, 9t
HVL. *See* half-value layer
hybridoma, 658, 660f
hydrocephalus, 412, 436–437, 437f
hydrogen
    electronic configuration of, 152t
    isotopes of, 148
hydrogen fluoride (HF), formation of, 156
hydrogen molecule, bonds of, 149
hydronephrosis, 556, 560–562
hydrophilic impurities, 319
hydrostatic pressure, and lung perfusion studies, 500–501
hydroxyapatite (HA), 571–572, 576
hydroxy ephedrine carbon C 11 (^{11}C-HED). *See* carbon C 11 hydroxy ephedrine
hydroxy ^{123}I-iodobenzyl propyl diamine (^{123}I-HIPDM), for brain imaging, 418
hydroxymethylene diphosphonate (HMDP), 573–574
hyperparathyroidism, 603–604
hypersensitivity reactions, 334t
hypertension
    essential, 556
    renovascular, 557–558, 558f, 560, 561f
hyperthyroidism, 450, 673
    FDG uptake in, 622
    imaging of, 453, 454f
    radioiodine therapy for, 453, 454
    treatment of, 3
hypertrophic pulmonary osteoarthropathy, 582, 585f
hyperuricemia, bone imaging in, 584, 586f
hypothyroidism, 450, 453, 455f
hypoxic cells, 136–137

I

I. *See* iodine
^{123}I. *See* iodine I 123
^{124}I. *See* iodine I 124
^{125}I. *See* iodine I 125
^{128}I. *See* iodine I 128
^{131}I. *See* iodine I 131
IBA Group–Eastern Isotopes, 9
IBD. *See* inflammatory bowel disease
ibritumomab tiuxetan
    In 111 (^{111}In-ibritumomab tiuxetan)
        dosing regimen for, 677, 677f
        labeling process for, 186, 675, 677t
        mixing order for, 321
        nomenclature for, 661
        for non-Hodgkin's lymphoma, 676–677
        radiochemical purity of, 286t, 288t
    Y 90 (^{90}Y-ibritumomab tiuxetan), 224, 666–667
        adverse effects of, 678
        dosing regimen for, 677–679, 677f, 678f, 679f
        labeling process for, 186, 674, 674f, 675, 677t
        mixing order for, 321
        nomenclature for, 661
        for non-Hodgkin's lymphoma, 3–4, 655, 666–667, 674, 676–680, 682
        radiation safety of, 679–680
        radiochemical purity of, 287t
        therapeutic uses of, 655
Ice, Rodney D., 9t
ICRP. *See* International Commission on Radiological Protection
IDA analogues. *See* iminodiacetic acid analogues
identification of radiochemicals, 275–284
    gas chromatography for, 275–278, 277f
    high-pressure liquid chromatography for, 275, 278–279, 278f
    paper chromatography for, 275, 279–283
    reference standard for, 275
    solid-phase extraction for, 275, 279–280, 280f
    thin-layer chromatography for, 275, 279–283, 292f
identification of radionuclides, 274–275
    by half-life, 274–275
    by radiation type, 274, 274f
idiosyncratic reactions, 334t
idiotypes, 658
IgA. *See* immunoglobulin A
IgD. *See* immunoglobulin D
IgE. *See* immunoglobulin E
IgG. *See* immunoglobulin G
IgM. *See* immunoglobulin M
^{123}I-IMP. *See* N-isopropyl-p-^{123}I-iodoamphetamine
IL-2 iodine I 123. *See* iodine I 123-labeled interleukin-2
IL-8 technetium Tc 99m. *See* technetium Tc 99m-labeled interleukin-8
imaging procedures, 2–3. *See also specific types*
    average effective dose in, 105t–106t
    diagnostic radiopharmaceuticals for, 158
    occupational exposure from, 113
    optimum effective half-life in, 159
imaging systems, 74–78. *See also specific types*
imaging target, selection of, 345
imciromab penetate indium In 111. *See* indium In 111 imciromab pentetate
^{123}I-MIBG. *See* iobenguane iodine I 123

^{131}I-MIBG. *See* iobenguane iodine I 131
iminodiacetic acid (IDA) analogues, 178, 179f, 183, 193–194
    biodistribution of, 523–524, 532, 533f
    for hepatobiliary imaging, 517, 523–524, 532–533
    physicochemical properties of, 523
    radiation dose of, 524, 524t
immune system, 656–661, 656f
immunoglobulin A (IgA), 657–658, 657t
immunoglobulin D (IgD), 657–658, 657t
immunoglobulin E (IgE), 657–658, 657t
immunoglobulin G (IgG), 657–658, 657t
    radiolabeled
        for infection imaging, 592
        localization mechanisms of, 596t
    structure of, 656–657, 656f, 657f
    subclassses of, 657–658, 658t
immunoglobulin M (IgM), 657–658, 657t
IMP iodine I 123. *See* N-isopropyl-p-^{123}I-iodoamphetamine
improper heating, 321–322
improper mixing order, 321
impurities, chemical, 293
impurities, radiochemical, 275–285, 319–320. *See also* radiochemical purity
    and distribution in body, 284, 284f
    hydrophilic, 319
    lipophilic, 319
impurities, radionuclide. *See* radionuclide purity
^{111}In. *See* indium In 111
^{113m}In. *See* indium In 113m
incapacitation syndrome, acute, 138, 139
incineration, 125
incubation/time delays, 322
indeterminate errors, 78–79
Indiana University Medical Center, 8
indirect effects, of radiation, 128–129, 128f
indium
    chemistry of, 215–220
    electronic configuration of, 215
    physical properties of, 216t
    production of, 215–216
indium In 111 (^{111}In), 5t, 27t
    applications of, 207
    autoradiography of, 353
    decay of, 22, 216, 216f
    detection efficiency of, 159t
    dose calibrator measurements of, 62–63, 64t, 287
    half-life of, 160
    impurities in, 320
    labeling with, 160, 661–662, 662t, 664–665, 676, 677t
    localization mechanisms of, 596t
    physical properties of, 216t
    production of, 215–216
    pulse-height energy spectrum of, 70, 71f
    radiolabeling approach for, 186
    radionuclidic purity of, 276t
    shipping guidelines for, 121t, 122t
    as traditional radionuclide, 346, 347t
indium In 111 capromab pendetide (^{111}In-capromab pendetide), 218, 666, 667–668
    indication for, 666
    mixing order for, 321
    for prostate imaging, 218, 666, 667–668, 668f
    radiochemical purity of, 286t, 288t
    radiolabeling of, 665
indium In 111 chloride (^{111}In-chloride), 217
    for bone marrow imaging, 610–611
    impurities in, 293
    indications for, 217
    pH of, 295
    radiochemical purity of, 286t
    solution, 217
indium In 111 ibritumomab tiuxetan (^{111}In-ibritumomab tiuxetan)
    dosing regimen for, 677, 677f
    labeling process for, 186, 675, 677t
    mixing order for, 321
    nomenclature for, 661
    for non-Hodgkin's lymphoma, 676–677
    preparation of, 676
    radiochemical purity of, 286t, 288t
indium In 111 imciromab pentetate (^{111}In-imciromab pentetate), 666
    for myocardial infarction imaging, 488–489, 490t
indium In 111-labeled antibodies, 218
    for tumor imaging, 596
indium In 111-labeled immunoglobulin G (^{111}In-IgG), 592
indium In 111-labeled platelets, 219–220, 220f
indium In 111-labeled white blood cells, 218–219
    for infection imaging, 591, 592f, 593t, 594–596
    normal scan with, 592f
    procedure for labeling, 201f, 218–219
    proposed mechanism for, 219, 219f
indium In 111 oxyquinoline (^{111}In-oxine), 217
    adverse reactions to, 338–339
    chemical structure of, 218f, 591f
    indications for, 217
    for platelet labeling, 219–220, 220f
    radiochemical purity of, 286t
    for white blood cell labeling, 217, 219, 219f, 591, 592f, 594
indium In 111 pentetate (^{111}In-DTPA), 217–218
    chemical structure of, 218f
    for CSF imaging, 158, 217–218, 412, 426–428, 426f, 427t
    of CSF leaks, 438
    of hydrocephalus, 437, 437f
    for gastric emptying studies, 527
    half-life of, 158
    indications for, 217–218
    intrathecal, safety considerations in, 427–428
    radiochemical purity of, 286t
indium In 111 pentetreotide (^{111}In-pentetreotide), 218, 668
    administration of, 599
    chemical structure of, 218f
    indications for, 218, 599
    normal scan with, 599, 599f
    pediatric dosing of, 325t
    preparation of, 599
    radiochemical purity of, 286t, 288t
    stability of, improving, 655–656
    for tumor imaging, 598, 599–600, 600f, 600t
indium In 111 satumomab pendetide (^{111}In-satumomab pendetide), 666
    indications for, 666
    radiochemical purity of, 286t
    radiolabeling of, 665
indium In 111 tropolone (^{111}In-tropolone)
    chemical structure of, 591f
    for platelet labeling, 219
    for white blood cell labeling, 591
indium In 113m (^{113m}In)
    ^{113}Sn–^{113m}In generator, 44
    for liver and spleen imaging, 517
indium In 113m hydroxide colloid, for liver imaging, 520, 520t
INDs. *See* investigational new drugs
inelastic collisions, 52
inertness, 148, 149
infants
    breast-feeding and radiopharmaceuticals for, 328–332
    close contact and exposure of, 332
    dosing for, 324–328
infarction, myocardial. *See* myocardial infarction
infection imaging
    bone, 582, 582f–583f
    clinical considerations in, 594–596
    gallium Ga 67 citrate for, 590–591, 590f, 593t
    localization mechanisms in, 596t
    nuclear medicine, indications for, 594
    PET, 596
    PET/CT, 650–651
    physiologic principles of, 589–590
    radiolabeled antibiotics for, 593–594
    radiolabeled antibody for, 668
    radiolabeled immunoglobulin G for, 592
    radiolabeled interleukins for, 594
    radiolabeled monoclonal antibodies for, 593
    radiolabeled white blood cells for, 591–592, 592f, 593f, 593t, 594–596
    total-body SPECT, 589–596
inflammatory bowel disease (IBD), 592, 595
inflammatory process, 589–590, 650–651
infrared radiation, 53t
inhalation, of radiation, 95–96
injections. *See also* specific types
    radiation exposure in, 112–113
    syringe and shielding for, 112–113, 112f, 113f
    technetium Tc 99m, 171–172, 172f
inspections, 119–120
instant thin-layer chromatography (ITLC), 280–283, 292f
Institute for Molecular Imaging, 254
Institute of Clinical PET, 254, 381
institutional review boards (IRBs), 111–112, 359, 385
instructions, safety, 119
instrumentation. *See also* specific instruments
    quality control for, 299–304
    radiation detection, 71–85, 262–265
    regulatory requirements on, 400–402, 406
insulinoma, indium In 111 pentetreotide imaging of, 600t

intake retention fraction (IRF), 117
integrated approach, to technetium-nonessential compounds, 185
integrins, 668–669
intensifying screen, for autoradiography, 352, 352*f*
interferons, 594
interleukins
    in inflammatory process, 589, 594
    radiolabeled
        for infection imaging, 594
        localization mechanisms of, 596*t*
interlobular artery, 547, 548*f*
intermediate neutrons, 51
intermediate nucleus, 31
internal conversion, 23–24, 24*f*, 159
International Commission on Radiological Protection (ICRP), 109–110, 128, 383
    ICRP 26
        cancer risk estimates in, 99
        effective dose in, 89, 89*t*
        tissue-weighting factor in, 104
    ICRP 60, 109–110
        cancer risk estimates in, 99, 100, 102, 102*t*
        deterministic effects in, 98
        detriment in, 104
        dose equivalent in, 89
        effective dose in, 89–90, 89*t*
        equivalent dose in, 88, 89
        hereditary risk estimates in, 103, 104
        radiation risk factors in, 89–90
        tissue-weighting factor in, 104
    ICRP 92, 110
    ICRP 103
        effective dose in, 89–90, 89*t*
        risk and detriment in, 104
International Committee for Standardization in Hematology Panel on Diagnostic Applications of Radionuclides, 219
International Isotope Society, 254
International Organization for Standardization (ISO), 259, 315–316
International Symposium on Radiopharmaceutical Sciences, 345
International System of Units, 4, 28
International Unit (IU), 4
interphase, 130, 130*f*
interstitial fluid, and brain imaging, 417
intrathecal administration, safety considerations in, 427–428
intrinsic efficiency, 81
inverse square law, 96–97, 112–113
investigational levels, 115, 115*t*
investigational new drugs (INDs), 357–359, 375, 376–377
    applicant or sponsor for, 382, 389
    application for, 384–385, 386*t*
    exploratory, 358–359, 384
    physician sponsorship of, 389
    preclinical studies for, 382
    process for, 385–387, 386*t*, 387*t*, 388*f*
    review process for, 391–394
        biostatistics, 393
        chemistry, 392–393
        clinical, 392, 393–394
        nonclinical, 392–393
        pharmacology and toxicity, 392

in vitro autoradiography, 351
in vitro binding assays, 348–350
    autoradiography *vs.*, 348, 349*t*, 350–351
    competitive binding, 349, 349*f*
    control studies of target-binding affinity in, 349
    direct, 349, 349*f*
    disadvantages of, 353–354
    procedure for, 349, 349*f*
    quantification of biomarker–target binding in, 350
    specific and nonspecific target binding in, 350, 350*f*
    target-binding selectivity in, 350, 350*f*
    variables in, 349, 350*t*
in vivo function studies, 3, 158, 695–703. *See also specific types*
iobenguane iodine I 123 (^{123}I-MIBG), 214, 601
    for cardiac neuronal imaging, 490–492
    chemistry of, 160, 215*f*
    FDA approval of, 214
    indications for, 214
    localization mechanism of, 601*f*
    pediatric dosing of, 327
    radiochemical purity of, 286*t*
    substrate specificity of, 169
    thyroid protection with, 214
    for tumor imaging, 596
iobenguane iodine I 131 (^{131}I-MIBG), 214, 691
    chemistry of, 160, 215*f*
    drugs affecting, 601–602, 601*t*
    FDA approval of, 214
    indications for, 214
    localization mechanism of, 601*f*
    normal scan with, 602, 602*f*
    pediatric dosing of, 602
    radiochemical purity of, 286*t*
    thyroid protection with, 214
    for tumor imaging, 596, 601–603, 603*f*, 604*f*
iodide trapping, 445
iodinated I 123 metaiodobenzylguanidine (^{123}I-MIBG). *See* iobenguane iodine I 123
iodinated I 125 albumin (^{125}I-albumin or ^{125}I-HSA), 208, 214
    diagnostic application of, 158
    labeling with, 161
    pediatric dosing of, 327
    for plasma volume measurement, 697–699, 699*f*, 700*t*
    radiochemical purity of, 286*t*
    for red blood cell–plasma measurement, combined, 699–700
iodinated I 131 albumin (^{131}I-HSA), 214
    adverse reactions to, 334–335
    for CSF imaging, 426, 427*t*
    for gastrointestinal imaging, 528
    radiochemical purity of, 286*t*
iodinated I 131 albumin aggregated (^{131}I-MAA), radiochemical purity of, 286*t*
iodinated I 131 albumin colloid (^{131}I-HSA colloid), for liver imaging, 520, 520*t*
iodinated I 131 metaiodobenzylguanidine (^{131}I-MIBG). *See* iobenguane iodine I 131
iodinated radiopharmaceuticals, 213–215. *See also specific types*

iodination (radioiodination), 157–158, 210–213, 347
    antibody, 662–663, 674
    electrophilic substitution for, 210, 210*f*, 662, 662*f*
    impurities in, 293
    isotope exchange for, 210–211, 211*f*
    oxidants for, 211–213, 662, 663*f*
    protein, 210, 210*f*, 212, 212*t*
    reagents for, 211–213, 662
    specific activities of, 161–162, 162*t*
    tyrosine, 210, 210*f*
iodine (I)
    chemistry of, 158–159, 207–213
    daily turnover in body, 445, 446*f*
    electronic configuration of, 157
    isotopes of, 446
    metabolism in thyroid, 445, 446*f*
    oxidation state of, 157
    physical properties of, 207, 208*t*
    redox reactions of, 155, 157–158
iodine, radioactive, uptake test with radioactive, 445, 451–452, 695
        interpretation of, 451–452, 451*f*
        procedure for, 451
        rationale for, 451
iodine escape peak, 70
iodine I 123 (^{123}I), 5*t*–6*t*, 27*t*
    applications of, 207
    for autoradiography, 352, 352*t*
    for brain imaging, 411
    chemistry of, 207
    decay of, 22
    dose calibrator measurements of, 62–63, 62*f*, 63*f*, 64*t*, 209–210, 287
    half-life of, 207
    imaging properties of, 207
    impurities in, 34
    labeling with, 160, 346–347, 662–663, 662*t*
    occupational exposure to, 117, 117*t*
    physical properties of, 208*t*
    production of, 209–210
    pulse-height energy spectrum of, 70, 70*f*
    radiation absorbed dose of, 449, 449*t*
    radionuclidic purity of, 276*t*
    redox reactions in, 157
    shipping guidelines for, 121*t*, 122*t*
    specific activity of, 162*t*
    in SPECT, 207
    for thyroid cancer scans, 456, 456*f*
    as traditional radionuclide, 346, 347*t*
iodine I 123 hydroxy iodobenzyl propyl diamine (^{123}I-HIPDM). *See* hydroxy ^{123}I-iodobenzyl propyl diamine (^{123}I-HIPDM)
iodine I 123 iobenguane. *See* iobenguane iodine I 123
iodine I 123 iodofiltic acid (^{123}I-BMIPP). *See* iodofiltic acid iodine I 123
iodine I 123 iodophenylpentadecanoic acid (^{123}I-IPPA), for heart imaging, 459
iodine I 123 iofetamine (^{123}I-IMP). *See* N-isopropyl-*p*-^{123}I-iodoamphetamine
iodine I 123-labeled interleukin-2 (^{123}I-IL-2), 594
iodine I 123 sodium iodide. *See* sodium iodide I 123

iodine I 124 (^{124}I), 27t
  applications of, 207
  as nontraditional radionuclide, 346, 347t
  physical properties of, 208t
  radiation absorbed dose of, 449, 449t
iodine I 125 (^{125}I), 6t, 27t
  applications of, 207, 208
  for autoradiography, 352, 352t, 353
  in binding assays, 349
  chemistry of, 207–208
  decay of, 22, 209, 209f
  dose calibrator measurements of, 303, 303t
  half-life of, 208
  labeling with, 160, 346–347, 662–663, 662t
  as nontraditional radionuclide, 346, 347t
  occupational exposure to, 117, 117t
  physical properties of, 208t
  production of, 209
  pulse-height energy spectrum of, 70, 70f
  radiation absorbed dose of, 449, 449t
  shipping guidelines for, 121t, 122t
  specific activity of, 162t
iodine I 125 albumin. See iodinated I 125 albumin
iodine I 125 human serum albumin. See iodinated I 125 albumin
iodine I 125 iothalamate sodium. See iothalamate sodium I 125 (^{125}I-iothalamate)
iodine I 128 (^{128}I), 446
iodine I 131 (^{131}I), 6t, 27t
  applications of, 4, 207
  biologic properties of, 447–449
  chemistry of, 207
  decay of, 208, 208f
  decay table for, 29–30
  detection efficiency of, 159t
  distribution in body, 447, 447f
  dose assay of, 292
  dose calibrator measurements of, 303, 303t
  excretion in breast milk, 448–449
  historical perspectives on, 446–447
  labeling with, 160, 661–663, 662t
  occupational exposure to, 117, 117t
  physical properties of, 208t
  production of, 208–209
  pulse-height energy spectrum of, 70f
  radiation absorbed dose of, 207, 449, 449t
  redox reactions in, 157–158
  release criteria, instructions, and records for, 456t
  shipping guidelines for, 121t, 122t
  specific activity of, 161–162, 162t
  in SPECT, 207
  for thyroid studies, 447–453, 450t
    radioactive iodine uptake test, 451–452, 451f
    thyroid scan, 452–453
  for thyroid therapy, 453–456
iodine I 131 albumin. See iodinated I 131 albumin
iodine I 131 human serum albumin. See iodinated I 131 albumin
iodine I 131 iobenguane. See iobenguane iodine I 131
iodine I 131 iodohippurate sodium. See iodohippurate sodium I 131
iodine I 131 ioflubenzamide (^{131}I-ioflubenzamide), 691
iodine I 131-labeled rose bengal (^{131}I-labeled rose bengal), for hepatobiliary imaging, 522
iodine I 131 LYM-1 (^{131}I-LYM-1)
  labeling process for, 681, 682f
  for non-Hodgkin's lymphoma, 674, 681
iodine I 131 sodium iodide. See sodium iodide I 131
iodine I 131 tositumomab. See tositumomab I 131
iodine monochloride, 211–212
iodine redox reactions, 156
Iodo-Beads, 211, 212–213, 212f
iodofiltic acid iodine I 123 (^{123}I-BMIPP), for heart imaging, 459, 472–473, 473f
Iodo-Gen, 211, 212–213, 212f, 662–663, 663f
iodohippurate sodium I 123 (^{123}I-OIH)
  radiochemical purity of, 286t
  chemical structure of, 211f
iodohippurate sodium I 131 (^{131}I-OIH), 214–215
  biologic properties of, 555, 555t
  chemical structure of, 211f
  kidney imaging, 214–215
  for renal imaging, 548, 555
iodophenylpentadecanoic acid iodine I 123. See iodine I 123 iodophenylpentadecanoic acid
iofetamine iodine I 123. See N-isopropyl-p-^{123}I-iodoamphetamine
ioflubenzamide iodine I 131. See iodine I 131 ioflubenzamide
^{123}I-OIH. See iodohippurate sodium I 123
^{131}I-OIH. See iodohippurate sodium I 131
ion chamber well geometry, 303, 303f
ion channels, 163, 163f
ion collection methods, 59–63
ionic bonds, 149
ionization chambers, 60, 304–305, 304f
  calibration of, 304–305
  in ion collection methods, 59–61
  in nuclear pharmacy, 262–263, 264, 264f
  in radionuclide dose calibrator, 60–61, 60f
ionization detector, 59–60, 60f
  current/voltage relationship in, 59–60, 60f
  Geiger region in, 59–60, 60f
  ionization region in, 59, 60f
  proportional region in, 59, 60f
  recombination region in, 59, 60f
ionization interactions, 47
  alpha particle, 47–48, 48f, 49f
  beta particle, 49–50, 49f, 50f
ionization potential, 149, 153t
ionization region, 59, 60f
iothalamate sodium I 125 (^{125}I-iothalamate), 215
  biologic properties of, 554t, 555
  chemical structure of, 211f
  diagnostic application of, 158
  radiochemical purity of, 215, 286t
  renal elimination of, 551
  for renal imaging, 555
    development of, 552
    in glomerular filtration rate measurement, 3, 158, 208, 215, 552, 569, 702–703
    properties of, 549–550
Iotrex, 208
IPA. See isopropyl alcohol
IPPA iodine I 123. See iodine I 123 iodophenylpentadecanoic acid
IRBs. See institutional review boards
IRF. See intake retention fraction
iron ascorbate technetium Tc 99m. See technetium Tc 99m iron ascorbate
iron Fe 52 (^{52}Fe), for bone marrow imaging, 610–611
ischemia, cerebral, imaging of, 433–435, 436f
ISO. See International Organization for Standardization
isobar(s), 13
isobaric transition, 22–23
isomer(s), 13, 166–168
  metastable, 23
isomeric transition, 23, 158
isopropyl alcohol (IPA), 314, 315
N-isopropyl-p-^{123}I-iodoamphetamine (^{123}I-IMP), for brain imaging, 418
isotone(s), 13
isotope(s), 13, 148, 160, 160t
  carbon, 18, 18t
isotope dilution analysis, 696, 696f
isotope exchange, for radioiodination, 210–211, 211f
isotope production reactors, 32
isotopic labeling, 160–161, 161t
isotypic markers, 658
ITLC. See instant thin-layer chromatography
IU. See International Unit

J

Japanese survivors
  acute effects in, 138
  cancer risk estimates for, 99, 101
  cancer types in, 140
  hereditary risk estimates for, 104
  radiation effects in pregnancy, 143
  radiation risk for, 89–90, 98
  risk assessment for, 99
Joint Commission, The (TJC), 253
Joliot, Frederic, 1, 31
Joliot, Irene Curie, 1, 31
juxtamedullary nephrons, 547, 548f

K

Kavula, Michael, 8, 9t
Kawada, Tom K., 9t
Keesee, Richard, 8, 9, 9t
Kefauver-Harris amendments, 375, 376
Kelsey, Frances, 375
keV. See kiloelectron volt
kidney(s)
  anatomy and physiology of, 547–548, 548f
  blood flow in, 547–548

clearance in, 548–549, 550t
drug elimination in, 550–551
excretion mechanisms of, 548–551
extraction efficiency or ratio in, 549–550, 550t, 551
FDG uptake in, 624–625, 624f
functioning unit of (nephron), 547–548, 548f
functions of, 547
kidney cancer, 562, 565f
kidney imaging, 547–569
agents for, 193, 194, 196, 198–199, 551–555
biologic properties of, 552–555
clearance agents, 552, 553t
development of, 551–552
ideal properties of, 549, 550t
imaging agents, 552, 553t
dynamic, 2
function, 555–556, 569
interpretation of, 559–569
in vivo function, 3, 158
iodine I 125, 208
iodohippurate sodium I 131, 214–215
nuclear medicine procedures for, 555–569
perfusion, 555–568
PET, 562, 567f
pharmacologic intervention in, 547
physiologic principles of, 547–548
rationale for, 556
unique feature of, 547
kidney transplant, evaluation of, 562, 563f
kiloelectron volt (keV), 16
Kinevac. *See* sincalide
kit, technetium. *See* technetium kit
kit preparation screen, 268, 268f
knockout mice, 357
Kowalsky, Richard J., 9t
81mKr. *See* krypton 81m
krypton 81m (81mKr), 26t
for lung ventilation studies, 505, 505t
K-shell electrons, 14–15, 15f
Kupffer's cells, 517, 518–519, 518f, 521

L

label(s)
PET, 309
radioactive material, 120, 122–123, 123f, 123t
radiopharmaceuticals, 269
regulatory requirements for, 401
labeling, 160–163
antibody, 186, 187f, 661–665, 674, 674f
radioiodine, 662–663, 662f, 663f, 664f, 674
radionuclide considerations in, 662, 662t
bifunctional chelating agents for, 179, 181
for antibodies, 186, 187f, 664–665, 664f, 665f
for peptides, 186, 187f, 656, 668–669
for technetium-nonessential compounds, 185–186, 186f
chemistry of, 346–348
choice of radiolabel, 346

indium In 111, 160, 186, 661–662, 662t, 664–665, 676, 677t
isotopic, 160, 161, 161t
molecular location of radiolabel, 348
nonisotopic, 160, 161
peptide, 186, 187f, 656
PET radionuclides, 160, 244–246, 244t
red blood cell, 700–701
strategy for, 346–348
technetium, 186, 186f
yttrium Y 90, 160, 186, 661–662, 662t, 664–665, 674, 674f, 676, 677t
labeling precursors, for PET radionuclides, 244–246, 244t
Lacks, Henrietta, 133
lacrimal glands
anatomy of, 608–610, 610f
dacryoscintigraphy of, 608–610, 611f
lactation. *See* breast-feeding; breast milk
lactoperoxidase, 211
LAFWs. *See* laminar-airflow workbenches
LAL. *See* limulus amebocyte lysate test
laminar-airflow workbenches (LAFWs), 259–260, 315
laminar-flow hood, 260, 261–262, 262f, 315
large bowel, FDG uptake in, 625
large-vessel hematocrit (LVH), 695–696
laryngeal cancer, PET/CT of, 622f, 626–627
latency phase, of acute syndrome, 138
Lathrop, Kathryn, 8
L configuration, 167–168, 167f, 168f
LDE. *See* lens dose equivalent
lead barrels, 260, 260f
lead bricks, 261
lead collimator, 76
lead development, 344–346
lead L-block, 261, 262f
lead shielding, 97, 98f, 98t
in nuclear pharmacy, 260–261
sleeves, for dose calibrator testing, 301–302, 301f, 302f
for syringes, 112–113, 112f, 113f, 260, 260f, 261f
for vials, 261
lead structures, identification of, 345–346
lead x-ray peak, 70
leaking source, report on, 407
left anterior descending artery, 460, 460f
left circumflex artery, 460, 460f
left coronary artery, 460, 460f
left ventricular ejection fraction (LVEF), 474, 480
lens dose equivalent (LDE), 91
LET. *See* linear energy transfer
leukemia
radiation-induced, 139–140, 143–144
treatment of, 223
leukocyte(s). *See* white blood cells
leukocyte-poor plasma (LPP), 201f, 202, 219
leukocyte-rich plasma (LRP), 201–202, 201f, 219
LeukoScan. *See* technetium Tc 99m sulesomab
Levine, Geoffrey, 9t
lexidronam samarium Sm 153. *See* samarium Sm 153 lexidronam
license(s)
for nuclear pharmacy, 265
records required for, 406–407

for radioactive material, 394–408
amendments of, 397–398
application for, 397
general administrative requirements for, 398–400
general information for, 395–398
general technical requirements for, 400
notification to NRC, 398, 407
renewal of, 397
report requirements for, 407
for radioiodine use, 456
lidofenin technetium Tc 99m (99mTc-HIDA). *See* technetium Tc 99m lidofenin
LiF. *See* lithium fluoride
ligands, 153–154, 154f
coordinating, for technetium, 174–175, 175f, 179, 180f
light
ultraviolet, 53t
visible, 53, 53t
limulus amebocyte lysate (LAL) test, 296–298, 297f, 315, 335
Linac. *See* linear accelerators
linear accelerators, 33–34, 33f
linear attenuation coefficient, 97, 98t
linear energy transfer (LET), 51, 51f, 52t, 88
and cell survival curve, 133, 134f
and direct effects, 128
and radiosensitivity, 134f, 136
linearity
of dose calibrators, 299t, 300–302
attenuation method for testing, 301–302, 301f, 302f
decay method for testing, 301, 301f
report on, 302t
of spectrometer, 66–67
linear nonthreshold (LNT) relationship, 100, 100f
linear–quadratic model, 134f, 135, 140
line of nuclear stability, 17–18, 18f
line of response, in PET, 77, 77f
lipophilic impurities, 319
liposomes, technetium Tc 99m-labeled, for lymphoscintigraphy, 608
liquid scintillation counter, 63–65, 71, 73–74, 74f
components of, 74f
quenching in, 73–74, 75f
liquid spills, 125–126
liquid targets, of cyclotrons, 243–244
lithium, electronic configuration of, 152t
lithium fluoride (LiF), formation of, 149, 156, 156f
liver
anatomy of, 517–520, 518f
blood flow in, 517, 519f, 530, 530f
cells of, 517–520, 518f, 519f
uptake in bone scans, 574, 574f
liver cancer
brachytherapy for, 688–690
PET/CT of, 641–643, 642f
liver imaging, 517–525, 529–530
agents for, 517–525
for hepatic artery infusion pump evaluation, 529, 531f
physiologic principles of, 517–520
procedure for, 530

radiocolloid, 2–3, 520–522, 520t, 521f, 529–530
    adverse reactions and toxicity in, 522
    biologic properties and, 520–522
    blood clearance and, 521
    blood flow and, 520–521, 521f
    disease state and, 521, 521f
    localization and metabolic fate in, 521–522, 521f
    opsonization and, 521
    particle size and, 521
    radiation dose in, 522, 522t
    rationale for, 529
    static, 2–3, 3f
liver metastases, brachytherapy for, 689–690
LLD. *See* lower-level energy discriminator
LNT. *See* linear nonthreshold relationship
lobules, liver, 517, 518f
loops of Henle, 547
lower-level energy discriminator (LLD), 66–68, 67f
LSD. *See* lung scanning dose
L-shell electrons, 15, 15f
LSO. *See* lutetium oxyorthosilicate
lung(s)
    anatomy of, 499, 500f
    circulation in, 499, 501f, 502
    tissue-weighting factor for, 104t
lung cancer
    differential lung perfusion studies of, 513, 514f
    indium In 111 pentetreotide imaging of, 600t
    PET/CT of, 629–634, 631f–632f, 633f
    radiation-induced, 140
    reimbursable PET procedures in, 239t, 240–241
lung perfusion studies
    adverse reactions in, 337
    agents for, 499, 501–504
    combined with ventilation studies, 504–505, 507–513. *See also* ventilation–perfusion lung scan
    hydrostatic pressure and, 500–501
    physiologic principles of, 499–501
    technetium Tc 99m aggregated albumin for, 189, 337, 501–504, 503f
lung scanning dose (LSD), 503–504
lung ventilation studies
    agents for, 499, 504–508
    combined with perfusion studies, 504–505, 507–513. *See also* ventilation–perfusion lung scan
    inhalation risk in, 95–96
    physiologic principles of, 499
    radioaerosols for, 507–508
    Technegas for, 508
    technetium Tc 99m pentetate for, 505, 507, 507f
    technetium Tc 99m sulfur colloid for, 508
    xenon Xe 133 for, 220, 505–507, 505t
Luschka, foramina of, 415–416
lutetium oxyorthosilicate (LSO), 78, 231, 233
LVEF. *See* left ventricular ejection fraction
LVH. *See* large-vessel hematocrit
LYM-1

labeling process for, 681, 682f
    for non-Hodgkin's lymphoma, 674, 681, 682–683
lymphoblastic lymphoma, 637f
lymphocytes, 656, 656f
lymphoma
    gallium Ga 67 citrate for imaging, 595f, 596, 597–598, 597f, 634
    Hodgkin's
        gallium Ga 67 citrate imaging of, 597–598, 597f, 634
        PET/CT of, 634–636, 635f, 637f
    lymphoblastic, 637f
    non-Hodgkin's
        gallium Ga 67 citrate imaging of, 597–598, 634
        PET/CT of, 634–636, 636f
        treatment of, 3–4, 207, 655, 666–667
            ^{111}In-ibritumomab tiuxetan for, 676–677
            ^{131}I-tositumomab for, 3–4, 207, 655, 666–667, 674, 676, 680–683
            LYM-1 for, 674, 681, 682–683
            myeloablative, 675–676, 682
            nonmyeloablative, 675–676, 682
            radioimmunotherapy, 3–4, 673–683
            ^{90}Y-epratuzumab for, 674, 681–682
            ^{90}Y-ibritumomab tiuxetan for, 3–4, 655, 666–667, 674, 676–680, 682
    reimbursable PET procedures in, 239t, 240–241
    response to treatment, assessing, 645f, 646f
lymphopenia, 138
lymphoscintigraphy, 606–608
    breast, 606, 607, 636
    melanoma, 606–607, 609f
    radiopharmaceuticals for, 607–608
    sentinel node concept in, 606
Lymphoseek. *See* technetium Tc 99m pentetate–mannosyl–dextran

M

MAA, ^{99m}Tc. *See* technetium Tc 99m aggregated albumin
MAA, ^{131}I. *See* iodinated I 131 albumin aggregated
macroaggregated albumin (MAA). *See* technetium Tc 99m aggregated albumin
magnetic resonance imaging (MRI), 343–344, 344t
    contrast-enhanced, of myocardial infarction, 489–490, 491f, 492f
    of head and neck cancer, 626–627
    hepatobiliary and spleen, 528–529
    renal, 547, 555
    of tumors, 598
major spill procedure, 126, 126t
malignant effusions, treatment of, 673, 687
mammary glands. *See also* breast; breast milk
    radioiodide concentration in, 448–449

mammography, positron emission, 636, 638f
Manhattan Project, 375
mass, of radiopharmaceuticals, and toxicity, 5–7, 147
Massachusetts College of Pharmacy, 9
Massachusetts Institute of Technology, 446
mass–activity relationship, 25
mass defect, 16
mass–energy relationship, 16
mass number, 13, 148
mass selective detectors (MSDs), 277
matter, radiation interactions with, 47–57
    charged-particle, 47–51
    electromagnetic, 52–56
    neutron, 51–52
maximum detectable activity, 82–83, 84f
maximum-intensity projections (MIPs), 620
maximum tolerated dose (MTD), 675–676, 683
MDP, ^{99m}Tc. *See* technetium Tc 99m medronate
mean absorbed dose ($D_{T,R}$), 91f
mean life, 29
measurement units, 4, 25–28, 87–91
mebrofenin technetium Tc 99m. *See* technetium Tc 99m mebrofenin
mechanical reactions, 334t
Meckel's diverticulum, imaging of, 528, 539–541, 542f
media-fill procedure, 314, 315f
mediated transport, 163, 165–166
medical event, 396, 407
Medical Internal Radiation Dose (MIRD), 92, 383, 675
medical officers, 393
medical physicist, authorized, 395
Medical University of South Carolina, 9
medical use, definition of, 396
medical use of byproduct material
    general administrative requirements for, 398–400
    general information for, 395–398
    general technical requirements for, 400
    NRC regulations on, 394–408, 395t
    records required for, 406–407
    report requirements for, 407
medical use of radiopharmaceuticals, NRC regulations on, 394–395
medronate technetium Tc 99m (^{99m}Tc-MDP). *See* technetium Tc 99m medronate
medulla, renal, 547
MedWatch, 334
megaelectron volt (MeV), 16
megamouse project, 142
melanocyte-stimulating hormone receptor analogues, 671
melanoma
    iodine I 131 ioflubenzamide for, 691
    lymphoscintigraphy of, 606–607, 609f
    PET/CT of, 638–640
    reimbursable PET procedures in, 239t, 240–241, 380
    staging of, 606, 606t, 638–640
melatonin, 424
membrane filter integrity, 296
meningitis, aseptic, 334t
mental retardation, 142, 143, 144f
Mercer University, 9

mercury 203 (^{203}Hg)
  decay of, 19, 20f, 23–24, 24f
  for renal imaging, 552
mercury Hg 197 chlormerodrin
  (^{197}Hg-chlormerodrin)
    for brain imaging, 416
    for renal imaging, 552
mercury Hg 203 chlormerodrin
  (^{203}Hg-chlormerodrin), for brain imaging, 416
meridional *(mer)* isomer, 168, 168f
mertiatide technetium Tc 99m
  (^{99m}Tc-MAG3). *See* technetium Tc 99m mertiatide
meso-tartaric acid, 166, 166f
metabolic bone disease, 584, 586f
metabolic processes, PET of, 234t, 236–237
metabolism agents
  for brain imaging, 411, 422–423
  for heart imaging, 459, 471–473
metaiodobenzylguanidine, iodinated. *See* iobenguane iodine I 123; iobenguane iodine I 131
metal(s), 347
  alkali, 148
  alkaline-earth, 223
metal chelate, 153–155, 347, 664–665, 664f, 665f
metal complex, 153–154
metal ions, solution chemistry of, 154–155
metal–ligand complex, 153–154, 154f
metaphase, 130, 130f
metaphosphoric acid, 222
metastable isomers, 23
metastable state, 13, 23
metastases
  bone, pain palliation for, 578, 582, 584f, 673, 683–686, 684t
  brain, PET/CT of, 628–629, 629f
methimazole, 445
metoclopramide, for gastroparesis, 544
MeV. *See* megaelectron volt
MI. *See* myocardial infarction
MIBG, iodinated I 123. *See* iobenguane iodine I 123
MIBG, iodinated I 131. *See* iobenguane iodine I 131
mice
  antibodies from
    adverse reaction to, 337–338, 658–659, 673, 674
    production of, 658–659, 659f, 660f
  gestational stages in, 142–143
  gonadal effects in, 142
  knockout, 357
  risk studies in, 99
microbial contamination risk levels, 313–314
microbiologic control, 295–296, 313–317
microbiologic tests, 315
microcuries, 4, 25–28, 161
microPET, 355–357, 355t
microscope, in nuclear pharmacy, 265
microSPECT, 355–357, 355t
microspheres, yttrium Y 90-labeled. *See* yttrium Y 90 microspheres
microwaves, 53, 53t
millicuries, 4, 25–28, 161
mineralocorticoids, 601

minimum detectable activity, 83–84
minimum dose, pediatric, 325–326
minimum sensitivity (MS), 83
minor spill procedure, 126
MIPs. *See* maximum-intensity projections
MIRD. *See* Medical Internal Radiation Dose
MIT. *See* monoiodotyrosine
mitosis, 129–130, 130f, 135
mixing order, improper, 321
^{99}Mo. *See* molybdenum Mo 99
mobile medical service, 402, 406
mobile phase
  in gas chromatography, 277
  in high-pressure liquid chromatography, 278
  in paper and thin-layer chromatography, 279–280
mole, 148
molecular imaging, 343–344
  concept of, 1, 343
  definition of, 343
  in drug discovery, 357
  methods of, 343–344, 344t
Molecular Insight Pharmaceuticals, 690, 690t
mole fraction, 320
moly-99 breakthrough test, 38, 38f, 275
molybdenum Mo 99 (^{99}Mo), 26t
  ^{99}Mo–^{99m}Tc generator, 34–44. *See also* technetium Tc 99m generator
  decay of, 23, 23f
  for liver imaging, 520
  shipping guidelines for, 121t, 122t
monitoring
  area, 118–119
  effluent, 118
  package, 120, 120t, 123
  personnel, 114–118
    devices for, 115–116, 117f
    investigational levels for, 115, 115t
  sealed-source, 118–119
monoclonal antibodies
  nomenclature for, 661
  radiolabeled
    for bone marrow imaging, 612
    for infection imaging, 593
    localization mechanisms of, 596t
monoiodotyrosine (MIT), 445
Monro, foramen of, 415
Moore, George, 416
morphine sulfate, for hepatobiliary imaging, 525
MOSE [di-β-(morpholinoethyl)-selenide], 164–165, 165f, 418
motion, and PET/CT, 618–619
mouse. *See* mice
M phase, 130, 130f, 137, 137f
MPI. *See* myocardial perfusion imaging
MRI. *See* magnetic resonance imaging
MS. *See* minimum sensitivity
MSDs. *See* mass selective detectors
MUGA. *See* equilibrium radionuclide ventriculography
multichannel analyzer (MCA), 68, 69f, 70f, 71f, 274, 274f
multicystic dysplastic kidney, 562, 567f
multifactorial effects, 104
multigated acquisition (MUGA). *See* equilibrium radionuclide ventriculography

multinodular goiter
  imaging of, 453, 453f, 454f
  radioiodine therapy for, 453, 454–455
multiple-dose therapy, 137
multiple myeloma, 139
multiplicative risk model, 99
multitarget theory, 134
murine antibodies
  adverse reaction to, 337–338, 658–659, 673, 674
  production of, 658–659, 659f, 660f
musculoskeletal system, FDG uptake in, 625, 626f
myc oncogene, 141
myeloablative therapy, 675–676, 682
myeloma, reimbursable PET procedures in, 239t
myocardial blood flow, 460–461
  resting, 460–461
  studies of. *See* myocardial perfusion imaging
myocardial infarction (MI), 459, 474
  imaging of, 488–490
    contrast-enhanced MRI, 489–490, 491f, 492f
    infarct-avid agents for, 488–489, 490t
  myocardial viability studies in, 484
  nontransmural, 484
  transmural, 484
myocardial perfusion imaging (MPI), 459, 473–479
  agents for, 197, 200, 220, 459, 460t, 467–471, 467t
  dual-isotope protocols for, 476–477, 479f
  exercise stress and, 461–462
  extraction and retention of tracers in, 466–467, 466f
  interpretation of, 479, 480f, 481f, 482f, 483f, 484f
  major goal of, 461
  PET, 237, 239–240, 239t, 380, 459
    agents for, 459, 460t, 467t, 470–471
    protocols for, 477–479
  PET/CT, 646–648, 648f, 649f
  pharmacologic stress and, 462–466
    precautions, adverse effects, contraindications, 464–466, 465t
    protocols for, 465t
  physiologic principles of, 460–467
  procedures for, 474–479
  rationale for, 474
  rest–stress principle in, 461, 461f
  risk stratification in, 492–494, 493f, 493t, 494t
  SPECT, 459, 460t, 467–470, 467t
  technetium protocols for, 476, 477f, 478f
  thallium protocol for, 474–476, 475f, 477f
myocardial stunning, 473, 484
myocardial viability imaging, 459, 484–488
  agents for, 459, 460t, 471–473
  PET, 239t, 240, 487–488
  PET/CT, 647–648, 648f, 649f
  procedures for and interpretation of, 484–488

rationale for, 484
technetium procedures for, 486–487
thallium procedures for, 484–486
myocardium, hibernating, 460–461, 484
myocytes
metabolic pathways in, 471, 472f
transport in, 468, 468f
Myoscint. See indium In 111 imciromab pentetate

N

^{13}N. See nitrogen N 13
n, γ nuclear reaction, 32
^{24}Na. See sodium Na 24
Nagasaki. See Japanese survivors
NaI scintillation counters. See sodium iodide scintillation counters
nanocuries, 25–28
nanotechnology, 1
NARM. See naturally occurring and accelerator-produced radioactive material
nasolacrimal system, 608–610, 610f
nasopharyngeal cancer, PET/CT of, 626–627, 627f
National Academy of Sciences, 109
National Association of Nuclear Pharmacies, 257
National Council on Radiation Protection and Measurements (NCRP), 110
on cancer risk estimates, 96
on dose calibrators, 299–300
on dose limits, 111
on exposure during pregnancy, 144
on exposure time, 96
on volunteer subjects, 112
national coverage decision (NCD), on PET procedures, 239–241
National Formulary, 235, 252
National Institute of Standards and Technology (NIST), 61
National Institutes of Health (NIH), 8, 9
National Oncologic PET Registry (NOPR), 240–241, 380
naturally occurring and accelerator-produced radioactive material (NARM), 377
natural sources of radiation, 112
nausea, 135, 138–139
NCA. See no carrier added
NCD. See national coverage decision
NCEs. See new chemical entities
N-chlorosuccinimide (NCS), 662, 663f
NCRP. See National Council on Radiation Protection and Measurements
NCS. See N-chlorosuccinimide
NDAs. See new drug applications
neck, FDG uptake in, 622, 623f, 624f
negative-ion cyclotrons, 33
negative predictive value (NPV), 484–485
negatron(s), 17
interactions with matter, 49–50, 49f, 50f
photon conversion to, 53, 55–56, 56f, 57t
properties of, 48t
negatron decay, 18, 19, 20f, 22–23, 51
NEO. See Nuclear Education Online
neon, electronic configuration of, 149, 152t
neonatal hepatitis, 537, 538f

NeoTect. See technetium Tc 99m depreotide
nephrons, 547–548, 548f
neural capillaries, brain, 412, 412f
neuroblastoma, imaging of, 601–602, 604f
neuroendocrine system, 598
neuroendocrine tumors, imaging of, 598–600
neurology, reimbursable PET procedures in, 239t, 240, 617
neuronal agents, for brain imaging, 411, 423–426
neuronal imaging, cardiac, 490–492
neurotransmission, 423–424, 424f
agents for imaging, 424–426
PET of, 234t, 237–238
neutral atom, 13, 148
neutrino, 19–23, 47, 48t
neutron(s), 13, 47, 147–148
binding energy of, 16–17, 17f
capture reactions of, 51–52
energy levels of, 15–16, 16f
fast, 51
free, decay of, 51
interactions with matter, 51–52
intermediate, 51
properties of, 48t
scatter reactions of, 51–52
thermal, 51–52
neutron activation, 52
neutron/proton (n/p) ratio, 17–18
in electron capture decay, 22–23
in negatron decay, 18, 19, 20f
in nuclear reaction, 31
in positron decay, 18, 21
neutropenia, 138
neutrophils, 589–590
NeutroSpec. See technetium Tc 99m fanolesomab
new chemical entities (NCEs), 346, 357, 359
new drug applications (NDAs), 253, 305, 358, 359, 376–377
goal of, 389
process of, 389–394, 391f
review process for, 391–394
biostatistics, 393
chemistry, 392–393
clinical, 392, 393–394
nonclinical, 392–393
pharmacology and toxicity, 392
submission of, 394
new products, adverse reactions to, 338
new radiopharmaceuticals development. See radiopharmaceutical development
NHL. See non-Hodgkin's lymphoma
NHS. See N-hydroxysuccinimide ester
N-hydroxysuccinimide ester (NHS), 663
NIH. See National Institutes of Health
NIST. See National Institute of Standards and Technology
nitrogen
coordinate chemical bonds of, 153
electronic configuration of, 152t, 153
nitrogen N 13 (^{13}N), 6t, 26t
for brain imaging, 422
decay of, 21, 21f
half-life of, 207, 275
labeling precursors for, 244–245, 244t
labeling with, 160, 161t, 346–347
on-site production of, 34

oxo anions of, 244–245
for PET, 207, 230t, 378
production of, 244t
radiochemical purity of, 286t
radionuclidic purity of, 276t
shipping guidelines for, 122t
as traditional radionuclide, 346, 347t
nitrogen N 13 ammonia. See ammonia N 13
no carrier added (NCA), 162, 178, 215
nodule, solitary pulmonary
PET of, reimbursable, 626
PET/CT of, 629–634, 630f
nodules, thyroid, 450–451
imaging of, 452–453, 453f
radioiodine therapy for, 453, 454–455
noise, in PET, 232, 232t
nondiffusible tracers, for brain imaging, 411, 416–418
capillary permeability and, 418
interstitial fluid and, 417–418
mechanisms of localization, 417
tumor cell uptake and, 418
vascularity and, 417
nonfunctioning thyroid nodules, 451, 453, 453f
non-Hodgkin's lymphoma
gallium Ga 67 citrate imaging of, 597–598, 634
PET/CT of, 634–636, 636f
treatment of, 3–4, 207, 655, 666–667
^{111}In-ibritumomab tiuxetan for, 676–677
^{131}I-tositumomab for, 3–4, 207, 655, 666–667, 674, 676, 680–683
LYM-1 for, 674, 681, 682–683
myeloablative, 675–676, 682
nonmyeloablative, 675–676, 682
radioimmunotherapy, 3–4, 673–683
^{90}Y-epratuzumab for, 674, 681–682
^{90}Y-ibritumomab tiuxetan for, 3–4, 655, 666–667, 674, 676–680, 682
nonisotopic labeling, 160, 161
nonmyeloablative therapy, 675–676, 682
nonneural capillaries, brain, 412, 412f
non-small cell lung cancer
PET/CT of, 629
reimbursable PET procedures in, 239t, 240–241
nonspecific target binding, 350, 350f
nonstochastic effects, 98. See also deterministic effects
nontraditional radionuclides, 346, 347t
nontransmural myocardial infarction, 484
NOPR. See National Oncologic PET Registry
norepinephrine, 424, 601
normal (Gaussian) distribution, 79–80, 80f
normal-pressure hydrocephalus, 436–437, 437f
NOTA (1,4,7-triazacyclononane-1,4,7-triyl) triacetic acid, 668–669, 669f
notices, safety, 119
NPT. See nuclear pharmacy technician
NPV. See negative predictive value
NRC. See Nuclear Regulatory Commission

NTA, ^{99m}Tc(CO)$_3$. *See* technetium Tc 99m(CO)$_3$nitrilotriacetic acid
nuclear binding energy, 16–17, 17f, 31
Nuclear Education Online (NEO), 10
nuclear force, 16–17
nuclear medicine. *See also specific agents and procedures*
    definitions of, 1
    history of, 1
    procedures in, 2–4
    radiation exposure in, 112–113
    radionuclides in, 26t–27t
    radiopharmaceuticals used in, 5t–7t, 147
nuclear medicine technologists, 4
nuclear pharmacist, authorized, 265, 395–396
nuclear pharmacy, 257–271
    centralized, 8–9, 257, 374
    companies, 9
    computer system in, 259–260, 259f
    development of, 257–258
    education programs, 9–10, 257
    equipment for, 258–262, 265
    facilities of, 258–262
    floor plan of, 258, 258f
    handling techniques in, 266, 266t
    historical perspectives on, 8–11
    licensing of, 265
    personnel of, 265–266
    pioneers in, 9t
    practice, 8–9, 266–271, 373–374
        APhA guidelines on, 266, 374
        compounding, 267–268, 267f, 268f
        dispensing, 269–270, 269f
        distribution, 270
        drug information provision, 270–271, 271t
        drug order provision, 266–270
        health and safety, 270
        procurement, 266–267
        records, 271, 271t
        standards, NRC requirements and, 408
    quality assurance in, 268–269
    quality control in, 260, 273–309
    radiation detection instrumentation in, 262–265
    radiation exposure in, 112–113
    sites of, 257–258, 374
    specialty certification in, 9–11, 10t, 11f, 266, 266t, 374, 405
    storage and shielding in, 260–261, 261t
    supplies, 265
Nuclear Pharmacy Inc., 9
Nuclear Pharmacy Practice Guidelines, 10
Nuclear Pharmacy Symposium, 10
nuclear pharmacy technician (NPT), 265
nuclear reactions, 31, 32–33
    ^{235}U(n,f)byproducts, 33
    A(n, γ)A*→B, 32
    A(n,p)B reaction, 32–33
    n, γ, 32
nuclear reactors, 31–32, 32f
Nuclear Regulatory Commission (NRC), 38, 110, 373
    on breast-feeding, 330–331
    on dose determination, 292–293
    on emergency procedures, 125–126
    on exposure time, 96
    historical perspective on, 375–376
    inspections by, 119–120
    on medical use of byproduct material, 394–408, 395t
    on medical use of radiopharmaceuticals, 394–395
    mission of, 394
    on monitoring of radioactive packages, 120, 120t
    notification to, 398, 407
    on nuclear pharmacy facilities, 258
    on nuclear pharmacy regulations, 265
    on PET, 253
    policy statement on regulation, 377–378
    on radioiodine use, 456
    on record-keeping, 271
    regulatory authority of, 377–378
    regulatory outlook for, 408–409
    requirements of, and nuclear pharmacy practice standards, 408
    on safety, 7, 270
    on shipment of radioactive material, 120–125
nuclear stability, line of, 17–18, 18f
nucleons
    binding energy of, 16–17, 17f
    energy levels of, 15–16, 16f
nucleus, 13, 147–148
    energy levels of, 15–16, 16f
    intermediate, 31
nuclides, 13–14, 147–148
    classification of, 13
    notation for, 13
    parent and daughter, 19
    on periodic table, 14, 14f
    stable, 14, 17–18, 18f, 148
    unstable or radioactive. *See* radionuclide(s)
Nutt, Ron, 626
N value, of nuclides, 13

O

^{15}O. *See* oxygen O 15
Oak Ridge Institute of Nuclear Studies, 9
Oak Ridge National Laboratory, 8, 375
obstructive uropathy, 556, 560–562, 562f
occupational dose, 111
occupational exposure, 96, 109, 111, 112–114. *See also* health care personnel
Occupational Safety and Health Administration (OSHA), 253, 270, 373, 391f
OctreoScan. *See* indium In 111 pentetreotide
octreotide, 598, 598f, 668
Oddi, sphincter of, 518f, 519–520, 532
OER. *See* oxygen enhancement ratio
Office of Federal and State Materials and Environmental Management Programs, 377
off-label uses
    chromic phosphate P 32 suspension, 222–223
    fluorodopa F 18, 237t, 238
    PET radiopharmaceuticals, 237, 237t
    sodium acetate C 11, 237, 237t

Ohio State University, 9
Oliver, Larry, 9, 9t
Onalta. *See* yttrium Y 90 edotreotide
oncogene, 132, 141
oncology. *See also* cancer
    PET procedures in, reimbursable, 239t, 240–241, 380, 617, 626
    PET/CT in, 617–618, 625–643
        for assessing tumor response to therapy, 628f, 644, 645f, 646f
Oncolym. *See* LYM-1
OncoScint. *See* indium In 111 satumomab pendetide
on-site production, 4, 34
opposite double-strand break, in DNA, 131, 131f
opsonization, of radiocolloids, 521
optical imaging, 343, 344t
optically active isomers, 166
optically stimulated luminescent dosimeter (OSLD), 116
optical radiation, 15, 15f
orbitals, electron, 149, 152f
order entry screen, 269, 269f
ordering, of radiopharmaceuticals, 266–270
organogenesis, 142
organs
    critical, 91
    radiation dose to, 91–92
    radiosensitivity of, 89, 135–137
    target, 91–92, 163
    tissue-weighting factors for, 89, 89t, 90t, 104, 104t
oropharyngeal cancer, PET/CT of, 626–627
Orphan Drug Act of 1983, 389
orthophosphoric acid, 221
OSHA. *See* Occupational Safety and Health Administration
OSLD. *See* optically stimulated luminescent dosimeter
osmolality, 295
osteoarthropathy, hypertrophic pulmonary, 582, 585f
osteoblasts, 572
osteoclasts, 572
osteocytes, 572
osteogenic activity, PET of, 236
osteomyelitis, 582, 582f–583f, 584t, 594, 596f, 651, 668
ovarian cancer
    PET/CT of, 643, 644f
    radiolabeled antibody for diagnosis of, 666
    reimbursable PET procedures in, 239t, 380
ovaries, radiation effects on, 142, 142t
Oxford Survey of Childhood Cancers, 144
oxidants
    for radioiodination, 211–213, 662, 663f
    radiolytic, 320–321
oxidation, 156, 156t
oxidation–reduction reactions, 155–158
    iodine, 155, 157–158
    lithium fluoride, 156, 156f
    technetium, 156–157, 157f, 172–177
    terminology in, 156, 156t
oxidation state, 150t–151t, 156
oxidative decomposition, 323

oxidative metabolism
  heart imaging of, 459
  PET of, 236–237, 237f
oxidizing agent, 156t
oxidronate technetium Tc 99m (^{99m}Tc-HDP). *See* technetium Tc 99m oxidronate
oxygen
  electronic configuration of, 152t
  isotopes of, 148
  and radiosensitivity, 136–137
oxygen enhancement ratio (OER), 136
oxygen O 15 (^{15}O), 6t, 26t
  for blood flow measurement, 234–235
  for brain imaging, 422
  decay of, 21, 21f
  gaseous, labeling precursor for, 244, 244t
  half-life of, 207, 246
  for heart imaging, 459, 460t, 466f, 470–471, 470t
  labeling with, 160, 161t
  on-site production of, 34
  for PET, 207, 230t, 234–235, 246–247, 378
  production of, 244, 244t
  radionuclidic purity of, 276t
oxygen O 15 carbon monoxide. *See* carbon monoxide O 15
oxygen O 15 water
  for heart imaging, 470–471, 470t
  for PET, 234–235, 246–247
  radiochemical purity of, 286t
  radiosynthesis of, 246–247, 247f
oxyquinoline indium I 111. *See* indium In 111 oxyquinoline

P

^{30}P. *See* phosphorus P 30
^{32}P. *See* phosphorus P 32
package inserts, information on, 273–274, 284–285, 286t–287t
package labeling, 120, 122–123, 123f, 123t
package markings, 122
packaging requirements, 121–122
Paget's disease, 584, 586f
PAH (para aminohippurate) clearance, 549–550
pair production, 53, 55–56, 56f, 57t
palmitate carbon C 11. *See* carbon C 11 palmitate
pancake probe, 263, 263f
pancreatic cancer
  PET/CT of, 640–641, 641f
  reimbursable PET procedures in, 239t
paper chromatography, 275, 279–283
para aminohippurate (PAH) clearance, 549–550
paraganglioma, imaging of, 600t, 604f
parathyroid gland
  anatomy and physiology of, 603–604
  hyperplasia of, 603–604
  tumors of, imaging of, 603–605, 605f
parathyroid hormone (PTH), 603–604
paratopes, 659
parent–daughter radionuclide generators, 34, 35t
parent nuclide, 19

Parker, Paul, 8
Parkinson's disease, 424–425, 425t, 435, 436f, 649–650
particle accelerators, 33–34, 33f, 242–244, 243t
particle aggregation, 319
particle number and size, in radiopharmaceuticals, 294, 294t
particle sedimentation, 324
particulate radiation, 17, 47, 158. *See also* alpha particle(s); beta particles
passive diffusion, 163–165, 163f, 548
PE. *See* pulmonary embolism
PEC. *See* primary engineering control
pediatric bone scan, 580, 580f
pediatric dosing, 324–328
  age-based adjustment, 326
  body surface area and, 326, 327t
  height-based adjustment, 326–327
  minimum dose, 325–326
  PET/CT, 327–328
  radiation absorbed dose, 325, 325t
  radiopharmaceuticals containing benzyl alcohol, 327
  technetium Tc 99m aggregated albumin, 325t, 327
  weight-based adjustment, 326
pelvis, renal, 547, 548f
penile cancer, lymphoscintigraphy in, 606
pentagastrin, for Meckel's diverticulum imaging, 540
pentetate, indium In 111. *See* indium In 111 pentetate
pentetate, technetium Tc 99m. *See* technetium Tc 99m pentetate
pentetate, tin Sn 117m. *See* tin Sn 117m pentetate
pentetate, ytterbium Yb 169. *See* ytterbium Yb 169 pentetate
peptides, 167
  arginine–glycine–aspartic acid (RGD) analogues, 669–670
  gastrin-releasing peptide analogues, 670–671
  melanocyte-stimulating hormone receptor analogues, 671
  radiolabeled, 655–656, 668–671
    advantages of, 655
    bifunctional chelates for, 186, 187f, 656, 668–669
    disadvantages of, 655–656
    stability of, improving, 655–656
    for tumor imaging, 596, 598–600
  somatostatin-receptor analogues, 655–656, 668
    chemical structure of, 598, 598f
    for neuroendocrine tumors, 598–600, 600f
  in technetium-nonessential compounds, 185–186
perchlorate, with thyroid agents, 445
perfusion imaging
  heart. *See* myocardial perfusion imaging
  lung. *See* lung perfusion studies
period(s), on periodic table, 148
periodic table, 14, 14f, 147–148, 150t–151t, 160, 160t, 172
peritoneal effusions, treatment of, 687
personnel. *See* health care personnel

personnel protective equipment (PPE), 314
PET. *See* positron emission tomography
PET/CT, 74, 233, 344, 380–381, 617–652
  acquisition process in, 618–620
  artifacts in, 618, 619–620, 620f, 627
  exercise, strenuous, and, 618, 619f
  "eyes to thighs," 618, 627
  false-positive and false-negative studies in, 651
  fasting for, 618, 618f
  FDG uptake in, 620–625
    in bladder, 624–625, 625f
    in brain, 621, 621f, 622f
    in brown fat, 618, 620f, 627
    in digestive system, 625
    in heart, 620, 622–624
    in kidneys, 624–625, 624f
    in musculoskeletal system, 625, 626f
    in neck, 622, 623f, 624f
    time for, 618
  future of, 651–652
  nononcologic, 646–651
    cardiac, 646–648, 648f, 649f
    dementia, 649–650, 651f
    epilepsy, 648–649, 650f
    infection, 650–651
  normal, 620–625, 621f
  oncology
    for assessing tumor response to therapy, 628f, 644, 645f, 646f
    bladder cancer, 625f
    brain cancer and metastases, 628–629, 629f
    breast cancer, 620f, 636–638, 639f
    cervical cancer, 643, 643f
    clinical use of, 625–643
    colorectal cancer, 633f, 638, 640f
    esophageal cancer, 634, 634f, 635f
    head and neck cancer, 626–627, 627f, 628f
    laryngeal cancer, 622f, 626–627
    liver cancer, 641–643, 642f
    lung cancer, 629–634, 631f–632f, 633f
    lymphoma, 634–636, 635f, 636f, 637f
    melanoma, 638–640
    ovarian cancer, 643, 644f
    pancreatic cancer, 640–641, 641f
    patient preparation for, 617–618
    thyroid cancer, 627–628, 628f
    whole-body, 618
  patient motion and, 618–619
  patient positioning for, 618
  pediatric, 327–328
  scanner for, 78, 78f
  total-body, 618
petechiae, 138
PETNET Solutions, 9
pH, of radiopharmaceuticals, 294–295
PHA. *See* pulse-height analyzer
phagocytosis, 166, 589
pharmacist, 396. *See also* nuclear pharmacist
pharmacologic stress
  for myocardial perfusion imaging, 462–466, 474
  precautions, adverse effects, contraindications, 464–466, 465t, 474
  protocols for, 465t

pharmacology review, 392
Pharmaco Nuclear, 9
pharmacy, nuclear. *See* nuclear pharmacy
Pharmacy Technician Certification Board, 265
Pharmatopes, 9
Phase 0 trial, 358–359
Phase 1 trial, 357, 358*f*, 359, 387–388, 392, 393
Phase 2 trial, 357, 358*f*, 359, 388, 392, 393–394
Phase 3 trial, 357, 358*f*, 359, 388–389, 392, 393–394
Phase 4 trial, 357, 358*f*, 359, 389
phenobarbital, for hepatobiliary imaging, 525
pheochromocytoma, imaging of, 600*t*, 601–602, 603*f*
phospholipid bilayer, of cell membrane, 163, 163*f*, 164*f*
phosphorus
    chemistry of, 221–224
    electronic configuration of, 221
phosphorus P 30 (^{30}P), production of, 1, 31
phosphorus P 32 (^{32}P), 6*t*, 26*t*
    applications of, 207
    for bone imaging, 573
    decay of, 222, 222*f*
    labeling with, 160, 161*t*
    production of, 222
    shipping guidelines for, 121*t*, 122*t*
phosphorus P 32 chromic phosphate suspension. *See* chromic phosphate P 32 suspension
phosphorus P 32 sodium phosphate. *See* sodium phosphate P 32 solution
photoelectric effect, 53–54, 53*f*, 54*t*, 56*f*, 57*t*, 128
photomultiplier (PM) tube, 65–66, 66*f*, 231
photon(s), 53
    colinear property of, 77
    Compton scatter and, 53, 54–55, 55*f*, 55*t*, 56*f*, 57*t*
    interactions with matter, 53–56
    pair production and, 53, 55–56, 56*f*, 57*t*
    photoelectric effect and, 53–54, 53*f*, 54*t*, 56*f*, 57*t*
    properties of, 48*t*
photon attenuation, 54, 54*f*
photon detection efficiencies, 159, 159*t*
photon energy, 52–53, 53*t*
photopeak, 69
photopeak energy resolution, 71
physician, regulatory definition of, 396
PIB. *See* Pittsburgh compound B
pi (π) bonds, 152–153, 152*f*
Pick's disease. *See* frontotemporal dementia
pinocytosis, 166
PIOPED. *See* Prospective Investigation of Pulmonary Embolism Diagnosis
PIPSE [di-β-(piperidinoethyl)-selenide], 164–165, 165*f*, 418
Pittsburgh compound B (PIB), 426
pituitary adenoma, indium In 111 pentetreotide imaging of, 600*t*
placarding, of transport vehicle, 123–124, 124*f*
placental transport
    of iodide, 449
    of sodium pertechnetate Tc 99m, 416

planar camera, 76
planar imaging, 3, 3*f*
plasmacrit, 695
plasma flow, renal, 549–550, 551, 555
plasma (cell) membrane
    structure of, 163, 163*f*, 164*f*
    transport across, 163–166
plasma volume studies, 158, 161, 208, 695–700, 697–699, 699*f*, 700*t*
platelet(s), 695
    indium In 111 labeled, 219–220, 220*f*
plating efficiency, 133
PLE. *See* protein-losing enteropathy
pleural effusions, treatment of, 687
Plummer's disease, 450, 453, 454–455
PM. *See* photomultiplier tube
PnAO, ^{99m}Tc. *See* technetium Tc 99m propyleneamine oxime
point mutation, 141
Poisson distribution, 79, 84–85
polycythemia vera, treatment of, 223, 673, 686–687
Polycythemia Vera Study Group, 223, 686–687
portal vein, 517, 519*f*
Porter, William C., 9*t*
positive-ion cyclotrons, 33, 215
positive predictive value (PPV), 484–485
positron(s), 17
    interactions with matter, 49–50, 49*f*, 50*f*
    photon conversion to, 53, 55–56, 56*f*, 57*t*
    properties of, 48*t*
positron annihilation reaction, 21, 21*f*
positron decay, 18, 21–23, 21*f*, 22*f*, 229–230, 231*f*
positron emission mammography (PEM), 636, 638*f*
positron emission tomography (PET), 3, 71, 74, 77–78, 229–254
    applications of, 229, 233–238
    attenuation in, 232, 232*t*
    bacterial endotoxin testing in, 298–299
    bone, 577*f*, 578, 579*f*, 580
    brain, 411–412. *See also* brain imaging
    of cardiovascular parameters, 234–236, 234*t*
        blood flow, 234–235, 234*t*
        blood volume, 234–236
    clinical research initiatives in, 359–362, 360*t*
    coincidence detection in, 77–78, 77*f*, 230–231, 230*f*, 231*f*
    CT with. *See* PET/CT
    dead time in, 232, 232*t*
    detector materials in, 78
    diagnostic efficacy of, 234
    future outlook for, 229, 253–254
    heart, 617. *See also* heart imaging
        cardiac neuronal, 490–492
        myocardial perfusion, 459, 460*t*, 467*t*, 470–471, 470*t*, 477–479
        myocardial viability, 239*t*, 240, 487–488
        risk stratification in, 492–494, 493*t*
    image quality in, factors affecting, 231–232, 232*t*

    infection, 594, 596
    line of response in, 77, 77*f*
    of metabolic processes, 234*t*, 236–237
    of neurotransmission, 234*t*, 237–238
    new drug applications for, 391
    noise in, 232, 232*t*
    nuclear pharmacy for, 257
    as preeminent molecular imaging method, 344
    protocol requirements for, 233–234
    protocol simplicity for, 234
    quality assurance in, 252–253, 305–309
        acceptance criteria, 308–309
        complaint handling, 309
        control of components, containers, and closures, 307
        distribution, 309
        execution of, 306
        facilities and equipment for, 306–307
        finished drug product controls, 308–309
        labeling and packaging, 309
        laboratory controls, 308
        oversight of, 306
        personnel resources for, 306
        production and process controls, 307–308
        records, 309
        stability testing, 308
    quality control in, 274, 299, 306
    radiochemical purity in, 283
    radionuclide formulation for, 251–252
    radionuclide identity in, 275
    radionuclide labeling for, 160, 244–246, 244*t*
    radionuclide production for, 34, 241–251
        automated modular systems for, 250–251
        remotely operated systems for, 250
        robotic systems for, 251
    radionuclide radiosynthesis for, 246–249
    radionuclides used in, 207, 229, 230*t*, 378, 617
    random events in, 232, 232*t*
    regulatory issues in, 253, 274, 305–306, 315, 378–381, 391, 408–409
    reimbursable procedures, 238–241, 239*t*, 253, 379–381, 617
        cardiology, 239–240, 239*t*, 380
        neurology, 239*t*, 240, 380
        oncology, 239*t*, 240–241, 380, 626
    renal, 562, 567*f*
    scatter in, 231–232, 232*f*, 232*t*
    scintillation crystals for, 230–231, 231*t*
    small-animal, 355–357, 355*t*
    spatial resolution in, 78, 78*t*, 232, 232*t*
    SPECT *vs.*, 77–78, 79*t*, 229
    tumor, 598
    two-dimensional *vs.* three-dimensional, 233, 233*f*
    USP monographs on, 390, 392*t*
positronium, 33
posting requirements, 119
postlabeling approach, 186, 187*f*, 664–665

potassium iodide, for thyroid protection, 448, 448t
potentiometers, 66
PPE. See personnel protective equipment
PPi, ^{99m}Tc. See technetium Tc 99m pyrophosphate
PPV. See positive predictive value
practice guidelines, APhA, 266, 374
precision, 79
preclinical studies, 382
predictive value, 484–485
pregnancy
    occupational exposure in, 117–118
    radiation effects in, 142–144
preimplantation period, 142
prelabeling approach, 186, 187f, 665, 666f
preparative manipulations, common problems associated with, 321–323
prescribed dosage, 396
prescribed dose, 396
preservatives, 295
pretargeted radioimmunotherapy, 683
primary engineering control (PEC), 313, 315–316
probes
    end-window, 263, 263f
    energy-compensated, 262–263, 264
    for Geiger-Müller detectors, 263, 263f
    pancake, 263, 263f
procurement, 266–267
prodromal phase, of acute syndrome, 138
programmed cell death, 673–674
proper supervision, 398–399
prophase, 130, 130f
proportional counters, 60
proportional region, 59, 60f
propyleneamine oxime technetium Tc 99m (^{99m}Tc-PnAO). See technetium Tc 99m propyleneamine oxime
propylthiouracil, 445
Prospective Investigation of Pulmonary Embolism Diagnosis (PIOPED), 509–513
ProstaScint. See indium In 111 capromab pendetide
prostate cancer
    indium In 111 capromab pendetide for diagnosis of, 218, 666, 667–668, 668f
    reimbursable PET procedures in, 239t
protection, 92–97. See also radiation safety
    agencies involved in, 109–110
    distance and, 96–97, 97t, 114
    dose system used in, 91f
    exposure time and, 96, 96t, 114
    of human research subjects, 397
    shielding and, 97, 98f, 98t, 114
protective techniques, 266, 266t
protein iodination, 210, 210f, 212, 212t
protein-losing enteropathy (PLE), 528
    in vitro procedure for, 528
    scintigraphic imaging of, 528, 529f
protein-mediated transport, 163, 165–166
proton(s), 13, 147–148
    binding energy of, 16–17, 17f
    energy levels of, 15–16, 16f
    ratio with neutrons, 17–18
        in electron capture decay, 22–23
        in negatron decay, 18, 19, 20f
        in nuclear reaction, 31
        in positron decay, 18, 21

proto-oncogenes, 141
proximal convoluted tubule, 547
PTH. See parathyroid hormone
Public Health Services Act of 1944, 376
pulmonary artery, 499–500, 500f
pulmonary circulation, 499–501, 500f. See also lung perfusion studies
pulmonary embolism (PE), 499
    blood flow effects of, 501
    computed tomography pulmonary angiography of, 511–513
    COPD vs., 509–513, 512f, 513f
    helical CT of, 508
    ventilation–perfusion imaging of, 504–505, 508–513, 511f, 513f
pulmonary nodule, solitary
    PET of, reimbursable, 626
    PET/CT of, 629–634, 630f
pulmonary osteoarthropathy, hypertrophic, 582, 585f
pulmonary vein, 499–500
pulse-height analyzer (PHA), 66–67, 67f, 76
pulse-height energy spectrum, 66–70, 68f, 69f, 70f, 71f, 72f
pulse-height spectrometry, 68–70
pump and damper, in HPLC, 278–279
Purdue University, 8, 9, 257
Pure Food and Drug Act of 1906, 375
purine neurotransmitters, 424
purity, chemical, 293
    in ^{99m}Tc generator, 38–39, 39f
purity, enantiomeric, 293
purity, radiochemical, 275–284, 275–285
    in ^{99m}Tc generator, 38–39, 39f
    in breast-milk excretion, 331
    corrective actions for, 283
    counting instruments for, 284
    definition of, 284
    identification methods for, 275–284, 288t–291t
        gas chromatography, 275–278, 277f
        high-pressure liquid chromatography, 275, 278–279, 278f
        paper chromatography, 275, 279–283
        reference standard, 275
        solid-phase extraction, 275, 279–280, 280f
        thin-layer chromatography, 275, 279–283, 292f
purity, radionuclidic, 274, 275
    in ^{99m}Tc generator, 38–39, 38f, 39f, 275, 277f
    of commonly used radionuclides, 276t
    definition of, 275
pyloric sphincter, 542
PYP, ^{99m}Tc. See technetium Tc 99m pyrophosphate
pyrogenic meningitis, 334t
pyrogens, testing for, 295, 296–299, 308, 315
pyrophosphate technetium Tc 99m. See technetium Tc 99m pyrophosphate
pyrophosphoric acid, 222

Q

QC. See quality control
QSARs. See quantitative structure–activity relationships

Quadramet. See samarium Sm 153 lexidronam
quality assurance
    in nuclear pharmacy, 268–269
    in PET, 252–253, 305–309
        acceptance criteria, 308–309
        complaint handling, 309
        control of components, containers, and closures, 307
        distribution, 309
        execution of, 306
        facilities and equipment for, 306–307
        finished drug product controls, 308–309
        labeling and packaging, 309
        laboratory controls, 308
        oversight of, 306
        personnel resources for, 306
        production and process controls, 307–308
        records, 309
        stability testing, 308
quality control (QC)
    dose calibrators, 263, 285–293, 299–304
    instrument, 299–304
    nuclear pharmacy, 260, 273–309
    PET, 274, 306
    radiopharmaceuticals, 273–299
        biologic considerations in, 295–299
        chemical considerations in, 293–294
        definition of, 273
        pharmaceutical considerations in, 294–295
        radiation considerations in, 274–293
        sources on information on, 273–274
    transport of radioactive material, 124
quality factors (Q), 89
quantitative structure–activity relationships (QSARs), 346
quenching, in liquid scintillation counter, 73–74, 75f

R

R. See roentgen
^{226}Ra. See radium 226
rabbit test, for pyrogens, 296–297
racemate/racemic mixture, 166
^{223}Ra-chloride. See radium Ra 223 chloride
raclopride C 11 (^{11}C-raclopride, ^{11}C-RAC)
    for brain imaging, 425
    formulation of, 251
    for neuroreceptor binding assessment, 238, 238f
    for PET, 234t, 237–238, 251
    radiochemical purity of, 286t
    radiosynthesis of, 249
rad. See radiation absorbed dose
radiation. See also specific types
    annihilation, 21, 21f
    dose of, 4
    electromagnetic, 17, 47, 158. See also gamma rays; x-ray(s)
    interactions with matter, 47–57

optical, 15, 15f
particulate, 17, 47, 158. *See also* alpha particle(s); beta particles
properties of, 48t
protection from. *See* protection
risk of. *See* risk
radiation absorbed dose (rad), 4, 88
radiation absorbed dose, in children, 325, 325t
radiation absorption, in air and water, 95, 95t
radiation biology, 127–145. *See also* Biologic effect(s)
radiation detection, 59–85
  counting, 78–85
    chi-square test of, 84–85, 85t
    efficiency in, 81–82
    errors in, 78–79, 80–81, 81t
    instrumentation for, 71–78
    statistics in, 78–80, 80f, 80t
  in imaging systems, 74–78
  instrumentation for, 71–85, 262–265
  ion collection methods of, 59–63
  maximum detectable activity in, 82–83, 84f
  minimum detectable activity in, 83–84
  in nuclear pharmacy, 262–265
  resolving time in, 82–83
  scintillation methods of, 63–71
radiation dose method, 675–676
radiation dosimetry, 90, 91–92
radiation effects
  as adverse reactions, 338
  biologic, 87, 127–145
    acute, 138–139
    carcinogenic, 98, 99–102, 139–141
    gastrointestinal, 135, 138–139
    hereditary (genetic), 98, 99, 102–104, 131–132, 141–142
    measurement of, 88
  deterministic, 98–99, 99f, 110, 127–128, 128f, 334t
  direct, 128–129, 128f
  indirect, 128–129, 128f
  in pregnancy, 142–144
  stochastic, 98–99, 99f, 110, 128, 128f
radiation measurement, 4, 25–28, 87–91
radiation safety, 7–8, 109–126
  agencies involved in, 109–110
  area monitoring for, 118–119
  caution signs for, 119, 119f
  definition of, 109
  dose limits in
    annual, 111, 111t
    occupational, 111
    volunteer subjects, 111–112
  emergency procedures in, 125–126
  federal regulations on, 109. *See also* specific agencies and regulations
  nuclear pharmacy, 270
  personnel monitoring for, 114–118
  shipping, 120–125
radiation safety committee (RSC), 398
radiation safety officer (RSO), 110, 112, 396, 399, 405
radiation safety program, 110
radiation synovectomy, 222–223, 673, 687–688, 688t
radiation-weighting factor, 88, 88t
radiation workers. *See* health care personnel
radioactive decay, 13, 17–30
  alpha particle, 24, 25f
  calculations of, 28–30
  definition of, 17
  diagnostic radiopharmaceuticals, 158–159
  dual-process, 21
  electron capture, 22–23, 22f, 158–159, 229
  negatron, 18, 19, 20f, 22–23, 51
  positron, 18, 21–23, 21f, 22f, 229–230, 231f
  rate of (activity), 25–28
  in storage, 125, 402, 406
radioactive decay law, 25
Radioactive Drug Research Committee (RDRC), 111, 358–359, 358t, 382–384
radioactive iodine uptake (RAIU) test, 445, 451–452, 695
  interpretation of, 451–452, 451f
  procedure for, 451
  rationale for, 451
radioactive material
  definition of, 121
  NRC monitoring requirements for, 120, 120t
  package labels for, 120, 122–123, 123f, 123t
  packaging requirements for, 121–122
  receipt of, 112–113, 120
  separation distance, from crew or handlers, 123, 123t
  threshold values for, 121, 121t
  transport of, 112–113, 120–125, 270, 271
radioactive material licenses, 394–408
radioactive waste
  disposal of, 125
  management in nuclear pharmacy, 260, 260f
  transfer to authorized recipient, 125
radioactivity, 25–30
  assessment of, 285–293
  calculations of, 28–30
  discovery of, 25
  dosage of, 4
radioactivity calibration, 37–38, 38f
radioaerosols
  components of device, 507f
  for lung ventilation studies, 507–508
radiobromination, 347
radiochemical, definition of, 147
radiochemical identity, 275–284. *See also* radiochemical purity
radiochemical purity, 275–285
  in ^{99m}Tc generator, 38–39, 39f
  in breast-milk excretion, 331
  corrective actions for, 283
  counting instruments for, 284
  definition of, 284
  identification methods for, 275–284, 288t–291t
    gas chromatography, 275–278, 277f
    high-pressure liquid chromatography, 275, 278–279, 278f
    paper chromatography, 275, 279–283
  reference standard, 275
  solid-phase extraction, 275, 279–280, 280f
  thin-layer chromatography, 275, 279–283, 292f
radiochemists, 4
radiochromatography. *See* chromatography
radiocolloids. *See also* specific colloids
  adverse reactions to, 337, 522
  for liver imaging, 2–3, 520–522, 520t, 521f, 529–530
  for lymphoscintigraphy, 607–608
  for radiation synovectomy, 687
  for spleen imaging, 526, 529–530
radiohalogenation, 347
radioimmunodiagnosis (RID), 655
  radionuclides for, 662, 662t
radioimmunotherapy (RIT), 3–4, 655
  antibody considerations in, 674–675
  dosing methods in, 675–676
  methods of, 675–676
  for non-Hodgkin's lymphoma, 3–4, 673–683
  pretargeted, 683
  radionuclides for, 662, 662t, 674–675
radioiodinated iodine I 131 serum albumin (RISA, ^{131}I-HSA), 426
radioiodination, 157–158, 210–213, 347
  antibody, 662–663, 674
  electrophilic substitution for, 210, 210f, 662, 662f
  impurities in, 293
  isotope exchange for, 210–211, 211f
  oxidants for, 211–213, 662, 663f
  protein, 210, 210f, 212, 212t
  reagents for, 211–213, 662
  specific activities of, 161–162, 162t
  tyrosine, 210, 210f
radioiodine(s). *See also* specific iodines
  adverse reactions to, 336–337
  biologic properties of, 447–449
  chemistry of, 207–213
  distribution in body, 447, 447f
  historical perspectives on, 446–447
  impurities in, 320
  labeling with, 210–213
  license for use, 456
  mass of, 336, 337t
  pH of, 294
  physical properties of, 207, 208t
  precautions in handling of, 157–158, 158t
  production of, 208–210
  radiation absorbed dose of, 449, 449t
  redox reactions in, 157–158
  for thyroid studies, 447–453
radioiodine therapy, 453–456
  procedures in, 454–456
  rationale for, 453–454
  safety considerations in, 456, 456t
radiolabeling, 160–163
  antibody, 186, 187f, 661–665, 674, 674f
    radioiodine, 662–663, 662f, 663f, 664f, 674
    radionuclide considerations in, 662, 662t
  bifunctional chelating agents for, 179, 181
    for antibodies, 186, 187f, 664–665, 664f, 665f

for peptides, 186, 187f, 656,
   668–669
 for technetium-nonessential
   compounds, 185–186, 186f
chemistry of, 346–348
choice of radiolabel, 346
indium In 111, 160, 186, 661–662,
   662t, 664–665, 676, 677t
isotopic, 160–161, 161t
molecular location of radiolabel, 348
nonisotopic, 160, 161
peptide, 186, 187f, 656, 668–669
PET radionuclides, 160, 244–246,
   244t
red blood cell, 700–701
strategy for, 346–348
technetium, 186, 186f
techniques of, requirements for,
   347–348
yttrium Y 90, 160, 186, 661–662, 662t,
   664–665, 674, 674f, 676, 677t
radiolytic decomposition, 323–324
radiolytic oxidants, 320–321
radiolytic reduction, 323
radionuclide(s), 14, 148
 chemistry of, 158–159
 half-life of, 28, 28f
 impurities in, 34
 line of nuclear stability and, 17–18, 18f
 in nuclear medicine, 26t–27t
 production of, 31–45
   accelerator methods of, 34,
     242–244
   automated modular systems for,
     250–251
   cyclotrons for, 33–34, 33f, 209,
     215, 220, 241, 242–244, 243f,
     244t
   generators for. See radionuclide
     generators
   linear accelerators for, 33–34, 33f
   nuclear reactors for, 31–32, 32f
   for PET, 241–251
   remotely operated systems for, 250
   robotic systems for, 251
 promising, clinical research initiatives
   on, 359–362, 360t, 361t
 traditional and nontraditional, 346,
   347t
radionuclide dose calibrators, 60–63
 adjustment of measurements in, 61–63,
   62f, 63f, 64t
 components of, 60–61, 60f
 copper filter in, 62–63, 62f, 63f, 64t
 primary standards for, 61
 secondary standards for, 61
 sensitivity of, 61, 61f
 syringe volume in, 63, 64t
radionuclide dose method, 675–676
radionuclide generators, 31, 34–45, 35t,
   241–242, 242t
 ^{68}Ge–^{68}Ga, 45, 242, 242t
 ^{82}Sr–^{82}Rb, 34, 41–42, 42f, 44, 207,
   241–242, 242f, 242t
 ^{90}Sr–^{90}Y, 45
 ^{113}Sn–^{113m}In, 44
 ^{188}W–^{188}Re, 44, 45
 disposal of, 44
 physics of, 39–44

record keeping on, 267–268, 267f
secular equilibrium, 41–42, 42f, 44
storage of, 260
technetium Tc 99m, 23, 34–44,
   171–172
transient equilibrium, 40–41, 40f, 44
radionuclide identity, 274–275
 by half-life, 274–275
 by radiation type, 274, 274f
radionuclidic purity, 274, 275
 in ^{99m}Tc generator, 38–39, 38f, 39f,
   275, 277f
 of commonly used radionuclides, 276t
 definition of, 275
radiopharmaceutical(s), 1–2. See also specific
   agents
 adverse reactions to, 333–339
 appearance and color of, 294, 294t
 chemical and physical forms of, 4, 4t,
   147
 chemistry of, 147–169
 commercialization of, 359
 definition of, 1, 147
 diagnostic, 158–160
 dosage forms of, 4–5, 4t
 first-generation, 168–169
 formulation problems for, 319–324
 half-lives of, 28, 28f, 147
 ideal properties of, 158–160
 in vivo localization of, 163–168
 iodinated, 213–215
 mass and toxicity of, 5–7, 147
 in nuclear medicine, 5t–7t, 147
 on-site production of, 4
 osmolality of, 295
 particle number and size of, 294, 294t,
   295f
 perspective on use, 4–8
 pH of, 294–295
 quality control of, 273–299
   biologic considerations in,
     295–299
   chemical considerations in,
     293–294
   definition of, 273
   pharmaceutical considerations in,
     294–295
   radiation considerations in,
     274–293
   sources of information on,
     273–274
 regulation of. See regulatory control
 requirements for safe use, 7–8
 routes of administration, 4–5, 4t
 second-generation, 169
 stabilizers and preservatives in, 295
 technetium, 178–179
 therapeutic, 2, 158, 160
 unique properties of, 5, 147
radiopharmaceutical development, 168–169,
   178–179, 343–362
 autoradiography in, 348, 349t,
   350–353
 clinical research initiatives in, 359–362,
   360t, 361t
 commercialization in, 359
 drug development vs., 345–346, 346t
 ex vivo tissue biodistribution in,
   353–355

flow chart of, 344–345, 344f
identification of lead structures in,
   345–346
in vitro binding assays in, 348–350,
   349t
molecular imaging concept and,
   343–344
as multidisciplinary effort, 343
preparation for human use in, 357–362,
   382–384
radiolabeling strategy in, 346–348
regulatory process in, 357–359, 358f,
   375, 376–377, 381–394
selection of imaging target in, 345
small-animal imaging in, 355–357
target and lead development in,
   344–346
radiopharmacists, 4
radiosensitivity, 89, 135–137
 cell cycle effects on, 137, 137f
 dose rate and, 133, 134f
 linear energy transfer and, 134f, 136
 oxygen effects on, 136–137
 relative, of cells, 135, 136t
radiotherapy, 673–691. See also specific agents
 for bone metastases or pain, 673,
   683–686
 for effusions, 673, 687
 for liver cancer, 688–690
 new agents for, 690–691
 for non-Hodgkin's lymphoma,
   673–683
 for polycythemia vera, 673, 686–687
 pretargeted, 683
 for synovectomy, 673, 687–688
radio waves, 53, 53t
radium, early use of, 1
radium 226 (^{226}Ra)
 decay of, 24, 25f
 decay rate based on, 25
radium jaw, 573
radium paint, 572–573
radium Ra 223 chloride (^{223}Ra-chloride)
 for bone pain palliation, 686
 clinical trials with, 690t
radon 222 (^{222}Rn), 24, 25f
radon gas, exposure to, 112
RAIU. See radioactive iodine uptake test
random coincidence events, 77
random errors, 78–79
random events, in PET, 232, 232t
rare earths, 224
^{82}Rb. See rubidium Rb 82
RBCs. See red blood cell(s)
RBE. See relative biologic effectiveness
R configuration, 167–168, 168f
RDRC. See Radioactive Drug Research
 Committee
^{186}Re. See rhenium Re 186
^{188}Re. See rhenium Re 188
reabsorption, renal, 548–549
reactions, adverse. See adverse reactions
reactive fluoride fluorine F 18, 244t,
   245–246, 246f
reactors, nuclear, 31–32, 32f
reagents, for radioiodination, 211–213, 662
reassortment, in radiobiology, 137
recall system, 309
receipt of radioactive material, 112–113, 120

recentness of professional training, 400
recertification, in nuclear pharmacy, 11
recipient, authorized, transfer to, 125
recoil energy, 24
recombinant thyroid-stimulating hormone (rhTSH), 455–456
recombination region, 59, 60f
record(s)
    maintaining applicable, 397
    nuclear pharmacy, 271, 271t
        compounding, 267–268, 267f, 268f
        computerized, 259–260, 259f
        generator, 267–268, 267f
    PET quality insurance, 309
    radiation safety, 110
    radioiodine therapy, 456t
    regulatory requirements for, 406–407
rectilinear scanners, 2, 74–76, 75f
red blood cell(s), 695
    cohort and random labeling of, 700–701
    technetium Tc 99m-labeled, 196–197
        activity in breast milk, 331
        excessive technetium Tc 99 in, 320
        excessive technetium Tc 99m in, 320
        for gastrointestinal imaging, 517
        for GI bleeding detection, 527–528, 527t, 539, 541f
        for heart imaging, 459, 460t, 471
        heat-denatured, for spleen imaging, 526, 526f, 529, 532f
        in vitro method for, 196–197
        in vivo method for, 197
        labeling mechanism of, 197
        mixing order for, 321
        modified in vivo method for, 197
        radiochemical purity of, 287t, 289t
        red blood cell concentration in, 322
red blood cell survival, 700–701, 701f
red blood cell volume, 158, 221, 695, 696–697, 697f, 698t, 699–700
red marrow, tissue-weighting factor for, 104t
redox (oxidation–reduction) reactions, 155–158
red pulp, of spleen, 525
reducing agent, 156t
    for technetium, 173–174, 175–176, 176f
reduction, 156, 156t. See also oxidation–reduction reactions
reductive decomposition, 323
reference standard (RS), 275
R_eff. See effective rate of loss
refrigeration, 260, 265
regadenoson
    chemical structure of, 462f
    for myocardial perfusion imaging, 462, 464–466, 478f
    precautions, adverse effects, contraindications, 464–466, 465t
regional cerebral blood flow (rCBS). See cerebral blood flow
regional mobile medical service, 402, 406

regulatory control, 373–409
    authority of FDA and NRC, 377–378
    federal vs. state, 373, 377
    history of, 375–377
    outlook for, 408–409
    overlap and inconsistencies in, 389, 390f
    PET, 253, 274, 305–306, 315, 378–381, 391, 408–409
    policy statement on, NRC, 377–378
    products and practice environments, 373–374
    radiopharmaceutical development, 357–359, 358f, 375, 376–377
    radiopharmaceuticals in breast milk, 330–331
regulatory tasks, daily, 270, 270f
reimbursable procedures, PET, 238–241, 239t, 253, 379–381, 617, 626
relative biologic effectiveness (RBE), 88
relative front, in chromatography, 279–280, 279f
relative risk model, 99
release, criteria for, 402
release of effluent, 125
release to sanitary sewer, 125
rem. See roentgen equivalent man
remotely operated systems, for radionuclide production, 250
renal artery stenosis, 557–558, 558f, 560, 561f
renal cell carcinoma, 562, 565f
renal clearance, 548–549, 550t
renal columns of Bertin, 547, 548f
renal excretion, 447–448, 548–551
renal imaging. See kidney imaging
renal pelvis, 547, 548f
renal plasma flow (RPF), 549–550, 551, 555
renal transplant, evaluation of, 562, 563f
renin–angiotensin system, 556–557, 557f
renography, 2, 552, 556–558
    captopril, 556–558, 558f, 560, 561f
    diuresis, 556, 557f, 560–562, 562f
    interpretation of, 559–562
    normal, 556, 556f, 559–560, 559f, 560f
    for transplant evaluation, 562, 563f
renovascular hypertension (RVH), 557–558, 558f
reoxygenation, in radiobiology, 137–138
repair, in radiobiology, 137
replication, DNA, 129–130
repopulation, in radiobiology, 137
reporting systems, for adverse reactions, 334, 335t
report requirements, 407
RES. See reticuloendothelial system
research subjects, dose limits for, 111–112
reserpine, 601
residual solvents, 293–294
resin purification, 245–246, 246f
resolving time, 82–83
respiration, 499, 500f. See also lung ventilation studies
respiratory zone, 499, 500f
resting blood flow, myocardial, 460–461
restricted area, 258f, 259–260
reticuloendothelial system (RES), 517. See also liver; spleen

review process, for new drugs/radiopharmaceuticals, 391–394
rhenium Re 186 (^{186}Re), 27t
    labeling with, 662t, 674
    shipping guidelines for, 122t
rhenium Re 186 etidronate (^{186}Re-HEDP), for bone pain palliation, 684, 684t, 686
rhenium Re 186 sulfide (^{186}Re-sulfide), for radiation synovectomy, 687–688, 688t
rhenium Re 188 (^{188}Re)
    ^{188}W–^{188}Re generator, 44, 45
    labeling with, 662t
    shipping guidelines for, 122t
rhTSH. See recombinant thyroid-stimulating hormone
Richards, Powell, 171
RID. See radioimmunodiagnosis
right coronary artery, 460, 460f
right-to-left shunt evaluation, 513, 515f
ring chromosome, 131, 132f
risk, 98–105
    definition of, 98
    key questions on, 98
    radiation, vs. other risks, 104–105
risk assessment, 99
risk estimates, 99
    for cancer, 99, 100–102, 104, 128
    for hereditary disorders, 99, 103–104, 103t
risk models
    for cancer, 99
    for hereditary disorders, 102–104
RIT. See radioimmunotherapy
rituximab (Rituxan), 673, 677–678, 682
^{222}Rn. See radon 222
RNV. See equilibrium radionuclide ventriculography
Robert A. Taft Sanitary Engineering Center, 9
robotic systems, for radionuclide production, 251
roentgen (R), 87–88, 90
roentgen equivalent man (rem), 87, 88
Roosevelt, Theodore, 375
rose bengal, iodine I 131-labeled, for hepatobiliary imaging, 522
routes of administration, 4–5, 4t
RPF. See renal plasma flow
RSC. See radiation safety committee
RSO. See radiation safety officer
rubidium chloride Rb 82
    for brain imaging, 411, 416
    extraction and retention of, 466f
    for heart imaging, 459, 460t, 467t, 470, 471
    interpretation of findings with, 479, 481f, 482f, 483f
    PET/CT, 648
    protocols for, 477–479, 479t
    for PET, 234–235, 234t
    blood flow, 234–235
    reimbursable procedures, 239–240, 239t
rubidium Rb 82 (^{82}Rb), 6t, 26t
    ^{82}Sr–^{82}Rb generator, 34, 41–42, 42f, 44, 207, 241–242, 242f, 242t
    decay of, 21, 22f
    half-life of, 207
    labeling with, 160

for PET, 207, 230*t*, 234–235, 234*t*, 378
radionuclidic purity of, 276*t*
shipping guidelines for, 122*t*
Rutherford, Ernest, 25, 28, 51

S

^{35}S. *See* sulfur 35
SA. *See* specific activity
SAAC. *See* single amino acid chelate
safety, intrathecal injection, 427–428
safety, nuclear pharmacy, 270
safety, radiation, 7–8, 109–126
    agencies involved in, 109–110
    area monitoring for, 118–119
    caution signs for, 119, 119*f*
    definition of, 109
    dose limits in
        annual, 111, 111*t*
        occupational, 111
        volunteer subjects, 111–112
    emergency procedures in, 125–126
    federal regulations on, 109. *See also specific agencies and regulations*
    personnel monitoring for, 114–118
    shipping, 120–125
safety, radioiodine therapy, 456, 456*t*
sagittal plane, 3, 3*f*
sagittal sinus, superior, 414–415, 436
Sakasitz, Richard, 9, 9*t*
salivary glands
    FDG uptake in, 622, 624*f*
    radioiodide secretion by, 448
samarium
    chemistry of, 224–225
    electronic configuration of, 224
samarium Sm 153 (^{153}Sm), 6*t*, 27*t*
    applications of, 207
    chemical structure of, 224, 224*f*
    decay of, 224, 225*f*
    labeling with, 160
    production of, 224
    shipping guidelines for, 121*t*, 122*t*
samarium Sm 153 lexidronam (^{153}Sm-lexidronam), 224–225
    for bone pain palliation, 684, 684*t*, 685–686
    indications for, 225
    radiochemical purity of, 286*t*
sample injector port
    in gas chromatography, 277
    in high-pressure liquid chromatography, 279
Sanchez, Robert, 9, 9*t*
Sandostatin. *See* octreotide
sanitary sewer, release to, 125
SAR. *See* structure–activity relationship
sarcoma, renal, 567*f*
SCA. *See* single-channel analyzer
scan(s), rectilinear, 74–76
scanner, radiochromatogram, 279, 280*f*, 284
scatter, in PET, 231–232, 232*f*, 232*t*
scattered coincidence events, 77
scatter reactions, neutron, 51–52
scintillation counters, 63–71
    liquid, 63–65, 71, 73–74, 74*f*, 75*f*
    maximum detectable activity for, 82–83

minimum detectable activity for, 83–84
in nuclear pharmacy, 262–263, 264–265
photopeak energy resolution in, 71
pulse-height analyzer in, 66–67, 67*f*
pulse-height energy spectrum in, 66–70, 68*f*, 69*f*, 70*f*, 71*f*, 72*f*
sample counting in, 71–74
solid-crystal, 63–66
scintillation crystals, for PET, 230–231, 231*t*
scintillation probe, 71, 73, 73*f*
scintillation well counter, 71, 72, 73*f*, 264–265, 265*f*, 284
S configuration, 167–168, 168*f*
scrotal scintigraphy, 559, 568, 568*f*
SDE. *See* shallow-dose equivalent
SDR. *See* structure–distribution relationship
^{75}Se. *See* selenium Se 75
sealed source, 396, 401, 406
sealed-source monitoring, 118–119
second-generation radiopharmaceuticals, 169
Section on Nuclear Pharmacy, APhA, 10
secular equilibrium generators, 41–42, 42*f*, 44
sedimentation, of particles, 324
seizures (epilepsy)
    PET/CT of, 629, 648–649, 650*f*
    SPECT of, 431–432, 432*f*
selective internal radiation therapy (SIRT), 688
selenium Se 75 (^{75}Se), diffusion in brain, 164–165, 165*f*
sensitivity
    of dose calibrators, 61, 61*f*
    minimum, 83
sentinel node(s)
    concept of, 606
    in melanoma, 606–607
separate double-strand break, in DNA, 131, 131*f*
Sep-Pak cartridge, 279–280, 280*f*
sequence rule, 168
serotonin, 424, 589
sestamibi technetium Tc 99m. *See* technetium Tc 99m sestamibi
shallow-dose equivalent (SDE, Hs), 91
Shaw, Stanley, 9, 9*t*
shells, electron, 14–15, 15*f*, 148–149, 152*f*
shielding, 97, 98*f*, 98*t*, 114
    emergency, 125–126
    in nuclear pharmacy, 260–261, 261*t*
    sleeves, for dose calibrator testing, 301–302, 301*f*, 302*f*
    syringe, 112–113, 112*f*, 113*f*, 260, 260*f*, 261, 261*f*
    for syringes, 112–113, 112*f*, 113*f*, 260, 260*f*, 261*f*
    for vials, 261
shin splints, 581–582, 581*f*
shipping of radioactive material, 120–125
    accident in, reporting of, 124
    liquid, absorbent material for, 124
    from nuclear pharmacy, 260, 270, 271
    package labeling for, 120, 122–123, 123*f*, 123*t*
    package markings for, 122
    package monitoring in, 120, 120*t*, 123
    packaging requirements for, 121–122
    personnel training for, 124

placarding of vehicle, 123–124, 124*f*
quality control measures for, 124
separation distance in, 123, 123*t*
shipping papers for, 124
threshold values for, 121, 121*t*
shrink-wrap, 260
shunt evaluation, 436, 438–440, 439*f*, 440*f*
SIB. *See* N-succinimidyl 3- or 4-^{131}I-iodobenzoate
side effects. *See* adverse reactions
side-window Geiger-Müller detectors, 63
sievert (Sv), 87, 88, 90
sigma (σ) bonds, 152–153, 152*f*
sigmoid cell survival curve, 133, 133*f*
signs, caution, 119, 119*f*
sincalide, for hepatobiliary imaging, 524–525, 532–533
single amino acid chelate (SAAC), 182, 182*f*
single carbon bonds, 152–153, 152*f*
single-channel analyzer (SCA), 68
single-dose therapy, 137
single-photon emission computed tomography (SPECT), 3, 71, 74, 76, 76*f*
    brain, 165, 411–412. *See also* brain imaging
    clinical research initiatives in, 359–362, 360*t*
    coincidence sum peak in, 70
    computed tomography with (SPECT/CT), 74
    heart. *See also* heart imaging
        cardiac neuronal, 490–492
        perfusion, 459, 460*t*, 467–470, 467*t*
        risk stratification in, 492–494, 493*f*, 493*t*
    PET *vs.*, 77–78, 79*t*, 229
    as preeminent molecular imaging method, 344
    radionuclides used in, 207
    small-animal, 355–357, 355*t*
    total-body, 589–612
        bone marrow, 611–612, 612*f*
        infection, 589–596
        tumor, 596–605
single-strand break, in DNA, 131, 131*f*
sinusoidal cells, hepatic, 517–519, 518*f*
SIR-Spheres. *See* yttrium Y 90-labeled resin microspheres
SIRT. *See* selective internal radiation therapy
skin cancer, radiation-induced, 140
skin erythema, 87, 127, 128
^{153}Sm. *See* samarium Sm 153
small-animal imaging, 355–357
    advantages of, 356
    limitations of, 356–357
small bowel, FDG uptake in, 625
small-cell lung cancer
    PET/CT of, 629–634, 633*f*
    reimbursable PET procedures in, 239*t*
Smith, Anne C., 9*t*
^{113}Sn. *See* tin Sn 113 (^{113}Sn)–indium 113m (^{113m}In) generator
^{117m}Sn. *See* tin Sn 117m
SNM. *See* Society of Nuclear Medicine
Society of Non-Invasive Imaging in Drug Development, 254
Society of Nuclear Medicine (SNM), 1, 92, 239, 334, 343, 345, 381

sodium acetate C 11 (^{11}C-sodium acetate)
    for heart imaging, 237, 459, 460t, 473
    off-label uses of, 237, 237t
    for oxidative metabolism measurement, 236–237, 237f
    for PET, 234t, 236–237, 248
    radiochemical purity of, 286t
    radiosynthesis of, 248, 248f
sodium borohydride, as reducing agent, 174
sodium chromate Cr 51 (^{51}Cr-sodium chromate), 221
    indications for, 221
    pediatric dosing of, 327
    radiochemical purity of, 286t
    for red blood cell–plasma measurement, combined, 699–700
    for red blood cell survival measurement, 700–701, 701f
    for red blood cell volume measurement, 696–697, 697f, 698t
sodium dithionite, as reducing agent, 174
sodium fluoride F 18
    for bone imaging, 573, 576f, 578, 579f, 580
        nononcologic, 646, 648f
        patient preparation for, 646
        PET/CT, 644–646, 647f
    for osteogenic activity measurement, 236
    for PET, 234t, 236
    radiochemical purity of, 286t
sodium iodide I 123 (^{123}I-sodium iodide)
    capsules or solutions, 213–214, 213t, 286t
    excretion in breast milk, 331
    pediatric dosing of, 325t
    radiochemical purity of, 286t
    for thyroid studies, 450t
sodium iodide I 131 (^{131}I-sodium iodide)
    activity in breast milk, 331–332
    adverse reactions to, 338
    appearance and color of, 294t
    biologic properties of, 447–449
    capsules or solutions, 213–214, 213t
    distribution in body, 447, 447f
    dose calibrator measurements of, 293
    excretion in breast milk, 448–449
    ionic bonding of, 149
    precautions in handling of, 157–158, 158t
    redox reactions in, 157
    relative bioavailability of, 214
    shipping of, 123
    specific activity of, 161–162
    substrate specificity of, 169
    for thyroid studies, 447–453, 450t
        radioactive iodine uptake test, 451–452, 451f
        thyroid scan, 452–453
    for thyroid therapy, 453–456
sodium iodide (NaI) scintillation counters, 65–66, 65f, 66f
    in rectilinear scanners, 74–76, 75f
    in scintillation probe, 73, 73f
    in scintillation well counter, 72, 73f
    in SPECT, 76
sodium Na 24 (^{24}Na), early use of, 1

sodium pertechnetate Tc 99m (^{99m}Tc-sodium pertechnate), 171–172, 172f, 187
    for brain imaging, 411, 416–417
    coordination geometry of, 155f
    for dacryoscintigraphy, 610, 611f
    excretion in breast milk, 416
    for first-pass radionuclide angiocardiography, 481–484, 486f
    as free technetium Tc 99m, 285
    impurities of, 284, 284f, 285, 293
    ionic bonding of, 149
    for Meckel's diverticulum imaging, 528, 539–541, 542f
    organ distribution of, 416, 417f
    oxidation state of, 184
    for parathyroid tumor imaging, 604–605, 605f
    pediatric dosing of, 325t
    radiochemical purity of, 287t, 289t
    for radionuclide cystography, 559
    radionuclidic purity of, 275
    for scrotal scintigraphy, 559, 568f
    for thyroid studies, 449–450, 452, 452f, 453f
sodium phosphate P 32 solution (^{32}P-sodium phosphate), 223
    appearance and color of, 294, 294t
    for bone pain palliation, 684, 684t
    indications for, 223
    for polycythemia vera, 686–687
    radiochemical purity of, 286t
sodium–potassium pump, 166
sodium trimetaphosphate, 222
soft-tissue sarcoma, reimbursable PET procedures in, 239t
soil, burial in, 125
Solazed. See iodine I 131 ioflubenzamide
solid-crystal scintillation counters, 63–66
solid-phase extraction, 275, 279–280, 280f
solid spills, 125–126
solid targets, of cyclotrons, 243–244
solitary pulmonary nodule
    PET of, reimbursable, 626
    PET/CT of, 629–634, 630f
Soloman, Arthur C., 9t
solution chemistry, of metal ions, 154–155
solvent front, in chromatography, 279, 279f
solvent reservoirs, in HPLC, 278
solvents, residual, 293–294
Solyom, Peter, 8
somatic cells, radiation damage to, 102, 141
somatostatin, 598, 598f
somatostatin receptor(s) (SSTRs), 598, 668
somatostatin-receptor imaging agents, 668
    chemical structure of, 598, 598f
    for neuroendocrine tumors, 598–600, 600f
    stability of, improving, 655–656
sources, of radiation exposure, 95–96, 112–114
spatial resolution, in PET, 78, 78t, 232, 232t
specialty certification, in nuclear pharmacy, 9–11, 10t, 11f, 266, 266t, 374, 405
Specialty Council on Nuclear Pharmacy, 10–11, 10t
specialty petition committee, 10, 10t
specific activity, 161–162, 162t, 293
specific ionization (SI), 50–51
    for alpha particles, 50, 51f
    particle velocity and, 50–51, 52t

specific target binding, 350, 350f
SPECT. See single-photon emission computed tomography
spectrometer
    calibration and linearity of, 66–67
    pulse-height, 68–70
    solid-crystal scintillation, 65–66, 66f
S phase, 130–131, 130f, 137
sphincter of Oddi, 518f, 519–520, 532
spill(s)
    emergency procedures for, 125–126
    gas, clearance time for, 118
    major, procedure for, 126, 126t
    minor, procedure for, 126
spleen imaging, 517, 525–526, 529–530
    agents for, 517, 526
    heat-denatured ^{99m}Tc red blood cells for, 526, 526f
    physiologic principles of, 525
    procedure for, 530
    rationale for, 529
spots, wet vs. dry, in chromatography, 282–283, 283f, 283t
^{82}Sr. See strontium Sr 82 (^{82}Sr)–rubidium 82 (^{82}Rb) generator
^{85}Sr. See strontium Sr 85
^{87m}Sr. See strontium Sr 87m
^{89}Sr. See strontium Sr 89
^{90}Sr. See strontium Sr 90 (^{90}Sr)–yttrium 90 (^{90}Y) generators
SSTRs. See somatostatin receptor(s)
stability
    drug, common problems associated with, 323–324
    nuclear, line of, 17–18, 18f
stability testing, PET, 308
stabilizers, 295
stable nuclides, 14, 17–18, 18f, 148
standard deviation, 80–81, 81t
standard operating procedures (SOPs), 315–316
stannous chloride, in technetium preparation, 154, 156–157, 174
stannous ion, inadequate, 321
stannous pyrophosphate kits, 196, 196t, 222
state boards of pharmacy, 258, 265, 268, 269
state regulation, 373, 377
static imaging studies, 2–3, 3f
stationary phase, in paper and thin-layer chromatography, 279–280
stem cell(s), 656, 656f
stem cell transplantation, autologous, 675
stereochemistry, 166–168
stereogenic center, 168
stereoisomers, 166–168
sterile preparation area, 259
sterile techniques, 313–317
sterility testing, 295–296, 306, 308, 315
stochastic effects, 98–99, 99f, 110, 128, 128f
    cancer as, 98, 128, 139
stomach
    emptying of, studies of, 517, 527, 542–544
    FDG uptake in, 625
    regions of, 542
    uptake in bone scans, 574, 574f
storage
    decay in, 125, 402, 406
    in nuclear pharmacy, 260–261, 261t

STPP, technetium Tc 99m. *See* technetium Tc 99m tripolyphosphate
streaking, in chromatography, 282–283, 283*f*
stress, exercise, for myocardial perfusion imaging, 461–462, 474
stress, pharmacologic
   for myocardial perfusion imaging, 462–466, 474
      precautions, adverse effects, contraindications, 464–466, 465*t*
      protocols for, 465*t*
stress fractures, 581–582, 581*f*
stress test, acetazolamide, 435, 436*f*
strontium
   chemistry of, 223–224
   electronic configuration of, 223
strontium chloride Sr 89 (^{89}Sr-chloride), 223–224
   for bone pain palliation, 684–685, 684*t*
   indications for, 223–224
strontium Sr 82 (^{82}Sr)–rubidium 82 (^{82}Rb) generator, 34, 41–42, 42*f*, 44, 241–242, 242*f*, 242*t*
strontium Sr 85 (^{85}Sr), applications of, 223
strontium Sr 87m (^{87m}Sr), applications of, 223
strontium Sr 87m citrate (^{87m}Sr-citrate), for bone imaging, 573
strontium Sr 89 (^{89}Sr), 6*t*, 26*t*
   for bone imaging, 573
   decay of, 222*f*, 223
   production of, 223
   shipping guidelines for, 122*t*
strontium Sr 90 (^{90}Sr)–yttrium 90 (^{90}Y) generators, 45
structure–activity relationship (SAR), 166, 346, 348
structure–distribution relationship (SDR), 166
subacute toxicity tests, 382
Subcommittee on Human Applications, AEC, 375–376
substance oxidized, 156*t*
substance reduced, 156*t*
substrate(s), 163
substrate-specific tracers, 169
succimer technetium Tc 99m (^{99m}T-DMSA). *See* technetium Tc 99m succimer
N-succinimidyl 3- or 4-^{131}I-iodobenzoate (SIB), 663, 664*f*
sulesomab technetium Tc 99m. *See* technetium Tc 99m sulesomab
sulfur 35 (^{35}S), labeling with, 161*t*
sulfur colloid technetium Tc 99m. *See* technetium Tc 99m sulfur colloid
Summers, Larry, 8
sum peak, coincidence, 70
superior sagittal sinus, 414–415, 436
superscan, bone, 582, 584*f*
supervision, proper, 398–399
suppressor genes, 141
surface contamination, control of, 314–315
survey instruments, 304–305, 304*f*. *See also specific instruments*
   regulatory requirements on, 400–401, 406
survival, red blood cell, 700–701, 701*f*

survival curves, cell, 132–135, 133*f*
   dose fractionation and, 134, 134*f*, 138
   linear energy transfer and, 133, 134*f*
   linear–quadratic model of, 134*f*, 135, 140
Sv. *See* sievert
S values, 92, 93*t*–94*t*, 675
Swanson, Dennis P., 9*t*
Syllabus for Nuclear Pharmacy Training, 10
syn configuration, 168, 168*f*
Syncor International Corporation, 9
synovectomy, radiation, 222–223, 673, 687–688, 688*t*
syringe(s)
   shielding for, 112–113, 112*f*, 113*f*, 260, 260*f*, 261, 261*f*
   spent, disposal of, 260
systematic errors, 78–79

T

Taplin, George, 522
target and lead development, 344–346
target binding, specific and nonspecific, 350, 350*f*
target-binding affinity, control studies of, 349
target-binding selectivity, 350, 350*f*
target-blocking studies, 354–355
targeted therapy, with alpha particle emitters, 690, 690*t*
targeting molecule–linker–BFCA–^{99m}Tc, 185
target organ, 91–92, 163
target selection, 345
target-to-nontarget (T:NT) ratio, 348, 354
tartaric acid, isomers of, 166, 166*f*
Task Analysis of Nuclear Pharmacy Practice, 10
Tc. *See* technetium
^{90}Tc. *See* technetium Tc 90
^{94m}Tc. *See* technetium Tc 94m
^{97}Tc. *See* technetium Tc 97
^{99}Tc. *See* technetium Tc 99
^{99m}Tc. *See* technetium Tc 99m
^{110}Tc. *See* technetium Tc 110
tear drainage, dacryoscintigraphy of, 608–610, 611*f*
teboroxime technetium 99m. *See* technetium Tc 99m teboroxime
Technegas, 508
technetium (Tc), 171–202
   carrier, 42–43, 43*f*, 177–178
   chemistry of, 156–157, 172–178
   discovery of, 171
   electronic configuration of, 172, 173*t*
   half-life of, 171
   isotopes of, 171
   origin of name, 171
   oxidation states of, 156, 172–174, 173*t*, 174*t*
   radiolabeling approaches for, 186, 186*f*, 187*f*
   radiopharmaceuticals, 178–179. *See also specific types*
   redox reactions of, 156–157, 157*f*, 172–177
   reduced form of, 172–173
technetium compounds, 179–186
   coordination geometries of, 154, 154*t*, 155*f*
   cores of, 179–184, 180*f*
      [O = Tc = O]$^+$, 183, 183*f*
      [Tc]$^+$, 181, 181*f*
      [Tc]$^{3+}$, 182, 183
      [Tc(CO)$_3$]$^+$, 181–182, 181*f*
      [Tc]$^{4+}$, 182, 183
      [Tc-HYNIC], 184, 184*f*
      [Tc = O]$^{3+}$, 180*f*, 183
      [Tc ≡ N]$^{2+}$, 183*f*, 184
   Tc(I), 179–182
   Tc(III), 182–183
   Tc(IV), 182–183
   Tc(V), 183–184
   Tc(VII), 184
   technetium-essential, 179, 184–185
   technetium-nonessential, 184, 185–186
   technetium-tagged, 184, 185–186
technetium-essential compounds, 179, 184–185
technetium kit, 174–178, 267–268
   ancillary chelating agent for, 175
   antioxidant stabilizers for, 176–177, 177*f*, 177*t*
   coordinating ligand for, 174–175, 175*f*
   reducing agents for, 175–176, 176*f*
technetium-nonessential compounds, 184, 185–186
technetium-tagged compounds, 184, 185–186
technetium Tc 90 (^{90}Tc), 171
technetium Tc 94m (^{94m}Tc), as nontraditional radionuclide, 347*t*
technetium Tc 97 (^{97}Tc), 171
technetium Tc 99 (^{99}Tc), 26*t*
   discovery of, 171
   half-life of, 171
   shipping guidelines for, 122*t*
technetium Tc 99m (^{99m}Tc), 6*t*–7*t*, 26*t*
   for autoradiography, 352–353, 352*t*
   for bone imaging, 573–578
      adverse reactions to agents, 577
      altered biodistribution of agents, 578
      biologic properties of agents, 575–576
      development of agents, 573–574
      dosimetry of agents, 576–577, 577*t*
      impurities in agents, 319, 574, 574*f*
      localization mechanisms of agents, 576, 577*f*
      preparation of agents, 574–575, 574*f*, 575*f*
   for brain imaging, 411, 416
   for infection imaging, 593–596
   chemistry of, 169
   decay of, 28*f*, 171, 172*f*
   detection efficiency of, 159, 159*t*
   development of, 178–179
   diagnostic application of, 159, 171–172
   discovery of, 171
   dose calibrator measurements of, 303, 303*t*
   excessive, problem of, 320
   excretion in breast milk, 331–333
   first-generation, 174, 178
   free, 285
   half-life of, 4, 28*f*, 171

736　INDEX

impurities in, 34, 319–320
labeling with, 160, 346, 661–662, 662t, 665, 666f
media-fill procedure for, 314, 315f
for myocardial perfusion imaging, 467, 468t
 in dual-isotope protocols, 476–477, 479f
 interpretation of findings with, 479, 480f, 482f, 484f
 protocols for, 476, 477f, 478f
for myocardial viability imaging, 486–487
preparation of, 174–178. See also technetium kit
preservatives in, 295
production of, 4, 23, 34–44, 171–172
pulse-height energy spectrum of, 71f
radiation absorbed dose of, 449t
radiochemical purity of, 279–280, 284–285
radionuclidic purity of, 275, 276t
second-generation, 169, 174, 178–179
shipping guidelines for, 121t, 122t
substrate nonspecificity of, 169
substrate specificity of, 169
S values of, 92, 93t–94t
as traditional radionuclide, 346, 347t
technetium Tc 99m(CO)$_3$nitrilotriacetic acid [^{99m}Tc(CO)$_3$-NTA], for renal imaging, 569
technetium Tc 99m aggregated albumin (^{99m}Tc-MAA), 188–189
adverse reactions to, 337, 503–504
appearance and color of, 294, 294t
biologic properties of, 188, 502
for differential lung perfusion studies, 513, 514f
excessive technetium Tc 99 in, 320
excretion in breast milk, 331–333, 332f, 333t
indications for, 189
for liver imaging, 529, 531f
localization in lung, 502, 503f
for lung perfusion studies, 189, 337, 501–504, 503f
lung scanning dose of, 503–504
particles of, 188, 189f
 cerebral microembolization of, 504
 clearance from lung, 502, 503f
 hardness and composition of, 502
 number of, 294, 294t, 295f, 501–502, 504
 reduced concentration of, 504, 504t
 sedimentation of, 324
 size of, 294, 294t, 295f, 501, 502f, 504
 toxicity of, 503–504
pediatric dosing of, 325t, 327, 504, 504t
preparation of, 188, 188t, 501
radioaerosols and, 507
radiochemical purity of, 287t, 288t
for right-to-left shunt evaluation, 513, 515f
specific activity of, 162
thyroid uptake of, 504, 505f

for ventilation–perfusion lung scan, 508–509, 510f, 511f, 512f, 513f
technetium Tc 99m albumin (^{99m}Tc-HSA), 188
for CSF imaging, 426
for gastrointestinal imaging, 528, 529f
for lymphoscintigraphy, 607–608
radiochemical purity of, 287t
technetium Tc 99m albumin colloid, for liver imaging, 520, 520t
technetium Tc 99m annexin V (^{99m}Tc-annexin V), for myocardial infarction imaging, 489, 490t
technetium Tc 99m antimony sulfide colloid
for liver imaging, 520
for lymphoscintigraphy, 607
technetium Tc 99m apcitide (^{99m}Tc-apcitide), 189, 669
chemical structure of, 190f
indications for, 189
radiochemical purity of, 287t, 288t, 292f
radiolabeling approach for, 186
stability of, improving, 656
technetium Tc 99m arcitumomab (^{99m}Tc-arcitumomab), 665–666
indication for, 666
radiochemical purity of, 287t, 288t
technetium Tc 99m bicisate (^{99m}Tc-ECD), 189–191
for brain imaging, 165, 189–191, 191f, 411, 418–419, 420–422, 421f, 422t
 in epilepsy, 430–432
chemical structure of, 190f
indications for, 189
L,L isomer of, 191, 191f, 420f
organ distribution of, 421f
radiochemical purity of, 287t
[Tc = O]$^{3+}$ core of, 183
as technetium-essential compound, 185
technetium Tc 99m bisaminethiol complexes (^{99m}Tc-BAT), for brain imaging, 418
technetium Tc 99m chelates, common problems associated with, 320–321
technetium Tc 99m ciprofloxacin (^{99m}Tc-ciprofloxacin), 593, 596t
technetium Tc 99m depreotide (^{99m}Tc-depreotide), 669
chemical structures of, 670f
radiochemical purity of, 287t, 288t
radiolabeling approach for, 186
technetium Tc 99m diethylenetriaminepentaacetic acid. See technetium Tc 99m pentetate
technetium Tc 99m disofenin (^{99m}Tc-DISIDA), 191
biodistribution of, 523–524, 532, 533f
booster dose of, 532
chemistry of, 178–179, 179f, 191
for hepatobiliary imaging, 523–524, 532–533, 533f
 of acute cholecystitis, 536f, 538f
 of biliary atresia, 539f
 of chronic cholecystitis, 536f
 of neonatal hepatitis, 538f
indications for, 191
physicochemical properties of, 523
radiation dose of, 524, 524t

radiochemical purity of, 287t, 289t
substrate specificity of, 169
technetium Tc 99m etidronate (^{99m}Tc-EHDP)
biologic properties of, 575–576
for bone imaging, 573–578
development of, 573–574
localization mechanisms of, 576, 577f
technetium Tc 99m exametazime (^{99m}Tc-HMPAO), 191–193
autoradiography of, 353
for brain imaging, 165, 190f, 192, 411, 419–420, 419f, 422t
 in brain death, 429, 431f
 of cerebrovascular reserve, 435, 436f
 in epilepsy, 430–432, 432f
chemical structure of, 192, 192f
critical organ of, 91
D,L isomer of, 419–420, 419f, 420f
excessive technetium Tc 99 in, 320
excessive technetium Tc 99m in, 320
indications for, 192
for liposome labeling, 608
meso-isomer of, 419, 419f
organ distribution of, 419f
pH of, 295
radiochemical purity of, 287t, 289t
reductive decomposition in, 323
[Tc = O]$^{3+}$ core of, 183
as technetium-essential compound, 185
for white blood cell labeling, 192, 201–202, 202f, 592, 593f, 594
with and without stabilizer, 192
technetium Tc 99m fanolesomab (^{99m}Tc-fanolesomab), 666
adverse reactions to, 338
indication for, 666
for infection imaging, 593
technetium Tc 99m furifosmin (^{99m}Tc-furifosmin)
[Tc]$^{3+}$ core of, 183
as technetium-essential compound, 185
technetium Tc 99m generator, 34–44, 171–172
^{99m}Tc–^{99}Mo relationship in, at various times, 41, 41t
chemical purity in, 38–39, 39f
commercial, 35, 36f
decay scheme in, 23, 23f
disposal of, 44
dry system of, 35–38, 37f
elution in, 35–36, 36f, 37f
expiration times for eluate in, 38, 39t
physics of, 39–44
production of, 35, 35f
quality control of eluate in, 37–39
radioactive calibration in, 37–38, 38f
radionuclidic purity in, 38–39, 38f, 39f, 275, 277f
shipping of, 123
technetium content in eluates in, 42–44, 42f, 43t, 44t
transient equilibrium in, 40–41, 40f
wet system of, 35–38, 37f
yield of, 36–37, 171
technetium Tc 99m glucarate (^{99m}Tc-glucarate), for myocardial infarction imaging, 489, 490t

technetium Tc 99m gluceptate (^{99m}Tc-GH), 193
    biologic properties of, 553–554, 555t
    for brain imaging, 193, 411, 416
        in brain death, 429, 429f, 430f
    chemical structure of, 193, 193f
    hepatobiliary excretion of, 554, 554f
    indications for, 193
    radiochemical purity of, 287t, 289t
    radiolabeling approach for, 186
    for renal imaging, 553–554
        cortical scintigraphy, 558–559
        development of, 552
technetium Tc 99m glucoheptonate (^{99m}Tc-glucoheptonate), 186
technetium Tc 99m iminodiacetic acid analogues (^{99m}Tc-IDA), 178, 179f, 183, 193–194
    biodistribution of, 523–524, 532, 533f
    for hepatobiliary imaging, 517, 523–524, 532–533
    physicochemical properties of, 523
    radiation dose of, 524, 524t
technetium Tc 99m iron ascorbate (^{99m}Tc-iron ascorbate), for renal imaging, 552
technetium Tc 99m-labeled antibodies, for tumor imaging, 596
technetium Tc 99m-labeled immunoglobulin G (^{99m}Tc-IgG), 592
technetium Tc 99m-labeled interleukin-8 (^{99m}Tc-IL-8), 594
technetium Tc 99m-labeled white blood cells, 192, 201–202
    for infection imaging, 591–592, 593f, 593t, 594–596
    localization mechanisms of, 596t
    procedure for labeling, 201, 201f
    proposed mechanism of, 202, 202f
    white blood cell concentration in, 322
technetium Tc 99m lidofenin (^{99m}Tc-HIDA)
    biodistribution of, 523–524
    chemical structure of, 523f
    chemistry of, 178–179, 179f
    for hepatobiliary imaging, 523–524
    physicochemical properties of, 523
    radiochemical purity of, 287t
technetium Tc 99m liposomes, for lymphoscintigraphy, 608
technetium Tc 99m mebrofenin (^{99m}Tc-BRIDA), 193–194
    biodistribution of, 523–524, 524f
    booster dose of, 532
    chemistry of, 178–179, 179f, 193
    for hepatobiliary imaging, 193, 523–524, 532–533
    indications for, 193
    labeling with, 161
    oxidation state of, 174
    physicochemical properties of, 523
    radiation dose of, 524, 524t
    radiochemical purity of, 287t, 289t
    substrate specificity of, 169
technetium Tc 99m medronate (^{99m}Tc-MDP)
    adverse reactions to, 338–339, 577
    biologic properties of, 575–576
    blood clearance of, 575, 576f
    for bone imaging, 194, 573–578, 579f, 580, 644–646
    chemical structure of, 175f
    dosimetry of, 576–577, 577t
    effective dose of, 89, 90t
    indications for, 194
    labeling with, 161
    localization mechanisms of, 576, 577f
    occupational exposure from, 114, 115t
    preparation of, 194, 194t, 575, 575f
    radiochemical purity of, 284f, 289t
technetium Tc 99m mertiatide (^{99m}Tc-MAG3)
    air in preparation of, 177
    biologic properties of, 554–555, 555t
    chemical structure of, 195f
    coordination geometries of, 154t
    excessive technetium Tc 99 in, 320
    excessive technetium Tc 99m in, 320
    hepatobiliary excretion of, 555
    indications for, 194
    oxidation state of, 174
    preparation of, 157, 157f, 194–195
    radiochemical purity of, 279–280, 280f, 287t, 289t
    renal elimination of, 551
    for renal imaging, 548, 554–555
        captopril renography, 558
        diuresis renography, 556
        normal renogram, 559f, 560f
        of renal tubular function, 569
        for transplant evaluation, 563f
    substrate specificity of, 169
    $[Tc=O]^{3+}$ core of, 183
    as technetium-essential compound, 185
technetium Tc 99m-N,N'-bis(mercaptoacetyl) ethylenediamine (^{99m}Tc-DADS), 552
technetium Tc 99m-N-(NOEt)$_2$, 183f, 184
technetium Tc 99m nofetumomab merpentan, 666
technetium Tc 99m oxidronate (^{99m}Tc-HDP), 194t, 195
    biologic properties of, 575–576
    for bone imaging, 195, 573–578, 580
        of stress fracture, 581f
    chemical structure of, 175f
    development of, 573–574
    dosimetry of, 576–577, 577t
    indications for, 195
    localization mechanisms of, 576, 577f
    radiochemical purity of, 287t, 289t
    radiochromatographic analysis of, 82–83, 83t
technetium Tc 99m pentetate (^{99m}Tc-DTPA), 195–196, 196t
    adverse reactions to, 337
    biologic properties of, 552–553, 555t
    for brain imaging, 196, 411, 416
        in brain death, 429
    chemical structure of, 175f
    for CSF imaging, 158, 195, 337, 427t, 428, 438–439, 438f, 439f, 440f
    as diagnostic radiopharmaceutical, 158
    extraction efficiency of, 549
    half-life or, 158
    indications for, 196
    for lung ventilation studies, 505, 507, 507f
    oxidation state of, 174
    pediatric dosing of, 325t
    radiochemical purity of, 287t, 289t
    for renal imaging, 552–553
        captopril renography, 558
        development of, 552
        in glomerular filtration rate measurement, 158, 196, 552, 553, 569, 702, 703
    $[Tc]^{4+}$ core of, 183
    for ventilation–perfusion lung scan, 508
technetium Tc 99m pentetate–mannosyl–dextran (^{99m}Tc-Lymphoseek), for lymphoscintigraphy, 608
technetium Tc 99m phytate, for liver imaging, 520
technetium Tc 99m propyleneamine oxime (^{99m}Tc-PnAO)
    for brain imaging, 191f, 418–419
    as technetium-essential compound, 185
technetium Tc 99m pyrophosphate (^{99m}Tc-PPi, ^{99m}Tc-PYP), 196, 196t
    adverse reactions to, 337, 577
    altered biodistribution of, 578
    biologic properties of, 575–576
    for bone imaging, 573–578
    development of, 573
    indications for, 196
    for myocardial infarction imaging, 488, 490t
    oxidation state of, 174
    pediatric dosing of, 325t
    radiochemical purity of, 287t, 289t, 574, 574f
    $[Tc]^{4+}$ core of, 183
technetium Tc 99m red blood cells, 196–197
    activity in breast milk, 331
    close contact and exposure to, 332
    excessive technetium Tc 99 in, 320
    excessive technetium Tc 99m in, 320
    for gastrointestinal imaging, 517
    for GI bleeding detection, 527–528, 527t, 539, 541f
    for heart imaging, 459, 460t, 471
    heat-denatured, for spleen imaging, 526, 526f
    in vitro method for, 196–197
    in vivo method for, 197
    labeling mechanism of, 197
    mixing order for, 321
    modified in vivo method for, 197
    radiochemical purity of, 287t, 289t
    red blood cell concentration in, 322
technetium Tc 99m sestamibi (^{99m}Tc-sestamibi), 197–198
    biochemistry of, 467t
    chemical bonds of, 153
    chemical structure of, 181f
    close contact and exposure to, 332
    coordination geometries of, 154t
    indications for, 197
    for myocardial perfusion imaging, 197, 459, 460t, 467t, 469
        extraction and retention of, 466–467, 466f
        interpretation of findings with, 479, 480f, 482f, 484f
        pharmacologic stress and, 463, 464
        protocol for, 476, 477f, 478f
    for myocardial viability imaging, 486–487

oxidation state of, 174
radiochemical purity of, 287t, 289t,
   292f
[Tc]⁺ core of, 181, 181f
as technetium-essential compound, 185
for tumor imaging, 596
   parathyroid, 605, 605f
for ventriculography, 471
technetium Tc 99m sodium pertechnetate.
   See sodium pertechnetate Tc 99m
technetium Tc 99m stannous albumin
   colloid kit, 520
technetium Tc 99m stannous phytate, 607
technetium Tc 99m succimer (⁹⁹ᵐTc-
   DMSA), 198–199
   biologic properties of, 553, 555t
   chemical structure of, 198f
   extraction efficiency of, 549
   indications for, 198
   oxidation state of, 174
   radiochemical purity of, 287t, 290t
   for renal imaging, 552, 553, 554f
      cortical scintigraphy, 558–559,
         564f–568f
      of horseshoe kidney, 568f
      of multicystic dysplastic kidney,
         567f
      of renal cell carcinoma, 565f
      of renal sarcoma, 567f
      of urinary tract infection, 566f
   [Tc]³⁺ core of, 183
technetium Tc 99m sulesomab
   (⁹⁹ᵐTc-sulesomab), 593, 668
technetium Tc 99m sulfur colloid
   (⁹⁹ᵐTc-SC), 172, 199
   adverse reactions to, 337, 338–339, 522
   appearance and color of, 294t
   biologic properties of, 520–522
   blood clearance of, 521
   for bone marrow imaging, 611–612,
      612f
   chemistry of, 199
   for dacryoscintigraphy, 610
   filtered, 608
   for gastrointestinal imaging, 517,
      527–528
      of gastric emptying, 527,
         542–544, 543f, 544f
      of gastroesophageal reflux, 527,
         541–542, 542f
      of gastrointestinal bleeding,
         527–528, 527t
   indications for, 199
   for liver imaging, 517, 520–522, 520t,
      521f, 529–530, 530f, 531f
      blood flow and, 520–521, 521f
      disease state and, 521, 521f
      particle size and, 521
   localization and metabolic fate of,
      521–522, 521f
   for lung ventilation studies, 508
   for lymphoscintigraphy, 606–608,
      609f, 636
      breast, 607
      melanoma, 606–607
   mixing order for, 321
   normal, no gelatin, no EDTA, 199,
      200f
   opsonization of, 521

oxidation state of, 174, 184
particle number of, 294, 294t, 295f
particle size of, 294, 294t, 295f, 521,
   608, 608t
pediatric dosing of, 325t
preparation of, 199, 608, 608t
radiation dose of, 522, 522t
radiochemical purity of, 284, 287t,
   290t
for spleen imaging, 517
stabilizing effect of EDTA in, 199, 200f
technetium Tc 99m teboroxime
   (⁹⁹ᵐTc-teboroxime), 182, 183
   for heart imaging, 466, 466f
technetium Tc 99m tetrofosmin
   (⁹⁹ᵐTc-tetrofosmin), 199–200
   air in preparation of, 177
   biochemistry of, 467t
   for heart imaging, 200, 459, 460t, 467t,
      469–470, 471
      extraction and retention of,
         466–467, 466f
      pharmacologic stress and, 463, 464
      protocol for, 476, 477f, 478f
      for viability studies, 486–487
   [O = Tc = O]⁺ core of, 183, 183f
   radiochemical purity of, 287t, 290t
   reaction sequence for labeling, 200, 201f
   as technetium-essential compound, 185
technetium Tc 99m tripolyphosphate
   (⁹⁹ᵐTc-STPP), 573
technetium Tc 99m TRODAT-1, 183
technetium Tc 99m TROTEC-1
   (⁹⁹ᵐTc-TROTEC-1), 181–182, 181f
technetium Tc 110 (¹¹⁰Tc), 171
technical requirements, regulatory, 400–402
TEDE. See total effective dose equivalent
Temple University, 9
terminology
   antibody, 661, 661t
   dose, 88–91, 91f
   oxidation–reduction, 156, 156t
terrestrial radiation, 112
testes
   radiation effects on, 142, 142t
   radiosensitivity of, 135
testicular cancer, reimbursable PET procedures
   in, 239t
testicular torsion, 568, 568f
TETA, 668–669, 669f
tetrofosmin technetium Tc 99m. See
   technetium Tc 99m tetrofosmin
Texatopes, 9
²³⁴Th. See thorium 234
thallium
   chemistry of, 215–217
   electronic configuration of, 215
   physical properties of, 216t
   production of, 215–217
   in scintillation counters, 66–67
thallium protocols
   for myocardial perfusion imaging,
      474–476, 475f, 477f
   for myocardial viability imaging,
      484–486
thallium Tl 201 (²⁰¹Tl), 7t, 27t
   applications of, 207
   autoradiography of, 353
   decay of, 22, 216–217, 217f

impurities in, 34
labeling with, 160
physical properties of, 216t
production of, 216–217
pulse-height energy spectrum of, 70f
radionuclidic purity of, 276t
shipping guidelines for, 121t, 122t
thallium Tl 201 thallous chloride
   (²⁰¹Tl-chloride), 220
   biochemistry of, 467t
   for brain imaging, 423
   indications for, 220
   for myocardial perfusion imaging, 220,
      459, 460t, 467–469, 467t, 468t
      in dual-isotope protocols,
         476–477, 479f
      extraction and retention of, 467
      interpretation of findings with,
         479, 482f
      pharmacologic stress and, 463
      protocol for, 474–476, 475f, 477f
   for myocardial viability imaging, 471,
      484–486
   pediatric dosing of, 327
   radiochemical purity of, 287t
   substrate specificity of, 169
   for tumor imaging, 201, 220, 596
      parathyroid, 604–605, 605f
The Joint Commission (TJC), 253
therapeutic applications, 673–691. See also
   specific agents
   for bone metastases or pain, 673,
      683–686
   for effusions, 673, 687
   for liver cancer, 688–690
   new agents for, 690–691
   for non-Hodgkin's lymphoma,
      673–683
   for polycythemia vera, 673, 686–687
   pretargeted, 683
   for synovectomy, 673, 687–688
therapeutic dosage, 396
therapeutic dose, 396
therapeutic index, 675
therapeutic procedures, 3–4. See also specific
   types
therapeutic radiopharmaceuticals, 2, 158, 160
TheraSpheres. See yttrium Y 90-labeled glass
   microspheres
thermal conductivity detectors, 277
thermal energy, 31–32
thermal neutrons, 51–52
thermoluminescent dosimeters, 116
thin-layer chromatography (TLC), 275,
   279–283, 292f
thiosulfate, for radioiodination, 157
three-dimensional imaging, 3, 3f
   PET, 233, 233f
   PET/CT, 620
three-phase bone scan, 578–586, 584t. See
   also bone imaging
threshold dose, for deterministic effects, 98,
   128
threshold level, in scintillation counter, 66
thrombocytopenia, 138
thyroglobulin, 445
thyroid cancer
   indium In 111 pentetreotide imaging
      of, 600t

PET/CT of, 627–628, 628f
radiation-induced, 139–140
reimbursable PET procedures in, 239t, 240–241
treatment of, 3, 446–447, 455–456, 455f, 456f, 673
thyroid gland
agenesis of, 453, 455f
FDG uptake in, 620, 622, 624f
function of, 445–446, 446f
in vivo function studies of, 3, 158, 695
pathophysiology of, 450–451
static imaging studies of, 3, 3f
tissue-weighting factor for, 104t
thyroid gland uptake, 448, 449t
in bone scans, 574, 574f
in lung perfusion studies, 504, 505f
thyroid hormones, 445–446
thyroiditis, 450
FDG uptake in, 622, 624f
thyroid nodules, 450–451
cold (nonfunctioning), 451, 453, 453f
hot (functioning), 450–451, 452–453, 453f, 454–455
imaging of, 452–453
radioiodine therapy for, 453, 454–455
warm, 453
thyroid-protecting agents, 214, 448, 448t
thyroid scan, 452–453
interpretation of, 452–453, 453f
procedure for, 452, 452f
rationale for, 452
thyroid-stimulating hormone (TSH), 445–446
recombinant, 455–456
thyroid studies, 445–453
historical perspectives on, 446–447
occupational exposure in, 117, 117t
physiologic principles of, 445–446
radioactive iodine uptake test, 445, 451–452, 451f
radiopharmaceuticals for, 447–450, 450t
sodium iodide I 131, 213, 213t
specific activity in, 161–162
thyroid scan, 452–453, 452f, 453f
thyroid therapy, 453–456
procedures in, 454–456
rationale for, 453–454
safety considerations in, 456
thyrotropin, 445–446
thyroxine, 445
TI. See transport index
TID. See transient ischemic dilation
time
activity vs., 28, 28f
calibration, 30
clearance, for radioactive gases, 118
of exposure, 96, 96t, 114
time/incubation delays, 322
tin Sn 113 (^{113}Sn)–indium 113m (^{113m}In) generator, 44
tin Sn 117m (^{117m}Sn), shipping guidelines for, 122t
tin Sn 117m pentetate (^{117m}Sn-DTPA), for bone pain palliation, 684t, 686
tissue biodistribution, ex vivo, 353–355
animal care in, 355
control studies of biomarker distribution in, 354

factors affecting, 353–354, 354t
target-blocking studies in, 354–355
tissue-weighting factor (W_T), 89, 89t, 90t, 104, 104t
TJC. See The Joint Commission
TL. See transfer ligand
^{201}Tl. See thallium Tl 201
TLC. See thin-layer chromatography
TLD. See tumor lethal dose
T-lymphocytes, 656, 656f
tonsils, FDG uptake in, 620, 622, 624f
tositumomab I 131 (^{131}I-tositumomab), 667
dosing regimen for, 680–681, 680f, 681f
maximum tolerated dose of, 676
nomenclature for, 661
for non-Hodgkin's lymphoma, 3–4, 207, 655, 666–667, 676, 680–683
radiation safety of, 681
therapeutic uses of, 655
total-body PET/CT, 618
total-body SPECT, 589–612
total effective dose equivalent (TEDE), 91, 96, 402, 406, 506
Townsend, David, 626
toxic goiter, 450
imaging of, 453, 453f, 454f
radioiodine therapy for, 453, 454–455
toxicity, mass and, 5–7, 147
toxicity review, 392
toxicity testing/studies, 382
trabecular bone, 571, 572f
"tracee," 695
trace metals, in indium In 111 chloride, 293
tracer(s), 695. See also radiopharmaceutical(s)
definition of, 158
effective, 158
tracer kinetics, 353
traditional radionuclides, 346, 347t
training requirements, 399–400, 405
tramtrack sign, in bone imaging, 582, 585f
transfer ligand (TL), 155
transient equilibrium generators, 40–41, 40f, 44
transient ischemic dilation (TID), 493
transition energy, 19
transition zone, of lungs, 499, 500f
translocations, chromosomal, 103t, 131–132, 132f, 141
transmural myocardial infarction, 484
trans-oxygen atoms, 183
transport across plasma membrane, 163–166
active, 164f, 165–166
mediated, 163, 165–166
passive, 163–165, 163f
transport index (TI), 120, 122, 123, 123t
transport in renal excretion, 548
transport of radioactive material, 120–125
accident in, reporting of, 124
in health care setting, 112–113
liquid, absorbent material for, 124
from nuclear pharmacy, 260, 270, 271
package labeling for, 120, 122–123, 123f, 123t
package markings for, 122
package monitoring in, 120, 120t, 123
packaging requirements for, 121–122
personnel training for, 124
placarding of vehicle, 123–124, 124f

quality control measures for, 124
separation distance in, 123, 123t
shipping papers for, 124
threshold values for, 121, 121t
transverse plane, 3, 3f
treadmill exercise, and myocardial perfusion imaging, 461–462
triethylenetetramine (TETA), 668–669, 669f
triiodothyronine, 445
trisomies, 103t
tritium (^{3}H)
in autoradiography, 352, 353
in binding assays, 349
tropolone indium I 111. See indium In 111 tropolone
TROTEC-1, technetium Tc 99m (^{99m}Tc-TROTEC-1). See technetium Tc 99m TROTEC-1
TSH. See thyroid-stimulating hormone
Tubis, Manuel, 8
tubular necrosis, acute, 562
tubular secretion, renal, 548–549, 569
tubules, renal, 547
tumor(s). See also cancer
imaging of
adrenal medullary, 600–603, 603f, 604f
gallium Ga 67 citrate for, 595f, 596, 597–598, 597f
indium In 111 pentetreotide, 598–600, 600f, 600t
neuroendocrine, 598–600
parathyroid, 603–605, 605f
PET, 598
principal objectives of, 596
radiolabeled peptides for, 596, 598–600
thallous chloride Tl 201, 220, 596
total-body SPECT, 596–605
radiation-induced, types of, 139–140
recurrence of, PET of, 433, 435f
response to treatment, assessing, 628f, 644, 645f, 646f
treatment of
dose fractionation in, 137–138
four Rs of, 137–138
iodine I 125 in, 208
multiple dose, 137
oxygen effects on, 136–137
single-dose, 137
tumor lethal dose (TLD), 137
tungsten, electron energy levels of, 14–15, 15f
tungsten shielding
for syringes, 113, 113f
for vials, 261
tungsten W 188 (^{188}W)–rhenium Re 188 (^{188}Re) generator, 44, 45
Tyco Healthcare/Mallinckrodt, 9
type A packages, 121–122, 122t
type B packages, 121
tyrosine, iodination of, 210, 210f

U

^{235}U. See uranium 235
^{235}U(n,f) byproducts, 33
ULD. See upper-level energy discriminator

ultrasound, 343–344, 344t
  for hepatobiliary and spleen imaging, 528–529
  of infections, 594
  renal, 555
  scrotal, 559
  of tumors, 598
UltraTag. See technetium Tc 99m red blood cells
ultraviolet light, 53t
uninodular thyroid nodules, 454–455
unit dosage, 397
unit dose
  determination of, 285
  preparation of, 260, 260f
United Nations Scientific Committee on the Effects of Atomic Radiation (UNSCEAR), 99, 109
United States Pharmacopeia (USP), 38–39, 235, 252, 266, 334
  on chromatography, 275–277
  on compounded sterile products, 313–317
  on endotoxin testing, 297, 298–299, 308, 315
  on PET, 305, 306–308, 381
  on PET radiopharmaceuticals, 390, 392t
  on quality control, 273–274
  on radiochemical purity, 284–285, 286t–287t
U.S. Public Health Service, 9
units of activity, 25–28, 161
university hospital nuclear pharmacy, 257
University of Arkansas, 9–10
University of Chicago, 8, 34, 171, 416
University of Cincinnati, 9
University of Indiana, 9
University of Michigan, 9, 257
University of Minnesota, 9, 416
University of Nebraska, 8, 9
University of New Mexico, 8–10, 257
University of North Carolina, 9
University of Oklahoma, 9
University of Pittsburgh, 9
University of Southern California, 8, 9, 257
University of Tennessee, 8, 9, 257
University of Toronto, 9
University of Utah, 8, 9
University of Washington, 8
University of Wisconsin, 9
unrestricted area, 258–259, 258f
UNSCEAR. See United Nations Scientific Committee on the Effects of Atomic Radiation
unsealed byproduct material
  dosages of, 400–401, 406
    regulatory requirements on, 400–401, 402–405
    written directive not required, 402–404
    written directive required, 404–405
upper-level energy discriminator (ULD), 66–68, 67f
uranium 235 (^{235}U)
  ^{235}U(n,f)byproducts, 33
  in nuclear reactor, 31, 32f
urea (^{14}C-urea)
  breath test, 703
  capsules, 225

urinary tract, FDG uptake in, 620
urinary tract infection (UTI), 559, 562, 566f
urine flow, 547
uropathy, obstructive, 556, 560–562, 562f
USP. See United States Pharmacopeia
UTI. See urinary tract infection

V

valence electrons, 148
vasovagal reactions, 334t
ventilation, 499, 500f. See also lung ventilation studies
ventilation–perfusion lung scan, 504–505, 507–513
  chronic obstructive pulmonary disease in, 509–511, 512f, 513f
  high probability of pulmonary embolism in, 509, 511f
  indications for, 508
  interpretation of, 509–513
  match in, 505, 509–511, 512f
  mismatch in, 505
  normal, 509, 510f
  rationale for, 508
ventriculography, cardiac, 459
  equilibrium radionuclide, 474, 479–480
    interpretation of, 481, 485f
    procedure for, 480–481, 485f
    rationale for, 480
  imaging agents for, 471
ventriculoperitoneal shunt, evaluation of, 438–440, 439f, 440f
Verluma. See technetium Tc 99m nofetumomab merpentan
vertebral artery, 413–414, 414f
vertical-flow hood, 262, 262f
vesicoureteral reflux, 559, 568, 568f
vial shields, 261
visible light, 53, 53t
vitamin B_{12} absorption, test of, 703
volume of distribution, 550–551
volume variation, in dose calibrators, 303–304
volunteer subjects, dose limits for, 111–112
vomiting, 135, 138–139
VQ. See ventilation–perfusion lung scan

W

^{188}W. See tungsten W 188 (^{188}W)–rhenium Re 188 (^{188}Re) generator
warm thyroid nodules, 453
waste, radioactive
  disposal of, 125
  management in nuclear pharmacy, 260, 260f
  transfer to authorized recipient, 125
water
  covalent bonds of, 149
  radiation absorption in, 95, 95t
water oxygen O 15 (^{15}O-water)
  for heart imaging, 470–471, 470t
  for PET, 234–235, 247
  radiochemical purity of, 286t
  radiosynthesis of, 247, 247f
wavelength, of electromagnetic radiation, 52–53, 53t

WBCs. See white blood cells
WBH. See whole-body hematocrit
weight-based dose adjustment, 326
weighting factors
  radiation-, 88, 88t
  tissue-, 89, 89t, 90t, 104, 104t
well-differentiated thyroid carcinoma, 455–456, 455f, 456f
wet system, of ^{99m}Tc generator, 35–38, 37f
white blood cells, 589–590, 650–651, 695
  indium In 111-labeled, 217, 218–219
    for infection imaging, 591, 591f, 593t, 594–596
    localization mechanisms of, 596t
    normal scan with, 592f
    procedure for labeling, 201f, 218–219
    proposed mechanism for, 219, 219f
  technetium Tc 99m-labeled, 192, 201–202
    for infection imaging, 591–592, 593f, 593t, 594–596
    localization mechanisms of, 596t
    procedure for labeling, 201, 201f
    proposed mechanism of, 202, 202f
  white blood cell concentration in, 322
white-I label, 120, 122–123, 123f, 123t
white pulp, of spleen, 525
whole blood volume, determination of, 3
whole-body badges, 114–116, 117f
whole-body bone scan, 578–586. See also bone imaging
whole-body hematocrit (WBH), 695
whole-body oncology studies, 618
whole-body radiation, acute effects of, 138–139
William Beaumont Hospital, 9
Willis, circle of, 414
window, in scintillation counter, 66, 67f
wipe test, 120, 120t, 260
Wolf, Walter, 8, 9t
workers, radiation. See health care personnel
World Health Organization, 333
World Molecular Imaging Congress, 254, 345
W_R. See radiation-weighting factor
written directive, 397, 399, 404–405, 406
W_T. See tissue-weighting factor

X

^{127}Xe. See xenon Xe 127
^{133}Xe. See xenon Xe 133
^{135}Xe. See xenon Xe 135
xenon Xe 127 (^{127}Xe), 220
  dose calibrator measurements of, 287
  for lung ventilation studies, 505, 505t
  radiochemical purity of, 287t
xenon Xe 133 (^{133}Xe), 7t, 27t, 220–221
  administration of, 505–506, 506f
  applications of, 159, 207, 220
  biologic distribution of, 506–507, 507t
  for brain imaging, 220
  decay of, 220–221, 221f
  detection efficiency of, 159, 159t

dose calibrator measurements of, 303
hydrostatic pressure and, 500–501
labeling with, 160
for lung ventilation studies, 220, 505–507, 505t
occupational exposure to, 117t
production of, 220
pulse-height energy spectrum of, 71f
radiochemical purity of, 287t
radionuclidic purity of, 276t
safety control of, 506
shipping guidelines for, 121t, 122t
for ventilation–perfusion lung scan, 508–509, 510f, 511f, 512f, 513f
xenon Xe 135 (^{135}Xe), for lung ventilation studies, 505, 505t
X-linked disorders, 103t, 104
x-ray(s), 17, 47
    interactions with matter, 52–56
    plain, for bone imaging, 571
    properties of, 52–53
    protection from, 92–97
x-ray peak
    characteristic, 70
    lead, 70
x-ray procedures, average effective dose in, 105t–106t

Y

^{86}Y. *See* yttrium Y 86
^{90}Y. *See* yttrium Y 90
^{169}Yb-DTPA. *See* ytterbium Yb 169 pentetate
yellow-III label, 120, 122–123, 123f, 123t
yellow-II label, 120, 122–123, 123f, 123t
Young's rule, 326
ytterbium Yb 169 pentetate (^{169}Yb-DTPA), for CSF imaging, 426, 427, 427t
yttrium
    chemistry of, 224
    electronic configuration of, 224
yttrium Y 86 (^{86}Y), as nontraditional radionuclide, 347t
yttrium Y 90 (^{90}Y), 7t, 26t
    ^{90}Sr–^{90}Y generator, 45
    applications of, 207
    decay of, 222f, 224
    dose calibrator measurements of, 63, 64t, 287–292
    labeling with, 160, 661–662, 662t, 664–665, 674, 674f, 676, 677t
    production of, 224
    radiolabeling approach for, 186
    shipping guidelines for, 121t, 122t
yttrium Y 90 chloride solution (^{90}Y-chloride), 224
yttrium Y 90 citrate (^{90}Y-citrate), for radiation synovectomy, 687–688, 688t
yttrium Y 90 DOTATOC (^{90}Y-DOTATOC), 691
yttrium Y 90 DOTA-tyr^3-octreotide, 691
yttrium Y 90 edotreotide (^{90}Y-edotreotide), 691
yttrium Y 90 epratuzumab (^{90}Y-hLL2), for non-Hodgkin's lymphoma, 674, 681–682
yttrium Y 90 ibritumomab tiuxetan (^{90}Y-ibritumomab tiuxetan), 224, 666–667
    adverse effects of, 678
    dosing regimen for, 677–679, 677f, 678f, 679f
    labeling process for, 186, 674, 674f, 675, 677t
    mixing order for, 321
    nomenclature of, 661
    for non-Hodgkin's lymphoma, 3–4, 655, 666–667, 674, 676–680, 682
    radiation safety of, 679–680
    radiochemical purity of, 287t
    therapeutic uses of, 655
yttrium Y 90-labeled antibodies, 224
yttrium Y 90-labeled glass microspheres
    dose calculation for, 690
    for liver cancer, 688–690
yttrium Y 90-labeled resin microspheres
    dose calculation for, 690
    for hepatic metastases, 689–690
yttrium Y 90 microspheres (^{90}Y-microspheres), 688–690
    dose calculation for, 690
    for liver cancer, 688–690
yttrium Y 90 silicate (^{90}Y-silicate), for radiation synovectomy, 687–688, 688t
yttrium Y 90 SMT487 (^{90}Y-SMT487), 691

Z

Zevalin. *See* ibritumomab tiuxetan
Zimmer, A. Michael, 9t
zirconium Zr 89 (^{89}Zr), as nontraditional radionuclide, 347t
^{89}Zr. *See* zirconium Zr 89
Z value, of nuclides, 13